HANDBOOK OF STATISTICS
VOLUME 16

Handbook of Statistics

VOLUME 16

General Editor
C. R. Rao

ELSEVIER
AMSTERDAM · LAUSANNE · NEW YORK · OXFORD · SHANNON · TOKYO

Order Statistics: Theory & Methods

Edited by

N. Balakrishnan
Department of Mathematics and Statistics
McMaster University
Hamilton, Ontario, Canada

C. R. Rao
Center for Multivariate Analysis
Department of Statistics, The Pennsylvania State University
University Park, PA, USA

1998

ELSEVIER
AMSTERDAM · LAUSANNE · NEW YORK · OXFORD · SHANNON · SINGAPORE · TOKYO

ELSEVIER SCIENCE B.V.
Sara Burgerhartstraat 25
P.O. Box 211, 1000 AE Amsterdam, The Netherlands

Library of Congress Cataloging-in-Publication Data
A catalog record from the Library of Congress has been applied for.

ISBN: 0-444-82091-4

© 1998 Elsevier Science B.V. All rights reserved.

No part of this publication may be reproduced, stored in a retrieval system or transmitted in any form or by any means, electronic, mechanical, photocopying, recording or otherwise, without the prior written permission of the publisher, Elsevier Science B.V., Copyright & Permissions Department, P.O. Box 521, 1000 AM Amsterdam, The Netherlands.

Special regulations for readers in the U.S.A.-This publication has been registered with the Copyright Clearance Center Inc. (CCC), 222 Rosewood Drive, Danvers, MA 01923. Information can be obtained from the CCC about conditions under which photocopies of parts of this publication may be made in the U.S.A. All other copyright questions, including photocopying outside the U.S.A., should be referred to the Publishers unless otherwise specified.

No responsibility is assumed by the publisher for any injury and/or damage to persons or property as a matter of products liability, negligence or otherwise, or from any use of operation of any methods, products, instructions or ideas contained in the material herein.

⊚ The paper used in this publication meets the requirements of ANSI/NISO Z39.48-1992 (Permanence of Paper).

Printed in The Netherlands.

Preface

The area of **Order Statistics** received a tremendous attention from numerous researchers during the past century. During this period, several major theoretical advances were made in this area of research. As a matter of fact, two of those have been adjudged as *breakthroughs* in the field of Statistics [see: Kotz, S. and Johnson, N. L. (1991). *Breakthroughs in Statistics*, Vols. 1 and 2, Springer-Verlag, New York]. In the course of these developments, order statistics have also found important applications in many diverse areas including life-testing and reliability, robustness studies, statistical quality control, filtering theory, signal processing, image processing, and radar target detection.

Based on this immense activity, we decided to prepare this Handbook on **Order Statistics and Their Applications**. We feel that we have successfully brought together theoretical researchers working on theoretical and methodological advancements on order statistics and applied statisticians and engineers developing new and innovative applications of order statistics. Altogether, there are 44 articles covering most of the important theoretical and applied aspects of order statistics. For the convenience of readers, the subject matter has been divided into two volumes, the first one (Handbook – **16**) focusing on Theory and Methods and the second one (Handbook – **17**) dealing primarily with Applications. Each volume has also been organized according to different parts, with each part specializing on one aspect of **Order Statistics**.

The articles in this volume have been classified into nine parts as follows:

Part I – Introduction and Basic Properties
Part II – Orderings and Bounds
Part III – Relations and Identities
Part IV – Characterizations
Part V – Extremes and Asymptotics
Part VI – Robust Methods
Part VII – Resampling Methods
Part VIII – Related Statistics
Part IX – Related Processes

We have also presented an elaborate Author Index as well as a Subject Index in order to facilitate an easy access to all the material included in the volume.

Part I contains three articles – the first one by N. Balakrishnan and C. R. Rao presents an introduction to order statistics, the second one by H. L. Harter and

N. Balakrishnan presents a historical perspective on the developments in the subject of order statistics, and the last article by P. R. Tadikamalla and N. Balakrishnan discusses the computer simulation of order statistics.

Part II contains three articles – the first one by B. C. Arnold and J. A. Villasenor discusses some results on the Lorenz ordering of order statistics, the second one by P. J. Boland, M. Shaked and J. G. Shanthikumar discusses results on the stochastic ordering of order statistics, and the last article by T. Rychlik reviews results on bounds for expectations of L-estimates.

Part III contains an exhaustive review article by N. Balakrishnan and K. S. Sultan on the recurrence relations and identities for moments of order statistics from arbitrary as well as many specific distributions.

Part IV contains three articles – the first one by C. R. Rao and D. N. Shanbhag discusses some recent methods used for characterizations results based on order statistics and record values, the second one by U. Gather, U. Kamps and N. Schweitzer reviews the characterization results based on identical distributions of functions of order statistics, and the last article by U. Kamps reviews the characterization results based on recurrence relations and identities for moments of order statistics.

Part V contains three articles – the first one by J. Galambos provides an exposure to the univariate extreme value theory and applications, the second article by P. K. Sen elaborates various applications of the asymptotic results, and the last article by R. J. Tomkins and H. Wang reviews the zero-one laws for large order statistics.

Part VI contains two articles – the first one by D. R. Jensen and D. E. Ramirez discussed some exact properties of Cook's distance while the other article by A. Childs and N. Balakrishnan presents some results on order statistics arising from independent and non-identically distributed Pareto random variables and illustrates their application to robust estimation of the location and scale parameters of the Pareto distribution.

Part VII contains two articles – the first one by R. L. Strawderman and D. Zelterman explains a semiparametric bootstrap method for simulation extreme order statistics while the other article by C. Ma and J. Robinson discusses some approximations to the distributions of sample quantiles.

Part VIII contains two articles – the first one by H. A. David and H. N. Nagaraja reviews the developments on concomitants of order statistics while the second article by V. Nevzorov and N. Balakrishnan provides an updated review on records.

Part IX contains two articles – the first one by B. Szyszkowicz elaborates on weighted sequential empirical type processes with their applications to change-point problems while the second article by M. Csörgö and B. Szyszkowicz discusses sequential quantile process and Bahadur-Kiefer process.

It needs to be mentioned here that the companion volume (Handbook – **17**), focusing on applications of order statistics, has been divided similarly into six parts.

Preface

While preparing this volume as well as the companion volume (Handbook – 17), we have made a very clear distinction between *order statistics* and *rank order statistics*, the latter being an integral part of the area of **Nonparametric Statistics**. Even though there is an overlap between the two and also that order statistics play a role in Nonparametric Statistics, one of the most important uses of order statistics is in the development of parametric inferential methods, as is clearly evident from this volume. Unfortunately, some researchers still view *Order Statistics* as part of **Nonparametric Statistics**. Strangely enough, this view is also present in *Mathematical Reviews*.

We express our sincere thanks to Mr. Gerard Wanrooy (North-Holland, Amsterdam) for his interest in this project and for providing constant support and encouragement during the course of this project. We also thank Mrs. Debbie Iscoe for helping us with the typesetting of some parts of this volume. Thanks are also due to the Natural Sciences and Engineering Research Council of Canada and the U.S. Army Research Office for providing individual research grants to the editors which facilitated the editorial work of this volume. Our final special thanks go to all the authors for showing interest in this project and for preparing fine expository articles in their respective topics of expertise.

We sincerely hope that theoretical researchers, applied scientists and engineers, and graduate students involved in the area of **Order Statistics** will all find this Handbook to be a useful and valuable reference in their work.

N. Balakrishnan
C. R. Rao

Table of contents

Preface v

Contributors xvii

PART I. INTRODUCTION AND BASIC PROPERTIES

Ch. 1. Order Statistics: An Introduction 3
N. Balakrishnan and C. R. Rao

1. Introduction 3
2. Marginal distributions of order statistics 4
3. Joint distributions of order statistics 5
4. Properties 7
5. Moments and product moments 7
6. Recurrence relations and identities 8
7. Bounds 10
8. Approximations 10
9. Characterizations 11
10. Asymptotics 12
11. Best linear unbiased estimation and prediction 12
12. Inference under censoring 14
13. Results for some specific distributions 16
14. Outliers and robust inference 17
15. Goodness-of-fit tests 18
16. Related statistics 18
17. Generalizations 20
    References 21

Ch. 2. Order Statistics: A Historical Perspective 25
H. Leon Harter and N. Balakrishnan

1. Introduction 25
2. Distribution theory and properties 26
3. Measures of central tendency and dispersion 27

4. Regression coefficients 28
5. Treatment of outliers and robust estimation 29
6. Maximum likelihood estimators 29
7. Best linear unbiased estimators 30
8. Recurrence relations and identities 32
9. Bounds and approximations 32
10. Distribution-free tolerance procedures 33
11. Prediction 35
12. Statistical quality control and range 36
13. Multiple comparisons and studentized range 36
14. Ranking and selection procedures 38
15. Extreme values 39
16. Plotting positions on probability paper 40
17. Simulation methods 41
18. Ordered characteristic roots 41
19. Goodness-of-fit tests 43
20. Characterizations 45
21. Moving order statistics and applications 45
22. Order statistics under non-standard conditions 46
23. Multivariate order statistics and concomitants 47
24. Records 47
    References 48

Ch. 3. Computer Simulation of Order Statistics 65
Pandu R. Tadikamalla and N. Balakrishnan

1. Introduction 65
2. Direct generation of order statistics 65
3. Generation of uniform (0, 1) ordered samples 65
4. Generation of progressive Type-II censored order statistics 68
5. Miscellaneous topics 69
   References 70

PART II. ORDERINGS AND BOUNDS

Ch. 4. Lorenz Ordering of Order Statistics and Record Values 75
Barry C. Arnold and Jose A. Villasenor

1. Introduction 75
2. The Lorenz order 75
3. Order statistics and record values 77
4. Lorenz ordering of order statistics 78
5. Lorenz ordering of record values 81
6. Remarks 86
   References 86

Ch. 5. Stochastic Ordering of Order Statistics   89
Philip J. Boland, Moshe Shaked and J. George Shanthikumar

1. Introduction   89
2. Stochastic orderings   90
3. Stochastic order for order statistics from one sample   94
4. Stochastic order for order statistics from two samples   99
   Acknowledgement   102
   References   102

Ch. 6. Bounds for Expectations of *L*-Estimates   105
Tomasz Rychlik

1. Introduction   105
2. Distribution bounds   106
3. Moment and support bounds   112
4. Moment bounds for restricted families   123
5. Quantile bounds for restricted families   135
   References   142

PART III. RELATIONS AND IDENTITIES

Ch. 7. Recurrence Relations and Identities for Moments of Order Statistics   149
N. Balakrishnan and K. S. Sultan

1. Introduction   149
2. Notations   153
3. Recurrence relations for single moments   154
4. Recurrence relations for product moments   161
5. Relations between moments of order statistics from two related populations   171
6. Normal and half normal distributions   172
7. Cauchy distribution   175
8. Logistic and related distributions   177
9. Gamma and related distributions   184
10. Exponential and related distributions   188
11. Power function and related distributions   190
12. Pareto and related distributions   193
13. Rayleigh distribution   199
14. Linear-exponential distribution   200
15. Lomax distribution   203
16. Log-logistic and related distributions   204
17. Burr and truncated Burr distributions   209
18. Doubly truncated parabolic and skewed distributions   211
19. Mixture of two exponential distributions   212
20. Doubly truncated Laplace distribution   212
21. A class of probability distributions   216
    Acknowledgement   221
    References   222

PART IV. CHARACTERIZATIONS

Ch. 8. Recent Approaches to Characterizations Based on Order Statistics and Record Values 231
C. R. Rao and D. N. Shanbhag

1. Introduction 231
2. Some basic tools 232
3. Characterizations based on order statistics 236
4. Characterizations involving record values and monotonic stochastic processes 249
   Acknowledgment 253
   References 254

Ch. 9. Characterizations of Distributions via Identically Distributed Functions of Order Statistics 257
Ursula Gather, Udo Kamps and Nicole Schweitzer

1. Introduction 257
2. Characterizations of exponential distributions based on normalized spacings 259
3. Related characterizations of other continuous distributions 268
4. Characterizations of uniform distributions 270
5. Characterizations of specific continuous distributions 272
6. Characterizations of geometric and other discrete distributions 280
   References 285

Ch. 10. Characterizations of Distributions by Recurrence Relations and Identities for Moments of Order Statistics 291
Udo Kamps

1. Introduction 291
2. Characterizations by sequences of moments and complete function sequences 293
3. Characterizations of exponential distributions 296
4. Related characterizations in classes of distributions 297
5. Characterizations based on a single identity 302
6. Characterizations of normal and other distributions by product moments 305
   References 308

PART V. EXTREMES AND ASYMPTOTICS

Ch. 11. Univariate Extreme Value Theory and Applications 315
Janos Galambos

1. Introduction 315
2. The classical models 317

3. Applications and statistical inference   324
4. Deviations from the classical models   329
   Acknowledgements   330
   References   331

Ch. 12. Order Statistics: Asymptotics in Applications   335
Pranab Kumar Sen

1. Introduction   335
2. Some basic results in order statistics   337
3. Some basic asymptotics in order statistics   341
4. Robust estimation and order statistics: asymptotics in applications   343
5. Trimmed LSE and regression quantiles   350
6. Asymptotics for concomitants of order statistics   352
7. Concomitant L-functionals and nonparametric regression   357
8. Applications of order statistics in some reliability problems   361
9. TTT asymptotics and tests for aging properties   365
10. Concluding remarks   370
    References   371

Ch. 13. Zero-One Laws for Large Order Statistics   375
R. J. Tomkins and Hong Wang

1. Introduction   375
2. Zero-One laws for the upper-case probability   376
3. Zero-one laws for the lower-case probability   379
4. Zero-One laws for the lower-case probability when ranks vary   382
   Acknowledgements   383
   References   384

PART VI. ROBUST METHODS

Ch. 14. Some Exact Properties Of Cook's D_I 387
D. R. Jensen and D. E. Ramirez

1. Introduction   387
2. Preliminaries   388
3. The structure of Cook's D_I 390
4. Normal-Theory properties   393
5. Modified versions of D_I 398
6. Summary   400
   References   401

Ch. 15. Generalized Recurrence Relations for Moments of Order Statistics from Non-Identical Pareto and Truncated Pareto Random Variables with Applications to Robustness   403
Aaron Childs and N. Balakrishnan

1. Introduction   403
2. Relations for single moments   405
3. Relations for product moments   407
4. Results for the multiple-outlier model (with a slippage of p observations)   412
5. Generalization to the truncated Pareto distribution   413
6. Robustness of the MLE and BLUE   415
7. Robustness of the censored BLUE   416
8. Conclusions   421
   Acknowledgements   426
   Appendix A   426
   Appendix B   432
   References   438

PART VII. RESAMPLING METHODS

Ch. 16. A Semiparametric Bootstrap for Simulating Extreme Order Statistics   441
Robert L. Strawderman and Daniel Zelterman

1. Introduction   441
2. A semiparametric bootstrap approximation to X_j 444
3. A saddlepoint approximation to the bootstrap distribution   448
4. Numerical implementation   451
5. Simulation results   453
6. Example: The British coal mining data   458
   Acknowledgements   460
   References   461

Ch. 17. Approximations to Distributions of Sample Quantiles   463
Chunsheng Ma and John Robinson

1. Introduction and definitions   463
2. Smirnov's lemma   467
3. Normal approximation   468
4. Saddlepoint approximation   475
5. Bootstrap approximation   479
   References   482

PART VIII. RELATED STATISTICS

Ch. 18. Concomitants of Order Statistics 487
H. A. David and H. N. Nagaraja

1. Introduction and summary 487
2. Finite-sample distribution theory and moments 488
3. Asymptotic theory 493
4. Estimation and tests of hypotheses 496
5. The rank of $Y_{[r:n]}$ 501
6. Selection through an associated variable 504
7. Functions of concomitants 506
   References 510

Ch. 19. A Record of Records 515
Valery B. Nevzorov and N. Balakrishnan

1. Introduction 515
2. Classical records 515
3. Definitions 517
4. Representations of record times and record values using sums of independent terms 518
5. Distributions and probability structure of record times 520
6. Moments of records times and numbers of records 523
7. Limit theorems for record times 525
8. Inter-Record times 525
9. Distributions and probability structure of record values in sequences of continuous random variables 527
10. Limit theorems for record values from continuous distributions 528
11. Record values from discrete distributions 528
12. Weak records 529
13. Bounds and approximations for moments of record values 530
14. Recurrence relations for moments of record values 531
15. Joint distributions of record times and record values 532
16. Generalizations of the classical record model 534
17. k^{th} record times 534
18. k^{th} inter-record times 537
19. k^{th} record values for the continuous case 538
20. k^{th} record values for the discrete case 540
21. Weak k^{th} record values 540
22. k_n-records 541
23. Records in sequences of dependent random variables 543
24. Random record models 545
25. Nonstationary record models 545
26. Multivariate records 558
27. Relations between records and other probabilistic and statistical problems 558
28. Nonclassical characterizations based on records 559
29. Processes associated with records 560
30. Diverse results 560

Acknowledgement 561
References 561

PART IX. RELATED PROCESSES

Ch. 20. Weighted Sequential Empirical Type Processes with Applications to Change-Point Problems 573
Barbara Szyszkowicz

1. Introduction 573
2. Weighted empirical processes based on observations 579
3. "Bridge-type" two-time parameter empirical processes 590
4. Weighted empirical processes based on ranks 599
5. Weighted empirical processes based on sequential ranks 604
6. "Bridge-type" empirical processes of sequential ranks 609
7. Contiguous alternatives 614
8. Weighted multi-time parameter empirical processes 621
   Acknowledgement 628
   References 628

Ch. 21. Sequential Quantile and Bahadur–Kiefer Processes 631
Miklós Csörgő and Barbara Szyszkowicz

1. Introduction: Basic notions, definitions and some preliminary results 631
2. Deviations between the general and uniform quantile processes and their sequential versions 649
3. Weighted sequential quantile processes in supremum and L_p-metrics 654
4. A summary of the classical Bahadur-Kiefer process theory via strong invariance principles 662
5. An extension of the classical Bahadur–Kiefer process theory via strong invariance principles 680
6. An outline of a sequential version of the extended Bahadur-Kiefer process theory via strong invariance principles 683
   Acknowledgement 686
   References 686

Author Index 689

Subject Index 701

Contents of Previous Volumes 715

Contributors

B. C. Arnold, *Department of Statistics, University of California at Riverside, Riverside, CA 92521-0138, USA* (Ch. 4)

N. Balakrishnan, *Department of Mathematics and Statistics, McMaster University, 1280 Main Street West, Hamilton, Ontario, Canada, L8S 4K1* (Chs. 1, 2, 3, 7, 15, 19)

P. J. Boland, *Department of Statistics, University College Dublin, Belfield, Dublin 4, Ireland* (Ch. 5)

A. Childs, *Department of Mathematics and Statistics, McMaster University, 1280 Main Street West, Hamilton, Ontario, Canada L8S 4K1* (Ch. 15)

M. Csörgő, *Department of Mathematics and Statistics, Carleton University, 4302 Herzberg Building, 1125 Colonel By Drive, Ottawa, Ontario, Canada K1S 5B6* (Ch. 21)

H. A. David, *Department of Statistics, Snedecor Hall, Iowa State University, Ames, IA 50011-1210, USA* (Ch. 18)

J. Galambos, *Department of Mathematics, TU 038-16, Temple University, Philadelphia, PA 19122, USA* (Ch. 11)

U. Gather, *Fachbereich Statistik, Lehrstuhl Mathematische Statistik und Industrielle Anwendungen, Univ. Dortmund, Vogelpothsweg 87, D-44221 Dortmund, Germany* (Ch. 9)

H. L. Harter, *203 N. Mckinley Avenue, Champaign, IL 61821-3251, USA* (Ch. 2)

D. R. Jensen, *Virginia Polytechnic Inst. and State University, Blacksburg, VA 24061, USA* (Ch. 14)

U. Kamps, *Institute of Statistics, Aachen Institute of Technology, D-52056 Aachen, Germany* (Ch. 9)

C. Ma, *School of Mathematics and Statistics, The University of Sydney, Sydney, NSW 2006, Australia* (Ch. 17)

H. N. Nagaraja, *Department of Statistics, Ohio State University, 1958 Neil Avenue, Columbus, OH 43210, USA* (Ch. 18)

V. B. Nevzorov, *Department of Mathematics and Mechanics, St. Petersburg State University, Bibliotchnaya Square 2, St. Petersburg 198904, Russia* (Ch. 19)

D. E. Ramirez, *Virginia Polytechnic Inst. and State University, Blacksburg, VA 24061, USA* (Ch. 14)

C. R. Rao, *The Pennsylvania State University, Center for Multivariate Analysis, Dept. of Statistics, 325 Classroom Bldg., University Park, PA 16802-6105, USA* (Chs. 1, 8)

J. Robinson, *School of Mathematics and Statistics, The University of Sydney, Sydney, NSW 2006, Australia* (Ch. 17)

T. Rychlik, *Lansjerów 1/1/34, 85617 Bydgoszez, Poland* (Ch. 6)

N. Schweitzer, *Institute of Statistics, Aachen Institute of Technology, D-52056 Aachen, Germany* (Ch. 9)

P. K. Sen, *Department of Statistics, University of North Carolina, Chapel Hill, NC 27599-3260, USA* (Ch. 12)

M. Shaked, *Department of Mathematics, University of Arizona, Tucson, AZ 85721, USA* (Ch. 5)

D. N. Shanbhag, *School of Mathematics and Statistics, University of Sheffield, Sheffield, Yorkshire S3 7RH, UK* (Ch. 8)

J. G. Shanthikumar, *School of Business Admin., University of California at Berkeley, Berkeley, CA 94720, USA* (Ch. 5)

R. L. Strawderman, *Department of Biostatistics, University of Michigan, 1420 Washington Heights, Ann Arbor, MI 48109-2029, USA* (Ch. 16)

K. S. Sultan, *Department of Mathematics, Al-Azhar Univ., Nasr City, Cairo 11884, Egypt* (Ch. 7)

B. Szyszkowicz, *Department of Mathematics and Statistics, Carleton University, 4302 Herzberg Building, 1125 Colonel By Drive, Ottawa, Ontario, Canada K1S 5B6* (Chs. 20, 21)

P. R. Tadikamalla, *Joseph M. Katz Graduate School of Business, University of Pittsburgh, Pittsburgh, PA 15260, USA* (Ch. 3)

R. J. Tomkins, *Department of Mathematics and Statistics, University of Regina, Regina, Saskatchewan, Canada S4S 0A2* (Ch. 13)

J. A. Villaseñor *Colegio de Postgraduados, ISEI, Carr Mexico-Texcoco km 35.5, Montecillo-Texcoco Edo. de Mex, Mexico CP 56230* (Ch. 4)

H. Wang, *Department of Mathematics and Statistics, Memorial University of Newfoundland, St. John's, Newfoundland, Canada A1C 5S7* (Ch. 13)

D. Zelterman, *Department of Biostatistics, Yale University, 208034 Yale Station, New Haven, CT 06520, USA* (Ch. 16)

PART I

Introduction and Basic Properties

Order Statistics: An Introduction

N. Balakrishnan and C. R. Rao

1. Introduction

Let $X_1, X_2, \ldots, X_n$ be n random variables. Then the corresponding order statistics are obtained by arranging these n X_i's in nondecreasing order, and are denoted by $X_{1:n}, X_{2:n}, \ldots, X_{n:n}$. Here, $X_{1:n}$ is the first order statistic denoting the smallest of the X_i's, $X_{2:n}$ is the second order statistic denoting the second smallest of the X_i's, ..., and $X_{n:n}$ is the nth order statistic denoting the largest of the X_i's. It is important to mention here that though this notation for order statistic is used by most authors, some other notations are also employed in the literature. For example, some authors use $X_{(i)}$ to denote the ith order statistic in a sample of size n. In situations where the sample size does not change, this notation will obviously cause no confusion. Some authors also use $X_{i,n}$ or $X_{n:i}$ or $X_{(i)n}$ to denote the ith order statistic. Because all these notations are prevalent in literature, we have deliberately allowed all these notations in the volume. Since any notation used by the author(s) in any specific chapter will remain consistent, and hence will not cause any confusion to the readers, we have left the notation exactly as used by all the authors.

Observe that the above definition of order statistics required neither the X_i's to be independent nor the X_i's to be identically distributed. However, a major portion of the remarkably large body of literature on order statistics has focussed on the case when the X_i's are independently and identically distributed. Of course, the common distribution may be continuous or discrete. Most of the early work on order statistics dealt with the continuous case assuming a probability density function $f(x)$ and a cumulative distribution function $F(x)$.

Developments on order statistics until 1962 were synthesized in the edited volume of Sarhan and Greenberg (1962) which also contained many valuable tables. The two volumes prepared by Harter (1970a,b) presented numerous tables which facilitate the use of order statistics in tests of hypotheses and in estimation methods based on complete as well as doubly Type-II censored samples from many different life-time distributions of interest. These two volumes have been recently revised and expanded by Harter and Balakrishnan (1996, 1997). An encyclopedic treatise on order statistics has been given by David (1970, 1981), while Arnold, Balakrishnan and Nagaraja (1992) presented a textbook on order statistics at an introductory level. Galambos (1978, 1987) prepared a volume dealing primarily with the asymptotic theory of extreme order statistics; applications of this extreme value theory in engineering problems have been highlighted in the volume by Castillo (1988). An excellent survey of all the developments concerning outliers has been made by Barnett and Lewis (1978, 1984, 1993). Arnold and Balakrishnan (1989) synthesized in their monograph the recurrence relations, bounds and approximations for order statistics. Finally, the book by Balakrishnan and Cohen (1991) elaborated various methods of estimation based on complete and censored samples. Naturally, these volumes should be consulted if one wishes to get an authoritative treatment to any of the topics mentioned above.

In this Chapter, we simply give an elementary introduction to order statistics which should be regarded as "a bare essential description" on the topic that would facilitate the reader to follow all the other chapters present in this handbook. We present the marginal distributions, joint distributions, moments and product moments of order statistics. We also present brief details on bounds and approximations for order statistics, exact distribution results for some specific distributions, asymptotic results, and on few related statistics such as concomitant order statistics and record values.

2. Marginal distributions of order statistics

The cumulative distribution function of $X_{i:n}$ ($1 \leq i \leq n$) is given by

$$F_{i:n}(x) = \Pr(X_{i:n} \leq x) = \sum_{r=i}^{n} \binom{n}{r} \{F(x)\}^r \{1 - F(x)\}^{n-r}$$

$$= \int_0^{F(x)} \frac{n!}{(i-1)!(n-i)!} t^{i-1} (1-t)^{n-i} \, dt \qquad (2.1)$$

for $-\infty < x < \infty$. Specifically, we find from (2.1) the cumulative distribution functions of $X_{1:n}$ and $X_{n:n}$ to be

$$F_{1:n}(x) = 1 - \{1 - F(x)\}^n, \qquad -\infty < x < \infty, \qquad (2.2)$$

and

$$F_{n:n}(x) = \{F(x)\}^n, \qquad -\infty < x < \infty. \qquad (2.3)$$

Instead of writing the cumulative distribution in a binomial form [as in (2.1)], one may express it in a negative binomial form as [see Pinsker, Kipnis and Grechanovsky (1986)]

$$F_{i:n}(x) = \sum_{r=0}^{n-i} \binom{n-1-r}{i-1} \{F(x)\}^i \{1 - F(x)\}^{n-i-r}, \qquad -\infty < x < \infty.$$

$$(2.4)$$

Observe that all the expressions given above hold for any arbitrary population whether continuous or discrete. For discrete populations, the probability mass function of $X_{i:n}$ $(1 \leq i \leq n)$ may be obtained from (2.1) by differencing as

$$f_{i:n}(x) = \Pr(X_{i:n} = x) = F_{i:n}(x) - F_{i:n}(x-)$$
$$= \int_{F(x-)}^{F(x)} \frac{n!}{(i-1)!(n-i)!} t^{i-1}(1-t)^{n-i} \, dt \ . \tag{2.5}$$

In particular, we also have

$$f_{1:n}(x) = \{1 - F(x-)\}^n - \{1 - F(x)\}^n \ , \tag{2.6}$$

and

$$f_{n:n}(x) = \{F(x)\}^n - \{F(x-)\}^n \ . \tag{2.7}$$

The review paper by Nagaraja (1992) lucidly accounts all the developments on discrete order statistics.

On the other hand, if the population is absolutely continuous, then the probability density function of $X_{i:n}$ can be obtained from (2.1) by differentiation as

$$f_{i:n}(x) = \frac{n!}{(i-1)!(n-i)!} \{F(x)\}^{i-1}\{1 - F(x)\}^{n-i} f(x), \quad -\infty < x < \infty \ . \tag{2.8}$$

In particular, we have

$$f_{1:n}(x) = n\{1 - F(x)\}^{n-1} f(x), \quad -\infty < x < \infty \ , \tag{2.9}$$

and

$$f_{n:n}(x) = n\{F(x)\}^{n-1} f(x), \quad -\infty < x < \infty \ . \tag{2.10}$$

3. Joint distributions of order statistics

The joint distributions of order statistics can be similarly derived and will naturally look a lot more complicated. For example, the joint cumulative distribution function of $X_{i:n}$ and $X_{j:n}$ $(1 \leq i < j \leq n)$ can be shown to be

$$F_{i,j:n}(x_i, x_j) = F_{j:n}(x_j) \quad \text{for} \quad x_i \geq x_j$$
$$= \sum_{s=j}^{n} \sum_{r=i}^{s} \frac{n!}{r!(s-r)!(n-s)!} \{F(x_i)\}^r \tag{3.1}$$
$$\times \{F(x_j) - F(x_i)\}^{s-r} \{1 - F(x_j)\}^{n-s} \quad \text{for } x_i < x_j \ .$$

This expression holds for any arbitrary population whether continuous or discrete.

For discrete populations, the joint probability mass function of $X_{i:n}$ and $X_{j:n}$ $(1 \leq i < j \leq n)$ may be obtained from (3.1) by differencing as

$$\begin{aligned} f_{i,j:n}(x_i, x_j) &= \Pr(X_{i:n} = x_i,\ X_{j:n} = x_j) \\ &= F_{i,j:n}(x_i, x_j) - F_{i,j:n}(x_i-, x_j) \\ &\quad - F_{i,j:n}(x_i, x_j-) + F_{i,j:n}(x_i-, x_j-)\ . \end{aligned} \quad (3.2)$$

On the other hand, if the population is absolutely continuous, then the joint probability density function of $X_{i:n}$ and $X_{j:n}$ $(1 \leq i < j \leq n)$ can be obtained from (3.1) by differentiation as

$$\begin{aligned} f_{i,j:n}(x_i, x_j) &= \frac{n!}{(i-1)!(j-i-1)!(n-j)!}\{F(x_i)\}^{i-1}\{F(x_j) - F(x_i)\}^{j-i-1} \\ &\quad \times \{1 - F(x_j)\}^{n-j} f(x_i) f(x_j), \quad x_i < x_j\ . \end{aligned} \quad (3.3)$$

This expression may also be derived directly by starting with the joint density function of all n order statistics given by

$$f_{1,2,\ldots,n:n}(x_1, x_2, \ldots, x_n) = n! \prod_{i=1}^n f(x_i), \quad x_1 < x_2 < \cdots < x_n\ , \quad (3.4)$$

and integrating out all the other variables. In fact, from (3.4) we can obtain the joint density function of $X_{i_1:n}, X_{i_2:n}, \ldots, X_{i_k:n}$ $(1 \leq i_1 < i_2 < \cdots < i_k \leq n)$ by integration as

$$\begin{aligned} f_{i_1,\ldots,i_k:n}(x_{i_1}, \ldots, x_{i_k}) &= \frac{n!}{(i_1-1)!(i_2-i_1-1)!\cdots(n-i_k)!}\{F(x_{i_1})\}^{i_1-1} \\ &\quad \times \{F(x_{i_2}) - F(x_{i_1})\}^{i_2-i_1-1} \times \cdots \times \{1 - F(x_{i_k})\}^{n-i_k} \\ &\quad \times f(x_{i_1})f(x_{i_2})\cdots f(x_{i_k}), \\ &\quad -\infty < x_{i_1} < x_{i_2} < \cdots < x_{i_k} < \infty\ . \end{aligned} \quad (3.5)$$

This joint density function, particularly for the cases of $k = 3$ and 4, become very useful in developing Edgeworth approximate inference for some distributions; see, for example, the article by Balakrishnan and Gupta (1998) in the companion volume.

This joint distribution will be far more complicated in the discrete case due to the possibility of ties. The joint probability mass function of $X_{i_1:n}, X_{i_2:n}, \ldots, X_{i_k:n}$ $(1 \leq i_1 < i_2 < \cdots < i_k \leq n)$ in this case can be shown to be

$$\begin{aligned} &f_{i_1,i_2,\ldots,i_k:n}(x_{i_1}, x_{i_2}, \ldots, x_{i_k}) \\ &= \frac{n!}{(i_1-1)!(i_2-i_1-1)!\cdots(n-i_k)!} \int_B u_{i_1}^{i_1-1}(u_{i_2} - u_{i_1})^{i_2-i_1-1}\cdots \\ &\quad \times (u_{i_k} - u_{i_{k-1}})^{i_k-i_{k-1}-1}(1 - u_{i_k})^{n-i_k} du_{i_1}\cdots du_{i_k}\ , \end{aligned} \quad (3.6)$$

where B is the k-dimensional space given by

$$B = \{(u_{i_1}, \ldots, u_{i_k}): u_{i_1} \leq u_{i_2} \leq \cdots \leq u_{i_k}, \ F(x_r-) \leq u_r \leq F(x_r)$$
$$\text{for } r = i_1, i_2, \ldots, i_k\} \ .$$

For more details on discrete order statistics, one may refer to Chapter 3 of Arnold, Balakrishnan and Nagaraja (1992) and the review article by Nagaraja (1992).

4. Properties

Order statistics from an arbitrary distribution possess some interesting properties. It is important to mention here that, though there are marked similarities between order statistics from continuous and discrete distributions, some important properties satisfied by order statistics from continuous distributions do not hold for order statistics from discrete distributions.

For example, in the continuous case, it is well-known that order statistics form a Markov chain. On the other hand, order statistics from discrete distributions do not form a Markov chain in general and do so under some conditions. Reference may be made to Nagaraja (1986a,b) and Rüschendorf (1985) in this regard.

For a review of some results concerning various stochastic orderings connected with order statistics, refer to the articles by Arnold and Villaseñor (1998) and Boland, Shaked and Shanthikumar (1998) in this volume.

Due to the fact that the cumulative distributions of order statistics are tail probabilities of a binomial distribution, and that binomial tail probabilities form a log-concave sequence, the distribution functions of order statistics form a log-concave sequence. This result has been further generalized to the case when the order statistics arise not necessarily from an i.i.d. sample; see Balasubramanian and Balakrishnan (1993a).

5. Moments and product moments

Let us denote the single moments of order statistics, $E(X_{i:n}^k)$, by $\mu_{i:n}^{(k)}$, $1 \leq i \leq n$. Clearly, these moments can be determined in the continuous case by

$$\mu_{i:n}^{(k)} = \int_{-\infty}^{\infty} x^k f_{i:n}(x) \, dx \qquad (5.1)$$

$$= \frac{n!}{(i-1)!(n-i)!} \int_{-\infty}^{\infty} x^k \{F(x)\}^{i-1} \{1 - F(x)\}^{n-i} f(x) \, dx \ ,$$

and in the discrete case by

$$\mu_{i:n}^{(k)} = \sum_x x^k f_{i:n}(x), \qquad 1 \leq i \leq n \ . \qquad (5.2)$$

Similarly, let us denote the product moments of order statistics, $E(X_{i:n}^k X_{j:n}^\ell)$, by $\mu_{i,j:n}^{(k,\ell)}$ for $1 \leq i < j < n$. Once again, these moments can be determined in the continuous case by

$$\mu_{i,j:n}^{(k,\ell)} = \iint_{x<y} x^k y^\ell f_{i,j:n}(x,y)\, dx\, dy$$

$$= \frac{n!}{(i-1)!(j-i-1)!(n-j)!} \iint_{x<y} x^k y^\ell \{F(x)\}^{i-1} \{F(y)-F(x)\}^{j-i-1}$$
$$\times \{1-F(y)\}^{n-j} f(x) f(y)\, dx\, dy, \qquad (5.3)$$

and in the discrete case by

$$\mu_{i,j:n}^{(k,\ell)} = \sum\sum_{x\leq y} x^k y^\ell f_{i,j:n}(x,y), \qquad 1 \leq i < j \leq n. \qquad (5.4)$$

For convenience, let us denote $\mu_{i:n}^{(1)}$ by $\mu_{i:n}$ and $\mu_{i,j:n}^{(1,1)}$ by $\mu_{i,j:n}$.

The variances of order statistics, $\mathrm{Var}(X_{i:n})$, are denoted by $\sigma_{i,i:n}$ ($1 \leq i \leq n$), and can be determined as

$$\sigma_{i,i:n} = \mathrm{Var}(X_{i:n}) = \mu_{i:n}^{(2)} - \mu_{i:n}^2, \qquad 1 \leq i \leq n. \qquad (5.5)$$

Similarly, the covariances of order statistics, $\mathrm{Cov}(X_{i:n}, X_{j:n})$, are denoted by $\sigma_{i,j:n}$ for $1 \leq i < j \leq n$, and can be determined as

$$\sigma_{i,j:n} = \mathrm{Cov}(X_{i:n}, X_{j:n}) = \mu_{i,j:n} - \mu_{i:n}\mu_{j:n}, \qquad 1 \leq i < j \leq n. \qquad (5.6)$$

These moments can all be derived in explicit form in the case of some distributions such as uniform, exponential, Pareto, power function, logistic, extreme value and Weibull. However, they have to be determined by numerical methods in most other cases. Reference may be made to the handbook of tables by Harter and Balakrishnan (1996) for tables of means, variances and covariances of order statistics for numerous distributions. See also the papers by Basu and Singh (1998), Balakrishnan and Chan (1998), Balakrishnan and Aggarwala (1998), and Balakrishnan and Lee (1998) in the companion volume.

By starting with the joint density function of three, four, ... order statistics and performing integration or summation, we can similarly determine third-order, fourth-order, ... moments as well. For example, see the paper by Balakrishnan and Gupta (1998) in the companion volume.

6. Recurrence relations and identities

From the basic identity

$$\left(\sum_{i=1}^n X_{i:n}^k\right)^\ell = \left(\sum_{i=1}^n X_i^k\right)^\ell, \qquad (6.1)$$

several identities for single and product moments of order statistics can be established. For example, by choosing $\ell = 1$ and taking expectations on both sides, we get the identity

$$\sum_{i=1}^{n} \mu_{i:n}^{(k)} = n\mathrm{E}(X^k) = n\mu_{1:1}^{(k)} \ . \tag{6.2}$$

Similarly, by taking $k = 1$ and $\ell = 2$ and then using binomial expansion, we obtain

$$\sum_{i=1}^{n} X_{i:n}^2 + 2\sum_{i=1}^{n-1} \sum_{j=i+1}^{n} X_{i:n} X_{j:n} = \sum_{i=1}^{n} X_i^2 + 2\sum_{i=1}^{n-1} \sum_{j=i+1}^{n} X_i X_j \ .$$

Now taking expectations on both sides, we get

$$\sum_{i=1}^{n} \mu_{i:n}^{(2)} + 2\sum_{i=1}^{n-1} \sum_{j=i+1}^{n} \mu_{i,j:n} = n\mathrm{E}(X^2) + \frac{2n(n-1)}{2} \{\mathrm{E}(X)\}^2$$

which, when used with (6.2) and simplified, yields an identity for product moments of order statistics as

$$\sum_{i=1}^{n-1} \sum_{j=i+1}^{n} \mu_{i,j:n} = \binom{n}{2}\{\mathrm{E}(X)\}^2 = \binom{n}{2}\mu_{1:1}^2 \ . \tag{6.3}$$

These and several more identities are reviewed in the article by Balakrishnan and Sultan (1998) in this volume.

Another topic in which considerable amount of work has been carried out is in the derivation of recurrence relations. For example, by starting from (2.1), one can establish the *triangle rule* for single moments of order statistics from any arbitrary distribution given by

$$r\mu_{r+1:n}^{(k)} + (n-r)\mu_{r:n}^{(k)} = n\mu_{r:n-1}^{(k)} \ . \tag{6.4}$$

A similar recurrence relation for the product moments of order statistics from any arbitrary distribution is given by

$$(r-1)\mu_{r,s:n}^{(k,\ell)} + (s-r)\mu_{r-1,s:n}^{(k,\ell)} + (n-s+1)\mu_{r-1,s-1:n}^{(k,\ell)} = n\mu_{r-1,s-1:n}^{(k,\ell)} \ . \tag{6.5}$$

These and many other recurrence relations satisfied by the product moments of order statistics from any arbitrary distribution are reviewed in the article by Balakrishnan and Sultan (1998) in this volume. These authors, in addition, also review recurrence relations satisfied by the single and the product moments of order statistics from several specific continuous distributions. The monograph by Arnold and Balakrishnan (1989) also provides an elaborate discussion on this topic.

In this context, it is important to mention here a duality result satisfied by recurrence relations and identities among order statistics established by Balasubramanian and Balakrishnan (1993b) in the most general case when X_i's are jointly arbitrarily distributed.

7. Bounds

For any continuous distribution with mean μ and variance σ^2, Hartley and David (1954) and Gumbel (1954) have shown that the universal bound for the mean of $X_{n:n}$ is given by

$$E(X_{n:n}) \leq \mu + \sigma(n-1)/\sqrt{2n-1}, \qquad n = 2, 3, \ldots, \tag{7.1}$$

and that this bound is attainable if and only if the population inverse cumulative distribution function is

$$F^{-1}(u) = \mu + \sigma \frac{\sqrt{2n-1}}{n-1}(nu^{n-1} - 1), \qquad 0 < u < 1. \tag{7.2}$$

Similarly, for any continuous symmetric distribution with mean μ and variance σ^2, a tighter universal bound can be obtained for the mean of $X_{n:n}$ and is given by

$$E(X_{n:n}) \leq \mu + \frac{\sigma n}{\sqrt{2(2n-1)}} \left\{ 1 - \frac{1}{\binom{2n-2}{n-1}} \right\}^{1/2}, \qquad n = 2, 3, \ldots. \tag{7.3}$$

This bound is attainable if and only if the population inverse cumulative distribution function is given by

$$F^{-1}(u) = \mu + \frac{\sigma}{\sqrt{2}} \left\{ \frac{2n-1}{1 - \binom{2n-2}{n-1}^{-1}} \right\}^{1/2} \{u^{n-1} - (1-u)^{n-1}\}, \qquad 0 < u < 1. \tag{7.4}$$

Several extensions and generalizations of these results are available in the literature. There are also numerous other bounds for moments of order statistics as well as for some functions of order statistics. For a review of this topic, one may refer to David (1981) and Arnold and Balakrishnan (1989); see also the review article by Rychlik (1998) in this volume.

8. Approximations

Since, as mentioned earlier, the moments of order statistics are hard to compute for many distributions and particularly so for large sample sizes, it becomes highly desirable to develop some approximations for these quantities. One such approximation is due to David and Johnson (1954) and is based on the fact that

$$X_{i:n} \stackrel{d}{=} F^{-1}(U_{i:n}), \tag{8.1}$$

where $U_{i:n}$ denotes the ith order statistic in a sample of size n from the uniform(0,1) distribution. Now, upon expanding $F^{-1}(U_{i:n})$ in (8.1) in a Taylor series around the value $E(U_{i:n}) = p_i = \frac{i}{n+1} = 1 - q_i$, and making use of the expressions for the central moments of uniform order statistics, we can develop the necessary series approximations for the single and the product moments of order statistics $X_{i:n}$. For example, we get the series approximation for $\mu_{i:n}$ as

$$\begin{aligned}\mu_{i:n} = {} & F^{-1}(p_i) + \frac{p_i q_i}{2(n+2)} F^{-1^{(2)}}(p_i) \\ & + \frac{p_i q_i}{(n+2)^2} \left\{ \tfrac{1}{3}(q_i - p_i) F^{-1^{(3)}}(p_i) + \tfrac{1}{8} p_i q_i F^{-1^{(4)}}(p_i) \right\} \\ & + \frac{p_i q_i}{(n+2)^3} \left[-\tfrac{1}{3}(q_i - p_i) F^{-1^{(3)}}(p_i) + \tfrac{1}{4}\{(q_i - p_i)^2 - p_i q_i\} F^{-1^{(4)}}(p_i) \right. \\ & \left. + \tfrac{1}{6} p_i q_i (q_i - p_i) F^{-1^{(5)}}(p_i) + \tfrac{1}{48} p_i^2 q_i^2 F^{-1^{(6)}}(p_i) \right] ,\end{aligned} \qquad (8.2)$$

where $F^{-1^{(1)}}(p_i)$, $F^{-1^{(2)}}(p_i)$, $F^{-1^{(3)}}(p_i)$, ... are the successive derivatives of $F^{-1}(u)$ evaluated at $u = p_i$.

David and Johnson (1954) have given similar series approximations for the first four cumulants and cross-cumulants of order statistics from an arbitrary continuous distribution $F(\cdot)$. Note that these approximations are all of order $O(1/n^3)$. Recently, Childs and Balakrishnan (1998a) have given a Maple program which can be used to derive such series approximations up to any specified order by accounting for as many terms as desired in the underlying Taylor series expansion in (8.1).

Some other methods of approximation are also available in the literature. One may refer to David (1981) and Arnold and Balakrishnan (1989) for elaborate reviews of these developments.

9. Characterizations

Over the years, several interesting characterizations of distributions have been established based on some properties of order statistics. The earliest result in this connection is the one by Hoeffding (1953) which states that the entire triangular array of means of order statistics, $\{\mu_{i:n}: 1 \leq i \leq n, n = 1, 2, \ldots\}$, characterizes the parent distribution. Chan (1967) established the same result with the array of just means of extreme order statistics. Several refinements on this result have been made and a good survey of all these results has been made by Huang (1989). There are numerous other characterization results based on different properties of order statistics and of some statistics based on order statistics such as spacings and records. For an authoritative treatment on all these developments, one may refer to Kagan, Linnik and Rao (1973), Galambos and Kotz (1978), Chapter 6 of Arnold, Balakrishnan and Nagaraja (1992), and Rao and Shanbhag (1993). The review articles by Gather, Kamps and Schweitzer (1998), Kamps (1998) and Rao and Shanbhag (1998) in this volume all provide updated reviews on different characterization problems involving order statistics.

10. Asymptotics

Following the early works of Fréchet (1927), Fisher and Tippett (1928) and von Mises (1936), Gnedenko (1943) laid a rigorous foundation for the limiting behavior of extremes by providing specifically necessary and sufficient conditions for the weak convergence of the extremes.

Assume that there exist sequences $\{a_n\}$ and $\{b_n > 0\}$ such that

$$\lim_{n \to \infty} F^n(a_n + b_n x) = G(x) \tag{10.1}$$

at all continuity points of $G(x)$. If such sequences $\{a_n\}$ and $\{b_n\}$ exist, then the distribution F is said to belong to the *domain of maximal attraction* of the nondegenerate distribution G. Under the condition in (10.1), the limiting cumulative distribution function G of the largest order statistic (appropriately normalized) is one of the following three types:

$$G_1(x) = e^{-x^{-\alpha}}, \quad 0 < x < \infty, \ \alpha > 0 \ ,$$
$$G_2(x) = e^{-(-x)^{\alpha}}, \quad -\infty < x < 0, \ \alpha > 0 \ ,$$

and

$$G_3(x) = e^{-e^{-x}}, \quad -\infty < x < \infty \ . \tag{10.2}$$

G_1, G_2 and G_3 are referred to as the Fréchet, Weibull and extreme value distributions, respectively.

Similarly, if there exist sequences $\{a_n^*\}$ and $\{b_n^* > 0\}$ such that $(X_{1:n} - a_n^*)/b_n^*$ converges in distribution to a nondegenerate random variable, then its distribution function is one of the following three types:

$$G_1^*(x) = 1 - e^{-(-x)^{-\alpha}}, \quad -\infty < x < 0, \ \alpha > 0 \ ,$$
$$G_2^*(x) = 1 - e^{-x^{\alpha}}, \quad 0 < x < \infty, \ \alpha > 0 \ ,$$

and

$$G_3^*(x) = 1 - e^{-e^{x}}, \quad -\infty < x < \infty \ . \tag{10.3}$$

There are numerous sophisticated results dealing with the asymptotic behavior of order statistics and of some functions of order statistics. For an elaborate treatment on this subject, interested readers may refer to Galambos (1987), Resnick (1987), Leadbetter, Lindgren and Rootzén (1983), Castillo (1988), Reiss (1989), Serfling (1980), and Shorack and Wellner (1986). The articles by Galambos (1998) and Sen (1998) in this volume will provide additional and up-to-date details on this topic.

11. Best linear unbiased estimation and prediction

Let X be the vector of order statistics obtained from either a complete sample of size n or from a Type-II censored sample obtained from n units. Then, if the underlying family of distributions is a scale-parameter family of distributions with density function $\frac{1}{\sigma} f(\frac{x}{\sigma})$, the Best Linear Unbiased Estimator (BLUE) of σ is given by

$$\sigma^* = \frac{\mu^T \Sigma^{-1}}{\mu^T \Sigma^{-1} \mu} X , \qquad (11.1)$$

where μ denotes the vector of means and Σ the matrix of variances and covariances of order statistics from the corresponding standard distribution. Further, the variance of this BLUE is given by

$$\text{Var}(\sigma^*) = \sigma^2/(\mu^T \Sigma^{-1} \mu) . \qquad (11.2)$$

Similarly, if X is the vector of order statistics obtained from a location-scale family of distributions with density function $\frac{1}{\sigma} f(\frac{x-\mu}{\sigma})$, then the BLUEs of μ and σ are given by [Lloyd (1952)]

$$\mu^* = -\frac{\mu^T \Sigma^{-1} (\mathbf{1}\mu^T - \mu\mathbf{1}^T) \Sigma^{-1}}{(\mu^T \Sigma^{-1} \mu)(\mathbf{1}^T \Sigma^{-1} \mathbf{1}) - (\mu^T \Sigma^{-1} \mathbf{1})^2} X \qquad (11.3)$$

and

$$\sigma^* = \frac{\mathbf{1}^T \Sigma^{-1} (\mathbf{1}\mu^T - \mu\mathbf{1}^T) \Sigma^{-1}}{(\mu^T \Sigma^{-1} \mu)(\mathbf{1}^T \Sigma^{-1} \mathbf{1}) - (\mu^T \Sigma^{-1} \mathbf{1})^2} X , \qquad (11.4)$$

where $\mathbf{1}$ is a column vector of 1's of the same dimension as X. Furthermore, the variances and covariances of these BLUEs are given by

$$\text{Var}(\mu^*) = \frac{\sigma^2(\mu^T \Sigma^{-1} \mu)}{(\mu^T \Sigma^{-1} \mu)(\mathbf{1}^T \Sigma^{-1} \mathbf{1}) - (\mu^T \Sigma^{-1} \mathbf{1})^2} , \qquad (11.5)$$

$$\text{Var}(\sigma^*) = \frac{\sigma^2(\mathbf{1}^T \Sigma^{-1} \mathbf{1})}{(\mu^T \Sigma^{-1} \mu)(\mathbf{1}^T \Sigma^{-1} \mathbf{1}) - (\mu^T \Sigma^{-1} \mathbf{1})^2} , \qquad (11.6)$$

and

$$\text{Cov}(\mu^*, \sigma^*) = \frac{-\sigma^2(\mu^T \Sigma^{-1} \mathbf{1})}{(\mu^T \Sigma^{-1} \mu)(\mathbf{1}^T \Sigma^{-1} \mathbf{1}) - (\mu^T \Sigma^{-1} \mathbf{1})^2} . \qquad (11.7)$$

These formulas may be used, along with the explicit expressions for the means, variances and covariances of order statistics from some specific distributions such as the uniform, exponential, power function and Pareto, in order to derive the BLUEs of the parameters of these distributions in an explicit form. In all other cases, the formulas in (11.3)–(11.7) may be used to determine the BLUEs by means of numerical methods. The papers by Balakrishnan and Aggarwala (1998), Balakrishnan and Chan (1998), and Balakrishnan and Lee (1998) in the companion volume all provide examples for this case.

Optimal properties of BLUEs and their large-sample approximations are all discussed in great length by Balakrishnan and Cohen (1991). Some other efficient linear estimation methods based on order statistics are also discussed by these authors. The papers by Sarkar and Wang (1998), Ali and Umbach (1998), Alimoradi and Saleh (1998), and Hosking (1998) in the companion volume all present additional details on this topic. Balakrishnan and Rao (1997, 1998) have recently established that the BLUEs are also trace-efficient as well as determinant-efficient linear unbiased estimators, and that they have complete covariance matrix dominance in the class of all linear unbiased estimators.

Having observed the first r order statistics, $(X_{1:n}, \ldots, X_{r:n})$, from a sample of size n, the Best Linear Unbiased Predictor (BLUP) of $X_{s:n}$ $(s > r)$ is given by [Goldberger (1962)]

$$X_{s:n}^* = \mu^* + \sigma^* \mu_{s:n} + \omega^T \Sigma^{-1}(X - \mu^* \mathbf{1} - \sigma^* \mu) \;, \tag{11.8}$$

where μ^* and σ^* are the BLUEs of μ and σ based on the first r order statistics, and $\omega^T = (\sigma_{1,s:n}, \ldots, \sigma_{r,s:n})$. Once again, this formula may be used along with the explicit expressions for the means, variances and covariances of order statistics from some specific distributions such as the uniform, exponential, power function and Pareto, in order to derive the BLUPs in an explicit form. For example, in the case of the exponential distribution with location parameter μ and scale parameter σ, the BLUP of $X_{s:n}$ in (11.8) reduces to

$$X_{s:n}^* = X_{r:n} + \left(\sum_{i=r+1}^{s} \frac{1}{n-i+1} \right) \sigma^* \;, \tag{11.9}$$

where

$$\sigma^* = \frac{1}{(r-1)} \sum_{i=2}^{r} (n-i+1)(X_{i:n} - X_{i-1:n}) \tag{11.10}$$

is the BLUE of σ based on the first r order statistics. In most other cases, the BLUP may be determined by employing numerical methods in the formula in (11.8).

Recently, Doganaksoy and Balakrishnan (1997) showed an interesting and useful connection between the BLUEs and the BLUPs. For an elaborate review on the prediction of order statistics, one may refer to the article by Kaminsky and Nelson (1998) in the companion volume.

12. Inference under censoring

In many life-testing experiments, it is quite common not to observe complete data and only to observe some form of censored data. This may be based on cost and/or time considerations. Of course, there are many kinds of censoring possible as described briefly below:

Type-I censoring. Suppose n identical units are placed on a life test, and the experimenter decides to observe all the failures up to pre-fixed time T and then stop the experimentation at time T with possibly some units still surviving. Here, the number of failures is a random variable (actually a binomial random variable), and this type of censoring is called *Type-I censoring*. More precisely, it is called Type-I right censoring as the censoring occurs only on the right side of the sample. However, it is easy to introduce Type-I censoring on both sides of the sample by assuming that the experimenter will observe all units that fail only in the pre-fixed time interval (T_L, T_U).

Type-II censoring. Suppose n identical units are placed on a life-test, and the experimenter decides to observe only a pre-fixed number of failures, say $n - s$, and then stop the experiment as soon as the $(n - s)^{\text{th}}$ failure occurs (thus censoring the last s units still surviving). Here, the number of failures is fixed but the termination time is random (actually the $(n - s)^{\text{th}}$ order statistic from a sample of size n), and this type of censoring is called *Type-II censoring*. Once again, this is only Type-II right censoring. But, it is easy to introduce Type-II censoring on both sides of the sample by assuming that the experimenter will not observe the first r items to fail and also the last s items still surviving.

Hybrid Censoring. Instead of choosing either Type-I or Type-II censoring, the experimenter may sometimes find it more convenient to combine the two forms of censoring and adopt a censoring method as follows. Pre-fix a time T and also the number of failures as $n - s$. Then, the experimenter may observe only $n - s$ failures if they all occur before time T and then terminate the experiment immediately after the $(n - s)^{\text{th}}$ failure; or observe all failures $(< n - s)$ that occur until time T and terminate the experiment at the pre-fixed time T. Thus, the time of termination in this case is $\min\{X_{n-s:n}, T\}$. This method of censoring is called *Hybrid censoring*.

Random Censoring. Suppose n subjects are placed on a life-test and with the ith subject there is a random censoring time C_i. In this case, the experimenter will observe $X_i = \min(Y_i, C_i)$, for $i = 1, \ldots, n$, where C_i is the censoring time and Y_i is the failure time. It is common to assume that Y_i and C_i are independent, but it is not necessary. Hence, each life-time is censored by an independent time and the experimenter will have the information whether the observation is the actual failure time or its censoring time. This type of censoring, which is commonly encountered in clinical trials, is called *Random censoring*.

Progressive Censoring. Suppose n identical units are placed on a life-test, and the experimenter decides to observe only m failures and censor the remaining $n - m$ units progressively as follows: At the time of the first failure, R_1 of the $n - 1$ surviving units are randomly removed from the life-test; at the time of the next failure, R_2 of the $n - 2 - R_1$ surviving units are randomly removed from the life-test, and so on; finally, at the time of the mth failure, all the remaining

$n - m - R_1 - \cdots - R_{m-1}$ surviving units are removed from the life-test. This type of censoring is called *Progressive censoring*. More precisely, this is called Progressive Type-II censoring. But, it is easy to introduce Progressive Type-I censoring by pre-fixing the times of censoring.

No matter what type of censoring is adopted by the experimenter, it is easy to construct the corresponding likelihood function and derive the maximum likelihood estimates of the parameters of the underlying life-time distribution. For example, in the case of Type-II right censoring described above, the likelihood function is given by

$$L = \frac{n!}{s!} \left\{ \prod_{i=1}^{n-s} f(x_{i:n}) \right\} \{1 - F(x_{n-s:n})\}^s, \quad x_{1:n} < \cdots < x_{n-s:n},$$

with the first part corresponding to the $n - s$ failures observed and the second part corresponding to the s censored units. Similarly, in the case of Progressive Type-II censoring described above, the likelihood function is

$$L = n(n - R_1 - 1) \cdots (n - R_1 - \cdots - R_{m-1} - m + 1)$$
$$\times \prod_{i=1}^{m} \left\{ f(x_i)\{1 - F(x_i)\}^{R_i} \right\}, \quad x_1 < x_2 < \cdots < x_m.$$

The maximum likelihood estimation of parameters based on all these different types of censored data has been discussed by numerous authors; see, for example, the books by Mann, Schafer and Singpurwalla (1974), Bain (1978), Nelson (1982), Lawless (1982), Cohen and Whitten (1988), Cohen (1991), Bain and Engelhardt (1991), Balakrishnan and Cohen (1991), and Harter and Balakrishnan (1996).

13. Results for some specific distributions

Though the distributional properties of order statistics and some statistics based on them are quite complicated and are impossible to derive in an explicit form in the case of most distributions, they do take on a nice and pleasing form in the case of some distributions like exponential and uniform. For example, if the order statistics $X_{1:n}, \ldots, X_{n:n}$ are from the standard exponential distribution, then the normalized spacings defined by

$$Z_1 = nX_{1:n}, \; Z_2 = (n-1)(X_{2:n} - X_{1:n}), \ldots, \quad (13.1)$$
$$Z_i = (n - i + 1)(X_{i:n} - X_{i-1:n}), \ldots, \; Z_n = X_{n:n} - X_{n-1:n}$$

are all i.i.d. standard exponential random variables [Sukhatme (1937)]. From (13.1), it readily follows that

$$X_{i:n} \stackrel{d}{=} \sum_{j=1}^{i} Z_j/(n - j + 1). \quad (13.2)$$

These results have found very important applications in developing exact inferential methods for the exponential distribution. For these and some other properties connected with order statistics from the exponential distribution, one may refer to the article by Basu and Singh (1998) in the companion volume.

Next, let us consider the order statistics $U_{1:n}, \ldots, U_{n:n}$ from the uniform(0,1) distribution. In this case, it can be shown that the random variables

$$V_1 = \frac{U_{1:n}}{U_{2:n}}, \; V_2 = \left(\frac{U_{2:n}}{U_{3:n}}\right)^2, \ldots, V_{n-1} = \left(\frac{U_{n-1:n}}{U_{n:n}}\right)^{n-1}, \; V_n = U_{n:n}^n \qquad (13.3)$$

are all i.i.d. uniform(0,1) random variables [Malmquist (1950)]. From (13.3), it readily follows that

$$U_{i:n} = \prod_{j=i}^{n} V_j^{1/j} . \qquad (13.4)$$

Once again, these results have found very important applications in developing exact inferential methods for the uniform distribution. This result has also been utilized to develop some efficient simulation algorithms for the machine generation of uniform order statistics; see the article by Tadikamalla and Balakrishnan (1998) in this volume.

If the order statistics $X_{1:n}, \ldots, X_{n:n}$ are from the normal $N(\mu, \sigma^2)$ population, then it can be shown that the statistic

$$T = \sum_{i=1}^{n} c_i X_{i:n} / S \qquad (13.5)$$

is independent of both $\overline{X}$ and S, where $\overline{X}$ denotes the sample mean, S^2 denotes the sample variance, and $\sum_{i=1}^{n} c_i = 0$. This result has found significant applications in the tests for potential outliers in normal samples; for example, see Barnett and Lewis (1993). Another interesting property of order statistics in this case is that the sum of the elements in any row or column of the variance-covariance matrix of all n order statistics is σ^2. In other words, when $\sigma^2 = 1$, the variance-covariance matrix is doubly stochastic. For more details on properties of order statistics from various distributions, one may refer to David (1981) and Arnold, Balakrishnan and Nagaraja (1992).

14. Outliers and robust inference

Order statistics enter into the areas of *Outliers* and *Robust Inference* in a very natural manner since outliers in statistical data are often expected to be a few extreme order statistics. Hence, test statistics based on extreme order statistics that measure the amount of departure of these observations from the rest of the data are used effectively to test whether those extreme order statistics are indeed outliers or not; see, for example, Barnett and Lewis (1993) for an exhaustive treatment on this topic. The article by Jensen and Ramirez (1998) in this volume is a good example of this kind as it illustrates the role of Cook's distance in regression problems.

Due to the suspicion (and also premise) that a few extreme order statistics are indeed the outliers, robust estimates and robust inferential methods can be produced by either trimming or downweighing those extreme order statistics. Examples of this kind of estimates include trimmed means, Winsorized means, and M-estimates. Their properties and usage in developing various inferential procedures form the area of *Robust Inference*. A good review of this topic may be found in the books by Andrews et al. (1972), Huber (1981), and Tiku, Tan and Balakrishnan (1986). The article by Childs and Balakrishnan (1998b) in this volume discusses the moments of order statistics arising from a Pareto sample containing multiple outliers and uses them to examine the bias and mean square error of some linear estimators of location and scale parameters under the presence of multiple outliers, and then presents some robust estimators.

15. Goodness-of-fit tests

There are many subjective as well as objective statistical methods available in the literature for checking the assumption of a particular distribution for the data at hand, and most of these make use of order statistics either explicitly or implicitly. An excellent source for this area of research is the book by D'Agostino and Stephens (1986).

One of the simplest and most commonly used methods is based on $Q - Q$ plots. All this does is to plot the sample order statistics $X_{i:n}$ against the corresponding expected values (from the standard distribution) $\mu_{i:n}$. A near straight line in the plot will suggest that the model assumed for the data is an appropriate one. While this is a subjective method, one could easily propose an objective method from it by using the correlation coefficient between the two sets of values as a test statistic.

Another well-known example for the use of order statistics in goodness-of-fit tests is Shapiro and Wilk's (1965) test for normality. These authors used the ratio of the best linear unbiased estimator of σ (the population standard deviation) and the sample standard deviation S as a test statistic. The empirical distribution function statistic is another classic example illustrating the use of order statistics in goodness-of-fit tests. The papers by Lockhart and Stephens (1998) and Shapiro (1998) in the companion volume will provide additional valuable details in this direction.

16. Related statistics

Let $X_1, X_2, \ldots$ be a sequence of i.i.d. random variables with common distribution function F. An observation X_j is called an *upper record* if it is at least as large as the maximum of all preceding observations. Similarly, an observation X_j is called a *lower record* if it is at most as large as the minimum of all preceding observations. These are referred to as *classical records* in the literature. In the case of absolutely continuous distributions, it is easy to show that for the nth upper record value R_n,

$$\Pr(R_n \geq r) = \{1 - F(r)\} \sum_{i=0}^{n} \{-\ln(1 - F(r))\}^k / k! \ , \qquad (16.1)$$

from which the density function of R_n can be readily obtained to be

$$f_{R_n}(r) = f(r)\{-\log(1 - F(r))\}^n / n! \ . \qquad (16.2)$$

Similar expressions can be given for the lower record values as well. Glick (1978), Nagaraja (1988), Nevzorov (1987), Arnold and Balakrishnan (1989), Arnold, Balakrishnan and Nagaraja (1992, 1998) and Ahsanullah (1995) all provide excellent reviews of all the developments on the theory and applications of records and related statistics.

Instead of keeping track of the largest X yet seen, one may keep track of the kth largest X yet seen which then gives rise to the so-called k^{th} *record values* [Grudzien and Szynal (1985)]. There are many other record models, extensions and generalizations proposed in the literature. The article by Nevzorov and Balakrishnan (1998) in this volume provides an exhaustive up-to-date review on the subject of *Records*.

Suppose $(X_1, Y_1), \ldots, (X_n, Y_n)$ are n i.i.d. bivariate observations from the distribution $F(x,y)$. Further, suppose the pairs are ordered by their X variates and the order statistics of X are denoted as usual by $X_{i:n}$, $1 \leq i \leq n$. Then, the Y variate associated with $X_{i:n}$ is called the *concomitant of the ith order statistic* and is usually denoted by $Y_{[i:n]}$ [David (1973)]. These concomitant order statistics become useful in some problems involving selection. The probability density function of $Y_{[i:n]}$ can be written as

$$f_{Y_{[i:n]}}(y) = \int_{-\infty}^{\infty} f(y|x) f_{i:n}(x) \, \mathrm{d}x \ , \qquad (16.3)$$

where $f_{i:n}(x)$ is the density function of the ith order statistic in a sample of size n from the marginal distribution of X. Similarly, the joint density function of $Y_{[i_1:n]}, \ldots, Y_{[i_k:n]}$ ($1 \leq i_1 < \cdots < i_k \leq n$) can be written as

$$f_{Y_{[i_1:n]}, \ldots, Y_{[i_k:n]}}(y_{i_1}, y_{i_2}, \ldots, y_{i_k})$$
$$= \int_{-\infty}^{\infty} \int_{-\infty}^{x_{i_k}} \cdots \int_{-\infty}^{x_{i_2}} \prod_{r=1}^{k} f(y_{i_r} | x_{i_r}) f_{i_1, \ldots, i_k : n}(x_{i_1}, \ldots, x_{i_k}) \, \mathrm{d}x_{i_1} \ldots \mathrm{d}x_{i_k} \ .$$
$$(16.4)$$

From this expression, for example, means, variances and covariances of concomitants of order statistics can all be determined.

Many distributional properties of concomitants and asymptotics have been studied. Inferential methods have been developed for the parameters of the underlying bivariate distribution. Some other models of introducing concomitants of order statistics have also been proposed in the literature. The article by David and Nagaraja (1998) in this volume reviews all these developments.

Kamps (1993) has introduced *generalized order statistics* which includes order statistics as well as record values as special cases.

17. Generalizations

Suppose X_i's are independent random variables distributed as F_i, $i = 1, 2, \ldots, n$. Then, the marginal density function of the ith order statistic $X_{i:n}$ can be written as [Vaughan and Venables (1972)]

$$f_{i:n}(x) = \frac{1}{(i-1)!(n-i)!} \left|^+ \begin{matrix} F_1(x) & \cdots & F_n(x) \\ \cdots & \cdots & \cdots \\ F_1(x) & \cdots & F_n(x) \\ f_1(x) & \cdots & f_n(x) \\ 1-F_1(x) & \cdots & 1-F_n(x) \\ \cdots & \cdots & \cdots \\ 1-F_x(x) & \cdots & 1-F_n(x) \end{matrix} \right|^+ \begin{matrix} \\ i-1 \text{ rows} \\ \\ \\ \\ n-i \text{ rows} \end{matrix}$$

$$-\infty < x < \infty, \quad (17.1)$$

where $^+|A|^+$ denotes the permanent of the matrix A. Similarly, the joint density function of $X_{i:n}$ and $X_{j:n}$ $(1 \le i < j \le n)$ can be written as

$$f_{i,j:n}(x, y) = \frac{1}{(i-1)!(j-i-1)!(n-j)!}$$

$$\times \left|^+ \begin{matrix} F_1(x) & \cdots & F_n(x) \\ \cdots & \cdots & \cdots \\ F_1(x) & \cdots & F_n(x) \\ f_1(x) & \cdots & f_n(x) \\ F_1(y)-F_1(x) & \cdots & F_n(y)-F_n(x) \\ \cdots & \cdots & \cdots \\ F_1(y)-F_1(x) & \cdots & F_n(y)-F_n(x) \\ f_1(y) & \cdots & f_n(y) \\ 1-F_1(y) & \cdots & 1-F_n(y) \\ \cdots & \cdots & \cdots \\ 1-F_1(y) & \cdots & 1-F_n(y) \end{matrix} \right|^+ \begin{matrix} \\ i-1 \text{ rows} \\ \\ \\ \\ j-i-1 \text{ rows} \\ \\ \\ n-j \text{ rows} \end{matrix},$$

$$-\infty < x < y < \infty. \quad (17.2)$$

Upon making use of these permanent expressions of order statistics and some known properties of permanents, most of the results available on the moments of order statistics from an i.i.d. sample have been generalized to the case when the order statistics arise from n independent and non-identically distributed variables. For example, such a generalization of the triangle rule in (6.4) is given by [Balakrishnan (1988)]

$$r\mu_{r+1:n}^{(k)} + (n-r)\mu_{r:n}^{(k)} = \sum_{i=1}^{n} \mu_{r:n-1}^{[i](k)}, \qquad (17.3)$$

where $\mu_{r:n-1}^{[i](k)}$ denotes the kth moment of the rth order statistic obtained from $n-1$ variables (with X_i having been removed from the original set of n variables). It is important to mention here that these results have been further generalized to the case when the order statistics arise from n arbitrarily jointly distributed random variables. In this respect, many properties (such as log-concavity, duality, etc.) of order statistics and recurrence relations for moments have all been generalized to this very general case. For this purpose, different approaches have been used to establish these results, including the method of induction, probabilistic approach, indicator method, operator method, and generating function approach.

References

Ahsanullah, M. (1995). *Record Statistics*. Nova Science Publishers, Commack, New York.
Ali, M. M. and D. Umbach (1998). Optimal linear inference using selected order statistics in location-scale models. In the companion volume.
Alimoradi, S. and A. K. Md. E. Saleh (1998). On some *L*-estimation in linear regression models. In the companion volume.
Andrews, D. F., P. J. Bickel, F. R. Hampel, P. J. Huber, W. H. Rogers and J. W. Tukey (1972). *Robust Estimates of Location*. Princeton University Press, Princeton, New Jersey.
Arnold, B. C. and N. Balakrishnan (1989). *Relations, Bounds and Approximations for Order Statistics*. Lecture Notes in Statistics – **53**, Springer-Verlag, New York.
Arnold, B. C., N. Balakrishnan and H. N. Nagaraja (1992). *A First Course in Order Statistics*. John Wiley & Sons, New York.
Arnold, B. C., N. Balakrishnan and H. N. Nagaraja (1998). *Records*. John Wiley & Sons, New York (to appear).
Arnold, B. C. and J. A. Villaseñor (1998). Lorenz ordering of order statistics and record values. In this volume.
Bain, L. J. (1978). *Statistical Analysis of Reliability and Life-Testing Models: Theory and Methods*. Marcel Dekker, New York.
Bain, L. J. and M. Engelhardt (1991). *Statistical Analysis of Reliability and Life-Testing Models*. Second edition, Marcel Dekker, New York.
Balakrishnan, N. (1988). Recurrence relations for order statistics from n independent and non-identically distributed random variables. *Ann. Inst. Statist. Math.* **40**, 273–277.
Balakrishnan, N. and R. Aggarwala (1998). Recurrence relations for single and product moments of order statistics from a generalized logistic distribution with applications to inference and generalizations to double truncation. In the companion volume.
Balakrishnan, N. and P. S. Chan (1998). Log-gamma order statistics and linear estimation of parameters. In the companion volume.

Balakrishnan, N. and A. C. Cohen (1991). *Order Statistics and Inference: Estimation Methods.* Academic Press, San Diego.

Balakrishnan, N. and S. S. Gupta (1998). Higher order moments of order statistics from exponential and right-truncated exponential distributions and applications to life-testing problems. In the companion volume.

Balakrishnan, N. and S. K. Lee (1998). Order statistics from the Type III generalized logistic distribution and applications. In the companion volume.

Balakrishnan, N. and C. R. Rao (1997). A note on the best linear unbiased estimation based on order statistics. *Amer. Statist.* **51**, 181–185.

Balakrishnan, N. and C. R. Rao (1998). Some efficiency properties of best linear unbiased estimators. *J. Statist. Plan. Infer.* (to appear).

Balakrishnan, N. and K. S. Sultan (1998). Recurrence relations and identities for moments of order statistics. In this volume.

Balasubramanian, K. and N. Balakrishnan (1993a). A log-concavity property of probability of occurrence of exactly r arbitrary events. *Statist. Prob. Lett.* **16**, 249–251.

Balasubramanian, K. and N. Balakrishnan (1993b). Duality principle in order statistics. *J. Roy. Statist. Soc. Ser. B* **55**, 687–691.

Barnett, V. and T. Lewis (1978, 1984, 1993). *Outliers in Statistical Data.* First edition, Second edition, Third edition, John Wiley & Sons, Chichester, England.

Basu, A. P. and B. Singh (1998). Order statistics in exponential distribution. In the companion volume.

Boland, P. J., M. Shaked and J. G. Shanthikumar (1998). Stochastic ordering of order statistics. In this volume.

Castillo, E. (1988). *Extreme Value Theory in Engineering.* Academic Press, San Diego.

Chan, L. K. (1967). On a characterization of distributions by expected values of extreme order statistics. *Amer. Math. Monthly* **74**, 950–951.

Childs, A. and N. Balakrishnan (1998a). Series approximations for moments of order statistics using MAPLE, *Submitted for publication.*

Childs, A. and N. Balakrishnan (1998b). Generalized recurrence relations for moments of order statistics from non-identical Pareto and truncated Pareto random variables with applications to robustness. In this volume.

Cohen, A. C. (1991). *Truncated and Censored Samples: Theory and Applications.* Marcel Dekker, New York.

Cohen, A. C. and B. J. Whitten (1988). *Parameter Estimation in Reliability and Life Span Models.* Marcel Dekker, New York.

D'Agostino, R. B. and M. A. Stephens (Eds.) (1986). *Goodness-of-fit Techniques.* Marcel Dekker, New York.

David, F. N. and N. L. Johnson (1954). Statistical treatment of censored data. I. Fundamental formulae. *Biometrika* **41**, 228–240.

David, H. A. (1970, 1981). *Order Statistics.* First edition, Second edition, John Wiley & Sons, New York.

David, H. A. (1973). Concomitants of order statistics. *Bull. Internat. Statist. Inst.* **45**, 295–300.

David, H. A. and H. N. Nagaraja (1998). Concomitants of order statistics. In this volume.

Doganaksoy, N. and N. Balakrishnan (1997). A useful property of best linear unbiased predictors with applications to life-testing. *Amer. Statist.* **51**, 22–28.

Fisher, R. A. and L. H. C. Tippett (1928). Limiting forms of the frequency distribution of the largest or smallest member of a sample. *Proc. Camb. Philos. Soc.* **24**, 180–190.

Frechét, M. (1927). Sur la loi de probabilité de l'écart maximum. *Ann. Soc. Polonaise Math.* **6**, 92–116.

Galambos, J. (1978). *The Asymptotic Theory of Extreme Order Statistics.* John Wiley & Sons, New York.

Galambos, J. (1987). *The Asymptotic Theory of Extreme Order Statistics.* Second edition, Kreiger, Malabar, Florida.

Galambos, J. (1998). Univariate extreme value theory and applications. In this volume.

Galambos, J. and S. Kotz (1978). *Characterizations of Probability Distributions.* Lecture Notes in Mathematics – **675**, Springer-Verlag, New York.

Gather, U., U. Kamps and N. Schweitzer (1998). Characterizations of distributions via identically distributed functions of order statistics. In this volume.

Glick, N. (1978). Breaking records and breaking boards. *Amer. Math. Monthly* **85**, 2–26.

Gnedenko, B. (1943). Sur la distribution limite du terme maximum d'une serie aleatoire. *Ann. Math.* **44**, 423–453.

Goldberger, A. S. (1962). Best linear unbiased prediction in the generalized linear regression model. *J. Amer. Statist. Assoc.* **57**, 369–375.

Grudzien, Z. and D. Szynal (1985). On the expected values of kth record values and associated characterizations of distributions. In *Probability and Statistical Decision Theory* (Eds., F. Konecny, J. Mogyorodi and W. Wertz), Vol. A, pp. 119–127, Reidel, Dordrecht, The Netherlands.

Gumbel, E. J. (1954). The maxima of the mean largest value and of the range. *Ann. Math. Statist.* **25**, 76–84.

Harter, H. L. (1970a,b). *Order Statistics and Their Use in Testing and Estimation*. Vols. 1 and 2, U.S. Government Printing Office, Washington, D.C.

Harter, H. L. and N. Balakrishnan (1996). *CRC Handbook of Tables for the Use of Order Statistics in Estimation*. CRC Press, Boca Raton, Florida.

Harter, H. L. and N. Balakrishnan (1997). *Tables for the Use of Range and Studentized Range in Tests of Hypotheses*. CRC Press, Boca Raton, Florida.

Hartley, H. O. and H. A. David (1954). Universal bounds for mean range and extreme observation. *Ann. Math. Statist.* **25**, 85–99.

Hoeffding, W. (1953). On the distribution of the expected values of the order statistics. *Ann. Math. Statist.* **24**, 93–100.

Hosking, J. R. M. (1998). L-estimation. In the companion volume.

Huang, J. S. (1989). Moment problem of order statistics: A review. *Internat. Statist. Rev.* **57**, 59–66.

Huber, P. J. (1981). *Robust Statistics*. John Wiley & Sons, New York.

Jensen, D. R. and D. E. Ramirez (1998). Some exact properties of Cook's D_I. In this volume.

Kagan, A. M., Yu. V. Linnik, and C. R. Rao (1973). *Characterization Problems in Mathematical Statistics*. John Wiley & Sons, New York (English translation).

Kaminsky, K. S. and P. I. Nelson (1998). Prediction of order statistics. In the companion volume.

Kamps, U. (1993). *A Concept of Generalized Order Statistics*. Teubner, Stuttgart, Germany.

Kamps, U. (1998). Characterizations of distributions by recurrence relations and identities for moments of order statistics. In the companion volume.

Lawless, J. F. (1982). *Statistical Models & Methods for Lifetime Data*. John Wiley & Sons, New York.

Leadbetter, M. R., G. Lindgren and H. Rootzén (1983). *Extremes and Related Properties of Random Sequences and Processes*. Springer-Verlag, New York.

Lloyd, E. H. (1952). Least-squares estimation of location and scale parameters using order statistics. *Biometrika* **39**, 88–95.

Lockhart, R. A. and M. A. Stephens (1998). The probability plot: Tests of fit based on the correlation. In the companion volume.

Malmquist, S. (1950). On a property of order statistics from a rectangular distribution. *Skandinavisk Aktuarietidskrift* **33**, 214–222.

Mann, N. R., R. E. Schafer and N. D. Singpurwalla (1974). *Methods for Statistical Analysis of Reliability and Life Data*. John Wiley & Sons, New York.

Nagaraja, H. N. (1986a). A note on conditional Markov property of discrete order statistics. *J. Statist. Plan. Infer.* **13**, 37–43.

Nagaraja, H. N. (1986b). Structure of discrete order statistics. *J. Statist. Plan. Infer.* **13**, 165–177.

Nagaraja, H. N. (1988). Record values and related statistics – A review. *Comm. Statist. – Theory Meth.*, **17**, 2223–2238.

Nagaraja, H. N. (1992). Order statistics from discrete distributions (with discussion). *Statistics* **23**, 189–216.

Nelson, W. (1982). *Applied Life Data Analysis*. John Wiley & Sons, New York.

Nevzorov, V. (1987). Records, *Teoriya Veroyatnostey i ee Primenenija*, **32** 219–251 (in Russian). Translated version in *Theory of Probability and its Applications* **32** (1988), 201–228.

Nevzorov, V. and N. Balakrishnan (1998). A record of records. In this volume.
Pinsker, I. Sh., V. Kipnis and E. Grechanovsky (1986). A recursive formula for the probability of occurrence of at least m out of N events. *Amer. Statist.* **40**, 275–276.
Rao, C. R. and D. N. Shanbhag (1993). *Choquet–Deny Type Functional Equations with Applications to Stochastic Models*. John Wiley & Sons, Chichester, England.
Rao, C. R. and D. N. Shanbhag (1998). Recent approaches to characterizations based on order statistics and record values. In this volume.
Reiss, R. D. (1989). *Approximate Distributions of Order Statistics: With Applications to Nonparametric Statistics*. Springer-Verlag, Berlin.
Resnick, S. I. (1987). *Extreme Values, Regular Variation, and Point Processes*. Springer-Verlag, New York.
Rüschendorf, L. (1985). Two remarks on order statistics. *J. Statist. Plan. Infer.* **11**, 71–74.
Rychlik, T. (1998). Bounds for expectations of L-estimates. In this volume.
Sarhan, A. E. and B. G. Greenberg (Eds.) (1962). *Contributions to Order Statistics*. John Wiley & Sons, New York.
Sarkar, S. K. and W. Wang (1998). Estimation of scale parameter based on a fixed set of order statistics. In the companion volume.
Sen, P. K. (1998). Order statistics: Asymptotic in applications. In this volume.
Serfling, R. J. (1980). *Approximation Theorems of Mathmetical Statistics*. John Wiley & Sons, New York.
Shapiro, S. S. (1998). The role of order statistics in distributional assessment. In the companion volume.
Shapiro, S. S. and M. B. Wilk (1965). An analysis of variance test for normality (complete samples). *Biometrika* **52**, 591–611.
Shorack, G. R. and J. A. Wellner (1986). *Empirical Processes with Applications to Statistics*. John Wiley & Sons, New York.
Sukhatme, P. V. (1937). Tests of significance for samples of the χ^2 population with two degrees of freedom. *Ann. Eugenics* **8**, 52–56.
Tadikamalla, P. R. and N. Balakrishnan (1998). Computer simulation of order statistics. In the companion volume.
Tiku, M. L., W. Y. Tan and N. Balakrishnan (1986). *Robust Inference*. Marcel Dekker, New York
Vaughan, R. J. and W. N. Venables (1972). Permanent expressions for order statistics densities. *J. Roy. Statist. Soc. Ser. B* **34**, 308–310.
von Mises, R. (1936). La distribution de la plus grande de n valeurs, *Rev. Math. Union Interbalcanique* **1**, 141–160. Reproduced in *Selected Papers of Richard von Mises*, American Mathematical Society, II (1964), 271–294.

Order Statistics: A Historical Perspective

H. Leon Harter and N. Balakrishnan

1. Introduction

Let $X_1, X_2, \ldots, X_n$ be n independent and identically distributed random variables from a specified or unspecified population which, when arranged in non-decreasing order of magnitude, are denoted by $X_{1:n} \leq X_{2:n} \leq \cdots \leq X_{n:n}$. Then, $X_{1:n}, X_{2:n}, \ldots, X_{n:n}$ are collectively called the *order statistics* of the sample and $X_{i:n}$ ($i = 1, 2, \ldots, n$) is called the ith order statistic of the sample. In another sense of the expression, *order statistics* is that branch of the subject of Statistics which deals with the mathematical properties of order statistics and with statistical methods based upon them.

We make a sharp distinction, as does David (1970, 1981), between *order statistics* and *rank order statistics*, which is a branch of *nonparametric statistics*. Though there is considerable overlap between order statistics and nonparametric (or distribution-free) statistics, the former is not (as some authors have claimed) a branch of the latter. This is readily evident from the fact that one of the most frequently encountered applications of order statistics is in the estimation of parameters of specified distributions. Examples include best linear unbiased estimation based on all or some of the order statistics of a complete or censored sample, and maximum likelihood estimation based on a censored sample. Order statistics are also used in estimation of regression coefficients. Most of the alternatives to the method of least squares (unlike that method itself) are based on order statistics, as are methods for the treatment of outliers. Order statistics are useful not only in point estimation but also in point prediction and in determining confidence intervals, tolerance intervals, and prediction intervals.

Other applications discussed include use of the range in statistical quality control; use of the studentized range in multiple comparison tests; use of order statistics in ranking and selection procedures; use of extreme values in hydrology, meteorology, seismology, econometrics, strength of materials, and other applied fields; plotting positions on probability paper; use of ordered characteristic roots in multivariate analysis; and use of the empirical cdf and spacings in goodness-of-fit tests.

Order statistics were discussed more or less extensively in a number of earlier books, but the first book devoted exclusively to order statistics [other than the book on extreme values by Gumbel (1958)] was a book edited by Sarhan and Greenberg (1962), containing contributions by the editors and several other authors. Other books on the subject were the two volumes by Harter (1970a,b) on order statistics and their use in testing and estimation, and the recent revision of these two volumes by Harter and Balakrishnan (1996, 1997).

The first textbook on order statistics was the one by David (1970) [second edition, 1981]. More recently Arnold, Balakrishnan and Nagaraja (1992) published a textbook for advanced undergraduate courses that does not assume an advanced mathematical or statistical background.

Harter (1978b [1983], 1983b, 1991a–d, 1992, 1993) published an eight-volume chronological annotated bibliography of order statistics. The first seven volumes give substantially complete coverage up through 1969. The first two volumes contain subject and author indices and citation lists, but those for Volumes III–VII are combined in Volume VIII with a list of errata and comments by N. Balakrishnan on developments on order statistics from 1970 to date.

In addition to the above mentioned volumes, a good number of other books and monographs have appeared in the literature discussing some specific issues relating to order statistics. We have cited these in the appropriate sections of this review article.

2. Distribution theory and properties

Distribution theory of order statistics has received considerable attention since the early 1900s. Numerous papers appeared dealing with general properties of order statistics from continuous as well as discrete populations. Order statistics (and their moments) from several specific continuous populations including the uniform, exponential, normal, logistic, Cauchy, lognormal, Gamma, Weibull, Pareto, extreme value, Laplace, t, and Beta have been studied by various authors. Similar work on order statistics from specific discrete populations including the uniform, geometric, binomial, Poisson, negative binomial, and multinomial has been carried out. The dependence structure of order statistics has also been studied in great detail. It has been proved, for example, that order statistics from discrete distributions, unlike in the case of continuous distributions, need not form a Markov chain. In addition, order statistics arising from finite populations under sampling schemes such as simple random sampling and stratified random sampling have also been discussed. Nagaraja (1992) has provided a detailed review of the distribution theory and properties of discrete order statistics. See also Chapters 2 and 5 of David (1981) and Chapters 2–4 of Arnold, Balakrishnan and Nagaraja (1992).

Barlow and Proschan (1975, 1981) established some reliability properties of order statistics. For example, they proved that if F is IFR (Increasing Failure Rate) or DFR (Decreasing Failure Rate) or IFRA (Increasing Failure Rate on

Average) or DFRA (Decreasing Failure Rate on Average) or NBU (New Better than Used) or NWU (New Worse than Used), then the distributions of order statistics from F also inherit the same reliability properties. Some extensions of these results have been provided by Takahasi (1988) and Nagaraja (1990).

The article by Balakrishnan and Rao (1998) in this volume reviews various developments that have taken place on the distribution theory and properties of order statistics, while the papers of Arnold and Villaseñor (1998) and Boland, Shaked and Shanthikumar (1998) review results on some stochastic orderings of distributions of order statistics.

3. Measures of central tendency and dispersion

Order statistics and functions of order statistics are often used as descriptive statistics for samples and for finite populations. The most common measures of central tendency are the mean and the mode (not based on order statistics), the median and the midrange. The median of n ordered observations is defined as the middle observation when n is odd (i.e. $X_{0.5(n+1):n}$), and as the average of the two middle observations when n is even (i.e. $(X_{0.5n:n} + X_{0.5(n+2):n})/2$). The midrange is defined as the average of the two extreme observations. Less commonly used are the quasi-midranges (quasi-medians), which are defined as the midranges of the observations remaining after discarding equal numbers of largest and smallest observations. The midrange was used by the ancient Greeks and Egyptians and by the Arabs during the middle ages. The median seems not to have come into use until early in the modern era. The quasi-midranges have been used, though less frequently, for over a century.

Common measures of dispersion include the standard deviation and the average deviation from the mean (both not based on order statistics), the range, the quasi-ranges, the quartile deviation (semi-interquartile range), and the average absolute deviation from the median. The range is defined as the difference between the largest and smallest observations, and the quasi-ranges as the differences between the largest and smallest observations remaining after discarding equal numbers of largest and smallest observations. The quartile deviation (semi-interquartile range) is defined as half the difference between the first and third quartiles, which are defined by $Q_1 = X_{0.25(n+1):n}$ and $Q_3 = X_{0.75(n+1):n}$, where a fractional subscript indicates interpolation between adjacent order statistics. The average absolute deviation from the median is the average of the absolute values of the deviations of the observations from their median. The range has been used since ancient times, and the average absolute deviation from the median and the quartile deviation for well over 200 and 150 years, respectively; the quasi-ranges apparently were first used in this century.

Some of the above described measures of central tendency and dispersion have since become so basic and important a part of descriptive statistics that even introductory texts in Statistics (for example, see Moore and McCabe, 1993) have now incorporated these details.

4. Regression coefficients

The problem of determining the coefficients in the equation of the straight line which best fits (in some specific sense) three or more non-collinear points in the (x, y)-plane whose coordinates are pairs of associated values of two related variables, x and y, dates back at least as far as Galileo Galilei (1632). This problem of linear regression in two variables can be generalized to non-linear regression and/or regression in more than two variables.

Over two hundred years ago, two methods based on order statistics were proposed for determining the best-fitting straight line through three or more points. Boscovich (1757) proposed a method based on two criteria:
- the sums of the positive and negative residuals (in the y-direction) shall be numerically equal;
- the sum of their absolute values shall be a minimum.

Boscovich (1760) presented a geometric method and Laplace (1793) an analytic method based on the above two criteria of Boscovich. Laplace (1786) gave a procedure for minimizing the maximum absolute deviation from the fitted line, a criterion suggested earlier by Euler (1749) and Lambert (1765). The method of least squares (not based on order statistics) was proposed independently by Legendre (1805), Adrain (1808) and Gauss (1809). Because of its computational simplicity and because of a mistaken belief in the universality of the normal law of error, on which it is based, it soon superseded the method of least absolute values and the minimax method in general practice. However, a few authors continued to study the two earlier methods. Fourier (1823, 1824) formulated both as what would now be called linear programming problems. Later authors, notably Edgeworth (1887), dropped Boscovich's first criterion, which constrains the line to pass through the mean point $(\bar{x}, \bar{y})$, and applied the second criterion without this constraint. The revival of interest in the method of least absolute values and the minimax method has continued in the present century, accelerated in the second half of the century by the development of powerful computers, which made computational simplicity less important. A summary of the history of these two methods of estimation is given in the Encyclopedia articles by Harter (1985a,b). Robustness considerations have also led to the development of many more methods (based on order statistics either explicitly or implicitly). A fuller account of the method of least squares and various alternatives, along with the related problem of the treatment of outliers, is given in a six-part article by Harter (1974–1976).

Of course, the problems of determining the best sample measures of central tendency and dispersion (for use in estimating the unknown population mean and standard deviation) and the best linear or non-linear regression equation are closely related. The solutions of all three problems depend upon the distribution of the errors or residuals (deviations of the observed values from those predicted by the regression equation). If the error distribution is normal, the maximum likelihood estimators are the sample mean and standard deviation (with n, not

$n - 1$, in the denominator) and the least squares regression coefficients. If the error distribution is double exponential (Laplace), the maximum likelihood estimators are the sample median, $\sqrt{2}$ times the mean absolute deviation from the median, and the least absolute values regression coefficients (see, for example, Johnson, Kotz and Balakrishnan, 1995). If the error distribution is uniform, the maximum likelihood estimators are the sample midrange, the sample semirange/$\sqrt{3}$, and the minimax regression coefficients.

5. Treatment of outliers and robust estimation

Another problem arises when the data may be contaminated by spurious observations (outliers) which come from distributions with different means and/or different standard deviations than the assumed distribution. Many methods have been proposed for testing outliers as well as for rejecting or modifying them (or their weights). Among the earliest criteria for rejection of outliers were those of Peirce (1852) and Chauvenet (1863). Newcomb (1886) proposed reducing the weights of extreme observations suspected of being spurious instead of discarding them entirely. Rider (1933) gave an excellent summary of methods proposed up to that time. Later methods include those of Grubbs (1950) and Dixon (1950), both of which make extensive use of extreme order statistics.

Robust estimation procedures have been developed by Huber (1964, 1972, 1973, 1977, 1981) and others for use when the underlying distribution is unknown or when the presence of spurious observations is suspected. These include trimmed and Winsorized procedures which are based on trimmed and Winsorized estimators, and also some adaptive procedures which make use of the sample data to decide what estimator to use. These procedures were developed by Tukey and McLaughlin (1963), Dixon and Tukey (1968), Hogg (1967, 1972, 1974), Harter (1974–1976, Part V), and Harter, Moore and Curry (1979), among many others.

Further information about the treatment of outliers and robust estimation is contained in several books including those by Tukey (1970), Andrews et al. (1972), Doornbos (1976), Launer and Wilkinson (1979), Hoaglin, Mosteller and Tukey (1985), Rousseeuw and Leroy (1987), Hawkins (1980), Tiku, Tan and Balakrishnan (1986), Hampel et al. (1986), Barnett and Lewis (1994), and Maddala and Rao (1997). Reference may also be made to a series of journal articles by Harter (1974–1976) and several Encyclopedia articles by Harter (1983a, 1985a,b). In the paper by Geisser (1998) in the companion volume, the Bayesian treatment of the outlier problem has been reviewed.

6. Maximum likelihood estimators

Fisher (1922) laid a firm mathematical foundation for the method of maximum likelihood and gave it that name for the first time, though it had previously been used by him [Fisher (1912)] and much earlier by Lambert (1760), Daniel Bernoulli

(1778), and Gauss (1809) [see Sheynin (1966)]. Maximum likelihood estimators (MLEs) are not in general based on order statistics, but they do involve order statistics in two important cases:

• When the domain of the population is limited in one or both directions and the limit(s) must be estimated from the data. For example, given a sample of size n from a uniform distribution over the interval (a, b), the smallest order statistic is the MLE of a and the largest order statistic is the MLE of b (assuming that a and b are unknown, of course);

• When Type II censoring (single or double) occurs. If the sample is Type II censored from below (from above), the smallest (largest) uncensored observation plays a special role.

Among the earliest authors to consider maximum likelihood estimators from censored samples were Cohen (1950, 1957, 1959) and Halperin (1952). During the 1960's, Harter and Moore (1965, 1966a,b, 1967a,b, 1968a,b, 1969) and Cohen (1961, 1963, 1965, 1966) published several papers on the maximum likelihood estimation, from censored samples, of the parameters of various populations. Each of these articles includes the derivation of the likelihood equations, a discussion of the numerical solution of these equations, and the asymptotic variances and covariances of the estimators. These results have been summarized in book form by Harter (1970b) and by Cohen (1991). A revision [Harter and Balakrishnan (1996)] of the first volume is also worth mentioning here. In addition to these volumes, numerous other books have appeared in the life-testing and reliability literature wherein details of the maximum likelihood estimation based on censored samples from different populations may be found; see, for example, Mann, Schafer and Singpurwalla (1974), Bain (1978), Kalbfleisch and Prentice (1980), Nelson (1982), Lawless (1982), Schneider (1986), Cohen and Whitten (1988), Bain and Engelhardt (1991), Balakrishnan and Cohen (1991), and Balakrishnan (1995).

7. Best linear unbiased estimators

The basic idea of using linear functions of sample order statistics to estimate population parameters (particularly, the location and scale parameters) goes back at least as far as papers by Daniell (1920) and Karl Pearson (1920). Interest in such procedures and others based on order statistics was heightened by landmark papers by Mosteller (1946) and Wilks (1948). Godwin (1949b) found best (minimum variance) unbiased estimators of the standard deviation of a normal population from complete samples of size $n \leq 10$. Ogawa (1951) and Lloyd (1952) made seminal contributions on the use of the Gauss-Markov theorem on least squares to derive best linear unbiased estimators (BLUEs), and applied the results to estimation of parameters of normal and other populations. Gupta (1952) found BLUEs of the mean and standard deviation of a normal population from censored samples of size $n \leq 10$. He also derived alternative linear estimators

(ALEs), which are unbiased, but not of minimum variance, and compared their efficiency with that of the BLUEs. The property of unbiasedness depends only on the expected values of order statistics, but the property of minimum variance depends on their variances and covariances. At that time, i.e. in 1952, the variances and covariances of normal order statistics were available only for $n \leq 10$ [Godwin (1949a)]; the tables were extended up through $n = 20$ by Teichroew (1956), $n = 30$ by Yamauti (1972), and $n = 50$ by Tietjen, Kahaner and Beckman (1977) [and more accurately by Parrish (1992)]. By making use of the tables for these extended sample sizes, Balakrishnan (1991) prepared tables of BLUEs (for complete as well as Type II censored samples) for sample sizes up to 40. A book edited by Sarhan and Greenberg (1962) summarized a wide variety of contributions to order statistics, including those of the editors and others on best linear unbiased estimators. One may also refer to the article in the companion volume by Sarkar and Wang (1998), which discusses the positivity of the BLUE of the scale parameter and also makes a comparison of this estimator with nonlinear ones.

Mann (1965, 1967a,b, 1968, 1969) advocated the use of best linear invariant estimators (minimum mean-square-error estimators which are invariant under transformations of location and scale) instead of best linear unbiased estimators. In cases of highly asymmetric censoring, the best linear invariant estimators, like the maximum likelihood estimators, may be strongly biased, but if so they also have substantially smaller mean square error than the BLUEs, of which they are simple linear functions.

Asymptotically optimum quantiles (spacings) for linear estimation and the corresponding weights (coefficients) have been tabulated by Ogawa (1951, 1962) and other authors for the normal population and by various authors for several other populations. These are especially useful for large samples, or whenever the expected values, variances, and covariances of the order statistics are not available. A summary of work on this subject has been given by Harter (1971) as part of a discussion of optimization problems in estimation of parameters. Also included in this article is a list of Air Force Institute of Technology M.S. Theses on best and nearly best linear unbiased and linear invariant estimation. Chan and Cheng (1988) provided a review on this topic of research. The book by Balakrishnan and Cohen (1991) has a chapter devoted to this topic in which work done on several other populations has been described and supplemented with a list of appropriate references. Ali and Umbach (1998), in their paper in the companion volume, have presented an elaborate review of various developments on this problem.

The book by Harter (1970b) includes tables of expected values of normal, exponential, Weibull and Gamma populations. Added in the revision [Harter and Balakrishnan (1996)] are tables of variances and covariances for those populations and of expected values, variances and covariances for several other populations, and a discussion of their use in best and nearly best linear unbiased and linear invariant estimation. Harter (1988) summarized results on Weibull, log-Weibull, and Gamma order statistics and their use in point and interval estimation of parameters and plotting on probability paper.

8. Recurrence relations and identities

In the 1960s, several papers were published dealing with recurrence relations and identities satisfied by single and product moments of order statistics; for example, see Sillitto (1964), Govindarajulu (1963a,b), Downton (1966), Srikantan (1962), Krishnaiah and Rizvi (1966), and Shah (1966, 1970). These results were extended, generalized and improved in a number of papers that followed. For example, by generalizing one of the results of Govindarajulu (1963a), Joshi (1971) established that for symmetric continuous distributions one needs to evaluate at most one single moment if n is even and at most one single moment and $(n-1)/2$ product moments if n is odd, in order to evaluate the means, variances and covariances of order statistics from a sample of size n (given these quantities for all sample sizes less than n). Similar improvement was made by Joshi and Balakrishnan (1982) who established that for an arbitrary continuous distribution one needs to evaluate at most two single moments and $(n-2)/2$ product moments if n is even and at most two single moments and $(n-1)/2$ product moments if n is odd. Balakrishnan (1986) extended this result to order statistics from discrete distributions. Balakrishnan and Malik (1986), by proving that the basic identities for the moments of order statistics will be satisfied automatically if some of the recurrence relations are used in the computation of moments of order statistics, issued a warning that the basic identities should not be used for checking the computations. Two simple identities established by Joshi (1973) also led to the derivation of several general identities for single and product moments of order statistics. Malik, Balakrishnan and Ahmed (1988) and Arnold and Balakrishnan (1989) have reviewed all the recurrence relations and identities for moments of order statistics from an arbitrary distribution.

Furthermore, following on the lines of Shah (1966, 1970), who established recurrence relations for logistic order statistics, several recurrence relations were also derived for single and product moments of order statistics from specific continuous distributions such as the exponential, normal, Gamma, power function, Pareto, half logistic, log-logistic, linear-exponential, generalized logistic, and generalized half logistic, and their truncated forms. For many of these cases, the recurrence relations are also complete in the sense that they could be used systematically in a simple recursive manner in order to compute all single and product moments of order statistics for all sample sizes. These results have all been synthesized in a review article by Balakrishnan, Malik and Ahmed (1988); see also Arnold and Balakrishnan (1989).

The article by Balakrishnan and Sultan (1998) in this volume provides an updated review of various recurrence relations and identities that hold for moments of order statistics from arbitrary as well as specific distributions.

9. Bounds and approximations

Universal bounds for moments of order statistics were first derived by Hartley and David (1954), Gumbel (1954b), Moriguti (1951, 1953, 1954), and Ludwig

(1959, 1960). Since then, numerous papers have appeared discussing bounds for moments of order statistics (or functions of them). Arnold (1985) used the Hölder inequality (instead of the Cauchy-Schwarz inequality) to derive p-norm bounds. David (1986) [see also David (1988)] derived some bounds for order statistics arising from X_i's, where $X_i = Y_i + Z_i$ ($i = 1, 2, \ldots, n$), which are in terms of the order statistics arising from Y_i's and Z_i's. Imputational-type bounds, assuming that the moments are known exactly for a few sample sizes, were derived by Balakrishnan (1990) which provide simple improvements over the Hartley-David-Gumbel bounds.

An orthogonal inversion expansion method was proposed by Sugiura (1962, 1964) to derive bounds and approximations for single and product moments of order statistics, which was extended by Mathai (1975, 1976) to the product moments of k order statistics. Joshi and Balakrishnan (1983) and Balakrishnan and Joshi (1985) improved on the results of Sugiura by deriving imputational-type bounds, while Joshi (1969) modified Sugiura's method to make it suitable for cases where the moments of extreme order statistics do not exist (like Cauchy).

Following the pioneering work of van Zwet (1964), inequalities for moments of order statistics in terms of quantiles of the distribution were discussed extensively by many authors including Barlow and Proschan (1975, 1981) and Gupta and Panchapakesan (1974). Dependence properties and stochastic comparisons of order statistics remains an active area of research. The paper by Boland, Shaked and Shanthikumar (1998) in this volume reviews this topic.

While most of the developments have focused on the independent case, some papers have appeared dealing with bounds for order statistics in the dependent case. Surprisingly simple and elegant results were derived in this case by Arnold and Groeneveld (1979). Since then, a number of papers have appeared on this problem; David (1988), Arnold (1988), and Arnold and Balakrishnan (1989) have provided useful reviews. More recent reviews of the various developments on this topic have been presented by Rychlik (1998) in this volume.

Series approximations for single and product moments of order statistics, which involve the derivatives of the inverse cumulative distribution function, were presented by David and Johnson (1954) and Clark and Williams (1958). Balakrishnan and Johnson (1998), in addition to reviewing these, have presented (in their article in this volume) improved series approximations for the moments of order statistics in the case of symmetric distributions. An Edgeworth-type expansion for distributions of order statistics has been discussed by Reiss (1983, 1989), who has also demonstrated its application in inference problems. The paper by Balakrishnan and Gupta (1998) in to the companion volume is also in a similar vein, but is restricted to the case of the exponential and truncated exponential distributions and utilizes exact moments and mixed moments of order up to four.

10. Distribution-free tolerance procedures

Wilks (1941) presented a method based on truncated sample ranges [quasi-ranges] for determining the sample size required for setting tolerance limits

$[L_1 = X_{r:n}, L_2 = X_{n-r+1:n}$, where $X_{i:n}$ $(i = 1, 2, \ldots, n)$ is the ith order statistic in a sample of size n] on a random variable X having any unknown continuous density function $f(x)$ and having a given degree of stability.

Wilks (1942) defined $100R_\alpha\%$ tolerance limits $L_1(x_1, x_2, \ldots, x_n)$ and $L_2(x_1, x_2, \ldots, x_n)$ for probability level α of a sample S_1 of size n from a population with density function $f(x)$ as two functions of the X's in S_1 such that the probability is α that at least $100R_\alpha\%$ of the X's of a further indefinitely large sample S_2 (i.e. the population) will lie between L_1 and L_2. He then noted that the same notion clearly applies if S_2 is a finite sample of size N, rather than an indefinitely large one, in which case we would be interested in the largest integer N_α such that the probability is at least α that at least $100\bar{R}_\alpha\% (\bar{R}_\alpha = N_\alpha/N)$ of the X's in S_2 would lie between L_1 and L_2. Assuming only that $f(x)$ is a probability density function (i.e. continuous), he used order statistics of the first sample S_1, in particular the smallest order statistic $X_{1:n}$ and the largest order statistic $X_{n:n}$, to set tolerance limits satisfying the above conditions. For general values of n, N, and α (e.g., 0.99 or 0.95), he tabulated N_α; also the limiting values of $N_\alpha/N = \bar{R}_\alpha$ as n increases indefinitely. Scheffé and Tukey (1945) validated Wilks's results, which had assumed a continuous probability density function, assuming only a continuous cumulative distribution function, and then modified his results so that they are valid for any distribution function.

Let X be a random variable having a density (or probability function) $f(x; \boldsymbol{\theta})$, with cdf $F(x; \boldsymbol{\theta})$ where $\boldsymbol{\theta}$ is a vector of parameters. Let $X_1, X_2, \ldots, X_n$ be a random sample from $f(x; \boldsymbol{\theta})$, and $X_{i:n}$, $i = 1, \ldots, n$, be the ith order statistic of the sample. Let β and γ be two constants such that $0 < \beta < 1$ and $0 < \gamma < 1$. If $\underline{L}$ and $\overline{L}$ are determined so that

$$P[P_X(\underline{L} \leq X \leq \overline{L}) \geq \beta] \geq \gamma$$

for all $\boldsymbol{\theta}$, then $[\underline{L}, \overline{L}]$ is called a β-content two-sided tolerance interval at confidence level γ. A two-sided (γ, β) tolerance interval with limits $\underline{L}$ and $\overline{L}$ based on order statistics can be obtained from

$$P[F(\overline{L}) - F(\underline{L}) \geq \beta] = \gamma$$

where $\underline{L} = X_{r:n}$, $\overline{L} = X_{n-s+1:n}$. A lower (γ, β) tolerance limit is given by $\underline{L} = X_{r:n}$ and an upper (γ, β) tolerance limit is given by $\overline{L} = X_{n-s+1:n}$. If the cdf is not continuous, the statement of the two-sided tolerance intervals must be modified as indicated by Scheffé and Tukey (1945).

For the two-sided tolerance intervals, one chooses any $r, s > 0$ such that $r + s = m$, and then finds the largest m for which

$$I_{1-\beta}(m, n - m + 1) \geq \gamma \;,$$

where $I_p(a, b)$ is the incomplete Beta-function ratio. Somerville (1958) tabulated the m values for selected values of β, γ and n, and also the γ values when $r = 1$, $s = 1$ for selected values of β and n. Murphy (1948) gave charts for selected values of β, m and n. When $r + s = 1$, the two-sided interval reduces to a one-sided interval $[-\infty, X_{n:n}]$ or $[X_{1:n}, \infty]$, which requires solving $\beta^n \leq 1 - \gamma$ for n. When $r = s = 1$, the interval becomes $[X_{1:n}, X_{n:n}]$, which requires solving $n\beta^{n-1} - (n-1)\beta^n \leq 1 - \gamma$. Owen (1962) tabulated the smallest n for selected values of β and γ when $m = 1$ or 2. Belson and Nakano (1963) gave tables and a nomograph for one-sided tolerance intervals, and Nelson (1974) gave a nomograph for two-sided tolerance intervals for the case $m = 2$. Govindarajulu (1977) gave methods for solving for any one of the four parameters m, n, β and γ in terms of the other three.

Wald (1943) extended Wilks' method for setting tolerance limits to the bivariate and multivariate cases. Tukey (1947, 1948) discussed statistically equivalent blocks and tolerance regions in the continuous and discontinuous cases, respectively. Murphy (1948) gave further results (mostly graphical) on nonparametric tolerance limits and regions. Both Wilks (1948) and Wolfowitz (1949), in survey papers, presented Wald's results on multivariate tolerance regions, and the former also dealt with Tukey's generalization. Further advances on multivariate tolerance regions were made by Fraser and Wormleighton (1951), Fraser (1951, 1953), Fraser and Guttman (1956), Kemperman (1956), Somerville (1958), Jílek and Likař (1960), Walsh (1962), and Quesenberry and Gessaman (1968).

Further information on distribution-free tolerance procedures is given in monographs by Guttman (1970) and Guenther (1977) and review papers by Guenther (1972) and Patel (1986). Some of these authors also considered tolerance procedures for specific continuous and discrete distributions and for restricted families of distributions. Jílek (1981) prepared an extensive bibliography on statistical tolerance regions, including distribution-free results.

11. Prediction

Considerable work has been done on the construction of prediction intervals and point predictions. Prediction intervals, unlike confidence intervals which provide intervals for parameters of the assumed distribution, provide intervals for some statistics concerning the experiment. Consequently, point predictions and prediction intervals have been used extensively in life-testing problems.

Work has focused primarily on two types of prediction problems. In the first, referred to as the "one-sample problem", one is concerned with predicting the lifetime of a surviving unit having observed a censored sample. In the second, referred to as the "two-sample problem", one is interested in predicting the lifetime of an order statistic (often the smallest or the largest) from a future sample having already observed an independent complete or censored sample. Distributions of the pivotal quantities have been discussed for a variety of parent populations, and tables of percentage points of these pivotal quantities have also been constructed. Some approximate prediction intervals have also been proposed for some distributions. Patel (1989) provided an exhaustive review of various developments in this area of research, while Aitchison and Dunsmore (1975) consolidated the progress until the mid 70s. Hahn and Meeker (1991) have recently discussed the usefulness of constructing such prediction intervals. Kaminsky and Nelson (1998), in a paper in the companion volume, have reviewed some developments on this topic.

12. Statistical quality control and range

Shewhart (1931), the father of statistical quality control, proposed the use of $\overline{X}$ and σ charts to determine whether or not a process is in a state of statistical control. These charts have center lines at $\overline{\overline{X}}$ and $\overline{\sigma}$, the means of the means and standard deviations, respectively, of a number of small samples (rational subgroups), and 3-sigma control limits at distances above and below the center lines depending upon $\overline{\sigma}$ and the subgroup size. The values of $\overline{X}$ and σ for successive subgroups are plotted on these charts, and if any of them fall outside the control limits that is an indication of lack of control and a signal that an assignable cause of variation must be found and eliminated to bring the process into control. Shewhart considered and tentatively rejected the use of the sample range R instead of the sample standard deviation σ, but Pearson (1935) justified the use of the range for small samples, and tabulated factors for the calculation of control limits, based on the mean range $\overline{R}$, for $\overline{X}$ and R charts. Pearson and Haines (1935) made a study of the use of range instead of standard deviation in quality control charts. They found that the mean subgroup standard deviation is slightly more powerful in detecting the presence of assignable causes than the mean range, but the difference is small for samples (subgroups) of size $n \leq 10$. Because of the greater ease of calculating the range, it soon replaced the standard deviation in most applications of quality control. The paper by Schneider and Barbera (1998) in the companion volume highlights the role of order statistics in quality control problems.

13. Multiple comparisons and studentized range

If an investigator wishes to compare the means of two groups of observations, the standard procedure is to perform a Student t test at an appropriate significance level α. Suppose, however, he wishes to compare the means of $m(\geq 3)$ groups. If he were to perform $m(m-1)/2$ t tests, one for each pair of means, each at significance level α, then the probability that a significant difference would be found between the largest and smallest means would be larger than α (much larger for large m), even if the groups all came from populations with equal means. This dilemma was recognized a century and a half ago by Cournot (1843). Some protection against too many significant differences is provided by first performing an analysis of variance on the m groups and then performing t tests only if an F test shows overall significance of difference among means. This procedure, called the protected LSD (least significant difference) test, has often been used in practice, but it is not clear that it provides sufficient protection. During the past fifty-odd years, a number of other procedures have been proposed, including several based on the studentized range, notably those of Newman (1939) and Keuls (1952), Tukey (1953), and Duncan (1955).

The differences LSD and WSD between any two out of m means of samples of size n required for significance (at the α level) by the LSD test and Tukey's studentized range test, respectively, are given by

$$\text{LSD} = t(\alpha, v)s_{\bar{d}} = q(\alpha, 2, v)s_{\bar{x}},$$
$$\text{WSD} = q(\alpha, m, v)s_{\bar{x}},$$

where v is the number of degrees of freedom for the error mean square, s^2; $t(\alpha, v)$ is the 2-tailed α point (upper $\alpha/2$ point) of Student's t with v degrees of freedom; $q(\alpha, r, v)$ is the upper α point of the studentized range of r observations with v degrees of freedom for s^2; $s_{\bar{x}}$ is the standard error of the mean, $s_{\bar{x}} = \sqrt{s^2/n}$; and $s_{\bar{d}}$ is the standard error of the difference between means, $s_{\bar{d}} = \sqrt{2s^2/n} = \sqrt{2}s_{\bar{x}}$.

The critical ranges ISD and SSD for p out of m ordered means of samples of size n required for significance (at the α level) by the Newman-Keuls test and Duncan's multiple range test, respectively, are given by

$$\text{ISD} = q(\alpha, p, v)s_{\bar{x}},$$
$$\text{SSD} = Q(p, \alpha, v)s_{\bar{x}},$$

where $Q(p, \alpha, v)$ is the studentized range of p observations with v degrees of freedom for s and protection level $\gamma_{p,\alpha} = (\gamma_{2,\alpha})^{p-1} = (1-\alpha)^{p-1}$, and where the other symbols are defined as above. In performing these tests, one first tests the range of all m means, then the ranges of $(m-1)$ ordered means, etc., stopping when significance is no longer found, since, by definition, if a set of means do not differ significantly, then no subset of that set differs significantly.

Harter (1957) made a study of error rates and sample sizes for the above multiple comparison tests and others, assuming an underlying normal distribution. Harter, Clemm and Guthrie (1959) published a technical report containing tables of the probability integral and percentage points of the studentized range and critical values for Duncan's multiple range test. The tables of critical values for Duncan's test and excerpts from the table of percentage points of the studentized range were published in journal articles [Harter (1960a,b)]. All of the above results (with revised tables for error rates and sample sizes) were reproduced in book form [Harter (1970a) and Harter and Balakrishnan (1997)], along with theory and tables for the range of samples from a normal population, including (1) percentage points of the ratio of two ranges and related tables and (2) probability integral, percentage points, and moments of the range. See also the revision by Harter and Balakrishnan (1997).

A fuller account of the early history of multiple comparison tests was published by Harter (1980). Volume VIII of the collected works of John W. Tukey (1994) is devoted to his work on multiple comparisons during the period 1948–1983, including his previously unpublished (but well-known and influential) 1953 memorandum on the problem of multiple comparisons. Also included are a biography and a bibliography of John W. Tukey, and extensive references to the work of other authors. Tukey argues cogently for the use of confidence procedures rather than significance procedures in almost all cases. However, multiple range procedures, which the authors prefer, are not adapted to the construction of confidence intervals.

14. Ranking and selection procedures

As an alternative to significance testing, multiple decision procedures called ranking and selection procedures were developed in the 1950's. Given k populations, the problem is to select the best one (usually defined as the one with the largest mean or the one with the smallest variance) or a subset containing the best one. Two different approaches have been proposed.

Under the indifference-zone approach, the object is to select the population with the largest mean (smallest variance) when the difference (ratio) of the two largest populations means (smallest population variances) lies outside the indifference zone comprising values near 0(1). The procedure is to draw a random sample of size n from each population and select the population corresponding to the sample with the largest mean (smallest variance), with ties broken by randomization. The selection is considered to be correct if the mean (variance) of the selected population lies within the indifference zone. The problem is to determine the common sample size n such that the probability of a correct selection is at least P^* whenever $\mu_{[k]} - \mu_{[k-1]} \geq \delta^*$ for specified $\delta^* > 0$, $[\sigma^2_{[2]}/\sigma^2_{[1]} \geq r^*$ for specified $r^* > 1]$, where the bracketed subscripts indicate ordering from smallest to largest. It is assumed that $P^* > 1/k$, the value corresponding to random selection. Bechhofer (1954) [Bechhofer, Dunnett and Sobel (1954)] developed a single-sample [two-sample] multiple decision procedure for ranking means of normal populations with known variances [unknown common variance]. Bechhofer and Sobel (1954) developed a single-sample multiple decision procedure for ranking variances of normal populations. Desu and Raghavarao (1990) summarized work on the indifference-zone approach to selecting the best normal, Bernoulli and exponential populations.

Under the subset-selection approach, the object is to select a subset containing the best population, usually defined as the one having the largest mean (or the smallest variance). Seal (1955) and Gupta (1956) developed decision procedures for selection, from a set of $(n+1)$ normal populations, of a subset containing the one with the largest mean. Gupta and Sobel (1962) developed a procedure for choosing a subset containing the population with the smallest variance.

Gupta and Panchapakesan (1988) summarized work on both the indifference-zone formulation of Bechhofer (1954) and the subset-selection approach of Gupta (1956), with emphasis on procedures that are relevant to reliability models. Gupta and Sobel (1960) and Gupta (1963) developed procedures for binomial and Gamma populations, respectively. Gupta (1965) summarized results up to that date on ranking and selection procedures, using mainly the subset-selection approach. The selection procedures select a non-empty, small, best subset such that the probability is at least equal to a specific value P^* that the best population is included in the selected subset. Selection of a subset to contain all populations better than a standard is also discussed. A thorough account of various developments on ranking and selection problems may be seen in the volumes by Gupta and Panchapakesan (1979) and Gibbons, Olkin and Sobel (1977).

15. Extreme values

While the theory of extreme values has been worked out systematically only within the past 75 years, a few probabilists, engineers, and statisticians dealt with specific extreme-value problems much earlier. Nicolas Bernoulli (1709) found the life expectancy of the last to die of n men, who will all die within a year and who are equally likely to die at any instant during that time, to be $n/(n+1)$, the expected value of the largest observation in a sample of size n from a uniform distribution on the interval (0,1). Chaplin (1880, 1882) considered the relation between the tensile strengths of long and short bars. He showed that if the probability is p that the tensile strength of a bar of given length and cross section exceeds a specified value, then the probability is p^n that the strength of the weakest of n such bars (or of a single bar of the same cross section but n times as long) exceeds the specified value. Galton (1902) and Pearson (1902) considered what ratio, in a competition, the first of two prizes should bear to the second one, assuming that the values of the prizes should be proportional to the respective excesses of merit of the two prize-winners over the third competitor, who receives no prize.

Gumbel (1954a) dated the modern history of extreme-value theory from a fundamental paper by von Bortkiewicz (1922) on the distribution of the range and on the mean range in samples from a normal distribution. Dodd (1923) was the first to study the largest value for other than normal distributions. Tippett (1925) tabulated the probability integral of the largest value in a sample of size $n = 1(1)1000$ from a normal population, and the mean range for $n = 2(1)1000$.

Fréchet (1927) introduced the concept of a class of initial distributions, and was the first to obtain an asymptotic distribution of the largest values. He introduced the stability postulate, which states that the distribution of the largest value should be equal to the initial distribution, except for a linear transformation. Fisher and Tippett (1928) used the stability postulate and found, in addition to Fréchet's asymptotic distribution of largest values, two others valid for other initial types, along with analogous results for smallest values. R. von Mises (1936) classified the initial distributions of the largest value, and gave sufficient conditions under which the three asymptotic distributions are valid. Gnedenko (1943) gave necessary and sufficient conditions, and pointed out that the results for largest values can easily be extended to smallest values.

Gumbel (1954a) gave a concise summary of extreme-value theory and its applications, including outlying observations, Galton's problem, floods and droughts, meteorological phenomena, gust loads, breaking strength of materials, quality control, duration of human life, extinction times for bacteria, radioactive decay, and the stock market. A fuller account was given in his book [Gumbel (1958)]. More recent advances have been presented in a book by Galambos (1978) [1987].

Extreme-value distributions, especially the Weibull distribution, which is the third asymptotic distribution of smallest values, are often used as models for device life. The Weibull distribution also plays an important role in studies of the size effect on material strength, based on the weakest-link theory proposed by Chaplin (1880, 1882). In fact, the Weibull distribution was developed in this context [Weibull (1939a,b)]. Harter (1977) published a study of statistical problems in the size effect on material strength. Harter (1978a) compiled a bibliography of over 600 publications on extreme-value theory, of which over 200 deal with applications to the size effect on material strength. In addition to this review article, the tract of de Haan (1970), the two editions of the book by Galambos (1978, 1987), and the volumes by Leadbetter, Lindgren and Rootzen (1983), Tiago de Oliveira (1984) and Resnick (1987) have provided an exhaustive coverage of this topic. Engineering applications of extreme-value theory have been highlighted, on the same lines as Gumbel (1958), by Castillo (1988). Considerable work on the asymptotic theory of central and intermediate order statistics, following the works of Daniell (1920), Chibisov (1964) and Stigler (1973a,b), has been synthesized in the books by Serfling (1980), Shorack and Wellner (1986) and Reiss (1989). Review papers by Galambos (1998) and Sen (1998) in this volume will provide an elaborate treatment of these topics.

Hydrologists, with their concern about such things as the return period of floods, have been among the chief users of extreme-value theory, and have been active in the development of various probability papers and in discussion of the proper plotting positions. While there is no consensus on the subject, Gumbel and many others contend that, at least for purposes of estimating return periods, the plotting position $i/(n+1)$ should be used for the ith order statistic of a sample of size n. This gives a return period of $(n+1)$ years for the largest of n annual floods (or the smallest of n annual droughts). This contrasts with a return period of $2n$ years corresponding to the plotting position $(i-1/2)/n$ used by earlier authors. Further discussion of plotting positions is given in the next section.

16. Plotting positions on probability paper

Probability paper was used as early as 1896, and was mentioned in the literature more than 50 times before 1950, mainly by hydrologists, many of whom used the plotting position $(i-0.5)/n$ proposed by Hazen (1914). Gumbel (1942) considered the modal position $(i-1)/(n-1)$ and the mean position $i/(n+1)$ [the latter proposed by Weibull (1939a,b)], and chose the latter. Lebedev (1952) and others proposed the use of $(i-0.3)/(n+0.4)$, which is approximately the median position advocated by Johnson (1951). Blom (1958) suggested $(i-\alpha)/(n+1-2\alpha)$, where α is a constant (usually $0 \leq \alpha \leq 1$), which includes all of the above plotting positions as special cases. Moreover, by proper choices of α, one can approximate

$F[E(x_i)]$, the position proposed by Kimball (1946), for any distribution of interest. Gumbel (1954a) stated some postulates which plotting positions should satisfy. Chernoff and Lieberman (1954) discussed the optimum choice of plotting positions in various situations. They showed that if the plotting position $i/(n+1)$ is used to plot sample data from a normal population with unknown standard deviation, the slope of the best-fitting (least squares) regression line seriously overestimates σ. The best plotting position for estimating σ appears to be $F[E(x_i)]$, which is not equal to $E[F(x_i)] = i/(n+1)$ for the normal distribution, or for any other distribution except the uniform. It is clear that the optimum plotting position depends upon the purpose of the investigation and may also depend upon the underlying distribution [see Harter (1984) and Harter and Wiegand (1985)]. Barnett (1975) has also discussed the critical choice of plotting positions. Fowlkes (1987) has presented an exhaustive folio of several important distributions through their theoretical quantile-quantile plots.

17. Simulation methods

As the role of order statistics became increasingly important in many different problems including the treatment of outliers and robustness issues, methods of simulation of order statistics also became of great interest. Schucany (1972), using the beta distribution property of uniform order statistics, suggested the "descending method" starting from the largest order statistic. Lurie and Hartley (1972) similarly presented the "ascending method" starting from the smallest order statistic, which was observed by Reeder (1972) and Lurie and Mason (1973) to be slightly slower in general than the descending method. By combining the two methods, Ramberg and Tadikamalla (1978) and Horn and Schlipf (1986) presented algorithms for the simulation of central order statistics. While Schmeiser (1978) provided a general survey of simulation of order statistics, Gerontidis and Smith (1982) discussed the "inversion method" and the "grouping method" for generating order statistics. The books by Kennedy and Gentle (1980) and Devroye (1986) present a good discussion on this topic. A simple and efficient algorithm for generating a progressively Type II censored sample has recently been given by Balakrishnan and Sandhu (1995). The article by Tadikamalla and Balakrishnan (1998) in this volume provides a review of various developments on this topic of research.

18. Ordered characteristic roots

Hotelling (1933), one of the pioneers in the field of multivariate analysis, was among the first to use methods based on order statistics in that field. He introduced the method of principal components, which are ordered roots of determinantal equations, in multiple factor analysis. Related work was performed by Thomson (1934), Girshick (1936), Hotelling (1936a), Aitken (1937), Bartlett

(1938), Lawley (1940) and Tintner (1945). Hotelling (1936b) studied relations between two sets of variates, and obtained distributions of canonical correlations, which are also roots of determinantal equations, and of functions of canonical correlations, arranged in order of magnitude. Other early contributors to the theory of canonical analysis were Girshick (1939), Bartlett (1947a,b, 1948), and Tintner (1946). Fisher (1938), in a fundamental paper on linear discriminant analysis, dealt with $s - 1$ orthogonal comparisons of s components, arranged in order of magnitude of their contributions. No fewer than four authors [Fisher (1939), Girshick (1939), Hsu (1939) and Roy (1939)] independently published fundamental results on the distribution of the ordered roots of determinantal equations, and during the next ten years further results were obtained by Roy (1940a,b, 1942a,b, 1945, 1946a,b), Hsu (1941), Wilks (1943), Anderson (1945, 1946, 1948), Bartlett (1947a,b), Geary (1948), Nanda (1948a,b), and Rao (1948). Tintner (1946) summarized some applications of four methods of multivariate analysis (discriminant analysis, principal components, canonical correlation, and weighted regression) to economic data. Tintner (1950) established formal relations among the four methods, all of which depend on ordered roots of determinantal equations.

During the decade 1950–1959, numerous authors contributed to distribution theory and/or computational methods for ordered roots and their applications to multivariate analysis. Test criteria for (i) multivariate analysis of variance, (ii) comparison of variance-covariance matrices, and (iii) multiple independence of variates when the parent population is multivariate normal were usually derived from the likelihood ratio principle until S. N. Roy (1953) formulated the union-intersection principle, on which Roy and Bose (1953) based their simultaneous test and confidence procedure. Roy and Bargmann (1958) used an alternative procedure, called the step-down procedure, in deriving a test for problem (iii), and J. Roy (1958) applied the step-down procedure to problems (i) and (ii). During the decade 1960–1969, further contributions were made to the use of ordered roots in multivariate analysis. Krishnaiah (1964) proposed an alternative test called the finite intersection test and investigated distribution problems associated with it. He showed that the confidence intervals associated with the finite intersection test are shorter than those associated with the step-down procedure. Further results on simultaneous test procedures were given by Srivastava (1966, 1969) and Krishnaiah (1969b).

Ordered roots were used in studies of serial correlation and regression analysis by Durbin and Watson (1950, 1951), Hannan (1955), Watson (1955), Watson and Hannan (1956), Theil and Nagar (1961), Henshaw (1966a,b) and Watson (1967). Zonal polynomials, which are homogeneous symmetric polynomials of the ordered characteristic roots of a symmetric matrix, were studied extensively by Constantine and James (1958), James (1960, 1961a,b), Constantine (1963), James (1964, 1966), and Constantine (1966), who applied them to a study of canonical correlations and other multivariate problems. Wigner (1959), Mehta (1960), and Dyson (1962a–c), among others, applied the theory of ordered roots to the study of the spacing of energy levels in nuclear spectra.

Further information on the theory and applications of ordered roots is given in books by Kendall (1957), Roy (1957), Anderson (1958), Pillai (1960), Lawley and Maxwell (1963), Rao (1965), Kendall and Stuart (1966), Morrison (1967), and Krishnaiah (1966, 1969a), and in review papers by Harter (1979, 1986).

19. Goodness-of-fit tests

Numerous methods have been proposed for testing the goodness of fit of a theoretical distribution to observed data. For many years the standard procedure was the chi-square test (not based on order statistics) introduced by Pearson (1900). This test involves grouping the observed data into classes with theoretical (expected) frequencies not too small (≥ 5, say) and comparing the observed frequencies with the theoretical ones. The test statistic is $\hat{\chi}^2 = \sum_{i=1}^{k}(O_i - E_i)^2/E_i$, where k is the number of classes and O_i and E_i are the observed and expected frequencies, respectively, of the ith class. The null hypothesis H_0 that the observed data came from the completely or partially specified theoretical distribution is rejected if $\hat{\chi}^2 > \chi^2_{\alpha,k-p-1}$, where α is the significance level and p is the number of parameters estimated in the specification of $F_0(x)$ in the null hypothesis $H_0 : F(x) = F_0(x)$.

Starting about 1930, several tests based on the empirical cdf were proposed. These include the Kolmogorov–Smirnov test [Kolmogorov (1933), Smirnov (1939)], the Cramér–von Mises test [Cramér (1928), von Mises (1931)], the Anderson-Darling test [Anderson and Darling (1952)], the Kuiper test [Kuiper (1960)], and the Watson test [Watson (1961)]. These tests tend to have somewhat greater power than the chi-square test, but they have a serious drawback when the population is not completely specified and one or more parameters must be estimated from the sample data. Whereas this situation can be handled very simply for the chi-square test by subtracting a degree of freedom for each parameter estimated from the data, it is much more troublesome in the case of tests based on the empirical cdf. While the critical values of such tests are distribution-free when the theoretical population is completely specified, this is no longer true when one or more parameters must be estimated from the data. In that case, it is necessary to compute separate tables of critical values for each population type, and modifications may be required to maximize the power of the tests. Lilliefors (1967) used Monte Carlo simulation to derive a modified test for normality using the Kolmorgorov–Smirnov (KS) test statistic and tabulated critical values. Lilliefors (1969) derived an analogous test for exponentiality. Green and Hegazy (1976) derived tables for modified tests for normal and other distributions using Cramér–von Mises (CvM) and Anderson-Darling (AD) statistics. Woodruff and Moore (1988) summarized work up to that time on modified tests based on the empirical cdf.

The Kolmogorov–Smirnov two-sided test statistic D_n is defined by

$$D_n = \max(D_n^+, D_n^-)$$

where the one-sided test statistics D_n^+ and D_n^- are given by

$$D_n^+ = l.u.b._{1 \leq i \leq n}[i/n - F_i],$$
$$D_n^- = l.u.b._{1 \leq i \leq n}[F_i - (i-1)/n] ,$$

where F_i is the theoretical (population) cdf $F(x_i)$ corresponding to the ith order statistic of a sample of size $n: x_{1:n} \leq x_{2:n} \leq \cdots \leq x_{n:n}$. The sample cdf is defined as a step function. If the sample cdf at $x_{i:n}$ is redefined as $(i - 0.5)/n$, the value midway through the jump from $(i-1)/n$ to i/n that takes place there, the analogous KS test statistics are

$$d_n^+ = l.u.b._{1 \leq i \leq n}[(i - .5)/n - F_i] = D_n^+ - .5/n,$$
$$d_n^- = l.u.b._{1 \leq i \leq n}[F_i - (i - .5)/n] = D_n^- - .5/n ,$$

and

$$d_n = l.u.b._{1 \leq i \leq n}|(i - .5)/n - F_i| = D_n - .5/n .$$

Pyke (1959) and Brunk (1962) proposed defining the sample cdf at the ith order statistic of a sample of size n as $i/(n+1)$, which is the unbiased estimate of the population cdf. If that is done, the test statistics analogous to the D's and d's are

$$C_n^+ = c_n^+ = l.u.b._{1 \leq i \leq n}[i/(n+1) - F_i],$$
$$C_n^- = c_n^- = l.u.b._{1 \leq i \leq n}[F_i - i/(n+1)] ,$$

and

$$C_n = c_n = \max(C_n^+, C_n^-) = l.u.b._{1 \leq i \leq n}|i/(n+1) - F_i| .$$

Other possible definitions include $(i - \delta)/(n + 1 - 2\delta)$, $0 \leq \delta \leq 0.5$, as suggested by Blom (1958). Miller (1956) tabulated critical values for the D's (corresponding to $\delta = 0.5$) and Durbin (1969) tabulated critical values for the C's (corresponding to $\delta = 0$). Harter, Khamis and Lamb (1984) tabulated more critical values of the C's and made a Monte Carlo study of the power of the tests based on the C's and D's for several hypothetical distributions. In a number of cases they found that tests based on the C's are more powerful than those based on the D's, especially when the standard deviation is greater under the alternative than under the null hypothesis. Khamis (1992, 1993) has summarized results on δ-corrected KS tests, including results for the case when one or more parameters are estimated. Little work has been done on the δ-correction for tests other than the KS test, but there is reason to believe that similar results hold for the Kuiper test (for which the two-sided test statistic is the sum, rather than the maximum, of the two one-sided test statistics) and perhaps others.

Other methods of testing goodness-of-fit include graphical techniques based on plotting on probability paper [see Wilk and Gnanadesikan (1968)], tests related to outlier tests [see Tiku (1975) and Barnett and Lewis (1978)], and tests based on spacings (differences between successive order statistics) [see Pyke (1965)]. Shapiro and Wilk (1965) introduced an analysis of variance test for normality for complete samples, and made remarks on extensions to incomplete samples and other distributional assumptions. Jackson (1967) proposed a comparable test for exponentiality. Shapiro, Wilk and Chen (1968) made a comparative study of various tests for normality. Mardia (1980) summarized results on tests of univariate and multivariate normality.

Comprehensive studies of goodness-of-fit tests are given in a volume by Shapiro (1980) and in a book edited by D'Agostino and Stephens (1986). The articles by Shapiro (1998) and Lockhart and Stephens (1998) in the companion volume provide more details on this area of research.

20. Characterizations

Following the fundamental papers of Hoeffding (1953), Fisz (1958), Rossberg (1960, 1972), Govindarajulu (1966), Ferguson (1967) and Desu (1971), numerous papers appeared discussing characterizations of several discrete and continuous distributions by making use of different properties of order statistics. Some of these characterizations are based on simple distributional properties which, therefore, become quite useful in the development of goodness-of-fit tests; see, for example, Csörgő, Seshadri and Yalovsky (1975). Kagan, Linnik and Rao (1973), Patil, Kotz and Ord (1975), Mathai and Pederzoli (1977), Galambos and Kotz (1978), Kakosyan, Klebanov and Melamed (1984), Ramachandran and Lau (1991), and Rao and Shanbhag (1994) all provide very elaborate and excellent discussions on various characterization results in addition to motivations and applications of these characterization results. While Galambos and Kotz (1978) deal with numerous characterization results based on properties of order statistics from exponential and some related distributions, the books by Kagan, Linnik and Rao (1973) and Mathai and Pederzoli (1977) mainly deal with characterizations of normal and stable distributions. The recent volume by Rao and Shanbhag (1994) illustrates nicely the role of Choquet–Deny type functional equations in characterization problems and has one of its chapters devoted to characterizations via properties of order statistics and record values.

Papers by Rao and Shanbhag (1998), Kamps (1998), Gather, Kamps and Schweitzer (1998) and Nevzorov and Balakrishnan (1998), in this volume, will all provide further details on this topic of research.

21. Moving order statistics and applications

Following the basic work of David (1955) on moving minima and moving maxima, distributions of moving order statistics were further discussed by Siddiqui (1970), Inagaki (1980), and David and Rogers (1983). Usage of the moving medians as current estimates of the location and as "robust" alternatives to the moving averages was discussed by Tukey (1970) in his classical book on *Empirical Data Analysis*. These median filters have been discussed further by Kuhlmann and

Wise (1981), Gallagher and Wise (1981), and Arce, Gallagher and Nodes (1986). Filters based on general moving order statistics and also on linear functions of them have been proposed and discussed by Bovik and Restrepo (1987), Palmieri and Boncelet (1988), and David (1992a).

Papers by Arce, Kim and Barner (1998), Barner and Arce (1998), Acton and Bovik (1998), Viswanathan (1998), in the companion volume, will all provide further details on developments in this area of research.

22. Order statistics under non-standard conditions

Some known results for order statistics in the i.i.d. case were extended, by Young (1967) and David and Joshi (1968), to the case when order statistics arise from exchangeable variables. In this case, Bhattacharyya (1970) and Dykstra, Hewett and Thompson (1973) also derived some inequalities for distributions of order statistics, while Maurer and Margolin (1976) presented an expression for the joint distribution of order statistics.

The case when order statistics arise from independent and non-identically distributed variables has been studied at great length. Sen (1970) established inequalities between the distributions of order statistics arising from F_i, $i = 1, \ldots, n$, and those arising from $\overline{F} = \frac{1}{n}\sum_{i=1}^{n} F_i$. Pledger and Proschan (1971) and Proschan and Sethuraman (1976) have provided extensions as well as some applications to reliability problems. Distributions of order statistics in this case were expressed by Vaughan and Venables (1972) in terms of permanents. These expressions were in turn used, along with some known properties of permanents, by Balakrishnan (1988, 1989), Bapat and Beg (1989), and Balakrishnan, Bendre and Malik (1992) to establish some relations for the moments of order statistics. David (1993) presented a simple probabilistic proof for some of these results. The permanent expression was also used by Bapat and Beg (1989) to prove the log-concavity of distributions of order statistics, a result for which simpler proofs as well as some extensions were given subsequently by Sathe and Bendre (1991) and Balasubramanian and Balakrishnan (1993a). Balakrishnan (1994) also discussed, in great detail, the order statistics arising from independent and non-identical exponential random variables, and illustrated their usefulness in the 'optimal' or 'robust' estimation of the exponential mean when possibly multiple outliers are present in the sample.

By relaxing even the assumption of independence, Sathe and Dixit (1990) presented a probabilistic argument to extend some of the recurrence relations to the case of order statistics arising from n arbitrarily distributed variables. Balasubramanian and Balakrishnan (1992) provided a simpler method of proof (using indicators) for this result. A duality principle satisfied by order statistics in this general case was established by Balasubramanian and Balakrishnan (1993b). The equivalence of linear relationships for order statistics from i.i.d. and i.ni.d. cases, exchangeable and arbitrary cases, and i.i.d. and exchangeable cases were established by Balasubramanian and Bapat (1991), Balasubramanian and

Balakrishnan (1994) and David (1995), respectively. The last two simply imply that every linear relationship for order statistics known in the i.i.d. case also holds for the arbitrary case (with appropriate changes).

23. Multivariate order statistics and concomitants

Barnett (1976) presented a survey of various attempts that were made to introduce multivariate order statistics. Bivariate and multivariate extremes, in particular, received considerable attention; for example, see Tiago de Oliveira (1975) and Galambos (1978, 1987).

Concomitants of order statistics, which arise when multivariate data are ordered by one of the components, were first introduced by David (1973); see also Bhattacharya (1974), who has discussed them under the name "induced order statistics". The asymptotic and small-sample theory of concomitants of order statistics has been discussed by David and Galambos (1974), Galambos (1978, 1987), and a number of authors since then. Extensive reviews of the developments on concomitants of order statistics have been prepared by Bhattacharya (1984) and David (1992b). The paper by David and Nagaraja (1998), in this volume, provides an updated review on this topic of research.

For the order statistics arising from a multivariate normal distribution, Siegel (1993) recently proved an interesting relationship among covariances [extending a result known in the i.i.d. case due to Govindarajulu (1966)]; see also Anderson (1993). This result (dealing with the minimum of multivariate normal variables) has been extended to the rth order statistic by Rinott and Samuel-Cahn (1994) and Wang, Sarkar and Bai (1996). Olkin and Viana (1998) have extended Siegel's result to the case of order statistics from elliptically contoured distributions.

24. Records

The first paper on this topic was written by Chandler (1952), who formulated the mathematical theory for the study of records. The asymptotic theory of records was developed in detail by Resnick (1973) and Shorrock (1973). Since then, numerous papers have appeared discussing diverse issues such as characterizations, bounds and approximations, relations, prediction and inference, and generalized concepts of records such as k^{th} record, records from improving populations, records from non-identical variables, and records from Markov and point processes. Review articles by Glick (1978), Nevzorov (1987), and Nagaraja (1988) have all discussed various developments on records. Further treatment has been given in the books by Galambos (1978, 1987), Resnick (1987), Ahsanullah (1988), Arnold and Balakrishnan (1989), and Arnold, Balakrishnan and Nagaraja (1992).

The paper by Nevzorov and Balakrishnan (1998), in this volume, provides an updated review of this topic of research.

References

Acton, S. T. and A.C. Bovik (1998). In the companion volume.
Adrain, R. (1808). Research concerning the probabilities of the errors which happen in making observations. *Analyst* (Philadelphia), **1**, 93–109.
Ahsanullah, M. (1988). *An Introduction to Record Statistics*. Ginn Press, Needham Heights, NJ.
Aitchison, J. and I. R. Dunsmore (1975). *Statistical Prediction Analysis*. Cambridge University Press, Cambridge, England.
Aitken, A. C. (1937). Studies in practical mathematics. II. The evaluation of the latent roots and latent vectors of a matrix. *J. Roy. Soc. Edinburgh* **57**, 269–304.
Ali, M. M. and D. Umbach (1998). In the companion volume.
Anderson, C. L. (1993). Extension of surprising covariances. *J. Amer. Statist. Assoc.* **88**, 1478.
Anderson, T. W. (1945). The non-central Wishart distribution and its application to problems in multivariate statistics. *Doctoral Dissertation*, Princeton University, Princeton, NJ. University Microfilms, Ann Arbor, MI.
Anderson, T. W. (1946). The non-central Wishart distribution and certain problems of multivariate statistics. *Ann. Math. Statist.* **17**, 409–431.
Anderson, T. W. (1948). The asymptotic distributions of the roots of certain determinantal equations. *J. Roy. Statist. Soc. Ser. B*, **10**, 132–136.
Anderson, T. W. (1958). *An Introduction to Multivariate Statistical Analysis*. John Wiley & Sons, New York.
Anderson, T. W. and D. A. Darling (1952). Asymptotic theory of certain "goodness of fit" criteria based on stochastic processes. *Ann. Math. Statist.* **23**, 193–212.
Andrews, D. F., P. J. Bickel, F. R. Hampel, P. J. Huber, W. H. Rogers and J. W. Tukey (1972). *Robust Estimates of Location: Survey and Advances*. Princeton University Press, Princeton, New Jersey.
Arce, G. R., Y.-T. Kim and K. E. Barner (1998). In the companion volume.
Arce, G. R., N. C. Gallagher and T. A. Nodes (1986). Median filters: Theory for one- and two-dimensional filters. *Advances in Computer Vision and Image Processing* **2**, 89–166.
Arnold, B. C. (1985). p-norm bounds on the expectation of the maximum of possibly dependent sample. *J. Multivar. Anal.* **17**, 316–332.
Arnold, B. C. (1988). Bounds on the expected maximum. *Commun. Statist. – Theory Meth.* **17**(7), 2135–2150.
Arnold, B. C. and J. A. Villaseñor (1998). In this volume.
Arnold, B. C. and N. Balakrishnan (1989). *Relations, Bounds and Approximations for Order Statistics*. Lecture Notes in Statistics No. **53**, Springer-Verlag, New York.
Arnold, B. C., N. Balakrishnan and H. N. Nagaraja (1992). *A First Course in Order Statistics*. John Wiley & Sons, New York.
Arnold, B. C. and R. A. Groeneveld (1979). Bounds on expectations of linear systematic statistics based on dependent samples. *Ann. Statist.* **7**, 220–223. Correction: **8** (1980), 1401.
Bain, L. J. (1978). *Statistical Analysis of Reliability and Life-Testing Models: Theory and Methods.*, Marcel Dekker, New York.
Bain, L. J. and M. Engelhardt (1991). *Statistical Analysis of Reliability and Life-Testing Models*. Second edition, Marcel Dekker, New York.
Balakrishnan, N. (1986). Order statistics from discrete distributions. *Commun. Statist. – Theory Meth.* **15**(3), 657–675.
Balakrishnan, N. (1988). Recurrence relations for order statistics from n independent and non-identically distributed variables. *Ann. Inst. Statist. Math.* **40**, 273–277.
Balakrishnan, N. (1989). Recurrence relations among moments of order statistics from two related sets of independent and non-identically distributed random variables. *Ann. Inst. Statist. Math.* **41**, 323–329.
Balakrishnan, N. (1990). Improving the Hartley-David-Gumbel bound for the means of extreme order statistics. *Statist. Probab. Lett.* **9**, 291–294.

Balakrishnan, N. (1991). Best linear unbiased estimates of the mean and standard deviation of normal distribution for complete and censored samples of sizes 21(1)30(5)40, *Unpublished Report*, McMaster University, Hamilton, Ontario, Canada.

Balakrishnan, N. (1994). Order statistics from non-identical exponential random variables and some applications (with discussion). *Comput. Statist. Data Anal.* **18**, 203–253.

Balakrishnan, N. (Ed.) (1998). *Recent Advances in Life-Testing and Reliability*. CRC Press, Boca Raton, FL.

Balakrishnan, N. and K. S. Sultan (1998). In this volume.

Balakrishnan, N., S. M. Bendre and H. J. Malik (1992). General relations and identities for order statistics from non-independent non-identical variables. *Ann. Inst. Statist. Math.* **44**, 177–183.

Balakrishnan, N. and A. C. Cohen (1991). *Order Statistics and Inference: Estimation Methods*. Academic Press, San Diego, CA.

Balakrishnan, N. and S. S. Gupta (1998). In the companion volume.

Balakrishnan, N. and P. C. Joshi (1985). Bounds for the mean of second largest order statistic in large samples. *Mathematische Operationsforschung und Statistik, Series Statistics* **16**, 457–464.

Balakrishnan, N. and H. J. Malik (1986). A note on moments of order statistics. *Amer. Statist.* **40**, 147–148.

Balakrishnan, N., H. J. Malik and S. E. Ahmed (1988). Recurrence relations and identities for moments of order statistics, II: Specific continuous distributions. *Commun. Statist. – Theory Meth.* **17**(8), 2657–2694.

Balakrishnan, N. and C. R. Rao (1998). In this volume.

Balakrishnan, N. and R. A. Sandhu (1995). A simple simulational algorithm for generating progressive Type-II censored samples. *Amer. Statist.* **49**, 229–230.

Balasubramanian, K. and N. Balakrishnan (1992). Indicator method for a recurrence relation for order statistics. *Statist. Probab. Lett.* **14**, 67–69.

Balasubramanian, K. and N. Balakrishnan, (1993a). A log-concavity property of probability of occurrence of exactly r arbitrary events. *Statist. Probab. Lett.* **16**, 249–251.

Balasubramanian, K. and N. Balakrishnan (1993b). Duality principle in order statistics. *J. Roy. Statist. Soc. Ser. B* **55**, 687–691.

Balasubramanian, K. and N. Balakrishnan (1994). Equivalence of relations for order statistics for exchangeable and arbitrary cases. *Statist. Probab. Lett.* **21**, 405–407.

Balasubramanian, K. and R. B. Bapat (1991). Identities for order statistics and a theorem of Rényi. *Statist. Probab. Lett.* **12**, 141–143.

Bapat, R. B. and M. I. Beg (1989). Order statistics for nonidentically distributed variables and permanents, *Sankhyā Ser. B* **51**, 79–93.

Barlow, R. E. and F. Proschan (1975). *Statistical Theory of Reliability and Life Testing: Probability Models*. Holt, Rinehart and Winston, New York. Second edition (1981), To Begin With, Silver Spring, MD.

Barner, K. E. and G. R. Arce (1998). In the companion volume.

Barnett, V. (1975). Probability plotting methods and order statistics. *Appl. Statist.* **24**, 95–108.

Barnett, V. (1976). The ordering of multivariate data (with discussion). *J. Roy. Statist. Soc. Ser. A* **139**, 318–354.

Barnett, V. and T. Lewis (1978). *Outliers in Statistical Data*. John Wiley & Sons, Chichester. Second edition, 1984. Third edition, 1994.

Bartlett, M. S. (1938). Further aspects of the theory of multiple regression. *Proc. Cambridge Philos. Soc. Suppl.* **34**, 33–40.

Bartlett, M. S. (1947a). The general canonical correlation distribution. *Ann. Math. Statist.* **18**, 1–17.

Bartlett, M. S. (1947b). Multivariate analysis. *J. Roy. Statist. Soc., Suppl.* **9**, 176–190; Discussion, 190–197.

Bartlett, M. S. (1948). Internal and external factor analysis. *British J. Psych. Statistical Section* **1**, 73–81.

Bechhofer, R. E. (1954). A single-sample multiple decision procedure for ranking means of normal populations with known variances. *Ann. Math. Statist.* **25**, 16–39.

Bechhofer, R. E., C. W. Dunnett and M. Sobel (1954). A two sample multiple decision procedure for ranking means of normal populations with a common unknown variance. *Biometrika* **41**, 170–176.

Bechhofer, R. E. and M. Sobel (1954). A single-sample multiple decision procedure for ranking variances of normal populations. *Ann. Math. Statist.* **25**, 273–289.

Belson, J. and K. Nakano (1965). Using single-sided nonparametric tolerance limits and percentiles. *Indust. Qual. Cont.* **21**, 566–569.

Bernoulli, D. (1778). Dijudicatio maxime probabilis plurium observationum discrepantium atque verisimillima inductio inde formanda. *Acta Academiae Scientiorum Petropolitanae* **1**(1), 3–23 (Memoirs).

Bernoulli, N. (1709). *Specimina Artis Conjectandi, ad quaestiones Juris Applicatae*, Basle. Reprinted in *Actorum Eruditorum quae Lipsiae Publicantur Supplementa* **4** (1711), 159–170.

Bhattacharya, P. K. (1974). Convergence of sample paths of normalized sums of induced order statistics. *Ann. Statist.* **2**, 1034–1039.

Bhattacharya, P. K. (1984). Induced order statistics: Theory and applications. In: *Handbook of Statistics, Vol. 4* (Eds., P. R. Krishnaiah and P. K. Sen), pp. 383–403, North-Holland, Amsterdam.

Bhattacharyya, B. B. (1970). Reverse submartingale and some functions of order statistics. *Ann. Math. Statist.* **41**, 2155–2157.

Blom, G. (1958). *Statistical Estimates and Transformed Beta-Variables*. John Wiley & Sons, New York; Almqvist & Wiksell, Stockholm.

Boland, P. J., M. Shaked and J. G. Shanthikumar (1998). In this volume.

Boscovich, R. J. (1757). De litteraria expeditione per pontificiam ditionem, et synopsis ampliorus operis, ac habentur plura ejus ex exemplaria etiam sensorum impressa. *Bononiensi Scientiarum et Artum Instituto Atque Academia Commentarii* **4**, 353–396.

Boscovich, R. J. (1760). De recentissimus graduum dimensionibus, et figura, ac magnitudine terrae inde derivanda, *Philosophiae Recentioris, a Benedicto Stay in Romano Archigynasis Publico Eloquentare Professore, versibus traditae, Libri X, cum adnotationibus et Supplementis P. Rogerii Boscovich S. J.*, Tomus II, pp. 406–426, esp. 420–425, Romae.

Bovik, A. C. and A. Restrepo (1987). Spectral properties of moving L-estimates of independent data. *J. Franklin Inst.* **324**, 125–137.

Brunk, H. D. (1962). On the range of the difference between hypothetical distribution function and Pyke's modified empirical distribution function. *Ann. Math. Statist.* **33**, 525–532.

Castillo, E. (1988). *Extreme Value Theory in Engineering*. Academic Press, San Diego, CA.

Chan, L. K. and Cheng, S. W. (1988). Linear estimation of the location and scale parameters based on selected order statistics. *Commun. Statist. – Theory Meth.* **17**(7), 2259–2278.

Chandler, K. N. (1952). The distribution and frequency of record values. *J. Roy. Statist. Soc. Ser. B* **14**, 220–228.

Chaplin, W. S. (1880). The relation between the tensile strengths of long and short bars. *van Nostrand's Engineering Magazine* **23**, 441–444.

Chaplin, W. S. (1882). On the relative tensile strengths of long and short bars. *Proceedings of the Engineers' Club, Philadelphia* **3**, 15–28.

Chauvenet, W. (1863). *A Manual of Spherical and Practical Astronomy*, Volume 2, J. B. Lippincott & Co., Philadelphia. (esp. Appendix, pp. 469–566, and Tables IX and X, pp. 593–599).

Chernoff, H. and G. J. Lieberman (1954). Use of normal probability paper. *J. Amer. Statist. Assoc.* **49**, 778–785.

Chibisov, D. M. (1964). On limit distributions for order statistics. *Theory Probab. Appl.* **9**, 142–148.

Clark, C. E. and G. T. Williams (1958). Distributions of the members of an ordered sample. *Ann. Math. Statist.* **29**, 862–870.

Cohen, A. C. (1950). Estimating the mean and variance of normal populations from singly and doubly truncated samples. *Ann. Math. Statist.* **21**, 557–569.

Cohen, A. C. (1957). On the solution of estimating equations for truncated and censored samples from normal populations. *Biometrika* **44**, 225–236.

Cohen, A. C. (1959). Simplified estimators for the normal distribution when samples are singly censored or truncated. *Technometrics* **1**, 217–237.

Cohen, A. C. (1961). Tables for maximum likelihood estimates: Singly truncated and singly censored samples. *Technometrics* **3**, 535–541.

Cohen, A. C. (1963). Progressively censored samples in life testing. *Technometrics* **5**, 237–239.

Cohen, A. C. (1965). Maximum likelihood estimation in the Weibull distribution based on complete and censored samples. *Technometrics* **7**, 579–588.

Cohen, A. C. (1966). Life testing and early failure. *Technometrics* **8**, 539–549.

Cohen, A. C. (1991). *Truncated and Censored Samples – Theory and Applications*. Marcel Dekker, New York.

Cohen, A. C. and B. J. Whitten (1988). *Parameter Estimation in Reliability and Life Span Models*. Marcel Dekker, New York.

Constantine, A. G. (1963). Some non-central distribution problems in multivariate analysis. *Ann. Math. Statist.* **34**, 1270–1285.

Constantine, A. G. (1966). The distribution of Hotelling's generalized T_0^2. *Ann. Math. Statist.* **37**, 215–225.

Constantine, A. G. and A. T. James (1958). On the general canonical correlation distribution. *Ann. Math. Statist.* **29**, 1146–1166.

Cournot, A. A. (1843). *Exposition de la Théorie des Chances et des Probabilités*. Librairie de L. Hachette, Paris.

Cramér, H. (1928). On the composition of elementary errors. *Skand. Aktuarietidska*. **14**, 13–74, 141–180.

Csörgő, M., V. Seshadri and M. Yalovsky (1975). Applications of characterizations in the area of goodness-of-fit. In: *Statistical Distributions in Scientific Work – Vol. 2: Model Building and Model Selection* (Eds., G. P. Patil, S. Kotz and J. K. Ord), pp. 79–90, D. Reidel, Dordrecht.

D'Agostino, R. B. and M. A. Stephens (Eds.) (1986). *Goodness-of-fit Techniques*. Marcel Dekker, New York.

Daniell, P. J. (1920). Observations weighted according to order. *Amer. J. Math.* **42**, 222–236.

David, F. N. and N. L. Johnson (1954). Statistical treatment of censored data. I. Fundamental formulae. *Biometrika*. **41**, 228–240.

David, H. A. (1955). A note on moving ranges. *Biometrika* **42**, 512–515.

David, H. A. (1970). *Order Statistics*. John Wiley & Sons, New York. Second edition, 1981.

David, H. A. (1973). Concomitants of order statistics. *Bull. Internat. Statist. Inst.* **45**, 295–300.

David, H. A. (1986). Inequalities for ordered sums. *Ann. Inst. Statist. Math.* **38**, 551–555.

David, H. A. (1988). General bounds and inequalities in order statistics. *Commun. Statist. – Theory Meth.* **17**(7), 2119–2134.

David, H. A. (1992a). Some properties of order-statistics filters. *Circuits Systems and Signal Processing* **11**, 109–114.

David, H. A. (1992b). Concomitants of order statistics: Review and recent developments. In: *Multiple Comparisons, Selection, and Applications in Biometry* (Ed., F. M. Hoppe), pp. 507–518, Marcel Dekker, New York.

David, H. A. (1993). A note on order statistics for dependent variates. *Amer. Statist.* **47**, 198–199.

David, H. A. (1995). On recurrence relations for order statistics. *Statist. Probabab. Lett.* **24**, 133–138.

David, H. A. and J. Galambos (1974). The asymptotic theory of concomitants of order statistics. *J. Appl. Probabab.* **11**, 762–770.

David, H. A. and H. N. Nagaraja (1998). In this volume.

David, H. A. and M. P. Rogers (1983). Order statistics in overlapping samples, moving order statistics and U-statistics. *Biometrika* **70**, 245–249.

De Haan, L. (1970). *On Regular Variation and its Application to the Weak Convergence of Sample Extremes*. Mathematical Centre Tracts No. 32, Mathematisch Centrum, Amsterdam.

Desu, M. M. (1971). A characterization of the exponential distribution by order statistics. *Ann. Math. Statist.* **42**, 837–838.

Desu, M. M. and D. Raghavarao (1990). *Sample Size Methodology*. Chapter 6. Academic Press, San Diego, CA.

Devroye, L. (1986). *Non-uniform Random Variate Generation*. Springer-Verlag, New York.

Dixon, W. J. (1950). Analysis of extreme values. *Ann. Math. Statist.* **21**, 488–506.

Dixon, W. J. and J. W. Tukey (1968). Approximate behavior of the distribution of Winsorized t (trimming/Winsorization 2). *Technometrics* **10**, 83–98.

Dodd, E. L. (1923). The greatest and the least variate under general laws of error. *Trans. Amer. Math. Soc.* **25**, 525–539.

Doornbos, R. (1976). *Slippage Tests*. Second edition. Mathematical Centre Tracts No. 15, Mathematisch Centrum, Amsterdam.

Downton, F. (1966). Linear estimates with polynomial coefficients. *Biometrika* **53**, 129–141.

Duncan, D. B. (1955). Multiple range and multiple F tests. *Biometrics* **11**, 1–42.

Durbin, J. (1969). Tests for serial correlation in regression analysis based on the periodogram of least-squares residuals. *Biometrika* **56**, 1–13.

Durbin, J. and G. S. Watson (1950). Testing for serial correlation in least squares regression. I. *Biometrika* **37**, 409–428.

Durbin, J. and G. S. Watson (1951). Testing for serial correlation in least squares regression. II. *Biometrika* **38**, 159–178.

Dykstra, R. L., J. E. Hewett and W. A. Thompson, Jr. (1973). Events which are almost independent. *Ann. Statist.* **1**, 674–681.

Dyson, F. J. (1962a–c). Statistical theory of the energy levels of complex systems. I–III. *J. Math. Phys.* **3**, 140–175.

Edgeworth, F. Y. (1887). On observations relating to several quantities. *Hermathena* **6**(13), 279–285.

Euler, L. (1749). *Pièce qui a Remporté le Prix de l'Académie Royale des Sciences en 1748, sur les Inégalités du Mouvement de Saturne et de Jupiter*. Martin, Coignard et Guerin, Paris.

Ferguson, T. S. (1967). On characterizing distributions by properties of order statistics. *Sankhyā Ser. A* **29**, 265–278.

Fisher, R. A. (1912). On an absolute criterion for fitting frequency curves. *Messenger of Mathematics* (2) **41**, 155–160.

Fisher, R. A. (1922). On the mathematical foundations of theoretical statistics. *Philos. Trans. Roy. Soc. London, Ser. A* **222**, 309–368.

Fisher, R. A. (1938). The statistical utilization of multiple measurements. *Ann. Eug.* **8**, 376–386.

Fisher, R. A. (1939). The sample distribution of some statistics obtained from non-linear equations. *Ann. Eug.* **9**, 238–249.

Fisher, R. A. and L. H. C. Tippett (1928). Limiting forms of the frequency distribution of the largest or smallest member of a sample. *Proc. Cambridge Philos. Soc.* **24**, 180–190.

Fisz, M. (1958). Characterization of some probability distributions. *Skand. Aktuarietidskr.* **41**, 65–70.

Fourier, J. B. J. (1823). Solution d'une question particulière au calcul des inégalités, premier extrait. *Histoire de l'Académie des Sciences* pour 1823, p. xxix.

Fourier, J. B. J. (1824). Solution d'une question particulière au calcul des inégalités, second extrait. *Histoire de l'Académie des Sciences* pour 1824, p. xlvii.

Fowlkes, E. B. (1987). *A Folio of Distributions: A Collection of Theoretical Quantile-Quantile Plots*. Marcel Dekker, New York.

Fraser, D. A. S. (1951). Sequentially determined statistically equivalent blocks. *Ann. Math. Statist.* **22**, 372–381.

Fraser, D. A. S. (1953). Nonparametric tolerance regions. *Ann. Math. Statist.* **24**, 44–55.

Fraser, D. A. S. and I. Guttman (1956). Tolerance regions. *Ann. Math. Statist.* **27**, 162–179.

Fraser, D. A. S. and R. Wormleighton (1951). Nonparametric estimation IV. *Ann. Math. Statist.* **22**, 294–298.

Fréchet, M. (1927). Sur la loi de probabilité de l'écart maximum. *Rocznik Polskie Towarzystwo Matematiczne [Annales de la Société Polonaise de Mathématique* (Cracow)] **6**, 93–116.

Galambos, J. (1978). *The Asymptotic Theory of Extreme Order Statistics*. John Wiley & Sons, New York. Second edition, 1987, Krieger, Melbourne, FL.

Galambos, J. (1998). In this volume.

Galambos, J. and S. Kotz (1978). *Characterizations of Probability Distributions*. Lecture Notes in Mathematics No. **675**, Springer-Verlag, New York.

Galilei, G. (1632). *Dialogo sopra i due massimi sistemi del mondo, Tolemaico, e Copernicano*. Landini, Florence. English translation, *Dialogue concerning the two chief world systems, Ptolemaic and Copernican*, by S. Drake (with foreword by A. Einstein). University of California Press, Berkeley (1953).

Gallagher, N. C. and G. L. Wise (1981). A theoretical analysis of the properties of median filters. *IEEE Transactions on Acoustics and Speech Signal Processing* **29**, 1136–1141.

Galton, F. (1902). The most suitable proportion between the values of first and second prizes. *Biometrika* **1**, 385–390.

Gather, U., U. Kamps and N. Schweitzer (1998). In this volume.

Gauss, C. F. (1809). *Theoria Motus Corporum Coelestium in Sectionibus Conicis Solem Ambientium*. Frid. Perthes et I. H. Besser, Hamburgi.

Geary, R. C. (1948). Studies in relations between economic time series. *J. Roy. Statist. Soc. Ser. B* **10**, 140–158.

Geisser, S. (1998). In the companion volume.

Gerontidis, I. and R. L. Smith (1982). Monte Carlo generation of order statistics from general distributions. *Appl. Statist.* **31**, 238–243.

Glick, N. (1978). Breaking records and breaking boards. *Amer. Math. Monthly* **85**, 2–26.

Gibbons, J. D., I. Olkin and M. Sobel (1977). *Selecting and Ordering Populations: A New Statistical Methodology*. John Wiley & Sons, New York.

Girshick, M. A. (1936). Principal components. *J. Amer. Statist. Assoc.* **31**, 519–528.

Girshick, M. A. (1939). On the sampling theory of roots of determinantal equations. *Ann. Math. Statist.* **10**, 203–224.

Gnedenko, B. (1943). Sur la distribution limite du terme maximum d'une série aléatoire. *Ann. Math.* (2) **44**(3), 423–453.

Godwin, H. J. (1949a). Some low moments of order statistics. *Ann. Math. Statist.* **20**, 279–285.

Godwin, H. J. (1949b). On the estimation of dispersion by linear systematic statistics. *Biometrika* **36**, 92–100.

Govindarajulu, Z. (1963a). On moments of order statistics and quasi-ranges from normal populations. *Ann. Math. Statist.* **34**, 633–651.

Govindarajulu, Z. (1963b). Relationships among moments of order statistics in samples from two related populations. *Technometrics* **5**, 514–518.

Govindarajulu, Z. (1966). Characterization of normal and generalized truncated normal distributions using order statistics. *Ann. Math. Statist.* **37**, 1011–1015.

Govindarajulu, Z. (1977). A note on distribution-free tolerance limits. *Naval Res. Logist. Quart.* **24**, 381–384.

Green, J. R. and Y. A. S. Hegazy (1976). Powerful modified goodness-of-fit tests. *J. Amer. Statist. Assoc.* **71**, 204–209.

Grubbs, F. E. (1950). Sample criteria for testing outlying observations. *Ann. Math. Statist.* **21**, 27–58.

Guenther, W. C. (1972). Tolerance intervals for univariate distributions. *Naval Res. Logist. Quart.* **19**, 309–333.

Guenther, W. C. (1977). *Sampling Inspection in Statistical Quality Control*. Chapter 4. Statistical Monographs and Courses No. 37, Charles Griffin and Co., London.

Gumbel, E. J. (1942). Simple tests for given hypotheses. *Biometrika* **32**, 317–333.

Gumbel, E. J. (1954a). *Statistical Theory of Extreme Values and Some Practical Applications*. National Bureau of Standards, Applied Mathematics Series No. 33, U.S. Government Printing Office, Washington, DC.

Gumbel, E. J. (1954b). The maxima of the mean largest value and of the range. *Ann. Math. Statist.* **25**, 76–84.

Gumbel, E. J. (1958). *Statistics of Extremes*. Columbia University Press, New York.

Gupta, A. K. (1952). Estimation of the mean and standard deviation of a normal population from a censored sample. *Biometrika* **39**, 260–273.

Gupta, S. S. (1956). On a decision rule for a problem in ranking means. *Ph.D. Thesis, Mimeograph Series No. 150*, Institute of Statistics, University of North Carolina, Chapel Hill, NC.

Gupta, S. S. (1963). On a selection and ranking procedure for gamma populations. *Ann. Inst. Statist. Math.* **14**, 199–216.

Gupta, S. S. (1965). On some multiple decision (selection and ranking) rules. *Technometrics* **7**, 225–245.

Gupta, S. S. and S. Panchapakesan (1974). Inference for restricted families. In: *Reliability and Biometry* (Eds., F. Proschan and R. J. Serfling), pp. 551–596, SIAM, Philadelphia.

Gupta, S. S. and S. Panchapakesan (1979). *Multiple Decision Procedures: Theory and Methodology of Selecting and Ranking Populations*. John Wiley & Sons, New York.

Gupta, S. S. and S. Panchapakesan (1988). Selection and ranking procedures in reliability models. In: *Handbook of Statistics – Volume 7: Quality Control and Reliability* (Eds., P. R. Krishnaiah and C. R. Rao), pp. 131–156, North-Holland, Amsterdam.

Gupta, S. S. and M. Sobel (1960). Selecting a subset containing the best of several binomial populations. In: *Contributions to Probability and Statistics: Essays in Honor of Harold Hotelling* (Eds., I. Olkin et al.), pp. 224–248, Stanford University Press, Stanford, CA.

Gupta, S. S. and M. Sobel (1962). On selecting a subset containing a population with the smallest variance. *Biometrika* **49**, 495–507.

Guttman, I. (1970). *Statistical Tolerance Regions: Classical and Bayesian*. Charles Griffin and Co., London.

Hahn, G. J. and W. Q. Meeker (1991). *Statistical Intervals: A Guide for Practitioners*. John Wiley & Sons, New York.

Halperin, M. (1952). Maximum likelihood estimation in truncated samples. *Ann. Math. Statist.* **23**, 226–238.

Hampel, F. R., E. M. Ronchetti, P. J. Rousseeuw, and W. A. Stahel (1986). *Robust Statistics: The Approach Based on Influence Functions*. John Wiley & Sons, New York.

Hannan, E. J. (1955). Exact tests for serial correlation. *Biometrika* **42**, 133–142.

Harter, H. L. (1957). Error rates and sample sizes for range tests in multiple comparisons. *Biometrics* **13**, 511–536.

Harter, H. L. (1960a). Critical values for Duncan's new multiple range test. *Biometrics* **16**, 671–685.

Harter, H. L. (1960b). Tables of range and studentized range. *Ann. Math. Statist.* **31**, 1122–1147.

Harter, H. L. (1970a). *Order Statistics and their Use in Testing and Estimation, Vol. 1: Tests Based on Range and Studentized Range of Samples from a Normal Population*. U.S. Government Printing Office, Washington, DC.

Harter, H. L. (1970b). *Order Statistics and their Use in Testing and Estimation, Vol. 2: Estimates Based on Order Statistics of Samples from Various Populations*. U.S. Government Printing Office, Washington, DC.

Harter, H. L. (1971). Some optimization problems in parameter estimation. In: *Optimizing Methods in Statistics* (Ed., J. S. Rustagi), pp. 36–62, Academic Press, New York.

Harter, H. L. (1974-76). The method of least squares and some alternatives. Parts I–VI. *Internat. Statist. Rev.* **42** (1974), 147–174, 235–264, 282; **43** (1975), 1–44, 125–190, 269–278; **44** (1976), 113–159.

Harter, H. L. (1977). Statistical problems in the size effect on material strength. In: *Application of Statistics* (Ed., P.R. Krishnaiah), pp. 201–221, North-Holland, Amsterdam.

Harter, H. L. (1978a). A bibliography of extreme-value theory. *Internat. Statist. Rev.* **46**, 279–306.

Harter, H. L. (1978b). *The Chronological Annotated Bibliography of Order Statistics, Volume I: Pre-1950*. U.S. Government Printing Office, Washington, DC. Revised edition, 1983, American Sciences Press, Columbus, OH.

Harter, H. L. (1979). Some early uses of order statistics in multivariate analysis. *Internat. Statist. Rev.* **47**, 267–282.

Harter, H. L. (1980). Early history of multiple comparison tests. In: *Handbook of Statistics, Volume 1: Analysis of Variance* (Ed., P. R. Krishnaiah), pp. 617–622, North-Holland, Amsterdam.

Harter, H. L. (1983a). Harter's adaptive robust method. In *Encyclopedia of Statistical Sciences, Vol. 3* (Eds., S. Kotz and N.L. Johnson), pp. 576–578, John Wiley & Sons, New York.

Harter, H. L. (1983b). *The Chronological Annotated Bibliography of Order Statistics.* Volume II: 1950–1959. American Sciences Press, Columbus, OH.

Harter, H. L. (1984). Another look at plotting positions. *Commun. Statist. – Theory Meth.* **13**, 1613–1633.

Harter, H. L. (1985a). Method of least absolute values. In *Encyclopedia of Statistical Sciences, Vol. 5* (Eds., S. Kotz and N. L. Johnson), pp. 462–464, John Wiley & Sons, New York.

Harter, H. L. (1985b). Minimax method. In *Encyclopedia of Statistical Sciences, Vol. 5* (Eds., S. Kotz and N. L. Johnson), pp. 514–516, John Wiley & Sons, New York.

Harter, H. L. (1986). Some applications of order statistics to multivariate analysis II. *Commun. Statist. – Theory Meth.*, **15**, 2609–2649.

Harter, H. L. (1988). Weibull, log-Weibull and Gamma order statistics. In: *Handbook of Statistics – Volume 7: Quality Control and Reliability* (Eds., P. R. Krishnaiah and C. R. Rao), pp. 433–466, North-Holland, Amsterdam.

Harter, H. L. (1991a). *The Chronological Annotated Bibliography of Order Statistics, Volume III: 1960–1961.* American Sciences Press, Columbus, OH.

Harter, H. L. (1991b). *The Chronological Annotated Bibliography of Order Statistics, Volume IV: 1962–1963.* American Sciences Press, Columbus, OH.

Harter, H. L. (1991c). *The Chronological Annotated Bibliography of Order Statistics, Volume V: 1964–1965.* American Sciences Press, Columbus, OH.

Harter, H. L. (1991d). *The Chronological Annotated Bibliography of Order Statistics, Volume VI: 1966–1967.* American Sciences Press, Columbus, OH.

Harter, H. L. (1992). *The Chronological Annotated Bibliography of Order Statistics, Volume VII: 1968–1969.* American Sciences Press, Columbus, OH.

Harter, H. L. (1993). *The Chronological Annotated Bibliography of Order Statistics, Volume VIII: Indices, with a Supplement (by N. Balakrishnan) on 1970–1992.* American Sciences Press, Columbus, OH.

Harter, H. L. and N. Balakrishnan (1996). *CRC Handbook of Tables for the Use of Order Statistics in Estimation.* CRC Press, Boca Raton, FL.

Harter, H. L. and N. Balakrishnan (1997). *Tables for the Use of Range and studentized Range in Tests of Hypotheses.* CRC Press, Boca Raton, FL.

Harter, H. L. and N. Balakrishnan (1997). *Tables for the Use of Range and Studentized Range in Tests of Hypotheses.* CRC Press, Boca Raton, FL.

Harter, H. L., D. S. Clemm and E. H. Guthrie (1959). The probability integrals of the range and of the studentized range: Probability integral and percentage points of the studentized range; Critical values for Duncan's new multiple range test. *WADC Technical Report 58-484, Volume II*, Wright Air Development Center, Wright-Patterson Air Force Base, OH.

Harter, H. L., H. J. Khamis and R. E. Lamb (1984). Modified Kolmogorov-Smirnov tests of goodness of fit. *Commun. Statist. – Simul. Comput.* **13**(3), 293–323.

Harter, H. L. and A. H. Moore (1965). Maximum-likelihood estimation of the parameters of Gamma and Weibull populations from complete and censored samples. *Technometrics* **7**, 639–643.

Harter, H. L. and A. H. Moore (1966a). Maximum-likelihood estimation of the parameters of a normal population from singly and doubly censored samples. *Biometrika* **53**, 205–213.

Harter, H. L. and A. H. Moore (1966b). Local-maximum-likelihood estimation of the parameters of three-parameter lognormal populations from complete and censored samples. *J. Amer. Statist. Assoc.* **61**, 842–851.

Harter, H. L. and A. H. Moore (1967a). Asymptotic variances and covariances of maximum-likelihood estimators, from censored samples, of the parameters of Weibull and Gamma populations. *Ann. Math. Statist.* **38**, 557–570.

Harter, H. L. and A. H. Moore (1967b). Maximum-likelihood estimation, from censored samples, of the parameters of a logistic distribution. *J. Amer. Statist. Assoc.* **62**, 675–684.

Harter, H. L. and A. H. Moore (1968a). Conditional maximum-likelihood estimation, from singly censored samples, of the scale parameters of Type II extreme-value distributions. *Technometrics* **10**, 349–359.

Harter, H. L. and A. H. Moore (1968b). Maximum-likelihood estimation, from doubly censored samples, of the parameters of the first asymptotic distribution of extreme values. *J. Amer. Statist. Assoc.* **63**, 889–901.

Harter, H. L. and A. H. Moore (1969). Conditional maximum-likelihood estimation, from singly censored samples, of the shape parameters of Pareto and limited distributions. *IEEE Trans. Reliability* **R-18**, 76–78.

Harter, H. L., A. H. Moore and T. F. Curry (1979). Adaptive robust estimation of location and scale paramters of symmetric populations. *Commun. Statist. – Theory Meth.* **8**, 1473–1491.

Harter, H. L. and R. P. Wiegand (1985). A Monte Carlo study of plotting positions. *Commun. Statist. – Simul. Comput.* **14**, 317–343.

Hartley, H. O. and H. A. David (1954). Universal bounds for mean range and extreme observation. *Ann. Math. Statist.* **25**, 85–99.

Hawkins, D. M. (1980). *Identification of Outliers*. Chapman and Hall, New York.

Hazen, A. (1914). Storage to be provided in impounding reservoirs for municipal water supply. *Trans. Amer. Soc. Civil Engineers* **77**, 1539–1640. Discussion, 1641–1649.

Henshaw, R. C., Jr. (1966a). Applications of the general linear model to seasonal adjustment of economic time series. *Econometrica* **34**, 381–395.

Henshaw, R. C., Jr. (1966b). Testing single-equation least squares regression models for autocorrelated disturbances. *Econometrica*, **34**, 646–660.

Hoaglin, D. C., F. Mosteller and J. W. Tukey (1985). *Understanding Robust and Exploratory Data Analysis*. John Wiley & Sons, New York.

Hoeffding, W. (1953). On the distribution of the expected values of the order statistics. *Ann. Math. Statist.* **24**, 93–100.

Hogg, R. V. (1967). Some observations on robust estimation. *J. Amer. Statist. Assoc.* **62**, 1179–1186.

Hogg, R. V. (1972). More light on the kurtosis and related statistics. *J. Amer. Statist. Assoc.* **67**, 422–424.

Hogg, R. V. (1974). Adaptive robust procedures: A partial review and some suggestions for future applications and theory. *J. Amer. Statist. Assoc.* **69**, 909–923; Discussion, 923–927.

Horn, P. S. and J. S. Schlipf (1986). Generating subsets of order statistics with applications to trimmed means and means of trimmings. *J. Statist. Comput. Simul.* **24**, 83–97.

Hotelling, H. (1933). Analysis of a complex of statistical variables into principal components. *J. Educ. Psych.* **24**, 417–441, 498–520.

Hotelling, H. (1936a). Simplified calculation of principal components. *Psychometrika* **1**, 27–35.

Hotelling, H. (1936b). Relations between two sets of variates. *Biometrika* **28**, 321–377.

Hsu, P. L. (1939). On the distribution of roots of certain determinantal equations. *Ann. Eug.* **9**, 250–258.

Hsu, P. L. (1941). On the limiting distribution of roots of a determinantal equation. *J. London Math. Soc.* **16**, 183–194.

Huber, P. J. (1964). Robust estimation of a location parameter. *Ann. Math. Statist.* **35**, 73–101.

Huber, P. J. (1972). Robust statistics: A review. *Ann. Math. Statist.* **43**, 1041–1067.

Huber, P. J. (1973). Robust regression: Asymptotics, conjectures and Monte Carlo. *Ann. Statist.* **1**, 799–821.

Huber, P. J. (1977). *Robust Statistical Procedures*. Society for Industrial and Applied Mathematics, Philadelphia.

Huber, P. J. (1981). *Robust Statistics*. John Wiley & Sons, New York.

Inagaki, N. (1980). The distribution of moving order statistics. In: *Recent Developments in Statistical Inference and Data Analysis* (Ed., K. Matusita), pp. 137–142, North-Holland, Amsterdam.

Jackson, O. A. Y. (1967). An analysis of departures from the exponential distribution. *J. Roy. Statist. Soc. Ser. B* **29**, 540–549.

James, A. T. (1960). The distribution of the latent roots of the covariance matrix. *Ann. Math. Statist.* **31**, 151–158.

James, A. T. (1961a). The distribution of noncentral means with known covariance. *Ann. Math. Statist.* **32**, 874–882.

James, A. T. (1961b). Zonal polynomials of the real positive definite symmetric matrices. *Ann. Math.* (2), **74**, 456–469.

James, A. T. (1964). Distributions of matrix variates and latent roots derived from normal samples. *Ann. Math. Statist.* **35**, 475–501.

James, A. T. (1966). Inference on latent roots by calculation of hypergeometric functions of matrix argument. In: *Multivariate Analysis* (Ed., P. R. Krishnaiah), pp. 209–235, Academic Press, New York.

Jílek, M. (1981). A bibliography of statistical tolerance regions. *Mathematische Operationsforschung und Statistik, Series Statistics* **12**, 441–456.

Jílek, M. and O. Likař (1960). Tables of random sample sizes, needed for obtaining non-parametric tolerance regions. *Zastos. Mat.* **5**, 155–160.

Johnson, L. G. (1951). The median ranks of sample values in their population with an application to certain fatigue studies. *Indust. Math.* **2**, 1–9.

Johnson, N. L., S. Kotz and N. Balakrishnan (1994). *Continuous Univariate Distributions – Volume 1*, Second edition. John Wiley & Sons, New York.

Johnson, N. L., S. Kotz and N. Balakrishnan (1995). *Continuous Univariate Distributions – Volume 2*, Second edition. John Wiley & Sons, New York.

Joshi, P. C. (1969). Bounds and approximations for the moments of order statistics. *J. Amer. Statist. Assoc.* **64**, 1617–1624.

Joshi, P. C. (1971). Recurrence relations for the mixed moments of order statistics. *Ann. Math. Statist.* **42**, 1096–1098.

Joshi, P. C. (1973). Two identities involving order statistics. *Biometrika* **60**, 428–429.

Joshi, P. C. and N. Balakrishnan (1982). Recurrence relations and identities for the product moments of order statistics. *Sankhyā Ser. B.* **44**, 39–49.

Joshi, P. C. and N. Balakrishnan (1983). Bounds for the moments of extreme order statistics for large samples. *Mathematische Operationsforschung und Statistik, Series Statistics* **14**, 387–396.

Kagan, A. M., Yu. V. Linnik and C. R. Rao (1973). *Characterization Problems in Mathematical Statistics*. John Wiley & Sons, New York (English Translation).

Kakosyan, A. V., L. B. Klebanov and J. A. Melamed (1984). *Characterization of Distributions by the Method of Intensively Monotone Operators*. Lecture Notes in Mathematics No. **1088**, Springer-Verlag, New York.

Kalbfleisch, J. D. and R. L. Prentice (1980). *The Statistical Analysis of Failure Time Data*. John Wiley & Sons, New York.

Kaminsky, K. S. and P. I. Nelson (1998). In the companion volume.

Kamps, U. (1998). In this volume.

Kemperman, J. H. B. (1956). Generalized tolerance limits. *Ann. Math. Statist.* **27**, 180–186.

Kendall, M. G. (1957). *A Course in Multivariate Analysis*. Hafner Publishing Co., New York; Charles Griffin and Co., London.

Kendall, M. G. and A. Stuart (1966). *The Advanced Theory of Statistics, Volume 3: Design and Analysis, and Time Series*. Charles Griffin & Co., London; Hafner Publishing Co., New York.

Kennedy, W. J. and J. E. Gentle (1980). *Statistical Computing*. Marcel Dekker, New York.

Keuls, M. (1952). The use of the "studentized range" in connection with an analysis of variance. *Euphytica* **1**, 112–122.

Khamis, H. J. (1992). The δ-corrected Kolmogorov-Smirnov test with estimated parameters. *J. Nonparametric Statist.* **2**, 17–27.

Khamis, H. J. (1993). A comparative study of the δ-corrected Kolmogorov-Smirnov test. *J. Appl. Statist.* **20**, 401–421.

Kimball, B. F. (1946). Assignment of frequencies to a completely ordered set of data. *Trans. Amer. Geophysical Union* **27**, 843–846; *Discussion.* **28** (1947), 951–953.

Kolmogorov, A. (1933). Sulla determinazione empirica di una legge di distribuzione. *Giorn Ist. Ital. Attuari* **4**, 83–91.

Krishnaiah, P. R. (1964). Multiple comparison tests in multivariate case. *Aerospace Research Laboratories Technical Report* 64–124, Wright-Patterson Air Force Base, OH.

Krishnaiah, P. R. (ed.) (1966). *Multivariate Analysis.* Academic Press, New York.

Krishnaiah, P. R. (ed.) (1969a). *Multivariate Analysis – II.* Academic Press, New York.

Krishnaiah, P. R. (1969b). Simultaneous test procedures under general MANOVA models. In: *Multivariate Analysis – II* (Ed., P. R. Krishnaiah), pp. 121–143, Academic Press, New York.

Krishnaiah, P. R. and M. H. Rizvi (1966). A note on recurrence relations between expected values of functions of order statistics. *Ann. Math. Statist.* **37**, 733–734.

Kuhlmann, F. and G. L. Wise (1981). On second moment properties of median filtered sequences of independent data. *IEEE Trans. Commun.* **29**, 1374–1379.

Kuiper, N. H. (1960). Tests concerning random points on a circle. *Konink. Nederl. Akad. Wetensch, Proc A* **63** = *Indag. Math.* **22**, 38–47.

Lambert, J. H. (1760). *Photometria, sive de Mensura et Gradibus Luminis, Colorum, et Umbrae.* Augustae Vindelicorum, Augsburg.

Lambert, J. H. (1765). Anmerkungen und Zusätze zur practischen Geometrie. *Beytrage zum Gebrauche der Mathematik und deren Anwendung*, Vol. I, pp. 1–313, Berlin. Second edition, 1792.

Laplace, P. S. (1786). Mémoire sur le figure de la terre. *Mémoires de l'Académie royale des Sciences de Paris*, **Année 1783**, 17–46.

Laplace, P. S. (1793). Sur quelques points du système du monde. *Mémoires de l'Académie royale des Sciences de Paris*, **Année 1789**, 1–87, esp. 18–43.

Launer, R. L. and G. N. Wilkinson (Eds.) (1979). *Robustness in Statistics.* Academic Press, New York.

Lawless, J. F. (1982). *Statistical Models and Methods for Lifetime Data.* John Wiley & Sons, New York.

Lawley, D. N. (1940). The estimation of factor loadings by the method of maximum likelihood. *Proc. Roy. Soc. Edinburgh* **60**, 64–82.

Lawley, D. N. and A. E. Maxwell (1963). *Factor Analysis as a Statistical Method.* Butterworths, London. Second edition, 1971, American Elsevier Publishing Co., New York.

Leadbetter, M. R., G. Lindgren and H. Rootzen (1983). *Extremes and Related Properties of Random Sequences and Processes.* Springer-Verlag, New York.

Lebedev, V. V. (1952). Gidrologiya i Gidrometriya w Zadăcah (*Hydrology and Hydrometry in Problems*). Gidrometeorologičeskoe Izdatel′stvo, Leningrad. Second edition, 1955. Third edition, 1961.

Legendre, A. M. (1805). *Nouvelles Méthodes pour la Détermination des Orbites des Comètes.* Courcier, Paris (esp. pp. 72–80).

Lilliefors, H. W. (1967). On the Kolmogorov–Smirnov test for normality with mean and variance unknown. *J. Amer. Statist. Assoc.*, **62**, 399–402.

Lilliefors, H. W. (1969). On the Kolmogorov–Smirnov test for the exponential distribution with mean unknown. *J. Amer. Statist. Assoc.* **64**, 387–389.

Lloyd, E. H. (1952). Least-squares estimation of location and scale parameters using order statistics. *Biometrika* **39**, 88–95.

Lockhart, R. A. and M. A. Stephens (1998). In the companion volume.

Ludwig, O. (1959). Ungleichungen für Extremwerte und andere Ranggrössen in Anwendung auf biometrische Probleme. *Biometrische Z.* **1**, 203–209.

Ludwig, O. (1960). Über Erwartungswerte und Varianzen von Ranggrössen in kleinen Stichproben. *Metrika*, **3**, 218–233.

Lurie, D. and H. O. Hartley (1972). Machine generation of order statistics for Monte Carlo computations. *Amer. Statist.* **26**(1), 26–27.

Lurie, D. and R. L. Mason (1973). Empirical investigation of general techniques for computer generation of order statistics. *Commun. Statist.* **2**, 363–371.

Maddala, G. S. and C. R. Rao (Eds.) (1997). Handbook of Statistics – Volume 15: *Robust Inference*. North-Holland, Amsterdam.

Malik, H. J., N. Balakrishnan and S. E. Ahmed (1988). Recurrence relations and identities for moments of order statistics, I: Arbitrary continuous distribution. *Commun. Statist. – Theory Meth.* **17**(8), 2632–2655.

Mann, N. R. (1965). Point and interval estimates for reliability parameters when the failure times have the two-parameter Weibull distribution. *Doctoral Dissertation* University of California at Los Angeles. University Microfilms, Ann Arbor, MI.

Mann, N. R. (1967a). Results on location and scale parameter estimation with application to the extreme-value distribution. *Aerospace Research Laboratories Technical Report ARL 67-0023*, Wright-Patterson Air Force Base, OH.

Mann, N. R. (1967b). Tables for obtaining the best linear invariant estimates of parameters of the Weibull distribution. *Technometrics* **9**, 629–643.

Mann, N. R. (1968). Point and interval estimation procedures for the two-parameter Weibull and extreme-value distributions. *Technometrics* **10**, 231–256.

Mann, N. R. (1969). Cramér-Rao efficiencies of best linear invariant estimators of parameters of the extreme-value distribution under Type II censoring from above. *SIAM J. Appl. Math.* **17**, 1150–1162.

Mann, N. R., R. E. Schafer and N. D. Singpurwalla (1974). *Methods for Statistical Analysis of Reliability and Life Data*. John Wiley & Sons, New York.

Mardia, K. V. (1980). Tests of univariate and multivariate normality. In: *Handbook of Statistics – Volume 1: Analysis of Variance* (Ed., P. R. Krishnaiah), pp. 279–320, North-Holland, Amsterdam.

Mathai, A. M. (1975). Bounds for moments through a general orthogonal expansion in a pre-Hilbert space I. *Canad. J. Statist.* **3**, 13–34.

Mathai, A. M. (1976). Bounds for moments through a general orthogonal expansion in a pre-Hilbert space II. *Canad. J. Statist.* **4**, 1–12.

Mathai, A. M. and G. Pederzoli (1977). *Characterizations of the Normal Probability Law*. John Wiley & Sons, New York.

Maurer, W. and B. H. Margolin (1976). The multivariate inclusion-exclusion formula and order statistics from dependent variates. *Ann. Statist.* **4**, 1190–1199.

Mehta, M. L. (1960). On the statistical properties of the level-spacings in nuclear spectra. *Nuclear Physics* **18**, 395–419.

Miller, L. H. (1956). Tables of percentage points of Kolmogorov statistics. *J. Amer. Statist. Assoc.* **51**, 111–121.

Mises, R. de [von Mises, R.] (1936). La distribution de la plus grande de n valeurs. *Revue Mathématique de l'Union Interbalkanique* (Athens) **1**, 141–160.

Mises, R. von (1931). *Vorlesungen aus dem Gebiete der angewandten Mathematik. I Band. Wahrscheinlichkeitsrechnung und ihre Anwendung in der Statistik und Theorischen Physik*. Franz Deuticke, Leipzig-Wien. Reprint, 1945: Rosenberg, New York.

Moore, D. S. and G. P. McCabe (1993). *Introduction to the Practice of Statistics*. Second edition. W. H. Freeman & Co., Salt Lake City, Utah.

Moriguti, S. (1951). Extremal properties of extreme value distributions. *Ann. Math. Statist.* **22**, 523–536.

Moriguti, S. (1953). A modification of Schwarz's inequality with applications to distributions. *Ann. Math. Statist.* **24**, 107–113.

Moriguti, S. (1954). Bounds for second moments of the sample range. *Rep. Statist Appl. Res. JUSE* **3**, 57–64.

Morrison, D. F. (1967). *Multivariate Statistical Methods*. McGraw-Hill, New York.

Mosteller, F. (1946). On some useful "inefficient" statistics. *Ann. Math. Statist.* **17**, 377–408.

Murphy, R. B. (1948). Non-parametric tolerance limits. *Ann. Math. Statist.* **19**, 581–589.

Nagaraja, H. N. (1988). Record values and related statistics – A review. *Commun. Statist. – Theory Meth.* **17**(7), 2223–2238.

Nagaraja, H. N. (1990). Some reliability properties of order statistics. *Commun. Statist. – Theory Meth.* **19**(1), 307–316.

Nagaraja, H. N. (1992). Order statistics from discrete distributions (with discussion) *Statistics.* **23**, 189–216.

Nanda, D. N. (1948a). Distribution of a root of a determinantal equation. *Ann. Math. Statist.* **19**, 47–57.

Nanda, D. N. (1948b). Limiting distribution of a root of a determinantal equation. *Ann. Math. Statist.* **19**, 340–350.

Nelson, L. S. (1974). Distribution-free tolerance limits. *J. Qual. Technol.* **6**, 163–164.

Nelson, W. (1982). *Applied Life Data Analysis.* John Wiley & Sons, New York.

Nevzorov, V. B. (1987). Records, *Teoriya Veroyatnostei i ee Premeneniya.* (*Theory Probab. Appl.*) **32**(2), 219–251.

Nevzorov, V. and N. Balakrishnan (1998). In this volume.

Newcomb, S. (1886). A generalized theory of the combination of observations so as to obtain the best result. *Amer. J. Math.* **8**, 343–366.

Newman, D. (1939). The distribution of range in samples from a normal population, expressed in terms of an independent estimate of standard deviation. *Biometrika* **31**, 20–30.

Ogawa, J. (1951). Contributions to the theory of systematic statistics, I. *Osaka Math. J.* **3**, 175–213.

Ogawa, J. (1962). Determination of optimum spacings in the case of the normal distribution. In: *Contributions to Order Statistics* (Eds., A. E. Sarhan and B. G. Greenberg), pp. 272–283, John Wiley & Sons, New York.

Olkin, I. and M. Viana (1995). Correlation analysis of extreme observations from a multivariate normal distribution. *J. Amer. Statist. Assoc.* **90**, 1373–1379.

Owen, D. B. (1962). *Handbook of Statistical Tables.* Addison-Wesley, Reading, MA.

Palmieri, F. and C. G. Boncelet, Jr. (1988). Design of order statistics filters I: L-Filters. *Technical Report 88-4-2*, Department of Electrical Engineering, University of Delaware.

Parrish, R. S. (1992). Computing variances and covariances of normal order statistics. *Commun. Statist. – Simul. Comput.* **21**(1), 71–101.

Patel, J. K. (1986). Tolerance limits – A review. *Commun. Statist. – Theory Meth.* **15**(9), 2719–2762.

Patel, J. K. (1989). Prediction intervals – A review. *Commun. Statist. – Theory Meth.* **18**(7), 2393–2466.

Patil, G. P., S. Kotz and J. K. Ord (Eds.) (1975). *Statistical Distributions in Scientific Work* – Vol. 3: *Characterizations and Applications.* D. Reidel, Dordrecht.

Pearson, E. S. (1935). *The Applications of Statistical Methods to Industrial Standardisation and Quality Control.* Publication No. 600, British Standards Association, London.

Pearson, E. S. and J. Haines (1935). The use of range in place of standard deviation in small samples. *J. Roy. Statist. Soc. Suppl.* **2**, 83–98.

Pearson, K. (1900). On the criterion that a given system of deviations from the probable in the case of a correlated system of variables is such that it can be reasonably supposed to have arisen from random sampling. *Philos. Mag.* (5) **50**, 157–175.

Pearson, K. (1902). Note on Francis Galton's problem. *Biometrika* **1**, 390–399.

Pearson, K. (1920). On the probable errors of frequency constants. Part III. *Biometrika* **13**, 113–132.

Peirce, B. (1852). Criterion for the rejection of doubtful observations. *Astron. J.* **2**, 161–163.

Pillai, K.C. S. (1960). *Statistical Tables for Tests of Multivariate Hypotheses.* Statistical Center, University of the Philippines, Manila.

Pledger, G. and F. Proschan (1971). Comparisons of order statistics from heterogeneous distributions. In: *Optimizing Methods in Statistics* (Ed., J. S. Rustagi), pp. 89–113, Academic Press, New York.

Proschan, F. and J. Sethuraman (1976). Stochastic comparisons of order statistics from heterogeneous populations, with applications in reliability. *J. Multivar. Anal.* **6**, 608–616.

Pyke, R. (1959). The supremum and infimum of the Poisson process. *Ann. Math. Statist.* **30**, 568–576.

Pyke, R. (1965). Spacings. *J. Roy. Statist. Soc. Ser. B* **27**, 395–436; Discussion, 437–449.

Quesenberry, C. P. and M. P. Gessaman (1968). Nonparametric discrimination using tolerance regions. *Ann. Math. Statist.* **39**, 664–673.

Ramachandran, B. and K. S. Lau (1991). *Functional Equations in Probability Theory*. Academic Press, San Diego, CA.

Ramberg, J. S. and P. R. Tadikamalla (1978). On generation of subsets of order statistics. *J. Statist. Comput. Simul.* **6**, 239–241.

Rao, C. R. (1948). Tests of significance in multivariate analysis. *Biometrika* **35**, 58–79.

Rao, C. R. (1965). *Linear Statistical Inference and its Applications*. John Wiley & Sons, New York. Second edition, 1973.

Rao, C. R. and D. N. Shanbhag (1994). *Choquet – Deny Type Functional Equations with Applications to Stochastic Models*. John Wiley & Sons, Chichester, England.

Rao, C. R. and D. N. Shanbhag (1998). In this volume.

Reeder, H. A. (1972). Machine generation of order statistics. *Amer. Statist.* **26**(4), 56–57.

Reiss, R. D. (1983). In *Probability and Statistical Decision Theory*. Proceedings of 4th Pannonian Symposium on Mathematical Statistics, Vol. A, pp. 293–300, Bad Tatzmannsdorf, Austria.

Reiss, R. D. (1989). *Approximate Distributions of Order Statistics: With Applications to Nonparametric Statistics*. Springer-Verlag, Berlin, Germany.

Resnick, S. I. (1973). Record values and maxima. *Ann. Probab.* **1**, 650–662.

Resnick, S. I. (1987). *Extreme Values, Regular Variation, and Point Processes*. Springer-Verlag, New York.

Rider, P. R. (1933). Criteria for rejection of observations. *Washington University Studies, New Series, Science and Technology*, No. 8.

Rinott, Y. and E. Samuel-Cahn (1994). Covariance between variables and their order statistics for multivariate normal variables. *Statist Probab. Lett.* **21**, 153–155.

Rossberg, H. J. (1960). Über die Verteilungsfunktionen der Differenzen und Quotienten von Ranggrössen. *Math. Nachr.* **21**, 37–79.

Rossberg, H. J. (1972). Characterization of the exponential and the Pareto distributions by means of some properties of the distributions which the differences and quotients of order statistics are subject to. *Mathematische Operationsforschung und Statistik* **3**, 207–216.

Rousseeuw, P. J. and A. M. Leroy (1987). *Robust Regression and Outlier Detection*. John Wiley & Sons, New York.

Roy, J. (1958). Step-down procedure in multivariate analysis. *Ann. Math. Statist.* **29**, 1177–1187.

Roy, S. N. (1939). p-statistics or some generalizations in analysis of variance appropriate to multivariate problems. *Sankhyā* **4**, 381–396.

Roy, S. N. (1940a). Distribution of p-statistics on the non-null hypothesis. *Science and Culture* **5**, 562–563.

Roy, S. N. (1940b). Distribution of certain symmetric functions of p-statistics on the null hypothesis. *Science and Culture* **5**, 563.

Roy, S. N. (1942a). The sampling distribution of p-statistics and certain allied statistics on the non-null hypothesis. *Sankhyā* **6**, 15–34.

Roy, S. N. (1942b). Analysis of variance for the multivariate normal populations: the sampling distribution of the requisite p-statistics on the null and non-null hypotheses. *Sankhyā* **6**, 35–50.

Roy, S. N. (1945). The individual sampling distribution of the maximum, the minimum and any intermediate of the p-statistics on the null hypothesis. *Sankhyā* **7**, 133–158.

Roy, S. N. (1946a). Multivariate analysis of variance: the sampling distribution of the numerically largest of the p-statistics on the non-null hypothesis. *Sankhyā* **8**, 15–32.

Roy, S. N. (1946b). A note on multi-variate analysis of variance when the number of variates is greater than the number of linear hypotheses per character. *Sankhyā* **8**, 53–66.

Roy, S. N. (1953). On a heuristic method of test construction and its use in multivariate analysis. *Ann. Math. Statist.* **24**, 220–238.

Roy, S. N. (1957). *Some Aspects of Multivariate Analysis*. John Wiley & Sons, New York; Indian Statistical Institute, Calcutta.

Roy, S. N. and R. E. Bargmann (1958). Tests of multiple independence and the associated confidence bounds. *Ann. Math. Statist.* **29**, 491–503.

Roy, S. N. and R. C. Bose (1953). Simultaneous confidence interval estimation. *Ann. Math. Statist.* **24**, 513–536.

Rychlik, T. (1998). In this volume.

Sarhan, A. E. and B. G. Greenberg (Eds.) (1962). *Contributions to Order Statistics*. John Wiley & Sons, New York.

Sarkar, S. K. and W. Wang (1998). In the companion volume.

Sathe, Y. S. and S. M. Bendre (1991). Log-concavity of probability of occurrence of at least r independent events. *Statist. Probab. Lett.* **11**, 63–64.

Sathe, Y. S. and U. J. Dixit (1990). On a recurrence relation for order statistics. *Statist. Probab. Lett.* **9**, 1–4.

Scheffé, H. and J. W. Tukey (1945). Non-parametric estimation, I. Validation of order statistics. *Ann. Math. Statist.* **16**, 187–192.

Schmeiser, B. W. (1978). Order statistics in digital computer simulation. A survey. *Proceedings of the 1978 Winter Simulation Conference*, pp. 136–140.

Schneider, H. (1986). *Truncated and Censored Samples from Normal Populations*. Marcel Dekker, New York.

Schneider, H. and F. Barbera (1998). In the companion volume.

Schucany, W. R. (1972). Order statistics in simulation. *J. Statist. Comput. Simul.* **1**, 281–286.

Seal, K. C. (1955). On a class of decision procedures for ranking means of normal populations. *Ann. Math. Statist.* **26**, 387–398.

Sen, P. K. (1970). A note on order statistics for heterogeneous distributions. *Ann. Math. Statist.* **41**, 2137–2139.

Sen, P. K. (1998). In this volume.

Sen, P. K. and I. A. Salama (Eds.) (1992). *Order Statistics and Nonparametrics: Theory and Applications*. North-Holland, Amsterdam.

Serfling, R. J. (1980). *Approximation Theorems of Mathematical Statistics*. John Wiley & Sons, New York.

Shah, B. K. (1966). On the bivariate moments of order statistics from a logistic distribution. *Ann. Math. Statist.* **37**, 1002–1010.

Shah, B. K. (1970). Note on moments of a logistic order statistics. *Ann. Math. Statist.* **41**, 2151–2152.

Shanbhag, D. N. (1998). In this volume.

Shapiro, S. S. (1980). *How to Test Normality and Other Distributional Assumptions*. Vol. 3, ASQC Basic References in Quality Control: Statistical Techniques, American Society for Quality Control, Milwaukee, WI.

Shapiro, S. S. (1998). In the companion volume.

Shapiro, S. S. and M. B. Wilk (1965). An analysis of variance test for normality (complete samples). *Biometrika* **52**, 591–611.

Shapiro, S. S., M. B. Wilk and (Mrs.) H. J. Chen (1968). A comparative study of various tests for normality. *J. Amer. Statist. Assoc.* **63**, 1343–1372.

Shewhart, W. A. (1931). *Economic Control of Quality of Manufactured Product*. D. van Nostrand Company, New York.

Sheynin, O. B. (1966). Origin of the theory of errors. *Nature* **211**, 1003–1004.

Shorack, G. R. and J. A. Wellner (1986). *Empirical Processes with Applications to Statistics*. John Wiley & Sons, New York.

Shorrock, R. W. (1973). Record values and inter-record times. *J. Appl. Probab.* **10**, 543–555.

Siddiqui, M. M. (1970). Order statistics of a sample and of an extended sample. In: *Nonparametric Techniques in Statistical Inference* (Ed., M. L. Puri), pp. 417–423, Cambridge University Press, Cambridge, England.

Siegel, A. F. (1993). A surprising covariance involving the minimum of multivariate normal variables. *J. Amer. Statist. Assoc.* **88**, 77–80.

Sillitto, G. P. (1964). Some relations between expectations of order statistics in samples of different sizes. *Biometrika* **51**, 259–262.

Smirnov, N. (1939). Sur les écarts de la courbe de distribution empirique (Russian, French summary). *Mat. Sb.* **48** (N.S. **6**), 3–26.
Somerville, P. R. (1958). Values for obtaining non-parametric tolerance limits. *Ann. Math. Statist.* **29**, 599–601.
Srikantan, K. S. (1962). Recurrence relations between the PDF's of order statistics, and some applications. *Ann. Math. Statist.* **33**, 169–177.
Srivastava, J. N. (1966). Some generalizations of multivariate analysis of variance. In: *Multivariate Analysis* (Ed., P. R. Krishnaiah), pp. 129–144, Academic Press, New York.
Srivastava, J. N. (1969). Some studies on intersection tests in multivariate analysis of variance. In: *Multivariate Analysis – II* (Ed., P. R. Krishnaiah), pp. 145–168, Academic Press, New York.
Stigler, S. M. (1973a). The asymptotic distribution of the trimmed mean. *Ann. Statist.* **1**, 472–477.
Stigler, S. M. (1973b). Simon Newcomb, Percy Daniell, and the history of robust estimation 1885–1920. *J. Amer. Statist. Assoc.* **68**, 872–879.
Sugiura, N. (1962). On the orthogonal inverse expansion with an application to the moments of order statistics. *Osaka Math. J.* **14**, 253–263.
Sugiura, N. (1964). The bivariate orthogonal inverse expansion and the moments of order statistics. *Osaka J. Math.* **1**, 45–59.
Tadikamalla, P. R. and N. Balakrishnan (1998). In this volume.
Takahasi, K. (1988). A note on hazard rates of order statistics. *Commun. Statist. – Theory Meth.* **17**, 4133–4136.
Teichroew, D. (Ed.) (1956). Tables of expected values of order statistics and products of order statistics for samples of size twenty and less from the normal distribution. *Ann. Math. Statist.* **27,** 410–426.
Theil, H. and A. L. Nagar (1961). Testing the independence of regression disturbances. *J. Amer. Statist. Assoc.* **56**, 793–806.
Thomson, G. H. (1934). Hotelling's method modified to give Spearman's g. *J. Educ. Psych.* **25**, 366–374.
Tiago de Oliveira, J. (1975). Bivariate extremes. Extensions. *40th Session of the Internat. Statist. Inst.* Warsaw, Poland.
Tiago de Oliveira, J. (Ed.) (1984). *Statistical Extremes and Applications*. D. Reidel, Dordrecht.
Tietjen, G. L., D. K. Kahaner and R. J. Beckman (1977). Variances and covariances of the normal order statistics for sample sizes 2 to 50. *Selected Tables in Mathematical Statistics* **5**, 1–73.
Tiku, M. L. (1975). A new statistic for testing suspected outliers. *Commun. Statist.* **4**, 737–752; Comment by D. M. Hawkins, **6** (1977), 435–438; Rejoinder, **6**, 1417–1422.
Tiku, M. L., W. Y. Tan and N. Balakrishnan (1986). *Robust Inference*. Marcel Dekker, New York.
Tintner, G. (1945). A note on rank, multicollinearity and multiple regression. *Ann. Math. Statist.* **16**, 304–308.
Tintner, G. (1946). Some applications of multivariate analysis to economic data. *J. Amer. Statist. Assoc.* **41**, 472–500.
Tintner, G. (1950). Some formal relations in multivariate analysis. *J. Roy. Statist. Soc. Ser. B* **12**, 95–101.
Tippett, L. H. C. (1925). On the extreme individuals and the range of samples taken from a normal population. *Biometrika* **17**, 364–387.
Tukey, J. W. (1947). Non-parametric estimation, II. Statistically equivalent blocks and tolerance regions – the continuous case. *Ann. Math. Statist.* **18**, 529–539.
Tukey, J. W. (1948). Nonparametric estimation, III. Statistically equivalent blocks and multivariate tolerance regions – the discontinuous case. *Ann. Math. Statist.* **19**, 30–39.
Tukey, J. W. (1953). The problem of multiple comparisons. *Unpublished Memorandum*, Princeton University, Princeton, NJ.
Tukey, J. W. (1970). *Exploratory Data Analysis*. (Limited Preliminary Edition). Addison-Wesley, Reading, MA.
Tukey, J. W. (1994). *Collected Works, Volume VIII: Multiple Comparisons: 1948–1983* (Ed., H. J. Braun). Chapman & Hall, New York.

Tukey, J. W. and D. H. McLaughlin (1963). Less vulnerable confidence and significance procedures for location based on a single sample: trimming/Winsorization, 1. *Sankhyā Ser. A.* **25**, 331–352.

Van Zwet, W. R. (1964). *Convex Transformations of Random Variables.* Mathematical Centre Tracts No. 7, Mathematisch Centrum, Amsterdam.

Vaughan, R. J. and W. N. Venables (1972). Permanent expressions for order statistics densities. *J. Roy. Statist. Soc. Ser. B* **34**, 308–310.

Von Bortkiewicz, L. (1922). Die Variationsbreite beim Gauss'schen Fehlergesetz. *Nordisk Statistisk Tidskrift* **1**, 11–38, 193–220.

Viswanathan, R. (1998). In the companion volume.

Wald, A. (1943). An extension of Wilks' method for setting tolerance limits. *Ann. Math. Statist.* **14**, 45–55.

Walsh, J. E. (1962). Some two-sided distribution-free tolerance intervals of a general nature. *J. Amer. Statist. Assoc.* **57**, 775–784.

Wang, W., S. K. Sarkar and Z. D. Bai (1996). Some new results on covariances involving order statistics from dependent random variables. *J. Multivar. Anal.* **59**, 308–316.

Watson, G. S. (1955). Serial correlation in regression analysis. I. *Biometrika* **42**, 327–341.

Watson, G. S. (1961). Goodness-of-fit tests on a circle. *Biometrika* **48**, 109–114.

Watson, G. S. (1967). Linear least squares regression. *Ann. Math. Statist.* **38**, 1679–1699.

Watson, G. S. and E. J. Hannan (1956). Serial correlation in regression analysis. II. *Biometrika* **43**, 436–448.

Weibull, W. (1939a). A statistical theory of the strength of materials. *Ingeniörs Vetenskaps Akademien Handlingar*, Number 151, Generalstabens Litografiska Anstalts Förlag, Stockholm.

Weibull, W. (1939b). The phenomenon of rupture in solids. *Ingeniörs Vetenskaps Akademien Handlingar*, Number 153, Generalstabens Litografiska Anstalts Förlag, Stockholm.

Wigner, E. P. (1959). Statistical properties of real symmetric matrices with many dimensions. *Proceedings of the 4th Canadian Mathematical Congress*, pp. 174–184, University of Toronto Press, Toronto.

Wilk, M. B. and R. Gnanadesikan (1968). Probability plotting methods for the analysis of data. *Biometrika* **55**, 1–17.

Wilks, S. S. (1941). Determination of sample sizes for setting tolerance limits. *Ann. Math. Statist.* **12**, 118–119.

Wilks, S. S. (1942). Statistical prediction with special reference to the problem of tolerance limits. *Ann. Math. Statist.* **13**, 400–409.

Wilks, S. S. (1943). *Mathematical Statistics.* Princeton University Press, Princeton, NJ.

Wilks, S. S. (1948). Order statistics. *Bull. Amer. Math. Soc.* **54**, 6–50.

Wolfowitz, J. (1949). Non-parametric statistical inference. In: *Proc. Berkeley Symposium on Math. Statist. Probab.* (Ed., J. Neyman), pp. 93–113, University of California Press, Berkeley, CA.

Woodruff, B. W. and A. H. Moore (1988). Applications of goodness-of-fit tests in reliability. In: *Handbook of Statistics – Volume 7: Quality Control and Reliability* (Eds., P. R. Krishnaiah and C. R. Rao), pp. 113–126, North-Holland, Amsterdam.

Yamauti, Z. (Ed.) (1972). *Statistical Tables and Formulas with Computer Applications.* Japan Standards Association, Tokyo (Table C2, pp. 38–49).

Young D. H. (1967). Recurrence relations between the P.D.F.'s of order statistics of dependent variables, and some applications. *Biometrika* **54**, 283–292

Computer Simulation of Order Statistics

Pandu R. Tadikamalla and N. Balakrishnan

1. Introduction

Recall from the earlier chapters that if $X_1, X_2, \ldots, X_n$ denote a random sample from a distribution with distribution function $F(x)$ and $X_{1:n} \leq X_{2:n} \leq \cdots \leq X_{n:n}$ are the corresponding ordered observations, then $X_{1:n}, X_{2:n}, \ldots, X_{n:n}$ are called the order statistics corresponding to $X_1, X_2, \ldots, X_n$. Since order statistics play an important role in statistical inference, there has been considerable interest in computer generation of order statistics for Monte Carlo simulation studies.

A straightforward method for generating these order statistics is to generate a random sample of size n from the corresponding distribution and then sort the sample to obtain the desired order statistics. For medium and large values of n, this sorting procedure can be very time-consuming. Hence, a direct method for generating order statistics is of great value.

2. Direct generation of order statistics

Suppose we are interested in generating order statistics from a continuous distribution with distribution function $F(\cdot)$. If the inverse transformation method is used to generate the random sample [see Tadikamalla and Johnson (1981)], there is a direct correspondence between the order statistics of $(X_1, X_2, \ldots, X_n)$ and the order statistics of the associated uniform sample $(U_1, U_2, \ldots, U_n)$. Since F^{-1} is a monotonic function, $Y_i = F^{-1}(U_{i:n})$, $i = 1, 2, \ldots, n$, represent the order statistics from the distribution with distribution function $F(\cdot)$, where $U_{1:n} \leq U_{2:n} \leq \cdots \leq U_{n:n}$ are the order statistics from the uniform$(0, 1)$ distribution. Thus, if the inverse transformation method is used to sample from the distribution $F(\cdot)$, the problem of generating order statistics reduces to the problem of generation of order statistics from the uniform$(0, 1)$ distribution.

3. Generation of uniform(0, 1) ordered samples

Balakrishnan and Cohen (1991), David (1981), and Arnold, Balakrishnan and Nagaraja (1992) all discuss the marginal and joint probability densities of order statistics from a continuous distribution with probability density function $f(x)$ and distribution function $F(x)$. These results take a simple form in the case of the uniform$(0, 1)$ distribution, yielding some simple methods for sequential generation of order statistics. The ith order statistic from uniform$(0, 1)$ has a beta density given by

$$f(u) = \frac{n!}{(i-1)!(n-i)!} u^{i-1}(1-u)^{n-i}, \qquad 0 \le u \le 1. \tag{3.1}$$

The joint density of the ith and j^{th} order statistics $(i < j)$ from uniform$(0, 1)$ is given by

$$f(u_{i:n}, u_{j:n}) = \frac{n!}{(i-1)!(j-i-1)!(n-j)!} u_{i:n}^{i-1}(u_{j:n} - u_{i:n})^{j-i-1}(1-u_{j:n})^{n-j},$$

$$0 \le u_{i:n} < u_{j:n} \le 1. \tag{3.2}$$

From (3.1), it can be seen that the distribution of the largest order statistic $(U_{n:n})$ from a sample of size n reduces to

$$f_{n:n}(u) = nu^{n-1}, \qquad 0 \le u \le 1. \tag{3.3}$$

Similarly, the distribution of smallest order statistic $(U_{1:n})$ from a simple of size n is given by

$$f_{1:n}(u) = n(1-u)^{n-1}, \qquad 0 \le u \le 1. \tag{3.4}$$

Based on the above results and some other results pertaining to uniform and exponential spacings [Devroye (1986)], the methods for generating uniform$(0, 1)$ order statistics can be summarized as follows.

3.1. Sorting

A naive approach is to generate i.i.d. uniform$(0, 1)$ random variates and sort them. Devroye (1986) discusses several efficient sorting algorithms. Unsophisticated approach to sorting based on pairwise comparisons of the numbers would be very time consuming. The worst case and expected times taken by these algorithms are $O(n \log n)$ [see Knuth (1973)]. In the case of uniform$(0, 1)$ variates, one can take advantage of truncation and bucket sort the u_i's in expected time $O(n)$. For example, see Devroye and Klincsek (1981) and Devroye (1986).

3.2. Ascending order

Lurie and Hartley (1972) gave a direct method for generating a complete set of n uniform order statistics in ascending order, as follows: Let $v_1, v_2, v_3, \ldots$ be observations from the uniform$(0, 1)$ distribution. Set

$$u_{1:n} = 1 - v_1^{1/n}$$
$$u_{2:n} = 1 - (1-u_{1:n})v_2^{1/(n-1)}$$
$$\vdots$$
$$u_{i+1:n} = 1 - (1-u_{i:n})v_{i+1}^{1/(n-i)}$$
$$\vdots$$

Then $u_{1:n}, u_{2:n}, \ldots, u_{n:n}$ are order statistics from the uniform$(0,1)$ distribution.

3.3. Descending order

Schucany (1972) used a similar approach to generate the order statistics in descending order, without sort, as follows:

$$u_{n:n} = v_1^{1/n}$$
$$\vdots$$
$$u_{n-i:n} = u_{n-i+1:n}v_{i+1}^{1/(n-i)}$$
$$\vdots$$

Reeder (1972) suggested that generating order statistics in *descending* order is faster than that in ascending order, and Lurie and Mason (1973) confirmed this claim.

3.4. Based on uniform spacing

If $E_1, E_2, \ldots, E_{n+1}$ are i.i.d. exponential variates and G is their sum, then

$$u_{1:n} = E_1/G$$
$$u_{2:n} = u_{1:n} + (E_2/G)$$
$$\vdots$$
$$u_{j:n} = u_{j-1:n} + (E_j/G)$$
$$\vdots$$
$$u_{n:n} = u_{n-1:n} + (E_n/G)$$

are n order statistics from uniform$(0,1)$ distribution. Tadikamalla and Balakrishnan (1995) discuss several methods for generating exponential variates; also see Marsaglia, MacLaren and Bray (1964).

3.5. Subsets of order statistics

Suppose that one is interested in generating a subset of k consecutive uniform$(0, 1)$ order statistics from a sample of size n. If the required k order statistics are at the upper extreme of the sample, they can be generated using Schucany's (1972) descending method. If the required subset is at the lower extreme, they can be generated using Lurie and Hartley's (1972) ascending method.

Neither of these methods is convenient if the required subset falls near the middle of the sample. For example, if we are interested in generating the middle two order statistics from a sample of size $n = 50$, then both of the previously described methods require the generation of twenty six order statistics. A more efficient approach may be to generate the *largest required order statistic* (26th in our example) directly; then generate the remaining set conditionally using the descending method. Ramberg and Tadikamalla (1978) suggest generating $u_{i:n}$ as a beta variate (with parameters i and $n - i + 1$) and generating $u_{i-1:n}, u_{i-2:n}, \ldots$ using the above recursion algorithm suggested by Schucany (1972). If fast algorithms for generating beta variates are used [Schmeiser and Babu (1980) and Johnson, Kotz and Balakrishnan (1995)], this procedure may be worth the effort. This approach will probably be faster in generating a subset of order statistics when the required subset is somewhere in the middle of the sample.

Horn and Schlipf (1986) similarly discussed the simulation of the central $n - 2i + 2$ order statistics from a sample of size n from the uniform$(0, 1)$ distribution. Their approach is first to generate $u_{n-i+1:n}$ directly as a Beta$(n - i + 1, i)$ variate, next to generate $u_{i:n}$ as $u_{n-i+1:n}v$ where v is a Beta$(i, n - 2i + 1)$ variate, and finally to generate $u_{i+1:n}, \ldots, u_{n-i:n}$ as order statistics from a sample of size $n - 2i$ from the uniform$(u_{i:n}, u_{n-i+1:n})$ distribution.

Instead, if one is interested in simulating the i largest and smallest order statistics from a sample of size n from the uniform$(0, 1)$ distribution, Horn and Schlipf (1986) proposed an algorithm by combining the ascending and descending algorithms. Specifically, their method is first to generate $u_{n-i+1:n}$ directly as a Beta$(n - i + 1, i)$ variate, next to generate $u_{i:n}$ as $u_{n-i+1:n}v$ where v is a Beta$(i, n - 2i + 1)$ variate, and finally to generate $(u_{1:n}, \ldots, u_{i-1:n})$ and $(u_{n-i+2:n}, \ldots, u_{n:n})$ as order statistics from samples of size $i - 1$ each from uniform$(0, u_{i:n})$ and uniform$(u_{n-i+1:n}, 1)$ distributions, respectively.

4. Generation of progressive Type-II censored order statistics

Let us first define a progressive Type-II censored sample. Under this censoring scheme, n identical items are placed on a life test; immediately after the first failure, R_1 surviving items are removed at random from the test; after the next failure, R_2 surviving items are removed at random from the test, and so on; finally, immediately after the mth failure, R_m remaining surviving items are removed from the test. Under this censoring scheme, we will observe in all m failures and $R_1 + R_2 + \cdots + R_m = n - m$ progressively censored items.

In the case of the uniform$(0, 1)$ distribution, let us denote these progressive Type-II censored order statistics by $u_{1:m:n}, u_{2:m:n}, \ldots, u_{m:m:n}$.

Then, with

$$v_i = \frac{1 - u_{m-i+1:m:n}}{1 - u_{m-i:m:n}} \quad \text{for} \quad i = 1, 2, \ldots, m-1 \quad \text{and} \quad v_m = 1 - u_{1:m:n},$$

and

$$w_i = v_i^{i+R_m+R_{m-1}+\cdots+R_{m-i+1}} \quad \text{for} \quad i = 1, 2, \ldots, m,$$

Balakrishnan and Sandhu (1995) have established that w_i $(i = 1, 2, \ldots, m)$ are all independent and identically distributed as uniform$(0, 1)$. Incidentally, this result generalizes a distributional property of uniform order statistics established by Malmquist (1950) and stated earlier in this article. Using this result, Balakrishnan and Sandhu (1995) proposed the following algorithm for generating a progressive Type-II censored sample:

1. Generate m independent uniform$(0, 1)$ variates denoted by $w_1, w_2, \ldots, w_m$.
2. Set $v_i = w_i^{1/(i+R_m+R_{m-1}+\cdots+R_{m-i+1})}$ for $i = 1, 2, \ldots, m$.
3. Set $u_{i:m:n} = 1 - v_m v_{m-1} \cdots v_{m-i+1}$ for $i = 1, 2, \ldots, m$.

Then, $u_{1:m:n}, u_{2:m:n}, \ldots, u_{m:m:n}$ is the required progressive Type-II censored sample from the uniform$(0, 1)$ distribution. Of course, upon setting $x_{i:m:n} = F^{-1}(u_{i:m:n})$ for $i = 1, 2, \ldots, m$, we can produce the required progressive Type-II censored sample from the distribution $F(\cdot)$.

Aggarwala and Balakrishnan (1998) have also proposed a similar algorithm for generating a general progressive Type-II censored sample with left censoring present.

5. Miscellaneous topics

5.1. Partitioning method for the maximum of n i.i.d. random variables

Schmeiser (1978) developed a technique for generating the maximum of n random variables $X_1, X_2, \ldots, X_n$. The technique is based on partitioning the (0,1) interval and the range of X so that some times not all X_i need to be generated. He presents optimal single, optimal double and infinite partition plans. The technique is applicable for both i.i.d. and non-i.i.d. cases and yields greater savings for non-identically distributed random variables.

5.2. Generating the largest order statistic

In generating $X_{n:n} = F^{-1}(u_{n:n})$, Devroye (1980) considers the case of $u_{n:n}$ when n is very large and $u_{n:n} = v_1^{1/n}$ may cause numerical problems. The problem is that for a large n, $v_1^{1/n}$ is close to 1, so that in regular wordsize arithmetic, there may be an accuracy problem. This problem can be alleviated if we use $G(x) = 1 - F(x)$ and the algorithm will be as follows:

1. Generate an exponential variate E and a gamma variate G_n (with shape parameter n) [see Tadikamalla and Johnson (1981)].
2. Then set $X_{n:n} = G^{-1}(E/(E+G_n))$.

Nagaraja (1979) has given an interesting probability result concerning this generation problem.

5.3. Methods for exponential order statistics

Newby (1979) presents the following algorithm based on the exponential spacings (Sukhatme, 1937). If $E_1, E_2, \ldots, E_n$ are n i.i.d. exponential variates, then

$$E_{1:n} = E_1/n$$
$$E_{2:n} = E_{1:n} + E_2/(n-1)$$
$$\vdots$$
$$E_{i:n} = E_{i-1:n} + E_i/(n-i+1)$$
$$\vdots$$
$$E_{n:n} = E_{n-1:n} + E_n$$

5.4. Generating order statistics from arbitrary distributions

Rabonowitz and Berenson (1974) and Gerontidis and Smith (1982) studied the so-called grouping method which is a hybrid version of the inversion method and the bucket sorting method.

The *grouping method* can be briefly described as follows. Given an integer k (a suggested value is $n/4$), divide the range of the distribution into k equal probability intervals. Next, generate a multinomial vector $(m_1, m_2, \ldots, m_k)$ corresponding to the division of n objects independently among k equally likely cells [see Fishman (1978), Ho, Gentle, and Kennedy (1979) and Johnson, Kotz and Balakrishnan (1997)]. Then, draw m_j variables (from the specified general distribution) from the jth interval for $1 \leq j \leq k$, and sort each group of m_j variables directly and put the k groups together to obtain a complete ordered sample. Gerontidis and Smith (1982) recommend the inversion method if F^{-1} exists in a simple closed form as is in the case of the exponential distribution.

References

Aggarwala, R. and N. Balakrishnan (1998). Some properties of progressive censored ordered statistics from arbitrary and uniform distributions, with applications to inference and simulation. *J. Statist. Plan. Inf.* (to appear).

Arnold, B. C., N. Balakrishnan and H. N. Nagaraja (1992). *A First Course in Order Statistics*. John Wiley & Sons, New York.

Balakrishnan, N. and A. C. Cohen (1991). *Order Statistics and Inference: Estimation Methods.* Academic Press, San Diego.

Balakrishnan, N. and R. A. Sandhu (1995). A simple simulational algorithm for generating progressive Type-II censored samples. *Amer. Statist.* **49**, 229–230.

David, H. A. (1981). *Order Statistics.* Second edition, John Wiley & Sons, New York.

Devroye, L. (1986). *Non-uniform Random Variate Generation.* Springer-Verlag, New York.

Devroye, L. and T. Klincsek (1981). Average time behavior of distributive sorting algorithms. *Computing* **26**, 1–7.

Fishman, G. S. (1978). Sampling from the multinomial distribution on a computer. *Technical Report #78–5*, Curriculum in Operations Research and Systems Analysis. University of North Carolina, Chapel Hill.

Gerontidis, I. and R. L. Smith (1982). Monte Carlo generation of order statistics from general distributions. *Appl. Statist.* **31**, 238–243.

Ho, F. C. M., J. E. Gentle and W. J. Kennedy (1979). Generation of random variates from the multinomial distribution. *Proceedings of the Statistical Computing Section, Amer. Statist. Assoc.* 336–339.

Horn, P. S. and J. S. Schlipf (1986). Generating subsets of order statistics with applications to trimmed means and means of trimmings. *J. Statist. Comput. Simul.* **24**, 83–97.

Johnson, N. L., S. Kotz and N. Balakrishnan (1995). *Continuous Univariate Distributions–Vol. 2*, Second edition. John Wiley & Sons, New York.

Johnson, N. L., S. Kotz and N. Balakrishnan (1997). *Discrete Multivariate Distributions.* John Wiley & Sons, New York.

Knuth, D. E. (1973). *The Art of Computer Programming, Vol. 3: Searching and Sorting.* Addison-Wesley, Reading, Massachusetts.

Lurie, D. and H. O. Hartley (1972). Machine generation of order statistics for Monte Carlo computations. *Amer. Statist.* **26**, 26–27.

Lurie, D. and R. L. Mason (1973). Empirical investigation of general techniques for computer generation of order statistics. *Comm. Statist.* **2**, 363–371.

Marsaglia, G., M. D. MacLaren and T. A. Bray (1964). A fast procedure for generating exponential random variables. *Communications of the ACM* **7**, 298–300.

Nagaraja, H. N. (1979). Some relations between order statistics generated by different methods. *Comm. Statist. – Simul. Comput.* **B8**, 369–377.

Newby, M. J. H. (1979). The simulation of order statistics from life distributions. *Appl. Statist.* **28**, 298–301.

Rabonowitz, M. and M. L. Berenson (1974). A comparison of various methods of obtaining random order statistics for Monte Carlo computations. *Amer. Statist.* **28**, 27–29.

Ramberg, J. S. and P. R. Tadikamalla (1978). On the generation of subsets of order statistics. *J. Statist. Comp. Simul.* **7**, 239–241.

Reeder, H. A. (1972). Machine generation of order statistics. *Amer. Statist.* **26**, 56–57.

Schmeiser, B (1978). Generation of the maximum (minimum) value in digital computer simulation. *J. Statist. Comp. Simul.* **8**, 103–115.

Schmeiser, B. W. (1978). Order statistics in digital computer simulation. A survey. *Proceedings of the 1978 Winter Simulation Conference* 136–140.

Schmeiser, B. W. and A. J. G. Babu (1980). Beta variate generation via exponential majorizing functions. *Oper. Res.* **28**, 917–926.

Schucany, W. R. (1972). Order statistics in simulation. *J. Statist. Comp. Simul.* **1**, 281–286.

Sukhatme, P. V. (1937). Tests of significance for samples of the chi square population with two degrees of freedom. *Ann. Eugenics* **8**, 52–56.

Tadikamalla, P. R. (1975). Modeling and generating stochastic methods for simulation studies. *Ph. D. Thesis*, University of Iowa, Iowa City.

Tadikamalla, P. R. (1984). Modeling and generating stochastic inputs for simulation studies. *Amer. J. Math. Mgmt. Sci.* **3 & 4**, 203–223.

Tadikamalla, P. R. and M. E. Johnson (1981). A complete guide to gamma variate generation. *Amer. J. Math. Mgmt. Sci.* **1**, 213–236.

Tadikamalla, P. R. and N. Balakrishnan (1995). Computer Simulation, Chapter 33 in *The Exponential Distribution: Theory, Methods and Applications* (eds., N. Balakrishnan and A. P. Basu), Gordon and Breach Publishers, Newark, New Jersey.

PART II

Orderings and Bounds

Lorenz Ordering of Order Statistics and Record Values

Barry C. Arnold and Jose A. Villasenor

1. Introduction

The Lorenz order is a natural tool for the comparison of variabilities of non-negative random variables. In certain contexts (e.g. reliability) variability comparison of order statistics is of concern. Available results in this area are surveyed. Parallel results are described for record values. The record value results are of mathematical interest but, for the moment, do not seem to enjoy ready applicability.

2. The Lorenz order

We begin with a brief review of key ideas relating to the Lorenz order (a convenient reference for further details is Arnold (1987)).

Let $\mathscr{L}$ denote the class of all non-negative random variables with finite positive expectations. For a random variable X in $\mathscr{L}$ with distribution function F_X, we define its inverse distribution function or quantile function F_X^{-1} by

$$F_X^{-1}(y) = \sup\{x: F_X(x) \leq y\} \ . \tag{2.1}$$

The Lorenz curve L_X associated with the random variable X is then defined by

$$L_X(u) = \int_0^u F_X^{-1}(y)\,dy \Big/ \int_0^1 F_X^{-1}(y)\,dy, \quad u \in [0,1] \tag{2.2}$$

(cf Gastwirth (1971)). The Lorenz partial order $\leq_L$ on $\mathscr{L}$ is defined by

$$X \leq_L Y \Leftrightarrow L_X(u) \geq L_Y(u) \quad \forall\, u \in [0,1] \ . \tag{2.3}$$

If $X \leq_L Y$ we say that X exhibits no more inequality than Y (in the Lorenz sense).

The following two characterizations of Lorenz ordering date back to Hardy, Littlewood and Polya (1929).

THEOREM 2.1. Suppose $X \geq 0, Y \geq 0$ and $E(X) = E(Y)$. We have $X \leq_L Y$ if and only if $E(h(X)) \leq E(h(Y))$ for every continuous convex function $h: \mathbf{R}^+ \to \mathbf{R}$.

THEOREM 2.2. Suppose $X \geq 0, Y \geq 0$ and $E(X) = E(Y)$. We have $X \leq_L Y$ if and only if $E[(X - c)^+] \leq [E(Y - c)^+]$ for every $c \in \mathbf{R}^+$.

A trivial but useful observation is that $X \leq_L Y$ implies and is implied by $cX \leq_L dY$ for any $c, d \in (0, \infty)$. Rather than compare two random variables X and Y with possibly different expectations, it is sometimes convenient to compare $E(Y)X$ and $E(X)Y$ which necessarily have equal expectations.

Another interpretation of Lorenz ordered random variables is that one of them is essentially an averaging of the other. This is made precise in the following theorem.

THEOREM 2.3. (Strassen (1965)). Suppose $X, Y \in \mathscr{L}$. $X \leq_L Y$ if and only if there exist random variables Y', Z' defined on some probability space such that $Y \stackrel{d}{=} Y'$ and $X \stackrel{d}{=} cE(Y'|Z')$ for some $c > 0$. (Here and henceforth $\stackrel{d}{=}$ denotes "has the same distribution as").

Explicit computation of Lorenz curves is frequently difficult. Consequently, Lorenz ordering is often not identified by direct use of its definition (equation (2.3)). Moreover Theorems 2.1 and 2.2 involve evaluation of expectations of a large class of functions and thus do not provide convenient tools for identifying situations in which Lorenz ordering obtains. Theorem 2.3 occasionally is just the right tool but in general the representation alluded to in the theorem, although it is guaranteed to exist, is rarely explicitly available. What is needed and what is available is a list of relatively simple to check sufficient conditions for Lorenz ordering. The following list is adequate for our present purposes.

DEFINITION 2.1. We say that X is star-shaped with respect to Y, and write $X \leq_* Y$ if $F_X^{-1}(u)/F_Y^{-1}(u)$ is a non-increasing function of u.

We can use $*$-ordering to verify that Lorenz ordering obtains as a consequence of the following theorem.

THEOREM 2.4. Suppose $X, Y \in \mathscr{L}$. If $X \leq_* Y$, then $X \leq_L Y$.

DEFINITION 2.2. We will say that X is sign-change ordered with respect to Y and write $X \leq_{s.c.} Y$, if $[F_X^{-1}(v)/E(X)] - [F_Y^{-1}(v)/E(Y)]$ has at most one sign change (from $+$ to $-$) as v ranges from 0 to 1.

Sign change ordering is implied by but does not imply star ordering. It does however imply Lorenz ordering, i.e.,

THEOREM 2.5. Suppose $X, Y \in \mathscr{L}$. If $X \leq_{s.c.} Y$ then $X \leq_L Y$.

If densities exist, a simple sufficient condition can be stated for sign change ordering.

THEOREM 2.6. Suppose that X and Y are absolutely continuous members of $\mathscr{L}$ with $E(X) = E(Y)$ and densities f_X and f_Y. A sufficient condition for $X \leq_{s.c.} Y$ and thus for $X \leq_L Y$ is that $f_X(x) - f_Y(x)$ changes sign twice on $(0, \infty)$ and the sequence of signs of $f_X - f_Y$ is $-, +, -$.

Finally we list a theorem identifying a class of random transformations which increase variability in the Lorenz sense.

THEOREM 2.7. Suppose that $g: \mathbf{R}^{+2} \to \mathbf{R}^+$ is such that $g(z, x)/x \uparrow$ as $x \uparrow$ for every z. Suppose that X and Z are independent random variables with $X \in \mathscr{L}$ and $Y = g(Z, X) \in \mathscr{L}$, then $X \leq_L Y$.

3. Order statistics and record values

We focus on the absolutely continuous case corresponding to independent identically distributed observations $X_1, X_2, \ldots$ with common distribution function F_X and density f_X. The corresponding $(i, n)^{\text{th}}$ order statistic will have density

$$f_{X_{i:n}}(x) = \frac{n!}{(i-1)!(n-i)!} \{F_X(x)\}^{i-1} \{1 - F_X(x)\}^{n-i} f_X(x) . \tag{3.1}$$

Order statistics from a uniform$(0, 1)$ distribution play a special role in the discussion and will be denoted by $U_{i:n}$, $i = 1, 2, \ldots, n$, $n = 1, 2, \ldots$. For such order statistics (3.1) simplifies and one recognizes that uniform order statistics have beta distributions. Thus

$$U_{i:n} \sim \text{Beta}(i, n - i + 1) . \tag{3.2}$$

If the X_i's have common distribution function $F_X(x)$ then a convenient representation of $X_{i:n}$ is available using the inverse distribution function (2.1). Thus

$$X_{i:n} \stackrel{d}{=} F_X^{-1}(U_{i:n}) \tag{3.3}$$

where $U_{i:n}$ is the corresponding order statistic from a uniform$(0, 1)$ sample.

In the case of record values there are two useful representations available, one involving uniform samples and one exponential samples. Consider a sequence $X_1, X_2, \ldots$ of independent identically distributed random variables with common distribution function F_X and density f_X. Denote the nth upper record from this sequence by $X^{(n)}$ and the nth lower record by $X_{(n)}$. By convention the zeroth upper and lower record is X_1, i.e., $X^{(0)} = X_{(0)} = X_1$. We reserve the notation $U^{(n)}$ and $U_{(n)}$ to denote record values (upper and lower) corresponding to uniform$(0, 1)$ sequences and the notation $X^{*(n)}, X^*_{(n)}$ to denote record values corresponding to a standard exponential sequence (i.e., $X_i^* \sim \Gamma(1, 1)$, $i = 1, 2, \ldots$). It is not difficult to verify the following representations of upper and lower uniform records. For each n,

$$U^{(n)} \stackrel{d}{=} 1 - \prod_{j=0}^{n} U_j \tag{3.4}$$

and
$$U_{(n)} \stackrel{d}{=} \prod_{j=0}^{n} U_j . \tag{3.5}$$

The lack of memory property of the exponential distribution yields a simple representation of upper records from a standard exponential sequence. One finds, for each n,

$$X^{*(n)} \stackrel{d}{=} \sum_{j=0}^{n} X_j^* \tag{3.6}$$

so that

$$X^{*(n)} \sim \Gamma(n+1, 1) . \tag{3.7}$$

The general lower record sequence (corresponding to i.i.d. X's with common distribution function F_X) admits a representation involving a transformation of corresponding uniform lower records, thus

$$X_{(n)} \stackrel{d}{=} F_X^{-1}(U_{(n)}) . \tag{3.8}$$

The general upper record sequence can be related to either uniform or exponential upper records. Thus one may write

$$X^{(n)} \stackrel{d}{=} F_X^{-1}(U^{(n)}) \tag{3.9}$$

and

$$X^{(n)} \stackrel{d}{=} \psi_X(X^{*(n)}) \tag{3.10}$$

where

$$\psi_X(u) = F_X^{-1}(1 - e^{-u}), \quad u \geq 0 . \tag{3.11}$$

4. Lorenz ordering of order statistics

Representation (3.3) of the ith order statistic $X_{i:n}$ suggests a special role for uniform order statistics. It is then appropriate to first consider the Lorenz ordering relations which hold among various uniform order statistics. By examination of the densities of normalized uniform order statistics (i.e., of random variables of the form $V_{i:n} = \frac{n+1}{i} U_{i:n}$, having mean 1) it is possible to confirm (via Theorem 2.6) that sign change ordering occurs between certain pairs of uniform order statistics and consequently, by Theorem 2.5, Lorenz ordering occurs also. In this fashion we may verify

$$U_{i+1:n} \leq_L U_{i:n}, \quad \forall \, i, n \tag{4.1}$$
$$U_{i:n} \leq_L U_{i:n+1}, \quad \forall \, i, n \tag{4.2}$$

$$U_{n-j+1:n+1} \leq_L U_{n-j:n}, \quad \forall\, j, n \tag{4.3}$$

and

$$U_{n+2:2n+3} \leq_L U_{n+1:2n+1}, \quad \forall\, n. \tag{4.4}$$

Alternative arguments based on Strassen's theorem (Theorem 2.3) are possible for (4.1), (4.2) and (4.3) (Arnold and Villaseñor (1991)). Since Lorenz ordering refers to the relative variability of scaled random variables, some of the relations (4.1)–(4.4) will not be intuitive. The last relation (4.4), which may be described as "sample medians exhibit less inequality as sample size increases", perhaps is the most intuitively plausible.

It may be observed that in all cases in (4.1)–(4.4) it is the order statistic with the smaller mean which exhibits the most inequality as measured by the Lorenz order. Unfortunately, the condition $i/(n+1) \geq j/(m+1)$ is not sufficient to ensure that $U_{i:n} \leq_L U_{j:m}$. It would be too much to hope that relations (4.1)–(4.4) would hold for order statistics from any parent distribution. In some cases they will not hold; in other cases they will be reversed(!). It is of interest identify non-uniform parent distributions for which relations (4.1)–(4.4) or their reversal will hold. Much remains to be done in this area but some results are available. For example power-function and Pareto order statistics behave well. To this end observe that the representation $X = cU^\delta, 0 < \delta < 1$ yields a random variable with a power function distribution while $X = cU^\delta$ with $-1 < \delta < 0$ yields a Pareto random variable (we have $\delta > -1$ here in order to have $E(X) < \infty$). Density crossing arguments may be used to prove:

THEOREM 4.1. If X has a power function distribution [i.e., $F_X(x) = (x/c)^\gamma$, $0 \leq x \leq c$, $\gamma > 0$] then

$$X_{i+1:n} \leq_L X_{i:n}, \quad \forall\, i, n \tag{4.5}$$
$$X_{i:n} \leq_L X_{i:n+1}, \quad \forall\, i, n \tag{4.6}$$

and

$$X_{n-j+1:n+1} \leq_L X_{n-j:n}, \quad \forall\, j, n. \tag{4.7}$$

If X has a Pareto distribution [i.e. $F(x) = 1 - (x/c)^{-\alpha}$, $x > 0$, $\alpha > 1$, then the Lorenz orderings in (4.5), (4.6) and (4.7) are reversed.

Density crossing arguments may be used to prove that (4.4) holds for a quite general class of distributions.

THEOREM 4.2. Suppose that X has a symmetric density supported on the interval $[0, c]$ for some $c > 0$, then

$$X_{n+2:2n+3} \leq_L X_{n+1:2n+1}, \quad \forall\, n.$$

There is one distribution for which it is possible to identify all pairs (i, n) and (j, m) for which $X_{j:m} \leq_L X_{i:n}$. It is the exponential distribution. The proof relies on the fact that it is possible to represent exponential order statistics as linear combinations of independent exponential variables. One finds

THEOREM 4.3. (Arnold and Nagaraja (1991)). If X has an exponential distribution (i.e., $F_X(x) = 1 - e^{-\lambda x}, x > 0, \lambda > 0$), then for $i \leq j$ the following are equivalent

(i) $X_{j:m} \leq_L X_{i:n}$
(ii) $(n - i + 1)E(X_{i:n}) \leq (m - j + 1)E(X_{j:m})$.

Thus for observations from an exponential distribution (4.3) and (4.4) hold, (4.2) is reversed while (4.1) holds for $i \leq (1 - e^{-1})n$.

Theorem 2.7 may be used to derive sufficient conditions for (4.2) to hold. We have

THEOREM 4.4. Suppose F, the common distribution of the X's satisfies

$$F^{-1}(uF(x))/x \uparrow \text{ as } x \uparrow \quad \forall u \in (0,1) . \tag{4.8}$$

It follows that

(i) $X_{i+1:n} \leq_L X_{i:n}$ and
(ii) $X_{i:n} \leq_L X_{i:n+1}$.

PROOF. Observe that

$$X_{i:n} \stackrel{d}{=} F^{-1}(U_{i:i}F(X_{i+1:n}))$$

where the two random variables on the right are independent, then apply Theorem 2.7. This proves (i). To prove (ii), use Strassen's theorem and the fact that $X_{i:n+1} \stackrel{d}{=} F^{-1}(U_{n+1:n+1}F(X_{i:n}))$, where the two random variables on the right are independent.

REMARK. If F is differentiable it is not difficult to verify that a necessary and sufficient condition for (4.8) to hold is that

$$xF'(x)/F(x) \uparrow \text{ as } x \uparrow \tag{4.9}$$

Condition (4.8) holds for distribution functions which increase relatively quickly on their support. For example it is satisfied by the distribution function

$$F(x) = e^x - 1, \quad 0 \leq x \leq \log 2 .$$

Finally we remark that Lorenz ordering is often inherited by order statistics from their parent distributions. Thus

THEOREM 4.5. (Shaked and Shanthikumar 1994, p. 108). If $X, X_1, \ldots, X_n$ are i.i.d. and $Y, Y_1, Y_2, \ldots, Y_n$ are i.i.d. and if $X \leq_* Y$ then for every i, n

$$X_{i:n} \leq_L Y_{i:n} .$$

Note that the hypothesis of Theorem 4.5 is stronger than the assumption $X \leq_L Y$. Star-ordering is however inherited by order-statistics and then the conclusion of Theorem 4.5 follows as a consequence of Theorem 2.4. The theorem does not guarantee that Lorenz ordering is inherited (unless it occurs in the stronger form of star ordering). The Shaked and Shanthikumar approach (cf Theorem 5.5, below) can also be used to argue that if for some (j, m) we have $X_{j:m} \leq_* Y_{j:m}$ then for every (i, n) we have $X_{i:n} \leq_L Y_{i:n}$.

5. Lorenz ordering of record values

First we consider two particularly well behaved record value sequences: upper exponential records and lower uniform records (for definitions refer again to Section 3). Note that for upper exponential records we have

$$X^{*(n)} \sim \Gamma(n+1, 1)$$

and via a density crossing argument we find

$$X^{*(n)} \leq_L X^{*(n-1)} . \tag{5.1}$$

Turning to lower uniform records we have available the representation

$$U_{(n)} \stackrel{d}{=} \prod_{i=0}^{n} U_i \tag{5.2}$$
$$\stackrel{d}{=} U_{(n-1)} U_n$$

where the random variables appearing on the right hand side are independent. It is then a straightforward consequence of Strassen's theorem (Theorem 2.3) that

$$U_{(n-1)} \leq_L U_{(n)} \tag{5.3}$$

To identify other parent distributions for which lower or upper records are Lorenz ordered, it is natural to turn to representations (3.8), (3.9) and (3.10). We are led to seek conditions on F_X to ensure that

$$F_X^{-1}\left(\prod_{i=0}^{n-1} U_i\right) \leq_L F_X^{-1}\left(\prod_{i=1}^{n} U_i\right) \tag{5.4}$$

which will guarantee that lower X records are Lorenz ordered (with inequality increasing as n increases). In parallel fashion we may seek conditions on F_X to ensure that either

$$F_X^{-1}\left(1 - \prod_{i=0}^{n-1} U_i\right) \geq_L F_X^{-1}\left(1 - \prod_{i=0}^{n} U_i\right) \tag{5.5}$$

or, equivalently,

$$\psi_X\left(\sum_{i=0}^{n-1} X_i^*\right) \geq_L \psi_X\left(\sum_{i=1}^{n} X_i^*\right). \tag{5.6}$$

Conditions (5.5) and (5.6) would guarantee that upper X records are Lorenz ordered (with inequality decreasing as n increases).

First we consider possible extensions of (5.2) (Lorenz ordering of lower uniform records). Recall that (5.2) was an obvious consequence of Strassen's theorem. It is reasonable to ask whether there exist forms for F_X^{-1} which will allow us to conclude Lorenz ordering of records via Strassen's theorem. Two simple examples are readily found; but only two.

THEOREM 5.1. (i) If $F_X(x) = x^\gamma, 0 < x < 1$ for some $\gamma > 0$ then

$$X_{(n-1)} \leq_L X_{(n)}$$

(ii) If $\overline{F}_X(x) = x^{-\delta}, x > 1$, for some $\delta > 1$ (to ensure that $E(X)$ exists), then

$$X^{(n-1)} \leq_L X^{(n)}.$$

PROOF. (i) Here $X \stackrel{d}{=} U^{1/\gamma}$ and clearly

$$X_{(n)} \stackrel{d}{=} \left(\prod_{i=0}^{n} U_i\right)^{1/\gamma} = \prod_{i=0}^{n} U_i^{1/\gamma} = X_{(n-1)} U_n^{1/\gamma}$$

where the random variables on the right are independent. From Strassen's theorem we have $X_{(n-1)} \leq_L X_{(n)}$.

(ii) Here $X \stackrel{d}{=} 1/U^{1/\delta}$ and

$$X^{(n)} = 1/(U_{(n)})^{1/\delta}$$
$$= \prod_{i=0}^{n} U_i^{-1/\delta} = X^{(n-1)} U_n^{-1/\delta}$$

and so by Theorem 2.3, again, $X^{(n-1)} \leq_L X^{(n)}$.

Thus we have Lorenz ordering for lower records from a power function distribution and for upper records from a Pareto distribution.

Next consider possible extension of (5.1), which dealt with Lorenz ordering of exponential records. It will be recalled that (5.1) was justified by a density crossing argument (though it could have been justified as a consequence of the fact that sample means are Lorenz ordered, see e.g. Theorem 4 of Arnold and Villasenor (1986)). It is reasonable to seek sufficient conditions on ψ_X to guarantee that (5.6) holds via a density crossing argument (i.e., via Theorem 2.6).

To this end it is convenient to present the form of the density of $X^{(n)}/E(X^{(n)})$ when $X^{(n)}$ is as defined in (3.10). (Note we must divide by $E(X^{(n)})$, to use the density crossing (or sign-change) ordering which assumes equal means, to conclude Lorenz ordering.) To simplify expressions we introduce the notation

$$\mu^{(n)} = E(X^{(n)}) \tag{5.7}$$

and observe that $\mu^{(n)} \leq \mu^{(n+1)}$, obviously. Since $X^{*(n)} \sim \Gamma(n+1,1)$ it follows that $X^{(n)}/\mu^{(n)}$ has density of the form

$$f_n(x) = [h(\mu^{(n)}x)]^n e^{-h(\mu^{(n)}x)} h'(\mu^{(n)}x)[\mu^{(n)}/n!], \quad 0 < x < F_X^{-1}(1)/\mu^{(n)} \tag{5.8}$$

where for convenience we have introduced the notation

$$h(x) = \psi_X^{-1}(x) \tag{5.9}$$

and have assumed h to be differentiable. In order to conclude that $X^{(n)} \leq_L X^{(n-1)}$, it will suffice to require conditions on h which will ensure that the function

$$\log f_n(x) - \log f_{n-1}(x)$$

changes sign twice on the interval $(0, F_X^{-1}(1)/\mu^{(n-1)})$ and that the sequence of signs is $-,+,-$. One case in which this is relatively straightforward is the case in which $h(x) = x^\delta, \delta > 0$ which corresponds to Weibull random variables.

THEOREM 5.2. If $\overline{F}_X(x) = \exp -(x/\sigma)^{1/\delta}$ for some $\sigma > 0$ and $\delta > 0$ (i.e., X has a Weibull distribution) then

$$X^{(n)} \leq_L X^{(n-1)} \ .$$

PROOF. Here $h(x) = \sigma x^\delta$ and $\log f_n(x) - \log f_{n-1}(x)$ is of the form $c + \gamma \log(x/b) - x/b$ for certain constants b and c and the sign sequence $-,+,-$ on $(0,\infty)$ will be evident.

If the support of X is bounded above, i.e. if $F_X^{-1}(1) = M < \infty$, then it is evident since $\mu^{(n)} \geq \mu^{(n-1)}$ that $\log f_n(x) - \log f_{n-1}(x)$ will be $-\infty$ on the interval $(M/\mu^{(n)}, M/\mu^{(n-1)})$ and it will suffice that on the interval $(0, M/\mu^{(n)})$ it exhibit either two sign changes $(-,+,-)$ or just one with sign sequence $-,+$. From this observation we may conclude that a sufficient condition for $X^{(n)} \leq_L X^{(n-1)}$ when $F_X^{-1}(1) = M < \infty$, is that $\log f_n(x) - \log f_{n-1}(x)$ be monotone increasing on $(0, M/\mu^{(n)})$. We may formalize this as follows.

THEOREM 5.3. (Arnold and Villasenor (1995)). If $F_X^{-1}(1) = M < \infty$ then a sufficient condition to ensure that $X^{(n)} \leq_L X^{(n-1)}$ is

$$\left[(n-1)\frac{h'(u)}{h(u)} - h'(u) + \frac{h''(u)}{h'(u)}\right] u \uparrow \text{ as } u \uparrow \tag{5.10}$$

on $(0, M)$ where $h(u) = \psi_X^{-1}(u) = -\log(1 - F_X(u))$.

PROOF. Consider $\frac{d}{dx}[\log f_n(x) - \log f_{n-1}(x)] \stackrel{\Delta}{=} \Lambda(x)$ where f_n is as given in (5.8). This will be always positive if $x\Lambda(x)$ is always positive. The given condition readily follows.

Perhaps the most important application of Theorem 5.3 involves uniform upper records. In order to have $X \sim \text{uniform}(0,1)$ we must have $\psi_X(u) = 1 - e^{-u}$ and consequently $h_X(u) = -\log(1-u)$. In such a case we have

$$h'_X(u) = \frac{1}{1-u}$$

and

$$h''_X(u) = \frac{1}{(1-u)^2}.$$

It follows that

$$\left[(n-1)\frac{h'(u)}{h(u)} - h'(u) + \frac{h''(u)}{h'(u)}\right]u = \frac{(n-1)u}{(1-u)[-\log(1-u)]}$$

and by differentiation we can readily verify that this is an increasing function on $(0,1)$. Thus, reverting to our $U^{(n)}$ notation for upper uniform records, we conclude from Theorem 5.3 that

$$U^{(n)} \leq_L U^{(n-1)} . \tag{5.11}$$

Analogous computations may be used to verify that (5.10) holds for both of the following examples

$$h_X(u) = -\log[(1-u)]^\delta, \quad 0 < u < 1 \text{ where } 0 < \delta \leq 1$$

and

$$h_X(u) = [-\log(1-u^\delta)], \quad 0 < u < 1 \text{ where } \delta > 0 .$$

The first of these corresponds to a distribution function of the form

$$F_X(x) = 1 - (1-x)^\delta, \quad 0 < x < 1 ,$$

where $\delta \in (0,1]$, while the second corresponds to a power function distribution, i.e.,

$$F_X(x) = x^\delta, \quad 0 < x < 1 .$$

For both such distributions we have Lorenz ordered upper records, i.e.

$$X^{(n)} \leq_L X^{(n-1)} .$$

It may be observed that conditions (such as those in Theorem 5.3) on h_X may be rephrased in terms of conditions on F_X and its corresponding density f_X since $h_X(x) = -\log(1 - F_X(x))$. Thus Theorem 5.3 admits the following alternative phrasing.

THEOREM 5.3A. (Arnold and Villasenor, 1995). If $F_X^{-1}(1) = M < \infty$ then a sufficient condition for $X^{(n)} \leq_L X^{(n-1)}$ is that

$$x\left[\frac{f_X'(x)}{f_X(x)} + \frac{nf_X(x)}{[1 - F_X(x)][-\log(1 - F_X(x))]}\right] \uparrow \text{ as } x \uparrow . \tag{5.12}$$

For this, a sufficient condition is that

$$xf_X'(x)/f_X(x) \uparrow \text{ as } x \uparrow \tag{5.13}$$

and

$$xf_X(x)/F_X(x) \uparrow \text{ as } x \uparrow . \tag{5.14}$$

PROOF. (5.10) and (5.12) are readily seen to be equivalent since $h_X(x) = -\log(1 - F_X(x))$. To verify that (5.13) and (5.14) imply (5.12) we need only observe that $a/\{(1-a)[-\log(1-a)]\} \uparrow$ as $a \uparrow$ and so consequently $F_X(x)/([1 - F_X(x)][-\log(1 - F_X(x))]) \uparrow$ as $x \uparrow$.

As an example in which conditions (5.13) and (5.14) are satisfied, consider

$$F_X(x) = \tfrac{1}{2}(x + x^\delta), \quad 0 < x < 1 \tag{5.15}$$

where $\delta > 1$. Upper records from this distribution are consequently Lorenz ordered.

It is of course possible to have upper records that are Lorenz ordered in the reverse direction, i.e. to have $X^{(n-1)} \leq_L X^{(n)}$. Theorem 2.7 provides us with the following sufficient condition for such Lorenz ordering of upper records.

THEOREM 5.4. Suppose that for each $z > 0$, $\psi_X(z + w)/\psi_X(w) \uparrow$ as $w \uparrow$ then $X^{(n-1)} \leq_L X^{(n)}$.

PROOF. Since we can write

$$X^{(n)} = \psi_X(X_n^* + \psi_X^{-1}(X^{(n-1)}))$$

where X_n^* and $X^{(n-1)}$ are independent, then a sufficient condition to ensure that $X^{(n-1)} \leq_L X^{(n)}$ is, by Theorem 2.7, that $\psi_X(z + \psi_X^{-1}(x))/x \uparrow$ as $x \uparrow$ for each z. But, writing $w = \psi_X^{-1}(x)$, this is equivalent to $\psi_X(z + w)/\psi_X(w) \uparrow$ as $w \uparrow$ for each z.

EXAMPLE. If we consider $\psi_X(x) = \exp\sqrt{1 + x}$ it is readily verified that $\psi_X(z + w)/\psi_X(w) \uparrow$ as $w \uparrow$ for each z.

REMARK. If we assume that F_X is differentiable then a sufficient condition for $\psi_X(z + w)/\psi_X(w) \uparrow$ as $w \uparrow \forall z$ and hence for $X^{(n-1)} \leq_L X^{(n)}$ is that $zF_X'(z)/\overline{F}_X(z) \downarrow$ as $z \uparrow$.

A result parallel to that described for order statistics at the end of section 4, is true for record values. Thus Lorenz ordering for record values is often inherited from the parent distributions.

THEOREM 5.5. If $X, X_1, X_2, \ldots$ are i.i.d. and $Y, Y_1, Y_2, \ldots$, are i.i.d. and if $X \leq_* Y$ then for every n

$$X^{(n)} \leq_L Y^{(n)} .$$

PROOF. Referring to Eq. (3.9) we may write

$$F_{X^{(n)}}^{-1}(u)/F_{Y^{(n)}}^{-1}(u) = F_X^{-1}(F_{U^{(n)}}^{-1}(u))/F_Y^{-1}(F_{U^{(n)}}^{-1}(u))$$

and conclude that this ratio is a non-increasing function of u since $F_{U^{(n)}}^{-1}(u)$ is an increasing function and, by hypothesis, $F_X^{-1}(u)/F_Y^{-1}(u)$ is a non-increasing function of u. Thus $X^{(n)} \leq_* Y^{(n)}$ and consequently $X^{(n)} \leq_L Y^{(n)}$.

Note that the hypotheses of the theorem is stronger than the assumption that $X \leq_L Y$. A parallel argument confirms that if for some $m, X^{(m)} \leq_* Y^{(m)}$ then for every $n, X^{(n)} \leq_* Y^{(n)}$ and hence $X^{(n)} \leq_L Y^{(n)}$.

6. Remarks

It will be observed that most of the explicit results obtained deal with exponential, Weibull, uniform, power-function and Pareto distributions. This is disappointing but perhaps not surprising. These are indeed the distributions on the positive half-line which have the most analytically tractable distribution functions. They reasonably are most easily dealt with. Open questions abound in this area. For example, it is intriguing that the non-standard example displayed in equation (5.15) for which Lorenz ordering of records obtained, is a mixture of two distributions for each of which the Lorenz ordering of records obtains. Further insight into the structure of the class of distributions for which $X^{(n)} \leq_L X^{(n-1)}$, is needed. Similarly it would be desirable to characterize the classes of distributions for which respectively (4.5), (4.6) or (4.7) (or their reversals) hold.

References

Arnold, B. C. (1987). *Majorization and the Lorenz Order: A Brief Introduction*. Lecture Notes in Statist., Vol. 43, Springer-Verlag, Berlin.

Arnold, B. C. and H. N. Nagaraja (1991). Lorenz ordering of exponential order statistics. *Statist. Probab. Lett.* **11**, 485–490.

Arnold, B. C. and J. A. Villaseñor (1986). Lorenz ordering of means and medians. *Statist. Probab. Lett.* **4**, 47–49.

Arnold, B. C. and J. A. Villaseñor (1991). Lorenz ordering of order statistics, in *Stochastic Orders and Decision Under Risk*. K. Mosler and M. Scarsini, Eds. Institute of Mathematical Statistics, Lecture Notes – Monograph Series, Vol. 19, 38–47.

Arnold, B. C. and J. A. Villaseñor (1995). Lorenz ordering of record values. Technical Report #228, Department of Statistics, University of California, Riverside.

Gastwirth, J. L. (1971). A general definition of the Lorenz curve. *Econometrica* **39**, 1037–1039.

Hardy, G. H., J. E. Littlewood and G. Polya (1929). Some simple inequalities satisfied by convex functions. *Messenger of Mathematics* **58**, 148–152.

Shaked, M. and J. G. Shanthikumar (1994). *Stochastic orders and their applications*. Academic Press, Boston, Mass.

Strassen, V. (1965). The existence of probability measures with given marginals. *Ann. Math. Statist.* **36**, 423–439.

Stochastic Ordering of Order Statistics

Philip J. Boland, Moshe Shaked and J. George Shanthikumar

1. Introduction

There are many ways in which one might say that the random variable X is smaller than the random variable Y. In the "usual" stochastic ordering, one says that X is stochastically smaller than Y (and write $X \leq_{st} Y$) if $F_X(t) \geq F_Y(t)$ for all t. That is $X \leq_{st} Y$ if the distribution function F_Y of Y is everywhere dominated by that of X. If $X_1, X_2, \ldots, X_n$ is a set of random variables, then $X_{(1)} \leq X_{(2)} \leq \cdots \leq X_{(n)}$ will denote the corresponding order statistics. When it is important to emphasize the number n of observations, then $X_{(1:n)} \leq X_{(2:n)} \leq \cdots \leq X_{(n:n)}$ will be used. It is clear from the definition of order statistics that they are ordered in the usual stochastic ordering, that is $X_{(i)} \leq_{st} X_{(j)}$ for any $i < j$. However there are many other stochastic orders which are of interest in various settings. The hazard rate (or failure rate) ordering is of particular interest in reliability and survival analysis situations. The likelihood ratio ordering is a powerful ordering of random variables which is possessed by many one parameter families of distributions. Other orders which are considered here are dispersive ordering, convex transform ordering, starshaped ordering and superadditive ordering. In this chapter an attempt is made to summarize many of the known results on stochastic ordering of order statistics in a compact way. The text "Stochastic Orders and Their Applications" by Shaked and Shanthikumar (1994) is a comprehensive treatment on stochastic orderings. Indeed further characterizations of the orders discussed here together with proofs of many of the results may be found in this treatise. In section 2 the definitions and some characterizations of the stochastic orders mentioned above are given. Various notions of positive dependence between two random variables X and Y are also reviewed.

In section 3 we give results on stochastic comparisons of order statistics based on one set of observations $X_1, X_2, \ldots, X_n$. Furthermore the bivariate dependence between order statistics $X_{(i)}$ and $X_{(j)}$ is treated, with implications for conditional ordering of order statistics. In section 4 the emphasis is on comparisons of order statistics from two independent sets of observations $X_1, X_2, \ldots, X_n$ and $Y_1, Y_2, \ldots, Y_n$.

2. Stochastic orderings

In this section we give definitions of many of the commonly used stochastic orders for comparing random variables. We also summarize some of the equivalent characterizations and properties of these orderings which we will find useful for our discussion of stochastic ordering of order statistics in sections 3 and 4.

F_X will denote the distribution function and $\overline{F}_X = 1 - F_X$ the survival function of the random variable X. If X is absolutely continuous (discrete), f_X will denote its density (mass function). We say that a function g is increasing if it is nondecreasing, and $a/0$ is interpreted as ∞ for $a > 0$.

The random variable X is smaller than the random variable Y in the *usual stochastic order* ($X \leq_{st} Y$) if $F_X(t) \geq F_Y(t)$ or equivalently $\overline{F}_X(t) \leq \overline{F}_Y(t)$ for all t. A very useful characterization of this stochastic order is $X \leq_{st} Y \iff E(g(X)) \leq E(g(Y))$ for all increasing g whenever the expectations exist. Although our interest here is principally on univariate orderings, we mention the natural extension of the usual stochastic order to random vectors given this characterization of the univariate order. The random vector $\boldsymbol{X} = (X_1, X_2, \ldots, X_n)$ is smaller than the random vector $\boldsymbol{Y} = (Y_1, Y_2, \ldots, Y_n)$ in (the usual) stochastic order (and written $\boldsymbol{X} \leq_{st} \boldsymbol{Y}$) if $E(g(\boldsymbol{X})) \leq E(g(\boldsymbol{Y}))$ for all increasing functions g whenever the expectations exist. It is easy to see that it is closed under mixtures in the sense that if $\boldsymbol{X}, \boldsymbol{Y}$, and Θ are random vectors where $[\boldsymbol{X}|\Theta = \theta] \leq_{st} [\boldsymbol{Y}|\Theta = \theta]$ for all θ, then $\boldsymbol{X} \leq_{st} \boldsymbol{Y}$. The usual stochastic order is closed under convolutions whereby if X_1, X_2, Y_1, and Y_2 are independent and $X_i \leq_{st} Y_i$ for $i = 1, 2$, then $X_1 + X_2 \leq_{st} Y_1 + Y_2$.

The hazard rate ordering is an ordering for random variables which compares lifetimes with respect to their hazard rate functions. If X is an absolutely continuous nonnegative random variable with density f_X, then we define $r_X(t) = f_X(t)/\overline{F}_X(t)$ to be its hazard rate (or failure rate) function. If X and Y with respective hazard rate functions r_X and r_Y satisfy $r_X(t) \geq r_Y(t)$ for all $t > 0$, then we say that X is smaller than Y in the hazard rate ordering and write $X \leq_{hr} Y$. Since the hazard rate function at time t represents the instantaneous rate of failure given survival to time t, $X \leq_{hr} Y$ means intuitively that given the lifetimes X and Y exceed t (for any fixed t), then X is more likely to fail than Y in the 'immediate' future. The hazard rate ordering is particularly useful in reliability theory and survival analysis due to the importance of the hazard rate function in these areas. The hazard rate ordering can of course be extended to nonnegative random variables. The condition $r_X(t) \geq r_Y(t)$ for all t is equivalent to the requirement that $\overline{F}_Y(t)/\overline{F}_X(t)$ is increasing in t, and this later condition can be used to define the hazard rate ordering in more generality (when for example densities don't exist). We will use this definition here, and hence say that X is smaller than Y in the *hazard rate ordering* (and write $X \leq_{hr} Y$) if $\overline{F}_Y(t)/\overline{F}_X(t)$ is increasing in t. Keilson and Sumita (1982) used this definition to define an order termed "uniform stochastic order in the positive direction". It is easy to verify that $\overline{F}_Y(t)/\overline{F}_X(t)$ is increasing in $t \iff P[X > t + s | X > t] \leq P[Y > t + s | Y > t]$ for all $s > 0$ and $t \iff [X|X > t] \leq_{st} [Y|Y > t]$ for all t. Hence $X \leq_{hr} Y \iff$ conditional on survival of both X and Y to t (for any t), X is less than Y in the usual stochastic order. This makes it clear that the hazard rate ordering is a stronger ordering than the usual stochastic ordering. A nonnegative random variable X with hazard function r_X is IFR (increasing failure or hazard rate) if r_X is increasing (equivalently $\overline{F}_X$ is log concave on its support). Similarly X is DFR (decreasing failure or hazard rate) if r_X is decreasing. The hazard rate ordering is not necessarily closed under convolutions since if X_1, X_2 are independent and Y_1, Y_2 are independent where $X_i \leq_{hr} Y_i$ for $i = 1, 2$, then it does not necessarily follow that $X_1 + X_2 \leq_{hr} Y_1 + Y_2$. However it can be shown that $X_1 + X_2 \leq_{hr} Y_1 + Y_2$ if X_1, X_2, Y_1 and Y_2 are IFR. See Keilson and Sumita (1982) and Shanthikumar and Yao (1991) for more on the hazard rate ordering and convolutions. Finally the hazard rate ordering is not necessarily closed under mixtures, although if X, Y, and Θ are random variables where $[X|\Theta = \theta] \leq_{hr} [Y|\Theta = \theta']$ for all θ and θ', then $X \leq_{hr} Y$ (Shaked and Shanthikumar (1994)).

If X and Y are continuous random variables with respective densities or mass functions f_X and f_Y, then we say X is smaller than Y in the *likelihood ratio ordering* ($X \leq_{lr} Y$) if $f_Y(t)/f_X(t)$ is increasing in t over the union of supports of X and Y. If $\{X_\theta : \theta \in \Theta \subseteq R\}$ is a family of random variables where $X_\theta \leq_{lr} X_{\theta'}$ for any $\theta < \theta'$, then we say $\{X_\theta : \theta \in \Theta\}$ has the monotone likelihood ratio property. This is equivalent to saying that $f_\theta(t_1)f_{\theta'}(t_2) \geq f_\theta(t_2)f_{\theta'}(t_1)$ whenever $t_1 < t_2$ and $\theta < \theta'$, or that the function $h(\theta, t) = f_\theta(t)$ is TP$_2$ (totally positive of order 2) in θ and t. The family of exponential distributions is but one with the monotone likelihood ratio property. It is well known that the likelihood ratio order is stronger than the hazard rate order. The likelihood ratio order is not in general closed under convolutions, but if X_1, X_2 are independent and Y_1, Y_2 are independent where $X_1 \leq_{lr} Y_1, X_2 \leq_{lr} Y_2$ and the densities or mass functions of X_2, Y_2 are log concave, then $X_1 + X_2 \leq_{lr} Y_1 + Y_2$ (Shanthikumar and Yao (1991)). The likelihood ratio order has a closure property for mixtures like that of the hazard rate order, that is if X, Y, and Θ are random variables where $[X|\Theta = \theta] \leq_{lr} [Y|\Theta = \theta']$ for all θ and θ', then $X \leq_{lr} Y$. The following Fig. 1 gives the implications between the three stochastic orders discussed up to this point.

$$X \leq_{lr} Y \Longrightarrow X \leq_{hr} Y \Longrightarrow X \leq_{st} Y$$

Fig. 1.

The dispersive ordering is a variability comparison between two random variables. Let F_X^{-1} and F_Y^{-1} be the right continuous inverses of F_X and F_Y respectively. We say that X is smaller than Y in the *dispersive order* ($X \leq_{disp} Y$) if $F_X^{-1}(\beta) - F_X^{-1}(\alpha) \leq F_Y^{-1}(\beta) - F_Y^{-1}(\alpha)$ for all $0 \leq \alpha < \beta \leq 1$. This is a variability or tail like ordering as it requires the difference of any two quantiles of X to be smaller than the difference of the corresponding quantiles of Y in order for $X \leq_{disp} Y$. Unlike any of the orderings so far discussed, the dispersive order is closed under shifts, that is $X \leq_{disp} Y \Longleftrightarrow X + c \leq_{disp} Y$ for any c. An equivalent definition of $X \leq_{disp} Y$ is to require that $F_Y^{-1}(F_X(t)) - t$ be increasing in t. It is easy to see that for example if U_i is uniform on (a_i, b_i) for $i = 1, 2$, then $U_1 \leq_{disp} U_2$ if

b_1 $a_1 \leq b_2 - a_2$. Also if X_λ is exponential with mean $1/\lambda$, then $\lambda_1 > \lambda_2 \implies X_{\lambda_1} \leq_{\text{disp}} X_{\lambda_2}$. The dispersive order is not closed under convolutions. A random variable Z is *dispersive* if $X + Z \leq_{\text{disp}} Y + Z$ whenever Z is independent of X, Y and $X \leq_{\text{disp}} Y$. It can be shown (Lynch, Mimmack and Proschan (1983), Lewis and Thompson (1981), and Shaked and Shanthikumar (1994)) that Z is dispersive $\iff Z$ has a log concave density (Z is strongly unimodal). Furthermore $X \leq_{\text{disp}} X + Y$ for any Y independent of $X \iff X$ has a log concave density (Droste and Wefelmeyer (1985)).

Finally we define three orderings which we refer to as the transform orders: the convex, the star, and the superadditive orders (see Shaked and Shanthikumar (1994)). They are also dealt with extensively in Barlow and Proschan (1981). Let us now suppose X and Y are nonnegative random variables where X has interval support. One says that X is less than Y in the *convex transform order* $(X \leq_c Y)$ if $F_Y^{-1} F_X(t)$ is convex on the support of X. If X and Y are Weibull random variables with the same scale parameter but with shape parameters α_X and α_Y respectively, then $X \leq_c Y \iff \alpha_Y \leq \alpha_X$. If Y is exponential with $F_Y(t) = 1 - e^{-\lambda t}$, then $F_Y^{-1} F_X(t) = -\frac{1}{\lambda} \ln \overline{F}_X(t)$ has derivative $r_X(t)/\lambda$. Hence X is IFR (has increasing failure rate) $\iff X \leq_c \text{Exp}(\lambda)$, and this observation is one motivation for the definition of the convex transform ordering (which was introduced by van Zwet (1964)). We say that X is less than Y in the *star order* $(X \leq_* Y)$ if $F_Y^{-1} F_X(t)$ is starshaped (that is $F_Y^{-1} F_X(t)/t$ is increasing in t). It is well known to reliabilists that $X \leq_* \text{Exp}(\lambda) \iff X$ is IFRA (increasing failure rate average). Shaked and Shanthikumar (1994) show that the star order and the dispersive order are related by: $X \leq_* Y \iff \ln X \leq_{\text{disp}} \ln Y$. We say that X is less than Y in the *superadditive order* $(X \leq_{\text{su}} Y)$ if $F_Y^{-1} F_X(s+t) \geq F_Y^{-1} F_X(s) + F_Y^{-1} F_X(t)$ for all $s, t > 0$ (that is $F_Y^{-1} F_X$ is superadditive on the support of X). Theorists in reliability know that $X \leq_{\text{su}} \text{Exp}(\lambda)$ is equivalent to saying that X is NBU (new better than used). Any increasing convex function on $[0, \infty)$ is starshaped, and every increasing starshaped function on $[0, \infty)$ is also superadditive. Figure 2 shows the implications between the last four stochastic orders which have been introduced.

$$X \leq_c Y \implies X \leq_* Y \implies X \leq_{\text{su}} Y$$

$$\Updownarrow$$

$$\ln X \leq_{\text{disp}} \ln Y$$

Fig. 2.

There are many notions of positive dependence between two random variables X and Y. The reader would do well to consult Jogdeo (1982), Barlow and Proschan (1981), and Tong (1980) for more extensive treatments of this subject. Perhaps the strongest notion of dependence between two random variables X and Y is TP_2 (totally positive of order 2) dependence. X and Y are TP_2 *dependent* if their joint density or mass function $f(x,y)$ is totally positive of order 2 in x and y, or more precisely if

$$\begin{vmatrix} f(x_1,y_1) & f(x_1,y_2) \\ f(x_2,y_1) & f(x_2,y_2) \end{vmatrix} \geq 0$$

for any $x_1 < x_2$, $y_1 < y_2$. We say that the random variables X and Y are *right corner set increasing* (RCSI) if for any fixed x and y, $P[X > x, Y > y | X > x', Y > y']$ is increasing in x' and y'. One says that Y is *stochastically increasing* in X if $P[Y > y | X = x]$ is increasing in x for all y, and write $SI(Y|X)$. Lehmann (1966) uses the term *positively regression dependent* to describe SI.

We say that Y is *right tail increasing* in X if $P[Y > y | X > x]$ is increasing in x for all y, and write $RTI(Y|X)$. Similarly Y is *left tail decreasing* in X if $P[Y > y | X < x]$ is decreasing in x for all y, and we denote this by $LTD(Y|X)$. The random variables X and Y are *associated* (written $A(X,Y)$) if $\text{Cov}[\Gamma(X,Y), \Delta(X,Y)] \geq 0$ for all pairs of increasing binary functions Γ and Δ. Finally we say X and Y are *positively quadrant dependent* if

$$P[X \leq x, Y \leq y] \geq P[X \leq x]P[Y \leq y]$$

for all x, y, and write **PQD(X,Y)**. The various implications between these notions of dependence, at least for continuous random variables are summarized in the following Fig. 3 (Barlow and Proschan (1981)).

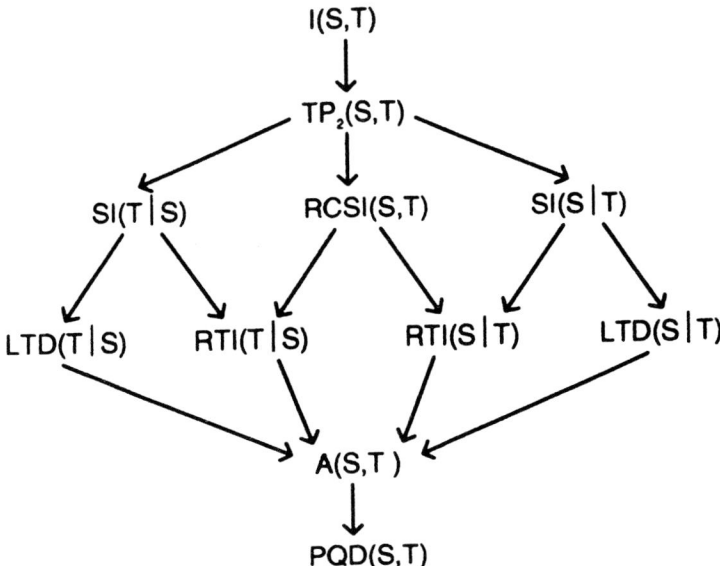

Fig. 3. Implications among notions of bivariate dependence.

3. Stochastic order for order statistics from one sample

In this section we discuss stochastic order relations between the order statistics $X_{(i)}$ and $X_{(j)}$ based on one set of observations $X_1, X_2, \ldots, X_n$. The first two results concerning the usual stochastic order and the hazard rate order hold in considerable generality. We next consider the case where $X_1, X_2, \ldots, X_n$ are independent and identically distributed, and then proceed to the situation where $X_1, X_2, \ldots, X_n$ are independent but otherwise arbitrarily distributed. The case when $X_1, X_2, \ldots, X_n$ represent a random sample taken without replacement from a finite population is also mentioned. Most of the results are for the situation where $X_1, X_2, \ldots, X_n$ are absolutely continuous with density functions.

It should be clear from the definition of the order statistics $X_{(i)}$ and $X_{(j)}$ (where $i < j$) that $P[X_{(j)} > t] \geq P[X_{(i)} > t]$ when $X_1, X_2, \ldots, X_n$ is *any* set of random variables. Therefore in general one has the following result.

THEOREM 3.1. Let $X_1, X_2, \ldots, X_n$ be random variables. Then for $i < j$, $X_{(i)} \leq_{\text{st}} X_{(j)}$.

The next Theorem shows that for *any independent* random variables $X_1, X_2, \ldots, X_n$, the order statistics are ordered with respect to the hazard rate ordering. The proof for the case when $X_1, X_2, \ldots, X_n$ are absolutely continuous was given in Boland, El-Neweihi and Proschan (1994). The proof given here is new but is similar in nature to that of the main result in Shaked and Shanthikumar (1995).

THEOREM 3.2. Let $X_1, X_2, \ldots, X_n$ be independent random variables. Then for $i < j$, $X_{(i)} \leq_{\text{hr}} X_{(j)}$.

PROOF. For any t, let $N_t = \sum_{i=1}^n I(X_i > t)$. Now $I(X_i > t)$ has log concave density, and it is easy to see that as t increases $I(X_i > t)$ decreases in the likelihood ratio ordering. Using the aforementioned result on preservation of the likelihood ratio order under convolutions (see also Theorem 1.C.5 of Shaked and Shanthikumar (1994), or Shanthikumar and Yao (1991)), it follows that N_t decreases in the likelihood ratio order as t increases. Since the likelihood ratio order is stronger than the hazard rate order, it follows that N_t is decreasing (with t) in the hazard rate order. Note now that $P(N_t > n - j + 1) = P(X_{(j)} > t)$. Hence for any $s > 0$ and $i < j$, one has that

$$\frac{P(N_t > n - j + 1)}{P(N_{t+s} > n - j + 1)} \leq \frac{P(N_t > n - i + 1)}{P(N_{t+s} > n - i + 1)}$$

or

$$\frac{P(X_{(j)} > t)}{P(X_{(j)} > t + s)} \leq \frac{P(X_{(i)} > t)}{P(X_{(i)} > t + s)}$$

or

$$\frac{P(X_{(i)} > t + s)}{P(X_{(i)} > t)} \leq \frac{P(X_{(j)} > t + s)}{P(X_{(j)} > t)}$$

or $X_{(i)} \leq_{\text{hr}} X_{(j)}$. □

Figure 4 illustrates the hazard rate functions of the order statistics $X_{(1)}$, $X_{(2)}$, $X_{(3)}$ and $X_{(4)}$ (denoted respectively by $r_{(1)}(t)$, $r_{(2)}(t)$, $r_{(3)}(t)$ and $r_{(4)}(t)$) when X_1, X_2, X_3, X_4 are independent exponential random variables with respective parameters 1, 2, 3 and 4.

Now suppose $X_1, X_2, \ldots, X_n$ are independent identically distributed random variables with common density f. The density function of the ith order statistic is given by

$$f_{(i)}(t) = \frac{n!}{(i-1)!(n-i)!} F^{i-1}(t) f(t) \overline{F}^{n-i}(t) .$$

Therefore the ratio of the density functions of $X_{(j)}$ and $X_{(i)}$ for $i < j$ is given by

$$\frac{f_{(j)}(t)}{f_{(i)}(t)} = c \left(\frac{F(t)}{\overline{F}(t)} \right)^{j-i} \quad \text{which is increasing in } t .$$

This yields the following result (see Chan, Proschan and Sethuraman (1991)).

THEOREM 3.3. Let $X_1, X_2, \ldots, X_n$ be independent identically distributed random variables with common density. Then $X_{(i)} \leq_{\text{lr}} X_{(j)}$ for $i < j$.

Another likelihood ratio order result for maximums and minimums is given by Shaked and Shanthikumar (1994). Their result compares the maximum (mini-

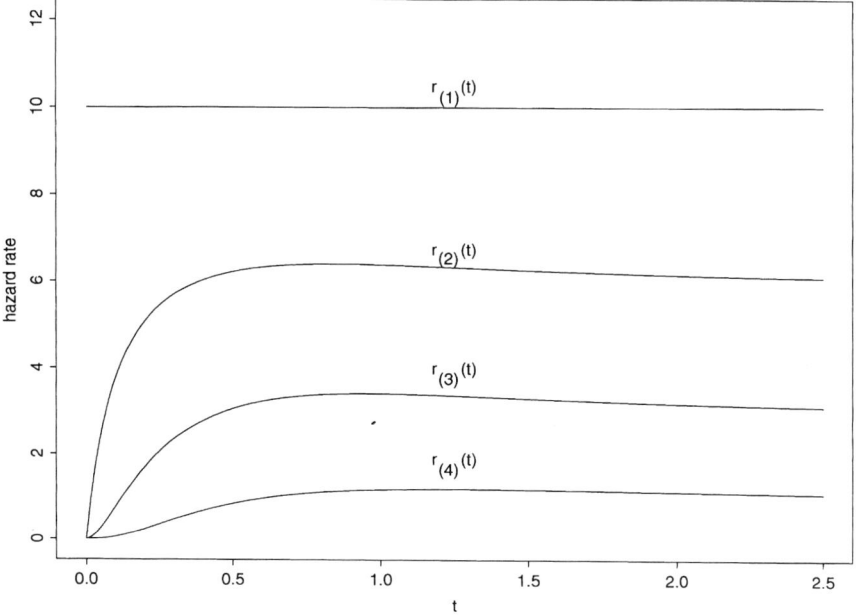

Fig. 4. Hazard rate functions of exponential order statistics.

mum) of $\{X_1, X_2, \ldots, X_n\}$ with the maximum (minimum) of $\{X_1, X_2, \ldots, X_n, X_{n+1}\}$ in the independent identically distributed (i.i.d.) case where X is either absolutely continuous or discrete.

THEOREM 3.4. Let $X_1, X_2, \ldots, X_n, X_{n+1}$ be independent identically distributed random variables which are either absolutely continuous or discrete. Then

$$X_{(n:n)} \leq_{\text{lr}} X_{(n+1:n+1)} \quad \text{and} \quad X_{(1:n)} \geq_{\text{lr}} X_{(1:n+1)} .$$

We consider now the dispersive ordering in the i.i.d. case. Although we expect the jth order statistic $X_{(j)}$ to be larger than $X_{(i)}$ ($i < j$) in many senses, it is not clear if we should expect $X_{(j)}$ to be more "dispersed" than $X_{(i)}$. It is true however in the case where $X_1, X_2, \ldots, X_n$ are exponentially distributed with parameter λ. In this situation it is well known that $X_{(j)} =_d X_{(i)} + (X_{(j)} - X_{(i)})$ where $X_{(i)}$ and $(X_{(j)} - X_{(i)})$ are independent, and $X_{(j)} - X_{(i)}$ is the sum of $j - i$ independent exponential random variables. $X_{(j)} - X_{(i)}$ has log concave density since the sum of independent random variables with log concave densities has log concave density, and therefore it follows from our discussion of the convolution properties of the dispersive order in Section 2 that $X_{(i)} \leq_{\text{disp}} X_{(j)}$. Bagai and Kochar (1986) (or see Theorem 2.B.13 in Shaked and Shanthikumar (1994)) have shown that if $X \leq_{\text{hr}} Y$ and either X or Y are DFR, then $X \leq_{\text{disp}} Y$. We know (Theorem 3.2) that when $X_1, X_2, \ldots, X_n$ are independent then $X_{(i)} \leq_{\text{hr}} X_{(j)}$ for any $i < j$, and hence if either $X_{(i)}$ or $X_{(j)}$ is DFR it follows that $X_{(i)} \leq_{\text{disp}} X_{(j)}$.

The following example shows that in spite of the above situation for exponential random variables, there is not much hope for the dispersive ordering of order statistics in general. In this example we see that the minimum $X_{(1)}$ and the maximum $X_{(2)}$ from a sample of size 2 of uniform random variables on $[0, 1]$ are not ordered in the dispersive order.

EXAMPLE 3.5. Let X_1 and X_2 be independent random variables uniformly distributed on $[0, 1]$. If $X_{(1)}$ and $X_{(2)}$ are the minimum and maximum of $\{X_1, X_2\}$, we have that $F_{(1)}(t) = 1 - (1-t)^2$ and $F_{(2)}(t) = t^2$. Hence

$$F_{(1)}^{-1}(\beta) - F_{(1)}^{-1}(\alpha) \leq F_{(2)}^{-1}(\beta) - F_{(2)}^{-1}(\alpha)$$
$$\iff \sqrt{1-\alpha} + \sqrt{\alpha} \leq \sqrt{\beta} + \sqrt{1-\beta} \quad \text{for } 0 \leq \alpha < \beta \leq 1 . \tag{3.1}$$

The function $g(t) = \sqrt{1-t} + \sqrt{t}$ on $[0, 1]$ is symmetric around $\frac{1}{2}$ and increasing on $[0, \frac{1}{2}]$. Hence (3.1) is valid if $\alpha < \beta \leq \frac{1}{2}$ and the reverse holds if $\frac{1}{2} \leq \alpha < \beta \leq 1$. In particular $X_{(1)}$ is not less than $X_{(2)}$ in the dispersive ordering, although one might be tempted to say that $X_{(1)}$ is less dispersed than $X_{(2)}$ for lower quantiles (and the reverse for higher quantiles).

It seems natural to expect some degree of positive dependence between the order statistics $X_{(i)}$ and $X_{(j)}$ from any set of observations $X_1, X_2, \ldots, X_n$. In the independent identically distributed case where $i < j$ and f is the common density function, then the joint density $f_{(i),(j)}$ of $X_{(i)}$ and $X_{(j)}$ is given by

$$f_{(i),(j)}(s,t) = \begin{cases} c(i,j,n)F^{i-1}(s)[F(t)-F(s)]^{j-i-1}[\overline{F}(t)]^{n-j}f(s)f(t) & s \leq t \\ 0 & \text{otherwise} \end{cases}$$

where $c(i,j,n)$ is a constant. Algebraically it is easy to see that $f_{(i),(j)}$ is TP$_2$ in s and t is equivalent to

$$[F(t_1) - F(s_1)][F(t_2) - F(s_2)] \geq [F(t_2) - F(s_1)][F(t_1) - F(s_2)] ,$$

which is always true. This proves the following result:

THEOREM 3.6. Let $X_1, X_2, \ldots, X_n$ be independent identically distributed random variables which are absolutely continuous. Then $X_{(i)}$ and $X_{(j)}$ are TP$_2$ dependent.

Note then in particular that for any $i < j$, $P[X_{(j)} > t | X_{(i)} = s]$ is increasing in s, and that as a result the family of conditional random variables $\{X_{(j)} | X_{(i)} = s : s \in \text{Support } X\}$ is increasing in the usual stochastic order.

The next two Theorems of Boland, El-Neweihi and Proschan (1994) provide results for the hazard rate ordering when comparing order statistics from $X_1, X_2, \ldots, X_n$ with those from $X_1, X_2, \ldots, X_n, X_{n+1}$. Hence we use the notation $X_{(i:n)}$ and $X_{(i:n+1)}$ to denote the ith order statistic from $X_1, X_2, \ldots, X_n$ and $X_1, X_2, \ldots, X_n, X_{n+1}$ respectively. In the first of these results we assume that $r_{n+1}(t) \leq r_k(t)$ for all t and $k = 1, \ldots, n$, while for the second we assume that $r_{n+1}(t) \geq r_k(t)$ for all t and $k = 1, \ldots, n$.

THEOREM 3.7. Let $X_1, X_2, \ldots, X_n, X_{n+1}$ be independent random variables where $X_k \leq_{\text{hr}} X_{n+1}$ for $k = 1, \ldots, n$. Then $X_{(i-1:n)} \leq_{\text{hr}} X_{(i:n+1)}$ for $i = 2, \ldots, n+1$.

THEOREM 3.8. Let $X_1, X_2, \ldots, X_n, X_{n+1}$ be independent random variables where $X_{n+1} \leq_{\text{hr}} X_k$ for $k = 2, \ldots, n$. Then $X_{(i:n)} \geq_{\text{hr}} X_{(i:n+1)}$ for all $i = 1, \ldots, n$.

Note that the continuous version of Theorem 3.4 is a special case of Theorems 3.7 and 3.8.

One might expect the order statistics to be ordered in the independently distributed case with respect to the (even stronger) likelihood ratio order. Surprisingly enough we will see in the following example that this is not the case in general (even for $n = 2$), although we know it is true if the X_i's are identically distributed. It will also provide us with an example of two random variables X and Y where $X \leq_{\text{hr}} Y$ but $X \not\leq_{\text{lr}} Y$.

EXAMPLE 3.9. Let f_i and $\overline{F}_i = 1 - F_i$ be the density and survival function of X_i, $i = 1, 2$. The densities of $X_{(2)}$ and $X_{(1)}$ are given respectively by: $f_1(t)(1 - \overline{F}_2(t)) + f_2(t)(1 - \overline{F}_1(t))$ and $f_1(t)\overline{F}_2(t) + f_2(t)\overline{F}_1(t)$. Now $X_{(2)}$ is greater in the likelihood ratio ordering than $X_{(1)} \iff (f_1(t)(1 - \overline{F}_2(t)) + f_2(t)(1 - \overline{F}_1(t)))/ (f_1(t)\overline{F}_2(t) + f_2(t)\overline{F}_1(t))$ is increasing in t. Equivalently by taking the derivative of the right-hand side and performing simple algebraic manipulations we must have

$$[f_1(t)f_2'(t) - f_2(t)f_1'(t)][\overline{F}_2(t) - \overline{F}_1(t)]$$
$$+ 2[f_1(t)f_2^2(t) + f_2(t)f_1^2(t)] \geq 0 \quad \text{for all } t \geq 0 \ . \tag{3.2}$$

Clearly one can construct f_1, f_2 such that the left-hand side of (3.2) is negative for at least one $t > 0$. Here is such a choice: let

$$f_1(t) = \begin{cases} \frac{3}{2} & 0 \leq t \leq \frac{2}{3} \\ 0 & \text{otherwise} \ , \end{cases}$$

and let

$$f_2(t) = \begin{cases} 1 & 0 \leq t \leq .50 \\ 1 - 25(t - .50) & .50 \leq t \leq .52 \\ .50 & .52 \leq t \leq 1.49 \ . \end{cases}$$

Now the right hand limit of the left side of (3.2) as t approaches .50 is equal to

$$\left[\tfrac{3}{2}(-25) - 0\right]\left[\tfrac{1}{2} - \tfrac{1}{4}\right] + 2\left(\tfrac{3}{2} + \tfrac{9}{4}\right) = -\tfrac{150}{16} + \tfrac{15}{2} < 0 \ ,$$

and hence $X_{(1)}$ is not smaller than $X_{(2)}$ in the likelihood ratio order.

Although the order statistics from an independent set of random variables are not necessarily increasing in the likelihood ratio ordering, Bapat and Kochar (1994) have proved that this is true under mild conditions if the independent random variables are themselves ordered according to the likelihood ratio order. Their result is the following theorem, which is proved using the theory of permanents. Note that Theorem 3.10 seems to be more general than Theorem 3.3 since identically distributed random variables are likelihood ratio ordered, however the proof of Theorem 3.10 requires *differentiable* densities with common interval support.

THEOREM 3.10. Let $X_1, X_2, \ldots, X_n$ be independent random variables with respective differentiable density functions $f_1, f_2, \ldots, f_n$ which have common interval support. If $X_1 \leq_{\text{lr}} X_2 \leq_{\text{lr}} \cdots \leq_{\text{lr}} X_n$, then

$$X_{(1)} \leq_{\text{lr}} X_{(2)} \leq_{\text{lr}} \cdots \leq_{\text{lr}} X_{(n)} \ .$$

Note that in particular if $X_1, X_2, \ldots, X_n$ are independent exponential random variables with respective parameters $\lambda_1, \lambda_2, \ldots, \lambda_n$, then it follows from the above that $X_{(1)} \leq_{\text{lr}} X_{(2)} \leq_{\text{lr}} \cdots \leq_{\text{lr}} X_{(n)}$. Another example is where $X_1, X_2, \ldots, X_n$ are independent normal random variables with variance σ^2 and respective means $\mu_1, \mu_2, \ldots, \mu_n$.

In the independent (but not necessarily identically distributed) case of random variables $X_1, X_2, \ldots, X_n$, it is not necessarily true that $X_{(i)}$ and $X_{(j)}$ are TP$_2$ dependent. In fact even in the case where X_1 and X_2 are independent exponentials with respective means 1 and .5, $X_{(1)}$ and $X_{(2)}$ are not TP$_2$ dependent (in fact they do not even satisfy the weaker condition of RCSI – see Boland, Hollander, Joag-Dev and Kochar (1994)). The following result (Boland et al., 1994) does show however that in general $X_{(j)}$ is right tail increasing in $X_{(i)}$ whenever $i < j$.

THEOREM 3.11. Let $X_1, X_2, \ldots, X_n$ be independently distributed random variables. Then for any $i < j$, $\text{RTI}(X_{(j)} | X_{(i)})$.

Theorem 3.11 implies that as s varies, the family of conditional order statistics $\{(X_{(j)} | X_{(i)} > s)\}$ is stochastically increasing in s in the usual stochastic order whenever $i < j$.

Now let us suppose that $X_1, X_2, \ldots, X_n$ represent a random sample of size n taken without replacement from a finite population. Then $X_1, X_2, \ldots, X_n$ are identically distributed discrete random variables which are clearly dependent. The following result of Boland et al. (1994) shows a strong dependence between order statistics in this situation.

THEOREM 3.12. Let $(X_1, X_2, \ldots, X_n)$ represent the observations of a random sample taken without replacement from a finite population. Then for any i and j, $X_{(i)}$ and $X_{(j)}$ have a TP_2 joint mass function.

A consequence of the above result is that in this finite population sampling situation, for any $i < j$, $\text{SI}(X_{(j)} | X_{(i)})$. Hence in particular the family of conditional random variables $\{X_{(j)} | X_{(i)} = s\}$ is increasing in s in the usual stochastic order.

Now suppose that $[X | \Theta = \theta]$ is a random variable for each value of the random variable Θ. Then X itself is a mixture with respect to Θ. Let $(X_1, X_2, \ldots, X_n)$ be a sample of size n from this mixture. Then although the X_i's are identically distributed they are dependent through Θ. Boland and El-Neweihi (1995) show the following dependence result for consecutive order statistics.

THEOREM 3.13. Let $X_1, X_2, \ldots, X_n$ be a sample of size n from a mixture distribution with respect to Θ, where the family of density functions $\{f_\theta(t): \theta \in \Theta\}$ has the monotone likelihood ratio property. Then for any i, $X_{(i)}$ and $X_{(i+1)}$ are TP_2 dependent.

4. Stochastic order for order statistics from two samples

In this section we assume that $X_1, X_2, \ldots, X_n$ and $Y_1, Y_2, \ldots, Y_n$ are two sets of random variables of size n, and we will make various stochastic comparisons between the order statistics $X_{(i)}$ and $Y_{(i)}$ for $i = 1, 2, \ldots, n$.

If $X_1, X_2, \ldots, X_n$ is a set of n independent random variables and likewise $Y_1, Y_2, \ldots, Y_n$ where $X_i \leq_{st} Y_i$ for $i = 1, \ldots, n$, then it is easy to verify that (Ross (1983)) $\mathbf{X} = (X_1, X_2, \ldots, X_n) \leq_{st} (Y_1, Y_2, \ldots, Y_n) = \mathbf{Y}$. Using the fact that $g_{(i)}(x_1, x_2, \ldots, x_n) = x_{(i)}$ is increasing, we have the following result for the usual stochastic order.

THEOREM 4.1. Let $X_1, X_2, \ldots, X_n$ be one set of independent random variables, and $Y_1, Y_2, \ldots, Y_n$ be another set where $X_i \leq_{st} Y_i$ for each $i = 1, 2, \ldots, n$. Then $X_{(i)} \leq_{st} Y_{(i)}$ for $i = 1, 2, \ldots, n$.

In order to give further results on stochastic comparisons of order statistics from two samples we introduce the concept of majorization of two vectors $a = (a_1, a_2, \ldots, a_n)$ and $b = (b_1, b_2, \ldots, b_n)$. We say a is majorized by b (and write $a \prec b$) if $\sum_{i=1}^{m} a_{(i)} \leq \sum_{i=1}^{m} b_{(i)}$ holds for $m = 1, 2, \ldots, n-1$ and $\sum_{i=1}^{n} a_{(i)} = \sum_{i=1}^{n} b_{(i)}$. If a is majorized by b, then in particular a is less dispersed or spread out than b is. The interested reader should consult Marshall and Olkin (1979), for more on majorization and its applications.

Now suppose $X_1, X_2, \ldots, X_n, Y_1, Y_2, \ldots, Y_n$ are independent random variables with proportional hazard (or failure rate) functions with $\lambda_1, \lambda_2, \ldots, \lambda_n$, $\lambda'_1, \lambda'_2, \ldots, \lambda'_n$ as the respective constants of proportionality. The situation where all of the random variables are exponentially distributed is but one example. The following result of Proschan and Sethuraman (1976) compares the vector of order statistics of X with the corresponding vector of order statistics of Y, and generalizes earlier work of Pledger and Proschan (1971), and Sen (1970) on individual order statistics.

THEOREM 4.2. Let $X_1, X_2, \ldots, X_n, Y_1, Y_2, \ldots, Y_n$ be independent random lifetimes with proportional hazard functions where $\lambda_1, \lambda_2, \ldots, \lambda_n, \lambda'_1, \lambda'_2, \ldots, \lambda'_n$ are the constants of proportionality. Suppose $\lambda = (\lambda_1, \lambda_2, \ldots, \lambda_n) \prec (\lambda'_1, \lambda'_2, \ldots, \lambda'_n) = \lambda'$. Then $(X_{(1)}, X_{(2)}, \ldots, X_{(n)}) \leq_{st} (Y_{(1)}, Y_{(2)}, \ldots, Y_{(n)})$ and in particular $X_{(i)} \leq_{st} Y_{(i)}$ for all $i = 1, \ldots, n$.

The following example shows that Theorem 4.1 for the usual stochastic order does not generalize to the hazard rate ordering.

EXAMPLE 4.3. If X_1, X_2 (Y_1, Y_2) are independent exponential random variables with parameters or hazards λ_1, λ_2 (λ'_1, λ'_2), then $X_{(2)}$ ($Y_{(2)}$) is the lifetime of the parallel system formed from X_1 and X_2 (Y_1 and Y_2). We let r_{λ_1, λ_2} ($r_{\lambda'_1, \lambda'_2}$) be the hazard rate function of $X_{(2)}$ ($Y_{(2)}$). If $(\lambda_1, \lambda_2) = (.5, 15)$ and $(\lambda'_1, \lambda'_2) = (.5, 5)$, then since $X_i \leq_{st} Y_i$ for $i = 1, 2$ it follows from Theorem 4.1 that $X_{(2)} \leq_{st} Y_{(2)}$. However it is not true (as seen from Fig. 5 below) that $X_{(2)}$ is smaller than $Y_{(2)}$ in the hazard rate ordering.

The following result (see Lynch, Mimmack and Proschan (1987) or Shaked and Shanthikumar (1994)) gives reasonable conditions under which the order statistics from two samples are comparable under the hazard rate ordering. A slightly more general result (with technical conditions) is given in Shaked and Shanthikumar (1995).

THEOREM 4.4. Let $X_1, X_2, \ldots, X_n$ and $Y_1, Y_2, \ldots, Y_n$ be two sets of independent random variables where $X_i \leq_{hr} Y_j$ for all i and j, $(i, j = 1, 2, \ldots, n)$. Then $X_{(i)} \leq_{hr} Y_{(i)}$ for $i = 1, 2, \ldots, n$.

Suppose now that $X_{(2)} = \max(X_1, X_2)$ where X_1 and X_2 are independent exponential random variables with respective parameters λ_1 and λ_2. The following result (see Boland, El-Neweihi and Proschan (1994)) shows that the more spread out (λ_1, λ_2) is in the sense of majorization, the greater $X_{(2)}$ is in the hazard rate ordering.

EXAMPLE 4.5. Let X_1, X_2, Y_1 and Y_2 be independent exponential random variables with respective parameters $\lambda_1, \lambda_2, \lambda_1'$ and λ_2'. Then $(\lambda_1, \lambda_2) \prec (\lambda_1', \lambda_2') \implies X_{(2)} \leq_{\text{hr}} Y_{(2)}$.

The above result suggests that perhaps an analogue of Theorem 4.2 for the hazard rate ordering might be valid. However Boland, El-Neweihi and Proschan (1994) show that this is not the case even when comparing the maxima of two samples of independent exponential random variables of size 3. In particular they show that if X_1, X_2, X_3, Y_1, Y_2, and Y_3 are independent exponentials with respective parameters $0.1, 4, 6, 0.1, 1$, and 9, then although $(0.1, 4, 6) \prec (0.1, 1, 9)$ it is not true that $X_{(3)}$ is smaller than $Y_{(3)}$ in the hazard rate ordering (although $X_{(3)} \leq_{\text{st}} Y_{(3)}$ from Theorem 4.2).

Chan, Proschan and Sethuraman (1991) prove the analogue of Theorem 4.4 for the likelihood ratio order, as given below.

THEOREM 4.6. Let $X_1, X_2, \ldots, X_n$ and $Y_1, Y_2, \ldots, Y_n$ be two sets of independent absolutely continuous random variables where $X_i \leq_{\text{lr}} Y_j$ for all i and j $(i, j = 1, 2, \ldots, n)$. Then $X_{(i)} \leq_{\text{lr}} Y_{(i)}$ for $i = 1, 2, \ldots, n$.

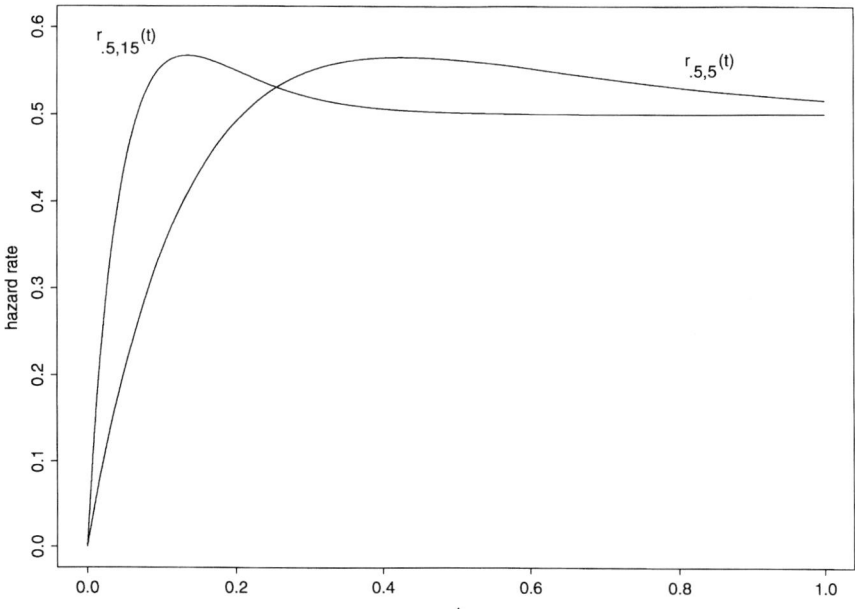

Fig. 5. Hazard rate functions for the maximum of 2 exponentials.

Now suppose $X_1, X_2, \ldots, X_n$ is a set of independent identically distributed random variables each with distribution equal to that of X. Then it is well known (see Barlow and Proschan (1981)) that the distribution function of the ith order statistic $X_{(i)}$ is given by

$$F_{X_{(i)}}(t) = B_{(i)}^n(F_X(t))$$

where

$$B_{(i)}^n(p) = \frac{n!}{(i-1)!(n-i)!} \int_0^p u^{i-1}(1-u)^{n-i} du \quad \text{for } 0 \leq p \leq 1 \ .$$

Suppose also that $Y_1, Y_2, \ldots, Y_n$ is another set of independent random variables with the same distribution as that of Y. Then it follows from the above representation that for any $i = 1, 2, \ldots, n$,

$$F_{Y_{(i)}}^{-1}(F_{X(i)}) = F_Y^{-1}(F_X) \ .$$

This observation allows one to prove without much difficulty the following stochastic comparisons for order statistics in the dispersive and transform orders.

THEOREM 4.7. Let $X_1, X_2, \ldots, X_n$ be independently distributed random variables with the same distribution as X, and similarly $Y_1, Y_2, \ldots, Y_n$ be independent random variables with the same distribution as Y.

(a) If $X \leq_{\text{disp}} Y$, then $X_{(i)} \leq_{\text{disp}} Y_{(i)}$ for $i = 1, 2, \ldots, n$.
(b) If $X \leq_c Y$, then $X_{(i)} \leq_c Y_{(i)}$ for $i = 1, 2, \ldots, n$.
(c) If $X \leq_* Y$, then $X_{(i)} \leq_* Y_{(i)}$ for $i = 1, 2, \ldots, n$.
(d) If $X \leq_{\text{su}} Y$, then $X_{(i)} \leq_{\text{su}} Y_{(i)}$ for $i = 1, 2, \ldots, n$.

Theorem 4.7 a is due to Bartoszewicz (1986), while the idea for Theorem 4.7b–d is in Barlow and Proschan (1981). As indicated earlier many of the results in this paper can be found in the extensive treatment of stochastic orders by Shaked and Shanthikumar (1994).

Acknowledgement

This work was supported by

1. NSF Grant DMS 930389 (M. Shaked)
2. NSF Grant DMS 9308149 (J.G. Shanthikumar)

References

Bagai, I. and S. C. Kochar (1986). On tail ordering and comparison of failure rates. *Commun. Statist. – Theory and Methods* **15**, 1377–1388.

Bapat, R. B. and S. C. Kochar (1994). On likelihood-ratio ordering of order statistics. *Linear Algebra and Its Applications* **199**, 281–291.

Barlow, R. E. and F. Proschan (1981). *Statistical Theory of Reliability and Life Testing: Probability Models*. To Begin With, Silver Spring, MD, USA.

Bartoszewicz, J. (1986). Dispersive ordering and the total time on test transformation. *Statist. Prob. Lett.* **4**, 285–288.

Boland, P. J. and E. El-Neweihi (1995). Dependence properties of consecutive order statistics from mixtures, preprint. Dept. of Statistics, University College Dublin.

Boland, P. J., E. El-Neweihi and F. Proschan (1994). Applications of the hazard rate ordering in reliability and order statistics. *J. Appl. Prob.* **31**, 180–192.

Boland, P.J., M. Hollander, K. Joag-Dev and S. Kochar (1994). Bivariate dependence properties of order statistics. *J. Multivar. Anal.*, **56** (1), 75–89.

Chan, W., F. Proschan and J. Sethuraman (1991). Convex ordering among functions, with applications to reliability and mathematical statistics. In: *Topics in Statistical Dependence*, ed. H. W. Block, A. R. Sampson and T. H. Savits. IMS Lecture Notes **16**, 121–134.

Droste, W. and W. Wefelmeyer (1985). A note on strong unimodality and dispersivity. *J. Appl. Prob.* **2**, 235–239.

Keilson, J. and U. Sumita (1982). Uniform stochastic ordering and related inequalities. *Canad. J. Statist.* **10**, 181–189.

Jogdeo, K. (1982). Concepts of dependence. In: *Encyclopedia of Statistical Sciences*. Eds S. Kotz and N. L. Johnson, **2**, 324–334.

Lewis, T. and J. W. Thompson (1981). Dispersive distributions and the connection between dispersivity and strong unimodality. *J. Appl. Prob.* **18**, 76–90.

Lehmann, E. (1966). Some concepts of dependence. *Ann. Math. Statist.* **37**, 1137–1153.

Lynch, J., G. Mimmack and F. Proschan (1987). Uniform stochastic orderings and total positivity. *Canad. J. Statist.* **15**, 63–69.

Marshall, A. W and I. Olkin (1979). *Inequalities: Theory of Majorization and its Applications*. Academic Press, New York, N.Y.

Pledger, G. and F. Proschan (1971). Comparisons of order statistics and of spacings from heterogeneous distributions. In *Optimizing Methods in Statistics*, Ed. J. S. Rustagi, Academic Press, New York.

Proschan, F. and J. Sethuraman (1976). Stochastic comparisons of order statistics from heterogeneous populations with applications in reliability. *J. Multivar. Anal.* **6**, 4, 608–616.

Ross, S. (1983). *Stochastic Processes*. Wiley, New York.

Sen, P. K. (1970). A note on order statistics for heterogeneous distributions. *Ann. Math. Statist.* **41**, 2137–2139.

Shaked, M. and J. G. Shanthikumar (1994). *Stochastic orders and their applications*. Academic Press, San Diego, CA.

Shaked, M. and J. G. Shanthikumar (1995). Hazard rate ordering of k-out-of-n systems, *Statist. Prob. Lett.* **23** (1), 1–8.

Shanthikumar, J. G. and D. D. Yao (1991). Bivariate characterization of some stochastic order relations. *Adv. Appl. Prob.* **23**, 642–659.

Tong, Y. L. (1980). *Probability Inequalities in Multivariate Distributions*. Academic Press, New York.

Van Zwet W. (1964). *Convex transforms of random variables*. Mathematische Centrum, Amsterdam.

Bounds for Expectations of *L*-Estimates

Tomasz Rychlik

1. Introduction

An original sample of a given size, say n, will be denoted by $X_1, \ldots, X_n$. The respective order statistics will be written as $X_{1:n} \leq \cdots \leq X_{n:n}$. For a given sequence $(c_i)_{i=1}^n \in \mathcal{R}^n$, the *L*-estimate will be defined as $\sum_{i=1}^n c_i X_{i:n}$. Our objective is to present accurate bounds on the expectation of the *L*-estimates under various assumptions on the distribution of the original sample, with indicating the distributions for which the bounds are attained. Usually, we focus the attention on deriving upper bounds. Some lower ones can be immediately obtained by simple transformations, e.g. $\inf \mathrm{E} \sum_{i=1}^n c_i X_{i:n} = -\sup \mathrm{E} \sum_{i=1}^n (-c_i) X_{i:n}$. The special attention will be devoted to the case of single order statistics. In principle, we consider two types of assumptions:

- $X_i, i = 1, \ldots, n$, are identically distributed, but can be dependent in an arbitrary way (i.d case),
- $X_i, i = 1, \ldots, n$, are identically distributed and independent (i.i.d case).

In Section 2 we shall study the expectations of *L*-estimates in the case of dependent random variables with a given common distribution function (for the i.i.d sample, there are well known formulas). For single order statistics, we present a method of obtaining extremes of the expectation and variance of an arbitrary integrable function of the order statistic. Also, we consider the case of nonidentically distributed samples. In Section 3 we present some bounds for both the i.d and i.i.d cases in terms of natural location and scale parameters of the marginal parent distribution, including the expectation and an absolute central moment, the left end point and the length of the support. The results for the i.d case enable us to find some sharp deterministic bounds which hold true for arbitrary samples. In Section 4 we obtain inequalities for the expectations of order statistics based on variables coming from restricted families of distributions: life distributions with monotone failure probability and rate, symmetric and symmetric unimodal. The inequalities will be expressed in terms of the first and second moments of the original variables. The bounds given in Sections 3 and 4 were obtained by means of projecting the quantile and related functions onto convex cones. In Section 5 we apply the Jensen inequality to derive bounds for expected order statistics coming from the restricted families in terms of quantiles of the parent marginal distribution.

Some bounds presented in the paper were discussed in the monographs by David (1981, Chapter 4, mainly the i.i.d case), Arnold and Balakrishnan (1989, Chapter 3), Arnold et al. (1992, Chapters 5, 6). However, a significant part of the paper content have been obtained quite recently and is reviewed for the first time here. Also, some results have not been published by now.

2. Distribution bounds

2.1. I.d case. Assume that $X_1, \ldots, X_n$ are arbitrarily dependent and have a common fixed distribution function F. Let $Q_F(x) = \sup\{t: F(t) \leq x\}$ denote the respective right continuous quantile function and let $\mu = EX_1 = \int_0^1 Q_F(x) \, dx$ be finite. Consider an arbitrary L-estimate with coefficients c_i, $i = 1, \ldots, n$, and define C as the greatest convex function on $[0, 1]$, vanishing at 0, and satisfying

$$C\left(\frac{j}{n}\right) \leq \sum_{i=1}^{j} c_i, \quad j = 1, \ldots, n \ .$$

THEOREM 1 (Rychlik, 1993a). *With D standing for the right derivative of C, we have*

$$E \sum_{i=1}^{n} c_i X_{i:n} \leq \int_0^1 Q_F(x) D(x) \, dx \ . \quad (1)$$

The bound is best possible.

Observe that both the C and D depend merely on the coefficients of L-estimates, which is ignored in notation, for brevity. Each D is a nondecreasing right continuous jump function which jumps at some multiplicites of $\frac{1}{n}$. This clearly allows us to write the right-hand side of (1) as a combination of subintegrals

$$\int_0^1 Q_F(x) D(x) \, dx = \sum_{i=1}^{n} d_i \int_{\frac{i-1}{n}}^{\frac{i}{n}} Q_F(x) \, dx \ , \quad (2)$$

where

$$d_i = D\left(\frac{i-1}{n}\right) = n\left(C\left(\frac{i}{n}\right) - C\left(\frac{i-1}{n}\right)\right), \quad i = 1, \ldots, n \ . \quad (3)$$

In fact, $(d_i)_{i=1}^n$ is the l_2-projection of $(c_i)_{i=1}^n$ onto the convex cone of nondecreasing sequences in $\mathscr{R}^n$. In the special case of single m^{th} order statistic $D = D_{m:n} = \frac{n}{n+1-m} \mathbf{1}_{[\frac{m-1}{n}, 1)}$, and, in consequence, (1) becomes

$$EX_{m:n} \leq \frac{n}{n+1-m} \int_{\frac{m-1}{n}}^{1} Q_F(x) \, dx \ . \quad (4)$$

We now turn to describing the conditions for which (1) becomes equality. We express them in terms of distribution functions of order statistics. It can be shown that $F_{i:n}$, $i=1,\ldots,n$, are the distribution functions of $X_{i:n}$, $i=1,\ldots,n$, respectively, based on some possibly dependent F-distributed random variables iff

$$\sum_{i=1}^{n} F_{i:n} = nF ,\tag{5}$$

and, clearly,

$$F_{i:n} \geq F_{i+1:n}, \quad i=1,\ldots,n-1 .\tag{6}$$

There are various ways of constructing ordered variables $X_{i:n}$, $i=1,\ldots,n$, with distributions satisfying (5) and (6) (the simplest one consists in taking $Q_{F_{i:n}}(U)$, $i=1,\ldots,n$, for U being a standard uniform random variable). Also, there are many ways of constructing identically distributed parent variables X_i, $i=1,\ldots,n$, whose order statistics have given distributions (the simplest one consists in the random reordering of $X_{1:n},\ldots,X_{n:n}$). Let $0=i_0<i_1<\cdots<i_v=n$ be the integers for which $(\frac{i_j}{n}, \sum_{i=1}^{i_j} c_i)$, $j=1,\ldots,v$, belong to the graph of C. Some of $\frac{i}{n}$ are the break points of C. Then (1) becomes equality iff

$$P\left(Q_F\left(\frac{i_{j'}}{n}\right) \leq X_{i_{j-1}+1:n} = X_{i_j:n} \leq Q_F\left(\frac{i_{j''}}{n}\right), \ j=1,\ldots,v\right) = 1 \tag{7}$$

holds together with (5) and (6). Here $[\frac{i_{j'}}{n}, \frac{i_{j''}}{n}]$ is the interval of linearity of C that contains both the $\frac{i_{j-1}}{n}$ and $\frac{i_j}{n}$. In particular, if all $\frac{i_j}{n}$, $j=1,\ldots,v$, are the break points of C, then the respective $F_{i:n}$, $i=1,\ldots,n$, are determined uniquely:

$$F_{i:n} = \max\left\{\min\left\{\frac{nF-i_{j-1}}{i_j-i_{j-1}},1\right\},0\right\},$$
$$i = i_{j-1}+1,\ldots,i_j, \ j=1,\ldots,v . \tag{8}$$

On the other hand, for the special case of sample mean, yields $C(x)=x$, $v=n$, and so (7) becomes trivial. Indeed, we see that $E\frac{1}{n}\sum_{i=1}^n X_{i:n} = \int_0^1 Q_F(x)\,dx = \mu$ for any type of interdependence among the variables.

Inequality (1) can be deduced from the representation

$$E\sum_{i=1}^{n} c_i X_{i:n} = \int_{-\infty}^{+\infty} x \sum_{i=1}^{n} c_i F_{i:n}(dx) ,$$

and the solution of the linear programming problem

$$\min \sum_{i=1}^{n} c_i F_{i:n}(x) = C(F(x)), \quad x \in \mathcal{R} ,\tag{9}$$

where the minimum is taken over all $(F_{i:n}(x))_{i=1}^n$ satisfying (5)–(6), and from a basic property of the stochastic order. This asserts that

$$\int_a^b g(x) G(dx) \leq \int_a^b g(x) H(dx) \tag{10}$$

for all nondecreasing functions g if $G \geq H$ on (a, b) and $G = H$ at the ends (see, e.g. Marshall and Olkin, 1979, p. 444). At each point x, the minimum in (9) is attained by a vector $(F^\star_{i:n}(x))_{i=1}^n$, whose coordinates form distribution functions, as x varies, with properties equivalent to (7).

As a consequence of Theorem 1, we get the best possible lower bounds

$$E \sum_{i=1}^n c_i X_{i:n} \geq \int_0^1 Q_F(x) \overline{D}(x) \, dx \;,$$

where $\overline{D} \colon [0, 1) \mapsto \mathcal{R}$ is the right derivative of the smallest concave function $\overline{C}$ satisfying $\overline{C}(\frac{j}{n}) \geq \sum_{i=1}^j c_i$, $j = 0, \ldots, n$.

EXAMPLE 1. Trimmed mean (single order statistic)

$$\frac{n}{k} \int_0^{\frac{k}{n}} Q_F(x) \, dx \leq E \frac{1}{k + 1 - j} \sum_{i=j}^k X_{i:n}$$

$$\leq \frac{n}{n + 1 - j} \int_{\frac{j-1}{n}}^1 Q_F(x) \, dx \;,\tag{11}$$

$1 \leq j \leq k \leq n$. The same bounds hold for the expectation of the Winsorized mean $E \frac{1}{n} [j X_{j:n} + \sum_{i=j+1}^{k-1} X_{i:n} + (n + 1 - k) X_{k:n}]$.

EXAMPLE 2. Difference of two order statistics

$$0 \leq E(X_{j:n} - X_{i:n}) \leq \frac{n}{n+1-j} \int_{\frac{j-1}{n}}^1 Q_F(x) \, dx - \frac{n}{i} \int_0^{\frac{i}{n}} Q_F(x) \, dx \;.$$

It is worth noting that

$$\int_0^1 h(Q_F(x)) \overline{D}(x) \, dx \leq E \sum_{i=1}^n c_i h(X_{i:n}) \leq \int_0^1 h(Q_F(x)) D(x) \, dx \tag{12}$$

for every nondecreasing function such that $E h(X_1)$ is finite, and equality in (1) implies that in (12).

Let us now concentrate on single order statistics $X_{m:n}$, $m = 1, \ldots, n$, (cf (11)). In this case C has a single break at $\frac{m-1}{n}$, $C(\frac{j}{n}) = \sum_{i=1}^j c_i$, $j = 1, \ldots, m-1, n$, and

$$P\left(X_{m-1:n} \leq Q_F\left(\frac{m-1}{n}\right) \leq X_{m:n} = X_{n:n}\right) = 1 \;,$$

and

$$F^\star_{m:n} = \max\left\{\frac{nF + 1 - m}{n + 1 - m}, 0\right\} \;,\tag{13}$$

by (7)–(8), respectively. Plugging $h_t = -\mathbf{1}_{(-\infty, t]}$ into (12) yields $P(X_{m:n} \leq t) \geq F^\star_{m:n}(t)$.

THEOREM 2 (Caraux and Gascuel, 1992a, Rychlik, 1992a). *For every sequence of dependent identically F-distributed random variables $X_1, \ldots, X_n$, we have*

$$\max\left\{\frac{nF(x) + 1 - m}{n + 1 - m}, 0\right\} \leq P(X_{m:n} \leq x) \leq \min\left\{\frac{n}{m}F(x), 1\right\}, \qquad (14)$$

For each lower and upper bound, for each $n \in \mathcal{N}$, and $m = 1, \ldots, n$, there exists a sequence of random variables satisfying the bound on the whole real axis.

Accordingly, the joint distributions, which maximize (minimize) the expectation of a given order statistic, minimize (maximize) uniformly the respective distribution function. There are numerous methods of constructing distributions with this property. The first example was presented by Mallows (1969) who constructed an n-dimensional density with uniform marginals for which $EX_{1:n}$ is minimal. Lai and Robbins (1976) extended the Mallows example to the case of arbitrary, possibly nonidentical marginals (the results were expressed in terms of the supremum of expected sample maximum), and noticed the existence of stochastically extreme distribution of the maximum. Lai and Robbins (1978) constructed a sequence of i.d variables (deterministic functions of a single uniform variable) such that every element of the sequence of maxima has the maximal expectation. They proved that these maxima and those of i.i.d sequence tend in distribution to the upper end-point of the support of F at the same rate. Distributions of other extreme order statistics exhibit the analogous asymptotic behavior (see Rychlik, 1992b).

We now describe a method of calculating the accurate lower and upper bounds for the expectation and variance of an arbitrary measurable function h of a given order statistic, presented in Rychlik (1994). To omit trivial solutions we assume that $Eh(X_1)$ and $Varh(X_1)$ are finite. A first useful result, which is of an intrinsic interest, is the characterization of the all distributions of a given order statistic for the all possible dependent F-distributed samples. Distribution function $F_{m:n}$ of some $X_{m:n}$ is characterized by two conditions: one is that the increase of $F_{m:n}$ does not exceed the increase of nF, and the other is (14). In other words, each $F_{m:n}$ has a density function $f_{m:n}$ with respect to F, not greater than n F-almost surely and such that its indefinite integral

$$F_{m:n}(x) = \int_{-\infty}^{x} f_{m:n}(t) F(dt) \quad \text{satisfies (14) for all } x \in \mathcal{R}. \qquad (15)$$

Therefore

$$Eh(X_{m:n}) = \int_{-\infty}^{+\infty} h(x) f_{m:n}(x) F(dx), \qquad (16)$$

$$Varh(X_{m:n}) = \int_{-\infty}^{+\infty} h^2(x) f_{m:n}(x) F(dx) - \left(\int_{-\infty}^{+\infty} h(x) f_{m:n}(x) F(dx)\right)^2, \qquad (17)$$

and, in consequence, the original problem can be replaced by calculating the extremes of the right-hand side functionals in (16) and (17) for $f_{m:n}$ from the convex set of functions, say $\mathscr{F}_{m:n}$, satisfying $0 \leq f_{m:n} \leq n$ and (15). Embedding $\mathscr{F}_{m:n}$ into a properly chosen topological space, we can prove that the extreme values of (16) and (17) are attained by some extreme points of $\mathscr{F}_{m:n}$. In the case of continuous F, a density $f_{m:n}$ is extreme in $\mathscr{F}_{m:n}$ iff

$$f_{m:n} = \text{either } 0 \text{ or } n \text{ on}$$

$$\left\{ x: \max\left\{ \frac{nF(x)+1-m}{n+1-m}, 0 \right\} < F_{m:n} < \min\left\{ \frac{n}{m}F(x), 1 \right\} \right\}, \tag{18}$$

which an open set, i.e. a countable, finite, or empty union of disjoint open intervals. For general F, extreme $f_{m:n}$ may have at most one value between 0 and n on each subinterval of (18) (for the precise definition, see Rychlik, 1994).

Being confined to such a simple class of densities, we can easily guess the form of solution for typical functions h. E.g., for h nondecreasing, the expectation is maximized by the density maximally moved to the right, i.e. $f^\star_{m:n} = \frac{n}{n+1-m}$ on $[Q_F(\frac{m-1}{n}), +\infty)$ (cf (12) and (13)). The variance can be maximized if we maximally disperse the density in the both directions so that we obtain $f^{\star\star}_{m:n} = \frac{n}{m}$ on $(-\infty, a]$ and $\frac{n}{n+1-m}$ on $[b, +\infty)$ for some $a < b$, determined by conditions $\int_{-\infty}^{+\infty} f^{\star\star}_{m:n}(x) F(dx) = 1$ and parametric maximization of (17).

EXAMPLE 3. (Rychlik, 1994). Let $X_1, \ldots, X_n$ be uniformly distributed on $[-1, 1]$. Then

$$\frac{m}{n} - 1 \leq EX_{m:n} \leq \frac{m-1}{n},$$

with the notation $k = \max\{m, n+1-m\}$

$$\frac{\left(\frac{2k}{n}-1\right)^3 - (2k-1)^{-2}}{6\frac{k}{n}} \leq EX^2_{m:n} \leq \left[1 - \frac{m(n+1-m)}{n(n+1)}\right]^2$$

$$+ \frac{1}{3}\left[\frac{m(n+1-m)}{n(n+1)}\right]^2,$$

$$\frac{1}{3n^2} \leq \text{Var}X_{m:n} \leq 1 + \frac{m^2 + (n+1-m)^2}{6n^2} - \frac{2n(n+1)}{3m(n+1-m)}$$

$$\times \left[\sqrt{\frac{4}{(n+1-2m)^2} + 1} + \sqrt{\frac{(n+1-2m)^2}{4} + 1} - \frac{2}{|n+1-2m|}\right].$$

2.2. *Related results.* We shortly mention the case of order statistics of possibly dependent, nonidentically distributed random variables $X_1, \ldots, X_n$, with distribution functions $F_1, \ldots, F_n$, respectively. It is easy to generalize (14) to

$$\max\left\{\frac{\sum_{i=1}^{n} F_i(x) + 1 - m}{n + 1 - m}, 0\right\} \leq P(X_{m:n} \leq x) \leq \min\left\{\frac{1}{m}\sum_{i=1}^{n} F_i(x), 1\right\} \tag{19}$$

so that

$$\frac{1}{n+1-m}\int_{m-1}^{n} Q_{\sum F_i}(x)\,dx \geq EX_{m:n} \geq \frac{1}{m}\int_{0}^{m} Q_{\sum F_i}(x)\,dx \ . \tag{20}$$

Generally, the bounds are not always attainable. Rychlik (1995) presented the necessary and sufficient conditions for F_i, $i = 1, \ldots, n$, so that each inequality in (19) and (20) becomes equality. E.g., the right-hand side bounds are attained if the all densities of F_i, $i = 1, \ldots, n$, with respect to $\sum_{i=1}^{n} F_i$ are not greater than $\frac{1}{m}$ on $(-\infty, Q_{\sum F_i}(m)]$. If $\sum_{i=1}^{n} F_i$ has a jump at $Q_{\sum F_i}(m)$, then the condition can be weakened for this point. Observe that for $m = 1$, the condition is satisfied and the bounds are sharp for arbitrary marginals, which was obtained by Lai and Robbins (1976). Tchen (1980) proved that there exists an infinite sequence of dependent random variables $(X_i)_{i=1}^{\infty}$ with arbitrarily chosen respective distribution functions $(F_i)_{i=1}^{\infty}$ such that $EX_{n:n} = \int_{n-1}^{n} Q_{\sum F_i}(x)\,dx$ (equivalently, $EX_{1:n} = \int_{0}^{1} Q_{\sum F_i}(x)\,dx$) for all $n \in \mathcal{N}$.

As we can see, various problems for the case arbitrarily dependent random variables have surprisingly nice solutions. Serious difficulties arise when we make some assumptions on the interdependence of variables. Kemperman (1993) considered bounds for the expectations of L-estimates of variables with given k-dimensional marginals, $1 < k < n$, and derived interesting lower and upper estimates of the bounds for expected order statistics from k-independent samples. In cooperation with Ott and Loh, he established that

$$0.213421 \approx \frac{297 + 5\sqrt{5}}{1444} \leq \inf\ EX_{1:3} < \frac{77}{360} \approx 0.213889$$

for the pairwise independent standard uniform variables X_1, X_2, X_3, which improved a result of Mallows (1969)

$$0.208333 \approx \tfrac{5}{24} \leq \inf\ EX_{1:3} \leq \tfrac{3}{14} \approx 0.214286 \ .$$

Gravey (1985) presented the following inequality for identically distributed $(n + 1 - m)$-exchangeable random variables:

$$EX_{m:n} \leq \binom{n}{m-1} \int_{1-\binom{n}{m-1}^{-1}}^{1} Q_{F_{1:n+1-m}}(x)\,dx \ .$$

2.3. I.i.d case.

In this case, we have explicit formulas for the distribution functions of order statistics

$$F_{m:n}(x) = \sum_{i=m}^{n} \binom{n}{i} F^i(x)(1 - F(x))^{n-i}$$

and for the expectation of arbitrary L-estimate

$$E \sum_{i=1}^{n} c_i X_{i:n} = \int_0^1 Q_F(x) \sum_{i=1}^{n} c_i N_{i:n}(x) \, dx \ . \tag{21}$$

where

$$N_{i:n}(x) = n \binom{n-1}{i-1} x^{i-1}(1-x)^{n-i}, \quad i = 1, \ldots, n, \tag{22}$$

is the Bernstein basis of the polynomials of degree not exceeding $n-1$. For abbreviation, we write N instead of a combination $\sum_{i=1}^{n} c_i N_{i:n}$, remembering the dependence of N on the coefficients of the L-estimate. For the mth order statistic $N = N_{m:n}$, which is the density function of mth order statistic of standard uniform sample of size n.

EXAMPLE 3 (continued). Let $X_1, \ldots, X_n$ be independent uniformly distributed on $[-1, 1]$. Then

$$EX_{m:n} = \frac{2m}{n+1} - 1 \ ,$$

$$EX_{m:n}^2 = 1 - 4\frac{m(n+1-m)}{(n+1)(n+2)} \ ,$$

$$\text{Var} X_{m:n} = \frac{4m(n+1-m)}{(n+1)^2(n+2)} \ .$$

3. Moment and support bounds

3.1. I.d case. Suppose now that X_i, $i = 1, \ldots, n$, are dependent and identically distributed, but we merely know a pair of location and scale parameters of the parent distribution. Because dropping either of the parameter constraints would enable us to increase trivially the expectation of L-estimates beyond any bounds, these are the minimal reasonable assumptions on the marginal distribution.

Suppose first that the expectation $EX_1 = \mu$ and the pth absolute central moment $E|X_1 - \mu|^p = \sigma_p^p$ are given, where $1 < p < \infty$ is fixed. Applying (1) and the Hölder inequality, we can write

$$E \sum_{i=1}^{n} c_i(X_{i:n} - \mu) \leq \int_0^1 (Q_F(x) - \mu) D(x) \, dx$$

$$= \int_0^1 (Q_F(x) - \mu)(D(x) - a) \, dx$$

$$\leq \|D - a\|_q \sigma_p \ ,$$

for an arbitrary real a, and $q = \frac{p}{p-1}$. If we choose a_q that minimizes $\|D - a\|_q$, (in other words, a_q is the coefficient of l_q-projection of $(d_i)_{i=1}^n$ onto vectors with identical coordinates (cf (3)), we obtain the sharp bound. Indeed, for the non-decreasing right continuous function

$$Q_{F^*} = \mu + \frac{\sigma_p}{\| \sqrt[p-1]{|D - a_q|} \|_p} \sqrt[p-1]{|D - a_q|} \, \text{sgn}(D - a_q) \tag{23}$$

yields

$$\int_0^1 (Q_{F^*}(x) - \mu) D(x) \, dx = \|D - a_q\|_q \sigma_p \, .$$

THEOREM 3 (Rychlik, 1993b). If X_i, $i = 1, \ldots, n$, are identically distributed, and $EX_1 = \mu$, and $E|X_1 - \mu|^p = \sigma_p^p$, $1 < p < +\infty$, then

$$E \sum_{i=1}^n c_i (X_{i:n} - \mu) \leq \|D - a_q\|_q \sigma_p = \|(d_i - a_q)_{i=1}^n\|_q \sigma_p \,, \tag{24}$$

(cf (3)), which is equality iff the joint distribution of $X_1, \ldots, X_n$ satisfies (5)–(7), and the common distribution function F^* has the quantile function defined in (23).

Observe that for the mean-variance bound ($p = q = 2$), formula (23) can be significantly simplified so more that $a_2 = \frac{1}{n} \sum_{i=1}^n d_i = \sum_{i=1}^n c_i$.

It was Arnold (1980) who adopted the application of the Hölder inequality for establishing bounds on expected order statistics from the i.i.d samples to the dependent case. Using the Lai–Robbins distribution bound (13) for $m = n$ and the Schwarz inequality, he determined the mean-variance bounds for the sample maximum and range (analogous results for $p \neq 2$ are given in Arnold, 1985, see also Arnold, 1988). Other special cases of Theorem 3 are the mean-variance bounds for the expectations of order statistics:

$$\mu - \sqrt{\frac{n-m}{m}} \sigma_2 \leq EX_{m:n} \leq \mu + \sqrt{\frac{m-1}{n+1-m}} \sigma_2 \,, \tag{25}$$

given in Caraux and Gascuel (1992b). Arnold and Groeneveld (1979) proved (25) as well as the respective bounds for trimmed means and differences of order statistics, using the Schwarz inequality to the left-hand side of (24).

Caraux and Gascuel (1992b) presented a version of (25) for nonidentically distributed variables, with μ and σ_2^2 replaced by $\bar{\mu} = \frac{1}{n} \sum_{i=1}^n EX_i$ and $\frac{1}{n} \sum_{i=1}^n [\text{Var} X_i - (EX_i - \bar{\mu})^2]$, respectively, where the equality cannot be reached for any mean and variance conditions, though. The necessary implicit condition is that all given moments come from marginal distributions satisfying the assumptions of Rychlik (1995). Aven (1985) proved two bounds on the expected maximum, dependent on means, variances, and variances of differences (or, equivalently, covariances):

$$EX_{n:n} \leq \bar{\mu} + \sqrt{\frac{n-1}{n}} \left(\sum_{i=1}^n \left[\text{Var}(X_i - X_j) + (EX_i - \bar{\mu})^2 \right] \right)^{\frac{1}{2}} ,$$

$$EX_{n:n} \leq \max_{1 \leq i \leq n} EX_i + \sqrt{\frac{n-1}{n}} \min_{1 \leq j \leq n} \left(\sum_{i=1}^n \text{Var}(X_i - X_j) \right)^{\frac{1}{2}} .$$

The former one was extented to all L-estimates by Lefèvre (1986)

$$\mathrm{E}\sum_{i=1}^n c_i(X_{i:n} - \mu) \leq \left[\sum_{j=1}^n \left(c_j - \frac{1}{n}\sum_{i=1}^n c_i\right)^2\right]^{\frac{1}{2}} \\ \times \min_{1 \leq j \leq n}\left[\sum_{i=1}^n \left[\mathrm{Var}(X_i - X_j) + (\mathrm{E}X_i - \bar{\mu})^2\right]\right]^{\frac{1}{2}}. \tag{26}$$

Repeating the proof from David (1988) and applying the fact that $\left(\frac{d_i}{n}\right)_{i=1}^n$ is the l_2-projection of $(c_i)_{i=1}^n$ onto nondecreasing sequences, we can essentially improve (26) for non-nondecreasing $(c_i)_{i=1}^n$ in the i.d case if we replace $\|(c_j - \frac{1}{n}\sum_{i=1}^n c_i)_{j=1}^n\|_2$ by $\frac{1}{n}\|(d_j - \sum_{i=1}^n c_i)_{j=1}^n\|_2$.

We now extend Theorem 3 to the case $p = 1$.

THEOREM 4 (Rychlik, 1993b). For the i.d case, with $\mathrm{E}X_1 = \mu$ and $\mathrm{E}|X_1 - \mu| = \sigma_1$, we have

$$\mathrm{E}\sum_{i=1}^n c_i(X_{i:n} - \mu) \leq \|D - a_1\|_\infty \sigma_1 = \tfrac{1}{2}(d_n - d_1)\sigma_1. \tag{27}$$

With $0 < \frac{i_j}{n} \leq \frac{i_k}{n} < 1$ denoting the first and last jump points of D, (27) becomes equality iff

$$\mathrm{P}(X_{i_j+1:n} = X_{i_k:n} = \mu) = 1,$$

$$\mu - \sum_{i=1}^{i_j} \mathrm{E}X_{i:n} = \mu + \sum_{i=i_k+1}^n \mathrm{E}X_{i:n} = \frac{n}{2}\sigma_1, \tag{28}$$

and (7) hold.

THEOREM 5 (Rychlik, 1993b). If X_i, $i = 1,\ldots,n$, are i.d., $\mathrm{E}X_1 = \mu$, and $\mathrm{P}(a \leq X_1 \leq b) = 1$, then

$$\mathrm{E}\sum_{i=1}^n c_i(X_{i:n} - \mu) \leq (b - \mu)\sum_{i=1}^n c_i - (b - a)C\left(\frac{b - \mu}{b - a}\right). \tag{29}$$

Let $\frac{i_j}{n}$ and $\frac{i_k}{n}$ be the nearest left and right to $\frac{b-\mu}{b-a}$ jump points of D, respectively. Then equality in (29) is attained iff

$$\mathrm{P}(X_{i_j:n} = a) = \mathrm{P}(X_{i_k+1:n} = b) = 1, \tag{30}$$

$$\sum_{i=i_j+1}^{i_k} \mathrm{P}(X_{i:n} = a) = n\frac{b - \mu}{b - a} - i_j, \tag{31}$$

$$\sum_{i=i_j+1}^{i_k} \mathrm{P}(X_{i:n} = b) = i_k - n\frac{b - \mu}{b - a}, \tag{32}$$

and (7) hold.

The case $p = \infty$ (ess sup norm) is derived for $\mu - a = b - \mu = \sigma_\infty$. Fixing $b - a$, and minimizing the right-hand side of (29) with respect to b, we derive an accurate bound under given expectation and support length.

COROLLARY 1. Under the hypotheses of Theorem 5 for some a, b satisfying $b - a = \lambda$, and for $\frac{i_k}{n}$ and $\frac{i_j}{n}$ being the smallest jump point of D satisfying $d_{i_k} \geq \sum_{i=1}^{n} c_i$, and the preceding one, respectively, yields

$$E \sum_{i=1}^{n} c_i (X_{i:n} - \mu) \leq \left[\frac{i_k}{n} \sum_{i=1}^{n} c_i - \sum_{i=1}^{i_k} c_i \right] \lambda . \tag{33}$$

If $d_{i_k} < \sum_{i=1}^{n} c_i$, we have the equality in (33) iff

$$P\left(X_{1:n} = X_{i_k:n} = \mu - \left(1 - \frac{i_k}{n}\right)\lambda, \ X_{i_k+1:n} = X_{n:n} = \mu + \frac{i_k}{n}\lambda \right) = 1 .$$

Otherwise, it suffices that (30)–(32) and (7) hold for some b, $\mu + \frac{i_j}{n}\lambda \leq b \leq \mu + \frac{i_k}{n}\lambda$ and $a = b - \lambda$.

EXAMPLE 4. In the notation of Theorems 4–5, and Corollary 1,

$$EX_{m:n} \leq \mu + \frac{n}{2(n+1-m)} \sigma_1 ,$$

$$EX_{m:n} \leq b + \mu - \frac{\max\{(b-\mu)n - (m-1)(b-a), 0\}}{n+1-m} , \tag{34}$$

$$EX_{m:n} \leq \mu + \frac{m-1}{n} \lambda .$$

We can replace the mean by another location parameter: either of the support endpoints (assumed finite). It would be natural and convenient to consider nonnegative random variables ($a = 0$), with given either an ordinary p^{th} moment $EX_1^p = \mu_p^p$, $p \geq 1$, or the right support endpoint μ_∞.

THEOREM 6. If X_i, $i = 1, \ldots, n$, are nonnegative i.i.d random variables, then

$$E \sum_{i=1}^{n} c_i X_{i:n} \leq \| \max\{D, 0\} \|_\infty \mu = \max\{d_n, 0\} \mu , \tag{35}$$

$$E \sum_{i=1}^{n} c_i X_{i:n} \leq \| \max\{D, 0\} \|_q \mu_p$$

$$= \sqrt[q]{\sum_{i=1}^{n} (\max\{d_i, 0\})^q} \mu_p, \quad 1 < p \leq \infty . \tag{36}$$

The bounds are the best possible.

We do not indicate here for which joint distributions bounds (35)–(36) are attained (see Rychlik, 1993b). Observe that we can similarly obtain the sharp inequalities for the expectations of L-estimates of arbitrary i.d X_i, $i = 1, \ldots, n$, with a known absolute p^{th} moment $m_p^p = \mathrm{E}|X_1|^p$, $p \geq 1$ or $m_\infty = \inf\{M : \mathrm{P}(|X_1| \leq M) = 1\}$:

$$\mathrm{E} \sum_{i=1}^{n} c_i X_{i:n} \leq \|(d_i)_{i=1}^n\|_q m_p, \quad 1 \leq p \leq \infty . \tag{37}$$

3.2. Deterministic bounds. We first analyze the quantile function (23). This is stepwise, $v(\leq n)$-valued, with jumps at some of $\frac{i}{n}$, $i = 1, \ldots, n-1$. It will be convenient to write the respective marginal distribution as

$$\mathrm{P}\left(X_1 = x_{i:n}^\star = \mu + \frac{\sqrt[p-1]{|d_i - a_q|}}{\|(\sqrt[p-1]{|d_i - a_q|})_{i=1}^n\|_q} \mathrm{sgn}(d_i - a_q) \sigma_p \right) = \frac{1}{n} \tag{38}$$

for $i = 1, \ldots, n$. If it happens that some $x_{i:n}^\star$ coincide, then we simply add their probabilities and obtain a well defined probability distribution. Combination of (38) with (5)–(7), ensuring the equality in (24), can be replaced by the following one:

$$\mathrm{P}(X_{i:n} = x_{i:n}^\star, \ i = 1, \ldots, n) = 1 . \tag{39}$$

On the other hand, let X_i, $i = 1, \ldots, n$, be defined as the ith coordinate of a random permutation of given numbers $x_1, \ldots, x_n$ (intuitively, this is the outcome of exhaustive drawing without replacement of n balls labelled $x_1, \ldots, x_n$ from an urn). It is easy to verify that X_i, $i = 1, \ldots, n$, are dependent identically distributed with the expectation $\bar{x} = \frac{1}{n} \sum_{i=1}^n x_i$, and p^{th} absolute central moment $s_p^p = \frac{1}{n} \sum_{i=1}^n |x_i - \bar{x}|^p$, and deterministic order statistics $x_{i:n}$, $i = 1, \ldots, n$. By Theorem 3,

$$\sum_{i=1}^n c_i(x_{i:n} - \bar{x}) \leq \|(d_i - a_q)_{i=1}^n\|_q s_p . \tag{40}$$

Setting $x_i = x_{i:n}^\star$, $i = 1, \ldots, n$, with μ and σ_p replaced by their sample counterparts, we obtain the random sample with the distribution satisfying (38) and (39), and so (40) becomes equality. From inequality (24), concerning the expectations of L-estimates of i.d samples, we have concluded its deterministic counterpart (40), which holds for samples from arbitrary populations.

It is easy to verify that for the other results obtained hitherto in this Section, except of Theorem 5, there exist jump marginal distribution functions with values in the set $\{\frac{i}{n}, i = 0, \ldots, n\}$ for which the respective upper bounds are attained. Repeating the above reasoning leading to (40), we obtain the sharp deterministic counterparts of (27), (33), (35), (36), and (37):

$$\sum_{i=1}^n c_i(x_{i:n} - \bar{x}) \leq \tfrac{1}{2}(d_n - d_1)s_1 , \tag{41}$$

$$\sum_{i=1}^n c_i(x_{i:n} - \bar{x}) \leq \left(\frac{i_k}{n} \sum_{i=1}^n c_i - \sum_{i=1}^{i_k} c_i \right)(x_{n:n} - x_{1:n}) , \tag{42}$$

(i_k is defined in Corollary 1),

$$\sum_{i=1}^{n} c_i(x_{i:n} - x_{1:n}) \leq \max\{d_n, 0\}(\bar{x} - x_{1:n}), \tag{43}$$

$$\sum_{i=1}^{n} c_i(x_{i:n} - x_{1:n}) \leq \|(\max\{d_i, 0\})_{i=1}^{n}\|_q \left[\frac{1}{n}\sum_{i=1}^{n}(x_i - x_{1:n})^p\right]^{\frac{1}{p}},$$

$$\sum_{i=1}^{n} c_i(x_{i:n} - x_{1:n}) \leq \sum_{d_i>0} d_i(x_{n:n} - x_{1:n}),$$

$$\sum_{i=1}^{n} c_i x_{i:n} \leq \max\{-d_1, d_n\} \sum_{i=1}^{n} |x_i|,$$

$$\sum_{i=1}^{n} c_i x_{i:n} \leq \|(d_i)_{i=1}^{n}\|_q \left(\sum_{i=1}^{n} |x_i|^p\right)^{\frac{1}{p}},$$

$$\sum_{i=1}^{n} c_i x_{i:n} \leq \sum_{i=1}^{n} |d_i| \max\{-x_{1:n}, x_{n:n}\}.$$

Rychlik (1992c) proved that the p-norms, $1 \leq p \leq +\infty$, can be replaced by any symmetric norm $\|\cdot\|$ on $\mathcal{R}^n$, so that,

$$\sum_{i=1}^{n} c_i(x_{i:n} - \bar{x}) \leq \|(d_i - a_\star)_{i=1}^{n}\|_\star \|(x_i - \bar{x})_{i=1}^{n}\| \tag{44}$$

is the best possible bound, where $\|\cdot\|_\star$ is the conjugate norm of $\|\cdot\|$, and $a_\star$ minimizes $\|(d_i - a)_{i=1}^{n}\|_\star$ in $a \in \mathcal{R}$. This is a generalization of so called the Samuelson inequality on general L-estimates and symmetric norms.

Samuelson (1968) raised and solved the problem of how much can a single observation deviate from the sample mean in the standard deviation units (the case of $c_1 = \cdots = c_{n-1} = 0$, $c_n = 1$, $p = 2$ in (40), answer: $\sqrt{n-1}$). Samuelson's paper stimulated intensive investigations of the problem and its modifications: alternative proofs, rediscoveries of earlier results, and extensions. Six different proofs were reviewed by Arnold and Balakrishnan (1989). The earliest proofs found in literature were due to Thompson (1935) and Scott (1936), mentioned by Wolkowicz and Styan (1980). We do not attempt to present a complete record of consecutive contributions, referring the reader to Arnold (1988) for a comprehensive bibliography, and Olkin (1992) for the most recent review, with yet another proof. We merely point out several partial results leading to (44), and do not mention generalizations of the Samuelson inequality in regression models.

For the Samuelson assumptions (deviations from the mean in the standard deviation units), Scott (1936) established the bound for $x_{n-1:n}$. The bounds for arbitrary order statistics follow directly from Mallows and Richter (1969), and were explicitly stated by Boyd (1971) and Hawkins (1971). Mallows and Richter (1969) established the inequalities for selection differentials $\frac{1}{j}\sum_{i=1}^{j} x_{i:n}$, and $\frac{1}{k}\sum_{i=n+1-k}^{n} x_{i:n}$, and their difference. The respective results for $x_{n:n} - x_{1:n}$, $x_{n-1:n} - x_{1:n}$, and the difference of arbitrary order statistics were derived by Nair

(1948), David et al. (1954), and Fahmy and Proschan (1981) (implicitly in Arnold and Groeneveld (1974)), respectively, and for the L-estimates with nondecreasing coefficients by David (1988).

Beesack (1973) derived (44) for the single order statistics and smooth symmetric Luxeburg norms. Arnold and Groeneveld (1981) proved (41) and (43) for the selection differentials, and Groeneveld (1982) obtained (42) and (43) for single order statistics and their differences. Mărgăritescu and Nicolae (1990) presented bounds on $x_{m:n} - \bar{x}$ in terms of differences of sample differentials, which cannot be deduced from (40)–(42). Boyd (1971) and Hawkins (1971) refined the trivial bound (40) on $x_{1:n}$ in the case $p = 2$ as follows

$$x_{1:n} - \bar{x} \leq -\frac{1}{\sqrt{n-1}} s_2 \ .$$

Also, Thompson (1955) noted that

$$x_{1:n} - x_{n:n} \leq -\frac{2\sqrt{2}n}{\sqrt{2n^2 - 1 + (-1)^n}} s_2 \ .$$

3.3. *I.i.d case.* If $X_1, \ldots, X_n$ are independent and identically F-distributed, with the expectation μ and variance σ_2, then by (21), (22) and the Schwarz inequality

$$\begin{aligned} EX_{n:n} - \mu &= \int_0^1 (Q_F(x) - \mu) N_{n:n}(x) \, dx \\ &= \int_0^1 (Q_F(x) - \mu)(N_{n:n}(x) - 1) \, dx \\ &\leq \|N_{n:n} - 1\|_2 \|Q_F - \mu\|_2 \\ &= \frac{n-1}{\sqrt{2n-1}} \sigma_2. \end{aligned} \quad (45)$$

We can subtract any real from $N_{n:n}$ above, but we minimize the norm by choosing $\int_0^1 N_{n:n}(x) \, dx = 1$. Bound (45) is tight, and the equality is attained iff

$$Q_{F^\star}(x) - \mu = \frac{N_{n:n}(x) - 1}{\|N_{n:n} - 1\|_2} \sigma_2 = \frac{\sqrt{2n-1}}{n-1} (nx^{n-1} - 1)\sigma_2 \ , \quad (46)$$

i.e., for the i.i.d sample with the common distribution function

$$F^\star(x) = \left[\frac{1}{n}\left(\frac{n-1}{\sqrt{2n-1}} \frac{x-\mu}{\sigma_2} + 1\right)\right]^{\frac{1}{n-1}} \ ,$$

$$\mu - \frac{\sqrt{2n-1}}{n-1}\sigma_2 \leq x \leq \mu + \sqrt{2n-1}\sigma_2 \ . \quad (47)$$

The result was presented independently by Gumbel (1954) and Hartley and David (1954). Observe that similar bounds can be derived for arbitrary L-estimates:

$$\mathrm{E}\sum_{i=1}^{n} c_i(X_{i:n} - \mu) \leq \left\| N - \sum_{i=1}^{n} c_i \right\|_2 \sigma_2 , \qquad (48)$$

specified by Hartley and David (1954) and Ludwig (1960) for the cases of single order statistics and differences of them, respectively. However, (48) is not sharp unless the argument of the right-hand side norm, generally a polynomial of degree $n-1$, is nondecreasing. This is actually nondecreasing for the sample range, and the selection differentials so that the bounds

$$\mathrm{E}(X_{n:n} - X_{1:n}) \leq \sqrt{\frac{2n^2}{2n-1} - \frac{2(n!)^2}{(2n-1)!}} \sigma_2 \qquad (49)$$

and

$$\mathrm{E}\frac{1}{k}\sum_{i=n+1-k}^{n} X_{i:n} - \mu \leq \binom{n}{k}\left[\frac{1}{2n-1}\sum_{i,j=n+1-k}^{n} \frac{\binom{n-1}{i-1}\binom{n-1}{j-1}}{\binom{2n-2}{i+j-2}} - 1\right]^{\frac{1}{2}} \sigma_2 ,$$

due to Plackett (1947) and Nagaraja (1981), respectively, are accurate. Applying the Hölder inequality, Gilstein (1981) provided the following sharp bound:

$$\mathrm{E}X_{n:n} \leq n\sqrt[q]{\frac{p-1}{np-1}} m_p, \quad 1 < p < \infty ,$$

attained for

$$Q_{F^*}(x) = n^{\frac{2-p}{p-1}} \sqrt[q]{\frac{np-1}{p-1}} m_p x^{\frac{n-1}{p-1}}, \quad 0 < x < 1 .$$

Likewise, Arnold (1985) obtained sharp bounds for the expected maximum and range of random variables with fixed expectation and p^{th} absolute central moment σ_p. The formulas are complicated and we omit presenting them here.

Another extension of (45) is due to Lin (1988)

$$|\mathrm{E}(X_{n:n} - \mu)^k - \alpha\sigma_k^k| \leq \sqrt{\frac{(n-1)^2}{2n-1} + (\alpha-1)^2 \sigma_{2k}^{2k}}, \quad \alpha \in \mathscr{R}, \ k \in \mathscr{N} ,$$

becoming equality for

$$F(x) = \sqrt[n-1]{\frac{\alpha}{n}} + \mathrm{const}\,(n, \alpha, \sigma_{2k})\, x^k . \qquad (50)$$

Rustagi (1957), using variational methods, derived the sharp bounds on the expectation of $X_{n:n}$, dependent on μ, σ_2, and $\sigma_\infty = \inf\{M: \mathrm{P}(|X_1 - \mu| \leq M) = 1\}$. The upper one is $\sqrt{2n-1}\sigma_2$, attained by a distribution function of the form $\sqrt[n]{ax+b}$, with possible jumps at the support ends (cf (47), (50)). The lower one is attained by a three-point distribution. Analogous results were presented earlier by Hartley and David (1954) for the sample range.

It was Moriguti (1953) who overcame the problems in establishing sharp bounds, caused by a generally nonmonotone polynomial N and went beyond special cases. He replaced the polynomials of the right hand side of (21) by properly constructed modifications such that their norms provide the best mean-variance bounds for the left-hand side. We present here a simpler version of Moriguti's Theorem 1:

LEMMA 1 (Moriguti, 1953). Let f be an integrable function in some $[a,b]$, and let Pf be the right derivative of the greatest convex function, say $G(x) = \int_a^x Pf(t)\,dt$, not greater than $F(x) = \int_a^x f(t)\,dt$, the indefinite integral of f. Then

$$\int_a^b g(x) f(x)\,dx \leq \int_a^b g(x) Pf(x)\,dx \quad (51)$$

for all nondecreasing functions g for which both the integrals are finite. Furthermore,

$$\int_a^b g(x) f(x)\,dx = \int_a^b g(x) Pf(x)\,dx \quad (52)$$

iff g is constant on every open interval where $F > G$.

Relation (51), rewritten as

$$\int_a^b g(x) F(dx) \leq \int_a^b g(x) G(dx)$$

is fulfilled, since F precedes G in the stochastic ordering (cf (10)). Both F and G are continuous and $\{F > G\}$ is a union of open intervals. If g is constant on an interval (c,d), where $F > G$ and $F = G$ at c and d, then

$$\int_c^d g(x)f(x)\,dx = g(c)[F(d) - F(c)]$$
$$= g(c)\int_c^d Pf(x)\,dx$$
$$= \int_c^d g(x)Pf(x)\,dx$$

and so (52) is justified.

Observe that $g = Pf$ is nondecreasing and satisfies (52), and, accordingly, this is the L_2-projection of f onto the convex cone of square integrable nondecreasing functions in $[a,b]$ (see, for example, Balakrishnan, 1981, Section 1.4). Applying (21), (51) and the Schwarz inequality, we obtain

$$\mathrm{E}\sum_{i=1}^{n} c_i(X_{i:n} - \mu) \le \int_0^1 (Q_F(x) - \mu)\left[PN(x) - \int_0^1 PN(t)\,dt\right] dx \tag{53}$$

$$\le \left\|PN - \sum_{i=1}^{n} c_i\right\|_2 \sigma_2,$$

which becomes equality for

$$Q_{F^*}(x) = \frac{PN(x) - \sum_{i=1}^n c_i}{\|PN - \sum_{i=1}^n c_i\|_2}\,\sigma_2, \tag{54}$$

by (52). It is easy to note that PN is defined by different formulas on the different elements of a partition of the domain. In some intervals, the projection coincides with the projected function, and in the other ones, this is the constant equal to the mean value of the original polynomial over the interval. However, the points of partition must be usually determined numerically. Moriguti (1953) described the form of PN for the cases of single order statistics, and their differences. He did not state (53) and (54) for general L-estimates, because this notion was not known yet. Balakrishnan (1993) derived explicit formulas for a few extreme order statistics. Ludwig (1973) presented a table of the sharp bounds for expected differences of order statistics in small samples.

Lemma 1 can be applied to calculating bounds more general than (53).

THEOREM 7. *For n i.d random variables with finite expectation μ and p^{th} absolute central moment σ_p^p, $1 \le p \le \infty$,*

$$\mathrm{E}\sum_{i=1}^{n} c_i(X_{i:n} - \mu) \le \|PN - a_q\|_q \sigma_p, \tag{55}$$

is the best possible bound, where a_q minimizes $\|PN - a\|_q$ in $a \in \mathscr{R}$.

If $p = 1$, then

$$\|PN - a_\infty\|_\infty = \tfrac{1}{2}(PN(1) - PN(0)),$$

and the equality in (55) is attained only if PN is constant on neighborhoods of the both endpoints, say on $[0, \alpha]$ and $[\beta, 1]$, and F^* is a three point distribution function, valued $\mu - c$, $\mu + d$, and μ, with some probabilities $a \le \alpha$, $b \le 1 - \beta$, and $1 - a - b$, respectively, and c and d satisfy:

$$bd - ac = 0,$$
$$bd + ac = \sigma_1.$$

Otherwise the equality in (55) holds for no marginal distribution.

If $1 < p < \infty$, then the equality in (55) holds iff

$$Q_{F^*} = \mu + \frac{\sqrt[p-1]{|PN - a_q|}\,\mathrm{sgn}(PN - a_q)}{\|\sqrt[p-1]{|PN - a_q|}\|_q}\,\sigma_p.$$

If $p = \infty$, then (55) becomes equality for any distribution such that

$$P(X_1 = \mu - \sigma_\infty) = \sup\{x\colon PN(x) < a_1\},$$
$$P(X_1 = \mu + \sigma_\infty) = \sup\{x\colon PN(1-x) > a_1\}.$$

Inequality (55) is an obvious extension of (53). It is sharp, because the functions providing the equality in the norm inequality satisfy conditions implying (52). Observe that the unique distribution function attaining the bound for some $1 < p < \infty$ is a mixture of a finitely valued discrete distribution with an absolutely continuous one, being the $(p-1)$st power of the inverse of N, a polynomial of degree $n-1$.

Sugiura (1962) noticed that (48) can be treated as the first term expansion in the Legendre orthonormal basis

$$\varphi_n = \frac{\sqrt{2n+1}\,n!}{(2n+1)!} N^{(n)}_{n+1:2n+1}, \quad n \geq 0.$$

Setting $Q_F = \sum_{j=0}^\infty a_j \varphi_j$, $N = \sum_{j=0}^\infty b_j \varphi_j$, where $a_j = \int Q_F \varphi_j$ and $b_j = \int N \varphi_j$, and applying the Schwarz inequality, we have

$$\left| \int_0^1 Q_F(x) N(x)\, dx - \sum_{j=0}^k a_j b_j \right| \leq \sqrt{\sum_{j=k+1}^\infty a_j^2} \sqrt{\sum_{j=k+1}^\infty b_j^2},$$

and so

$$\left| E \sum_{i=1}^n c_i (X_{i:n} - \mu) - \sum_{j=1}^k a_j b_j \right| \leq \sqrt{\|N\|_2^2 - \sum_{j=0}^k b_j^2} \sqrt{\sigma_2^2 - \sum_{j=0}^k a_j^2}. \tag{56}$$

Observe that for a given L-estimate, the coefficients b_j, $j \geq 0$, are uniquely determined. Since each φ_j is a polynomial of degree j, each a_j can be considered as a combination of expected order statistics from subsamples of size $l \leq j$, (e.g. subsequent sample maxima $X_{l:l}$). For single order statistics, Joshi (1969) multiplied $Q_F(x)$ by a polynomial factor $x^k(1-x)^l$ taken from $N_{m:n}(x)$, and following Sugiura's arguments derived a counterpart of (56) with μ and σ_2 replaced by the respective moments of order statistics of smaller samples, which is useful when $\sigma_2 = \infty$.

Numerous bounds based on expectations of order statistics in smaller samples are presented in literature. For instance, Balakrishnan (1990), and Balakrishnan and Bendre (1993) refined inequality (48) for order statistics by making use of expectations of two maxima in smaller samples. Lin (1988) applied the Schwarz inequality to prove for nonnegative parent variables

$$EX_{m:n} \leq \left[\frac{mn}{(m-1)(n+1)} EX_{m-1:n-1}^2 \right]^{\frac{1}{2}}, \tag{57}$$

which becomes equality for the uniform marginal distribution. By means of Hölder inequality, Kamps (1991) derived the following extension:

$$\mathrm{E} X_{m:n}^r \leq \frac{n!}{(m-1)!} \sqrt[p]{\frac{(m+k-n-1)!}{k!}} \frac{\Gamma\left(m+\frac{n-k}{p-1}\right)}{\Gamma\left(\frac{(n+1)p-k-1}{p-1}\right)} \sqrt[p']{\mathrm{E} X_{m+k-n:k}^{rp}} \; ,$$

$n+1-m \leq k \leq n$, attainable by the power distribution

$$F^\star(x) = \mathrm{const}(p, n-k) x^{\frac{(p-1)r}{n-k}} \; ,$$

where $r > 0$ and $p > 1$. Further generalization of (57) were presented by Gajek and Gather (1991), and Gajek and Lenic (1993).

4. Moment bounds for restricted families

4.1. Symmetric random variables. Bounds for the case of symmetric (about μ) random variables are obtained by a standard trick, making use of the property

$$\mu - Q_F(x) = Q_F(1-x-0) - \mu \; . \tag{58}$$

In consequence, for the dependent case, (1) can be rewritten as

$$\begin{aligned}
\mathrm{E} \sum_{i=1}^n c_i(X_{i:n} - \mu) &\leq \int_{\frac{1}{2}}^1 (Q_F(x) - \mu) D^s(x) \, dx \\
&= \sum_{i=1}^{[\frac{n}{2}]} (d_{n+1-i} - d_i) \int_{1-\frac{i}{n}}^{1-\frac{i-1}{n}} (Q_F(x) - \mu) \, dx \; ,
\end{aligned} \tag{59}$$

where $D^s(x) = D(x) - D(1-x-0)$. Applying further norm inequalities, we conclude

THEOREM 8 (Rychlik, 1993b). *If X_i, $i = 1, \ldots, n$, are dependent symmetrically identically distributed about μ, and $\mathrm{E}|X_1 - \mu|^p = \sigma_p^p$ for some $1 < p < \infty$, then*

$$\begin{aligned}
\mathrm{E} \sum_{i=1}^n c_i(X_{i:n} - \mu) &\leq \frac{1}{\sqrt[p]{2}} \left[\int_{\frac{1}{2}}^1 (D^s(x))^q \, dx \right]^{\frac{1}{q}} \sigma_p \\
&= \frac{1}{\sqrt[p]{2}} \left[\frac{1}{n} \sum_{i=1}^{[\frac{n}{2}]} (d_{n+1-i} - d_i)^q \right]^{\frac{1}{q}} \sigma_p \; .
\end{aligned} \tag{60}$$

Inequality (60) is tight and becomes equality iff

$$P\left(\mu - X_{j:n} = X_{n+1-j:n} - \mu = \frac{d_{n+1-j} - d_j}{2p\left[\frac{1}{n}\sum_{i=1}^{[\frac{n}{2}]}(d_{n+1-i} - d_i)^q\right]^{\frac{1}{p}}}\sigma_p,\right.$$

$$\left. j = 1, \ldots, \left[\frac{n}{2}\right]\right) = 1 \ .$$

For the cases $p = 1$ and $p = +\infty$, the best bounds immediately follow from the general ones presented in Theorems 4 and 5, respectively. For $p = 1$, bound (27) is also the best possible for symmetric samples. Conditions of equality in (27) are analogous to those of Theorem 4, with (28) strengthened by

$$\sum_{i=1}^{i_j} P(X_{i:n} \leq \mu - x) = \sum_{i=i_k+1}^{n} P(X_{i:n} \geq \mu + x) \quad \text{for all } x > 0 \ .$$

For $p = +\infty$, it suffices to replace a and b by $\mu - \sigma_\infty$ and $\mu + \sigma_\infty$, respectively, in the statement of Theorem 5. In particular, the bounds (34), and

$$EX_{m:n} - \mu \leq \sqrt[p]{\frac{n}{2}} \frac{\sqrt[q]{\min\{m-1, n+1-m\}}}{n+1-m} \sigma_p, \tag{61}$$

$$EX_{m:n} - \mu \leq \min\left\{\frac{m-1}{n+1-m}, 1\right\}\sigma_\infty \tag{62}$$

are sharp. The bounds for the expectations of quasiranges $X_{m:n} - X_{n+1-m:n}$, $m > \frac{n+1}{2}$, are twice of those in (61)–(62) for the respective m^{th} order statistics. Arnold (1985) determined the p-norm bounds for the sample maximum and range (the case $p = 2$ was treated in Arnold, 1980).

In the i.i.d case, we obtain general bounds applying (58), and Moriguti's projection P (see Lemma 1).

THEOREM 9. *If X_i, $i = 1, \ldots, n$, are symmetric i.i.d random variables, with expectation μ and p^{th} absolute central moment σ_p^p, $1 \leq p \leq +\infty$, then*

$$E\sum_{i=1}^{n} c_i(X_{i:n} - \mu) \leq \frac{1}{\sqrt[p]{2}}\|PN^s\|_q \sigma_p \tag{63}$$

is an accurate bound, where

$$N^s(x) = N(x) - N(1-x) = \sum_{i=1}^{n}(c_i - c_{n+1-i})N_{i:n}(x), \quad \tfrac{1}{2} \leq x \leq 1 \ .$$

We leave to the reader to verify (63) and describe the conditions of becoming equality. As in the asymmetric case, the main difficulty lies in the precise determination of the greatest convex minorant of a given polynomial. Generally, we can merely derive numerical approximations. Certainly, if the polynomial N^s is actually increasing on $[\frac{1}{2}, 1)$, it is its norm that gives the infimum of the bound in

(63). This occurs, e.g., for the polynomials corresponding with the sample maximum and range, and selection differential. Plackett (1947) derived the equality in (49) for a symmetric distribution. Moriguti (1951) proved that

$$EX_{n:n} - \mu \leq \frac{n}{\sqrt{2}} \sqrt{\frac{1}{2n-1} - B(n,n)} \sigma_2 , \tag{64}$$

generalized by Arnold (1985) to

$$EX_{n:n} - \mu \leq \frac{n}{\sqrt[p]{2}} \left(\int_{\frac{1}{2}}^{1} [x^{n-1} - (1-x)^{n-1}]^q \, dx \right)^{\frac{1}{q}} \sigma_p, \quad 1 < p < \infty ,$$

(half of the respective bound for the sample range). These bounds are attained by the respective quantile functions

$$Q_{F^\star}(x) = \mu + \mathrm{const}(p) \sqrt[p-1]{x^{n-1} - (1-x)^{n-1}} \, \mathrm{sgn}(x - \tfrac{1}{2}) .$$

The bound for the selection differentials is a huge formula. On the other hand, analogous inequality for the single order statistics

$$EX_{m:n} - \mu \leq \left[\frac{B(2m-1, 2n+1-2m) - B(n,n)}{2B^2(m, n+1-m)} \right]^{\frac{1}{2}} \sigma_2 , \tag{65}$$

given in Sugiura (1962) is not sharp for $1 < m < n$, because the function

$$Q(x) = \mu + \mathrm{const}(m)[x^{m-1}(1-x)^{n-m} - x^{n-m}(1-x)^{m-1}] \tag{66}$$

providing the equality in (65) is nonmonotone. In fact, the sharp bounds are

$$EX_{m:n} - \mu \leq \left[\int_{\frac{1}{2}}^{x_0} (N^s_{m:n}(x))^q \, dx + (N^s_{m:n}(x_0))^q (1-x_0) \right]^{\frac{1}{q}} \frac{\sigma_p}{\sqrt[p]{2}} , \tag{67}$$

where $m > \frac{n+1}{2}$, and $1 \leq p \leq \infty$, and x_0 is defined by

$$N^s_{m:n}(x_0)(1-x_0) = \int_{x_0}^{1} N^s_{m:n}(x) \, dx .$$

Equality in (67) holds iff the distribution function $F^\star$ is proportional to the $(p-1)^{\mathrm{st}}$ power of the inverse of $N^s_{m:n}$ on the interval $(-\sqrt[p-1]{Q_{F^\star}(x_0)}, +\sqrt[p-1]{Q_{F^\star}(x_0)})$, and has two jumps at the support endpoints of height $1 - x_0$. This problem is closely related to that of establishing bounds on m^{th} quasirange of possibly asymmetric random variables, for which the bound, twice as large as (67), is attained for the same distribution. The solution was primarily derived for $p = 2$ by Moriguti (1953, Example 4).

Moriguti (1951) derived two sharp lower bounds on the variance of sample maximum. With the notation

$$M_n(\lambda) = \int_{\frac{1}{2}}^{1} \frac{n^2[x^{n-1} - (1-x)^{n-1}]^2}{n[x^{n-1} + (1-x)^{n-1}] - 2\lambda} \, dx ,$$

we have

$$\text{Var} X_{n:n} \geq \left(\frac{1}{M_n(0)} - 1 \right) (EX_{n:n})^2 ,$$

and

$$\text{Var} X_{n:n} \geq \lambda_n \sigma_2^2 ,$$

where $0 \leq \lambda_n \leq \frac{n}{2^{n-1}}$ is uniquely determined by $M_n(\lambda_n) = 1$.

4.2. Projection method. We describe a method of determining sharp bounds on the expectations of L-estimates in terms of the first two moments of parent distribution that comes from a given restricted family. The method, proposed recently by Gajek and Rychlik (1996), applies projections onto convex cones of functions. We actually apply the following lemma.

LEMMA 2. *Let $\mathscr{C}$ be a convex cone in a real inner product space $\mathscr{X}$. If for a given $f \in \mathscr{X}$ there exists its projection $P_\mathscr{C} f$ onto $\mathscr{C}$ (i.e., an element of $\mathscr{C}$ closest to f), then $P_\mathscr{C} f$ is unique and satisfies*

$$(f, g) \leq (P_\mathscr{C} f, g) \quad \text{for all } g \in \mathscr{C} , \tag{68}$$

$$(f, P_\mathscr{C} f) = \|P_\mathscr{C} f\|^2 . \tag{69}$$

A more popular stronger version of Lemma 2 for the closed convex cones of the Hilbert spaces asserts that $P_\mathscr{C} f$ does exist and is uniquely characterized by (68)–(69) (see, e.g., Balakrishnan, 1981, Section 1.4).

EXAMPLE 5. Suppose that $\mathscr{X}$ is the space of right continuous square integrable functions on $[0, 1)$, with the inner product defined by $(f, g) = \int_0^1 f(x) g(x) \, dx$. Let $\mathscr{C}$ be the family of nondecreasing functions, orthogonal to constants in $\mathscr{X}$. In other words, this is the class of all $Q_F - \mu$, where Q_F is the quantile of a distribution with a finite second moment and μ is the respective expectation. By Lemma 1, $PN - (PN, 1) = P \sum_{i=1}^{n} c_i (N_{i:n} - 1)$ is actually the projection of N onto $\mathscr{C}$. Therefore, by (21), (68), and the Schwarz inequality, we have for the i.i.d case

$$E \sum_{i=1}^{n} c_i (X_{i:n} - \mu) = (Q_F - \mu, N)$$

$$\leq \left(Q_F - \mu, P_\mathscr{C} N - \sum_{i=1}^{n} c_i \right)$$

$$\leq \left\| P_\mathscr{C} N - \sum_{i=1}^{n} c_i \right\| \sigma_2 .$$

By (69), the equality holds for

$$Q_{F^*} = \mu + \frac{P_{\mathscr{C}}N - \sum_{i=1}^{n} c_i}{\|P_{\mathscr{C}}N - \sum_{i=1}^{n} c_i\|} \sigma_2 .$$

In fact, the majority of results in Section 3 follows from implicit applications of the projection method. The classes of all quantile functions and the all anti-symmetric about $\frac{1}{2}$ quantiles are convex cones. These applications are not apparent for the i.␣d case, since D itself is nondecreasing. However, the deterministic bounds in Subsection 3.2 hold due to the fact that $(d_i)_{i=1}^n$ is the projection of $(nc_i)_{i=1}^n$ onto nondecreasing sequences.

The projection method provides a possibility of deriving sharp bounds for distributions from a restricted family, say $\mathscr{F}$, if the corresponding family of quantile functions $\mathscr{Q} = \{Q_F\colon F \in \mathscr{F}\}$ is closed under positive combinations. It suffices to find the projections of functions N and D onto the family of quantiles, and calculate their norms. Then

$$E \sum_{i=1}^{n} c_i X_{i:n} \leq \|P_{\mathscr{Q}} N\| m_2 \qquad (70)$$

in the i.i.d case, and

$$E \sum_{i=1}^{n} c_i X_{i:n} \leq \|P_{\mathscr{Q}} D\| m_2 \qquad (71)$$

in the dependent one. If $\mathscr{Q} - \mu = \{Q_F - \mu\colon F \in \mathscr{F}\}$ is a convex cone, we can derive more subtle evaluations

$$E \sum_{i=1}^{n} c_i (X_{i:n} - \mu) \leq \|P_{\mathscr{Q}-\mu} N\| \sigma_2 \qquad (72)$$

and

$$E \sum_{i=1}^{n} c_i (X_{i:n} - \mu) \leq \|P_{\mathscr{Q}-\mu} D\| \sigma_2 , \qquad (73)$$

respectively. Bounds (70)–(73) are attained by quantiles proportional to respective projections.

Below we show applications of the projection method for deriving bounds on the expectations of L-estimates from families determined by partial orderings of distributions. Other techniques for these families will be discussed in Section 5. First we consider the convex partial ordering (c-ordering) introduced by van Zwet (1964) for life distributions: $F \prec_c G$ iff $Q_G F$ is convex on the support of F. The ordering allows to compare the skewness of life distributions (cf also Oja, 1981). Consider the family $\mathscr{F}_{\succ_c G}$ of all life distributions with finite second moments, succeeding G in the convex ordering. We assume that a fixed known distribution function G is continuous and strictly increasing on its support and has a finite second moment. Two standard examples of such families are the ones induced by the uniform and exponential distributions. The elements of the former are the life distributions with decreasing failure probability (possible jump at 0 and a non-increasing density on $\mathscr{R}_+$), the latter consists of distributions with decreasing failure rate. We concentrate here on the dependent case. Note that

$$\int_0^\infty Q_F^2 G(x) G(\mathrm{d}x) = \int_0^1 Q_F^2(x) \, \mathrm{d}x = \mu_2^2 < \infty, \quad F \in \mathscr{F}_{\succ_c G},$$

and

$$\int_0^\infty Q_F G(x) DG(x) G(\mathrm{d}x) = \int_0^1 Q_F(x) D(x) \, \mathrm{d}x \, .$$

It follows that the compositions $Q_F G$, $\mathscr{F}_{\succ_c G}$, form the convex cone

$$\mathscr{Q}_{\succ_c G} = \left\{ f: [0, \infty) \mapsto \mathscr{R}_+ : f(0) = 0, \quad \int_0^\infty f^2(x) G(\mathrm{d}x) < \infty, \right.$$

$$\left. f - \text{nondecreasing and convex} \right\} . \quad (74)$$

in the space of the right continuous square G-integrable functions with the inner product $(f, g) = \int_0^\infty f(x) g(x) G(\mathrm{d}x)$. The problem of finding the accurate sharp bound on the expectation of a given L-estimate of i.d variables with a distribution function $F \in \mathscr{F}_{\succ_c G}$ and a given second moment will be solved once we determine the projection of the respective function DG onto (74). The latter problem is not trivial, because there are no general methods of constructing projections onto convex functions like for projecting onto monotone ones (cf Swetits et al., 1989, Ubhaya, 1989, 1990).

Fortunately, DG has a nice form: this is a nondecreasing jump function, taking $v (\leq n)$ values. For the case of single order statistics, $D_{m:n} G$ takes only two values: zero near the origin and a positive value further. These specific properties enable us to prove that the bounds on the expectations of the all L-estimates of F-distributed variables, $F \succ_c G$, are attained for distribution functions from a parametric subclass of $\mathscr{F}_{\succ_c G}$. The bounds for the expectations of single order statistics are explicitly derived. In description of distributions attaining the bounds, we confine ourselves to the presentation of the respective marginals. In each case, the construction of the joint distribution which satisfies

$$E \sum_{i=1}^n c_i X_{i:n} = \int_0^1 Q_F(x) D(x) \, \mathrm{d}x$$

is omitted (we refer to Section 2.1).

LEMMA 3 (Gajek and Rychlik, 1996). (a) If f is a jump function in $[0, \infty)$ and has v values, then its projection $P_{\succ_c G} f$ onto (74) is a piecewise linear function, with v pieces at most.

(b) If $f = 0$ in $[0, a)$, and $M > 0$ in $[a, \infty)$, then

$$P_{\succ_c G} f(x) = \max\{\lambda^*(x - \alpha^*), 0\} \quad G - \text{a.e.} ,$$

where α^* maximizes

$$A(\alpha) = \frac{\left[\int_a^\infty (x - \alpha) G(dx)\right]^2}{\int_\alpha^\infty (x - \alpha)^2 G(dx)}$$

for $0 \le \alpha < a$ and

$$\lambda^* = M \frac{\int_a^\infty (x - \alpha^*) G(dx)}{\int_{\alpha^*}^\infty (x - \alpha^*)^2 G(dx)} > 0 .$$

THEOREM 10. (a) For every L-estimate of a dependent sample with a distribution function $F \in \mathscr{F}_{\succ_c G}$ and a second moment μ_2^2, there exists $F^\star \in \mathscr{F}^0_{\succ_c G} \subset \mathscr{F}_{\succ_c G}$,

$$\mathscr{F}^0_{\succ_c G} = \Bigg\{ F : F(x) = G\Bigg(y_0 + \alpha_j(x - x_{j-1}) + \sum_{i=1}^{j-1} \alpha_i(x_i - x_{i-1}) \Bigg),$$

$$x \in [x_{j-1}, x_j] \quad \text{for some } y_0 \ge 0, \alpha_1 \ge \cdots \ge \alpha_n \ge 0, \quad (75)$$

$$0 = x_0 < \cdots < x_n = +\infty \Bigg\}$$

such that

$$\mathrm{E}_F \sum_{i=1}^n c_i X_{i:n} \le \mathrm{E}_{F^\star} \sum_{i=1}^n c_i X_{i:n} = \|P_{\succ_c G} DG\|_{\mu_2}, \quad F \in \mathscr{F}_{\succ_c G} , \quad (76)$$

and

$$Q_{F^\star} G = \frac{P_{\succ_c G} DG}{\|P_{\succ_c G} DG\|} \mu_2 . \quad (77)$$

(b) In particular, for $\alpha^* \in [0, Q_G(\frac{m-1}{n}))$ maximizing

$$A(\alpha) = \frac{\left[\int_{Q_G(\frac{m-1}{n})}^\infty (x - \alpha) G(dx)\right]^2}{\int_\alpha^\infty (x - \alpha)^2 G(dx)} ,$$

yields

$$\mathrm{E}_F X_{m:n} \le \frac{n}{n+1-m} \sqrt{1 - \frac{m-1}{n} - A(\alpha^*)} \mu_2, \quad F \in \mathscr{F}_{\succ_c G} . \quad (78)$$

The equality in (78) holds for

$$F^\star(x) = G\Bigg(\alpha^* + x \frac{\sqrt{1 - \frac{m-1}{n} - A(\alpha^*)}}{A(\alpha^*) \mu_2} \int_{Q_G(\frac{m-1}{n})}^\infty (t - \alpha^*) G(dt) \Bigg) .$$

Below we specify the results for $G(x) = x$ and $G(x) = 1 - e^{-x}$. Observe that in the former case (75) is the family of mixtures of a possible pole at the origin with not more than n uniform distributions with the left endpoint at 0.

COROLLARY 2 (Gajek and Rychlik, 1996). Let X_i, $i = 1, \ldots, n$, be i.d with a decreasing probability life distribution and $EX_1^2 = \mu_2^2$.
    If $\frac{m-1}{n} \leq \frac{1}{3}$, then

$$EX_{m:n} \leq \frac{\sqrt{3}}{2}\left(1 + \frac{m-1}{n}\right)\mu_2,$$

where the equality holds iff X_i, $i = 1, \ldots, n$, are uniformly distributed on $[0, \sqrt{3}\mu_2]$.
    Otherwise

$$EX_{m:n} \leq \frac{2}{3}\sqrt{\frac{2n}{n+1-m}}\mu_2,$$

which becomes equality iff X_i, $i = 1, \ldots, n$, are $[0, \sqrt{2n/(n+1-m)}\mu_2]$-uniformly distributed with probability $\frac{3}{2}(1 - \frac{m-1}{n})$ and have an atom at 0 with probability $\frac{3}{2}\frac{m-1}{n} - \frac{1}{2}$.

COROLLARY 3 (Gajek and Rychlik, 1996). Suppose that $X_1, \ldots, X_n$ have a common DFR distribution and $EX_1^2 = \mu_2^2$. Then

$$EX_{m:n} \leq \begin{cases} (\ln\frac{n}{n+1-m} + 1)\frac{\mu_2}{\sqrt{2}}, & \text{if } \frac{m-1}{n} \leq 1 - e^{-1}, \\ \sqrt{2n/(n+1-m)e}\,\mu_2, & \text{otherwise}. \end{cases}$$

The former bound is attained by the exponential distribution with the scale parameter $\frac{\mu_2}{\sqrt{2}}$. The latter holds iff F^* is a mixture of the exponential distribution with the scale $\sqrt{n/2e(n+1-m)}\mu_2$ and the degenerate one at zero, with probabilities $(1 - \frac{m-1}{n})e$ and $1 - (1 - \frac{m-1}{n})e$, respectively.

We can derive analogous results for the independent variables, if we determine projections of NG, a polynomial in G, onto (74). One can guess that projecting such functions is a more difficult task than for step functions and we cannot expect deriving simple analytic formulas. This means that, paradoxically, it is easier to establish bounds for all possible dependent samples than for that with the given precisely defined product distribution. The independent case was examined in Gajek and Rychlik (1998), where various moment bounds for the i.d and i.i.d cases were also numerically compared. We note here that projections of general NG onto (74) are defined by different formulas on different intervals: the pieces where $P_{\succ_c G}NG = NG$ alternate with pieces of linearity. Also, $P_{\succ_c G}N_{m:n}G$ may first coincide with $N_{m:n}G$, and then becomes linear.

We proceed to present the bounds for i.d life variables with distributions from the restricted family

$$\mathscr{F}_{\prec_c G} = \{F : F \prec_c G, \int_0^1 Q_F^2(x)\,dx < \infty\}$$

for a given G. The problem lies now in projecting DG onto

$$\mathscr{D}_{\prec_c G} = \Big\{f : [0,\infty) \mapsto \mathscr{R}_+ : f(0) = 0,\ f - \text{nondecreasing},$$
$$\text{concave},\ \int_0^\infty f^2(x) G(dx) < \infty\Big\}\ .$$

Omitting details, we present merely the final results: a counterpart of Theorem 10, and respective implications for distributions with an increasing failure probability (*J*-shaped, for short), and for the IFR distributions.

THEOREM 11. (a) For every *L*-estimate there exists $F^\star \in \mathscr{F}^0_{\prec_c G}$,

$$\mathscr{F}^0_{\prec_c G} = \Big\{F : F(x) = G\Big(\alpha_j(x - x_{j-1}) + \sum_{i=1}^{j-1}\alpha_i(x_i - x_{i-1})\Big),$$
$$x \in [x_{j-1}, x_j]\ \text{for some}\ 0 \le \alpha_1 \le \cdots \le \alpha_n, \tag{79}$$
$$0 = x_0 < \cdots < x_n = +\infty\Big\} \subset \mathscr{F}_{\prec_c G}$$

such that for all $F \in \mathscr{F}_{\prec_c G}$

$$E_F \sum_{i=1}^n c_i X_{i:n} \le E_{F^\star} \sum_{i=1}^n c_i X_{i:n} = \|P_{\prec_c G} DG\|_{\mu_2}\ .$$

(b) If there exists $\alpha = \alpha(m, n, G) \ge Q_G\big(\tfrac{m-1}{n}\big)$ satisfying

$$\frac{1}{\alpha}\int_0^\alpha x^2 G(dx) = \int_{Q_G(\frac{m-1}{n})}^\alpha x G(dx)\ ,$$

then

$$E_F X_{m:n} \le \frac{n}{n+1-m}\sqrt{1 - G(\alpha) + \frac{1}{\alpha^2}\int_0^\alpha x^2 G(dx)}\mu_2,\quad F \in \mathscr{F}_{\prec_c G}\ .$$

The equality holds iff

$$F^\star(x) = G\Big(\frac{\alpha}{\mu_2}\sqrt{1 - G(\alpha) + \frac{1}{\alpha^2}\int_0^\alpha t^2 G(dt)}x\Big),$$
$$0 < x < \frac{\mu_2}{\sqrt{1 - G(\alpha) + \frac{1}{\alpha^2}\int_0^\alpha t^2 G(dt)}}$$

Otherwise

$$E_F X_{m:n} \le \frac{n \int_{Q_G(\frac{m-1}{n})}^\infty x G(dx)}{(n+1-m)\sqrt{\int_0^\infty x^2 G(dx)}}\mu_2\ ,$$

attained for a properly rescaled G.

COROLLARY 4. Suppose that X_i, $i = 1, \ldots, n$, are dependent i.i.d with a J-shaped distribution and $EX_1^2 = \mu_2^2$.
If $\frac{m-1}{n} < \frac{1}{\sqrt{3}}$, then

$$EX_{m:n} \leq \frac{n}{n+1-m}\sqrt{1 - \frac{2(m-1)}{\sqrt{3}n}}\mu_2 .$$

The equality holds for a mixture of the $\left[0, \mu_2/\sqrt{1 - \frac{2(m-1)}{\sqrt{3}n}}\right]$-uniform and the degenerate, concentrated at $\mu_2/\sqrt{1 - \frac{2(m-1)}{\sqrt{3}n}}$, distributions with probabilities $\sqrt{3}\frac{m-1}{n}$ and $1 - \sqrt{3}\frac{m-1}{n}$, respectively.
If $\frac{m-1}{n} > \frac{1}{\sqrt{3}}$, then

$$EX_{m:n} \leq \frac{\sqrt{3}}{2}\left(1 + \frac{m-1}{n}\right)\mu_2 ,$$

attained by random variables with the uniform distribution on the interval $[0, \sqrt{3}\mu_2]$.

COROLLARY 5. If X_i, $i = 1, \ldots, n$, are i.i.d with an IFR distribution and a given μ_2, then for $\alpha = \alpha(m, n)$ uniquely defined by

$$\frac{2}{\alpha}(1 - e^{-\alpha}) - e^{-\alpha} = \left(1 - \frac{m-1}{n}\right)\ln\frac{ne}{n+1-m} ,$$

we have

$$EX_{m:n} \leq \frac{n}{n+1-m}\frac{\sqrt{2}}{\alpha}e^{-\frac{\alpha}{2}}\sqrt{e^\alpha - 1 - \alpha}\,\mu_2 .$$

The equality holds iff

$$F^*(x) = 1 - \exp\left(-\frac{\sqrt{2}}{\mu_2}e^{-\frac{\alpha}{2}}\sqrt{e^\alpha - 1 - \alpha}\,x\right), \quad 0 < x < \frac{\alpha e^{\frac{\alpha}{2}}\mu_2}{\sqrt{e^\alpha - 1 - \alpha}} ,$$

which is a combination of a right truncated exponential distribution with a pole at the truncation point.

Combining the projection method for families restricted by the convex ordering relations and the standard transformation of Subsection 4.1, we can establish analogous results for families of symmetric distributions defined by means of s-ordering. We recall the definition: $F \prec_s G$ iff $Q_G F$ is concave and convex on the nonpositive and nonnegative parts of the support of F, respectively. The s-ordering was defined and investigated by van Zwet (1964) (see also Lawrence, 1975). This is a comparison of distribution peakedness, and the s-ordering of a pair is inherited by the respective kurtoses.

Let us consider the distributions symmetric about an arbitrary point $\mu \in \mathcal{R}$, say F_μ, such that

$$F_\mu(x) = F(x - \mu), \quad \text{and} \quad F(x) = 1 - F(-x), \quad x < 0 . \tag{80}$$

Then, for a given G_μ, $F_\mu \succ_s G_\mu$ ($F_\mu \prec_s G_\mu$) iff $Q_F G_{|\mathscr{R}_+} \in \mathscr{Q}_{\succ_c G}$ ($\mathscr{Q}_{\prec_c G}$, respectively) and (80) hold. This and (59) imply that the problem of determining the bound for an expected L-estimate of dependent variables with a given variance and a distribution function F_μ such that $F_\mu \succ_s G_\mu$, (G – fixed) consists in finding the projection of the composition $D^s G$ onto $\mathscr{Q}_{\succ_c G}$, and calculating its norm. In contrast with bound (76), the norm is multiplied by $\frac{1}{\sqrt{2}} \sigma_2$, which follows from $\int_0^\infty Q_F G(x) G(\mathrm{d}x) = \frac{1}{2} \sigma_2^2$. The bound is attained by $F_\mu^\star$ defined by

$$Q_{F^\star} G = \frac{P_{\succ_c G} D^s G}{\|P_{\succ_c G} D^s G\|} \frac{\sigma_2}{\sqrt{2}}$$

(cf (77)) on the positive axis. According to (80), $F^\star$ is extended on $\mathscr{R}_-$ and shifted by μ. In the same manner we can obtain results for distributions $F_\mu \prec_s G_\mu$, the only difference being in choosing the cone $\mathscr{Q}_{\prec_s G}$ instead of $\mathscr{Q}_{\succ_s G}$.

Since $D^s G$ is also a step function, and

$$D^s_{m:n} = \frac{n}{n+1-m} \mathbf{1}_{[\max\{\frac{m-1}{n}, 1-\frac{m-1}{n}\}, 1)} \, ,$$

we can directly apply Lemma 2. In addition, every D^s has $[\frac{n+1}{2}]$ values at most, and so we reduce nearly twice the number of parameters in representations (75) and (79). We shall not present here theorems for the classes of distributions determined by the s-comparisons with a general G. These are similar to Theorems 10(a) and 11(a). We describe only the bounds for the special cases of symmetric unimodal and U-shaped distributions, which follow and precede the uniform distribution in the s-ordering.

THEOREM 12 (Gajek and Rychlik, 1996). Suppose that $X_1, \ldots, X_n$ are i.d random variables, with a symmetric about μ and unimodal distribution and variance σ_2^2.
If $0 < \frac{m-1}{n} < \frac{1}{3}$, then

$$EX_{m:n} \leq \mu + \frac{2n}{3(n+1-m)} \sqrt{\frac{m-1}{n}} \sigma_2 \, .$$

This becomes equality iff $X_1 = \mu$ with probability $1 - 3\frac{m-1}{n}$ and is uniformly distributed on $\left[\mu - \sqrt{\frac{n}{m-1}} \sigma_2, \mu + \sqrt{\frac{n}{m-1}} \sigma_2\right]$ with probability $3\frac{m-1}{n}$.
If $\frac{1}{3} \leq \frac{m-1}{n} \leq \frac{2}{3}$, then

$$EX_{m:n} \leq \mu + \sqrt{3} \frac{m-1}{n} \sigma_2 \, ,$$

which is equality iff X_1 is uniformly distributed on $[\mu - \sqrt{3}\sigma_2, \mu + \sqrt{3}\sigma_2]$.
If $\frac{m-1}{n} > \frac{2}{3}$, then

$$EX_{m:n} \leq \mu + \frac{2}{3} \sqrt{\frac{n}{n+1-m}} \sigma_2 \, .$$

Here the equality is attained by the mixture of the degenerate distribution at μ and the $[\mu - \sqrt{\frac{n}{n+1-m}}\sigma_2, \mu + \sqrt{\frac{n}{n+1-m}}\sigma_2]$-uniform distribution with probabilities $3\frac{m-1}{n} - 2$ and $3(1 - \frac{m-1}{n})$, respectively.

THEOREM 13. Suppose that X_i, $i = 1, \ldots, n$, are dependent random variables with a common symmetric about μ and U-shaped distribution and a finite given variance σ_2^2.

If $|\frac{m-1}{n} - \frac{1}{2}| < \frac{1}{\sqrt{3}}$, then

$$EX_{m:n} \leq \mu + \frac{n}{n+1-m}\sqrt{\frac{1}{4} - \frac{1}{\sqrt{3}}\left|\frac{m-1}{n} - \frac{1}{2}\right|}\,\sigma_2\;.$$

The equality holds iff the distribution of X_1 has the symmetric uniform density on $[\mu - \sigma_2/\sqrt{1 - 4\sqrt{3}|\frac{m-1}{n} - \frac{1}{2}|}, \mu + \sigma_2/\sqrt{1 - 4\sqrt{3}|\frac{m-1}{n} - \frac{1}{2}|}]$, and two atoms at the ends of the interval, with identical probabilities $\frac{1}{2} - \sqrt{3}|\frac{m-1}{n} - \frac{1}{2}|$.

If $\frac{1}{\sqrt{3}} < |\frac{m-1}{n} - \frac{1}{2}| < \frac{1}{2}$, then

$$EX_{m:n} \leq \mu + \sqrt{3}\frac{m-1}{n}\sigma_2,$$

which becomes equality for X_i, $i = 1, \ldots, n$, uniformly distributed on $[\mu - \sqrt{3}\sigma_2, \mu + \sqrt{3}\sigma_2]$.

We have not considered the trivial case of sample minimum, for which $D_{1:n}^s \equiv 0$. In fact, $EX_{1:n}$ attains the upper bound μ iff X_i, $i = 1, \ldots, n$, are identical whatever the parent distribution is. For the other order statistics of symmetric variables, there are joint distributions such that $EX_{m:n} > \mu$ (even for $m = 2$ and large n). This contrasts with the i.i.d case, in which we derive the trivial upper bound on $EX_{m:n}$ if $m \leq [\frac{n+1}{2}]$.

The s-ordering is also well defined for pairs of asymmetric distributions. The families of compositions $Q_F G$ for all F being in either of s-relations with G are closed under nonnegative combinations, and applying the approach developed in this Section we can derive tight bounds for these families. Barlow and Proschan (1966) defined the starshaped ordering of life distributions: $F \succ_\star G$ iff $\frac{Q_F G(x)}{x}$ is nondecreasing, which is implied by $F \succ_c G$. We say that $Q_F G$ is superadditive (subadditive) iff $Q_F G(x + y) \geq (\leq) Q_F G(x) + Q_F G(y)$ for all x and y. The relation allows to define another more general partial ordering (see Marshall and Proschan, 1972). The properties of being in this relation as well as in the $\star$-relation with the exponential distribution have a natural meaning in the reliability theory (see Barlow and Proschan, 1975). The counterpart of the starshaped ordering for the symmetric distributions, the r-ordering, was introduced by Lawrence (1975). For all these relations, with G fixed, the resulting families of compositions are convex cones, which, theoretically, makes it possible to determine a number of sharp bounds on L-estimates.

5. Quantile bounds for restricted families

Another way of deriving bounds on expectations of L-estimates, with distributions from restricted families determined by partial orderings makes use of the Jensen inequality. The bounds are expressed in terms of quantiles of the parent distribution. In this Section, we change the convention of presentation, assumed in the previous ones. We first discuss the bounds in the i.i.d case, which were presented in literature before. The basic results were published in the sixties. Then we derive the respective bounds for the dependent case.

5.1. *I.i.d case.* Consider the sample $X_1, \ldots, X_n$ with a common distribution $F \in \mathscr{F}_{\prec_c G}$, G is fixed and known. Since $Q_F G$ is concave, and $N_{m:n} G(x) G(dx)$ is a probability measure,

$$\begin{aligned} EX_{m:n} &= \int_0^1 Q_F(x) N_{m:n}(x) \, dx \\ &= \int_0^\infty Q_F G(x) N_{m:n} G(x) G(dx) \\ &\leq Q_F G \left(\int_0^\infty x N_{m:n} G(x) G(dx) \right) \qquad (81) \\ &= Q_F G \left(\int_0^1 Q_G(x) N_{m:n}(x) \, dx \right) \\ &= Q_F G(EY_{m:n}) \;, \end{aligned}$$

where $Y_{m:n}$ is the mth order statistic of the independent G-distributed sample. If $F \in \mathscr{F}_{\succ_c G}$, we obtain

$$EX_{m:n} \geq Q_F G(EY_{m:n}). \qquad (82)$$

The equalities in (81)–(82) hold if $F = G$ up to a location-scale transformation. The inequalities were obtained by van Zwet (1964). For the special case of uniform G, we have the bounds derived earlier by Blom (1958)

$$EX_{m:n} \geq Q_F\left(\frac{m}{n+1}\right) \qquad (83)$$

for F with a decreasing probability on its support and the reversed one for J-shaped distributions.

Kamps (1991) used analogous arguments to establish bounds on moments of order statistics

$$EX_{m:n}^r \leq (\geq) Q_F^r\left(\frac{m}{n+1}\right), \qquad r > 0 \;, \qquad (84)$$

provided that Q_F^r is concave (convex), and becoming equality iff F is a power distribution with the exponent r. Inequalities (84) follow from the special case (83) if we replace the original variables by their r^{th} powers. An intermediate generalization of (84) is

$$EX^r_{m:n} \leq (\geq, =) Q^r_F(EY^r_{m:n}), \qquad r > 0,$$

which is satisfied if $Q^r_F G$ is concave (convex, and linear, respectively) on the support of G, and G is the distribution function of Y_i, $i = 1, \ldots, n$.

We also write the implications of (81) and (82) for the IFR and DFR random variables, respectively. If X_i, $i = 1, \ldots, n$, are IFR, then

$$EX_{m:n} \leq Q_F\left(1 - \exp\left(-\sum_{i=1}^{m} \frac{1}{n+1-i}\right)\right), \tag{85}$$

and for the DFR variables the inequality in (85) is reversed. The integral approximations

$$\int_{n+1-m}^{n+1} \frac{dx}{x} < \sum_{i=1}^{m} \frac{1}{n+1-i} < \int_{n+\frac{1}{2}-m}^{n+\frac{1}{2}} \frac{dx}{x}$$

allow to simplify (85) by the evaluations

$$EX_{m:n} < Q_F\left(\frac{m}{n+\frac{1}{2}}\right)$$

for the IFR variables and

$$EX_{m:n} > Q_F\left(\frac{m}{n+1}\right)$$

for the DFR ones (cf (83)). Barlow and Proschan (1966) established the following sharp inequalities for the mth order statistic of IFR distributed random variables with a given p^{th} quantile $Q_F(p)$:

$$\sum_{j=0}^{m-1} \binom{n}{j} \int_0^{Q_F(p)} \left[1 - e^{-x \ln \frac{Q_F(p)}{1-p}}\right]^j e^{x(j-n) \ln \frac{Q_F(p)}{1-p}} dx$$

$$\leq EX_{m:n} \leq \max\left\{Q_F(p), \frac{p}{-\ln(1-p)} \sum_{j=1}^{m} \frac{1}{n+1-m}\right\}.$$

Van Zwet (1964) used c-comparisons with distributions, for which the formulas for the expectations of order statistics have simple analytic expressions, to approximate the respective expectations for the c-related distributions, which do not have such nice formulas. Comparisons with the Pareto and negative Pareto distributions $G(x) = 1 - \frac{1}{x}$ and $-\frac{1}{x}$, respectively, yield: if $\frac{1}{1-F(x)}$ is convex (concave), then

$$EX_{m:n} \leq (\geq) Q_F\left(\frac{m}{n}\right),$$

and if $\frac{1}{F(x)}$ is concave (convex), then

$$EX_{m:n} \leq (\geq) Q_F\left(\frac{m-1}{n}\right).$$

For instance, for the normal distribution function F both $\frac{1}{1-F(x)}$ and $\frac{1}{F(x)}$ are convex, and so we have

$$Q_F\left(\frac{m-1}{n}\right) \leq EX_{m:n} \leq Q_F\left(\frac{m}{n}\right). \tag{86}$$

Note that any convex combination $\sum_{i=1}^{n} c_i N_{i:n}$ is a probability density function. Replacing a single $N_{i:n}$ by the combination in (81), we derive

$$E\sum_{i=1}^{n} c_i X_{i:n} \leq Q_F G\left(E\sum_{i=1}^{n} c_i Y_{i:n}\right), \quad F \in \mathscr{F}_{\prec_c G} \tag{87}$$

(cf Arnold and Balakrishnan, 1989, Section 3.4). It follows that

$$E\sum_{i=1}^{n} c_i X_{i:n} \leq Q_F\left(\frac{1}{n+1}\sum_{i=1}^{n} ic_i\right), \quad F - J\text{-shaped}, \tag{88}$$

and

$$E\sum_{i=1}^{n} c_i X_{i:n} \leq Q_F\left(1 - \exp\left(-\sum_{j=1}^{n}\frac{1}{j}\sum_{i=n+1-j}^{n} c_i\right)\right), \quad F - \text{IFR}. \tag{89}$$

In particular, Nagaraja (1981) derived

$$E\frac{1}{k}\sum_{i=n+1-k}^{n} X_{i:n} \leq Q_F\left(\frac{2n+1-k}{2(n+1)}\right),$$

and

$$E\frac{1}{k}\sum_{i=n+1-k}^{n} X_{i:n} \leq Q_F\left(1 - \exp\left(1 - \sum_{j=k+1}^{n}\frac{1}{j}\right)\right)$$

$$< Q_F\left(1 - \frac{2k+1}{(2n+1)e}\right),$$

respectively. Barlow and Proschan (1966) noted that (87)–(89) hold for nonnegative combinations with $\sum_{i=1}^{n} c_i \leq 1$, also. More generally, the inequalities are satisfied if some c_i are negative, but $0 < \sum_{i=1}^{n} c_i \leq 1$, and $\sum_{i=1}^{n} c_i N_{i:n} \geq 0$ on $[0, 1]$. Indeed, by (87), vanishing of $Q_F G$ at the origin, and its concavity, we have

$$E\sum_{i=1}^{n} c_i X_{i:n} \leq \sum_{i=1}^{n} c_i Q_F G\left(\frac{E\sum_{i=1}^{n} c_i Y_{i:n}}{\sum_{i=1}^{n} c_i}\right)$$

$$= \sum_{i=1}^{n} c_i Q_F G\left(\frac{E\sum_{i=1}^{n} c_i Y_{i:n}}{\sum_{i=1}^{n} c_i}\right) + \left(1 - \sum_{i=1}^{n} c_i\right) Q_F G(0) \tag{90}$$

$$\leq Q_F G\left(E\sum_{i=1}^{n} c_i Y_{i:n}\right).$$

Necessary and sufficient conditions on the coefficients $c_1, \ldots, c_n$ such that (87) holds were obtained by van Zwet (cf Lawrence, 1975, p. 423).

Since $N^s_{m:n}$, $m > \frac{n+1}{2}$, is positive on $(\frac{1}{2}, 1)$, and integrates to a number less than 1, we can repeat the arguments of (81) and (90) to deduce

$$\mathrm{E} X_{m:n} - \mu \leq \int_{\frac{1}{2}}^{1} Q_F(x) N^s_{i:n}(x)\, \mathrm{d}x \qquad (91)$$
$$\leq Q_F G(\mathrm{E}(Y_{m:n} - Y_1))$$

for all $F \prec_s G$ and $m > \frac{n+1}{2}$ (van Zwet, 1964). Since $Q_F G$ is antisymmetric, (91) is reversed for $m < \frac{n+1}{2}$. For brevity, we assume further that F and G are symmetric about 0. If $F \in \mathcal{F}_{\succ_s G}$, then

$$|\mathrm{E} X_{m:n}| \geq |Q_F G(\mathrm{E} Y_{m:n})|\ .$$

For the special case of s-comparisons with the symmetric uniform distribution, we have

$$|\mathrm{E} X_{m:n}| \leq (\geq) \left| Q_F\left(\frac{m}{n+1}\right) \right| \qquad (92)$$

for symmetric U-shaped (unimodal) parent distribution functions F (see also Ali and Chan, 1965). Kabir and Rahman (1974) improved (92) assuming the stronger condition that $\frac{Q_F(x)}{x - \frac{1}{2}}$ is convex (or concave) on $(\frac{1}{2}, 1)$.

Van Zwet (1964) proved that normal $F \prec_s$ logistic G, and applying (91), the fact that $\mathrm{E} Y_{m:n} = \sum_{j=n-m}^{m} \frac{1}{j-1}$, and an integral approximation, he derived

$$\mathrm{E} X_{m:n} < Q_F\left(\left[1 + \exp\left(-\sum_{j=n-m}^{m} \frac{1}{j-1} \right) \right]^{-1} \right) < Q_F\left(\frac{m - \frac{1}{2}}{n} \right),$$

for all $m > \frac{n+1}{2}$, which improves (86). Analogous bounds for the selection differential of the normal population were presented by Nagaraja (1981, p. 444). In this case, the inequality

$$\frac{1}{k} \sum_{i=n+1-k}^{n} \mathrm{E} X_{i:n} < Q_F G\left(\frac{1}{k} \sum_{i=n+1-k}^{n} \mathrm{E} Y_{m:n} \right)$$

follows from the fact that

$$\frac{1}{k} \sum_{i=n+1-k}^{n} N^s_{i:n} = \frac{1}{k} \sum_{i=\max\{k, n-k\}+1}^{n} N^s_{i:n}$$

is a positive density of a substochastic measure on $(\frac{1}{2}, 1)$. Similar inequalities for general L-estimates can be verified if $c_i = 0$ for $i < \frac{n+1}{2}$, and $c_i \geq 0$ for $i > \frac{n+1}{2}$, and $\sum_{i > \frac{n+1}{2}} c_i \leq 1$ (cf Arnold and Balakrishnan, 1989). Lawrence (1975) proved that the necessary and sufficient condition for

$$E\sum_{i=1}^{n} c_i X_{i:n} \leq (\geq) Q_F G\left(E\sum_{i=1}^{n} c_i Y_{m:n}\right)$$

for all $F \prec_s (\succ_s) G$ is

$$\sum_{i=1}^{n} c_i - \sum_{j=1}^{n}\sum_{i=1}^{j} c_i \binom{n}{j}[x^{n-j}(1-x)^j + x^j(1-x)^{n-j}] \in [0,1]([-1,0])$$

for every $\frac{1}{2} \leq x \leq 1$. Lawrence (1975) also presented the respective conditions for asymmetric distribution functions.

5.2. I.d case. Suppose that X_i, $i=1,\ldots,n$, and Y_i, $i=1,\ldots,n$, are two sequences of dependent random variables with life distributions F and G, respectively. We assume that $F \prec_c G$ and G is known. Since $Q_F G$ is a concave function on $[0,\infty)$, and $D_{m:n} = \frac{n}{n+1-m}\mathbf{1}_{[\frac{m-1}{n},1)}$ is a probability density function on $[0,1)$,

$$EX_{m:n} \leq \int_0^1 Q_F(x) D_{m:n}(x)\,dx$$

$$= \int_0^\infty Q_F G(x) D_{m:n} G(x) G(dx)$$

$$\leq Q_F G\left(\int_0^\infty x D_{m:n} G(x) G(dx)\right) \quad (93)$$

$$= Q_F G\left(\int_0^1 Q_G(x) D_{m:n}(x)\,dx\right)$$

$$= Q_F G(\sup\ EY_{m:n})\ ,$$

where the supremum is taken over the all possible joint distributions of $Y_1,\ldots,Y_n$. If $F \succ_c G$, then for $\overline{D}_{m:n} = \frac{n}{m}\mathbf{1}_{[0,\frac{m}{n})}$, we can similarly prove

$$EX_{m:n} \geq Q_F G\left(\int_0^1 Q_G(x) \overline{D}_{m:n}(x)\,dx\right) \quad (94)$$

$$= Q_F G(\inf\ EY_{m:n})\ ,$$

with the infimum taken over the same set as in (93). For specific choices of G, we obtain

$$EX_{m:n} \leq Q_F\left(\frac{1}{2} + \frac{m-1}{2n}\right)\ ,$$

if F is J-shaped,

$$EX_{m:n} \geq Q_F\left(\frac{m}{2n}\right)\ ,$$

if F has a decreasing failure probability on $\mathcal{R}_+$, and, furthermore,

$$EX_{m:n} \leq Q_F\left(1 - \frac{n+1-m}{ne}\right)\ ,$$

if F has an increasing failure rate, and

$$EX_{m:n} \geq Q_F\left(1 - e^{-1}\left(\frac{n}{n-m}\right)^{\frac{n}{m}-1}\right),$$

if F is a DFR distribution.

Observe that the single order statistics in (93) and (94) can be replaced by L-estimates such that the respective jump functions D and $\overline{D}$ are nonnegative and integrate to one. The latter condition can be further weakened when $Q_F G(0) = 0$ (cf (90)). Let us concentrate on the case of L-estimates such that D is a substochastic density function, i.e.,

$$\sum_{i=1}^{j} c_i \geq 0, \quad j = 1, \ldots, n-1, \quad 0 < \sum_{i=1}^{n} c_i = \int_0^1 D(x)\,dx \leq 1. \tag{95}$$

Then $F \prec_c G$ implies

$$E \sum_{i=1}^{n} c_i X_{i:n} \leq \sum_{i=1}^{n} c_i \int_0^1 Q_F G(x) \frac{DG(x)}{\sum_{i=1}^{n} c_i} G(dx)$$

$$\leq \sum_{i=1}^{n} c_i Q_F G\left(\frac{1}{\sum_{i=1}^{n} c_i} \int_0^1 Q_G(x) D(x)\,dx\right)$$

$$+ \left(1 - \sum_{i=1}^{n} c_i\right) Q_F G(0)$$

$$\leq Q_F G\left(\int_0^1 Q_G(x) D(x)\,dx\right)$$

$$= Q_F G\left(\sup E \sum_{i=1}^{n} c_i Y_{i:n}\right)$$

If we admit $\sum_{i=1}^{n} c_i > 1$ in (95), we would merely have

$$E \sum_{i=1}^{n} c_i X_{i:n} \leq \sum_{i=1}^{n} c_i Q_F G\left(\frac{\sup E \sum_{i=1}^{n} c_i Y_{i:n}}{\sum_{i=1}^{n} c_i}\right).$$

Likewise, under the assumption $F \succ_c G$, $\sum_{i=1}^{n} c_i > 0$, and $\sum_{i=j}^{n} c_i \geq 0$, $j = 2, \ldots, n$, we derive

$$E \sum_{i=1}^{n} c_i X_{i:n} \geq \max\left\{\sum_{i=1}^{n} c_i, 1\right\} Q_F G\left(\frac{\int_0^1 Q_G(x)\overline{D}(x)\,dx}{\max\{\sum_{i=1}^{n} c_i, 1\}}\right).$$

If F and G are symmetric about 0, and $F \prec_s G$, then

$$EX_{m:n} \leq \int_{\frac{1}{2}}^{1} Q_F(x) D_{m:n}^s(x) \, dx$$

$$\leq Q_F G \left(\int_{\frac{1}{2}}^{1} Q_G(x) D_{m:n}^s(x) \, dx \right) \quad (96)$$

$$= Q_F G(\sup E Y_{m:n}) ,$$

because $D_{m:n}^s$ is a substochastic density on $[\frac{1}{2}, 1]$. By similar arguments we conclude for $F \succ_s G$

$$EX_{m:n} \geq Q_F G \left(\int_{\frac{1}{2}}^{1} Q_G(x) \overline{D}_{m:n}^s(x) \, dx \right) \quad (97)$$

$$= Q_F G(\inf E Y_{m:n}) .$$

Note that the right-hand side of (96) is positive for all $m = 2, \ldots, n$, in contrast with the i.i.d case, where the respective upper bound (91) is nontrivial for $m > \frac{n+1}{2}$, only. Similarly, the lower bound in (97) is negative unless $m = n$. Plugging G being symmetric uniform into (96) and (97), yields

$$EX_{m:n} \leq Q_F \left(\frac{1}{2} + \frac{m-1}{2n} \right)$$

for the symmetric U-shaped distributions F, and

$$EX_{m:n} \geq Q_F \left(\frac{m}{2n} \right)$$

for the symmetric unimodal ones.

Certainly, we can verify

$$E \sum_{i=1}^{n} c_i X_{i:n} \leq Q_F G \left(\sup E \sum_{i=1}^{n} c_i Y_{i:n} \right), \quad F \prec_s G , \quad (98)$$

provided that

$$\int_{\frac{1}{2}}^{1} D^s(x) \, dx = C(1) - 2C(\tfrac{1}{2}) \leq 1 \quad (99)$$

(condition $D^s(x) \geq 0$, $x \in [\frac{1}{2}, 1]$, holds trivially). The necessary and sufficient condition for (99) is

$$\sum_{i=1}^{j} c_i + \frac{n - 2j}{2(k - j)} \sum_{i=j+1}^{k} c_i \geq \frac{1}{2} \sum_{i=1}^{n} c_i - \frac{1}{2} \quad \text{for all } 0 \leq j \leq \frac{n}{2} \leq k \leq n$$

(cf Rychlik, 1993a). Otherwise

$$E \sum_{i=1}^{n} c_i X_{i:n} \leq \left[\sum_{i=1}^{n} c_i - 2C\left(\frac{1}{2} \right) \right] Q_F G \left(\frac{\sup E \sum_{i=1}^{n} c_i Y_{i:n}}{\sum_{i=1}^{n} c_i - 2C(\frac{1}{2})} \right) . \quad (100)$$

A counterpart of (98) and (100) for $F \succ_s G$ is

$$\mathrm{E}\sum_{i=1}^{n} c_i X_{i:n} \geq \max\left\{2\overline{C}\left(\frac{1}{2}\right) - \sum_{i=1}^{n} c_i, 1\right\}$$
$$\times Q_F G\left(\frac{\inf \mathrm{E}\sum_{i=1}^{n} c_i Y_{i:n}}{\max\{2\overline{C}(\frac{1}{2}) - \sum_{i=1}^{n} c_i, 1\}}\right),$$

Also, $2\overline{C}(\frac{1}{2}) - \sum_{i=1}^{n} c_i \leq 1$ iff

$$\sum_{i=1}^{j} c_i + \frac{n - 2j}{2(k - j)} \sum_{i=j+1}^{k} c_i \leq \frac{1}{2}\sum_{i=1}^{n} c_i + \frac{1}{2} \quad \text{for all } 0 \leq j \leq \frac{n}{2} \leq k \leq n.$$

We finally remark that the bounds presented in this Subsection are sharp. For instance, (93) becomes equality iff $F \sim_c G$ and the joint distribution of X_i, $i = 1, \ldots, n$, satisfies the conditions described in Section 2. It is also worth noting that since it is usually easier to determine the extremes of expected L-estimates in the i.d case than the respective value for the independent sample, the bounds presented here may be used for approximations in the i.i.d case.

References

Ali, M. M. and L. K. Chan (1965). Some bounds for expected values of order statistics. *Ann. Math. Statist.* **36**, 1055–1057.

Arnold, B. C. (1980). Distribution-free bounds on the mean of the maximum of a dependent sample. *SIAM J. Appl. Math.* **38**, 163–167.

Arnold, B. C. (1985). *p*-Norm bounds on the expectation of the maximum of possibly dependent sample. *J. Multivar. Anal.* **17**, 316–332.

Arnold, B. C. (1988). Bounds on the expected maximum. *Commun. Statist. – Theor. Meth.* **17**, 2135–2150.

Arnold, B. C. and N. Balakrishnan (1989). *Relations, Bounds and Approximations for Order Statistics*. Lecture Notes in Statistics, Vol. 53, Springer-Verlag, New York.

Arnold, B. C., N. Balakrishnan and H. N. Nagaraja (1992). *A First Course in Order Statistics*. Wiley, New York.

Arnold, B. C. and R. A. Groeneveld (1974). Bounds for deviations between sample population statistics. *Biometrika* **61**, 387–389.

Arnold, B. C. and R. A. Groeneveld (1979). Bounds on expectations of linear systematic statistics based on dependent samples. *Ann. Statist.* **7**, 220–223. Erratum in *Ann. Statist.* **8**, 1401.

Arnold, B. C. and R. A. Groeneveld (1981). Maximal deviation between sample and population means in finite populations. *J. Amer. Statist. Assoc.* **76**, 443–445.

Aven, T. (1985). Upper (lower) bounds on the mean of the maximum (minimum) of a number of random variables. *J. Appl. Probab.* **22**, 723–728.

Balakrishnan, A. V. (1981). *Applied Functional Analysis*. 2nd Edition. Springer-Verlag, New York.

Balakrishnan, N. (1990). Improving the David-Hartley-Gumbel bound for the mean of extreme order statistics. *Statist. Probab. Lett.* **9**, 291–294.

Balakrishnan, N. (1993). A simple application of binomial-negative binomial relationship in the derivation of sharp bounds for moments of order statistics based on greatest convex minorants. *Statist. Probab. Lett.* **18**, 301–305.

Balakrishnan, N. and S. M. Bendre (1993). Improved bounds for expectations of linear functions of order statistics. *Statistics* **24**, 161–165.

Barlow, R. E. and F. Proschan (1966). Inequalities for linear combinations of order statistics from restricted families. *Ann. Math. Statist.* **37**, 1574–1591.

Barlow, R. E. and F. Proschan (1975). *Statistical Analysis of Reliability and Life Testing*. Holt, Rinehart and Winston, New York.

Beesack, P. R. (1973). On bounds for the range of ordered variates. *Publ. of the Electrotechnical Faculty of Belgrade Univ., Math. Phys. Series* **428**, 93–96.

Blom, G. (1958). *Statistical Estimates and Transformed Beta Variables*. Almqvist and Wiksells, Uppsala.

Boyd, A. V. (1971). Bound for order statistics. *Publ. of the Electrotechnical Faculty of Belgrade Univ., Math. Phys. Series* **365**, 31–32.

Caraux, G. and O. Gascuel (1992a). Bounds on distribution functions of order statistics for dependent variates. *Statist. Probab. Lett.* **14**, 103–105.

Caraux, G. and O. Gascuel (1992b). Bounds on expectations of order statistics via extremal dependences. *Statist. Probab. Lett.* **15**, 143–148.

David, H. A. (1981). *Order Statistics*. 2nd Edition. Wiley, New York.

David, H. A. (1988). General bounds and inequalities in order statistics. *Commun. Statist. – Theor. Meth.* **17**, 2119–2134.

David, H. A., H. O. Hartley and E. S. Pearson (1954). The distribution of the ratio, in a single normal sample, of range to standard deviation. *Biometrika* **41**, 482–493.

Fahmy, S. and F. Proschan (1981). Bounds on differences of order statistics. *Amer. Statist.* **35**, 46–47.

Gajek, L. and U. Gather (1991). Moment inequalities for order statistics with applications to characterizations of distributions. *Metrika* **38**, 357–367.

Gajek, L. and E. Lenic (1993). Moment inequalities for order and record statistics under restrictions on their distributions. *Ann. Univ. Mariae Curie-Skłodowska, Sect. A* **47**, 27–34.

Gajek, L. and T. Rychlik (1996). Projection method for moment bounds on order statistics from restricted families. I. Dependent case. *J. Multivar. Anal.* **57**, 156–174.

Gajek, L. and T. Rychlik (1998). Projection method for moment bounds on order statistics from restricted families. II. Independent case. *J. Multivar. Anal.* **64**, to appear.

Gilstein, C. Z. (1981). Bounds on expectations of linear combinations of order statistics (preliminary report). Abstract number 177–100. *Inst. Math. Statist. Bull.* **10**, 253.

Gravey, A. (1985). A simple construction of an upper bound for the mean of the maximum of n identically distributed random variables. *J. Appl. Prob.* **22**, 844–851.

Groeneveld, R. A. (1982). Best bounds for order statistics and their expectations in range and mean units with applications. *Commun. Statist. – Theor. Meth.* **11**, 1809–1815.

Gumbel, E. J. (1954). The maxima of the mean largest value and of the range. *Ann. Math. Statist.* **25**, 76–84.

Hartley, H. O. and H. A. David (1954). Universal bounds for mean range and extreme observation. *Ann. Math. Statist.* **25**, 85–99.

Hawkins, D. M. (1971). On the bounds of the range of order statistics. *J. Amer. Statist. Assoc.* **66**, 644–645.

Joshi, P. C. (1969). Bounds and approximatons for the moments of order statistics. *J. Amer. Statist. Assoc.* **64**, 1617–1624.

Kabir, A. B. M. L. and M. Rahman (1974). Bounds for expected values of order statistics. *Commun. Statist. – Theor. Meth.* **3**, 557–566.

Kamps, U. (1991). Inequalities for moments of order statistics and characterizations of distributions. *J. Statist. Plan. Infer.* **27**, 397–404.

Kemperman, J. H. B. (1993). Personal communication.

Lai, T. L. and H. Robbins (1976). Maximally dependent random variables. *Proc. Nat. Acad. Sci. U.S.A.* **73**, 286–288.

Lai, T. L. and H. Robbins (1978). A class of dependent random variables and their maxima. *Z. Wahrsch. Verw. Gebiete* **42**, 89–111.

Lawrence, M. J. (1975). Inequalities for s-ordered distributions. *Ann. Statist.* **3**, 413–428.

Lefèvre, C. (1986). Bounds on the expectation of linear combinations of order statistics with application to Pert networks. *Stoch. Anal. Appl.* **4**, 351–356.

Lin, G. D. (1988). Characterizations of uniform distributions and of exponential distributions. *Sankhyā Ser. A* **50**, 64–69.

Ludwig, O. (1960). Über Erwartungswerte und Varianzen von Ranggrössen in kleinen Stichproben. *Metrika* **3**, 218–233.

Ludwig, O. (1973). Differenzen der Erwartungswerte von Ranggrössen in kleinen Stichproben. In: B. Bereanu, M. Iosifescu, T. Postelnicu and P. Tăutu, eds., *Proc. 4th Conf. on Probab. Theory.* Editura Acad., pp. 299–303.

Mallows, C. L. (1969). Extrema of expectations of uniform order statistics. *SIAM Rev.* **11**, 410–411.

Mallows, C. L. and D. Richter (1969). Inequalities of Chebyshev type involving conditional expectations. *Ann. Math. Statist.* **40**, 1922–1932.

Mărgăritescu, E. and T. Nicolae (1990). Best linear bounds for order statistics in terms of the mean and the generalized range. *Stud. Cerc. Mat.* **42**, 41–45.

Marshall, A. W. and I. Olkin (1979). *Inequalities: Theory of Majorization and Its Applications,* Academic Press, New York.

Marshall, A. W. and F. Proschan (1972). Classes of distributions applicable in replacement, with renewal theory implications. In: L. Le Cam, J. Neyman and E. L. Scott, eds., *Proc. 6th Berkeley Symp. Math. Statist. Prob.*, Vol. 1. Univ. of California Press, Berkeley, pp. 395–415.

Moriguti, S. (1951). Extremal properties of extreme value distributions. *Ann. Math. Statist.* **22**, 523–536.

Moriguti, S. (1953). A modification of Schwarz's inequality with applications to distributions. *Ann. Math. Statist.* **24**, 107–113.

Nagaraja, H. N. (1981). Some finite sample results for the selection differential. *Ann. Inst. Statist. Math.* **33**, 437–448.

Nair, K. R. (1948). The distribution of the extreme deviate from the sample mean and its studentized form. *Biometrika* **35**, 118–144.

Oja, H. (1981). On location, scale, skewnwess and kurtosis of univariate distributions. *Scand. J. Statist.* **8**, 154–168.

Olkin, I. (1992). A matrix formulation on how deviant can an observation be. *Amer. Statist.* **46**, 205–209.

Plackett, R. L. (1947). Limits of the ratio of mean range to standard deviation. *Biometrika* **34**, 120–122.

Rustagi, J. S. (1957). On minimizing and maximizing a certain integral with statistical applications. *Ann. Math. Statist.* **28**, 309–328.

Rychlik, T. (1992a). Stochastically extremal distributions of order statistics for dependent samples. *Statist. Probab. Lett.* **13**, 337–341.

Rychlik, T. (1992b). Weak limit theorems for stochastically largest order statistics. In: I. A. Salama, P. K. Sen, eds., *Order Statistics and Nonparametrics. Theory and Applications.* North-Holland, pp. 141–154.

Rychlik, T. (1992c). Sharp inequalities for linear combinations of elements of monotone sequences. *Bull. Polish Acad. Sci. Math.* **40**, 247–254.

Rychlik, T. (1993a). Bounds for expectation of L-estimates for dependent samples. *Statistics* **24**, 1–7.

Rychlik, T. (1993b). Sharp bounds on L-estimates and their expectations for dependent samples. *Commun. Statist. – Theory Meth.* **22**, 1053–1068. Erratum in *Commun. Statist. – Theory Meth.* **23**, 305–306.

Rychlik, T. (1994). Distributions and expectations of order statistics for possibly dependent random variables. *J. Multivar. Anal.* **48**, 31–42.

Rychlik, T. (1995). Bounds for order statistics based on dependent variables with given nonidentical distributions. *Statist. Probab. Lett.* **23**, 351–358.

Samuelson, P. A. (1968). How deviant can you be? *J. Amer. Statist. Assoc.* **63**, 1522–1525.

Scott, J. M. C. (1936). Appendix to paper by Pearson and Chandra Sekar. *Biometrika* **28**, 319–320.

Sugiura, N. (1962). On the orthogonal inverse expansion with an application to the moments of order statistics. *Osaka Math. J.* **14**, 253–263.

Swetits, J. J., S. E. Weinstein and Yuesheng Xu (1990). Approximation in $L_p[0, 1]$ by n-convex functions. *Numer. Funct. Anal. and Optimiz.* **11**, 167–179.

Tchen, A. (1980). Inequalities for distributions with given marginals. *Ann. Probab.* **8**, 814–827.

Thompson, W. R. (1935). On a criterion for the rejection of observations and the distribution of the ratio of deviation to sample standard deviation. *Ann. Math. Statist.* **6**, 214–219.

Thompson, G. W. (1955). Bounds on the ratio of range to standard deviation. *Biometrika* **42**, 268–269.

Ubhaya, V. A. (1989). L_p approximation by quasi-convex and convex functions. *J. Math. Anal. Appl.* **139**, 574–585.

Ubhaya, V. A. (1990). L_p approximation by subsets of convex functions of several variables. *J. Approx. Theor.* **63**, 144–155.

Van Zwet, W. R. (1964). *Convex Transformations of Random Variables*. Math. Centre Tracts, Vol. 7, Mathematisch Centrum, Amsterdam.

Wolkowicz, H. and G. P. H. Styan (1980). Reply to the letter of Kabe. *Amer. Statist.* **34**, 250–251.

PART III

Relations and Identities

Recurrence Relations and Identities for Moments of Order Statistics

N. Balakrishnan and K. S. Sultan

1. Introduction

The theory of statistical inference has been mostly developed assuming that the samples are drawn from some specific (known) probability distribution. These developments broadly fall into two main classes, viz., (i) statistical estimation, and (ii) tests of hypotheses. Yet there are numerous situations in practice wherein it may not be possible to assume a specific functional form for the underlying population distribution function. This led to the development of the so-called "nonparametric" or "distribution-free" methods in statistics that are based on relatively mild assumptions regarding the underlying population; see, for example, Gibbons and Chakraborty (1994), Hollander and Wolfe (1973) and Lehmann (1975). In the beginning [e.g., Scheffe (1943)], order statistics were also considered as a part of the nonparametric statistics and one was simply concerned with the properties of a random sample whose observations when arranged in ascending or descending order of magnitude. Order statistics and their moments quickly gained their importance in many statistical problems. Linear functions of order statistics were found to be extremely useful in the estimation of parameters and also in testing of hypotheses problems. The application of Gauss–Markov theorem of least squares by Lloyd (1952) [also see Sarhan and Greenberg (1962) and David (1981)] to derive linear functions of order statistics (termed as "linear estimators") for estimating the location and scale parameters of distributions, is one fine example. Blom's (1958, 1962) "nearly best linear estimators", Jung's (1955, 1962) "asymptotically best linear estimators", Dixon's (1960) "Winsorized estimators", and Tukey and McLaughlin's (1963) "trimmed estimators" are all based on order statistics and they simply demonstrate the importance of order statistics in statistical inference. Knowledge of the moments of order statistics, in particular their means, variances and covariances, allows us to find the expected value and variance of a linear function of order statistics, and hence permits us to obtain estimators and their efficiencies. There are both theoretical and practical reasons for this sound development of the theory of order statistics. From a theoretical point of view, it is often desirable to develop methods of inference which remain valid under a wide class of population distributions. It is also desirable from a practical point of view to make a statistical procedure as simple and broadly applicable as possible. Order statistics are most applicable in many engineering fields since in these cases the smallest, the largest or an intermediate future realization of a random variable is more important than the mean or the median of the distribution. For example, the lowest strength value of a critical structural component is an important factor in assessing the value of a structural design and the largest value is also important for some similar assessments. One could refer to Gumbel (1958) for many such practical examples where the usefulness of order statistics is stressed upon. In particular, applications of extreme order statistics in many practical situations have been mentioned by Gumbel (1958) and Castillo (1988), and the asymptotic theory of extremes and of related statistics has been developed at length by Galambos (1978, 1987).

It is of interest to mention here that life tests provide an ideal illustration of the advantages of order statistics in the case of censored samples. It is often desirable to stop the experiment after the failure of a certain number of items under test instead of waiting for all the items to fail since such life-testing experiments may take a long time to complete (Mann, Schafer and Singpurwalla, 1974; Lawless, 1982; Nelson, 1982; Bain and Engelhardt, 1991; Cohen, 1991; Cohen and Whitten, 1988, and Balakrishnan and Cohen, 1991). Note that in this situation, unlike in most other cases, the data arrives already in a naturally ordered way by the method of experimentation. A nice historical note on the history and different roles of order statistics has been prepared by Harter (1988).

In the recent years, order statistics have been used quite extensively in outlier detection (Barnett and Lewis, 1993; Hawkins, 1980), goodness-of-fit tests (D'Agostino and Stephens, 1986) and robustness studies. Several robust estimators and robust tests have been successfully developed using censored samples based on the fact that the lowest and highest few observations in a sample are the most likely to be the results of departure from assumed distribution of contamination; for example, see David (1979, 1981), David and Shu (1978), Dixon (1960), Dixon and Tukey (1968), Hogg (1967, 1974), Huber (1972, 1982), Mosteller and Tukey (1977), Shu (1978), Yuen (1971), Andrews et al. (1972) and Tiku, Tan and Balakrishnan (1986). Another important application of order statistics is data compression (Eisenberger and Posner, 1965) wherein the sample may be replaced by enough order statistics to allow both satisfactory estimation of parameters and also a test of the assumed underlying distribution.

The moments of order statistics have assumed considerable interest in recent years and, in fact, have been tabulated quite extensively for many distributions. Many authors have investigated and derived several recurrence relations and identities satisfied by the single as well as product moments of order statistics primarily to reduce the amount of direct computations. However, one could list the following three main reasons why these recurrence relations and identities for the moments of order statistics are important:

(i) they reduce the amount of direct computations quite considerably,

(ii) they usefully express the higher order moments of order statistics in terms of the lower order moments and hence make the evaluation of higher order moments easy,

(iii) they are very useful in checking the computation of the moments of order statistics.

In addition, Joshi (1973) and Joshi and Balakrishnan (1981b) have demonstrated a very interesting application of these recurrence relations and identities among order statistics in establishing some combinatorial identities. For the normal distribution, Davis and Stephens (1978) have illustrated a very good application of an identity between moments of normal order statistics in obtaining an improved approximation of the variance-covariance matrix of normal order statistics.

Exact lower order moments of order statistics in small samples from the normal distribution, along with some relations satisfied by these moments, were first obtained by Jones (1948). Godwin (1949) recognized some more recurrence relations and extended Jones's results for larger sample sizes. A simple recurrence formula between moments of order statistics, commonly known as the "normalized moments" was derived by Cole (1951). Sillitto (1951, 1964) established some recurrence relations for an arbitrary distribution and also used them to obtain some relations for means of range from different sample sizes. Srikantan (1962) derived some recurrence relations between probability density functions of order statistics and extended them for the moments of order statistics; he also investigated the numerical error propagation in such recursive computations. Some additional relations between probability density functions of order statistics were recognized by Young (1967). Recurrence relations which are closely related to Srikantan's results, between expected values of functions of order statistics, were obtained by Krishnaiah and Rizvi (1966). While Melnick (1964) showed that Cole's result is also valid for samples drawn from discrete populations, Arnold (1977) presented an alternative proof for Cole's result covering mixtures of continuous and discrete distributions and also obtained some additional relations exactly on the same lines. Some more relations for both single and product moments of order statistics were established by Downton (1966). David and Joshi (1968) noted that many of these results hold even for order statistics obtained from exchangeable variates. David and Joshi's (1968) result has been further used by Balakrishnan (1987b) in proving some additional results for order statistics obtained from exchangeable variates. Balakrishnan and Malik (1986a) displayed that two well-known identities among moments of order statistics follow directly from some basic recurrence relations satisfied by these moments. As a result, Balakrishnan and Malik (1986a) showed that it is not meaningful to apply these identities for the purpose of checking the computations of these moments whenever these recurrence relations are used in the computational procedure. In an interesting paper, Govindarajulu (1963a) derived some relations between the moments of order statistics from a symmetric distribution and the moments of order statistics from its folded distribution (folded at the center) and also investigated the cumulative rounding error committed by using these relations. In another article, Govindarajulu (1963b) summarized many of these results and established some more recurrence relations and identities satisfied by the single and product moments of order statistics and used these results in order to determine the number of single and double integrals to be evaluated for the calculation of means, variances and covariances of order statistics in a sample of size n, assuming these quantities for all sample sizes less than n to be known, for an arbitrary continuous distribution symmetric about zero. He showed that in this case, one has to evaluate at most one single integral and $(n-2)/2$ double integrals if n is even, and one single integral and $(n-1)/2$ double integrals if n is odd. By a simple generalization of one of the results of Govindarajulu (1963b), Joshi (1971) displayed that for distributions symmetric about zero the number of double integrals to be evaluated for even values of n is in fact zero. Joshi and Balakrishnan (1982) established similar results for an arbitrary continuous distribution and showed that, in that case, one has to evaluate at most two single integrals and $(n-2)/2$ double integrals if n is even, and two single integrals and $(n-1)/2$ double integrals if n is odd. Some more relations and identities satisfied by variances and covariances of order statistics are also given by Joshi and Balakrishnan (1982). For even values of n, in addition, Balakrishnan (1982) proved that the sums $\sum_{r=1}^{n-i} \mu_{r,r+i:n}$, $(1 \leq i \leq n-1)$, of sub-diagonal product moments of order statistics from a sample of size n can all be calculated from the single and product moments of order statistics from samples of size less than n.

Joshi (1973) proved two interesting identities involving order statistics and applied them to obtain some combinatorial identities. Following a similar method, Joshi and Balakrishnan (1981b) used many recurrence relations and identities between order statistics in exhibiting different methods of deriving such combinatorial results. By this process, they gave alternate proofs to several identities given in Riordan (1968). Joshi's (1973) identities have been generalized by Balakrishnan and Malik (1985). Similar results for the joint distributions of two order statistics have been derived by Balakrishnan and Malik (1987b). These results also have their usefulness in establishing several interesting combinatorial identities. By making use of Khatri's (1962) integral representation for the joint probability mass function of two order statistics from an arbitrary discrete population, Balakrishnan (1986) showed that most of these results among moments of order statistics hold for the case of discrete distributions as well. Using these results then, Balakrishnan (1986) established the above described bounds of Joshi and Balakrishnan (1982) for any arbitrary discrete distribution. These bounds can be further improved for the case of symmetric distributions as mentioned above. These recurrence relations (ones for the product moments in particular) are highly useful as the computation of the product moments is very difficult and expensive too; see Barnett (1966). In addition, for some specific distributions like the normal, logistic, Cauchy, double exponential, exponential, Pareto, power function, etc., some more recurrence relations and identities can be derived for the moments of order statistics by making use of the functional relationship between the probability density function and the cumulative distribution function of the underlying population. In many of these cases, the recurrence relations are also complete in the sense that they will enable one to compute all the single and product moments of all order statistics for all sample sizes in a simple recursive manner.

In this review article, we update the reviews of Malik, Balakrishnan and Ahmed (1988) and Balakrishnan, Malik and Ahmed (1988). Bearing in mind the importance and usefulness of recurrence relations and identities for moments of order statistics, we list and analyze all these results. We present the recurrence relations and identities satisfied by the single moments of order statistics for any arbitrary distribution in Section 3, and the recurrence relations and identities satisfied by the product moments and covariances of order statistics in Section 4. We also mention the interrelationships between many of these results. In addition to all these results, several recurrence relations satisfied by the moments of order statistics from some specific continuous distributions are also available and these are presented in Sections 5–19. For the normal distribution, some identities satisfied by the covariances of order statistics have been used by Davis and Stephens (1978) in obtaining improved approximation for the variance-covariance matrix of normal order statistics. Joshi and Balakrishnan (1981a) have applied some such identities among the covariances of normal order statistics in obtaining an expression for the variance of standardized and studentized selection differential. Besides, for distributions like the logistic, exponential, double exponential, power function, Pareto, Lomax, log-logistic, Burr, mixture of two exponential, parabolic and skewed distributions and their truncated forms, these recurrence relations can be used effectively to compute all the single and the product moments of order statistics for all sample sizes. It must be remarked here that only simple algebraic operations are performed for the evaluation of both single and product moments when using these recurrence relations and therefore, the rounding errors can be kept negligible at least for small and moderately large sample sizes by performing the necessary calculations at a high precision on a computer. Many of the results mentioned above have been synthesized in the monograph by Arnold and Balakrishnan (1989); see also Balakrishnan (1988).

2. Notations

Let X be a continuous random variable having a cumulative distribution function (cdf) $F(x)$ and probability density function (pdf) $f(x)$. Let $X_1, X_2, \ldots, X_n$ be a random sample of size n from this distribution and $X_{1:n} \leq X_{2:n} \leq \cdots \leq X_{n:n}$ be the corresponding order statistics obtained by arranging the $X_i's$ in ascending order of magnitude. Then the pdf of $X_{r:n}$ $(1 \leq r \leq n)$ is given by [see David (1981, p. 9) and Arnold, Balakrishnan and Nagaraja (1992, p. 10)]

$$f_{r:n}(x) = \{B(r, n-r+1)\}^{-1}[F(x)]^{r-1}[1-F(x)]^{n-r}f(x), \quad -\infty < x < \infty,$$

(2.1)

and the joint density of $X_{r:n}$ and $X_{s:n}$ $(1 \leq r < s \leq n)$ is given by [see David (1981, p. 10) and Arnold, Balakrishnan and Nagaraja (1992, p. 16)]

$$f_{r,s:n}(x,y) = \{B(r, s-r, n-s+1)\}^{-1}[F(x)]^{r-1}[F(y) - F(x)]^{s-r-1} \\ \times [1 - F(y)]^{n-s} f(x) f(y), \quad -\infty < x < y < \infty, \quad (2.2)$$

where $B(a,b) = \Gamma(a)\Gamma(b)/\Gamma(a+b)$ is the complete beta function and $B(a,b,c) = \Gamma(a)\Gamma(b)\Gamma(c)/\Gamma(a+b+c)$, $(a,b,c > 0)$, is the generalized beta function.

We shall denote the single moment $E(X_{r:n}^k)$ by $\mu_{r:n}^{(k)}$, $(1 \leq r \leq n, k \geq 1)$, the product moment $E(X_{r:n}^j X_{s:n}^k)$ by $\mu_{r,s:n}^{(j,k)}$, $(1 \leq r < s \leq n)$, and the covariance between $X_{r:n}$ and $X_{s:n}$ by $\sigma_{r,s:n}$. Note that for $1 \leq r < s \leq n$, $\mu_{r,s:n} = \mu_{s,r:n}$ and $\sigma_{r,s:n} = \sigma_{s,r:n}$. For simplicity, we shall also use $\mu_{r:n}$ for $\mu_{r:n}^{(1)}$, $\mu_{r,r:n}$ for $\mu_{r:n}^{(2)}$, $\mu_{r,s:n}$ for $\mu_{r,s:n}^{(1,1)}$, and $\sigma_{r,r:n}$ for variance of $X_{r:n}$. We shall also assume that all these quantities exist.

Also, the distribution function of $X_{r:n}$ $(1 \leq r \leq n)$ is [see David (1981, p. 8) and Arnold, Balakrishnan and Nagaraja (1992, p. 12)]

$$F_{r:n}(x) = \sum_{i=r}^{n} \binom{n}{i} \{F(x)\}^i \{1 - F(x)\}^{n-i} \\ = I_{F(x)}(r, n-r+1), \quad -\infty < x < \infty, \quad (2.3)$$

where

$$I_p(a,b) = \{B(a,b)\}^{-1} \int_0^p u^{a-1}(1-u)^{b-1} \, du, \quad a, b > 0 \quad (2.4)$$

is the incomplete beta ratio.

3. Recurrence relations for single moments

With the density function of $X_{r:n}$ as given in Eq. (2.1), the single moment $\mu_{r:n}^{(k)}$ $(1 \leq r \leq n, k \geq 1)$ is given by

$$\mu_{r:n}^{(k)} = \{B(r, n-r+1)\}^{-1} \int_{-\infty}^{\infty} x^k [F(x)]^{r-1} [1 - F(x)]^{n-r} f(x) \, dx. \quad (3.1)$$

Then the single moments $\mu_{r:n}^{(k)}$ satisfy the following recurrence relations and identities.

RELATION 3.1. For any arbitrary distribution,

$$\sum_{r=1}^{n} \mu_{r:n}^{(k)} = nE(X^k) = n\mu_{1:1}^{(k)}. \quad (3.2)$$

For example, for $k = 1$ we have $\sum_{r=1}^{n} \mu_{r:n} = nE(X) = n\mu_{1:1}$, and for $k = 2$ we have $\sum_{r=1}^{n} \mu_{r:n}^{(2)} = nE(X^2) = n\mu_{1:1}^{(2)}$, as noted by Hoeffding (1953).

RELATION 3.2. For any arbitrary distribution,

$$r\mu_{r+1:n}^{(k)} + (n-r)\mu_{r:n}^{(k)} = n\mu_{r:n-1}^{(k)}, \quad 1 \le r \le n . \tag{3.3}$$

Relation 3.2 was derived by Cole (1951) and it just requires the value of the kth moment of any one order statistic in a sample of size n in order to compute k^{th} moment of the remaining $n-1$ order statistics, assuming that these moments in samples of size less than n are known. This relation has been extended to the case of discrete distributions by Melnick (1964); also see Abdel-Aty (1954) and David (1981). An alternative proof which covers mixtures of discrete and continuous cases has been given by Arnold (1977); also see Balakrishnan and Malik (1986a).

RELATION 3.3. For n even, say $n = 2m$, and $k \ge 1$,

$$\frac{1}{2}\left(\mu_{m+1:2m}^{(k)} + \mu_{m:2m}^{(k)}\right) = \mu_{m:2m-1}^{(k)} . \tag{3.4}$$

This relation follows directly from Relation 3.2. Note that for $k=1$, in particular, this relation simply implies that the expected value of the median in a sample of size $n = 2m$ (even) is equal to the expected value of the median in a sample of size $n - 1 = 2m - 1$ (odd).

RELATION 3.4. For any arbitrary distribution symmetric about zero,

$$\mu_{r:n}^{(k)} = (-1)^k \mu_{n-r+1:n}^{(k)}, \quad r = 1, 2, \ldots, [n/2] . \tag{3.5}$$

This result was first derived by Jones (1948) for the special case of the standard normal distribution.

RELATION 3.5. For any arbitrary distribution symmetric about zero, and n even, say $n = 2m$,

$$\mu_{m:n-1}^{(k)} = \begin{cases} \mu_{m:n}^{(k)}, & k \text{ even} \\ 0, & k \text{ odd} \end{cases} . \tag{3.6}$$

This follows immediately if we use Relation 3.4 in Relation 3.3.

RELATION 3.6. For $m = 1, 2, \ldots, n-r$ and $k \ge 1$,

$$(n-r)^{(m)}\mu_{r:n}^{(k)} = \sum_{i=0}^{m}(-1)^{(i)}(n)^{(m-i)}\binom{m}{i}\mu_{r+i:n-m+i}^{(k)} , \tag{3.7}$$

where $(n)^{(m)}$ denotes $n(n-1)(n-2)\cdots(n-m+1)$.

RELATION 3.7. For $1 \le r \le n-1$ and $k \ge 1$,

$$\mu_{r:n}^{(k)} = \sum_{j=r}^{n}(-1)^{j-r}\binom{j-1}{r-1}\binom{n}{j}\mu_{j:j}^{(k)} . \tag{3.8}$$

This relation follows directly from Relation 3.6 if we set $m = n - r$. Note that this relation expresses the kth moment of the rth order statistic in a sample of size n in terms of the kth moment of the largest order statistic in samples of size $r, r+1, \ldots, n$. In fact, this relation is the solution for Relation 3.2 in terms of the largest order statistic in samples up to size n. Refer to Balakrishnan and Malik (1986a) for additional comments on the use of Relation 3.7.

RELATION 3.8. For $m = 1, 2, \ldots, r-1$ and $k \geq 1$,

$$(r-1)^{(m)} \mu_{r:n}^{(k)} = \sum_{j=n-m}^{n} (m-n+j)^{(n-m+j)} (n)^{(n-j)} \binom{m}{n-j} \mu_{r-m:j}^{(k)}. \tag{3.9}$$

RELATION 3.9. For $2 \leq r \leq n$,

$$\mu_{r:n}^{(k)} = \sum_{j=n-r+1}^{n} (-1)^{j-n+r-1} \binom{j-1}{n-r} \binom{n}{j} \mu_{1:j}^{(k)}. \tag{3.10}$$

This relation follows directly from Relation 3.8 if we set $m = r - 1$. This relation expresses the kth moment of the rth order statistic in a sample of size n in terms of the kth moment of the smallest order statistic in samples of size $n-r+1, n-r+2, \ldots, n$. As a matter of fact, this relation is the solution for Relation 3.2 in terms of the smallest order statistic in samples up to size n. One has to be careful, however, while using Relations 3.7 and 3.9 as increasing values of n result in large combinatorial terms and hence in large error. A detailed discussion of this has been made by Srikantan (1962); also see Balakrishnan and Malik (1986a).

Stating these results also in terms of distribution functions of order statistics, Lange (1996) has recently illustrated the usefulness of Relations 3.7 and 3.9 through two applications to waiting time problems in urn sampling.

RELATION 3.10. For $1 \leq r \leq m < n$ and $k \geq 1$,

$$\binom{n}{m} \mu_{r:m}^{(k)} = \sum_{i=0}^{n-m} \binom{r+i-1}{i} \binom{n-r-i}{m-r} \mu_{r+i:n}^{(k)}. \tag{3.11}$$

This result was derived by Sillitto (1964). Note that this relation expresses the kth moment of the rth order statistic in a sample of size m in terms of the kth moment of $r, r+1, \ldots, n-m+r$ order statistics from a sample of size n larger than m.

RELATION 3.11. For $j + \ell \leq n - 1$ and $k \geq 1$,

$$\sum_{i=j+1}^{n-\ell} (i-1)^{(j)} (n-i)^{(\ell)} \mu_{i:n}^{(k)} = j! \ell! \binom{n}{j+\ell+1} \mu_{j+1:j+\ell+1}^{(k)}. \tag{3.12}$$

This result, due to Downton (1966), follows immediately from Relation 3.10 if we set $r = j + 1$ and $m = j + \ell + 1$.

RELATION 3.12. For $n \geq 2$ and $k \geq 1$,

$$\sum_{r=1}^{n} \frac{1}{r} \mu_{r:n}^{(k)} = \sum_{r=1}^{n} \frac{1}{r} \mu_{1:r}^{(k)} . \qquad (3.13)$$

RELATION 3.13. For $n \geq 2$ and $k \geq 1$,

$$\sum_{r=1}^{n} \frac{1}{(n-r+1)} \mu_{r:n}^{(k)} = \sum_{r=1}^{n} \frac{1}{r} \mu_{r:r}^{(k)} . \qquad (3.14)$$

Relations 3.12 and 3.13 were derived by Joshi (1973). These two results are quite useful for checking the computations of the single moments of order statistics and also in establishing some interesting combinatorial identities. These results have been generalized by Balakrishnan and Malik (1985) and their general results are given in the following four relations, where we denote

$$C_i = \begin{cases} 1, & \text{for } i=1 \\ (n+1)(n+2)\cdots(n+i-1), & \text{for } i=2,3,\ldots. \end{cases}$$

RELATION 3.14. For $i \geq 1$ and $k \geq 1$,

$$\sum_{r=1}^{n} \mu_{r:n}^{(k)} / \{r(r+1)\cdots(r+i-1)\} = \frac{1}{C_i} \sum_{r=1}^{n} \binom{r+i-2}{i-1} \mu_{1:r}^{(k)}/r . \qquad (3.15)$$

Note that for $i = 1$ this result reduces to Relation 3.12.

RELATION 3.15. For $i \geq 1$ and $k \geq 1$,

$$\sum_{r=1}^{n} \mu_{r:n}^{(k)} / \{(n-r+1)(n-r+2)\cdots(n-r+i)\}$$
$$= \frac{1}{C_i} \sum_{r=1}^{n} \binom{r+i-2}{i-1} \mu_{r:r}^{(k)}/r . \qquad (3.16)$$

Note that for $i = 1$, this result reduces to Relation 3.13.

RELATION 3.16. For $i \geq 1$ and $k \geq 1$,

$$\sum_{r=1}^{n} \mu_{r:n}^{(k)} / \{r(r+1)(r+2)\cdots(r+i-1)(n-r+1)$$
$$\times (n-r+2)\cdots(n-r+i)\} \qquad (3.17)$$
$$= \frac{1}{C_{2i}} \sum_{r=1}^{n} \binom{r+2i-2}{i-1} \left(\mu_{1:r}^{(k)} + \mu_{r:r}^{(k)} \right)/r .$$

RELATION 3.17. For $i, j \geq 1$ and $k \geq 1$,

$$\sum_{r=1}^{n} \mu_{r:n}^{(k)} / \{r(r+1)(r+2) \cdots (r+i-1)(n-r+1)$$

$$\times (n-r+2) \cdots (n-r+j)\} \qquad (3.18)$$

$$= \frac{1}{C_{i+j}} \sum_{r=1}^{n} \frac{1}{r} \left\{ \binom{r+i+j-2}{i-1} \mu_{1:r}^{(k)} + \binom{r+i+j-2}{j-1} \mu_{r:r}^{(k)} \right\}.$$

Note that for $i = j$, this result reduces to Relation 3.16.

RELATION 3.18. For an arbitrary distribution,

$$\frac{1}{(n+1) \cdots (n+r)} \sum_{i=r+1}^{n+r} \mu_{i:n+r} \sum_{j=r+1}^{i} \frac{1}{n+r+1-j} \binom{r-j+i-1}{r-1}$$

$$= \sum_{i=r+1}^{n+r} \frac{1}{i(i-1) \cdots (i-r)} \mu_{i:i}, \quad r \geq 1, \qquad (3.19)$$

$$\frac{1}{(n+1) \cdots (n+r)} \sum_{i=1}^{n} \mu_{i:n+r} \sum_{j=i}^{n} \frac{1}{j} \binom{j+r-i-1}{r-1}$$

$$= \sum_{i=r+1}^{n+r} \frac{1}{i(i-1) \cdots (i-r)} \mu_{1:i}, \quad r \geq 1. \qquad (3.20)$$

RELATION 3.19. For an arbitrary distribution and $m = 1, 2, \ldots$,

$$\frac{1}{(n+1) \cdots (n+r)} \sum_{i=r+1}^{n+r} \mu_{i:n+r} \sum_{j=r+1}^{i} \frac{1}{(n+r+1-j) \cdots (n+r+m-j)}$$

$$\times \binom{r+i-j-1}{r-1} = \frac{1}{c_m} \sum_{i=r+1}^{n+r} \binom{i+m-r-2}{m-1} \frac{1}{i(i-1) \cdots (i-r)} \mu_{i:i},$$

$$\text{for } r = 1, 2, \ldots, \qquad (3.21)$$

$$\frac{1}{(n+1) \cdots (n+r)} \sum_{i=1}^{n} \mu_{i:n+r} \sum_{j=i}^{n} \frac{1}{j(j+1) \cdots (j+m-1)} \binom{j+r-i-1}{r-1}$$

$$= \frac{1}{c_m} \sum_{i=r+1}^{n+r} \binom{i+m-r-2}{m-1} \frac{1}{i(i-1) \cdots (i-r)} \mu_{1:i}, \text{ for } r = 1, 2, \ldots, \qquad (3.22)$$

where $c_1 = 1$ and $c_m = (n+1) \cdots (n+m-1)$ for $m = 1, 2, 3, \ldots$.

RELATION 3.20. For an arbitrary distribution,

$$\sum_{j=1}^{n} \binom{n}{j} \frac{(k+j-1)!(n+k-j)!}{(n+2k)!} \mu_{k+j:n+2k} \qquad (3.23)$$

$$= \sum_{j=1}^{n} \frac{(k)!(j+k-1)!}{(j+2k)!} \mu_{k+1:j+2k}, \quad k \geq 1,$$

$$\sum_{j=1}^{n} \binom{n}{j-1} \frac{(k+j-1)!(n+k-j)!}{(n+2k)!} \mu_{k+j:n+2k} \qquad (3.24)$$

$$= \sum_{j=1}^{n} \frac{(k)!(j+k-1)!}{(j+2k)!} \mu_{j+k:j+2k}, \quad k \geq 1.$$

Relations 3.18, 3.19 and 3.20 were derived by Joshi and Shubha (1991). Note that these relations are generalizations of the results given by Joshi (1973) and Balakrishnan and Malik (1985) and presented in Relations 3.12–3.17.

Now let us denote $\chi_{n:r} = \mu_{r+1:n} - \mu_{r:n}$ $(r = 1, 2, \ldots, n-1)$ and $w_n = \mu_{n:n} - \mu_{1:n}$, $n \geq 2$. Note that $\chi_{r:n}$ is the expected value of the difference between $(r+1)^{\text{th}}$ and r^{th} order statistics in a sample of size n, while w_n is the expected value of the range $w_n = \mu_{n:n} - \mu_{1:n}$ in a sample of size n. We present here some relations satisfied by these quantities.

RELATION 3.21. For $n \geq 3$,

$$nw_{n-1} - (n-1)w_n = \mu_{n-1:n} - \mu_{2:n} . \qquad (3.25)$$

This is easily derived from Relation 3.2.

RELATION 3.22. For $n \geq 3$,

$$w_n = w_{n-1} = \frac{1}{n}(\chi_{n:1} + \chi_{n:n-1}) . \qquad (3.26)$$

This is obtained directly from Relation 3.21 by noting that

$$\mu_{n-1:n} - \mu_{2:n} - w_n = -(\chi_{n:1} + \chi_{n:n}) .$$

These identities were proved by Sillitto (1951) and Cadwell (1953).

RELATION 3.23. For $1 \leq r \leq n - 1$,

$$\binom{n}{r} \sum_{i=0}^{r} (-1)^{i+1} \binom{r}{i} w_{n-r+i} = \chi_{n:r} + \chi_{n:n-r} . \qquad (3.27)$$

In particular, if we set $r = 1$ in the above result, we deduce Relation 3.22.

RELATION 3.24. For $n \geq 3$,

$$\{1 - (-1)^n\} w_n = w_{n-1} + \sum_{i=1}^{n-2} (-1)^{i+1} \binom{n-1}{i} w_{i+1} ; \qquad (3.28)$$

for odd values of n, we have

$$2w_n = w_{n-1} + \sum_{i=1}^{n-2} (-1)^{i+1} \binom{n-1}{i} w_{n-i} . \qquad (3.29)$$

This result follows from Relation 3.23 by setting $r = n - 1$ and then making use of Relation 3.22.

RELATION 3.25. For $n \geq 3$,

$$\{1 - (-1)^n\} w_n = \sum_{j=2}^{n-1} (-1)^j \binom{n}{j} w_j ; \qquad (3.30)$$

for odd values of n, we have

$$2w_n = \sum_{j=2}^{n-1} (-1)^j \binom{n}{j} w_j . \qquad (3.31)$$

These relations are due to Romanovsky (1933) and Sillitto (1951).

RELATION 3.26. For any arbitrary distribution,

$$n\chi_{n-1:r-1} - (n - r + 1)\chi_{n:r-1} = r\chi_{n:r}, \quad 2 \leq r \leq n - 1 . \qquad (3.32)$$

This relation was derived by Sillitto (1951). Note that this relation expresses the expectation of the difference between the $(r + 1)^{\text{th}}$ and r^{th} order statistics in a sample of size n in terms of the expected values of the differences between the rth and $(r - 1)^{\text{th}}$ order statistics in samples of size n and $n - 1$, respectively. The distribution of these differences was first discussed by Galton (1902) and was investigated further by Pearson (1902). Relation 3.26 follows easily from Relation 3.2.

RELATION 3.27. For $v \leq r - 1$,

$$\chi_{n:r} = \frac{(n)^{(v)}}{(r)^{(v)}} \sum_{i=0}^{v} (-1)^i \binom{v}{i} \frac{(n-r+i)^{(i)}}{(n-v+i)^{(i)}} \chi_{n-v+i:r-v} , \qquad (3.33)$$

where, as before, $(n)^{(v)} = n(n-1)(n-2) \cdots (n-v+1)$. This relation was established by Sillitto (1951) and it follows directly by a repeated application of Relation 3.26. Note that it expresses $\chi_{n:r}$ in terms of the χ's in samples of size less than or equal to n and of order less than or equal to r. In particular, setting $v = 1$ in the above result, we deduce Relation 3.26.

More generally, denoting the ith quasi-range $X_{n-i:n} - X_{i+1:n}$, $(i = 0, 1, 2, \ldots, [(n-2)/2])$, by $W_{i,n}$ and its expected value $\mathrm{E}(W_{i,n}) = \mu_{n-i:n} - \mu_{i+1:n}$ by $w_{i,n}$, we immediately have the following relations. Note that $W_{0,n}$ will be the sample range and $w_{0,n}$ will be its expected value according to this notation.

RELATION 3.28. For any arbitrary distribution symmetric about zero,

$$w_{i,n} = 2\mu_{n-i:n}, \quad i = 0, 1, 2, \ldots, [(n-2)/2] \;. \tag{3.34}$$

This follows directly from Relation 3.4.

RELATION 3.29. For any arbitrary distribution,

$$iw_{i,n} + (n-i)w_{i-1,n} = nw_{i-1,n-1}, \quad i = 0, 1, 2, \ldots, [(n-2)/2] \;. \tag{3.35}$$

These results were obtained by Govindarajulu (1963b). Note that Relation 3.29, after dividing both sides by n, can be used for working downwards in numerical evaluation of the expected values of the sample quasi-ranges, without serious accumulation of rounding errors. The above result follows easily from Relation 3.2.

4. Recurrence relations for product moments

With the joint density of $X_{r:n}$ and $X_{s:n}$ as given in (2.2), the product moment $\mu_{r,s:n}$ $(1 \le r < s \le n)$ is given by

$$\mu_{r,s:n} = \{B(r, s-r, n-r+1)\}^{-1} \int_{-\infty}^{\infty} \int_{x}^{\infty} xy[F(x)]^{r-1}[F(y) - F(x)]^{s-r-1}$$

$$\times [1 - F(y)]^{n-s} f(x) f(y) \, dy \, dx \;. \tag{4.1}$$

Then, the product moments $\mu_{r,s:n}$ satisfy the following recurrence relations and identities.

RELATION 4.1. For any arbitrary distribution,

$$\sum_{r=1}^{n} \sum_{s=1}^{n} \mu_{r,s:n} = n\mathrm{E}(X^2) + n(n-1)\{\mathrm{E}(X)\}^2 = n(\sigma^2 + n\mu^2) \;. \tag{4.2}$$

RELATION 4.2. For any arbitrary distribution,

$$\sum_{r=1}^{n-1}\sum_{s=r+1}^{n}\mu_{r,s:n} = n(n-1)\{E(X)\}^2/2 = \binom{n}{2}\mu^2 . \tag{4.3}$$

This result follows immediately upon using Relation 3.1 in Relation 4.1.

RELATION 4.3. For any arbitrary distribution,

$$\sum_{r=1}^{n}\sum_{s=1}^{n}\sigma_{r,s:n} = n\mathrm{Var}(X) = n\sigma^2 . \tag{4.4}$$

This result follows immediately upon using Relation 3.1 in Relation 4.1.

RELATION 4.4. For any arbitrary distribution symmetric about zero,

$$\mu_{r,s:n} = \mu_{n-s+1,n-r+1:n} . \tag{4.5}$$

RELATION 4.5. For any arbitrary distribution symmetric about zero,

$$\sigma_{r,s:n} = \sigma_{n-s+1,n-r+1:n} . \tag{4.6}$$

This result follows immediately upon using Relation 3.4 in Relation 4.4.

RELATION 4.6. For any arbitrary distribution,

$$(r-1)\mu_{r,s:n} + (s-r)\mu_{r-1,s:n} + (n-s+1)\mu_{r-1,s-1:n}$$
$$= n\mu_{r-1,s-1:n-1}, \quad 2 \leq r < s \leq n . \tag{4.7}$$

This result was established by Govindarajulu (1963b). Note that, due to this relation, it will be enough if we compute $n-1$ suitably chosen product moments, e.g. the immediate upper-diagonal product moments $\mu_{r,r+1:n}$ $(1 \leq r \leq n-1)$ in order to determine all the product moments since the remaining product moments, viz. $\mu_{r,s:n}$ $(1 \leq r < s \leq n, s-r \geq 2)$, can all be obtained by making use of Relation 4.6. Also refer to Balakrishnan and Malik (1986a) for additional comments on the use of Relation 4.6.

Using Relations 4.6 and 3.2, Balakrishnan (1989) has established the following relation satisfied by the covariances of order statistics.

RELATION 4.7. For $2 \leq r < s \leq n$,

$$(n-r)\sigma_{r,s:n} + (s-r)\sigma_{r-1,s:n} + (n-s+1)\sigma_{r-1,s-1:n}$$
$$= n\left\{\sigma_{r-1,s-1:n-1} + (\mu_{r-1:n-1} - \mu_{r-1:n})(\mu_{s-1:n-1} - \mu_{s:n})\right\} . \tag{4.8}$$

RELATION 4.8. For any arbitrary distribution,

$$\sum_{i=1}^{n-1}\sum_{j=i+1}^{n} \mu_{i,j:n} = \tfrac{1}{2}n(n-1)\mu_{1,2:2} \; . \tag{4.9}$$

This result is due to Govindarajulu (1963b).

RELATION 4.9. For any arbitrary distribution and even n,

$$\mu_{1,n:n} = \sum_{i=1}^{(n-2)/2} (-1)^{i-1}\binom{n}{i}\mu_{i:i}\mu_{n-i:n-i} + \frac{1}{2}(-1)^{(n-2)/2}\binom{n}{n/2}\mu_{n/2:n/2}^2 \; . \tag{4.10}$$

This result, due to Govindarajulu (1963b), has been generalized by Joshi (1971) as follows.

RELATION 4.10. For any arbitrary distribution and $1 \leq k \leq n-1$,

$$B(1,n-k,k)\mu_{k,n:n} + \sum_{i=0}^{k-1}(-1)^{n-i}\binom{k-1}{i}B(1,n-k,k-i)\mu_{1,n-k+1:n-i}$$

$$= \sum_{i=1}^{n-k}(-1)^{n-k-i}\binom{n-k-1}{i-1}\frac{1}{i(n-i)}\mu_{i:i}\mu_{n-i:n-i} \; . \tag{4.11}$$

It must be noted that Relation 4.10 contains both the product moments $\mu_{1,n-k+1:n}$ and $\mu_{k,n:n}$. Hence, for any arbitrary parent distribution, Relation 4.10 is useful only for $k=1$ and in this case it simply becomes Relation 4.9.

RELATION 4.11. For any arbitrary distribution symmetric about zero and $1 \leq k \leq n-1$,

$$\{1+(-1)^n\}B(1,n-k,k)\mu_{1,n-k+1:n}$$

$$= \sum_{i=1}^{k-1}(-1)^{n-i+1}\binom{k-1}{i}B(1,n-k,k-i)\mu_{1,n-k+1:n-i}$$

$$+ \sum_{i=2}^{n-k}(-1)^{n-k-i}\binom{n-k-1}{i-1}\frac{1}{i(n-i)}\mu_{i:i}\mu_{n-i:n-i} \; . \tag{4.12}$$

This result follows easily from Relation 4.10 upon using Relation 4.4. From Relation 4.11, it follows that if the parent distribution is symmetric about zero, then for even values of n all the product moments $\mu_{1,s:n}$ ($s = 2, 3, \ldots, n$) can be obtained provided that all the first and product moments in samples of sizes less than n are available.

RELATION 4.12. For any arbitrary distribution symmetric about zero and n even,

$$2\mu_{1,2:n} = \sum_{i=1}^{n-2}(-1)^{i-1}\binom{n}{i}\mu_{1,2:n-i} . \tag{4.13}$$

This result follows immediately from Relation 4.11 by setting $k = n - 1$ and is due to Joshi (1971); see also Govindarajulu (1963b) who proved this result for the normal case.

RELATION 4.13. For any arbitrary distribution symmetric about zero,

$$B(1, 2m - 1, 1)\mu_{1,2m:2m} = \sum_{i=2}^{2m-1}(-1)^{i-1}\binom{2m-2}{i-1} \\ \times \frac{1}{i(2m+1-i)}\mu_{i:i}\mu_{2m+1-i:2m+1-i} , \tag{4.14}$$

and

$$2B(1, 2m - 1, 1)\mu_{1,2m:2m} = \sum_{i=2}^{2m-2}(-1)^{i-1}\binom{2m-2}{i-1}\frac{1}{i(2m-i)}\mu_{i:i}\mu_{2m-i:2m-i} . \tag{4.15}$$

These results have been established by Joshi (1971). The first result is obtained by setting $k = 2$ and $n = 2m + 1$ in Relation 4.11 and noting that the LHS becomes zero. The second result follows in a straightforward way by setting $k = 1$ and $n = 2m$ in Relation 4.11.

In addition, making use of Relations 3.2, 3.5, 4.4, 4.6 and 4.11, Joshi (1971) also arrived at the following result.

THEOREM 4.1. In order to find the first, second and product moments of order statistics in a sample size n drawn from an arbitrary continuous distribution symmetric about zero, given these moments for all sample sizes less than n, one has to evaluate at most one single integral if n is even, and one single integral and $(n-1)/2$ double integrals if n is odd.

RELATION 4.14. For $1 \leq r < s \leq n$,

$$\mu_{r,s:n} + \sum_{i=0}^{r-1}\sum_{j=0}^{n-s}(-1)^{n-i-j}\binom{n}{j}\binom{n-j}{i}\mu_{n-s-j+1,n-r-j+1:n-i-j} \\ = \sum_{i=1}^{s-r}(-1)^{s-r-i}\binom{n}{s-i}\binom{s-i-1}{r-1}\mu_{s-i:s-i}\mu_{i:n-s+i} . \tag{4.16}$$

This result, due to Joshi and Balakrishnan (1982), is a generalization of Joshi's (1971) result. Note that if we set $s = n$ in the above result, we deduce Relation 4.10.

RELATION 4.15. For $1 \leq r \leq n-1$,

$$\mu_{r,r+1:n} + (-1)^n \mu_{n-r,n-r+1:n}$$
$$= \sum_{i=0}^{r-1} \sum_{j=1}^{n-r-1} (-1)^{n-i-j+1} \binom{n}{j} \binom{n-j}{i} \mu_{n-r-j,n-r-j+1:n-i-j} \quad (4.17)$$
$$+ \sum_{i=1}^{r-1} (-1)^{n-i+1} \binom{n}{i} \mu_{n-r:n-r+1:n-i} + \binom{n}{r} \mu_{r:r} \mu_{1:n-r} .$$

This result follows simply by setting $s = r+1$ in Relation 4.14. Note that when n is odd, we only need to calculate $(n-1)/2$ product moments $\mu_{r,r+1:n}$ ($1 \leq r \leq (n-1)/2$) since the remaining moments $\mu_{r,r+1:n}$ $\{(n+1)/2 \leq r \leq n-1\}$ can all be obtained by using Relation 4.15. Similarly, when n is even, say $n = 2m$, we only need to calculate $(n-2)/2 = (m-1)$ product moments $\mu_{r,r+1:n}$ ($1 \leq r \leq m-1$) since the remaining moments $\mu_{r,r+1:n}$ ($m \leq r \leq n-1$) can all be obtained by using Relation 4.15.

RELATION 4.16. For n even (say $n = 2m$) and $m \geq 1$,

$$2\mu_{m,m+1:2m} = \sum_{i=0}^{m-1} \sum_{j=1}^{m-1} (-1)^{i+j-1} \binom{2m}{j} \binom{2m-j}{i} \mu_{m-j,m-j+1:2m-i-j}$$
$$+ \sum_{i=1}^{m-1} (-1)^{i-1} \binom{2m}{i} \mu_{m,m+1:2m-i} + \binom{2m}{m} \mu_{m:m} \mu_{1:m} . \quad (4.18)$$

This follows directly from Relation 4.15 by setting $n = 2m$ and $r = m$. Note that this result will enable us to compute the moment $\mu_{m,m+1:2m}$ from the single and product moments in samples of sizes less than $2m$.

RELATION 4.17. For $r = 1, 2, \ldots, [n/2]$,

$$\{1 + (-1)^n\} \mu_{r,n-r+1:n} = \sum_{j=1}^{r-1} \sum_{k=0}^{r-1} (-1)^{n-j-k+1} \binom{n}{j} \binom{n-j}{k} \mu_{r-j,n-r-j+1:n-j-k}$$
$$+ \sum_{k=1}^{r-1} (-1)^{n-k+1} \binom{n}{k} \mu_{r,n-r+1:n-k} \quad (4.19)$$
$$+ \sum_{i=1}^{n-2r+1} (-1)^{n-i+1} \binom{n}{r+i-1} \binom{n-r-i}{r-1}$$
$$\times \mu_{n-r-i+1:n-r-i+1} \mu_{i:r+i-1} .$$

Relation 4.17 shows that if n is even, then the product moments $\mu_{r,n-r+1:n}$ for $1 \leq r \leq [n/2]$ can all be obtained from the moments in samples of sizes $n-1$ and less. For even values of n, if we set $r = 1$ in Relation 4.17 we deduce Relation 4.9.

These results have been established by Joshi and Balakrishnan (1982). Making use of these relations along with Relation 3.2, they also arrived at the following result.

THEOREM 4.2. *In order to find the first, second and product moments of order statistics in a sample of size n drawn from an arbitrary continuous distribution, given these moments in samples of sizes $n-1$ and less, one has to evaluate at most two single integrals and $(n-2)/2$ double integrals if n is even, and two single integrals and $(n-1)/2$ double integrals if n is odd.*

RELATION 4.18. For an arbitrary distribution symmetric about zero and n even,

$$2\mu_{r,s:n} = 2\mu_{n-s+1,n-r+1:n}$$

$$= \sum_{i=1}^{s-r}(-1)^{s-r-i+1}\binom{s-i-1}{r-1}\binom{n}{s-i}\mu_{1:s-i}\mu_{i:n-s+i}$$

$$+ \sum_{i=1}^{r-1}(-1)^{i-1}\binom{n}{i}\mu_{n-s+1,n-r+1:n-i} \qquad (4.20)$$

$$+ \sum_{j=1}^{n-s}\sum_{i=0}^{r-1}(-1)^{i+j-1}\binom{n}{j}\binom{n-j}{i}\mu_{n-s-j+1,n-r-j+1:n-i-j} \; .$$

Relation 4.18, due to Joshi and Balakrishnan (1982), gives an explicit expression for the product moments $\mu_{r,s:n}$ for even values of n, in terms of the single and product moments of order statistics in samples of sizes $(n-1)$ and less. Hence, for distributions symmetric about zero and even values of n, all the product moments $\mu_{r,s:n}$ can be evaluated from moments in samples of sizes less than n and note that this is in accordance with Theorem 4.1. In addition, setting $r=1$ and $s=n-k+1$ in Relation 4.18, we obtain Relation 4.11.

Thomas and Samuel (1996) have obtained the following comparatively simpler alternate forms for Relation 4.15 and Relation 4.17 when n is even.

RELATION 4.19. For $1 \leq r \leq n-1$,

$$\mu_{r,r+1:n} + (-1)^n \mu_{n-r,n-r+1:n} = \sum_{j=1}^{n-r-1}(-1)^{j+1}\binom{n}{j}\mu_{r,r+1:n-j}$$

$$+ \phi_r \sum_{j=1}^{r-1}(-1)^{n-j+1}\binom{n}{j}\mu_{n-r:n-r+1:n-j} \qquad (4.21)$$

$$+ (-1)^{n-r+1}\binom{n}{r}\mu_{r:r}\mu_{n-r:n-r} \; ,$$

where

$$\phi_r = \begin{cases} 0 & \text{if } r = 1, \\ 1 & \text{otherwise}. \end{cases} \quad (4.22)$$

RELATION 4.20. For $r = 1, 2, \ldots, [n/2]$,

$$\mu_{r,n-r+1:n} = \phi_r \sum_{j=1}^{r-1}(-1)^{j+1}\binom{n}{j}\mu_{r,n-r+1:n-j} + \frac{1}{2}\sum_{j=0}^{n-2r}(-1)^{r+j-1}\binom{r+j-1}{r-1}$$
$$\times \binom{n-r-j-1}{r-1}\binom{n}{r+j}\mu_{r+j,r+j}\mu_{n-r-j:n-r-j},$$
$$(4.23)$$

where ϕ_r is defined as in (4.22).

RELATION 4.21. For an arbitrary continuous distribution and for $1 \leq r \leq n-2$,

$$(n-i)(n-i-1)\sum_{j=0}^{i}\frac{\binom{i}{j}}{\binom{n-2}{j}}\sum_{k=0}^{j}\mu_{n-j-1,n-k:n}k = n(n-1)\mu_{n-i-1,n-i:n-i}.$$
$$(4.24)$$

This result for the case $i = 1$ is a special case of Relation 4.6 for $r = n-1$ and $s = n$. Note that if we set $i = n-2$ in the above result, we deduce Relation 4.8, which is the same as Relation 4.2 since $\mu_{1,2:2} = \mu_{1:1}^2 = \mu^2$.

RELATION 4.22. For an arbitrary distribution,

$$\sum_{r=1}^{n-1}\mu_{r,r+1:n} + \sum_{j=2}^{n}\binom{n}{j}\mu_{1,j:j} = \sum_{j=1}^{n-1}\binom{n}{j}\mu_{j:j}\mu_{1:n-j}. \quad (4.25)$$

RELATION 4.23. For $1 \leq r \leq n-1$ and $1 \leq k \leq n-r$,

$$\sum_{s=r+1}^{n-k+1}\binom{n-s}{k-1}\mu_{r,s:n} + \sum_{i=1}^{r}\sum_{s=r+1}^{r+k}\binom{s-i-1}{s-r-1}\binom{n-s}{n-k-r}\mu_{i,s:n} \quad (4.26)$$
$$= \binom{n}{k}\mu_{1:k}\mu_{r:n-k}.$$

RELATION 4.24. For $1 \leq k \leq n-1$,

$$\sum_{s=2}^{n-k+1}\binom{n-s}{k-1}\mu_{1,s:n} + \sum_{s=2}^{k+1}\binom{n-s}{n-k-1}\mu_{1,s:n} = \binom{n}{k}\mu_{1:k}\mu_{1:n-k}. \quad (4.27)$$

The above results are due to Joshi and Balakrishnan (1982). Relation 4.24 follows directly from Relation 4.23 if we set $r = 1$. Note that in Relation 4.24, the equation for k is same as the equation for $n - k$, and so there are only $[n/2]$ distinct equations. Also note that Relation 4.22 contains $\mu_{1,s:n}$ ($2 \leq s \leq n$) and first order moments only. Thus, for even values of n, knowledge of $(n-2)/2$ of these, e.g., $\mu_{1,2:n}, \mu_{1,3:n}, \ldots, \mu_{1,n/2:n}$, is enough to calculate all the product moments provided the first moment in samples of sizes $n - 1$ and less are known. Similarly, for odd values of n, it is sufficient to know $(n-1)/2$ moments, e.g., $\mu_{1,2:n}, \mu_{1,3:n}, \ldots, \mu_{1,(n+1)/2:n}$. Note that these numbers are precisely the same as the numbers given in Theorem 4.2 for the double integrals to be evaluated for the calculation of all the product moments. This is quite expected since the moments $\mu_{1,s:n}$ ($2 \leq s \leq n$) along with Relation 4.6 are also sufficient for the evaluation of all the product moments.

RELATION 4.25. For $1 \leq r \leq n - 1$,

$$\sum_{s=r+1}^{n} \mu_{r,s:n} + \sum_{i=1}^{r} \mu_{i,r+1:n} = n\mu_{1:1}\mu_{r:n-1} . \tag{4.28}$$

This result follows immediately by setting $k = 1$ in Relation 4.22.

David and Balakrishnan (1996) have recently given a new interpretation for Relation 4.25 and applied it to derive a convenient computational formula for the variance of a lightly trimmed mean. Balakrishnan and David (1998) have generalized this result and have derived a convenient computation formula for the variance of a lightly trimmed mean when multiple outliers are possibly present in the sample.

RELATION 4.26. For $1 \leq r \leq n - 1$,

$$\sum_{s=r+1}^{n} \sigma_{r,s:n} + \sum_{i=1}^{r} \sigma_{i,r+1:n} = \left(r\mu_{1:1} - \sum_{i=1}^{r} \mu_{i:n} \right) (\mu_{r+1:n} - \mu_{r:n}) . \tag{4.29}$$

Relations 4.25 and 4.26, established by Joshi and Balakrishnan (1982), give extremely simple and useful results for checking the computations of product moments and covariances of order statistics in a random sample of size n. In particular, setting $r = 1$ and $r = n - 1$ in Relation 4.26, we derive the relations

$$2\sigma_{1,2:n} + \sum_{s=3}^{n} \sigma_{1,s:n} = (\mu_{1:1} - \mu_{1:n})(\mu_{2:n} - \mu_{1:n}) \tag{4.30}$$

and

$$2\sigma_{n-1,n:n} + \sum_{i=1}^{n-2} \sigma_{i,n:n} = (\mu_{n:n} - \mu_{1:1})(\mu_{n:n} - \mu_{n-1:n}) , \tag{4.31}$$

respectively.

RELATION 4.27. For an arbitrary distribution symmetric about zero and for $1 \leq r \leq n-1$,

$$\sum_{s=r+1}^{n} \mu_{r,s:n} + \sum_{i=1}^{r} \mu_{i,r+1:n} = 0 \ . \tag{4.32}$$

This result follows directly from Relation 4.25 upon using the fact that $\mu_{1:1} = 0$ in this case.

RELATION 4.28. For an arbitrary distribution symmetric about zero and even n, say $n = 2m$,

$$\sum_{i=1}^{m} \mu_{i,m+1:2m} = 0 \ . \tag{4.33}$$

This result follows from Relation 4.25 by setting $n = 2m$ and $r = m$, and then using Relation 4.4.

RELATION 4.29. For an arbitrary distribution and for $1 \leq i \leq n-1$,

$$\begin{aligned}\sum_{r=1}^{n-i} \mu_{r,r+i:n} &= \sum_{r=1}^{n-i} \sum_{j=1}^{i} (-1)^{j-1} \binom{n}{r+i-j} \\ &\quad \times \binom{n-r-i+j-1}{j-1} \mu_{1:n-r-i+j} \mu_{r:r+i-j} \\ &\quad + (-1)^{i} \sum_{r=1}^{n-i} \binom{r+i-2}{i-1} \binom{n}{r+i} \mu_{1,r+i:r+i} \ . \end{aligned} \tag{4.34}$$

This result has been derived by Balakrishnan (1982).

RELATION 4.30. For $n \geq 2$,

$$\sum_{r=1}^{n-1} \mu_{r,r+1:n} = \sum_{r=1}^{n-1} \binom{n}{r} \mu_{1:n-r} \mu_{r:r} - \sum_{r=2}^{n} \binom{n}{r} \mu_{1,r:r} \ . \tag{4.35}$$

This result is obtained simply by setting $i = 1$ in Relation 4.29.

RELATION 4.31. For any arbitrary distribution and even n,

$$\mu_{1,n:n} = \sum_{i=1}^{(n-2/2)} (-1)^{i-1} \binom{n}{i} \mu_{1:i} \mu_{1:n-i} + \frac{1}{2} (-1)^{(n-2)/2} \binom{n}{n/2} \mu^{2}_{1:n/2} \ . \tag{4.36}$$

This result is obtained by setting $i = n - 1$ in Relation 4.29.

Note that Relation 4.31 expresses the product moments $\mu_{1,n:n}$ for even values of n in terms of means of smallest order statistics in samples of sizes $(n-1)$ and less, while Relation 4.9 similarly expresses it in terms of the means of largest order statistics in samples of sizes $(n-1)$ and less.

For even values of n, it can be easily seen from Relation 4.29 that sums of the sub-diagonal product moments of order statistics in a sample of size n, that is, sums of the form $\sum_{r=1}^{n-i} \mu_{r,r+1:n}$ $(1 \leq i \leq n-1)$, can all be calculated from the single and product moments of order statistics from samples of sizes $(n-1)$ and less. This result has been noted by Balakrishnan (1982).

Now we present the identities established by Balakrishnan and Malik (1987b). These could be regarded as extensions to the identities involving the single moments that have been derived by Joshi (1973) and Balakrishnan and Malik (1985), which have been presented earlier as Relations 3.13–3.18.

RELATION 4.32. For an arbitrary distribution and $n \geq 2$,

$$\sum_{s=2}^{n}\sum_{r=1}^{s-1} \mu_{r,s:n}/r = n\sum_{s=2}^{n}\sum_{r=1}^{s-1} \mu_{1,r+1:s}/\{(s-1)s\}$$

$$= n\sum_{s=2}^{n}(-1)^s \binom{n-1}{s-1} \mu_{s-1,s:s}/\{(s-1)s\} \qquad (4.37)$$

$$= n\sum_{s=1}^{n-1}\sum_{r=0}^{s-1}(-1)^r \binom{s-1}{r} \mu_{r+1,r+2:r+2}/\{(r+1)(r+2)\} \ .$$

RELATION 4.33. For $n \geq 2$,

$$\sum_{s=2}^{n}\sum_{r=1}^{s-1} \mu_{r,s:n}/(s-r) = n\sum_{s=2}^{n}\sum_{r=1}^{s-1} \mu_{r,r+1:s}/\{(s-1)s\}$$

$$= n\sum_{s=2}^{n}(-1)^s \binom{n-1}{s-1} \mu_{1,s:s}/\{(s-1)s\} \qquad (4.38)$$

$$= n\sum_{s=1}^{n-1}\sum_{r=0}^{s-1}(-1)^r \binom{s-1}{r} \mu_{1,r+2:r+2}/\{(r+1)(r+2)\} \ .$$

RELATION 4.34. For $n \geq 2$,

$$\sum_{s=2}^{n}\sum_{r=1}^{s-1} \mu_{r,s:n}/(n-s+1) = n\sum_{s=2}^{n}\sum_{r=1}^{s-1} \mu_{r,s:s}/\{(s-1)s\}$$

$$= n\sum_{s=2}^{n}(-1)^s \binom{n-1}{s-1} \mu_{1,2:s}/\{(s-1)s\} \qquad (4.39)$$

$$= n\sum_{s=1}^{n-1}\sum_{r=0}^{s-1}(-1)^r \binom{s-1}{r} \mu_{1,2:r+2}/\{(r+1)(r+2)\} \ .$$

RELATION 4.35. For $n \geq 2$,

$$\sum_{s=2}^{n}\sum_{r=1}^{s-1} \mu_{r,s:n}/\{r(s-r)\} = \sum_{r=1}^{n-1}\sum_{s=r+1}^{n}\left(1+(-1)^{r+1}\frac{(s)_r}{(n-s+r+1)_r}\right)$$
$$\times \mu_{r,r+1:n-s+r+1}/\{r(n-s+r+1)\}, \quad (4.40)$$

where $(n)_r$ denotes $(n-1)(n-2)\cdots(n-r)$, for $r \geq 1$.

RELATION 4.36. For $n \geq 2$,

$$\sum_{s=2}^{n}\sum_{r=1}^{s-1} \mu_{r,s:n}/\{r(n-s+1)\} = \sum_{r=1}^{n-1}\sum_{s=r+1}^{n}\left(1+(-1)^{r+1}\frac{(s)_r}{(n-s+r+1)_r}\right)$$
$$\times \mu_{r,n-s+r+1:n-s+r+1}/\{r(n-s+r+1)\} .$$
$$(4.41)$$

RELATION 4.37. For $n \geq 2$,

$$\sum_{s=2}^{n}\sum_{r=1}^{s-1} \mu_{r,s:n}/\{(s-r)(n-s+1)\}$$
$$= \sum_{r=1}^{n-1}\sum_{s=r+1}^{n}\left(1+(-1)^{r+1}\frac{(s)_r}{(n-s+r+1)_r}\right) \quad (4.42)$$
$$\times \mu_{n-s+1,n-s+r+2:ns+r+1}/\{r(n-s+r+1)\} .$$

5. Relations between moments of order statistics from two related populations

Let $X_{1:n} \leq X_{2:n} \leq \cdots \leq X_{n:n}$ be the order statistics obtained from a random sample of size n drawn from a symmetric population (symmetric about zero) with cdf $F(x)$ and pdf $f(x)$. Further, let $Y_{1:n} \leq Y_{2:n} \leq \cdots \leq Y_{n:n}$ be the order statistics obtained from a random sample of size n drawn from the population having pdf $f^*(x)$ and cdf $F^*(x)$, where

$$F^*(x) = 2F(x) - 1 \quad \text{and} \quad f^*(x) = 2f(x), \quad x > 0 .$$

That is, the distribution of Y's is obtained by folding the distribution of X's at zero. Denoting $E(X_{r:n}^k)$ by $\mu_{r:n}^{(k)}$, $E(Y_{r:n}^k)$ by $v_{r:n}^{(k)}$ ($1 \leq r \leq n$, $k > 0$), $E(X_{r:n}^i X_{s:n}^j)$ by $\mu_{r,s:n}^{(i,j)}$ and $E(Y_{r:n}^i Y_{s:n}^j)$ by $v_{r,s:n}^{(i,j)}$ ($1 \leq r < s \leq n$, $i,j > 0$), Govindarajulu (1963a) obtained the following relations satisfied by these moments:

For $1 \leq r \leq n$ and $k > 0$,

$$\mu_{r:n}^{(k)} = 2^{-n}\left[\sum_{m=0}^{r-1}\binom{n}{m}v_{r-m:n-m}^{(k)} + (-1)^k\sum_{m=r}^{n}\binom{n}{m}v_{m-r+1:m}^{(k)}\right] ; \quad (5.1)$$

for $1 < r < s < n$ and $i, i > 0$,

$$\mu_{r,s:n}^{(i,j)} = 2^{-n}\left[\sum_{m=0}^{r-1}\binom{n}{m}v_{r-m,s-m:n-m}^{(i,j)} + (-1)^i\sum_{m=r}^{s-1}\binom{n}{m}v_{m-r+1:m}^{(i)}\right.$$
$$\left.\times v_{s-m:n-m}^{(j)} + (-1)^{i+j}\sum_{m=s}^{n}\binom{n}{m}v_{m-s+1,m-r+1:m}^{(i,j)}\right]; \quad (5.2)$$

for $1 \le r \le n$ and $k > 0$,

$$\left(1+(-1)^{n+k-1}\right)v_{r:n}^{(k)} = 2^{(n-r+1)}\sum_{m=0}^{r-1}(-1)^m\binom{n-r+m}{m}\mu_{r-m:n}^{(k)}$$
$$+ (-1)^k\sum_{m=r}^{n-1}(-1)2^{n-m}\binom{n}{m}v_{r:m}^{(k)}; \quad (5.3)$$

and for $1 \le r < s \le n$ and $i, j > 0$,

$$v_{r,s:n}^{(i,j)} = 2^n\sum_{m=0}^{r-1}(-1)2^{-m}\binom{n}{m}\mu_{r-m,s-m:n-m}^{(i,j)} + (-1)^{r+i}\binom{n}{r}$$
$$\times \sum_{m=1}^{s-r}r(s-m)^{-1}\binom{n-r}{s-r-m}v_{s-m:s-m}^{(i)}v_{m:n-s+m}^{(j)} \quad (5.4)$$
$$+ (-1)^{r+i+j}\sum_{m=1}^{r}\sum_{m'=1}^{n-s+1}(-1)^{m-1}\binom{n}{s+m'-1}\binom{s+m'-1}{r-m}$$
$$\times v_{m',s-r+m':s-r+m+m'-1}^{(j,i)}.$$

Relations (5.1) and (5.2) express the μ's explicity in terms of the v's; hence, both single and product moments of order statistics from the symmetric distribution (symmetric about zero) can all be obtained with a knowledge of the single and product moments of order statistics from its folded distribution. Making use of the explicit expressions available for the moments of order statistics from the exponential distribution, Govindarajulu (1966) has applied these relations to compute both single and product moments of order statistics from the double exponential (Laplace) distribution. Similar work has been carried out for the double Weibull distribution by Balakrishnan and Kocherlakota (1985). Balakrishnan, Govindarajulu and Balasubramanian (1993) have also given an interesting probabilistic interpretation for relations (5.1) and (5.2).

6. Normal and half normal distributions

For any differentiable function $g(x)$ such that differentiation of $g(x)$ with respect to its argument and expectation of $g(X)$ with respect to an absolutely continuous distribution are interchangeable, Govindarajulu (1963b) has shown that

$$Eg'(X_{r:n}) = -\sum_{s=1}^{n} E[g(X_{r:n})f'(X_{s:n})/f(X_{s:n})] , \tag{6.1}$$

for $1 \leq r \leq n$, where f denotes the probability density function of the distribution. For example, for $g(x) = x$, relation (6.1) gives

$$-\sum_{s=1}^{n} E[X_{r:n}f'(X_{s:n})/f(X_{s:n})] = 1, \quad 1 \leq r < s \leq n . \tag{6.2}$$

For the standard normal distribution with pdf

$$f(x) = \frac{1}{\sqrt{2\pi}} e^{-x^2/2}, \quad -\infty < x < \infty ,$$

it is easy to see that $f'(x) = -xf(x)$, and hence relation (4.2) immediately gives the identity

$$\sum_{s=1}^{n} \mu_{r,s:n} = 1, \quad 1 \leq r \leq n ; \tag{6.3}$$

also see Seal (1956). Also since $\sum_{s=1}^{n} \mu_{s:n} = nE(X) = 0$, Eq. (6.3) yields the identity

$$\sum_{s=1}^{n} \sigma_{r,s:n} = 1, \quad 1 \leq r \leq n . \tag{6.4}$$

In other words, each row (or column) in the variance-covariance matrix of normal order statistics add up to 1. These identities have been applied by Davis and Stephens (1978) for checking the accuracy of the approximation for variances and covariances of normal order statistics while improving David and Johnson's (1954) approximations for these quantities. Note that identities (6.3) and (6.4) also hold for the half normal (chi) distribution with pdf

$$f(x) = \sqrt{2/\pi} e^{-x^2/2}, \quad x > 0 ,$$

as in this case also we have $f'(x) = -xf(x)$.

For normal and half normal distributions, Govindarajulu (1963b) has also established the relation

$$\mu_{r:n}^{(2)} = 1 + r\binom{n}{r} \sum_{m=0}^{n-r} (-1)^m \frac{1}{m+r} \binom{n-r}{m} \mu_{m+r-1, m+r:m+r} , \tag{6.5}$$

for $1 \leq r \leq n$. Proceeding on similar lines, Joshi and Balakrishnan (1981a) have established the following identities satisfied by the moments of order statistics from normal and half normal distributions:

$$\sum_{s=1}^{n} \mu_{1,s:n} = 1 ; \tag{6.6}$$

for $2 \leq r \leq n$,

$$\sum_{s=r}^{n} \mu_{r,s:n} = 1 + \sum_{s=r}^{n} \mu_{r-1,s:n} \; ; \tag{6.7}$$

for $1 \leq r \leq n-1$,

$$\sum_{s=r+1}^{n} \mu_{r,s:n} = \sum_{s=r+1}^{n} \mu_{s:n}^{(2)} - (n-r) \; ; \tag{6.8}$$

for $2 \leq r \leq n$,

$$\sum_{s=1}^{n} \mu_{r,s:n} = 1 + n\mu_{1:1}\mu_{r-1:n-1} \; ; \tag{6.9}$$

for $n \geq 1$,

$$\sum_{s=1}^{n} \sigma_{1,s:n} = 1 - \mu_{1:1}\mu_{1:n} \; ; \tag{6.10}$$

and for $2 \leq r \leq n$,

$$\sum_{s=1}^{n} \sigma_{r,s:n} = 1 - (n-r+1)\mu_{1:1}(\mu_{r:n} - \mu_{r-1:n}) \; . \tag{6.11}$$

Note that identities (6.6), (6.9), (6.10) and (6.11) reduce to the identities (6.3) and (6.4) for the normal distribution by making use of the fact that $\mu_{1:1} = E(X) = 0$. Identities (6.6)–(6.11) have been applied by Joshi and Balakrishnan (1981b) for obtaining an expression for the variance of standardized and studentized selection differentials of reach statistics. See also Schaeffer, Van Vleck and Velasco (1970) and Burrows (1972, 1975) for some related work on the selection differential for the normal distribution.

Balakrishnan and Malik (1987c) have established the following identities satisfied by the moments of normal order statistics:

For $n \geq 2$,

$$\sum_{s=2}^{n} \sum_{r=1}^{s-1} \mu_{r,s:n}/(n-s+1) = (n-1) - n \sum_{r=2}^{n} \mu_{1:r}^{(2)}/\{(r-1)r\}$$

$$= (n-1) - n \sum_{r=2}^{n} \mu_{r:r}^{(2)}/\{(r-1)r\} \; ; \tag{6.12}$$

for $n \geq 2$,

$$\sum_{s=2}^{n} \sum_{r=1}^{s-1} \mu_{r,s:n}/\{(n-s+1)(n-s+2)\}$$

$$= (1/2 + 1/3 + \cdots + 1/n) - \sum_{r=2}^{n} \mu_{r:r}^{(2)}/r \; ; \tag{6.13}$$

and for $i = 3, 4, \ldots,$

$$\sum_{s=2}^{n} \sum_{r=1}^{s-1} \mu_{r,s:n} / \{(n-s+1)(n-s+1) \cdots (n-s+i)\}$$
$$= \frac{C_i}{(i-1)(i-2)} \left[\left\{ \binom{n+i-2}{i-2} - (i-1) \right\} - \sum_{r=2}^{n} \binom{r+i-3}{i-3} \mu_{r:r}^{(2)} \right], \tag{6.14}$$

where $C_i = 1/\{(n+1)(n+2)\cdots(n+i-2)\}$.

7. Cauchy distribution

For the Cauchy distribution with density function

$$f(x) = \frac{1}{\pi(1+x^2)}, \quad -\infty < x < \infty,$$

Barnett (1966) has obtained the following recurrence relation between the first and second order single moments of order statistics:
For $3 \le r \le n-2$,

$$\mu_{r:n}^{(2)} = \frac{n}{\pi}(\mu_{r:n-1} - \mu_{r-1:n-1}) - 1, \tag{7.1}$$

so that the variances are given by

$$\mathrm{Var}(X_{r:n}) = \sigma_{r,r:n} = \frac{n}{\pi}(\mu_{r:n-1} - \mu_{r-1:n-1}) - 1 - \mu_{r:n}^{(2)} \tag{7.2}$$

for $3 \le r \le n-2$. More generally, Barnett (1966) has obtained the following recurrence relation for the higher order moments:

$$\mu_{r:n}^{(k)} = \frac{n}{(k-1)\pi} \left(\mu_{r:n-1}^{(k-1)} - \mu_{r-1:n-1}^{(k-1)} \right) - \mu_{r:n}^{(k-2)} \tag{7.3}$$

for $k+1 \le r \le n-k$.

Relations (7.1)–(7.3) provide a very reasonable way of obtaining the variances and also higher order moments since the expected values of order statistics could be easily calculated to any required accuracy. Similar results have been obtained for the truncated Cauchy distribution by Khan, Yaqub and Parvez (1983).

Recently, Vaughan (1994) has derived general recurrence relations for the single moments of order statistics from the Cauchy distribution in terms of infinite series:

$$\mu_{r:n}^{(k)} = \sum_{i=0}^{j-1} \frac{(-1)^i n}{\pi(2j-2i-1)} \left(\mu_{r:n-1}^{(2j-2i-1)} - \mu_{r-1:n-1}^{(2j-2i-1)} \right) + (-1)^j, \quad \text{if } k = 2j \tag{7.4}$$

and

$$\mu_{r:n}^{(k)} = \frac{n}{\pi} \sum_{a=1}^{j} \frac{(-1)^{j-a}}{2a} \binom{j}{a} \left[\sum_{b=0}^{a} \binom{a}{b} \left[\mu_{r:n-1}^{(2b)} - \mu_{r-1:n-1}^{(2b)} \right] \right] + (-1)^{j} \mu_{r:n},$$

$$\text{if } k = 2j+1 \ . \quad (7.5)$$

In particular, if $k = 2$, Eq. (7.4) reduces to (7.1). One advantage of such formulae is that the calculation of $\mu_{r:n}^{(k)}$ requires lower order moments of lower order statistics for smaller sample sizes [i.e., $(n-1)$].

Infinite series expression for mean $\mu_{r:n}^{(1)} = \mu_{r:n}$ of $X_{r:n}$ has been derived recently by Joshi and Chakraborty (1996a) in terms of Riemann zeta functions (see Abramowitz and Stegun, 1964)

$$\zeta(s) = \sum_{k=1}^{\infty} k^{-s}, \quad s \geq 1 \ .$$

This is given by

$$\mu_{r:n} = \sum_{m=1}^{[(n-2)/2]} d_{2m+1} \frac{\zeta(2m+1)}{\pi^{2m+1}}, \quad (7.6)$$

where

$$d_{2m+1} = (-1)^{m} \frac{n!(2m+1)!h_m}{(m-r)!(r-1)!2^{2m}},$$

and

$$h_m = \sum_{j=\max(0,2m-r+2)}^{n-r} (-1)^{j} \binom{n-r}{j} \binom{r+j}{2m+1} \frac{1}{r+j} \ . \quad (7.7)$$

They have also tabulated the coefficients d_{2m+1} for $4 \leq n \leq 10$ and $[(n+3)/2] \leq r \leq n-1$.

Joshi and Chakraborty (1996b) have applied the techniques given in Joshi and Chakraborty (1996a) for deriving expressions for the single and product moments as:

$$\mu_{r:n}^{(k)} = \frac{-n}{\pi(r-1)} \mu_{r-1:n-1}^{(k-1)} + \frac{2(n)!}{\pi(r-1)!} \sum_{i=1}^{\infty} \frac{(r+2i-2)!}{(n+2i-1)!} \zeta(2i) \mu_{r+2i-1:n+2i-1}^{(k-1)} ,$$

$$(7.8)$$

and

$$\mu_{r:n}^{(k)} = \frac{n}{\pi(n-r)} \mu_{r:n-1}^{(k-1)} - \frac{2(n)!}{\pi(n-r)!} \sum_{i=1}^{\infty} \frac{(n-r+2i-1)!}{(n+2i-1)!} \zeta(2i) \mu_{r:n+2i-1}^{(k-1)} \ .$$

$$(7.9)$$

In approximating $\mu_{r:n}^{(k)}$ by first few terms of the series expansions given above, it is observed that Eq. (7.9) gives faster convergence than Eq. (7.8) for $k+1 \leq r \leq [(n+1)/2]$.

For the product moments $\mu_{r,s:n}$, Joshi and Chakraborty (1996a) have shown that

$$\mu_{r,s:n} = -\frac{n}{\pi(r-1)}\mu_{s-1:n-1} + \frac{2(n)!}{\pi(r-1)!}\sum_{i=1}^{\infty}(r+2i-2)!b_{n-1(i)}\mu_{s+2i-1:n+2i-1} \quad (7.10)$$

$$= \frac{n}{\pi(n-s)}\mu_{r:n-1} - \frac{2(n)!}{\pi(n-s)!}\sum_{i=1}^{\infty}(n-s+2i-1)!b_{n-1(i)}\mu_{r:n+2i-1}, \quad (7.11)$$

where $b_{t(i)} = \frac{\zeta(2i)}{(2i+t)!}$, $t = 0, 1, 2, \ldots$, with Eq. (7.10) giving better results than Eq. (7.11) for $r + s < n$ in the sense of faster convergence.

Again, Joshi and Chakraborty (1996b) have derived another form for $\mu_{r,s:n}$ as

$$\mu_{r,s:n} = \frac{n(n-1)}{\pi^2(r-1)(n-s)} + \frac{2(n)!}{\pi^2(r-1)(n-s)!}\sum_{i=1}^{\infty}(n-s+2i-1)!b_{n-2(i)}$$

$$+ \frac{2(n)!}{\pi^2(r-1)!(n-s)}\sum_{i=1}^{\infty}(r+2i-2)!b_{n-2(i)} \quad (7.12)$$

$$- \frac{4(n)!}{\pi^2(r-1)!(n-s)!}\sum_{i=1}^{\infty}\sum_{j=1}^{\infty}(r+2i-2)!(n-s+2j-1)!a_{n-2(i,j)},$$

where $a_{t(i,j)} = \frac{\zeta(2i)\zeta(2j)}{(2i+2j+t)!}$, for $t = 0, 1, 2, \ldots$.

8. Logistic and related distributions

With the density function given by [for more details, see Balakrishnan (1992); Johnson, Kotz and Balakrishnan (1995)]

$$f(x) = e^{-x}/(1+e^{-x})^2, \quad -\infty < x < \infty,$$

and the cumulative distribution function given by

$$F(x) = 1/(1+e^{-x}), \quad -\infty < x < \infty,$$

it is easy to see that

$$f(x) = F(x)\{1 - F(x)\}, \quad -\infty < x < \infty.$$

Making use of this relation, Shah (1966, 1970) has derived the following recurrence relations:

For $1 \leq r \leq n$ and $k = 1, 2, \ldots,$

$$\mu_{r+1:n+1}^{(k)} = \mu_{r:n}^{(k)} + \left(\frac{k}{r}\right)\mu_{r:n}^{(k-1)} ; \tag{8.1}$$

for $1 \leq r \leq n-1$,

$$\mu_{r,r+1:n+1} = \frac{n+1}{n-r+1}\left[\mu_{r,r+1:n} - \left(\frac{r}{n+1}\right)\mu_{r+1:n+1}^{(2)} - \frac{1}{n-r}\mu_{r:n}\right] ; \tag{8.2}$$

and for $1 \leq r < s \leq n$, $s - r \geq 2$,

$$\mu_{r,s:n+1} = \frac{n+1}{n-s+2}\left[\mu_{r,s:n} - \mu_{r,s-1:n} + \left(\frac{n-s+2}{n+1}\right)\mu_{r,s-1:n+1} - \frac{1}{n-s+1}\mu_{r:n}\right] . \tag{8.3}$$

Due to the symmetry of the distribution, we also have the relations $\mu_{r:n}^{(k)} = (-1)^k \mu_{n-r+1:n}^{(k)}$, $\mu_{1,2:2} = 0$ and $\mu_{r,s:n} = \mu_{n-s+1,n-r+1:n}$ as noted earlier in Sections 3 and 4. Making use of these relations along with recurrence relations (8.1)–(8.3), one could compute all the single and product moments of order statistics for all sample sizes. Explicit expressions for these moments are also available; see, for example, Birnbaum and Dudman (1963), Gupta and Shah (1965), Gupta, Qureishi and Shah (1967), and Balakrishnan (1992).

By considering the half logistic distribution with cumulative distribution function $F(x)$ and density function

$$f(x) = 2e^{-x}/(1 + e^{-x})^2, \quad 0 \leq x < \infty ,$$

and noting that

$$f(x) = F(x)\{1 - F(x)\} + (1/2)\{1 - F(x)\}^2$$
$$= \{1 - F(x)\} - (1/2)\{1 - F(x)\}^2$$
$$= (1/2)\{1 - F^2(x)\} ,$$

Balakrishnan (1985) has established the following recurrence relations satisfied by the single and product moments of order statistics:

For $n \geq 1$ and $k = 0, 1, 2, \ldots,$

$$\mu_{1:n+1}^{(k+1)} = 2\left[\mu_{1:n}^{(k+1)} - \left(\frac{k+1}{n}\right)\mu_{1:n}^{(k)}\right] ; \tag{8.4}$$

for $n \geq 1$ and $k = 0, 1, 2, \ldots,$

$$\mu_{2:n+1}^{(k+1)} = \frac{(n+1)(k+1)}{n}\mu_{1:n}^{(k)} - \left(\frac{n-1}{2}\right)\mu_{1:n+1}^{(k+1)} ; \tag{8.5}$$

for $2 \leq r \leq n$ and $k = 0, 1, 2, \ldots$,

$$\mu_{r+1:n+1}^{(k+1)} = \frac{1}{r}\left[\frac{(n+1)(k+1)}{(n-r+1)}\mu_{r:n}^{(k)} + \left(\frac{n+1}{2}\right)\mu_{r-1:n}^{(k+1)} - \frac{n-2r+1}{2}\mu_{r:n+1}^{(k+1)}\right] ; \tag{8.6}$$

for $1 \leq r \leq n-1$,

$$\mu_{r,r+1:n+1}^{(2)} = \mu_{r,n+1}^{(2)} + \frac{2(n+1)}{n-r+1}\left[\mu_{r,r+1:n} - \mu_{r:n}^{(2)} - \left(\frac{1}{n-r}\right)\mu_{r:n}\right] ; \tag{8.7}$$

for $n \geq 2$,

$$\mu_{2,3:n+1} = \mu_{3:n+1}^{(2)} + (n+1)\left[\mu_{2:n} - \left(\frac{n}{2}\right)\mu_{1:n-1}^{(2)}\right] ; \tag{8.8}$$

for $2 \leq r \leq n-1$,

$$\mu_{r+1,r+2:n+1} = \mu_{r+2:n+1}^{(2)} + \frac{n+1}{r(r+1)}\left[2\mu_{r+1:n} + n\left(\mu_{r-1,r:n-1} - \mu_{r:n-1}^{(2)}\right)\right] ; \tag{8.9}$$

for $1 \leq r \leq n-2$ and $s - r \geq 2$,

$$\mu_{r,s:n+1} = \mu_{r,s-1:n+1} + \frac{2(n+1)}{n-s+2}\left[\mu_{r,s:n} - \mu_{r,s-1:n} - \left(\frac{1}{n-s+1}\right)\mu_{r:n}\right] ; \tag{8.10}$$

for $3 \leq s \leq n$,

$$\mu_{2,s+1:n+1} = \mu_{3,s+1:n+1} + (n+1)\left[\mu_{s:n} - \left(\frac{n}{2}\right)\mu_{1,s-1:n-1}\right] ; \tag{8.11}$$

and for $2 \leq r \leq n-2$ and $s - r \geq 2$,

$$\mu_{r+1,s+1:n+1} = \mu_{r+2,s+1:n+1} + \frac{n+1}{r(r+1)}\left[2\mu_{s:n} - n\left(\mu_{r,s-1:n-1} - \mu_{r-1,s-1:n-1}\right)\right] . \tag{8.12}$$

Starting with $\mu_{1:1} = \log 4$ and $\mu_{1:1}^{(2)} = \pi^2/3$, recurrence relations (8.4)–(8.12) could be systematically applied in order to compute all the single and product moments of order statistics for all sample sizes in a simple recursive way. This recursive evaluation of the means, variances and covariances of order statistics from a half logistic distribution has been successfully employed by Balakrishnan and Puthenpura (1986) while determining the coefficients for the best linear unbiased estimators of the location and scale parameters of a half logistic distribution.

These results have been extended by Balakrishnan and Kocherlakota (1986) to the doubly truncated logistic distribution with cumulative distribution function $F(x)$ and density function

$$f(x) = \frac{1}{P-Q} e^{-x}/(1+e^{-x})^2, \qquad Q_1 \leq x \leq P_1 ,$$

where Q and $1-P$ ($Q < P$) are, respectively, the proportions of truncation on the left and right of the standard logistic distribution, and $Q_1 = \log(\frac{Q}{1-Q})$ and $P_1 = \log(\frac{P}{1-P})$. Denoting $Q_2 = Q(1-Q)/(P-Q)$ and $P_2 = P(1-P)/(P-Q)$, it can be seen that

$$\begin{aligned} f(x) &= (1-2Q)F(x) - (P-Q)\{1-F(x)\}^2 + Q_2 \\ &= P_2 + (2P-1)\{1-F(x)\} - (P-Q)\{1-F(x)\}^2 \\ &= P_2 + (2P-1)F(x)\{1-F(x)\} + (P+Q-1)\{1-F(x)\}^2 . \end{aligned}$$

Making use of these relations, Balakrishnan and Kocherlakota (1986) have established the following recurrence relations:
For $k = 0, 1, 2, \ldots$,

$$\mu_{1:2}^{(k+1)} = Q_1^{k+1} + \frac{1}{P-Q} \left[P_2(P_1^{k+1} - Q_1^{k+1}) + (2P-1) \right. \\ \left. \times \left(\mu_{1:1}^{(k+1)} - Q_1^{k+1} \right) - (k+1)\mu_{1:1}^{(k)} \right] ; \tag{8.13}$$

for $n \geq 2$ and $k = 0, 1, 2, \ldots$,

$$\mu_{1:n+1}^{(k+1)} = Q_1^{k+1} + \frac{1}{P-Q} \left[P_2 \left(\mu_{1:n-1}^{(k+1)} - Q_1^{k+1} \right) + (2P-1) \right. \\ \left. \times \left(\mu_{1:1}^{(k+1)} - Q_1^{k+1} \right) - \left(\frac{k+1}{n} \right) \mu_{1:n}^{(k)} \right] ; \tag{8.14}$$

for $k = 0, 1, 2, \ldots$,

$$\mu_{2:2}^{(k+1)} = P_1^{k+1} - \frac{1}{P-Q} \left[Q_2(P_1^{k+1} - Q_1^{k+1}) + (1-2Q) \right. \\ \left. \times \left(P_1^{k+1} - \mu_{1:1}^{(k+1)} \right) - (k+1)\mu_{1:1}^{(k)} \right] ; \tag{8.15}$$

for $n \geq 3$ and $k = 0, 1, 2, \ldots$,

$$\mu_{2:n+1}^{(k+1)} = \mu_{1:n+1}^{(k+1)} + \frac{n+1}{P-Q} \left[\frac{P_2}{n-1} \left(\mu_{2:n-1}^{(k+1)} - \mu_{1:n-1}^{(k+1)} \right) + \frac{2P-1}{n} \right. \\ \left. \times \left(\mu_{2:n}^{(k+1)} \mu_{1:n}^{(k+1)} \right) - \frac{(k+1)}{n(n-1)} \mu_{2:n}^{(k)} \right] ; \tag{8.16}$$

for $2 \leq r \leq n-1$ and $k = 0, 1, 2, \ldots,$

$$\mu_{r+1:n+1}^{(k+1)} = \frac{n+1}{r(2P-1)} \left[\frac{k+1}{n-r+1} \mu_{r:n}^{(k)} - \frac{nP_2}{n-r+1} \left(\mu_{r:n-1}^{(k+1)} - \mu_{r-1:n-1}^{(k+1)} \right) \right.$$
$$- \frac{1}{n+1} \{(n+1)(P+Q-1) - r(2P-1)\} \mu_{r:n+1}^{(k+1)} \qquad (8.17)$$
$$\left. + (P+Q-1) \mu_{r-1:n}^{(k+1)} \right] ;$$

for $n \geq 2$ and $k = 0, 1, 2, \ldots,$

$$\mu_{n+1:n+1}^{(k+1)} = \frac{n+1}{n(2P-1)} \left[(k+1) \mu_{n:n}^{(k)} - nP_2 \left(P_1^{k+1} - \mu_{n-1:n-1}^{(k+1)} \right) \right.$$
$$- \frac{1}{n+1} \{(n+1)(P+Q-1) - n(2P-1)\} \qquad (8.18)$$
$$\left. \times \mu_{n:n+1}^{(k+1)} + (P+Q-1) \mu_{n-1:n}^{(k+1)} \right] ;$$

$$\mu_{1,2:3} = \mu_{1:3}^{(2)} + \frac{3}{2(P-Q)} \left[2P_2 \left(P_1 \mu_{1:1} - \mu_{1:1}^{(2)} \right) \right.$$
$$\left. + (2P-1) \left(\mu_{1,2:2} - \mu_{1:2}^{(2)} \right) - \mu_{1:2} \right] ; \qquad (8.19)$$

$$\mu_{2,3:3} = \mu_{3:3}^{(2)} + \frac{3}{2(P-Q)} \left[\mu_{2:2} - 2Q_2 \left(\mu_{1:1}^{(2)} - Q_1 \mu_{1:1} \right) \right.$$
$$\left. + (2Q-1) \left(\mu_{2:2}^{(2)} - \mu_{1,2:2} \right) \right] ; \qquad (8.20)$$

for $1 \leq r \leq n-2,$

$$\mu_{r,r+1:n+1} = \mu_{r:n+1}^{(2)} + \frac{n+1}{(n-r+1)(P-Q)} \left[\frac{nP_2}{n-r} \left(\mu_{r,r+1:n-1} - \mu_{r:n-1}^{(2)} \right) \right.$$
$$\left. + (2P-1) \left(\mu_{r,r+1:n} - \mu_{r:n}^{(2)} \right) - \frac{1}{n-r} \mu_{r:n} \right] ; \qquad (8.21)$$

for $n \geq 2,$

$$\mu_{n-1,n:n+1} = \mu_{n-1:n+1}^{(2)} + \frac{n+1}{2(P-Q)} \left[nP_2 \left(P_1 \mu_{n-1:n-1} - \mu_{n-1:n-1}^{(2)} \right) \right.$$
$$\left. + (2P-1) \left(\mu_{n-1,n:n} - \mu_{n-1:n}^{(2)} \right) - \mu_{n-1:n} \right] ; \qquad (8.22)$$

for $n \geq 2$,

$$\mu_{2,3:n+1} = \mu_{3:n+1}^{(2)} + \frac{n+1}{2(P-Q)} \left[\mu_{2:n} - nQ_2 \left(\mu_{1:n-1}^{(2)} - Q_1 \mu_{1:n-1} \right) \right.$$
$$\left. - (1 - 2Q) \left(\mu_{2:n}^{(2)} - \mu_{1,2:n} \right) \right] ; \qquad (8.23)$$

for $2 \leq r \leq n-1$,

$$\mu_{r+1,r+2:n+1} = \mu_{r+2:n+1}^{(2)} + \frac{n+1}{(r+1)(P-Q)} \left[\frac{1}{r} \mu_{r+1:n} - \frac{nQ_2}{r} \left(\mu_{r:n-1}^{(2)} - \mu_{r-1,r:n-1} \right) \right.$$
$$\left. - (1 - 2Q) \left(\mu_{r+1:n}^{(2)} - \mu_{r,r+1:n} \right) \right] ; \qquad (8.24)$$

for $1 \leq r \leq n-2$,

$$\mu_{r,n:n+1} = \mu_{r,n-1:n+1} + \frac{n+1}{2(P-Q)} \left[nP_2 \left(P_1 \mu_{r:n-1} - \mu_{r,n-1:n-1} \right) \right.$$
$$\left. + (2P - 1)(\mu_{r,n:n} - \mu_{r,n-1:n}) - \mu_{r:n} \right] ; \qquad (8.25)$$

for $s \leq n-1$ and $s - r \geq 2$,

$$\mu_{r,s:n+1} = \mu_{r,s-1:n+1} + \frac{n+1}{(n-s+2)(P-Q)} \left[\frac{nP_2}{n-s+1} \left(\mu_{r,s:n-1} - \mu_{r,s-1:n-1} \right) \right.$$
$$\left. + (2P - 1)(\mu_{r,s:n} - \mu_{r,s-1:n}) - \frac{1}{n-s+1} \mu_{r:n} \right] ; \qquad (8.26)$$

for $s \geq 3$,

$$\mu_{2,s+1:n+1} = \mu_{3,s+1:n+1} + \frac{n+1}{2(P-Q)} \left[\mu_{s:n} - nQ_2 \left(\mu_{1,s-1:n-1} - Q_1 \mu_{s-1:n-1} \right) \right.$$
$$\left. + (2Q - 1)(\mu_{2,s:n} - \mu_{1,s:n}) \right] ; \qquad (8.27)$$

and for $r \geq 2$, $s - r \geq 2$,

$$\mu_{r+1,s+1:n+1} = \mu_{r+2,s+1:n+1} + \frac{n+1}{(r+1)(P-Q)} \left[\frac{1}{r} \mu_{s:n} - \frac{nQ_2}{r} \right.$$
$$\left. \times (\mu_{r,s-1:n-1} - \mu_{r-1,s-1:n-1}) + (2Q - 1)(\mu_{r+1,s:n} - \mu_{r,s:n}) \right] . \qquad (8.28)$$

Now starting with the values of $\mu_{1:1}$ and $\mu_{1:1}^{(2)}$, for example, recurrence relations (8.13)–(8.28) could be systematically applied in a simple recursive way in order to compute the first two single moments and product moments of all order statistics for all sample sizes; also see Balakrishnan and Joshi (1983). Exact and explicit expressions for the first two single moments and product moments of order statistics from the doubly truncated logistic distribution have been derived by Tarter (1966).

The pdf of a generalized half logistic distribution is given by

$$f(x) = \frac{2(1-kx)^{(1/k)-1}}{\left[1+(1-kx)^{1/k}\right]^2}, \quad 0 \le x \le 1/k, \quad k \ge 0,$$

and the cdf is given by

$$F(x) = \frac{1-(1-kx)^{1/k}}{\left[1+(1-kx)^{1/k}\right]}, \quad 0 \le x \le 1/k, \quad k \ge 0.$$

It is easy to note that

$$(1-kx)f(x) = 1 - F(x) - (1/2)\{1 - F(x)\}^2.$$

Using the above differential equation, Balakrishnan and Sandhu (1995) have established the following recurrence relations:

For $n \ge 1$ and $i = 0, 1, 2, \ldots$,

$$\mu_{1:n+1}^{(i+1)} = 2\left[\mu_{1:n}^{(i+1)} - \frac{i+1}{n}\left(\mu_{1:n}^{(i)} - \mu_{1:n}^{(i+1)}\right)\right]; \tag{8.29}$$

for $1 \le r \le n$ and $i = 0, 1, 2, \ldots$,

$$\mu_{r+1:n+1}^{(i+1)} = \mu_{r:n+1}^{(i+1)} + \frac{n+1}{r}\left[\frac{2(i+1)}{n-r+1}\left(\mu_{r:n}^{(i)} - k\mu_{r:n}^{(i+1)}\right) - \left(\mu_{r:n}^{(i+1)} - \mu_{r-1:n}^{(i+1)}\right)\right],$$

$$\mu_{0:n}^{(i)} = 0, \quad \text{for } n \ge 1 \text{ and } i = 0, 1, 2, \ldots; \tag{8.30}$$

for $1 \le r \le n-1$,

$$\mu_{r,r+1:n+1} = \mu_{r:n+1}^{(2)} + \frac{2(n+1)}{n-r+1}\left[\mu_{r,r+1:n} - \mu_{r:n}^{(2)} + \frac{1}{n-r}\left(k\mu_{r,r+1:n} - \mu_{r:n}\right)\right]; \tag{8.31}$$

for $1 \le r < s \le n$ and $s - r \ge 2$,

$$\mu_{r,s:n+1} = \mu_{r,s:n+1} + \frac{2(n+1)}{n-s+2}\left[\mu_{r,s:n} - \mu_{r,s-1:n} + \frac{1}{n-s+1}\left(k\mu_{r,s:n} - \mu_{r:n}\right)\right]; \tag{8.32}$$

for $2 \le r \le n-1$,

$$\mu_{r+1,r+2:n+1} = \mu_{r+2:n}^{(2)} + \frac{2(n+1)}{r(r+1)}$$
$$\times \left[\mu_{r+1,:n} - k\mu_{r,r+1:n} - \frac{n}{2}\left(\mu_{r:n-1}^{(2)} - \mu_{r-1,r:n-1}\right)\right]; \tag{8.33}$$

for $2 \leq r < s \leq n$ and $s - r \geq 2$,

$$\mu_{r+1,s+1:n+1} = \mu_{r+2,s+1:n+1} + \frac{2(n+1)}{r(r+1)}\left[\mu_{r:n} - k\mu_{r,s:n}\right. \quad (8.34)$$

$$\left. - \frac{n}{2}(\mu_{r,s-1:n-1} - \mu_{r-1,s-1:n-1})\right];$$

and for $n \geq 3$,

$$\mu_{1,n+1:n+1} = \frac{2(n+1)}{n(n-1)}\left[(n-1+k)\mu_{1,n:n} + \mu_{1,n-1:n} - \frac{n-1}{n+1}\mu_{1,n,n+1}\right.$$

$$\left. - \frac{1}{n+1}\mu_{1,n-1:n+1} - \mu_{r:n}\right]. \quad (8.35)$$

Letting the shape parameter $k \to 0$ in the relations (8.29)–(8.35), we obtain the relations for the half logistic distribution presented in Eqs. (8.4)–(8.12).

The recurrence relations in (8.29) and (8.30) will enable one to compute all the single moments of all order statistics for all sample sizes. Also, the recurrence relations (8.31)–(8.35) are complete in the sense that they will enable one to compute all the product moments of all order statistics for all sample sizes in a simple recursive manner. This can be done for any choice of the shape parameter k, and the required recursive computational algorithm is explained in detail by Balakrishnan and Sandhu (1995).

Moreover, recurrence relations between moment generating functions as well as factorial moment generating functions of order statistics from the doubly truncated logistic distribution have been established by Mohie El-Din, Mahmoud and Abu-Youssef (1992) and Mohie El-Din and Sultan (1995), respectively.

9. Gamma and related distributions

For the gamma distribution with probability density function

$$f(x) = e^{-x}x^{m-1}/\Gamma(m), \quad x > 0, \quad m > 0, \quad (9.1)$$

explicit expressions for the moments of order statistics have been derived by Gupta (1960) for integral values of m, and by Krishnaiah and Rizvi (1967) for a general value of m; see also Breiter and Krishnaiah (1968). Note that for the gamma random variable X with density function given by (9.1), the kth moment

$$E(X^k) = \Gamma(m+k)/\Gamma(m)$$

exists for all $k > -m$ and consequently, $\mu_{r:n}^{(k)}$ also exists for $k > -m$ (David, 1981, p. 34). For integral values of m, Joshi (1979a) has established the following recurrence relations satisfied by the single moments of order statistics:

For $k = 1, 2, \ldots$,

$$\mu_{1:n}^{(k)} = (k/n)\Gamma(m) \sum_{t=0}^{m-1} \mu_{1:n}^{(t+k-m)}/t! , \qquad (9.2)$$

and for $2 \leq r \leq n$ and $k = 1, 2, \ldots$,

$$\mu_{r:n}^{(k)} = \mu_{r-1:n-1}^{(k)} + (k/n) \sum_{t=0}^{m-1} \mu_{r:n}^{(t+k-m)}/t! . \qquad (9.3)$$

Relation (9.2) expresses the kth order moment of $X_{1:n}$ in terms of the lower order moments of $X_{1:n}$. In particular, it expresses the mean of $X_{1:n}$ in terms of moments of orders $-(m-1), -(m-2), \ldots, -1$ of $X_{1:n}$. Similarly, relation (9.3) expresses the kth order moment of $X_{r:n}$ in terms of the kth order moment of $X_{r-1:n-1}$ and lower order moments of $X_{r:n}$. Hence, from the relation

$$\mu_{r:n}^{(k)} = \sum_{j=n-r+1}^{n} (-1)^{j-n+r-1} \binom{j-1}{n-r} \binom{n}{j} \mu_{1:j}^{(k)}$$

presented earlier in Section 3 and the recurrence relations (9.2) and (9.3), it is clear that if the negative moments of orders $-(m-1), -(m-2), \ldots, -1$ of the smallest order statistic in samples of size $j \leq n$ are known, then one could calculate all the moments $\mu_{r:n}^{(k)}$ for $1 \leq r \leq n$ and $k = 1, 2, \ldots$.

Young (1971) has also established a simple relation between moments of order statistics from the symmetrical inverse multinomial distribution and the order statistics of independent standardized gamma variables with integer parameter m.

Thomas and Moothathu (1991) have obtained a recurrence relation for the moments of different orders of the largest order statistic from a gamma distribution with shape parameter m, which is

$$\sum_{j=1}^{n} A_{n-1,j} \mu_{n:n}^{(k-n+j)} = \frac{\Gamma(k+nm-n+1)}{(\Gamma(m))^n} n^{-k-(m-1)n} , \qquad (9.4)$$

where

(i) $n \geq 2$ is an integer and k is a real number such that $k > \max(-m, -n(m-1)-1)$,
(ii) for $r = 1, 2, 3, \ldots, n-1$, $A_{0,1} = 1$,
$A_{r,1} = -r^{-1}[k-r+1+(n-r)(m-1)]A_{r-1,1}$,
(iii) for $j = 2, 3, \ldots, r$,
$A_{r,j} = r^{-1}(n-r)A_{r-1,j-1} - r^{-1}[k-r+j+(n-r)(m-1)]A_{r-1,j}$,
$A_{r,r+1} = r^{-1}(n-r)A_{r-1,r}$

and $\mu_{n:n}^{(0)} = 1$. Thomas and Moothathu (1991) have presented a numerical example for illustrating the application of recurrence relation (9.4).

Based on the functional relationship between the distribution functions of gamma random variables with shape parameter m and $m-1$, $m > 1$, Thomas (1993) has derived the following recurrence relation:

For $n \geq 2$ and r and m real numbers, $r \geq 0$, $m > 1$,

$$\mu_{n,n:m}^{(r)} = \sum_{k=0}^{n-1} \binom{n-1}{k}(m-1)^{-k-1}(-1)^k \sum_{j=1}^{k+1} A_{k+r+1,n-k-1,k+1:m-1}^{(k,j)} \mu_{n,n:m-1}^{(r+j)},$$

(9.5)

where

(i) $\mu_{r,n:m}^{(k)} = \mu_{r:n}^{(k)}$ with shape parameter m,
(ii) for $k = 1, 2, \ldots, t-1$, $A_{r,s,t:m}^{(0,1)} = 1$,
(iii) $A_{r,s,t:m}^{(k,1)} = -(s+k)^{-1}[r - k + 1 + (t-k)(m-1)]A_{r,s,t:m}^{(k-1,1)}$,
(iv) $A_{r,s,t:m}^{(k,j)} = (s+k)^{-1}\left[(t-k)A_{r,s,t:m}^{(k-1,j-1)} - [r - k + j + (t-k)(m-1)]A_{r,s,t:m}^{(k-1,j)}\right]$,
(v) and $A_{r,s,t:m}^{(k,k+1)} = (s+k)^{-1}(t-k)A_{r,s,t:m}^{(k-1,k)}$.

Certain applications of relation (9.5) are presented by Thomas (1993).

Recently, for the generalized gamma distribution

$$g(x; a, b, c) = bc^{-ab}\{\Gamma(a)\}^{-1}x^{ab-1}e^{-(x/c)^b}, \quad x > 0,$$

(9.6)

where $a > 0$, $b > 0$ and $c > 0$, Thomas (1996) has proved the following two theorems based on the moments of the largest order statistic.

For $1 \leq r \leq n$ and $k = 1, 2, \ldots$,

$$\mu_{r:n;a,b,c}^{(k)} = r\binom{n}{r}\int_0^\infty x^k\{G(x; a, b, c)\}^{r-1}$$

$$\times \{1 - G(x; a, b, c)\}^{n-r}g(x; a, b, c)\,dx,$$

(9.7)

where $\mu_{r:n;a,b,c}^{(0)} = 1$ and $G(x; a, b, c)$ is the corresponding cdf. Also, he has proved the following two Lemma which are used to prove the two theorems.

LEMMA 9.1. If $M(u, v, w; a, b, c)$ is defined as

$$M(u, v, w; a, b, c) = \int_0^\infty x^u\{G(x; a, b, c)\}^v\{g(x; a, b, c)\}^w\,dx,$$

(9.8)

where u is a non-negative real number, v and w are integers such that $v \geq 1$ and $w \geq 1$, then the integral in (9.8) is finite for those values of u and w for which $u + (ab - 1)w + 1 > 0$.

LEMMA 9.2. If $M(r, s, t; a, b, c)$ is defined as in (9.8), where r is a positive real number, s and t are integers such that $r \geq t - 1$, $s \geq 0$ and $t \geq 2$, then for $k = 1, 2, \ldots, t-1$ we have the following:

$$M(r,s,t;a,b,c) = \sum_{j=1}^{k+1} J^{(k,j)}_{r,s,t;a,b,c} M(r-k+jb-b, s+k, t-k; a,b,c) ,$$
(9.9)

where

$$J^{(0,1)}_{r,s,t;a,b,c} = 1 ;$$

$$J^{(k,1)}_{r,s,t;a,b,c} = -(s+k)^{-1}\{ab(t-k)+r+1-t\}J^{(k-1,1)}_{r,s,t;a,b,c} ;$$

$$J^{(k,j)}_{r,s,t;a,b,c} = (s+k)^{-1}(t-k)bc^{-b}J^{(k-1,j-1)}_{r,s,t;a,b,c}$$
$$- (s+k)^{-1}\{b(at+j-ak-1)+r+1-t\}J^{(k-1,j)}_{r,s,t;a,b,c}$$

for $j = 2, 3, \ldots, k$, and

$$J^{(k,k+1)}_{r,s,t;a,b,c} = (s+k)^{-1}(t-k)bc^{-b}J^{(k-1,k)}_{r,s,t;a,b,c} .$$

THEOREM 9.1. *If $X_{n:n;a,b,c}$ denotes the largest order statistic in a random sample of size $n \geq 2$ drawn from the generalized gamma distribution (9.6) and for any $p \geq 0$, let*

$$\mu^{(p)}_{n:n;a,b,c} = E\left(X^p_{n:n;a,b,c}\right) ;$$

then for any finite real $r \geq n-1$, we have

$$\mu^{(r+(n-1)(b-1))}_{n:n;a,b,c} = \left\{J^{(n-1,n)}_{r,0,n;a,b,c}\right\}^{-1} \left\{ \frac{b^{n-1}c^{r-n+1}\Gamma((abn-n+r+1)/b)}{\{\Gamma(a)\}^n n^{(abn-n+r+1-b)/b}} \right.$$
$$\left. - \sum_{j=1}^{n-1} J^{(n-1,j)}_{r,0,n;a,b,c} \mu^{(r-n+jb-b+1)}_{n:n;a,b,c} \right\} ,$$
(9.10)

where the J's are constants as defined in Lemma 9.2.

THEOREM 9.2. *Let $n \geq 2$ be an integer, k be a real number such that $k \geq -b$, and let*

$$\mu^{(k)}_{n:n;a,b,c} = E\left(X^k_{n:n;a,b,c}\right) .$$

Then for every $\alpha > 1$, we have

$$\mu^{(k)}_{n:n;a,b,c} = c^{-b} \sum_{i=0}^{n-1} \left\{ \binom{n-1}{i}(-b)^{-i}(a-i)^{-i-1} \right.$$
$$\left. \times \sum_{j=1}^{i+1} J^{(i,j)}_{i+k+b,n-i-1,i+1;a-1,b,c} \mu^{(k+bj)}_{n:n;a-1,b,c} \right\} ,$$
(9.11)

where

$$J^{(0,1)}_{k+b,n-1,1;a-1,b,c} = 1 , \qquad (9.12)$$

and the other J's are constants as defined in Lemma 9.2.

10. Exponential and related distributions

With the probabililty density function being

$$f(x) = e^{-x}, \quad 0 < x < \infty ,$$

and the cumulative distribution function being

$$F(x) = 1 - e^{-x}, \quad 0 < x < \infty ,$$

it is easy to see that

$$f(x) = 1 - F(x), \quad 0 < x < \infty .$$

Making use of this relation, Joshi (1978, 1982) has established the following recurrence relations:

For $n \geq 1$ and $k = 1, 2, \ldots$,

$$\mu^{(k)}_{1:n} = (k/n)\mu^{(k-1)}_{1:n} ; \qquad (10.1)$$

for $2 \leq r \leq n$ and $k = 1, 2, \ldots$,

$$\mu^{(k)}_{r:n} = \mu^{(k)}_{r-1:n-1} + (k/n)\mu^{(k-1)}_{r:n} ; \qquad (10.2)$$

for $1 \leq r \leq n - 1$,

$$\mu_{r,r+1:n} = \mu^{(2)}_{r:n} + \mu_{r:n}/(n-r) ; \qquad (10.3)$$

and for $1 \leq r < s \leq n$ and $s - r \geq 2$,

$$\mu_{r,s:n} = \mu_{r,s-1:n} + \mu_{r:n}/(n-s+1) . \qquad (10.4)$$

Note that the relations in (10.1)–(10.4) could be used in a simple recursive way in order to compute all the single and product moments of order statistics for all sample sizes.

These results have been extended by Joshi (1979b) and Balakrishnan and Joshi (1984) for the doubly truncated exponential distribution with cumulative distribution function $F(x)$ and probabilitiy density function

$$f(x) = \frac{1}{P-Q}e^{-x}, \quad Q_1 \leq x \leq P_1 ,$$

where Q and $1 - P$ $(Q < P)$ are the proportions of truncation on the left and right of the standard exponential distribution, respectively, and

$Q_1 = -\log(1-Q)$ and $P_1 = -\log(1-P)$. Denoting $(1-Q)/(P-Q)$ by Q_2 and $(1-P)/(P-Q)$ by P_2, it is easy to see that

$$f(x) = Q_2 - F(x)$$
$$= P_2 + \{1 - F(x)\} \ .$$

Making use of these two differential equations, Joshi (1979b) and Balakrishnan and Joshi (1984) have derived the following recurrence relations:

For $n \geq 2$ and $k = 1, 2, \ldots$,

$$\mu_{1:n}^{(k)} = Q_1^k Q_2 - \mu_{1:n-1}^{(k)} P_2 + (k/n)\mu_{1:n}^{(k-1)} \ ; \qquad (10.5)$$

for $2 \leq r \leq n-1$ and $k = 1, 2, \ldots$,

$$\mu_{r:n}^{(k)} = \mu_{r-1:n-1}^{(k)} Q_2 - \mu_{r:n-1}^{(k)} P_2 + (k/n)\mu_{r:n}^{(k-1)} \ ; \qquad (10.6)$$

for $n \geq 2$ and $k = 1, 2, \ldots$,

$$\mu_{n:n}^{(k)} = \mu_{n-1:n-1}^{(k)} Q_2 - P_1^k P_2 + (k/n)\mu_{n:n}^{(k-1)} \ ; \qquad (10.7)$$

for $2 \leq r \leq n$ and $k = 1, 2, \ldots$,

$$\mu_{r:n}^{(k)} = \mu_{r-1:n}^{(k)} + \binom{n}{r-1}\left[k\sum_{j=r}^{n}(1-Q_2)^{n-j}B(r, j-r+1)\mu_{r:j}^{(k-1)} \right.$$
$$\left. + (1-Q_2)^{n-r+1}\left(P_1^k - \mu_{r-1:r-1}^{(k)}\right)\right] \ ; \qquad (10.8)$$

for $2 \leq r \leq n$ and $k = 1, 2, \ldots$,

$$\mu_{r:n}^{(k)} = \mu_{r-1:n}^{(k)} + \binom{n}{r-1}\left[Q_2^{r-1}\left(\mu_{1:n-r+1}^{(k)} - Q_1^{(k)}\right) \right.$$
$$\left. - k\sum_{j=2}^{r}Q_2^{r-j}B(j-1, n-r+2)\mu_{j-1:n-r+j}^{(k-1)}\right] \ ; \qquad (10.9)$$

for $1 \leq r \leq n-2$,

$$\mu_{r,r+1:n} = \mu_{r:n}^{(2)} + \frac{1}{n-r}\left[\mu_{r:n} - nP_2\left(\mu_{r,r+1:n-1} - \mu_{r:n-1}^{(2)}\right)\right] \ ; \qquad (10.10)$$

for $n \geq 2$,

$$\mu_{n-1,n:n} = \mu_{n-1:n}^{(2)} + \mu_{n-1:n} - nP_2\left[P_1\mu_{n-1:n-1} - \mu_{n-1:n-1}^{(2)}\right] \ ; \qquad (10.11)$$

for $1 \leq r < s \leq n$ and $s - r \geq 2$,

$$\mu_{r,s:n} = \mu_{r,s-1:n} + \frac{1}{n-s+1}\left[\mu_{r:n} - nP_2\left(\mu_{r,s:n-1} - \mu_{r,s-1:n-1}\right)\right] \ ; \qquad (10.12)$$

and for $1 \leq r \leq n-2$,

$$\mu_{r,n:n} = \mu_{r,n-1:n} + \mu_{r:n} - nP_2\left[P_1\mu_{r:n-1} - \mu_{r,n-1:n-1}\right] . \tag{10.13}$$

Starting with the values of $\mu_{1:1}$ and $\mu_{1:1}^{(2)}$, for example, the recurrence relations (10.5)–(10.13) would enable one to evaluate the means, variances and covariances of all order statistics for all sample sizes in a simple and systematic recursive way for arbitrary values of the proportions of truncation Q and P. This computation of the moments could be carried out by means of a simple computer program without introducing serious rounding errors. Explicit finite series expressions for the first and second order single moments and product moments of order statistics for the special case of the right truncated exponential distribution have been derived by Saleh, Scott and Junkins (1975).

Similar simple recursive computational procedures have been suggested by Balakrishnan and Joshi (1981a) and Khan, Yaqub and Parvez (1983) for the evaluation of the single moments of order statistics from the Weibull distribution.

Recurrence relations between the moment generating functions of order statistics from the doubly truncated exponential distribution have been established by Mohie El-Din, Mahmoud and Abu-Youssef (1992). Again, recurrence relation between factorial moment generating functions of order statistics from the doubly truncated exponential distribution have been derived by Mohie El-Din and Sultan (1995).

11. Power function and related distributions

With the probability density function being

$$f(x) = va^{-v}x^{v-1}, \quad 0 \leq x \leq a, \quad a, v \geq 0,$$

and the cumulative distribution function being

$$F(x) = a^{-v}x^v, \quad 0 \leq x \leq a ,$$

it is easy to see that

$$F(x) = xf(x)/v, \quad 0 \leq x \leq a .$$

Making use of this relation, Balakrishnan and Joshi (1981b) have established the following recurrence relations for the single and product moments of order statistics:

For $1 \leq r \leq n-1$ and $k = 1, 2, \ldots,$

$$\mu_{r:n}^{(k)} = \{nv/(nv+k)\}\mu_{r,n-1}^{(k)} ; \tag{11.1}$$

for $n \geq 1$ and $k = 1, 2, \ldots,$

$$\mu_{n:n}^{(k)} = \frac{a^k nv}{(nv+k)} ; \tag{11.2}$$

for $n \geq 3$ and $1 \leq r \leq n-2$,

$$\mu_{r,r+1:n} = \frac{v}{v(n-r)+1}\left[(n-r)\mu_{r:n}^{(2)} + n\left(\mu_{r,r+1:n-1} - \mu_{r:n-1}^{(2)}\right)\right]; \quad (11.3)$$

for $n \geq 2$,

$$\mu_{n-1,n:n} = \frac{v}{v+1}\left[\mu_{n-1:n}^{(2)} + n\left(a\mu_{n-1:n-1} - \mu_{n-1:n-1}^{(2)}\right)\right]; \quad (11.4)$$

for $s - r \geq 2$ and $1 \leq r < s \leq n$,

$$\mu_{r,s:n} = \frac{v}{v(n-s+1)+1}\left[(n-s+1)\mu_{r,s-1:n} + n\left(\mu_{r,s:n-1} - \mu_{r,s-1:n-1}\right)\right]; \quad (11.5)$$

and for $1 \leq r \leq n-2$,

$$\mu_{r,n:n} = \frac{v}{v+1}\left[\mu_{r,n-1:n} + n\left(a\mu_{r:n-1} - \mu_{r,n-1:n-1}\right)\right]. \quad (11.6)$$

Note that the recurrence relations (11.1)–(11.6) could be used in a simple recursive way in order to compute all the single and product moments of order statistics for all sample sizes. Explicit expressions for these moments are also available and are due to Malik (1967a).

These results have also been extended by Balakrishnan and Joshi (1981b) for the doubly truncated power function distribution with cdf $F(x)$ and the density function

$$f(x) = \frac{1}{P-Q}va^{-v}x^{v-1}, \quad Q_1 \leq x \leq P_1, \quad a, v > 0 ,$$

where Q and $1 - P$ are, respectively, the proportions of truncation on the left and right of the power function distribution, and $Q_1 = aQ^{1/v}$ and $P_1 = aP^{1/v}$. Denoting $Q/(P-Q)$ by Q_2 and $P/(P-Q)$ by P_2, it is easy to see that

$$xf(x) = v\{Q_2 + F(x)\}$$
$$= v[P_2 - \{1 - F(x)\}], \quad Q_1 \leq x \leq P_1 .$$

Making use of these two differential equations, Balakrishnan and Joshi (1981b) have derived the following recurrence relations:
For $k = 1, 2, \ldots$,

$$\mu_{1:1}^{(k)} = \frac{v}{v+k}\left(P_2 P_1^k - Q_2 Q_1^k\right); \quad (11.7)$$

for $n \geq 2$ and $k = 1, 2, \ldots$,

$$\mu_{1:n}^{(k)} = \frac{nv}{nv+k}\left(P_2\mu_{1:n-1}^{(k)} - Q_2 Q_1^k\right); \quad (11.8)$$

for $n \geq 3$, $2 \leq r \leq n-1$, and $k = 1, 2, \ldots$,

$$\mu_{r:n}^{(k)} = \frac{nv}{nv+k}\left(P_2\mu_{r:n-1}^{(k)} - Q_2\mu_{r-1:n-1}^{(k)}\right) ; \tag{11.9}$$

for $n \geq 2$ and $k = 1, 2, \ldots$,

$$\mu_{n:n}^{(k)} = \frac{nv}{nv+k}\left(P_2 P_1^k - Q_2\mu_{n-1:n-1}^{(k)}\right) ; \tag{11.10}$$

for $n \geq 3$ and $1 \leq r \leq n-2$,

$$\mu_{r,r+1:n} = \frac{v}{v(n-r)+1}\left[(n-r)\mu_{r:n}^{(2)} + nP_2\left(\mu_{r,r+1:n-1} - \mu_{r:n-1}^{(2)}\right)\right] ; \tag{11.11}$$

for $n \geq 2$,

$$\mu_{n-1,n:n} = \frac{v}{v+1}\left[\mu_{n-1:n}^{(2)} + nP_2\left(P_1\mu_{n-1:n-1} - \mu_{n-1:n-1}^{(2)}\right)\right] ; \tag{11.12}$$

for $1 \leq r < s \leq n$ and $s - r \geq 2$,

$$\mu_{r,s:n} = \frac{v}{v(n-s+1)+1}\left[(n-s+1)\mu_{r,s-1:n} + nP_2\left(\mu_{r,s:n-1} - \mu_{r,s-1:n-1}\right)\right] ; \tag{11.13}$$

and for $1 \leq r \leq n-2$,

$$\mu_{r,n:n} = \frac{v}{v+1}\left[\mu_{r,n-1:n} + nP_2\left(P_1\mu_{r:n-1} - \mu_{r,n-1:n-1}\right)\right] . \tag{11.14}$$

Recurrence relations (11.7)–(11.14) can be used systematically to obtain all the single and product moments of order statistics from the doubly truncated power function distribution for any sample size.

Recently, Mohie El-Din, Mahmoud and Sultan (1996) have used the properties of the hypergeometric function and its contiguous functions to derive the following recurrence relations between the moments of order statistics from the doubly truncated power function distribution:

$$\left(n-r+1+\frac{k}{v}\right)\mu_{r:n}^{(k)} = \frac{k}{v}a^v P\mu_{r:n}^{(k-v)} + (n-r+1)\mu_{r+1:n}^{(k)} ; \tag{11.15}$$

$$\left(\frac{k}{v}+n\right)\mu_{r:n}^{(k)} = \frac{k}{v}a^v P\mu_{r:n}^{(k-v)} + n\mu_{r-1:n-1}^{(k)} ; \tag{11.16}$$

$$(n+1)\left(\frac{k}{v}+r\frac{1}{P_2}\right)\mu_{r:n}^{(k)} = \frac{k}{v}a^v Q\mu_{r:n}^{(k-v)} + r\left(n+1+\frac{k}{v}\right)\frac{1}{P_2}\mu_{r+1:n+1}^{(k)} ; \tag{11.17}$$

$$(n+1)Q\mu_{r:n}^{(k)} = (n+1)Pa^v\mu_{r:n}^{(k-v)} - rP\frac{1}{P_2}\mu_{r+1:n+1}^{(k)} ; \tag{11.18}$$

$$P\left[\frac{2k}{v}+n+1-\frac{1}{P_2}\left(n+1-r+\frac{k}{v}\right)\right]\mu_{r:n}^{(k)}$$
$$=\frac{k}{v}PQa^v\mu_{r:n}^{(k-v)}+\left(n+1+\frac{k}{v}\right)a^{-v}\mu_{r:n}^{(k+v)}\ ; \quad (11.19)$$

$$\left(\frac{k}{v}+r\right)\mu_{r:n}^{(k)}=\frac{k}{v}Qa^v\mu_{r:n}^{(k-v)}+r\mu_{r+1:n}^{(k)}\ ; \quad (11.20)$$

$$P\left(r+\frac{k}{v}\right)\mu_{r:n}^{(k)}=\left(n+1+\frac{k}{v}\right)a^{-v}\mu_{r:n}^{(k+v)}-Q(n+1-r)\mu_{r-1:n}^{(k)}\ ; \quad (11.21)$$

$$\left(n+1-r+\frac{k}{v}\right)Q\mu_{r:n}^{(k)}=\left(n+1+\frac{k}{v}\right)a^{-v}\mu_{r:n}^{(k+v)}-rP\mu_{r+1:n}^{(k)}\ ; \quad (11.22)$$

$$P\left[1+\frac{k}{v}+(r-1)\frac{1}{P_2}\right]\mu_{r:n}^{(k)}=\left(n+1+\frac{k}{v}\right)a^{-v}\mu_{r:n}^{(k+v)}-nQ\mu_{r-1:n-1}^{(k)}\ ; \quad (11.23)$$

$$P\left[n+1-2r-\frac{1}{P_2}\left(\frac{k}{v}+n+1-r\right)\right]\mu_{r:n}^{(k)}$$
$$=Q(n+1-r)\mu_{r-1:n}^{(k)}-Pr\mu_{r+1:n}^{(k)}\ ; \quad (11.24)$$

$$P(n+1)\left[n+1-r-\frac{1}{P_2}\left(\frac{k}{v}+n+1\right)\right]\mu_{r:n}^{(k)}$$
$$=Q(n+1)(n+1-r)\mu_{r-1:n}^{(k)}-rP\frac{1}{P_2}\left(n+1+\frac{k}{v}\right)\mu_{r+1:n+1}^{(k)}\ ; \quad (11.25)$$

$$P\left[r-n+\left(n+\frac{k}{v}\right)\frac{1}{P_2}\right]\mu_{r:n}^{(k)}=rP\mu_{r+1:n}^{(k)}-nQ\mu_{r-1:n-1}^{(k)}\ ; \quad (11.26)$$

and

$$P(n+1)\left[n-\frac{1}{P_2}\left(\frac{k}{v}+r+n\right)\right]\mu_{r:n}^{(k)}$$
$$=n(n+1)Q\mu_{r-1:n-1}^{(k)}-rP\frac{1}{P_2}\left(n+1+\frac{k}{v}\right)\mu_{r+1:n+1}^{(k)}\ . \quad (11.27)$$

12. Pareto and related distributions

With the probability density function as

$$f(x)=va^v x^{-(v+1)},\quad x\geq a,\quad a,v>0\ ,$$

and the cumulative distribution function as

$$F(x) = 1 - a^v x^{-v}, \quad x \geq a,$$

it is easy to see that

$$1 - F(x) = xf(x)/v, \quad x \geq 0. \tag{12.1}$$

For the Pareto distribution, single and product moments of order statistics have been discussed by several authors; for example, see Downton (1954), Malik (1966), Kabe (1972) and Huang (1975).

In particular, Huang (1975) has shown that $\mu_{r:n}^{(k)}$ exists for $v > k/(n-r+1)$, and is given by

$$\mu_{r:n}^{(k)} = a^k \frac{n!}{(n-r)!} \frac{\Gamma(n-r+1-k/v)}{\Gamma(n+1-k/v)},$$

where $\Gamma(\cdot)$ is the complete gamma function. Making use of the relation in (12.1), Balakrishnan and Joshi (1982) have derived the following recurrence relations:

For $n \geq 1$ and $k = 1, 2, \ldots$,

$$(nv - k)\mu_{1:n}^{(k)} = nva^k; \tag{12.2}$$

for $2 \leq r \leq n$ and $k = 1, 2, \ldots$,

$$(nv - k)\mu_{r:n}^{(k)} = nv\mu_{r-1:n-1}^{(k)}; \tag{12.3}$$

for $1 \leq r \leq n - 1$,

$$\{(n-r)v - 1\}\mu_{r,r+1:n} = v(n-r)\mu_{r:n}^{(2)}; \tag{12.4}$$

and for $1 \leq r < s \leq n$ and $s - r \geq 2$,

$$\{v(n-s+1) - 1\}\mu_{r,s:n} = v(n-s+1)\mu_{r,s-1:n}. \tag{12.5}$$

Note that when $v > k$, relations (12.2)–(12.5) could be used to compute the first k single moments and also product moments of all order statistics for all sample sizes. In particular, if $v > 2$, these relations would enable one to evaluate the means, variances and covariances of all order statistics for any sample size.

However, in the case of the doubly truncated Pareto distribution, range of X is finite and so $\mu_{r:n}^{(k)}$ exists for all r, n and k. Without loss of any generality, taking the scale parameter $a = 1$ since the random variable $Y = X/a$ has a one-parameter Pareto distribution, Balakrishnan and Joshi (1982) have considered the doubly truncated Pareto distribution with cdf $F(x)$ and density function

$$f(x) = \frac{1}{P-Q} v x^{-(v+1)}, \quad Q_1 \leq x \leq P_1,$$

where Q and $1 - P$ ($P < Q$) are, respectively, the proportions of truncation on the left and right of the standard Paetro distribution, and $Q_1 = (1-Q)^{-1/v}$ and

$P_1 = (1-P)^{-1/v}$. Now denoting $(Q-1)/(P-Q)$ by Q_2 and $(P-1)/(P-Q)$ by P_2, Balakrishnan and Joshi (1982) have established the following recurrence relations for the single and product moments of order statistics:

For $n \geq 2$ and $k = 1, 2, \ldots,$

$$(nv - k)\mu_{n:n}^{(k)} = nv\left(P_2 P_1^k - Q_2 \mu_{n-1:n-1}^{(k)}\right) ; \qquad (12.6)$$

for $n \geq 2$ and $k = 1, 2, \ldots,$

$$(nv - k)\mu_{1:n}^{(k)} = nv\left(P_2 \mu_{1:n-1}^{(k)} - Q_2 Q_1^k\right) ; \qquad (12.7)$$

for $n \geq 3$, $2 \leq r \leq n-1$ and $k = 1, 2, \ldots,$

$$(nv - k)\mu_{r:n}^{(k)} = nv\left(P_2 \mu_{r:n-1}^{(k)} - Q_2 \mu_{r-1:n-1}^{(k)}\right) ; \qquad (12.8)$$

for $n \geq 3$, $2 \leq r \leq n-1$ and $k = 1, 2, \ldots,$

$$\{(n-r+1)v - k\}\mu_{r:n}^{(k)} = v\left[(n-r+1)\mu_{r-1:n}^{(k)} + nP_2\left(\mu_{r:n-1}^{(k)} - \mu_{r-1:n-1}^{(k)}\right)\right] ; \qquad (12.9)$$

for $n \geq 2$,

$$(v-1)\mu_{n-1,n:n} = v\left[\mu_{n-1:n}^{(2)} + nP_2\left(P_1 \mu_{r-1:n-1} - \mu_{n-1:n-1}^{(2)}\right)\right] ; \qquad (12.10)$$

for $n \geq 3$ and $1 \leq r \leq n-2$,

$$\{v(n-r) - 1\}\mu_{r,r+1:n} = v\left[(n-r)\mu_{r:n}^{(2)} + nP_2\left(\mu_{r,r+1:n-1} - \mu_{r:n-1}^{(2)}\right)\right] ; \qquad (12.11)$$

for $n \geq 2$,

$$(v-1)\mu_{1,2:n} = v\left[\mu_{2:n}^{(2)} + nQ_2\left(\mu_{1:n-1}^{(2)} - Q_1 \mu_{1:n-1}\right)\right] ; \qquad (12.12)$$

for $n \geq 3$ and $2 \leq r \leq n-1$,

$$(rv - 1)\mu_{r,r+1:n} = v\left[r\mu_{r+1:n}^{(2)} + nQ_2\left(\mu_{r:n-1}^{(2)} - \mu_{r-1,r:n-1}\right)\right] ; \qquad (12.13)$$

for $s - r \geq 2$ and $1 \leq r < s \leq n$,

$$\{v(n-s+1) - 1\}\mu_{r,s:n} = v\left[(n-s+1)\mu_{r,s-1:n} + nP_2\left(\mu_{r,s:n-1} - \mu_{r,s-1:n-1}\right)\right] ; \qquad (12.14)$$

and for $1 \leq r \leq n-2$,

$$(v-1)\mu_{r,n:n} = v\left[\mu_{r,n-1:n} + nP_2\left(P_1 \mu_{r:n-1} - \mu_{r,n-1:n-1}\right)\right] ; \qquad (12.15)$$

if $v(n - s + 1) = 1$, then for $1 \leq r < s \leq n$ and $s - r \geq 2$,

$$(s - r - 1)\mu_{r,s:n} = n\left[P_2\mu_{r,s:n-1} - Q_2\mu_{r,s-1:n-1}\right] - r\mu_{r+1,s:n}, \qquad (12.16)$$

and for $1 \leq r \leq n - 2$,

$$(n - r - 1)\mu_{r,n:n} = n\left[P_2 P_1 \mu_{r:n-1} - Q_2\mu_{r,n-1:n-1}\right] - r\mu_{r+1,n:n}. \qquad (12.17)$$

Thus, starting from

$$\mu_{1:1}^{(k)} = \log(Q_2/P_2)/(P - Q), \qquad \text{if } k = v$$
$$= \frac{v}{v - k}(P_2 P_1^k - Q_2 Q_1^k), \qquad \text{if } k \neq v$$

obtained by direct integration, recurrence relations (12.6)–(12.8) would enable one to obtain all the single moments of order statistics, except when $k = nv$. Note that it is possible to have a situation with $k = nv$. For example, when $n = 6$, $v = 1/6$ and $k = 1$. For this case, relations (12.6)–(12.8) immediately yield

$$\mu_{r-1:n-1}^{(k)} = P_2\mu_{r:n-1}^{(k)}/Q_2, \quad 2 \leq r \leq n - 1, \qquad (12.18)$$

and

$$\mu_{n-1:n-1}^{(k)} = P_2 P_1^k/Q_2. \qquad (12.19)$$

Noting now that relations (12.6)–(12.9) along with relations (12.18) and (12.19) would enable one to evaluate all the single moments except $\mu_{n:n}^{(k)}$ and $\mu_{1:n}^{(k)}$, Balakrishnan and Joshi (1982) have obtained the following explicit finite series expressions for these moments when $nv = k$ is an integer:

For $n \geq 1$,

$$\mu_{n:n}^{(k)} = n(-1)^{n-1}(P - Q)^{-n}\left[\log(Q_2/P_2) + \sum_{j=0}^{n-2} P_2^{-(j+1)}/(j+1)\right]$$

and

$$\mu_{1:n}^{(k)} = n(P - Q)^{-n}\left[\log(Q_2/P_2) + \sum_{j=0}^{n-2}(-1)^j Q_2^{-(j+1)}/(j+1)\right].$$

Further, note that relations (12.10) and (12.11) could be applied to evaluate $\mu_{r,r+1:n}$ ($1 \leq r \leq n - 1$) recursively except when $v(n - r) = 1$ and that relations (12.12) and (12.13) could be applied to evaluate $\mu_{r,r+1:n}$ ($1 \leq r \leq n - 1$) recursively except when $vr = 1$. Hence, the recurrence relations (12.10)–(12.13) give the product moments $\mu_{r,r+1:n}$ ($1 \leq r \leq n - 1$) for all values of r, n and v except for the case when $vr = v(n - r) = 1$, that is, when $n = 2m$, $r = m$ and $v = 1/m$. However, for any arbitrary continuous distribution, we also have the relation

$$2\mu_{m,m+1:2m} = \sum_{j=1}^{m-1}\sum_{k=0}^{m-1}(-1)^{j+k-1}\binom{2m}{k}\binom{2m-j}{k}\mu_{m-j,m-j+1:2m-j-k}$$
$$+ \sum_{k=1}^{m-1}(-1)^{k+1}\binom{2m}{k}\mu_{m,m+1:2m-k} + \binom{2m}{m}\mu_{1:m}\mu_{m:m},$$

mentioned earlier in Section 4, which will enable one to determine $\mu_{m,m+1:2m}$ when $v = 1/m$, using the single and product moments of order statistics from smaller sample sizes. Note that in this way, one could systematically compute all the product moments $\mu_{r,r+1:n}$ ($1 \le r \le n-1$) for all choices of r, n and v, and the remaining product moments $\mu_{r,s:n}$ ($1 \le r < s \le n$, $s - r \ge 2$) could all be determined by using Relation 4.6.

Khurana and Jha (1991) have used the properties of the hypergeometric function and its contiguous functions [see Rainville (1960)] to obtain some recurrence relations between moments of order statistics from the Pareto distribution. They are as follows:

$$(k - vr)\mu_{r:n}^{(k)} = k(1 - Q)\mu_{r:n}^{(k+v)} - vr\mu_{r+1:n}^{(k)} ; \qquad (12.20)$$

$$(nv - k)\mu_{r:n}^{(k)} = nv\mu_{r:n-1}^{(k)} - k(1 - Q)\mu_{r:n}^{(k+v)} ; \qquad (12.21)$$

$$(n+1)[kQ_2 + v(n-r+1)]\mu_{r:n}^{(k)} = v(n+1-k/v)(n+1-r)\mu_{r:n+1}^{(k)}$$
$$+ kQ_2(n+1)(1-P)\mu_{r:n}^{(k+v)} ; \qquad (12.22)$$

$$(1 - P)\mu_{r:n}^{(k)} = \mu_{r:n}^{(k-v)} - \left(1 - \frac{r}{n+1}\right)(P - Q)\mu_{r:n+1}^{(k)} ; \qquad (12.23)$$

$$(1 - P)\mu_{r:n}^{(k)} = (1 - Q)\mu_{r-1:n}^{(k)} - \left(\frac{n+1-k/v}{n+1}\right)(P - Q)\mu_{r:n+1}^{(k)} ; \qquad (12.24)$$

$$\left[(n+1-2k/v) - \frac{k/v - r}{Q_2}\right]\mu_{r:n}^{(k)} = \frac{n+1-k/v}{1-Q}\mu_{r:n}^{(k-v)} - \frac{k}{v}(1-P)\mu_{r:n}^{(k+v)} ;$$
$$(12.25)$$

$$(n+1-r-k/v)\mu_{r:n}^{(k)} = (n+1-r)\mu_{r-1:n}^{(k)} - \frac{k}{v}(1-P)\mu_{r:n}^{(k+v)} ; \qquad (12.26)$$

$$(1-Q)(n+1-r-k/v)\mu_{r:n}^{(k)} = (n+1-k/v)\mu_{r:n}^{(k-v)} - r(1-P)\mu_{r+1:n}^{(k)} ; \qquad (12.27)$$

$$(r-k/v)(1-P)\mu_{r:n}^{(k)} = (n+1-k/v)\mu_{r:n}^{(k-v)} - (n+1-r)(1-Q)\mu_{r-1:n}^{(k)} ; \qquad (12.28)$$

$$[(1-Q)(1-k/v) + (n-r)(P-Q)]\mu_{r:n}^{(k)}$$
$$= (n+1-k/v)\mu_{r:n}^{(k-v)} - n(1-P)\mu_{r:n-1}^{(k)} ; \tag{12.29}$$

$$\left[n+1-2r-\frac{(r-k/v)}{Q_2}\right]\mu_{r:n}^{(k)} = (n+1-r)\mu_{r-1:n}^{(k)} - r\frac{1-P}{1-Q}\mu_{r+1:n}^{(k)} ; \tag{12.30}$$

$$\left[\frac{n+1-k/v}{Q_2}-r\right]\mu_{r:n}^{(k)} = \frac{(n+1-k/v)(n+1-r)}{(n+1)Q_2}\mu_{r:n+1}^{(k)} + r\frac{1-P}{1-Q}\mu_{r+1:n}^{(k)} ; \tag{12.31}$$

$$\left[1-r-\frac{(n+k/v)}{Q_2}\right]\mu_{r:n}^{(k)} = (n+1-r)\mu_{r-1:n}^{(k)} - n\frac{1-P}{1-Q}\mu_{r:n+1}^{(k)} ; \tag{12.32}$$

and

$$\left[n - \frac{(k/v+r-2n-1)}{Q_2}\right]\mu_{r:n}^{(k)} = \frac{n(1-P)}{1-Q}\mu_{r:n-1}^{(k)}$$
$$+ \frac{(n+1-k/v)(n+1-r)}{(n+1)Q_2}\mu_{r:n+1}^{(k)} . \tag{12.33}$$

Recurrence relations for moments of order statistics for untruncated Pareto distribution are readily obtained from (12.20)–(12.33) on setting $Q = 1 - P = 0$. These are:

$$\{(k/v) - r\}\mu_{r:n}^{(k)} = (k/v)\mu_{r-1:n}^{(k+v)} - r\mu_{r+1:n}^{(k)} ; \tag{12.34}$$

$$(n+1-r-k/v)\mu_{r:n}^{(k)} = \frac{(n+1-k/v)(n+1-r)}{n+1}\mu_{r:n+1}^{(k)} ; \tag{12.35}$$

$$\mu_{r:n}^{(k-v)} = \frac{n+1-r}{n+1}\mu_{r:n+1}^{(k)} ; \tag{12.36}$$

$$\mu_{r-1:n}^{(k)} = \frac{n+1-k/v}{n+1}\mu_{r:n+1}^{(k)} ; \tag{12.37}$$

$$(n+1-r-k/v)\mu_{r:n}^{(k)} = (n+1-k/v)\mu_{r:n}^{(k-v)} ; \tag{12.38}$$

and

$$(n+1-r-k/v)\mu_{r:n}^{(k)} = (n+1-r)\mu_{r-1:n}^{(k)} . \tag{12.39}$$

13. Rayleigh distribution

With the density function as

$$f(x) = xe^{-x^2/2}, \quad x \geq 0,$$

and the cumulative distribution function as

$$F(x) = 1 - e^{-x^2/2}, \quad x \geq 0,$$

it is easy to see that

$$f(x) = x\{1 - F(x)\}, \quad x \geq 0.$$

Making use of this relation, it is easy to establish the following recurrence relations for the single and product moments of order statistics:

For $n \geq 1$ and $k = 0, 1, 2, \ldots$,

$$\mu_{1:n}^{(k+2)} = \frac{k+2}{n} \mu_{1:n}^{(k)} ; \tag{13.1}$$

for $2 \leq r \leq n$ and $k = 0, 1, 2, \ldots$,

$$\mu_{r:n}^{(k+2)} = \mu_{r-1:n}^{(k+2)} + \frac{k+2}{n-r+1} \mu_{r:n}^{(k)} ; \tag{13.2}$$

for $1 \leq r \leq n-1$ and $i, j \geq 0$,

$$\mu_{r,r+1:n}^{(i,j+2)} = \mu_{r:n}^{(i+j+2)} + \frac{j+2}{n-r} \mu_{r,r+1:n}^{(i,j)} ; \tag{13.3}$$

for $1 \leq r < s \leq n$, $s - r \geq 2$, and $i, j \geq 0$,

$$\mu_{r,s:n}^{(i,j+2)} = \mu_{r,s-1:n}^{(i,j+2)} + \frac{j+2}{n-s+1} \mu_{r,s:n}^{(i,j)} ; \tag{13.4}$$

for $n \geq 2$, and $i, j \geq 0$,

$$\mu_{1,2:n}^{(i+2,j)} = (i+2)\mu_{1,2:n}^{(i,j)} + (n-1)\mu_{1:n}^{(i+j+2)} ; \tag{13.5}$$

for $2 \leq r \leq n-1$, and $i, j \geq 0$,

$$\mu_{r,r+1:n}^{(i+2,j)} = \frac{i+2}{r} \mu_{r,r+1:n}^{(i,j)} - \frac{1}{r}\left((n-r)\mu_{r:n}^{(i+j+2)} - n\mu_{r-1,r:n-1}^{(i+2,j)}\right) ; \tag{13.6}$$

and for $1 \leq r < s \leq n$, $s - r \geq 2$, and $i, j \geq 0$,

$$\mu_{r,s:n}^{(i+2,j)} = \mu_{r+1,s:n}^{(i+2,j)} - \frac{n}{r}\left(\mu_{r,s-1:n-1}^{(i+2,j)} - \mu_{r-1,s-1:n-1}^{(i+2,j)}\right) + \frac{i+2}{r} \mu_{r,s:n}^{(i,j)} . \tag{13.7}$$

Exact and explicit expressions for the first two single moments and also the product moments of order statistics have been given by Dyer and Whisenand

(1973a,b). After computing these moments from the exact and explicit expressions, the recurrence relations (13.1)–(13.7) could then be systematically applied in a simple recursive way in order to compute the higher order single and product moments of order statistics for all sample sizes from the Rayleigh distribution.

14. Linear-exponential distribution

With the density function of the linear-exponential distribution with increasing hazard rate as

$$f(x) = (\lambda + vx)e^{-(\lambda x + vx^2/2)}, \quad 0 \leq x < \infty, \quad \lambda, v > 0,$$

and the cumulative distribution function as

$$F(x) = 1 - e^{-(\lambda x + vx^2/2)}, \quad 0 \leq x < \infty,$$

it is easy to note that

$$f(x) = (\lambda + vx)\{1 - F(x)\}$$
$$= (\lambda + vx) - (\lambda + vx)F(x), \quad x \geq 0.$$

Making use of these relations, Balakrishnan and Malik (1986b) have established the following recurrence relations for the single and product moments of order statistics:

For $n \geq 1$ and $k = 0, 1, 2, \ldots$,

$$\mu_{1:n}^{(k+2)} = \frac{k+2}{v}\left(\frac{1}{n}\mu_{1:n}^{(k)} - \frac{\lambda}{k+1}\mu_{1:n}^{(k+1)}\right); \tag{14.1}$$

for $2 \leq r \leq n$ and $k = 0, 1, 2, \ldots$,

$$\mu_{r:n}^{(k+2)} = \mu_{r-1:n}^{(k+2)} + \frac{k+2}{v}\left[\frac{1}{n-r+1}\mu_{r:n}^{(k)} - \frac{\lambda}{k+1}\left(\mu_{r:n}^{(k+1)} - \mu_{r-1:n-1}^{(k+1)}\right)\right]; \tag{14.2}$$

for $1 \leq r \leq n-1$ and $i, j \geq 0$,

$$\mu_{r,r+1:n}^{(i,j+2)} = \mu_{r:n}^{(i+j+2)} + \frac{j+2}{v}\left[\frac{1}{n-r}\mu_{r,r+1:n}^{(i,j)} - \frac{\lambda}{j+1}\left(\mu_{r,r+1:n}^{(i,j+1)} - \mu_{r:n}^{(i+j+1)}\right)\right]; \tag{14.3}$$

for $1 \leq r < s \leq n$, $s - r \geq 2$ and $i, j \geq 0$,

$$\mu_{r,s:n}^{(i,j+2)} = \mu_{r,s-1:n}^{(i,j+2)} + \frac{j+2}{v}\left[\frac{1}{n-s+1}\mu_{r,s:n}^{(i,j)} - \frac{\lambda}{j+1}\left(\mu_{r,s:n}^{(i,j+1)} - \mu_{r,s-1:n}^{(i,j+1)}\right)\right]; \tag{14.4}$$

for $n \geq 2$ and $i, j \geq 0$,

$$\mu_{1,2:n}^{(i+2,j)} = \frac{i+2}{v}\left[\mu_{1,2:n}^{(i,j)} - \frac{\lambda}{i+1}\left\{(n-1)\mu_{1,:n}^{(i+j+1)} + \mu_{1,2:n}^{(i+1,j)}\right\}\right]$$
$$- (n-1)\mu_{1:n}^{(i+j+2)} ; \qquad (14.5)$$

for $2 \leq r \leq n-1$ and $i, j \geq 0$,

$$\mu_{r,r+1:n}^{(i+2,j)} = \frac{i+2}{rv}\left[\mu_{r,r+1:n}^{(i,j)} - \frac{\lambda}{i+1}\left\{(n-r)\mu_{r:n}^{(i+j+1)} - n\mu_{r-1,r:n-1}^{(i+1,j)} + r\mu_{r,r+1:n}^{(i+1,j)}\right\}\right]$$
$$- \frac{1}{r}\left\{(n-r)\mu_{r:n}^{(i+j+2)} - n\mu_{r-1,r:n-1}^{(i+2,j)}\right\} ; \qquad (14.6)$$

and for $1 \leq r < s \leq n$, $s-r \geq 2$ and $i, j \geq 0$,

$$\mu_{r,s:n}^{(i+2,j)} = \mu_{r+1,s:n}^{(i+2,j)} - \frac{n}{r}\left(\mu_{r,s-1:n-1}^{(i+2,j)} - \mu_{r-1,s-1:n-1}^{(i+2,j)}\right)$$
$$+ \frac{i+2}{rv}\left[\mu_{r,s:n}^{(i,j)} - \frac{\lambda}{i+1}\left\{n\left(\mu_{r,s-1:n-1}^{(i+1,j)} - \mu_{r-1,s-1:n-1}^{(i+1,j)}\right)\right.\right. \qquad (14.7)$$
$$\left.\left. - r\left(\mu_{r+1,s:n}^{(i+1,j)} - \mu_{r,s:n}^{(i+1,j)}\right)\right\}\right] .$$

Letting $v \to 0$ in these recurrence relations, we deduce the recurrence relations presented earlier in Section 8 for the single and product moments of order statistics from exponential distribution.

Setting $\lambda = 0$ and $v = 1$ in these recurrence relations, we deduce the recurrence relations presented earlier in Section 11 for the single and product moments of order statistics from the Rayleigh distribution.

Exact and explicit expressions for the means and product moments of order statistics have been derived by Balakrishnan and Malik (1986b). The recurrence relations (14.1)–(14.7) could then be systematically applied in a simple recursive way in order to compute the higher order single and product moments of order statistics for all sample sizes from the linear exponential distribution with increasing hazard rate.

The doubly truncated linear-exponential pdf is

$$f(x) = \frac{\lambda + vx}{P - Q}e^{-(\lambda x + vx^2/2)}, \quad Q_1 \leq x \leq P_1, \quad \lambda, v > 0, \qquad (14.8)$$

where $P = 1 - e^{-(\lambda P_1 + vP_1^2/2)}$ and $Q = 1 - e^{-(\lambda Q_1 + vQ_1^2/2)}$.

Letting $Q_2 = \frac{1-Q}{P-Q}$ and $P_2 = \frac{1-P}{P-Q}$, the cumulative distribution function (cdf) is

$$F(x) = Q_2 - \frac{f(x)}{(\lambda + vx)}, \qquad (14.9)$$

and hence

$$f(x) = P_2(\lambda + vx) + (\lambda + vx)(1 - F(x)) \ . \tag{14.10}$$

Making use of the relation (14.10), Mohie El-Din et al. (1997) have derived the following recurrence relations:

For $n \geq 1$ and $k = 1, 2, 3, \ldots,$

$$\mu_{1:n}^{(k)} = \frac{n\lambda}{k+1} \left[\mu_{1:n}^{(k+1)} + P_2 \mu_{1:n-1}^{(k+1)} - Q_2 Q_1^{(k+1)} \right]$$
$$+ \frac{nv}{k+2} \left[\mu_{1:n}^{(k+2)} + P_2 \mu_{1:n-1}^{(k+2)} - Q_2 Q_1^{(k+2)} \right] \ ; \tag{14.11}$$

for $2 \leq r \leq n$,

$$\mu_{r:n}^{(k)} = \frac{n\lambda}{k+1} \left[\mu_{r:n}^{(k+1)} + P_2 \mu_{r:n-1}^{(k+1)} - Q_2 \mu_{r-1:n-1}^{(k+1)} \right]$$
$$+ \frac{nv}{k+2} \left[\mu_{r:n}^{(k+2)} + P_2 \mu_{r:n-1}^{(k+2)} - Q_2 \mu_{r-1:n-1}^{(k+2)} \right] \ ; \tag{14.12}$$

for $1 \leq r \leq n-1$ and $i, j \geq 0$,

$$\mu_{r,r+1:n}^{(i,j)} = \frac{n\lambda P_2}{j+1} \left[\mu_{r,r+1:n-1}^{(i,j+1)} - \mu_{r,n-1}^{(i+j+1)} \right] + \frac{nv P_2}{j+2} \left[\mu_{r,r+1:n-1}^{(i,j+2)} - \mu_{r,n-1}^{(i+j+2)} \right]$$
$$+ \frac{\lambda(n-r)}{j+1} \left[\mu_{r,r+1:n}^{(i,j+1)} - \mu_{r,n}^{(i+j+1)} \right] + \frac{v(n-r)}{j+2} \left[\mu_{r,r+1:n}^{(i,j+2)} - \mu_{r,n}^{(i+j+2)} \right] \ ; $$
$$\tag{14.13}$$

for $1 \leq r < s \leq n$, $n - r \geq 2$ and $i, j \geq 0$,

$$\mu_{r,s:n}^{(i,j)} = \frac{n\lambda P_2}{j+1} \left[\mu_{r,s:n-1}^{(i,j+1)} - \mu_{r,s-1,n-1}^{(i,j+1)} \right] + \frac{nv P_2}{j+2} \left[\mu_{r,s:n-1}^{(i,j+2)} - \mu_{r,s-1,n-1}^{(i,j+2)} \right]$$
$$+ \frac{\lambda(n-s+1)}{j+1} \left[\mu_{r,s:n}^{(i,j+1)} - \mu_{r,s-1,n}^{(i,j+1)} \right]$$
$$+ \frac{v(n-s+1)}{j+2} \left[\mu_{r,s:n}^{(i,j+2)} - \mu_{r,s-1,n}^{(i,j+2)} \right] \ ; \tag{14.14}$$

for $n \geq 2$ and $i, j \geq 0$,

$$\mu_{1,2:n}^{(i,j)} = \frac{\lambda}{i+1} \left[nQ_2 \mu_{1:n-1}^{(i+j+1)} - \mu_{2:n}^{(i+j+1)} + \mu_{1,2:n}^{(i+1,j)} \right]$$
$$+ \frac{v}{i+2} \left[nQ_2 \mu_{1:n-1}^{(i+j+2)} - \mu_{2:n}^{(i+j+2)} + \mu_{1,2:n}^{(i+2,j)} \right] \tag{14.15}$$
$$- n \left[\frac{\lambda Q_2 Q_1^{i+1}}{i+1} + \frac{v Q_2 Q_1^{i+2}}{i+2} \right] \mu_{1:n-1}^{(j)} \ ;$$

for $2 \leq r \leq n-1$ and $i, j \geq 0$,

$$\mu_{r,r+1:n}^{(i,j)} = \frac{\lambda}{i+1} \left[nQ_2 \left(\mu_{r,n-1}^{(i+j+1)} - \mu_{r-1,r:n-1}^{(i+1,j)} \right) - r \left(\mu_{r+1:n}^{(i+j+1)} - \mu_{r,r+1:n}^{(i+1,j)} \right) \right]$$
$$+ \frac{v}{i+2} \left[nQ_2 \left(\mu_{r:n-1}^{(i+j+2)} - \mu_{r-1,r:n-1}^{(i+2,j)} \right) - r \left(\mu_{r+1:n}^{(i+j+2)} - \mu_{r,r+1:n}^{(i+2,j)} \right) \right] ;$$

(14.16)

and for $1 \leq r < s \leq n$, $s - r \geq 2$ and $i, j \geq 0$,

$$\mu_{r,s:n}^{(i+2,j)} = \mu_{r+1,s:n}^{(i+2,j)} - \frac{nQ_2}{r} \left(\mu_{r,s-1,n-1}^{(i+2,j)} - \mu_{r-1,s-1,n-1}^{(i+2,j)} \right)$$
$$+ \frac{i+2}{rv} \left\{ \mu_{r,s:n}^{(i,j)} - \frac{\lambda}{i+1} \left[Q_2 n \left(\mu_{r,s-1,n-1}^{(i+1,j)} - \mu_{r-1,s-1,n-1}^{(i+1,j)} \right) \right. \right.$$
$$\left. \left. - r \left(\mu_{r+1,s:n}^{(i+1,j)} - \mu_{r,s:n}^{(i+1,j)} \right) \right] \right\} .$$

(14.17)

Letting $Q_1 = 0, Q_2 = 1$ and $P_2 = 0$ in the relations (14.11)–(14.17), we obtain the results of Balakrishnan and Malik (1986b) for the untruncated case.

15. Lomax distribution

With the density function and cumulative distribution function as

$$f(x) = \alpha(1+x)^{-(\alpha+1)}, \quad x \geq 0, \quad \alpha > 0 ,$$
$$F(x) = 1 - (1+x)^{-\alpha}, \quad x \geq 0, \quad \alpha > 0,$$

it is easily noted that

$$(1+x)f(x) = \alpha\{1 - F(x)\}, \quad x \geq 0 . \tag{15.1}$$

A good discussion on some properties of this distribution, with special reference to the distributions of sample median and range, is given by Burr (1968) and Burr and Cislak (1968). For this distribution, we have $\mu_{1:1}^{(k)} = E(X^k) = \alpha B(k+1, \alpha - k)$, $B(\cdot, \cdot)$ being the complete beta function, which exists for all $k < \alpha$. Hence, the single moments of order statistics $\mu_{r:n}^{(k)}$ exist for all $k < \alpha$. Making use of the relation in (15.1), Balakrishnan (1987a) has established the following recurrence relations for the single and product moments of order statistics:

For $n \geq 1$ and $k = 1, 2, \ldots$,

$$\mu_{1:n}^{(k)} = k\mu_{1:n}^{(k-1)}/(n\alpha - k) ; \tag{15.2}$$

for $2 \leq r \leq n$ and $k = 1, 2, \ldots$,

$$\mu_{r:n}^{(k)} = \left(k\mu_{r:n}^{(k-1)} + (n-r+1)\alpha\mu_{r-1:n}^{(k)} \right)/\{(n-r+1)\alpha - k\} ; \tag{15.3}$$

for $1 \leq r \leq n-1$,

$$\mu_{r,r+1:n} = \left((n-r)\alpha\mu_{r:n}^{(2)} + \mu_{r:n}\right)/\{(n-r)\alpha - 1\} ; \tag{15.4}$$

and for $1 \leq r \leq n$ and $s - r \geq 2$,

$$\mu_{r,s:n} = \{(n-s+1)\alpha\mu_{r,s-1:n} + \mu_{r:n}\}/\{(n-s+1)\alpha - 1\} . \tag{15.5}$$

Note that the recurrence relations (15.2)–(15.5) would enable one to obtain all the single and product moments of all order statistics for any sample size in a simple recursive way.

16. Log-logistic and related distributions

With the density function as

$$f(x) = \beta x^{\beta-1}/(1+x^\beta)^2, \quad x \geq 0, \quad \beta > 0 ,$$

and the cumulative distribution function as

$$F(x) = x^\beta/(1+x^\beta), \quad x \geq 0 ,$$

it is easy to note that

$$xf(x) = \beta F(x)\{1 - F(x)\} \tag{16.1}$$
$$= \beta\left[F(x) - F^2(x)\right] \tag{16.2}$$
$$= \beta\left[\{1 - F(x)\} - \{1 - F(x)\}^2\right] . \tag{16.3}$$

Note that the log-logistic distribution is a particular member of the family XII of distributions of Burr (1942) (also see Johnson, Kotz and Balakrishnan, 1994). Some further discussion on properties of Burr's family XII of distributions, with special reference to the distributions of sample median and range, have been made by Burr (1968) and Burr and Cislak (1968); also see Malik (1967b), Block and Rao (1973), O'Quigley and Struthers (1982) and Bennet (1983). Note that, for this distribution, we have $\mu_{1:1}^{(k)} = E(X^k) = \Gamma(1 + k/\beta)\Gamma(1 - k/\beta)$ and it exists only for $k < \beta$. Hence, $\mu_{r:n}^{(k)}$ exists for all $1 \leq r \leq n$ only if $k < \beta$.

Making use of the relations (16.1)–(16.3), Balakrishnan and Malik (1987a) have established the following recurrence relations for the single and product moments of order statistics:
For $n \geq 1$ and $k = 1, 2, \ldots$,

$$\mu_{1:n+1}^{(k)} = \left(1 - \frac{k}{n\beta}\right)\mu_{1:n}^{(k)} ; \tag{16.4}$$

for $1 \leq r \leq n$ and $k = 1, 2, \ldots$,

$$\mu_{r+1:n+1}^{(k)} = \left(1 + \frac{k}{r\beta}\right)\mu_{r:n}^{(k)} ; \tag{16.5}$$

for $1 \leq r \leq n-1$,

$$\mu_{r,r+1:n+1} = \frac{1}{n-r+1}\left[(n+1)\left(1-\frac{1}{(n-r)\beta}\right)\mu_{r,r+1:n} - r\mu_{r+1:n+1}^{(2)}\right] ; \quad (16.6)$$

for $1 \leq r \leq n-1$,

$$\mu_{r+1,r+2:n+1} = \frac{1}{r+1}\left[(n+1)\left(1+\frac{1}{r\beta}\right)\mu_{r,r+1:n} - (n-r)\mu_{r+1:n+1}^{(2)}\right] ; \quad (16.7)$$

for $1 \leq r < s \leq n$ and $s - r \geq 2$,

$$\mu_{r,s:n+1} = \mu_{r,s-1:n+1} + \frac{n+1}{n-s+2}\left[\left(1-\frac{1}{(n-s+1)\beta}\right)\mu_{r,s:n} - \mu_{r,s-1:n}\right] ; \quad (16.8)$$

and for $1 \leq r < s \leq n$ and $s - r \geq 2$,

$$\mu_{r+1,s+1:n+1} = \mu_{r+2,s+1:n+1} + \frac{n+1}{r+1}\left[\left(1+\frac{1}{r\beta}\right)\mu_{r,s:n} - \mu_{r+1,s:n+1}\right] . \quad (16.9)$$

Starting with the first k moments of X (assume $\beta > k$), the recurrence relations (16.4)–(16.9) could be used recursively in order to evaluate the first k single moments and also the product moments of all order statistics for any sample size. Thus, for example, starting with the values of $E(X) = \Gamma(1+1/\beta)\Gamma(1-1/\beta)$ and $E(X^2) = \Gamma(1+2/\beta)\Gamma(1-2/\beta)$, one could evaluate the means, variances and covariances of all order statistics for any sample size. The necessary values of the gamma function could be obtained either from the detailed tables of the complete gamma function given in Abramowitz and Stegun (1964) or by using the algorithem due to Pike and Hill (1966) or by using GAMMA subroutine from IMSL.

By considering the doubly truncated log-logistic distribution with cumulative distribution function $F(x)$ and density function

$$f(x) = \frac{1}{P-Q}\frac{\beta x^{\beta-1}}{(1+x^\beta)^2}, \quad Q_1 \leq x \leq P_1, \ \beta > 0 ,$$

and noting that

$$xf(x) = \beta\Big[QQ_2 + (1-2Q)F(x) - (P-Q)(F(x))^2\Big]$$
$$= \beta\Big[PP_2 - (1-2P)(1-F(x)) - (P-Q)(1-F(x))^2\Big] ,$$

where Q and $1-P$ ($Q < P$) are, respectively, the proportions of truncation on the left and right of the log-logistic distribution, $Q_1 = \left(\frac{Q}{1-Q}\right)^{1/\beta}$, $P_1 = \left(\frac{P}{1-P}\right)^{1/\beta}$, $Q_2 = \left(\frac{1-Q}{P-Q}\right)^{1/\beta}$, and $P_2 = \left(\frac{1-P}{P-Q}\right)^{1/\beta}$, Balakrishnan and Malik (1987b) have established the following recurrence relations for the single and product moments of order statistics:

For $k = 1, 2, \ldots,$

$$\mu_{1:2}^{(k)} = \frac{1}{P - Q}\left[PP_2 P_1^k - \{(1 - 2P) + k/\beta\}\mu_{1:1}^{(k)} - QQ_2 Q_1^k\right] ; \qquad (16.10)$$

for $n \geq 2$ and $k = 1, 2, \ldots,$

$$\mu_{1:n+1}^{(k)} = \frac{1}{P - Q}\left[PP_2 \mu_{1:n-1}^{(k)} - \{(1 - 2P) + k/n\beta\}\mu_{1:n}^{(k)} - QQ_2 Q_1^k\right] ; \qquad (16.11)$$

for $n \geq 2$ and $k = 1, 2, \ldots,$

$$\mu_{2:n+1}^{(k)} = \frac{1}{P - Q}\Big[\{(1 - 2Q) + k/\beta\}\mu_{1:n}^{(k)} - (1 - Q - P)\mu_{2:n}^{(k)}$$
$$- nQQ_2\left(\mu_{1:n-1}^{(k)} - Q_1^k\right)\Big] ; \qquad (16.12)$$

for $2 \leq r \leq n - 1$ and $k = 1, 2, \ldots,$

$$\mu_{r+1:n+1}^{(k)} = \frac{1}{P - Q}\Big[\{(1 - 2Q) + k/n\beta\}\mu_{r:n}^{(k)} - (1 - Q - P)\mu_{r+1:n}^{(k)}$$
$$- \frac{nQQ_2}{r}\left(\mu_{r:n-1}^{(k)} - \mu_{r-1:n-1}^{(k)}\right)\Big] ; \qquad (16.13)$$

for $k = 1, 2, \ldots,$

$$\mu_{2:2}^{(k)} = \frac{1}{P - Q}\left[QQ_2 Q_1^k + \{(1 - 2Q) + k/\beta\}\mu_{1:1}^{(k)} - PP_2 P_1^k\right] ; \qquad (16.14)$$

for $n \geq 2$ and $k = 1, 2, \ldots,$

$$\mu_{n+1:n+1}^{(k)} = \frac{1}{P - Q}\left[QQ_2 \mu_{n-1:n-1}^{(k)} + \{(1 - 2Q) + k/n\beta\}\mu_{n:n}^{(k)} - PP_2 P_1^k\right] ; \qquad (16.15)$$

for $n \geq 2$,

$$\mu_{2,3:n+1} = \mu_{3,n+1}^{(2)} + \frac{(n + 1)}{2(P - Q)}\Big[\{(1 - 2Q) + 1/\beta\}\mu_{1,2:n} - (1 - 2Q)\mu_{2:n}^{(2)}$$
$$- nQQ_2\left(\mu_{1:n-1}^{(2)} - Q_1\mu_{1:n-1}\right)\Big] ; \qquad (16.16)$$

for $2 \leq r \leq n - 1$,

$$\mu_{r+1,r+2:n+1} = \mu_{r+2:n+1}^{(2)} + \frac{n + 1}{r(r + 1)(P - Q)}\Big[\{r(1 - 2Q) + 1/\beta\}\mu_{r,r+1:n}$$
$$- r(1 - 2Q)\mu_{r+1:n}^{(2)} - nQQ_2\left(\mu_{r:n-1}^{(2)} - \mu_{r-1,r:n-1}\right)\Big] ; \qquad (16.17)$$

for $1 \leq r \leq n-2$,

$$\mu_{r,r+1:n+1} = \mu_{r:n+1}^{(2)} + \frac{n+1}{(n-r)(n-r+1)(P-Q)}$$
$$\times \left[(n-r)(1-2P)\mu_{r:n}^{(2)} - \{(n-r)(1-2P)+1/\beta\}\mu_{r,r+1:n} \right.$$
$$\left. + nPP_2\left(\mu_{r,r+1:n-1} - \mu_{r:n-1}^{(2)}\right) \right] ; \qquad (16.18)$$

for $n \geq 2$,

$$\mu_{n-1,n:n+1} = \mu_{n-1:n+1}^{(2)} + \frac{n+1}{2(P-Q)} \left[(1-2P)\mu_{n-1:n}^{(2)} - \{(1-2P)+1/\beta\} \right.$$
$$\left. \times \mu_{n-1,n:n} + nPP_2\left(P_1\mu_{n-1:n-1} - \mu_{n-1:n-1}^{(2)}\right) \right] ; \qquad (16.19)$$

for $s-r \geq 2$ and $1 \leq r < s \leq n-1$,

$$\mu_{r,s:n+1} = \mu_{r,s-1:n+1} + \frac{n+1}{(n-s+1)(n-s+2)(P-Q)}$$
$$\times \left[nPP_2\left(\mu_{r,s:n-1} - \mu_{r,s-1:n-1}\right) + (n-s+1)(1-2P)\mu_{r,s-1:n} \right.$$
$$\left. - \{(n-s+1)(1-2P)+1/\beta\}\mu_{r,s:n} \right] ; \qquad (16.20)$$

for $1 \leq r \leq n-2$,

$$\mu_{r,n:n+1} = \mu_{r,n-1:n+1} + \frac{n+1}{2(P-Q)} \left[nPP_2\left(P_1\mu_{r:n-1} - \mu_{r,n-1:n-1}\right) \right.$$
$$\left. + (1-2P)\mu_{r,n-1:n} - \{(1-2P)+1/\beta\}\mu_{r,n:n} \right] ; \qquad (16.21)$$

for $3 \leq s \leq n$,

$$\mu_{2,s+1:n+1} = \mu_{3,s+1:n+1} + \frac{n+1}{2(P-Q)} \left[\{(1-2Q)+1/\beta\}\mu_{1,s:n} \right.$$
$$\left. - nQQ_2\left(\mu_{1,s-1:n-1} - Q_1\mu_{s-1:n-1}\right) - (1-2Q)\mu_{2,s:n} \right] ; \qquad (16.22)$$

and for $r \geq 2$, $s \leq n$ and $s-r \geq 2$,

$$\mu_{r+1,s+1:n+1} = \mu_{r+2,s+1:n+1} + \frac{n+1}{r(r+1)(P-Q)}$$
$$\times \left[\{r(1-2Q)+1/\beta\}\mu_{r,s:n} - nQQ_2(\mu_{r,s-1:n-1} \right. \qquad (16.23)$$
$$\left. - \mu_{r-1,s-1:n-1}) - r(1-2Q)\mu_{r+1,s:n} \right].$$

Starting from $\mu_{1:1}^{(k)}$, the recurrence relations (16.10)–(16.23) along with the relation $\mu_{1,2:2} = \mu_{1:1}^{(2)}$ would enable one to obtain the first k single moments and also the product moments of all order statistics for any sample size in a simple recursive way. Thus, for example, starting with the value of

$$\mu_{1:1}^{(1)} = E(X) = \frac{1}{P-Q} \int_{1-P}^{1-Q} u^{-1/\beta}(1-u)^{1/\beta} \, du \;,$$

and

$$\mu_{1:1}^{(2)} = E(X^2) = \frac{1}{P-Q} \int_{1-P}^{1-Q} u^{-2/\beta}(1-u)^{2/\beta} \, du \;,$$

the recurrence relations (16.10)–(16.23) could be used systematically in a simple recursive manner in order to evaluate the means, variances and covariances of all order statistics for all sample sizes. Note that for $\beta \geq 2$, the above expressions could be rewritten as

$$\mu_{1:1}^{(1)} = E(X) = \frac{1}{P-Q} B(1-1/\beta, 1+1/\beta)\{I_{1-Q}(1-1/\beta, 1+1/\beta) - I_{1-P}(1-1/\beta, 1+1/\beta)\} \;,$$

and

$$\mu_{1:1}^{(2)} = E(X^2) = \frac{1}{P-Q} B(1-2/\beta, 1+2/\beta) \\ \times \{I_{1-Q}(1-2/\beta, 1+2/\beta) - I_{1-P}(1-2/\beta, 1+2/\beta)\} \;,$$

where $I_\alpha(a,b)$ is the incomplete beta ratio defined by

$$I_\alpha(a,b) = \frac{1}{B(a,b)} \int_0^\alpha t^{a-1}(1-t)^{b-1} dt, \quad a,b > 0, \quad 0 < \alpha < 1 \;,$$

whose values could be taken either from the extensive tables of Pearson (1968) or from Harvard Computation Laboratory (1955) or by using BETAI subroutine from IMSL.

The probability density function of a log-logistic distribution with scale parameter θ is

$$f(x) = \beta \theta x^{\beta-1}(1 + \theta x^\beta)^{-2}, \quad 0 \leq x < \infty \;, \tag{16.24}$$

where β is the shape parameter. This is a special case of Burr type XII distribution (Tadikamalla, 1980).

It is easy to note that

$$F(x)\{1 - F(x)\} = \frac{x}{\beta} f(x) \;, \tag{16.25}$$

$$(1 - F(x))^2 = \frac{x^{1-\beta}}{\beta \theta} f(x) \;, \tag{16.26}$$

where $F(x) = 1 - (1 + \theta x^\beta)^{-1}$, $0 \leq x < \infty$.

Ali and Khan (1987) have used (16.25) and (16.26) to derive the following recurrence relations for the single as well as product moments of order statistics:

For $2 \leq r \leq n$,

$$\mu_{r:n}^{(k-\beta)} = \frac{(n-r+1)\theta}{r-1} \mu_{r-1:n}^{(k)} ; \tag{16.27}$$

for $n \geq 1$,

$$\mu_{1:n}^{(k-\beta)} = \frac{np\theta}{k} \mu_{1:n+1}^{(k)} ; \tag{16.28}$$

for $1 \leq r \leq n$ and $\theta = 1$,

$$\mu_{r:n}^{(-k)} = \mu_{n-r:n}^{(k)} ; \tag{16.29}$$

for $k < \beta$,

$$E(X^k) = \mu_{1:1}^{(k)} = \theta^{-k/\beta} \Gamma(1 - k/\beta)\Gamma(1 + k/\beta) ; \tag{16.30}$$

for $1 \leq r < s \leq n-1$,

$$\mu_{r,s:n}^{(j,k)} = \mu_{r,s-1:n}^{(j,k)} + \frac{n}{n-s+1} \{1 - k/\beta(n-s)\}$$
$$\times \mu_{r,s:n-1}^{(j,k)} - \frac{n}{n-s+1} \mu_{r,s-1:n-1}^{(j,k)} ; \tag{16.31}$$

for $1 \leq r \leq n-2$,

$$\mu_{r,n:n}^{(j,k)} = \mu_{r,n-1:n}^{(j,k)} + \frac{nk}{\beta(n-r-1)} \mu_{r,n-1:n-1}^{(j,k)}$$
$$- \frac{r}{n-r+1} \left(\mu_{r+1,n:n}^{(j,k)} - \mu_{r+1,n-1:n}^{(j,k)} \right) ; \tag{16.32}$$

and for $1 \leq r < s \leq n$,

$$\mu_{r,s:n}^{(j,k-p)} = \frac{\beta\theta(n-s+2)(n-s+1)}{k(n+1)} \left(\mu_{r,s:n+1}^{(j,k)} - \mu_{r,s-1:n+1}^{(j,k)} \right) . \tag{16.33}$$

Some of these are identical to the results presented earlier.

17. Burr and truncated Burr distributions

The pdf of the doubly truncated Burr distribution is

$$f(x) = \frac{mp\theta}{P-Q} x^{p-1}(1+\theta x^p)^{-(m+1)}, \quad Q_1 \leq x \leq P_1 ,$$

where Q and $1-P$ ($Q < P$) are the proportions of truncation on the left and right of the Burr distribution.

As convention, Khan and Khan (1987) have taken

$$\mu_{0:n}^{(k)} = Q_1^k, \quad n \geq 0 \text{ and } \mu_{n:n-1}^{(k)} = P_1^k, \quad n \geq 1, \tag{17.1}$$

and

$$\mu_{r,r:n}^{(j,k)} = \mu_{r:n}^{(j+k)}, \quad 1 \leq r \leq n \text{ and } \mu_{n-1,n:n-1}^{(j,k)} = P_1^k \mu_{n-1:n-1}^{(j)}. \tag{17.2}$$

They then derived the following recurrence relations for the single moments of order statistics and well as product moments of order statistics:
For $2 \leq r \leq n-1$ and $k \neq mnp$,

$$(1 - k/mnp)\mu_{r:n}^{(k)} = Q_2 \mu_{r-1:n-1}^{(k)} - P_2 \mu_{r:n-1}^{(k)} + \frac{k}{mnp\theta} \mu_{r:n}^{(k-p)}; \tag{17.3}$$

and for $1 \leq r < s \leq n-1$, $s - r \geq 2$ and $k \neq (n-s+1)mp$,

$$\{1 - k/mp(n-s+1)\}\mu_{r,s:n}^{(j,k)} = \frac{-nP_2}{n-s+1}\left[\mu_{r,s:n-1}^{(j,k)} - \mu_{r,s-1:n-1}^{(j,k)}\right]$$

$$+ \mu_{r,s-1:n}^{(j,k)} + \frac{k}{mp\theta(n-s+1)}\mu_{r,s:n}^{(j,k-p)}, \tag{17.4}$$

where $Q_2 = (1-Q)/(P-Q)$ and $P_2 = (1-P)/(P-Q)$.
The important deductions for $k \neq mnp$ in view of (17.1)–(17.4) are:

$$(1 - k/mp)\mu_{1:1}^{(k)} = Q_2 Q_1^k - P_2 P_1^k + \frac{k}{mp\theta}\mu_{1:1}^{(k-p)}; \tag{17.5}$$

$$(1 - k/mp)\mu_{1:n}^{(k)} = Q_2 Q_1^k - P_2 \mu_{1:n-1}^{(k)} + \frac{k}{mpn\theta}\mu_{1:n}^{(k-p)}, \quad n \geq 1; \tag{17.6}$$

$$(1 - k/mnp)\mu_{n:n}^{(k)} = Q_2 \mu_{n-1:n-1}^{(k)} - P_2 P_1^k + \frac{k}{mpn\theta}\mu_{n:n}^{(k-p)}, \quad n \geq 1. \tag{17.7}$$

If $k = mnp$, one can write from (17.3)

$$\mu_{r:n-1}^{(k)} = \frac{Q_2}{P_2}\mu_{r-1:n-1}^{(k)} + \frac{1}{P_2\theta}\mu_{r:n}^{(k-p)}, \quad 2 \leq r \leq n-1. \tag{17.8}$$

The results which are presented in this section are also true for Lomax distribution $(p = 1, \theta = 1/a)$, Weibull–Gamma $(\theta = 1/a)$, Weibull-Exponential $(m = 1, \theta = 1/a)$ and Log-Logistic distribution $(m = 1, \theta = a^{-p})$.

18. Doubly truncated parabolic and skewed distributions

The doubly truncated parabolic distribution has its pdf as

$$f(x) = \frac{6x(1-x)}{P-Q}, \quad 0 < Q_1 \le x \le P_1 < 1, \tag{18.1}$$

where $P = P_1^2(3 - 2P_1)$ and $Q = Q_1^2(3 - 2Q_1)$.

Mohie El-Din, Mahmoud and Abu-Youssef (1991) have derived the following recurrence relations for the single and product moments of order statistics:

For $2 \le r \le n-1$,

$$\left(1 + \frac{k}{3n}\right)\mu_{r:n}^{(k)} = Q_2 \mu_{r-1:n-1}^{(k)} - P_2 \mu_{r:n-1}^{(k)} + \frac{k}{6n}\left(\mu_{r:n}^{(k-1)} + \mu_{r:n}^{(k-2)}\right); \tag{18.2}$$

for $1 \le r \le n-2$,

$$\left(1 + \frac{k}{3(n-r)}\right)\mu_{r,r+1:n}^{(j,k)} = \mu_{r:n}^{(j+k)} - \frac{np_2}{n-r}\left(\mu_{r,r+1:n-1}^{(j,k)} - \mu_{r:n-1}^{(j+k)}\right)$$
$$+ \frac{k}{6(n-r)}\left(\mu_{r,r+1:n}^{(j,k-1)} + \mu_{r,r+1:n}^{(j,k-2)}\right); \tag{18.3}$$

and

$$\left(1 + \frac{k}{3}\right)\mu_{n-1,n:n}^{(j,k)} = \mu_{n-1:n}^{(j+k)} - np_2\left(P_1^k \mu_{n-1:n-1}^{(j)} - \mu_{n-1:n-1}^{(j+k)}\right)$$
$$+ \frac{k}{6}\left(\mu_{n-1,n:n}^{(j,k-1)} + \mu_{n-1,n:n}^{(j,k-2)}\right), \tag{18.4}$$

where $P_1^k = \mu_{n:n-1}^{(k)}$, $P_2 = (1-P)/(P-Q)$ and $Q_2 = (1-Q)/(P-Q)$.

The doubly truncated skewed population has its pdf as

$$f(x) = \frac{12x^2(1-x)}{P-Q}, \quad 0 < Q_1 \le x \le P_1 < 1, \tag{18.5}$$

where $P = P_1^3(4 - 3P_1)$ and $Q = Q_1^3(4 - 3Q_1)$.

Mohie El-Din, Mahmoud and Abu-Youssef (1991) have derived the following recurrence relations for the single and product moments of order statistics in this case:

For $2 \le r \le n-1$,

$$\left(1 + \frac{k}{4n}\right)\mu_{r:n}^{(k)} = Q_2 \mu_{r-1:n-1}^{(k)} - P_2 \mu_{r:n}^{(k)} + \frac{k}{12n}\left(\mu_{r:n}^{(k-1)} + \mu_{r:n}^{(k-2)} + \mu_{r:n}^{(k-3)}\right); \tag{18.6}$$

and for $1 \le r \le n-1$ and $s - r \ge 2$,

$$\left(1 + \frac{k}{4(n-s+1)}\right)\mu_{r,s:n}^{(j,k)} = -\frac{P_2(n-1)}{n-s+1}\left(\mu_{r,s:n-1}^{(j,k)} - \mu_{r,s-1:n-1}^{(j,k)}\right)$$
$$+ \frac{k}{12n}\left(\mu_{r,s:n}^{(j,k-1)} + \mu_{r,s:n}^{(j,k-2)} + \mu_{r,s:n}^{(j,k-3)}\right), \quad (18.7)$$

where $P_2 = (1-P)/(P-Q)$ and $Q_2 = (1-Q)/(P-Q)$.

19. Mixture of two exponential distributions

The pdf of a mixture of two exponential distributions is

$$f(x) = \lambda f_1(x) + (1-\lambda)f_2(x), \quad x \geq 0, \ 0 < \lambda < 1, \quad (19.1)$$

where

$$f_j(x) = \alpha_j e^{-\alpha_j x}, \quad j = 1, 2; \quad (19.2)$$

the corresponding cdf is

$$F(x) = \lambda F_1(x) + (1-\lambda)F_2(x), \quad x \geq 0, \ 0 < \lambda < 1, \quad (19.3)$$

where

$$F_j(x) = 1 - e^{-\alpha_j x}, \quad j = 1, 2. \quad (19.4)$$

Nassar and Mahmoud (1985) have used (19.1)–(19.4) to obtain the following recurrence relations for the moments of order statistics:
For $r = 0, 1, 2, \ldots,$ and $n = r+1, r+2, \ldots,$

$$\mu_{r+1:n}^{(2)} - \mu_{r:n}^{(2)} = \{(1/\alpha_1) + (1/\alpha_2)\}\frac{1}{n-r}\mu_{r+1:n} + \{(1/\alpha_1) + (1/\alpha_2)\}\frac{1}{n-r}$$
$$\times \left(\alpha_1 \frac{\partial}{\partial \alpha_1}\mu_{r+1:n} - \alpha_2 \frac{\partial}{\partial \alpha_2}\mu_{r+1:n}\right), \quad (19.5)$$

and for $r = 1, 2, \ldots,$ and $n = r+1, r+2, \ldots,$

$$n\left(\mu_{r:n}^{(2)} - \mu_{r-1:n-1}^{(2)}\right) = \{(1/\alpha_1) + (1/\alpha_2)\}\mu_{r:n} + \{(1/\alpha_1) + (1/\alpha_2)\}$$
$$\times \left(\alpha_1 \frac{\partial}{\partial \alpha_1}\mu_{r:n} - \alpha_2 \frac{\partial}{\partial \alpha_2}\mu_{r:n}\right). \quad (19.6)$$

20. Doubly truncated Laplace distribution

The pdf of the doubly truncated Laplace distribution is

$$g(x) = \frac{e^{-|x|}}{2(1-P-Q)}, \quad Q_1 \leq x \leq P_1, \quad (20.1)$$

where $Q_1 = \log(2Q) < 0$ and $P_1 = -\log(2P) > 0$, and the corresponding distribution function is

$$G(x) = \begin{cases} 0, & x < Q_1 \\ \frac{e^x - 2Q}{2(1-P-Q)}, & Q_1 \leq x \leq 0 \\ \frac{2 - 2Q - e^{-x}}{2(1-P-Q)}, & 0 \leq x \leq P_1 \\ 1, & \text{otherwise} . \end{cases} \qquad (20.2)$$

Lien, Balakrishnan and Balasubramanian (1992) have proved the following theorem for the single moments of order statistics.

THEOREM 20.1. The kth moment of the ith order statistic from a random sample of size n drawn from a (Q,P)-truncated Laplace distribution in (20.1) can be written as:

$$\mu_{i:n}^{(k)} = \Delta_{Q,P}^{(k)}(i,n) + (-1)^k \Delta_{P,Q}^{(k)}(n-i,n) , \qquad (20.3)$$

where $\Delta_{Q,P}^{(k)}(i,n)$ satisfies the following recurrence relation:

$$\Delta_{Q,P}^{(k)}(i,n) = (-k/n)\Delta_{Q,P}^{(k-1)}(i,n) - \{Q/(1-P-Q)\}\Delta_{P,Q}^{(k)}(i-1,n-1)$$
$$+ \{(1-P)/(1-P-Q)\}\Delta_{P,Q}^{(k)}(i,n-1), \quad k \geq 1 , \qquad (20.4)$$

$$\Delta_{Q,P}^{(0)}(i,n) = -(1-P-Q)^{-n}(1-2Q)^{i-1}(1-2P)^{n-i+1}2^{-n}C(n,i-1)$$
$$+ \Delta_{P,Q}^{(0)}(i-1,n) , \qquad (20.5)$$

$$C(n, i-1) = \frac{n!}{(i-1)!(n-i+1)!}, \quad \Delta_{Q,P}^{(k)}(0,n) = Q_1^k, \text{ and}$$

$$\Delta_{Q,P}^{(k)}(n, n-1) = 0 .$$

Similarly, for the product moments of order statistics of doubly truncated Laplace distribution, Lien, Balakrishnan and Balasubramanian (1992) have proved the following theorem.

THEOREM 20.2. The product moment of the ith and $(i+1)^{\text{th}}$ order statistics from a random sample of size n drawn from a (Q,P)-truncated Laplace distribution in (20.1) can be written as

$$\mu_{i,i+1:n} = A_{Q,P}(i,n) + A_{Q,P}(n-i,n) - \Delta_{Q,P}^{(1)}(i,i)\Delta_{Q,P}^{(1)}(n-i,n-i)C(n,i) , \qquad (20.6)$$

where $\Delta_{Q,P}^{(1)}(i,i)$ and $\Delta_{P,Q}^{(1)}(n-i,n-i)$ can be calculated by using Eqs. (20.4) and (20.5), and $A_{Q,P}(i,n)$ satisfies the following relations:

$$A_{Q,P}(i,n) = \Delta_{Q,P}^{(2)}(i+1,n) - \frac{1}{i}\Delta_{Q,P}^{(1)}(i+1,n) + \frac{nQ}{i(1-P-Q)} \\ \times \left[\Delta_{Q,P}^{(2)}(i,n-1) - A_{Q,P}(i-1,n-1)\right], \quad (20.7)$$

and

$$A_{Q,P}(1,n) = \frac{nQ(1-Q_1)}{(1-P-Q)}\Delta_{Q,P}^{(1)}(1,n-1) - (n-1)\left[\Delta_{Q,P}^{(2)}(1,n) - \Delta_{Q,P}^{(1)}(1,n)\right] \\ + \frac{n(1-P)}{(1-P-Q)}\left[\Delta_{Q,P}^{(2)}(1,n-1) - \Delta_{Q,P}^{(1)}(1,n-1)\right]. \quad (20.8)$$

20.1. Non-overlapping mixture model

Let X be a two component mixed random variable,

$$X \stackrel{D}{=} \begin{cases} Z_1 & \text{with probability} \quad \pi \\ Z_2 & \text{with probability} \quad 1-\pi, \end{cases} \quad (20.9)$$

where $Z_1 \stackrel{D}{=} f_1(z_1)$ and $Z_2 \stackrel{D}{=} f_2(z_2)$, and supports $R_1 = (a_1, b_1)$ and $R_2 = (a_2, b_2)$, $a_1 > -\infty$, $b_1 < \infty$ and $b_1 < a_2$.

From (20.9), we have

$$F(x) = \begin{cases} 0 & \text{for } x < a_1 \\ \pi F_1(x) & \text{for } x \in R_1 \\ \pi & \text{for } b_1 \leq x \leq a_2 \\ \pi + (1-\pi)F_2(x) & \text{for } x \in R_2 \\ 1 & \text{for } x > b_2. \end{cases} \quad (20.10)$$

Let us denote the single and product moments of order statistics corresponding to the components Z_1 and Z_2 by $\mu_{i:n}^{(k)}[1]$, $\mu_{i:n}^{(k)}[2]$ and $\mu_{i,j:n}[1]$, $\mu_{i,j:n}[2]$, respectively. Then, Lien, Balakrishnan and Balasubramanian (1992) have proved the following two theorems.

THEOREM 20.3. For $1 \leq i \leq n$ and $k = 1, 2, \ldots,$

$$\mu_{i:n}^{(k)} = \sum_{t=0}^{i-1}\binom{n}{t}\pi^t(1-\pi)^{n-t}\mu_{i-k:n-t}^{(k)}[2] + \sum_{t=i}^{n}\binom{n}{t}\pi^t(1-\pi)^{n-t}\mu_{i:t}^{(k)}[1].$$

$$(20.11)$$

THEOREM 20.4. For $1 \leq i < j \leq n$,

$$\mu_{i,j:n} = \sum_{t=0}^{i-1}\binom{n}{t}\pi^t(1-\pi)^{n-t}\mu_{i-t,j-t:n-t}[2] + \sum_{t=j}^{n}\binom{n}{t}\pi^t(1-\pi)^{n-t}\mu_{i,j:t}[1] \\ + \sum_{t=i}^{j-1}\binom{n}{t}\pi^t(1-\pi)^{n-t}\mu_{j-t:n-t}[2]. \quad (20.12)$$

The results given in Theorems 20.3 and 20.4 can be extended to the case when X is a m-component mixed random variable defined by

$$X \stackrel{D}{=} Z_t \text{ with probability } \pi_t, \quad \sum_{t=1}^{m} \pi_t = 1, \tag{20.13}$$

where $Z_1, Z_2, \ldots, Z_m$ are random variables with pdf's $f_1, f_2, \ldots, f_m$, cdf's $F_1, F_2, \ldots, F_m$ and supports $R_1 = (a_1, b_1)$, $R_2 = (a_2, b_2), \ldots, R_m = (a_m, b_m)$, respectively, with $b_{t-1} \leq a_t$ for $t = 1, 2, \ldots, m$, $a_1 > -\infty$ and $b_m < \infty$.

From (20.13), it is clear that the cumulative distribution function of X is

$$F(x) = \begin{cases} 0 & \text{for } x < a_1, \\ \pi_{(t-i)} + \pi_t F_t(x) & \text{for } x \in R_t, \ t = 1, 2, \ldots, m, \\ \pi_{(t)} & \text{for } b_t \leq x \leq a_{t_1+1}, \ t = 1, 2, \ldots, m-1, \\ 1 & \text{for } x > b_m, \end{cases} \tag{20.14}$$

where $\pi_{(0)} = 0$ and $\pi_{(t)} = \sum_{j=1}^{t} \pi_j$. As before, let us denote the single moments of order statistics by $\mu_{i:n}^{(k)}$ and the product moments of order statistics by $\mu_{i,j:n}$ and the corresponding quantities for the individual component Z_t by $\mu_{i:n}^{(k)}[t]$ and $\mu_{i,j:n}[t]$, for $t = 1, 2, \ldots, m$. Then, by proceeding exactly on the same lines as in Theorems 20.3 and 20.4, Lien, Balakrishnan and Balasubramanian (1992) have established the following two theorems.

THEOREM 20.5. For $1 \leq i \leq n$, and $k = 1, 2, \ldots$,

$$\mu_{i:n}^{(k)} = \sum_{t=1}^{m} \sum_{p=0}^{i-1} \sum_{q=i}^{n} \binom{n}{p, q-p, n-q} \pi_{(t-1)}^{p} \pi_t^{q-p} \left(1 - \pi_{(t)}\right)^{n-q} \mu_{i-p:q-p}^{(k)}[t]. \tag{20.15}$$

THEOREM 20.6. For $1 \leq i < j \leq n$,

$$\mu_{i,j:n} = \sum_{t=1}^{m} \sum_{p=0}^{i-1} \sum_{q=j}^{n} \binom{n}{p, q-p, n-q} \pi_{(t-1)}^{p} \pi_t^{q-p} \left(1 - \pi_{(t)}\right)^{n-q} \mu_{i-p,j-p:q-p}[t]$$

$$+ \sum_{t=1}^{m-1} \sum_{t'=t+1}^{m} \sum_{p=0}^{i-1} \sum_{q=j}^{n} \sum_{r=i}^{j-1} \sum_{s=r}^{j-1} \binom{n}{p, i-p+s-r, r-i, q-s, n-q}$$

$$\times \pi_{(t-1)}^{p} \pi_t^{i-p+s-r} \left(\pi_{(t'-1)} - \pi_{(t)}\right)^{r-i} \pi_{t'}^{q-s}$$

$$\times \left(1 - \pi_{(t')}\right)^{n-q} \mu_{i-p:s-r+i-p}[t] \mu_{j-s:q-s}[t']. \tag{20.16}$$

It may be noted that Theorems 20.5 and 20.6 reduce to Theorems 20.3 and 20.4 when $m = 2$.

Furthermore, recurrence relations between factorial moment generating functions from the doubly truncated Laplace distribution have been derived by Mohie El-Din and Sultan (1995).

21. A class of probability distributions

Consider a class of probability distributions F given by [Kamps (1991)]

$$\frac{d}{dt}F^{-1}(t) = \frac{1}{d}t^p(1-t)^{q-p-1}, \quad t \in (0,1), \tag{21.1}$$

where p and q are integers and $d > 0$.

From (21.1), we have four cases with respect to p and q, $c \in R$:

(i) Putting $p = 0$ and $q = 0$ in (21.1), we get

$$\frac{d}{dt}F^{-1}(t) = \frac{1}{d(1-t)}, \text{ and hence, } F(x) = 1 - e^{-d(x-c)}, \quad x \in (c, \infty)$$

which is the exponential distribution.

(ii) Putting $p = 0$ and $q \neq 0$ in (21.1), we get

$$\frac{d}{dt}F^{-1}(t) = \frac{1}{d}(1-t)^{q-1}, \text{ and hence,}$$

$$F(x) = 1 - [dq(c-x)]^{1/q}, \quad x \in \begin{cases} (c - 1/dq, c), & q > 0 \\ (c - 1/dq, \infty), & q < 0 \end{cases};$$

$q < 0$: Pareto, special Burr XII (Lomax) distributions.

(iii) Putting $p \neq -1$ and $q = p + 1$ in (21.1), we have

$$\frac{d}{dt}F^{-1}(t) = \frac{1}{d}t^p, \text{ and hence,}$$

$$F(x) = 1 - [dq(x-c)]^{1/q}, \quad x \in \begin{cases} (c, c + 1/dq), & q > 0 \\ (-\infty, c + 1/dq), & q < 0 \end{cases};$$

$q > 0$: power function distributions.

(iv) Putting $p = -1$ and $q = -1$ in (21.1), we get

$$\frac{d}{dt}F^{-1}(t) = \frac{1}{dt(1-t)}, \text{ and hence } F(x) = \left[1 + e^{-d(x-c)}\right]^{-1},$$

$$x \in (-\infty, \infty),$$

which is the logistic distribution.

Kamps (1991, 1992) has proved two theorems for obtaining general recurrence relations between moments of order statistics based on the class of probability distributions (21.1). They are as follows.

THEOREM 21.1. Let X be a random variable with distribution function F, $m \geq 1$ a constant, and F satisfying (21.1) with $F^{-1}(t) \geq 0$, if $m \notin N$. Then for all $k, n \in N$, $2 \leq r \leq n$, satisfying $1 \leq k + p \leq n + q$, and $X_{r:n}$ the rth order statistic with $-\infty < \mathrm{E}(X_{k:n}^m)$, $\mathrm{E}(X_{k-1:n}^m)$, $\mathrm{E}(X_{k+p:n+q}^{m-1}) < \infty$, the identity

$$\mathrm{E}(X_{k:n}^m) - \mathrm{E}(X_{k-1:n}^m) = mC(k,n,p,q)\mathrm{E}\left(X_{k+p:n+q}^{m-1}\right), \tag{21.2}$$

with the constant $C(k,n,p,q)$ as

$$C(k,n,p,q) = \mathrm{E}(X_{k:n}) - \mathrm{E}(X_{k-1:n}) = \frac{1}{d} \frac{\binom{n}{k-1}}{(k+p)\binom{n+q}{k+p}}, \tag{21.3}$$

holds true.

In Table 1, some examples as special cases are presented for the relation (21.2). From Table 1, we observe that:

(i) The recurrence relations for moments of order statistics from exponential distributions have been obtained by Joshi (1978) with $d = 1$, Khan, Yaqub and Parvez (1983) with $d = 1$, Azlarov and Volodin (1986) with $m = 2$, and Lin (1988a,b) with $d = 1$.

(ii) The recurrence relations for moments of order statistics from Pareto and Lomax distributions have been derived by Khan and Khan (1987) and Lin (1988a,b).

(iii) The recurrence relations for moments of order statistics from power function distribution have been derived by Lin (1988a,b) with $d = 1$ and $q = 1$.

(iv) The recurrence relations for moments of order statistics from logistic distribution with $d = 1$ have been derived by Shah (1970).

THEOREM 21.2. Let $\alpha \in R$, $\beta \geq 0$ with $\alpha + \beta \neq 0$, and a function h_1 on $(0,1)$, with

$$\frac{d}{dt} h_1(t) = \frac{1}{d} t^p (1-t)^{q-p-1}, \quad t \in (0,1), \tag{21.4}$$

and constants $d > 0$, $p, q \in Z$ be given, such that the expression $\{\beta h_1(t)\}^{1/\beta}$ is defined, if $\beta > 0$.

Furthermore, let F be given by

$$F^{-1}(t) = \begin{cases} \exp\{h_1(t)\}, & \beta = 0, \ t \in (0,1) \\ \{\beta h_1(t)\}^{1/\beta}, & \beta > 0, \end{cases} \tag{21.5}$$

and let the moments $\mathrm{E}(X_{r:n}^{\alpha+\beta})$, $\mathrm{E}(X_{r-1:n}^{\alpha+\beta})$, and $\mathrm{E}(X_{r+p:n+q}^{\alpha})$ of order statistics exist for some integers r, n with $2 \leq r \leq n$, and $1 \leq r + p \leq n + q$. Then the recurrence relation

$$\mathrm{E}(X_{r:n}^{\alpha+\beta}) - \mathrm{E}\left(X_{r-1:n}^{\alpha+\beta}\right) = (\alpha+\beta)C_1 \mathrm{E}\left(X_{r+p:n+q}^{\alpha}\right) \tag{21.6}$$

Table 1
Some examples based on Theorem 21.1

p,q	$F(x)$	Distribution	$dC(k,n,p,q)$
$0,0$	$1 - \exp[-d(x-c)]$, $x \in (c, \infty)$	Exponential	$\frac{1}{n-k+1}$
$0, > 0$	$1 - [dq(c-x)]^{1/q}$, $x \in (c-1/dq, c)$		$\frac{n!(n-k+q)!}{(n+q)!(n-k+1)!}$
$0, < 0$	$1 - [dq(c-x)]^{1/q}$, $x \in (c-1/dq, \infty)$	Pareto Lomax	$\frac{n!(n-k+q)!}{(n+q)!(n-k+1)!}$
$-1, 0$	$\exp[d(x-c)]$, $x \in (-\infty, c)$		$\frac{1}{k-1}$
$>-1, p+1$	$[dq(x-c)]^{1/q}$, $x \in (c, c+1/dq)$	Power function	$\frac{n!(k+p-1)!}{(n+p+1)!(k-1)!}$
$<-1, p+1$	$[dq(x-c)]^{1/q}$, $x \in (-\infty, c+1/dq)$		$\frac{n!(k+p-1)!}{(n+p+1)!(k-1)!}$
$-1, -1$	$[1 + \exp(-d(x-c))]^{-1}$, $x \in (-\infty, \infty)$	Logistic	$\frac{n}{(k-1)(n-k+1)}$

with $C_1 = \frac{1}{d} \frac{\binom{n}{r-1}}{(r+p)\binom{n+q}{r+p}}$ holds.

Table 2 presents some examples of probability distributions included with respect to the integers p, q and the parameter β, wherein

D1: Pareto distribution [Malik (1966)].

D2: Weibull distribution [Khan, Yaqub and Parvez (1983)], especially Rayleigh distribution and exponential distribution when $\beta = 1$ [Joshi (1978), Khan, Yaqub and Parvez (1983), Azlarov and Volodin (1986) and Lin (1988a)].

D3: power function distributions ($\beta > 0, q = 1, c = 1/d$) ($\beta = 1$ in Lin (1988a)). Burr XII distributions when $q < 0$ [Khan and Khan (1987)], Lomax distributions when $\beta = 1$ [Lin (1988a)], and log-logistic distributions [$\beta > 0$, $p = 0, q = -1, d = \beta, c = -1/\beta$].

D4: logistic distribution [Shah (1970), Khan, Yaqub and Parvez (1983) and Lin (1988a)].

D5: power function distributions [Malik (1967a)].

Kamps and Mattner (1993) have generailized the above results of Kamps (1991) by proving the following theorem.

THEOREM 21.3. Let F satisfy (21.1), and let k be an integer with $2 \leq k \leq n$ and $1 \leq k + p \leq n + q$. Then

$$E[g(X_{k:n})] - E[g(X_{k-1:n})] = C\, E[g'(X_{k+p:n+q})] , \qquad (21.7)$$

where $C = \frac{\binom{n}{k-1}}{d(k+p)\binom{n+q}{k+p}}$ holds for every absolutely continuous function g provided that the right hand side in (21.7) exists.

Table 2
Some examples based on Theorem 21.2

p, q, β	$F(x)$	Distribution	$(\alpha + \beta)C_1$
$0, 0, 0$	$1 - \exp(cd)x^{-d}$, $x \in (e^c, \infty)$ and $c \in R$	D1	$\frac{\alpha}{d(n-r+1)}$
$0, 0, > 0$	$1 - \exp(-d[x^\beta/\beta - c])$, $x \in ((\beta c)^{1/\beta}, \infty)$, $c > 0$	D2	$\frac{\alpha+\beta}{d(n-r+1)}$
$0, \neq 0, 0$	$1 - [dq(c - \log x)]^{1/q}$, $x \in \begin{cases} (e^{c-1/dq}, e^c), & q > 0 \\ (e^{c-1/dq}, \infty), & q < 0 \end{cases}$		$\frac{\alpha(n+q-r)!n!}{d(n+q)!(n-r+1)!}$
$0, \neq 0, > 0$	$1 - [dq(c - x^\beta/\beta]^{1/q}$, $x \in \begin{cases} \left([\beta(c - 1/dq)]^{1/\beta}, [\beta c]^{1/\beta}\right), & q > 0 \\ \left([\beta(c - 1/dq)]^{1/\beta}, \infty\right), & q < 0 \end{cases}$ $c \geq 1/dq$	D3	$\frac{n!(\alpha+\beta)(n+q-r)!}{d(n-r+1)!(n+q)!}$
$-1, -1, 1$	$1 + [\exp(-d(x - c))]^{-1}$, $x \in (-\infty, \infty)$	D4	$\frac{n(\alpha+1)}{d(r-1)!(n-r+1)}!$
$-1, 0, 0$	$e^{-cd}x^d$, $x \in (0, e^c)$	D5	$\frac{\alpha}{d(r-1)}$

Recently, Mohie El-Din, Abu-Youssef and Sultan (1996) have proved the following theorem for obtaining a general identity for product moments of order statistics in a class of distribution functions, including Pareto, Weibull, exponential, Rayleigh and Burr distributions.

THEOREM 21.4. Let $\alpha \in R$, $\beta \geq 0$ with $\alpha + \beta \neq 0$, $h(t)$ be a function on $(0, 1)$ with

$$\frac{d}{dt}h(t) = \frac{1}{d}(1 - t)^{q-1}, \quad d > 0, \; q \in Z, \tag{21.8}$$

F be given by

$$F^{-1}(t) = \begin{cases} \exp\{h(t)\}, & \beta = 0 \\ (\beta h(t))^{1/\beta}, & \beta > 0 \end{cases} \tag{21.9}$$

where $(\beta h(t))^{1/\beta} \in R$, and let

$$-\infty < E\left(X_{r:n}^j X_{s:n}^{\alpha+\beta}\right), \; E\left(X_{r:n}^j X_{s-1:n}^{\alpha+\beta}\right), \; E\left(X_{r:n+q}^j X_{s:n+q}^{\alpha}\right) < \infty, \quad j > 0,$$

for some integers r, s and n with $1 \leq r < s \leq \min(n, n + q)$, $n + q \geq 2$. Then the recurrence relation

$$E\left(X_{r:n}^j X_{s:n}^{\alpha+\beta}\right) - E\left(X_{r:n}^j X_{s-1:n}^{\alpha+\beta}\right) = (\alpha + \beta)C(n, q, s)E\left(X_{r:n+q}^j X_{s:n+q}^{\alpha}\right), \tag{21.10}$$

is valid, where

$$C(n,q,s) = \frac{n!(n+q-s)!}{d(n-s+1)!(n+q)!} \ . \tag{21.11}$$

Note that Theorem 21.4 corresponds to Theorem 21.1 in the case of single order statistics.

Some examples of probability distributions based on the above theorem are presented and the corresponding values of the integer q and parameter β.

EXAMPLES. (i) $q = 0, \beta = 0$ and $c \in R$,

$$\frac{d}{dt}h(t) = \frac{1}{d(1-t)} \ .$$

Then,

$$F(x) = 1 - e^{cd}x^{-d}, \quad x \in (e^c, \infty)$$

i.e., the Pareto distribution.

(ii) $q = 0, \beta > 0$ and $c \geq 0$,

$$h(t) = \frac{1}{d} \log \frac{1}{(1-t)} + c \ .$$

Then,

$$F(x) = 1 - \exp\left\{-d\left(\frac{1}{\beta}x^\beta - c\right)\right\}, \quad x \in \left((\beta c)^{\frac{1}{\beta}}, \infty\right)$$

i.e., the Weibull distribution, exponential distribution with $\beta = 1$, and Rayleigh distribution with $\beta = 2$.

(iii) $q < 0, \beta > 0$ and $c \geq \frac{1}{dq}$,

$$h(t) = -\frac{1}{dq}(1-t)^q + c \ .$$

Then,

$$F(x) = 1 - \left[dq\left(c - \frac{1}{\beta}x^\beta\right)\right]^{\frac{1}{q}}, \quad x \in \left(\beta\left(c - \frac{1}{dq}\right)^{\frac{1}{\beta}}, \infty\right)$$

i.e., the Burr distribution.

(iv) $q \neq 0, \beta = 0$ and $c \in R$,

$$h(t) = -\frac{1}{dq}(1-t)^q + c \ .$$

Then,

$$F(x) = 1 - [dq(c - \log x)]^{\frac{1}{q}}, \quad \begin{cases} x \in \left(e^{c-\frac{1}{dq}}, e^c\right), & q > 0 \\ x \in \left(e^{c-\frac{1}{dq}}, \infty\right), & q < 0 \ . \end{cases}$$

Table 3
Some Examples Based on Theorem 21.4

q, β	$F(x)$	Distribution	$(\alpha + \beta)C(n, q, s)$
$0, 0$	$1 - e^{cd}x^{-d}, c \in R,$ $x \in (e^c, \infty)$	Pareto	$\frac{\alpha}{d(n-s+1)}$
$0, > 0$	$1 - \exp\{-d(\frac{1}{\beta}x^\beta - c)\},$ $x \in ((\beta c)^{\frac{1}{\beta}}, \infty), c \geq 0$	Weibull	$\frac{\alpha+\beta}{d(n-s+1)}$
$0, 1$	$1 - \exp(-d(x - c)),$ $x \in (c, \infty)$	exponential	$\frac{\alpha+1}{d(n-s+1)}$
$0, 2$	$1 - \exp(-d(x^2/2 - c)),$ $x \in ((2c)^{\frac{1}{2}}, \infty)$	Rayleigh	$\frac{\alpha+2}{d(n-s+1)}$
$< 0, > 0$	$1 - (dq(c - \frac{1}{\beta}x^\beta))^{\frac{1}{q}},$ $x \in ((c - \frac{1}{dq})^{\frac{1}{\beta}}, \infty)$	Burr	$\frac{n!(\alpha+\beta)(n+q-s)!}{d(n-s+1)!(n+q)!}$
$> 0, 0$	$1 - (dq(c - \log x))^{\frac{1}{q}},$ $x \in (e^{c - \frac{1}{qd}}, e^c)$		
$< 0, 0$	$1 - (dq(c - \log x))^{\frac{1}{q}},$ $x \in (e^{c - \frac{1}{qd}}, \infty)$		

Table 3 shows the particular cases of the results which are known from the literature.

From Table 3, we note the following:

(1) The recurrence relation of order statistics for Pareto distribution which is special case from (2.15) have been reviewed by Balakrishnan, Malik and Ahmed (1988).
(2) The product moments of order statistics from Weibull distribution have been derived by Khan, Parvez and Yaqub (1983).
(3) Khan and Khan (1987) have derived the recurrence relation between the product moments of order statistics from Burr distribution.
(4) Balakrishnan, Malik and Ahmed (1988) have presented the recurrence relation for the product moments of order statistics for exponential distribution.

Acknowledgement

This work was done while the second author was visiting McMaster University as a Channel Scholar funded by the Government of Egypt.

References

Abdel-Aty, S. H. (1954). Ordered variables in discontinuous distributions. *Statistica Neerlandica* **8**, 61–82.

Abramowitz, M. and I. A. Stegun (Eds.) (1964). *Handbook of Mathematical Functions with Formulas, Graphs and Mathematical Tables*. U.S. Government Printing Office, Washington.

Ali, M. M. and A. H. Khan (1987). On order statistics from the log-logistic distribution. *J. Statist. Plann. Inf.* **17**, 103–108.

Andrews, D. F., P. J. Bickel, F. R. Hampel, P. J. Huber, W. H. Rogers and J. W. Tukey (1972). *Robust Estimates of Location*. Princeton University Press, Princeton.

Arnold, B. C. (1977). Recurrence relations between expectations of functions of order statistics. *Scand. Actuar. J.* 169–174.

Arnold, B. C. and N. Balakrishnan (1989). *Relations, Bounds and Approximations for Order Statistics*. Lecture Notes in Statistics, No. **53**, Springer-Verlag, New York.

Arnold, B. C., N. Balakrishnan and H. N. Nagaraja (1992). *A First Course in Order Statistics*. John Wiley & Sons, New York.

Azlarov, T. A. and N. A. Volodin (1986). *Characterization Problems Associated with the Exponential Distribution*. Springer-Verlag, New York.

Bain, L. J. and M. Engelhardt (1991). *Statistical Analysis of Reliability and Lif-Testing Models*, Second edition. Marcel Dekker, New York.

Balakrishnan, N. (1982). A note on sum of the sub-diagonal product moments of order statistics. *J. Statist. Res.* **16**, 37–42.

Balakrishnan, N. (1985). Order statistics from the half logistic distribution. *J. Statist. Comput. Simul.* **20**, 287–309.

Balakrishnan, N. (1986). Order statistics from discrete distributions. *Commun. Statist. – Theor. Meth.* **15**, 657–675.

Balakrishnan, N. (1987a). Recurrence relations for moments of order statistics from Lomax distribution. *Technical Report*, McMaster University, Hamilton, Canada.

Balakrishnan, N. (1987b). A note on moments of order statistics from exchange variates. *Commun. Statist. – Theor. Meth.* **16**, 885–891.

Balakrishnan, N. (Ed.) (1988). Order statistics and Applications. A special edition of *Commun. Statist. – Theor. Meth.* **17**.

Balakrishnan, N. (1989). A relation for the covariances of order statistics from n independent and non-identically distributed random variables. *Statist. Hefte* **30**, 141–146.

Balakrishnan, N. (Ed.) (1992). *Handbook of the Logistic Distribution*. Marcel Dekker, New York.

Balakrishnan, N. and A. C. Cohen (1991). *Order Statistics and Inference: Estimation Methods*. Academic Press, San Diego.

Balakrishnan, N. and H. A. David (1998). On the variance of a lightly trimmed mean when multiple outliers are present in the sample. *Amer. Statist.* (submitted).

Balakrishnan, N., Z. Govindarajulu and K. Balasubramanian (1993). Relationships between moments of two related sets of order statistics and some extensions. *Ann. Inst. Statist. Math.* **45**, 243–247.

Balakrishnan, N. and P. C. Joshi (1981a). A note on order statistics from Weibull distribution. *Scand. Actuar. J.* 121–122.

Balakrishnan, N. and P. C. Joshi (1981b). Moments of order statistics from doubly truncated power function distribution. *Aligarh J. Statist.* **1**, 98–105.

Balakrishnan, N. and P. C. Joshi (1982). Moments of order statistics from doubly truncated Pareto distribution. *J. Indian Statist. Assoc.* **20**, 109–117.

Balakrishnan, N. and P. C. Joshi (1983). Single and product moments of order statistics from symmetrically truncated logistic distribution. *Demonstratio Mathematica* **16**, 833–841.

Balakrishnan, N. and P. C. Joshi (1984). Product moments of order statistics from doubly truncated exponential distribution. *Naval Res. Logist. Quart.* **31**, 27–31.

Balakrishnan, N. and S. Kocherlakota (1985). On the double Weibull distribution: Order statistics and estimation. *Sankhyā Ser. B* **47**, 161–178.

Balakrishnan, N. and S. Kocherlakota (1986). On the moments of order statistics from doubly truncated logistic distribution. *J. Statist. Plann. Inf.* **13**, 117–129.

Balakrishnan, N. and H. J. Malik (1985). Some general identities involving order statistics. *Commun. Statist. – Theor. Meth.* **14**, 333–339.

Balakrishnan, N. and H. J. Malik (1986a). A note on moments of order statistics. *Amer. Statist.* **40**, 147–148.

Balakrishnan, N. and H. J. Malik (1986b). Order statistics from the linear-exponential distribution, Part I: Increasing hazard rate case. *Commun. Statist. – Theor. Meth.* **15**, 179–203.

Balakrishnan, N. and H. J. Malik (1987a). Moments of order statistics from truncated log-logistic distribution. *J. Statist. Plann. Inf.* **16**, 251–267.

Balakrishnan, N. and H. J. Malik (1987b). Some identities involving product moments of order statistics. *Technical Report*, University of Guelph, Guelph, Canada.

Balakrishnan, N. and H. J. Malik (1987c). Some general identities involving the joint distribution of two order statistics. *Technical Report*, University of Guelph, Guelph, Canada.

Balakrishnan, N., H. J. Malik and S. E. Ahmed (1988). Recurrence relations and identities for moments of order statistics, II: Specific continuous distributions. *Commun. Statist.–Theor. Meth.* **17**, 2657–2694.

Balakrishnan, N. and S. Puthenpura (1986). Best linear unbiased estimation of location and scale parameters of the half logistic distribution. *J. Statist. Comput. Simul.* **25**, 193–204.

Balakrishnan, N. and R. A. Sandhu (1995). Recurrence relations for single and product moments of order statistics from a generalized half logistic distribution with applications to inference. *J. Statist. Comput. Simul.* **52**, 385–398.

Barnett, V. D. (1966). Order statistics estimators of the location of the Cauchy distribution. *J. Amer. Statist. Assoc.* **61**, 1205–1218. Correction **63**, 383–385.

Barnett, V. and T. Lewis (1993). *Outliers in Statistical Data*, Third edition. John Wiley & Sons, Chichester, U.K.

Bennet, S. (1983). Log-logistic regression model for survival data. *Appl. Statist.* **32**, 165–171.

Birnbaum A. and J. Dudman (1963). Log-logistic order statistics. *Ann. Math. Statist.* **34**, 658–663.

Block, H. W. and B. R. Rao (1973). A beta warning-time distribution and a distended beta distribution. *Sankhyā Ser. B* **35**, 79–84.

Blom, G. (1958). *Statistical Estimates and Transformed Beta-Variables*. Almqvist and Wiksell, Uppsala, Sweden.

Blom, G. (1962). Nearly best linear estimates of location and scale parameters. In *Contributions to Order Statistics* (Eds., A. E. Sarhan and B. G. Greenberg), pp. 34–46. John Wiley & Sons, New York.

Breiter, M. C. and P. R. Krishnaiah (1968). Tables for the moments of gamma order statistics. *Sankhyā Ser. B* **30**, 59–72.

Burr, I. W. (1942). Cumulative frequency functions. *Ann. Math. Statist.* **13**, 215–232.

Burr, I. W. (1968). On general system of distributions, III. The sample range. *J. Amer. Statist. Assoc.* **62**, 636–643.

Burr, I. W. and P. J. Cislak (1968). On a general system of distributions, I. Its curve-shape characteristics, II. The sample median. *J. Amer. Statist. Assoc.* **63**, 627–635.

Burrows, P. M. (1972). Expected selection differentials for directional selection. *Biometrics* **28**, 1091–1100.

Burrows, P. M. (1975). Variances of selection differentials in normal samples. *Biometrics* **31**, 125–133.

Cadwell, J. H. (1953). The distribution of quasi-ranges in samples from a normal population. *Ann. Math. Statist.* **24**, 603–613.

Castillo, E. (1988). *Extreme Value Theory in Engineering*. Academic Press, San Diego.

Cohen, A. C. (1991). *Truncated and Censored Samples: Theory and Applications*. Marcel Dekker, New York.

Cohen, A. C. and B. J. Whitten (1988). *Parameter Estimation in Reliability and Life Span Models*. Marcel Dekker, New York.

Cole, R. H. (1951). Relations between moments of order statistics. *Ann. Math. Statist.* **22**, 308–310.

D'Agostino, R. B. and M. A. Stephens (Eds.) (1986). *Goodness-of-Fit Techniques*. Marcel Dekker, New York.

David, H. A. (1979). Robust estimation in the presence of outliers. In *Robustness in Statistics* (Eds., R. L. Launer and G. N. Wilkinson), 61–74. Academic Press, New York.

David, H. A. (1981). *Order Statistics*, Second Edition. John Wiley & Sons, New York.

David, H. A. and N. Balakrishnan (1996). Product moments of order statistics and the variance of a lightly trimmed mean. *Statist. Prob. Lett.* **29**, 85–87.

David, H. A. and P. C. Joshi (1968). Recurrence relations between moments of order statistics for exchangeable variates. *Ann. Math. Statist.* **39**, 272–274.

David, F. N. and N. L. Johnson (1954). Statistical treatment of censored data. I. Fundamental formulae. *Biometrika* **41**, 228–240.

David, H. A. and V. S. Shu (1978). Robustness of location estimators in the presence of an outlier. In *Contributions to Survey Sampling and Applied Statistics: Papers in Honor of H. O. Hartley* (Ed., H. A. David), 235–250. Academic Press, New Yok.

Davis, C. S. and M. A. Stephens (1978). Approximating the covariance matrix of normal order statistics. Algorithm AS 128. *Appl. Statist.* **27**, 206–212.

Dixon, W. J. (1960). Simplified estimation from censored normal samples. *Ann. Math. Statist.* **31**, 385–391.

Dixon, W. J. and J. W. Tukey (1968). Approximate behavior of the distribution of Winsorized t (trimming/Winsorization 2). *Technometrics* **10**, 83–98.

Downton, F. (1954). Least-squares estimates using ordered observations. *Ann. Math. Statist.* **25**, 303–316.

Downton, F. (1966). Linear estimates with polynomial coefficients. *Biometrika* **53**, 129–141.

Dyer, D. D. and C. W. Whisenand (1973a). Best linear unbiased estimator of the parameter of the Rayleigh distribution – I: Small sample theory for censored order statistics. *IEEE Trans. on Reliab.* **22**, 27–34.

Dyer, D. D. and C. W. Whisenand (1973b). Best linear unbiased estimator of the parameter of the Rayleigh distribution – II: Optimum theory for selected order statistics. *IEEE Trans. on Reliab.* **22**, 229–231.

Eisenberger, I. and E. C. Posner (1965). Systematic statistics used for data compression in space telemetry. *J. Amer. Statist. Assoc.* **60**, 97–133.

Galambos, J. (1978). *The Asymptotic Theory of Extreme Order Statistics*. John Wiley & Sons, New York; Second edition, 1987. Krieger, Malabar, FL.

Galton, F. (1902). The most suitable proportion between the values of first and second prizes. *Biometrika* **1**, 385–390.

Gibbons, J. D. and S. Chakraborty (1994). *Nonparametric Statistical Inference*, Second edition. Marcel Dekker, New York.

Godwin, H. J. (1949). Some low moments of order statistics. *Ann. Math. Statist.* **20**, 279–285.

Govindarajulu, Z. (1963a). Relationships among moments of order statistics in samples from two related populations. *Technometrics* **5**, 514–518.

Govindarajulu, Z. (1963b). On moments of order statistics and quasi-ranges from normal populations. *Ann. Math. Statist.* **34**, 633–651.

Govindarajulu, Z. (1966). Best linear estimates under symmetric censoring of the parameters of a double exponential population. *J. Amer. Statist. Assoc.* **61**, 248–258.

Gumbel, E. J. (1958). *Statistics of Extremes*. Columbia University Press, New York.

Gupta, S. S. (1960). Order statistics from the gamma distribution. *Technometrics* **2**, 243–262.

Gupta, S. S., A. S. Qureishi and B. K. Shah (1967). Best linear unbiased estimators of the parameters of the logistic distribution using order statistics. *Technometrics* **9**, 43–56.

Gupta, S. S. and B. K. Shah (1965). Exact moments and percentage points of the order statistics and the distribution of the range from the logistic distribution. *Ann. Math. Statist.* **36**, 907–920.

Harter, H. L. (1988). History and role of order statistics. *Commun. Statist. – Theor. Meth.* **17**, 2091–2107.

Harvard Compuation Laboratory (1955). *Tables of the Cumulative Binomal Probability istribution*, Harvard University Press, Cambridge, Mass.

Hawkins, D. M. (1980). *Identification of Outliers*. Chapman and Hall, London.

Hoeffding, W. (1953). On the distribution of the expected values of the order statistics. *Ann. Math. Statist.* **24**, 93–100.

Hogg, R. V. (1967). Some observations on robust estimation. *J. Amer. Statist. Assoc.* **62**, 1179–1186.

Hogg, R. V. (1974). Adaptive robust procedures: a partial review and some suggestions for future applications and theory. *J. Amer. Statist. Assoc.* **69**, 909–923.

Hollander, M. and D. A. Wolfe (1973). *Nonparameteric Statistical Methods*. John Wiley & Sons, New York.

Huang, J. S. (1975). A note on order statistics from Pareto distribution. *Scand Actuar. J.* 187–190.

Huber, P. J. (1972). Robust statistics: a review. *Ann. Math. Statist.* **43**, 1041–1067.

Huber, P. J. (1982). *Robust Statistics*. John Wiley & Sons, New York.

Johnson, N. L., S. Kotz and N. Balakrishnan (1994). *Continuous Univariate Distributions*, Vol. 1, Second edition. John Wiley & Sons, New York.

Johnson, N. L., S. Kotz and N. Balakrishnan (1995). *Continuous Univariate Distributions*, Vol. 2, Second edition. John Wiley & Sons, New York.

Jones, H. L. (1948). Exact lower moments of order statistics in small samples from a normal distribution. *Ann. Math. Statist.* **19**, 270–273.

Joshi, P. C. (1971). Recurrence relations for the mixed moments of order statistics. *Ann. Math. Statist.* **42**, 1096–1098.

Joshi, P. C. (1973). Two identities involving order statistics. *Biometrika* **60**, 428–429.

Joshi, P. C. (1978). Recurrence relations between moments of order statistics from exponential and truncated exponential distributions. *Sankhyā Ser. B* **39**, 362–371.

Joshi, P. C. (1979a). On the moments of gamma order statistics. *Naval Res. Logist. Quart.* **26**, 675–679.

Joshi, P. C. (1979b). A note on the moments of order statistics from doubly truncated exponential distribution. *Ann. Inst. Statist. Math.* **31**, 321–324.

Joshi, P. C. (1982). A note on the mixed moments of order statistics from exponential and truncated exponential distributions. *J. Statist. Plann. Inf.* **6**, 13–16.

Joshi, P. C. and N. Balakrishnan (1981a). An identity for the moments of normal order statistics with applications. *Scand. Actuar. J.* 203–213.

Joshi, P. C. and N. Balakrishnan (1981b). Applications of order statistics in combinatorial identities. *J. Comb. Infor. System Sci.* **6**, 271–278.

Joshi, P. C. and N. Balakrishnan (1982). Recurrence relations and identities for the product moments of order statistics. *Sankhyā Ser. B* **44**, 39–49.

Joshi, P. C. and S. Chakraborty (1996a). Moments of Cauchy order statistics via Riemann zeta function. In *Statistical Theory and Applications*: Papers in Honor of Herbert A. David (Eds., H. N. Nagaraja, Pranab K. Sen and Donald F. Morrison), 117–127. Springer-Verlag, New York.

Joshi, P. C. and S. Chakraborty (1996b). Single and product moments of Cauchy order statistics. *Comun. Statist. – Theor. Meth.* **25**, 1837–1844.

Joshi, P. C. and Shubha (1991). Some identities among moments of order statistics. *Commun. Statist. – Theor. Meth.* **20**, 2837–2843.

Jung, J. (1955). On linear estimates defined by a continuous weight function. *Ark. Math.* **3**, 199–209.

Jung, J. (1962). Approximation of least-squares estimates of location and scale parameters. In *Contributions to Order Statistics* (Eds., A. E. Sarhan and B. G. Greenberg), 28–33, John Wiley & Sons, New York.

Kabe, D. G. (1972). On moments of order statistics from the Pareto distribution. *Skand. Aktuarietdskr.* **55**, 179–181.

Kamps, U. (1991). A general recurrence relation for moments of order statistics in a class of probability distributions and characterizations. *Metrika* **38**, 215–225.

Kamps, U. (1992). Identities for the difference of moments of successive order statistics and record values. *Metron* **50**, 179–187.

Kamps, U. and L. Mattner (1993). An identity for expectations of order statistics. *Metrika* **40**, 361–365.

Khan, A. H. and I. A. Khan (1987). Moments of order statistics from Burr distribution and its characterizations. *Metron* **45**, 21–29.

Khan, A. H., S. Parvez and M. Yaqub (1983). Recurrence relations between product moments of order statistics. *J. Statist. Plann. Inf.* **8**, 175–183.

Khan, A. H., M. Yaqub and S. Parvez (1983). Recurrence relations between moments of order statistics. *Naval Res. Logist. Quart.* **30**, 419–441.

Khatri, C. G. (1962). Distributions of order statistics for discrete case. *Ann. Inst. Statist. Math.* **14**, 167–171.

Khurana, A. P. and V. D. Jha (1991). Recurrence relations between moments of order statistics from a doubly truncated Pareto distribution. *Sankhyā Ser. B* **21**, 11–16.

Krishnaiah, P. R. and M. H. Rizvi (1966). A note on recurrence relations between expected values of functions of order statistics. *Ann. Math. Statist.* **37**, 733–734.

Krishnaiah, P. R. and M. H. Rizvi (1967). A note on the moments of gamma order statistics. *Technometrics* **9**, 315–318.

Lange, K. (1996). Illustration of some moment identities for order statistics. *Statist. Prob. Lett.* **29**, 245–249.

Lawless, J. F. (1982). *Statistical Models and Methods for Lifetime Data*. John Wiley & Sons, New York.

Lehmann, E. L. (1975). *Nonparametric Statistical Methods Based on Ranks*. McGraw-Hill, New York.

Lin G. D. (1988a). Characterizations of distributions via relationships between two moments of order statistics. *J. Statist. Plann. Inf.* **19**, 73–80.

Lin G. D. (1988b). Characterizations of uniform distributions and exponential distributions. *Sankhyā Ser. B* **50**, 64–69.

Lien, D. D., N. Balakrishnan and K. Balasubramanian (1992). Moments of order statistics from a non-overlapping mixture model with applications to truncated Laplace distribution. *Commun. Statist. – Theor. Meth.* **21**, 1909–1928.

Lloyd, E. H. (1952). Least-squares estimation of location and scale parameters using order statistics. *Biometrika* **39**, 88–95.

Malik, H. J. (1966). Exact moments of order statisics from the Pareto distribution. *Skandinavisk Aktuarietidskrift* **49**, 144–157.

Malik, H. J. (1967a). Exact moments of order statistics from a power function distribution. *Skandinavisk Aktuarietidskrift* 64–69.

Malik, H. J. (1967b). Exact distribution of the quotient of independent generalized gamma variables. *Canad. Math. Bull.* **10**, 463–466.

Malik, H. J., N. Balakrishnan and S. E. Ahmed (1988). Recurrence relations and identities for moments of order statistics, I: Arbitrary continuous distributions. *Commun. Statist. – Theor. Meth.* **17**, 2623–2655.

Mann, N. R., R. E. Schafer and N. D. Singpurwalla (1974). *Methods for Statistical Analysis of Reliability and Lifetime Data*. John Wiley & Sons, New York.

Melnick, E. L. (1964). *Moments of Ranked Poisson Variates*, M. S. Thesis, Virginia Polytechnic Institute.

Mosteller, F. and J. W. Tukey (1977). *Data Analysis and Regression*. Addison-Wesley, Reading, Massachusetts.

Mohie El-Din, M. M., M. A. W. Mahmoud and S. E. Abu-Youssef (1991). Moments of order statistics from parabolic and skewed distributions and characterization of Weibull distribution. *Commun. Statist. – Simul. Comput.* **20**, 639–645.

Mohie El-Din, M. M., M. A. W. Mahmoud and S. E. Abu-Youssef (1992). Recurrence relations between moment generating function of order statistics from doubly truncated continuous distributions. *Egypt. Statist. J.* **36**, 82–94.

Mohie El-Din, M. M., S. E. Abu-Youssef and K. S. Sultan (1996). An identity for the product moments of order statistics. *Metrika* **44**, 95–100.

Mohie El-Din, M. M., M. A. W. Mahmoud and K. S. Sultan (1995). On moment generating function of order statistics for doubly truncated exponential distribution. *Metron* **53**, 171–183.

Mohie El-Din, M. M., M. A. W. Mahmoud and K. S. Sultan (1996). On order statistics of doubly truncated power function distribtion. *Metron* **54**, 83–93.

Mohie El-Din, M. M., M. A. W. Mahmoud, S. E. Abu-Youssef and K. S. Sultan (1997). Order statistics from the doubly truncated linear-exponential distribution and its characterizations. *Commun. Statist. – Simul. Comput.* **26**, 281–290.

Mohie El-Din, M. M. and K. S. Sultan (1995). Recurrence relations for expectations of functions of order statistics for doubly truncated distributions and their applications. *Commun. Statist. – Theor. Meth.* **24**, 997–1010.

Nassar, M. M. and M. R. Mahmoud (1985). On characterization of a mixture of exponential distributions. *IEEE Trans. on Reliab.* **34**, 484–488.

Nelson, W. (1982). *Applied Life Data Analysis*. John Wiley & Sons, New York.

O'Quigley, J. and L. Struthers (1982). Several models based upon the logistic and log-logistic distributions. *Computer Programs in Biomedicine* **15**, 3–12.

Pearson, K. (1902). Note on Francis Galton's difference problem. *Biometrika* **1**, 390–399.

Pearson, K. (1968). *Tables of the Incomplete Beta Function*, Second Edition. Cambridge University Press, London.

Pike, M. C. and I. D. Hill (1966). Logarithm of gamma function. Algorithm 291. *Commun. of ACM* **9**, 684.

Rainville, E. D. (1960). *Special Functions*. Macmillan, New York.

Riordan, J. (1968). *Combinatorial Identities*. John Wiley & Sons, New York.

Romanovsky, V. (1933). On a property of the mean ranges in samples from a normal population and on some integrals of Professor T. Hojo. *Biometrika* **25**, 195–197.

Saleh, A. K. Md. E., C. Scott and D. B. Junkins (1975). Exact first and second order moments of order statistics from the truncated exponential distribution. *Naval Res. Logist. Quart.* **22**, 65–77.

Sarhan, A. E. and B. G. Greenberg (Eds) (1962). *Contributions to Order Statistics*. John Wiley & Sons, New York.

Schaeffer, L. R., L. D. Van Vleck and J. A. Velasco (1970). The use of order statistics with selected records. *Biometrics* **26**, 854–859.

Scheffé, H. (1943). Statistical inference in the non-parametric case. *Ann. Math. Statist.* **14**, 305–332.

Shu, V. S. (1978). *Robust Estimation of a Location Parameter in the Presence of Outliers*. Ph.D. Thesis, Iowa State University, Ames, Iowa.

Shah, B. K. (1966). On the bivariate moments of order statistics from a logistic distribution. *Ann. Math. Statist.* **37**, 1002–1010.

Shah, B. K. (1970). Note on moments of a logistic order statistics. *Ann. Math. Statist.* **41**, 2150–2152.

Sillitto, G. P. (1951). Interrelations between certain linear systematic statistics of samples from any continuous population. *Biometrika* **38**, 377–382.

Sillitto, G. P. (1964). Some relations between expectations of order statistics in samples of different sizes. *Biometrika* **51**, 259–262.

Srikantan, K. S. (1962). Recurrence relations between the PDF's of order statistics, and some applications. *Ann. Math. Statist.* **33**, 169–177.

Tadikamalla, P. R. (1980). A look at the Burr and related distributions. *Internat. Statist. Rev.* **48**, 337–344.

Tarter, M. E. (1966). Exact moments and product moments of order statistics from the truncated logistic distribution. *J. Amer. Statist. Assoc.* **61**, 514–525.

Tiku, M. L., W. Y. Tan and N. Balakrishnan (1986). *Robust Inference*. Marcel Dekker, New York.

Thomas, P. Y. (1993). On some identities involving the moments of extremes from gamma distribution. *Commun. Statist. – Theor. Meth.* **22**, 2321–2326.

Thomas, P. Y. (1996). On the moments of extremes from generalized gamma distribution. *Commun. Statist. – Theor. Meth.* **25**, 1825–1836.

Thomas, P. Y. and T. S. K. Moothathu (1991). Recurrence relations for different order statistics of extremes from gamma distribution. *Commun. Statist. – Theor. Meth.* **20**, 945–950.

Thomas, P. and P. Samuel (1996). A note on recurrence relations for the product moments of order statistics. *Statist. Prob. Lett.* **29**, 245–249.

Tukey, J. W. and D. H. McLaughlin (1963). Less vulnerable confidence and significance procedures for location based on a single sample: Trimming/Winsorization, 1. *Sankhyā Ser. A* **25**, 331–352.

Vaughan, D. C. (1994). The exact values of the expected values, variances and covariances of the order statistics from the Cauchy distribution. *J. Statist. Comput. Simul.* **49**, 21–32.

Young, D. H. (1967). Recurrence relations between the P.D.F.'s of order statistics of dependent variables, and some applications. *Biometrika* **24**, 283–292.

Young, D. H. (1971). Moment relations for order statistics of the standardized gamma distribution and the inverse multinomial distribution. *Biometrika* **58**, 637–640.

Yuen, K. K. (1971). A note on Winsorized t. *Appl. Statist.* **20**, 297–303.

PART IV

Characterizations

Recent Approaches to Characterizations Based on Order Statistics and Record Values

C. R. Rao and D. N. Shanbhag

1. Introduction

Ferguson (1964, 1965) and Crawford (1966) were amongst the earliest authors who characterized geometric and exponential distributions via properties of order statistics. In various papers, Ahsanullah, Govindarajulu, Arnold, Galambos and many others have since obtained interesting characterization results based on order statistics. Galambos and Kotz (1978), David (1981), Azlarov and Volodin (1986), Arnold, Balakrishnan and Nagaraja (1992), Rao and Shanbhag (1994), Aly (1988), and Kamps (1995) among others have reviewed the existing literature on important properties of order statistics.

Shorrock (1972a,b, 1973), Nagaraja (1977), Gupta (1984), Dallas (1981), Nayak (1981), and Rao and Shanbhag (1986) have observed properties of, or established characterizations based on, record values. The monographs of Galambos and Kotz (1978), Azlarov and Volodin (1986), and Rao and Shanbhag (1994) have reviewed some of the major results on the topic.

Many of the characterization results based either on order statistics or record values have implicit links with the integrated Cauchy functional equation or its variants. The monograph of Rao and Shanbhag (1994) gives the relevant details as regards this and shows that the recent advances on the functional equation lead us, in places, to improved and unified versions of the results in the existing literature. (A slightly restrictive coverage of the link that we have referred to here appears in Ramachandran and Lau (1991).)

The purpose of the present paper is to review characterization results based on order statistics and record values, having links with the integrated Cauchy functional equation or its variants, and make further observations on these going beyond what Rao and Shanbhag (1994) have already pointed out. In the process of doing this, we show that many of the cited results could be arrived at via the strong memoryless property characterization of the exponential and geometric distributions. We also give some statistical applications of the characterization results discussed.

2. Some basic tools

The following are some basic tools that are to be referred to in the present discussion.

THEOREM 1. Let f be a non-negative real locally integrable Borel measurable function on $\mathbf{R}_+$, other than a function which is identically 0 almost everywhere $[L]$, such that it satisfies

$$f(x) = \int_{\mathbf{R}_+} f(x+y)\mu(dy) \quad \text{for almost all } [L]x \in \mathbf{R}_+ \tag{2.1}$$

for some σ-finite measure μ on (the Borel σ-field of) $\mathbf{R}_+$ with $\mu(\{0\}) < 1$ (yielding trivially that $\mu(\{0\}^c) > 0$), where L corresponds to Lebesgue measure. Then, either μ is arithmetic with some span λ and

$$f(x+n\lambda) = f(x)b^n, \quad n = 0, 1, \ldots \quad \text{for almost all } [L]x \in \mathbf{R}_+$$

with b such that

$$\sum_{n=0}^{\infty} b^n \mu(\{n\lambda\}) = 1$$

or μ is nonarithmetic and

$$f(x) \propto \exp(\eta x) \quad \text{for almost all } [L]x \in \mathbf{R}_+$$

with η such that

$$\int_{\mathbf{R}_+} \exp(\eta x) \mu(dx) = 1 \ .$$

The theorem is due to Lau and Rao (1982) and it has several interesting proofs including that based on exchangeability, given by Alzaid, Rao and Shanbhag (1987); see Ramachandran and Lau (1991) or Rao and Shanbhag (1994) for more details.

COROLLARY 1. Let $\{(v_n, w_n): n = 0, 1, \ldots\}$ be a sequence of vectors with non-negative real components such that $v_n \neq 0$ for at least one n, $w_0 < 1$, and the largest common divisor of the set $\{n: w_n > 0\}$ is unity. Then

$$v_m = \sum_{n=0}^{\infty} v_{m+n} w_n, \quad m = 0, 1, \ldots \tag{2.2}$$

if and only if

$$v_n = v_0 b^n, \quad n = 0, 1, 2, \ldots, \quad \text{and} \quad \sum_{n=0}^{\infty} w_n b^n = 1$$

for some $b > 0$.

PROOF. The "if" part is trivial, and to have the "only if" part, apply the theorem considering $f: \mathbf{R}_+ \to \mathbf{R}_+$ such that

$$f(x) = v_{[x]}, \quad x \in \mathbf{R}_+ ,$$

where $[\cdot]$ is the integral part, and μ as a measure concentrated on $\{0, 1, \ldots\}$ such that $\mu(\{n\}) = w_n$ for $n = 0, 1, \ldots$.

COROLLARY 2. Let X be a nonnegative random variable with $P\{X = 0\} < 1$ and h be a monotonic right continuous function on $\mathbf{R}_+$ such that $E(|h(X)|) < \infty$ and $E(h(X)) \neq h(0)$. Then

$$E\{h(X - x)|X \geq x\} = E(h(X)), \quad x \in \mathbf{R}_+ \quad \text{with} \quad P\{X \geq x\} > 0 \quad (2.3)$$

if and only if either h^* is nonarithmetic and X is exponential, or for some $\lambda > 0$, h^* is arithmetic with span λ and $P\{X \geq n\lambda + x\} = P\{X \geq x\} \ (P\{X \geq \lambda\})^n$, $n \in \mathbf{N}_0, x \in \mathbf{R}_+$, where

$$h^*(x) = \begin{cases} (h(x) - h(0))/(E(h(X)) - h(0)), & x \geq 0 \\ 0, & x < 0 \end{cases}$$

(We define h^* to be arithmetic or nonarithmetic according to whether the measure determined by it on $\mathbf{R}_+$ is arithmetic or nonarithmetic.)

PROOF. Note, writing $\overline{F}(x) = P\{X \geq x\}, x \in \mathbf{R}_+$, that (2.3) is equivalent to

$$\int_{\mathbf{R}_+} \overline{F}(x + y) \mu_{h^*}(dy) = \overline{F}(x), \quad x \in \mathbf{R}_+ ,$$

where μ_{h^*} is the measure determined by h^*. Theorem 1 then establishes the "only if" part of the assertion. As the "if" part of the assertion is trivial, we then have the corollary.

Corollary 1 is essentially given in Lau and Rao (1982) and is a slight generalization of a lemma established earlier by Shanbhag (1977). (Shanbhag takes $w_1 > 0$ in place of the condition that the largest common divisor of the set of n for which $w_n > 0$ is 1, even though he does not assume a priori $w_0 < 1$.) Corollary 2, in the form that we have presented here, has appeared earlier in Rao and Shanbhag (1994, p. 108); a somewhat different version of this result has been given by Klebanov (1980). If Y is a nonnegative random variable and we take

$$h(x) = P\{Y \leq x\}, \quad x \in \mathbf{R}_+ ,$$

then it follows that for a nonnegative random variable X which is independent of Y with $P\{X \geq Y\} > P\{Y = 0\}$, (2.3) is equivalent to

$$P\{X \geq Y + x | X \geq Y\} = P\{X \geq x\}, \quad x \in \mathbf{R}_+ .$$

Consequently, we have that the characterization based on the strong memoryless property, for the exponential and geometric distributions, given by Shimizu (1978) and Ramachandran (1979) follows essentially as a by-product of Corollary 2. (It is a simple exercise to see that the characterization based on the version of the strong memoryless property appearing above gives as a corollary that based on the version of the property with ">" in place of "≥").

THEOREM 2. Let f be a nonnegative real locally integrable Borel measurable function on $\mathbf{R}$, other than a function which is identically zero almost everywhere $[L]$, such that it satisfies

$$f(x) = \int_{\mathbf{R}} f(x+y)\mu(dy) \quad \text{for almost all } [L]x \in \mathbf{R} \qquad (2.4)$$

for some σ-finite measure μ on $\mathbf{R}$ satisfying $\mu(\{0\}) < 1$ or equivalently $\mu(\{0\}^c) > 0$ (with L as Lebesgue measure). Then, either μ is nonarithmetic and

$$f(x) = c_1 \exp\{\eta_1 x\} + c_2 \exp\{\eta_2 x\} \quad \text{for almost all } [L]x \in \mathbf{R} \ ,$$

or μ is arithmetic with some span λ and

$$f(x) = \xi_1(x) \exp\{\eta_1 x\} + \xi_2(x) \exp\{\eta_2 x\} \quad \text{for almost all } [L]x \in \mathbf{R} \ ,$$

with c_1 and c_2 as nonnegative real numbers, ξ_1 and ξ_2 as periodic nonnegative Borel measurable functions having period λ, and $\eta_i, i = 1, 2$ as real numbers such that $\int_{\mathbf{R}} \exp\{\eta_i x\}\mu(dx) = 1$. (For the uniqueness of the representations but for the ordering of the terms, one may assume for example that $c_2 = 0$ and $\xi_2 \equiv 0$ if $\eta_1 = \eta_2$.)

Theorem 2 is a corollary to a general theorem of Deny (1961). For the details of various proofs for this theorem and other related results, see Ramachandran and Lau (1991) and Rao and Shanbhag (1994). It is interesting to note here that if in Theorem 1 f is bounded and $\mu(\mathbf{R}_+) < \infty$, then the theorem follows easily as a corollary to Theorem 2 (see Fosam, Rao and Shanbhag (1993) for more details). Also, even when it is not of much relevance to characterization problems discussed in the following sections, it is worth pointing out in this place that the next result is a corollary to Theorem 2 and is in the same spirit as Corollary 1 of Theorem 1. (Use an argument analogous to that used in the proof of Corollary 1 to see the validity of this result.)

COROLLARY 3. Let $\{(v_n, w_n): n = 0, \pm 1, \ldots\}$ be a sequence of two-component vectors with nonnegative real components such that $w_0 < 1$ and at least one $v_n \neq 0$. Then

$$v_m = \sum_{n=-\infty}^{\infty} w_n v_{n+m}, \quad m = 0, \pm 1, \ldots$$

if and only if

$$v_m = B(m)b^m + C(m)c^m, \quad m = 0, \pm 1, \ldots$$

and $\sum_{m=-\infty}^{\infty} w_m b^m = \sum_{m=-\infty}^{\infty} w_m c^m = 1$ for some $b, c > 0$ and non-negative periodic functions B, C with the largest common divisor of $\{m: w_m > 0\}$ as their common period.

Corollary 3 was proved via a direct proof by Ramachandran (1984).

Davies and Shanbhag (1987), Shanbhag (1991), and Rao and Shanbhag (1994) have given general results subsuming versions of Deny's theorem as well as the Lau-Rao theorem, via arguments based on exchangeability, amongst other things. Although these arguments turn out to be involved due to technical difficulties one has to encounter in the general cases, one could illustrate a key idea appearing in these through a proof for Corollary 1 via de Finetti's theorem.

As the "if" part of the result is trivial, it is sufficient if we establish the "only if" part of the result. There is no loss of generality in assuming that $w_0 = 0$. The functional equation in the lemma implies that

$$v_m = \sum_{n=1}^{\infty} v_{m+n} w_n^*, \quad m = 0, 1, \ldots \tag{2.5}$$

with

$$w_n^* = \sum_{k=1}^{\infty} 2^{-k} w_n^{(k)}, \quad n = 1, 2, \ldots,$$

where $\{w_n^{(k)}\}$ is the k-fold convolution of $\{w_n\}$ with itself. There exists then some $n_0 > 0$ such that $w_n^* > 0$ for all $n \geq n_0$. Substituting for $v_{m+n}, n = 1, 2, \ldots, n_0 - 1$ in (2.5) successively (when $n_0 > 1$), we can get from (2.5)

$$v_m = \sum_{n=n_0}^{\infty} v_{m+n} \hat{w}_n, \quad m = 0, 1, \ldots \tag{2.6}$$

with $\hat{w}_n > 0, n = n_0, n_0 + 1, \ldots$ (2.6) implies that $v_n > 0$ for all n. (Note that here $v_m = 0 \Leftrightarrow v_n = 0$ for all $n > m$.) Define a sequence of exchangeable random variables $\{X_n: n = 1, 2, \ldots\}$ such that

$$P\{X_1 = x_1, \ldots, X_n = x_n\} = \frac{v_{x_1 + \cdots + x_n}}{v_0} \hat{w}_{x_1} \ldots \hat{w}_{x_n}, \quad x_i \geq n_0,$$

$$i = 1, 2, \ldots, n; \quad n = 1, 2, \ldots.$$

We have then, in view of de Finetti's theorem, that for all $x, y \in \{n_0, n_0 + 1, \ldots\}$,

$$0 = \frac{1}{v_0} \{v_{(x+y)+(x+y)} - 2v_{(x+y)+x+y} + v_{x+y+x+y}\}$$

$$= E\{((P\{X_1 = x+y|\mathcal{T}\}/\hat{w}_{x+y})$$

$$- (P\{X_1 = x|\mathcal{T}\}/\hat{w}_x)(P\{X_1 = y|\mathcal{T}\}/\hat{w}_y))^2\}$$

where $\mathcal{T}$ is the tail σ-field of $\{X_n\}$. Hence, it follows that

$$(P\{X_1 = x|\mathcal{T}\}/\hat{w}_x)(P\{X_1 = y|\mathcal{T}\}/\hat{w}_y)$$
$$= P\{X_1 = x+y|\mathcal{T}\}/\hat{w}_{x+y}, \quad x,y = n_0, n_0+1, \ldots, \text{a.s.},$$

which implies that there exists a positive real number b such that

$$P\{X_1 = x|\mathcal{T}\}/\hat{w}_x = b^x, \quad x = n_0, n_0+1, \ldots, \text{a.s.} \tag{2.7}$$

with $\sum_{n=n_0}^{\infty} \hat{w}_n b^n = 1$. (Note that b is unique.) From (2.6) and (2.7), we get that

$$\frac{v_x}{v_0} = b^x, \quad x = 0, 1, \ldots,$$

which implies, in view of (2.2), that $\sum_{n=0}^{\infty} w_n b^n = 1$; consequently we have the required result.

The above proof simplifies slightly if the assumptions of Shanbhag's lemma are met. With minor alterations in the proof, one could also produce a proof based on de Finetti's theorem for Corollary 3; further details in this respect will be available from a forthcoming article of the authors.

The next result that we need is Theorem 3 given below; in the statement of the theorem, we assume the following definition.

DEFINITION. Let X be a real-valued random variable with $E(X^+) < \infty$. Define a real-valued Borel measurable function s on $\mathbf{R}$ satisfying $s(x) = E\{X - x | X \geq x\}$ for all x such that $P\{X \geq x\} > 0$. This function is called the mean remaining life function (m.r.l. function for short).

The restriction of the m.r.l. to $(-\infty, b)$ where b is the right extremity of the distribution of X is clearly left continuous and hence is determined by its knowledge on a dense subset of $(-\infty, b)$.

THEOREM 3. Let $b (\leq \infty)$ denote the right extremity of the distribution function (d.f.) F of a random variable X with $E(X^+) < \infty$ and let s be its m.r.l. function. Further, let $A = \{y : \lim_{x \uparrow y} s(x) \text{ exists and equals } 0\}$. Then $b = \infty$ if A is empty and $b = \inf\{y : y \in A\}$ if A is non-empty. Moreover, for every $-\infty < y < x < b$

$$\frac{1 - F(x-)}{1 - F(y-)} = \frac{s(y)}{s(x)} \exp\left\{-\int_y^x \frac{dz}{s(z)}\right\}, \tag{2.8}$$

and for every $-\infty < x < b$, $1 - F(x-)$ is given by the limit of the right-hand side of (2.8) as $y \to -\infty$.

For a proof for Theorem 3 as well as some other interesting properties of the m.r.l. function, see Kotz and Shanbhag (1980) or Rao and Shanbhag (1994).

3. Characterizations based on order statistics

Ferguson (1964, 1965) and Crawford (1966) were among the earliest authors who characterized geometric and exponential distributions via properties of order statistics. They showed that if X and Y are independent nondegenerate random variables, then $\min\{X, Y\}$ is independent of $X - Y$ if and only if for some $\alpha > 0$ and $\beta \in \mathbf{R}$, we have $\alpha(X - \beta)$ and $\alpha(Y - \beta)$ to be either both exponential or both geometric (in the usual sense). Using effectively the strong memoryless property characterization of the geometric and exponential distributions, Rao and Shanbhag (1994; pp. 196–197) have essentially established the following extended version of the Ferguson–Crawford result.

THEOREM 4. Let X and Y be as in the Ferguson–Crawford result and y_0 be a point such that there are at least two support points of the distribution of $\min\{X, Y\}$ in $(-\infty, y_0]$. Let ϕ be a real-valued Borel measurable function on $\mathbf{R}$ such that its restriction to $(-\infty, y_0]$ is nonvanishing and strictly monotonic. Then $X - Y$ and $\phi(\min\{X, Y\})I_{\{\min\{X,Y\} \leq y_0\}}$ are independent if and only if for some $\alpha \in (0, \infty)$ and $\beta \in \mathbf{R}$, $\alpha(X - \beta)$ and $\alpha(Y - \beta)$ are both exponential, or geometric on $\mathbf{N}_0$, in which case $X - Y$ and $\min\{X, Y\}$ are independent.

The "if" part (of the theorem) and the result that $\min\{X, Y\}$ and $X - Y$ are independent if X and Y are as in the "if" part are trivial. To prove the "only if" part here, one could follow Rao and Shanbhag to show first that if $P\{X \geq Y\} > 0$, then the assertion implies that $l \in (-\infty, y_0)$[1], $P\{Y = l | X \geq Y, Y \leq y_0| < 1$, and

$$P\{X \geq Y + x | X \geq Y, Y \leq y_0\}$$
$$= P\{X \geq^* l + x | X \geq^* l\}, \quad x \in \mathbf{R}_+ ,$$

where l is the left extremity of the distribution of Y and "$\geq^*$" denotes "$\geq$" if l is a discontinuity point of the distribution of Y and it denotes "$>$" otherwise. Observe now that if $P\{Y = l\} = 0$, then, unless $P\{X = l\} = 0$, we have

$$P\{X - Y < 0\} = P\{X - Y < 0 | \min\{X, Y\} = l\}$$
$$= P\{X - Y < 0 | X = l\} = P\{Y > l\} = 1 ,$$

contradicting the condition that $P\{X \geq Y\} > 0$. In view of the observation that we have made in Section 2 on Corollary 2, we may then appeal to Corollary 2 to have that the conditional distribution of $X - l$ given that $X \geq l$ is exponential if the conditional distribution of $Y - l$ given that $Y \leq y_0$ is nonarithmetic, and that of $\lambda[(X - l)/\lambda]$ given that $X \geq l$, where $[\cdot]$ denotes the integral part, is geometric on $\{0, \lambda, 2\lambda, \ldots\}$ if the conditional distribution of $Y - l$ given that $Y \leq y_0$ is arithmetic with span λ. This, in turn, implies, because of the "independence" condition in the assertion, that the left extremity of the distribution of X is less than or equal to that of Y and that $P\{Y \geq X\} > 0$. Hence, by symmetry, a further result with the places of X and Y interchanged (and the obvious notational change in l) follows, and one is then led to the result sought.

---

[1] indeed, it now follows trivially that $l < y_0$ since $P\{\min\{X, Y\} \leq y\} = cP\{X \geq Y, Y \leq y\}$ for each $y \in (-\infty, y_0]$ and some $c > 0$.

Rao and Shanbhag (1994, p. 197) have effectively observed the following two simple corollaries of Theorem 4; the latter of these two results essentially extends a result of Fisz (1958).

COROLLARY 4. If in Theorem 4, X and Y are additionally assumed to be identically distributed, then the assertion of the theorem holds with $|X - Y|$ in place of $X - Y$.

PROOF. The corollary follows on noting that, under the assumptions, for any $y \in \mathbf{R}$, $|X - Y|$ and $\phi(\min\{X, Y\})I_{\{\min\{X,Y\} \leq y\}}$ are independent if and only if $X - Y$ and $\phi(\min\{X, Y\})I_{\{\min(X,Y) \leq y\}}$ are independent.

COROLLARY 5. Let X and Y be two i.i.d. nondegenerate positive random variables and y_0 be as defined in Theorem 4. Then $\min\{X, Y\}/\max\{X, Y\}$ and $\min\{X, Y\}I_{\{\min(X,Y) \leq y_0\}}$ are independent if and only if for some $\alpha > 0$ and $\beta \in \mathbf{R}$, $\alpha(\log X - \beta)$ is either geometric or exponential.

PROOF. Define $X^* = \log X$, $Y^* = \log Y$ and $y_0^* = \log y_0$. (Note that y_0 here has to be positive.) Noting that $-\log(\min\{X, Y\}/\max\{X, Y\}) = |X^* - Y^*|$ and
$$\min\{X, Y\}I_{\{\log(\min\{X,Y\}) \leq \log y_0\}} = \exp\{\min\{X^*, Y^*\}\}I_{\{\min\{X^*,Y^*\} \leq y_0^*\}}$$
we can hence get the result from Corollary 4.

REMARKS 1. (i) If we replace in Corollary 5, the condition on the existence of y_0 by that there exists a point y_0' such that there are at least two support points of the distribution of $\max\{X, Y\}$ in $[y_0', \infty)$, then the assertion of the corollary with $\min\{X, Y\}I_{\{\min\{X,Y\} \leq y_0\}}$ replaced by $\max\{X, Y\}I_{\{\max\{X,Y\} \geq y_0'\}}$ and $\log X$ replaced by $-\log X$ holds. This follows because $\min\{X^{-1}, Y^{-1}\} = (\max\{X, Y\})^{-1}$ and $\max\{X^{-1}, Y^{-1}\} = (\min\{X, Y\})^{-1}$. The result that is observed here is indeed a direct extension of Fisz's (1958) result, and it is yet another result mentioned in Rao and Shanbhag (1994). (Fisz characterizes the distribution in question via the independence of $\max\{X, Y\}/\min\{X, Y\}$ and $\max\{X, Y\}$.)

(ii) Under the assumptions in Theorem 4, the condition that $X - Y$ and $\phi(\min\{X, Y\})I_{\{\min\{X,Y\} \leq y_0\}}$ be independent is clearly equivalent to that for each $y \in (-\infty, y_0]$, $X - Y$ be independent of $I_{\{\min\{X,Y\} \leq y\}}$. (The remark with $X - Y$ replaced by $|X - Y|$ applies to Corollary 4, i.e. when we have the assumptions as in the corollary.)

(iii) Theorem 4 remains valid if the "independence" condition appearing in the assertion is replaced by that conditional upon $\min\{X, Y\} \in (-\infty, y_0]$, $X - Y$ and $\min\{X, Y\}$ are independent. (The corresponding remark in the case of Corollary 4 is now obvious.)

(iv) If the assumptions in Theorem 4 are met with $P\{X \geq Y\} > 0$, then on modifying slightly, the Rao–Shanbhag argument that we have referred to in the proof of the theorem proves that conditionally upon $\min\{X, Y\} \in (-\infty, y_0]$, $(I_{\{X=Y\}}, (X - Y)^+)$ and $\min\{X, Y\}$ are independent only if $l \in (-\infty, y_0)$, the conditional distribution of $X - l$ given that $X \geq l$ is exponential if the conditional distribution of $Y - l$ given that $Y \leq y_0$ is nonarithmetic, and that of $\lambda[\frac{X-l}{\lambda}]$ given that $X \geq l$ is geometric on $\{0, \lambda, 2\lambda, \ldots\}$ if the conditional distribution of $Y - l$ given that $Y \leq y_0$ is arithmetic with span λ, where l is the left extremity of the distribution of Y and $[\cdot]$ denotes the integral part.

(v) The version of Theorem 4 with $\min\{X, Y\}$ in place of $\phi(\min\{X, Y\})$ holds if in place of "two support points" we take "two nonzero support points" or in place of "there are ... in $(-\infty, y_0]$" we take "the left extremity of the distribution of $\min\{X, Y\}$ is nonzero and is less than y_0". The result in (iii) above and that mentioned here are essentially variations of Theorem 8.2.1 of Rao and Shanbhag (1994). (Incidentally, the cited result of Rao and Shanbhag requires a minor notational alteration such as the one where "$(1, \min\{X, Y\})I_{\{\min\{X,Y\} \leq y_0\}}$" appears in place of "$\min\{X, Y\}I_{\{\min\{X,Y\} \leq y_0\}}$".)

There is an interesting variant of Theorem 4; Rossberg (1972), Ramachandran (1980), Rao (1983), Lau and Ramachandran (1991), and Rao and Shanbhag (1994) among others have produced versions of this theorem. A special case of this result for $n = 2$ was given in a somewhat restricted form by Puri and Rubin (1970); see also Lau and Rao (1982), Stadje (1994) and Rao and Shanbhag (1995a) for comments and extensions on the Puri-Rubin (1970) result. We give this variant as our next theorem and show that it is also linked with the strong memoryless property characterization of the exponential and geometric distributions.

THEOREM 5. Let $n \geq 2$ and $X_1, \ldots, X_n$ be i.i.d. random variables with d.f. F that is not concentrated on $\{0\}$. Further, let $X_{1:n} \leq \cdots \leq X_{n:n}$ denote the corresponding order statistics. Then, for some $1 \leq i < n$,

$$X_{i+1:n} - X_{i:n} \stackrel{d}{=} X_{1:n-i} , \quad (3.1)$$

where $X_{1:n-i} = \min\{X_1, \ldots, X_{n-i}\}$, if and only if one of the following two conditions holds:

(i) F is exponential.

(ii) F is concentrated on some semilattice of the form $\{0, \lambda, 2\lambda, \ldots\}$ with $F(0) = \alpha$ and $F(j\lambda) - F((j-1)\lambda) = (1-\alpha)(1-\beta)\beta^{j-1}$ for $j = 1, 2, \ldots$ for some $\alpha \in (0, \binom{n}{i}^{-1/i}]$ and $\beta \in [0, 1)$ such that $P\{X_{i+1:n} > X_{i:n}\} = (1-\alpha)^{n-i}$ (which holds with $\alpha = \binom{n}{i}^{-1/i}$ or $\beta = 0$ if and only if

$$F(0) - F(0-) = \binom{n}{i}^{-1/i}$$

and

$$F(\lambda) - F(\lambda-) = 1 - \binom{n}{i}^{-1/i}$$

for some $\lambda > 0$). (The existence of cases $\beta > 0$ can easily be verified.)

Note that (3.1) is equivalent to the assertion that $P\{X_{i+1:n} > X_{i:n}\} = P\{X_{1:n-i} > 0\}$ and we have independent nonnegative nondegenerate random variables Y and Z such that

$$P\{Y > Z + y | Y > Z\} = P\{Y > y | Y > 0\}, \quad y \in \mathbf{R}_+$$

with $Y \stackrel{d}{=} X_{1:n-i}$ and $Z \stackrel{d}{=} X_{i:i}$ (in obvious notation). For nonnegative nondegenerate independent random variables Y_1 and Y_2 with supports of their distributions to be equal, we have

$$P\{Y_1 > Y_2 + y | Y_1 > Y_2\} = P\{Y_1 > y | Y_1 > 0\}, \quad y \in \mathbf{R}_+$$

if and only if the conditional distribution of Y_1 given that $Y_1 > 0$ is either exponential or geometric on $\{\lambda, 2\lambda, \ldots\}$ for some $\lambda > 0$ or degenerate at some positive point. If F is of the form

$$F = \alpha F_1 + (1 - \alpha) F_2$$

with F_1 as degenerate at the origin and F_2 as either exponential or geometric on $\{\lambda, 2\lambda, \ldots\}$ or degenerate at a positive point, then it is easily seen that $P\{X_{i+1:n} > X_{i:n}\} = P\{X_{1:n-i} > 0\}$ if and only if either $\alpha = 0$ and F is exponential or $\alpha \in (0, (\binom{n}{i})^{-1/i}]$ and F is as in (ii) in the statement of the theorem. In view of what we have observed here, the theorem is then obvious.

The following two remarks are taken from Rao and Shanbhag (1995a). These explain respectively as to how the existence of $\beta > 0$ in the theorem above (i.e., Theorem 5) follows and how the recent result of Stadje (1994) is a corollary to the theorem.

REMARK 2. Suppose we consider a family of the distributions of the form in (ii), but not necessarily satisfying the condition that $P\{X_{i+1:n} > X_{i:n}\} = (1 - \alpha)^{n-i}$. Then, if we take a fixed $\beta \in (0, 1)$ and allow α to vary, we get for a sufficiently small α, $P\{X_{i+1:n} > X_{i:n}\} < (1 - \alpha)^{n-i}$, and for $\alpha = \binom{n}{i}^{-1/i}$, we get $P\{X_{i+1:n} > X_{i:n}\} > (1 - \alpha)^{n-i}$; since we have now $P\{X_{i+1:n} > X_{i:n}\}$ to be a continuous function of α, we have the existence of an α value such that $P\{X_{i+1:n} > X_{i:n}\} = (1 - \alpha)^{n-i}$. This proves that the claim made by us under brackets immediately after the statement of the theorem is justified.

REMARK 3. If $n = 2$ and $i = 1$, we get $\binom{n}{i}^{-1/i} = 1/2$. In this case, if neither $\alpha = 1/2$ nor $\beta = 0$, we get $P\{X_{i+1:n} > X_{i:n}\} = 1 - P\{X_1 = X_2\} = \{2(1 - \alpha)(\alpha + \beta)\}/(1 + \beta)$; consequently we have here $P\{X_{i+1:n} > X_{i:n}\} = (1 - \alpha)^{n-i}$, i.e. the probability to be equal to $1 - \alpha$, if and only if $\beta = 1 - 2\alpha$. One can hence see as to how Stadje's result follows as a corollary to Theorem 5.

The sketch of the argument that we have produced above to see the validity of Theorem 5 tells us further that the following theorem holds. Arnold and Ghosh (1976) and Arnold (1980) have dealt with specialized versions of this result; see, also, Zijlstra (1983) and Fosam et al. (1993) for further specialized versions and some comments on the earlier literature.

THEOREM 6. Let $n \geq 2$ and $X_1, \ldots, X_n$ be nondegenerate i.i.d. random variables with d.f. F. Also, let $X_{1:n}, \ldots, X_{n:n}$ be order statistics as in Theorem 5. Then, for some $i \geq 1$, the conditional distribution of $X_{i+1:n} - X_{i:n}$ given that $X_{i+1:n} - X_{i:n} > 0$ is the same as the distribution of $X_{1:n-i}$, where $X_{1:n-i}$ is as defined in Theorem 5 if and only if F is either exponential, or, for some $\lambda > 0$, geometric on $\{\lambda, 2\lambda, \ldots\}$.

The next two theorems are in the same spirit as Theorem 6 and extend slightly the results given in Fosam and Shanbhag (1994). Once again these results follow as corollaries to the strong memoryless property characterization of the exponential and geometric distributions. The results given in Fosam and Shanbhag (1994) and hence so also the theorems given here, in turn, subsume the specialized results given by Liang and Balakrishnan (1992, 1993).

THEOREM 7. Let $n \geq 2$ and $1 \leq k \leq n - 1$ be integers and $Y_1, Y_2, \ldots, Y_n$ be independent positive random variables such that $P\{Y_1 > Y_2 > \cdots > Y_n\} > 0$ and for each $i = 1, 2, \ldots, k$, the conditional distribution of Y_{i+1} given that $Y_{i+1} > Y_{i+2} > \cdots > Y_n$ be nonarithmetic. (The condition on Y_i's is clearly met if Y_i's are independent positive random variables such that for each $i = 2, \ldots, n$ and $y > 0, P\{Y_i > y\} > 0$.) Then

$$P\{Y_i - Y_{i+1} > y | Y_1 > Y_2 > \cdots > Y_n\} = P\{Y_i > y | Y_1 > Y_2 > \cdots > Y_i\},$$
$$y > 0; \quad i = 1, 2, \ldots, k \quad (3.2)$$

(where the right-hand side of the identity is to be read as $P\{Y_1 > y\}$ for $i = 1$) if and only if $Y_i, i = 1, 2, \ldots, k$, are exponential random variables. (The result also holds if ">" in (3.2) is replaced by "$\geq$" with "$Y_1 > Y_2 > \cdots > Y_n$" and "$Y_{i+1} > Y_{i+2} > \cdots > Y_n$" in the assumptions replaced respectively by "$Y_1 \geq Y_2 \geq \cdots \geq Y_n$" and "$Y_{i+1} \geq Y_{i+1} \geq \cdots \geq Y_n$".)

PROOF. Defining for each $i = 1, 2, \ldots, k, X^{(i)}$ and $Y^{(i)}$ to be independent positive random variables with distribution functions $P\{Y_i \leq x | Y_1 > Y_2 > \cdots > Y_i\}$, $x \in \mathbf{R}_+$ and $P\{Y_{i+1} \leq x | Y_{i+1} > Y_{i+2} > \cdots > Y_n\}$, $x \in \mathbf{R}_+$, we see that (3.2) can be rewritten as

$$P\{X^{(i)} > Y^{(i)} + x | X^{(i)} > Y^{(i)}\} = P\{X^{(i)} > x\}, \quad x > 0; \quad i = 1, 2, \ldots, k \;.$$

Consequently, in view of the strong memoryless characterization of the exponential distributions, it follows that (3.2) is valid if and only if the distribution functions $P\{Y_i \leq x | Y_1 > Y_2 > \cdots > Y_i\}, x \in \mathbf{R}_+$, are those corresponding to exponential random variables for $i = 1, 2, \ldots, k$. It is easy to see inductively that we have the distribution functions $P\{Y_i \leq x | Y_1 > Y_2 > \ldots > Y_i\}, x \in \mathbf{R}_+$, for $i = 1, 2, \ldots, k$ as those corresponding to exponential distributions if and only if the random variables $Y_1, \ldots, Y_k$ are exponential. Hence we have the theorem.

THEOREM 8. Let $n \geq 2$ and $1 \leq k \leq n - 1$ be integers and $Y_1, Y_2, \ldots, Y_n$ be independent nonnegative integer-valued random variables such that

$P\{Y_1 \geq Y_2 \geq \cdots \geq Y_n\} > 0$ and for each $i = 1, 2, \ldots, k$, the conditional distribution of Y_{i+1} given $Y_{i+1} \geq Y_{i+2} \geq \cdots \geq Y_n$ be arithmetic with span 1. Also, let

$$P\{Y_{i+1} = 0 | Y_1 \geq Y_2 \geq \cdots \geq Y_n\} < 1, \quad i = 1, 2, \ldots, k \ .$$

(The conditions on Y_i's are clearly met if Y_i's are independent nonnegative integer-valued random variables such that $P\{Y_1 \geq 1\} > 0, P\{Y_i = 1\} > 0$ for $2 \leq i \leq k + j$ and $P\{Y_i = 0\} > 0$ for $k + j < i \leq n$ for some $j \geq 1$.) Then

$$P\{Y_i - Y_{i+1} \geq y | Y_1 \geq Y_2 \geq \cdots \geq Y_n\} = P\{Y_i \geq y | Y_1 \geq Y_2 \geq \cdots \geq Y_i\}$$
$$y = 0, 1, \ldots; \quad i = 1, 2, \ldots, k \quad (3.3)$$

(where the right-hand side of the identity is to be read as $P\{Y_1 \geq y\}$ for $i = 1$) if and only if $Y_i, i = 1, 2, \ldots, k$, are geometric random variables.

Theorem 8 follows essentially via the argument in the proof of Theorem 7 but with "geometric" in place of "exponential".

REMARKS 4. (i) As observed by Fosam and Shanbhag (1994), the specialized version of Theorem 7 given by them subsumes the "only if" part (i.e. the major part) of the Liang-Balakrishnan (1992) theorem; note that if X and Y are independent positive random variables such that 0 is a cluster point of the distribution of Y, then, conditionally upon $X > Y$, the random variables $X - Y$ and Y are independent only if

$$P\{X > Y + x | X > Y\}(= \lim_{y \downarrow 0} P\{X > Y + x | X > Y, Y \leq y\})$$
$$= P\{X > x\}, \quad x \in (0, \infty) \ .$$

Consequently, it follows that under the weaker assumption in the Fosam-Shanbhag result in place of its original assumption, the Liang-Balakrishnan theorem holds. This improved theorem also holds if A is replaced by $A^* = \{Y_1 \geq Y_2 \geq \cdots \geq Y_n\}$.

(ii) If X and Y are independent nonnegative integer-valued random variables such that $P\{Y = 0\} > 0$, then, conditionally upon $X \geq Y$, the random variables $X - Y$ and Y are independent only if

$$P\{X \geq Y + x | X \geq Y\}(= P\{X \geq Y + x | X \geq Y, Y = 0\})$$
$$= P\{X \geq x\}, \quad x = 0, 1, \ldots \ .$$

In view of this, we have that comments analogous to those on the Liang–Balakrishnan (1992) theorem (but with Theorem 8 in place of Theorem 7) also apply to the Liang–Balakrishnan (1993) theorem. (Note that in this latter case, we restrict ourselves to the independence conditionally upon A^*, where A^* is as in (i)).

(iii) Under a somewhat more complicated assumption, it can be shown that the equation (3.3) with "$\geq$" replaced by "$>$" leads us to characterizations of shifted geometric distributions.

Before discussing further results that are linked with Corollary 2, let us give the next general result. This latter result could be viewed as one of the important tools in the remainder of the present study:

THEOREM 9. Let Y and Z be independent random variables with distributions such that the corresponding supports are equal and Y is continuous. Further, let ϕ be a nonarithmetic (or nonlattice) real monotonic function on $\mathbf{R}_+$ such that $E(|\phi(Y-Z)|) < \infty$. Then, for some constant $c \neq \phi(0+)$,

$$E\{\phi(Y-Z)|Y \geq Z, Z\} = c \quad \text{a.s.} \tag{3.4}$$

if and only if Y is exponential, upto a change of location. (By the conditional expectation in (3.4), we really mean the one with $I_{\{Y \geq Z\}}$ in place of $Y \geq Z$; the assertion of the theorem also holds if "$Y \geq Z$" is replaced by "$Y > Z$.")

Under the stated assumptions in Theorem 9, we have (3.4) to be equivalent to

$$E\{\phi((Y-z)+)|Y \geq z\} = c \quad \text{for each } z \in \text{supp}[G] \text{ with } P\{Y \geq z\} > 0 , \tag{3.5}$$

where G is the d.f. of Y. If $z_1, z_2 \in \text{supp}[G]$ such that $z_1 < z_2$ with $P\{Y \geq z_1\} = P\{Y \geq z_2\} > 0$, then, from (3.5), it easily follows that the equation in it holds for each $z \in [z_1, z_2]$; consequently, we see that (3.5) is equivalent to the assertion obtained from it by deleting "$\in \text{supp}[G]$" and we get Theorem 9 as a consequence of Corollary 2. (One could also arrive at the result directly without appealing to Corollary 2, from Theorem 1.)

COROLLARY 6. Let F be continuous and, as before, let $X_{1:n}, \ldots, X_{n:n}$ for $n \geq 2$ be n ordered observations based on a random sample of size n from F. Further, let i be a fixed positive integer less than n and ϕ be a nonarithmetic (or nonlattice) real monotonic function on $\mathbf{R}_+$ such that $E(|\phi(X_{i+1:n} - X_{i:n})|) < \infty$. Then, for some constant $c \neq \phi(0+)$,

$$E(\phi(X_{i+1:n} - X_{i:n})|X_{i:n}) = c \quad \text{a.s.} \tag{3.6}$$

if and only if F is exponential, within a shift.

We can express (3.6) as (3.4) with Y and Z as independent random variables such that $Z \stackrel{d}{=} X_{i:i}$ and $Y \stackrel{d}{=} X_{1:n-i}$; consequently, we get Corollary 6 as a corollary to Theorem 9.

One could now raise a question as to how crucial is the assumption of continuity of Y for the validity of Theorem 9. The continuity assumption (when taken in conjunction with other assumptions in the theorem) implies that

$$E\{\phi(Y-z)|Y > z\} = c \quad \text{for a.a.}[G]z \in \mathbf{R} \tag{3.7}$$

is equivalent to

$$E\{\phi((Y-z)-)|Y>z\} = c \quad \text{for each } z \in \mathbf{R} \text{ with } P\{Y>z\}>0, \quad (3.8)$$

where G is the d.f. of Y. The equivalence mentioned here (or any relevant alternative version of it) is the reason as to why one is able to get Theorem 9 via Corollary 2 or Theorem 1. Suppose we now have the assumptions in Theorem 9 met with Y nondegenerate in place of continuous. Then, if ϕ is left continuous and satisfies the condition that

$$G(x+\cdot) = G((y+\cdot)-) \text{ a.e. } [|\phi(\cdot+)-\phi(0+)|]$$
$$\text{whenever } 0 < G(x) = G(y-) < G(y), \quad (3.9)$$

then it easily follows that (3.7) is equivalent to (3.8); thus, we have cases other than those met in Theorem 9 under which (3.7) and (3.8) are equivalent.

Taking a clue from the observations made above and using essentially the same arguments as those that led us to Theorem 9 and Corollary 6 respectively, we can now give the following theorem and corollary. The theorem given here answers the question that we have raised above partially.

THEOREM 10. Let Y and Z be independent nondegenerate random variables such that the corresponding distributions have the same support and the same set of discontinuity points. Let ϕ be a monotonic real left continuous nonconstant function on $\mathbf{R}_+$ for which (3.9) is met (or, more generally, a monotonic real nonconstant function for which (3.7) and (3.8) are equivalent) and $E(|\phi(Y-Z)|) < \infty$, where G is the d.f. of Y. Then, for some constant $c \neq \phi(0+)$,

$$E\{\phi(Y-Z)|Y>Z,Z\} = c \quad \text{a.s.} \quad (3.10)$$

if and only if the left extremity l, of the distribution of Y is finite, and either ϕ is nonarithmetic (or nonlattice) and the conditional distribution of $Y-l$ given that $Y>l$ is exponential, or for some $\lambda > 0$, ϕ is arithmetic (or lattice) with span λ and the conditional survivor function, $\overline{G}_l$, of $Y-l$ given that $Y>l$ satisfies for some $\beta \in (0,1)$

$$\overline{G}_l(x+n\lambda) = \beta^n \overline{G}_l(x), \quad x>0; \; n=0,1,\ldots$$

COROLLARY 7. Let $X_{1:n},\ldots,X_{n:n}$ be ordered observations based on a random sample of size $n(\geq 2)$ from a nondegenerate distribution with d.f. F. Let $1 \leq i \leq n-1$ be a given integer and ϕ be a monotonic real left continuous nonconstant function on $\mathbf{R}_+$ such that $E(|\phi(X_{i+1:n}-X_{i:n})|) < \infty$ and (3.9) met with F in place of G. Then, for some constant $c \neq \phi(0+)$

$$E(\phi(X_{i+1:n}-X_{i:n})|X_{i+1:n}>X_{i:n},X_{i:n}) = c \quad \text{a.s.}$$

if and only if the left extremity, l, of F is finite, and either ϕ is nonarithmetic (or nonlattice) and (with $X_1 \sim F$) the conditional distribution of X_1 given that $X_1 > l$ is exponential, within a shift, or for some $\lambda > 0$, ϕ is arithmetic (or lattice) with span λ and the conditional survivor function, $\overline{F}_l$, of X_1 given that $X_1 > l$ satisfies for some $\beta \in (0, 1)$

$$\overline{F}_l(x + n\lambda) = \beta^n \overline{F}_l(x), \quad x > l; \; n = 0, 1, 2, \ldots.$$

REMARKS 5. (i) Suppose now that the assumptions in Theorem 10 are met, but, with (3.7) and (3.8) such that "$Y \geq z$" appears in place of "$Y > z$" in both of them and "$\phi(\cdot+)$" appears in place of "$\phi(\cdot-)$" in (3.8), and with "right continuous" in place of "left continuous". Then (3.10) with "$Y > Z$" replaced by "$Y \geq Z$" holds if and only if l, the left extremity of Y, is finite, and ϕ is either nonarithmetic and $Y - l$ is exponential, or for some $\lambda > 0$, ϕ is arithmetic with span λ and for some $\beta \in (0, 1)$

$$P\{Y - l \geq x + n\lambda\} = \beta^n P\{Y - l \geq x\}, \quad x \in \mathbf{R}_+; \; n = 0, 1, \ldots.$$

(This follows essentially via the same argument as in the earlier case.)

(ii) As a further corollary of Corollary 7, it follows that if $X_{1:n}, \ldots, X_{n:n}$ are as defined in Corollary 7, then conditionally upon $\{X_{i+1:n} > X_{i:n}\}$, the random variables $X_{i+1:n} - X_{i:n}$ and $X_{i:n}$ are independent if and only if conditionally upon $X_1 > l$, the random variable $X_1 - l$ is either exponential or geometric on $\{\lambda, 2\lambda, \ldots\}$ for some $\lambda > 0$, where $X_1 \sim F$. This latter result gives as a corollary Rogers's (1963) extension of Fisz's result.

(iii) In view of Shanbhag's (1977) lemma, it follows that the following variant of Corollary 7 holds:

THEOREM 11. Let X_1 and X_2 be i.i.d. nondegenerate integer-valued random variables with support of the type $I \cap \mathbf{Z}$ with I as an interval and $\phi: \mathbf{N}_0 \to \mathbf{R}$ a function such that $E(|\phi(|X_1 - X_2|)|) < \infty$, $\phi(1) > \phi(0)$ and $\phi(n+2) - 2\phi(n+1) + \phi(n) \geq 0$ for all $n \in \mathbf{N}_0$ (i.e., the second differences of ϕ are nonnegative on $\mathbf{N}_0$). Then, for some c,

$$E(\phi(|X_1 - X_2|) | \min\{X_1, X_2\}) = c \quad \text{a.s.}$$

if and only if X_1 is geometric, but for a shift. (For a proof of the theorem, see Rao and Shanbhag (1994, pp. 200–201).)

(iv) Specialized versions of Corollary 6 and Theorem 11 have appeared in Beg and Kirmani (1979), and Kirmani and Alam (1980) respectively. (See, also Rao and Shanbhag (1986).) Corollary 7 is essentially due to Rao and Shanbhag (1994).

(v) Let $X_{1:n} \leq \cdots \leq X_{n:n}$ denote the n ordered observations in a random sample of size $n(\geq 2)$ from a nondegenerate d.f. F concentrated on $\mathbf{N}_0$. Arnold (1980) effectively raised the question as to whether the independence of $X_{2:n} - X_{1:n}$ and the event $\{X_{1:n} = m\}$ for a fixed $m \geq 1$, when obviously $F(m) - F(m-) > 0$, implies that F is geometric (possibly within a shift or a change of scale). Some partial results on the conjecture have appeared in Sreehari (1983) and Alzaid et al. (1988).

However, that the conjecture in its existing form is false is shown by the following example.

EXAMPLE. Let m be a positive integer and for each $c \in (0, 1)$, let $f^{(c)}: (0, 1) \to (0, \infty)$ such that

$$f^{(c)}(q) = q^{-m}\left\{\frac{1}{1-q^2} + \frac{1}{1-c}\right\}, \quad q \in (0, 1) .$$

Note that for each c,

$$\lim_{q \downarrow 0} f^{(c)}(q) = \lim_{q \uparrow 1} f^{(c)}(q) = \infty .$$

Also, we have for each c, $f^{(c)}(q)$ to be strictly decreasing when $q \in (0, \sqrt{\frac{m}{m+2}}]$, and for each $c \in (0, 1/2]$, $f^{(c)}(q)$ to be strictly increasing in q when $q \in [\sqrt{\frac{m}{m+1}}, 1)$; this could be verified by obtaining the derivative of $f^{(c)}(q)$ with respect to q and observing that this is uniformly negative and (for $c \in (0, \frac{1}{2}]$) uniformly positive respectively, on the intervals in question. Consequently, for a sufficiently small c,

$$\lim_{q \downarrow c} f^{(c)}(c/q) > f^{(c)}\left(c/\sqrt{\frac{m}{m+1}}\right) > f^{(c)}\left(\sqrt{\frac{m}{m+1}}\right) .$$

As $f^{(c)}$ is continuous for each c, it then follows that for a sufficiently small c, we have a $q \in (c, c/\sqrt{\frac{m}{m+1}}]$ such that

$$f^{(c)}(q) = f^{(c)}(c/q)$$

with $q \neq c/q$. Hence, it follows that we can find distinct $q_1, q_2 \in (0, 1)$ such that

$$q_1^{-m}\left\{\frac{1}{1-q_1^2} + \frac{1}{1-q_1q_2}\right\} = q_2^{-m}\left\{\frac{1}{1-q_2^2} + \frac{1}{1-q_1q_2}\right\} .$$

Suppose now that X_1 and X_2 are independent identically distributed random variables such that

$$P\{X_1 = j\} = (q_1^j + q_2^j)/\{(1 - q_1)^{-1} + (1 - q_2)^{-1}\}, \quad j = 0, 1, \ldots .$$

It is then a simple exercise to check that

$$P\{|X_1 - X_2| = j, \min\{X_1, X_2\} = m\}$$
$$= P\{|X_1 - X_2| = j\}P\{\min\{X_1, X_2\} = m\}, \quad j = 0, 1, \ldots ,$$

implying that the conjecture is false.

There are characterization results based on order statistics, which are arrived at in the literature via techniques different from those we have used so far in the present study. We give below three such results. These results follow respectively via Theorem 2, Theorem 3 and essentially the technique that we have used immediately above the statement of Theorem 3, to prove Corollary 1 (together with a certain uniqueness property appearing in Rao and Shanbhag (1995b). The details of the existing literature on these results could be found in Kagan, Linnik and Rao (1973), Rao and Shanbhag (1994), and Rao and Shanbhag (1995b); note, in particular, that Theorem 12 is essentially due to Rao and Shanbhag (1994) and it extends a result of Shimizu, and Theorem 14 is due to Rao and Shanbhag (1995b) and it deals with a conjecture of Dufour. (A special case of Theorem 14 was established earlier by Leslie and van Eeden (1993).)

THEOREM 12. Let $X_1, \ldots, X_n, n \geq 2$, be i.i.d. positive random variables and $a_1, \ldots, a_n$ be positive real numbers not equal to 1, such that the smallest closed subgroup of $\mathbf{R}$ containing $\log a_1, \ldots, \log a_n$ equals $\mathbf{R}$ itself. Then, in obvious notation, for some $m \geq 1$

$$\min\{X_1 a_1, \ldots, X_n a_n\} \stackrel{d}{=} X_{1:m} \qquad (3.11)$$

if and only if the survivor function of X_1 is of the form

$$\overline{F}(x) = \exp\{-\lambda_1 x^{\alpha_1} - \lambda_2 x^{\alpha_2}\}, \quad x \in \mathbf{R}_+ \qquad (3.12)$$

with $\lambda_1, \lambda_2 \geq 0, \lambda_1 + \lambda_2 > 0$ and $\alpha_r (r = 1, 2)$ as positive numbers such that $\sum_{i=1}^{n} a_i^{-\alpha_r} = m$. (If $\alpha_1 = \alpha_2$, the distribution corresponding to (3.12) is Weibull.)

The theorem is an obvious consequence of Theorem 2 since, defining $\overline{F}$ to be the survivor function of X_1, we have (3.11) to be equivalent to

$$\prod_{i=1}^{n} \overline{F}(x/a_i) = (\overline{F}(x))^m, x \in \mathbf{R}_+ \ ,$$

which implies, in view of the assumption in the theorem (i.e. in Theorem 12) on a_i's and X_i's, that $\overline{F}$ is nonvanishing on $\mathbf{R}_+$ with $\overline{F}(0+)(= \overline{F}(0)) = 1$; note that the "if" part of the result here is trivial and hence it is sufficient if we prove its "only if" part.

THEOREM 13. Let F be a continuous d.f. with finite mean and $X_{1:n}, \ldots, X_{n:n}$ be the ordered observations based on a random sample of size $n (\geq 2)$ from it. Then, for some $1 \leq i < n$

$$E(X_{i:n}|X_{i+1:n} = x) = ax - b \quad \text{for almost all } [F]x \in \mathbf{R} \ . \qquad (3.13)$$

only if $a > 0$ and the d.f. has the following form, to within a shift and a change of scale:

(i) $F(x) = e^x$ for $x \leq 0$ if $a = 1$.
(ii) $F(x) = x^\theta$ for $x \in [0, 1]$ if $a \in (0, 1)$.
(iii) $F(x) = (-x)^\theta$ for $x \leq -1$ if $a > 1$, where $\theta = a[i(1-a)]^{-1}$.

As (3.13) implies that

$$\frac{\int_{-\infty}^{x} yF^i(\mathrm{d}y)}{(F(x))^i} = ax - b, \quad x \in (l, \infty) \cap \mathrm{supp}[F] \,, \tag{3.14}$$

where l is the left extremity of F, and the left hand side of the equation in (3.14) is an increasing nonconstant function on $(l, \infty) \cap \mathrm{supp}[F]$, it follows that $a > 0$. This, in turn, implies that the right-hand side of and, hence, the left-hand side of the equation in (3.14) is strictly increasing on $(l, \infty) \cap \mathrm{supp}[F]$, giving us $(l, \infty) \cap \mathrm{supp}[F]$ as an interval. As the relation (3.14) can then be translated into that corresponding to the mean residual life for the random variable $-X_{i:i}$, we then get Theorem 13 as a corollary to Theorem 3.

THEOREM 14. Let r and n be positive integers greater than or equal to 3 such that either $r, n \in \{3, 4\}$ with $r = n$, or $r, n \geq 5$. Also, let $X_1, \ldots, X_n$ be i.i.d. positive random variables and $X_{1:n}, \ldots, X_{n:n}$ be the corresponding order statistics. Define $X_{0:n} = 0$,

$$D_{i,n} = (n - i + 1)(X_{i:n} - X_{i-1:n}), \quad i = 1, 2, \ldots, n \,,$$

and

$$S_{i,n} = \sum_{j=1}^{n} D_{j,n}, \quad i = 1, 2, \ldots, n \,.$$

Then, if $(S_{1,n}/S_{r,n}, S_{2,n}/S_{r,n}, \ldots, S_{r-1,n}/S_{r,n})$ is distributed as the vector of order statistics relative to a random sample of size $r - 1$ from the uniform distribution on (0,1), we have X_1 to be an exponential random variable.

Essentially using the theme involved in the proof based on exchangeability, given in Section 2 for Corollary 1, but without involving de Finetti's theorem, Rao and Shanbhag (1995b) have proved Theorem 14 when $r, n \geq 5$; the result for $r = n = 3, 4$ follows from the certain uniqueness theorem on the problem, established by Rao and Shanbhag (1995b). (The validity of the result for $r, n = 3, 4$ also follows from the existing literature, see, Leslie and van Eeden (1993) for the relevant references.) Theorem 14 for $r, n \geq 5$ has also been proved independently via a different argument by Xu and Yang (1995; private communication).

REMARKS 6. (i) The Proof of Theorem 13 sketched here provides a link between Theorem 13 and the uniqueness theorem corresponding to the mean residual life function (i.e. Theorem 3), and reveals the analogy between Theorem 13 and a characterization result given in Hall and Wellner (1981). (Incidentally, Theorem 13 is a result of Ferguson.)

(ii) We have not dealt with in this review characterization results of the type considered in Chan (1967).

4. Characterizations involving record values and monotonic stochastic processes

Some characterization results essentially of the type met in the last section for order statistics also hold for record values:

DEFINITION. Let $\{X_n: n = 1, 2, \ldots\}$ be a sequence of i.i.d. random variables such that the right extremity of the common distribution of X_j's is not one of its discontinuity points.

Then, if we define $\{L(i): i = 1, 2, \ldots\}$ such that $L(1) = 1$ and for $i > 1$

$$L(i) = \inf\{j: j > L(i-1), X_j > X_{L(i-1)}\} \, ,$$

the sequence $\{X_{L(i)}: i = 1, 2, \ldots\}$ is called the sequence of record values and the sequence $\{L(i): i = 1, 2, \ldots\}$ as that of record times.

To give us some idea regarding characterization results based on record values, we shall touch upon here briefly three characterization results. These are respectively due to Witte (1988), Rao and Shanbhag (1986, 1994) and Dallas (1981). (The result given by Dallas though is slightly weaker than that given here; the present version is due to Rao and Shanbhag (1994). We also present the Rao–Shanbhag result in a somewhat different form with a different proof.) Lau and Rao (1982), Rao and Shanbhag (1986) and several others have given specialized versions of Witte's result. Gupta (1984), and Huang and Li (1993) among others have given specialized versions of the Rao-Shanbhag result (i.e. Theorem 16).

THEOREM 15. Let $\{R_i: i = 1, 2, \ldots\}$ be a sequence of record values corresponding to a d.f. F (as in the definition above). For some $k \geq 1$, $R_{k+1} - R_k \stackrel{d}{=} X_1$ where X_1 is a random variable with d.f. F, if and only if X_1 is exponential or, for some $a > 0$, X_1 is geometric on $\{a, 2a, \ldots\}$ (i.e. $a^{-1}X_1 - 1$ is geometric in the usual sense).

The following condensed version of essentially Witte's proof is given in Rao and Shanbhag (1994, pp. 205–206):

PROOF. Clearly the condition that $R_{k+1} - R_k \stackrel{d}{=} X_1$ is equivalent to that $X_1 > 0$ a.s. (i.e. $F(0) = 0$) and

$$\int_{(0,\infty)} \frac{\overline{F}(x+y)}{\overline{F}(y)} F_k(dy) = \overline{F}(x), \quad x \in (0, \infty) \, , \tag{4.1}$$

where F_k is the d.f. of R_k and $\overline{F}(\cdot) = 1 - F(\cdot)$. Note that if (4.1) holds with $F(0) = 0$ then given any point $s_0 \in \mathrm{supp}[F]$ there exists a point $s_1 \in \mathrm{supp}[F_k]$ such that $s_0 + s_1 \in \mathrm{supp}[F]$ and hence $\in \mathrm{supp}[F_k]$. Consequently, from the condition, we get that the smallest closed subgroup of $\mathbf{R}$ containing $\mathrm{supp}[F_k]$ equals that containing $\mathrm{supp}[F]$. In view of Theorem 1, we have then immediately that if (4.1) holds with $F(0) = 0$, then either X_1 is exponential or for some $a > 0$, $a^{-1}X_1 - 1$ is geometric (in the usual sense). The converse of the assertion is trivial and hence we have the theorem.

The above proof is in effect that corresponding to a version of the strong memoryless property characterization of the exponential and geometric distributions. Note that for F concentrated on $(0, \infty)$, we can express (4.1) as

$$\mathscr{E}_{\mu_1 \times \mu_2}(I_{\{(x,y) \in \mathbf{R}^2 : x > y+z\}})/\mathscr{E}_{\mu_1 \times \mu_2}(I_{\{(x,y) \in \mathbf{R}^2 : x > y\}}) = \mathscr{E}_{\mu_1}(I_{\{x \in \mathbf{R} : x > z\}}), z > 0,$$

where $\mathscr{E}_\mu$ denotes the integral with respect to μ, and μ_1 and μ_2 are measures concentrated on $(0, \infty)$ such that for every Borel subset B of $(0, \infty)$,

$$\mu_1(B) = P_F(B)$$

and

$$\mu_2(B) = \int_B (\overline{F})^{-1} F_k(dy),$$

where P_F is the measure determined on (the Borel σ-field of) $\mathbf{R}$ by F.

THEOREM 16. Let $\{R_i : i = 1, 2, \ldots\}$ be as in Theorem 15, k be a positive integer, and ϕ be a nonconstant real monotonic left continuous function on $\mathbf{R}_+$ such that $E(|\phi(R_{k+1} - R_k)|) < \infty$ and (3.9) is met with G replaced by F_k, where F_k is the d.f. of R_k. Then, for some $c \neq \phi(0+)$,

$$E\{\phi(R_{k+1} - R_k) | R_k\} = c \quad \text{a.s.} \tag{4.2}$$

if and only if the left extremity, l_k, of the distribution of R_k is finite, and either ϕ is nonarithmetic and the conditional distribution of $X_1 - l_k$ given that $X_1 > l_k$ is exponential, or for some $\lambda > 0$, ϕ is arithmetic with span λ and for some $\beta \in (0, 1)$

$$P\{X_1 - l_k \geq x + n\lambda\} = \beta^n P\{X_1 - l_k \geq x\}, \quad x > 0; \; n = 0, 1, \ldots,$$

where X_1 is a random variable distributed with d.f. F.

(4.2) can be expressed as (3.10) with Y such that its distribution is given by the conditional distribution of X_1 given that $X_1 \geq l_k$ and $Z \stackrel{d}{=} R_k$, and hence the theorem follows easily as a Corollary to Theorem 10. (Note that we allow here the case with $P\{Y = l_k\} > 0$ and $P\{Z = l_k\} = 0$.)

COROLLARY 8. Let the assumptions in Theorem 16 be met. Then the following assertions hold:

(i) If F is continuous or has its left extremity as one of its continuity points and ϕ is nonarithmetic, then, for some $c \neq \phi(0+)$, (4.2) is met if and only if F is exponential, within a shift.

(ii) If ϕ is arithmetic with span a and F is arithmetic with span greater than or equal to a, then, for some $c \neq \phi(0+)$, (4.2) is valid if and only if F has a finite left extremity and the conditional distribution of the residual value of X_1 over the kth support point of F given that this is positive is geometric on $\{a, 2a, \ldots\}$, where X_1 is a random variable with d.f. F.

(The corollary follows trivially from Theorem 16.)

THEOREM 17. Let $\{R_i\}$ be as defined in the two previous theorems, but with F continuous. Let $k_2 > k_1 \geq 1$ be fixed integers. Then, on some interval of the type $(-\infty, a)$, with $a >$ the left extremity of the distribution of R_{k_1}, the conditional distribution of $R_{k_2} - R_{k_1}$ given $R_{k_1} = x$ is independent of x for almost all x if and only of F is exponential, within a shift.

It easily follows that the independence in question is equivalent to the condition that for almost all $[F]c \in (-\infty, \min\{a, b\})$, where b is the right extremity of F, the r.v. $R^{(c)}_{k_2-k_1} - c$, where $R^{(c)}_{k_2-k_1}$ is the $(k_2 - k_1)^{\text{th}}$ record value corresponding to an i.i.d. sequence with d.f. F_c such that

$$F_c(x) = \begin{cases} \frac{F(x)-F(c)}{1-F(c)} & \text{if } x > c \\ 0 & \text{otherwise,} \end{cases}$$

is distributed independently of c. As F is continuous, the latter condition is seen to be equivalent to the condition that the left extremity, l, of F is finite and for some $a > l$

$$(1 - F(c + x)) = (1 - F(l + x))(1 - F(c)),$$
$$c \in (-\infty, a) \cap \text{supp}[F], x \in \mathbf{R}_+ .$$

(note that the last equation implies that $b = \infty$). In view of the Marsaglia-Tubilla (1975) result, the assertion of the theorem then follows. (Incidentally, the Marsaglia-Tubilla result referred to here could be arrived at as a corollary to either of Theorems 1 and 2.)

REMARKS 7. (i) Downton's (1969) result (with the correction as in Fosam et al. (1993) may be viewed as a specialized version of Theorem 15 for $k = 1$. Moreover, if we assume F to be concentrated on $\{0, 1, 2, \ldots\}$ with $F(1) - F(1-) > 0$ (and the right extremity condition met), then as a corollary to the theorem it follows that, for some $k \geq 1, R_{k+1} - R_k \stackrel{d}{=} X_1 + 1$ if and only if F is geometric.

(ii) When F is continuous, Theorem 16 holds without the left continuity assumption of ϕ. (This is also so of Corollary 7 of the last section.)

(iii) A version of Dallas's (1981) result has also appeared in Nayak (1981). With obvious alterations in its proof, one could easily see that Theorem 17 holds with R_{k_2} and R_{k_1} replaced respectively by $X_{k_2:n}$ and $X_{k_1:n}$ (assuming of course that $n \geq k_2 > k_1 \geq 1$). A variant of this latter result appears in Gather (1989); the result in this case is that if F is a continuous d.f. with support equal to $\mathbf{R}_+$, then F is exponential if and only if $X_{j_r:n} - X_{i:n} \stackrel{d}{=} X_{j_r-i:n-i}, r = 1, 2$ holds for fixed $1 \leq i < j_1 < j_2 \leq n$ and $n \geq 3$.

Finally, in this section, we mention a theorem that leads us to a characteristic property of Yule and Poisson processes. Rao and Shanbhag (1989, 1994) have shown implicitly that there is a link between a version of Theorem 1 (revealed first in Rao and Shanbhag (1986)) and this result:

Suppose that $\{X_n: n = 1, 2, \ldots\}$ is a sequence of independent positive random variables. Define $N(t) = \sup\{n: X_1 + \cdots + X_n \leq t\}, t > 0$. Let $n_0 (\geq 2)$ be a fixed positive integer and let $\{X_n: n = 1, 2, \ldots\}$ be such that it satisfies additionally $P\{N(t) = n\} > 0$ for $n = 1, 2, \ldots, n_0$ and all $t > 0$. We have then the following theorem.

THEOREM 18. The conditional distribution of $N(y)$ given $N(t) = n$ for each $0 < y < t, t > 0$ and $n = 1, 2, \ldots, n_0$ is nondegenerate binomial with index n and success probability parameter independent of n if and only if for some $\lambda_0 > 0$ and $\lambda \neq 0$ such that $\lambda_0 + n_0 \lambda > 0$,

$$P\{X_i > x\} = \exp\{-(\lambda_0 + (i-1)\lambda)x\}, \quad x \in \mathbf{R}_+, \quad i = 1, 2, \ldots, n_0 + 1$$
(4.3)

(For a proof of this theorem, see Rao and Shanbhag (1994; pp. 218–219).)

As a corollary of Theorem 18, we have the characterization given below:

COROLLARY 9. If we assume $P\{N(t) = n\} > 0$ for every $n \geq 1$ and every $t > 0$, then the conditional distribution of $N(y)$ given $N(t) = n$ is nondegenerate binomial with parameters as stated in Theorem 18 for every $0 < y < t < \infty$ and every $n \geq 1$ if and only if the process $\{N(t)\}$ is Yule. (The process constructed with intervals such that $P\{X_n > x\} = e^{-\{\lambda_0 + (n-1)\lambda\}x}, x \in \mathbf{R}_+$, with $\lambda_0 > 0$ and $\lambda \geq 0$, is referred to as Yule; the process reduces to a Poisson process if $\lambda = 0$.)

REMARKS 8. (i) In Theorem 18 and Corollary 9 the success probability parameter corresponding to the binomial distribution equals $(e^{\lambda y} - 1)/(e^{\lambda t} - 1)$ if $\lambda \neq 0$ and y/t if $\lambda = 0$.

(ii) From the proof of Theorem 18 in Rao and Shanbhag (1994), it is clear that both Theorem 18 and Corollary 9 remain valid even when in each case, the requirement of the conditional distribution is replaced by that $P\{N(y) = 0 | N(t) = n\}$ and $P\{N(y) = n | N(t) = n\}$ are as in the conditional distribution.

(iii) Corollary 9 gives us a characterization of a Poisson process if it is assumed additionally that for some i, j with $i \neq j$ and some $x > 0, P\{X_i \leq x\} = P\{X_j \leq x\}$. It also gives us a version of Liberman's (1985) characterization of a Poisson process in the class of renewal processes as a special case. (Suppose that we have a renewal process generated by a positive i.i.d. sequence, with index set $[0, \infty)$ and that $t > 0$ and n is a positive integer. Then for the process, we have the conditional distribution of the epochs at which the events during $(0, t]$ occur given that during the interval there are n events, to be the same as that of n ordered observations from the uniform distribution on $(0, t]$ only if for each $0 < y < t$, the conditional distribution of the number of events during $(0, y]$ given that there are n events during $(0, t]$ is binomial $(n, y/t)$. Also, if we have a renewal process on $[0, \infty)$, then, under the assumptions in question, the criterion of Theorem 18 with $n_0 = 2$ characterizes a Poisson process.

(iv) Even when the a priori conditions that $P\{N(t) = n\} > 0$ for $n = 1, 2, \ldots, n_0$ and all $t > 0$ in Theorem 18 and that $P\{N(t) = n\} > 0$ for every $n \geq 1$ and every $t > 0$ in Corollary 9 are not assumed, the respective results still hold, provided we understand by the conditional distributions their versions selected such that they are as stated whenever $P\{N(t) = n\} = 0$.

Characterization problems arise naturally in areas such as reliability, statistical inference and model building where one is interested in knowing whether a particular hypothesis or model is equivalent to some other hypothesis or model that is appealing in some sense. For example, the problem of whether the uniform-order-statistics distribution of the vector in Dufour's conjecture, characterizes exponential distributions appears when one bases a test of the hypothesis that the r.v. X_1 is exponentially distributed on the vector in question. (This is also true of the analogous characterization of the Poisson process met in Remarks 8, iii).) There are several tests of uniformity available and the result relative to the Dufour conjecture suggests a possible way of testing the exponentiality of a sequence of ordered observations before the complete set is observed. The strong memoryless property characterization of exponential distributions has interesting and important applications in queuing theory and other areas (see, for example, Rao and Shanbhag (1994).) The property in the characterization result given by Liang and Balakrishnan (1992) was shown to be of relevance in estimation theory by Sackrowitz and Samuel-Cahn (1984); it is now natural to ask whether the property in question is valid for nonexponential $Y_1, \ldots, Y_k$ so that one could explore the possibility of using it for other distributions. However, that it is a characterization property of exponential distributions tells us that the possibility does not arise.

The results that we have listed here mostly concern exponential or geometric distributions in one form or another. Although these are theoretical results, we expect these to be of potential importance in applications. Some illustrations of these are provided above. Also, we have made, in this article, an effort to unify a certain set of results in characterization theory via techniques cited in Section 2 of the article.

Acknowledgment

Research sponsored by the Army Research Office under Grant DAAHO4-93-G-0030. The United States Government is authorized to reproduce and distribute reprints for governmental purposes notwithstanding any copyright notation hereon.

References

Ahsanullah, M. (1975). A characterization of the exponential distribution. *Statistical Distributions in Scientific Work*, Vol. 3, pp. 131–135, eds. Patil, G. P., Kotz, S. and Ord, J. K. Dortrecht, Reidel.

Ahsanullah, M. (1978). A characterization of the exponential distribution by spacing of order statistics. *J. Appl. Prob.* **15**, 650–653.

Ahsanullah, M. (1984). A characterization of the exponential distribution by higher order gap. *Metrika* **31**, 323–326.

Ahsanullah, M. (1987). Record statistics and the exponential distribution. *Pak. J. Statist.* **3**, 17–40.

Ahsanullah, M. (1989). On characterization of the uniform distribution based on functions of order statistics. *Aligarh. J. Statist.* **9**, 1–6.

Ahsanullah, M. and B. Holland (1984). Record values and the geometric distribution. *Statistiche Hefte* **25**, 319–327.

Aly, M. A. H. (1988). Some contributions to characterization theory with applications in stochastic processes. Ph.D. Thesis, University of Sheffield.

Alzaid, A. A. (1983). Some contributions to characterization theory. Ph.D. Thesis. University of Sheffield.

Alzaid, A. A., C. R. Rao and D. N. Shanbhag (1987). Solution of the integrated Cauchy equation using exchangeability, *Sankhyā Ser. A* **49**, 189–194.

Alzaid, A. A., K. S. Lau, C. R. Rao and D. N. Shanbhag (1988). Solution of Deny convolution equation restricted to a half line via a Random Walk approach. *J. Multivar Anal.* **24**, 309–329.

Arnold, B. C. (1980). Two characterizations of the geometric distribution. *J. Appl. Prob.* **17**, 570–573.

Arnold, B. C., N. Balakrishnan and H. N. Nagaraja (1992). *A first course in order statistics*. J. Wiley and Sons, Inc., New York.

Arnold, B. C. and M. Ghosh (1976). A characterization of the geometric distribution by properties of order statistics. *Scand. Actuar. J.* **58**, 232–234.

Azlarov, T. A. and N. A. Volodin (1986). *Characterization Problems Associated with the Exponential Distribution*. Springer-Verlag.

Beg, M. I. and S. N. U. A. Kirmani (1979). On characterizing the exponential distribution by a property of truncated spacing. *Sankhyā Ser. A* **41**, 278–284.

Chan, L. K. (1967). On a characterization of distributions by expected values of extreme order statistics. *Amer. Math. Monthly* **74**, 950–951.

Crawford, G. B. (1966). Characterization of geometric and exponential distributions. *Ann. Math. Statist.* **37**, 1790–1795.

Dallas, A. C. (1981). Record values and exponential distributions. *J. Appl. Prob.* **18**, 949–951.

David H. A. (1981). *Order Statistics*. Wiley, New York, 2nd edn.

Davies, P. L. and D. N. Shanbhag (1987). A generalization of a theorem of Deny with applications in characterization theory. *J. Math. Oxford* (2) **38**, 13–34.

Deny, J. (1961). Sur l'equation de convolution $\mu = \mu * \sigma$. Semin. Theory Potent. M. Brelot. Fac. Sci. Paris, 1959–1960, 4 ann.

Downton, F. (1969). An integral equation approach to equipment failure. *J. Roy. Statist, Soc. Ser. B* **31**, 335–349.

Dufour, R. (1982). Tests d'ajustement pour des échantillons tronqués, Ph.D. dissertation, Montréal.

Ferguson, T. S. (1964). A characterization of the exponential distribution. *Ann. Math. Statist.* **35**, 1199–1207.

Ferguson, T. S. (1965). A characterization of the geometric distribution. *Amer. Math. Monthly* **72**, 256–260.

Ferguson, T. S. (1967). On characterizing distributions by properties of order statistics. *Sankhyā Ser A* **29**, 265–277.

Fisz, M. (1958). Characterization of some probability distributions. *Scand. Aktiarict.* 65–67.

Fosam, E. B. (1993). Characterizations and structural aspects of probability distributions. Ph.D. thesis, Sheffield University.

Fosam, E. B., C. R. Rao and D. N. Shanbhag (1993). Comments on some papers involving the integrated Cauchy functional equation. *Statist. Prob. Lett.* **17**, 299–302.

Fosam, E. B. and D. N. Shanbhag (1994). Certain characterizations of exponential and geometric distributions. *J.R.S.S. Ser. B* **56**, 157–160.

Galambos, J. (1975a). Characterizations in terms of properties of the smaller of two observations. *Commun. Statist.* **4**(3), 239–244.

Galambos, J. (1975b). Characterizations of probability distributions by properties of order statistics I and II. *Statistical Distributions in Scientific Work*, Vol 3, pp. 71–101, eds. Patil, G. P., Kotz, S. and Ord, J. K. Dortrecht, Reidel.

Galambos, J. and S. Kotz (1978). Characterizations of probability distributions. *Lecture Notes in Mathematics*, 675, Springer-Verlag, Berlin.

Gather, U. (1989). On a characterization of the exponential distribution by properties of order statistics. *Statist. Prob. Lett.* **7**, 93–96.

Govindarajulu, Z. (1975). Characterization of the exponential distribution using lower moments of order statistics. *Statistical Distributions in Scientific Work*, Vol 3, pp. 117–129, eds. Patil, G. P., Kotz, S. and Ord, J. K., Dortrecht, Reidel.

Govindarajulu, Z. (1980). Characterization of the geometric distribution using properties of order statistics. *J. Statist.Plan. Inf.* **4**, 237–247.

Gupta, R. C. (1984). Relationships between order statistics and record values and some characterization results. *J. Appl. Prob.* **21**, 425–430.

Hall, W. J. and J. A. Wellner (1981). Mean residual life. *Statistics and Related Topics* pp. 169–184, eds. Csorgo, M., Dawson, D. A., Rao, J. N. K. and Saleh, M. K. Md. E. Amsterdam, North-Holland.

Huang, J. S. (1978). On a "lack of memory" property, *Stat. Tech. Rept.* Univ. of Guelph, Canada.

Huang, W. L. and S. H. Li (1993). Characterization results based on record values. *Statistica Sinica* **3**, 583–589.

Kagan, A. M., Yu. V. Linnik and C. R. Rao (1973). *Characterization Problems in Mathematical Statistics*. J. Wiley and Sons, New York.

Kamps, U. (1995). *A concept of generalized order statistics*. B. G. Teubner, Stuttgart.

Kirmani, S. N. U. A. and S. N. Alam (1980). Characterization of the geometric distribution by the form of a predictor. *Comm. Statist. A*, **9**, 541–548.

Kirmani, S. N. U. A. and M. I. Beg (1984). On characterization of distributions by expected records. *Sankhyā Ser. A* **46**, 463–465.

Klebanov, L. B. (1980). Some results connected with characterizations of the exponential distribution. *Theor. Veoj. i Primenen.* **25**, 628–633.

Kotlarski, I. I. (1967). On characterizing the normal and gamma distributions. *Pacific J. Math.* **20**, 69–76.

Kotz, S. and D. N. Shanbhag (1980). Some new approaches to probability distributions. *Adv. Appl. Prob.* **12**, 903–921.

Lau, K. S. and C. R. Rao (1982). Integrated Cauchy functional equation and characterizations of the exponential law. *Sankhyā Ser. A* **44**, 72–90.

Lau, K. and C. R. Rao (1984). Integrated Cauchy functional equation on the whole line. *Sankhyā Ser. A* **46**, 311–319.

Leslie, J. R. and C. van Eeden (1993). On a characterization of the exponential distribution on a type 2 right censored sample. *Ann. Stat.* **21**, 1640–1647.

Liang, T. C. and N. Balakrishnan (1992). A characterization of exponential distributions through conditional independence. *J. Roy. Statist. Soc. Ser. B* **54**, 269–271.

Liang, T. C. and N. Balakrishnan (1993). A characterization of geometric distributions through conditional independence. *Austral. J. Statist.* **35**, 225–228.

Liberman, U. (1985). An order statistic characterization of the Poisson renewal process. *J. Appl. Prob.* **22**, 717–722.

Marsaglia, G. and A. Tubilla (1975). A note on the lack of memory property of the exponential distribution. *Ann. Prob.* **3**, 352–354.

Mohan, N. R. and S. S. Nayak (1982). A characterization based on the equidistribution of the first two spacings of record values. *Z. Wahrsh. Verw. Gebiete* **60**, 219–221.

Nagaraja, H. N. (1975). Characterization of some distributions by conditional moments. *J. Indian Statist. Assoc.* **13**, 57–61.

Nagaraja, H. N. (1977). On a characterization based on record values. *Austral. J. Statist.* **19**, 70–73.

Nagaraja, H. N. and R. C. Srivastava (1987). Some characterizations of geometric type distributions based on order statistics. *J. Statist. Plan. Inf.* **17**, 181–191.

Nayak, S. S. (1981). Characterizations based on record values. *J. Indian Stat. Assoc.* **19**, 123–127.

Puri, P. S. and H. Rubin (1970). A characterization based on the absolute difference of two i.i.d. random variables. *Ann. Math. Statist.* **41**, 251–255.

Ramachandran, B. (1979). On the strong memoryless property of the exponential and geometric laws. *Sankhyā Ser. A* **41**, 244–251.

Ramachandran, B. (1984). Renewal-type equations on **Z**. *Sankhyā Ser. A* **46**, 319–325.

Ramachandran, B. and K. S. Lau (1991). *Functional equations in probability theory*. Academic Press, Inc., New York.

Rao, C. R. and D. N. Shanbhag (1986). Recent results on characterization of probability distributions: A unified approach through extensions of Deny's theorem. *Adv. Applied Prob.* **18**, 660–678.

Rao, C. R. and D. N. Shanbhag (1989). Recent advances on the integrated Cauchy functional equation and related results in applied probability. Papers in honor of S. Karlin (eds. T. W. Anderson, K. B. Athreya and D. L. Iglehart). Academic Press, 239–253.

Rao, C. R. and D. N. Shanbhag (1994). *Choquet-Deny type functional equations with applications to stochastic models*. J. Wiley and Sons, Ltd, Chichester.

Rao, C. R. and D. N. Shanbhag (1995a). A note on a characteristic property based on order statistics. *Proc. Amer. Math. Soc.* (to appear).

Rao, C. R. and D. N. Shanbhag (1995b). A conjecture of Dufour on a characterization of the exponential distributions. Center for Multivariate Analysis, Penn Stat Univ., Tech. Report 95–105.

Rogers, G. S. (1963). An alternative proof of the characterization of the density Ax^B. *Amer. Math. Monthly* **70**, 857–858.

Rossberg, H. J. (1972). Characterization of the exponential and the Pareto distributions by means of some properties of the distributions which the difference and quotients of the order statistics are subject to. *Math. Operatonsforch Statist.* **3**, 207–216.

Sackrowitz, H. and E. Samuel-Cahn (1984). Estimation of the mean of a selected negative exponential population. *J. Roy. Statist. Ser. B* **46**, 242–249.

Shanbhag, D. N. (1977). An extension of the Rao–Rubin characterization of the Poisson distribution. *J. Appl. Prob.* **14**, No. 3, 640–646.

Shanbhag, D. N. (1991). Extended versions of Deny's theorem via de Finetti's theorem. *Comput. Statist. Data Analysis* **12**, 115–126.

Shimizu, R. (1978). Solution to a functional equation and its applications to some characterization problems. *Sankhyā Ser. A* **40**, 319–332.

Shimizu, R. (1979). On a lack of memory of the exponential distribution. *Ann. Inst. Statist. Math.* **39**, 309–313.

Shorrock, R. W. (1972a). A limit theorem for inter-record times. *J. Appl. Prob.* **9**, 219–233; Correction, *J. Appl. Prob.* **9**, 877.

Shorrock, R. W. (1972b). On record values and record times. *J. Appl. Prob.* **9**, 316–326.

Shorrock, R. W. (1973). Record values and inter-record times. *J. Appl. Prob.* **10**, 543–555.

Sreehari, M. (1983). A characterization of the geometric distribution. *J. Appl. Prob.* **20**, 209–212.

Srivastava, R. C. (1974). Two characterizations of the geometric distribution. *J. Amer. Statist. Assoc.* **69**, 267–269.

Srivastava, R. C. (1979). Two characterizations of the geometric distribution by record values. *Sankhyā Ser. B* **40**, 276–278.

Stadje, W. (1994). A characterization of the exponential distribution involving absolute difference of i.i.d. random variables. *Proc. Amer. Math. Society* **121**, 237–243.

Witte, H.-J. (1988). Some characterizations of distributions based on the ICFE. *Sankhyā Ser. A* **50**, 59–63.

Zijlstra, M. (1983). Characterizations of the geometric distribution by distribution properties. *J. Appl. Prob.* **20**, 843–850.

Characterizations of Distributions via Identically Distributed Functions of Order Statistics

Ursula Gather, Udo Kamps and Nicole Schweitzer

1. Introduction

The theory of order statistics provides a variety of useful distributional equations for specific underlying distributions. The question arises as to whether, under suitable regularity conditions, such an identity is a characteristic property of the corresponding distribution. This paper reviews corresponding characterization results.

Related characterizations using order statistics are based, e.g., on inequalities for moments (see Rychlik, Chapter 6), recurrence relations for moments (see Kamps, Chapter 10), conditional moments, and the independence of functions of order statistics (cf Rao and Shanbhag, Chapter 8). For earlier surveys of this material and additional results we refer to Kotz (1974), Galambos (1975a,b), Galambos, Kotz (1978), Azlarov, Volodin (1986), Arnold et al. (1992, Chapter 6.4), Rao, Shanbhag (1994) and Johnson et al. (1994, 1995).

Throughout this paper, let $X_1, \ldots, X_n$, $n \geq 2$, denote i.i.d. random variables each with distribution function F. Let $X_{1,n} \leq \cdots \leq X_{n,n}$ be the corresponding order statistics. Let $\mathbb{N}$ denote the set of positive integers. Moreover, let F^{-1} be the pseudo-inverse of F defined by $F^{-1}(y) = \inf\{x; F(x) \geq y\}$, $y \in (0,1)$.

There is a large number of publications on characterizations of distributions via moments of order statistics initiated by Hoeffding (1953) who shows that the triangular array $(EX_{r,n})_{1 \leq r \leq n,\, n \in \mathbb{N}}$ of expected values of order statistics characterizes the underlying distribution if the first absolute moment exists. We do not consider results of this type here, but refer to Galambos, Kotz (1978, Chapter 3.4), Kamps (1995, Chapter II.2) and to the review articles of Hwang, Lin (1984), Huang (1989) and Lin (1989).

It should be noted however that characterization results where moment conditions are involved are certainly not necessarily stronger than a result where a corresponding distributional identity is assumed. In theorems based on moment conditions the existence of these moments is often implicitly assumed and this is of course restrictive.

The starting point for many characterizations of exponential distributions via identically distributed functions of order statistics is the well-known result of Sukhatme (1937): The normalized spacings

(1.1) $\quad D_{1,n} = nX_{1,n} \quad \text{and} \quad D_{r,n} = (n-r+1)(X_{r,n} - X_{r-1,n}), \quad 2 \le r \le n$

from an exponential distribution with parameter λ, i.e., $F(x) = 1 - e^{-\lambda x}$, $x \ge 0$, $\lambda > 0$ ($F \sim \text{Exp}(\lambda)$ for short), are again independent and identically exponentially distributed (cf David 1981, p. 20). Thus, we have

(1.2) $\quad F \sim \text{Exp}(\lambda)$ implies that $D_{1,n}, \ldots, D_{n,n}$ are i.i.d. $\sim \text{Exp}(\lambda)$.

This property is also valid for several other models of ordered random variables. In Kamps (1995, p. 81) it is derived for so-called generalized order statistics.

Malmquist (1950) considers ratios of order statistics from a standard uniform distribution $U[0,1]$ and obtains

(1.3) $\quad F \sim U[0,1]$ implies that $(X_{r,n}/X_{r+1,n})^r$, $1 \le r \le n-1$, are i.i.d. $\sim U[0,1]$.

Malmquist points out that for exponential distributions (1.2) follows by transformation. Motivated by this work, Renyi (1953) proves (1.2) by direct arguments, and notes that

$$X_{r,n} = \sum_{i=1}^{r} D_{i,n}/(n-i+1) .$$

Thus, an order statistic from an exponential distribution can be represented by a weighted sum of i.i.d. exponential random variables, i.e.,

(1.4) $\quad F \sim \text{Exp}(\lambda)$ implies $X_{r,n} = \sum_{i=1}^{r} Y_i/(n-i+1)$, $1 \le r \le n$,

with $Y_1, \ldots, Y_n$ i.i.d. $\sim \text{Exp}(\lambda)$.

Independently from Malmquist and Renyi, (1.2) is derived by Epstein, Sobel (1953). They also show that

(1.5) $\quad F \sim \text{Exp}(\lambda)$ implies $X_{r+1,n} - X_{r,n} \sim X_{1,n-r}$ for all $1 \le r \le n-1$.

We note that the early work of Sukhatme (1937) is not referred to in Malmquist (1950), Epstein, Sobel (1953) and Renyi (1953).

The distributional relation (1.5) can be stated in a more general form as

(1.6) $\quad F \sim \text{Exp}(\lambda)$ implies $X_{s,n} - X_{r,n} \sim X_{s-r,n-r}$ for all $1 \le r < s \le n$,

since, applying (1.4), we have

$$X_{s,n} - X_{r,n} = \sum_{i=r+1}^{s} Y_i/(n-i+1) = \sum_{i=1}^{s-r} Y_{i+r}/((n-r)-i+1) \sim X_{s-r,n-r} .$$

Characterizations of special distributions by distributional identities are often related to those of the exponential distribution. On the one hand, the ideas of the proofs may be applied in the discrete case to characterize the geometric distribution. Considering (1.6), then the condition $X_{r+1,n} > X_{r,n}$ is imposed to avoid ties, i.e., equality of adjacent order statistics (cf Section 6). On the other hand, transformations of the exponential distribution yield characterizations, e.g., of power function, Pareto and Weibull distributions (cf Section 3).

The following sections discuss step by step the development of characterizations corresponding to the above types of results, namely via spacings, in Section 2 for exponential distributions and by transformations for other continuous distribution functions in Section 3. With respect to general distributional characterizing properties, uniform distributions are treated in Section 4 whereas other specific continuous distributions are discussed in Section 5.

We do not present proofs in the present collection of results. However, it is evident that many of the characterizing distributional properties of order statistics are reduced to simple defining functional equations of distributions. Under weak assumptions, the main part of some proofs consists in the solution of more complicated and general functional equations such as the integrated Cauchy functional equation (ICFE) (cf Ramachandran, Lau 1991, Rao, Shanbhag 1994).

Finally we note that though aiming at a complete survey, we might have overlooked some contributions to the field.

2. Characterizations of exponential distributions based on normalized spacings

From Section 1 we summarize the following properties for exponential distributions. Let $1 \leq r < s \leq n \in \mathbb{N}$.

Normalized spacings are exponentially distributed

(2.1) $\quad D_{r,n} \sim \text{Exp}(\lambda) \quad (\text{cf } (1.2))$

and hence distributed according to the underlying distribution function

(2.2) $\quad D_{r,n} \sim F \quad (\text{cf } (1.2))$.

Different normalized spacings are identically distributed

(2.3) $\quad D_{r,n} \sim D_{s,n} \quad (\text{cf } (1.2))$.

Each order statistic is distributed as a weighted sum of i.i.d. exponential random variables $Y_1, \ldots, Y_r$

(2.4) $\quad X_{r,n} \sim \sum_{i=1}^{r} Y_i/(n-i+1) \quad (\text{cf } (1.4))$.

Moreover, there are several characterizations based on

(2.5) $\quad X_{s,n} - X_{r,n} \sim X_{s-r,n-r} \quad (\text{cf } (1.6))$.

Often, the above properties are replaced by corresponding relations for expectations, densities or failure rates. For $r = 1$, (2.2) and (2.4) are connected and then read

(2.6) $\quad nX_{1,n} \sim X_1 \quad \text{or} \quad nX_{1,n} \sim F$.

The proofs of several results need certain aging properties of the underlying distributions to be fulfilled. Those properties are introduced first.

(2.7) F is called IFR (DFR) (more precisely: F possesses the IFR (DFR) property) (IFR: increasing failure rate; DFR: decreasing failure rate) if $[1 - F(x + z)]/[1 - F(x)]$ decreases (increases) with respect to x on the support of F (supp(F) for short) for all $z \geq 0$.

If F is absolutely continuous with density function f, then F is IFR (DFR) if its failure rate $f/(1 - F)$ increases (decreases) on the support of F.

The following conditions are weaker than the above ones.

(2.8) F is called NBU (NWU) (NBU: new better than used; NWU: new worse than used), if

$$1 - F(x + y) \underset{(\geq)}{\leq} (1 - F(x)) \cdot (1 - F(y)) \quad \text{for all } x, y, x + y \in \text{supp}(F) \ .$$

(2.9) F is called DMRL (IMRL) (DMRL: decreasing mean residual life; IMRL: increasing mean residual life) if

$$\tfrac{1}{1-F(x)} \int_x^\infty (1 - F(y)) \, dy \text{ decreases (increases) with respect to } x \text{ on supp}(F).$$

2.1. Results based on $D_{r,n} \sim \text{Exp}(\lambda)$

The following result is due to Rossberg (1972) with a proof based on complex analysis.

(2.1.1) (Rossberg 1972) Let $\alpha(F) = \inf\{x; F(x) > 0\} > -\infty$. Suppose that for some fixed $r, n \in \mathbb{N}$, $2 \leq r \leq n$, the Laplace transform $\int_{\alpha(F)}^\infty e^{-sx} \, dF^{r-1}(x)$ is nonzero for all $s \in \mathbb{C}$ with $\text{Re}(s) > 0$. Then $D_{r,n} \sim \text{Exp}(1)$ iff $X_1 - \alpha(F) \sim \text{Exp}(1)$.

The assumption concerning the Laplace transform can not be dropped since (2.1) is satisfied by other distributions than the exponential. Rossberg (1972) gives the following example. If

$$F(x) = F_R(x) = 1 - e^{-x}(1 + \tfrac{4}{a^2}(1 - \cos ax)), \quad a \geq 2\sqrt{2}, \ x \geq 0 \ ,$$

then (2.1) is fulfilled, but the corresponding Laplace transform has zeros in $\{s \in \mathbb{C}; \ \text{Re}(s) > 0\}$. A related stability result is shown by Shimizu (1980). For specific distributions it may be difficult to verify the assumption concerning the Laplace transform. Therefore, in Pudeg (1990) it is replaced by an aging property.

(2.1.2) (Pudeg 1990) Let F be IFR (or DFR) and let $\alpha(F) > -\infty$. Then (2.1) holds for a pair (r, n), $2 \leq r \leq n$, iff $X_1 - \alpha(F) \sim \text{Exp}(\lambda)$ for some $\lambda > 0$.

Obviously, the assumption "F is IFR (or DFR)" can be replaced by "$1 - (1 - F)^{n-r+1}$ is IFR (or DFR)". Pudeg points out that this condition in (2.1.2) can be weakened to "$1 - (1 - F)^{n-r+1}$ is DMRL (or IMRL)". Moreover, she shows that Rossberg's example still serves as a counterexample. F_R does not possess the DMRL or IMRL property and hence is neither IFR nor DFR.

In a recent paper, Riedel and Rossberg (1994) treat the problem of characterizing exponential distributions by a distributional property of a contrast $X_{r+s,n} - X_{r,n}$ (see also Section 2.5). Their main assumption concerns the asymptotic behaviour of the survival function of the contrast. For further results and details as well as for a discussion we refer to Riedel, Rossberg (1994).

(2.1.3) (Riedel, Rossberg 1994) Let F be absolutely continuous with a continuous and bounded density f on $[0, \infty)$, and let $\lambda > 0$. Then $F \sim \text{Exp}(\lambda)$ if one of the following conditions is satisfied.

a) There exists a triple (r, s, n), $1 \leq s \leq n - r$, such that

$$P(X_{r+s,n} - X_{r,n} \geq x) - e^{-\lambda(n-r)x} = o(x^s), \quad x \to 0,$$

and $f(x)/[1 - F(x)] - \lambda$ does not change its sign for any $x \geq 0$.

b) There exists a quadruple (r, s_1, s_2, n), $1 \leq s_1 < s_2 \leq n - r$, such that

$$P(X_{r+s_i,n} - X_{r,n} \geq x) - e^{-\lambda(n-r)x} = o(x^{s_i}), \quad x \to 0, \text{ for } i = 1, 2.$$

Seshadri et al. (1969) are interested in applications of characterizations to goodness of fit tests. They assume for instance that all normalized spacings (cf (2.1.1)) are exponentially distributed to conclude that the underlying distribution is exponential. Other results of Seshadri et al. (1969) are presented in Section 5.

(2.1.4) (Seshadri, Csörgö, Stephens 1969) Let $F^{-1}(0+) \geq 0$, $\lambda > 0$ and $n \geq 2$. Then $F \sim \text{Exp}(\lambda)$ iff $D_{j,n} \sim \text{Exp}(\lambda)$ for all $1 \leq j \leq n$.

Fang and Fang (1989) give a multivariate characterization result which can be seen as an extension of (2.1.4). In a previous paper Fang, Fang (1988) introduce the class τ_n of multivariate l_1-norm symmetric distributions which are versions of multivariate exponential distributions. More precisely, τ_n is the class of all multivariate distribution functions of non-negative random variables $X_1, \ldots, X_n$ for which $P(X_1 > x_1, \ldots, X_n > x_n) = h\left(\sum_{i=1}^n x_i\right)$ for all $(x_1, \ldots, x_n) \in [0, \infty)^n$ and some function h on $[0, \infty)$.

Obviously, the joint distribution function of n i.i.d. exponential random variables is a member of τ_n.

(2.1.5) (Fang, Fang 1989) Let $X_1, \ldots, X_n \geq 0$ be exchangeable random variables with a joint distribution function G and a joint continuous density function. Then $G \in \tau_n$ iff $(X_1, \ldots, X_n) \sim (D_{1,n}, \ldots, D_{n,n})$.

2.2. Results based on $D_{r,n} \sim F$ and related characterizations

For the particular case $r = 1$, Desu (1971) proves a characterization result based on $nX_{1,n} \sim F$ (i.e., (2.6)) for all $n \geq 2$ (see also Bell, Sarma 1980). There is a link here to extreme value theory where limits of minima and maxima with respect to n

are considered (see Galambos 1978, p. 188 and Galambos 1975a, p. 78). The requirement "for all $n \in \mathbb{N}$" has been modified in three ways. The distributional identity (2.6) needs to hold only for some special n or for two different values of n. Moreover, the equality of expectations $nEX_{1,n} = EX_1$ for all $n \in \mathbb{N}$ suffices to characterize the exponential distribution. This is shown in Chan (1967). It is also possible to choose certain subsequences of positive integers (see the remarks in Section 1). Property (2.6) is analyzed more generally by Shimizu (1979).

(2.2.1) (Shimizu 1979) Let $m, n_1, \ldots, n_m \in \mathbb{N}$, $c, a_1, \ldots, a_m > 0$ with

$$\begin{cases} c > \max_{1 \leq k \leq m} a_k, & \text{if } m > 1 \\ c = a_1, & \text{if } m = 1 \end{cases} \quad \text{and let}$$

$$\begin{cases} \alpha \text{ be the uniquely determined positive real number with } \sum_{k=1}^{m} a_k^\alpha = c^\alpha, & \text{if } m > 1 \\ \alpha > 0 \text{ arbitrary, if } m = 1. \end{cases}$$

Let the trivial case $m = n_1 = a_1 = 1$ be excluded. Let $(X_j^{(k)})_{1 \leq j \leq n_k, 1 \leq k \leq m}$ be i.i.d. random variables with distribution function F satisfying $0 = F(0) < F(x) < 1$ for some $x > 0$, and $X_{1,n_k}^{(k)} = \min_{1 \leq j \leq n_k} X_j^{(k)}$. Then

$$\min_{1 \leq k \leq m} \left(\frac{cn_k^{1/\alpha}}{a_k} X_{1,n_k}^{(k)} \right) \sim F$$

iff

$$F(x) = \begin{cases} 1 - \exp(-x^\alpha H(-\ln x)), & x \geq 0 \\ 0, & x < 0 \end{cases}$$

where H is a positive, bounded function with periods $A_k = \ln(cn_k^{1/\alpha}/a_k)$, $1 \leq k \leq m$.

The above general setting reduces to the case $c = \alpha = 1$ by means of the monotone transformation $X_j^{(k)} \to (cX_j^{(k)})^\alpha$. Davies, Shanbhag (1987) point out that the proof can be simplified applying the integrated Cauchy functional equation (ICFE; cf Lau, Rao 1982, Rao, Shanbhag 1986, Alzaid et al. 1988, Ramachandran, Lau 1991, Rao et al. 1994, Rao, Shanbhag 1994). As corollaries, characterizations of exponential distributions result without assuming continuity of F. Galambos (1975b, p. 92) notes that therefore there is no discrete distribution satisfying (2.6). However in the discrete case a similar condition can be considered (see Galambos 1975b, p. 92, Bagchi 1989, (6.13), (6.14)).

For the special case $m = a_1 = c = \alpha = 1$, $n_1 = n$, we obtain the following corollary from (2.2.1).

(2.2.2) (cf Shimizu 1979) Let F be as in (2.2.1). Then (2.6) is satisfied iff F is given by $F(x) = \begin{cases} 1 - e^{-xH(-\ln x)}, & x > 0 \\ 0, & x \leq 0 \end{cases}$ with some positive, bounded function H having period $\ln n$.

Under additional conditions, known characterization results for exponential distributions are obtained.

(2.2.3) (Arnold 1971, Shimizu 1979) Let $\operatorname{supp}(F) = (0, \infty)$. Then $F \sim \operatorname{Exp}(\lambda)$ iff $n_i X_{1,n_i} \sim F$ for $1 < n_1 < n_2$ with $\ln n_1 / \ln n_2$ irrational.

Arnold (1971) requires (2.6) for two different values of n and Gupta (1973) requires "$\lim_{x \to 0} F(x)/x = \lambda$ for some $0 < \lambda < \infty$". Desu's (1971) result is obviously contained in Arnold's (1971) result. Dallas (1977) (not cited by Shimizu 1979) states a general solution of the distributional identity (2.6) which is equivalent to

$$(1 - F(x))^n = 1 - F(nx)$$

(see also Bosch 1977, Riedel 1981, Galambos, Kotz 1983 and (5.2.5)).

The relation between the periodic function A in Dallas (1977) and the function H in (2.2.1) is given by

$$H(-\ln x) = n^{A(\log_n x)} .$$

Because of the above representation of F, additional conditions are of interest which guarantee that H is constant. Gupta (1973) and Dallas (1977) point out that such conditions must restrict the behaviour of F at the origin, since the function $A(\log_n x)$ oscillates infinitely often in neighbourhoods of zero. This motivates Gupta's (1973) condition requiring $0 < \lim_{x \to 0+} F(x)/x = \lambda < \infty$.

We also refer to Arnold et al. (1992, p. 146) for a discussion of the functional equation. The particular case $n = 2$ in the above equation appears in Kamps (1990) in a different context. Marshall, Olkin (1991) work on a multivariate version of the above functional equation leading to a multivariate characterization result analogous to Desu's (1971) theorem.

Arnold, Isaacson (1976) discuss solutions to $\min(X, Y) \sim aX$ and $\min(X, Y) \sim aX \sim bY$ for positive constants a and b and independent non-negative random variables X and Y.

Galambos (1975a) points out that prior to Desu (1971) there was a stronger result of Sethuraman (1965) which is quoted by Galambos as follows. We present Sethuraman's result in detail in (5.3.2) and (5.3.3).

(2.2.4) (Sethuraman 1965, Galambos 1975a) Let $\operatorname{supp}(F) = (0, \infty)$. If we have for $\alpha_1, \alpha_2 \in \mathbb{R}$, $\ln \alpha_1 / \ln \alpha_2$ irrational, and for $n_1, n_2 \in \mathbb{N}$ that $\alpha_1 X_{1,n_1} \sim \alpha_2 X_{1,n_2} \sim F$, then there exist $\alpha > 0$ and $\lambda > 0$ such that $X^\alpha \sim \operatorname{Exp}(\lambda)$. The constants α_1 and α_2 are necessarily of the form $\alpha_i = n_i^{1/\alpha}$, $i = 1, 2$.

For further details see Galambos (1975a). Another corollary of Shimizu's result is the following.

(2.2.5) (Shimizu 1979) Let F be as in (2.2.1), $a_1, \ldots, a_n > 0$ with $\sum_{k=1}^{n} a_k = 1$ and $\ln a_i / \ln a_j$ irrational for some $i, j \in \{1, \ldots, n\}$. Then $F \sim \operatorname{Exp}(\lambda)$ iff $\min_{1 \le k \le n} X_k / a_k \sim F$.

Using failure rates, a weaker condition that (2.2) is sufficient to characterize exponential distributions.

(2.2.6) (cf Ahsanullah 1981) Let F be absolutely continuous, $F(0) = 0$ and strictly increasing on $(0, \infty)$. Let the failure rates r_{X_1} and $r_{D_{r,n}}$ of X_1 and $D_{r,n}$, respectively, be continuous on the right at zero. Moreover, assume that r_{X_1} attains its maximum or minimum at zero. Then $F \sim \text{Exp}(\lambda)$ iff there exists a pair (r, n), $2 \leq r \leq n$, such that $r_{D_{r,n}}(0+) = r_{X_1}(0+)$.

In Ahsanullah's (1981) paper, the equation $r_{D_{r,n}}(x) = r_{X_1}(x)$ is required for all $x \geq 0$, whereas the validity for $x = 0$ is sufficient as Gajek, Gather (1989) point out. Moreover, the IFR or DFR property of F is assumed which can be replaced by the condition that zero is an extremal point of r_{X_1}.

In Ahsanullah (1977) the weaker condition of NBU or NWU is assumed. However, there is a gap in the proof (cf Gather 1989) and we only have the desired result in the particular case $r = n$.

(2.2.7) (Ahsanullah 1977) Let F be absolutely continuous, strictly increasing on $(0, \infty)$ and NBU or NWU. Then $F \sim \text{Exp}(\lambda)$ iff $D_{n,n} \sim F$ for some $n \in \mathbb{N}$.

The related characterization (2.2.8) is shown in Ahsanullah (1987) where the characteristic property reduces to (2.6) in the case $r = 1$. Kakosyan et al. (1984) and Ahsanullah (1988a,b) also deal with random sums.

(2.2.8) (Ahsanullah 1987) Let F be absolutely continuous with density function f, strictly increasing on $(0, \infty)$ and let f be continuous on the right at zero. Then $F \sim \text{Exp}(\lambda)$ iff there exists a pair (r, n), $2 \leq r \leq n - 1$, such that $\sum_{i=1}^{r} D_{i,n} \sim \sum_{j=1}^{r} Y_j$ with $Y_1, \ldots, Y_r$ i.i.d. $\sim F$.

We now turn to a random sample size. Let $(X_i)_{i \in \mathbb{N}}$ be a sequence of i.i.d random variables with an absolutely continuous distribution function F. We consider a sample $X_1, \ldots, X_N$ of random sample size N. The random variable N is assumed to be independent of $(X_i)_{i \in \mathbb{N}}$. In (2.2.10) and (2.2.11) below, N is also assumed to be geometrically distributed with $P(N = k) = (1 - p)p^{k-1}$, $k \in \mathbb{N}$, $p \in (0, 1)$. Let $X_{1,N} \leq \cdots \leq X_{N,N}$ be the corresponding order statistics. Kakosyan et al. (1984, p. 77) characterize exponential distributions via identical distributions of $NX_{1,N}$ and X_1 which is a corollary of (5.3.6).

(2.2.9) (Kakosyan, Klebanov, Melamed 1984) Let $(X_i)_{i \in \mathbb{N}}$ be a sequence of i.i.d. random variables with distribution function F, $X_1 > 0$, F continuous for $x \geq 0$, and let $\lim_{x \to 0} F(x)/x$ exist and be finite. Moreover, let $N \geq 2$ be an integer-valued random variable independent of $(X_i)_{i \in \mathbb{N}}$. Then $F \sim \text{Exp}(\lambda)$ for some $\lambda > 0$ iff $NX_{1,N} \sim X_1$.

Kakosyan et al. (1984) conjecture that the distributional identity

$$(1 - p) \sum_{i=1}^{N} X_i \sim nX_{1,n}$$

for some fixed $n \in \mathbb{N}$ where N is geometrically distributed characterizes an exponential distribution. Under the additional assumption that F is IFR or DFR, Ahsanullah (1988a) deals with this assertion and points out that $nX_{1,n}$ can be replaced by $NX_{1,N}$. In Ahsanullah (1988b) the following more general result (with respect to r) is shown.

(2.2.10) (Ahsanullah 1988b) Let $(X_i)_{i \in \mathbb{N}}$ be a sequence of i.i.d. random variables with distribution function F, F absolutely continuous, $F(0+) \geq 0$, let F be IFR or DFR, $EX_1 < \infty$, and $0 < \lim_{x \to 0+} F(x)/x = \lambda < \infty$. Moreover, let N be a geometrically distributed random variable independent of $(X_i)_{i \in \mathbb{N}}$. Then $F \sim \text{Exp}(\lambda)$ iff there exists a pair (r, n), $1 \leq r \leq n$, such that $(1-p) \sum_{i=1}^{N} X_i \sim D_{r,n}$.

(2.2.11) (Ahsanullah 1988a) Let $(X_i)_{i \in \mathbb{N}}$ be a sequence of i.i.d. random variables with distribution function F, $F^{-1}(0+) \geq 0$, $F(x) < 1$ for all $x > 0$, F IFR or DFR, $EX_1 < \infty$, and $0 < \lim_{x \to 0+} F(x)/x = \lambda < \infty$. Moreover, let N be a geometrically distributed random variable independent of $(X_i)_{i \in \mathbb{N}}$. Then $F \sim \text{Exp}(\lambda)$ iff $(1-p) \sum_{i=1}^{N} X_i \sim NX_{1,N}$.

2.3. Results based on $D_{r,n} \sim D_{s,n}$

Normalized spacings from an exponential distribution are identically distributed. For the class of IFR (DFR) distributions, Ahsanullah (1976, 1978a) shows that (2.3) with $r = 1$ and $s = r+1$, respectively, is sufficient to characterize exponential distributions. In Ahsanullah (1978b) we find that (2.3) is a characteristic property of the exponential distribution if the underlying distribution function is absolutely continuous.

(2.3.1) (Ahsanullah 1978b) Let F be absolutely continuous, strictly increasing on $(0, \infty)$, and IFR or DFR. Then $F \sim \text{Exp}(\lambda)$ iff there exists a triple (r, s, n), $2 \leq r < s \leq n$, with (2.3).

Gajek, Gather (1989) do not use (2.3) as a distributional identity, but only require the equality of the corresponding densities or of the failure rates at zero.

(2.3.2) (Gajek, Gather 1989) Let F be absolutely continuous, $F^{-1}(0+) = 0$, strictly increasing on $(0, \infty)$, and IFR or DFR. Moreover, the densities $f^{D_{r,n}}$ and $f^{D_{s,n}}$ of $D_{r,n}$ and $D_{s,n}$ are assumed to be continuous on the right at zero. Then $F \sim \text{Exp}(\lambda)$ iff there exists a triple (r, s, n), $1 \leq r < s \leq n$, such that $f^{D_{r,n}}(0) = f^{D_{s,n}}(0)$.

We also find a result using expectations of $D_{r,n}$ and $D_{s,n}$. Ahsanullah (1981) considers the case $s = r + 1$.

(2.3.3) (Ahsanullah 1981) Let F be absolutely continuous, $\text{supp}(F) = (0, \infty)$, and IFR or DFR. Moreover, let $EX_1 < \infty$. Then $F \sim \text{Exp}(\lambda)$ iff there exists a pair (r, n), $2 \leq r \leq n$, such that $ED_{r,n} = ED_{r-1,n}$.

As pointed out in Gather, Szekely (1989) it is an open problem whether the fact that certain linear statistics are identically distributed characterizes exponential distributions. The answer can of course not be affirmative in general, since $X_{r,n} - X_{r-1,n}$ and $X_{s,n} - X_{s-1,n}$, $2 \leq r < s \leq n$, are identically distributed if F is uniform (see Section 4.1).

2.4. Results based on $X_{r,n} \sim \sum_{i=1}^{r} Y_i(n-i+1)$

As shown above, an order statistic from an exponential distribution can always be represented as a weighted sum of i.i.d. random variables. In characterization results such sums are assumed to have the same underlying distribution as some order statistic. Ahsanullah, Rahman (1972) assume the validity of (2.4) for all $n \in \mathbb{N}$ and $r \leq n$ to characterize the exponential distribution. For $r = 1$ this is the result of Desu (1971) (see 2.2.).

(2.4.1) (Ahsanullah, Rahman 1972) Let F be continuous, $\mathrm{supp}(F) \subseteq (0, \infty)$ and $n \geq 2$. Then $F \sim \mathrm{Exp}(\lambda)$ iff (2.4) is satisfied for all $1 \leq r \leq n$ with $Y_1, \ldots, Y_n$ i.i.d. $\sim F$.

This result and possible other assumptions are investigated in Huang (1974a). We note that it suffices to consider equality of corresponding moment equations for certain sequences of indices (cf Huang 1974b, Kamps 1992b), if the existence of the expectations involved is ensured. Huang (1974b) discusses Desu's (1971) result which corresponds to the case $r = 1$. Then, e.g., the condition that $\mathrm{E}(nX_{1,n}) = \mathrm{E}X_1$ is valid for all $n \geq 2$ can be weakened by using Müntz's theorem and requiring equality for a sequence $(n_i)_{i \in \mathbb{N}} \subset \mathbb{N}$ with a divergent sum of reciprocals $\left(\sum_{i=1}^{\infty} n_i^{-1} = \infty\right)$.

2.5. Results based on $X_{s,n} - X_{r,n} \sim X_{s-r,n-r}$

Several papers deal with characterizations of exponential distributions based on equation (2.5):

$$X_{s,n} - X_{r,n} \sim X_{s-r,n-r} \ .$$

The first result is due to Puri, Rubin (1970) for the case $n = 2$. Rossberg (1972) gives a more general result where $s = r + 1$ which is also proved in Rao (1983) using the ICFE (see also Ramachandran 1982).

(2.5.1) (Rossberg 1972) Let F be continuous and $\alpha(F) = 0$. Then $F \sim \mathrm{Exp}(\lambda)$ iff there exists a pair (r, n), $1 \leq r < n$, with (2.5) where $s = r + 1$.

Continuity of F is required in (2.5.1) as pointed out by Becker (1984, Chapter 4). There has to be an assumption on the support of F to characterize exponential distributions. Related results are also available for discrete distributions (cf (6.8)).

For arbitrary s, either an aging property of F is assumed or (2.5) is required for two different values of s. We first quote a result of Gajek, Gather (1989) where

(2.5) is replaced by the corresponding moment equation. The moments are assumed to exist (see the remarks in Section 1).

(2.5.2) (Iwińska 1986, Gajek, Gather 1989) Let F be absolutely continuous, $F^{-1}(0+) = 0$, strictly increasing on $(0, \infty)$ and NBU or NWU. Then $F \sim \text{Exp}(\lambda)$ iff there exists a triple (r, s, n), $1 \leq r < s \leq n$, with

$$EX_{s,n} - EX_{r,n} = EX_{s-r,n-r} .$$

A similar result is proven earlier by Ahsanullah (1984) using the stronger condition that (2.5) is fulfilled and that F is IFR or DFR. Moreover, Gajek, Gather (1989) show that, under suitable smoothness conditions, relation (2.5) can be replaced by the equality of $(s-r-1)^{\text{th}}$ derivatives of the corresponding failure rates evaluated at zero. Under the conditions of (2.5.2) and based on (2.5), exponential distributions were characterized before by Iwińska (1985).

(2.5.3) (Gather 1988) Let F be continuous and strictly increasing on $(0, \infty)$. Then $F \sim \text{Exp}(\lambda)$ iff there exists a quadruple (r, s_1, s_2, n), $1 \leq r < s_1 < s_2 \leq n$, such that (2.5) holds for $s = s_1$ and $s = s_2$.

This result, which is proved using Jensen's inequality, is already stated in Ahsanullah (1975). However, in the proof the NBU/NWU property of F is implicitly used as pointed out by Gather (1988). The discrete analogue is shown in (6.5).

Without any further assumption, the equation

$$EX_{r+1,n} - EX_{r,n} = EX_{1,n-r}$$

valid for one pair (r, n), $1 \leq r \leq n - 1$, does not characterize exponential distributions. For every choice of r and n there is a distribution different from the exponential with the above property as shown in Kamps (1991, 1992a). E.g., the distribution given by

$$F(x) = \left(1 + e^{cd} x^{-d}\right)^{-1}, \quad c > 0, \quad d = \frac{n}{n-1} \text{ and } x > 0$$

satisfies the moment relation in the case $r = 2$.

Similar to (2.1.3), Riedel, Rossberg (1994) prove characterizations of exponential distributions by comparing the survival functions on both sides of (2.5). For further results and comments we refer to their paper.

(2.5.4) (Riedel, Rossberg 1994) Let F be absolutely continuous with a continuous and bounded density f on $[0, \infty)$ and $F(0) = 0$. Then $F \sim \text{Exp}(\lambda)$ for some $\lambda > 0$ if one of the following conditions is fulfilled.

a) There exists a triple (r, s, n), $1 \leq s \leq n - r$, such that

$$P(X_{r+s,n} - X_{r,n} \geq x) - P(X_{s,n-r} \geq x) = o(F^s(x)), \quad x \to 0 ,$$

and $f(x)/[1 - F(x)] - f(0)$ does not change sign for $x \geq 0$.

b) There exists a quadruple (r, s_1, s_2, n), $1 \leq s_1 < s_2 \leq n - r$, such that

$$P(X_{r+s_i,n} - X_{r,n} \geq x) - P(X_{s_i,n-r} \geq x) = o(F^{s_i}(x)), \quad x \to 0, \; i = 1, 2.$$

3. Related characterizations of other continuous distributions

By transformations we immediately obtain similar results (as in Section 2) for other continuous distributions. Let G be the underlying continuous distribution function of $Y_1, \ldots, Y_n$ with order statistics $Y_{1,n}, \ldots, Y_{n,n}$. Moreover, let F be the exponential distribution function with parameter $\lambda > 0$, $X_1, \ldots, X_n$ i.i.d $\sim F$ and let $X_{1,n}, \ldots, X_{n,n}$ be the corresponding order statistics. Then we can use the relations

(3.1) $\quad G^{-1}F(X_{r,n}) \sim Y_{r,n}, \quad F^{-1}G(Y_{r,n}) \sim X_{r,n}, \quad 1 \leq r \leq n,$

to establish results similar to those in Section 2.

The following table shows some continuous distribution functions, their pseudo-inverse functions and explicit transformations $(Y_{r,n} = a, \; Y_{r-1,n} = b, \; Y_1 = c)$ concerning the distributional identity (2.2).

	$G(x)$		$x \in$	distribution	$G^{-1}(y)$, $y \in (0,1)$
1	$(x/v)^\lambda$	$\lambda, v > 0$	$(0, v)$	power fct	$vy^{1/\lambda}$
2	$1 - (v/x)^\lambda$	$\lambda, v > 0$	(v, ∞)	Pareto	$v(1-y)^{-1/\lambda}$
3	$1 - e^{-\lambda x^v}$	$\lambda, v > 0$	$(0, \infty)$	Weibull	$\left(\frac{1}{\lambda}\ln\frac{1}{1-y}\right)^{1/v}$
4	$(1 + e^{-\lambda x})^{-1}$	$\lambda > 0$	$(-\infty, \infty)$	logistic	$\frac{1}{\lambda}\ln\frac{y}{1-y}$
5	$1 - (1 + x^v)^{-\lambda}$	$\lambda, v > 0$	$(0, \infty)$	Burr XII	$((1-y)^{-1/\lambda} - 1)^{1/v}$

	$G^{-1}F(x)$	$F^{-1}G(x)$	$X_{r,n} - X_{r-1,n}$, $r \geq 2$
1	$v(1 - e^{-\lambda x})^{1/\lambda}$	$-\frac{1}{\lambda}\ln\left(1 - \left(\frac{x}{v}\right)^\lambda\right)$	$\frac{1}{\lambda}\ln((v^\lambda - b^\lambda)/(v^\lambda - a^\lambda))$
2	ve^x	$\ln\frac{x}{v}$	$\ln\frac{a}{b}$
3	$x^{1/v}$	x^v	$a^v - b^v$
4	$\frac{1}{\lambda}\ln(e^{\lambda x} - 1)$	$\frac{1}{\lambda}\ln(1 + e^{\lambda x})$	$\frac{1}{\lambda}\ln((1 + e^{\lambda a})/(1 + e^{\lambda b}))$
5	$(e^x - 1)^{1/v}$	$\ln(1 + x^v)$	$\ln((1 + a^v)/(1 + b^v))$

	distributional identity (cf (2.2))	
1	$\left(\frac{v^\lambda - b^\lambda}{v^\lambda - a^\lambda}\right)^{n-r+1}$	$\sim \frac{v^\lambda}{v^\lambda - c^\lambda}$
2	$\left(\frac{a}{b}\right)^{n-r+1}$	$\sim \frac{c}{v}$
3	$(n - r + 1)(a^v - b^v)$	$\sim c^v$
4	$\left(\frac{1 + e^{\lambda a}}{1 + e^{\lambda b}}\right)^{n-r+1}$	$\sim 1 + e^{\lambda c}$
5	$\left(\frac{1 + a^v}{1 + b^v}\right)^{n-r+1}$	$\sim 1 + c^v$

Since

$$X_{r,n} - X_{r-1,n} \sim \frac{X_1}{n-r+1} \sim \frac{F^{-1}G(Y_1)}{n-r+1}$$

according to (1.2) and (3.1), the last table shows distributional identities for several distributions.

If G is the standard uniform distribution ($\lambda = v = 1$ in case 1) with order statistics $U_{1,n}, \ldots, U_{n,n}$, we obtain

$$\left(\frac{1-U_{r-1,n}}{1-U_{r,n}}\right)^{n-r+1} \sim \frac{1}{1-U}, \text{ with } U \sim U[0,1] .$$

Observing that

$$1 - U_{r,n} \sim U_{n-r+1,n}, \quad 1 \leq r \leq n ,$$

we get (1.3).

Several authors use the above transformations to find results for power function and Pareto distributions (see Renyi 1953, Desu 1971, Rossberg 1972, Ahsanullah 1989).

It has been noted in (1.3) that Malmquist (1950) derives (1.2) by applying his result to the uniform distribution. Rossberg (1972) deduces from (2.5.1) that G is the distribution function of a Pareto distribution with parameters $\lambda > 0$ and $v = 1$ iff there exists a pair (r,n), $1 \leq r < n$, such that $Y_{r+1,n}/Y_{r,n} \sim Y_{1,n-r}$. Rossberg also states a result analogous to (2.1.1). Desu (1971) transforms his result (see Section 2.2) with respect to power function distributions. As a corollary to (2.2.7), Ahsanullah (1989) considers an absolutely continuous distribution function G with $\text{supp}(G) = [0,1]$. The assumption of the NBU/NWU property of the underlying distribution in (2.2.7) is replaced by

$$G(x \cdot y) \underset{(\leq)}{\geq} G(x) \cdot G(y) .$$

Then G is the distribution function of a power function distribution iff $Y_{1,n}/Y_{2,n} \sim G$ for some $n \in \mathbb{N}$ (see (4.2.1)).

Gupta (1979) shows a result concerning the independence of functions of order statistics from exponential distributions and obtains analogous results for power function and Pareto distributions via transformations. In the same context, Shah, Kabe (1981) consider also Burr XII and logistic distributions.

In two papers, Janardan, Taneja (1979a,b) deal with Weibull distributions. In the first, they are concerned with (2.5.3) for Weibull distributions in the same way as Ahsanullah (1975). Since the proof of the latter fails without assuming the NBU/NWU property of the underlying distribution function, a similar additional condition is needed. However, with the above transformation for Weibull variables, Gather's (1988) result can be utilized. In Janardan, Taneja (1979b), analogues to Desu's (1971) and Gupta's (1973) result are shown in a direct way.

Applying Shimizu's (1979) result (for α arbitrary), we obtain characterizations of Weibull distributions without using a transformation.

Dimaki, Xekalaki (1993) present a characterization of Pareto distributions via identical distributions of $X_{s,n}/X_{r,n}$ and $X_{s-r,n-r}$ for $s \in \{s_1, s_2\}$, $1 \le r < s_1 < s_2 \le n$. This assertion can directly be obtained via transformation from Gather's (1988) result (see (2.5.3)) which is not cited, however. The authors also restate Desu's (1971) result in terms of Pareto distributions.

4. Characterizations of uniform distributions

In the previous section we mentioned results for uniform distributions which are obtained from characterizations of exponential distributions by a simple transformation. We now gather together several other results dealing with uniform distributions.

4.1. Characterizations based on spacings

Normalized spacings play an important role in characterization results for exponential distributions as shown in Section 2. In particular, the distributional identity

$$X_{s,n} - X_{r,n} \sim X_{s-r,n-r}$$

is valid for $1 \le r < s \le n$, if $F \sim \text{Exp}(\lambda)$ (cf (2.5)).

A similar relation holds if $F \sim U[0, a]$ for some $a > 0$:

(4.1.1) $X_{s,n} - X_{r,n} \sim X_{s-r,n}$

(cf Hajós, Rényi 1954). Several authors deal with characterizations of uniform distributions based on (4.1.1). In the exponential case, aging properties of the underlying distribution function are used as assumptions in some of the theorems (cf Section 2). In characterization results for the uniform distribution, super-additivity or sub-additivity of F are appropriate conditions.

(4.1.2) The distribution function F is called super-additive (sub-additive) if $F(x+y) \underset{(\le)}{\ge} F(x) + F(y)$ for all $x, y, x+y \in \text{supp}(F)$.

Many interesting distribution functions have such a property. E.g., power function distributions with $F(x) = x^\alpha$, $x \in (0, 1)$, are super-additive if $\alpha \ge 1$ and sub-additive if $0 < \alpha \le 1$. Moreover, any NWU-distribution function is sub-additive. Huang et al. (1979) present a corresponding result.

(4.1.3) (Huang, Arnold, Ghosh 1979) Let F be continuous, strictly increasing on $\text{supp}(F)$, and let F be super-additive or sub-additive. Then $F \sim U[0, a]$ for some $a > 0$ iff there exists a pair (r, n), $1 \le r \le n-1$, such that $X_{r+1,n} - X_{r,n} \sim X_{1,n}$.

By considering (4.1.1) for $s = r + 1$, the question arises whether the uniform distribution is the only one satisfying

$$X_{i,n} - X_{i-1,n} \sim X_{j,n} - X_{j-1,n}$$

for some $i \neq j$. A partial answer to this is due to Ahsanullah (1989).

(4.1.4)   (Ahsanullah 1989) Let F be absolutely continuous with density function f, $F(0) = 0$, $F(1) = 1$, and either $f(x) \geq f(y)$ or $f(x) \leq f(y)$ for all $x, y \in (0, 1)$, $x \geq y$. Then $F \sim U[0, 1]$ iff there exists a pair (r, n), $2 \leq r \leq n$, such that $X_{r,n} - X_{r-1,n} \sim X_{r-1,n} - X_{r-2,n}$.

The assumption "$r \neq (n + 1)/2$" in Ahsanullah (1989) can be dropped since monotonicity of f excludes the case that f is an arbitrary symmetric density.

Moreover, Ahsanullah (1989) proves a characterization result for the standard uniform distribution using (4.1.1) for $s = n$ and $r = 1$.

(4.1.5)   (Ahsanullah 1989) Let F be absolutely continuous, symmetric, either super-additive or sub-additive, $F^{-1}(0+) = 0$, $F(1) = 1$. Then $F \sim U[0, 1]$ iff $X_{n,n} - X_{1,n} \sim X_{n-1,n}$ for some $n \geq 2$.

Without requiring super-additivity, Huang et al. (1979) prove the following result based on the first spacing in a sample of size n.

(4.1.6)   (Huang, Arnold, Ghosh 1979) Let F be strictly increasing on $\mathrm{supp}(F) = [0, a]$, $0 < a < \infty$, and let F be absolutely continuous with continuous density f on $(0, a)$, $f(0+) < \infty$, $f(a-) < \infty$. Then $F \sim U[0, a]$ iff $X_{2,n} - X_{1,n} \sim X_{1,n}$.

Under the conditions of (4.1.6), the fact that $a - X_{n,n}$ and $X_{n,n} - X_{n-1,n}$ have the same distribution characterizes the $U[0, a]$-distribution, too. A refinement of (4.1.6) is shown in Shimizu, Huang (1983) stating that, for an absolutely continuous distribution function F being strictly increasing on $\mathrm{supp}(F)$, $F \sim U[0, a]$ for some $0 < a < \infty$ iff $X_{2,n} - X_{1,n} \sim X_{1,n}$ for some $n \geq 2$.

4.2. Miscellaneous results

As in the exponential case (cf (2.1.1) and (2.1.2)), Ahsanullah (1989) shows by transformation and applying (2.2.7) that if the ratio of the first two order statistics is uniformly distributed then the underlying distribution is uniform.

However, $X_{1,n}/X_{2,n} \sim F$ is not a characterizing property of the standard uniform distribution. For all power function distributions with $F(x) = x^\alpha$, $\alpha > 0$, $x \in (0, 1)$, we have the same property.

(4.2.1)   (Ahsanullah 1989) Let F be absolutely continuous, $F^{-1}(0+) = 0$, $F(1) = 1$, and let either $F(x \cdot y) \geq F(x) \cdot F(y)$ or $F(x \cdot y) \leq F(x) \cdot F(y)$ for all $x, y \in \mathrm{supp}(F)$. Then $F(x) = x^\alpha$, $x \in (0, 1)$, for some $\alpha > 0$ iff $X_{1,n}/X_{2,n} \sim F$.

We now quote two results of Madreimov, Petunin (1983) where the forms of expected (contrasts of) order statistics are used as characterizing properties.

(4.2.2) (Madreimov, Petunin 1983) Let F be continuous and let $X_1, \ldots, X_n$, $X \sim F$ be independent random variables.

a) Then $F \sim U[0,1]$ iff $E(X_{n,n} - X_{i,n}) = P(X \in (X_{i,n}, X_{n,n}))$ for all $i \in \mathbb{N}$ and $n \geq i$.
b) Then $F \sim U[0,1]$ iff there exists a pair (i,j), $1 \leq i < j \leq n$, such that $E(X_{i,n}) = P(X \in (X_{j-i,n}, X_{j,n}))$ for all $n \geq 2$.

We refer to (5.2.7) for a simultaneous characterization of the standard uniform and of exponential distributions and to Section 5.6 for results of Ghurye (1960).

Finally, we mention a characterization based on moments. It is well known that the property $EX_{1,n} = 1/(n+1)$ for all $n \in \mathbb{N}$ implies that the underlying distribution is standard uniform (cf Galambos, Kotz 1978, p. 55). A simple corollary of this is the following which is of interest for goodness of fit tests.

(4.2.3) (Galambos, Kotz 1978) Let $F^{-1}(0+) \geq 0$, $S_r = \sum_{i=1}^{r} X_i$, $1 \leq r \leq n+1$, and $V_r = S_r/S_{n+1}$, $1 \leq r \leq n$. Then $F \sim U[0,1]$ iff $(V_1, \ldots, V_n) \sim (X_{1,n}, \ldots, X_{n,n})$.

5. Characterizations of specific continuous distributions

In this section we review some characterization results for specific continuous distributions such as normal, exponential, Weibull and logistic distributions. We do not consider further general characterizations of distributions such as the following by Kotlarski, Sasvári (1992) where for independent random variables X_1, X_2, X_3 the joint distributions of $\max(X_1, X_3)$ and $\max(X_2, X_3)$ as well as of $\max(X_1, X_3)$ and $\min(X_2, X_3)$ determine the distributions of X_1, X_2, X_3 (see also Kotlarski 1978 and (5.6.1)).

Several of the results in this section are based on ratios of partial sums of random variables which behave like uniform order statistics.

Throughout this section, $U_1, \ldots, U_m$, $m \in \mathbb{N}$, are i.i.d. random variables from a standard uniform distribution with order statistics $U_{1,m} \leq \cdots \leq U_{m,m}$.

5.1. Characterizations of normal and logarithmic normal distributions

(5.1.1) (Csörgö, Seshadri 1971a) Let F have mean μ and variance σ^2, $|\mu| < \infty$, $0 < \sigma < \infty$, and let $n = 2k + 3$, $k \geq 2$. Moreover, let

$$Z_1 = (X_1 - X_2)/\sqrt{2}, \quad Z_2 = (X_1 + X_2 - 2X_3)/\sqrt{6}, \ldots ,$$

$$Z_{n-1} = (X_1 + \cdots + X_{n-1} - (n-1)X_n)/\sqrt{n(n-1)}, \quad Z_n = \sum_{i=1}^{n} X_i/\sqrt{n} ,$$

$$W_r = Z_{2r-1}^2 + Z_{2r}^2, \quad S_r = \sum_{i=1}^{r} W_i, \quad 1 \leq r \leq k+1, \text{ and}$$

$$V_r = S_r/S_{k+1}, \quad 1 \le r \le k \ .$$

Then $F \sim N(\mu, \sigma^2)$ iff $(V_1, \ldots, V_k) \sim (U_{1,k}, \ldots, U_{k,k})$.

The result (5.1.3) below for logarithmic normal distributions is a corollary of (5.1.1). If the mean μ is known, then the following theorem may be used.

(5.1.2) (Csörgö, Seshadri 1971a) Let F have mean μ, $|\mu| < \infty$, let F be symmetric about μ and let $n = 2k$, $k \ge 3$. Moreover, let

$$Z_i = X_i - \mu, \ 1 \le i \le n, \ W_r = Z^2_{2r-1} + Z^2_{2r}, \ S_r = \sum_{i=1}^{r} W_i, \ 1 \le r \le k, \text{ and}$$

$$V_r = S_r/S_k, \ 1 \le r \le k - 1 \ .$$

Then $F \sim N(\mu, \sigma^2)$ for some $0 < \sigma < \infty$ iff $(V_1, \ldots, V_{k-1}) \sim (U_{1,k-1}, \ldots, U_{k-1,k-1})$.

Two further characterizations of normal distributions are shown by the same authors in Csörgö, Seshadri (1971b) based on two independent samples of random variables. This result is related to the Behrens-Fisher problem.

(5.1.3) (Csörgö, Seshadri 1971a) Let F be absolutely continuous with finite mean μ and variance σ^2, $|\mu| < \infty$, $0 < \sigma < \infty$, $F^{-1}(0+) > 0$, and let $Y_i = \ln X_i$, $1 \le i \le n$, $n = 2k + 3$, $k \ge 2$. Moreover, let

$$Z_1 = (Y_1 - Y_2)/\sqrt{2}, \ Z_2 = (Y_1 + Y_2 - 2Y_3)/\sqrt{6}, \ldots \ ,$$

$$Z_{n-1} = (Y_1 + \cdots + Y_{n-1} - (n-1)Y_n)/\sqrt{n(n-1)}, \quad Z_n = \sum_{i=1}^{n} Y_i/\sqrt{n} \ ,$$

$$W_r = Z^2_{2r-1} + Z^2_{2r}, \quad S_r = \sum_{i=1}^{r} W_i, \ 1 \le r \le k+1, \text{ and}$$

$$V_r = S_r/S_{k+1}, \ 1 \le r \le k \ .$$

Then F is the distribution function of the logarithmic normal distribution with density

$$f(x) = x^{-1}(2\pi\sigma^2)^{-1/2} \exp\{-(2\sigma^2)^{-1}(\ln x - \mu)^2\}, \quad x > 0, \text{ iff}$$

$$(V_1, \ldots, V_k) \sim (U_{1,k}, \ldots, U_{k,k}) \ .$$

Klebanov (1972) states a result based on identical distributions of the range and the range of orthogonal transformations of random variables.

(5.1.4) (Klebanov 1972) Let F be absolutely continuous with continuous density function and $n \ge 4$. Let $Y_{1,n} \le \cdots \le Y_{n,n}$ be the order statistics based on $Y_1, \ldots, Y_n$ where $(Y_1, \ldots, Y_n)' = A(X_1, \ldots, X_n)'$ for some orthogonal transformation $A: \mathbb{R}^n \to \mathbb{R}^n$. Then $F \sim N(0, \sigma^2)$ for some $0 < \sigma < \infty$ iff $X_{n,n} - X_{1,n} \sim Y_{n,n} - Y_{1,n}$ for any orthogonal transformation $A: \mathbb{R}^n \to \mathbb{R}^n$.

Ahsanullah, Hamedani (1988) contribute two results concerning the distribution of the square of the minimum of two i.i.d. random variables. The minimum $X_{1,2}$ can be replaced by the maximum $X_{2,2}$.

(5.1.5) (Ahsanullah, Hamedani 1988) Let F be absolutely continuous and symmetric about zero. Then $F \sim N(0,1)$ iff $X_{1,2}^2 \sim \chi^2(1)$ where $\chi^2(1)$ is the chi-square distribution with one degree of freedom.

Without requiring a symmetric underlying distribution we find

(5.1.6) (Ahsanullah, Hamedani 1988) Let F be absolutely continuous. Then $F \sim N(0,1)$ iff $X_{1,2}^2 \sim \chi^2(1)$ and $X_1/X_2 \sim C(0)$ where $C(0)$ is the Cauchy distribution with median zero.

5.2. Characterizations of exponential distributions

In this section we also consider two-parameter exponential distributions with density function

$$f(x) = \lambda \exp\{-\lambda(x-a)\}, \quad x > a, \; \lambda > 0 \;,$$

which we denote by $\text{Exp}(\lambda, a)$ so that $\text{Exp}(\lambda) \equiv \text{Exp}(\lambda, 0)$.

The following theorems (5.2.1) and (5.2.2) are quoted in Csörgö, Seshadri (1971a). They are used to obtain two characterizations of the Poisson process. Several other characterizations of exponential distributions can be found in Section 5.3 on Weibull distributions.

(5.2.1) (Seshadri, Csörgö, Stephens 1969) Let F have mean $1/\lambda$, $0 < \lambda < \infty$, and $X_1 > 0$, $n \geq 3$. Let $S_r = \sum_{i=1}^{r} X_i$, $1 \leq r \leq n$, and $V_r = S_r/S_n$, $1 \leq r \leq n-1$. Then $F \sim \text{Exp}(\lambda)$ iff $(V_1, \ldots, V_{n-1}) \sim (U_{1,n-1}, \ldots, U_{n-1,n-1})$.

Csörgö et al. (1975) and Menon, Seshadri (1975) point out that the original proof of (5.2.1) is incorrect and present a new proof for $n \geq 3$. In Csörgö et al. (1975) an additional result for two-parameter exponential distributions is stated. A correct proof of the following characterization of $\text{Exp}(\lambda, a)$-distributions is given in Dufour et al. (1984).

(5.2.2) (Seshadri, Csörgö, Stephens 1969) Let $n \geq 3$ and let F have mean $a + 1/\lambda$, $\lambda > 0$, $X_1 > a$. Define $S_r = \sum_{i=1}^{r} D_{i,n}$, $1 \leq r \leq n$, $X_{0,n} = a$, and $V_r = S_r/S_n$, $1 \leq r \leq n-1$. Then $F \sim \text{Exp}(\lambda, a)$ iff $(V_1, \ldots, V_{n-1}) \sim (U_{1,n-1}, \ldots, U_{n-1,n-1})$.

Dufour (1982) conjectured that $(V_1, \ldots, V_{r-1}) \sim (U_{1,r-1}, \ldots, U_{r-1,r-1})$, $V_i = S_i/S_r$, $1 \leq i \leq r-1$, $S_i = \sum_{j=1}^{i} D_{j,n}$, $1 \leq i \leq r$, for some $2 \leq r < n$ is a characteristic property of exponential distributions. Seshadri et al. (1969) and Dufour et al. (1984) (see (5.2.2)) prove this result for the uncensored case $n = r \geq 3$. It should be noted that the distributional identity $D_{1,2}/(D_{1,2} + D_{2,2}) \sim U[0,1]$ does not characterize exponential distributions (cf Menon, Seshadri 1975). Leslie, van Eeden (1993) prove Dufour's conjecture for the case $r \geq \frac{2}{3}n + 1$ and they point out the use of such results in goodness of fit testing.

(5.2.3) (Leslie, van Eeden 1993) Let $F^{-1}(0+) \geq 0$, $\frac{2}{3}n+1 \leq r \leq n-1$. Let $S_i = \sum_{j=1}^{i} D_{j,n}$, $1 \leq i \leq r$, and $V_i = S_i/S_r$, $1 \leq i \leq r-1$. Then $F \sim \text{Exp}(\lambda)$ for some $\lambda > 0$ iff $(V_1, \ldots, V_{r-1}) \sim (U_{1,r-1}, \ldots, U_{r-1,r-1})$.

For further characterization results used in the proof of (5.2.3) we refer to van Eeden (1991) and Leslie, van Eeden (1993).

In Xu, Yang (1995) it is shown that Dufour's conjecture is true for all $5 \leq r \leq n$. The cases $r = 2, 3, 4$ are still not determined. If, however, the distribution of X_1 is restricted to either the class of NBU or NWU distributions, Xu, Yang (1995) show that Dufour's conjecture is true if $r \geq 2$.

In Seshadri et al. (1969) we also find a related result based on spacings of uniform order statistics.

(5.2.4) (Seshadri, Csörgö, Stephens 1969) Let $X_1 > 0$, $EX_1 = 1$, $V_r = X_r/\sum_{i=1}^{n} X_i$, $1 \leq r \leq n-1$. Then $F \sim \text{Exp}(1)$ iff $(V_1, \ldots, V_{n-1}) \sim (U_{1,n-1}, \ldots, U_{n-1,n-1} - U_{n-2,n-1})$.

Galambos, Kotz (1983) point out that Desu's (1971) result (see Section 2.2) is related to an assertion for the distribution of the integer part of a random variable. They quote a result based on a conditional distribution and give further details.

(5.2.5) (Galambos, Kotz 1983) Let $F^{-1}(0+) \geq 0$, $X_i^{(t)} = [X_i/t] + 1$, $t > 0$, $i = 1, 2$, where $[x]$ denotes the integer part of $x \in \mathbb{R}$. If the distribution of

$$\min\left(X_1^{(t)}, X_2^{(t)}\right) | X_1^{(t)} + X_2^{(t)} = 2m + 1$$

is uniform on $1, \ldots, m$ for every $m \in \mathbb{N}$, then $F \sim \text{Exp}(\lambda)$ for some $\lambda > 0$.

Not truncating the X_i's to integers, Galambos (1975a) shows that the underlying distribution with continuous density is exponential iff the distribution of $2X_{1,2}$ given $X_1 + X_2 = s$ is uniform on $[0, s]$ (see also Berk 1977, Patil, Seshadri 1964).

For characterizations of exponential distributions related to those of logistic distributions we refer to Section 5.5 and to a result of Ghurye (1960) which is mentioned in Section 5.6.

Huang et al. (1979) restate (5.2.1) as follows and present a simultaneous characterization of exponential distributions and the standard uniform distribution.

(5.2.6) (Huang, Arnold, Ghosh 1979) Under the assumptions of (5.2.1) we have $F \sim \text{Exp}(\lambda)$ iff $(V_1, V_2) \sim (U_{1,n-1}, U_{2,n-1})$.

(5.2.7) (Huang, Arnold, Ghosh 1979) Let $V_1, \ldots, V_{n-1}$ be as in (5.2.1). Moreover, let $Y_{1,n-1}, \ldots, Y_{n-1,n-1}$ be order statistics of i.i.d. random variables $Y_1, \ldots, Y_n$ with distribution function G and continuous density function. Then $F \sim \text{Exp}(\lambda)$ and $G \sim U[0, 1]$ iff $(V_1, V_2) \sim (Y_{1,n-1}, Y_{2,n-1})$.

5.3. Characterizations of Weibull distributions

As a corollary to their results, Csörgö, Seshadri (1971a) state the following characterization of Weibull distributions based on ratios of partial sums.

(5.3.1) (Csörgö, Seshadri 1971a) Let F be absolutely continuous with finite mean, $X_1 > 0$ and $n \geq 3$. Define $Y_i = X_i^2$, $1 \leq i \leq n$, $S_r = \sum_{i=1}^{r} Y_i$, $1 \leq r \leq n$, $V_r = S_r/S_n$, $1 \leq r \leq n-1$. Then F is the distribution function of a Weibull distribution with density function $f(x) = 2\lambda x \exp\{-\lambda x^2\}$, $x > 0$, $\lambda > 0$, iff $(V_1, \ldots, V_{n-1}) \sim (U_{1,n-1}, \ldots, U_{n-1,n-1})$.

Sethuraman (1965) shows a characterization result for distributions which, in extreme value theory, are well known as limiting distributions of normalized minima of i.i.d. random variables. For details on characterizations based on asymptotic properties of extremes we refer to Galambos (1978). The random variables $X_1, \ldots, X_n$ in (5.3.2) are not assumed to be identically distributed but only compatible which means that $P(X_i > X_j) > 0$ for all $i \neq j$, $1 \leq i, j \leq n$. A preliminary theorem is the following.

(5.3.2) (Sethuraman 1965) Let $X_1, \ldots, X_n$ be compatible and $X_i \sim F_i$, $1 \leq i \leq n$. Then there exist constants $p_2, \ldots, p_n > 0$ with

$$(1 - F_i(x))^{p_i} = 1 - F_1(x), \quad 2 \leq i \leq n,$$

iff $\quad X_{1,n} \sim X_{1,n} | X_{1,n} = X_i \quad$ for all $1 \leq i \leq n$.

Obviously, if $p_i \in \mathbb{N}$, then the equation $(1 - F_i(x))^{p_i} = 1 - F_1(x)$ yields that F_1 is the distribution of the minimum of a number of p_i i.i.d. random variables with distribution function F_i. Such an interpretation is also possible when $1/p_i \in \mathbb{N}$. Sethuraman's main theorem reads as follows and characterizes Weibull, reflected Weibull distributions and the double exponential distribution. For a related characterization of the latter we also refer to Dubey (1966).

(5.3.3) (Sethuraman 1965) Let $X_1, \ldots, X_n$ be compatible, $X_i \sim F_i$, $1 \leq i \leq n$, and let $X_{1,n} \sim X_{1,n} | X_{1,n} = X_i$ for all $1 \leq i \leq n$. Let $F_1, \ldots, F_n$ be of the same type, i.e., there exist constants $a_2 \geq 1, \ldots, a_n \geq 1$, $b_2, \ldots, b_n \in \mathbb{R}$ such that $F_i(x) = F_1(a_i x + b_i)$, $2 \leq i \leq n$. For $a_2 > 1$ one has $a_i > 1$, $b_2/(1 - a_2) = b_i/(1 - a_i)$, $3 \leq i \leq n$. If further $\ln a_i / \ln a_j$ is irrational for some pair (i,j), $i, j \geq 2$, then

$$\begin{cases} F \text{ is of the type } \Phi_{1,\alpha}, & p_2 > 1 \\ F \text{ is of the type } \Phi_{2,\alpha}, & p_2 < 1 \end{cases}.$$

For $a_1 = 1$, one has $a_i = 1$, $3 \leq i \leq n$. If further b_i/b_j is irrational for some pair (i,j), $i, j \geq 2$, then F is of the type Δ. $\Phi_{1,\alpha}$, $\Phi_{2,\alpha}$ and Δ denote the limiting distributions of the sample minimum:

$$\Phi_{1,\alpha}(x) = \begin{cases} 1 - \exp\{-(-x)^{-\alpha}\}, & x < 0 \\ 1, & x \geq 0 \end{cases}, \quad \alpha > 0,$$

$$\Phi_{2,\alpha}(x) = \begin{cases} 0, & x \leq 0 \\ 1 - \exp\{-x^{\alpha}\}, & x > 0 \end{cases}, \quad \alpha > 0,$$

$$\Delta(x) = 1 - \exp\{-e^x\}, \quad x \in \mathbb{R}.$$

In Shimizu, Davies (1981) and Kakosyan et al. (1984) several characterizations of distributions are shown which are based on relations similar to (2.6) (see Section 2.2 for characterizations of exponential distributions). In Shimizu, Davies (1981) we find two characterizations of Weibull distributions based on a modified condition (2.6). A solution of a general functional equation leads to (5.3.4) for order statistics from a sample with random sample size. Another characterization deals with order statistics of ratios of independent random variables.

Let Wei(λ, α) denote the Weibull distribution with distribution function $F(x) = 1 - \exp\{-\lambda x^{\alpha}\}$, $\alpha > 0$, $\lambda > 0$, $x > 0$.

(5.3.4) (Shimizu, Davies 1981) Let $(X_i)_{i \in \mathbb{N}}$ be a sequence of i.i.d. random variables with distribution function F. Let F be non-degenerate, $\alpha > 0$, and N an integer-valued random variable independent of $(X_i)_{i \in \mathbb{N}}$ such that $P(N \geq 2) = 1$, $\ln N$ has finite expectation, and is not concentrated on a lattice $(kp)_{k \in \mathbb{N}}$ for any $p > 0$. Then $F \sim \text{Wei}(\lambda, \alpha)$ for some $\lambda > 0$ iff $N^{1/\alpha} X_{1,N} \sim X_1$.

(5.3.5) (Shimizu, Davies 1981) Let F be non-degenerate and $\alpha > 0$. Let further $Y_1, \ldots, Y_n > 0$ be random variables independent of $X_1, \ldots, X_n$ such that $P\left(\sum_{i=1}^n Y_i^{\alpha} = 1\right) = 1$ and $P(\ln Y_i / \ln Y_j$ is irrational for some i and $j) > 0$. Moreover, let $Z_i = X_i/Y_i$, $1 \leq i \leq n$. Then $F \sim \text{Wei}(\lambda, \alpha)$ for some $\lambda > 0$ iff $Z_{1,n} \sim X_1$.

In their book, Kakosyan et al. (1984, Chapter 3.1) consider characterizations by properties of order statistics associated with non-linear statistics. The following is a corollary of a more general theorem where $X_{1,n}$ is replaced by the infimum of a countable number of random variables.

(5.3.6) (Kakosyan, Klebanov, Melamed 1984) Let $(X_i)_{i \in \mathbb{N}}$ be a sequence of i.i.d. random variables with distribution function F. Let $X_1 > 0$, F be continuous on $[0, \infty)$ and non-degenerate. Let $(a_i)_{i \in \mathbb{N}}$ be a sequence of positive constants satisfying $\sum_{i=1}^{\infty} a_i^{\alpha} = 1$ for some $\alpha > 0$ and let $0 < \lim_{x \to 0+} F(x)/x^{\alpha} = \lambda < \infty$. Then $F \sim \text{Wei}(\lambda, \alpha)$ iff $X_1 \sim \inf_{i \in \mathbb{N}} X_i/a_i$.

Gupta's (1973) result (see Section 2.2) is a consequence of (5.3.6) for $a_1 = \cdots = a_n = \frac{1}{n}, \alpha = 1$. The theorem can also be extended to conditionally independent random variables (see Kakosyan et al. 1984, p. 75). Moreover, the numbers $a_i, i \in \mathbb{N}$, can be replaced by random variables independent of $(X_i)_{i \in \mathbb{N}}$, (see Kakosyan et al. 1984, p. 77).

For related characterizations of exponential, logistic and other distributions and for more details we refer to Chapter 3.1 in Kakosyan et al. (1984).

5.4. Characterizations of gamma distributions

Csörgö Seshadri (1971a) also use the fact that certain ratios of partial sums are distributed as uniform order statistics to obtain the following characterizations for gamma distributions of order $1/n$ the densities of which are given by

$$f(x) = (\pi/\lambda)^{-1/n}(x-a)^{-(n-1)/n}\exp\{-\lambda(x-a)\}, \quad \lambda > 0, \; x > a .$$

These distributions are denoted by $\text{Gam}(\frac{1}{n}, \lambda, a)$.

(5.4.1) (Csörgö, Seshadri 1971a) Let $n = 2k$, $k \geq 3$, $X_1 > 0$. Let $Y_i = X_{2i-1} + X_{2i}$, $1 \leq i \leq k$, $S_r = \sum_{i=1}^{r} Y_i$, $V_r = S_r/S_k$, $1 \leq r \leq k-1$. Then $F \sim \text{Gam}(\frac{1}{2}, \lambda, 0)$ for some $\lambda > 0$ iff $(V_1, \ldots, V_{k-1}) \sim (U_{1,k-1}, \ldots, U_{k-1,k-1})$.

In terms of normalized spacings the same authors obtain

(5.4.2) (Csörgö, Seshadri 1971a) Let $n = 2k$, $k \geq 3$, $X_1 > a$. Let $Y_i = X_{2i-1} + X_{2i}$, $1 \leq i \leq k$, $D_{i,k}^Y = (k-i+1)(Y_{i,k} - Y_{i-1,k})$, $1 \leq i \leq k$, $Y_{0,k} = 2a$, $S_r = \sum_{i=1}^{r} D_{i,k}^Y$, $1 \leq r \leq k$, $V_r = S_r/S_k$, $1 \leq r \leq k-1$. Then $F \sim \text{Gam}(\frac{1}{2}, \lambda, a)$ for some $\lambda > 0$ iff $(V_1, \ldots, V_{k-1}) \sim (U_{1,k-1}, \ldots, U_{k-1,k-1})$.

These results can easily be generalized to obtain characterizations of $\text{Gam}(\frac{1}{n}, \lambda, a)$-distributions as pointed out by Csörgö, Seshadri (1971a).

5.5. Characterizations of the logistic distribution

In George, Mudholkar (1981a) we find the following theorem on related characterizations of the standard exponential and the logistic distribution by properties of the minimum and maximum of two random variables. A random variable X has a (standard) logistic distribution, if its distribution function is given by $F(x) = (1 + e^{-x})^{-1}$, $x \in \mathbb{R}$, briefly $F \sim \text{Lgc}$.

(5.5.1) (George, Mudholkar 1981a) Let $F(0) = 1/2$ and $t\,\varphi(t)$ be integrable where φ is the characteristic function of F. Let a random variable Z with distribution function G be independent of $X_1, X_2 \sim F$.

a) If $G \sim \text{Exp}(1)$, we have

$F \sim \text{Lgc}$ iff $X_{1,2} + Z \sim X_1$ iff $X_{2,2} - Z \sim X_1$.

b) If $F \sim \text{Lgc}$, we have

$G \sim \text{Exp}(1)$ iff $X_{1,2} + Z \sim X_1$ iff $X_{2,2} - Z \sim X_1$.

It is also possible to obtain a simultaneous characterization of exponential and logistic distributions by the same arguments as in (5.5.1). For further details we refer to George, Mudholkar (1982) and Galambos (1992).

(5.5.2) (George, Mudholkar 1981a) Let Z_1, Z_2 be non-negative random variables with some non-lattice distribution function G and independent of X_1 and X_2 which have distribution function F. Then $F \sim \text{Lgc}$ and $G \sim \text{Exp}(1)$ iff $X_{1,2} + Z_1 \sim X_1$ and $X_{2,2} - Z_2 \sim X_1$.

George, Mudholkar (1981a) also present a characterization based on the characteristic function of the sample median and herewith generalize a result in George, Mudholkar (1981b).

(5.5.3) (George, Mudholkar 1981a) Let F be absolutely continuous with density function f and characteristic function φ, let $n = 2m - 1$ for some $m \in \mathbb{N}$ and let $\varphi_{m,n}$ be the characteristic function of the sample median $X_{m,n}$. Moreover, let $F(0) = 1/2$,

$$f(x) = o(e^{kx}), \; x \to \infty, \text{ for every } k \in \mathbb{N},$$
$$f(x) = o(e^{-kx}), \; x \to -\infty, \text{ for every } k \in \mathbb{N},$$

and $t^k \varphi(t)$ be integrable for every $k \in \mathbb{N}$.

Then $F \sim \text{Lgc}$ iff $\varphi_{m,n}(t) = \prod_{j=1}^{m-1}(1 + t^2/j^2)\varphi(t)$ for all $t \in \mathbb{R}$.

As a consequence of the above theorem, George, Mudholkar (1981a) get the following theorem which again connects logistic and exponential distributions.

(5.5.4) (George, Mudholkar 1981a) Let $Z_1, \ldots, Z_{n-1}$ be independent Laplace-distributed random variables with densities $g_1, \ldots, g_{n-1}$ where $g_j(z) = \frac{j}{2}\exp\{-j|z|\}$, $1 \leq j \leq n-1$. Let $X_1, \ldots, X_{2n-1}$ be independent of $Z_1, \ldots, Z_{n-1}$, F be absolutely continuous and let the same conditions as in (5.5.2) be fulfilled. Then $F \sim \text{Lgc}$ iff $X_{n,2n-1} + \sum_{j=1}^{n-1} Z_j \sim X_1$.

George and Mudholkar (1981a,b) mention as special case of their result and under the above regularity conditions with respect to F that if Z is a Laplace variable with density $g(z) = \frac{1}{2}e^{-|z|}, z \in \mathbb{R}$, and independent of X_1, X_2, X_3, then $F \sim \text{Lgc}$ iff $X_{2,3} + Z \sim X_1$. Since Z may be represented as a difference of two i.i.d. exponential random variables Y_1 and Y_2, the above result can also be stated in this way. Let Y_1 and Y_2 be i.i.d. standard exponentially distributed random variables which are independent of X_1, X_2 and X_3. Then $F \sim \text{Lgc}$ iff $X_{2,3} + Y_1 - Y_2 \sim X_1$.

An open problem regarding characterizations is mentioned in Arnold et al. (1992, p. 150). It is questioned whether a logistic distribution can be characterized by its property that the mid-range is distributed as the median in a sample of size 3.

5.6. Miscellaneous results

Kagan et al. (1973) give some remarks on order statistics. Actually they deal with linear statistics $\sum_{i=1}^n a_i X_i$ which correspond to M-statistics of the form $\max((X_1 - a)/b_1, \ldots, (X_n - a)/b_n)$ with $a \in \mathbb{R}$ and $b_1, \ldots, b_n > 0$. They state a result where, under several regularity conditions, identical distributions of two M-statistics characterize the underlying distribution function of the form $F(x) = \exp\{-\sum_{i=1}^k e_i/(x-a)^i\}$. Moreover, a theorem of Ghurye (1960) is cited (see Kagan et al. 1973, p. 443) where in the case that $W = \sum_{i=1}^n (X_i - X_{1,n})$, the uniform distribution of $((X_1 - X_{1,n})/W, \ldots, (X_n - X_{1,n})/W)$ over some surface characterizes a two-parameter exponential distribution and a uniform distribution in the case that $W = X_{n,n} - X_{1,n}$. One further characterization of uniform distributions is shown in Ghurye (1960) based on a uniform distribution of $(X_1/X_{n,n}, \ldots, X_n/X_{n,n})$.

Kotlarski (1979) presents the following theorem by using maxima of a random number of random variables. The underlying distributions of the random variables involved are uniquely determined but not explicitly given (see the introductory remarks to this section).

(5.6.1)   (Kotlarski 1979) Let the random variables N, $(X_i)_{i \in \mathbb{N}}$, $(Y_i)_{i \in \mathbb{N}}$ be independent, $\text{supp}(N) \subset \mathbb{N}_0, P(N = 1) > 0$,

$(X_i)_{i \in \mathbb{N}} \sim F$, F continuous, $F(a) = 0$, $F(b) = 1$, $0 < F(x) < 1$ for $a < x < b$, $-\infty \le a < b \le \infty$,

$(Y_i)_{i \in \mathbb{N}} \sim G$, G continuous, $G(c) = 0$, $G(d) = 1$, $0 < G(y) < 1$ for $c < y < d$, $-\infty \le c < d \le \infty$.

Moreover, let

$$U = a \quad \text{for } N = 0, \text{ and } U = X_{N,N} \quad \text{for } N > 0,$$
$$V = c \quad \text{for } N = 0, \text{ and } V = Y_{N,N} \quad \text{for } N > 0.$$

Then the joint distribution of the two-dimensional random variable (U, V) uniquely determines the distributions of N, X_1 and Y_1.

Under the additional assumption that X_1 and Y_1 have positive densities on the interiors of their supports, Kotlarski (1985) shows a procedure to obtain the distributions of N, X_1 and Y_1 if the joint distribution of U and V is given. He gives an example where $H(x,y) = (1 + x^2 y^2)/2$, $x, y \in [0, 1]$, is the joint distribution function of U and V which leads to standard uniform distributions of X_1 and Y_1.

6. Characterizations of geometric and other discrete distributions

A random variable X with distribution function F is said to be geometrically distributed with parameter $p \in (0, 1)$ and with the positive integers $\mathbb{N}$ or $\mathbb{N}_0 = \mathbb{N} \cup \{0\}$ as its support if

$$P(X = k) = (1 - p)p^{k-1} \quad \text{for all} \quad k \in \mathbb{N} \quad (F \sim \text{Geo}(p) \text{ for short}),$$

or if $P(X = k) = (1 - p)p^k \quad \text{for all} \quad k \in \mathbb{N}_0 \quad (F \sim \text{Geo}_{\mathbb{N}_0}(p) \text{ for short})$,

respectively

We will also consider arbitrary lattice supports.

There is a variety of characterization results based on the independence of functions of order statistics. For details and a review we refer to Becker (1984), Galambos (1975b) and Srivastava (1986). Some other characterizations are also based on a relation similar to (1.6). Since for non-continuous distributions ties may occur with positive probability, often conditional distributions are considered. E.g., we find that if $F \sim \text{Geo}(p)$, $p \in (0, 1)$, then

(6.1) $X_{s,n} - X_{r,n} | X_{r+1,n} > X_{r,n} \sim X_{s-r,n-r}$ for all $1 \leq r < s \leq n$.

For certain applications it may be useful to define a geometric distribution with an arbitrary lattice as support. Assuming condition (6.1), the lattice structure of the support follows as shown by Becker (1984, p. 62).

Concerning similarities and distinctions of the results and their proofs for exponential and geometric distributions we refer to the remarks in Galambos (1975b) and Arnold et al. (1984). For instance, there is no discrete distribution satisfying $nX_{1,n} \sim F$ (i.e., (2.6)). A related condition is considered in Galambos (1975b, p. 92) and in Bagchi (1989) (cf (6.12)–(6.14)). Arnold, Ghosh (1976) show that (6.1) for $s = n = 2$ and $r = 1$ characterizes geometric distributions within the class of non-degenerate distributions with $\text{supp}(F) \subset \mathbb{N}_0$ and $P(X_1 = 1) > 0$, which is explicitly used in the proof. They conjecture that (6.1) for $s = r + 1$ and arbitrary $n \geq s$ is also a characteristic property which indeed is proven by Arnold (1980) using Shanbhag's (1977) lemma. For a different proof of (6.2) we refer to Zijlstra (1983).

(6.2)   (Arnold 1980) Assume that $\text{supp}(F) \subset \mathbb{N}_0$ and $0 < P(X_1 = 1) < 1$. Then $F \sim \text{Geo}(p)$ for some $p \in (0,1)$ iff there exists a pair (r,n), $1 \leq r < n$, such that

$$X_{r+1,n} - X_{r,n} | X_{r+1,n} > X_{r,n} \sim X_{1,n-r} .$$

Fosam et al. (1993) point out that the assumption $P(X_1 = 1) > 0$ is also implicitly made in Arnold, Ghosh (1976). Without this assumption, a modified theorem can be shown by using a Lau–Rao theorem where the geometric distributions are defined on some lattice:

$$P(X_1 = \lambda k) = (1-p)p^{k-1}, \quad k \in \mathbb{N} ,$$

for some $p \in (0,1)$ and some positive integer λ.

Arnold, Ghosh (1976) and Arnold (1980) ask whether

$$X_{s,n} - X_{r,n} | X_{s,n} > X_{r,n} \sim X_{s-r,n-r}$$

for some r,s with $s > r + 1$ characterizes geometric distributions. However, Zijlstra (1983) points out that this is not a property of geometric distributions. Conditioning on $X_{r+1,n} > X_{r,n}$ Zijlstra states the following theorem.

(6.3)   (Zijlstra 1983) Let $\text{supp}(F) \subset \mathbb{N}_0$ and $(1 - F(i+1))/(1 - F(i)) \geq 1 - F(0) > 0$ for all $i \in \mathbb{N}_0$. Then $F \sim \text{Geo}_{\mathbb{N}_0}(p)$ for some $p \in (0,1)$ iff there exists a triple (r,s,n), $2 \leq r + 1 < s \leq n$, such that

$$P(X_{s,n} - X_{r,n} > j | X_{r+1,n} > X_{r,n}) = P(X_{s-r,n-r} > j - 1) \quad \text{for all } j \in \mathbb{N}_0 .$$

The assumption of (6.3) is obviously fulfilled for NWU-distributions.

Similar to the result (2.5.2) for exponential distributions, Arnold's (1980) result is generalized by Becker (1984) with respect to arbitrary spacings under the IFR/DFR-assumption and by Schweitzer (1995) under the NBU/NWU-assumption (cf (2.5.2) for exponential distributions).

(6.4) (Becker 1984, Schweitzer 1995) Let F be discrete and NBU or NWU. Then $F \sim \text{Geo}(p)$ with support $\{mt; m \in \mathbb{N}\}$ for some $t \in \mathbb{R}$ iff there exists a triple (r,s,n), $1 \leq r < s \leq n$, satisfying

$$P(X_{s,n} - X_{r,n} \leq t | X_{r+1,n} > X_{r,n}) = P(X_{s-r,n-r} \leq t) .$$

Arnold et al. (1984) consider the particular case $s = n = 3$ and $r = 1$. Becker (1984) proves a characterization result of geometric distributions without an aging condition assuming the distributional identity for two different values of s and again applying Jensen's inequality. This result is similar to (2.5.3).

(6.5) (Becker 1984) Let F be discrete and let $\alpha(F) = t > -\infty$. Then $F \sim \text{Geo}(p)$ with support $\{mt; m \in \mathbb{N}\}$ iff there exists a quadruple (r, s_1, s_2, n), $1 \leq r < s_1 < s_2 \leq n$, such that

$$X_{s_1,n} - X_{r,n} | X_{r+1,n} > X_{r,n} \sim X_{s_1-r,n-r}$$

and

$$P(X_{s_2,n} - X_{r,n} > t | X_{r+1,n} > X_{r,n}) = P(X_{s_2-r,n-r} > t) .$$

As a corollary, Becker (1984) points out that, assuming a support with lattice structure, the validity of

$$P(X_{s,n} - X_{r,n} > t | X_{r+1,n} > X_{r,n}) = P(X_{s-r,n-r} > t)$$

for $s \in \{s_1, s_2\}$, $1 \leq r < s_1 < s_2 \leq n$, is sufficient to characterize geometric distributions. If in addition we condition on $X_{1,n}$, we obtain a characterization using only one distributional identity.

(6.6) (Nagaraja, Srivastava 1987) Let F be non-degenerate, $\text{supp}(F) \subset \mathbb{N}$ and $P(X_1 = 1) > 0$. Then $F \sim \text{Geo}(p)$ for some $p \in (0,1)$ iff there exists a pair (s,n), $1 \leq s \leq n$, such that

$$P(X_{s,n} - X_{1,n} = j | X_{1,n} = 1, X_{2,n} > X_{1,n}) = P(X_{s-1,n-1} = j)$$
for all $j \in \text{supp}(F)$

By considering the equality of the corresponding survival functions in (6.6), Schweitzer (1995) characterizes geometric distributions by means of the more general equation ($r = 1$ in (6.6)):

$$P(X_{s,n} - X_{r,n} > j | X_{r,n} = 1, X_{r+1,n} > X_{r,n}) = P(X_{s-r,n-r} > j) .$$

If the condition $X_{1,n} = 1$ in (6.6) is changed into $X_{1,n} = i$ for some $i > 1$, then there are distributions other than geometric satisfying the distributional identity. Nagaraja, Srivastava (1987) also show a result in this framework for $\text{Geo}_{\mathbb{N}_0}(p)$ requiring this identity for two different values of i.

(6.7) (Nagaraja, Srivastava 1987) Let F be non-degenerate, $\text{supp}(F) \subset \mathbb{N}_0$ and $P(X_1 = 0) > 0$. Then $F \sim \text{Geo}_{\mathbb{N}_0}(p)$ for some $p \in (0,1)$ iff there exists a triple (s,n,i), $1 \leq s \leq n$, $i \in \mathbb{N}$, such that

$$P(X_{s,n} - X_{1,n} = j | X_{1,n} = i, X_{2,n} > X_{1,n}) = P(X_{s-1,n-1} = j - 1)$$
$$\text{for all } j \in \mathbb{N} \ ,$$

and there exists some $k \in \mathbb{N}$, $k > i$, $k+1$ and $i+1$ relatively prime, such that

$$P(X_{s,n} - X_{1,n} = j | X_{1,n} = k, X_{2,n} > X_{1,n}) = P(X_{s-1,n-1} = j - 1)$$
$$\text{for all } 1 \le j \le i \ .$$

For discrete distributions, Puri, Rubin (1970, $r=1$, $n=2$), Ramachandran (1982), Zijlstra (1983, supp$(F) \subset \mathbb{N}_0$) and Becker (1984) deal with the following general characterization result based on (2.5) with $s = r+1$. Here ϵ_a denotes a degenerate distribution with mass in $a \in \mathbb{R}$. Theorem (6.8) gives the answer to the question found in Galambos (1975b, p. 93).

(6.8) (Puri, Rubin 1970, Ramachandran 1982, Zijlstra 1983, Becker 1984) Let F be some discrete distribution (i.e. supp(F) is countable). Then there exists a pair (r,n), $1 \le r < n$, such that
$$X_{r+1,n} - X_{r,n} \sim X_{1,n-r}$$

iff either

i) $F \sim \epsilon_0$ (F is degenerate)

or ii) $F \sim p_0 \epsilon_0 + (1-p_0)\epsilon_a$ with $p_0 = \binom{n}{r}^{-1/r}$, $a > 0$ (F has two mass points)

or iii) $F \sim p_0 \epsilon_0 + (1-p_0)(1-p) \sum_{m=1}^{\infty} p^{m-1} \epsilon_{tm}$ with $t > 0$, $0 < p_0 < \binom{n}{r}^{-1/r}$ and p is determined by $\binom{n}{r} \sum_{i=0}^{r}(-1)^i \binom{r}{i}(1-p_0)^i(1-p^{n-r})/(1-p^{n-r+i}) = 1$. For every pair (r,n), $1 \le r < n$, there exist $p_0, p \in (0,1)$ satisfying this equation. (F is a mixture of a degenerate distribution and a geometric distribution).

Puri, Rubin (1970) require an additional assumption for the support of F. Rossberg (1972) (see (2.5.1)) does not give the complete solution in the discrete case. However, it can be obtained by analytical methods. This is shown by Ramachandran (1982) using the Wiener–Hopf technique. Rao (1983) (see (2.5.1)) applies the ICFE in the continuous case. Zijlstra (1983) and Becker (1984) consider discrete distributions and make use of Shanbhag's (1977) lemma.

As a corollary, Becker (1984) obtains explicit representations of the distribution in (6.8) iii) for special cases.

(6.9) (Becker 1984) Let F be discrete with more than two mass points.

a) Then $X_{2,n} - X_{1,n} \sim X_{1,n-1}$ iff

$$F \sim \left(1 - \frac{n-1}{n} \frac{1-p^n}{1-p^{n-1}}\right) \epsilon_0 + \sum_{m=1}^{\infty} \frac{n-1}{n} \frac{1-p^n}{1-p^{n-1}} (1-p) p^{m-1} \epsilon_{tm}$$

for some $p \in (0,1)$ and $t > 0$.

b) Then $X_{3,n} - X_{2,n} \sim X_{1,n-2}$ $(n \ge 3)$ iff $F \sim \sum_{m=0}^{\infty} p_m \epsilon_{tm}$, $t > 0$ with

$$p_0 = 1 - \frac{1-p^n}{1-p^{n-1}} + \left(\left(\frac{1-p^n}{1-p^{n-1}}\right)^2 - \left(1 - \frac{2}{n(n-1)}\right)\frac{1-p^n}{1-p^{n-2}}\right)^{1/2}$$

and $p_m = (1-p_0)(1-p)p^{m-1}$, $m \in \mathbb{N}$, $p \in (0,1)$.

We now quote a result due to Puri (1966) characterizing geometric distributions by the identical distribution of $X_{2,2} - X_{1,2}$ and a sum of two independent random variables.

(6.10) (Puri 1966) Let $\text{supp}(F) \subset \mathbb{N}_0$, $0 < 1 - p = q = F(0) < 1$. Then $F \sim \text{Geo}_{\mathbb{N}_0}(p)$ for some $p \in (0,1)$ iff

$$X_{2,2} - X_{1,2} \sim Y_1 + Y_2,$$

where Y_1 and Y_2 are independent, $Y_1 \sim \frac{1}{1+p}\epsilon_0 + \frac{p}{1+p}\epsilon_1$ (Bernoulli distribution) and $Y_2 \sim \text{Geo}_{\mathbb{N}_0}(p)$.

Another result is based on two i.i.d. random variables characterizing discrete distributions by means of the distribution of the minimum conditioned on the sum of random variables.

(6.11) (Galambos 1975b) Let F be non-degenerate, $\text{supp}(F) \subset \mathbb{N}_0$ such that if $P(X_1 = k) = 0$ then $P(X_1 = m) = 0$ for all $m \geq k$. Moreover, let g and c be functions with $g(m) \geq 0$, $c(m) > 0$ for all $m \in \mathbb{N}_0$. Then $P(X_1 = k) = c\,g(k)v^k$ for all $k \in \mathbb{N}_0$, for some $v > 0$ and a norming constant c iff

$$P(X_{1,2} = k \mid X_1 + X_2 = m) = c(m)\,g(k)\,g(m-k), \quad 0 \leq k \leq \frac{m-1}{2}$$

for all $m \in \mathbb{N}$, m odd satisfying $P(X_1 + X_2 = m) > 0$.

Galambos (1975b) discusses some special cases. E.g., taking $c(m) = 2/(m+1)$ and $g(k) = 1$ leads to a characterization of geometric distributions by the property that $X_{1,2}$ conditioned on $X_1 + X_2 = m$ is uniformly distributed. Other choices yield characterizations of binomial, Poisson and discrete Pareto distributions.

There are also characterizations of geometric distributions similar to results for exponential distributions based on (2.6). It is easily seen that the condition

(6.12) $P(X_{1,n} \geq k) = P(X_1 \geq kn)$ for all $n \in \mathbb{N}$ and $k = 1$

characterizes geometric distributions.

(6.13) (Galambos 1975b) Let $\text{supp}(F) \subset \mathbb{N}_0$. Then $F \sim \text{Geo}_{\mathbb{N}_0}(p)$ for some $p \in (0,1)$ iff $P(X_{1,n} \geq 1) = P(X_1 \geq n)$ for all $n \geq 2$.

Bagchi (1989) requires (6.12) for two values of n and for all $k \in \mathbb{N}$.

(6.14) (Bagchi 1989) Let $\text{supp}(F) \subset \mathbb{N}_0$ and $F(0) < 1$. Then $F \sim \text{Geo}_{\mathbb{N}_0}(p)$ for some $p \in (0,1)$ iff $P(X_{1,n} \geq k) = P(X_1 \geq kn)$ for all $k \in \mathbb{N}$ and for two incommensurable values $1 < n_1 < n_2$ of n (i.e., $\log_{n_1} n_2$ is irrational).

Neither assuming (6.12) for all k and a single value of n nor for all n and some $k > 1$ is sufficient to characterize geometric distributions. Bagchi (1989) considers some examples.

Aly (1988) contributes the following theorem dealing with the distribution of the minimum in a sample of size n.

(6.15) (Aly 1988) Let $\mathrm{supp}(F) = \mathbb{N}_0$ (i.e., $P(X_1 = j) > 0$ for all $j \in \mathbb{N}_0$) and let $c > 0$. Then $F \sim \mathrm{Geo}_{\mathbb{N}_0}(p)$ with p being that root of the equation $(c - n)x^n = c - nx^{n-1}$ which lies in $(0, 1)$ iff $P(X_{1,n} = j, X_{2,n} - X_{1,n} \geq 1) = cP(X_{1,n} = j)$ for all $j \in \mathbb{N}_0$.

Finally we summarize some characterization results of geometric distributions with support $\mathbb{N}_0$ due to Nagaraja, Srivastava (1987) which are usually referred to as characterizations by means of independence and conditional independence (see also Nagaraja 1992).

(6.16) (Nagaraja, Srivastava 1987) Let $\mathrm{supp}(F) \subset \mathbb{N}_0$, $P(X_1 = 0) > 0$, $P(X_1 = 1) > 0$. Then $F \sim \mathrm{Geo}_{\mathbb{N}_0}(p)$ for some $p \in (0, 1)$ iff there exists a pair (r, n), $1 \leq r \leq n$, such that

$$P(X_{r,n} - X_{1,n} = j | X_{1,n} = 0) = P(X_{r,n} - X_{1,n} = j | X_{1,n} = 1)$$

for all $j \in \mathrm{supp}(F^{X_{r,n} - X_{1,n}})$.

In the same paper we also find a theorem characterizing modified geometric type distributions based on

$$P(X_{s,n} - X_{r,n} = 0 | X_{r,n} = x, X_{r,n} > X_{r-1,n})$$
$$= P(X_{s,n} - X_{r,n} = 0 | X_{r,n} > X_{r-1,n})$$

for a triple (r, s, n), $2 \leq r < s \leq n$, as well as a theorem characterizing modified geometric distributions based on

$$P(X_{s,n} - X_{r,n} = j | X_{r,n} = 1, X_{r,n} > X_{r-1,n})$$
$$= P(X_{s,n} - X_{r,n} = j | X_{r,n} = 2, X_{r,n} > X_{r-1,n})$$

for a triple (r, s, n), $2 \leq r < s \leq n$, and for all $j \in \mathrm{supp}(F^{X_{s,n} - X_{r,n}})$.
For more details we refer to Nagaraja, Srivastava (1987).

References

Ahsanullah, M. (1975). A characterization of the exponential distribution. In: G. P. Patil et al., eds., Statistical Distributions in Scientific Work, Vol. 3. Reidel, Dordrecht, 131–135.

Ahsanullah, M. (1976). On a characterization of the exponential distribution by order statistics. *J. Appl. Prob.* **13**, 818–822.

Ahsanullah, M. (1977). A characteristic property of the exponential distribution. *Ann. Statist.* **5**, 580–582.

Ahsanullah, M. (1978a). A characterization of the exponential distribution by spacings. *J. Appl. Prob.* **15**, 650–653.

Ahsanullah, M. (1978b). A characterization of the exponential distribution by spacings. *Ann. Inst. Statist. Math.* **30** A, 163–166.

Ahsanullah, M. (1981). On characterizations of the exponential distribution by spacings. *Statist. Hefte* **22**, 316–320.

Ahsanullah, M. (1984). A characterization of the exponential distribution by higher order gap. *Metrika* **31**, 323–326.

Ahsanullah, M. (1987). Two characterizations of the exponential distribution. *Commun. Statist. – Theory Meth.* **16**, 375–381.

Ahsanullah, M. (1988a). Characteristic properties of order statistics based on random sample size from an exponential distribution. *Statistica Neerlandica* **42**, 193–197.

Ahsanullah, M. (1988b). On a conjecture of Kakosyan, Klebanov and Melamed. *Statistical Papers* **29**, 151–157.

Ahsanullah, M. (1989). On characterizations of the uniform distribution based on functions of order statistics. *Aligarh J. Statist.* **9**, 1–6.

Ahsanullah, M. and G. G. Hamedani (1988). Some characterizations of normal distribution. *Calcutta Statist. Assoc. Bull.* **37**, 95–99.

Ahsanullah, M. and M. Rahman (1972). A characterization of the exponential distribution. *J. Appl. Prob.* **9**, 457–461.

Aly, M. A. H. (1988). Some Contributions to Characterization Theory with Applications in Stochastic Processes. Ph.D. Thesis, University of Sheffield.

Alzaid, A. A., K. S. Lau, C. R. Rao and D. N. Shanbhag (1988). Solution of Deny convolution equation restricted to a half line via a random walk approach. *J. Multivar. Anal.* **24**, 309–329.

Arnold, B. C. (1971). Two characterizations of the exponential distribution using order statistics. Unpublished manuscript.

Arnold, B. C. (1980). Two characterizations of the geometric distribution. *J. Appl. Prob.* **17**, 570–573.

Arnold, B. C. and M. Ghosh (1976). A characterization of geometric distributions by distributional properties of order statistics. *Scand. Actuarial J.* 232–234.

Arnold, B. C. and D. Isaacson (1976). On solutions to $\min(X, Y) \stackrel{d}{=} aX$ and $\min(X, Y) \stackrel{d}{=} aX \stackrel{d}{=} bY$. *Z. Wahrscheinlichkeitstheorie verw. Gebiete* **35**, 115–119.

Arnold, B. C., N. Balakrishnan and H. N. Nagaraja (1992). *A First Course in Order Statistics*. Wiley, New York.

Arnold, B. C., A. Becker, U. Gather and H. Zahedi (1984). On the Markov property of order statistics. *J. Statist. Plan. Inf.* **9**, 147–154.

Azlarov, T. A. and N. A. Volodin (1986). *Characterization Problems Associated with the Exponential Distribution*. Springer, New York.

Bagchi, S. N. (1989). Characterisations of the geometric distribution using distributional properties of the order statistics. *Prob. and Math. Statist.* **10**, 143–147.

Becker, A. (1984). Charakterisierungen diskreter Verteilungen durch Verteilungs eigenschaften von Ordnungsstatistiken. Dissertation, Aachen University of Technology.

Bell, C. B. and Y. R. K. Sarma (1980). A characterization of exponential distributions based on order statistics. *Metrika* **27**, 263–269.

Berk, R. H. (1977). Characterizations via conditional distributions. *J. Appl. Prob.* **14**, 806–816.

Bosch, K. (1977). Eine Charakterisierung der Exponentialverteilungen. *ZAMM* **57**, 609–610.

Chan, L. K. (1967). On a characterization of distributions by expected values of extreme order statistics. *Amer. Math. Monthly* **74**, 950–951.

Csörgö, M. and V. Seshadri (1971a). Characterizing the Gaussian and exponential laws via mappings onto the unit interval. *Z. Wahrscheinlichkeitstheorie verw. Geb.* **18**, 333–339.

Csörgö, M. and V. Seshadri (1971b). Characterizations of the Behrens-Fisher and related problems (A goodness of fit point of view). *Theory Prob. Appl.* **16**, 23–35.

Csörgö, M., V. Seshadri and M. Yalovsky (1975). Applications of characterizations in the area of goodness of fit. In: G. P. Patil et al., eds., Statistical Distributions in Scientific Work, Vol. 2. Reidel, Dordrecht, 79–90.

Dallas, A. (1977). On the minimum of a random sample. *Math. Operationsforsch. Statist., Ser. Statistics* **8**, 511–513.

David, H. A. (1981). *Order Statistics*. 2nd ed. Wiley, New York.

Davies, P. L. and D. N. Shanbhag (1987). A generalization of a theorem of Deny with applications in characterization theory. *Quart. J. Math.* Oxford (2) **38**, 13–34.

Desu, M. M. (1971). A characterization of the exponential distribution by order statistics. *Ann. Math. Statist.* **42**, 837–838.

Dimaki, C. and E. Xekalaki (1993). Characterizations of the Pareto distribution based on order statistics. In: V. V. Kalashnikov and V. M. Zolotarev, eds., *Stability Problems for Stochastic Models*. Springer, Berlin, 1–16.

Dubey, S. D. (1966). Characterization theorems for several distributions and their applications. *J. Industrial Mathematics* **16**, 1–22.

Dufour, R. (1982). Tests d' ajustement pour des échantillons tronqués ou censurés. Ph.D. Thesis, Université de Montréal.

Dufour, R., U. R. Maag and C. van Eeden (1984). Correcting a proof of a characterization of the exponential distribution. *J. Roy. Statist. Soc. B* **46**, 238–241.

Epstein, B. and M. Sobel (1953). Life testing. *J. Amer. Statist. Assoc.* **48**, 486–502.

Fang, B. Q. and K. T. Fang (1989). A characterization of multivariate l_1-norm symmetric distributions. *Statist. Prob. Lett.* **7**, 297–299.

Fang, K. T. and B. Q. Fang (1988). Some families of multivariate symmetric distributions related to exponential distribution. *J. Multivar. Anal.* **24**, 109–122.

Fosam, E. B., C. R. Rao and D. N. Shanbhag (1993). Comments on some papers involving the integrated Cauchy functional equation. *Statist. Prob. Lett.* **17**, 299–302.

Gajek, L. and U. Gather (1989). Characterizations of the exponential distribution by failure rate and moment properties of order statistics. In: J. Hüsler and R. D. Reiss, eds., *Extreme Value Theory*. Springer, Berlin, 114–124.

Galambos, J. (1975a). Characterizations of probability distributions by properties of order statistics I. In: G. P. Patil et al., eds., *Statistical Distributions in Scientific Work*, Vol. **3**. Reidel, Dordrecht, 71–88.

Galambos, J. (1975b). Characterizations of probability distributions by properties of order statistics II. In: G. P. Patil et al., eds., *Statistical Distributions in Scientific Work*, Vol. **3**. Reidel, Dordrecht, 89–101.

Galambos, J. (1978). *The Asymptotic Theory of Extreme Order Statistics*. Wiley, New York.

Galambos, J. (1992). Characterizations (Chapter 7). In: N. Balakrishnan, ed., *Handbook of the Logistic Distribution*. Dekker, New York, 169–188.

Galambos, J. and S. Kotz (1978). *Characterizations of Probability Distributions*. Springer, Berlin.

Galambos, J. and S. Kotz (1983). Some characterizations of the exponential distribution via properties of the geometric distribution. In: P. K. Sen, ed., *Essays in Honour of Norman L. Johnson*. North-Holland, Amsterdam, 159–163.

Gather, U. (1988). On a characterization of the exponential distribution by properties of order statistics. *Statist. Prob. Lett.* **7**, 93–96.

Gather, U. (1989). Personal correspondence with M. Ahsanullah (11.4.1989), author's reply (6.7.1989).

Gather, U. and G. Szekely (1989). Characterizations of distributions by linear forms of order statistics. Technical Report, Department of Statistics, University of Dortmund.

George, E. O. and G. S. Mudholkar (1981a). Some relationships between the logistic and the exponential distributions. In: C. Taillie et al., eds., *Statistical Distributions in Scientific Work*, Vol. 4. Reidel, Dordrecht, 401–409.

George, E. O. and G. S. Mudholkar (1981b). A characterization of the logistic distribution by a sample median. *Ann. Inst. Statist. Math.* **33** A, 125–129.

George, E. O. and G. S. Mudholkar (1982). On the logistic and exponential laws. *Sankhyā A* **44**, 291–293.

Ghurye, S. G. (1960). Characterization of some location and scale parameter families of distributions. In: I. Olkin et al., eds., Contributions to Probability and Statistics, Essays in Honor of Harold Hotelling. Stanford University Press, Stanford, 203–215.

Gupta, R. C. (1973). A characteristic property of the exponential distribution. *Sankhyā B* **35**, 365–366.

Gupta, R. C. (1979). The order statistics of exponential, power function and Pareto distributions and some applications. *Math. Operationsforsch. Statist., Ser. Statistics* **10**, 551–554.

Hajós, G. and A. Rényi (1954). Elementary proofs of some basic facts concerning order statistics. *Acta Math. Acad. Sci. Hungar.* **5**, 1–6.

Hoeffding, W. (1953). On the distribution of the expected values of the order statistics. *Ann. Math. Statist.* **24**, 93–100.

Huang, J. S. (1974a). On a theorem of Ahsanullah and Rahman. *J. Appl. Prob.* **11**, 216–218.

Huang, J. S. (1974b). Characterizations of the exponential distribution by order statistics. *J. Appl. Prob.* **11**, 605–609.

Huang, J. S. (1989). Moment problem of order statistics: A review. *Internat. Statist. Rev.* **57**, 59–66.

Huang, J. S., B. C. Arnold and M. Ghosh (1979). On characterizations of the uniform distribution based on identically distributed spacings. *Sankhyā B* **41**, 109–115.

Hwang, J. S. and G. D. Lin (1984). Characterizations of distributions by linear combinations of moments of order statistics. *Bull. Inst. Math., Acad. Sinica* **12**, 179–202.

Iwińska, M. (1985). On a characterization of the exponential distribution by order statistics. In: Numerical Methods and Their Applications, Proc. 8th Sess. Poznan Circle Zesz. Nauk. Ser I, Akad. Ekon. Poznan **132**, 51–54.

Iwińska, M. (1986). On the characterizations of the exponential distribution by order statistics and record values. *Fasciculi Mathematici* **16**, 101–107.

Janardan, K. G. and V. S. Taneja (1979a). Characterization of the Weibull distribution by properties of order statistics. *Biom. J.* **21**, 3–9.

Janardan, K. G. and V. S. Taneja (1979b). Some theorems concerning characterization of the Weibull distribution. *Biom. J.* **21**, 139–144.

Johnson, N. L., S. Kotz and N. Balakrishnan (1994). *Continuous Univariate Distributions.* Volume 1, 2nd ed. Wiley, New York.

Johnson, N. L., S. Kotz and N. Balakrishnan (1995). *Continuous Univariate Distributions.* Volume 2, 2nd ed. Wiley, New York.

Kagan, A. M., Y. V. Linnik and C. R. Rao (1973). *Characterization Problems in Mathematical Statistics.* Wiley, New York.

Kakosyan, A. V., L. B. Klebanov and J. A. Melamed (1984). *Characterization of Distributions by the Method of Intensively Monotone Operators.* Springer, Berlin.

Kamps, U. (1990). Characterizations of the exponential distribution by weighted sums of i.i.d random variables. *Statistical Papers* **31**, 233–237.

Kamps, U. (1991). A general recurrence relation for moments of order statistics in a class of probability distributions and characterizations. *Metrika* **38**, 215–225.

Kamps, U. (1992a). Identities for the difference of moments of successive order statistics and record values. *Metron* **50**, 179–187.

Kamps, U. (1992b). Characterizations of the exponential distribution by equality of moments. *Allg. Statist. Archiv* **76**, 122–127.

Kamps, U. (1995). A *Concept of Generalized Order Statistics.* Teubner, Stuttgart.

Klebanov, L. B. (1972). A characterization of the normal distribution by a property of order statistics. *Math. Notes* **13**, 71–73.

Kotlarski, I. I. (1978). On some characterizations in probability by using minima and maxima of random variables. *Aequationes Mathematicae* **17**, 77–82.

Kotlarski I.I. (1979). On characterizations of probability distributions by using maxima of a random number of random variables. *Sankhyā A* **41**, 133–136.

Kotlarski, I. I. (1985). Explicit formulas for characterizations of probability distributions by using maxima of a random number of random variables. *Sankhyā A* **47**, 406–409.

Kotlarski, I. I. and Z. Sasvári (1992). On a characterization problem of statistics. *Statistics* **23**, 85–93.

Kotz, S. (1974). Characterizations of statistical distributions: a supplement to recent surveys. *Internat. Statist. Rev.* **42**, 39–65.

Lau, K. S. and C. R. Rao (1982). Integrated Cauchy functional equation and characterizations of the exponential law. *Sankhyā A* **44**, 72– 90.

Leslie, J. and C. van Eeden (1993). On a characterization of the exponential distribution based on a type 2 right censored sample. *Ann. Statist.* **21**, 1640–1647.

Lin, G. D. (1989). Characterizations of distributions via moments of order statistics: a survey and comparison of methods. In: Y. Dodge, ed., Statistical Data Analysis and Inference. Elsevier, Amsterdam, 297–307.

Madreimov, I. and Petunin, Y. I. (1983). A characterization of the uniform distribution with the aid of order statistics. *Theor. Prob. and Math. Statist.* **27**, 105–110.

Malmquist, S. (1950). On a property of order statistics from a rectangular distribution. *Skand. Aktuarietidskrift* **33**, 214–222.

Marshall, A. W. and I. Olkin (1991). Functional equations for multivariate exponential distributions. *J. Multivar. Anal.* **39**, 209–215.

Menon, M. V. and V. Seshadri (1975). A characterization theorem useful in hypothesis testing. In: Contributed Papers, 40th Session of the *Internat. Statist. Inst.,* Voorburg, 586–590.

Nagaraja, H. N. (1992). Order statistics from discrete distributions. *Statistics* **23**, 189–216.

Nagaraja, H. N. and R. C. Srivastava (1987). Some characterizations of geometric type distributions based on order statistics. *J. Statist. Plan. Inf.* **17**, 181–191.

Patil, G. P. and V. Seshadri (1964). Characterization theorems for some univariate probability distributions. *J. Roy. Statist. Soc. B* **26**, 286–292.

Pudeg, A. (1990). Charakterisierung von Wahrscheinlichkeitsverteilungen durch Verteilungseigenschaften der Ordnungsstatistiken und Rekorde. Dissertation, Aachen University of Technology.

Puri, P. (1966). Probability generating functions of absolute difference of two random variables. *Proc. National Acad. of Sciences* **56**, 1059–1061.

Puri, P. and H. Rubin (1970). A characterization based on the absolute difference of two i.i.d. random variables. *Ann. Math. Statist.* **41**, 2113–2122.

Ramachandran, B. (1982). An integral equation in probability theory and its applications In: G. Kallianpur et al., eds., Statistics and Probability: Essays in Honor of C. R. Rao. North-Holland, Amsterdam, 609–616.

Ramachandran, B. and K. S. Lau (1991). *Functional Equations in Probability Theory.* Academic Press, Boston.

Rao, C. R. (1983). An extension of Deny's theorem and its application to characterizations of probability distributions. In: P. J. Bickel et al., eds., A Festschrift for Erich L. Lehmann. Wadsworth, Belmont, 348–366.

Rao, C. R. and D. N. Shanbhag (1986). Recent results on characterization of probability distributions: a unified approach through extensions of Deny's theorem. *Adv. Appl. Prob.* **18**, 660–678.

Rao, C. R. and D. N. Shanbhag (1994). *Choquet–Deny Type Functional Equations with Applications to Stochastic Models.* Wiley, Chichester.

Rao, C. R., T. Sapatinas and D. N. Shanbhag (1994). The integrated Cauchy functional equation: some comments on recent papers. *Adv. Appl. Prob.* **26**, 825–829.

Rényi, A. (1953). On the theory of order statistics. *Acta Math. Acad. Sci. Hungar.* **4**, 191–227.

Riedel, M. (1981). On Bosch's characterization of the exponential distribution function. *ZAMM* **61**, 272–273.

Riedel, M. and H. J. Rossberg (1994). Characterization of the exponential distribution function by properties of the difference $X_{k+s:n} - X_{k:n}$ of order statistics. *Metrika* **41**, 1–19.

Rossberg, H. J. (1972). Characterization of the exponential and the Pareto distributions by means of some properties of the distributions which the differences and quotients of order statistics are subject to. *Math. Operationsforsch. Statist.* **3**, 207–216.

Schweitzer, N. (1995). Charakterisierungen von Wahrscheinlichkeitsverteilungen durch identische Verteilung von Funktionen von Ordnungsstatistiken. Master Thesis, Aachen University of Technology.

Seshadri, V., M. Csörgö and M. A. Stephens (1969). Tests for the exponential distribution using Kolmogorov-type statistics. *J. Roy. Statist. Soc. B* **31**, 499–509.

Sethuraman, J. (1965). On a characterization of the three limiting types of the extreme. *Sankhyā A* **27**, 357–364.

Shah, S. M. and D. G. Kabe (1981). Characterizations of exponential, Pareto, power function, Burr and logistic distributions by order statistics. *Biom. J.* **23**, 141–146.

Shanbhag, D. N. (1977). An extension of the Rao–Rubin characterization of the Poisson distribution. *J. Appl. Prob.* **14**, 640–646.

Shimizu, R. (1979). A characterization of the exponential distribution. *Ann. Inst. Statist. Math. A* **31**, 367–372.

Shimizu, R. (1980). Functional equation with an error term and the stability of some characterizations of the exponential distribution. *Ann. Inst. Statist. Math. A* **32**, 1–16.

Shimizu, R. and L. Davies (1981). General characterization theorems for the Weibull and the stable distributions. *Sankhyā A* **43**, 282–310.

Shimizu, R. and J. S. Huang (1983). On a characteristic property of the uniform distribution. *Ann. Inst. Statist. Math. A* **35**, 91–94.

Srivastava, R. C. (1986). On characterizations of the geometric distribution by independence of functions of order statistics. *J. Appl. Prob.* **23**, 227–232.

Sukhatme, P. V. (1937). Tests of significance for samples of the χ^2-population with two degrees of freedom. *Ann. Eugenics* **8**, 52–56.

Van Eeden, C. (1991). On a conjecture concerning a characterization of the exponential distribution. *CWI Quarterly* **4**, 205–211.

Xu, J. L. and G. L. Yang (1995). A note on a characterization of the exponential distribution based on a type II censored sample. *Ann. Statist.* **23**, 769–773.

Zijlstra, M. (1983). Characterizations of the geometric distribution by distributional properties. *J. Appl. Prob.* **20**, 843–850.

Characterizations of Distributions by Recurrence Relations and Identities for Moments of Order Statistics

Udo Kamps

1. Introduction

Recurrence relations and identities for moments of order statistics are often helpful in numerical computations as well as for theoretical purposes. They have been extensively investigated and we find a variety of results for arbitrary and specific distributions.

The most important recurrence relation for moments of order statistics from arbitrary distributions is given by Cole (1951) in the continuous case and by Melnick (1964) in the discrete case and it is frequently used:

(1.1) $\quad (n-r)\,\mathrm{E}\,X_{r,n}^\alpha + r\,\mathrm{E}\,X_{r+1,n}^\alpha = n\,\mathrm{E}\,X_{r,n-1}^\alpha, \quad 1 \leq r \leq n-1$.

Let, throughout this paper, $X, X_1, \ldots, X_n$, $n \geq 2$, be independent and identically distributed random variables with distribution function F, and let $X_{1,n} \leq \cdots \leq X_{n,n}$ denote the order statistics based on $X_1, \ldots, X_n$.

Using the integral representation of a moment via the pseudo inverse of the underlying distribution function

$$\mathrm{E}\,X_{r,n}^\alpha = r\binom{n}{r} \int_0^1 (F^{-1}(t))^\alpha t^{r-1}(1-t)^{n-r}\,dt \ ,$$

(1.1) obviously holds true for arbitrary distributions. Moreover, the assumption of independence of $X_1, \ldots, X_n$ can be weakened; requiring exchangeable random variables turns out to be sufficient (David, Joshi 1968).

(1.1) has the following interpretation. If all moments of order α are known in a sample of size $n-1$ and if $\mathrm{E}\,X_{i,n}^\alpha$ is known for any i, $1 \leq i \leq n$, then all moments of order α in a sample of size n can be computed.

Here, the term "recurrence relation" is not used in the strict sense which means that it is not necessarily possible to reconstruct a whole system of moments via some given set of moments.

Identities for arbitrary distributions are shown in Govindarajulu (1963), David (1981), Arnold, Balakrishnan (1989), Balakrishnan, Cohen (1991), Arnold et al. (1992) and in the detailed review by Malik et al. (1988). Moreover, we point out the important results on recurrence relations and identities for the moments and distribution functions of order statistics from dependent random variables which can be found, e.g., in Young (1967), David, Joshi (1968), Balakrishnan (1987), Sathe, Dixit (1990), Balasubramanian, Bapat (1991), Balakrishnan et al. (1992), Balasubramanian, Balakrishnan (1993), David (1993) and Balasubramanian et al. (1994).

The relation (1.1) can be modified to obtain

(1.2)
$$(n-r)\left(\mathrm{E}X_{r+1,n}^{\alpha} - \mathrm{E}X_{r,n}^{\alpha}\right) = n\left(\mathrm{E}X_{r+1,n}^{\alpha} - \mathrm{E}X_{r,n-1}^{\alpha}\right),$$
$$n\left(\mathrm{E}X_{r,n}^{\alpha} - \mathrm{E}X_{r,n-1}^{\alpha}\right) = r\left(\mathrm{E}X_{r,n}^{\alpha} - \mathrm{E}X_{r+1,n}^{\alpha}\right), \quad 1 \leq r \leq n-1.$$

The relations (1.2) are helpful to modify several identities shown in the sequel. For reviews on recurrence relations and identities for moments of order statistics from specific distributions, we again refer to David (1981), Arnold, Balakrishnan (1989), Balakrishnan, Cohen (1991), Arnold et al. (1992) and to the detailed account of Balakrishnan et al. (1988).

In the present article we focus on characterizations of distributions by identities and recurrence relations for moments of order statistics. If not explicitly stated, the appearing expectations are always assumed to exist.

In Section 2 we mention but do not review characterizations by sequences of moments and moment differences. Results of this type as well as characterizations by identities and recurrence relations are mainly based on complete sequences of functions. Several of these sequences are cited. Section 3 contains characterizations of exponential distributions which are due to Govindarajulu (1975). Recurrence relations which are valid in classes of distributions and corresponding characterizations of distributions are subject matter of Section 4. There are also several characterization results which are based on a single identity in contrast to the ones deduced by means of a complete function sequence. Some of these results are cited in Section 5, and we refer to the literature on inequalities for moments of order statistics. Finally, we review some results based on product moments of order statistics including Govindarajulu's (1966) results for normal distributions.

Other characterizations by means of order statistics are based on, e.g., identically distributed functions of order statistics (see Gather, Kamps and Schweitzer, Chapter 9), inequalities for moments (see Rychlik, Chapter 6), conditional moments and on the independence of functions of order statistics (cf Rao and Shanbhag, Chapter 8).

Some characterization results based on identities for moments of order statistics are also shown in Chapter 9 (Gather, Kamps and Schweitzer.) of this volume. Occasionally, the assumption of identically distributed functions of order statistics can be weakened to a corresponding moment condition.

It should be noted that there are many characterization results based on conditional moments of order statistics. We do not review these results but refer to, e.g., Ferguson (1967), Beg, Kirmani (1974), Galambos, Kotz (1978), Khan, Beg (1987), Khan, Khan (1987), Khan, Abu-Salih (1989), Rauhut (1989), Beg, Balasubramanian (1990), Mohie El-Din et al. (1991) and Balasubramanian, Beg (1992).

Several of the results in this chapter, e.g., results via recurrence relations and inequalities for moments of order statistics, can also be shown for generalized order statistics. In Kamps (1995) a concept of generalized order statistics is proposed as a unified approach to a variety of models of ordered random variables including, e.g., ordinary order statistics and k^{th} record values. Well known results for ordinary order statistics and record values can be subsumed, generalized, and integrated within a general framework. Hence, these results are also valid in other models of ordered random variables such as sequential order statistics, Pfeifer's records and k_n-records from non-identical distributions.

2. Characterizations by sequences of moments and complete function sequences

Hoeffding (1953) showed that the expected values $(EX_{r,n})_{1 \leq r \leq n, n \in \mathbb{N}}$ characterize the underlying distribution function, if the first absolute moment exists. The assertion remains valid, if only the sequence of minima $(EX_{1,n})_{n \in \mathbb{N}}$ or if the sequence of maxima $(EX_{n,n})_{n \in \mathbb{N}}$ is known (Chan 1967, Konheim 1971). Pollak (1973) assumes knowledge of some subsequence $(EX_{r(n),n})_{n \in \mathbb{N}}$, choosing for each n some $r(n)$ with $1 \leq r(n) \leq n$.

However, by this the original assumption of Hoeffding is not really weakened. The whole triangular array of expectations of the order statistics can be reconstructed by the cited sequences of moments via relation (1.1) as pointed out by Mallows (1973) and Kadane (1974).

Interesting results can easily be obtained. The following examples are shown in Galambos (1975).

(2.1)
    i) If $EX_{1,n} = \dfrac{1}{n}$ $\forall n \in \mathbb{N}$, then $F(x) = 1 - e^{-x}$, $x > 0$.

    ii) If $EX_{1,n} = \dfrac{1}{n+1}$ $\forall n \in \mathbb{N}$, then $F(x) = x$, $x \in (0,1)$.

The connection between characterizations of distributions by moments of order statistics and the completeness of certain function sequences is indicated by the following. Let the random variables X and Y be distributed according to F and G, respectively. Thus, if the expectations $EX_{r,n}$ and $EY_{r,n}$ of order statistics coincide for some sequence of indices, then

$$\int_0^1 \left(F^{-1}(t) - G^{-1}(t)\right) t^{r-1}(1-t)^{n-r} \, dt = 0$$

yields the equality of F and G via a complete function sequence.

(2.2) (cf Hwang, Lin 1984a) Let $L_\delta(A)$ be the space of δ-integrable functions on a measurable set $A \subset \mathbb{R}$. A sequence $(f_n)_{n \in \mathbb{N}}$ of functions in $L_\delta(A)$ is called complete on $L_\delta(A)$, if for all functions $g \in L_\delta(A)$ the condition

$$\int_A g(x) f_n(x)\, dx = 0 \quad \forall n \in \mathbb{N}$$

implies

$$g(x) = 0 \quad \text{a.e. on } A\ .$$

Notations: $L_\delta(A) = L_\delta(a,b)$, if $A = (a,b)$; $L(a,b) = L_1(a,b)$.

We now list some complete sequences of functions. The Müntz–Szász lemma permits the choice of a real subsequence of Hoeffding's triangular scheme; this result dates back to Müntz (1914), Szász (1916) and it is cited, e.g., in Boas (1954), Hwang, Lin (1984a) and Lin (1989a).

(2.3) (Müntz 1914, Szász 1916) Let $(n_j)_{j \in \mathbb{N}} \subset \mathbb{N}$ with $n_1 < n_2 < \cdots$. Then the sequence $(x^{n_j})_{j \in \mathbb{N}}$ of polynomials is complete on $L(0,1)$, iff $\sum_{j=1}^\infty n_j^{-1} = \infty$.

In the sequel, a sequence $(n_j)_{j \in \mathbb{N}} \subset \mathbb{N}$ with $n_1 < n_2 < \cdots$ and $\sum_{j=1}^\infty n_j^{-1} = \infty$ is called a Müntz–Szász sequence. As an example it may be used to modify the assertions (2.1). A generalization of this lemma is given in

(2.4) (Hwang 1983) Let f be an absolutely continuous function on a bounded interval $[a,b]$ with $f(a)f(b) \geq 0$ and $|f'(x)| \geq k > 0$ a.e. on $[a,b]$. Moreover, let $(n_j)_{j \in \mathbb{N}} \subset \mathbb{N}$ be a subsequence of $\mathbb{N}$ with $n_1 < n_2 < \cdots$ and $\sum_{j=1}^\infty n_j^{-1} = \infty$. Then the function sequence $(f^{n_j}(x))_{j \in \mathbb{N}}$ is complete on $L(a,b)$, iff f strictly increases on $[a,b]$.

The assertion (2.4) remains valid, if the assumption $|f'(x)| \geq k > 0$ a.e. on $[a,b]$ is replaced by $f'(x) \neq 0$ a.e. on (a,b) (cf Hwang, Lin 1984a,b). In order to modify the condition imposed on the function f, a theorem of Zaretzki (cf Natanson 1961, Lin 1988b) can be applied.

(2.5) (Zaretzki) If f is a continuous and strictly increasing function on a bounded interval $[a,b]$, then f^{-1} is absolutely continuous on $[f(a), f(b)]$, iff $f'(x) \neq 0$ a.e. on (a,b).

In (2.6) we show some results on the completeness of sequences

$$(x^{r-1}(1-x)^{n-r})_{(r,n) \in I_j}\ .$$

(2.6) For any set I_j of pairs of indices (r,n), $1 \leq r \leq n$, the sequence of polynomials $(x^{r-1}(1-x)^{n-r})_{(r,n) \in I_j}$ is complete on $L(0,1)$:

$I_1 = \{(r,n)|$ let $\mu \in \mathbb{N}$ fixed; for each $n \geq \mu$ choose some $r = r_n$, $1 \leq r_n \leq n$, with $r_\mu \leq r_n \leq r_\mu + n - \mu\}$ (Huang 1975),

$I_2 = \{(r,n)|$ for each $n \geq 2$ choose some $r = r_n$ with $1 \leq r_n \leq n\}$ (Huang, Hwang 1975, see Hwang, Lin 1984a, p. 187),

$I_3 = \{(r,n)|$ each $n \in (n_j)_{j\in\mathbb{N}}$ with $n_j \to \infty, j \to \infty$, is combined with all $r = r_j$, $r_j \in \{1,\ldots,n_j\}\}$ (Hwang 1978),

$I_4 = \{(r,n)|$ for given sequences $(n_j)_{j\in\mathbb{N}}, (\mu_j)_{j\in\mathbb{N}} \subset \mathbb{N}$ satisfying $\mu_{j+1} > n_j \geq \mu_j > 1, j \in \mathbb{N}$ and $\sum_{j=1}^{\infty} \sum_{\nu=\mu_j}^{n_j} \frac{1}{\nu-1} = \infty$, each $n \in (n_j)_{j\in\mathbb{N}}$ is combined with all $r = r_j$, $r_j \in \{\mu_j,\ldots,n_j\}\}$ (Hwang, Lin 1984a).

When deriving characterization results via a complete function sequence argument, different sequences of polynomials can be chosen as shown above. In all such results shown in the sequel it then has to be ensured that the corresponding function g in the sense of (2.2) is integrable. In the following, the appropriate conditions are not always stated explicitly (cf (4.3)). Since such integrability conditions are obvious, they are omitted and implicitly assumed.

Characterizations of distributions by sequences of moments, results on the completeness of certain function sequences and many other facts and investigations may be found in Arnold, Meeden (1975), Galambos (1975), Galambos, Kotz (1978) and in a series of papers by Huang, Hwang and Lin (see e.g., Huang 1975, Hwang 1978). The articles by Hwang, Lin (1984a), Huang (1989) and Lin (1989a) are reviews on this topic.

In the above references we also find characterizations by sequences of moment differences.

(2.7) (Lin 1988b) Let F^{-1} be absolutely continuous on $(0,1)$, $E|X| < \infty$, and $(n_j)_{j\in\mathbb{N}} \subset \mathbb{N}$ a Müntz–Szász sequence. Then the sequence $\left(E(X_{1,n_j} - X_{1,n_j+1})\right)_{j\in\mathbb{N}}$ characterizes the distribution function F up to a location parameter.

As corollaries from (2.7), Lin (1988b) shows characterizations of exponential and uniform distributions similar to those in (2.1).

(2.8) (Lin 1988b) Under the conditions of (2.7) we find

i) $E(X_{1,n_j} - X_{1,n_j+1}) = \frac{1}{n_j(n_j+1)}$ for all $j \in \mathbb{N}$ iff $F(x) = 1 - \exp(-(x-\mu))$, $x \in (\mu, \infty)$ for some $\mu \in \mathbb{R}$.

ii) $E(X_{1,n_j} - X_{1,n_j+1}) = \frac{1}{(n_j+1)(n_j+2)}$ for all $j \in \mathbb{N}$ iff $F(x) = x - c, x \in (c, c+1)$ for some $c \in \mathbb{R}$.

Lin (1988b) also presents an analogue of (2.7) concerning a sequence of expected spacings.

(2.9) (Lin 1988b) Let $r \in \mathbb{N}$ and let the conditions of (2.7) be fulfilled. Then the sequence $\left(E(X_{r+1,n_j} - X_{r,n_j})\right)_{n_j \geq r+1}$ characterizes the distribution function F up to a location parameter.

For earlier results and more details we refer to, e.g., Govindarajulu et al. (1975), Saleh (1976) and Madreimov, Petunin (1983).

Furthermore, there are characterizations of exponential distributions by identical expectations of

$$X_{r_j,n_j} \quad \text{and} \quad \sum_{i=1}^{r_j} \frac{1}{n_j - i + 1} X_i$$

for suitable sequences $(r_j, n_j)_{j \in \mathbb{N}}$ of indices (cf Gather, Kamps and Schweitzer, Chapter 9, Section 2.4, Ahsanullah, Rahman 1972, Kotz 1974, Huang 1974a,b and Kamps 1990, 1992b).

3. Characterizations of exponential distributions

Govindarajulu (1975) presents characterizing recurrence relations which are shown in this section (see also Azlarov, Volodin 1986, Chapter 6). The characterizing properties are assumed to hold true for all sufficiently large sample sizes $n \in \mathbb{N}$. Obviously, the results can be weakened by applying complete sequences of functions as introduced in the previous section.

Let, throughout this section, $(X_i)_{i \in \mathbb{N}}$ be a sequence of i.i.d random variables with some non-degenerate distribution function F, $F(0) = 0$ and $EX_1^2 < \infty$. Moreover, let $X_{0,n} = 0$ and $F \sim \exp(\lambda)$ denote that F is the distribution function of an exponential distribution:

$$F(x) = 1 - e^{-\lambda x}, \quad x > 0, \; \lambda > 0 .$$

(3.1) (Govindarajulu 1975) $F \sim \exp(\lambda)$ iff

$$EX_{i+1,n}^2 - EX_{i,n}^2 = \frac{2}{(n-i)\lambda} EX_{i+1,n} \quad \text{for some } i \in \mathbb{N}_0 \text{ and for all } n \geq i+1 .$$

Applying relation (1.2), this result can also be stated as follows.

(3.2) (Govindarajulu 1975) $F \sim \exp(\lambda)$ iff

$$EX_{i,n}^2 - EX_{i-1,n-1}^2 = \frac{2}{n\lambda} EX_{i,n} \quad \text{for some } i \in \mathbb{N} \text{ and for all } n \geq i .$$

(3.3) (Govindarajulu 1975) $F \sim \exp(\lambda)$ iff

$$\text{Var } X_{i+1,n} - \text{Var } X_{i,n} = (EX_{i+1,n} - EX_{i,n})^2$$

for some $i \in \mathbb{N}_0$ and for all $n \geq i+1$.

The condition in (3.3) can be rewritten as

(3.4) $\quad EX_{i+1,n}^2 - EX_{i,n}^2 = 2EX_{i+1,n}(EX_{i+1,n} - EX_{i,n}) .$

Applying (1.2), the assertion (3.3) can also be stated as follows.

(3.5) (Govindarajulu 1975) $F \sim \exp(\lambda)$ iff

$$\text{Var } X_{i,n} - \text{Var } X_{i-1,n-1} = (EX_{i,n} - EX_{i-1,n-1})^2$$

for some $i \in \mathbb{N}$ and for all $n \geq i$.

For $i \geq 2$, the relation in (3.5) can be rewritten as

(3.6) $\quad EX_{i+1,n}^2 - EX_{i,n-1}^2 = 2EX_{i+1,n}(EX_{i+1,n} - EX_{i,n-1})$.

Applying (1.1) to both sides we obtain (3.4). Other characterizations make use of covariances between order statistics.

(3.7) $\quad$ (Govindarajulu 1975) $F \sim \exp(\lambda)$ iff

$$\text{Var } X_{i,n} = \text{Cov}(X_{i,n}, X_{i+1,n}) \quad \text{for some } i \in \mathbb{N} \text{ and for all } n \geq i \ .$$

(3.8) $\quad$ (Govindarajulu 1975) $F \sim \exp(\lambda)$ iff

$$\text{Var } X_{i,n} = \frac{1}{n-i} \sum_{j=i+1}^{n} \text{Cov}(X_{i,n}, X_{j,n})$$

$$\text{for some } i \in \mathbb{N}_0 \text{ and for all } n \geq i+1 \ .$$

(3.9) $\quad$ (Govindarajulu 1975) $F \sim \exp(\lambda)$ iff

$$\text{Cov}(X_{i,n}, X_{k,n}) = \text{Cov}(X_{i,n}, X_{k+1,n})$$

$$\text{for some } i \in \mathbb{N}_0, k \geq i, \text{ and for all } n \geq k \ .$$

If the rhs of the characterizing identity in (3.10) is replaced by the constant 1, we obtain a characterization of the standard normal distribution (see (6.6)).

(3.10) $\quad$ (Govindarajulu 1975) Let $EX_1 = 1/\lambda$. $F \sim \exp(\lambda)$ iff

$$\sum_{j=1}^{n} \text{Cov}(X_{i,n}, X_{j,n}) = \frac{1}{\lambda} EX_{i,n} \quad \text{for some } i \in \mathbb{N} \text{ and for all } n \geq i \ .$$

4. Related characterizations in classes of distributions

Numerous articles on recurrence relations for moments of order statistics from specific distributions are found in the literature. For detailed surveys we refer to Balakrishnan et al. (1988) and Arnold, Balakrishnan (1989). The results can often be described as rather isolated; explicit expressions for the moments of some distribution lead to an identity. A step towards a systematic treatment is shown in Khan et al. (1983) and Lin (1988b). They derive a representation for the difference of moments of successive order statistics. Putting in special distributions leads to similar recurrence relations. Govindarajulu (1975) and Lin (1988b) state corresponding characterization results assuming the validity of some identity for a certain sequence of order statistics.

Lin (1988b) considers relations for uniform, Pareto, exponential and logistic distributions of the form

$$EX_{r,n}^\alpha - EX_{r-1,n}^\alpha = \alpha \, c(r,n,p,q) EX_{r+p,n+q}^{\alpha-1}$$

with integers p, q and certain constants $c(r, n, p, q)$, and presents corresponding characterization results by applying a Müntz–Szász sequence (cf (2.3)).

Motivated by the fact that there are similarly structured relations, a unified approach to several identities is shown in this section (cf Kamps 1991b). We proceed as follows. The starting point is a parametrized recurrence relation. Following, a characterization set-up may lead to a corresponding family of distributions applying an appropriate complete sequence of functions. Going backwards, the strong assumptions are dropped and the relation is verified within this class of distributions under mild conditions. This approach provides an insight into structural properties and relationships of several probability distributions. Moreover, isolated results can be subsumed and well known results can be generalized with respect to the parametrization of the underlying distribution and to moments of non-integral orders. This method is demonstrated for the class $\mathscr{F}$ of distributions introduced in Kamps (1991b).

More generally, results of this type can be shown for generalized order statistics including the assertions for ordinary order statistics as well as identities for records and other models of ordered random variables (cf Kamps 1995, Chapter III).

Let $\mathscr{F}$ be the class of distribution functions F, where F is given by the first derivative of its pseudo inverse function:

(4.1) $\quad (F^{-1})'(t) = \frac{1}{d} t^p (1-t)^{q-p-1}, \quad t \in (0,1)$,

with a constant $d > 0$ and integers p, q.

All possible pseudo inverse functions with (4.1) are shown in Kamps (1995, pp. 119–121) and particular distribution functions out of $\mathscr{F}$ are given by, e.g., $(c \in \mathbb{R})$

$F(x) = 1 - \exp\{-d(x-c)\}, \quad x \in (c, \infty)$ (exponential distributions),

$F(x) = 1 - (dq(c-x))^{1/q}$,

$\begin{cases} x \in \left(c - \frac{1}{dq}, c\right), & q > 0 \\ x \in \left(c - \frac{1}{dq}, \infty\right), & q < 0 \end{cases}$ ($q < 0$: Pareto, Burr XII distributions),

$F(x) = (dq(x-c))^{1/q}$,

$\begin{cases} x \in \left(c, c + \frac{1}{dq}\right), & q > 0 \\ x \in \left(-\infty, c + \frac{1}{dq}\right), & q < 0 \end{cases}$ ($q > 0$: power function distributions),

$F(x) = (1 + \exp\{-d(x-c)\})^{-1}, \quad x \in (-\infty, \infty)$ (logistic distributions).

For any $F \in \mathscr{F}$, (4.2) shows a corresponding recurrence relation for the moments of order statistics. The constant $c(r, n, p, q)$ appearing in the equations turns out to be the expectation of a certain spacing. We restrict ourselves to positive moments. In the case of negative moments, regularity conditions concerning the support have to be made. Using the representation

$$\mathrm{E} X_{r,n}^\alpha - \mathrm{E} X_{r-1,n}^\alpha = \alpha \binom{n}{r-1} \int_0^1 (F^{-1}(t))^{\alpha-1} (F^{-1})'(t) t^{r-1} (1-t)^{n-r+1} \, dt,$$

we derive

(4.2) (Kamps 1991b) Let the appearing order statistics be based on some distribution function $F \in \mathscr{F}$, let $\alpha \geq 1$ be a constant and $F^{-1}(0) \geq 0$, if $\alpha \notin \mathbb{N}$. Then for all $r, n \in \mathbb{N}, 2 \leq r \leq n$, satisfying

$$1 \leq r + p \leq n + q$$

and

$$-\infty < \mathrm{E} X^{\alpha}_{r,n}, \mathrm{E} X^{\alpha}_{r-1,n}, \mathrm{E} X^{\alpha-1}_{r+p,n+q} < \infty \; ,$$

the identity

$$\mathrm{E} X^{\alpha}_{r,n} - \mathrm{E} X^{\alpha}_{r-1,n} = \alpha \, c(r, n, p, q) \mathrm{E} X^{\alpha-1}_{r+p,n+q}$$

is valid where the constant $c(r, n, p, q)$ is given by

$$c(r, n, p, q) = \mathrm{E} X_{r,n} - \mathrm{E} X_{r-1,n} = \frac{1}{d} \frac{\binom{n}{r-1}}{(r+p)\binom{n+q}{r+p}} \; .$$

Assuming the validity of the recurrence relation in (4.2) for an appropriate sequence of pairs of indices $((r_j, n_j))_{j \in \mathbb{N}}$, the correspondingly parametrized distribution function in the family $\mathscr{F}$ can be characterized. The sequence has to be chosen such that the sequence of polynomials

$$\left\{ t^{r_j - 1} (1-t)^{n_j - r_j + 1} \right\}_j \left(\text{or } \{ (1-t)^{n_j} \}_j \right)$$

is complete on the space $L(0, 1)$. Appropriate sequences are shown in Section 2. Such a characterization of the exponential distribution is given by Govindarajulu (1975) (see (3.1)) in the case $\alpha = 2$. Lin (1988b) obtains characterizations of exponential, uniform, Pareto and logistic distributions applying the Müntz–Szász Lemma (see (2.3)).

The results are contained in

(4.3) (Kamps 1991b) Let the order statistics be based on the distribution function F, let F^{-1} be absolutely continuous on the interval $(0, 1)$ and let $\alpha \geq 1$ be a constant with

$$\begin{cases} |\{t \in (0,1); F^{-1}(t) = 0\}| \in \{0, 1\}, & \alpha \in \mathbb{N}, \quad \alpha \geq 2 \\ F^{-1}(t) > 0 \quad \text{for all } t \in (0,1), \quad \alpha \notin \mathbb{N} \; . \end{cases}$$

Moreover, let integers p, q and a sequence $((r_j, n_j))_{j \in \mathbb{N}}$ according to the above remark are given satisfying the conditions

$$2 \leq r_j \leq n_j, 1 \leq r_j + p \leq n_j + q$$

and

$$-\infty < \mathrm{E}X^{\alpha}_{r_j,n_j}, \mathrm{E}X^{\alpha}_{r_j-1,n_j}, \mathrm{E}X^{\alpha-1}_{r_j+p,n_j+q} < \infty,$$

for all $j \in \mathbb{N}$.

Then for some constant $d > 0$ we have

$$(F^{-1})'(t) = \frac{1}{d} t^p (1-t)^{q-p-1} \quad \text{a.e. on } (0,1),$$

iff

$$\mathrm{E}X^{\alpha}_{r_j,n_j} - \mathrm{E}X^{\alpha}_{r_j-1,n_j} = \alpha\, c(r_j,n_j,p,q) \mathrm{E}X^{\alpha-1}_{r_j+p,n_j+q}$$

with

$$c(r_j,n_j,p,q) = \frac{1}{d} \frac{\binom{n_j}{r_j-1}}{(r_j+p)\binom{n_j+q}{r_j+p}}$$

for all elements of the sequence $((r_j,n_j))_{j \in \mathbb{N}}$ and assuming

$$(F^{-1}(t))^{\alpha-1}\left(d(F^{-1})'(t) - t^p(1-t)^{q-p-1}\right) \in L(0,1).$$

If $r \in \mathbb{N}$ is fixed (e.g., using sequences according to the lemma of Müntz, Szász, cf (2.3)) the constant d may depend on r. Applying another complete function sequence (see (2.6)), an appropriately modified integrability condition is required. The characterization result itself is less important in view of the strong assumptions. However, this approach leads to a class of distributions the elements of which are related by a similar recurrence relation for moments of order statistics. In this regard, the characterization set-up works as a method for a systematic treatment of recurrence relations.

The relation in (4.2) involves moments of orders α and $\alpha - 1$. A more general relation is shown in Kamps (1992a) with moments of order $\alpha + \beta$ on the lhs and a moment of order α on the rhs of the identity choosing $\alpha \in \mathbb{R}$, $\beta \geq 0$ and $\alpha + \beta \neq 0$. Characterizations can be derived by analogy with (4.3) using appropriate complete sequences of functions. Several distributions possess recurrence relations of this type such as Weibull, power function, Pareto, Burr XII, logistic distributions and logarithmic and reflected versions of them. Some equations are shown earlier in Khan et al. (1983) for Weibull distributions and in Khan, Khan (1987) for Burr XII distributions. For more details and examples we refer to Kamps (1992a) and to Kamps (1995, pp. 129–133).

A general identity for expectations of functions of order statistics is shown in Kamps, Mattner (1993). Let $F \in \mathscr{F}, r, n, p$ and q be integers with $2 \leq r \leq n$ and $1 \leq r + p \leq n + q$. Then the relation

(4.4) $\quad \mathrm{E}\big(g(X_{r,n}) - g(X_{r-1,n})\big) = c(r,n,p,q) \mathrm{E}g'(X_{r+p,n+q})$

holds true for every absolutely continuous function g provided that the rhs in (4.4) exists and with $c(r,n,p,q)$ as in (4.2).

The choice $g(x) = x^\alpha$ yields the equations in (4.2). Putting $g(x) = x^{\alpha/\beta+1}$ with $\alpha \in \mathbb{R}, \beta > 0, \alpha + \beta \neq 0$, and rewriting the results in terms of $Y = (\beta X)^{1/\beta}$ leads to relations for moments of orders $\alpha + \beta$ and α valid in a transformed, and by this enlarged, class of distributions as shown in Kamps (1992a). Putting similarly $g(x) = e^{\alpha x}$ and $Y = e^X$ yields identities involving moments of order α only. Choosing $g(x) = e^{itx}$, $t \in \mathbb{R}$, we obtain

$$\varphi_{r,n}(t) - \varphi_{r-1,n}(t) = cit\varphi_{r+p,n+q}(t) ,$$

where $\varphi_{j,n}$ denotes the characteristic function of $X_{j,n}$. Such identities for characteristic functions may be useful since simple explicit expressions are not available in general. Explicit expressions are available for power function distributions (Malik 1967), Pareto distributions (Malik 1966), and for the logistic distribution (Balakrishnan, Cohen 1991, p. 38).

Characterization results based on (4.4) can easily be obtained. Let F be strictly increasing on its support and p, q and d fixed. If (4.4) holds for some fixed function g, with $g'(x) \neq 0$ except for isolated points, and all r and n, then F is necessarily contained in $\mathscr{F}$. In fact, not all r and n are needed, applying certain complete sequences of functions as in (4.3) (in (4.3): $g(x) = x^\alpha$). If instead (4.4) holds for a fixed pair (r,n) and for a sufficiently rich class of functions g, then $F \in \mathscr{F}$ follows again.

Another parametrized recurrence relation for a class of distributions including Cauchy and doubly truncated Cauchy distributions as well as a corresponding characterization result are shown in Kamps (1995, pp. 134–137).

Finally, we cite further results of Lin (1988b) and Dimaki, Xekalaki (1993). For power function distributions, Lin (1988b) derives three characterizations by identities for moments of order statistics.

(4.5)   (Lin 1988b) Let $r, p, q \in \mathbb{N}, E|X|^\alpha < \infty$ for some $\alpha \in \mathbb{N}, \alpha \geq p$, and let $(n_j)_{j \in \mathbb{N}}$ be a Müntz–Szász sequence (cf (2.3)). Then for a given constant $\lambda > 0$, the following statements are equivalent:

i) $F(x) = (x/\lambda)^{p/q}, x \in (0, \lambda)$,
ii) $EX_{r,n_j}^\alpha = \lambda^p EX_{r+q,n_j+q}^{\alpha-p}$ for all $n_j \geq r$,
iii) $EX_{r,n_j}^\alpha = \lambda^p \left(EX_{r+q-1,n_j+q-1}^{\alpha-p} - EX_{r+q-1,n_j+q}^{\alpha-p} \right)$ for all $n_j \geq r$,
iv) $EX_{r,n_j}^\alpha - EX_{r,n_j+1}^\alpha = \lambda^p EX_{r+q+1,n_j+q+1}^{\alpha-p}$ for all $n_j \geq r$.

Dimaki, Xekalaki (1993) show characterizations of Pareto distributions with distribution functions

$$F(x) = 1 - vx^{-\alpha}, \quad v \leq x < \infty, \quad v > 0, \quad \alpha > 0$$

via

(4.6) i) $EX_{i+1,n}^2 - EX_{i,n}^2 = \frac{2}{\alpha(n-i)} EX_{i+1,n}^2$ for all $i \in \mathbb{N}$ and $n \geq i+1$, and
ii) $EX_{i+1,n}^2 - EX_{i,n-1}^2 = \frac{2}{n\alpha} EX_{i+1,n}^2$ for all $i \in \mathbb{N}$ and $n \geq i+1$.

The identity in (4.6) i) is shown in Kamps (1992a) for moments of arbitrary order. The relation (4.6) ii) is obtained from (4.6) i) by using (1.2).

5. Characterizations based on a single identity

In the previous sections, complete function sequences have been used to obtain characterization results based on identities for moments of order statistics. In other words, the validity of infinitely many identities has been required. Other characterizations can be found which are based on a single relation. We cite a result of Too, Lin (1989) and refer to characterizing relations which arise from characterizations via identically distributed functions of order statistics. Most important, there is a variety of results on inequalities for moments of order statistics. Arnold, Balakrishnan (1989) present an excellent annotated compendium of such results. We also refer to Rychlik's contribution to this volume (cf Chapter 4). If it is possible to characterize equality in some inequality for moments, we obtain a characterization of a probability distribution. In contrast to the ones deduced by means of complete function sequences, characterization results are derived under mild conditions and, simultaneously, recurrence relations are obtained for those distributions characterized by equality. We now cite only some of these characterizations.

First results on bounds for moments of order statistics date back to Plackett (1947) and Moriguti (1951) and are then generalized by Gumbel (1954) and Hartley, David (1954). We have:

Let $EX_1 = 0$ and $EX_1^2 = 1$.
Then

$$EX_{n,n} \leq \frac{n-1}{(2n-1)^{1/2}},$$

and we find equality iff F is a special power function distribution:

$$F(x) = \left(\frac{1+bx}{n}\right)^{1/(n-1)}, \quad b = \frac{n-1}{(2n-1)^{1/2}},$$

$$x \in \left(-\frac{(2n-1)^{1/2}}{n-1}, (2n-1)^{1/2}\right).$$

Lin (1988a) applies the Cauchy–Schwarz inequality to representations of moments of order statistics and record values to obtain characterization theorems for uniform and exponential distributions, respectively.

(5.1)   (Lin 1988a) If $EX^2 < \infty$ and $2 \leq r \leq n$, then

$$(EX_{r,n})^2 \leq \frac{rn}{(r-1)(n+1)} EX_{r-1,n-1}^2$$

with equality iff F is the distribution function of a degenerate distribution at 0 or of a uniform distribution on $(0, c)$ for some $c > 0$.

Thus, results of this type, i.e., characterizing equality in an inequality for moments, lead to bounds for certain moments of order statistics as well as to new recurrence relations for moments with respect to the distributions characterized by equality.

In particular, we observe that fixing only two or three moments is sufficient to determine the corresponding distributions uniquely; this fact is pointed out by Lin (1988a) (cf (5.1) and (5.5)).

Theorem (5.1) directly implies the example (5.5) of Too, Lin (1989) which can also be deduced from (5.4). The result is remarkable in view of the variety of characterization theorems (see Sections 2, 3, 4) in which assumptions are imposed on a sequence of moments.

The results of Lin (1988a) are taken up in Gajek, Gather (1991) and Kamps (1991a) and are generalized with respect to appearing powers and indices of order statistics and records applying Hölder's inequality and its inverse version (see Mitrinović 1970, p. 54, Beckenbach, Bellman 1961, p. 21/2).

(5.2)   (Gajek, Gather 1991) Let F be non-degenerate, $\alpha_1, \alpha_2, \in \mathbb{R}, \alpha = \alpha_1 + \alpha_2$, $p_1, p_2 \in \mathbb{R}\setminus\{0, 1\}, n, n_1, n_2, r, r_1, r_2 \in \mathbb{N}, 1 \leq r_1 \leq n_1, 1 \leq r_2 \leq n_2, 1 \leq r \leq n$ such that $p_1^{-1} + p_2^{-1} = 1, n = n_1 p_1^{-1} + n_2 p_2^{-1}$ and $r = r_1 p_1^{-1} + r_2 p_2^{-1}$. Then we have for $p_1 > 1$:

$$\left(r\binom{n}{r}\right)^{-1} E|X_{r,n}|^\alpha \leq \left(\left(r_1\binom{n_1}{r_1}\right)^{-1} E|X_{r_1,n_1}|^{\alpha_1 p_1}\right)^{1/p_1}$$

$$\times \left(\left(r_2\binom{n_2}{r_2}\right)^{-1} E|X_{r_2,n_2}|^{\alpha_2 p_2}\right)^{1/p_2}.$$

The inequality holds with the reverse sign if $0 < p_1 < 1$ or $p_1 < 0$ and it is generally assumed that the upper bound is finite.

The equality sign holds iff for some constant $c > 0$

$$|F^{-1}(t)|^{\alpha_1 p_1 - \alpha_2 p_2} = c t^{r_2 - r_1}(1-t)^{n_2 - r_2 - n_1 + r_1} \quad \text{for all } t \in (0, 1).$$

For several examples and particular cases of (5.2), e.g., $r_1 = r_2$ and $n_2 - r_2 = n_1 - r_1$, we refer to Gajek, Gather (1991). A similar result involving two moments of order statistics is shown in Kamps (1991a).

(5.3)   (Kamps 1991a) Let F be non-degenerate, $\alpha > 0, p > 1, 1 \leq r \leq n, 1 \leq i \leq j \leq n$ satisfying $n - r = j - i, F^{-1}(0+) \geq 0$ and $EX^{\alpha p} < \infty$. Then we have

$$EX_{r,n}^\alpha \leq \frac{n!}{(r-1)!}\left(\frac{j!}{(i-1)!}\right)^{-1/p}$$

$$\times \left(\Gamma\left(\frac{rp-i}{p-1}\right)\left(\Gamma\left(\frac{(n+1)p-(j+1)}{p-1}\right)\right)^{-1}\right)^{1-1/p} (EX_{i,j}^{\alpha p})^{1/p}.$$

In the case $i < r$ we have equality iff F is the distribution function of a power function distribution:

$$F(x) = c^{-(p-1)/(p(r-i))} x^{\alpha(p-1)/(r-i)}, \quad x \in (0, c^{1/\alpha p}) \text{ for some } c > 0 .$$

More details and further results using Jensen's inequality and inequalities of Diaz, Metcalf and Póeya, Szegö can be found in Kamps (1991a) and Kamps (1995, Chapter IV). Moreover, results of this type are valid for generalized order statistics (cf Kamps 1995, Chapter IV) such that related inequalities hold true in several models of ordered random variables including order statistics and record values as particular cases.

Too, Lin (1989) show characterizing recurrence relations for moments of order statistics and record values. If the existence of the appearing moments is ensured, then we have for $p \in \mathbb{N}$:

(5.4)   (Too, Lin 1989)

$$\left(r\binom{n}{r}\right)^{-1} EX_{r,n}^2 - 2\left((r+p)\binom{n+p}{r+p}\right)^{-1} EX_{r+p,n+p}$$
$$+ \left((r+2p)\binom{n+2p}{r+2p}\right)^{-1} = 0$$

iff F is the distribution function of a power function distribution with $F(x) = x^{1/p}, x \in (0, 1)$.

The above theorem states that, without any further assumptions, two particular moments of order statistics characterize the underlying distribution function, if the identity in (5.4) is satisfied. The interesting and remarkable special case (5.5) is a simple consequence of (5.4).

(5.5)   (Too, Lin 1989) $2EX^2 = EX_{2,2} = \frac{2}{3}$ iff $F(x) = x$, $x \in (0, 1)$.

Finally, we point out that characterizing moment relations can also be found in Chapter 9 of this volume (cf Gather, Kamps and Schweitzer, Chapter 9 (2.3.3), (2.5.2)). Occasionally it is possible to weaken the assumption in characterization results based on identically distributed functions of order statistics. It may be sufficient to consider corresponding moment equations. $F \sim \exp(\lambda)$ denotes that F is the distribution function of an exponential distribution with parameter $\lambda > 0$: $F(x) = 1 - e^{-\lambda x}, x > 0$.

(5.6)   (Ahsanullah 1981, cf Chapter 9 (2.3.3)) Let F be absolutely continuous, $\text{supp}(F) = (0, \infty)$ and IFR or DFR. Moreover, let $EX_1 < \infty$. Then $F \sim \exp(\lambda)$ iff there exists a pair (r, n), $2 \leq r \leq n$, such that $ED_{r,n} = ED_{r-1,n}$ with $D_{1,n} = nX_{1,n}$ and $D_{r,n} = (n - r + 1)(X_{r,n} - X_{r-1,n}), 2 \leq r \leq n$.

(5.7)   (Iwińska 1986, Gajek, Gather 1989, cf Chapter 9 (2.5.2)) Let F be absolutely continuous, $F^{-1}(0+) = 0$, strictly increasing on $(0, \infty)$ and NBU or NWU. Then $F \sim \exp(\lambda)$ iff there exists a triple (r, s, n), $1 \leq r < s \leq n$, with

$$EX_{s,n} - EX_{r,n} = EX_{s-r,n-r} .$$

6. Characterizations of normal and other distributions by product moments

In the previous sections we dealt with recurrence relations and identities for single moments of order statistics. There is also a variety of relations for product moments of order statistics from arbitrary and specific distributions. Surveys of such assertions are given in Balakrishnan et al. (1988), Malik et al. (1988) and Arnold, Balakrishnan (1989). We now review related characterization results of Govindarajulu (1966) and Lin (1989b). Some of Govindarajulu's (1966) characterizations of normal distributions are generalized by Lin (1989b). For further details on moments of order statistics from normal distributions we refer to David (1981, Chapter 3.2) and Arnold et al. (1992, Chapter 4.9).

Let F be non-degenerate, $EX^2 < \infty$ and let

$$\Phi(x) = (2\pi)^{-1/2} \int_{-\infty}^{x} \exp(-y^2/2)\, dy, \quad x \in \mathbb{R},$$

be the distribution function of the standard normal distribution.

(6.1) (Govindarajulu 1966)

$$EX_{n,n}^2 - E(X_{n-1,n}X_{n,n}) = 1 \quad \text{for all } n \geq 2$$

iff there exists $A \in [-\infty, \infty)$ such that $F(x) = \frac{\Phi(x) - \Phi(A)}{1 - \Phi(A)}$, $x \in (A, \infty)$.

Replacing X_i by $-X_i$ in (6.1), $1 \leq i \leq n$, we obtain

(6.2) (Govindarajulu 1966)

$$EX_{1,n}^2 - E(X_{1,n}X_{2,n}) = 1 \text{ for all } n \geq 2$$

iff there exists $B \in (-\infty, \infty]$ such that $F(x) = \frac{\Phi(x)}{\Phi(B)}$, $x \in (-\infty, B)$.

(6.3) (Govindarajulu 1966) Let $F(0) = 0$.

$$\sum_{j=1}^{n} E(X_{1,n}X_{j,n}) = 1 \quad \text{for all } n \geq 2 \quad \text{iff} \quad F(x) = 2\Phi(x) - 1, \quad x \in (0, \infty).$$

Replacing X_i by $-X_i$ in (6.3), $1 \leq i \leq n$, leads to

(6.4) (Govindarajulu 1966) Let $F(0) = 1$.

$$\sum_{j=1}^{n} E(X_{j,n}X_{n,n}) = 1 \quad \text{for all } n \geq 2 \quad \text{iff} \quad F(x) = 2\Phi(x)\ 1, \quad x \in (-\infty, 0).$$

(6.5) (Govindarajulu 1966) Let $EX = 0$.

$$\sum_{j=1}^{n} E(X_{i,n}X_{j,n}) = 1 \quad \text{for some } i \in \mathbb{N} \text{ and all } n \geq i$$

$$\text{iff} \quad F(x) = \Phi(x), \quad x \in \mathbb{R}.$$

The assertion (6.5) can be formulated in terms of covariances since $EX = 0$.

(6.6) (Govindarajulu 1966) Let $EX = 0$.
$$\sum_{j=1}^{n} \text{Cov}(X_{i,n}, X_{j,n}) = 1 \text{ for some } i \in \mathbb{N} \text{ and all } n \geq i$$
$$\text{iff} \quad F(x) = \Phi(x), \quad x \in \mathbb{R}.$$

On applying the representation
$$E(X_{r,n}^j X_{s,n}^k) = \frac{n!}{(r-1)!(s-r-1)!(n-s)!}$$
$$\times \int_0^1 \int_u^1 (F^{-1}(u))^j (F^{-1}(v))^k u^{r-1}(v-u)^{s-r-1}(1-v)^{n-s}\,dv\,du$$
$$1 \leq r < s \leq n, \quad j,k \in \mathbb{N}_0 \,,$$

for product moments of order statistics from arbitrary distributions, Lin (1989b) derives characterizations of uniform, exponential and normal distributions. The results use the Müntz–Szász theorem (cf (2.3)); thus the following assumption is made.

(6.7) Let $(n_i)_{i \in \mathbb{N}}$ be a sequence of integers satisfying
$$2 \leq n_1 < n_2 < \cdots \quad \text{and} \quad \sum_{i=1}^{\infty} \frac{1}{n_i} = \infty \,.$$

The first result of Lin (1989b) characterizes uniform distributions via a relation between single and product moments.

(6.8) (Lin 1989b) Let $E|X|^\alpha < \infty$ for some $\alpha \geq 1$, $F^{-1}(x) > 0$ for $x \in (0,1)$ and $F^{-1}(0+) = 0$. Moreover, let (6.7) be given and let $\lambda > 0$. Then $F \sim R[0, \lambda]$ (i.e., $F(x) = x/\lambda$, $x \in (0, \lambda)$) iff there exists a tuple (j,k), $j,k \in \mathbb{N}_0, j+k+1 \leq \alpha$, such that
$$E\left(X_{1,n_i}^j X_{2,n_i}^k\right) = \frac{1}{\lambda(j+1)} n_i E X_{1,n_i-1}^{j+k+1} \quad \text{for all } i \in \mathbb{N} \,.$$

In the special case $j = k = 0$ the characterizing property in (6.8) reduces to $EX_{1,n_i-1} = \lambda/n_i$ for all $i \in \mathbb{N}$ which is a well known characterization of uniform distributions (see (2.1), Galambos, Kotz 1978, p. 55, and Gather, Kamps and Schweitzer, Chapter 9).

(6.9) (Lin 1989b) Let F^{-1} be absolutely continuous on $(0,1)$, $E|X|^\alpha < \infty$ for some $\alpha \geq 1$, $F^{-1}(0) = 0, F^{-1}(x) > 0$ for $x \in (0,1)$. Moreover, let (6.7) be given and let $\lambda > 0$. Then $F \sim R[0, \lambda]$ iff there exists a triple $(r,j,k), r,k \in \mathbb{N}$, $j \in \mathbb{N}_0, j+k \leq \alpha$ such that
$$E\left(X_{r,n_i}^j X_{r+1,n_i}^k\right) - E X_{r,n_i}^{j+k} = \frac{k\lambda}{n_i+1} E\left(X_{r,n_i+1}^j X_{r+1,n_i+1}^{k-1}\right) \quad \text{for all } n_i \geq r+1 \,.$$

With the same lhs as in (6.9), Lin (1989b) also obtains a characterization of exponential distributions.

(6.10) (Lin 1989b) Let the conditions of (6.9) be given. Then $F \sim \exp(\lambda)$ iff there exists a triple (r, j, k), $r, k \in \mathbb{N}, j \in \mathbb{N}_0, j + k \leq \alpha$ such that

$$E\left(X_{r,n_i}^j X_{r+1,n_i}^k\right) - EX_{r,n_i}^{j+k} = \frac{k}{\lambda(n_i - r)} E\left(X_{r,n_i}^j X_{r+1,n_i}^{k-1}\right) \text{ for all } n_i \geq r + 1 \ .$$

In the special case $j = 0$, (6.9) and (6.10) simplify to the relations between single moments of order statistics as shown in Lin (1988b) (cf Section 4). Moreover, Lin deals with an extended normal distribution with distribution function

$$\Phi_k(x) = \left(\int_{-\infty}^{\infty} \exp\left(-\frac{t^{2k}}{2k}\right) dt\right)^{-1} \int_{-\infty}^{x} \exp\left(-\frac{t^{2k}}{2k}\right) dt, \quad x \in \mathbb{R}, \ k \in \mathbb{N} \ .$$

The following theorem extends Govindarajulu's (1966) characterization of truncated normal distributions (cf (6.1)).

(6.11) (Lin 1989b) Let $EX^{2k} < \infty$ for some $k \in \mathbb{N}$ and let (6.7) be given. Then $EX_{n_i,n_i}^{2k} - E(X_{n_i-1,n_i} X_{n_i,n_i}^{2k-1}) = 1$ for all $i \in \mathbb{N}$ iff

$$F(x) = \frac{\Phi_k(x) - \Phi_k(A)}{1 - \Phi_k(A)}, \quad x \in (A, \infty) \text{ for some } A \geq -\infty \ .$$

The following result generalizes a characterization of the standard normal distribution due to Govindarajulu (1966) (cf. 6.5)).

(6.12) (Lin 1989b) Let $EX^{2k} < \infty$ for some $k \in \mathbb{N}$ and let (6.7) be given. Moreover, let $EX^{2k-1} = 0, F^{-1}$ differentiable on $(0, 1)$ and $r \in \mathbb{N}$. Then

$$\sum_{s=1}^{n_i} E\left(X_{r,n_i} X_{s,n_i}^{2k-1}\right) = 1 \quad \text{for all } n_i \geq r + 1 \quad \text{iff} \quad F(x) = \Phi_k(x), \quad x \in \mathbb{R} \ .$$

In addition, Lin (1989b) presents a modification and extension of Govindarajulu's (1966) result (6.3) based on the identity in (6.12) for normal distributions truncated on the left at zero.

(6.13) (Lin 1989b) Let $EX^{2k} < \infty$ for some $k \in \mathbb{N}$ and let (6.7) be given. Moreover, let $F^{-1}(0+) = 0$ and F^{-1} differentiable on $(0, 1)$. Then

$$\sum_{s=1}^{n_i} E\left(X_{1,n_i} X_{s,n_i}^{2k-1}\right) = 1 \quad \text{for all } i \in \mathbb{N} \quad \text{iff} \quad F(x) = 2\Phi_k(x) - 1, \quad x > 0 \ .$$

Finally, we refer to two recent papers. Khan (1995) applies a Müntz–Szász sequence to derive characterizations of power function and Pareto distributions via recurrence relations for product moments of order statistics. By analogy with (4.2), Mohie El-Din et al. (1996) present a general identity for the product moments of order statistics where characterizations may be obtained as in (4.3).

References

Ahsanullah, M. (1981). On characterizations of the exponential distribution by spacings. *Statist. Hefte* **22**, 316–320.

Ahsanullah, M. and M. Rahman (1972). A characterization of the exponential distribution. *J. Appl. Prob.* **9**, 457–461.

Arnold, B. C. and N. Balakrishnan (1989). *Relations, Bounds and Approximations for Order Statistics.* Springer, Berlin.

Arnold, B. C., N. Balakrishnan and H. N. Nagaraja (1992). *A First Course in Order Statistics.* Wiley, New York.

Arnold, B. C. and G. Meeden (1975). Characterization of distributions by sets of moments of order statistics. *Ann. Statist.* **3**, 754–758.

Azlarov, T. A. and N. A. Volodin (1986). *Characterization Problems Associated with the Exponential Distribution.* Springer, New York.

Balakrishnan, N. (1987). A note on moments or order statistics from exchangeable variates. *Commun. Statist. – Theory Meth.* **16**, 855–861.

Balakrishnan, N., S. M. Bendre and H. J. Malik (1992). General relations and identities for order statistics from non-independent non-identical variables. *Ann. Inst. Statist. Math.* **44**, 177–183.

Balakrishnan, N. and A. C. Cohen (1991). *Order Statistics and Inference: Estimation Methods.* Academic Press, Boston.

Balakrishnan, N., H. J. Malik and S. E. Ahmed (1988). Recurrence relations and identities for moments of order statistics, II: Specific continuous distributions. *Commun. Statist. – Theory Meth.* **17**, 2657–2694.

Balasubramanian, K. and N. Balakrishnan (1993). Duality principle in order statistics. *J. Roy. Statist. Soc. B* **55**, 687–691.

Balasubramanian, K., N. Balakrishnan and H. J. Malik (1994). Identities for order statistics from non-independent non-identical variables. *Sankhyā B* **56**, 67–75.

Balasubramanian, K. and R. B. Bapat (1991). Identities for order statistics and a theorem of Renyi. *Statist. & Prob. Lett.* **12**, 141–143.

Balasubramanian, K. and M. I. Beg (1992). Distributions determined by conditioning on a pair of order statistics. *Metrika* **39**, 107–112.

Beckenbach, E. F. and R. Bellman (1961). *Inequalities.* Springer, Berlin.

Beg, M. I. and K. Balasubramanian (1990). Distributions determined by conditioning on a single order statistic. *Metrika* **37**, 37–43.

Beg, M. I. and S. N. U. A. Kirmani (1974). On a characterization of exponential and related distributions. *Austral. J. Statist.* **16**, 163–166. Correction (1976), *Austral. J. Statist.* **18**, 85.

Boas, R. P. (1954). *Entire Functions.* Academic Press, New York.

Chan, L. K. (1967). On a characterization of distributions by expected values of extreme order statistics. *American Math. Monthly* **74**, 950–951.

Cole, R. H. (1951). Relations between moments of order statistics. *Ann. Math. Statist.* **22**, 308–310.

David, H. A. (1981). *Order Statistics.* 2nd ed., Wiley, New York.

David, H. A. (1993). A note on order statistics for dependent variates. *Amer. Statist.* **47**, 198–199.

David, H. A. and P. C. Joshi (1968). Recurrence relations between moments of order statistics for exchangeable variates. *Ann. Math. Statist.* **39**, 272–274.

Dimaki, C. and E. Xekalaki (1993). Characterizations of the Pareto distribution based on order statistics. In: V. V. Kalashnikov and V.M. Zolotarev, eds., *Stability Problems for Stochastic Models.* Springer, Berlin, 1–16.

Ferguson, T. S. (1967). On characterizing distributions by properties of order statistics. *Sankhyā Ser. A* **29**, 265–278.

Gajek, L. and U. Gather (1989). Characterizations of the exponential distribution by failure rate and moment properties of order statistics. In: J. Hüsler and R. D. Reiss, eds., *Extreme Value Theory.* Springer, Berlin, 114–124.

Gajek, L. and U. Gather (1991). Moment inequalities for order statistics with applications to characterizations of distributions. *Metrika* **38**, 357–367.

Galambos, J. (1975). Characterization of probability distributions by properties of order statistics I. In: Patil, G. P. et al., eds., *Statistical Distributions in Scientific Work*, Vol. 3. Reidel, Dordrecht, 71–88.

Galambos, J. and S. Kotz (1978). *Characterizations of Probability Distributions*. Springer, New York.

Govindarajulu, Z. (1963). On moments of order statistics and quasi-ranges from normal populations. *Ann. Math. Statist.* **34**, 633–651.

Govindarajulu, Z. (1966). Characterization of normal and generalized truncated normal distributions using order statistics. *Ann. Math. Statist.* **37**, 1011–1015.

Govindarajulu, Z. (1975). Characterization of the exponential distribution using lower moments of order statistics. In: Patil, G. P. et al., eds., *Statistical Distributions in Scientific Work*, Vol. 3. Reidel, Dordrecht, 117–129.

Govindarajulu, Z., J. S. Huang and A. K. Md. E. Saleh (1975). Expected value of the spacings between order statistics. In: Patil, G. P. et al., eds., *Statistical Distributions in Scientific Work*, Vol. 3. Reidel Dordrecht, 143–147.

Gumbel, E. J. (1954). The maxima of the mean largest value and of the range. *Ann. Math. Statist.* **25**, 76–84.

Hartley, H. O. and H. A. David (1954). Universal bounds for mean range and extreme observation. *Ann. Math. Statist.* **25**, 85–99.

Hoeffding, W. (1953). On the distribution of the expected values of the order statistics. *Ann. Math. Statist.* **24**, 93–100.

Huang, J. S. (1974a). On a theorem of Ahsanullah and Rahman. *J. Appl. Prob.* **11**, 216–218.

Huang, J. S. (1974b). Characterizations of the exponential distribution by order statistics. *J. Appl. Prob.* **11**, 605–608.

Huang, J. S. (1975). Characterization of distributions by the expected values of the order statistics. *Ann. Inst. Statist. Math.* **27**, 87–93.

Huang, J. S. (1989). Moment problem of order statistics: A review. *Internat. Statist. Rev.* **57**, 59–66.

Huang, J. S. and J. S. Hwang (1975). L_1-completeness of a class of beta densities. In: Patil, G. P. et al., eds., *Statistical Distributions in Scientific Work*, Vol. 3. Reidel, Dordrecht, 137–141.

Hwang, J. S. (1978). A note on Bernstein and Müntz–Szasz theorems with applications to the order statistics. *Ann. Inst. Statist. Math.* **30** A, 167–176.

Hwang, J. S. (1983). On a generalized moment problem. *Proc. American. Math. Soc.* **87**, 88–89.

Hwang, J. S. and G. D. Lin (1984a). Characterizations of distributions by linear combinations of moments of order statistics. *Bulletin of the Institute of Mathematics, Academia Sinica* **12**, 179–202.

Hwang, J. S. and G. D. Lin (1984b). On a generalized moment problem II. *Proc. Amer. Math. Soc.* **91**, 577–580.

Iwińska, M. (1986). On the characterizations of the exponential distribution by order statistics and record values. *Fasciculi Mathematici* **16**, 101–107.

Kadane, J. B. (1974). A characterization of triangular arrays which are expectations of order statistics. *J. Appl. Prob.* **11**, 413–416.

Kamps, U. (1990). Characterizations of the exponential distribution by weighted sums of i.i.d random variables. *Statistical Papers* **31**, 233–237.

Kamps, U. (1991a). Inequalities for moments of order statistics and characterizations of distributions. *J. Statist. Plan. Inf.* **27**, 397–404.

Kamps, U. (1991b). A general recurrence relation for moments of order statistics in a class of probability distributions and characterizations. *Metrika* **38**, 215–225.

Kamps, U. (1992a). Identities for the difference of moments of successive order statistics and record values. *Metron* **50**, 179–187.

Kamps, U. (1992b). Characterizations of the exponential distribution by equality of moments. *Allgemeines Statistisches Archiv* **76**, 122–127.

Kamps, U. (1995). *A Concept of Generalized Order Statistics*. Teubner, Stuttgart.

Kamps, U. and L. Mattner (1993). An identity for expectations of functions of order statistics. *Metrika* **40**, 361–365.

Khan, A. H. (1995). Characterizations of the power function and the Pareto distributions through moments of order statistics. Preprint.

Khan, A. H. and M. S. Abu-Salih (1989). Characterizations of probability distributions by conditional expectation of order statistics. *Metron* **47**, 171–181.

Khan, A. H. and M. I. Beg (1987). Characterization of the Weibull distribution by conditional variance. *Sankhyā Ser. A* **49**, 268–271.

Khan, A. H. and I. A. Khan (1987). Moments of order statistics from Burr distribution and its characterizations. *Metron* **45**, 21–29.

Khan, A. H., M. Yaqub and S. Parvez (1983). Recurrence relations between moments of order statistics. *Nav. Res. Log. Quart.* **30**, 419–441. Corrigendum **32**, 693 (1985).

Konheim, A. G. (1971). A note on order statistics. *Amer. Math. Monthly* **78**, 524.

Kotz, S. (1974). Characterizations of statistical distributions: a supplement to recent surveys. *Internat. Statist. Rev.* **42**, 39–65.

Lin, G. D. (1988a). Characterizations of uniform distributions and of exponential distributions. *Sankhyā Ser. A* **50**, 64–69.

Lin, G. D. (1988b). Characterizations of distributions via relationships between two moments of order statistics. *J. Statist. Plan. Inf.* **19**, 73–80.

Lin, G. D. (1989a). Characterizations of distributions via moments of order statistics: A survey and comparison of methods. In: Y. Dodge, ed., *Statistical Data Analysis and Inference*. North-Holland, Amsterdam, 297–307.

Lin, G. D. (1989b). The product moments of order statistics with applications to characterizations of distributions. *J. Statist. Plan. Inf.* **21**, 395–406.

Madreimov, I. and Y. I. Petunin (1983). A characterization of the uniform distribution with the aid of order statistics. *Theor. Prob. and Math. Statist.* **27**, 105–110.

Malik, H. J. (1966). Exact moments of order statistics from the Pareto distribution. *Skand. Aktuarietidskrift* **49**, 144–157.

Malik, H. J. (1967). Exact moments of order statistics from a power-function distribution. *Skand. Aktuarietidskrift* **50**, 64–69.

Malik, H. J., N. Balakrishnan and S. E. Ahmed (1988). Recurrence relations and identities for moments of order statistics, I: Arbitrary continuous distribution. *Commun. Statist. – Theory Meth.* **17**, 2623–2655.

Mallows, C. L. (1973). Bounds on distribution functions in terms of expectations of order statistics. *Ann. Prob.* **1**, 297–303.

Melnick, E. L. (1964). Moments of ranked Poisson variates. M.S. Thesis, Virginia Polytechnic Institute.

Mitrinović, D. S. (1970). *Analytic Inequalities*. Springer, Berlin.

Mohie El-Din, M. M., S. E. Abu-Youssef and K. S. Sultan (1996). An identity for the product moments of order statistics. *Metrika* **44**, 95–100.

Mohie El-Din, M. M., M. A. W. Mahmoud and S.E. Abo-Youssef (1991). Moments of order statistics from parabolic and skewed distributions and a characterization of Weibull distribution. *Commun. Statist. – Simul.* **20**, 639–645.

Moriguti, S. (1951). Extremal property of extreme value distributions. *Ann. Math Statist.* **22**, 523–536.

Müntz, C. H. (1914). Über den Approximationssatz von Weierstraß. Festschrift für H. A. Schwarz, Berlin, 303–312.

Natanson, I. P. (1961). *Theorie der Funktionen einer reellen Veränderlichen*. Akademie-Verlag, Berlin.

Plackett, R. L. (1947). Limit of the ratio of mean range to standard deviation. *Biometrika* **34**, 120–122.

Pollak, M. (1973). On equal distributions. *Ann. Statist.* **1**, 180–182.

Pólya, G. and G. Szegö (1964). *Aufgaben und Lehrsätze aus der Analysis, Erster Band*. Springer, Berlin.

Rauhut, B. (1989). Characterization of probability distribution by conditioned moments of spacings. *Methods Oper. Res.* **58**, 571–579.

Saleh, A. K. MD. E. (1976). Characterization of distributions using expected spacings between consecutive order statistics. *J. Statist. Res.* **10**, 1–13.

Sathe, Y. S. and U. J. Dixit (1990). On a recurrence relation for order statistics. *Statist. & Prob. Lett.* **9**, 1–4.

Szász, O. (1916). Über die Approximation stetiger Funktionen durch lineare Aggregate von Potenzen. *Math. Ann.* **77**, 482–496.

Too, Y. H. and G. D. Lin (1989). Characterizations of uniform and exponential distributions. *Statist. & Prob. Lett.* **7**, 357–359.

Young, D. H. (1967). Recurrence relations between the P.D.F.'s of order statistics of dependent variables, and some applications. *Biometrika* **54**, 283–292.

PART V

Extremes and Asymptotics

Univariate Extreme Value Theory and Applications

Janos Galambos

1. Introduction

For a long period of time, probability theory was meant to interpret the laws of averages but even within this theory occurrences now associated with extreme values had been viewed as accidents or surprises without regular laws. This view has changed only quite recently, and the present article is devoted to describe the present stage of extreme value theory both from the mathematical and practical points of view. Because the subject matter described by the extremes is very sensitive to slight errors in approximations, more controversies can occur in interpretation of results than when dealing with averages. This fact will become clear in the paper. We shall also see that a number of mathematical results developed independently of extreme value theory gain significant practical applications; I would mention the fields of characterization of probability distribution and extensions of some Bonferroni-type inequalities. Although the present paper is not dealing with the above mentioned associated fields in detail, their significance will be demonstrated in a number of ways through examples.

Extreme value theory is mainly a model building tool, but it can also be utilized in statistical evaluations. It concerns the largest or the smallest in a set of random variables where the random variables in question are either actual observations or just hypothetical quantities for describing a model. Hence, extreme value theory is more than the study of the largest or smallest order statistics since values other than the extremes may become meaningless in certain situations: a spacecraft may be destroyed by the first failure of essential components. The following examples well demonstrate the variety of applied problems and the mathematical difficulties faced in extreme value theory.

EXAMPLE 1.1. (Fatigue failure). Let S be the random time to the failure of a sheet of metal used in constructing the body of an aircraft. Let us hypothetically divide the sheet into n smaller pieces and let X_j denote the strength (time to failure) of the jth piece in this division, where labelling is made by some predetermined rule. Then, by the weakest link principle (a chain breaks at its weakest link),

$$S = \min(X_1, X_2, \ldots, X_n)$$

where the X_j are similar in nature to S, n is chosen by us, but we have no control over the interdependence of the X_j. If n could be chosen arbitrarily large and if an acceptable model can be developed for the X_j, then a limiting distribution for the minimum in such a model should be the exact distribution of S.

EXAMPLE 1.2 (Warranty period). The manufacturer of a piece of equipment with a large number n of components wants to determine the warranty period in such a manner that, with high probability, no (expensive) component should fail during this period. That is, warranty period is determined by the distribution of the minimum of the time to failure of the components. Here, the components can have a variety of underlying distributions, and their interdependence may be very strong. Therefore, a single model cannot be expected to cover all structures.

EXAMPLE 1.3 (Statistical outliers). Assume that each of n terminally ill patients is diagnosed to have an expected life of one year. Can it be justified that one of these patients is still alive five years later? Indeed, and this is what is expected from extreme value theory. Here we face n independent and identically distributed random variables with expectation one, and we want to determine the distribution of their maximum. Since 'terminally ill' will lead to the unique underlying distribution $F(x) = 1 - e^{-x}, x \geq 0$, (a characterization theorem), the exact distribution of the maximum of n observations can easily be computed.

There are very few applied problems when the population distribution can accurately be determined via a characterization theorem. In all such cases an approximation to the population distribution should be avoided. Rather, an asymptotic distribution theory should replace exact formulas and statistical inference should be based on the asymptotic model for the extremes. This will be made very clear in the following sections.

In extreme value theory we usually face two sequences of random variables. The first sequence, which we denote by $X_1, X_2, \ldots, X_n$, is the one on which the extreme value model is based. These are some times statistical observations as in Example 1.3 and in other cases explicity present but unknown variables such as component lives in Example 1.2 or just hypothetical variables as in Example 1.1. By specifying an extreme value model we mean that we make distributional assumptions on the X_j (univariate and multivariate), and we develop asymptotic distributions for the extremes

$$W_n = \min(X_1, X_2, \ldots, X_n) \quad \text{and} \quad Z_n = \max(X_1, X_2, \ldots, X_n)$$

after some normalization. The distribution function of X_j is denoted by $F_j(x)$ but if they are known to be identically distributed then the common distribution function is $F(x)$. We set

$$L_n(x) = P(W_n \leq x) \quad \text{and} \quad H_n(x) = P(Z_n \leq x) \ .$$

If more variables than just W_n or Z_n become of interest, we use the standard notation of the present book for order statistics: $X_{1:n} \leq X_{2:n} \leq \cdots \leq X_{n:n}$ for

$X_1, X_2, \ldots, X_n$ in a nondecreasing order. In such cases, evidently $W_n = X_{1:n}$ and $Z_n = X_{n:n}$.

A second sequence $Y_1, Y_2, \ldots, Y_N$ of random variables in extreme value theory is a set of (usually independent) statistical observations on an extreme of a model, and their common distribution function therefore is an extreme value distribution (of the model in question). These observations are usually utilized to test the model (its extreme value distribution) and to estimate parameters.

EXAMPLE 1.4. Let Y be the age of the longest living European individual at the time of death of all those currently living. Let n be a large number and split the population of Europe into disjoint groups of n persons in each group. Assume that the number of groups is large as well. In a group, let $X_1, X_2, \ldots, X_n$ be the ages (at death) of the individuals of the group. Now, whatever model we impose on the X_j, if a limiting distribution of Z_n after some normalization exists then Z_n in a randomly selected group is an observation on Y. Since n is assumed large, the distribution of Z_n can be replaced by its asymptotic distribution (which we call an extreme value distribution), and upon selecting N groups at random, N observations $Y_1, Y_2, \ldots, Y_N$ obtain on Y. (Since the values Y_j are future values, one can only test the model in the form of $P(Y > A)$ for a variety of values of A, or, by adding some stationarity assumptions in time, values from the past are used for Z_n as Y_j.)

We now turn to a systematic description of the mathematical results of extreme value theory and their practical applications.

2. The classical models

In a classical model, $X_1, X_2, \ldots, X_n$ are independent and identically distributed (i.i.d.) random variables and their common distribution function is denoted by $F(x)$. We assume that there are sequences a_n and $b_n > 0$ such that, as $n \to +\infty$,

$$H_n(a_n + b_n z) = F^n(a_n + b_n z) \to H(z) \tag{2.1}$$

where $H(z)$ is a proper nondegenerate distribution function (since $H(z)$ turns out to be continuous, convergence is pointwise). Such a model is called a classical model for the maximum, and $H(z)$ is called an extreme value distribution (function) for the maximum (in a classical model). The distribution function $F(x)$ satisfying (2.1) is said to be in the domain of attraction of $H(z)$ which fact is expressed by the notation $F \in D(H)$.

Similar definitions apply to the minimum W_n. However, since

$$W_n = \min(X_1, X_2, \ldots, X_n) = -\max(-X_1, -X_2, \ldots, -X_n) \tag{2.2}$$

we make mathematical statements on Z_n only. We use (2.2) for transforming such statements to W_n when necessary.

Before proceeding further, note that, upon replacing z by $A + Bz$ and setting $a_n^* = a_n + b_n A$ and $b_n^* = b_n B$, where A and $B > 0$ are arbitrary constants, (2.1) becomes

$$H_n(a_n^* + b_n^* z) \to H(A + Bz) \qquad (2.1a)$$

That is, if $H(z)$ is an extreme value distribution (in a classical model) then so is the whole parametric family $H(A + Bz)$ and their domains of attraction are identical. The distribution functions $H(z)$ and $H(A + Bz)$ are said to have the same type, and the convergence in (2.1) is always understood to be a convergence to the type of $H(z)$.

From a classical theorem on the convergence to types (see Lemma 2 on p. 188 of Galambos (1995)) it follows that if $H(z)$ and $H^*(z)$ are two extreme value distributions which are not of the same type than $D(H)$ and $D(H^*)$ are disjoint. Furthermore, upon writing up (2.1) in the following two ways for a fixed m

$$F^{nm}(a_n + b_n z) \to H^m(z) \quad \text{and} \quad F^{nm}(a_{nm} + b_{nm} z) \to H(z)$$

we get from the cited Lemma 2 that an extreme value distribution $H(z)$ in (2.1) must satisfy the functional equation

$$H^m(A_m + B_m z) = H(z) \quad \text{all } z \text{ and all } m \geq 1 \qquad (2.3)$$

where A_m and $B_m > 0$ are suitable constants. Depending whether $B_m > 1$ or $B_m < 1$ or $B_m = 1$ for one $m > 1$ (and then for all $m > 1$) the solution of (2.3) is necessarily of the same type as

$$H_{1,\gamma}(z) = \begin{cases} \exp(-z^{-\gamma}) & \text{if } z > 0 \\ 0 & \text{otherwise} \end{cases} \qquad (2.4)$$

or

$$H_{2,\gamma}(z) = \begin{cases} 1 & \text{if } z > 0 \\ \exp(-(-z)^\gamma) & \text{if } z \leq 0 \end{cases} \qquad (2.5)$$

or

$$H_{3,0}(z) = \exp(-e^{-z}) \quad \text{all } z \qquad (2.6)$$

respectively. The parameter $\gamma > 0$ in both cases above. In turn, each of the above distributions is of the same type as

$$H_{(c)}(z) = \exp\left\{-(1 + cz)^{-1/c}\right\} \quad \text{if } 1 + cz > 0 \qquad (2.7)$$

if we adopt the convention that $H_{(0)}(z) = \lim H_{(c)}(z)$ as $c \to 0$ through values $c \neq 0$. This way, $H_{(c)}(z)$ is of the type of (2.6) if $c = 0$, while (2.4) or (2.5) obtains depending whether $c > 0$ or $c < 0$. We use alternatively (2.7) and (2.4) through (2.6) whichever is convenient for a particular result. The form $H_{(c)}(z)$ of (2.7) appears particularly convenient in statistical inference since a decision among the forms (2.4) through (2.6) is a matter of estimating the parameter c, more precisely, we just have to decide whether $c > 0$ or $c < 0$ or $c = 0$. This, however, is not as easy as it sounds and, in fact, this 'simple task' generates more controversy to extreme value theory than any other methodology. The problem and controversy stem from the fact that $H_{2,\gamma}(z)$ represents a random variable which is bounded from above while the other two types are unbounded. The reduction of this very critical problem of boundedness to deciding whether a parameter (c in (2.7)) is negative or not made the problem relatively simple. On the other hand, the conclusions of boundedness are not always acceptable which leads to the rejection of the classical model as appropriate in such circumstances. We shall return to specifics of this problem in the next section. Here, we continue the mathematical analysis of the classical models.

First, we define the endpoints of a distribution function as

$$\alpha(F) = \inf\{x\colon F(x) > 0\} \quad \text{and} \quad \omega(F) = \sup\{x\colon F(x) < 1\}$$

The following theorems on domains of attraction are due to Gnedenko (1943) who unified and generalized the scattered results of von Bortkiewicz (1922), Dodd (1923), von Mises (1923), Tippett (1925), Fréchet (1927), Fisher and Tippett (1928) and von Mises (1936). Gnedenko's (1943) work is very thorough and generally accepted as the foundation of extreme value theory. For a review of Gnedenko's work and its influence on extreme value theory over the past half century, see Galambos (1994).

THEOREM 2.1. $F \in D(H_{1,\gamma})$ if and only if, $\omega(F) = +\infty$ and for all $x > 0$, as $t \to +\infty$,

$$\lim \frac{1 - F(tx)}{1 - F(t)} = x^{-\gamma} \tag{2.8}$$

The normalizing constants a_n and b_n of (2.1) can always be chosen as $a_n = 0$ and $b_n = \inf\{x\colon 1 - F(x) \le 1/n\}$.

THEOREM 2.2. $F \in D(H_{2,\gamma})$ if and only if, $\omega(F) < +\infty$ and the distribution function $F^*(x) = F(\omega(F) - 1/x), x > 0$, belongs to $D(H_{1,\gamma})$. The normalizing constants a_n and b_n of (2.1) can be chosen as $a_n = \omega(F)$ and

$$b_n = \omega(F) - \inf\{x\colon 1 - F(x) \le 1/n\}$$

THEOREM 2.3. $F \in D(H_{3,0})$ if and only if, there is a function $u(t) > 0$ such that, for all real x, as $t \to \omega(F)$ with $t < \omega(F)$,

$$\lim \frac{1 - F(t + xu(t))}{1 - F(t)} = e^{-x} \tag{2.9}$$

The normalizing constants a_n and b_n of (2.1) can be chosen as $a_n = \inf\{x\colon 1 - F(x) \le 1/n\}$ and $b_n = u(a_n)$.

These three theorems, when combined with the earlier quoted result on the possible types of limiting distribution $H(z)$ in (2.1), forms the foundation of extreme value theory for the classical models. Recall (2.2) in order to restate all the above results for the minimum.

The theory is not complete in the above stated form. Theorem 2.3 has an unspecified function $u(t)$ in its formulation, there is no mention of the quality of approximation (speed of convergence, convergence of densities when exist) in any of the theorems, and the choice of the normalizing constants might not be the best or most convenient. Gnedenko himself raised these questions but only partial solutions were provided by him through examples rather than general statements. Gnedenko also initiated the extension of weak convergence of (2.1) to weak laws and strong laws which mean that, with some constants A_n, either Z_n/A_n converges to one or $Z_n - A_n$ converges to zero in probability and almost surely, respectively. All these problems have been addressed in the literature in the past three decades which lifted the classical theory to a well developed subject matter.

Before we can formulate newer results we have to introduce two concepts. The conditional expectation

$$R(t) = E(X - t | X > t), \alpha(F) < t < \omega(F) \tag{2.10}$$

where X is a random variable with distribution function $F(x)$, is called the expected residual life at age t. One can compute $R(t)$ by the formula

$$R(t) = \frac{1}{1 - F(t)} \int_t^{\omega(F)} [1 - F(y)] \, dy \tag{2.11}$$

Next, assume that $F(x)$ is differentiable. Then we define the failure rate or hazard rate of X or $F(x)$ by the limiting conditional instantaneous failure

$$r(t) = \lim \frac{1}{\Delta t} P(X \leq t + \Delta t | X > t) = \frac{F'(t)}{1 - F(t)} \tag{2.12}$$

where $\Delta t > 0$ and $\Delta t \to 0$. Let $\alpha(F) \leq m < \omega(F)$ be a fixed number, and let us integrate (2.12) starting at m. We get

$$g(x) = -\log(1 - F(x)) = \int_m^x r(t) \, dt + C \tag{2.13}$$

where $C = -\log(1 - F(m))$. The function $g(x)$ is known as the cumulative hazard function which, of course, is meaningful even if $F(x)$ is not differentiable. Evidently,

$$F(x) = 1 - \exp(-g(x)), \quad \alpha(F) \leq x < \omega(F) \tag{2.14}$$

The relations (2.13) and (2.14) entail that $r(t)$ uniquely determines $F(x)$. A similar statement is true in regard to $R(t)$ as well. The following argument is instructive in which we assume that $F(x)$ is continuous but only for the sake of allowing us to use Riemann integration and ordinary differentiation. The final conclusion is valid for all distribution functions. Now, for $F(x)$, define

$$F_{\text{int}}(x; C) = 1 - C \int_x^{\omega(F)} [1 - F(y)] \, dy \tag{2.15}$$

If we assume that $R(t)$ is finite for F, then C can be chosen so that $F_{\text{int}}(x; C)$ is a distribution function (we refer to F_{int} as an integral distribution function of F). If F is continuous, then F_{int} is differentiable, and we have

$$1/R_F(t) = r_{F_{\text{int}}}(t) \tag{2.16}$$

where the subscripts in R and r emphasize the respective underlying distribution. From the uniqueness statement at (2.14) we thus have that $R_F(t)$ uniquely determines F_{int} via (2.16). However, upon differentiating (2.15) the uniqueness of F itself follows. The mathematically oriented reader may want to note that the characterization of $F(x)$ by the second conditional residual moment $R_2(t) = E[(X - t)^2 | X > t]$ is much more difficult to prove but true. See Galambos and Hagwood (1992) for a proof as well as for further references on the subject matter.

With the introduction of $R(t)$ and $r(t)$ we can discuss the choice of $u(t)$ in (2.9). In all early examples in the literature (2.9) was demonstrated to be valid with $u(t) = 1/r(t)$. These examples were centered at the normal, exponential, gamma and some special cases of what is now known as Weibull distributions. One remarkable theorem of von Mises (1936) justified to view $u(t) = 1/r(t)$ as the main choice in (2.9). It is established in von Mises (1936) that if, for all x close to $\omega(F)$, the second derivative $F''(x)$ exists and $f(x) = F'(x) \neq 0$ then the limit

$$\lim \frac{d(1/r(x))}{dx} = 0 \quad (x \to \omega(F)) \tag{2.17}$$

entails (2.9) with $u(t) = 1/r(t)$ (in all conditions above, evidently $x < \omega(F)$). Let us record that (2.17), with the differentiation carried out, becomes ($x < \omega(F)$ and $x \to \omega(F)$)

$$\lim \frac{f'(x)[1 - F(x)]}{f^2(x)} = -1 \tag{2.17a}$$

Since $r(t)$ is not defined for all distribution functions $F(x)$, one may expect that perhaps the hazard function $g(x)$ of (2.13) and (2.14) may lead to a universal choice of $u(t)$ in (2.9). An attempt in this direction is the work of Marcus and Pinsky (1969) but their ultimate conclusion is that even if $r(t)$ is defined one cannot always choose $u(t) = 1/r(t)$ in (2.9). That is, (2.9) can be valid with some $u(t)$ but not with $1/r(t)$.

The universal choice of $u(t)$ by $R(t)$ was discovered by de Haan (1970). The proof of the theorem of von Mises, stated at (2.17), implicitly contains the fact that $R(t)$ could have been used in place of $1/r(t)$ in (2.9). That is, it follows from (2.17) that $R(t)$ is finite and

$$R(t)r(t) \to 1 \quad \text{as} \quad t \to \omega(F) \tag{2.18}$$

This is sufficient for replacing $1/r(t)$ by $R(t)$ in (2.9) since (2.9) is a convergence of distributions to exponentiality. The recognition that (2.18) follows from (2.17) suggests that the universal choice of $u(t)$ in (2.9) might be $R(t)$. This is indeed the case. The following result is due to de Haan (1970).

THEOREM 2.4. $F \in D(H_{3,0})$ if and only if, $R(t)$ is finite and (2.9) holds with $u(t) = R(t)$.

For special classes of distributions, covering most widely applied distributions in practice, Galambos and Xu (1990) established

THEOREM 2.5. Assume that $R(t)$ is regularly varying at infinity (implying that $\omega(F) = +\infty$). Then $F \in D(H_{3,0})$ if, and only if, $R(t)/t \to 0$ as $t \to +\infty$.

We give the simple form of definition of regular variation. $R(t)$ is said to be regularly varying at infinity with index δ if $R(t) = t^\delta s(t)$, where $s(t)$ satisfies

$$\lim \frac{s(\lambda t)}{s(t)} = 1, \quad \lambda > 0 \quad \text{fixed and} \quad t \to +\infty$$

In other words, $s(t)$ is slowly varying. The theorem of Galambos and Xu thus states that if $R(t)$ is regularly varying then either $\delta < 1$ or $\delta = 1$ and $s(t) \to 0$ as $t \to +\infty$ characterize $D(H_{3,0})$. Note that (2.9) does not have to be tested. In many cases, one simply gets $R(t) \sim t^\delta$ (such is the case of the normal distribution with $\delta = -1$, and for the exponential distribution $R(t) = $ constant, i.e., $\delta = 0$). A variant of Theorem 2.5 is also proved by Galambos and Xu (1990) for $\omega(F) < +\infty$. Since Gnedenko has shown that, whatever $u(t)$ in (2.9), $u(t)/t \to 0$ as $t \to +\infty$ for $\omega(F) = +\infty$, only the sufficiency part of Theorem 2.5 is new.

There is more to the limit relation (2.18) than just its allowing us to choose between $R(t)$ and $1/r(t)$ as a normalization at (2.9). It turns out (see Sweeting (1985)) that (2.18) is both necessary and sufficient for the densities to converge:

$$nb_n F^{n-1}(a_n + b_n z) f(a_n + b_n z) \to H'_{3,0}(z)$$

locally uniformly, assuming that (2.1) holds with $H = H_{3,0}$. Sweeting (1985) establishes similar results for the other two domains of attraction as well. Since a result of Galambos and Obretenov (1987) states that if $F \in D(H_{3,0})$ and if $r(t)$ is monotonic then (2.18) holds, we have that for monotonic $r(t)$ we can use $1/r(t)$ as normalization in (2.9) and we can freely switch from convergence of distribution functions to convergence of densities. The investigations concerning (2.18) led to a better understanding of the von Mises condition (2.17) as well. Pickands (1986) showed that (2.17) is both necessary and sufficient for (2.1) and its first and second derivative variants to hold with $H = H_{3,0}$.

There is a large number of other results on domains of attraction in which $r(t)$ plays an important role. See Section 2.7 in Galambos (1987); several of the results appeared there for the first time. See also this book by the present author for a very extensive bibliography.

While Theorems 2.1–2.3 give specific instructions on finding normalizing constants a_n and b_n for (2.1) to hold, these choices are not the only possibilities. However, there is nothing particular to extreme value theory on the extent to which one can modify one set of normalizing constants without violating (2.1). From a general rule on (weak) convergence of probability distributions we have that if (a_n, b_n) and (a_n^*, b_n^*) are two sets of normalizing constants in (2.1) then we must have

$$\lim(a_n - a_n^*)/b_n = 0 \quad \text{and} \quad \lim b_n^*/b_n = 1 \, (n \to +\infty) \qquad (2.19)$$

Conversely, if (2.19) holds and (2.1) holds with (a_n, b_n) then (2.1) remains to hold with (a_n^*, b_n^*). There are two significant consequences of the relative freedom of choosing normalizing constants. First, we have the rule in (2.19) which tells us how accurately we have to solve the equation $1 - F(x) = 1/n$ for a continuous distribution $F(x)$ in order to get an a_n or b_n by the cited instructions. Second, different choices of the normalizing constants result in different speeds of convergence in (2.1). This is evident by looking at numerical comparisons between $F^n(a_n + b_n z)$ and $H_{3,0}(z)$, say, when F is known to be in $D(H_{3,0})$. By changing a_n or b_n for a fixed (large) n but not changing z will result in a different $F^n(a_n + b_n z)$ but $H_{3,0}(z)$ remains unchanged. The first generally applicable estimate of the speed of convergence appeared in Section 2.10 of Galambos (1978), which estimate is a clear demonstration of the influence of the normalizing constants as well as of the population distribution on the speed of convergence. Namely, the two main terms in estimating (from above) the absolute difference

$$| P(Z_n \leq a_n + b_n z) - H(z) |$$

are $r_{1,n}^* = \dfrac{2 z_n^2(z)}{n}$ and $r_{2,n}^* = | z_n(z) + \log H(z) | \qquad (2.20)$

where $z_n(z) = n[1 - F(a_n + b_n z)]$. Now, by taking logarithm in (2.1), Taylor's expansion yields that $z_n(z) \sim -\log H(z)$ entailing that $z_n(z)$ is bounded for fixed z and $r_{2,n}^* \to 0$ as $n \to +\infty$ and z is kept fixed. Hence, $r_{1,n}^*$ is always of the magnitude of $1/n$, regardless of F, a_n and b_n, while no uniform estimate can be given for $r_{2,n}^*$ since it strongly depends on F as well as a_n and b_n. In fact, if we choose a_n and b_n by the instructions of Theorem 2.3 for a standard normal population, one finds from the estimates of Galambos (1978), Section 2.10, that the difference in (2.1) is asymptotically equal to $(\log \log n)^2 / \log n$. Hall (1979) and (1980) established that this speed in the normal case can be improved by modified choices of a_n and b_n, but it always stays above $c/(\log n)$, where $c > 0$ is a positive constant. On the other hand, W.J. Hall and J.A. Wellner (1979) show that, for the exponential distribution, the uniform speed of convergence is of the magnitude of $1/n$. We have already seen in the discussion at (2.20) that such an estimate cannot be improved for any F and for any choice of a_n and b_n. However, improvements in the speed of convergence can be achieved by modifying (2.1) itself. Since we know that $H(z)$ in (2.1) is one of three types of distribution, let us insist that we approximate F^n by $H_{1,\gamma}$ regardless whether F is in its domain of attraction or not.

That is, let us consider the difference

$$F^n(a_n + b_n z) - H_{1,\gamma_n}(C_n + D_n z) \qquad (2.21)$$

where a_n and $b_n > 0$ are computed by the rules of Theorems 2.1 through 2.3 and then γ_n, C_n and D_n are computed by some new rule based on F, a_n and b_n, where this new rule is aimed at obtaining a smaller value in (2.21) than one would have for the difference with the proper H in (2.1). Under the assumption of the von Mises condition (2.17a), Gomes (1978) (or see its published form Gomes (1984)) worked out the details of such an approximation, known as penultimate approximation. One of the major results of Gomes (1978) is presented in the book Galambos (1987), Section 2.10, which provides an opportunity for the reader to compare penultimate approximation with a variety of estimates on the speed of convergence.

A further deviation from (2.1) is when one chooses some monotonic transformation of Z_n rather than the linear one of (2.1). One hopes with such a modification that a better approximation can be achieved as well as more types of H will result as limiting distribution. A systematic theory of this nature has been initiated by Pancheva (1985), which has already reached quite a high level of development. See Pancheva (1994) for a survey on nonlinear normalization.

3. Applications and statistical inference

We present practical problems which will be approximated by a classical model. Even though the independence of the underlying random variables will not always be justified the approximation by a classical model will be, due to the mathematical generalization of the results on classical models to so called weakly dependent or graph-dependent random variables. The discussion of such models is postponed to the next section.

Let us start with the longest living terminally ill patient in a group of n, each having an expected life of one year (Example 1.3). If X_j is the actual (remaining) life of patient j, the no-aging property of the exponential distribution entails (see Galambos and Kotz (1978), p.8) that we may assume that $P(X_j \leq x) = 1 - e^{-x}, x \geq 0$, and the X_j are quite evidently independent. Then

$$P(Z_n \leq \log n + z) = (1 - e^{-\log n - z})^n \to \exp(-e^{-z})$$

from which we have that if there are in the viewing range of a TV station about $n = 200$ such patients, then, with probability $1 - e^{-1} = 0.63$ (we took $z = 0$ above), $Z_{200} > \log 200 > 5$, i.e., a patient will turn up at the TV station "who was condemned by a doctor to die in one year" but still alive after 5 years. Since news like this are taken over by TV stations in the USA, n is in fact much larger, and thus from the strong law $P(Z_n/\log n \to 1) = 1$ (see p. 262 in Galambos (1987)) one can conclude that, almost surely, someone will turn up in the news who was declared terminally ill but still lives 8, 10, or even more years later.

The example above is a special example for statistical outliers. We have n independent observations and we assume (justified or not) that the underlying distribution is $F(x)$. The data nicely fit $F(x)$ but a few observations seem to be questionable. These can be outliers (mismeasurements, etc.), but it may equally be true that the assumption $F(x)$ is wrong and extreme observations are the good signs for it. For example, if $F(x)$ is assumed standard normal, and the bulk of the data seem to confirm this (and in most cases this is true due to the asymptotic normality of quantiles), then the strong law $P(Z_n/(2 \log n)^{1/2} \to 1) = 1$ or the more informative strong law $P(Z_n - a_n \to 0) = 1$, where a_n is the solution of $1 - F(x) = 1/n$ (see Galambos (1987), p. 265–266) can be used to determine that how large the largest observation should be.

The two examples above differ in the way in which $F(x)$ is chosen but both utilize an exact model. The case of outliers requires the model to be exact, but the case of terminally ill patients can be modified to just ill patients in a life-threatening disease to conclude that, with high probability, someone gets into the news for seemingly beating the odds. The only exception would be if life distributions $F(x)$ for any group of people would have a very limited $\omega(F)$ implying that no one in the group can survive beyond the age $\omega(F)$. Can $\omega(F)$ be infinity? We got such a case with the exponential distribution. But note that with $F(x) = P(X \le x) = 1 - e^{-x}, X > 20$, say, is still possible but $P(X > 20) = 2/10^9$ would require, via the strong law of large numbers, to have hundreds of millions of people terminally ill with $E(X) = 1$ in order to observe one of them living more than 20 years after getting ill. This is significant to understand when we discuss the next model.

We, in fact, return to Example 1.4 and try to estimate the age of the longest living person in Europe. We assume that the random life lengths of individuals are independent and identically distributed but we do not use or attempt to identify the common distribution $F(x)$. Rather, we decompose the total population into disjoint large subgroups, and we estimate the highest age in the population from the maxima in the subgroups. But the subgroups can further be decomposed into disjoint large subgroups, so the maximum in a subgroup is the maximum of other maxima, and so on. This process of decomposition can go down in a large number of steps, and when we move back up we see that at every stage the maximum has the same type of distribution as the one whose maximum is taken. If this process would never terminate we would get that the distribution at every stage is an extreme value distribution. The Decomposition Principle, that we now adopt, adds that the number of steps in the above process is so large that a 'jump' to the limit is justified. With this we eliminated the most troubling parts we face in extreme value theory: we do not need the form of the population distribution and we will not argue with the very sensitive n^{th} power of distributions. Rather, we take the largest value from subgroups as observations whose underlying distribution is $H_{(c)}(A + Bz)$ of (2.7), and we want to decide whether $c > 0, c < 0$ or $c = 0$, and to estimate the parameters A and B. Since all observations are assumed to come from $H_{(c)}$ very little extreme value theory is remaining in the statistical inference. Indeed, Tiago de Oliveira (1984) suggests the use of loglikelihood functions and the Neyman-Pearson theory for developing the critical regions for choosing between any two of the possibilities for $c(c > 0$ versus $c = 0$ or $c < 0$ versus $c = 0$ or $c > 0$ versus $c < 0$). Tiago de Oliveira himself points out that the above method is not very effective if the sample size is not very large which is the case in our longest living problem. However, the method is applicable to large data sets. Two other test statistics are recommended in Tiago de Oliveira and Gomes (1984), one of which is refinement of a test first investigated by Gumbel (1958). This test statistic is based on four quantiles only, and thus its simplicity is quite appealing. Usually one has larger sample sizes in hydrology where for flood levels The Decomposition Principle can again be applied (annual flood is the highest monthly floods, the monthly floods in turn can be decomposed into weekly floods, daily floods, etc). Thus, assuming that flood levels are distributed with $H_{(c)}$ and annual floods are sufficiently weakly dependent so a weak dependence model of the next section justifies the approximation by classical models, one can apply the cited methods. If the set of data is not large, it appears that The Decomposition Principle should be replaced by the full extent of the limiting extreme value theory and statistical inference should be based on the 'original' observations rather than on a single selected value from subgroups. However, when a characterization theorem is not available for the population distribution, it must not be attempted to be identified or even approximated. Instead, a domain of attraction is to be selected on a statistical basis and then inference is to proceed with the appropriate limiting distribution. In this direction, three basic methods have been developed: (i) the threshold method, (ii) method by probability papers, and (iii) decisions based on limit results for domains of attraction. Before describing these methods, one should emphasize two very important rules.

Rule 1. For making decisions on a maximum, use always upper extremes only. That is, if $X_1, X_2, \ldots, X_n$ are observations, decide on an $r = r(n)$, and use only the order statistics $X_{n-r+1:n} \leq X_{n-r+2:n} \leq \cdots \leq X_{n-1:n} \leq X_{n:n}$. Contrary to the views of some 'outsider statisticians', this does not lead to loss of information (see Janssen (1989)). More importantly, low indexed order statistics may simply mislead the observer. Castillo, Galambos and Sarabia (1989) give the following warning example. Let $F_1 \in D(H_{1,\gamma})$ and $F_2 \in D(H_{3,0})$. Assume that the parameters of F_1 and F_2 are such that $F_1(z_0) = F_2(z_0) = 1/2$. Then observations on

$$F(x) = \begin{cases} F_1(x) & \text{if } x < z_0 \\ F_2(x) & \text{if } x > z_0 \end{cases}$$

will produce 50% of the data from F_1 and 50% from F_2, but the maximum will be among those which come from F_2 and ultimately the observations generated by F_1 will become irrelevant for the maximum. Clearly, $F \in D(H_{3,0})$, but any statistical decision based on all observations would reject such a hypothesis.

Rule 2. When in an applied problem a decision has been made on c and the parameters A and B of $H_{(c)}(A + Bz)$ have been estimated, the result must fully be utilized for future actions by observing that records (maxima) are guaranteed to be broken (at least for continuous variables). That is, preventive measures protecting against the level of a past disaster – such as a previous flood – is an open invitation for a new disaster. One must compute from $H_{(c)}(A+Bz)$ the likely and unlikely levels of new disasters, together with their return periods, and only then should there be a decision on the magnitude of 'preventive investments'.

Let us return to the cited three statistical methods.

(i) The Threshold Method. Assume that it has been decided by some logical argument that $\omega(F) = +\infty$, and the maximum satisfies (2.1). Hence, the problem is to decide whether $F \in D(H_{3,0})$ or $F \in D(H_{1,\gamma})$. Apply Theorem 2.4. Note that (2.9) with $u(t) = R(t)$ means that in order for $F \in D(H_{3,0})$, the normalized variable $(X - t)/R(t)$ must be asymptotically exponential, given that $X > t$. Hence, if we choose a large value t, select those observations which exceed t, say $Y_1 = X_{i_1}$, $Y_2 = X_{i_2}, \ldots, Y_N = X_{i_N}$. Estimate $R(t)$ by the arithmetical mean $R_N(t)$ of $Y_j - t, 1 \leq j \leq N$, and test the exponentiality of $(Y_j - t)/R_N(t)$. Acceptance means the acceptance of $F \in D(H_{3,0})$ and rejection means accepting $F \in D(H_{1,\gamma})$. In the form just described the method is formulated in Galambos (1980), which is enhanced by the analysis of the method by Tiago de Oliveira (1984) and Castillo (1988), p. 218. In a somewhat different formulation and in the context of hydrology, Todorovic (1978) and (1979) applied such a model, which has then been extended by a number of investigators. Smith (1984) and (1994) made the most extensive studies on this line.

(ii) Method with Probability Papers. Plot the empirical distribution function of the observations on a Gumbel probability paper, defined as a coordinate system (v, w) for the points (x, p), $0 < p < 1$, using the scales $v = x$ and $w = -\log(-\log p)$. Compare the shape of the upper tail of this plot with the shape of $H_{(c)}(A + Bz)$, also drawn on a Gumbel paper. Whichever shape is imitated by the empirical distribution function $F_n(x)$, F is in the domain of attraction of that H_c. More precisely, since

$$w = -\log(-\log H_c(A + Bz)) = (1/c)\log[1 + c(A + Bz)]$$

is the straight line $w = A + Bz$ for $c = 0$, otherwise a logarithmic curve with coefficient $1/c$, w above is concave, convex (concave up) or straight line according as $c > 0$, $c < 0$ or $c = 0$. That is, the shape of w is sufficient to make distinction among the possible three types of H_c. Now, Castillo and Galambos (1986) and with more details, Castillo et al. (1989), proved that the upper tail, $x \geq (n-r)/n$, of $F_n(x)$ has the same shape as H_c if $F \in D(H_c)$. This can very effectively be utilized in one of two ways. Either the shape of the upper tail of F_n is very clearly concave or convex or a line, in which case visual inspection is sufficient to conclude that $F \in D(H_c)$. (Here, and throughout this section, $D(H_c)$ is an abbreviation for one of the three domains $D(H_c: c > 0) = D(H_{1,\gamma}), D(H_c: c < 0) = D(H_{2,\gamma})$ and $D(H_c, c = 0) = D(H_{3,0})$ of attraction.) When visual inspection is not convincing, then the very accurate statistical method is available: choose a value r, and fit a straight line to the points (x_j, w_j), where $x_j = X_{n-r+j:n}$, and

$w_j = -\log(-\log F_n(x_j))$, $1 \leq j \leq r/2$, using the method of least squares, and fit another line to (x_j, w_j) for $r/2 < j \leq r$. From the angle between these two lines, decide on the shape of the upper tail of $F_n(x)$, using tables developed in Castillo (1988), pp. 219–226.

(iii) Decisions from Limit Results. The results stated in Theorems 2.1–2.4 has been unified and extended by de Haan and his colleagues into a single form using $D(H_c)$ and inverse functions of $F(x)$. These results give specific forms, based once again on the upper extremes, which converge to c if $F \in D(H_c)$, and the accuracy of these limit theorems are also established. Such limit theorems are then used as estimators, both as point estimators and confidence bounds, for c. de Haan (1994) gives a detailed presentation of this method, including earlier references. We therefore omit further details.

All these methods have widely been used in practice. We have mentioned the works of Todorovic in hydrology. The Method with Probability Papers (Castillo calls it The Curvature Method in his book, p. 219) has been applied to a wide variety of data in Castillo (1988), using the accurate rather than the visual method. This includes human life length, and the conclusion by Castillo is that the data are not sufficient to make a statistical decision between $D(H_{2,\gamma})$ and $D(H_3, 0)$. From another set of data on length of human life, de Haan (1994), however, concludes, using method (iii), that $F \in D(H_{2,\gamma})$, meaning that human life is bounded. The present author sides with Castillo on this matter.

In a large number of applied works, $D(H_{3,0})$ is assumed without any statistical considerations. The reason for this is that $\omega(F) < +\infty$ can some times be rejected on practical grounds, and $D(H_{1,\gamma})$ may not be reasonable for some practical problems due to the implication that if $F \in D(H_{1,\gamma})$ then the moments $E(X_j^k)$, $k > \gamma$, are not finite. This can be deduced from (2.8), or see Theorem 2.7.11 in Galambos (1987).

Let us look at one general problem when the minimum W_n of the components $X_j, 1 \leq j \leq n$, is our interest. The simplest one of such cases is an instrument which breaks down as soon as any one of its components breaks down. Such an instrument is equivalent to a series system for which an example is n electric bulbs connected to a single line, and thus if one of them burns out then the electricity stops flowing. Another example is the case of a sheet of metal discussed in Section 1. The components now are hypothetical, and the strength $S = W_n$. In both of these last examples an approximation by classical models is justified (in the case of the metal, a weakly dependent model of the next section is to be utilized). By an appeal to (2.2), Theorem 2.1–2.4 can be applied in the same ways in which Z_n is treated. In particular, if we know apriori that the random variable approximated by W_n is positive, then, via (2.2)

$$L(z) = 1 - H_{2,\gamma}(-z) = 1 - \exp(-z^\gamma), \quad z > 0$$

provides the underlying distribution, that is, we get a Weibull model. Although the Weibull family of distributions is used in engineering models quite frequently for no other reason than its possible use to fit a large variety of data, stemming from the influence of the shape parameter on the densities $L'(z)$, here we provided a theoretical justification for adopting Weibull models.

For another approach to material strength, applying extreme value theory and arriving at the Weibull model, see Taylor (1994) and its references.

There are several ways of placing the statistical aspects of extreme value theory into a larger framework. See Falk and Marohn (1992), Falk, Hüsler and Reiss (1994), Marohn (1994), and Reiss (1989).

There are a number of conference proceedings on the extremes (Galambos (1981), Tiago de Oliveira (1984), Hüsler and Reiss (1989) and Galambos, Lechner and Simiu (1994)) and two basic books with emphasis on applications (Gumbel (1958) and Castillo (1988)).

4. Deviations from the classical models

There are two directions in which one can extend a classical model: (i) one can drop the assumption of stationarity of distribution but keeps independence, or (ii) one develops dependent models. Both come up in practice in a natural way, even within the topic of instrument failure. Even if we take the simplest models of instruments: parallel or series systems, it is against engineering experience that each such system's failure rate function is accurately a power function (Weibull distributions). It turns out that if we model such systems with independent component life but without the assumption of identical distributions, then the asymptotic distributions of the normalized extremes will include all distributions with monotonic failure rate and practically only these (additional mathematical niceties are required). This is one of the most pleasing results in extreme value theory: mathematical conclusions and practical experience come together in such a nice way. The first such result goes back to Mejzler (1949). For extensions of Mejzler's result, together with surveys of this subject, see Mejzler (1965) and Weissman (1994). See also Section 3.10 in Galambos (1987).

The following graph-dependent model extends the result of Mejzler. For a graph $G = (V, E)$ with vertex-set $V = \{1, 2, \ldots, n\}$ and arbitrary edge-set E, the random variables $X_1, X_2, \ldots, X_n$ are called G-dependent (for the maximum), if, as $x \to \min\{\omega(F_j) : 1 \leq j \leq n\}$, (i) disjoint subsets of the events $\{X_j > x\}$ are asymptotically independent if there is no edge connecting these subsets, (ii) every subset of the events $\{X_j > x\}$ with a single edge in the subset satisfies a boundedness condition compared with the independent case, and (iii) the number $N(E)$ of the edges of G is of smaller magnitude than the total possible number of edges n^2, then the asymptotic distribution of the properly normalized maximum Z_n is the same as for independent variables.

Note that the dependence structure of those subsets in which there are two or more edges have no influence on Z_n. On the other hand, the model is accurate for the maximum in the sense that the assumptions do not guarantee the convergence of the distribution of $X_{n-1:n}$, even when normalized. In this regard the model is a significant deviation from independence.

The model with accurate translation of the assumptions into mathematical formulas is introduced in Galambos (1972) and developed further in Galambos (1988). See also Section 3.9 in Galambos (1987).

If one adds to the graph-dependent models that the sequence $X_1, X_2, \ldots, X_n$ is stationary and the edge set E is restricted to $E = \{(i,j): 1 \leq i \leq n-s, i+1 \leq j \leq i+s\}$, where s is an integer with $s = s(n) \to +\infty$ and $s(n)/n \to 0$ as $n \to +\infty$, we get the dependence model on which the book Leadbetter et al. (1983) is based.

It appears that the most natural dependent model, that of exchangeable variables, is the most difficult to handle. If it is assumed that $X_1, X_2, \ldots, X_n$ is a finite segment of an infinite sequence of exchangeable variables, and thus de Finetti's theorem entailing conditional independence when properly conditioned applies, then quite general results follow (Berman (1964)). However, for finitely exchangeable variables the present author's result, collected in Chapter 3 of Galambos (1987), is too abstract for practical applications. It would be of great interest and value to analyze exchangeable models with some further assumptions in order to obtain specific rather than abstract limit theorems.

Both for the graph dependent models and for exchangeability Bonferroni-type inequalities provide the most effective tools. Define the events $A_j = A_j(x) = \{X_j > x\}$, $1 \leq j \leq n$, and let $m_n(x)$ be the number of those A_j which occur. Then

$$P(m_n(x) = 0) = P(Z_n \leq x) .$$

Hence, by applying Bonferroni-type inequalities to the left hand side, inequalities on the distribution of Z_n obtain. Because the same kinds of inequalities do not apply to $P(m_n(x) \geq r)$ if $r \neq 1$ as to the case of $m_n(x) = 0$, we can now understand why Z_n is special among all other order statistics. It should be noted that the classical Bonferroni-type inequalities do not suffice for obtaining meaningful bounds in the case of graph dependence. However, the theory of Bonferroni-type bounds is very rich, it has gone through a very fast development in recent years, although an extension of the classical Bonferroni-type bounds by Rényi (1961) suffices for the graph dependent model. Since we shall deal with Bonferroni-type bounds elsewhere, we do not go into further details.

Acknowledgements

The present work is a slightly modified version of the author's Inaugural Address to the Hungarian Academy of Sciences, delivered in June, 1994. The author's election to membership in the Hungarian Academy was held at the Annual Meeting of the Academy.

My sincere thanks go to the three academicians Imre Kátai, Zoltán Daróczy and András Prékopa who acted as my 'Floor Managers' at the election.

Just as my Inaugural Address, the present paper is dedicated to the Memory of My Parents. They would have been very proud of the moment of my receiving the Certificate of Membership. But more importantly, their devotion to my upbringing and their setting of the standards for everything in my life were the most important contribution to my achievements. At different stages of my education, Lajos Barna, János Knoll and the late József Masszi, Attila Gyüre and Alfréd Rényi influenced me the most. I am very grateful to them. For her understanding the way of life of a scientist and for her spirited arguments about interpreting a mathematical result in practice, I thank my wife, Éva Galambos.

References

Berman, S. M. (1964). Limit theorems for the maximum term in stationary sequences. *Ann. Math. Statist.* **35**, 502–516.
Castillo, E. (1988). *Extreme Value Theory in Engineering*. Academic Press, New York.
Castillo, E. and J. Galambos (1986). Determining the domain of attraction of an extreme value distribution. Technical Report, Department of Mathematics, Temple University.
Castillo, E., J. Galambos and J. M. Sarabia (1989). The selection of the domain of attraction of an extreme value distribution from a set of data. In: Hüsler, J. and R.-D. Reiss, eds., *Extreme Value Theory*. Springer, Heidelberg, 181–190.
De Haan, L. (1970). *On Regular Variation and its Application to the Weak Convergence of Sample Extremes*. Math. Centre Tracts, Vol. 32, Amsterdam.
De Haan, L. (1994). Extreme value statistics. In: Galambos, J. et al., eds., *Extreme Value Theory and Applications*, Vol. I. Kluwer, Dordrecht, 93–122.
Dodd, E. L. (1923). The greatest and least variate under general laws of error. *Trans. Amer. Math. Soc.* **25**, 525–539.
Falk, M. and F. Marohn (1992). Laws of small numbers: Some applications to conditional curve estimation. In: Galambos, J. and Kátai, eds., *Probability Theory and Applications*. Kluwer, Dordrecht, 257–278.
Falk, M., J. Hüsler and R.-D. Reiss (1994). *Laws of Small Numbers: Extremes and Rare Events*. Birkhäuser, Basel.
Fisher, R. A. and L. H. C. Tippett (1928). Limiting forms of the frequency distributions of the largest or smallest member of a sample. *Proc. Cambridge Philos. Soc.* **24**, 180–190.
Fréchet, M. (1927). Sur la loi de probabilite de l'écart maximum. *Ann. de la Soc. polonaise de Math* (Cracow) **6**, 93–116.
Galambos, J. (1972). On the distribution of the maximum of random variables. *Ann. Math. Statist.* **43**, 516–521.
Galambos, J. (1978). *The Asymptotic Theory of Extreme Order Statistics*. Wiley, New York.
Galambos, J. (1980). A statistical test for extreme value distributions. In: Gnedenko, B. V and I. Vincze, eds., *Nonparametric Statistical Inference* (Vol. 32, Colloquia Math. Soc. János Bolyai), North Holland, Amsterdam, 221–230.
Galambos, J. (ed.) (1981). Extreme Value Theory and Application. 43rd Session ISI, Invited Papers, 837–902.
Galambos, J. (1987). *The Asymptotic Theory of Extreme Order Statistics*. 2nd edition. Krieger, Malabar, Florida.
Galambos, J. (1988). Variants of the graph dependent model in extreme value theory. *Commun. Statist. –Theory Meth.* **17**, 2211–2221.
Galambos, J. (1994). The development of the mathematical theory of extremes in the past half century (in Russian). *Teor. Veroyatnost. i Primen.* **39**, 272–293.
Galambos, J. (1995). *Advanced Probability Theory*. 2nd edition. Marcel Dekker, New York.
Galambos, J. and Ch. Hagwood (1992). The characterization of a distribution function by the second moment of the residual life. *Commun. Statist. –Theory Meth.* **21**, 1463–1468.
Galambos, J. and S. Kotz (1978). *Characterizations of Probability Distributions*. Lecture Notes in Math., 675. Springer, Heidelberg.

Galambos, J. and A. Obretenov (1987). Restricted domains of attraction of $\exp(-e^{-x})$. *Stochastic Proc. Appl.* **25**, 265–271.

Galambos, J. and Y. Xu (1990). Regularly varying expected residual life and domains of attraction of extreme value distributions. *Ann. Univ. Sci. Budapest, Sectio Math.* **33**, 105–108.

Galambos, J., J. Lechner and E. Simiu (eds.) (1994). *Extreme Value Theory and Applications*, Vol. I - III. Vol. I: Kluwer, Dordrecht. Vol. II: J. of Research NIST (their volume 99) Vol. III: NIST Special Publication 866.

Gnedenko, B. V. (1943). Sur la distribution limite du terme maximum d'une serie aléatioire. *Ann. Math.* **44**, 423–453.

Gomes, M. I. (1978). Some probabilistic and statistical problems in extreme value theory. Thesis for Ph.D., University of Sheffield.

Gomes, M. I. (1984). Penultimate limiting forms in extreme value theory. *Ann. Inst. Statist. Math.* **36**, 71–85.

Gumbel, E. J. (1958). *Statistics of Extremes*. Columbia Univ. Press, New York.

Hall, P. (1979). On the rate of convergence of normal extremes. *J. Appl. Probab.* **16**, 433–439.

Hall, P. (1980). Estimating probabilities for normal extremes. *Adv. Appl. Probab.* **12**, 491–500.

Hall, W. J. and J. A. Wellner (1979). The rate of convergence in law of the maximum of an exponential sample. *Statist. Neerlandica* **33**, 151–154.

Hüsler, J. and R.-D. Reiss, (eds.) (1989). *Extreme Value Theory*. Lecture Notes in Statist. Vol. 51. Springer, Heidelberg.

Janssen, A. (1989). The role of extreme order statistics for exponential families. In: Hüsler, J. and R.-D. Reiss, eds. *Extreme Value Theory*, Springer, Heidelberg, 204–221.

Leadbetter, M. R., Lindgren, G. and H. Rootzen (1983). *Extremes and Related Properties of Random Sequences and Processes*. Springer, New York.

Marcus, M. B. and M. Pinsky (1969). On the domain of attraction of $\exp(-e^{-x})$. *J. Math. Anal. Appl.* **28**, 440–449.

Marohn, F. (1994). On testing the exponential and Gumbel distributions. In: Galambos, J. et al., eds. *Extreme Value Theory and Applications*, Vol. I. Kluwer, Dordrecht, 159–174.

Mejzler, D. G. (1949). On a theorem of B.V. Gnedenko (in Russian). *Sb. Trudov. Inst. Mat. Akad. Nauk. Ukrain. SSR* **12**, 31–35.

Mejzler, D. G. (1965). On a certain class of limit distributions and their domain of attraction. *Trans. Amer. Math. Soc.* **117**, 205–236.

Pancheva, E. (1985). Limit theorems for extreme order statistics under nonlinear normalization. In: V.M. Zolotarev et al., eds., *Stability Problems for Stochastic Models*, Lecture Notes in Math., Vol. 1155, Springer, Heidelberg, 284–309.

Pancheva, E. (1994). Extreme value limit theory with nonlinear normalization. In: Galambos, J. et al., eds., *Extreme Value Theory and Applications*, Vol. I. Kluwer, Dordrecht, 305–318.

Pickands, J. (1986). The continuous and differentiable domains of attraction of the extreme value distributions. *Ann. Probab.* **14**, 996–1004.

Reiss, R.-D. (1989). *Approximate Distributions of Order Statistics*. Springer, Heidelberg.

Rényi, A. (1961). A general method for proving theorems in probability theory and some of its applications. Original in Hungarian. Translated into English in Selected Papers of A. Rényi, Vol. 2, Akademiai Kiado, Budapest, 1976, 581–602.

Smith, R. L. (1984). Threshold methods for sample extremes. In: J. Tiago de Oliveira, ed., *Statistical Extremes and Applications*, Reidel, Dordrecht, 623–638.

Smith, R. L. (1994). Multivariate threshold methods. In: J. Galambos, et al., eds., *Extreme Value Theory and Applications*, Kluwer, Dordrecht, 225–248.

Sweeting, T. J. (1985). On domains of uniform local attraction in extreme value theory. *Ann. Probab.* **13**, 196–205.

Taylor, H. M. (1994). The Poisson-Weibull flaw model for brittle fiber strength. In: J. Galambos, et al., eds., *Extreme Value Theory and Applications*. Kluwer, Dordrecht, 43–59.

Tiago de Oliveira, J., ed. (1984). *Statistical Extremes and Applications*. Reidel, Dordrecht.

Tiago de Oliveira, J. (1984). Univariate extremes; Statistical choice. In: the preceding edited book, 91–107.

Tiago de Oliveira, J. and M. I. Gomes (1984). The test statistics for choice of univariate extreme models. In: J. Tiago de Oliveira, ed., *Statistical Extremes and Applications*. Reidel, Dordrecht, 653–668.

Todorovic, P. (1978). Stochastic models of floods. *Wat. Res.* **14**, 345–356.

Todorovic, P. (1979). A probabilistic approach to analysis and prediction of floods. 42nd Session I.S.I., Invited Papers, 113–124.

Von Bortkiewicz, L. (1922). Variationsbreite und mittlerer Fehler. *Stizungsberichte Berliner Math. Ges.* **21**.

Von Mises, R. (1923). Über die Variationsbreite einer Beobachtungsreihe. *Sitzungsberichte Berlin. Math. Ges.* **22.**

Von Mises, R. (1936). La distribution de la plus grande de n valeurs. Reprinted in Selected Papers II, Amer. Math. Soc., Providence, R.I., 1954, 271–294.

Weissman, I. (1994). Extremes for independent nonstationary sequences. In: J. Galambos et al., eds., *Extreme Value Theory and Applications*, Vol. III. NIST Special Publication **866**, 211–218.

Order Statistics: Asymptotics in Applications

Pranab Kumar Sen

1. Introduction

The very definition of *order statistics* is somewhat confined to univariate setups where ordering of the observations can be made unambiguously, although with some modifications, such interpretations can be made in some multivariate cases as well. To set the things in a proper perspective, let us consider a sample $X_1, \ldots, X_n$ of n independent and identically distributed (i.i.d.) (real valued) random variables (r.v.), and assume that they come from a distribution F, defined on the real line $\mathcal{R}$. We arrange these X_i in an ascending order of magnitude, and obtain the so called order statistics

$$X_{n:1} \leq X_{n:2} \leq \cdots \leq X_{n:n} , \tag{1.1}$$

where the strict inequality signs hold with probability one if F is continuous. If F admits jump discontnuities, ties among the X_i and hence $X_{n:i}$, may occur with a positive probability, and in statistical analysis, these are to be taken into account. Conventionally, we let $X_{n:0} = -\infty$ and $X_{n:n+1} = \infty$; if the X_i are nonnegative r.v.'s, as is the case in *reliability* and *survival analysis*, we have $F(0) = 0$, so the $X_{n:i}$ are all nonnegative, and hence, we let $X_{n:0} = 0$. Similarly, if the d.f. F has a finite upper endpoint $b(< \infty)$, we would let $X_{n:n+1} = b$. An optimal, unbiased estimator of F is the sample (or *empirical*) distribution function F_n, defined by

$$F_n(x) = n^{-1} \sum_{i=1}^{n} I\{X_i \leq x\}, \quad x \in \mathcal{R} , \tag{1.2}$$

where $I(A)$ stands for the indicator function of the set A. In the case of a continuous F, we have from the above two expressions,

$$F_n(x) = k/n, \quad \text{for } X_{n:k} \leq x < X_{n:k+1}, \ k = 0, 1, \ldots, n . \tag{1.3}$$

A simple modification of this relation holds for the discrete case where the jumps of F_n occur at the distinct order statistics but their magnitudes depend on the number of ties at those values. Note that even when F is continuous, F_n is a step function, so the continuity property is not preserved by the estimator. Nevertheless, the sample order statistics and the empirical distribution have a one-to-one relation, although both are stochastic in nature. For this reason, in the literature, order statistics and empirical distributions are often blended in a broader tone which allows the percolation of general asymptotics under appropriate regularity conditions; we may refer to Shorack and Wellner (1986), Koul (1992), Sen and Singer (1993) and Jurečková and Sen (1996), among others, for some comprehensive treatise of related asymptotics. Mostly there is a theoretical flavor in such developments where sophisticated probabilistic tools have been incorporated, often justifiably, with deep coatings of abstractions. For this reason, Sen and Singer (1993, Chapter 4) dealt with sound methodology at an intermediate level with due emphasis on potential applications in a variety of fields, including biostatistics.

The primary objective of the current study is to focus on the basic role played by the asymptotics in order statistics in various applications. In order to put the things in a proper perspective, in Section 2, we start with some of the basic results in order statistics which provide the access to the general asymptotics to be presented here. These asymptotic counterparts are then provided in Section 3. One of the most useful application areas of order statistics is *robust* and *efficient estimation* in *location-scale* and *regression-scale* families of densities, and the related asymptotics for the so called *L-estimators* are presented in Section 4. In a parametric mold, estimators of the scale parameters are generally obtained by using *L*-estimators as well. But in a semiparametric or nonparametric setup, asymptotic distribution (mostly, normal) of such *L*-estimators (or related test statistics) involve some nuisance (variance) functionals which are needed to be estimated from the sample as well. *Jackknifing* techniques play an important role in this context, and this is discussed briefly in the same section. Some emphasis has also been placed on *functional jackknifing* and other variants of the classical jackknife methodology. *Censoring* (*truncation*) schemes of various types arising in practice calls for some modification in the formulation of *L*-statistics, and these are briefly included in this presentation. Asymptotics for *trimmed least squares* and *regression quantiles* are presented in Section 5. Asymptotics for *induced order statistics* or *concomitants of order statistics* are considered in Section 6. These results are of considerable use in analysis of association or dependence pattern in bivariate models and in survival analysis and some related areas. In a multivariate situation, the conditional distribution of one of the variables, given the others, leads to a formulation of *nonparametric regression models,* where the *linearity of regression* or *homoscedasticity* condition may not be that crucial. In this context, *conditional L-functionals* are often used, and their methodology depends on a somewhat different type of asymptotics entailing a possibly slower rate of convergence; these are treated in Section 7. Applications to statistical analysis of mixed-effects models are also considered in this section. Interesting applications of order statistics asymptotics relate to *statistical strength of a bundle of parallel filaments, system availability* and to some other measures in multicomponent systems arising in problems in reliability theory, and Section 8 deals with the relevant methodology. The *total time on testing* (TTT) concept, intimately associated with order statistics for nonnegative random variables, occupies a prominent stand in the main avenue of reliability and *life-testing* models (where the failure time or length of life is typically nonnegative). In this context, characterizations of different forms of *aging* properties depend of such TTT transformations, and related TTT-asymptotics having genesis in asymptotics for order statistics are of basic importance from the application point of view; these are presented in a systematic manner in Section 9. The concluding section is devoted to some general observations on asymptotics in order statistics.

2. Some basic results in order statistics

An order statistic $X_{n:k}$ is classified as a sample *quantile* or an *extreme value* according as k/n is bounded away from 0 and 1 or not. Thus, $X_{n:1}$ is the lower extreme value, $X_{n:n}$ is the upper one, while $X_{n:[np]}$ (for $0 < p < 1$) is the sample *p*-quantile. We denote the distribution function (d.f.) of $X_{n:k}$ by $G_{nk}(\cdot)$. Then for every n, k and F, not necessarily continuous, we have

$$G_{nk}(x) = \sum_{r=k}^{n} \binom{n}{r} [F(x)]^r [1 - F(x)]^{n-r}, \quad x \in \mathcal{R} . \tag{2.1}$$

In particular, for $k = 1$, we have $G_{n1}(x) = 1 - [1 - F(x)]^n$, and for $k = n$, $G_{nn}(x) = [F(x)]^n$. These results do not require the (absolute) continuity of the underlying d.f. F, but the i.i.d. structure of the X_i is presumed here. Sans independence, the relevance of the Binomial law in (2.1) is lost, while for independent but not necessarily i.d.r.v.'s, (2.1) provides some useful bounds on the actual d.f.; we may refer to Sen (1970) for some details. Galambos (1984) reviewed some interesting results in the dependent case.

For nonnegative random variables for which the support of the d.f. F is $\mathcal{R}^+ = [0, \infty)$, closely related to the order statistics are the *sample spacings* $\{l_{n1}, \ldots, l_{nn}\}$, which are defined by letting

$$l_{nj} = X_{n:j} - X_{n:j-1}, \quad \text{for } j = 1, \ldots, n , \tag{2.2}$$

where we set conventionally $X_{n:0} = 0$. Then l_{nj} is defined as the jth spacing in a sample of size n from the d.f. F, $j = 1, \ldots, n$. Keeping in mind the simple exponential density for which the spacings were considered first, we may as well define the *normalized spacings* by letting

$$d_{nj} = (n - j + 1) l_{nj}, \quad \text{for } j = 1, \ldots, n . \tag{2.3}$$

In the aforementioned exponential case, the d_{nj} are i.i.d.r.v.'s having the same exponential distribution. However this characterization of the exponential distribution may not generally hold for other distributions, and much of the reliability theory deals with this aspect of the spacings. The concept of total time on test upto the kth failure point has its genesis in this setup. If we let

$$D_{nk} = \sum_{j \leq k} d_{nj}, \quad \text{for } k = 1, \ldots, n . \tag{2.4}$$

then, for any timepoint $t \in [X_{n:k}, X_{n:k+1})$, in a without replacement scheme, the total lifetime of these n units upto the time t, when all of them enter into the scheme at a common timepoint, say 0, is given by

$$D_n(t) = D_{nk} + (n-k)(t - X_{n:k}), \quad \text{for } k = 0, \ldots, n , \tag{2.5}$$

where conventionally, we let $D_{n0} = 0$, so that $D_n(t) = nt$, for all $t < X_{n:1}$. This concept will be incorporated in a later section to provide related asymptotic results which are very useful in the context of life testing and reliability models. For d.f.'s other than exponential ones, the study of the exact distributional properties of the (normalized) spacings (including their i.i.d. structure) may encounter considerable difficulties, although their asymptotic counterparts can be studied under fairly general regularity conditions. Asymptotics for order statistics play a basic role in this context too.

Let us next consider a *stochastic process* $Q_n(\cdot) = Q_n(t), t \in (0, 1)$, by letting

$$Q_n(t) = F_n^{-1}(t) = \inf\{x : F_n(x) \geq t\}, \quad t \in (0, 1) . \tag{2.6}$$

Then Q_n is termed the sample *quantile function*. Side by side, we may introduce the population quantile function $Q(\cdot)$ by letting

$$Q(t) = F^{-1}(t) = \inf\{x : F(x) \geq t\}, \quad t \in (0, 1) . \tag{2.7}$$

Intuitively, at least, Q_n estimates Q, although a more precise study of this estimation problem requires a good deal of asymptotics. While such quantiles are defined by a single order statistic, in general, functions of order statistics, such as an L-statistic, are estimators of a functional, say, $\theta(F)$ of the d.f. F. Such functionals may be location, scale, regression ones, and may even be more general in nature. Following Hoeffding (1948), we term $\theta(F)$ a *regular functional* or *estimable parameter* if there exist a sample size $m(\geq 1)$ and a statistic $T(X_1, \ldots, X_m)$, such that

$$\begin{aligned}\theta(F) &= E_F\{T(X_1, \ldots, X_m)\} \\ &= \int \cdots \int T(x_1, \ldots, x_m) \, dF(x_1) \cdots dF(x_m) ,\end{aligned} \tag{2.8}$$

for all F belonging to a class $\mathcal{F}$, called the *domain* of F. If m is the smallest sample size for which this estimability holds, it is called the *degree* of the parameter, and the corresponding $T(\cdot)$ is called the *kernel* which we may take to be a symmetric one without any loss of generality. A symmetric, unbiased estimator of $\theta(F)$ is given by

$$U_n = \binom{n}{m}^{-1} \sum_{\{1 \leq i_1 < \cdots < i_m \leq n\}} T(X_{i_1}, \ldots, X_{i_m}), \quad n \geq m . \tag{2.9}$$

It is easy to verify that

$$U_n = \binom{n}{m}^{-1} \sum_{\{1 \leq i_1 < \cdots < i_m \leq n\}} T(X_{n:i_1}, \ldots, X_{n:i_m}) , \qquad (2.10)$$

so that U_n is a function of the order statistics as well. This formulation of U-statistics has largely a nonparametric flavor wherein the d.f. F is allowed to be a member of a bigger class, and continuity properties are not that essential. Such U-statistics possess nonparametric optimality properties, and by virtue of the above characterization, order statistics share such optimality properties as well. As a matter of fact, the conditional distribution of the sample observations given the set of order statistics is (discrete) uniform over the set of all possible permutations, and hence, the vector of order statistics is (jointly) *sufficient* for the underlying d.f. F; it is *complete* under fairly general regularity conditions [see, Chapter 3 of Puri and Sen (1971)]. In a parametric setup, particularly for the exponential family of densities, *minimal sufficient statistics* exist and are appropriate subsets of this vector (or lower dimensional functions), but in a nonparametric setup, such characterization of minimal sufficiency may not be universal. It is interesting to note that the above formulation of estimable parameters is tied-down to unbiasedness, so that in a nonparametric setup, the population quantiles may not be estimable in the same sense. For this reason, sample quantiles are generally not expressible as U-statistics, and the treatment of statistical methodology differs from one setup to another. Nevertheless, the above formulation has paved the way for the definition of location, scale or regression functionals for which order statistics are very appropriate, and we shall present the relevant methodology in a later section dealing with the so-called L-statistics; such L-statistics are related to linear functionals of F and often may be expressed as U-statistics as well.

Although order statistics are well defined for all univariate F continuous or not, there may be tied observations (with a positive probability) if F admits jump discontinuities. For this reason, often, in practice, it is tacitly assumed that F is continuous, although due to rounding up or grouping, we may have in reality a discrete d.f. even if F is continuous. For a continuous F, if we denote by

$$Y_i = F(X_i) \text{ and } Y_{n:i} = F(X_{n:i}), \quad i = 1, \ldots, n , \qquad (2.11)$$

then the Y_i are i.i.d. with the uniform(0, 1) distribution, and therefore the $Y_{n:i}$ are the order statistics of a sample of size n from the uniform(0, 1) distribution. These are termed the reduced or *uniform order statistics* and the corresponding empirical distribution, denoted by $G_n(t), t \in (0, 1)$, is called the reduced or *uniform empirical* d.f. This reduction may not work out conveniently for F having jump discontinuities. Much of the asymptotics in the case of continuous F, to be dealt with in later sections, can be studied conveniently by appealing to this reduction to the uniform case.

Let us look into the infrastructure of order statistics and the so called *ranks antiranks*, and for simplicity of presentation, we again consider the case of continuous F. Let R_i be the rank of X_i among the n sample observations $X_1, \ldots, X_n$, for $i = 1, \ldots, n$. Then, by definition we have

$$X_i = X_{n:R_i}, \quad \text{for } i = 1, \ldots, n, \tag{2.12}$$

and the stochastic vector $\mathbf{R}_n = (R_1, \ldots, R_n)'$, termed the *rank-vector*, takes on each permutation of $(1, \ldots, n)$ with the common probability $(n!)^{-1}$. Likewise, we may define the *anti-ranks* S_i by letting

$$X_{n:i} = X_{S_i}, \quad \text{for } i = 1, \ldots, n, \tag{2.13}$$

so that the vector of anti-ranks $\mathbf{S}_n = (S_1, \ldots, S_n)'$ also takes on each permutation of the numbers $(1, \ldots, n)$ with the common probability $(n!)^{-1}$. Note that by definition

$$R_{S_i} = i = S_{R_i}, \quad \text{for } i = 1, \ldots, n. \tag{2.14}$$

This basic relationship between the vectors of order statistics, ranks and anti-ranks provides a key to the (asymptotic) distribution theory of various statistics based on such vectors.

The very definition of order statistics rests on an ordering of the sample observations. In the bivariate (or more generally, multivariate) case, such an ordering of the sample observations may not exist. For example, we may order the sample observations with respect to each of the coordinates, but then these orderings may not be concordant. We may also order the sample observations by choosing an appropriate *cutting function* reducing the dimension to one; but such an ordering may depend heavily on the choice of this otherwise arbitrary cutting function. Therefore, it is clear that order statistics in the multivariate case may not be defined uniquely, and moreover, may not possess the *affine invariance* property which is usually shared by the classical linear statistics for linear models. In practice, in a multivariate setup, usually the coordinatewise order statistics vectors are incorporated, and this may lead to some complications as will be discussed later on. In this setup, we will also encounter the so called *induced order statistics* or *concomitant of order statistics* which are defined as follows. Let $(X_i, Y_i), i = 1, \ldots, n$, be n i.i.d. bivariate observations, and we denote the order statistics corresponding to the X-coordinate values by $X_{n:i}, i = 1, \ldots, n$. By the definition of the anti-ranks, made earlier, we have $X_{n:i} = X_{S_i}, i = 1, \ldots, n$. Let us denote by

$$Y_{[i]} = Y_{S_i}, \quad \text{for } i = 1, \ldots, n. \tag{2.15}$$

Then the $Y_{[i]}$ are termed the concomitant of (or induced) order statistics, where ordering is with respect to the X-coordinate values and the concomitance is with respect to the Y-coordinate values. In passing, it may be recalled that whereas by definition the $X_{n:i}$ are ordered, the induced order statistics may not necessarily be ordered; in this sense, the term concomitant of order statistics, coined by David (1973), is more meaningful than its counterpart; induced order statistics, due to Bhattacharya (1974). It may also be noted that by virtue of the Bhattacharya (1974) lemma, these induced order statistics are conditionally (given the order statistics $X_{n:k}$) independent, but are not identically distributed in general. Thus, the $Y_{[i]}$ may not be generally exchangeable r.v.'s, although marginally, the Y_i are i.i.d.r.v.'s.

Another approach to multivariate models is to formulate the conditional distribution of one coordinate given the others, and quantile functions relating to such a conditional d.f. may then be used as suitable regression functionals. This formulation, however, has an uncomfortable feature: the definition of the sample counterpart of the conditional d.f. may entail some arbitrariness, and in view of that, some complications may arise in the definition of the sample conditional quantiles and in the study of their properties. Generally, slower rates of convergence hold in such a case, and we shall refer to these in a later section.

Some other definitions and notions will be introduced as and when they are relevant in the subsequent sections.

3. Some basic asymptotics in order statistics

We look back at (2.1) and make use of the DeMoivre–Laplace Central Limit Theorem on the binomial distribution. This leads us to the following basic result for sample quantiles: whenever the d.f. F has a continuous and positive density $f(x)$ at the population p-quantile ξ_p,

$$n^{1/2}(X_{n:[np]} - \xi_p) \rightsquigarrow \mathcal{N}(0, p(1-p)/f^2(\xi_p)). \tag{3.1}$$

It is not necessary to choose particularly the rank $k = [np]$, and for the above asymptotic normality result to hold, it suffices to choose a k, such that $|k - [np]| = o(n^{1/2})$. In this setup, it is not necessary to assume that the D.F. F has a finite second moment (as is needed for the sample mean), but the positivity of $f(\xi_p)$ is needed here but not for the sample mean. If we assume further that the density $f(\cdot)$ is absolutely continuous with a bounded derivative at ξ_p, then following Bahadur (1966), we may obtain the following asymptotic representation, termed the *Bahadur representation* for sample quantiles:

$$X_{n:[np]} = \xi_p - [f(\xi_p)]^{-1} \sum_{i=1}^{n} \{I(X_i \leq \xi_p) - p\} + R_n(p) , \tag{3.2}$$

where $I(A)$ stands for the indicator function of a set A, and

$$R_n(p) = O(n^{-3/4}(\log n)^{1/2}) \text{ almost surely (a.s.), as } n \to \infty . \tag{3.3}$$

Ghosh (1971) managed to show that sans the differentiability condition on f, (3.2) holds with $R_n(p) = o_p(n^{-1/2})$. Various asymptotic results on sample quantiles emerge from (3.1), (3.2) and its weaker version due to Ghosh (1971). First, the result extends directly to the case of a vector $[X_{n:k_1}, \ldots, X_{n:k_q}]'$ of sample quantiles, where $k_j = [np_j]; 0 < p_1 < \cdots < p_q < 1$. Parallel to (3.1), here we would have a q-variate normal distribution with null mean vector and dispersion matrix $\Gamma = ((\gamma_{jl}))$, where $\gamma_{jl} = (p_j \wedge p_l - p_j p_l)/[f(\xi_{p_j})f(\xi_{p_l})]$, for $j, l = 1, \ldots, q$. The Bahadur representation extends to this vector case under parallel regularity conditions. Another important extension relates to multivariate observations where the

$\mathbf{X}_i = (X_{i1}, \ldots, X_{im})'$ are m-vectors, and for each of the m coordinates, we have the ordered observations. For the jth coordinate, the sample p_j-quantile is denoted by $X_{n:[np_j]}^{(j)}$ and its population counterpart by $\xi_{p_j}^{(j)}$, for $j = 1, \ldots, m$, where all the p_j belong to the open interval $(0, 1)$, but they need not be all distinct or ordered. In this case, using the weaker version of (3.2) for each coordinate, we arrive at an asymptotic multinormal distribution with null mean vector and dispersion matrix Γ having the elements γ_{jl}, $j, l = 1, \ldots, m$, where γ_{jj} has the same form as in before, but related to the jth marginal density $f_j(\cdot)$, while for $j \neq l$, $\gamma_{jl} = (p_{jl} - p_j p_l)/[f_j(\xi_{p_j}^{(j)})][f_l(\xi_{p_l}^{(l)})]$, and $p_{jl} = P\{X_{ij} \leq \xi_{p_j}^{(j)}, X_{il} \leq \xi_{p_l}^{(l)}\}$, for $j \neq l = 1, \ldots, m$. Finally, an extension of this result for more than one quantile for each coordinate also emerges from the same Bahadur representation in its weaker form.

We have noted earlier that the empirical d.f. and order statistics are interrelated, and this feature provides an asymptotic resolution of one via the other. As in after (2.11), we denote the uniform empirical d.f. by $G_n(\cdot)$, and consider the so called *uniform empirical distributional process* $W_n^o = \{W_n^o(t), t \in [0, 1]\}$, defined by

$$W_n^o(t) = n^{1/2}(G_n(t) - t), \quad t \in [0, 1] \ . \tag{3.4}$$

Also, let $W^o = \{W^o(t), t \in [0, 1]\}$ be a Gaussian function on the unit interval $[0, 1]$, where $EW^o(t) = 0$ and $E\{W^o(s)W^o(t)\} = \min(s, t)$, for all $s, t \in [0, 1]$. In the literature, W^o is referred to as a *Brownian bridge* or a *tied-down Wiener process*. Then, we have the basic weak convergence result: As n increases,

$$W_n^o \text{ converges in law to } W^o \ . \tag{3.5}$$

It is well known [viz., Billingsley (1968)] that for the Brownian bridge, we have, for every $\lambda(> 0)$,

$$P\{W^o(t) \leq \lambda, \forall t \in [0, 1]\} = 1 - e^{-2\lambda^2} \ , \tag{3.6}$$

and

$$P\{|W^o(t)| \leq \lambda, \forall t \in [0, 1]\} = 1 - \sum_{k \geq 1}(-1)^{k-1} e^{-2k^2\lambda^2} \ . \tag{3.7}$$

The last two equations provide simple asymptotics for the classical *Kolmogorov–Smirnov* type statistics. Specifically, for every continuous univariate d.f. F,

$$\lim_{n \to \infty} P\{\sup_x n^{1/2}[F_n(x) - F(x)] \geq \lambda\} = e^{-2\lambda^2} \ ; \tag{3.8}$$

$$\lim_{n \to \infty} P\{\sup_x n^{1/2}|F_n(x) - F(x)| \geq \lambda\} = \sum_{k \geq 1}(-1)^{k-1} e^{-2k^2\lambda^2} \tag{3.9}$$

For finite sample sizes, the equality signs can be replaced by $\geq$ signs, and also for F admitting jump discontinuities, we have a similar $\geq$ sign holding in this setup.

These provide access to one or two-sided (asymptotic) *condifence bands* for F, and by inversion, one also gets asymptotic confidence bands for the population quantile function $Q(\cdot) = \{Q(t) = F^{-1}(t), t \in (0,1)\}$. We may also define the *uniform empirical quantile process* $W_n^*(\cdot) = \{W_n^*(t), t \in (0,1)\}$, by letting

$$W_n^*(t) = n^{1/2}\{G_n^{-1}(t) - t\}, \quad t \in (0,1) \ . \tag{3.10}$$

Then the same asymptotic distributions (as in the case of W_n^o) hold here. However, for a d.f. F different from the uniform(0,1), we need to incorporate the density function in the definition of W_n^*, and for the related asymptotics, appropriate regularity conditions are also needed. We shall discuss these briefly later on.

These asymptotic results may not in general apply to the sample extreme values. Since the details of such extreme value asymptotics are provided in some other chapters of this volume, we shall refrain ourselves from a detailed discussion of them. Other asymptotic results cropping up in the subsequent sections will be presented as they arise.

4. Robust estimation and order statistics: asymptotics in applications

In the fifties, the *generalized least squares* (GLS) methodology has been successfully adopted for the *location-scale* family of (univariate) densities to derive BLUE (*best unbiased linear estimators*) of location and scale parameters based on order statistics; the methodology immediately covered the case of *censored* observations (including right and left truncation or censoring models) as well as the case of *selected few order statistics*; an excellent treatise of this subject matter is due to Sarhan and Greenberg (1962), believed to be the first one of its kind in a series of books, and some other monographs which are in citation in other chapters of this volume also deal with this important topic. The related developments on linear models have taken place mostly during the past twenty-five years or so. In this setup, the primary task has been the compilation of extensive tables for the variance-covariance matrix and expectation vector of order statistics in a finite sample setting, and for various parent distributions, this formidable task was accomplished within a short span of time. Nevertheless, it came as no surprise that asymptotic expressions for the expectations and variance-covariances of order statistics (as presented in the last section) can be profitably used to simplify the procedure, at least, for large sample sizes, and in this manner, the BLUE theory laid down the foundation of asymptotically (A-)BLUE estimators. In this quest, there has been even a shift from a purely parametric formulation to more comprehensive nonparametric ones wherein considerable emphasis has been placed on the so called *robustness* properties (against outliers, error contamination, gross errors and other forms of model departures).

Typically, (A-)BLUE can be expressed as

$$L_n = c_{n1}X_{n:1} + \cdots + c_{nn}X_{n:n} \ , \tag{4.1}$$

where the coefficients $\{c_{ni}\}$ depend on the argument i/n as well as the underlying d.f. F. The sample quantiles correspond to the particular case where only one of the c_{ni} is different from 0, although for finite sample sizes often an average of two or three consequtive order statistics is taken as a smooth quantile estimator. In this setup, we may as well conceive of a suitable *score function* $J_n(\cdot) = \{J_n(t), t \in (0,1)\}$, and express

$$c_{ni} = n^{-1}J_n(i/n), \quad \text{for } i = 1,\ldots,n \ . \tag{4.2}$$

From the last two equations, it follows that L_n may be equivalently written as

$$L_n = \int_{-\infty}^{\infty} xJ_n(F_n(x)) \, dF_n(x) \ . \tag{4.3}$$

In passing we may remark that in a Type I (right) censoring scheme, for a prefixed point $\xi < \infty$, the observations having larger values are censored. Thus, there are a (random) number (M) of the c_{ni} corresponding to these censored observations which are to be taken as equal to 0, while ordering of the uncensored observations yields the coefficients c_{ni} as in the uncensored case. Thus, for L_n we have a similar representation as in above, though $J_n(\cdot)$ will be taken to be equal to 0 at the upper end. A similar case arises with Type II (right) censoring, where there is an additional simplification that the number of censored observations is prefixed. Modifications for the left or both sided censoring schemes are straightforward. Bearing these features in mind, in an asymptotic setup, often, we justify that there exists a smooth score function $J(\cdot) = \{J(t), t \in (0,1)\}$, such that

$$J_n(t) \to J(t) \quad \text{as} \quad n \to \infty, \quad \forall t \in (0,1) \ , \tag{4.4}$$

where $J(\cdot)$ may vanish at the tail(s). Further noting that the empirical d.f. F_n converges almost surely to the true d.f. F, we may conceive of a function

$$\theta(F) = \int_{-\infty}^{\infty} xJ(F(x)) \, dF(x) = \int_0^1 F^{-1}(t)J(t) \, dt \ , \tag{4.5}$$

which is a linear functional of the underlying d.f. F. Much of the asymptotics deals with the behavior of $L_n - \theta(F)$ with special emphasis on its stochastic (as well as almost sure) convergence, asymptotic normality and related invariance principles. Most of these details can be found in the books of Serfling (1980) and Sen (1981c) as well as in some later ones in this area. For an up-to-date coverge at an advanced mathematical level, we may refer to Chapter 4 of Jurečková and Sen (1996). It follows from the above developments that an L-statistic is a linear functional of the empirical d.f. F_n, and hence, it belongs to the general class of (differentiable) statistical functionals, for which general asymptotics have been developed in a systematic manner. As such, we find it convenient to introduce such functionals and their related asymptotic theory, and incorporate them in the display of general asymptotic properties of a general class of L-estimators. In the von Mises–Hoeffding approach, a regular functional or estimable parameter is conceived as one which admits a finite-degree kernel that estimates it unbiasedly for all F belonging to a nonempty class, termed the domain of the functional. However, this finite-degree clause may not be always tenable; a very simple case is the population quantile or percentile which for an arbitrary F may not have a finite degree kernel. In the current state of art of these developments, basically the aim is to express a functional $T(F_n)$ in a Taylor's type expansion:

$$T(F_n) = T(F) + T'_F(F_n - F) + \text{rem}(F_n - F; T) \ , \tag{4.6}$$

where T'_F is the derivative of the functional, defined suitably, and the remainder term has some nice and manageable properties. Therefore there is a need to introduce suitable mode of differentiability (in functional spaces) so that the above representation works out well. The Frechet differentiability considered earlier in this context appears to be more stringent than generally needed in statistical applications, and at the present time, Hadamard or compact differentiability appears to be more convenient for statistical analysis.

Let $\mathscr{V}$ and $\mathscr{W}$ be two topological vector spaces and let $\mathscr{L}_1(\mathscr{V}, \mathscr{W})$ be the set of continuous linear transformations from $\mathscr{V}$ to $\mathscr{W}$. Let $\mathscr{A}$ be an open set of $\mathscr{V}$. Then a functional $\tau: \mathscr{A} \to \mathscr{W}$ is said to be Hadamard differentiable at $F \in \mathscr{A}$ if there exists a functional $\tau'_F \in \mathscr{L}_1(\mathscr{V}, \mathscr{W})$, termed the Hadamard derivative of τ at F, such that for any compact set $\mathscr{K}$ of $\mathscr{V}$,

$$\lim_{t \to 0} t^{-1}[\tau(F + tH) - \tau(F) - \tau'_F(tH)] = 0, \quad \text{uniformly for any } H \in \mathscr{K} \ . \tag{4.7}$$

Thus, we have from the last two equations,

$$\text{Rem}(tH, \tau) = \tau(F + tH) - \tau(F) - \tau'_F(tH) \ . \tag{4.8}$$

The concept of second-order Hadamard-differentiability has also been introduced in the literature, and we may refer to Jurečková and Sen (1996) for some detailed discussion; other pertinent references are all cited there. For L-functionals there are certain additional simplicities in the smooth case at least, and we shall refer to those later on. In this context we may note that though F_n estimates F unbiasedly, for a general nonlinear functional, $T(F_n)$ may not be an unbiased estimator of $T(F)$. Moreover, by virtue of the fact that $T'_F(F_n - F; T)$ is a linear functional, its general asymptotic properties (such as stochastic or a.s. convergence and asymptotic normality) can be established by standard methods, but its asymptotic variance may depend on the unknown F in a rather involved manner. Therefore there is a three front task endowed to the statisticians in using such functionals for drawing statistical conclusions:

(a) To eliminate or reduce the asymptotic bias of $T(F_n)$.
(b) To estimate the asymptotic mean squared error of $T(F_n)$ in a robust and efficient manner.
(c) To exhibit that the remainder term is negligible in the particular context.

Jackknifing and functional jackknifing play an important role in this context. To motivate jackknifing, we let $T(F_n) = T_n$ and suppose that

$$\mathrm{E}_F[T_n] = T(F) + n^{-1}a(F) + n^{-2}b(F) + \cdots, \tag{4.9}$$

where $a(F), b(F)$ etc., are possibly unknown functionals of the d.f. F. Let T_{n-1} be the same functional based on a sample of size $n-1$, so that $\mathrm{E}_F[T_{n-1}] = T(F) = (n-1)^{-1}a(F) + (n-1)^{-2}b(F) + \cdots$, and hence,

$$\mathrm{E}_F[nT_n - (n-1)T_{n-1}] = T(F) - [n(n-1)]^{-1}b(F) + O(n^{-3}). \tag{4.10}$$

Therefore, the order of the asymptotic bias is reduced to n^{-2} from n^{-1}. Motivated by this feature, from the base sample $(X_1, \ldots, X_n)$, we drop the ith observation, and denote the resulting estimator by $T_{n-1}^{(i)}$, for $i = 1, \ldots, n$. In the next step, we define the pseudovariables as

$$T_{n,i} = nT_n - (n-1)T_{n-1}^{(i)}, \quad i = 1, \ldots, n. \tag{4.11}$$

Then the jackknifed version T_n^J of the original estimator T_n is defined as

$$T_n^J = n^{-1} \sum_{i=1}^n T_{n,i}. \tag{4.12}$$

It follows from the last two equations that

$$\mathrm{E}_F[T_n^J] = T(F) + O(n^{-2}), \tag{4.13}$$

so that jackknifing effectively reduces the order of the asymptotic bias. Although this was the primary reason for introducing jackknifing (nearly half a century ago), these pseudovariables serve some other important purposes too. Toward this end, we define the jackknifed variance estimator V_n^J by

$$\begin{aligned}V_n^J &= (n-1)^{-1} \sum_{i=1}^n (T_{n,i} - T_n^J)^2 \\ &= (n-1) \sum_{i=1}^n (T_{n-1}^{(i)} - T_n^*)^2,\end{aligned} \tag{4.14}$$

where $T_n^* = n^{-1} \sum_{i=1}^n T_{n-1}^{(i)}$. We define a (nonincreasing) sequence of sub-sigma fields $\{\mathscr{C}_n\}$, by letting $\mathscr{C}_n = \mathscr{C}(X_{n:1}, \ldots, X_{n:n}: X_{n+j}, j \geq 1)$, for $n \geq 1$. Then proceeding as in Sen (1977), we may note that $T_n^* = \mathrm{E}\{T_{n-1}|\mathscr{C}_n\}$, and as a result,

$$T_n^J = T_n + (n-1)\mathrm{E}\{(T_n - T_{n-1})|\mathscr{C}_n\}, \tag{4.15}$$

so that jackknifing essentially adds the adjustment term based on the classical concept of conditional expectations. For a reverse martingale sequence, this adjustment is null. Similarly, we have

$$V_n^J = n(n-1)\mathrm{var}(T_{n-1}|\mathscr{C}_n), \tag{4.16}$$

whenever these conditional moments are defined properly. Again under suitable regularity conditions, the right hand side of the last equation converges a.s. to σ_T^2, the asymptotic mean squared error of $n^{1/2}[T_n - T(F)]$, so the second task is also accomplished by the jackknife technique. Finally, we note that Hadamard differentiability ensures that

$$|\text{Rem}(F_n - F; T)| = o(||F_n - F||) \;, \tag{4.17}$$

on the set where $||F_n - F|| \to 0$, and hence, the weak convergence results on the empirical d.f. imply the asymptotic negligibility of this remainder term. A better order of representation holds under second-order Hadamard-differentiability. In Jurečková and Sen (1996), considerable emphasis has been laid down to suitable first-order asymptotic distributional representation (FOADR) results for various nonlinear statistics. Typically, we have the following: As n increases,

$$T_n - T(F) = n^{-1} \sum_{i=1}^{n} \phi(X_i; F) + R_n \;, \tag{4.18}$$

where the score function ϕ is so normalized that

$$E_F \phi(X_i : F) = 0 \text{ and } E_F[\phi(X_i : F)]^2 = \sigma_T^2 \;, \tag{4.19}$$

and the remainder term R_n satisfies some smoothness properties. In such a case, even without the Hadamard differentiability of $T(F)$, the bias reduction and consistency of the variance estimator in the jackknifing method have been established. There are some other variants of the classical jackknife method, and delete k-jackknifing and functional jackknifing are noteworthy in this perspective. We will present a brief outline of these variants of the jackknife and we refer to Sen (1988a,b) and Jurečková and Sen (1996) for some discussion with good emphasis on the robustness aspects of these estimates.

As a simple example, we may consider the case of sample quantiles where the classical jackknife may not work out well. However, if instead of taking all possible subsamples of size $n-1$ from the base sample of size n, we take all possible $[\binom{n}{k}]$ subsamples of size $n-k$, where k is (moderately) large but k/n is small, then the delete-k jackknife works out well. On the other hand, in a regular case where the functional is sufficiently smooth, such a delete-k jackknife yields variance estimators that are asymptotically stochastically equivalent to the one provided by the classical jackknife [see, Sen (1989)]. This suggests that even in a regular case, instead of the classical jackknife, one may adopt a delete-k jackknife with a moderate value of k. For nonlinear statistics this may add more robustness flavor to the derived variance estimator. Along the same lines, we consider the functional jackknife. The pseudovariables $T_{n,i}$ defined by (4.11) are most likely to be less robust than the original T_n; this can easily be verified by considering a simple nonlinear estimator, such as the sample variance. Moreover, the jackknifed version in (4.12) being the simple average of these pseudovariables inherit all the nonrobustness properties of sample means. Thus, in this setup, error contaminations or outliers in the sample observations may have more noticeable impact on the pseudovariables in (4.11) and thereby the jackknifed estimator in (4.12) may lose its robustness prospects considerably. One possible way of recovering such robustness properties would be to lay a bit less emphasis on the bias-reduction role and jack-up the stability of variance estimators through the construction of other functionals. For statistical functionals admitting FOADR or Hadamard differentiability, it can be shown that the pseudovariables in (4.11) are exchangeable random variables (in a triangular scheme), so that a suitable measure of their central tendency can be advocated as an alternative jackknifed version. Such measures of central tendency are themselves typically L-functionals, so that instead of the simple mean in (4.12), we may consider an L-statistic based on the psdeudovariables in (4.11). In this context, to emphasize on the robustness aspects, we may consider analogues of the trimmed mean or Winsorized mean of these pseudovariables, or some other L-statistics for which the coefficients in the two tails are taken as equal to 0. Another possibility is to use a rank-weighted mean [Sen (1964)] of these pseudovariables as a jackknifed version. If $T_{n,(1)}, \ldots, T_{n,(n)}$ stand for the order statistics corresponding to the unordered pseudovariables in (4.11), then as in Sen (1964), we may define a kth order rank-weighted average as

$$T_{nJ,k} = \binom{n}{2k+1}^{-1} \sum_{j=k+1}^{n-k} \binom{j-1}{k}\binom{n-j}{k} T_{n,(j)}, \qquad (4.20)$$

where for $k = 0$, we have the mean of the pseudovariables (most nonrobust), while for $k = [(n+1)/2]$, we have their median (most robust). A choice of k as small as 2 or 3 may induce considerable robustness without much sacrifice of the efficiency aspect. Moreover, such rank weighted means are smooth L-functionals, and hence, their robustness properties are retained to a greater extent. In fact, they are also expressible as U-statistics, so that variance estimation by the jackknifing methodology can be easily accomplished. We refer to Sen (1988a,b) for various asymptotic properties of such functional jackknifed estimators and their related variance estimators.

It follows from the above discussion that under appropriate regularity conditions, for a statistical functional $T(F_n)$ there exist a suitable jackknifed version T_n^J and a jackknifed variance estimator V_n^J, such that as n becomes large,

$$n^{1/2}(T_n^J - T(F))/[V_n^J]^{1/2} \to_{\mathscr{D}} \mathscr{N}(0,1) . \qquad (4.21)$$

The last equation provides the desired tool for attaching a confidence interval for $T(F)$ based on $\{T_n^J, V_n^J\}$ and also for testing suitable hypotheses on the $T(F)$. In both the cases, due emphasis can be placed on the underlying robustness aspects. A similar picture holds for Efron's (1982) bootstrap methods whenever the asymptotic normality property holds; we refer to Sen (1988b) for some detailed discussion.

We conclude this section with some pertinent discussion on various types of censoring schemes arising in statistical applications, and on the scope of asymptotics already presented in the uncensored case. As has been discussed before, in a Type I (right) censoring case, observations beyond a truncation point are censored. This leads to an L-statistics where a (random) number of coefficients of the extreme order statistics are taken to be 0. This situation is similar to the case of the trimmed L-estimators, and we will discuss this in the next section. Secondly, consider the case of Type II (right) censoring where a given number of the extreme c_{ni} are taken to be equal to 0. This results in a greater simplification of the boundedness and differentiability conditions for the associated L-functionals, so that the asymptotic theory for the uncensored case treated earlier remains applicable in this context too. The situation is somewhat more complex in the case of random censoring schemes, and we present here briefly the necessary modifications.

In random censoring, we conceive of a set of censoring variables $C_1, \ldots, C_n$ which are i.i.d. according to a d.f. G, such that C_i and X_i are independent. Then the observable random elements are

$$Z_i = \min(X_i, C_i) \quad \text{and} \quad \delta_i = I(X_i = Z_i), \quad i = 1, \ldots, n \ . \tag{4.22}$$

Let us denote the order statistics for the Z_i by $Z_{n:1} \leq \cdots \leq Z_{n:n}$. Further, if $\overline{F}$ and $\overline{G}$ stand for the survival functions corresponding to the d.f.s F and G respectively, then the survival function for the Z_i is given by

$$\overline{H}(x) = 1 - H(x) = \overline{F}(x)\overline{G}(x), \quad x \in \mathcal{R} \ , \tag{4.23}$$

so that the $Z_{n:i}$ are intricately related to the empirical d.f. related to $H = 1 - \overline{H}$. For the estimation of the percentile points or other measures of the d.f. F, we therefore need first to estimate the d.f. F itself. A very popular estimator, known as the Kaplan–Meier (1958) *product-limit* (PL-)estimator, can be formulated as follows. Let

$$\begin{aligned} N_n(y) &= \sum_{i=1}^n I(Z_i > y), \quad y \in \mathcal{R}; \\ \alpha_i(y) &= I(Z_i \leq y, \delta_i = 1), \quad i = 1, \ldots, n; \ y \in \mathcal{R}, \\ \tau_n &= \max\{Z_i : 1 \leq i \leq n\} \ . \end{aligned} \tag{4.24}$$

Then the PL-estimator of $\overline{F}(y)$ is given by

$$\begin{aligned} \overline{P}_n(y) &= \prod_{i=1}^n \{N_n(Z_i)/[N_n(Z_i) + 1]\}^{\alpha_i(y)} I(y \leq \tau_n) \\ &= \prod_{\{i : Z_i \leq y\}} \left\{ \frac{n\overline{H}_n(Z_i)}{n\overline{H}_n(Z_i) + 1} \right\}^{\delta_i}, \quad y \leq \tau_n \ , \end{aligned} \tag{4.25}$$

where $\overline{H}_n = 1 - H_n$ is defined by

$$\overline{H}_n(y) = n^{-1} N_n(y) = n^{-1} \sum_{i=1}^n I(Z_i > y), \quad y \in \mathcal{R} \ . \tag{4.26}$$

As such, it is possible to replace the empirical d.f. F_n by the product-limit estimator P_n and define an L-statistic in this random censoring scheme as a suitable (linear) functional of P_n. The weak as well as strong convergence properties of the

PL-estimator have been extensively studied in the literature [viz., Shorack and Wellner (1986)], and as such, Hadamard differentiability and other tools adapted in the uncensored case may also be used in the random censoring case to yield parallel results. Finally, censoring in practice may be considerably different from that in theory and methodology [viz., Sen (1995c)], and there are certain basic issues that merit a careful examination. In a later section, we will discuss more about these findings.

5. Trimmed LSE and regression quantiles

L-estimators introduced in earlier sections work out well for the location-scale models. However, for the regression model

$$\mathbf{Y}_n = \mathbf{X}_n \boldsymbol{\beta} + \mathbf{e}_n ,\qquad(5.1)$$

where $\mathbf{Y}_n$ is the (n-)vector of sample observations, $\mathbf{X}_n$ is an $n \times p$ matrix of known regression constants, known as the design matrix, and $\mathbf{e}_n$ is an n-vector of i.i.d.r.v.'s with a d.f. F having location 0, the concept of sample quantiles deems a regression-equivariance property which is not apparent in general. Koenker and Bassett (1978) came up with a very clever idea of regression quantiles which satisfy such a basic requirement and thereby pave the way for further fruitful developments of robust estimation theory in linear models. Consider a fixed α ($0 < \alpha < 1$). Then their proposed α-regression quantile $\hat{\boldsymbol{\beta}}_n(\alpha)$ is defined by

$$\hat{\boldsymbol{\beta}}_n(\alpha) = \arg\min\left\{\sum_{i=1}^n \rho_\alpha(Y_i - \mathbf{x}'_i \mathbf{t}): \mathbf{t} \in \mathscr{R}_p\right\} ,\qquad(5.2)$$

where $\mathbf{Y}_n = (Y_1, \ldots, Y_n)'$, $\mathbf{x}'_i$ is the ith row of $\mathbf{X}_n$, for $i = 1, \ldots, n$, and

$$\rho_\alpha(x) = |x|\{(1 - \alpha)I(x < 0) + \alpha I(x > 0)\}, \quad x \in \mathscr{R} .\qquad(5.3)$$

Koenker and Bassett (1978) also managed to show that their regression quantile can also be characterized as an optimal solution ($\boldsymbol{\beta}$) of the following linear programming problem:

$$\begin{aligned}
&\alpha \sum_{i=1}^n r_i^+ + (1-\alpha)\sum_{i=1}^n r_i^- = \min ;\\
&\sum_{j=1}^p x_{ij}\beta_j + r_i^+ - r_i^- = Y_i, \quad i = 1,\ldots,n ;\\
&\beta_j \in \mathscr{R}, j = 1,\ldots,p;\ r_i^+ \geq 0,\ r_i^- \geq 0,\ i = 1,\ldots,n ;
\end{aligned}\qquad(5.4)$$

where r_i^+ (r_i^-) is the positive (negative) part of the residual $Y_i - \mathbf{x}'_i \boldsymbol{\beta}$, $i = 1,\ldots,n$. The Koenker–Bassett regression quantiles also lead to trimmed least squares estimators of the regression parameters. We choose two values $0 < \alpha_1 < \alpha_2 < 1$, and define the regression quantiles $\hat{\boldsymbol{\beta}}_n(\alpha_1)$ and $\hat{\boldsymbol{\beta}}_n(\alpha_2)$ as in before. Let then

$$c_{ni} = I\left\{\mathbf{x}_i' \hat{\boldsymbol{\beta}}_n(\alpha_1) < Y_i < \mathbf{x}_i' \hat{\boldsymbol{\beta}}_n(\alpha_2)\right\}, \quad i = 1, \ldots, n , \tag{5.5}$$

and let $\mathbf{C}_n = \text{Diag}(c_{n1}, \ldots, c_{nn})$. Then the (α_1, α_2)–trimmed least squares estimator $\mathbf{T}_n(\alpha_1, \alpha_2)$ is expressible as

$$\mathbf{T}_n(\alpha_1, \alpha_2) = (\mathbf{X}_n' \mathbf{C}_n \mathbf{X}_n)^- \mathbf{X}_n' \mathbf{C}_n \mathbf{Y}_n , \tag{5.6}$$

where $\mathbf{D}^-$ stands for a generalized inverse of $\mathbf{D}$. For the simple location model the above equation reduces to that for the classical trimmed mean. In this context it may be noted that the centering constant for this trimmed least squares estimator may differ from the true $\boldsymbol{\beta}$ when α_1 and α_2 are not complementary to each other or the d.f. F is not symmetric. Hence, in practice, it is tacitly assumed that the error distribution is symmetric about 0, and we take $\alpha_1 = 1 - \alpha_2 = \alpha$, for some $0 < \alpha < 1/2$.

Jurečková and Sen (1996, Chapter 4) have considered the asymptotic theory of regression quantiles and trimmed least squares estimators in general linear models in a unified and systematic manner. It follows from their general results that under mild regularity assumptions, denoting by $\mathbf{T}_n(\alpha) = \mathbf{T}_n(\alpha, 1 - \alpha)$, we have the FOADR:

$$\mathbf{T}_n(\alpha) - \boldsymbol{\beta} = (1 - 2\alpha)^{-1}(\mathbf{X}_n' \mathbf{X}_n)^{-1} \sum_{i=1}^n \mathbf{x}_i \psi(e_i) + o_p(n^{-1/2}) , \tag{5.7}$$

where $\psi(e) = F^{-1}(\alpha)$, e, or $F^{-1}(1 - \alpha)$, according as e is $< F^{-1}(\alpha)$, $F^{-1}(\alpha) \le e \le F^{-1}(1 - \alpha)$, or $> F^{-1}(1 - \alpha)$. The above representation also yields the following asymptotic (multi-)normality result:

> Under mild regularity assumptions, as n increases, $n^{1/2}[\mathbf{T}_n(\alpha) - \boldsymbol{\beta}]$ has asymptotically a p-variate normal distribution with null mean vector and dispersion matrix $\sigma^2(\alpha, F)\mathbf{Q}^{-1}$, where
>
> $$\sigma^2(\alpha, F) = (1 - 2\alpha)^{-1}\left\{\int_\alpha^{1-\alpha} (F^{-1}(u))^2 \, du + 2\alpha(F^{-1}(\alpha))^2\right\} , \tag{5.8}$$
>
> and it is assumed that $n^{-1}\mathbf{X}_n'\mathbf{X}_n$ converges to a positive definite limit $\mathbf{Q}$.

It may be noted that for the location model, Q reduces to 1 and (5.8) specifies the asymptotic mean squared error of the classical trimmed mean. Moreover higher-order asymptotics for such estimators in the general case, where α_1 and $1 - \alpha_2$ are not necessarily equal or F is possibly asymmetric, have been considered in detail in Section 4.7 of Jurečková and Sen (1996). We therefore omit these details.

We conclude this section with some remarks on robustness properties of regression quantiles and trimmed least squares estimators in linear model. The main motivation for such estimators is to borrow the robustness properties of sample quantiles through the regression-equivariance of these estimators, and at the sametime, by choosing α small enough, to retain their asymptotic efficiency to a greater extent. Depending on the largeness of the sample size (n), one may choose α sufficiently small to achieve this dual goal. Finally, we may note that such regression quantiles have also led to the development of another important class of robust estimators termed the regression rank scores estimators (Gutenbrunner and Jurečková, 1992) which have close affinity to the classical R-estimators of regression parameters. Since these have been presented in a unified manner in Chapter 6 of Jurečková and Sen (1996), we refrain ourselves from going over their details.

6. Asymptotics for concomitants of order statistics

For a sample of bivariate observations (X_i, Y_i), $i = 1, \ldots, n$, we define the concomitants of the order statistics $(X_{n:i})$ by $Y_{n[i]}$, $i = 1, \ldots, n$ as in (2.15), and note that they are conditionally (given the $X_{n:i}$) independent but not necessarily identically distributed random variables. If, however, X and Y are stochastically independent then the $Y_{n[i]}$ are i.i.d., and hence, in testing for stochastic independence of (X, Y), such concomitants of order statistics can be effectively used. Bhattacharya (1974, 1976) and Sen (1976a) considered general asymptotics for the partial sum sequence

$$S_{nk} = \sum_{j \leq k} \{Y_{n[j]} - m(X_{n:j})\}, \quad k = 1, \ldots, n; \ S_{n0} = 0 , \tag{6.1}$$

where we set

$$m(X_{n:k}) = \mathrm{E}(Y_{n[k]} | X_{n:k}), \quad k = 1, \ldots, n , \tag{6.2}$$

so that they represent the regression function of Y on X. There is also an intricate relationship between the concomitant of order statistics and mixed-rank statistics for testing bivariate independence [viz., Sen (1981b)]. Towards this, we define the marginal ranks of the X_i and Y_i by R_{ni} and S_{ni} respectively, so that

$$R_{ni} = \sum_{j=1}^{n} I(X_j \leq X_i), \quad S_{ni} = \sum_{j=1}^{n} I(Y_j \leq Y_i), \quad i = 1, \ldots, n . \tag{6.3}$$

Then a typical rank statistic for testing the hypothesis of independence of (X, Y) is of the following type:

$$M_n = \sum_{i=1}^{n} a_n(R_{ni}) b_n(S_{ni}) , \tag{6.4}$$

where the $a_n(k), b_n(k)$, $k = 1, \ldots, n$ are suitable scores. Side by side, we may define a mixed-rank statistic as

$$Q_n = \sum_{i=1}^{n} a_n(R_{ni}) b(Y_i) , \tag{6.5}$$

where $b(\cdot)$ is a suitable score function. It is clear that we may rewrite Q_n equivalently as

$$Q_n = \sum_{j=1}^{n} a_n(j) b(Y_{n[j]}) , \qquad (6.6)$$

so that Q_n is a linear combination (of functions) of the concomitants of order statistics. In that way the general asymptotic results considered by Sen (1981b) extend the results for partial sums, mentioned earlier, to more general functions. Ghosh and Sen (1971) considered a general class of rank order tests for regression with partially informed stochastic predictors. These statistics are all mixed (linear) rank statistics, developed before the concept of concomitants of order statistics surfaced in the statistical literature, and hence, the general asymptotics presented in Ghosh and Sen (1971) may as well be streamlined along the lines of Sen (1981b). An accompanying article by H. A. David and H. N. Nagaraja (in this volume) deals with some other aspects of these induced order statistics, and hence, we omit the details to avoid duplications to a greater extent.

There is, however, an important area where the concept of concomitants of order statistics has mingled with a more general concept of nonparametric regression function, and we will discuss this in more detail in a later section. In this section, we provide only an outline of this linkage and stress the role of asymptotic theory in that context. Suppose that we aim to estimate the unknown regression function

$$m(x) = E\{Y|X = x\}, \quad x \in \mathscr{C}, \text{ a compact set of } \mathscr{R} . \qquad (6.7)$$

Although in the normal case this regression function is linear and hence is describable by a finite dimensional parameter, in a general nonnormal setup, the regression function may not be linear or even be describable in terms of a finite number of parameters. For this reason, a regression functional approach is more appealing from practical point of view. But, like the sample mean in the univariate case, $m(x)$ in this conditional model may be quite sensitive (even to a greater extent) to error contaminations or outliers, and sans its plausible linearity, its appeal is greatly lost in a nonparametric setup. Therefore instead of defining the regression function in terms of the conditional mean, we may also define this in terms of the conditional median or some other robust measure of the central tendency of the conditional distribution of Y, given $X = x$. In either case, we need to estimate this unknown conditional d.f. (at a given x) from the sample data. Among various possibilities, the nearest neighbor (NN-) method and the kernel-smoothing method have emerged as the most popular ones. In a kernel method, we choose (i) some known density $K(x)$ possessing some smoothness properties (viz., unimodality, symmetry, differentiability upto a certain order and bounded support etc.) and (ii) a sequence $\{h_n\}$ of positive numbers converging to 0 (as $n \to \infty$), so that defining by $F_n(x, y)$ the sample (empirical) (bivariate) d.f., we may set for an $a \in \mathscr{C}$,

$$\hat{m}_n(a) = \frac{\int\int yK((a-x)/h_n)\,dF_n(x,y)}{\int\int K((a-x)/h_n)\,dF_n(x,y)}, \tag{6.8}$$

and propose this as a nonparametric estimator of $m(a)$, $a \in \mathscr{C}$. It is clear that such an estimator becomes a linear combination of the induced order statistics with stochastic weights depending on the order statistics of the X values lying in the vicinity of the chosen point a. A somewhat simpler situation is encountered in a NN-method. There we define the pseudovariables (Z_i) by letting

$$Z_i = |X_i - a|, \quad i = 1, \ldots, n, \tag{6.9}$$

which depend on the pivot a and are nonnegative r.v.'s. We denote the corresponding order statistics by $Z_{n:1} \leq \cdots \leq Z_{n:n}$. Consider then a nondecreasing sequence $\{k_n\}$ of positive integers, such that k_n goes to ∞, but $n^{-1}k_n \to 0$ as $n \to \infty$. Denote the concomitants of these order statistics by $Y_{n[j]}(a)$, for $j = 1, \ldots, n$. We define an empirical d.f. $F_n^*(\cdot, a)$ by letting

$$F_n^*(y; a) = k_n^{-1} \sum_{j \leq k_n} I(Y_{n[j]}(a) \leq y), \quad y \in \mathscr{R}. \tag{6.10}$$

Then, we may consider an estimator

$$\tilde{m}_n(a) = \int y\,dF_n^*(y; a) = k_n^{-1} \sum_{j \leq k_n} Y_{n[j]}(a), \quad a \in \mathscr{C}. \tag{6.11}$$

The simplicity of (6.11) over (6.8) is evident from the constant weight (i.e., k_n^{-1}) against stochastic ones which depend on the choice of the kernel density as well as the order statistics for the X characteristics in the neighborhood of a. Nevertheless, generally both the estimators are biased, and letting $k_n \sim nh_n$, it can be shown that both the methods possess similar bias and mean squared error properties. We refer to Bhattacharya and Gangopadhyay (1990) for some deeper asymptotic studies relating to such estimators. In a general framework, we may consider suitable (L-)functionals of the conditional d.f. $F(y|x)$ and estimate the same by considering their sample counterparts based on $F_n^*(\cdot|x)$. We refer to Gangopadhyay and Sen (1992) where other references are also cited. From the point of view of applications, asymptotic normality and consistency properties of these estimators suffice, while the other deeper asymptotic results are mostly of academic interest only.

In the area of survival analysis (with covariates), the Cox (1972) regression model or the so-called proportional hazard model (PHM) occupies a focal point. This is essentially a semi-parametric model where the regression of the survival time on the covariates is formulated in a parametric form, but the (baseline) hazard function is treated as arbitrary. We consider the simple model where n subjects have simultaneous entry into the study-plan: the ith subject (having survival time Y_i and a set of possibly stochastic concomitant variates $\mathbf{Z}_i = (Z_{i1}, \ldots, Z_{iq})'$, for some $q \geq 1$) has the (conditional) hazard rate (given $\mathbf{Z}_i = \mathbf{z}_i$)

$$h_i(y|\mathbf{z}_i) = h_o(y) \cdot \exp(\boldsymbol{\beta}'\mathbf{z}_i), \quad i = 1, \ldots, n, \ y \in \mathcal{R}^+ , \tag{6.12}$$

where the baseline hazard function $h_o(y)$ is an unknown, arbitrary nonnegative function, and $\boldsymbol{\beta} = (\beta_1, \ldots, \beta_q)'$ parameterizes the regression (on the covariates). In survival analysis and life testing problems, the Y_i are nonnegative r.v.'s, and they are observed in a time-sequential setup, namely, the smallest one $(Y_{n:1})$ is observed first, the second smallest next, and so on, the largest one $(Y_{n:n})$ emerges last. Thus, just before the ith failure point $(Y_{n:i} - o)$, the risk set $\mathcal{R}_i$ consists of the $n - i + 1$ subjects which have not failed uptil that timepoint, for $i = 1, \ldots, n$. Then in the case of no censoring, proceeding as in Cox (1972), we obtain the partial likelihood function

$$L_n^*(\boldsymbol{\beta}) = \prod_{i=1}^n \left\{ \frac{e^{\boldsymbol{\beta}'\mathbf{Z}_{n[i]}}}{\sum_{j \in \mathcal{R}_i} e^{\boldsymbol{\beta}'\mathbf{Z}_j}} \right\} , \tag{6.13}$$

where we note that by definition $\mathbf{Z}_{n[i]}$ is the concomitant of the ith order statistic $Y_{n:i}$, $i = 1, \ldots, n$, and this definition is adapted to the vector case without any modification. For testing the null hypothesis $H_0: \boldsymbol{\beta} = \mathbf{0}$ against the alternative $H_1: \boldsymbol{\beta} \neq \mathbf{0}$, Cox (1972) considered the following test statistic (based on the scores)

$$\mathscr{L}_n^* = \mathbf{U}_n^{*\prime} \mathbf{J}_n^{*-} \mathbf{U}_n^* , \tag{6.14}$$

where

$$\begin{aligned} \mathbf{U}_n^* &= (\partial/\partial\boldsymbol{\beta}) \log L_n^*(\boldsymbol{\beta})|_{\boldsymbol{\beta}=\mathbf{0}} ; \\ \mathbf{J}_n^* &= (\partial^2/\partial\boldsymbol{\beta}\partial\boldsymbol{\beta}') \log L_n^*(\boldsymbol{\beta})|_{\boldsymbol{\beta}=\mathbf{0}} , \end{aligned} \tag{6.15}$$

and $\mathbf{A}^-$ stands for the generalized inverse of $\mathbf{A}$. Let us write

$$\overline{\mathbf{Z}}_{n[i]}^* = (n - i + 1)^{-1} \sum_{j \in \mathcal{R}_i} \mathbf{Z}_j, \quad i = 1, \ldots, n . \tag{6.16}$$

Then by some standard arguments [viz., Sen (1981a)] it follows that

$$\mathbf{U}_n^* = \sum_{i=1}^n \left\{ \mathbf{Z}_{n[i]} - \overline{\mathbf{Z}}_{n[i]}^* \right\} , \tag{6.17}$$

and

$$\mathbf{J}_n^* = \sum_{i=1}^n (n - i + 1)^{-1} \sum_{j \in \mathcal{R}_i} \left(\mathbf{Z}_j - \overline{\mathbf{Z}}_{n[i]}^* \right) \left(\mathbf{Z}_j - \overline{\mathbf{Z}}_{n[i]}^* \right)' . \tag{6.18}$$

Therefore $\mathbf{U}_n^*$ is a linear combination of the concomitants of the failure order statistics, and $\mathbf{J}_n^*$ is also completely expressible in terms of these concomitants of order statistics. In this context, the asymptotics for partial sums of concomitants of order statistics developed by Bhattacharya (1974, 1976) and Sen (1976a) may not be totally adoptable, and general asymptotics based on suitable martingale constructions developed by Sen (1981a) provide a better alternative for the study of the asymptotic properties of the Cox (1972) test.

In Section 4 we have mentioned about various censoring schemes as may arise in practice. In the context of survival analysis, such censoring schemes are more commonly encountered. For Type I or II censoring, the modifications for the Cox procedure are straightforward. Instead of the sample size n, in the expression for the score statistic and the second derivative matrix, the sum (over i) extends only over the observable failure points. In the case of random censoring, the problem becomes more complex, and the Cox (1972) ingenuity leads to a very clever construction of the partial likelihood function where the risk sets are only adapted to the failure points (but not the censored ones). Thus the cardinality of the risk set $\mathcal{R}_i$ will be a number, say, r_i which is the number of subjects that have not failed or dropped out before the timepoint $Y_{n:i}^*$, and $Y_{n:i}^*$ is defined as the ith failure point allowing possible censoring due to drop outs. Therefore if out of n units, m results in failure and the remaining $n - m$ are censored, we would have m terms in the likelihood score statistic. Again these terms are a (random) subset of the totality of n terms, and hence, the characterization in terms of concomitant of order statistics remains in tact, but some extra manipulations are needed to handle the asymptotic theory in a manageable way. The martingale approach of Sen (1981a) remains adoptable without any major alteration.

In survival analysis, time-sequential tests based on progressive censoring schemes have received due attention in the recent past [viz., Sen (1981c, Chapter. 11)]. In the current context, this relates to a repeated significance testing (RST) of the same hypotheses (H_0 vs. H_1) on accumulating data. Basically when the dataset relating to the flow of events upto the ith failure is acquired, one can construct a partial likelihood score test statistic as in the Cox case, for every $i \geq 1$. Let us denote this partial sequence of test statistics by

$$\{L_{n,i}^* ; i = 1, \ldots, m_n\} , \qquad (6.19)$$

where m_n stands for the number of failure points among the n units in the study-plan. Marginally, for each $L_{n,i}^*$, we may argue as in the case of total sample, and using the inherent permutational invariance structure (under H_0) of the anti-ranks, we obtain a distribution-free test statistic. Therefore, the task is to construct a suitable stopping rule related to the partial sequence in (6.19), so that the resulting sequential test has some desirable properties. The invariance principles for concomitants of order statistics studied by Sen (1981a) in the context of the Cox (1972) model enables us to provide suitable Gaussian process (for $q = 1$) or Bessel process (for $q \geq 2$) approximations, so that the usual sequential testing methodology can be easily adopted here when n is large. There is an additional simplification here: The stopping variable (say, K_n) cannot be greater than m_n which in turn cannot be larger than n, so that $n^{-1}K_n$ is a bounded nonnegative random variable. For details, we refer to Chapter 11 in Sen (1981c). Motivated by this feature of progressive censoring schemes in survival analysis, Sen (1979) has considered a general class of quantile (history) processes based on suitable functions of concomitants of order statistics and incorporate them in the construction of suitable time-sequential tests. We start with a triangular array of constants (vectors)

$$\{\mathbf{c}_{ni};\ i=1,\ldots,n\}\ .\tag{6.20}$$

These vectors may contain nonstochastic design variates and stochastic concomitant variates too. Then in a RST setting, we construct a partial sequence of induced linear order statistics:

$$\mathbf{T}_{nk}^{*}=\sum_{i=1}^{k}g(Y_{n:i})\left[\mathbf{c}_{n[i]}+(n-k)^{-1}\sum_{j=1}^{k}\mathbf{c}_{n[i]}\right],\quad k=1,\ldots,n\ ,\tag{6.21}$$

where we take without any loss of generality $\sum_{i=1}^{n}\mathbf{c}_{ni}=\mathbf{0}$, and conventionally, we let $\mathbf{T}_{n0}^{*}=\mathbf{0}$; for $k=n$, we have $\mathbf{T}_{nn}^{*}=\mathbf{T}_{n}^{*}=\sum_{i=1}^{n}g(Y_{n:i})\mathbf{c}_{n[i]}$. In the above setting, the functional form of $g(\cdot)$ is assumed to be given. In the Cox model, we let $g(y)=1,\forall\, y$, so that we have an extension of the Cox statistics where additional information on the survival function can be acquired through nonconstant $g(\cdot)$. However, in the case of a nonconstant $g(\cdot)$, we essentially endup with a mixed rank statistics, and hence, the basic distribution-free property prevailing in the Cox model may no longer be tenable here. Nevertheless, incorporating suitable martingale characterizations, invariance principles for the partial sequence in (6.21) have been studied by Sen (1979), and results parallel to the Cox model have also been derived.

7. Concomitant L-functionals and nonparametric regression

In parametric linear models, apart from the basic linearity of the regressors, independence, homoscedasticity and normality of the error component constitute the basic regularity assumptions. Yet in practice, there may not be sufficient incentives to take these regularity conditions for granted, and plausible departures from the model based assumptions can affect, often seriously, the performance characteristics of classical statistical tools. Moreover, in analysis of covariance (ANOCOVA) or other mixed-effects models, there are, in addition to the fixed-effect components, some other random-effects component(s) which may even violate these regularity assumptions to a greater extent. For example, the regression of the primary variate on these stochastic (concomitant) regressors may not be closely linear, and even so, the conditional variance of the primary variate, given the covariates, may not be a constant at all levels of these concomitant variates. Therefore from robustness and validity considerations it seems more appropriate to formulate suitable nonparametric regression models wherein the above regularity assumptions are relaxed to a certain extent.

It will be more convenient for us to start with the simple nonparametric regression model, introduced in the last section, and then consider the more general case of mixed-effects models with some semi-parametric flavor. The formulation in the bivariate case sketched in (6.7) extends directly to the case where the concomitant variables $\mathbf{X}_i$ are q-vectors, for some $q\geq 1$. However, statistical manipulations become cumbersome when q becomes large, and the rate of convergence of estimators of the type in (6.8) becomes slower. Moreover, the conditional mean, defined in (6.7)–(6.8), may generally be highly nonrobust. Hence, we may find it more appealing to take recourse to suitable robust location functionals of the conditional distribution function, and in this respect, L-functionals including conditional quantiles are very appropriate. In our presentation here we mainly deal with the NN-methodology and note that parallel results hold for the kernel method as well.

Consider a set $(\mathbf{X}_i, Y_i)$, $i = 1, \ldots, n$ of n i.i.d. random $(q+1)$ vectors, and let $F(y|\mathbf{x})$ stand for the conditional d.f. of Y, given $\mathbf{X} = \mathbf{x}$. A conditional functional is a function of this conditional d.f., and hence, depends on $\mathbf{x}$ as well. In the general case of q-dimensional stochastic X's, we conceive of a suitable metric $\rho: \mathcal{R}^q \times \mathcal{R}^q \to \mathcal{R}^+$, and for a chosen pivot $\mathbf{x}_o$, define the nonnegative random variables

$$D_i^o = \rho(\mathbf{X}_i, \mathbf{x}_o), \quad i = 1, \ldots, n. \tag{7.1}$$

We denote the ordered values of the D_i^o by $(0 \leq) D_{n:1}^o \leq \cdots \leq D_{n:n}^o$, where the superscript indicates the dependence of these variables on the chosen pivot $\mathbf{x}_o$. Also, we choose a sequence of positive integers $\{k_n\}$, such that as n increases k_n also does so, but $n^{-1}k_n$ goes to 0. Typically, we choose $k_n \sim O(n^{q/(q+4)})$. Next corresponding to the chosen metric ρ and the pivot $\mathbf{x}_o$, we define the anti-ranks S_i^o, $i = 1, \ldots, n$ by letting $D_{n:i}^o = D_{S_i^o}^o$, for $i = 1, \ldots, n$. Then the k_n-NN empirical d.f. at the pivot $\mathbf{x}_o$ is defined as

$$\hat{F}_{n,k_n}(y) = k_n^{-1} \sum_{i=1}^{k_n} I(Y_{S_i} \leq y), \quad y \in \mathcal{R}. \tag{7.2}$$

In the particular case of a conditional p-quantile function, we may set $\theta(F(\cdot|\mathbf{x}_o)) = \inf\{y: F(y|\mathbf{x}_o) \geq p\}$, so the corresponding k_n-NN estimator is

$$\hat{\theta}_{n,k_n}(\mathbf{x}_o) = \inf\{y: \hat{F}_{n,k_n}(y|\mathbf{x}_o) \geq [pk_n]/k_n\}, \quad p \in (0,1). \tag{7.3}$$

In the sameway for a linear functional $\int J(F(y|\mathbf{x}_o))a(y)\,dF(y|\mathbf{x}_o)$ with suitable score functions $J(\cdot)$ and $a(\cdot)$, we may consider the plug-in estimator

$$\hat{\theta}_{n,k_n}(\mathbf{x}_o) = \int J(\hat{F}_{n,k_n}(y|\mathbf{x}_o))a(y)\,d\hat{F}_{n,k_n}(y|\mathbf{x}_o), \tag{7.4}$$

which is typically a L-statistic in the set $\{Y_{S_i^o}, i \leq k_n\}$. Typically such L-functionals are location functional, so they are measures of central tendency of the conditional d.f. $F(\cdot|\mathbf{x}_o)$. This explains the relevance of order and concomitants of order statistics in the study of nonparametric regression. In this context, we need to allow the pivot $\mathbf{x}_o$ to vary over a (possibly compact) set $\mathcal{C} \in \mathcal{R}^q$, and formulate a functional cloud $\{\theta(F(\cdot|\mathbf{x}); \mathbf{x} \in \mathcal{C}\}$. In that way, we need to have deeper weak convergence results for some multidimensional time-parameter stochastic processes. Some general asymptotics in this vein are considered by Sen (1993b), where other pertinent references are also cited.

Next, we proceed to examine the role of conditional (L-)functionals in mixed-effects models. In a conventional (normal theory) model, we denote the primary, design and (stochastic) concomitant variates for the ith observation by Y_i, $\mathbf{t}_i$ and $\mathbf{Z}_i$ respectively, for $i = 1, \ldots, n$. Then conditional on $\mathbf{Z}_i = \mathbf{z}_i$, we have

$$Y_i = \boldsymbol{\beta}'\mathbf{t}_i + \boldsymbol{\gamma}'\mathbf{z}_i + e_i, \quad i = 1, \ldots, n, \tag{7.5}$$

where $\boldsymbol{\beta}$ and $\boldsymbol{\gamma}$ are the regression parameter vectors for the fixed and random effects components, and the errors e_i are i.i.d. normal with null mean and a positive (unknown) variance σ^2. The e_i are assumed to be independent of the concomitant variates $\mathbf{Z}_i$, and the assumed joint normality of $(\mathbf{Z}_i, e_i)$ yields homoscedasticity, linearity of regression as well as the normality in the conditional setup. However, sans this joint normality (which may often be very questionable), a breakdown may occur in each of these three basic postulations. As such, two different models have been proposed to enhance robustness properties of statistical analysis tools. First, with respect to the linear model, assume that the d.f. $F(e|\mathbf{z})$ of e_i, given $\mathbf{Z}_i = \mathbf{z}$, is independent of $\mathbf{z}$ and is continuous. Thus only the normality part of the basic assumptions is relaxed here. In this still linear setup, the classical procedure works out well in an asymptotic setup when F has a finite second moment. However, it remains vulnerable to plausible departures from linearity as well as homoscedasticity. For this reason, in a nonparametric formulation [viz., Puri and Sen (1985, Chapter 8)] it is generally assumed that $F_i(y|\mathbf{z}) = P\{Y_i \leq y|\mathbf{Z}_i = \mathbf{z}\} = F(y - \boldsymbol{\beta}'\mathbf{t}_i|\mathbf{z})$, $i = 1, \ldots, n$, where F is arbitrary and continuous. Thus here we conform to a parametric (linear) form for the fixed-effects variables but to a nonparametric one for the stochastic covariates. In order to quantifying further this model in terms of appropriate regression functionals, we define a translation-invariant functional $\theta(F(\cdot|\mathbf{z}))$ (typically a measure of location of the conditional d.f. $F(\cdot|\mathbf{z})$), and consider the following quasi-parametric model:

$$\theta(F_i(\cdot|\mathbf{z})) = \theta(F(\cdot|\mathbf{z})) + \boldsymbol{\beta}'\mathbf{t}_i, \quad i = 1, \ldots, n. \tag{7.6}$$

This model has a finite dimensional regression parameter for the design variates but a regression functional for the random effect components. From robustness considerations this model appears to be more appropriate than the others. However, in this formulation the finite dimensional regression parameter $\boldsymbol{\beta}$ is estimable with the conventional $\sqrt{n}$ rate of convergence, while the estimators of the regression-functional $\theta(F(\cdot|\mathbf{z}))$ have a slower rate of convergence. In this way we end up with a robust estimator of the regression functional without compromising much on the efficiency of the estimator of $\boldsymbol{\beta}$.

The basic idea is simple. Recall that the $\mathbf{Z}_i$ qualify as genuine concomitant variates if their distribution is unaffected by the design variates. Thus, if we consider the model in (7.6) and integrating over the concomitant variate $\mathbf{z}$, we obtain the marginal model (sans the $\mathbf{Z}_i$) where

$$Y_i = \boldsymbol{\beta}'\mathbf{t}_i + e_i^*, \quad i = 1, \ldots, n, \tag{7.7}$$

where the e_i^* are i.i.d.r.v.'s with a continuous d.f., say, F^*. This is the classical linear model for which L-estimators of $\boldsymbol{\beta}$ have nice properties (we may refer to Section 5 where regression quantiles and trimmed least squares estimators have been considered in the same setup). We denote such an L-estimator of $\boldsymbol{\beta}$ by $\widetilde{\boldsymbol{\beta}}_n$, and note that under appropriate regularity assumptions, as n increases,

$$n^{1/2}\{\widetilde{\boldsymbol{\beta}}_n - \boldsymbol{\beta}\} \to_{\mathscr{D}} \mathscr{N}_p(\mathbf{0}, \mathbf{Q}^{-1} \cdot \xi^2) , \tag{7.8}$$

where $\mathbf{Q}$ has been defined in Section 5 and ξ^2 is a positive scalar constant depending on the particular form of the L-estimator and the d.f. F^*. Note that the $\sqrt{n}$-consistency property of this estimator is a consequence of the above result. In the next step, we consider the aligned observations (residuals)

$$\widetilde{Y}_{ni} = Y_i - \widetilde{\boldsymbol{\beta}}_n' \mathbf{t}_i, \quad i = 1, \ldots, n , \tag{7.9}$$

and observe that by virtue of (7.8) and the bounded nature of the (fixed-)$\mathbf{t}_i$, the perturbations of the residuals in (7.9) (around their true values) are $O_p(n^{-1/2})$. Next we consider the set of aligned vectors $(\widetilde{Y}_{ni}, \mathbf{Z}_i)$, $i = 1, \ldots, n$. On this set we incorporate the methodology of conditional functionals as has been presented before. However, we need to keep certain features in mind. First, these aligned stochastic vectors are not necessarily independent or even marginally identically distributed (though the $\mathbf{Z}_i$ are i.i.d.). Second, in view of the stochastic nature of the $\mathbf{Z}_i$, we need to formulate a set $\mathscr{C} \in \mathscr{R}^q$ (usually a compact one), and allowing $\mathbf{z}$ to vary over $\mathscr{C}$ and pointwise defining a conditional functional $\theta(\mathbf{z}) = \theta(F(\cdot|\mathbf{z}))$, $\mathbf{z} \in \mathscr{C}$, we obtain a functional process:

$$\Theta(\mathscr{C}) = \{\theta(\mathbf{z}): \mathbf{z} \in \mathscr{C}\} . \tag{7.10}$$

Thus our primary task is to construct a functional estimator (process) to estimate the functional in (7.10). Third, in this venture, we need to pay due attention to the apparently contradictory outcomes: bias due to possible oversmoothing and slower rate of convergence due to the infinite-dimensional nature of the parameter. Finally, a prescribed solution should be reasonably adoptable in actual practice when the sample size may not be enormously large. This last requirement may often preclude most of the contemporary refined local smoothing techniques based on pure asymptotic considerations.

A linearity theorem based approach has been considered by Sen (1995b, 1996b) and it works out reasonably well in this respect. If we denote by $Y_i^o = Y_i - \boldsymbol{\beta}' \mathbf{t}_i$, $i = 1, \ldots, n$, then we note that the $(Y_i^o, \mathbf{Z}_i)$ are i.i.d.r. vectors, so that the formulation of conditional quantiles presented earlier remains valid in this setup. In the next step, we consider a compact $\mathscr{K} \in \mathscr{R}^p$ and define

$$Y_{ni}^o(\mathbf{b}) = Y_i^o - n^{-1/2}\mathbf{b}'\mathbf{t}_i, \quad i = 1, \ldots, n, \quad \mathbf{b} \in \mathscr{K} . \tag{7.11}$$

Note that the concomitant variates are not affected by this regression-translation, and hence, corresponding to a pivot $\mathbf{z}_o \in \mathscr{C}$, we can define the k_n-NN order statistics $(D_{n:i}^o)$ as in before (7.2). With this definition, replacing the Y_i in (7.2) by the corresponding $Y_{ni}^o(\mathbf{b})$ in (7.11), we define the k_n-NN empirical d.f. by

$$\hat{F}_{n,k_n}(y|\mathbf{z}_o; \mathbf{b}), \quad \mathbf{z}_o \in \mathscr{C}, \quad \mathbf{b} \in \mathscr{K} \ . \tag{7.12}$$

Note that for $\mathbf{b} = \mathbf{0}$, the asymptotics presented before hold. Then Sen (1995b, 1996b) managed to show that under appropriate regularity conditions, as $n \to \infty$,

$$\sup_{\mathbf{z} \in \mathscr{C}} \sup_{\mathbf{b} \in \mathscr{K}} \sup_{y \in \mathscr{R}} k_n^{1/2} |\hat{F}_{n,k_n}(y|\mathbf{z}; \mathbf{b}) - \hat{F}_{n,k_n}(y|\mathbf{z}; \mathbf{0})| \to_P 0 \ . \tag{7.13}$$

This last result in turn implies that for smooth L-functionals (viz., Hadamard differentiable ones), the perturbation does not affect the asymptotics upto the order $k_n^{-1/2}$. Therefore conditional functionals based on the aligned residuals have the same (first order) asymptotic properties as the ones based on the true residuals. This methodology also suggests that we may improve the estimator of β based on the marginal model by gridding the compact set $\mathscr{C}$ into a number of buckets, estimating the parameters from each bucket by suitable L-functionals, and then combining these estimators by a version of the weighted least squares method. For details, we refer to Sen (1996b).

8. Applications of order statistics in some reliability problems

There are various problems cropping up in reliability and life testing models where order statistics play a vital role. The related asymptotics, mostly adopted from Sen (1995a), are outlined here with a view to foster more applications in practice. We consider four basic reliability models:

(a) Statistical strength of a bundle of parallel filaments;
(b) Reliability of K-component system – in series;
(c) Reliability of K-component system – in parallel;
(d) Systems availability, under spare and repair.

First, we consider Daniels' (1945) formulation of (a). Consider a bundle of n parallel filaments. Assume that a load to which the bundle may be subjected to is shared uniformly by the n filaments, whose individual strengths are denoted by $X_1, \ldots, X_n$ respectively, and further that the X_i are i.i.d. nonnegative r.v.'s with d.f. F defined on $\mathscr{R}^+$. Let $X_{n:1} < \cdots < X_{n:n}$ be the order statistics associated with the X_i. Then Daniels (1945) defined the bundle strength as

$$B_n = \max\{(n - k + 1)X_{n:k} : 1 \le k \le n\} \ , \tag{8.1}$$

and by very elaborate analysis he established the consistency and asymptotic normality of the perunit bundle strength ($Z_n = n^{-1}B_n$). Note that B_n is a well defined function of the sample order statistics, but it is not a linear combination of the order statistic, nor an extreme order statistic. Hence the classical asymptotic theory of order statistics may not be of much use in this context. Nevertheless, denoting by F_n the sample d.f., we may write, as in Sen, Bhattacharyya and Suh (1973),

$$Z_n = \sup\{x[1 - F_n(x)] : x \in \mathscr{R}^+\} \ . \tag{8.2}$$

This immediately suggests that the centering constant for Z_n should be

$$\theta(F) = \sup\{x[1 - F(x)] : x \in \mathscr{R}^+\} = x_o[1 - F(x_o)], \tag{8.3}$$

where we assume that $x_o = x_o(F)$ is a unique point where this supremum is attained. If the d.f. F is of known functional form (involving possibly some unknown parameters), $\theta(F)$ can be expressed as a function of these parameters (often quite simple in form), so that optimum estimation of $\theta(F)$ can be achieved through optimum parametric estimation of these parameters. However, such estimates are usually highly nonrobust [Sen, 1995a], and hence, we advocate the use of nonparametric asymptotics for this model. In this context the one-to-one relationship of the vector of sample order statistics and empirical d.f. plays a basic role in the formulation of general asymptotic results presented below.

First, by establishing a reverse sub-martingale property of $\{Z_n; n \geq 1\}$, Sen et al. (1973) were able to study (almost sure) convergence properties of Z_n and the order of its asymptotic bias term too. But more remarkably, the weak convergence of the empirical process $W_n = \{n^{1/2}[F_n(x) - F(x)], x \in \mathscr{R}^+\}$ to a tied-down Wiener process provides a vastly simpler and shorter proof of the asymptotic normality of $n^{1/2}\{Z_n - \theta(F)\}$ under minimal regularity assumptions on the d.f. F. It follows from this weak convergence result that the asymptotic mean squared error of $n^{1/2}(Z_n - \theta(F))$ is equal to

$$\gamma^2(F) = \theta^2(F)\{F(x_o)/\overline{F}(x_o)\}, \tag{8.4}$$

where $\theta(F) = x_o\overline{F}(x_o)$, so that x_o also depends on the underlying d.f. F, and can be estimated consistently. Second, this alternative approach enables one to encompass a large class of functionals (expressible as extrema of certain sample functions) and to establish their asymptotic properties under simpler regularity assumptions. Third, it also shows that there are possibly nonlinear and extremal type of functions of order statistics which are attracted by appropriate normal laws. Finally, this asymptotic normality and a related FOADR enable one to incorporate jackknife or bootstrap methods for estimating the (unknown) asymptotic mean squared error of such seemingly nonlinear estimators. We refer to Sen (1993a, 1994b) for some of these details.

Consider next Problem (b). A simple example is a chain with K loops whose individual breaking strengths are denoted by $X_1, \ldots, X_n$ respectively. Note that the chain breaks when at least one of the loops is broken, so that the breaking strength of the chain is defined as

$$C_K = \min\{X_i : 1 \leq i \leq K\} = X_{K:1}. \tag{8.5}$$

If independent copies of the system lifetime are available, one may proceed directly to incorporate the corresponding empirical d.f. in the estimation of the d.f. of C_K as well as suitable functionals of that d.f. In that respect, there is no additional complication involved in statistical modeling and analysis for this type of reliability models. A more interesting statistical problem arises when independent copies of the strength of the loops are available, and they provide additional statistical information too. It may be plausible to assume that the X_i are i.i.d.r.v.'s with a d.f. F defined on $\mathcal{R}^+$. Thus if we denote the d.f. of C_K by G_K and the corresponding reliability function by $\overline{G}_K = 1 - G_K$, then we have

$$\overline{G}_K(x) = [\overline{F}(x)]^K = \{1 - F(x)\}^K, \quad x \in \mathcal{R}^+ . \tag{8.6}$$

The last equation provides a convenient means for estimating the reliability function $\overline{G}_K$ from the sample d.f. F_n when a number of independent copies of the X_i are available. We may consider the plug-in estimator of G as follows:

$$\hat{G}_n(x) = 1 - \{1 - F_n(x)\}^K, \quad x \in \mathcal{R}^+ , \tag{8.7}$$

although this estimator may not be strictly unbiased when $K > 1$. An optimal unbiased estimator of this reliability function is obtained by using Hoeffding's (1948) U-statistics theory. Define a kernel (of degree K) by letting

$$\phi(x_1, \ldots, x_K) = \min\{x_1, \ldots, x_K\} , \tag{8.8}$$

and incorporate this in the following U-process:

$$G_{U,n}(x) = \binom{n}{K}^{-1} \sum_{n,K} I(\phi(X_{i_1}, \ldots, X_{i_K}) \leq x), \quad x \in \mathcal{R}^+ , \tag{8.9}$$

where the summation $\sum_{n,K}$ extends over all possible $\binom{n}{K}$ $1 \leq i_1 < \cdots < i_K \leq n$. Apart from its unbiasedness, the U-process $\{G_{U,n}(x), x \in \mathcal{R}^+\}$ possesses the Glivenko–Cantelli type convergence property, the Hoeffding-decomposition and projection properties, and weak as well as strong invariance principles also hold for them.

Suppose now that we want to estimate $\theta(G_K)$, the mean lifetime of the system. By definition, we have

$$\theta(G_K) = \int_0^\infty \{1 - G_K(x)\} \, dx . \tag{8.10}$$

Then the Hoeffding (1948) U-statistic for this estimator is given by

$$\begin{aligned} U_n &= \binom{n}{K}^{-1} \sum_{n,K} \phi(X_{i_1}, \ldots, X_{i_K}) \\ &= \binom{n}{K}^{-1} \sum_{i=1}^n \binom{n-i}{K-1} X_{n:i} , \end{aligned} \tag{8.11}$$

where $X_{n:1}, \ldots, X_{n:n}$ stand for the order statistics for a collection of n X_i, $i = 1, \ldots, n$. Thus U_n is an L-estimator and its asymptotic properties can be studied by an appeal to the standard U-statistics theory or the results presented in Sections 3 and 4. In general, we may consider a regular functional of the d.f. G_K (such as a percentile or a measure of location) and employ appropriate L-estimators based on $G_{U,n}$. In the literature such estimators are often termed generalized L-estimators.

The case of the K-component system – in parallel follows parallel to the 'in series' case where instead of the $\min\{X_1, \ldots, X_K\}$, we have to work with their maximum (as the system survives as long as at least one of the component is alive). Here $G_K(x) = F^K(x)$, $x \in \mathcal{R}^+$, and hence, similar changes are to be made in the definition of the kernel ϕ as well as the U-statistics constructed above. The parallel expression for the last equation in this case is the following:

$$U_n = \binom{n}{K}^{-1} \sum_{i=1}^{n} \binom{i-1}{K-1} X_{n:i} . \tag{8.12}$$

Therefore the relevance of order statistics asymptotics remains in tact in this case as well. In the same vein we may consider a k-out of-K multicomponent system and express the reliability function as well as regular functionals for this system in terms of the original d.f. F, so that similar order statistics based statistical analysis can be made.

In a single unit system supported by a single spare and a repair facility, when the operating unit fails it is instantaneously replaced by the spare and sent to the repair shop. Upon repair it goes to the spare-box for its subsequent use. Thus the system fails when an operating unit fails but the spare box is empty. This occurs when the lifetime (X) of the operating unit is shorter than the repairing time (Y) of the last failed unit. Since to start with there is a spare with the same d.f. F as the original unit, if we denote the d.f. of the repair times by G and if N is the number of operating units failure culminating in a system failure, then

$$P\{N = k+1\} = \alpha^{k-1}(1-\alpha), \quad k = 1, 2, 3, \ldots , \tag{8.13}$$

where $\alpha = P\{X > Y\} = \int_0^\infty G(x)\, dF(x)$ is a positive fraction. Note that ED, the expected system downtime is

$$ED = \int_0^\infty \int_0^\infty \{\overline{G}(x+t)/\overline{G}(x)\}\, dF(x)\, dt . \tag{8.14}$$

The meantime until the first system failure, measuring from a regeneration point, is given by

$$ET = (1-\alpha)^{-1}\theta(F); \qquad \theta(F) = \int_0^\infty \overline{F}(x)\, dx . \tag{8.15}$$

Assuming that (i) the repair of a failed unit restores it to its new condition, (ii) the original and spare units both have the same d.f., and (iii) X and Y are independent [viz., Barlow and Proschan (1991)], the limiting average availability of the system is defined by

$$A_{FG} = ET\{ET + ED\}^{-1}$$
$$= \theta(F)\{\theta(F) + (1-\alpha)ED\}^{-1} . \tag{8.16}$$

In a strict parametric mold, this model has been treated extensively in the literature, and reported in Barlow and Proschan (1991). A parallel nonparametric treatment is due to Sen and Bhattacharjee (1986) and Sen (1995a), among others.

Note that $\theta(F)$ has the natural nonparametric estimator $\theta(F_n) = \bar{X}_n$, the sample average of n lifetimes of the operating units. Moreover, a plug-in estimator of ED is given by

$$D_n = \int_0^\infty \int_0^\infty \{\bar{G}_n(x+t)/\bar{G}_n(x)\} \, dF_n(x) \, dt$$

$$= n^{-1} \sum_{i=1}^n n_i^{-1} \sum_{j=1}^n (Y_{n:j} - X_{n:i}) I(Y_{n:j} > X_{n:i}) , \qquad (8.17)$$

where the $X_{n:i}$ and $Y_{n:j}$ stand for the order statistics of the X and Y sample values respectively, and

$$n_i = \sum_{j=1}^n I(Y_{n:j} > X_{n:i}), \quad i = 1, \ldots, n . \qquad (8.18)$$

Thus a plug-in estimator of A_{FG} can be obtained by substituting these estimators in the form in (8.15). By definition this is a bounded and nonnegative random variable, and is expressible in terms of a (nonlinear) function of the order statistics $X_{n:i}$ and $Y_{n:j}$. Asymptotics for order statistics, albeit in a functional mode, play a basic role in the study of consistency and asymptotic normality of the plug-in estimator. Jackknifing has also been incorporated to reduce the order of the leading bias term and to obtain the jackknife variance estimator of this estimator. Some alternative estimators are also considered along the same lines.

It is clear that this order statistics asymptotics based approach can also be adopted in other related problems in such reliability models.

9. TTT asymptotics and tests for aging properties

We have introduced the sample spacings and their normalized versions in (2.2) through (2.4), and the TTT function in (2.5). In reliability theory and survival analysis, such TTT statistics play a basic role, specially in the context of testing for some aging properties. We start with a nonnegative r.v. X having a d.f. F and survival function $\bar{F}$, defined on $\mathcal{R}^+$. Thus $\bar{F}(0) = 1$. $\bar{F}(x)$ is nonincreasing in $x \in \mathcal{R}^+$, and $\bar{F}(\infty) = 0$. Moreover we assume that F admits a continuous density f, and define the hazard function as

$$h_F(x) = -(d/dx) \log \bar{F}(x) = f(x)/\bar{F}(x), \quad x \in \mathcal{R}^+ . \qquad (9.1)$$

Then $h_F(x)$ is nonnegative and defining the cumulative hazard function as $H_F(x) = \int_0^x h_F(t) \, dt$, we obtain that

$$\bar{F}(x) = \exp\{-H_F(x)\}, \quad x \in \mathcal{R}^+ . \qquad (9.2)$$

Aging properties of life distributions are then formulated in terms of the (cumulative) hazard function, density function and other associated functionals. Among these the mean residual life (MRL) function $e_F(x)$ is specially noteworthy. Note that by definition

$$e_F(x) = \left\{\int_x^\infty \overline{F}(y)\,\mathrm{d}y\right\}\big/\overline{F}(x) = E(X - x|X > x), \quad x \in \mathcal{R}^+. \quad (9.3)$$

By contrast, the mean life time is $e_F(0) = E(X) = \mu$, say. Moreover corresponding to a d.f. F with finite mean μ, we may define the first derived d.f. $TF(\cdot)$ by letting

$$TF(y) = \mu^{-1}\int_0^y \overline{F}(x)\,\mathrm{d}x;$$

$$\overline{TF}(y) = 1 - TF(y) = \mu^{-1}\int_y^\infty \overline{F}(x)\,\mathrm{d}x, \quad y \in \mathcal{R}^+. \quad (9.4)$$

Before we proceed to introduce the characterizations of various aging properties, we may point out the role of order statistics and the TTT transformation in the formulation of the sample counterparts of these measures.

We define the order statistics $X_{n:i}$ as in (1.1), F_n as in (1.2)–(1.3), the spacings l_{nj} as in (2.2), the normalized spacings d_{nj} as in (2.3), their cumulative entries D_{nk} as in (2.4), and the total time on test upto the point t by $D_n(t)$ as in (2.5). then the sample counterpart of μ is the sample mean

$$\hat{\mu}_n = \overline{X}_n = n^{-1}D_{nn}. \quad (9.5)$$

Likewise the plug-in estimator of $TF(y)$ is

$$TF_n(y) = (\hat{\mu}_n)^{-1}D_n(y)/n = D_n(y)/D_{nn}, \quad y \le X_{n:n}. \quad (9.6)$$

Then a plug-in estimator of the MRL function $e_F(y)$ is

$$\hat{e}_n(y) = \{\overline{F}_n(y)\}^{-1}\int_y^\infty \overline{F}_n(x)\,\mathrm{d}x$$
$$= (n-k)^{-1}\{D_{nn} - D_n(y)\}, \quad \text{for } X_{n:k} \le y < X_{n:k+1},\ k = 0,\ldots,n. \quad (9.7)$$

All these involve the TTT statistics at various timepoints, and as a result, are functions of the sample order statistics.

At the base of the characterizations of aging properties lies the simple exponential model $\overline{F}(x) = \exp\{-x/\mu\}I(x \ge 0)$ for which we have the following characterizations:

(i) $P\{X > x + y | X > x\} = \overline{F}(x+y)/\overline{F}(x) = \overline{F}(y) = P\{X > y\}$, for all $x, y \ge 0$.
(ii) The hazard function $h_F(y) = \mu^{-1}$ is a constant for all $y \ge 0$.
(iii) The MRL $e_F(x) = e_F(0)$, for all $x \in \mathcal{R}^+$.
(iv) The d_{nj} are i.i.d.r.v.'s having the same exponential d.f.

The different aging concepts are related to the negation of such properties of the survival or reliability function. We mention here only the most commonly used ones.

(i) *NB(W)U* Class. A d.f. F is new better (worse) than used if

$$\overline{F}(y) \ge (\le)\overline{F}(x+y)/\overline{F}(x), \quad \forall x, y \ge 0. \quad (9.8)$$

(ii) *I(D)FR* Class. A d.f. F has increasing (decreasing) failure rate if

$$h_F(y) \text{ is } \nearrow (\searrow) \text{ or } H_F(y) \text{ is convex (concave) in } y \in \mathcal{R}^+ . \qquad (9.9)$$

(iii) *I(D)FRA* Class. A d.f. F has increasing (decreasing) failure rate average if

$$\overline{F}(cx) \text{ is } \geq (\leq) [\overline{F}(x)]^c, \quad \forall x, c \geq 0 . \qquad (9.10)$$

(iv) *D(I)MRL* Class. A d.f. F has decreasing (increasing) mean residual life if

$$e_F(x) \text{ is } \searrow (\nearrow) \text{ in } x \geq 0 . \qquad (9.11)$$

(v) *NB(W)UE* Class. A d.f. F belongs to the new better (worse) than used in expectation if

$$e_F(0) \text{ is } \geq (\leq) e_F(x), \quad \forall x \geq 0 . \qquad (9.12)$$

(vi) *NB(W)RUE* Class. Let $\{N(t), t \geq 0\}$ be a counting process relating to the number of renewals under instantaneous replacement (perfect repair) upto the time point $t (\geq 0)$ so that $N(0) = 0$ with probability 1 and $N(t)$ is $\nearrow$, nonnegative and integer valued (r.v.). Under such repeated renewals it is well known that the remaining life

$$L(t) = \sum_{i=1}^{N(t)+1} X_i - t \ (t \geq 0) \qquad (9.13)$$

of the item in use at time t converges in law to a nondegenerate r.v. X_o (as $t \to \infty$) which has the d.f. $TF(y)$, the first derived d.f. for F, already introduced in (9.4). Thus the corresponding MRL function is defined as

$$e_{TF}(y) = E(X_o - y | X_o > y), \quad y \in \mathcal{R}^+ . \qquad (9.14)$$

Then a d.f. F belongs to the new better (worse) than renewal used in expectation if

$$e_F(0) \text{ is } \geq (\leq) e_{TF}(y), \quad \forall y \in \mathcal{R}^+ . \qquad (9.15)$$

There are some other concepts of aging (including the well known HNB(W)E property) which will not be included here. However, these are discussed in detail in a separate chapter in this volume, and hence the results to follow would apply to them as well.

It is well known that there is a partial ordering of these classes of life distributions. While the exponential family is the pivot of this complex, the IFR class contains the exponential ones as a vertex point and it is contained in the IFRA class, which is a subclass of the NBU class, and the NBUE class contains the NBU class; finally, the NBRUE class contains the NBUE class. A similar picture holds when the NBRUE class is replaced by the HNBUE class, although the

NBRUE class is neither contained in or contains the HNBUE class. Also a parallel implication diagram holds for the NWU etc. Therefore in testing for exponentiality of a d.f. F, it is of natural interest to consider a more general class of alternatives, and in this respect, the NBUE and NBRUE alternatives therefore appear to be more appropriate than the others.

Let us first consider the NBUE alternatives. Hollander and Proschan (1972) derived a test statistic for this class which can be expressed as

$$T_{n1} = n^{-1} \sum_{i=1}^{n} \left\{ \frac{3(n-i)}{2n} - \frac{i-1}{2n} \right\} X_{n:i} , \qquad (9.16)$$

which is clearly an L-statistic with smooth weights. A more general class of L-statistics for this testing problem has been considered by Koul (1978), and for the asymptotic normality of such test statistics, we may make use of the general asymptotics presented in Sections 3 and 4. Let us look into this testing problem from a slightly different angle. Let

$$\xi_F(y) = \mu^{-1} \left\{ \mu \bar{F}(y) - \int_y^\infty \bar{F}(x) \, dx \right\}, \quad y \geq 0 . \qquad (9.17)$$

Then for the exponential class, $\xi_F(y) = 0, \forall y \geq 0$, while under the NBUE class, it is nonnegative everywhere, and positive for the strict NBUE class. As such, as a measure of divergence from exponentiality along the NBUE avenue, we may consider the following:

$$\Delta(F) = \sup_{y \geq 0} \xi_F(y) . \qquad (9.18)$$

In the above definitions replacing the d.f. F by its sample counterpart F_n, we may define the sample counterparts of $\xi_F(\cdot)$ and $\Delta(F)$ respectively as

$$\hat{\xi}_n(y) = \xi_{F_n}(y), \, y \geq 0; \quad \text{and} \quad \hat{\Delta}_n = \Delta(F_n) . \qquad (9.19)$$

Further recalling the piecewise linearity of the $\hat{\xi}_n(y)$, we may proceed as in Bhattacharjee and Sen (1995) and show that

$$\hat{\Delta}_n = \max\{D_{nn}^{-1} D_{nk} - n^{-1}k : k = 0, 1, \ldots, n\} , \qquad (9.20)$$

where the D_{nk} have been defined in (2.4). Again the above equation relates to a (nonlinear) function of the order statistics. The same statistic was proposed by Koul (1978) from a somewhat different consideration. Using a multivariate beta distributional characterization of the $D_{nn}^{-1} d_{nk}, k = 1, \ldots, n$, a simpler derivation of the asymptotic distribution of $(n-1)^{1/2} \hat{\Delta}_n$ under exponentiality is given in Bhattacharjee and Sen (1995). This paper also contains some extension of this distribution theory under various types of censoring; in the case of random censoring a similar functional of the Kaplan–Meier product limit estimator has been used along with suitable versions of the classical jackknifing and bootstrapping technique to provide a workable solution to the large sample distribution theory (as is needed for the actual testing problem).

Let us next consider the case of NBRUE alternatives. The situation is a bit more complex. Using the convex-ordering ($\prec_c$) of TF and F, we may write

$$F \text{ is NBRUE} \iff TF \prec_c F , \qquad (9.21)$$

so that if we let

$$J_F(y) = \int_y^\infty \{\overline{F}(x) - \overline{TF}(x)\} \, dx, \quad y \in \mathscr{R}^+ , \qquad (9.22)$$

then $J_F(y) = 0, \forall y \geq 0$ when F is exponential, and it is nonnegative for the NBRUE class. Thus as a measure of divergence from the exponentiality along the NBRUE avenue, we may consider a linear functional

$$J_F^o = \int_0^\infty J_F(y) \, d\Omega(y) \equiv \int_0^\infty \xi_F(y)\Omega(y) \, dy , \qquad (9.23)$$

or a sup-norm functional

$$\Delta(F) = \sup\{\omega(y) J_F(y) : y \in \mathscr{R}^+\} , \qquad (9.24)$$

where $\omega(\cdot)$ is a nonnegative weight function, and $\Omega(y)$ is nonnegative and nondecreasing in $y (\geq 0)$. In particular, we may choose $\partial \Omega(y)/\partial y = \overline{F}(y), y \geq 0$, and $\omega(\cdot) \equiv 1$.

It is quite natural to consider the plug-in estimators of these functionals wherein we replace the unknown F by the sample counterpart F_n. Proceeding as in Sen and Bhattacharjee (1996), we may then consider a rescaled statistic

$$\begin{aligned}
L = \sum_{i=1}^{n-1} &\Big\{ D_{nk} d_{n,k+1}/D_{nn}^2 + d_{n,k+1}^2/(2D_{nn}^2) \\
&- nD_{nk}d_{n,k+1}/((n-k)D_{nn}^2) + nd_{n,k+1}^2/(2(n-k)D_{nn}^2) \\
&+ n(n-k)^{-1}\left[D_{nn}^2 d_{n,k+1} + D_{nk} d_{n,k+1}^2 + d_{n,k+1}^3/3\right]/D_{nn}^3 \Big\} .
\end{aligned} \qquad (9.25)$$

Similarly, if we define

$$\hat{J}_n(y) = n^{-1}\{D_{nn} - D_n(y)\} - D_{nn}^{-1} \int_y^\infty \{D_{nn} - D_n(x)\} \, dx, \quad y \geq 0 , \qquad (9.26)$$

then we have for each $k(= 0, 1, \ldots, n)$,

$$\begin{aligned}
\hat{J}_{nk} = \hat{J}_n(X_{n:k}) &= \sum_{j=k+1}^n u_{nj}\left\{\frac{n}{n-j+1}\left[U_{n,j-1} - \frac{j-1}{n}\right]\right\} \\
&+ \frac{1}{2}\sum_{j=k+1}^n \frac{nu_{nj}^2}{n-j+1} ,
\end{aligned} \qquad (9.27)$$

where $u_{nj} = d_{nj}/D_{nn}$ and $U_{nk} = \sum_{j \leq k} u_{nj}$. Then the sup-norm test statistic is

$$\hat{\Delta}_n = \max\{\hat{J}_{nk} : k \leq n\} . \tag{9.28}$$

Sen and Bhattacharjee (1996) employed the multivariate beta distribution of the $u_{nj}, j = 1, \ldots, n$ (under exponentiality of F) to provide simple L-statistic approximation to L_n and also managed to show that the sup-norm statistic has asymptotically (under the null hypothesis) the same distribution as of the supremum of a Wiener process on the unit interval $(0,1)$. Again the general asymptotics discussed in earlier sections provide the necessary tools in this investigation. This functional approach to testing the null hypothesis of exponentiality against various alternatives based on aging properties opens the doors for the TTT statistics, and these in turn are liked to the original sample order statistics. Therefore, the general asymptotics for order statistics provide the desired tools for the study of asymptotic properties of related tests.

10. Concluding remarks

Granted the sufficiency and completeness of sample order statistics prevailing in a large class of statistical models, the relevance of these statistics in statistical inference problems is quite apparent. On the top of that order statistics crop up in many situations as handy tools for summarizing the statistical information, and hence, from practical applications point of view, they are generally appealing even in a broader setup. Censoring schemes in survival and reliability analysis are the most notable areas of such applications. The interrelations of order statistics and the empirical processes have also added to the convenience of incorporating the order statistics asymptotics in some other areas of research. There are, however, certain other fields where parallel developments require more research on order statistics. One of the most noteworthy areas relates to multivariate analysis. There may not be a complete ordering of points on a two or higher dimensional space, and hence, order statistics may not be properly defined. Typically if the observations are themselves p vectors, for some $p > 1$, then for each coordinate, a set of order statistics can be defined as in the univariate case, and hence, these vectors of coordinatewise order statistics can be incorporated in drawing statistical conclusions. However, such a matrix of order statistics would not necessarily be affine-invariant (a property possessed by linear estimators based on the unordered variates). This drawback is not of that major concern when the different coordinate variates are not that linearly conformable. Nevertheless, in many problems in design of experiments, the vector of residuals may not have the full rank, and hence, a choice of a subset of variates having the full-rank property may not be unique. In this case, the coordinatewise order statistics may lose some of their natural appeal. Even in most simple cases, such as the bivariate normal distributions, the exact distribution theory of coordinatewise order statistics may become unmanageable when the sample size is not small. The asymptotics for coordinatewise order statistics are potentially usable in a much broader setup, although motivations for their adoptations need to be initiated on other grounds.

Concomitants of order statistics and conditional quantile functions are important developments in this context. We expect more research work in this direction in the near future.

References

Bahadur, R. R. (1966). A note on quantiles in large samples. *Ann. Math. Statist.* **37**, 577–580.

Barlow, R. E., and F. Proschan (1991). *Statistical Reliability Theory and Applications: Probability Models*. To Begin With, Silver Spring, MD.

Bhattacharya, P. K. (1974). Convergence of sample paths of normalized sums of induced order statistics. *Ann. Statist.* **2**, 1034–1039.

Bhattacharya, P. K. (1976). An invariance principle in regression analysis. *Ann. Statist.* **4**, 621–624.

Bhattacharya, P. K. and A. K. Gangopadhyay (1990). Kernel and nearest neighbor estimation of a conditional quantile. *Ann. Statist.* **17**, 1400–1415.

Bhattacharjee, M. C. and P. K. Sen (1995). On Kolmogorov–Smirnov type tests for NB(W)UE alternatives under some censoring schemes. In *Analysis of Censored Data* (eds. H. L. Koul, and J. V. Deshpande), IMS Mono. Lect. Notes Ser. No 27, Hayward, Calif., pp. 25–38.

Billingsley, P. (1968). *Convergence of Probability Measures*. John Wiley, New York.

Cox, D. R. (1972). Regression models and life tables (with discussion). *J. Roy. Statist. Soc. Ser. B* **74**, 187–220.

Daniels, H. A. (1945). The statistical theory of the strength of bundles of threads. *Proc. Roy. Soc. Ser. A* **183**, 405–435.

David, H. A. (1973). Concomitants of order statistics. *Bull. Internat. Statist. Inst.* **45**, 295–300.

David, H. A. (1981). *Order Statistics*. 2nd ed., John Wiley, New York.

Efron, B. (1982). *The Jackknife, the Bootstrap, and Other Resampling Plans*. SIAM, Philadelphia.

Galambos, J. (1984). Order Statistics. In *Handbook of Statistics, Vol. 4: Nonparametric Methods* (eds. P. R. Krishnaiah and P. K. Sen), North Holland, Amsterdam, pp. 359–382.

Gangopadhyay, A. K. and P. K. Sen (1990). Bootstrap confidence intervals for conditional quantile functions. *Sankhyā Ser. A* **52**, 346–363.

Gangopadhyay, A. K. and P. K. Sen (1992). Contiguity in nonparametric estimation of a conditional functional. In *Nonparametric Statistics and Related Topics* (ed. A. K. M. E. Saleh), North Holland, Amsterdam, pp. 141–162.

Gangopadhyay, A. K. and P. K. Sen (1993). Contiguity in Bahadur-type representation of a conditional quantile and application in conditional quantile process. In *Statistics and Probability, a Raghu Raj Bahadur Festschrift* (eds. J. K. Ghosh et al.) Wiley Eastern, New Delhi, pp. 219–231.

Ghosh, J. K. (1971). A new proof of the Bahadur representation and an application. *Ann. Math. Statist.* **42**, 1957–1961.

Ghosh, M. and P. K. Sen (1971). On a class of rank order tests for regression with partially informed stochastic predictors. *Ann. Math. Statist.* **42**, 650–661.

Gutenbrunner, C., and J. Jurečková (1992). Regression rank scores and regression quantiles. *Ann. Statist.* **20**, 305–330.

Hoeffding, W. (1948). A class of statistics with asymptotically normal distribution. *Ann. Math. Statist.* **19**, 293–325.

Hollander, M. and F. Proschan (1972). Testing whether new is better than used. *Ann. Math. Statist.* **43**, 1136–1146.

Hollander, M. and F. Proschan (1975). Testing for the mean residual life. *Biometrika* **62**, 585–593.

Jurečková, J. and P. K. Sen (1996). *Robust Statistical Procedures: Asymptotics and Interrelations*. John Wiley, New York.

Kaplan, E. L. and P. Meier (1958). Nonparametric estimation from incomplete observations. *J. Amer. Statist. Assoc.* **53**, 457–481.

Koenker, R. and G. Bassett (1978). Regression quantiles. *Econometrica* **46**, 33–50.

Koul, H. L. (1978). Testing for new better than used in expectation. *Commun. Statist. Theor. Meth. A* **7**, 685–701.

Koul, H. L. (1992). *Weighted Empiricals and Linear Models*. IMS Mono. Lect. Notes Ser. No 21, Hayward, Calif.

Puri, M. L. and P. K. Sen (1971). *Nonparametric Methods in Multivariate Analysis*. John Wiley, New York.

Puri, M. L. and P. K. Sen (1985). *Nonparametric Methods in General Linear Models*. John Wiley, New York.

Sarhan, A. E. and B. G. Greenberg (eds.) (1962) *Contributions to Order Statistics*. John Wiley, New York.

Sen, P. K. (1960). On some convergence properties of U-statistics. *Calcutta Statist. Assoc. Bull.* **10**, 1–18.

Sen, P. K. (1964). On some properties of the rank weighted means. *J. Ind. Soc. Agri. Statist.* **16**, 51–61.

Sen, P. K. (1970). A note on order statistics from heterogeneous distributions. *Ann. Math. Statist.* **41**, 2137–2139.

Sen, P. K. (1976). A note on invariance principles for induced order statistics. *Ann. Probab.* **4**, 1247–1257.

Sen, P. K. (1979). Weak convergence of some quantile processes arising in progressively censored tests. *Ann. Statist.* **7**, 414–431.

Sen, P. K. (1981a). The Cox regression model, invariance principles for some induced quantile processes and some repeated significance tests. *Ann. Statist.* **9**, 109–121.

Sen, P. K. (1981b). Some invariance principles for mixed rank statistics and induced order statistics and some applications. *Commun. Statist. Theor. Meth. A* **10**, 1691–1718.

Sen, P. K. (1981c). *Sequential Nonparametrics: Invariance Principles and Statistical Inference*. John Wiley, New York.

Sen, P. K. (1988a). Functional jackknifing: Rationality and general asymptotics. *Ann. Statist.* **16**, 450–469.

Sen, P. K. (1988b). Functional approaches in resampling plans: A review of some recent developments. *Sankhyā Ser. A* **50**, 394–435.

Sen, P. K. (1989). Whither delete-k jackknifing for smooth statistical functionals. In: *Statistical Data analysis and Inference* (ed. Y. Dodge), North Holland, Amsterdam, pp. 269–279.

Sen, P. K. (1993a). Resampling methods for the extrema of certain sample functions. In *Probability and Statistics* (eds. Basu, S. K. and Sinha, B. K.), Narosa Publ., Delhi, pp. 66–79.

Sen, P. K. (1993b). Perspectives in multivariate nonparametrics: Conditional functionals and ANOCOVA models. *Sankhyā Ser. A* **55**, 516–532.

Sen, P. K. (1994a). Regression quantiles in nonparametric regression. *J. Nonparamet. Statist.* **3**, 237–253.

Sen, P. K. (1994b). Extreme value theory for fibre bundles. In *Extreme Value Theory and Applications* (ed. Galambos, J.), Kluwer, Mass., pp. 77–92.

Sen, P. K. (1995a). Statistical Analysis of some reliability models: Parametrics, semi-parametrics and nonparametrics. *J. Statist. Plan. Infer.* **44**, 41–66.

Sen, P. K. (1995b). Robust and nonparametric methods in linear models with mixed effects. *Tetra Mount. Math. J.* **7**, 331–343.

Sen, P. K. (1995c). Censoring in theory and practice: statistical controversies and perspectives. In *Analysis of Censored Data* (eds. H. L. Koul and J. V. Deshpande), IMS Mono. Lect. Notes Ser. No 27, Hayward, Calif. pp. 175–192.

Sen, P. K. (1996a). Statistical functionals, Hadamard differentiability and martingales. In *A Festschrift for J. Medhi* (eds. Borthakur, A. C. and Chaudhury, H.), New Age Press, Delhi, pp. 29–47.

Sen, P. K. (1996b). Regression rank scores estimation in ANOCOVA. *Ann. Statist.* **24**, 1586–1602.

Sen, P. K. and M. C. Bhattacharjee (1986). Nonparametric estimators of avaialability under provisions of spare and repair, I. In *Reliability and Quality Control* (ed. A. P. Basu), North Holland, Amsterdam, pp. 281–296.

Sen, P. K. and M. C. Bhattacharjee (1996). Testing for a property of aging under renewals: Rationality and general asymptotics. In *Proc. Sec. Internat. Trien. Calcutta Symp. Prob. Statist.* (eds. S. K. Basu et al.) Narosa Pub., Calcutta, pp 283–295.

Sen, P. K., B. B. Bhattacharyya and M. W. Suh (1973). Limiting behavior of the extrema of certain sample functions. *Ann. Statist.* **1**, 297–311.

Sen, P. K. and J. M. Singer (1993). *Large Sample Methods in Statistics: An Introduction with Applications*. Chapman and Hall, New York.

Serfling, R. J. (1980). *Approximation Theorems of Mathematical Statistics*. John Wiley, New York.

Serfling, R. J. (1984). Generalized *L*-, *M*- and *R*-statistics. *Ann. Statist.* **12**, 76–86.

Shorack, G. R. and J. A. Wellner (1986). *Empirical Processes with Applications to Statistics*. John Wiley, New York.

Von Mises, R. (1947). On the asymptotic distribution of differentiable statistical functionals. *Ann. Math. Statist.* **18**, 309–348.

Zero-One Laws for Large Order Statistics

R. J. Tomkins and Hong Wang

1. Introduction

This article will present a survey of zero-one laws involving large order statistics of independent, identically distributed (i.i.d.) random variables. The importance of laws of this type in the study of the limiting behaviour of large order statistics has been established in the literature over the past forty years. In particular, these zero-one laws can be used to establish almost sure stability theorems for large order statistics.

Let $(\Omega, \mathcal{F}, P)$ be a probability space, and let $X_1, X_2, \ldots$ be a sequence of i.i.d. random variables defined on that space. For any given positive integer r, and for each $n \geq r$, define $Z_{r,n}$ to be the rth largest value in $\{X_1, \ldots, X_n\}$. Then $\{Z_{r,n}, n \geq r\}$ is called the rth *maximum sequence* of $\{X_n, n \geq 1\}$. If $i \leq n$ satisfies $X_i = Z_{r,n}$, then r is said to be the *rank* of X_i among $\{X_1, \ldots, X_n\}$.

Let F be the common distribution function (d.f.) of the sequence $\{X_n, n \geq 1\}$, i.e., $F(x) = P\{X \leq x\}$ for each real x.

This article will focus on probabilities of the form

$$P\{Z_{r,n} > u_n \text{ i.o.}\} \quad \text{and} \quad P\{Z_{r,n} \leq u_n \text{ i.o.}\}, \tag{1}$$

where $r \geq 1$ and $\{u_n\}$ is a given real sequence. (The abbreviation "i.o." is used for "infinitely often".) It is an easy consequence of the Hewitt–Savage Zero-One Law (see, for instance, Breiman (1967)) that the two probabilities in (1) can assume no values other than zero or one.

Typically, as will be seen, a zero-one law for the large order statistics $\{Z_{r,n}, n \geq r\}$, $r \geq 1$, presents a series whose terms depend on F and $\{u_n\}$ such that $P\{Z_{r,n} > u_n \text{ i.o.}\}$ (or $P\{Z_{r,n} \leq u_n \text{ i.o.}\}$) equals zero if the series converges, and equals one if it diverges. Such a series is called a *criterion series*.

Zero-one laws for $P\{Z_{r,n} > u_n \text{ i.o.}\}$, the so-called *upper-case probability*, are generally easier to derive than those for $P\{Z_{r,n} \leq u_n \text{ i.o.}\}$, the *lower-case probability*. An overview of upper-case results will appear in Section 2. The fundamental results in this case are due to Geffroy (1958/59) when $r = 1$, to Mori (1976) for general $r \geq 1$, and to Deheuvels (1986) and Wang (1991) when the ranks vary with n.

Zero-one laws for the lower-class probability will be presented in Section 3. Barndorff–Nielsen (1961) and Klass (1984, 1985) did pioneering work for $r = 1$, while fundamental results for a general rank $r \geq 1$ are due to Frankel (1972, 1976), Shorack and Wellner (1978), and Wang and Tomkins (1992).

Finally, Section 4 will be devoted to zero-one laws for probabilities of the form $P\{Z_{r_n,n} \leq u_n \text{ i.o.}\}$, where $\{r_n, n \geq 1\}$ is a non-decreasing sequence of integers obeying $1 \leq r_n \leq n$ for every n. Such a sequence $\{r_n\}$ is called a *rank sequence*. Since $n \geq r_n$, the random sequence $\{Z_{r_n,n}\}$ is well-defined. The study of this case was initiated by Deheuvels (1986) and extended by Wang (1991).

It should be noted that the results to follow can easily be restated to produce zero-one laws for the rth minimum sequence $\{X_{r,n}, n \geq r\}$, where $X_{1,n} \leq X_{2,n} \leq \cdots \leq X_{n,n}$ are the order statistics of $\{X_1, X_2, \ldots, X_n\}$, since $-X_{r,n}$ is the rth largest of $\{-X_1, -X_2, \ldots, -X_n\}$.

2. Zero-One laws for the upper-case probability

Let $\{x_n\}$ and $\{u_n\}$ be real sequences such that u_n is non-decreasing and $u_n \to \infty$ (under which circumstances we write "$u_n \uparrow \infty$"). Then it is a straightforward task to show that $\max\{x_1, \ldots, x_n\} \leq u_n$ for all large values of n iff (if and only if) $x_n \leq u_n$ for all large n. It follows easily that, if $u_n \uparrow \infty$,

$$P\{Z_{1,n} > u_n \text{ i.o.}\} = P\{X_n > u_n \text{ i.o.}\}$$

for *every* sequence of random variables $\{X_n\}$. In particular, if $X_1, X_2, \ldots$ are i.i.d. with common d.f. F, then it is a simple consequence of the Borel Zero-One Law (see, e.g. Chow and Teicher (1987), p. 61) that

$$P\{Z_{1,n} > u_n \text{ i.o.}\} = 0 \text{ or } 1$$

according as

$$\sum_{n=1}^{\infty} \{1 - F(u_n)\} \text{ converges or diverges},$$

for any real sequence $u_n \uparrow \infty$. This result appears to have been derived first by Geffroy (1958/59).

Tomkins (1996) produced a counter-example to show that the preceding results may fail if $\{u_n\}$ is not non-decreasing. Furthermore, he proved that

$$P\{Z_{1,n} > u_n \text{ i.o.}\} = P\left\{Z_{1,n} > \inf_{k \geq n} u_k \text{ i.o.}\right\}$$

when $u_n \to \infty$, for every random sequence $\{X_n\}$. It follows that, for any independent sequence $\{X_n\}$, $P\{Z_{1,n} > u_n \text{ i.o.}\} = 0$ iff

$$\sum_{n=1}^{\infty} \left[1 - F\left(\inf_{k \geq n} u_k\right)\right] < \infty$$

whenever $u_n \to \infty$. This result can also be deduced from Theorem 1 of Rothmann and Russo (1991), who defined the sequence $\{M_n\}$ by

$$M_n = \max\{X_i : n - a_n < i \leq n\}$$

for an integer sequence $\{a_n\}$ such that $1 \leq a_n \leq n$, and proved that $P\{M_n > u_n \text{ i.o.}\} = 0$ iff

$$\sum_{n=1}^{\infty} [1 - F(v_n)] < \infty \; ,$$

where $v_n = \inf\{u_i : i - a_i \leq n \leq i\}$, whenever $u_n \to \infty$. If $a_n = n$, clearly $M_n = Z_{1,n}$ and the result of Tomkins (1995) follows. (Rothmann and Russo's Theorem 1 is stated for i.i.d. uniformly-distributed random variables. However, the foregoing statement of their result for a general independent sequence follows via a standard transformation argument.)

Kiefer (1972) was the first author to tackle the upper-class probability for a general (fixed) $r \geq 1$. His result was refined by Mori (1976), as follows:

THEOREM 2.1. Let $\{X_n\}$ be an i.i.d. sequence with d.f F. Then, for any integer $r \geq 1$ and any real sequence $u_n \uparrow \infty$,

$$P\{Z_{r,n} > u_n \text{ i.o.}\} = 0 \quad \text{iff} \quad \sum_{n=1}^{\infty} n^{r-1}\{1 - F(u_n)\}^r < \infty \; .$$

Kiefer's (1972) version of Theorem 2.1 contained the superfluous hypothesis that $\{n[1 - F(u_n)]\}$ be non-decreasing. Note that Theorem 2.1 reduces to Geffroy's result when $r = 1$.

As one might anticipate, the variable-rank case is more complicated. Indeed, complete results are not yet available for the upper-class probability in this case. We will present some sufficient conditions which guarantee that $P\{Z_{r_n,n} > u_n \text{ i.o.}\} = 0$ for a given rank sequence $\{r_n\}$. First, here is a result of Deheuvels (1986).

THEOREM 2.2. Let $X_1, X_2, \ldots$ be an i.i.d. sequence with d.f. F. Let $\{r_n, n \geq 1\}$ be a rank sequence such that $r_n \to \infty$ and $\limsup_{n \to \infty} (r_n/n) < 1$. Let $\{u_n\}$ be a non-decreasing sequence such that

$$\lim_{n \to \infty} \frac{r_n - n[1 - F(u_n)]}{\sqrt{r_n}} = \infty \; . \tag{2}$$

Then $P\{Z_{r_n,n} > u_n \text{ i.o.}\} = 0$ if

$$\sum_{n=1}^{\infty} \frac{\sqrt{r_n}}{n} \left\{\frac{ne}{r_n}[1 - F(u_n)]\right\}^{r_n} \exp\{-n[1 - F(u_n)]\} < \infty \; .$$

Deheuvels (1986) also proved that Theorem 2.2 remains true for a sequence $\{u_n\}$, not necessarily monotone, if $\limsup_{n \to \infty} r_n^2/n < \infty$, $u_n \leq u_{n+1}$ whenever $r_n = r_{n+1}$, and $\lim_{n \to \infty} \{r_n - n[1 - F(u_n)]\}/\sqrt{r_n \log r_n} = \infty$.

More recently, Wang (1991) demonstrated that the conclusion of Theorem 2.2 still holds under the hypotheses $n[1 - F(u_n)] \to \infty$, $\limsup_{n\to\infty} r_n[1 - F(u_n)] < \infty$ and

$$\lim_{n\to\infty} \frac{r_n - n[1 - F(u_n)]}{\sqrt{n[1 - F(u_n)]}} = \infty . \tag{3}$$

It is easy to check that $r_n/n \to 0$ under Wang's assumptions, so his result cannot imply Theorem 2.2 (which assumes only that $\limsup_{n\to\infty} r_n/n < 1$). However, (3) implies (2), so neither Theorem 2.2 nor Wang's result can be deduced from the other.

Finding conditions under which $P\{Z_{r_n,n} > u_n \text{ i.o.}\} = 1$ is an open problem.

REMARK 1. If $\{r_n\}$ is a rank sequence such that $r_n \to r$ for some integer $r \geq 1$, then $r_n = r$ for all large n, so that $P\{Z_{r_n,n} > u_n \text{ i.o.}\} = P\{Z_{r,n} > u_n \text{ i.o.}\}$. Thus Theorem 2.1 can be applied in this case.

REMARK 2. Let $X_1, X_2, \ldots$ be i.i.d. with d.f. F. Suppose $\{u_n\}$ is non-decreasing, but that $u_n \to u < \infty$. Define $x_0 = \sup\{x: F(x) < 1\}$. Then Theorem 2.1 remains true. If $u < x_0$, then $P\{X_n > u_n \text{ i.o.}\} \geq P\{X_n > u \text{ i.o.}\} = 1$ by the Borel Zero-One law. It follows easily that $P\{Z_{r,n} > u_n \text{ i.o.}\} = 1, r \geq 1$. The same conclusion holds if $P\{X_1 = x_0\} > 0, u = x_0$ and $u_n < x_0, n \geq 1$, since $P\{X_n = x_0 \text{ i.o.}\} = 1$ in this case. In either case, $\sum_{n=1}^{\infty} n^{r-1}[1 - F(u_n)]^r = \infty$.

On the other hand, if $u > x_0$ then $u_n > x_0$ for all large n, say $n \geq N$. Hence $P\{X_n \leq u_n\} = 1$ for $n \geq N$. It follows easily that $P\{Z_{r,n} > u_n \text{ i.o.}\} = 0, r \geq 1$. The same conclusion holds if $u_n = x_0$ for all large n (so that $u = x_0$). In either case, $\sum_{n=1}^{\infty} n^{r-1}[1 - F(u_n)]^r < \infty$, since all but a finite number of terms equal zero.

Finally, suppose that $u = x_0, u_n < x_0$ for all n and $P\{X_1 = x_0\} = 0$. Define $Y_n = (x_0 - X_n)^{-1}$ and $v_n = (x_0 - u_n)^{-1}$. Both $\{Y_n\}$ and $\{v_n\}$ are well-defined, and $Y_n > v_n$ iff $X_n > u_n$. Hence, for $r \geq 1$,

$$P\{Z_{r,n} > u_n \text{ i.o.}\} = 0 \quad \text{iff} \quad P\{r^{\text{th}} \max\{Y_1, \ldots, Y_n\} > v_n \text{ i.o.}\} = 0$$

$$\text{iff} \quad \sum_{n=1}^{\infty} n^{r-1}[P\{Y_1 > v_n\}]^r < \infty$$

$$\text{iff} \quad \sum_{n=1}^{\infty} n^{r-1}[1 - F(u_n)]^r < \infty$$

by Theorem 2.1, which applies since $v_n \uparrow \infty$. It follows that Theorem 2.1 is true under the assumption that $\{u_n\}$ be non-decreasing, but not necessarily divergent. This observation appears to be new.

3. Zero-one laws for the lower-case probability

Barndorff–Nielsen (1961) proved that

$$P\{Z_{1,n} \leq u_n \text{ i.o.}\} = 0 \text{ or } 1$$

according as the series

$$\sum_{n=1}^{\infty} F^n(u_n) \frac{\log \log n}{n} \qquad (4)$$

converges or diverges, provided that $\{u_n\}$ is non-decreasing and $\{[F(u_n)]^n\}$ is non-increasing. Further, he showed by means of a counter-example that, in general, the hypothesis that $\{[F(u_n)]^n\}$ be non-increasing cannot be dropped. However, that hypothesis is not needed to establish that

$$P\{Z_{1,n} \leq u_n \text{ i.o.}\} = 0 \qquad (5)$$

when the series (4) converges. The proof of this part of Barndorff–Nielsen's results hinges on the following clever refinement of the Borel–Cantelli Lemma.

LEMMA 3.1 (Barndorff–Nielsen (1961)). Let $\{A_n\}$ be a sequence of events. If $P(A_n) \to 0$ as $n \to \infty$ and $\sum_{n=1}^{\infty} P(A_n A_{n+1}^c) < \infty$, where A_{n+1}^c denotes the complement of A_{n+1}, then $P\{A_n \text{ i.o.}\} = 0$.

An easy consequence of Lemma 3.1 is the following result: if $F^n(u_n) \to 0$ (or, equivalently, $n[1 - F(u_n)] \to \infty$) as $n \to \infty$ and

$$\sum_{n=1}^{\infty} F^n(u_n)[1 - F(u_n)] < \infty , \qquad (6)$$

then (5) holds, provided that $\{u_n\}$ is non-decreasing. Moreover, Wang and Tomkins (1992) proved that (6) implies $F^n(u_n) \to 0$ and (4) if $\{n[1 - F(u_n)]\}$ is non-decreasing.

Barndorff–Nielsen (1961) proved the "divergence part" of his theorem by means of a delicate analysis of the random sequence $\{Z_{1,m_n}, n \geq 1\}$, where $m_n = \exp\{4n/\log n\}$, $n \geq 1$; the sequence $\{m_n\}$ was used earlier (for different purposes) by Erdös (1942). Barndorff–Nielsen proved that the divergence of the series (4), together with the monotonicity of $\{[F(u_n)]^n\}$, yields $P\{Z_{1,m_n} \leq u_{m_n} \text{ i.o.}\} > 0$, so that $P\{Z_{1,n} \leq u_n \text{ i.o.}\} = 1$ by the Hewitt–Savage Zero-One Law.

Suppose that $\{n[1 - F(u_n)]\}$ is non-decreasing and divergent. Then, using Barndorff–Nielsen's approach, it can be shown that $P\{Z_{1,n} \leq u_n \text{ i.o.}\} = 0$ or 1 according as

$$\sum_{n=3}^{\infty} \frac{\log \log n}{n} \exp\{-n[1 - F(u_n)]\} \qquad (7)$$

converges or diverges (cf Galambos (1987), p. 252, and Wang and Tomkins (1992)). Under the same hypotheses, Robbins and Siegmund (1972) proved that $P\{Z_{1,n} \leq u_n \text{ i.o.}\} = 0$ or 1 according as the series

$$\sum_{n=1}^{\infty} [1 - F(u_n)] \exp\{-n[1 - F(u_n)]\} \tag{8}$$

converges or diverges.

Using a different tack, Klass (1984, 1985) proved that if $n[1 - F(u_n)] \to \infty$, then $P\{Z_{1,n} \leq u_n \text{ i.o.}\} = 0$ or 1 according as the series (8) converges or diverges. Notice that Klass's theorem places no monotonicity hypotheses on $\{n[1 - F(u_n)]\}$ or $\{(F(u_n))^n\}$. Klass's approach also involved the study of the behaviour of $\{Z_{1,n}\}$ along a subsequence $\{n_k\}$ but, unlike Barndorff–Nielsen (1961), Klass defined $\{n_k\}$ so as to depend on the given sequence $\{u_n\}$ and the d.f. F. Indeed, Klass presented a number of such sequences, called *monitoring sequences*. For instance, one monitoring sequence is given by: $n_1 = 1$ and, for $k \geq 1$,

$$n_{k+1} = \min\{j > n_k : (j - n_k)[1 - F(u_{n_k})] \geq \lambda\},$$

where $\lambda > 0$. Klass (1984) proved that $P\{Z_{1,n} \leq u_n \text{ i.o.}\} = 0$ iff

$$\sum_{n=1}^{\infty} \exp\{-n_k[1 - F(u_{n_k})]\} < \infty, \tag{9}$$

and then (Klass (1985)) demonstrated that (9) holds iff the series in (8) converges.

Godbole (1987) provided a shorter proof of Klass's result using a martingale argument.

Rothmann and Russo (1991, 1993) studied circumstances under which $P\{M_n \leq u_n \text{ i.o.}\} = 0$, where $M_n = \max\{X_i : n - a_n < i \leq n\}$ and $\{a_n\}$ are integers with $1 \leq a_n \leq n$, and presented criterion series for the validity of this equation for several families of sequences $\{a_n\}$, all obeying $\limsup_{n \to \infty} a_n/n < 1$.

A natural consequence of the appearance of Barndorff–Nielsen's (1961) zero-one law for $\{Z_{1,n}\}$ was a quest for similar results for $\{Z_{r,n}\}, r > 1$. Compared to the $r = 1$ case, the derivation of zero-one laws involving $\{Z_{r,n}\}$ is more complicated, mainly because the distribution function of $Z_{r,n}$, that is,

$$P\{Z_{r,n} \leq x\} = \sum_{j=0}^{r-1} \binom{n}{j} [1 - F(x)]^j F^{n-j}(x),$$

is much more complex than that of $Z_{1,n}$ (i.e., $P\{Z_{1,n} \leq x\} = F^n(x)$) when $r > 1$.

The first result in the general r case is due to Frankel (1972, 1976) who used techniques based on empirical processes to prove that, for $r \geq 1$, $P\{Z_{r,n} \leq u_n \text{ i.o.}\} = 0$ iff

$$\sum_{n=1}^{\infty} \frac{\{n[1 - F(u_n)]\}^r}{n} \exp\{-n[1 - F(u_n)]\} \tag{10}$$

converges, provided that both $\{u_n\}$ and $\{n[1 - F(u_n)]\}$ are non-decreasing and that $n[1 - F(u_n)] \to \infty$. Later, Shorack and Wellner (1978) demonstrated that Frankel's result remains valid if the monotonicity assumption about $\{n[1 - F(u_n)]\}$ is replaced by:

$$\liminf_{n\to\infty} \frac{n[1 - F(u_n)]}{\log\log n} \geq 1 \ .$$

Clearly, Frankel's criterion series (10) reduces to (8) when $r = 1$. It is natural to ask whether the other three criterion series (4), (6) and (7) from the $r = 1$ case have analogues for a general $r \geq 1$. An affirmative answer is given by the following theorem due to Wang and Tomkins (1992). (See Wang (1997) for refinements of this result.)

THEOREM 3.2. Let $\{X_n, n \geq 1\}$ be an i.i.d. sequence with d.f. F. Let $n[1 - F(u_n)] \to \infty$ for a non-decreasing real sequence $\{u_n\}$. Fix any integer $r \geq 1$ and let $Z_{r,n}$ be the rth largest of $\{X_1, \ldots, X_n\}$, $n \geq r$.

(i) $P\{Z_{r,n} \leq u_n \text{ i.o.}\} = 0$ if any of the following series converges:

$$\sum_{n=3}^{\infty} F^n(u_n) \frac{(\log\log n)^r}{n} \ ;$$

$$\sum_{n=1}^{\infty} n^{r-1} [1 - F(u_n)]^r F^n(u_n) \ ;$$

$$\sum_{n=r}^{\infty} P\{Z_{r,n} \leq u_n\} [1 - F(u_n)] \ ;$$

$$\sum_{n=1}^{\infty} n^{r-1} [1 - F(u_n)]^r \exp\{-n[1 - F(u_n)]\} \ ;$$

$$\sum_{n=3}^{\infty} \frac{(\log\log n)^r}{n} \exp\{-n[1 - F(u_n)]\} \ .$$

(ii) If $\{n[1 - F(u_n)]\}$ is non-decreasing, then $P\{Z_{r,n} \leq u_n \text{ i.o.}\} = 1$ if any of the five series above diverges.

REMARKS 1. If $\liminf_{n\to\infty} n[1 - F(u_n)] < \infty$, then $P\{Z_{r,n} \leq u_n \text{ i.o.}\} = 1$ for every $r \geq 1$ (cf. Wang and Tomkins (1992), Klass (1984)). Therefore, it makes sense to assume that $n[1 - F(u_n)] \to \infty$ in Theorem 3.2.

2. Wang and Tomkins (1992) showed that $\{n[1 - F(u_n)]\}$ is non-decreasing if $\{[F(u_n)]^n\}$ is non-increasing, but that the converse is false, in general. Thus, Theorem 3.2 contains Barndorff–Nielsen's (1961) result for the case $r = 1$.

3. It is evident from Theorem 3.2 that, under the condition that $\{n[1 - F(u_n)]\}$ is non-decreasing, all of the series listed in Theorem 3.2 (i) converge if any one of them converges.

4. Zero-One laws for the lower-case probability when ranks vary

This section will focus on circumstances under which

$$P\{Z_{r_n,n} \le u_n \text{ i.o.}\} = 0 \text{ or } 1$$

for a given rank sequence $\{r_n\}$ and a given real sequence $\{u_n\}$.

By definition, $1 \le r_n \le n$ and $r_n \le r_{n+1}, n \ge 1$, so $\lim_{n\to\infty} r_n$ exists. If $\lim_{n\to\infty} r_n = r < \infty$, then r is a positive integer and $r_n = r$ for all large n. Hence,

$$P\{Z_{r_n,n} \le u_n \text{ i.o.}\} = P\{Z_{r,n} \le u_n \text{ i.o.}\}$$

in this case, so Theorem 3.2 applies. Therefore, we will restrict our attention to the case where $\lim_{n\to\infty} r_n = \infty$.

The seminal work for variable ranks was done by Deheuvels (1986), who employed a methodology similar to the classical approach of Barndorff–Nielsen (1961) in the case $r = 1$. Deheuvels produced results for the lower order statistics from a uniform distribution on $(0,1)$, but also observed that his results can be readily translated into zero-one laws for $\{Z_{r_n,n}, n \ge 1\}$ for a general i.i.d sequence $\{X_n\}$.

THEOREM 4.1 (Deheuvels (1986)). Let $X_1, X_2, \ldots$ be an i.i.d. sequence with d.f. F, and let $\{u_n\}$ be a non-decreasing real sequence. Let $\{r_n, n > 1\}$ be a divergent rank sequence such that $\limsup_{n\to\infty} r_n/n < 1$ and

$$\lim_{n\to\infty} \frac{n[1 - F(u_n)] - r_n}{\sqrt{r_n}} = \infty \ . \tag{11}$$

If

$$\sum_{n=1}^{\infty} \frac{\sqrt{r_n}}{n} \left\{\frac{ne}{r_n}[1 - F(u_n)]\right\}^{r_n} \exp\{-n[1 - F(u_n)]\} \tag{12}$$

converges, then

$$P\{Z_{r_n,n} \le u_n \text{ i.o.}\} = 0 \ . \tag{13}$$

Deheuvels also showed that Theorem 4.1 remains true for non-monotone sequences $\{u_n\}$ if $u_n \le u_{n+1}$ whenever $r_n = r_{n+1}$ and either

$$0 < \liminf_{n\to\infty} \frac{n[1 - F(u_n)] - r_n}{\log \log n} \le \limsup_{n\to\infty} \frac{n[1 - F(u_n)] - r_n}{\log \log n} < \infty$$

or $\lim_{n\to\infty} \log r_{n+1}/\log r_n = 1$,

$$\limsup_{n\to\infty} \frac{n[1 - F(u_n)]}{(n+1)[1 - F(u_{n+1})]} < \infty \quad \text{and} \quad \liminf_{n\to\infty} \frac{n[1 - F(u_n)]}{r_n} \ge 1 \ .$$

THEOREM 4.2 (Deheuvels (1986)). Let $X_1, X_2, \ldots$ be a sequence of i.i.d. random variables with d.f. F. Let $\{r_n\}$ be a divergent rank sequence such that $r_n/(np_n) \to 1$ for some non-increasing sequence $\{p_n\}$. Let $\{u_n\}$ be a real sequence such that $n[1 - F(u_n)] \leq (n+1)[1 - F(u_{n+1})]$ and $u_n \leq u_{n+1}$ whenever $r_n = r_{n+1}$. Suppose that either (a) $n[1 - F(u_n)]/\log\log n \to A$ for some $A > 0$, $r_n = o(\log\log n)$ and $\liminf_{n\to\infty} r_n/\log\log n > 0$, or (b) $r_n = O(\log\log n)$ and

$$0 < \liminf_{n\to\infty} \frac{n[1 - F(u_n)]}{\log\log n} \leq \limsup_{n\to\infty} \frac{n[1 - F(u_n)]}{\log\log n} < \infty .$$

If the series (12) diverges, then

$$P\{Z_{r_n,n} \leq u_n \text{ i.o.}\} = 1 .$$

Wang (1991) adapted techniques of Klass (1984, 1985) to establish necessary and sufficient conditions for (13) for a particular class of rank sequences.

THEOREM 4.3 (Wang (1991)). Let $X_1, X_2, \ldots$ be i.i.d random variables with common d.f. F. Let $\{r_n\}$ be a divergent rank sequence and $\{u_n\}$ a non-decreasing sequence such that $n[1 - F(u_n)] \to \infty, F(u_n) \to 1$,

$$\limsup_{n\to\infty} r_n[1 - F(u_n)] < \infty \text{ and } \limsup_{n\to\infty} r_n/(n[1 - F(u_n)]) < 1 . \tag{14}$$

Then $P\{Z_{r_n,n} \leq u_n \text{ i.o.}\} = 0$ or 1 according as the series (12) converges or diverges.

Notice that Wang's hypotheses in (14) imply that $\limsup_{n\to\infty} r_n^2/n < \infty$ and, for some $\delta < 1$,

$$\lim_{n\to\infty} \frac{n[1 - F(u_n)] - r_n}{\sqrt{r_n}} = \lim_{n\to\infty} \sqrt{r_n} \left(\frac{n[1 - F(u_n)]}{r_n} - 1 \right)$$
$$\geq \lim_{n\to\infty} \sqrt{r_n}(\delta^{-1} - 1) = \infty .$$

Consequently, (11) holds and $r_n/n \to 0$, so Theorem 4.1 can be invoked to show that (13) holds when the series (12) converges. However, unlike Theorem 4.2, Theorem 4.3 does not require that $r_n = O(\log\log n)$ and places no monotonicity restrictions on $\{n[1 - F(u_n)]\}$.

Finally, we note that it follows readily from Stirling's formula that the series (12) converges if and only if

$$\sum_{n=1}^{\infty} \binom{n}{r_n} \frac{[1 - F(u_n)]^{r_n}}{n} \exp\{-n[1 - F(u_n)]\} < \infty .$$

Notice that, if $r_n = r < \infty$ for all $n \geq 1$, it is now easy to show that (12) is equivalent to the convergence of the series (10).

Acknowledgements

Both authors are grateful for research grants from the Natural Sciences and Engineering Research Council of Canada.

References

Barndorff–Nielsen, O. (1961). On the rate of growth of the partial maxima of a sequence of independent identically distributed random variables. *Math Scand.* **9**, 383–394.

Breiman, L. (1967). *Probability*. Addison-Wesley, Reading, Mass.

Chow, Y. S. and H. Teicher (1987). *Probability Theory: Independence, Interchangeability, Martingales.* 2nd ed. Springer, New York.

Deheuvels, P. (1986). Strong laws for the kth order statistics when $k \leq \log_2 n$. *Probab. Theory Rel. Fields* **72**, 133–154.

Erdös, P. (1942). On the law of the iterated logarithm. *Ann. Math.* **43**, 419–436.

Frankel, J. (1972). On the Law of the Iterated Logarithm for Order Statistics. Ph. D. dissertation, Columbia University.

Frankel, J. (1976). A note on downcrossings for extremal processes. *Ann. Probab.* **4**, 151–152.

Galambos, J. (1987). *The Asymptotic Theory of Extreme Order Statistics.* 2nd ed. Robert Briger Publishing Co., Malabat, Fla.

Geffroy, J. (1958/59). Contributions à la théorie des valeurs extrêmes. *Publ. Inst. Statist. Univ. Paris* **718**, 37–185.

Godbole, A. (1987). On Klass' series criterion for the minimal growth rate of partial maxima. *Statist. Probab. Lett.* **5**, 235–238.

Kiefer, J. (1972). Iterated logarithm analogues for sample quantities when $p_n \downarrow 0$. *Proc. 6th Berkeley Sympos. Math. Statist. Probab.*, 227–244. Univ. of California Press.

Klass, M. J. (1984). The minimal growth rate of partial maxima. *Ann. Probab.* **12**, 380–389.

Klass, M. J. (1985). The Robbins–Siegmund series criterion for partial maxima. *Ann. Probab.* **13**, 1369–1370.

Mori, T. (1976). The strong law of large numbers when extreme terms are excluded from sums. *Z. Wahrscheinlichkeitstheorie verw. Geb.* **36**, 189–194.

Robbins, H. and D. Siegmund (1972). On the law of the iterated logarithm for maxima and minima. *Proc. 6th Berkeley Sympos. Math. Statist. Probab.*, 51–70. Univ. of California Press.

Rothmann, M. D. and R. P. Russo (1991). Strong limiting bounds for a sequence of moving maxima. *Statist. Probab. Lett.* **11**, 403–410.

Rothmann, M. D. and R. P. Russo (1993). A series criterion for moving maxima. *Stochastic Process. Appl.* **46**, 241–247.

Shorack, G. R. and J. Wellner (1978). Linear bounds on the empirical distribution function. *Ann. Probab.* **6**, 349–353.

Tomkins, R. J. (1996). Refinement of a zero-one law for maxima. *Statist. Probab. Lett.* **27**, 67–69.

Wang, Hong (1991). Zero-one laws for extreme order statistics. Ph.D. dissertation, University of Regina.

Wang, Hong (1997). Generalized zero-one laws for large-order statistics. *Bernoulli* **3**, 429–444.

Wang, Hong and R. J. Tomkins (1992). A zero-one law for large order statistics, *Can. J. Statist.* **20**, 323–334.

PART VI

Robust Methods

Some Exact Properties Of Cook's D_I

D. R. Jensen and D. E. Ramirez

1. Introduction

Cook's (1977) D_I statistics are used widely for assessing influence of design points in regression diagnostics. These statistics typically contain a leverage component and a standardized residual component. Subsets having large D_I are said to be *influential*, reflecting high leverage for these points or outliers in the data.

In particular, consider a linear model $\mathbf{Y} = \mathbf{X}_0\boldsymbol{\beta} + \boldsymbol{\varepsilon}$, with $(\mathbf{Y}, \boldsymbol{\varepsilon}) \in \mathbb{R}^N$, $\mathbf{X}_0 \in F_{N \times k}$, and with $\boldsymbol{\beta} \in \mathbb{R}^k$ unknown. Partition $\mathbf{Y} = [\mathbf{Y}_1', \mathbf{Y}_2']'$, $\mathbf{X}_0 = [\mathbf{X}', \mathbf{Z}']'$, and $\boldsymbol{\varepsilon} = [\boldsymbol{\varepsilon}_1', \boldsymbol{\varepsilon}_2']'$ conformably, with $(\mathbf{Y}_1, \boldsymbol{\varepsilon}_1) \in \mathbb{R}^n$, $(\mathbf{Y}_2, \boldsymbol{\varepsilon}_2) \in \mathbb{R}^r$, $\mathbf{X} \in F_{n \times k}$ of rank $k < n$, and $\mathbf{Z} \in F_{r \times k}$, where $N = n + r$ and $r \leq k$. To assess the joint influence of design points comprising the rows of $\mathbf{Z}$, let $\hat{\boldsymbol{\beta}} = (\mathbf{X}_0'\mathbf{X}_0)^{-1}\mathbf{X}_0'\mathbf{Y}$ be the least-squares estimator for $\boldsymbol{\beta}$ in the full data, with $\hat{\boldsymbol{\beta}}_I = (\mathbf{X}'\mathbf{X})^{-1}\mathbf{X}'\mathbf{Y}_1$ from the reduced data. Versions of Cook's D_I statistics, patterned after the generalized distance of Mahalanobis (1936), are obtained on specializing

$$D_1(\hat{\boldsymbol{\beta}}, \mathbf{M}, c\hat{\sigma}^2) = (\hat{\boldsymbol{\beta}}_I - \hat{\boldsymbol{\beta}})'\mathbf{M}(\hat{\boldsymbol{\beta}}_I - \hat{\boldsymbol{\beta}})/c\hat{\sigma}^2 , \qquad (1.1)$$

where $\mathbf{M}(k \times k)$ is positive definite and $\hat{\sigma}^2$ is some estimator for the variance. Commonly used choices for $\mathbf{M}$ are $\mathbf{X}'\mathbf{X}$ and $\mathbf{X}_0'\mathbf{X}_0 = (\mathbf{X}'\mathbf{X} + \mathbf{Z}'\mathbf{Z})$, and for $\hat{\sigma}^2$ are the residual mean square S^2 from the full data and S_I^2 from the reduced data, and for c is $c = k$. Related constructs are the *DFBETA*'s to gauge the influence of row $I = i$ of $\mathbf{X}_0$ in single-case diagnostics, as given by $\{DF\beta_{j,i} = (\hat{\beta}_{ij} - \hat{\beta}_j)/S_i(C_{jj})^{1/2}; 1 \leq j \leq k\}$ for elements of $(\hat{\boldsymbol{\beta}}_i - \hat{\boldsymbol{\beta}}) \in \mathbb{R}^k$, with C_{jj} as a diagonal element of $\mathbf{C} = (\mathbf{X}_0'\mathbf{X}_0)^{-1}$. We return to these subsequently.

In practice, benchmarks are essential for gauging whether D_I is large enough to be deemed "influential." Since D_I is random, its stochastic properties are crucial in setting guidelines for its use. Exact properties of D_I in any of its forms essentially are unknown, despite an impressive list of references including books by Belsley, Kuh, and Welsch (1980), Chatterjee and Hadi (1988), Cook and Weisberg (1982), Fox (1991), and Rousseeuw and Leroy (1987). Thus benchmarks as currently prescribed at best are imprecise. Various authors refer $D_I(\hat{\boldsymbol{\beta}}, \mathbf{X}_0'\mathbf{X}_0, kS^2)$ to the Snedecor-Fisher distribution $F(k, N - k)$, but with some distrust. Cook and Weisberg (1982) refer $D_I(\hat{\boldsymbol{\beta}}, \mathbf{X}_0'\mathbf{X}_0, kS^2)$ to the 50th percentile of $F(k, N - k)$, for example, whereas Gray (1993) has questioned the appropriateness of scaling by k in the denominator. We return to these issues later.

Here we seek structural parameters that drive D_I stochastically, thereby accounting for diagnostic trends against random scatter in the data. At the same time, we seek to dispel some commonly held misconceptions. We show that (i) the k-dimensional distribution of $(\hat{\boldsymbol{\beta}}_I - \hat{\boldsymbol{\beta}})$ is degenerate of rank $r < k$ unless $r = k$; (ii) the numerator and denominator of $D_I(\hat{\boldsymbol{\beta}}, \mathbf{X}_0'\mathbf{X}_0, kS^2)$ are dependent, and the denominator is inflated stochastically by outliers resulting from shifted means; (iii) for $r = 1$ the distribution of D_I on scaling is $F(1, v)$ with $\hat{\sigma}^2$ chosen suitably; (iv) the exact distribution of D_I derives generally from a weighted sum, the weights depending on leverages and on variances before and after a shift; and (v) small values for D_I may indicate *inliers*, i.e., data points having error variance *smaller* than the remaining data. It is seen that design points having high leverage may mask the effects of a shift in either the mean or variance, whether the shift be up or down.

An outline of the paper follows. Section 2 contains conventions for notation, essentials regarding certain nonstandard distributions, and steps for the simultaneous reduction of rectangular matrices to quasidiagonal forms. The latter supports a canonical reduction of the model for subset diagnostics as in Section 3, thereby unmasking the detailed structure of D_I. Exact distributions are studied in Section 4; effects of leverage and shifts in the mean or variance are examined; and stochastic bounds are given for distributions of these types. Modifications to D_I are considered in Section 5, and Section 6 concludes with a brief summary.

2. Preliminaries

We first set conventions for notation; we next review essentials of some nonstandard distributions; and we then undertake the simultaneous reduction of rectangular matrices of different orders to quasidiagonal forms.

2.1. Notation. Spaces of note include the Euclidean k-space $\mathbb{R}^k$; its positive orthant $\mathbb{R}_+^k$; the collection $F_{n \times k}$ of real $(n \times k)$ matrices; and the symmetric (S_k), positive semidefinite (S_k^0), and positive definite (S_k^+) real $(k \times k)$ matrices. The transpose and inverse of $\mathbf{A}$ are denoted by $\mathbf{A}'$ and $\mathbf{A}^{-1}$ as appropriate; $\mathbf{A}^{1/2} \in S_k^+$ is the symmetric root of $\mathbf{A} \in S_k^+$; and $O(n)$ is the group of real orthogonal $(n \times n)$ matrices. Special arrays include the unit vector $\mathbf{1}_n = [1, \ldots, 1]' \in \mathbb{R}^n$, the identity matrix $\mathbf{I}_k$ of order $(k \times k)$, and the block-diagonal form $\text{Diag}(\mathbf{A}_1, \ldots, \mathbf{A}_r)$. Probability density and cumulative distribution functions are abbreviated as pdf and cdf; $\mathscr{L}(\mathbf{Y})$ refers to the law of distribution of $\mathbf{Y} \in \mathbb{R}^n$; and $N_n(\boldsymbol{\mu}, \boldsymbol{\Sigma})$ designates the Gaussian law on $\mathbb{R}^n$ having mean $E(\mathbf{Y}) = \boldsymbol{\mu} \in \mathbb{R}^n$ and dispersion matrix $V(\mathbf{Y}) = \boldsymbol{\Sigma} \in S_n^+$. Standard distributions on $\mathbb{R}_+^1$ include the cdf $G(\cdot; v, \lambda)$ of the chi-squared distribution $(\chi^2(v, \lambda))$ having v degrees of freedom, and the cdf $F(\cdot; k, v, \lambda)$ of the F-distribution having (k, v) degrees of freedom, both having noncentrality parameter λ. The latter is omitted in the case of central distributions. Further classes emerge as follows.

2.2. Special distributions. Suppose that elements of $\mathbf{U} = [U_1, U_2, \ldots, U_r]'$ are independent $\{N_1(\omega_i, 1); 1 \leq i \leq r\}$ random variates; let $\{\alpha_1, \alpha_2, \ldots, \alpha_r\}$ be positive weights; identify $T = \alpha_1 U_1^2 + \cdots + \alpha_r U_r^2$; and denote by $g_r(t; \alpha_1, \ldots, \alpha_r, \omega_1 \ldots, \omega_r)$ its pdf, and by $G_r(t; \alpha_1, \ldots, \alpha_r, \omega_1, \ldots, \omega_r)$ its cdf. These are written succinctly on occasion as $g_r(t; \boldsymbol{\alpha}', \boldsymbol{\omega}')$ and $G_r(t; \boldsymbol{\alpha}', \boldsymbol{\omega}')$ with $\boldsymbol{\alpha}' = [\alpha_1, \alpha_2, \ldots, \alpha_r]$ and $\boldsymbol{\omega}' = [\omega_1, \omega_2, \ldots, \omega_r]$. If in addition $\mathscr{L}(V) = G(\cdot; v)$ independently of $\mathbf{U}$, then the pdf of $W = vT/rV = v(\alpha_1 U_1^2 + \cdots + \alpha_r U_r^2)/rV$ is denoted by $f_r(w; \alpha_1, \ldots, \alpha_r, \omega_1, \ldots, \omega_r, v)$, and its cdf by $F_r(w; \alpha_1, \ldots, \alpha_r, \omega_1, \ldots, \omega_r, v)$. Series expansions for $g_r(t; \alpha_1, \ldots, \alpha_r, \omega_1, \ldots, \omega_r)$ and $G_r(t; \alpha_1, \ldots, \alpha_r, \omega_1, \ldots, \omega_r)$, and bounds on errors due to truncating each series, are developed for central and noncentral distributions in Kotz, Johnson, and Boyd (1967a,b); see also Johnson and Kotz (1970) and Mathai and Provost (1992). Building on the work of Gurland (1955) and Kotz et al. (1967a), Ramirez and Jensen (1991) derived series expansions for the central pdf and cdf of $\mathscr{L}(T/V)$, from which $f_r(w; \alpha_1, \ldots, \alpha_r, v)$ and $F_r(w; \alpha_1, \ldots, \alpha_r, v)$ follow directly on rescaling. Connections to standard distributions are that $G_r(t; \alpha, \ldots, \alpha, \omega_1, \ldots, \omega_r) = G(t/\alpha; r, \lambda)$ and $F_r(w; \alpha, \ldots, \alpha, \omega_1, \ldots, \omega_r, v) = F(w/\alpha; r, v, \lambda)$, with $\lambda = \boldsymbol{\omega}'\boldsymbol{\omega}/\alpha$.

Basic stochastic ordering properties of these distributions are summarized in the following.

LEMMA 1. *Given the foregoing developments, consider the distributions $G_r(t; \alpha_1, \ldots, \alpha_r, \omega_1, \ldots, \omega_r)$ and $F_r(w; \alpha_1, \ldots, \alpha_r, \omega_1, \ldots, \omega_r, v)$. Then*

(i) $G_r(t; \alpha_1, \ldots, \alpha_r, \omega_1, \ldots, \omega_r)$ *increases stochastically in each* $\{\alpha_i; 1 \leq i \leq r\}$, *i.e., for each fixed t and* $\boldsymbol{\omega}' = [\omega_1, \omega_2, \ldots, \omega_r]$, $G_r(t; \alpha_1, \ldots, \alpha_r, \boldsymbol{\omega}')$ *is a decreasing function of each* $\{\alpha_i; 1 \leq i \leq r\}$, *and similarly for* $F_r(t; \alpha_1, \ldots, \alpha_r, \omega_1, \ldots, \omega_r, v)$.

(ii) $G_r(t; \alpha_1, \ldots, \alpha_r, \omega_1, \ldots, \omega_r)$ *increases stochastically in each* $\{|\omega_i|; 1 \leq i \leq r\}$, *i.e., for each fixed t and* $\boldsymbol{\alpha}' = [\alpha_1, \alpha_2, \ldots, \alpha_r]$, $G_r(t; \boldsymbol{\alpha}', \omega_1, \ldots, \omega_r)$ *is a decreasing function of each* $\{|\omega_i|; 1 \leq i \leq r\}$, *and similarly for* $F_r(t; \alpha_1, \ldots, \alpha_r, \omega_1, \ldots, \omega_r, v)$.

PROOF. First consider $G_r(t; \alpha_1, \ldots, \alpha_r, \omega_1, \ldots, \omega_r)$. From the assumption that $[U_1, U_2, \ldots, U_r]$ are independent $\{N_1(\omega_i, 1); 1 \leq i \leq r\}$ random variates, it follows directly that $\{\mathscr{L}(U_i^2) = \chi^2(1, \omega_i^2); 1 \leq i \leq r\}$ are independent. It is well known that $G(\cdot; v, \lambda)$ increases stochastically with λ for any v, so that $\mathscr{L}(U_i^2)$ increases stochastically with $|\omega_i|$ for each $1 \leq i \leq r$. In a similar manner $\mathscr{L}(\alpha_i U_i^2)$ increases stochastically with α_i for fixed $|\omega_i|$ for each $1 \leq i \leq r$. The first parts of conclusions (i) and (ii) now follow since the convolution of nonnegative random variables is itself a stochastically increasing operator. Conclusions pertaining to $F_r(w; \alpha_1, \ldots, \alpha_r, \omega_1, \ldots, \omega_r, v)$ now follow on applying conclusions (i) and (ii) for $G_r(t; \alpha_1, \ldots, \alpha_r, \omega_1, \ldots, \omega_r)$ conditionally given $V = v$, and observing that the orderings hold independently of v and thus unconditionally as well. This concludes our proof. □

2.3. Rectangular matrix reductions. We undertake the simultaneous reduction of rectangular matrices $\mathbf{X} \in F_{n \times k}$ and $\mathbf{Z} \in F_{r \times k}$ to quasidiagonal forms. Details follow.

LEMMA 2. *Consider* $\mathbf{X} \in F_{n \times k}$ *of rank* $k < n$, *and let* $\mathbf{Z} \in F_{r \times k}$ *with* $r \leq k$. *Then there are orthogonal matrices* $\mathbf{Q}_1 \in O(n)$ *and* $\mathbf{Q}_2 \in O(r)$, *and a nonsingular matrix* $\mathbf{G} \in F_{k \times k}$, *such that*

$$\mathbf{Q}_1 \mathbf{X} \mathbf{G} = [\mathbf{I}_k, \mathbf{0}]' \text{ and } \mathbf{Q}_2 \mathbf{Z} \mathbf{G} = [\mathbf{D}_\gamma, \mathbf{0}] \quad (2.1)$$

for $r < k$, *whereas* $\mathbf{Q}_2 \mathbf{Z} \mathbf{G} = \mathbf{D}_\gamma$ *for* $r = k$, *where the elements* $\{\gamma_1 \geq \gamma_2 \geq \cdots \geq \gamma_r\}$ *of* $\mathbf{D}_\gamma = \text{Diag}(\gamma_1, \gamma_2, \ldots, \gamma_r)$ *comprise square roots of the ordered eigenvalues of* $\mathbf{H} = \mathbf{Z}(\mathbf{X}'\mathbf{X})^{-1}\mathbf{Z}'$.

PROOF. We offer a constructive proof using the expression $\mathbf{Q}_2 \mathbf{Z} \mathbf{G} = [\mathbf{D}_\gamma, \mathbf{0}]$ for $r < k$, with the understanding that $\mathbf{Q}_2 \mathbf{Z} \mathbf{G} = \mathbf{D}_\gamma$ for $r = k$. Take $(\mathbf{X}, \mathbf{Z})$ into $(\mathbf{X}(\mathbf{X}'\mathbf{X})^{-1/2}, \mathbf{Z}(\mathbf{X}'\mathbf{X})^{-1/2})$, and postulate that orthogonal matrices $\mathbf{P} \in O(k)$, $\mathbf{Q}_1 \in O(n)$, and $\mathbf{Q}_2 \in O(r)$ exist such that $\mathbf{Q}_1 \mathbf{X}(\mathbf{X}'\mathbf{X})^{-1/2}\mathbf{P}' = [\mathbf{I}_k, \mathbf{0}]'$ and $\mathbf{Q}_2 \mathbf{Z}(\mathbf{X}'\mathbf{X})^{-1/2}\mathbf{P}' = [\mathbf{D}_\gamma, \mathbf{0}]$. Existence of the postulated $\{\mathbf{P}, \mathbf{Q}_1, \mathbf{Q}_2\}$ may be seen as follows. Clearly $\mathbf{X}(\mathbf{X}'\mathbf{X})^{-1/2}\mathbf{P}'\mathbf{P}(\mathbf{X}'\mathbf{X})^{-1/2}\mathbf{X}' = \mathbf{X}(\mathbf{X}'\mathbf{X})^{-1}\mathbf{X}'$ is idempotent of order $(n \times n)$ and rank k for any $\mathbf{P} \in O(k)$, so that $\mathbf{Q}_1 \in O(n)$ can be found giving the spectral form $\mathbf{Q}_1 \mathbf{X}(\mathbf{X}'\mathbf{X})^{-1}\mathbf{X}'\mathbf{Q}_1' = \text{Diag}(\mathbf{I}_k, \mathbf{0})$. Moreover, since $\mathbf{P}(\mathbf{X}'\mathbf{X})^{-1/2}\mathbf{X}'\mathbf{X}(\mathbf{X}'\mathbf{X})^{-1/2}\mathbf{P}' = \mathbf{I}_k$ for any $\mathbf{P} \in O(k)$, it follows that $\mathbf{Q}_1 \mathbf{X}(\mathbf{X}'\mathbf{X})^{-1/2}\mathbf{P}' = [\mathbf{I}_k, \mathbf{0}]'$ achieves the singular decomposition of $\mathbf{X}(\mathbf{X}'\mathbf{X})^{-1/2}$, i.e., $\mathbf{X}(\mathbf{X}'\mathbf{X})^{-1/2} = \mathbf{Q}_1'[\mathbf{I}_k, \mathbf{0}]'\mathbf{P}$, with singular values $\mathbf{1}_k$. Now choose $(\mathbf{Q}_2, \mathbf{P})$ such that $\mathbf{Q}_2 \mathbf{Z}(\mathbf{X}'\mathbf{X})^{-1/2}\mathbf{P}' = [\mathbf{D}_\gamma, \mathbf{0}]$ from the singular decomposition $\mathbf{Z}(\mathbf{X}'\mathbf{X})^{-1/2} = \mathbf{Q}_2'[\mathbf{D}_\gamma, \mathbf{0}]\mathbf{P}$. Here $\mathbf{D}_\gamma = \text{Diag}(\gamma_1, \gamma_2, \ldots, \gamma_r)$ contains the ordered singular values of $\mathbf{Z}(\mathbf{X}'\mathbf{X})^{-1/2}$, these being square roots of the nonvanishing eigenvalues of $(\mathbf{X}'\mathbf{X})^{-1/2}\mathbf{Z}'\mathbf{Z}(\mathbf{X}'\mathbf{X})^{-1/2}$, or equivalently, of $\mathbf{H} = \mathbf{Z}(\mathbf{X}'\mathbf{X})^{-1}\mathbf{Z}'$. The proof is now complete on identifying $\mathbf{G}$ as $(\mathbf{X}'\mathbf{X})^{-1/2}\mathbf{P}'$. $\square$

3. The structure of Cook's D_I

We first express the model in a canonical form making transparent the essential features of D_I. We initially apply the method of least squares to uncover the structure of D_I, requiring no assumptions on the errors beyond their joint nonsingularity and centering at zero. We then impose specific requirements on the error distributions to support more detailed conclusions.

3.1. The canonical form. Since the original model $\mathbf{Y} = \mathbf{X}_0 \boldsymbol{\beta} + \boldsymbol{\varepsilon}$ and its canonical form are related one-to-one, it suffices to consider the latter. To consider effects of a shift in $E(\mathbf{Y}_2)$, we begin with the model

$$\begin{bmatrix} \mathbf{Y}_1 \\ \mathbf{Y}_2 \end{bmatrix} = \begin{bmatrix} \mathbf{X} \\ \mathbf{Z} \end{bmatrix} \boldsymbol{\beta} - \begin{bmatrix} \xi_1 \\ \xi_2 \end{bmatrix} + \begin{bmatrix} \varepsilon_1 \\ \varepsilon_2 \end{bmatrix} \quad (3.1)$$

for $r \leq k$, where $\xi' = [\xi_1', \xi_2']$ such that $\xi_1 = \mathbf{0} \in \mathbb{R}^n$ and $\xi_2 \in \mathbb{R}^r$. We first apply $\mathbf{Q}_1 \in O(n)$ as in Lemma 1 to $\mathbf{Y}_1$ and ε_1. For later reference partition these as $\mathbf{Q}_1 \mathbf{Y}_1 = [\mathbf{U}_1', \mathbf{U}_2', \mathbf{U}_3']'$ and $\mathbf{Q}_1 \varepsilon_1 = [\boldsymbol{\eta}_1', \boldsymbol{\eta}_2', \boldsymbol{\eta}_3']'$, with $(\mathbf{U}_1, \boldsymbol{\eta}_1) \in \mathbb{R}^r, (\mathbf{U}_2, \boldsymbol{\eta}_2) \in \mathbb{R}^s$, and $(\mathbf{U}_3, \boldsymbol{\eta}_3) \in \mathbb{R}^t$, such that $r + s = k$ and $t = n - k$. Further let $\mathbf{Q}_2 \mathbf{Y}_2 = \mathbf{U}_4$ and $\mathbf{Q}_2 \varepsilon_2 = \boldsymbol{\eta}_4$, with $(\mathbf{U}_4, \boldsymbol{\eta}_4) \in \mathbb{R}^r$. Now choose $\mathbf{G}$ as in Lemma 1; take $\mathbf{X} \to \mathbf{Q}_1 \mathbf{X} \mathbf{G}, \mathbf{Z} \to \mathbf{Q}_2 \mathbf{Z} \mathbf{G}$, and $\boldsymbol{\beta} \to \mathbf{G}^{-1} \boldsymbol{\beta} = \boldsymbol{\theta} \in \mathbb{R}^k$; and partition the latter as $\boldsymbol{\theta} = [\boldsymbol{\theta}_1', \boldsymbol{\theta}_2']'$, with $\boldsymbol{\theta}_1 \in \mathbb{R}^r$ and $\boldsymbol{\theta}_2 \in \mathbb{R}^s$ such that $r + s = k$. In summary, the model (3.1) may be written equivalently in canonical form as

$$\begin{bmatrix} \mathbf{U}_1 \\ \mathbf{U}_2 \\ \mathbf{U}_3 \\ \mathbf{U}_4 \end{bmatrix} = \begin{bmatrix} \mathbf{I}_r & \mathbf{0} \\ \mathbf{0} & \mathbf{I}_s \\ \mathbf{0} & \mathbf{0} \\ \mathbf{D}_\gamma & \mathbf{0} \end{bmatrix} \begin{bmatrix} \boldsymbol{\theta}_1 \\ \boldsymbol{\theta}_2 \end{bmatrix} - \begin{bmatrix} \mathbf{0} \\ \mathbf{0} \\ \mathbf{0} \\ \boldsymbol{\delta} \end{bmatrix} + \begin{bmatrix} \boldsymbol{\eta}_1 \\ \boldsymbol{\eta}_2 \\ \boldsymbol{\eta}_3 \\ \boldsymbol{\eta}_4 \end{bmatrix}, \quad (3.2)$$

i.e., as $\mathbf{U} = \mathbf{W}\boldsymbol{\theta} - \boldsymbol{\delta}_N + \boldsymbol{\eta}$ with $\boldsymbol{\delta}_N' = [\mathbf{0}', \mathbf{0}', \mathbf{0}', \boldsymbol{\delta}']$ and $\boldsymbol{\delta} = \mathbf{Q}_2 \xi_2 \in \mathbb{R}^r$. To proceed we suppose for now that $\boldsymbol{\delta} = \mathbf{0}$, i.e., that $\mathbf{Y} = \mathbf{X}_0 \boldsymbol{\beta} + \varepsilon$ and $\mathbf{U} = \mathbf{W}\boldsymbol{\theta} + \boldsymbol{\eta}$ are appropriate. Then solutions for (3.1) and (3.2) under the full and reduced data are related one-to-one, so that $(\hat{\boldsymbol{\beta}}_I - \hat{\boldsymbol{\beta}}) = \mathbf{G}(\hat{\boldsymbol{\theta}}_I - \hat{\boldsymbol{\theta}})$. The cases $r = k$ and $r < k$ often can be consolidated on letting $\mathbf{D}_\gamma^2 = \mathrm{Diag}(\gamma_1^2, \gamma_2^2, \ldots, \gamma_k^2) \in S_k^0$, with the understanding that $\mathbf{D}_\gamma^2 = \mathrm{Diag}(\gamma_1^2, \gamma_2^2, \ldots, \gamma_r^2, 0, \ldots, 0)$ for $r < k$. We reinstate $\boldsymbol{\delta} \neq \mathbf{0}$ subsequently.

Recall from Lemma 1 that elements of $\mathbf{D}_\gamma^2 = \mathrm{Diag}(\gamma_1^2, \gamma_2^2, \ldots, \gamma_r^2)$ for $r \leq k$ comprise the ordered eigenvalues of $\mathbf{H} = \mathbf{Z}(\mathbf{X}'\mathbf{X})^{-1}\mathbf{Z}'$. This is now seen to be the predictive dispersion matrix in predicting at r points comprising the rows of $\mathbf{Z}$, based on $\hat{\boldsymbol{\beta}}_I$ from the reduced data. If instead we predict at $\mathbf{Z}$ based on $\hat{\boldsymbol{\beta}}$ from the full data, then the predictive dispersion matrix is given by $\mathbf{H}_0 = \mathbf{Z}(\mathbf{X}'\mathbf{X} + \mathbf{Z}'\mathbf{Z})^{-1}\mathbf{Z}'$, with eigenvalues to be designated as $\{\lambda_1 \geq \lambda_2 \geq \cdots \geq \lambda_r\}$. The latter are clearly related to those of $\mathbf{H}$ through $\{\lambda_i = \gamma_i^2/(\gamma_i^2 + 1); 1 \leq i \leq r\}$. As these eigenvalues assume a critical role in the developments following, for later reference we set $\boldsymbol{\Lambda}_r = \mathrm{Diag}(\lambda_1, \lambda_2, \ldots, \lambda_r) = \mathbf{D}_\gamma(\mathbf{I}_r + \mathbf{D}_\gamma^2)^{-1}\mathbf{D}_\gamma$.

It should be noted that an elementary change of variables transfers $D_I(\hat{\boldsymbol{\beta}}, \mathbf{X}_0'\mathbf{X}_0, c\hat{\sigma}^2)$, in terms of $(\hat{\boldsymbol{\beta}}_I - \hat{\boldsymbol{\beta}})$, directly into $D_I(\hat{\boldsymbol{\theta}}, \mathbf{M}(\gamma), c\hat{\sigma}^2)$ in terms of $(\hat{\boldsymbol{\theta}}_I - \hat{\boldsymbol{\theta}})$, i.e.,

$$(\hat{\boldsymbol{\beta}}_I - \hat{\boldsymbol{\beta}})'(\mathbf{X}'\mathbf{X} + \mathbf{Z}'\mathbf{Z})(\hat{\boldsymbol{\beta}}_I - \hat{\boldsymbol{\beta}}) = (\hat{\boldsymbol{\theta}}_I - \hat{\boldsymbol{\theta}})'(\mathbf{I}_k + \mathbf{D}_\gamma^2)(\hat{\boldsymbol{\theta}}_I - \hat{\boldsymbol{\theta}}), \quad (3.3)$$

where for later reference we let $\mathbf{M}(\gamma) = (\mathbf{I}_k + \mathbf{D}_\gamma^2)$ for $r = k$, and $\mathbf{M}(\gamma) = \mathrm{Diag}((\mathbf{I}_r + \mathbf{D}_\gamma^2), \mathbf{I}_s)$ for $r < k$. This may be seen on writing $(\mathbf{X}'\mathbf{X} + \mathbf{Z}'\mathbf{Z}) = (\mathbf{X}'\mathbf{X})^{1/2} \mathbf{P}'(\mathbf{I}_k + \mathbf{D}_\gamma^2) \mathbf{P}(\mathbf{X}'\mathbf{X})^{1/2} = \mathbf{G}'^{-1}(\mathbf{I}_k + \mathbf{D}_\gamma^2)\mathbf{G}^{-1}$ as in Section 2.3, then combining terms to get the expression $(\hat{\boldsymbol{\beta}}_I - \hat{\boldsymbol{\beta}})' \mathbf{G}'^{-1}(\mathbf{I}_k + \mathbf{D}_\gamma^2)\mathbf{G}^{-1}(\hat{\boldsymbol{\beta}}_I - \hat{\boldsymbol{\beta}})$ and substituting $(\hat{\boldsymbol{\theta}}_I - \hat{\boldsymbol{\theta}})$ for $\mathbf{G}^{-1}(\hat{\boldsymbol{\beta}}_I - \hat{\boldsymbol{\beta}})$.

3.2. Basic properties of D_I. Essential properties of D_I follow directly. For the case $r = k$, let $\mathbf{U}_0 = [\mathbf{U}_1', \mathbf{U}_2']' = \mathbf{U}_1$ since $\mathbf{U}_2$ is now void, and observe that $\mathbf{W}' = [\mathbf{I}_k, \mathbf{0}, \mathbf{D}_\gamma]$. It follows easily that $(\hat{\boldsymbol{\theta}}_I - \hat{\boldsymbol{\theta}}) = (\mathbf{I}_k + \mathbf{D}_\gamma^2)^{-1}\mathbf{D}_\gamma(\mathbf{D}_\gamma \mathbf{U}_0 - \mathbf{U}_4)$. From the nondegeneracy on $\mathbb{R}^N$ of $\mathcal{L}(\mathbf{Y})$, and thus of $\mathcal{L}(\mathbf{U})$, we infer that the distribution of $(\hat{\boldsymbol{\theta}}_I - \hat{\boldsymbol{\theta}})$ is nonsingular on $\mathbb{R}^k$, as is $(\hat{\boldsymbol{\beta}}_I - \hat{\boldsymbol{\beta}}) = \mathbf{G}(\hat{\boldsymbol{\theta}}_I - \hat{\boldsymbol{\theta}})$. The residual vector in canonical form for the full data is $\mathbf{R}' = [(\mathbf{U}_0 - \hat{\boldsymbol{\theta}})', \mathbf{U}_3', (\mathbf{U}_4 - \mathbf{D}_\gamma \hat{\boldsymbol{\theta}})']$, and the corresponding sum of squares is given by

$$(N - k)S^2 = (\hat{\boldsymbol{\theta}}_I - \hat{\boldsymbol{\theta}})' \mathbf{D}_\gamma^{-1}(\mathbf{I}_k + \mathbf{D}_\gamma^2) \mathbf{D}_\gamma^{-1}(\hat{\boldsymbol{\theta}}_I - \hat{\boldsymbol{\theta}}) + \mathbf{U}_3' \mathbf{U}_3 , \qquad (3.4)$$

as may be seen on noting that $\hat{\boldsymbol{\theta}}_I = \mathbf{U}_0$ and combining the first and last terms of $\mathbf{R}'\mathbf{R}$.

The case $r < k$ is somewhat more delicate but proceeds as follows. From the reduced data in canonical form it is seen that $\hat{\boldsymbol{\theta}}_I' = [\hat{\boldsymbol{\theta}}_{I1}', \hat{\boldsymbol{\theta}}_{I2}'] = [\mathbf{U}_1', \mathbf{U}_2']$, whereas the corresponding error mean square is given by $(n - k)S_I^2 = \mathbf{U}_3' \mathbf{U}_3$. For the full data in canonical form we have for $r < k$ that $\mathbf{W}'\mathbf{W} = \mathrm{Diag}((\mathbf{I}_r + \mathbf{D}_\gamma^2), \mathbf{I}_s)$, and elements of $\hat{\boldsymbol{\theta}}' = [\hat{\boldsymbol{\theta}}_1', \hat{\boldsymbol{\theta}}_2']$ are given in partitioned form as $\hat{\boldsymbol{\theta}}_1 = (\mathbf{I}_r + \mathbf{D}_\gamma^2)^{-1}(\mathbf{U}_1 + \mathbf{D}_\gamma \mathbf{U}_4)$ and $\hat{\boldsymbol{\theta}}_2 = \mathbf{U}_2$, whereas the residual vector is given by

$$\mathbf{R}' = [(\mathbf{U}_1 - \hat{\boldsymbol{\theta}}_1)', (\mathbf{U}_2 - \hat{\boldsymbol{\theta}}_2)', \mathbf{U}_3', (\mathbf{U}_4 - \mathbf{D}_\gamma \hat{\boldsymbol{\theta}}_1)'] . \qquad (3.5)$$

The second subvector consists entirely of zeros since $\hat{\boldsymbol{\theta}}_2 = \mathbf{U}_2$.

We next summarize essential features of $(\hat{\boldsymbol{\beta}}_I - \hat{\boldsymbol{\beta}})$, S_I^2, and S^2 under the full and reduced data sets, as they stem from straightforward least-squares analyses.

THEOREM 1. Let $\hat{\boldsymbol{\beta}} = (\mathbf{X}_0'\mathbf{X}_0)^{-1}\mathbf{X}_0'\mathbf{Y}$ and $\hat{\boldsymbol{\beta}}_I = (\mathbf{X}'\mathbf{X})^{-1}\mathbf{X}'\mathbf{Y}_1$ be least-squares estimators for $\boldsymbol{\beta}$ in the full and reduced data under $\mathbf{Y} = \mathbf{X}_0\boldsymbol{\beta} + \boldsymbol{\varepsilon}$ as in (3.1), where $r \leq k$ and $\mathscr{L}(\mathbf{Y})$ is nonsingular on $\mathbb{R}^N$. Then

(i) $(\hat{\boldsymbol{\beta}}_I - \hat{\boldsymbol{\beta}})$ has a joint distribution on $\mathbb{R}^k$ of rank $r \leq k$, singular for $r < k$ and non-singular for $r = k$.

(ii) The residual sum of squares under the reduced data is given in canonical form as $(n - k)S_I^2 = \mathbf{U}_3'\mathbf{U}_3$.

(iii) The vector $(\hat{\boldsymbol{\theta}}_I - \hat{\boldsymbol{\theta}})$ is a component of the residual sum of squares $(N - k)S^2$ under the full data with $r = k$, whereas $(\hat{\boldsymbol{\theta}}_{I1} - \hat{\boldsymbol{\theta}}_1)$ is a component of $(N - k)S^2$ for $r < k$.

(iv) If $E(\boldsymbol{\varepsilon}) = \mathbf{0} = E(\boldsymbol{\eta})$ in (3.1) and (3.2) with $\boldsymbol{\xi}_2$ and $\boldsymbol{\delta}$ not necessarily null, then $E(\hat{\boldsymbol{\theta}}_I - \hat{\boldsymbol{\theta}}) = (\mathbf{I}_k + \mathbf{D}_\gamma^2)^{-1} \mathbf{D}_\gamma \boldsymbol{\delta}$ and thus $E(\hat{\boldsymbol{\beta}}_I - \hat{\boldsymbol{\beta}}) = \mathbf{G}(\mathbf{I}_k + \mathbf{D}_\gamma^2)^{-1} \mathbf{D}_\gamma \boldsymbol{\delta}$ for $r = k$, whereas $E(\hat{\boldsymbol{\theta}}_{I1} - \hat{\boldsymbol{\theta}}_1) = (\mathbf{I}_r + \mathbf{D}_\gamma^2)^{-1} \mathbf{D}_\gamma \boldsymbol{\delta}$ and thus $E(\hat{\boldsymbol{\beta}}_I - \hat{\boldsymbol{\beta}}) = \mathbf{G}_1 (\mathbf{I}_r + \mathbf{D}_\gamma^2)^{-1} \mathbf{D}_\gamma \boldsymbol{\delta}$ for $r < k$, where $\mathbf{G}_1 \in F_{k \times r}$.

PROOF. That conclusion (i) holds for $r = k$ was shown in developments preceding expression (3.4). For $r < k$, choose $\mathbf{G}$ as in Lemma 1 and Section 3.1, to be partitioned as $\mathbf{G} = [\mathbf{G}_1, \mathbf{G}_2]$ with $\mathbf{G}_1 \in F_{k \times r}$, and observe that $(\hat{\boldsymbol{\beta}}_I - \hat{\boldsymbol{\beta}}) = \mathbf{G}(\hat{\boldsymbol{\theta}}_I - \hat{\boldsymbol{\theta}}) = \mathbf{G}_1(\hat{\boldsymbol{\theta}}_{I1} - \hat{\boldsymbol{\theta}}_1)$ since $(\hat{\boldsymbol{\theta}}_{I2} - \hat{\boldsymbol{\theta}}_2) = \mathbf{0}$ identically. Moreover, from the nonsingularity of $\mathscr{L}(\mathbf{Y})$ on $\mathbb{R}^N$ we infer that $\mathscr{L}((\hat{\boldsymbol{\theta}}_{I1} - \hat{\boldsymbol{\theta}}_1))$ is nonsingular on $\mathbb{R}^r$, so that the joint distribution of $(\hat{\boldsymbol{\beta}}_I - \hat{\boldsymbol{\beta}})$ on $\mathbb{R}^k$ is singular of rank $r < k$. Conclusion (ii) is a direct consequence of the canonical form as noted earlier. Conclusion (iii) is apparent from (3.4) for $r = k$, and for $r < k$ it follows on noting that

$$(N-k)S^2 = (\hat{\boldsymbol{\theta}}_{I1} - \hat{\boldsymbol{\theta}}_1)'\mathbf{D}_\gamma^{-1}(\mathbf{I}_r + \mathbf{D}_\gamma^2)\mathbf{D}_\gamma^{-1}(\hat{\boldsymbol{\theta}}_{I1} - \hat{\boldsymbol{\theta}}_1) + \mathbf{U}_3'\mathbf{U}_3 , \qquad (3.6)$$

where $\mathbf{D}_\gamma$ is now of order $(r \times r)$, together with the fact that $(\hat{\boldsymbol{\theta}}_{I2} - \hat{\boldsymbol{\theta}}_2) = \mathbf{0}$ identically. To see conclusion (iv), we find from earlier developments with $r = k$ that $(\hat{\boldsymbol{\theta}}_I - \hat{\boldsymbol{\theta}}) = (\mathbf{I}_k + \mathbf{D}_\gamma^2)^{-1}\mathbf{D}_\gamma(\mathbf{D}_\gamma\mathbf{U}_0 - \mathbf{U}_4)$, so that $E(\hat{\boldsymbol{\theta}}_I - \hat{\boldsymbol{\theta}}) = (\mathbf{I}_k + \mathbf{D}_\gamma^2)^{-1}\mathbf{D}_\gamma[\mathbf{D}_\gamma\boldsymbol{\theta} - (\mathbf{D}_\gamma\boldsymbol{\theta} - \boldsymbol{\delta})] = (\mathbf{I}_k + \mathbf{D}_\gamma^2)^{-1}\mathbf{D}_\gamma\boldsymbol{\delta}$ under the model (3.2), and similarity for $r < k$. This completes our proof. □

The foregoing developments all follow from the least-squares principle assuming only nonsingularity and centering of the errors. In the next sections we examine D_I for assessing influence under specific error assumptions, with reference both to leverage and to outliers.

4. Normal-Theory properties

Properties of D_I are given next under Gaussian errors. The focus is on $(\hat{\boldsymbol{\beta}}_I - \boldsymbol{\beta})$ under a block-diagonal dispersion matrix $V(\mathbf{Y})$, to be specialized as circumstances demand.

4.1. Basic results. Properties of $(\hat{\boldsymbol{\beta}}_I - \boldsymbol{\beta})$ are developed through $(\hat{\boldsymbol{\theta}}_I - \hat{\boldsymbol{\theta}})$. Let $\mathbf{Q} = \text{Diag}(\mathbf{Q}_1, \mathbf{Q}_2)$ with $\mathbf{Q}_1$ and $\mathbf{Q}_2$ as in Section 3.1; suppose that $\mathbf{Y}_1$ and $\mathbf{Y}_2$ are mutually uncorrelated with dispersion matrices $\Omega_1 \in S_n^+$ and $\Omega_2 \in S_r^+$, so that $V(\mathbf{Y}) = \Omega = \text{Diag}(\Omega_1, \Omega_2)$; and let $\mathbf{Q}\Omega\mathbf{Q}' = \Sigma = \text{Diag}(\mathbf{Q}_1\Omega_1\mathbf{Q}_1', \mathbf{Q}_2\Omega_2\mathbf{Q}_2')$, so that $V(\mathbf{U}) = \Sigma$ with $\mathbf{U}$ as in (3.2). Now partition $\Sigma = [\Sigma_{ij}]$ conformably with $\mathbf{U} = [\mathbf{U}_1', \mathbf{U}_2', \mathbf{U}_3', \mathbf{U}_4']'$, so that $\mathbf{Q}_2\Omega_2\mathbf{Q}_2' = \Sigma_{44}$; observe that $\{\Sigma_{14}, \Sigma_{24}, \Sigma_{34}\}$ all vanish owing to the block-diagonal structure of Σ; and write $\Sigma_0 = [\Sigma_{ij}; i, j = 1, 2] = \Sigma_{11}$ corresponding to $\mathbf{U}_0 = [\mathbf{U}_1', \mathbf{U}_2']'$, since $\mathbf{U}_2$ and hence Σ_{22} are now void for $r = k$.

Proceed as in Section 3.2 and observe that $(\hat{\boldsymbol{\theta}}_I - \hat{\boldsymbol{\theta}}) = (\mathbf{I}_k + \mathbf{D}_\gamma^2)^{-1}\mathbf{D}_\gamma(\mathbf{D}_\gamma\mathbf{U}_0 - \mathbf{U}_4)$ for $r = k$. For $r < k$ it was seen that $(\hat{\boldsymbol{\theta}}_{I2} - \hat{\boldsymbol{\theta}}_2) = \mathbf{0}$ identically, whereas $(\hat{\boldsymbol{\theta}}_{I1} - \hat{\boldsymbol{\theta}}_1) = (\mathbf{I}_r + \mathbf{D}_\gamma^2)^{-1}\mathbf{D}_\gamma(\mathbf{D}_\gamma\mathbf{U}_1 - \mathbf{U}_4)$. Essential properties of $(\hat{\boldsymbol{\beta}}_I - \boldsymbol{\beta})$ may be studied through $(\hat{\boldsymbol{\theta}}_I - \hat{\boldsymbol{\theta}})$ under Gaussian errors as follows, where we allow for a mean shift of ξ_2 in $E(\mathbf{Y}_2)$, namely, $E(\mathbf{Y}_2) = \mathbf{Z}\boldsymbol{\beta} - \xi_2$ as in (3.1), and similarly for (3.2). To these ends define special arrays as

$$\boldsymbol{\mu}_r(\boldsymbol{\delta}) = (\mathbf{I}_r + \mathbf{D}_\gamma^2)^{-1}\mathbf{D}_\gamma\boldsymbol{\delta} , \qquad (4.1)$$

where $\boldsymbol{\delta} = \mathbf{Q}_2\xi_2$, and

$$\boldsymbol{\Xi}_r(\gamma) = (\mathbf{I}_r + \mathbf{D}_\gamma^2)^{-1}\mathbf{D}_\gamma(\mathbf{D}_\gamma\Sigma_{11}\mathbf{D}_\gamma + \Sigma_{44})\mathbf{D}_\gamma(\mathbf{I}_r + \mathbf{D}_\gamma^2)^{-1} . \qquad (4.2)$$

Both expressions apply for each $r \leq k$, where $\Sigma_{11} = \Sigma_0$ for $r = k$ as noted earlier. A central result is the following.

THEOREM 2. Suppose that $\mathscr{L}(\mathbf{Y}) = N_N(\mathbf{X}_0\boldsymbol{\beta} - \boldsymbol{\xi}, \boldsymbol{\Omega})$, with $\boldsymbol{\xi}' = [\boldsymbol{\xi}_1', \boldsymbol{\xi}_2']$ such that $\boldsymbol{\xi}_1 = \mathbf{0}$ and $\boldsymbol{\xi}_2 \in \mathbb{R}^r$, and with $\boldsymbol{\Omega} = \text{Diag}(\boldsymbol{\Omega}_1, \boldsymbol{\Omega}_2)$, and consider $(\hat{\boldsymbol{\beta}}_I - \hat{\boldsymbol{\beta}})$, $(\hat{\boldsymbol{\theta}}_I - \hat{\boldsymbol{\theta}})$ for $r = k$, and $(\hat{\boldsymbol{\theta}}_{I1} - \hat{\boldsymbol{\theta}}_1)$ for $r < k$. Then for $r = k$:

(i) The distribution of $(\hat{\boldsymbol{\theta}}_I - \hat{\boldsymbol{\theta}})$ is nonsingular on $\mathbb{R}^k$ and is given by $N_k(\boldsymbol{\mu}_k(\boldsymbol{\delta}), \boldsymbol{\Xi}_k(\gamma))$, where $\boldsymbol{\delta} = \mathbf{Q}_2\boldsymbol{\xi}_2$, with $\boldsymbol{\mu}_k(\boldsymbol{\delta})$ as in (4.1) and $\boldsymbol{\Xi}_k(\gamma)$ as in (4.2).

(ii) The distribution of $(\hat{\boldsymbol{\beta}}_I - \hat{\boldsymbol{\beta}})$ is nonsingular on $\mathbb{R}^k$ and is given by $N_k(\mathbf{G}\boldsymbol{\mu}_k(\boldsymbol{\delta}), \mathbf{G}\boldsymbol{\Xi}_k(\gamma)\mathbf{G}')$ with $\mathbf{G}$ as defined in Section 3.1. Moreover, for $r < k$:

(iii) The distribution of $(\hat{\boldsymbol{\theta}}_{I1} - \hat{\boldsymbol{\theta}}_1)$ is nonsingular on $\mathbb{R}^r$ and is given by $N_r(\boldsymbol{\mu}_r(\boldsymbol{\delta}), \boldsymbol{\Xi}_r(\gamma))$, where $\boldsymbol{\mu}_r(\boldsymbol{\delta})$ and $\boldsymbol{\Xi}_r(\gamma)$ are given in (4.1) and (4.2).

(iv) The distribution of $(\hat{\boldsymbol{\beta}}_I - \hat{\boldsymbol{\beta}})$ on $\mathbb{R}^k$ is the singular Gaussian distribution $N_k(\mathbf{G}_1\boldsymbol{\mu}_r(\boldsymbol{\delta}), \mathbf{G}_1\boldsymbol{\Xi}_r(\gamma)\mathbf{G}_1')$, where $\mathbf{G} = [\mathbf{G}_1, \mathbf{G}_2]$ is partitioned as in the proof for Theorem 1.

PROOF. These conclusions all follow from standard arguments pertaining to linear transformations of Gaussian variates. In particular, expected values follow from Theorem 1 (iv), to complete our proof. □

We turn next to structural parameters that drive D_I stochastically, in partial explanation for the occurrence of extreme values in practice. To these ends we specialize the matrix $\boldsymbol{\Omega}$ to include the cases $\boldsymbol{\Omega} = \sigma^2\mathbf{I}_N$ and $\boldsymbol{\Omega} = \boldsymbol{\Xi}(\sigma_1^2, \sigma_2^2) = \text{Diag}(\sigma_1^2\mathbf{I}_n, \sigma_2^2\mathbf{I}_r)$.

4.2. The scaling of D_I. The matter of scaling in the denominator of $D_I(\hat{\boldsymbol{\beta}}, \mathbf{M}, c\hat{\sigma}^2)$ surfaces immediately. Prospects for a known distributional form of D_I are predicated on a suitable scaling constant c and choice for $\hat{\sigma}^2$. Recall from Theorem 1 (iii) that $(\hat{\boldsymbol{\theta}}_I - \hat{\boldsymbol{\theta}})$ is a component of $(N - k)S^2$ for $r = k$, as is $(\hat{\boldsymbol{\theta}}_{I1} - \hat{\boldsymbol{\theta}}_1)$ for $r < k$. Moreover, Theorem 1 (iv) shows under Gaussian errors that the first quadratic form on the right of (3.4) has a noncentral distribution unless $\boldsymbol{\delta} = \mathbf{0}$. We conclude that $((\hat{\boldsymbol{\theta}}_I - \hat{\boldsymbol{\theta}}), S^2)$ are dependent, as are $((\hat{\boldsymbol{\theta}}_{I1} - \hat{\boldsymbol{\theta}}_1), S^2)$ and thus $((\hat{\boldsymbol{\beta}}_I - \hat{\boldsymbol{\beta}}), S^2)$, even under the Gaussian model $\mathscr{L}(\mathbf{Y}) = N_N(\mathbf{X}_0\boldsymbol{\beta}, \sigma^2\mathbf{I}_N)$. The conventional use of $D_I(\hat{\boldsymbol{\beta}}, \mathbf{X}_0'\mathbf{X}_0, kS^2)$ now may be seen to be flawed on three counts: (i) Its numerator has rank $r \leq k$, with rank k if and only if $r = k$. Based on extensive numerical studies, routine scaling by k was questioned by Gray (1993), who suggested scaling by subset size. On structural grounds we now concur that scaling by r, the number rows of $\mathbf{Z}$ and hence the rank of $\mathscr{L}(\hat{\boldsymbol{\beta}}_I - \hat{\boldsymbol{\beta}})$, now appears to be appropriate, giving $D_I(\hat{\boldsymbol{\beta}}, \mathbf{X}_0'\mathbf{X}_0, r\hat{\sigma}^2)$ for each $r \leq k$. (ii) The numerator and denominator of $D_I(\hat{\boldsymbol{\beta}}, \mathbf{X}_0'\mathbf{X}_0, kS^2)$ are not independent, thus precluding the emergence of known distributions for ratios of mean squares. (iii) Not only are $((\hat{\boldsymbol{\theta}}_I - \hat{\boldsymbol{\theta}}), S^2)$ dependent, but the denominator of $D_I(\hat{\boldsymbol{\beta}}, \mathbf{X}_0'\mathbf{X}_0, cS^2)$ will be inflated stochastically owing to its noncentral distribution when $\boldsymbol{\delta} \neq \mathbf{0}$. A mean shift therefore may mask evidence of influence otherwise apparent in the numerator of $D_I(\hat{\boldsymbol{\beta}}, \mathbf{X}_0'\mathbf{X}_0, c\hat{\sigma}^2)$ if scaled properly.

On the other hand, it is clear from Theorem 2 that $((\hat{\boldsymbol{\theta}}_I - \hat{\boldsymbol{\theta}}), S_I^2)$ are independent under Gaussian errors with dispersion matrix $\boldsymbol{\Xi}(\sigma_1^2, \sigma_2^2)$, and that $(n - k)S_I^2 = \mathbf{U}_3'\mathbf{U}_3$ has a central distribution regardless of the value of $\boldsymbol{\delta}$. To avoid anomalies of the foregoing types, henceforth we consider only versions of D_I of the type $D_I(\hat{\boldsymbol{\beta}}, \mathbf{M}, rS_I^2)$, including the case $r = k$.

4.3. Effects of leverage and outliers.
Two versions of leverages are germane. The eigenvalues $\{\gamma_1^2, \ldots, \gamma_r^2\}$ of $\mathbf{H} = \mathbf{Z}(\mathbf{X}'\mathbf{X})^{-1}\mathbf{Z}'$ serve as canonical leverages in predicting at $\mathbf{Z}$ based on $\hat{\boldsymbol{\beta}}_I$. Similarly, $\{\lambda_i = \gamma_i^2/(\gamma_i^2 + 1); 1 \leq i \leq r\}$ apply in predicting at $\mathbf{Z}$ based on $\hat{\boldsymbol{\beta}}$ from the full data as noted. Both figure prominently in the distribution of $D_I(\hat{\boldsymbol{\beta}}, \mathbf{M}, rS_I^2)$ under two commonly used choices of $\mathbf{M}$ to be studied next, followed by a comparative assessment of their diagnostic capabilities.

Several possibilities emerge for modeling what might be deemed to be "outliers." Here we consider (i) a possible shift in means, namely, $E(\mathbf{Y}_2) = \mathbf{Z}\boldsymbol{\beta} - \boldsymbol{\xi}_2$ as in (3.1) and the corresponding shift in (3.2), as well as (ii) a possible shift in variance between $\mathbf{Y}_1 \in \mathbb{R}^n$ and $\mathbf{Y}_2 \in \mathbb{R}^r$ in (3.1), or both. To model the latter, let $V(\mathbf{Y}_1) = \sigma_1^2 \mathbf{I}_n$ and $V(\mathbf{Y}_2) = \sigma_2^2 \mathbf{I}_r$, so that $V(\mathbf{Y}) = \boldsymbol{\Xi}(\sigma_1^2, \sigma_2^2) = \mathrm{Diag}(\sigma_1^2 \mathbf{I}_n, \sigma_2^2 \mathbf{I}_r)$ with $r \leq k$.

We first seek the distribution of $D_I(\hat{\boldsymbol{\beta}}, \mathbf{X}_0' \mathbf{X}_0, rS_I^2)$ through the canonical form $D_I(\hat{\boldsymbol{\theta}}, \mathbf{M}(\gamma), rS_I^2)$ with $\mathbf{M}(\gamma) = (\mathbf{I}_k + \mathbf{D}_\gamma^2)$ as defined following (3.3). To study $\mathscr{L}((\hat{\boldsymbol{\theta}}_I - \hat{\boldsymbol{\theta}})'(\mathbf{I}_k + \mathbf{D}_\gamma^2)(\hat{\boldsymbol{\theta}}_I - \hat{\boldsymbol{\theta}})/\sigma_1^2)$, we specialize Theorem 2, under the dispersion structure $\boldsymbol{\Xi}(\sigma_1^2, \sigma_2^2)$, to infer for $r = k$ that $\mathscr{L}((\hat{\boldsymbol{\theta}}_I - \hat{\boldsymbol{\theta}})/\sigma_1) = N_k(\boldsymbol{\mu}_k(\sigma_1), \boldsymbol{\Gamma}(\sigma_1^2, \sigma_2^2))$. Here $\boldsymbol{\mu}_k(\sigma_1) = \boldsymbol{\mu}_k(\delta)/\sigma_1$ as in (4.1), and $\boldsymbol{\Gamma}(\sigma_1^2, \sigma_2^2) = \boldsymbol{\Xi}_k(\gamma)/\sigma_1^2 = (\mathbf{I}_k + \mathbf{D}_\gamma^2)^{-2} \mathbf{D}_\gamma^2 (\mathbf{D}_\gamma^2 + \sigma_2^2/\sigma_1^2 \mathbf{I}_k)$ with $\boldsymbol{\Xi}_k(\gamma)$ as in (4.2), since diagonal matrices commute. To express $\mathscr{L}((\hat{\boldsymbol{\theta}}_1 - \hat{\boldsymbol{\theta}})'(\mathbf{I}_k + \mathbf{D}_\gamma^2)(\hat{\boldsymbol{\theta}}_I - \hat{\boldsymbol{\theta}})/\sigma_1^2)$ in terms of $G_k(t; \alpha', \omega')$ as in Section 2.2, we apply standard theory for distributions of quadratic forms in Gaussian variables, first constructing $\omega = [\boldsymbol{\Gamma}(\sigma_1^2, \sigma_2^2)]^{-1/2} \boldsymbol{\mu}_k(\sigma_1) = (\mathbf{D}_\gamma^2 + \sigma_2^2/\sigma_1^2 \mathbf{I}_k)^{-1/2} \boldsymbol{\delta}/\sigma_1$ with typical elements $\{\omega_i = \delta_i/[\sigma_1(\gamma_i^2 + \sigma_2^2/\sigma_1^2)^{1/2}]; 1 \leq i \leq k\}$. The weights $\{\alpha_1, \ldots, \alpha_k\}$ in turn are eigenvalues of $\boldsymbol{\Gamma}(\sigma_1^2, \sigma_2^2) \mathbf{M}(\gamma)$, to be recovered directly from

$$\boldsymbol{\Gamma}(\sigma_1^2, \sigma_2^2) \mathbf{M}(\gamma) = (\mathbf{I}_k + \mathbf{D}_\gamma^2)^{-2} \mathbf{D}_\gamma^2 (\mathbf{D}_\gamma^2 + \sigma_2^2/\sigma_1^2 \mathbf{I}_k)(\mathbf{I}_k + \mathbf{D}_\gamma^2) \quad (4.3)$$

which reduces to $\boldsymbol{\Gamma}(\sigma_1^2, \sigma_2^2) \mathbf{M}(\gamma) = (\mathbf{I}_k + \mathbf{D}_\gamma^2)^{-1} \mathbf{D}_\gamma^2 (\mathbf{D}_\gamma^2 + \sigma_2^2/\sigma_1^2 \mathbf{I}_k)$ since the product is diagonal and diagonal matrices commute. Essential properties of $D_I(\hat{\boldsymbol{\beta}}, \mathbf{X}_0' \mathbf{X}_0, rS_I^2)$ and its canonical form may be summarized as follows, where elements of $\{\gamma, \omega, \alpha, \sigma_2^2/\sigma_1^2\}$ emerge as structural parameters that drive $D_I(\hat{\boldsymbol{\beta}}, \mathbf{M}, rS_I^2)$ stochastically.

THEOREM 3. Suppose that $\mathscr{L}(\mathbf{Y}) = N_N(\mathbf{X}_0 \boldsymbol{\beta} - \boldsymbol{\xi}, \boldsymbol{\Xi}(\sigma_1^2, \sigma_2^2))$, with $\boldsymbol{\Xi}(\sigma_1^2, \sigma_2^2) = \mathrm{Diag}(\sigma_1^2 \mathbf{I}_n, \sigma_2^2 \mathbf{I}_r)$. Then

(i) For each $r \leq k$, $\mathscr{L}((\hat{\boldsymbol{\theta}}_I - \hat{\boldsymbol{\theta}})' \mathbf{M}(\gamma)(\hat{\boldsymbol{\theta}}_I - \hat{\boldsymbol{\theta}})/\sigma_1^2) = \mathscr{L}(\sum_{i=1}^r \alpha_i U_i^2)$, such that $\{U_1, U_2, \ldots, U_r\}$ are independent random $\{N_1(\omega_i, 1); 1 \leq i \leq r\}$ variates, where $\{\omega_i = \delta_i/[\sigma_1(\gamma_i^2 + \sigma_2^2/\sigma_1^2)^{1/2}]; 1 \leq i \leq r\}$, and $\{\alpha_i = \gamma_i^2(\gamma_i^2 + \sigma_2^2/\sigma_1^2)/(\gamma_i^2 + 1); 1 \leq i \leq r\}$. Its cdf thus is given by $G_r(t; \alpha_1, \ldots, \alpha_r, \omega_1, \ldots, \omega_r)$.

(ii) The distribution of $D_I(\hat{\boldsymbol{\theta}}, \mathbf{M}(\gamma), rS_I^2)$, and thus of $D_I(\hat{\boldsymbol{\beta}}, \mathbf{X}_0' \mathbf{X}_0, rS_I^2)$, is given by $F_r(w; \alpha_1, \ldots, \alpha_r, \omega_1, \ldots, \omega_r, n-k)$ for each $r \leq k$.

(iii) For the case of outliers of neither type, the distribution of $D_I(\hat{\boldsymbol{\beta}}, \mathbf{X}_0'\mathbf{X}_0, rS_I^2)$ is given by the central cdf $F_r(w; \gamma_1^2, \gamma_2^2, \ldots, \gamma_r^2, n-k)$.

PROOF. Arguments supporting conclusion (i) were outlined in the foregoing paragraphs. To continue, note that $\mathscr{L}(\mathbf{U}_3'\mathbf{U}_3/\sigma_1^2) = \chi^2(n-k, 0)$ independently of $(\hat{\boldsymbol{\theta}}_I - \hat{\boldsymbol{\theta}})$, so that standard developments yield the distribution $\mathscr{L}(D_I(\hat{\boldsymbol{\theta}}, \mathbf{M}(\gamma), kS_I^2))$ for the case $r=k$ as in Section (2.2). Parallel steps for $r<k$ give $\mathscr{L}(D_I(\hat{\boldsymbol{\theta}}, \mathbf{M}(\gamma), rS_I^2))$ without difficulty. Claims for the distribution $\mathscr{L}(D_I(\hat{\boldsymbol{\beta}}, \mathbf{X}_0'\mathbf{X}_0, rS_I^2))$ then follow as in (3.3) for each $r \leq k$, to give conclusion (ii). Conclusion (iii) now follows on specializing (ii), to complete our proof. □

We next consider $D_I(\hat{\boldsymbol{\beta}}, \mathbf{X}'\mathbf{X}, rS_I^2)$ as another version in common use, where our reasons for scaling by rS_I^2 persist. To these ends we proceed precisely as in the proof for Theorem 3, with the one exception that $\mathbf{M}(\gamma) = (\mathbf{I}_r + \mathbf{D}_\gamma^2)$, as used there, is now supplanted by $\mathbf{I}_r$ for each $r \leq k$. The principle findings may be summarized as follows.

THEOREM 4. Suppose that $\mathscr{L}(\mathbf{Y}) = N_N(\mathbf{X}_0 \boldsymbol{\beta} - \boldsymbol{\xi}, \boldsymbol{\Xi}(\sigma_1^2, \sigma_2^2))$, with $\boldsymbol{\Xi}(\sigma_1^2, \sigma_2^2) = \text{Diag}(\sigma_1^2 \mathbf{I}_n, \sigma_2^2 \mathbf{I}_r)$, and let $\{\lambda_i = \gamma_i^2/(\gamma_i^2 + 1); 1 \leq i \leq r\}$ be canonical leverages in predicting at $\mathbf{Z}$ based on $\hat{\boldsymbol{\beta}}$ from the full data. Then

(i) $\mathscr{L}((\hat{\boldsymbol{\theta}}_I - \hat{\boldsymbol{\theta}})'(\hat{\boldsymbol{\theta}}_I - \hat{\boldsymbol{\theta}})/\sigma_1^2) = \mathscr{L}(\sum_{i=1}^r \alpha_i U_i^2)$, such that $\{U_1, U_2, \ldots, U_r\}$ are independent random $\{N_1(\omega_i, 1); 1 \leq i \leq r\}$ variates, where $\{\omega_i = \delta_i/[\sigma_1(\gamma_i^2 + \sigma_2^2/\sigma_1^2)^{1/2}]; 1 \leq i \leq r\}$ as before, and $\{\alpha_i = \lambda_i(\gamma_i^2 + \sigma_2^2/\sigma_1^2)/(\gamma_i^2 + 1); 1 \leq i \leq r\}$. Its cdf thus is given by $G_r(t; \alpha_1, \ldots, \alpha_r, \omega_1, \ldots, \omega_r)$.

(ii) The distribution of $D_I(\hat{\boldsymbol{\theta}}, \mathbf{I}_r, rS_I^2)$, and thus of $D_I(\hat{\boldsymbol{\beta}}, \mathbf{X}'\mathbf{X}, rS_I^2)$, is given by $F_r(w; \alpha_1, \ldots, \alpha_r, \omega_1, \ldots, \omega_r, n-k)$ for each $r \leq k$.

(iii) For the case $\sigma_2^2/\sigma_1^2 = 1$, the distribution of $D_I(\hat{\boldsymbol{\theta}}, \mathbf{I}_r, rS_I^2)$, and thus of $D_I(\hat{\boldsymbol{\beta}}, \mathbf{X}'\mathbf{X}, rS_I^2)$, is given by $F_r(w; \lambda_1, \ldots, \lambda_r, \omega_1, \ldots, \omega_r, n-k)$ for each $r \leq k$, where $\{\omega_i; 1 \leq i \leq r\}$ are as before and $\{\lambda_i; 1 \leq i \leq r\}$ are canonical leverages in predicting at $\mathbf{Z}$ based on $\hat{\boldsymbol{\beta}}$ from the full data.

(iv) For the case of outliers of neither type, the distribution of $D_I(\hat{\boldsymbol{\beta}}, \mathbf{X}'\mathbf{X}, rS_I^2)$ is given by the central cdf $F_r(w; \lambda_1, \ldots, \lambda_r, n-k)$ for each $r \leq k$, and this function is bounded below by $F_r(w; \lambda_1, \ldots, \lambda_r, n-k) \geq F(w; r, n-k)$ pointwise for each $w \geq 0$.

PROOF. Except as noted, the bulk of the proof runs parallel to that of Theorem 3. That the cdf in conclusion (iv) is bounded below by a standard F-distribution is a consequence of the bound $\{\lambda_i \leq 1; 1 \leq i \leq r\}$ together with the orderings of Lemma 1. □

The foregoing developments now support comparison of the diagnostic capabilities of $D_I(\hat{\boldsymbol{\beta}}, \mathbf{X}_0'\mathbf{X}_0, rS_I^2)$ and $D_I(\hat{\boldsymbol{\beta}}, \mathbf{X}'\mathbf{X}, rS_I^2)$. This may be garnered by examining in detail the shift $\{\omega_i; 1 \leq i \leq r\}$ and scale $\{\alpha_i; 1 \leq i \leq r\}$ parameters of Theorems 3 and 4, when coupled with Lemma 1.

In the absence of outliers resulting from shifted means or variance, Theorems 3 and 4 and Lemma 1 confirm that $D_I(\hat{\boldsymbol{\beta}}, \mathbf{X}_0'\mathbf{X}_0, rS_I^2)$ and $D_I(\hat{\boldsymbol{\beta}}, \mathbf{X}'\mathbf{X}, rS_I^2)$ are driven stochastically by the canonical leverages $\{\gamma_i^2; 1 \leq i \leq r\}$ and $\{\lambda_i; 1 \leq i \leq r\}$, respectively. These may be very small in either case or, in the case of $D_I(\hat{\boldsymbol{\beta}}, \mathbf{X}_0'\mathbf{X}_0, rS_I^2)$, very large, but are bounded above by 1.0 for $D_I(\hat{\boldsymbol{\beta}}, \mathbf{X}_0'\mathbf{X}_0, rS_I^2)$. Large outcomes thereby associate stochastically with high subset leverages, small outcomes with low leverages, and intermediate outcomes with moderate, or with a mix of high and low leverages. In particular, if all leverages are equal for a given subset, say $\{\gamma_1^2 = \cdots = \gamma_r^2 = \gamma^2\}$ as in first-order orthogonal designs, then the distributions of $D_I(\hat{\boldsymbol{\beta}}, \mathbf{X}_0'\mathbf{X}_0, rS_I^2)$ and $D_I(\hat{\boldsymbol{\beta}}, \mathbf{X}'\mathbf{X}, rS_I^2)$ relate to standard F-distributions through scaling as $F_r(w; \gamma^2, \ldots, \gamma^2, n-k) = F(w/\gamma^2; r, n-k)$ and $F(w/\lambda; r, n-k)$, respectively, as in Section 2.2.

Nonetheless, since leverages can be found numerically from $\mathbf{X}_0$ independently of $\mathbf{Y}$, we find no diagnostic merit whatever in gauging subset leverages vicariously through a chance outcome of either $D_I(\hat{\boldsymbol{\beta}}, \mathbf{X}_0'\mathbf{X}_0, rS_I^2)$ or $D_I(\hat{\boldsymbol{\beta}}, \mathbf{X}'\mathbf{X}, rS_I^2)$. This point of view in turn obviates the need for benchmarks on these statistics for assessing subset influence due to leverages.

For diagnosing shifts in means or variance, the use of either version of D_I is clouded by a complexity of interrelationships among $\{\boldsymbol{\alpha}, \boldsymbol{\omega}, \gamma, \sigma_2^2/\sigma_1^2\}$. Expressions for $\{\omega_i; 1 \leq i \leq r\}$ and $\{\alpha_i; 1 \leq i \leq r\}$ from Theorems 3 and 4 link with Lemma 1 to gauge the stochastic behavior of both statistics as follows. Specifically, an increase in each scale factor $\{\alpha_i; 1 \leq i \leq r\}$ often is offset in part by a decrease in the corresponding noncentrality component $\{|\omega_i|; 1 \leq i \leq r\}$. For example, if $\boldsymbol{\delta}$ and γ are held fixed, then $|\omega_i|$ decreases, whereas α_i increases, with increasing σ_2^2/σ_1^2. Similarly, with $\boldsymbol{\delta}$ and σ_2^2/σ_1^2 fixed, it is seen that $|\omega_i|$ decreases and α_i increases with increasing leverage γ_i^2 for both statistics. Further such comparisons follow directly.

Perhaps most striking is the multiplier effect exerted on each of the weights $\{\alpha_i = \gamma_i^2(\gamma_i^2 + \sigma_2^2/\sigma_1^2)/(\gamma_i^2 + 1); 1 \leq i \leq r)$ in Theorem 3 by the extra leverage factor γ_i^2, dilating α_i for $\gamma_i^2 > 1$ and contracting α_i for $\gamma_i^2 < 1$. However, extreme dilation effects are preempted in Theorem 4 owing to the upper bound $\{\lambda_i \leq 1; 1 \leq i \leq r\}$.

Stochastic effects of variance outliers $(\sigma_2^2/\sigma_1^2 > 1)$ are masked by leverage, whether large or small, for both versions of D_I. On the other hand, a ratio σ_2^2/σ_1^2 near zero, if coupled with low subset leverages, will further depress $\{\alpha_i; 1 \leq i \leq r\}$. In consequence, small observed values for either statistic may indicate a downward shift in variance from $\mathbf{Y}_1$ to $\mathbf{Y}_2$ if coupled with a subset having small leverages. It follows that benchmarks thus are needed for gauging not only whether each statistic is *large* enough to be deemed 'influential,'' but whether it is *small* enough as well.

The diagnostic properties of both $D_I(\hat{\boldsymbol{\beta}}, \mathbf{X}_0'\mathbf{X}_0, rS_I^2)$ and $D_I(\hat{\boldsymbol{\beta}}, \mathbf{X}'\mathbf{X}, rS_I^2)$ at best are obscure, bringing into focus their practical merits as tools in subset diagnostics. The occurrence of small, intermediate, or large values has no clear link to either type of outlier, and therefore the practical value of each is less than definitive. These difficulties stem in part from the somewhat *ad hoc* choices for $\mathbf{M}$ in the statistic $D_I(\hat{\boldsymbol{\beta}}, \mathbf{M}, c\hat{\sigma}^2)$. In the next section we reexamine these choices in light of results provided in Theorem 2.

5. Modified versions of D_I

Difficulties pertaining to $D_I(\hat{\boldsymbol{\beta}}, \mathbf{M}, rS_I^2)$, as noted for $\mathbf{M} = \mathbf{X}_0'\mathbf{X}_0$ and $\mathbf{M} = \mathbf{X}'\mathbf{X}$, stem in part from the choice for $\mathbf{M}$. We next embrace a more natural choice in keeping with conclusions of Theorem 2. Further developments then yield exact normal-theory inferences for shifted outliers as modeled in (3.1) and (3.2).

5.1. Rescaling through M. Temporarily assuming an ideal model with no outliers, we choose $\mathbf{M}$ to achieve a central F- distribution as reference, not depending on leverages. We then reinstate outliers and examine diagnostic properties of the revised D_I statistic. The main focus is the canonical form of (3.2), but those findings are lifted on occasion back to (3.1).

The technical details run parallel to those of Section 4.3, but now using standard theory for distributions of quadratic forms in Gaussian variables to determine $\mathbf{M}$. Specifically, we take $\mathbf{M}$ as the inverse $\mathbf{B}(\gamma) = \mathbf{D}_\gamma^{-1}(\mathbf{I}_r + \mathbf{D}_\gamma^2)\mathbf{D}_\gamma^{-1}$ from $V(\hat{\boldsymbol{\theta}}_{I1} - \hat{\boldsymbol{\theta}}_1)$ for $r < k$, and from $V(\hat{\boldsymbol{\theta}}_I - \hat{\boldsymbol{\theta}})$ for $r = k$, as would be appropriate under homogeneous variance. Invertibility is assured by Theorem 2. Expressions for $\{\omega_i; 1 \leq i \leq r\}$ carry over intact, as these depend only on $\mathscr{L}(\hat{\boldsymbol{\theta}}_I - \hat{\boldsymbol{\theta}})$ and not $\mathbf{M}$. In contrast to Section 4.3, the weights $\{\alpha_1, \ldots, \alpha_r\}$ under a possible shift in variance are now eigenvalues of $\boldsymbol{\Gamma}(\sigma_1^2, \sigma_2^2)\mathbf{B}(\gamma)$ as given by

$$\boldsymbol{\Gamma}(\sigma_1^2, \sigma_2^2)\mathbf{B}(\gamma) = (\mathbf{I}_k + \mathbf{D}_\gamma^2)^{-2}\mathbf{D}_\gamma^2(\mathbf{D}_\gamma^2 + \sigma_2^2/\sigma_1^2 \mathbf{I}_k)\mathbf{D}_\gamma^{-2}(\mathbf{I}_k + \mathbf{D}_\gamma^2) \quad (5.1)$$

which reduces to $\boldsymbol{\Gamma}(\sigma_1^2, \sigma_2^2)\mathbf{B}(\gamma) = (\mathbf{I}_k + \mathbf{D}_\gamma^2)^{-1}(\mathbf{D}_\gamma^2 + \sigma_2^2/\sigma_1^2 \mathbf{I}_k)$ since the product is diagonal. Further let $\mathbf{V} = [(\mathbf{X}'\mathbf{X})^{-1} - (\mathbf{X}'\mathbf{X} + \mathbf{Z}'\mathbf{Z})^{-1}]$, and observe that $V(\hat{\boldsymbol{\beta}}_I - \hat{\boldsymbol{\beta}}) = \sigma^2 \mathbf{V}$ for the case that $\sigma_1^2 = \sigma_2^2$. With these preliminaries in place, we may stipulate as before the distribution of $D_I(\hat{\boldsymbol{\theta}}, \mathbf{B}(\gamma), rS_I^2)$ pertaining to (3.2). A corresponding version for (3.1) gives $D_I(\hat{\boldsymbol{\beta}}, \mathbf{V}^-, rS_I^2) = (\hat{\boldsymbol{\beta}}_I - \hat{\boldsymbol{\beta}})'\mathbf{V}^-(\hat{\boldsymbol{\beta}}_I - \hat{\boldsymbol{\beta}})/rS_I^2$ with $\mathbf{V}^-$ to be stipulated. Basic properties of these forms are summarized in the following.

THEOREM 5. *Suppose that* $\mathscr{L}(\mathbf{Y}) = N_N(\mathbf{X}_0\boldsymbol{\beta} - \xi, \boldsymbol{\Xi}(\sigma_1^2, \sigma_2^2))$, *with* $\boldsymbol{\Xi}(\sigma_1^2, \sigma_2^2) = \mathrm{Diag}(\sigma_1^2 \mathbf{I}_n, \sigma_2^2 \mathbf{I}_r)$, *and let* $\mathbf{B}(\gamma) = \mathbf{D}_\gamma^{-1}(\mathbf{I}_r + \mathbf{D}_\gamma^2)\mathbf{D}_\gamma^{-1}$. *Then*

(i) *For each* $r \leq k$, *the representation* $\mathscr{L}((\hat{\boldsymbol{\theta}}_{I1} - \hat{\boldsymbol{\theta}}_1)'\mathbf{B}(\gamma) \, (\hat{\boldsymbol{\theta}}_{I1} - \hat{\boldsymbol{\theta}}_1)/\sigma_1^2) = \mathscr{L}(\sum_{i=1}^r \alpha_i U_i^2)$ *holds, such that* $\{U_1, U_2, \ldots, U_r\}$ *are independent* $\{N_1(\omega_i, 1); 1 \leq i \leq r\}$ *random variables, where the translation parameters are* $\{\omega_i = \delta_i/[\sigma_1(\gamma_i^2 + \sigma_2^2/\sigma_1^2)^{1/2}]; 1 \leq i \leq r\}$ *as before, and* $\{\alpha_i = (\gamma_i^2 + \sigma_2^2/\sigma_1^2)/(\gamma_i^2 + 1); 1 \leq i \leq r\}$. *Its cdf thus is given by* $G_r(t; \alpha_1, \ldots, \alpha_r, \omega_1, \ldots, \omega_r)$.

(ii) *For* $r = k$, *the distribution of* $D_I(\hat{\boldsymbol{\theta}}, \mathbf{B}(\gamma), kS_I^2)$, *and thus of* $D_I(\hat{\boldsymbol{\beta}}, \mathbf{V}^{-1}, kS_I^2)$, *is given by* $F_k(w; \alpha_1, \ldots, \alpha_k, \omega_1, \ldots, \omega_k, n - k)$, *where* $\mathbf{V} = [(\mathbf{X}'\mathbf{X})^{-1} - (\mathbf{X}'\mathbf{X} + \mathbf{Z}'\mathbf{Z})^{-1}]$ *as defined following* (5.1).

(iii) Suppose that $\sigma_2^2/\sigma_1^2 = 1$. Then for $r = k$, the distribution of $D_I(\hat{\boldsymbol{\beta}}, \mathbf{V}^{-1}, kS_I^2)$ is the noncentral cdf $F(w; k, n - k, \lambda_k(\xi_2))$ with noncentrality parameter $\lambda_k(\xi_2) = \boldsymbol{\mu}_k(\xi_2)'\mathbf{V}^{-1}\boldsymbol{\mu}_k(\xi_2)/\sigma^2$, where $\boldsymbol{\mu}_k(\xi_2) = E(\hat{\boldsymbol{\beta}}_I - \hat{\boldsymbol{\beta}}) = \mathbf{G}(\mathbf{I}_k + \mathbf{D}_\gamma^2)^{-1}\mathbf{D}_\gamma\mathbf{Q}_2\xi_2$.

(iv) For $\sigma_2^2/\sigma_1^2 = 1$ and $r < k$, the distribution of $D_I(\hat{\boldsymbol{\beta}}, \mathbf{V}^-, rS_I^2)$ is the noncentral cdf $F(w; r, n - k, \lambda_r(\xi_2))$, where $\mathbf{V}^-$ is any reflexive symmetric g-inverse of $\mathbf{V}$, with noncentrality parameter $\lambda_r(\xi_2) = \boldsymbol{\mu}_r(\xi_2)'\mathbf{V}^-\boldsymbol{\mu}_r(\xi_2)/\sigma^2$, where $\boldsymbol{\mu}_r(\xi_2) = E(\hat{\boldsymbol{\beta}}_I - \hat{\boldsymbol{\beta}}) = \mathbf{G}_1(\mathbf{I}_r + \mathbf{D}_\gamma^2)^{-1}\mathbf{D}_\gamma\mathbf{Q}_2\xi_2$.

PROOF. Conclusions (i)–(iii) follow along the lines of the proof for Theorem 4. In particular, the expressions for $\boldsymbol{\mu}_k(\xi_2)$ as given in conclusion (iii), and $\boldsymbol{\mu}_r(\xi_2)$ in conclusion (iv), follow from Theorem 1 (iv) and the relation $\boldsymbol{\delta} = \mathbf{Q}_2\xi_2$. Conclusion (iv) requires lifting those results to include $(\hat{\boldsymbol{\beta}}_I - \hat{\boldsymbol{\beta}})$ on $\mathbb{R}^k$ when its distribution is singular of rank $r < k$. However, since $\mathscr{L}(\hat{\boldsymbol{\theta}}_{I1} - \hat{\boldsymbol{\theta}}_1) = N_r(\boldsymbol{\mu}_r(\boldsymbol{\delta}), \Omega_r)$ with $\Omega_r = (\mathbf{I}_r + \mathbf{D}_\gamma^2)^{-2}\mathbf{D}_\gamma^2(\sigma_1^2\mathbf{D}_\gamma^2 + \sigma_2^2\mathbf{I}_r)$ as in (4.2), we observe as before that $(\hat{\boldsymbol{\beta}}_I - \hat{\boldsymbol{\beta}}) = \mathbf{G}_1(\hat{\boldsymbol{\theta}}_{I1} - \hat{\boldsymbol{\theta}}_1)$ with $\mathbf{G} = [\mathbf{G}_1, \mathbf{G}_2]$ as in Theorem 1. It follows that $\mathscr{L}(\hat{\boldsymbol{\beta}}_I - \hat{\boldsymbol{\beta}}) = N_k(\boldsymbol{\mu}_r(\xi_2), \mathbf{G}_1\Omega_r\mathbf{G}_1')$ with $\boldsymbol{\mu}_r(\xi_2) = \mathbf{G}_1(\mathbf{I}_r + \mathbf{D}_\gamma^2)^{-1}\mathbf{D}_\gamma^2\mathbf{Q}_2\xi_2$. Theorem 9.2.3 of Rao and Mitra (1971) now assures that $\mathscr{L}((\hat{\boldsymbol{\beta}}_I - \hat{\boldsymbol{\beta}})'\mathbf{V}^-(\hat{\boldsymbol{\beta}}_I - \hat{\boldsymbol{\beta}})/\sigma^2) = \chi^2(r, \lambda_r(\xi_2))$, where $\mathbf{V}^-$ is any symmetric reflexive g-inverse of $\mathbf{V} = \mathbf{G}_1\Omega_r\mathbf{G}_1'$, where $\lambda_r(\xi_2) = \boldsymbol{\mu}_r(\xi_2)'\mathbf{V}^-\boldsymbol{\mu}_r(\xi_2)$. In particular, we may proceed constructively using $\mathbf{V}_0^- = \mathbf{G}_1\Omega_r^{-1}\mathbf{G}_1'$, to complete our proof. □

5.2. *Exact inferences for* $\boldsymbol{\delta}$. Here we assume that $E(\boldsymbol{\varepsilon}) = \mathbf{0}$ and $V(\boldsymbol{\varepsilon}) = \boldsymbol{\Xi}(\sigma_1^2, \sigma_2^2) = \text{Diag}(\sigma_1^2\mathbf{I}_n, \sigma_2^2\mathbf{I}_r)$ in (3.1) as before. It is remarkable that $(\hat{\boldsymbol{\theta}}_I - \hat{\boldsymbol{\theta}})$, and thus $(\hat{\boldsymbol{\beta}}_I - \hat{\boldsymbol{\beta}})$, contain quantifiable information about a mean shift. We focus on $\boldsymbol{\delta}$ as in (3.2), since ξ_2 in (3.1) is related one-to-one through $\boldsymbol{\delta} = \mathbf{Q}_2\xi_2$. We thus consider $(\hat{\boldsymbol{\theta}}_{I1} - \hat{\boldsymbol{\theta}}_1)$ for $r \leq k$, with the understanding that this becomes $(\hat{\boldsymbol{\theta}}_I - \hat{\boldsymbol{\theta}})$ at $r = k$.

Recall that $E(\hat{\boldsymbol{\theta}}_{I1} - \hat{\boldsymbol{\theta}}_1) = \boldsymbol{\mu}_r(\boldsymbol{\delta}) = (\mathbf{I}_r + \mathbf{D}_\gamma^2)^{-1}\mathbf{D}_\gamma\boldsymbol{\delta}$ under $E(\boldsymbol{\eta}) = \mathbf{0}$ as in Theorem 1 (iv), with $V(\hat{\boldsymbol{\theta}}_{I1} - \hat{\boldsymbol{\theta}}_1)$ as in (4.2) under a block-diagonal dispersion structure. Moreover, under Gaussian errors, $\mathscr{L}(\hat{\boldsymbol{\theta}}_{I1} - \hat{\boldsymbol{\theta}}_1)$ is Gaussian as in Theorem 2. If we now let $\hat{\boldsymbol{\delta}} = (\mathbf{I}_r + \mathbf{D}_\gamma^2)\mathbf{D}_\gamma^{-1}(\hat{\boldsymbol{\theta}}_{I1} - \hat{\boldsymbol{\theta}}_1)$ and specialize (4.2) for the case $V(\boldsymbol{\varepsilon}) = \boldsymbol{\Xi}(\sigma_1^2, \sigma_2^2)$, then the following conclusions are immediate. (i) The statistic $\hat{\boldsymbol{\delta}}$ is unbiased for estimating $\boldsymbol{\delta}$; (ii) its dispersion matrix is $V(\hat{\boldsymbol{\delta}}) = (\sigma_1^2\mathbf{D}_\gamma^2 + \sigma_2^2\mathbf{I}_r)$; and (iii) under Gaussian errors its distribution is $\mathscr{L}(\hat{\boldsymbol{\delta}}) = N_r(\boldsymbol{\delta}, (\sigma_1^2\mathbf{D}_\gamma^2 + \sigma_2^2\mathbf{I}_r))$. If we now fix $\boldsymbol{\delta}_0 = [\delta_{01}, \ldots, \delta_{0r}]' \in \mathbb{R}^r$ and then consider $D_I((\hat{\boldsymbol{\delta}} - \boldsymbol{\delta}_0), (\mathbf{I}_r + \mathbf{D}_\gamma^2)^{-1}, rS_I^2)$, then its properties are found directly as before without further difficulty. These may be summarized as follows.

THEOREM 6. Suppose that $\mathscr{L}(\mathbf{Y}) = N_N(\mathbf{X}_0\boldsymbol{\beta} - \boldsymbol{\xi}, \boldsymbol{\Xi}(\sigma_1^2, \sigma_2^2))$, and consider the statistic $D_I((\hat{\boldsymbol{\delta}} - \boldsymbol{\delta}_0), (\mathbf{I}_r + \mathbf{D}_\gamma^2)^{-1}, rS_I^2) = (\hat{\boldsymbol{\delta}} - \boldsymbol{\delta}_0)'(\mathbf{I}_r + \mathbf{D}_\gamma^2)^{-1}(\hat{\boldsymbol{\delta}} - \boldsymbol{\delta}_0)/rS_I^2$.

(i) Then the distribution of $D_I((\hat{\boldsymbol{\delta}} - \boldsymbol{\delta}_0), (\mathbf{I}_r + \mathbf{D}_\gamma^2)^{-1}, rS_I^2)$ has the cdf $F_r(w; \alpha_1, \ldots, \alpha_r, \omega_1, \ldots, \omega_r, n - k)$ with $\{\omega_i = (\delta_i - \delta_{0i})/[\sigma_1(\gamma_i^2 + \sigma_2^2/\sigma_1^2)^{1/2}]; 1 \leq i \leq r\}$ and $\{\alpha_i = (\gamma_i^2 + \sigma_2^2/\sigma_1^2)/(\gamma_i^2 + 1); 1 \leq i \leq r\}$, for each $r \leq k$.

(ii) If $\sigma_2^2/\sigma_1^2 = 1$, then the cdf of $D_I((\hat{\boldsymbol{\delta}} - \boldsymbol{\delta}_0), (\mathbf{I}_r + \mathbf{D}_\gamma^2)^{-1}, rS_I^2)$ is given by $F(w; r, n - k, \lambda(\boldsymbol{\delta}_0))$, with noncentrality parameter $\lambda(\boldsymbol{\delta}_0) = (\boldsymbol{\delta} - \boldsymbol{\delta}_0)'(\mathbf{I}_r + \mathbf{D}_\gamma^2)^{-1}(\boldsymbol{\delta} - \boldsymbol{\delta}_0)/\sigma^2 = \sum_{i=1}^r (\delta_i - \delta_{0i})^2/[\sigma^2(\gamma_i^2 + 1)]$ for each $r \leq k$.

PROOF. The proof proceeds step-by-step as in the proofs for Theorems 4 and 5, as no new difficulties are encountered. □

Theorem 6 supports exact normal-theory inferences for $\boldsymbol{\delta}$. To test $H : \boldsymbol{\delta} = \boldsymbol{\delta}_0$ against $A : \boldsymbol{\delta} \neq \boldsymbol{\delta}_0$, an exact test at level α rejects H whenever $D_I((\hat{\boldsymbol{\delta}} - \boldsymbol{\delta}_0), (\mathbf{I}_r + \mathbf{D}_\gamma^2)^{-1}, rS_I^2) > F_\alpha(r, n - k)$, the $100(1 - \alpha)^{\text{th}}$ percentile, under the assumption that $\sigma_2^2/\sigma_1^2 = 1$. Otherwise Theorem 6 (i) provides the distribution theory needed for a thorough study of disturbances in level and power of this test under a given ratio $\sigma_2^2/\sigma_1^2 \neq 1$. Indeed, it is clear from Lemma 1 that the actual level will be greater than the nominal level whenever $\sigma_2^2/\sigma_1^2 > 1$, and less for $\sigma_2^2/\sigma_1^2 < 1$. Regarding power of the test, it is seen from $\lambda_r(\boldsymbol{\delta}_0) = \sum_{i=1}^r (\delta_i - \delta_{0i})^2/[\sigma^2(\gamma_i^2 + 1)]$ that leverages tend to mask a given shift $(\boldsymbol{\delta} - \boldsymbol{\delta}_0)$ from the null hypothesis. In particular, a high leverage value γ_i^2 may suppress even a moderate component $|\delta_i - \delta_{0i}|$ of $(\boldsymbol{\delta} - \boldsymbol{\delta}_0)$.

In addition to hypothesis testing, Theorem 6 supports an exact normal-theory confidence region for $\boldsymbol{\delta}$, with confidence coefficient $1 - \alpha$, as given by

$$R(\boldsymbol{\delta}) = \{\boldsymbol{\delta} \in \mathbb{R}^r : (\hat{\boldsymbol{\delta}} - \boldsymbol{\delta})'(\mathbf{I}_r + \mathbf{D}_\gamma^2)^{-1}(\hat{\boldsymbol{\delta}} - \boldsymbol{\delta}) \leq rS_I^2 F_\alpha(r, n - k)\} \ . \quad (6.1)$$

It is of some interest that the principal axes of this ellipsoid are parallel to coordinate axes in the parameter space $\mathbb{R}^r$. This is a consequence of the canonical form (3.2). These hypothesis tests and confidence regions all carry over directly to include ξ_2 owing to the one-to-one relation $\boldsymbol{\delta} = \mathbf{Q}_2 \xi_2$.

6. Summary

In this paper, we have adapted the theory of singular decompositions to transform a linear model into canonical form. The design matrix in canonical form is a partitioned matrix whose nonzero block entries are either identity or diagonal matrices. The quadratic form determining the numerator of Cook's D_I statistic is preserved under this canonical transformation, and we have studied transformed versions of Cook's D_I in detail.

With standard Gaussian assumptions, and using an independent estimator for the variance from the reduced data, we have determined the structure of $D_I(\hat{\boldsymbol{\beta}}, \mathbf{M}, c\hat{\sigma}^2)$ statistics under each of two choices for $\mathbf{M}$ in current use, namely, $\mathbf{X}_0'\mathbf{X}_0$ and $\mathbf{X}'\mathbf{X}$, corresponding to the full and reduced data, respectively. The numerators of these statistics are shown to be represented in distribution as finite sums of weighted noncentral chi-squared variables, and the statistic D_I itself as a weighted sum of F-distributions when more than one row is deleted, and it is a scaled F-distribution when only one row has been deleted. In the presence of a mean shift or a shift in variance, the noncentralities of the factors depend on both the shifted means and variance, whereas the scaling parameters depend only on the variances. Both types of parameters depend on subset leverages, which tend to mask the effects of outliers of both types. These results are significant in both theory and practice, as they give the first derivation of the exact distributions of these statistics.

We have introduced modified versions of D_I on utilizing the inverse dispersion matrix for $(\hat{\boldsymbol{\beta}}_I - \hat{\boldsymbol{\beta}})$ when its distribution is nonsingular, and a generalized inverse otherwise. Under the usual Gaussian assumptions, these modified statistics therefore have F-distributions as reference when there are no shifts in either means or variance. Under a shift in means, a further modification yields statistics having noncentral F-distributions, which in turn support exact normal-theory tests at level α for hypotheses regarding the shifts, as well as the construction of exact confidence sets with coefficient $1 - \alpha$.

Our findings have profound implications regarding some of the bewildering array of diagnostic tools now available to users. As a case in point, the *DFBETA*'s are often presented in the literature as genuinely distinct gauges of influence in single-case diagnostics. How many *DFBETA*'s are there? We claim for each fixed row of X_0 that there is essentially only one. To support this claim, we apply Theorem 1 to infer for $r = 1$ that the joint distribution of $[DF\beta_{1,i}, \ldots, DF\beta_{k,i}]$ is singular on $\mathbb{R}^k$ of unit rank. In fact, the proof for Theorem 1 shows that these statistics all are scalings of the single random variable $(\hat{\theta}_{i1} - \hat{\theta}_1)/S_i$. The *DFBETA*'s accordingly are in fixed ratios determined beforehand by the structure of $\mathbf{X}_0$, independently of the chance value taken by $\mathbf{Y}$ in a random experiment. Each collection $[DF\beta_{1,i}, \ldots, DF\beta_{k,i}]$ thereby is an equivalence class to be represented by an arbitrary member as may be designated by the user.

Numerical studies have been undertaken by the authors to examine the effects of leverage and outliers in the use of D_I in regression diagnostics. We have developed the computer software to calculate the cdfs for these nonstandard distributions in linear models under the usual Gaussian assumptions. We are thus enabled to answer the commonly asked questions regarding benchmarks for various versions of Cook's D_I in gauging the stability of a linear model under small perturbations. These numerical studies will be reported elsewhere.

References

Belsley, D. A., E. Kuh and R. E. Welsch (1980). *Regression Diagnostics: Identifying Influential Data and Sources of Collinearity*. New York, John Wiley & Sons.
Chatterjee, S. and A. S. Hadi (1988). *Sensitivity Analysis in Linear Regression*. New York, John Wiley & Sons.
Cook, R. D. (1977). Detection of influential observations in linear regression. *Technometrics* **19**, 15–18.
Cook, R. D. and S. Weisberg (1982). *Residuals and Influence in Regression*. London, Chapman and Hall.
Fox, J. (1991). *Regression Diagnostics*. Newbury Park, California, Sage Publications, Inc.
Gray, J. B. (1993). Approximating the internal norm influence measure in linear regression. *Commun. Statist. – Simul. Comput.* **22**, 117–135.

Gurland, J. (1955). Distributions of definite and indefinite quadratic forms. *Ann. Math. Statist.* **26**, 122–127.

Johnson, N. L. and S. Kotz (1970). *Continuous Univariate Distributions-2*. Boston, Houghton Mifflin Co.

Kotz, S., N. L. Johnson and D. W. Boyd (1967a). Series representations of distributions of quadratic forms in normal variables. I. Central case. *Ann. Math. Statist.* **38**, 823–837.

Kotz, S., N. L. Johnson and D. W. Boyd (1967b). Series representations of distributions of quadratic forms in normal variables. II. Non-central case. *Ann. Math. Statist.* **38**, 838–848.

Mahalanobis, P. C. (1936). On the generalized distance in statistics. *Proceedings of the National Institute of Science (India)* **12**, 49–55.

Mathai, A. M. and S. B. Provost (1992). *Quadratic Forms in Random Variables: Theory and Applications*. New York, Marcel Dekker, Inc.

Ramirez, D. E. and D. R. Jensen (1991). Misspecified T^2 tests. II. Series expansions. *Commun. Statist. – Simul. Comput.* **20**, 97–108.

Rao, C. R. and S. K. Mitra (1971). *Generalized Inverse of Matrices and its Applications*. New York, John Wiley & Sons.

Rousseeuw, P. J. and A. M. Leroy (1987). *Robust Regression and Outlier Detection*. New York, John Wiley & Sons.

Generalized Recurrence Relations for Moments of Order Statistics from Non-Identical Pareto and Truncated Pareto Random Variables with Applications to Robustness

Aaron Childs and N. Balakrishnan

1. Introduction

Let $X_1, X_2, \ldots, X_n$ be independent random variables having cumulative distribution functions $F_1(x), F_2(x), \ldots, F_n(x)$ and probability density functions $f_1(x)$, $f_2(x), \ldots, f_n(x)$, respectively. Then the X_i's are said to be independent and non-identically distributed (I.NI.D.) random variables. Let $X_{1:n} \leq X_{2:n} \leq \cdots \leq X_{n:n}$ denote the order statistics obtained by arranging the n X_i's in increasing order of magnitude. Then the density function of $X_{r:n}$ ($1 \leq r \leq n$) can be written as (David, 1981, p. 22)

$$f_{r:n}(x) = \frac{1}{(r-1)!(n-r)!} \sum_p \prod_{a=1}^{r-1} f_{i_a}(x) f_{i_r}(x) \prod_{b=r+1}^{n} \{1 - F_{i_b}(x)\}, \tag{1.1}$$

where $\sum_p$ denotes the summation over all $n!$ permutations $(i_1, i_2, \ldots, i_n)$ of $(1, 2, \ldots, n)$. Similarly, the joint density function of $X_{r:n}$ and $X_{s:n}$ ($1 \leq r < s \leq n$) can be written as

$$f_{r,s:n}(x,y) = \frac{1}{(r-1)!(s-r-1)!(n-s)!} \sum_p \prod_{a=1}^{r-1} F_{i_a}(x) f_{i_r}(x)$$
$$\times \prod_{b=r+1}^{s-1} \{F_{i_b}(y) - F_{i_b}(x)\} f_{i_s}(y) \prod_{c=s+1}^{n} \{1 - F_{i_c}(y)\}, \quad x < y.$$
(1.2)

Alternatively, the densities in (1.1) and (1.2) can be written in terms of permanents of matrices; see Vaughan and Venables (1972).

In recent years, many of the recurrence relations for order statistics from I.I.D. samples have been generalized to the I.NI.D. case. This work was initiated by Balakrishnan (1994a,b) for the exponential and right-truncated exponential distributions. Childs and Balakrishnan (1995a,b) generalized the I.I.D. recurrence relations for doubly-truncated exponential and logistic models, while Balakrishnan and Balasubramanian (1995) generalized results for the power function distribution. All of these results were obtained by exploiting a basic differential equation satisfied by the distributions under consideration.

In this paper, we generalize the I.I.D. results for the Pareto and doubly-truncated Pareto models established by Balakrishnan and Joshi (1982). We first consider the case when the variables X_i's are independent having Pareto distributions with density functions

$$f_i(x) = v_i x^{-(v_i+1)}, \quad x \geq 1, \quad v_i > 0 \tag{1.3}$$

and cumulative distribution functions

$$F_i(x) = 1 - x^{-v_i}, \quad x \geq 1, \quad v_i > 0 \tag{1.4}$$

for $i = 1, 2, \ldots, n$. For a detailed discussion of various aspects of the Pareto distribution, one may refer to Arnold (1983) or Johnson, Kotz, and Balakrishnan (1994). The basic differential equations satisfied in this situation can be seen from (1.3) and (1.4) to be

$$F_i(x) = 1 - \frac{x}{v_i} f_i(x), \quad i = 1, 2, \ldots, n \ . \tag{1.5}$$

In order to guarantee the existence of all the means, variances, and covariances of order statistics for all sample sizes n, it is sufficient to assume that $v_i > 2$ for $i = 1, 2, \ldots, n$. However, without this assumption certain moments may still exist. Let us denote the single moments $E(X_{r:n}^k)$ by $\mu_{r:n}^{(k)}$, $1 \leq r \leq n$ and $k = 1, 2, \ldots$, and the product moments $E(X_{r:n} X_{s:n})$ by $\mu_{r,s:n}$ for $1 \leq r < s \leq n$. Let us also use $\mu_{r:n-1}^{[i](k)}$ and $\mu_{r,s:n-1}^{[i]}$ to denote the single and the product moments of order statistics arising from $n-1$ variables obtained by deleting X_i from the original n variables $X_1, X_2, \ldots, X_n$.

In Sections 2 and 3 we make use of the differential equations in (1.5) to establish several recurrence relations satisfied by the single and the product moments of order statistics. These relations will enable one to compute all the single and the product moments of all order statistics in a simple recursive manner. In Section 4 we will use the results of Sections 2 and 3 to deduce a set of recurrence relations for the multiple-outlier model. We will then generalize to the case when the variables X_i's are independent having doubly-truncated Pareto distributions in Section 5. In Section 6, we will use the multiple-outlier results from Section 4 to examine the robustness of the MLE and BLUE of the scale parameter σ of a one-parameter Pareto distribution. Finally, in Section 7, we will examine the robustness of the censored BLUE for the location and scale parameters of a two-parameter Pareto distribution, in the presence of multiple shape outliers.

2. Relations for single moments

In this section, we use the differential equations in (1.5) to establish the following recurrence relations for single moments.

RELATION 2.1. For $n \geq 1$ and $k = 0, 1, 2, \ldots$,

$$\mu_{1:n}^{(k)} = \frac{\sum_{i=1}^{n} v_i}{\sum_{i=1}^{n} v_i - k} .$$

RELATION 2.2. For $2 \leq r \leq n$ and $k = 0, 1, 2, \ldots$,

$$\mu_{r:n}^{(k)} = \frac{\sum_{i=1}^{n} v_i \mu_{r-1:n-1}^{[i](k)}}{\sum_{i=1}^{n} v_i - k} .$$

PROOF OF RELATION 2.1. For $n \geq 1$ and $k = 0, 1, 2, \ldots$, we use (1.1) and then (1.5) to write

$$\mu_{1:n}^{(k)} = \frac{1}{(n-1)!} \sum_p \int_1^\infty x^k f_{i_1}(x) \prod_{b=2}^{n} \{1 - F_{i_b}(x)\} \, dx$$

$$= \frac{1}{(n-1)!} \sum_p v_{i_1} \int_1^\infty x^{k-1} \prod_{b=1}^{n} \{1 - F_{i_b}(x)\} \, dx .$$

Integrating now by parts treating x^{k-1} for integration, we obtain

$$k \mu_{1:n}^{(k)} = \frac{1}{(n-1)!} \sum_p \left[-v_{i_1} + v_{i_1} \int_1^\infty x^k \sum_{j=1}^{n} f_{i_j}(x) \prod_{b=1, b \neq j}^{n} \{1 - F_{i_b}(x)\} \, dx \right]$$

$$= -\sum_{i=1}^{n} v_i + \left(\sum_{i=1}^{n} v_i \right) \mu_{1:n}^{(k)}$$

which, when rewritten, yields Relation 2.1.

PROOF OF RELATION 2.2. For $2 \leq r \leq n$ and $k = 0, 1, 2, \ldots$, we use (1.1) and then (1.5) to write

$$\mu_{r:n}^{(k)} = \frac{1}{(r-1)!(n-r)!} \sum_p \int_1^\infty x^k \prod_{a=1}^{r-1} F_{i_a}(x) f_{i_r}(x) \prod_{b=r+1}^{n} \{1 - F_{i_b}(x)\} \, dx$$

$$= \frac{1}{(r-1)!(n-r)!} \sum_p v_{i_r} \int_1^\infty x^{k-1} \prod_{a=1}^{r-1} F_{i_a}(x) \prod_{b=r}^{n} \{1 - F_{i_b}(x)\} \, dx .$$

Integrating now by parts treating x^{k-1} for integration gives

$$k\mu_{r:n}^{(k)} = \frac{1}{(r-1)!(n-r)!} \sum_p \left[-v_{i_r} \int_1^\infty x^k \sum_{j=1}^{r-1} f_{i_j}(x) \prod_{\substack{a=1 \\ a \neq j}}^{r-1} F_{i_a}(x) \right.$$

$$\times \prod_{b=r}^n \{1 - F_{i_b}(x)\} \, dx + v_{i_r} \int_1^\infty x^k \prod_{a=1}^{r-1} F_{i_a}(x) \sum_{j=r}^n f_{i_j}(x)$$

$$\left. \times \prod_{\substack{b=r \\ b \neq j}}^n \{1 - F_{i_b}(x)\} \, dx \right].$$

We then split the first term in the above sum into two through the term $\{1 - F_{i_r}(x)\}$ to obtain

$$k\mu_{r:n}^{(k)} = \frac{1}{(r-1)!(n-r)!} \sum_p \left[-v_{i_r} \int_1^\infty x^k \sum_{j=1}^{r-1} f_{i_j}(x) \prod_{\substack{a=1 \\ a \neq j}}^{r-1} F_{i_a}(x) \right.$$

$$\times \prod_{b=r+1}^n \{1 - F_{i_b}(x)\} \, dx + v_{i_r} \int_1^\infty x^k \sum_{j=1}^{r-1} f_{i_j}(x) \prod_{\substack{a=1 \\ a \neq j}}^r F_{i_a}(x)$$

$$\times \prod_{b=r+1}^n \{1 - F_{i_b}(x)\} dx + v_{i_r} \int_1^\infty x^k \prod_{a=1}^{r-1} F_{i_a}(x) \sum_{j=r}^n f_{i_j}(x)$$

$$\left. \times \prod_{\substack{b=r \\ b \neq j}}^n \{1 - F_{i_b}(x)\} \, dx \right]$$

$$= -\sum_{i=1}^n v_i \mu_{r-1:n-1}^{[i](k)} + \left(\sum_{i=1}^n v_i\right) \mu_{r:n}^{(k)}.$$

Relation 2.2 is derived simply by rewriting the above equation.

The recurrence relations presented in Relations 2.1 and 2.2 will enable one to compute all the single moments of all order statistics that exist in a simple recursive manner for any specified values of $v_i > 2$ $(i = 1, 2, \ldots, n)$.

Alternatively Relation 2.1 could be used, along with a general relation established by Balakrishnan (1988) which expresses $\mu_{r:n}^{(k)}$ in terms of the kth moment of the smallest order statistic in sample sizes up to n, to compute all single moments of all order statistics in a simple recursive manner. Another possibility is to use Relation 2.2 (with $r = n$) along with Balakrishnan's (1988) general relation which expresses $\mu_{r:n}^{(k)}$ in terms of the kth moment of the largest order statistic in samples of size up to n, to compute all the single moments of all order statistics.

We conclude this section by setting $v_1 = v_2 = \cdots = v_n = v$ in Relations 2.1 and 2.2 to obtain the following recurrence relations for I.I.D. Pareto random variables originally derived by Balakrishnan and Joshi (1982):

$$\mu_{1:n}^{(k)} = \frac{nv}{nv - k}$$

and

$$\mu_{r:n}^{(k)} = \frac{nv}{nv - k} \mu_{r-1:n-1}^{(k)} .$$

3. Relations for product moments

In this section we again use the differential equations in (1.5), but this time to establish the following four recurrence relations for the product moments of order statistics.

RELATION 3.1. For $n \geq 2$,

$$\mu_{1,2:n} = \frac{\sum_{i=1}^{n} v_i \mu_{1:n-1}^{[i]}}{\sum_{i=1}^{n} v_i - 2} .$$

RELATION 3.2. For $2 \leq r \leq n - 1$,

$$\mu_{r,r+1:n} = \frac{\sum_{i=1}^{n} v_i \mu_{r-1,r:n-1}^{[i]}}{\sum_{i=1}^{n} v_i - 2} .$$

RELATION 3.3. For $3 \leq s \leq n$,

$$\mu_{1,s:n} = \frac{\sum_{i=1}^{n} v_i \mu_{s-1:n-1}^{[i]}}{\sum_{i=1}^{n} v_i - 2} .$$

RELATION 3.4. For $2 \leq r < s \leq n$ and $s - r \geq 2$,

$$\mu_{r,s:n} = \frac{\sum_{i=1}^{n} v_i \mu_{r-1,s-1:n-1}^{[i]}}{\sum_{i=1}^{n} v_i - 2} .$$

PROOF OF RELATION 3.1. From Equation (1.2), let us consider for $n \geq 2$

$$\mu_{1,2:n} = \frac{1}{(n-2)!} \sum_{p} \int_{1}^{\infty} \int_{x}^{\infty} xy f_{i_1}(x) f_{i_2}(y) \prod_{c=3}^{n} \{1 - F_{i_c}(y)\} \, dy \, dx$$

$$= \frac{1}{(n-2)!} \sum_{p} \int_{1}^{\infty} x f_{i_1}(x) I_1(x) \, dx , \qquad (3.1)$$

where

$$I_1(x) = \int_x^\infty y f_{i_2}(y) \prod_{c=3}^n \{1 - F_{i_c}(y)\} \, dy$$

$$= v_{i_2} \int_x^\infty \prod_{c=2}^n \{1 - F_{i_c}(y)\} \, dy$$

upon using (1.5). Integrating now by parts yields

$$I_1(x) = v_{i_2} \left[-x \prod_{c=2}^n \{1 - F_{i_c}(x)\} + \int_x^\infty y \sum_{j=2}^n f_{i_j}(y) \prod_{\substack{c=2 \\ c \neq j}}^n \{1 - F_{i_c}(y)\} \, dy \right]$$

which, when substituted in (3.1), gives

$$\mu_{1,2:n} = \frac{1}{(n-2)!} \sum_P v_{i_2} \left[-\int_1^\infty x^2 f_{i_1}(x) \prod_{c=2}^n \{1 - F_{i_c}(x)\} \, dx \right.$$

$$\left. + \int_1^\infty \int_x^\infty xy f_{i_1}(x) \sum_{j=2}^n f_{i_j}(y) \prod_{\substack{c=2 \\ c \neq j}}^n \{1 - F_{i_c}(y)\} \, dy \, dx \right] . \quad (3.2)$$

Alternatively, we may write

$$\mu_{1,2:n} = \frac{1}{(n-2)!} \sum_P \int_1^\infty \int_1^y xy f_{i_1}(x) f_{i_2}(y) \prod_{c=3}^n \{1 - F_{i_c}(y)\} \, dx \, dy$$

$$= \frac{1}{(n-2)!} \sum_P \int_1^\infty y f_{i_2}(y) \prod_{c=3}^n \{1 - F_{i_c}(y)\} I_2(y) \, dy , \quad (3.3)$$

where

$$I_2(y) = \int_1^y x f_{i_1}(x) \, dx$$

$$= v_{i_1} \int_1^y \{1 - F_{i_1}(x)\} \, dx$$

upon using (1.5). Integrating now by parts yields

$$I_2(y) = v_{i_1} \left[y\{1 - F_{i_1}(y)\} - 1 + \int_1^y x f_{i_1}(x) \, dx \right]$$

which, when substituted in (3.3), gives

$$\mu_{1,2:n} = \frac{1}{(n-2)!} \sum_P v_{i_1} \left[\int_1^\infty y^2 f_{i_2}(y)\{1 - F_{i_1}(y)\} \prod_{c=3}^n \{1 - F_{i_c}(y)\} \, dy \right.$$

$$- \int_1^\infty y f_{i_2}(y) \prod_{c=3}^n \{1 - F_{i_c}(y)\} \, dy \quad (3.4)$$

$$\left. + \int_1^\infty \int_1^y xy f_{i_1}(x) f_{i_2}(y) \prod_{c=3}^n \{1 - F_{i_c}(y)\} \, dx \, dy \right] .$$

We now add the expressions for $\mu_{1,2:n}$ in (3.2) and (3.4) and simplify the resulting equation to get

$$2\mu_{1,2:n} = \left(\sum_{i=1}^{n} v_i\right)\mu_{1,2:n} - \sum_{i=1}^{n} v_i \mu_{1:n-1}^{[i]}.$$

Relation 3.1 is derived simply by rewriting the above equation.

PROOF OF RELATION 3.2. From Equation (1.2), let us consider for $2 \leq r \leq n-1$

$$\mu_{r,r+1:n} = \frac{1}{(r-1)!(n-r-1)!}$$

$$\times \sum_{p} \int_{1}^{\infty}\int_{x}^{\infty} xy \prod_{a=1}^{r-1} F_{i_a}(x) f_{i_r}(x) f_{i_{r+1}}(y) \prod_{c=r+2}^{n} \{1 - F_{i_c}(y)\} \, dy \, dx$$

$$= \frac{1}{(r-1)!(n-r-1)!} \sum_{p} \int_{1}^{\infty} x \prod_{a=1}^{r-1} F_{i_a}(x) f_{i_r}(x) I_1(x) \, dx,$$

(3.5)

where

$$I_1(x) = \int_{x}^{\infty} y f_{i_{r+1}}(y) \prod_{c=r+2}^{n} \{1 - F_{i_c}(y)\} \, dy$$

$$= v_{i_{r+1}} \int_{x}^{\infty} \prod_{c=r+1}^{n} \{1 - F_{i_c}(y)\} \, dy$$

upon using (1.5). Integrating now by parts yields

$$I_1(x) = v_{i_{r+1}}\left[-x \prod_{c=r+1}^{n} \{1 - F_{i_c}(x)\} + \int_{x}^{\infty} y \sum_{j=r+1}^{n} f_{i_j}(y) \right.$$

$$\left. \times \prod_{\substack{c=r+1 \\ c \neq j}}^{n} \{1 - F_{i_c}(y)\} \, dy\right]$$

which, when substituted in (3.5), gives

$$\mu_{r,r+1:n} = \frac{1}{(r-1)!(n-r-1)!}$$

$$\times \sum_{p} v_{i_{r+1}}\left[-\int_{1}^{\infty} x^2 \prod_{a=1}^{r-1} F_{i_a}(x) f_{i_r}(x) \prod_{c=r+1}^{n} \{1 - F_{i_c}(x)\} \, dx \right.$$

$$+ \int_{1}^{\infty}\int_{x}^{\infty} xy \prod_{a=1}^{r-1} F_{i_a}(x) f_{i_r}(x) \sum_{j=r+1}^{n} f_{i_j}(y)$$

$$\left. \times \prod_{\substack{c=r+1 \\ c \neq j}}^{n} \{1 - F_{i_c}(y)\} \, dy \, dx\right].$$

(3.6)

Alternatively, we may write

$$\mu_{r,r+1:n} = \frac{1}{(r-1)!(n-r-1)!}$$
$$\times \sum_P \int_1^\infty \int_1^y xy \prod_{a=1}^{r-1} F_{i_a}(x) f_{i_r}(x) f_{i_{r+1}}(y) \prod_{c=r+2}^n \{1 - F_{i_c}(y)\} \, dx \, dy$$
$$= \frac{1}{(r-1)!(n-r-1)!} \sum_P \int_1^\infty y f_{i_{r+1}}(y) \prod_{c=r+2}^n \{1 - F_{i_c}(y)\} I_2(y) \, dy \,,$$

(3.7)

where

$$I_2(y) = \int_1^y x \prod_{a=1}^{r-1} F_{i_a}(x) f_{i_r}(x) \, dx$$
$$= v_{i_r} \int_1^y \prod_{a=1}^{r-1} F_{i_a}(x) \{1 - F_{i_r}(x)\} \, dx$$

upon using (1.5). Integrating now by parts yields

$$I_2(y) = v_{i_r} \Bigg[y \prod_{a=1}^{r-1} F_{i_a}(y) \{1 - F_{i_r}(y)\} - \int_1^y x \sum_{j=1}^{r-1} f_{i_j}(x)$$
$$\times \prod_{\substack{a=1 \\ a \neq j}}^{r-1} F_{i_a}(x) \{1 - F_{i_r}(x)\} \, dx + \int_1^y x f_{i_r}(x) \prod_{a=1}^{r-1} F_{i_a}(x) \, dx \Bigg]$$

which, when substituted in (3.7), gives

$$\mu_{r,r+1:n} = \frac{1}{(r-1)!(n-r-1)!} \sum_P v_{i_r}$$
$$\times \Bigg[\int_1^\infty y^2 \prod_{a=1}^{r-1} F_{i_a}(y) f_{i_{r+1}}(y) \{1 - F_{i_r}(y)\} \prod_{c=r+2}^n \{1 - F_{i_c}(y)\} \, dy$$
$$- \int_1^\infty \int_1^y xy \sum_{j=1}^{r-1} f_{i_j}(x) \prod_{\substack{a=1 \\ a \neq j}}^{r-1} F_{i_a}(x) \{1 - F_{i_r}(x)\} f_{i_{r+1}}(y)$$
$$\times \prod_{c=r+2}^n \{1 - F_{i_c}(y)\} \, dx \, dy$$
$$+ \int_1^\infty \int_1^y xy \prod_{a=1}^{r-1} F_{i_a}(x) f_{i_r}(x) f_{i_{r+1}}(y) \prod_{c=r+2}^n \{1 - F_{i_c}(y)\} \, dx \, dy \Bigg].$$

We then split the second term in the above sum through the term $\{1 - F_{i_r}(x)\}$ to get

$$\mu_{r,r+1:n} = \frac{1}{(r-1)!(n-r-1)!} \sum_p v_{i_r}$$

$$\times \left[\int_1^\infty y^2 \prod_{a=1}^{r-1} F_{i_a}(y)\{1 - F_{i_r}(y)\} f_{i_{r+1}}(y) \prod_{c=r+2}^n \{1 - F_{i_c}(y)\} \, dy \right.$$

$$- \int_1^\infty \int_1^y xy \sum_{j=1}^{r-1} f_{i_j}(x) \prod_{\substack{a=1 \\ a \neq j}}^{r-1} F_{i_a}(x) f_{i_{r+1}}(y) \prod_{c=r+2}^n \{1 - F_{i_c}(y)\} \, dx \, dy$$

$$+ \int_1^\infty \int_1^y xy \sum_{j=1}^{r-1} f_{i_j}(x) \prod_{\substack{a=1 \\ a \neq j}}^{r} F_{i_a}(x) f_{i_{r+1}}(y) \prod_{c=r+2}^n \{1 - F_{i_c}(y)\} \, dx \, dy$$

$$\left. + \int_1^\infty \int_1^y xy \prod_{a=1}^{r-1} F_{i_a}(x) f_{i_r}(x) f_{i_{r+1}}(y) \prod_{c=r+2}^n \{1 - F_{i_c}(y)\} \, dx \, dy \right] .$$

(3.8)

We now add the expressions for $\mu_{r,r+1:n}$ in (3.6) and (3.8) and simplify the resulting equation to get

$$2\mu_{r,r+1:n} = \left(\sum_{i=1}^n v_i \right) \mu_{r,r+1:n} - \sum_{i=1}^n v_i \mu_{r-1,r:n-1}^{[i]} .$$

Relation 3.2 is derived simply by rewriting the above equation.

The proofs of Relations 3.3 and 3.4 are similar to the above, and have been relegated to Appendix A.

The recurrence relations presented in Relations 3.1–3.4 along with Relations 2.1 and 2.2 will enable one to compute all the product moments, and hence the covariances, of all order statistics in a simple recursive manner for any specified values of $v_i > 2$ ($i = 1, 2, \ldots, n$).

Alternatively, two of the general relations established by Balakrishnan, Bendre and Malik (1992) could also be used. One of these relations expresses $\mu_{r,s:n}$ in terms of product moments of the form $\mu_{i,i+1:m}$ for m up to n, and hence could be used in conjunction with Relations 3.1 and 3.2 to compute all the product moments of all order statistics in a simple recursive manner. The other relation that can be used expresses $\mu_{r,s:n}$ in terms of product moments of the form $\mu_{1,j:m}$ for m up to n, and so could be used in conjunction with Relations 3.1 and 3.3 to compute all of the product moments.

We conclude this section by setting $v_1 = v_2 = \cdots = v_n = v$ in Relations 3.1–3.4, to obtain the following results I.I.D. Pareto random variables:

$$\mu_{1,2:n} = \frac{nv}{nv - 2} \mu_{1:n-1}, \quad n \geq 2 ,$$

$$\mu_{r,r+1:n} = \frac{nv}{nv - 2} \mu_{r-1,r:n-1}, \quad 2 \leq r \leq n - 1 ,$$

$$\mu_{1,s:n} = \frac{nv}{nv - 2} \mu_{s-1:n-1}, \quad 3 \leq s \leq n ,$$

and

$$\mu_{r,s:n} = \frac{nv}{nv-2}\mu_{r-1,s-1:n-1}, \quad 2 \leq r < s \leq n, \ s-r \geq 2 \ .$$

These relations may be derived from the results of Balakrishnan and Joshi (1982).

4. Results for the multiple-outlier model (with a slippage of p observations)

In this section we consider the special case where $X_1, X_2, \ldots, X_{n-p}$ are independent Pareto random variables with parameter v, while $X_{n-p+1}, \ldots, X_n$ are independent Pareto random variables with parameter v^* (and independent of $X_1, X_2, \ldots, X_{n-p}$). This situation is known as the multiple-outlier model with a slippage of p observations; see David (1979) and Barnett and Lewis (1994, pp. 66–68). Here, we denote the single moments by $\mu_{r:n}^{(k)}[p]$ and the product moments by $\mu_{r,s:n}[p]$.

In this situation, the results established in Sections 2 and 3 reduce to the following recurrence relations:

(a) for $n \geq 1$ and $k = 0, 1, 2, \ldots,$

$$\mu_{1:n}^{(k)}[p] = \frac{(n-p)v + pv^*}{(n-p)v + pv^* - k} \ ; \tag{4.1}$$

(b) for $2 \leq r \leq n$ and $k = 0, 1, 2, \ldots,$

$$\mu_{r:n}^{(k)}[p] = \frac{(n-p)v\mu_{r-1:n-1}^{(k)}[p] + pv^*\mu_{r-1:n-1}^{(k)}[p-1]}{(n-p)v + pv^* - k} \ ; \tag{4.2}$$

(c) for $n \geq 2$,

$$\mu_{1,2:n}[p] = \frac{(n-p)v\mu_{1:n-1}[p] + pv^*\mu_{1:n-1}[p-1]}{(n-p)v + pv^* - 2} \ ; \tag{4.3}$$

(d) for $2 \leq r \leq n-1$,

$$\mu_{r,r+1:n}[p] = \frac{(n-p)v\mu_{r-1,r:n-1}[p] + pv^*\mu_{r-1,r:n-1}[p-1]}{(n-p)v + pv^* - 2} \ ; \tag{4.4}$$

(e) for $3 \leq s \leq n$,

$$\mu_{1,s:n}[p] = \frac{(n-p)v\mu_{s-1:n-1}[p] + pv^*\mu_{s-1:n-1}[p-1]}{(n-p)v + pv^* - 2} \ ; \tag{4.5}$$

and

(f) for $2 \leq r < s \leq n$ and $s - r \geq 2$,

$$\mu_{r,s:n}[p] = \frac{(n-p)v\mu_{r-1,s-1:n-1}[p] + pv^*\mu_{r-1,s-1:n-1}[p-1]}{(n-p)v + pv^* - 2} . \tag{4.6}$$

Note that these recurrence relations reduce (by setting $p = 0$) to those presented in Sections 2 and 3 for I.I.D. Pareto random variables. Thus by starting with Equations (4.1)–(4.6) for $p = 0$, all of the single and product moments for the I.I.D. case can be determined. These same relations could then be used again, this time with $p = 1$, to determine all of the single and product moments of all order statistics from a sample containing a single outlier. Continuing in this manner, the relations in (4.1)–(4.6) could be used to compute all the single and product moments (and hence covariances) of all order statistics from a p-outlier model in a simple recursive manner.

5. Generalization to the truncated Pareto distribution

We now generalize all of the preceding results by considering the case when the variables X_i's are independent having doubly-truncated Pareto distributions with density functions

$$f_i(x) = \frac{v_i x^{-(v_i+1)}}{L^{-v_i} - U^{-v_i}}, \quad L \leq x \leq U, \ v_i > 0 \tag{5.1}$$

and cumulative distribution functions

$$F_i(x) = \frac{L^{-v_i} - x^{-v_i}}{L^{-v_i} - U^{-v_i}}, \quad L \leq x \leq U, \ v_i > 0 \tag{5.2}$$

for $i = 1, 2, \ldots, n$. The differential equations in this case are

$$F_i(x) = C_i - \frac{x}{v_i} f_i(x), \quad i = 1, 2, \ldots, n , \tag{5.3}$$

where

$$C_i = \frac{L^{-v_i}}{L^{-v_i} - U^{-v_i}} . \tag{5.4}$$

We first point out though, that in a discussion of Balakrishnan (1994a), Arnold (1994) presented an alternative method for deriving the single and product moments of order statistics arising from I.NI.D. exponential random variables. This alternative method uses the fact that I.NI.D. exponential random variables are closed under minima, i.e., the minimum of a set of I.NI.D. exponential random variables is again an exponential random variable. He points out that the same is true for Pareto random variables, and that his method is also applicable to distributions close under maxima. Thus his method could possibly be used as an alternative to the recurrence relations presented in Sections 2 and 3 to derive all of the single and product moments of order statistics arising from I.NI.D. Pareto random variables. However, truncated I.NI.D. Pareto random variables are not closed under minima or maxima. Hence, the method of Arnold (1994) does not apply to the truncated Pareto distribution case.

On the other hand, the differential equation technique used in Sections 2 and 3 easily handles the truncated Pareto case. We now use this method to generalize the recurrence relations established in Section 2 and 3 to the doubly-truncated Pareto model in (5.1).

The recurrence relations derived in Section 2 for single moments generalize as follows:

(a) for $n \geq 2$ and $k = 0, 1, 2, \ldots$,

$$\mu_{1:n}^{(k)} = \frac{L^k \sum_{i=1}^n C_i v_i + \sum_{i=1}^n (1-C_i) v_i \mu_{1:n-1}^{[i](k)}}{\sum_{i=1}^n v_i - k} ;$$

(b) for $2 \leq r \leq n-1$ and $k = 0, 1, 2, \ldots$,

$$\mu_{r:n}^{(k)} = \frac{\sum_{i=1}^n C_i v_i \mu_{r-1:n-1}^{[i](k)} + \sum_{i=1}^n (1-C_i) v_i \mu_{r:n-1}^{[i](k)}}{\sum_{i=1}^n v_i - k} ;$$

and

(c) for $n \geq 2$ and $k = 0, 1, 2, \ldots$,

$$\mu_{n:n}^{(k)} = \frac{U^k \sum_{i=1}^n (1-C_i) v_i + \sum_{i=1}^n C_i v_i \mu_{n-1:n-1}^{[i](k)}}{\sum_{i=1}^n v_i - k} .$$

We will provide the proof of recurrence relation (b) in Appendix B. Relations (a) and (c) may be proved on similar lines.

The recurrence relations established in Section 3 for product moments can similarly be generalized and are given below (we will only provide the proof of relation (i) in Appendix B; the other proofs are similar):

(d) for $n \geq 3$,

$$\mu_{1,2:n} = \frac{L \sum_{i=1}^n C_i v_i \mu_{1:n-1}^{[i]} + \sum_{i=1}^n (1-C_i) v_i \mu_{1,2:n-1}^{[i]}}{\sum_{i=1}^n v_i - 2} ;$$

(e) for $2 \leq r \leq n-2$,

$$\mu_{r,r+1:n} = \frac{\sum_{i=1}^n C_i v_i \mu_{r-1,r:n-1}^{[i]} + \sum_{i=1}^n (1-C_i) v_i \mu_{r,r+1:n-1}^{[i]}}{\sum_{i=1}^n v_i - 2} ;$$

(f) for $n \geq 3$,

$$\mu_{n-1,n:n} = \frac{\sum_{i=1}^{n} C_i v_i \mu_{n-2,n-1:n-1}^{[i]} + U \sum_{i=1}^{n} (1 - C_i) v_i \mu_{n-1:n-1}^{[i]}}{\sum_{i=1}^{n} v_i - 2} ;$$

(g) for $3 \leq s \leq n - 1$,

$$\mu_{1,s:n} = \frac{L \sum_{i=1}^{n} C_i v_i \mu_{s-1:n-1}^{[i]} + \sum_{i=1}^{n} (1 - C_i) v_i \mu_{1,s:n-1}^{[i]}}{\sum_{i=1}^{n} v_i - 2} ;$$

(h) for $n \geq 3$,

$$\mu_{1,n:n} = \frac{L \sum_{i=1}^{n} C_i v_i \mu_{n-1:n-1}^{[i]} + U \sum_{i=1}^{n} (1 - C_i) v_i \mu_{1:n-1}^{[i]}}{\sum_{i=1}^{n} v_i - 2} ;$$

(i) for $2 \leq r < s \leq n - 1$ and $s - r \geq 2$,

$$\mu_{r,s:n} = \frac{\sum_{i=1}^{n} C_i v_i \mu_{r-1,s-1:n-1}^{[i]} + \sum_{i=1}^{n} (1 - C_i) v_i \mu_{r,s:n-1}^{[i]}}{\sum_{i=1}^{n} v_i - 2} ;$$

and

(j) for $2 \leq r \leq n - 2$,

$$\mu_{r,n:n} = \frac{\sum_{i=1}^{n} C_i v_i \mu_{r-1,n-1:n-1}^{[i]} + U \sum_{i=1}^{n} (1 - C_i) v_i \mu_{r:n-1}^{[i]}}{\sum_{i=1}^{n} v_i - 2} ;$$

Relations (a)–(j) are generalizations of the results of Balakrishnan and Joshi (1982) to the I.NI.D case. Further, if $L \to 1$ and $U \to \infty$, relations (a)–(c) reduce to Results 2.1 and 2.2, while relations (d)–(j) reduce to Results 3.1–3.4. The results for the right-truncated case (left-truncated case) can be deduced from relations (a)–(j) by letting $L \to 1 (U \to \infty)$.

6. Robustness of the MLE and BLUE

In this section, we introduce a scale parameter to the Pareto distribution in (1.3) and consider the p-outlier model described in Section 4. Specifically we consider the situation where $X_1, X_2, \ldots, X_{n-p}$ have a Pareto distribution with shape parameter v and scale parameter σ,

$$f(x) = v\sigma^v x^{-(v+1)}, \quad x \geq \sigma, \ v > 0, \ \sigma > 0 \tag{6.1}$$

while $X_{n-p+1}, \ldots, X_n$ have the same distribution but with shape parameter v^*

$$f(x) = v^* \sigma^{v^*} x^{-(v^*+1)}, \quad x \geq \sigma, \ v^* > 0, \ \sigma > 0 . \tag{6.2}$$

We will use the single and product moments obtained from the recurrence relations in Section 4 to examine the robustness of the maximum likelihood estimator (MLE) and best linear unbiased estimator (BLUE) of the scale parameter σ to the presence of multiple shape outliers. We will see how their bias and mean square error are affected by the presence of (possibly) multiple shape outliers.

For the I.I.D. case $(p = 0)$ the MLE for σ is

$$\hat{\sigma} = X_{1:n}$$

while the best linear unbiased estimator (BLUE) is given by

$$\sigma^* = \frac{vn-1}{vn\sigma} X_{1:n} \ . \tag{6.3}$$

This expression for the BLUE may be obtained in the usual way (as described in Arnold, Balakrishnan, and Nagaraja (1992)) by inverting the covariance matrix of the standardized I.I.D. Pareto random variables. Alternatively, it may be obtained by observing that $X_{1:n}$ is a complete sufficient statistic for the parameter σ. Thus the minimum variance unbiased estimator for σ is a function of $X_{1:n}$. Since $\frac{vn-1}{vn\sigma} X_{1:n}$ is unbiased, and a linear function of order statistics, it must also be the BLUE. See also Likeš (1969), and Kulldorff and Vännman (1973).

In Table 1 we present the bias and mean square error of the MLE and BLUE for $n = 10, 20$, and 30, $p = 0(1)5$, and various values of v and v^*. Whenever $v^* > v$, the observations with shape parameter v^* are actually likely to be smaller than those with shape parameter v. This case is therefore referred to as the 'inlier situation'. The actual outliers occur whenever $v^* < v$.

From Table 1 we see that in the outlier situation $(v^* < v)$ and the I.I.D. case $(v^* = v)$, the bias and mean square error of the BLUE are considerably smaller than those of the MLE. As v^* increases, the mean square error of both estimators decreases. However, the mean square error of the MLE decreases more rapidly than that of the BLUE. The result is that for some of the larger values of v^*, for example $v = 3$, $v^* = 18, 23$, and 28, the mean square error of the MLE becomes smaller than that of the BLUE. However, when the MLE has smaller mean square error, it is usually only slightly smaller than the BLUE, except in the extreme inlier situation $(v = 3, v^* = 23$ and $28)$.

Therefore, for estimation of the scale parameter σ of the Pareto distribution in (6.1) in (possibly) the presence of outliers, we recommend use of the BLUE given in (6.3) since it is significantly more efficient in the outlier situation, and usually only slightly less efficient than the MLE in the inlier situation.

7. Robustness of the censored BLUE

In this section, we introduce a location and scale parameter to the Pareto distribution in (1.3) and consider the p-outlier model described in Section 4. Specifically we consider the situation where $X_1, X_2, \ldots, X_{n-p}$ have a Pareto distribution with shape parameter v, location parameter μ, and scale parameter σ,

Table 1
10^2 (Bias of estimators of σ)/σ and 10^4(mean square error)/σ^2 in the presence of multiple shape outliers

$n = 10$

$v = 3$	Bias								
	v^*	2.1	2.5	3^+	8	13	18	23	28
$p = 1$	BLUE	0.1068	0.0585	0.0000	−0.4902	−0.8547	−1.1364	−1.3605	−1.5432
	MLE	3.5587	3.5088	3.4483	2.9412	2.5641	2.2727	2.0408	1.8519
$p = 2$	BLUE	0.2206	0.1190	0.0000	−0.8547	−1.3605	−1.6949	−1.9324	−2.1097
	MLE	3.6765	3.5714	3.4483	2.5641	2.0408	1.6949	1.4493	1.2658
$p = 3$	BLUE	0.3422	0.1818	0.0000	−1.1364	−1.6949	−2.0270	−2.2472	−2.4038
	MLE	3.8023	3.6364	3.4483	2.2727	1.6949	1.3514	1.1236	0.9615

$v = 3$	Mean Square Error								
	v^*	2.1	2.5	3^+	8	13	8	23	28
$p = 1$	BLUE	12.7190	12.3445	11.9048	8.8136	7.1975	6.3425	5.9051	5.7070
	MLE	26.2636	25.5184	24.6305	17.8253	13.4953	10.5708	8.5034	6.9881
$p = 2$	BLUE	13.6432	12.8160	11.9048	7.1975	5.9051	5.6497	5.7545	5.9865
	MLE	28.0647	26.4550	24.6305	13.4953	8.5034	5.8445	4.2626	3.2457
$p = 3$	BLUE	14.6946	13.3219	11.9048	6.3425	5.6497	5.8620	6.2564	6.6592
	MLE	30.0576	27.4443	24.6305	10.5708	5.8445	3.7023	2.5536	1.8671

$v = 5$	Bias								
	v^*	2.1	2.5	3	4	5^+	10	15	20
$p = 1$	BLUE	0.1258	0.1075	0.0851	0.0417	0.0000	−0.1852	−0.3390	−0.4688
	MLE	2.1692	2.1505	2.1277	2.0833	2.0408	1.8519	1.6949	1.5625
$p = 2$	BLUE	0.2685	0.2273	0.1778	0.0851	0.0000	−0.3390	−0.5797	−0.7595
	MLE	2.3148	2.2727	2.2222	2.1277	2.0408	1.6949	1.4493	1.2658
$p = 3$	BLUE	0.4318	0.3614	0.2791	0.1304	0.0000	−0.4688	−0.7595	−0.9574
	MLE	2.4814	2.4096	0.3256	2.1739	2.0408	1.5625	1.2658	1.0638

$v = 5$	Mean Square Error								
	v^*	2.1	2.5	3^+	4	5^+	10	15	20
$p = 1$	BLUE	4.7353	4.6485	4.5439	4.3475	4.1667	3.4521	2.9690	2.6389
	MLE	9.6195	9.4529	9.2507	8.8652	8.5034	6.9881	5.8445	4.9603
$p = 2$	BLUE	5.4622	5.2431	4.9899	4.5439	4.1667	2.9690	2.4126	2.1551
	MLE	10.9707	10.5708	10.1010	9.2507	8.5034	5.8445	4.2626	3.2457
$p = 3$	BLUE	6.4008	5.9824	5.5194	4.7575	4.1667	2.6389	2.1551	2.0270
	MLE	12.6729	11.8994	11.0742	9.6618	8.5034	4.9603	3.2457	2.2878

$n = 20$

$v = 3$	Bias								
	v^*	2.1	2.5	3^+	8	13	18	23	28
$p = 1$	BLUE	0.0258	0.0142	0.0000	−0.1302	−0.2415	−0.3378	−0.4219	−0.4960
	MLE	1.7212	1.7094	1.6949	1.5625	1.4493	1.3514	1.2658	1.1905
$p = 2$	BLUE	0.0524	0.0287	0.0000	−0.2415	−0.4219	−0.5618	−0.6734	−0.7645
	MLE	1.7483	1.7241	1.6949	1.4493	1.2658	1.1236	1.0101	0.9174
$p = 3$	BLUE	0.0799	0.0435	0.0000	−0.3378	−0.5618	−0.7212	−0.8403	−0.9328
	MLE	1.7762	1.7391	1.6949	1.3514	1.1236	0.9615	0.8403	0.7463
$p = 4$	BLUE	0.1083	0.0585	0.0000	−0.4219	−0.6734	−0.8403	−0.9592	−1.0482
	MLE	1.8051	1.7544	1.6949	1.2658	1.0101	0.8403	0.7194	0.6289

Table 1 (Contd.)

$v=3$	Mean Square Error								
	v^*	2.1	2.5	3^+	8	13	18	23	28
$p=1$	BLUE	2.9655	2.9240	2.8736	2.4526	2.1490	1.9283	1.7671	1.6495
	MLE	6.0286	5.9458	5.8445	4.9603	4.2626	3.7023	3.2457	2.8686
$p=2$	BLUE	3.0633	2.9761	2.8736	2.1490	1.7671	1.5641	1.4602	1.4134
	MLE	6.2216	6.0496	5.8445	4.2626	3.2457	2.5536	2.0614	1.6989
$p=3$	BLUE	3.1673	3.0300	2.8736	1.9283	1.5641	1.4314	1.4006	1.4168
	MLE	6.4239	6.1562	5.8445	3.7023	2.5536	1.8671	1.4243	1.1222
$p=4$	BLUE	3.2781	3.0858	2.8736	1.7671	1.4602	1.4006	1.4278	1.4861
	MLE	6.6362	6.2657	5.8445	3.2457	2.0614	1.4243	1.0426	0.7961
$v=5$	Bias								
	v^*	2.1	2.5	3	4	5^+	10	15	20
$p=1$	BLUE	0.0302	0.0259	0.0206	0.0102	0.0000	−0.0481	−0.0917	−0.1316
	MLE	1.0406	1.0363	1.0309	1.0204	1.0101	0.9615	0.9174	0.8772
$p=2$	BLUE	0.0622	0.0532	0.0421	0.0206	0.0000	−0.0917	−0.1681	−0.2326
	MLE	1.0730	1.0638	1.0526	1.0309	1.0101	0.9174	0.8403	0.7752
$p=3$	BLUE	0.0963	0.0820	0.0645	0.0313	0.0000	−0.1316	−0.2326	−0.3125
	MLE	1.1074	1.0929	1.0753	1.0417	1.0101	0.8772	0.7752	0.6944
$p=4$	BLUE	0.1327	0.1124	0.0879	0.0421	0.0000	−0.1681	−0.2878	−0.3774
	MLE	1.1442	1.1236	1.0989	1.0526	1.0101	0.8403	0.7194	0.6289
$v=5$	Mean Square Error								
	v^*	2.1	2.5	3	4	5^+	10	15	20
$p=1$	BLUE	1.0845	1.0752	1.0638	1.0417	1.0204	0.9261	0.8486	0.7848
	MLE	2.1884	2.1702	2.1478	2.1039	2.0614	1.8671	1.6989	1.5526
$p=2$	BLUE	1.1567	1.1359	1.1109	1.0638	1.0204	0.8486	0.7321	0.6523
	MLE	2.3275	2.2878	2.2396	2.1478	2.0614	1.6989	1.4243	1.2112
$p=3$	BLUE	1.2382	1.2032	1.1620	1.0868	1.0204	0.7848	0.6523	0.5769
	MLE	2.4802	2.4152	2.3375	2.1930	2.0614	1.5526	1.2112	0.9713
$p=4$	BLUE	1.3304	1.2781	1.2176	1.1109	1.0204	0.7321	0.5974	0.5350
	MLE	2.6485	2.5536	2.4420	2.2396	2.0614	1.4243	1.0426	0.7961
$n=30$									
$v=3$	Bias								
	v^*	2.1	2.5	3^+	8	13	18	23	28
$p=1$	BLUE	0.0114	0.0063	0.0000	−0.0591	−0.1122	−0.1603	−0.2039	−0.2437
	MLE	1.1351	1.1299	1.1236	1.0638	1.0101	0.9615	0.9174	0.8772
$p=2$	BLUE	0.0229	0.0126	0.0000	−0.1122	−0.2039	−0.2801	−0.3445	−0.3997
	MLE	1.1468	1.1364	1.1236	1.0101	0.9174	0.8403	0.7752	0.7194
$p=3$	BLUE	0.0348	0.0190	0.0000	−0.1603	−0.2801	−0.3731	−0.4474	−0.5081
	MLE	1.1587	1.1429	1.1236	0.9615	0.8403	0.7463	0.6711	0.6098
$p=4$	BLUE	0.0468	0.0255	0.0000	−0.2039	−0.3445	−0.4474	−0.5260	−0.5879
	MLE	1.1710	1.1494	1.1236	0.9174	0.7752	0.6711	0.5917	0.5291
$p=5$	BLUE	0.0592	0.0321	0.0000	−0.2437	−0.3997	−0.5081	−0.5879	−0.6490
	MLE	1.1834	1.1561	1.1236	0.8772	0.7194	0.6098	0.5291	0.4673
$v=3$	Mean Square Error								
	v^*	2.1	2.5	3^+	8	13	18	23	28
$p=1$	BLUE	1.2890	1.2771	1.2626	1.1340	1.0307	0.9474	0.8799	0.8252
	MLE	2.6064	2.5827	2.5536	2.2878	2.0614	1.8671	1.6989	1.5526
$p=2$	BLUE	1.3164	1.2920	1.2626	1.0307	0.8799	0.7807	0.7155	0.6732
	MLE	2.6608	2.6123	2.5536	2.0614	1.6989	1.4243	1.2112	1.0426

Table 1 (Contd.)

$p=3$	BLUE	1.3450	1.3072	1.2626	0.9474	0.7807	0.6920	0.6466	0.6262
	MLE	2.7169	2.6425	2.5536	1.8671	1.4243	1.1222	0.9069	0.7482
$p=4$	BLUE	1.3748	1.3227	1.2626	0.8799	0.7155	0.6466	0.6231	0.6223
	MLE	2.7748	2.6731	2.5536	1.6989	1.2112	0.9069	0.7044	0.5629
$p=5$	BLUE	1.4059	1.3386	1.2626	0.8252	0.6732	0.6262	0.6223	0.6368
	MLE	2.8346	2.7042	2.5536	1.5526	1.0426	0.7482	0.5629	0.4388
$v=5$	Bias								
	v^*	2.1	2.5	3	4	5$^+$	10	15	20
$p=1$	BLUE	0.0132	0.0114	0.0091	0.0045	0.0000	−0.0216	−0.0419	−0.0610
	MLE	0.6845	0.6826	0.6803	0.6757	0.6711	0.6494	0.6289	0.6098
$p=2$	BLUE	0.0270	0.0231	0.0184	0.0091	0.0000	−0.0419	−0.0789	−0.1117
	MLE	0.6983	0.6944	0.6897	0.6803	0.6711	0.6289	0.5917	0.5587
$p=3$	BLUE	0.0413	0.0353	0.0280	0.0137	0.0000	−0.0610	−0.1117	−0.1546
	MLE	0.7128	0.7067	0.6993	0.6849	0.6711	0.6098	0.5587	0.5155
$p=4$	BLUE	0.0563	0.0480	0.0378	0.0184	0.0000	−0.0789	−0.1411	−0.1914
	MLE	0.7278	0.7194	0.7092	0.6897	0.6711	0.5917	0.5291	0.4785
$p=5$	BLUE	0.0719	0.0611	0.0480	0.0231	0.0000	−0.0958	−0.1675	−0.2232
	MLE	0.7435	0.7326	0.7194	0.6944	0.6711	0.5747	0.5025	0.4464
$v=5$	Mean Square Error								
	v^*	2.1	2.5	3	4	5$^+$	10	15	20
$p=1$	BLUE	0.4688	0.4662	0.4630	0.4566	0.4505	0.4220	0.3970	0.3751
	MLE	0.9434	0.9383	0.9319	0.9193	0.9069	0.8488	0.7961	0.7482
$p=2$	BLUE	0.4887	0.4830	0.4762	0.4630	0.4505	0.3970	0.3558	0.3239
	MLE	0.9822	0.9713	0.9579	0.9319	0.9069	0.7961	0.7044	0.6277
$p=3$	BLUE	0.5102	0.5011	0.4901	0.4695	0.4505	0.3751	0.3239	0.2888
	MLE	1.0233	1.0060	0.9849	0.9447	0.9069	0.7482	0.6277	0.5342
$p=4$	BLUE	0.5335	0.5204	0.5048	0.4762	0.4505	0.3558	0.2991	0.2647
	MLE	1.0672	1.0426	1.0132	0.9579	0.9069	0.7044	0.5629	0.4601
$p=5$	BLUE	0.5588	0.5411	0.5204	0.4830	0.4505	0.3388	0.2797	0.2482
	MLE	1.1139	1.0813	1.0426	0.9713	0.9069	0.6644	0.5076	0.4004

$^+$ This is the I.I.D. case ($p=0$).

$$f(x) = v\sigma^v(x-\mu)^{-(v+1)}, \quad x \geq \sigma + \mu, \quad v > 0, \quad \sigma > 0 \tag{7.1}$$

while $X_{n-p+1}, \ldots, X_n$ have the same distribution but with shape parameter v^*.

$$f(x) = v^*\sigma^{v^*}(x-\mu)^{-(v^*+1)}, \quad x \geq \sigma + \mu, \quad v^* > 0, \quad \sigma > 0. \tag{7.2}$$

We will use the single and product moments obtained from the recurrence relations in Section 4 to examine the robustness of the full sample and censored BLUE of the location parameter μ and the scale parameter σ to the presence of multiple shape outliers. We will see how their bias and mean square error are affected by the presence of (possibly) multiple shape outliers.

Explicit expressions for the full sample BLUE may be obtained from the results of Kulldorff and Vännman (1973). To compute the censored BLUE's, we note that the covariance matrix of the standardized Pareto order statistics $(\sigma_{i,j:n})$ is of the form $(a_i b_j)$ where

$$a_i = \frac{\Gamma(n-i+1-2/v)}{\Gamma(n-i+1-1/v)\Gamma(n+1-2/v)} - \frac{\Gamma(n-i+1-1/v)\Gamma(n+1)}{\Gamma(n-i+1)\Gamma(n+1-1/v)^2} \tag{7.3}$$

and

$$b_j = \frac{\Gamma(n+1)\Gamma(n-j+1-1/v)}{\Gamma(n-j+1)}. \tag{7.4}$$

These may be obtained from the results of Huang (1975). We are therefore able to invert the covariance matrix $(\sigma_{i,j:n})$ and obtain explicit expressions for the censored BLUE's $\mu^*(r)$ and $\sigma^*(r)$ of μ and σ respectively, their variances $\text{var}(\mu^*(r))$ and $\text{var}(\sigma^*(r))$, and their covariance $\text{cov}(\mu^*(r), \sigma^*(r))$, as described, for example, in Arnold, Balakrishnan, and Nagaraja (1992). We have,

$$\mu^*(r) = \left(\left\{ \frac{a_2 - a_1}{a_1(a_2 b_1 - a_1 b_2)} - \frac{1}{a_1 b_1} \right\} X_{1:n} \right.$$
$$+ \sum_{i=1}^{n-r-2} \frac{a_i(b_{i+1} - b_{i+2}) + a_{i+1}(b_{i+2} - b_i) + a_{i+2}(b_i - b_{i+1})}{(a_{i+1} b_i - a_i b_{i+1})(a_{i+2} b_{i+1} - a_{i+1} b_{i+2})} X_{i+1:n}$$
$$\left. + \frac{b_{n-r-1} - b_{n-r}}{b_{n-r}(a_{n-r} b_{n-r-1} - a_{n-r-1} b_{n-r})} X_{n-r:n} \right) \bigg/ \left(\sum_{i,j} \sigma^{ij} - \frac{1}{a_1 b_1} \right),$$

$$\sigma^*(r) = \left(\left\{ \sum_{i,j} \sigma^{ij} - \frac{a_2 - a_1}{a_1(a_2 b_1 - a_1 b_2)} \right\} X_{1:n} \right.$$
$$- \sum_{i=1}^{n-r-2} \frac{a_i(b_{i+1} - b_{i+2}) + a_{i+1}(b_{i+2} - b_i) + a_{i+2}(b_i - b_{i+1})}{(a_{i+1} b_i - a_i b_{i+1})(a_{i+2} b_{i+1} - a_{i+1} b_{i+2})} X_{i+1:n}$$
$$\left. - \frac{b_{n-r-1} - b_{n-r}}{b_{n-r}(a_{n-r} b_{n-r-1} - a_{n-r-1} b_{n-r})} X_{n-r:n} \right) \bigg/ \left(\frac{b_1}{c} \sum_{i,j} \sigma^{ij} - \frac{1}{c a_1} \right),$$

$$\text{var}(\mu^*(r)) = \sigma^2 \bigg/ \left(\sum_{ij} \sigma^{ij} - \frac{1}{a_1 b_1} \right), \tag{7.5}$$

$$\text{var}(\sigma^*(r)) = \sigma^2 \bigg/ \left(\frac{b_1}{a_1 c^2} - \frac{1}{c^2 a_1^2 \sum_{i,j} \sigma^{ij}} \right), \tag{7.6}$$

and

$$\text{cov}(\mu^*(r), \sigma^*(r)) = -\sigma^2 \bigg/ \left(\frac{b_1}{c} \sum_{i,j} \sigma^{ij} - \frac{1}{c a_1} \right),$$

where a_i and b_i are given in (7.3) and (7.4) respectively,

$$c = \Gamma(n+1-1/v) \ ,$$

and $\sum_{i,j} \sigma^{ij}$ is the sum of all of the elements of the inverse matrix of the covariance matrix $(\sigma_{i,j:n})$ and is given by

$$\sum_{i,j} \sigma^{ij} = \sum_{i=2}^{n-r-1} \frac{a_{i-1}(2b_i - b_{i+1}) - 2a_i b_{i-1} + a_{i+1} b_{i-1}}{(a_i b_{i-1} - a_{i-1} b_i)(a_{i+1} b_i - a_i b_{i+1})}$$
$$+ \frac{a_2 - 2a_1}{a_1(a_2 b_1 - a_1 b_2)} + \frac{b_{n-r-1}}{b_{n-r}(a_{n-r} b_{n-r-1} - a_{n-r-1} b_{n-r})} \ .$$

In Tables 2 and 3, we present the bias and mean square error of the BLUE's $\mu^*(r)$ and $\sigma^*(r)$, respectively, with $r = 0, 10$, and 20% of n, for $n = 10$ and 20, $p = 0(1)4$, and various values of v and v^*. Note that the mean square error for the I.I.D. case $(p = 0)$ given in Tables 2 and 3 may be obtained using (7.5) and (7.6) respectively.

From Table 2, we see that when $v^* \geq v$ (the inlier situation) the bias and mean square error for the censored forms of the BLUE for v are only slightly larger than those of the full sample BLUE. But in the outlier situation $(v^* < v)$, as the outliers become more pronounced, the bias and mean square error increase for each form of the BLUE. However, they increase much less for the censored forms of the BLUE than they do for the full sample BLUE, especially for the larger values of p. The result is that as the outliers become more pronounced the bias and mean square error of the censored forms of the BLUE start to become smaller than those of the full sample BLUE. The difference becomes quite significant as p increases. The same observations remain true for the BLUE's for σ in Table 3.

Therefore, since the loss of efficiency due to censoring when there are no outliers present is minimal as compared with the possible gain in efficiency when outliers are present, we recommend use of the BLUE with 10% censoring for the estimation of both the location parameter μ and scale parameter σ.

8. Conclusions

We have established in this paper several recurrence relations for the single and the product moments of order statistics arising from n independent non-identically distributed Pareto random variables. These recurrence relations are simple in nature and could be applied systematically in order to compute all the single and the product moments of order statistics arising from I.NI.D. Pareto random variables for all values of n in a simple recursive manner, as long as the values of $v_i > 2$ $(i = 1, 2, \ldots, n)$ are known. The results for the case when the order statistics arise from a multiple-outlier model (with a slippage of p observations) from a Pareto population are deduced as special cases. We have also generalized all of the results to the doubly-truncated Pareto distribution. We have then applied the

Table 2
(Bias of estimators of μ/σ and $(MSE)/\sigma^2$ in the presence of multiple shape outliers

$n = 10$

$v = 3$	Bias								
	v^*	2.1	2.5	3^+	8	13	18	23	28
$p = 1$	BLUE0	−0.0548	−0.0258	0.0000	0.0812	0.0910	0.0891	0.0845	0.0793
	BLUE1	−0.0503	−0.0246	0.0000	0.0824	0.0925	0.0906	0.0859	0.0806
	BLUE2	−0.0467	−0.0234	0.0000	0.0859	0.0973	0.0955	0.0906	0.0850
$p = 2$	BLUE0	−0.1120	−0.0521	0.0000	0.1640	0.1939	0.2024	0.2048	0.2050
	BLUE1	−0.1036	−0.0498	0.0000	0.1665	0.1972	0.2059	0.2083	0.2086
	BLUE2	−0.0963	−0.0474	0.0000	0.1731	0.2073	0.2168	0.2196	0.2199
$p = 3$	BLUE0	−0.1717	−0.0789	0.0000	0.2458	0.2981	0.3185	0.3288	0.3348
	BLUE1	−0.1598	−0.0756	0.0000	0.2494	0.3031	0.3240	0.3345	0.3405
	BLUE2	−0.1488	−0.0720	0.0000	0.2586	0.3181	0.3411	0.3524	0.3590
$v = 3$	Mean Square Error								
	v^*	2.1	2.5	3^+	8	13	18	23	28
$p = 1$	BLUE0	0.2624	0.2278	0.2128	0.1886	0.1903	0.1923	0.1938	0.1948
	BLUE1	0.2470	0.2302	0.2165	0.1916	0.1934	0.1955	0.1970	0.1981
	BLUE2	0.2567	0.2416	0.2283	0.2010	0.2033	0.2057	0.2075	0.2087
$p = 2$	BLUE0	0.3227	0.2449	0.2128	0.1783	0.1842	0.1874	0.1888	0.1895
	BLUE1	0.2876	0.2460	0.2165	0.1810	0.1873	0.1906	0.1922	0.1928
	BLUE2	0.2938	0.2568	0.2283	0.1893	0.1969	0.2010	0.2028	0.2036
$p = 3$	BLUE0	0.3947	0.2642	0.2128	0.1833	0.2020	0.2122	0.2180	0.2216
	BLUE1	0.3393	0.2639	0.2165	0.1863	0.2060	0.2166	0.2226	0.2263
	BLUE2	0.3408	0.2741	0.2283	0.1945	0.2177	0.2303	0.2373	0.2416
$v = 5$	Bias								
	v^*	2.1	2.5	3	4	5^+	10	15	20
$p = 1$	BLUE0	−0.1820	−0.1253	−0.0814	−0.0301	0.0000	0.0583	0.0732	0.0771
	BLUE1	−0.1227	−0.0961	−0.0687	−0.0280	0.0000	0.0603	0.0764	0.0806
	BLUE2	−0.1061	−0.0851	−0.0624	−0.0266	0.0000	0.0633	0.0818	0.0870
$p = 2$	BLUE0	−0.3797	−0.2585	−0.1662	−0.0607	0.0000	0.1171	0.1510	0.1643
	BLUE1	−0.2772	−0.2074	−0.1434	−0.0567	0.0000	0.1211	0.1574	0.1718
	BLUE2	−0.2331	−0.1821	−0.1304	−0.0539	0.0000	0.1265	0.1681	0.1849
$p = 3$	BLUE0	−0.5911	−0.3988	−0.2539	−0.0916	0.0000	0.1758	0.2299	0.2545
	BLUE1	−0.4567	−0.3313	−0.2234	−0.0861	0.0000	0.1813	0.2396	0.2659
	BLUE2	−0.3850	−0.2919	−0.2040	−0.0820	0.0000	0.1886	0.2549	0.2856
$v = 5$	Mean Square Error								
	v^*	2.1	2.5	3	4	5^+	10	15	20
$p = 1$	BLUE0	0.7934	0.2871	0.2192	0.1812	0.1674	0.1534	0.1534	0.1546
	BLUE1	0.2498	0.2294	0.2108	0.1877	0.1752	0.1599	0.1601	0.1614
	BLUE2	0.2519	0.2371	0.2224	0.2022	0.1898	0.1722	0.1723	0.1740
$p = 2$	BLUE0	1.5365	0.4576	0.2905	0.1973	0.1674	0.1464	0.1499	0.1532
	BLUE1	0.4231	0.3292	0.2649	0.2026	0.1752	0.1522	0.1564	0.1601
	BLUE2	0.3694	0.3159	0.2700	0.2168	0.1898	0.1630	0.1677	0.1725
$p = 3$	BLUE0	2.4114	0.6838	0.3829	0.2159	0.1674	0.1467	0.1594	0.1685
	BLUE1	0.7149	0.4815	0.3393	0.2199	0.1752	0.1524	0.1667	0.1770
	BLUE2	0.5762	0.4378	0.3356	0.2337	0.1898	0.1622	0.1787	0.1918

Table 2 (Contd.)

$n = 20$

$v = 3$ Bias

v^*		2.1	2.5	3^+	8	13	18	23	28
$p = 1$	BLUE0	−0.0257	−0.0124	0.0000	0.0433	0.0517	0.0532	0.0526	0.0512
	BLUE2	−0.0237	−0.0118	0.0000	0.0441	0.0527	0.0543	0.0537	0.0522
	BLUE4	−0.0221	−0.0112	0.0000	0.0460	0.0553	0.0571	0.0565	0.0549
$p = 2$	BLUE0	−0.0520	−0.0249	0.0000	0.0864	0.1048	0.1105	0.1122	0.1123
	BLUE2	−0.0481	−0.0238	0.0000	0.0879	0.1068	0.1126	0.1143	0.1144
	BLUE4	−0.0449	−0.0226	0.0000	0.0916	0.1121	0.1184	0.1203	0.1204
$p = 3$	BLUE0	−0.0789	−0.0375	0.0000	0.1290	0.1581	0.1690	0.1738	0.1762
	BLUE2	−0.0732	−0.0358	0.0000	0.1313	0.1612	0.1722	0.1772	0.1796
	BLUE4	−0.0684	−0.0342	0.0000	0.1366	0.1692	0.1811	0.1864	0.1890
$p = 4$	BLUE0	−0.1065	−0.0502	0.0000	0.1710	0.2111	0.2273	0.2356	0.2404
	BLUE2	−0.0990	−0.0481	0.0000	0.1740	0.2151	0.2317	0.2401	0.2450
	BLUE4	−0.0925	−0.0459	0.0000	0.1807	0.2258	0.2437	0.2526	0.2578

$v = 3$ Mean Square Error

v^*		2.1	2.5	3^+	8	13	18	23	28
$p = 1$	BLUE0	0.1019	0.0969	0.0939	0.0882	0.0885	0.0890	0.0893	0.0896
	BLUE2	0.1018	0.0985	0.0957	0.0899	0.0902	0.0906	0.0910	0.0913
	BLUE4	0.1065	0.1034	0.1007	0.0943	0.0947	0.0953	0.0957	0.0960
$p = 2$	BLUE0	0.1117	0.1003	0.0939	0.0865	0.0886	0.0898	0.0904	0.0907
	BLUE2	0.1096	0.1017	0.0957	0.0881	0.0903	0.0916	0.0922	0.0925
	BLUE4	0.1137	0.1066	0.1007	0.0924	0.0950	0.0964	0.0972	0.0975
$p = 3$	BLUE0	0.1235	0.1041	0.0939	0.0887	0.0945	0.0976	0.0992	0.1001
	BLUE2	0.1190	0.1053	0.0957	0.0924	0.0965	0.0997	0.1014	0.1024
	BLUE4	0.1224	0.1100	0.1007	0.0947	0.1019	0.1056	0.1076	0.1086
$p = 4$	BLUE0	0.1372	0.1084	0.0939	0.0945	0.1063	0.1125	0.1160	0.1182
	BLUE2	0.1302	0.1093	0.0957	0.0964	0.1088	0.1153	0.1190	0.1212
	BLUE4	0.1327	0.1138	0.1007	0.1012	0.1155	0.1230	0.1271	0.1297

$v = 5$ Bias

v^*		2.1	2.5	3	4	5^+	10	15	20
$p = 1$	BLUE0	−0.0803	−0.0570	−0.0380	−0.0145	0.0000	0.0305	0.0400	0.0437
	BLUE2	−0.0563	−0.0447	−0.0324	−0.0136	0.0000	0.0315	0.0416	0.0455
	BLUE4	−0.0494	−0.0400	−0.0297	−0.0129	0.0000	0.0330	0.0445	0.0489
$p = 2$	BLUE0	−0.1651	−0.1162	−0.0769	−0.0292	0.0000	0.0608	0.0803	0.0887
	BLUE2	−0.1183	−0.0924	−0.0662	−0.0273	0.0000	0.0629	0.0836	0.0924
	BLUE4	−0.1031	−0.0826	−0.0607	−0.0260	0.0000	0.0658	0.0893	0.0993
$p = 3$	BLUE0	−0.2543	−0.1776	−0.1168	−0.0439	0.0000	0.0910	0.1208	0.1343
	BLUE2	−0.1865	−0.1434	−0.1013	−0.0412	0.0000	0.0940	0.1257	0.1400
	BLUE4	−0.1614	−0.1278	−0.0929	−0.0392	0.0000	0.0982	0.1341	0.1503
$p = 4$	BLUE0	−0.3473	−0.2410	−0.1575	−0.0588	0.0000	0.1210	0.1612	0.1801
	BLUE2	−0.2609	−0.1974	−0.1377	−0.0553	0.0000	0.1248	0.1677	0.1877
	BLUE4	−0.2247	−0.1758	−0.1264	−0.0527	0.0000	0.1301	0.1785	0.2014

$v = 5$ Mean Square Error

v^*		2.1	2.5	3	4	5^+	10	15	20
$p = 1$	BLUE0	0.1612	0.0970	0.0865	0.0794	0.0765	0.0731	0.0731	0.0733
	BLUE2	0.0943	0.0907	0.0871	0.0825	0.0797	0.0761	0.0761	0.0763
	BLUE4	0.0985	0.0957	0.0928	0.0886	0.0859	0.0818	0.0817	0.0821
$p = 2$	BLUE0	0.2649	0.1263	0.1002	0.0828	0.0765	0.0717	0.0729	0.0739
	BLUE2	0.1195	0.1077	0.0975	0.0856	0.0797	0.0746	0.0759	0.0771

Table 2 (Contd.)

	BLUE4	0.1188	0.1102	0.1021	0.0917	0.0859	0.0799	0.0815	0.0830
$p = 3$	BLUE0	0.3897	0.1651	0.1177	0.0868	0.0765	0.0721	0.0760	0.0787
	BLUE2	0.1586	0.1319	0.1110	0.0892	0.0797	0.0750	0.0793	0.0824
	BLUE4	0.1485	0.1303	0.1142	0.0953	0.0859	0.0802	0.0853	0.0891
$p = 4$	BLUE0	0.5374	0.2142	0.1393	0.0912	0.0765	0.0745	0.0824	0.0878
	BLUE2	0.2145	0.1643	0.1281	0.0934	0.0797	0.0774	0.0863	0.0923
	BLUE4	0.1900	0.1568	0.1292	0.0993	0.0859	0.0826	0.0931	0.1005

[+] This is the I.I.D. case ($p = 0$).

Table 3
(Bias of estimators of $\sigma)/\sigma$ and (MSE)$/\sigma^2$ in the presence of multiple shape outliers

$n = 10$

$v = 3$	Bias								
	v^*	2.1	2.5	3[+]	8	13	18	23	28
$p = 1$	BLUE0	0.0540	0.0255	0.0000	−0.0834	−0.0965	−0.0975	−0.0952	−0.0921
	BLUE1	0.0497	0.0244	0.0000	−0.0846	−0.0980	−0.0990	−0.0967	−0.0934
	BLUE2	0.0462	0.0232	0.0000	−0.0880	−0.1026	−0.1037	−0.1012	−0.0976
$p = 2$	BLUE0	0.1105	0.0516	0.0000	−0.1671	−0.2011	−0.2126	−0.2173	−0.2193
	BLUE1	0.1023	0.0494	0.0000	−0.1695	−0.2043	−0.2160	−0.2207	−0.2227
	BLUE2	0.0953	0.0470	0.0000	−0.1759	−0.2140	−0.2266	−0.2316	−0.2336
$p = 3$	BLUE0	0.1694	0.0781	0.0000	−0.2490	−0.3051	−0.3282	−0.3403	−0.3477
	BLUE1	0.1579	0.0749	0.0000	−0.2524	−0.3099	−0.3335	−0.3458	−0.3532
	BLUE2	0.1472	0.0715	0.0000	−0.2613	−0.3244	−0.3500	−0.3632	−0.3711
$v = 3$	Mean Square Error								
	v^*	2.1	2.5	3[+]	8	13	18	23	28
$p = 1$	BLUE0	0.2467	0.2142	0.2001	0.1777	0.1799	0.1822	0.1838	0.1849
	BLUE1	0.2323	0.2164	0.2035	0.1806	0.1829	0.1852	0.1869	0.1880
	BLUE2	0.2414	0.2271	0.2145	0.1894	0.1921	0.1949	0.1968	0.1980
$p = 2$	BLUE0	0.3037	0.2304	0.2001	0.1698	0.1776	0.1822	0.1846	0.1860
	BLUE1	0.2708	0.2314	0.2035	0.1724	0.1806	0.1853	0.1879	0.1892
	BLUE2	0.2766	0.2415	0.2145	0.1803	0.1899	0.1954	0.1982	0.1998
$p = 3$	BLUE0	0.3719	0.2487	0.2001	0.1771	0.1990	0.2112	0.2185	0.2232
	BLUE1	0.3200	0.2484	0.2035	0.1799	0.2028	0.2155	0.2231	0.2279
	BLUE2	0.3213	0.2579	0.2145	0.1878	0.2142	0.2290	0.2376	0.2431
$v = 5$	Bias								
	v^*	2.1	2.5	3	4	5[+]	10	15	20
$p = 1$	BLUE0	0.1796	0.1239	0.0807	0.0300	0.0000	−0.0590	−0.0752	−0.0803
	BLUE1	0.1215	0.0952	0.0681	0.0279	0.0000	−0.0610	−0.0783	−0.0837
	BLUE2	0.1052	0.0845	0.0620	0.0265	0.0000	−0.0639	−0.0836	−0.0899
$p = 2$	BLUE0	0.3748	0.2556	0.1646	0.0603	0.0000	−0.1182	−0.1537	−0.1687
	BLUE1	0.2744	0.2055	0.1423	0.0565	0.0000	−0.1220	−0.1601	−0.1759
	BLUE2	0.2312	0.1807	0.1296	0.0537	0.0000	−0.1274	−0.1706	−0.1888
$p = 3$	BLUE0	0.5836	0.3944	0.2516	0.0911	0.0000	−0.1769	−0.2329	−0.2590
	BLUE1	0.4519	0.3283	0.2218	0.0857	0.0000	−0.1824	−0.2424	−0.2701
	BLUE2	0.3817	0.2897	0.2027	0.0817	0.0000	−0.1895	−0.2574	−0.2894

Table 3 (Contd.)

$v = 5$	Mean Square Error								
	v^*	2.1	2.5	3	4	5^+	10	15	20
$p = 1$	BLUE0	0.7630	0.2766	0.2111	0.1745	0.1612	0.1478	0.1481	0.1494
	BLUE1	0.2408	0.2211	0.2031	0.1807	0.1687	0.1541	0.1545	0.1560
	BLUE2	0.2427	0.2284	0.2142	0.1947	0.1827	0.1659	0.1662	0.1681
$p = 2$	BLUE0	1.4786	0.4414	0.2803	0.0901	0.1612	0.1416	0.1459	0.1497
	BLUE1	0.4086	0.3178	0.2555	0.1951	0.1687	0.1472	0.1522	0.1565
	BLUE2	0.3568	0.3049	0.2604	0.2088	0.1827	0.1576	0.1631	0.1686
$p = 3$	BLUE0	2.3223	0.6606	0.3699	0.2081	0.1612	0.1427	0.1566	0.1667
	BLUE1	0.6916	0.4658	0.3279	0.2119	0.1687	0.1482	0.1638	0.1751
	BLUE2	0.5577	0.4235	0.3242	0.2252	0.1827	0.1577	0.1756	0.1897

$n = 20$									
$v = 3$	Bias								
	v^*	2.1	2.5	3^+	8	13	18	23	28
$p = 1$	BLUE0	0.0255	0.0123	0.0000	−0.0439	−0.0532	−0.0557	−0.0560	−0.0553
	BLUE2	0.0236	0.0117	0.0000	−0.0447	−0.0542	−0.0567	−0.0570	−0.0563
	BLUE4	0.0220	0.0112	0.0000	−0.0465	−0.0568	−0.0595	−0.0597	−0.0590
$p = 2$	BLUE0	0.0517	0.0247	0.0000	−0.0874	−0.1072	−0.1142	−0.1170	−0.1181
	BLUE2	0.0478	0.0236	0.0000	−0.0889	−0.1092	−0.1163	−0.1191	−0.1202
	BLUE4	0.0447	0.0225	0.0000	−0.0925	−0.1145	−0.1221	−0.1250	−0.1261
$p = 3$	BLUE0	0.0784	0.0373	0.0000	−0.1303	−0.1611	−0.1733	−0.1793	−0.1826
	BLUE2	0.0728	0.0357	0.0000	−0.1325	−0.1641	−0.1765	−0.1826	−0.1859
	BLUE4	0.0680	0.0340	0.0000	−0.1377	−0.1720	−0.1853	−0.1917	−0.1952
$p = 4$	BLUE0	0.1058	0.0500	0.0000	−0.1724	−0.2143	−0.2319	−0.2413	−0.2469
	BLUE2	0.0984	0.0479	0.0000	−0.1753	−0.2183	−0.2362	−0.2457	−0.2514
	BLUE4	0.0921	0.0457	0.0000	−0.1819	−0.2288	−0.2480	−0.2580	−0.2640

$v = 3$	Mean Square Error								
	v^*	2.1	2.5	3^+	8	13	18	23	28
$p = 1$	BLUE0	0.0988	0.0940	0.0911	0.0857	0.0860	0.0866	0.0870	0.0873
	BLUE2	0.0988	0.0956	0.0928	0.0872	0.0876	0.0882	0.0886	0.0889
	BLUE4	0.1033	0.1003	0.0976	0.0916	0.0920	0.0927	0.0931	0.0935
$p = 2$	BLUE0	0.1084	0.0973	0.0911	0.0843	0.0867	0.0882	0.0890	0.0895
	BLUE2	0.1063	0.0987	0.0928	0.0858	0.0883	0.0899	0.0908	0.0913
	BLUE4	0.1103	0.1034	0.0976	0.0899	0.0929	0.0947	0.0957	0.0962
$p = 3$	BLUE0	0.1199	0.1011	0.0911	0.0867	0.0933	0.0969	0.0989	0.1002
	BLUE2	0.1155	0.1022	0.0928	0.0884	0.0952	0.0990	0.1011	0.1024
	BLUE4	0.1188	0.1067	0.0976	0.0927	0.1005	0.1048	0.1072	0.1086
$p = 4$	BLUE0	0.1333	0.1052	0.0911	0.0929	0.1057	0.1127	0.1168	0.1194
	BLUE2	0.1265	0.1061	0.0928	0.0948	0.1082	0.1154	0.1197	0.1224
	BLUE4	0.1289	0.1105	0.0976	0.0995	0.1148	0.1231	0.1278	0.1308

$v = 5$	Bias								
	v^*	2.1	2.5	3	4	5^+	10	15	20
$p = 1$	BLUE0	0.0798	0.0567	0.0378	0.0145	0.0000	−0.0306	−0.0405	−0.0445
	BLUE2	0.0560	0.0445	0.0323	0.0135	0.0000	−0.0317	−0.0421	−0.0464
	BLUE4	0.0492	0.0399	0.0296	0.0129	0.0000	−0.0332	−0.0449	−0.0498
$p = 2$	BLUE0	0.1641	0.1156	0.0766	0.0291	0.0000	−0.0611	−0.0812	−0.0901
	BLUE2	0.1177	0.0920	0.0660	0.0272	0.0000	−0.0632	−0.0845	−0.0939
	BLUE4	0.1027	0.0823	0.0605	0.0259	0.0000	−0.0661	−0.0901	−0.1006
$p = 3$	BLUE0	0.2527	0.1767	0.1163	0.0438	0.0000	−0.0914	−0.1219	−0.1361
	BLUE2	0.1856	0.1427	0.1010	0.0411	0.0000	−0.0944	−0.1268	−0.1417

Table 3 (Contd.)

		2.1	2.5	3	4	5	10	15	20
	BLUE4	0.1608	0.1274	0.0926	0.0392	0.0000	−0.0985	−0.1350	−0.1519
p = 4	BLUE0	0.3452	0.2398	0.1568	0.0586	0.0000	−0.1215	−0.1624	−0.1821
	BLUE2	0.2596	0.1965	0.1372	0.0551	0.0000	−0.1253	−0.1689	−0.1896
	BLUE4	0.2238	0.1752	0.1260	0.0526	0.0000	−0.1305	−0.1796	−0.2031
v = 5		Mean Square Error							
	v^*	2.1	2.5	3	4	5^+	10	15	20
p = 1	BLUE0	0.1581	0.0952	0.0849	0.0779	0.0750	0.0718	0.0717	0.0720
	BLUE2	0.0926	0.0890	0.0855	0.0809	0.0783	0.0747	0.0747	0.0750
	BLUE4	0.0967	0.0939	0.0911	0.0870	0.0843	0.0803	0.0802	0.0806
p = 2	BLUE0	0.2600	0.1240	0.0984	0.0813	0.0750	0.0704	0.0717	0.0729
	BLUE2	0.1174	0.1058	0.0957	0.0840	0.0783	0.0733	0.0747	0.0760
	BLUE4	0.1167	0.1082	0.1003	0.0900	0.0843	0.0785	0.0802	0.0818
p = 3	BLUE0	0.3826	0.1623	0.1157	0.0852	0.0750	0.0710	0.0751	0.0780
	BLUE2	0.1559	0.1296	0.1091	0.0876	0.0783	0.0738	0.0784	0.0816
	BLUE4	0.1460	0.1280	0.1122	0.0935	0.0843	0.0789	0.0842	0.0883
p = 4	BLUE0	0.5278	0.2107	0.1370	0.0896	0.0750	0.0735	0.0818	0.0875
	BLUE2	0.2111	0.1617	0.1259	0.0917	0.0783	0.0764	0.0856	0.0919
	BLUE4	0.1870	0.1542	0.1270	0.0975	0.0743	0.0815	0.0923	0.1000

$^+$ This is the I.I.D. case ($p = 0$).

multiple-outlier results to conclude that the BLUE is more efficient than the MLE of the scale parameter σ of a one-parameter Pareto distribution. And we have also found that the censored BLUE's (based on 10% censoring) of the location parameter μ and the scale parameter σ of a two-parameter Pareto distribution are both quite robust to the presence of multiple outliers.

Acknowledgements

The authors would like to thank the Natural Sciences and Engineering Research Council of Canada for funding this research.

Appendix A

PROOF OF RELATION 3.3. From Equation (1.2), let us consider for $3 \leq s \leq n$

$$\mu_{1,s:n} = \frac{1}{(s-2)!(n-s)!} \sum_p \int_1^\infty \int_x^\infty x y f_{i_1}(x) \prod_{b=2}^{s-1} \{F_{i_b}(y) - F_{i_b}(x)\} f_{i_s}(y)$$

$$\times \prod_{c=s+1}^n \{1 - F_{i_c}(y)\} \, dy \, dx$$

$$= \frac{1}{(s-2)!(n-s)!} \sum_p \int_1^\infty x f_{i_1}(x) I_1(x) \, dx \,, \tag{A.1}$$

where

$$I_1(x) = \int_x^\infty y \prod_{b=2}^{s-1}\{F_{i_b}(y) - F_{i_b}(x)\} f_{i_s}(y) \prod_{c=s+1}^n \{1 - F_{i_c}(y)\} \, dy$$

$$= v_{i_s} \int_x^\infty \prod_{b=2}^{s-1}\{F_{i_b}(y) - F_{i_b}(x)\} \prod_{c=s}^n \{1 - F_{i_c}(y)\} \, dy$$

upon using (1.5). Integrating now by parts yields

$$I_1(x) = v_{i_s} \left[-\int_x^\infty y \sum_{\substack{j=2 \\ b \neq j}}^{s-1} f_{i_j}(y) \prod_{b=2}^{s-1}\{F_{i_b}(y) - F_{i_b}(x)\} \prod_{c=s}^n \{1 - F_{i_c}(y)\} \, dy \right.$$

$$\left. + \int_x^\infty y \prod_{b=2}^{s-1}\{F_{i_b}(y) - F_{i_b}(x)\} \sum_{j=s}^n f_{i_j}(y) \prod_{\substack{c=s \\ c \neq j}}^n \{1 - F_{i_c}(y)\} \, dy \right]$$

which, when substituted in (A.1), gives

$$\mu_{1,s:n} = \frac{1}{(s-2)!(n-s)!} \sum_p v_{i_s} \left[-\int_1^\infty \int_1^y xy f_{i_1}(x) \sum_{j=2}^{s-1} f_{i_j}(y) \right.$$

$$\times \prod_{\substack{b=2 \\ b \neq j}}^{s-1} \{F_{i_b}(y) - F_{i_b}(x)\} \prod_{c=s}^n \{1 - F_{i_c}(y)\} \, dy \, dx + \int_1^\infty \int_1^y xy f_{i_1}(x)$$

$$\left. \times \prod_{b=2}^{s-1}\{F_{i_b}(y) - F_{i_b}(x)\} \sum_{j=s}^n f_{i_j}(y) \prod_{\substack{c=s \\ c \neq j}}^n \{1 - F_{i_c}(y)\} \, dy \, dx \right]. \quad (A.2)$$

Alternatively, we may write

$$\mu_{1,s:n} = \frac{1}{(s-2)!(n-s)!} \sum_p \int_1^\infty \int_1^y xy f_{i_1}(x)$$

$$\times \prod_{b=2}^{s-1}\{F_{i_b}(y) - F_{i_b}(x)\} f_{i_s}(y) \prod_{c=s+1}^n \{1 - F_{i_c}(y)\} \, dx \, dy$$

$$= \frac{1}{(s-2)!(n-s)!} \sum_p \int_1^\infty y f_{i_s}(y) \prod_{c=s+1}^n \{1 - F_{i_c}(y)\} I_2(y) \, dy, \quad (A.3)$$

where

$$I_2(y) = \int_1^y x f_{i_1}(x) \prod_{b=2}^{s-1} \{F_{i_b}(y) - F_{i_b}(x)\} \, dx$$

$$= v_{i_1} \int_1^y \{1 - F_{i_c}(x)\} \prod_{b=2}^{s-1} \{F_{i_b}(y) - F_{i_b}(x)\} \, dx$$

upon using (1.5). Integrating now by parts yields

$$I_2(y) = v_{i_1}\left[-\prod_{b=2}^{s-1} F_{i_b}(y) + \int_1^y x f_{i_1}(x) \prod_{b=2}^{s-1}\{F_{i_b}(y) - F_{i_b}(x)\}\,dx\right.$$
$$\left.+ \int_1^y x\{1 - F_{i_1}(x)\} \sum_{j=2}^{s-1} f_{i_j}(x) \prod_{\substack{b=2\\b\neq j}}^{s-1}\{F_{i_b}(y) - F_{i_b}(x)\}\,dx\right]$$

which, when substituted in (A.3), gives

$$\mu_{1,s:n} = \frac{1}{(s-2)!(n-s)!} \sum_p v_{i_1}$$

$$\times \left[-\int_1^\infty y \prod_{b=2}^{s-1} F_{i_b}(y) f_{i_s}(y) \prod_{c=s+1}^n \{1 - F_{i_c}(y)\}\,dy\right.$$

$$+ \int_1^\infty \int_1^y xy f_{i_1}(x) \prod_{b=2}^{s-1}\{F_{i_b}(y) - F_{i_b}(x)\} f_{i_s}(y) \prod_{c=s+1}^n \{1 - F_{i_c}(y)\}\,dx\,dy$$

$$+ \int_1^\infty \int_1^y xy\{1 - F_{i_1}(x)\} \sum_{j=2}^{s-1} f_{i_j}(x)$$

$$\left.\times \prod_{\substack{b=2\\b\neq j}}^{s-1}\{F_{i_b}(y) - F_{i_b}(x)\} f_{i_s}(y) \prod_{c=s+1}^n \{1 - F_{i_c}(y)\}\,dx\,dy\right].$$

We then split the third term in the above sum through the term $\{1 - F_{i_1}(x)\} = \{F_{i_1}(y) - F_{i_1}(x)\} + \{1 - F_{i_1}(y)\}$ to get

$$\mu_{1,s:n} = \frac{1}{(s-2)!(n-s)!} \sum_p v_{i_1} \times \left[-\int_1^\infty y \prod_{b=2}^{s-1} F_{i_b}(y) f_{i_s}(y) \prod_{c=s+1}^n \{1 - F_{i_c}(y)\}\,dy\right.$$

$$+ \int_1^\infty \int_1^y xy f_{i_1}(x) \prod_{b=2}^{s-1}\{F_{i_b}(y) - F_{i_b}(x)\} f_{i_s}(y) \prod_{c=s+1}^n \{1 - F_{i_c}(y)\}\,dx\,dy$$

$$+ \int_1^\infty \int_1^y xy \sum_{j=2}^{s-1} f_{i_j}(x) \prod_{\substack{b=1\\b\neq j}}^{s-1}\{F_{i_b}(y) - F_{i_b}(x)\} f_{i_s}(y) \prod_{c=s+1}^n \{1 - F_{i_c}(y)\}\,dx\,dy$$

$$+ \int_1^\infty \int_1^y xy\{1 - F_{i_1}(y)\} \sum_{j=2}^{s-1} f_{i_j}(x)$$

$$\left.\times \prod_{\substack{b=2\\b\neq j}}^{s-1}\{F_{i_b}(y) - F_{i_b}(x)\} f_{i_s}(y) \prod_{c=s+1}^n \{1 - F_{i_c}(y)\}\,dx\,dy\right].$$

(A.4)

We now add the expressions for $\mu_{1,s:n}$ in (A.2) and (A.4) and simplify the resulting equation to get

$$2\mu_{1,s:n} = \left(\sum_{i=1}^{n} v_i\right)\mu_{1,s:n} - \sum_{i+1}^{n} v_i \mu_{s-1:n-1}^{[i]}.$$

Relation 3.3 is derived simply by rewriting the above equation.

PROOF OF RELATION 3.4. From Equation (1.2), let us consider for $2 \leq r < s \leq n$ and $s - r \geq 2$

$$\mu_{r,s:n} = \frac{1}{(r-1)!(s-r-1)!(n-s)!} \sum_p \int_1^\infty \int_x^\infty xy \prod_{a=1}^{r-1} F_{i_a}(x) f_{i_r}(x)$$

$$\times \prod_{b=r+1}^{s-1} \{F_{i_b}(y) - F_{i_b}(x)\} f_{i_s}(y) \prod_{c=s+1}^{n} \{1 - F_{i_c}(y)\}\, dy\, dx$$

$$= \frac{1}{(r-1)!(s-r-1)!(n-s)!} \sum_p \int_1^\infty x \prod_{a=1}^{r-1} F_{i_a}(x) f_{i_r}(x) I_1(x)\, dx,$$

(A.5)

where

$$I_1(x) = \int_x^\infty y \prod_{b=r+1}^{s-1} \{F_{i_b}(x) - F_{i_b}(x)\} f_{i_s}(y) \prod_{c=s+1}^{n} \{1 - F_{i_c}(y)\}\, dy$$

$$= v_{i_s} \int_x^\infty \prod_{b=r+1}^{s-1} \{F_{i_b}(y) - F_{i_b}(x)\} \prod_{c=s}^{n} \{1 - F_{i_c}(y)\}\, dy$$

upon using (1.5). Integrating now by parts yields

$$I_1(x) = v_{i_s}\left[-\int_x^\infty y \sum_{\substack{j=r+1 \\ b \neq j}}^{s-1} f_{i_j}(y) \prod_{b=r+1}^{s-1} \{F_{i_b}(y) - F_{i_b}(x)\} \prod_{c=s}^{n} \{1 - F_{i_c}(y)\}\, dy\right.$$

$$\left.+ \int_x^\infty y \prod_{b=r+1}^{s-1} \{F_{i_b}(y) - F_{i_b}(x)\} \sum_{j=s}^{n} f_{i_j}(y) \prod_{\substack{c=s \\ c \neq j}}^{n} \{1 - F_{i_c}(y)\}\, dy\right]$$

which, when substituted in (A.5), gives

$$\mu_{r,s:n} = \frac{1}{(r-1)!(s-r-1)!(n-s)!} \sum_p v_{i_s} \left[-\int_1^\infty \int_x^\infty xy \prod_{a=1}^{r-1} F_{i_a}(x) f_{i_r}(x) \right.$$

$$\times \sum_{j=r+1}^{s-1} f_{i_j}(y) \prod_{\substack{b=r+1 \\ b \neq j}}^{s-1} \{F_{i_b}(y) - F_{i_b}(x)\} \prod_{c=s}^{n} \{1 - F_{i_c}(y)\} \, dy \, dx$$

$$+ \int_1^\infty \int_x^\infty xy \prod_{a=1}^{r-1} F_{i_a}(x) f_{i_r}(x) \prod_{b=r+1}^{s-1} \{F_{i_b}(y) - F_{i_b}(x)\}$$

$$\left. \times \sum_{j=s}^{n} f_{i_j}(y) \prod_{\substack{c=s \\ c \neq j}}^{n} \{1 - F_{i_c}(y)\} \, dy \, dx \right]. \tag{A.6}$$

Alternatively, we may write

$$\mu_{r,s:n} = \frac{1}{(r-1)!(s-r-1)!(n-s)!} \sum_p \times \int_1^\infty \int_1^y xy \prod_{a=1}^{r-1} F_{i_a}(x) f_{i_r}(x)$$

$$\times \prod_{b=r+1}^{s-1} \{F_{i_b}(y) - F_{i_b}(x)\} f_{i_s}(y) \prod_{c=s+1}^{n} \{1 - F_{i_c}(y)\} \, dx \, dy$$

$$= \frac{1}{(r-1)!(s-r-1)!(n-s)!}$$

$$\times \sum_p \int_1^\infty y f_{i_s}(y) \prod_{c=s+1}^{n} \{1 - F_{i_c}(y)\} I_2(y) \, dy, \tag{A.7}$$

where

$$I_2(y) = \int_1^y x \prod_{a=1}^{r-1} F_{i_a}(x) f_{i_r} \prod_{b=r+1}^{s-1} \{F_{i_b}(y) - F_{i_b}(x)\} \, dx$$

$$= v_{i_r} \int_1^y \prod_{a=1}^{r-1} F_{i_a}(x) \{1 - F_{i_r}(x)\} \prod_{b=r+1}^{s-1} \{F_{i_b}(y) - F_{i_b}(x)\} \, dx$$

upon using (1.5). Integrating now by parts yields

$$I_2(y) = v_{i_r} \left[-\int_1^y x \sum_{j=1}^{r-1} f_{i_j}(x) \prod_{\substack{a=1 \\ a \neq j}}^{r-1} F_{i_a}(x) \{1 - F_{i_r}(x)\} \prod_{b=r+1}^{s-1} \{F_{i_b}(y) - F_{i_b}(x)\} \, dx \right.$$

$$+ \int_1^y x \prod_{a=1}^{r-1} F_{i_a}(x) f_{i_r}(x) \prod_{b=r+1}^{s-1} \{F_{i_b}(y) - F_{i_b}(x)\} \, dx$$

$$\left. + \int_1^y x \prod_{a=1}^{r-1} F_{i_a}(x) \{1 - F_{i_r}(x)\} \sum_{j=r+1}^{s-1} f_{i_j}(x) \prod_{\substack{b=r+1 \\ b \neq j}}^{s-1} \{F_{i_b}(y) - F_{i_b}(x)\} \, dx \right]$$

which, when substituted in (A.7), gives

$$\mu_{r,s:n} = \frac{1}{(r-1)!(s-r-1)!(n-2)!} \sum_{p} v_{i_r} \left[-\int_1^\infty \int_1^y xy \sum_{j=1}^{r-1} f_{i_j}(x) \right.$$

$$\times \prod_{\substack{a=1 \\ a \neq j}}^{r-1} F_{i_a}(x)\{1-F_{i_r}(x)\} \prod_{b=r+1}^{s-1} \{F_{i_b}(y)-F_{i_b}(x)\}f_{i_s}(y) \prod_{c=s+1}^{n} \{1-F_{i_c}(y)\} \, dx \, dy$$

$$+ \int_1^\infty \int_1^y xy \prod_{a=1}^{r-1} F_{i_a}(x)f_{i_r}(x) \prod_{b=r+1}^{s-1} \{F_{i_b}(y) - F_{i_b}(x)\}f_{i_s}(y)$$

$$\times \prod_{c=s+1}^{n} \{1-F_{i_c}(y)\} \, dx \, dy + \int_1^\infty \int_1^y xy \prod_{a=1}^{r-1} F_{i_a}(x)\{1-F_{i_r}(x)\} \sum_{j=r+1}^{s-1} f_{i_j}(x)$$

$$\times \prod_{\substack{b=r+1 \\ b \neq j}}^{s-1} \{F_{i_b}(y) - F_{i_b}(x)\}f_{i_s}(y) \prod_{c=s+1}^{n} \{1 - F_{i_c}(y)\} \, dx \, dy \right].$$

We then split the first term in the above sum through $\{1 - F_{i_r}(x)\}$ and the third term through $\{1 - F_{i_r}(x)\} = \{F_{i_r}(y) - F_{i_r}(x)\} + \{1 - F_{i_r}(y)\}$ to get

$$\mu_{r,s:n} = \frac{1}{(r-1)!(s-r-1)!9n-s)!} \sum_{p} v_{i_r}$$

$$\times \left[-\int_1^\infty \int_1^y xy \sum_{j=1}^{r-1} f_{i_j}(x) \prod_{\substack{a=1 \\ a \neq j}}^{r-1} F_{i_a}(x) \prod_{b=r+1}^{s-1} \{F_{i_b}(y)-F_{i_b}(x)\}f_{i_s}(y) \right.$$

$$\times \prod_{c=s+1}^{n} \{1-F_{i_c}(y)\} \, dx \, dy + \int_1^\infty \int_1^y xy \sum_{j=1}^{r-1} f_{i_j}(x) \prod_{\substack{a=1 \\ a \neq j}}^{r-1} F_{i_a}(x)$$

$$\times \prod_{b=r+1}^{s-1} \{F_{i_b}(y) - F_{i_b}(x)\}f_{i_s}(y) \prod_{c=s+1}^{n} \{1 - F_{i_c}(y)\} \, dx \, dy$$

$$+ \int_1^\infty \int_1^y xy \prod_{a=1}^{r-1} F_{i_a}(x)f_{i_r}(x) \prod_{b=r+1}^{s-1} \{F_{i_b}(y) - F_{i_b}(x)\}f_{i_s}(y)$$

$$\times \prod_{c=s+1}^{n} \{1-F_{i_c}(y)\} \, dx \, dy + \int_1^\infty \int_1^y xy \prod_{a=1}^{r-1} F_{i_a}(x) \sum_{j=r+1}^{s-1} f_{i_j}(x)$$

$$\times \prod_{\substack{b=r \\ b \neq j}}^{s-1}\{F_{i_b}(y) - F_{i_b}(x)\}f_{i_s}(y) \prod_{c=s+1}^{n} \{1 - F_{i_c}(y)\} \, dx \, dy$$

$$+ \int_1^\infty \int_1^y xy \prod_{a=1}^{r-1} F_{i_a}(x)\{1 - F_{i_r}(y)\} \sum_{j=r+1}^{s-1} f_{i_j}(x)$$

$$\left. \times \prod_{\substack{b=r+1 \\ b \neq j}}^{s-1} \{F_{i_b}(y) - F_{i_b}(x)\}f_{i_s}(y) \prod_{c=s+1}^{n} \{1 - F_{i_c}(y)\} \, dx \, dy \right]. \quad (A.8)$$

We now add the expressions for $\mu_{r,s:n}$ in (A.6) and (A.8) and simplify the resulting equation to get

$$2\mu_{r,s:n} = \left(\sum_{i=1}^{n} v_i\right)\mu_{r,s:n} - \sum_{i=1}^{n} v_i \mu_{r-1,s-1:n-1}^{[i]}.$$

Relation 3.4 is derived simply by rewriting the above equation.

Appendix B

PROOF OF RELATION (b). For $2 \leq r \leq n-1$ and $k = 0, 1, 2, \ldots$, we use (1.1) and then (5.3) to write

$$\mu_{r:n}^{(k)} = \frac{1}{(r-1)!(n-r)!}\sum_{p}\int_{L}^{U} x^k \prod_{a=1}^{r-1} F_{i_a}(x)f_{i_r}(x) \prod_{b=r+1}^{n} \{1 - F_{i_b}(x)\}\,dx$$

$$= \frac{1}{(r-1)!(n-r)!}\sum_{p} v_{i_r}\int_{L}^{U} x^{k-1}\prod_{a=1}^{r-1} F_{i_a}(x)\{C_{i_r} - F_{i_r}(x)\}$$

$$\times \prod_{b=r+1}^{n} \{1 - F_{i_b}(x)\}\,dx.$$

Integrating now by parts treating x^{k-1} for integration gives

$$k\mu_{r:n}^{(k)} = \frac{1}{(r-1)!(n-r)!}\sum_{p}\left[-v_{i_r}\int_{L}^{U} x^k \sum_{j=1}^{r-1} f_{i_j}(x)\prod_{\substack{a=1 \\ a\neq j}}^{r-1} F_{i_a}(x)\{C_{i_r} - F_{i_r}(x)\}\right.$$

$$\times \prod_{b=r+1}^{n}\{1 - F_{i_b}(x)\}\,dx + v_{i_r}\int_{L}^{U} x^k \prod_{a=1}^{r-1} F_{i_a}(x)f_{i_r}(x)\prod_{b=r+1}^{n}\{1 - F_{i_b}(x)\}\,dx$$

$$\left. + v_{i_r}\int_{L}^{U} x^k \prod_{a=1}^{r-1} F_{i_a}(x)\{C_{i_r} - F_{i_r}(x)\}\sum_{j=r+1}^{n} f_{i_j}(x))\prod_{\substack{b=r+1 \\ b\neq j}}^{n}\{1 - F_{i_b}(x)\}\,dx\right].$$

We then split the first term in the above sum into two through $\{C_{i_r} - F_{i_r}(x)\}$ and the third term through $\{C_{i_r} - F_{i_r}(x)\} = \{1 - F_{i_r}(x)\} + \{C_{i_r} - 1\}$ to get

$$k\mu_{r:n}^{(k)} = \frac{1}{(r-1)!(n-r)!} \sum_P \left[-C_{i_r} v_{i_r} \int_L^U x^k \sum_{j=1}^{r-1} f_{i_j}(x) \right.$$

$$\times \prod_{\substack{a=1 \\ a \neq j}}^{r-1} F_{i_a}(x) \prod_{b=r+1}^{n} \{1 - F_{i_b}(x)\} \, dx.$$

$$+ v_{i_r} \int_L^U x^k \sum_{j=1}^{r-1} f_{i_j}(x) \prod_{\substack{a=1 \\ a \neq j}}^{r} F_{i_a}(x) \prod_{b=r+1}^{n} \{1 - F_{i_b}(x)\} \, dx$$

$$+ v i_r \int_L^U x^k \prod_{a=1}^{r-1} F_{i_a}(x) f_{i_r}(x) \prod_{b=r+1}^{n} \{1 - F_{i_b}(x)\} \, dx$$

$$+ v_{i_r} \int_L^U x^k \prod_{a=1}^{r-1} F_{i_a}(x) \sum_{j=r+1}^{n} f_{i_j}(x) \prod_{\substack{b=r \\ b \neq j}}^{n} \{1 - F_{i_b}(x)\} \, dx$$

$$+ \{C_{i_r} - 1\} v_{i_r} \int_L^U x^k \prod_{a=1}^{r-1} F_{i_a}(x) \sum_{j=r+1}^{n} f_{i_j}(x) \prod_{\substack{b=r+1 \\ b \neq j}}^{n} \{1 - F_{i_b}(x)\} \, dx \right]$$

$$= -\sum_{i=1}^{n} C_i v_i \mu_{r-1:n-1}^{[i](k)} + \left(\sum_{i=1}^{n} v_i \right) \mu_{r:n}^{(k)} - \sum_{i=1}^{n} v_i (1 - C_i) \mu_{r:n-1}^{[i](k)} .$$

Relation (b) is derived simply by rewriting the above equation.

PROOF OF RELATION (i). From Equation (1.2), let us consider for $2 \leq r < s \leq n-1$ and $s - r \geq 2$

$$\mu_{r,s:n} = \frac{1}{(r-1)!(s-r-1)!(n-s)!} \sum_P$$

$$\times \int_L^U \int_x^U xy \prod_{a=1}^{r-1} F_{i_a}(x) f_{i_r}(x) \prod_{b=r+1}^{s-1} \{F_{i_b}(y) - F_{i_b}(x)\} f_{i_s}(y)$$

$$\times \prod_{c=s+1}^{n} \{1 - F_{i_c}(y)\} \, dy \, dx$$

$$= \frac{1}{(r-1)!(s-r-1)!(n-s)!} \sum_P \int_L^U x \prod_{a=1}^{r-1} F_{i_a}(x) f_{i_r}(x) I_1(x) \, dx , \quad \text{(B.1)}$$

where

$$I_1(x) = \int_x^U y \prod_{b=r+1}^{s-1} \{F_{i_b}(y) - F_{i_b}(x)\} f_{i_s}(y) \prod_{c=s+1}^{n} \{1 - F_{i_c}(y)\} \, dy$$

$$= v_{i_s} \int_x^U \prod_{b=r+1}^{s-1} \{F_{i_b}(y) - F_{i_b}(x)\} \{C_{i_s} - F_{i_s}(y)\} \prod_{c=s+1}^{n} \{1 - F_{i_c}(y)\} \, dy$$

upon using (5.3). Integrating now by parts yields

$$I_1(x) = v_{i_s}\Bigg[-\int_x^U y \sum_{j=r+1}^{s-1} f_{i_j}(y) \prod_{\substack{b=r+1\\b\neq j}}^{s-1} \{F_{i_b}(y) - F_{i_b}(x)\}\{C_{i_s} - F_{i_s}(y)\}$$

$$\times \prod_{c=s+1}^n \{1 - F_{i_c}(y)\}\,dy$$

$$+ \int_x^U y \prod_{b=r+1}^{s-1} \{F_{i_b}(y) - F_{i_b}(x)\} f_{i_s}(y) \prod_{c=s+1}^n \{1 - F_{i_c}(y)\}\,dy$$

$$+ \int_x^U y \prod_{b=r+1}^{s-1} \{F_{i_b}(y) - F_{i_b}(x)\}\{C_{i_s} - F_{i_s}(y)\}$$

$$\times \sum_{j=s+1}^n f_{i_j}(y) \prod_{\substack{c=s+1\\c\neq j}}^n \{1 - F_{i_c}(y)\}\,dy\Bigg]$$

which, when substituted in (B.1), gives

$$\mu_{r,s:n} = \frac{1}{(r-1)!(s-r-1)!(n-s)!}$$

$$\times \sum_P v_{i_s}\Bigg[-\int_L^U \int_x^U xy \prod_{a=1}^{r-1} F_{i_a}(x) f_{i_r}(x) \sum_{j=r+1}^{s-1} f_{i_j}(y)$$

$$\times \prod_{\substack{b=r+1\\b\neq j}}^{s-1} \{F_{i_b}(y) - F_{i_b}(x)\}\{C_{i_s} - F_{i_s}(y)\} \prod_{c=s+1}^n \{1 - F_{i_c}(y)\}\,dy\,dx$$

$$+ \int_L^U \int_x^U xy \prod_{a=1}^{r-1} F_{i_a}(x) f_{i_r}(x)$$

$$\times \prod_{b=r+1}^{s-1} \{F_{i_b}(y) - F_{i_b}(x)\} f_{i_s}(y) \prod_{c=s+1}^n \{1 - F_{i_c}(y)\}\,dy\,dx$$

$$+ \int_L^U \int_x^U xy \prod_{a=1}^{r-1} F_{i_a}(x) f_{i_r}(x)$$

$$\times \prod_{b=r+1}^{s-1} \{F_{i_b}(y) - F_{i_b}(x)\}\{C_{i_s} - F_{i_s}(y)\}$$

$$\times \sum_{j=s+1}^n f_{i_j}(y) \prod_{\substack{c=s+1\\c\neq j}}^n \{1 - F_{i_c}(y)\}\,dy\,dx\Bigg].$$

We then split the third term in the above sum through the term $\{C_{i_s} - F_{i_s}(y)\} = \{1 - F_{i_s}(y)\} + \{C_{i_s} - 1\}$ to get

$$\mu_{r,s:n} = \frac{1}{(r-1)!(s-r-1)!(n-s)!}$$

$$\times \sum_p v_{i_s} \left[-\int_L^U \int_x^U xy \prod_{a=1}^{r-1} F_{i_a}(x) f_{i_r}(x) \sum_{j=r+1}^{s-1} f_{i_j}(y) \right.$$

$$\times \prod_{\substack{b=r+1 \\ b \neq j}}^{s-1} \{F_{i_b}(y) - F_{i_b}(x)\} \{C_{i_s} - F_{i_s}(y)\} \prod_{c=s+1}^n \{1 - F_{i_c}(y)\} \, dy \, dx$$

$$+ \int_L^U \int_x^U xy \prod_{a=1}^{r-1} F_{i_a}(x) f_{i_r}(x)$$

$$\times \prod_{b=r+1}^{s-1} \{F_{i_b}(y) - F_{i_b}(x)\} f_{i_s}(y) \prod_{c=s+1}^n \{1 - F_{i_c}(y)\} \, dy \, dx$$

$$+ \int_L^U \int_x^U xy \prod_{a=1}^{r-1} F_{i_a}(x) f_{i_r}(x)$$

$$\times \prod_{b=r+1}^{s-1} \{F_{i_b} - F_{i_b}(x)\} \sum_{j=s+1}^n f_{i_j}(y) \prod_{\substack{c=s \\ c \neq j}}^n \{1 - F_{i_c}(y)\} \, dy \, dx$$

$$+ \{C_{i_s} - 1\} \int_L^U \int_x^U xy \prod_{a=1}^{r-1} F_{i_a}(x) f_{i_r}(x)$$

$$\times \prod_{b=r+1}^{s-1} \{F_{i_b}(y) - F_{i_b}(x)\} \sum_{j+s+1}^n f_{i_j}(y) \prod_{\substack{c=s+1 \\ c \neq j}}^n \{1 - F_{i_c}(y)\} \, dy \, dx \right]. \quad (B.2)$$

Alternatively, we may write

$$\mu_{r,s:n} = \frac{1}{(r-1)!(s-r-1)!(n-s)!} \sum_p \int_L^U \int_x^U xy \prod_{a=1}^{r-1} F_{i_a}(x) f_{i_r}(x)$$

$$\times \prod_{b=r+1}^{s-1} \{F_{i_b}(y) - F_{i_b}(x)\} f_{i_s}(y) \prod_{c=s+1}^n \{1 - F_{i_c}(y)\} \, dx \, dy$$

$$= \frac{1}{(r-1)!(s-r-1)!(n-s)!} \sum_p \int_L^U y f_{i_s}(y) \prod_{c=s+1}^n \{1 - F_{i_c}(y)\} I_2(y) \, dy,$$

$$(B.3)$$

where

$$I_2(y) = \int_L^y x \prod_{a=1}^{r-1} F_{i_a}(x) f_{i_r}(x) \prod_{b=r+1}^{s-1} \{F_{i_b}(y) - F_{i_b}(x)\} \, dx$$

$$= v_{i_r} \int_L^y \prod_{a=1}^{r-1} F_{i_a}(x) \{C_{i_r} - F_{i_r}(x)\} \prod_{b=r+1}^{s-1} \{F_{i_b}(y) - F_{i_b}(x)\} \, dx$$

upon using (5.3). Integrating now by parts yields

$$I_2(y) = v_{i_r} \Bigg[-\int_L^y x \sum_{\substack{j=1 \\ a \neq j}}^{r-1} f_{i_j}(x) \prod_{a=1}^{r-1} F_{i_a}(x) \{C_{i_r} - F_{i_r}(x)\} \prod_{b=r+1}^{s-1} \{F_{i_b}(y) - F_{i_b}(x)\} \, dx$$

$$+ \int_L^y x \prod_{a=1}^{r-1} F_{i_a}(x) f_{i_r}(x) \prod_{b=r+1}^{s-1} \{F_{i_b}(y) - F_{i_b}(x)\} \, dx$$

$$+ \int_L^y x \prod_{a=1}^{r-1} F_{i_a}(x) \{C_{i_r} - F_{i_r}(x)\} \sum_{j=r+1}^{s-1} f_{i_j}(x) \{F_{i_b}(y) - F_{i_b}(x)\} \, dx \Bigg]$$

which, when substituted in (B.3), gives

$$\mu_{r,s:n} = \frac{1}{(r-1)!(s-r-1)!(n-s)!} \sum_p v_{i_r}$$

$$\Bigg[-\int_L^U \int_L^y xy \sum_{\substack{j=1 \\ a \neq j}}^{r-1} f_{i_j}(x) \prod_{a=1}^{r-1} F_{i_a}(x) \{C_{i_r} - F_{i_r}(x)\} \cdot$$

$$\times \prod_{b=r+1}^{s-1} \{F_{i_b}(y) - F_{i_b}(x)\} f_{i_s}(y) \prod_{c=s+1}^{n} \{1 - F_{i_c}(y)\} \, dx \, dy$$

$$+ \int_L^U \int_L^y xy \prod_{a=1}^{r-1} F_{i_a}(x) f_{i_r}(x) \prod_{b=r+1}^{s-1} \{F_{i_b}(y) - F_{i_b}(x)\} f_{i_s}(y)$$

$$\times \prod_{c=s+1}^{n} \{1 - F_{i_c}(y)\} \, dx \, dy$$

$$+ \int_L^U \int_L^y xy \prod_{a=1}^{r-1} F_{i_a}(x) \{C_{i_r} - F_{i_r}(x)\} \sum_{j=r+1}^{s-1} f_{i_j}(x)$$

$$\times \prod_{\substack{b=r+1 \\ b \neq j}}^{s-1} \{F_{i_b}(y) - F_{i_b}(x)\} f_{i_s}(y) \prod_{c=s+1}^{n} \{1 - F_{i_c}(y)\} \, dx \, dy \Bigg].$$

We then split the first term in the above sum through $\{C_{i_r} - F_{i_r}(y)\}$ and the third term through $\{C_{i_r} - F_{i_r}(x)\} = \{F_{i_r}(y) - F_{i_r}(x)\} + \{C_{i_r} - F_{i_r}(y)\}$ to get

$$\mu_{r,s:n} = \frac{1}{(r-1)!(s-r-1)!(n-s)!} \sum_p v_{i_r}$$

$$\times \left[-C_{i_r} \int_L^U \int_L^y xy \sum_{\substack{j=1 \\ a \neq j}}^{r-1} f_{i_j}(x) \prod_{a=1}^{r-1} F_{i_a}(x) \prod_{b=r+1}^{s-1} \{F_{i_b}(y) \right.$$

$$-F_{i_b}(x)\} f_{i_s}(y) \prod_{c=s+1}^{n} \{1 - F_{i_c}(y)\} \, dx \, dy$$

$$+ \int_L^U \int_L^y xy \sum_{\substack{j=1 \\ a \neq j}}^{r-1} f_{i_j}(x) \prod_{a=1}^{r} F_{i_a}(x) \prod_{b=r+1}^{s-1} \{F_{i_b}(y) - F_{i_b}(x)\} f_{i_s}(y)$$

$$\prod_{c=s+1}^{n} \{1 - F_{i_c}(y)\} \, dx \, dy$$

$$+ \int_L^U \int_L^y xy \prod_{a=1}^{r-1} F_{i_a}(x) f_{i_r}(x) \prod_{\substack{b=r+1 \\ b \neq j}}^{s-1} \{F_{i_b}(y) - F_{i_b}(x)\} f_{i_s}(y)$$

$$\prod_{c=s+1}^{n} \{1 - F_{i_c}(y)\} \, dx \, dy$$

$$+ \int_L^U \int_L^y xy \prod_{a=1}^{r-1} F_{i_a}(x) \{C_{i_r} - F_{i_r}(y)\} \sum_{j=r+1}^{s-1} f_{i_j}(x)$$

$$\prod_{\substack{b=r+1 \\ b \neq j}}^{s-1} \{F_{i_b}(y) - F_{i_b}(x)\} f_{i_s}(y) \prod_{c=s+1}^{n} \{1 - F_{i_c}(y)\} \, dx \, dy$$

$$+ \int_L^U \int_L^y xy \prod_{a=1}^{r-1} F_{i_a}(x) \{C_{i_r} - F_{i_r}(y)\} \sum_{j=r+1}^{s-1} f_{i_j}(x)$$

$$\left. \times \prod_{\substack{b=r+1 \\ b \neq j}}^{s-1} \{F_{i_b}(y) - F_{i_b}(x)\} f_{i_s}(y) \prod_{c=s+1}^{n} \{1 - F_{i_c}(y)\} \, dx \, dy \right].$$

(B.4)

We now add the expression for $\mu_{r,s:n}$ in (B.2) and (B.4) and simplify the resulting equation to get

$$2\mu_{r,s:n} = \left(\sum_{i=1}^{n} v_i\right) \mu_{r,s:n} - \sum_{i=1}^{n} C_i v_i \mu_{r-1,s-1:n-1}^{[i]} - \sum_{i=1}^{n} (1 - C_i) v_i \mu_{r,s:n-1}^{[i]}.$$

Relation (i) is derived simply by rewriting the above equation.

References

Arnold, B. C. (1983). *Pareto Distributions*. International Cooperative Publishing House, Fairland, MD.

Arnold, B. C. (1994). Discussion of Balakrishnan, N. *Comput. Statist. Data Anal.* **18**, 203–253.

Arnold, B. C., N. Balakrishnan and H. N. Nagaraja (1992). *A First Course in Order Statistics*. John Wiley & Sons, New York.

Balakrishnan, N. (1988). Recurrence relations for order statistics from n independent and non-identically distributed random variables. *Ann. Inst. Statist. Math.* **40**, 273–277.

Balakrishnan, N. (1994a). Order statistics from non-identical exponential random variables and some applications (with discussion). *Commun. Statist. Data Anal.* **18**, 203–253.

Balakrishnan, N. (1994b). On order statistics from non-identical right-truncated exponential random variables and some applications. *Commun. Statist. – Theory Meth.* **23**, 3373–3393.

Balakrishnan, N. and K. Balasubramanian (1995). Order statistics from non-identical power function random variables. *Computn. Statist. – Theory Meth.* **24**, 1443–1454.

Balakrishnan, N., S. M. Bendre and H. J. Malik (1992). General relations and identities for order statistics from non-independent non-identical variables. *Ann. Inst. Statist. Math.* **44**, 177–183.

Balakrishnan, N. and P. C. Joshi (1982). Moments of order statistics from doubly truncated Pareto distribution. *J. Ind. Statist. Assoc.* **20**, 109–117.

Barnett, V. and T. Lewis (1994). *Outliers in Statistical Data*, Third edition. John Wiley & Sons, Chichester, England.

Childs, A. and N. Balakrishnan (1995a). Generalized recurrence relations for moments of order statistics from non-identical doubly-truncated exponential random variables. *Preprint*.

Childs, A. and N. Balakrishnan (1995b). Relations for single moments of order statistics from non-identical logistic random variables and assessment of the effect of multiple outliers on the bias of linear estimators of location and scale. *Preprint*.

David, H. A. (1979). Robust estimation in the presence of outliers. In *Robustness in Statistics* (Eds., R. L. Launer and G. N. Wilkinson). Academic Press, New York.

David, H. A. (1981). *Order Statistics*, Second edition. John Wiley & Sons, New York.

Huang, J. S. (1975). A note on order statistics from the Pareto distribution. *Scand. Act. J.* **2**, 187–190.

Johnson N. L., S. Kotz and N. Balakrishnan (1994). *Continuous Univariate Distributions*, Vol. 1, Second edition. John Wiley & Sons, New York.

Kulldroff, G. and K. Vännman (1973). Estimation of the location and scale parameters of a Pareto distributions by linear function of order statistics. *J. Amer. Statist. Assoc.* **68**, 218–227.

Likeš, J. (1969). Minimum variance unbiased estimates of the parameters of power-function and Pareto's distribution. *Statistische Hefte* **10**, 104–110.

Vaughan, R. J. and W. N. Venables (1972). Permanent expressions for order statistics densities. *J. Roy. Statist. Soc. Ser. B* **34**, 308–310.

PART VII

Resampling Methods

A Semiparametric Bootstrap for Simulating Extreme Order Statistics

Robert L. Strawderman and Daniel Zelterman

1. Introduction

The study of extreme order statistics has long been a concern of statisticians. Recently, it has become increasingly important with its application to environmental and public health issues. A high rate of cancer or a major flood are headlines that attract public attention, not the average rainfall or typical tumor risk. The Dutch government, for example, has legislated that levees and sea dikes must have a one in 10,000 chance of failure. This has provoked a series of statistical studies about what height is adequate protection (Dekkers and de Haan, 1989). Unusually high rates of childhood leukemia or birth defects are cause for alarm (Lagakos, Wessen, and Zelen, 1986). Other recent examples are the estimation of the maximum concentration of airborne pollutants in a metropolitan area (Smith, 1989) or the maximum exposure to radioactivity released by a nuclear power plant (Davison and Smith, 1990).

Exact distributions of extreme order statistics are typically difficult to obtain, and consequently the associated asymptotic theory has been both extensively investigated and employed in practice. David (1981), Serfling (1980), and Reiss (1989) are excellent general references on order statistics. Let $X_{1n} \geq X_{2n} \geq \cdots \geq X_{nn}$ denote the descending order statistics of a random sample of size n from a population with cumulative distribution function (CDF) $F(\cdot)$. Hereafter, we shall suppress the subscript n, its presence throughout being implied. The CDF $F(x)$, is said to be in the *domain of attraction* of the distribution function $G(x)$ if there exist sequences of real numbers $a_n > 0$, b_n, $n = 1, 2, \ldots$ such that for all real x,

$$\lim_{n \to \infty} F^n((x - b_n)/a_n) = G(x) \ . \tag{1}$$

The functional form of the limit $G(\cdot)$ (up to location and scale parameters) is either degenerate, or a member of the Type I, II, or III Gumbel family of distribution functions (Gnedenko, 1943). The Type I Gumbel family is generated from the base distribution $G(x) = \exp\{-e^{-x}\}, x \in \mathbb{R}$, and plays an important role in this paper; hereafter, we denote this family of distributions by $\mathscr{G}$.

Specification of the functional form of $G(\cdot)$ in (1) places a restriction on the extreme tail of $F(\cdot)$ only, and thus constitutes a *semiparametric* restriction on $F(\cdot)$. For example, many distributions, such as the normal, lognormal, gamma, Pareto, Gompertz, Weibull, Gumbel, and the logistic distributions, are all in the domain of attraction of $\mathcal{G}$. In addition, if a random variable X has a distribution $F(\cdot)$ in the domain of attraction of either the Type I Gumbel or Type II Gumbel distributions, then the distribution function of $\log X$ is also a member of $\mathcal{G}$ (cf. Davis and Resnick, 1984). In fact, it can be shown that each of the Gumbel families are contained within the richer family of *Generalized Extreme Value* distributions $\{G_\gamma : \gamma \in \mathbb{R}\}$ (cf. Dekkers and de Haan, 1989), where

$$G_\gamma(x) = \exp\left\{-(1+\gamma x)^{-1/\gamma}\right\} \quad x \in \mathbb{R} ; \tag{2}$$

$G_\gamma(x)$ is also known as the Generalized Pareto distribution (GPD; cf. Pickands, 1975). The GPD $G_\gamma(\cdot)$ is often used to model high level exceedances (e.g., Smith, 1989), but can also be used to model the behavior of X_1; Smith (1989) is a good introduction to this distribution. Other models for the extreme tail of a distribution are given in Hill (1975), Davis and Resnick (1984), Zelterman (1992), and Hsing (1993).

The use of *spacings*, or differences between adjacent order statistics, has been prevalent in the study of order statistics; Pyke (1965) is a seminal reference. Weissman (1978) used the sample spacings to motivate an estimator for extreme quantiles. Define

$$d_i = i(X_i - X_{i+1}), \quad i = 1, \ldots, k \tag{3}$$

as the *normalized sample spacings* between the k largest order statistics from a sample of size n. The spacings $\mathbf{d} = \{d_1, \ldots, d_k\}$ are independent for any n and choice of k if and only if $\{X_1, \ldots, X_k\}$ are the order statistics from an exponential distribution (Sukhatme, 1937; Pyke, 1965). For arbitrary distributions in $\mathcal{G}$, Weissman (1978, Theorem 3) proved that $\{a_n^{-1}d_1, \ldots, a_n^{-1}d_k\}$ are asymptotically independent unit exponential random variables for a suitably chosen sequence $a_n > 0$ and n much larger than k. Weissman (1978, Theorem 2) also showed that for some sequence $\{b_n, n \geq 1\}$, the random variables

$$\hat{M}_j = a_n^{-1}(X_j - b_n), \quad j = 1, \ldots, k$$

converge jointly in distribution to a k-dimensional extremal variate; the associated (asymptotic) marginal densities are

$$g_j(m) = \frac{\exp\{-e^{-m} - jm\}}{(j-1)!} \tag{4}$$

for $j = 1, \ldots, k$ and all real m. For any distribution in $\mathcal{G}$, Weissman (1978) proposes to estimate the $100(1 - c/n)^{\text{th}}$ percentile of $F(\cdot)$ by

$$\hat{\eta}_c = \tilde{b} - \tilde{a} \log c , \tag{5}$$

where $c > 0$ is much smaller than n,

$$\tilde{a} = \overline{X}_k - X_{k+1} \quad \text{and} \quad \tilde{b} = \tilde{a}\Psi(k+2) + X_{k+1} , \tag{6}$$

$\overline{X}_k$ is the sample mean of the k largest order statistics, and $\Psi(\cdot)$ is the digamma function (cf. Abramowitz and Stegun, 1972). In (6), $\tilde{a}$ and $\tilde{b}$ are respectively the minimum variance unbiased estimators for a_n and b_n assuming $(\hat{M}_1, \ldots, \hat{M}_k)'$ to be a k–dimensional extremal variate (Weissman, 1978, Theorem 5). This essentially corresponds to using

$$\tilde{X}_j = \tilde{a}M_j + \tilde{b} , \quad j = 1, \ldots, k , \tag{7}$$

as a stochastic approximation to X_j, where M_j has density (4). Boos (1984) compared the percentile estimator (5) to those available in SAS and empirically demonstrated that (5) could provide a substantial improvement in the estimation of percentiles beyond $p = 0.95$ when the tails of the parent distribution $F(\cdot)$ behave approximately as an exponential distribution. Boos (1984) also discussed empirical criteria for determining values of k for which the spacings **d** behave similarly to those from an exponential distribution. His study of "where the tail begins" was limited to large sample sizes (e.g., $n \geq 500$), and compared the approximation $P\{\tilde{X}_j \geq x | \tilde{a}, \tilde{b}\}$ to the true tail probability $P\{X_j \geq x\}$. This approximation may fail for moderate n, and we will consider alternative methods which rely less upon these parametric asymptotic approximations.

The methods to be discussed here are simulation-based and make use of bootstrap techniques. The bootstrap, introduced by Efron (1979), is a flexible technique that can be applied to a wide variety of problems. Introductions to the bootstrap are the monograph by Efron (1982) or Efron and Gong (1983). More recent reviews include DiCiccio and Romano (1988), Hinkley (1988), Efron and Tibshirani (1993), and Young (1994). The major appeal of the nonparametric bootstrap is that approximate samples from the distribution of a random variable can be obtained without specifying a parametric form for its distribution. The usefulness of this ability cannot be overestimated: the distribution, significance levels, bias, variance, etc… can all be approximated in a nonparametric framework. However, a drawback is that the bootstrap distribution of a statistic is a *theoretical* quantity that usually must be approximated. Generally, this is done via Monte Carlo simulation, although there has been some recent work on the application of saddlepoint methods (Davison and Hinkley, 1988). The primary appeal of the saddlepoint method is that Monte Carlo simulation is largely avoided. We will elaborate further on this approach in Section 3.

Recent work on quantile estimation using bootstrap techniques includes Davison (1988), Johns (1988), and Do and Hall (1991). None of these references give special attention to extreme quantiles, which necessarily involve the extreme order statistics of the sample. It is reasonable to expect that the usual nonparametric bootstrap will fare poorly as a method for estimating extreme quantiles of

$F(\cdot)$ since resampling the observed data will result in pseudo-samples with values no larger than X_1 or smaller than X_n. More specifically, suppose $X_1^*, \ldots, X_n^*$ is a bootstrap sample taken with replacement from the (ordered) data $X_1 \geq \cdots \geq X_n$, where each X_i has probability n^{-1} of appearing in the bootstrap sample. Then, under the bootstrap probability measure it follows that

$$P\left\{\max_i X_i^* = X_1\right\} = 1 - (1 - n^{-1})^n \ ,$$

which converges to $1 - \mathrm{e}^{-1} \doteq 0.63$ as $n \to \infty$. That is to say, even for large sample sizes, the naive bootstrap distribution of the largest order statistic will have a large point-mass at X_1. For additional discussion on problems associated with bootstrapping the largest order statistic, see Bickel and Freedman (1981, Section 6) and Loh (1984).

This chapter examines a semiparametric bootstrap approximation to the marginal distributions of the k largest order statistics in a sample of size n from $F(\cdot)$, where k is much smaller than n. The distribution $F(\cdot)$ is assumed unknown; however, it is also assumed that $F(\cdot)$ is a member of $\mathscr{G}$, hence the name semiparametric. The bootstrap approximation, proposed by Zelterman (1993) and summarized in Section 2, involves resampling the *normalized sample spacings* **d**. For a fixed value of k, the bootstrap selects k values from **d** with replacement, "un-normalizes" these bootstrapped values, and then adds their sum to X_{k+1} to simulate the behavior of $X_1, \ldots, X_k$. The motivation behind this bootstrap method is that when properly normalized the sample spacings **d** behave approximately as independent and identically distributed exponential random variables when the sample size n is large. Further details on the bootstrap technique are given in Section 2, where issues regarding the choice of k are also discussed. In Section 3, we propose a saddlepoint approximation to the bootstrap distribution. Traditional saddlepoint methods in statistics (e.g., Daniels, 1954) rely on the standard normal distribution, and do not really apply here since the limiting distribution of the normalized bootstrap statistic is extreme value, not standard normal. Instead, we employ the results of Wood et al. (1993) and develop a tail probability approximation that is appropriate for random variables having this non-normal limit distribution. In Section 4, we show how to implement the approximation using S-plus. In Section 5, we investigate the expansion empirically and demonstrate that this bootstrap approximation can be an accurate approximation to $P\{X_1 \leq x\}$ for an appropriately chosen value of k. We end in Section 6 with an investigation of the British Coal Mining data (Andrews and Herzberg, 1985, pp. 51–56).

2. A semiparametric bootstrap approximation to X_j

Zelterman (1993) describes a way to approximate the distribution of X_j via the nonparametric bootstrap. In order to understand how this is done, it is useful to write the jth largest order statistic X_j as a linear function of the normalized sample spacings $\mathbf{d} = \{d_1, \ldots, d_k\}$ defined in (3) and the order statistic X_{k+1}. Specifically, note that

$$X_j = X_{k+1} + \sum_{i=j}^{k} i^{-1} d_i$$

for any $k \geq 1$ and $j > k+1$. Based on this identity, Zelterman (1993) proposes

$$X_j^* = X_{k+1} + \sum_{i=j}^{k} i^{-1} d_i^*$$

as a stochastic approximation to X_j, where $\{d_1^*, \ldots, d_k^*\}$ is a bootstrap sample taken with replacement from $\{d_1, \ldots, d_k\}$. The validity of this bootstrap relies on the property that $\{d_1, \ldots, d_k\}$ are approximately independent and identically distributed. For distributions in $\mathcal{G}$, the normalized spacings $\mathbf{d}$ are approximately independent and exponentially distributed when the value of k is appropriately small relative to n.

The moment generating function for the bootstrap variate X_j^* (conditionally on $\mathbf{d}$ and X_{k+1}) is

$$\mathrm{E}\left(\exp(tX_j^*) \big| \mathbf{d}, X_{k+1}\right) = \exp(tX_{k+1}) \prod_{i=j}^{k} k^{-1} \sum_{r=1}^{k} \exp(td_r/i) , \qquad (8)$$

and is derived in Zelterman (1993). Using (8),

$$\mu_j^* = \mathrm{E}\left(X_j^* \big| \mathbf{d}, X_{k+1}\right) = \tilde{a} S_{j:k} + X_{k+1} \qquad (9)$$

and

$$(\sigma_j^*)^2 = \mathrm{Var}\left(X_j^* \big| \mathbf{d}, X_{k+1}\right) = s_d^2 (1 - k^{-1}) S_{j:k}^{(2)} , \qquad (10)$$

where $S_{m:k}^{(w)} = \sum_{r=m}^{k} r^{-w}$, $S_{m:k} = S_{m:k}^{(1)}$, $\tilde{a} = \overline{X}_k - X_{k+1}$, and s_d^2 is the sample variance of $\mathbf{d}$. It can also be shown that

$$\mathrm{Cov}\left(X_j^*, X_{j'}^* \big| \mathbf{d}, X_{k+1}\right) = \mathrm{Var}\left(X_{j \wedge j'}^* \big| \mathbf{d}, X_{k+1}\right) ;$$

that is, the proposed bootstrap approximation reflects the well-known Markovian relationship between X_j and $X_{j'}$ (cf David, 1981, Section 2.7).

Weissman (1978) uses $\tilde{X}_j = \tilde{a} M_j + \tilde{b}$ as a parametric model for estimating extreme quantiles of $F(\cdot)$, where M_j is a random variable with density function given by (4) and $\tilde{a}$ and $\tilde{b}$ are defined in (6). Conditional on the values of the fitted parameters $\tilde{a}$ and $\tilde{b}$, the first two moments of $\tilde{X}_j$ are

$$\mathrm{E}\left(\tilde{X}_j \big| \tilde{a}, \tilde{b}\right) = \tilde{a} S_{j:k} + X_{k+1} ,$$

and

$$\mathrm{Var}\left(\widetilde{X}_j \big| \tilde{a}, \tilde{b}\right) = \tilde{a}^2 \sum_{i=j}^{\infty} i^{-2} = \tilde{a}^2 \left(\pi^2/6 - \sum_{i=1}^{j-1} i^{-2}\right).$$

Since $\mathrm{E}(\widetilde{X}_j|\tilde{a},\tilde{b}) = \mu_j^*$, the parametric model and bootstrap approximation for X_j have the same conditional mean. The variance $\mathrm{Var}(\widetilde{X}_j|\tilde{a},\tilde{b})$ is not equal to $(\sigma_j^*)^2$, but the difference becomes negligible as $n, k \to \infty$ since the spacings become independent exponential random variables. In addition, if the normalized spacings are exponentially distributed with mean $a = \mathrm{E}(\tilde{a})$, then the order of the difference can be expressed in terms of a and k, and is given by

$$\mathrm{Var}\left(\widetilde{X}_j \big| X_{k+1}\right) - \mathrm{Var}\left(X_j^* \big| X_{k+1}\right) = O(a^2 k^{-1}).$$

Zelterman (1993) proves that the unconditional asymptotic distributions of X_j^* and $\widetilde{X}_j$ (properly normalized) coincide as $k, n \to \infty$. This motivates the use of $P\{X_j^* \le x | \mathbf{d}, X_{k+1}\}$ and $P\{\widetilde{X}_j \le x | \tilde{a}, \tilde{b}\}$ as approximations for $P\{X_j \le x\}$ when n is large and $k \ll n$. We show via simulation in Section 5 that $P\{X_j^* \le x | \mathbf{d}, X_{k+1}\}$ can be an excellent approximation to $P\{X_j \le x\}$ for relatively small n when k is chosen appropriately.

The asymptotic validity of the bootstrap method relies upon the normalized spacings $\mathbf{d}$ in (3) being an approximately *iid* sample from an exponential distribution. The quality of the bootstrap approximation is therefore sensitive to the choice of k. Finding the "optimal" value of k is very difficult since it is equivalent to determining where the tail of the distribution begins. We propose four methods of choosing k, and compare them in Section 5 using simulated data. The four methods are summarized in Table 1.

Two of the four proposed methods of choosing k use the *Gini statistic*

$$G_k = G_k(\mathbf{d}) = [k(k-1)\bar{d}]^{-1} \sum_{i=1}^{k-1} i(k-i)\left(d_{(i)k} - d_{(i+1)k}\right)$$

where $d_{(j)k}$ is the jth largest of the k normalized spacings $\mathbf{d}$ (Gail and Gastwirth, 1978). This statistic was proposed as way to test the null hypothesis "the data are exponentially distributed", and exploits the fact that the spacings between exponentially distributed observations are themselves exponentially distributed (cf. Pyke, 1965, Section 2.3). Under the null hypothesis, G_k is approximately dis-

Table 1
Methods used to estimate k

Method	Description
I	$k = \min\{m : p_{m+1} < 0.25\}$
II	$k = \mathrm{argmax}_j\{p_j\}$
III	$k = \min\{m : |\tilde{\gamma}_{m+1}|\sqrt{m+1} > 1.96\}$
IV	$k = \mathrm{argmin}_j\{(\tilde{\gamma}_j - R_j^{(1)})^2\}$

tributed as normal with mean $\frac{1}{2}$ and variance $[12(k-1)]^{-1}$ when $k > 5$; the exponential model is therefore rejected for large values of G_k.

Our first proposed method of choosing k is to use this test sequentially, examining successive values of k until the test rejects. The second method is to pick the value of k which maximizes the significance level

$$p_k = 1 - \Phi\{(G_k - 0.5)[12(k-1)]^{1/2}\} ,$$

where $\Phi(x)$ is the standard normal cumulative distribution function.
The Generalized Pareto distribution (GPD) defined in (2) can also be used to motivate the choice of k. The limit of

$$G_\gamma(x) = \exp\{-(1+\gamma x)^{-1/\gamma}\}$$

as $\gamma \to 0$, say $G_0(x)$, is the Type I Gumbel distribution. Under sufficient regularity conditions, any consistent estimator for γ converges to zero when $F(\cdot)$ is in $\mathscr{G}$. Dekkers et al. (1989) propose

$$\tilde{\gamma}_k = R_k^{(1)} + 1 - \frac{1}{2}\left\{1 - \frac{(R_k^{(1)})^2}{R_k^{(2)}}\right\}^{-1} ,$$

where

$$R_k^{(j)} = k^{-1}\sum_{i=1}^{k}(\log X_i - \log X_{k+1})^j .$$

They prove that (i) $\tilde{\gamma}_k$ is strongly consistent for γ when $k = o(n)$ and $k(\log n)^{-\delta} \to \infty$ for some $\delta > 0$ (Theorem 2.1); and, (ii) $\tilde{\gamma}_k \stackrel{.}{\sim} N(0, k^{-1})$ when $F(\cdot)$ is in $\mathscr{G}$ (Corollary 3.2). Our third proposal for estimating k will be to choose the first value of k for which $|\tilde{\gamma}_{k+1}|\sqrt{k+1} > 1.96$. This corresponds to choosing the largest k for which the exponential model (i.e., $\gamma = 0$) is not rejected by the data.

Our fourth method for choosing k is based upon the relationship between $\tilde{\gamma}_k$ and $R_k^{(1)}$. The statistic $R_k^{(1)}$ has the interpretation of being the empirical mean residual life of the log-transformed data. Hill (1975) proposed this estimator as part of the tail probability approximation

$$\check{F}(x) = 1 - \frac{k}{n}(x/X_{k+1})^{-1/R_k^{(1)}} \tag{11}$$

for $x \geq X_{k+1}$. The right-hand side of (11) is derived as the MLE assuming that the data follow (2) for $x \geq X_{k+1}$. The estimator in (11) was also motivated by Davis and Resnick (1984) from a less parametric point of view, where they prove that $\check{F}(x)$ is consistent as $n, k = o(n) \to \infty$. The results of Davis and Resnick (1984, Theorem 5.1) also imply that $R_k^{(1)} \stackrel{P}{\to} \gamma$. Combining these results with those of Dekkers et al. (1989), we see that

$$\tilde{\gamma}_k - R_k^{(1)} \xrightarrow{P} 0$$

whenever $F(\cdot)$ is in $\mathcal{G}$. This suggests choosing the value of k such that $(\tilde{\gamma}_k - R_k^{(1)})^2$ is minimized, or equivalently when $R_k^{(2)}/(R_k^{(1)})^2$ is closest to 2.

This fourth criteria has another important theoretical justification. It is known that the distribution function $F(\cdot)$ of a random variable X is in $\mathcal{G}$ if and only if

$$\lim_{t \to t_o} \frac{E[(X-t)^2 | X > t]}{(E[(X-t)|X > t])^2} = 2 \qquad (12)$$

where $t_o = \sup\{t: F(t) < 1\}$ (Balkema and de Haan, 1974). The ratio $R_k^{(2)}/(R_k^{(1)})^2$ is the empirical version of (12) for the log-transformed random variable, and converges in probability to 2 for any random variable in the domain of attraction of the Type I Gumbel family (Dekkers et al., 1989, p. 1840). This criteria cannot be used as a *test* of whether $F(\cdot)$ is in $\mathcal{G}$ since it can be shown that $R_k^{(2)}/(R_k^{(1)})^2 \xrightarrow{P} 2$ for distributions in the Type II Gumbel family as well. However, the fact that we are already assuming that $F(\cdot)$ is a member of $\mathcal{G}$ allows us to use it as way to choose the value of k "most consistent" with the model corresponding to $\gamma = 0$. We investigate these four methods of choosing k further in Section 5.

3. A saddlepoint approximation to the bootstrap distribution

In order to approximate $P\{X_j \leq x\}$ via $P\{X_j^* \leq x | \mathbf{d}, X_{k+1}\}$, we must find or approximate the distribution of the latter. The bootstrap distribution of a statistic is a theoretical quantity based upon an infinitely large number of resamples from the original data set. Most practical applications of the bootstrap involve Monte Carlo simulation, done occasionally for convenience but almost always because of the intractable functional form of the exact distribution function. It is easy to let the computer generate the bootstrap distribution, but the final result may be unwieldy if functionals of the distribution are required as well.

In situations where something is known about functionals of the true bootstrap distribution, we may take advantage of that information. Suppose that the statistic of interest, say T_n, has the general representation

$$T_n = t(\hat{F}) = t(F) + n^{-1} \sum_i L(X_i) , \qquad (13)$$

where $\hat{F}(\cdot)$ is the empirical distribution function, the form of $t(\cdot)$ doesn't depend upon n, and $L(\cdot)$ is a linear function of $X_1, \ldots, X_n$ which is not identically zero. Statistics satisfying (13) are known as *linear statistical functionals* and include the sample mean, M-estimators, and any other statistic having the general representation $t(G) = \int w(x) \, dG(x)$. The corresponding bootstrap version of T_n, say T_n^*, has the representation $t(F^*) = t(\hat{F}) + n^{-1} \sum_i L(X_i^*)$. One of the strengths of the bootstrap is that the distribution of $T_n - t(F)$ is often well-approximated by the bootstrap distribution of $T_n^* - t(\hat{F})$. Desirable features of the latter are that the probability mechanism generating the data (i.e., $\hat{F}$) is *known* and that the statistic $T_n^* - t(\hat{F})$ has the form of a sample mean. Davison and Hinkley (1988) exploit these facts to derive a simple method for accurately approximating the true bootstrap distribution of $T_n^* - t(\hat{F})$ using the theory of saddlepoint approximations developed for sums of i.i.d random variables (Daniels, 1954).

Daniels (1954) used the method of steepest descent to derive the saddlepoint approximation to the density of a sample mean. Lugannani and Rice (1980) take a similar approach in deriving an approximation for the *CDF*. A more "statistical" method for deriving such expansions which achieves the same end result is the *tilted Edgeworth expansion* (cf Barndorff–Nielsen and Cox, 1989, Section 4.3). In all cases, the derivation of these expansions incorporate an underlying assumption that the distribution of the statistic in question is asymptotically normal. This assumption is implicit in Davison and Hinkley (1988) as well. Traditional saddlepoint methods for approximating the distribution of X_j^* are not applicable here since the limiting distribution of X_j^* is extreme value, not normal. Wood et al. (1993) consider expansions in which the base distribution is not the standard normal distribution function. They do not, however, consider applications of their method in the context of the bootstrap. We will apply their results to construct an approximation to the conditional CDF of X_j^* given **d** and X_{k+1}.

Let $Z_1, Z_2, \ldots$ denote a sequence of random variables having the limit distribution $\Upsilon(x)$, possibly up to location and scale parameters. Suppose $K(t) = K_n(t)$ is the cumulant generating function (CGF) of Z_n and let $G(\cdot)$ denote the CGF corresponding to $\Upsilon(\cdot)$. Then, the following expansion, derived by Wood et al. (1993), may be used to approximate $P\{Z_n \geq x\}$:

$$P\{Z_n \geq x\} \approx 1 - \Upsilon(\hat{\xi}) + v(\hat{\xi})\left\{\hat{u}^{-1} - w_{\hat{\xi}}^{-1}\right\}, \tag{14}$$

where $v(\cdot)$ is the density function corresponding to $\Upsilon(\cdot)$. Although not explicit in the notation, it is important to realize that each of $\hat{u}$, $w_{\hat{\xi}}$, and $\hat{\xi}$ depend upon x; we describe below how to determine the value of each for every x.

The dependence on x of $\hat{u}$, $w_{\hat{\xi}}$, and $\hat{\xi}$ will be through the *saddlepoint* $\hat{t}_x$, which we define as the solution to $K'(\hat{t}_x) = x$. For any x, $\hat{t}_x$ maximizes $K(t) - tx$, and $K(\hat{t}_x) - x\hat{t}_x \leq 0$. Now, for any real ξ, let w_ξ be defined as the unique solution to $G'(w_\xi) = \xi$; thus, w_ξ is a function of ξ. Given this functional relationship, let $\hat{\xi}$ solve

$$G(w_{\hat{\xi}}) - G'(w_{\hat{\xi}})w_{\hat{\xi}} = K(\hat{t}_x) - x\hat{t}_x. \tag{15}$$

The left-hand side, or $G(w_\xi) - G'(w_\xi)w_\xi$, is the negative of the *Legendre–Fenchel transformation* of $G(\cdot)$ (cf McCullagh, 1987, p. 174), and is consequently a non-positive concave function of ξ. Since $K(\hat{t}_x) - x\hat{t}_x \leq 0$, it follows that (15) has at least one ($\xi = G'(0)$) and at most two solutions ($\xi_- < G'(0)$ and $\xi_+ > G'(0)$). More specifically, if

$$x < K'(0) \quad \text{then set} \quad \hat{\xi} = \xi_-$$
$$x = K'(0) = \mathrm{E}(Z_n) \quad \text{then set} \quad \hat{\xi} = G'(0)$$
$$x > K'(0) \quad \text{then set} \quad \hat{\xi} = \xi_+ .$$

The value of $w_{\hat{\xi}}$ is then determined by the equation $G'(w_{\hat{\xi}}) = \hat{\xi}$. Given $\hat{t}_x$, $\hat{\xi}$, and $w_{\hat{\xi}}$, set

$$\hat{u} = \hat{t}_x [K''(\hat{t}_x)]^{1/2} \left[G''(w_{\hat{\xi}}) \right]^{-1/2} \tag{16}$$

in (14). The expansion (14) is valid for $x \neq \mathrm{E}(Z_n)$; there is an analogous formula for $x = \mathrm{E}(Z_n)$ (Wood et al., 1993, Eq. 9). If $\Upsilon(x) = \Phi(x)$, then (14) reduces to the usual Lugannani–Rice tail probability formula. Further details of this inter-relationship may be found in Wood et al. (1993).

To approximate the distribution of X_j^* using (14), we need the CGF of X_j^*. From (8), the conditional CGF of X_j^* given **d** and X_{k+1} is

$$K(t) = t X_{k+1} + \sum_{i=j}^{k} \log \left(k^{-1} \sum_{r=1}^{k} \exp\{t d_r / i\} \right) . \tag{17}$$

Although not explicit in the notation, it is important to note that $K(t)$ depends upon the k largest observations in a sample of size n. The first and second derivatives of $K(t)$ are respectively

$$K'(t) = X_{k+1} + \sum_{i=j}^{k} \frac{h_{1i}(t)}{h_{0i}(t)}$$

and

$$K''(t) = \sum_{i=j}^{k} \left[\frac{h_{2i}(t)}{h_{0i}(t)} - \left(\frac{h_{1i}(t)}{h_{0i}(t)} \right)^2 \right] ,$$

where

$$h_{si}(t) = \sum_{r=1}^{k} \left(\frac{d_r}{i} \right)^s \exp\{t d_r / i\} .$$

The higher-order derivatives of $K(t)$ have similar representations. The saddlepoint $\hat{t}_x$ must be found numerically since no closed form solution exists to the equation $K'(t) = x$. Due to the asymptotic relationship between X_j^* and $\widetilde{X}_j = \tilde{b} + \tilde{a} M_j$, the relevant base distribution $\Upsilon_j(\cdot)$ to be used in the expansion (14) is the CDF of $\widetilde{X}_j$. Using (4), it can be shown that

$$\Upsilon_j(x) = \frac{1}{\Gamma(j)} \int_{e^{-v(x)}}^{\infty} z^{j-1} e^{-z} dz ,$$

where $v(x) = (x - \tilde{b})/\tilde{a}$. For $j = 1$, this integral simplifies considerably. In particular, we have

$$\Upsilon_1(x) = \exp\left\{-\exp\left(-\frac{x - \tilde{b}}{\tilde{a}}\right)\right\},$$

where $\Upsilon_1(x)$ is the CDF of $\tilde{a}M_1 + \tilde{b}$ and M_1 has the Type I Gumbel (extreme value) distribution.

The CGF corresponding to $\Upsilon_j(x)$ is

$$G_j(t) = \log \mathrm{E}\left(\exp\{t\tilde{X}_j\}\big|\tilde{a}, \tilde{b}\right) = t\tilde{b} + \log \mathrm{E}\left(\exp\{t\tilde{a}M_j\}\big|\tilde{a}, \tilde{b}\right).$$

Since

$$\mathrm{E}(\exp\{\beta M_j\}) = \Gamma(j - \beta)/\Gamma(j)$$

for $j - \beta > 0$, it follows that

$$G_j(t) = t\tilde{b} + \log \Gamma(j - \tilde{a}t) - \log \Gamma(j) \tag{18}$$

for $t < j/\tilde{a}$. The functions $\Upsilon_j(x)$ and $G_j(t)$ play the role of $\Upsilon(x)$ and $G(t)$ in the calculation of (14). To compute (14) at any value of x, the only remaining task is to determine $\hat{\xi}$, $w_{\hat{\xi}}$, and $\hat{u}$. These must obtained numerically, and an algorithm for doing so is described in Section 4.

REMARK. The manner in which X_j^* is constructed forces it to have a distribution with finite support. More specifically, $P\{X^* \leq x | \mathbf{d}, X_{k+1}\}$ is equal to zero for x less than $x_{min} = X_{k+1} + S_{(j+1):k} d_{min}$ and equal to one for x greater than $x_{max} = X_{k+1} + S_{(j+1):k} d_{max}$, where d_{min} and d_{max} are respectively the smallest and largest elements of $\mathbf{d}$ (Zelterman, 1993). The limits of the saddlepoint approximation are defined accordingly.

4. Numerical implementation

We concentrate here on describing the algorithm used for approximating $P\{X_1^* \leq x | \mathbf{d}, X_{k+1}\}$ via the saddlepoint method of Section 3. The same algorithm applies for approximating the CDF's of other order statistics with obvious modifications. Software for computing the saddlepoint approximation described in Section 3 has been written using a combination of the statistical software language S-plus and FORTRAN 77. Algorithms used for calculating the log-gamma function and its first two derivatives are due to Lanczos (1964), Bernardo (1976), and Schneider (1978) respectively, and are available from *statlib*, an electronic statistical software archive. Further information can be found in the Acknowledgments.

Let $P_W(x)$ denote the saddlepoint approximation to $P\{X_1^* \le x|\mathbf{d}, X_{k+1}\}$. At μ_1^* (the conditional mean of X_1^*; see Eq. 9), finding $P_W(\mu_1^*)$ requires no iteration. From Wood et al. (1993, eqn. 9),

$$P_W(\mu_1^*) = \Upsilon(\hat{\xi}) - \frac{v(\hat{\xi})G''(0)}{6}\left\{\frac{G'''(0)}{[G''(0)]^{3/2}} - \frac{K'''(0)}{[K''(0)]^{3/2}}\right\},$$

where $K(t)$ and $G(t) = G_1(t)$ are defined in (17) and (18) respectively,

$$\Upsilon(x) = \Upsilon_1(x) = \exp\left\{-\exp\left(-\frac{x-\tilde{b}}{\tilde{a}}\right)\right\},$$

and $v(x) = \frac{d}{ds}\Upsilon(s)\big|_{s=x}$ is the density function corresponding to $\Upsilon(x)$.

For $x \in [x_{min}, x_{max}]$ and $x \ne \mu_1^*$, the algorithm used for calculating $P_W(x)$ is necessarily more complicated, and is described below in an annotated outline format:

Given; $k, \mathbf{d}, X_{k+1}, \tilde{a}$, and $\tilde{b}$:

- SOLVE $K'(t) = x$ for $\hat{t}_x$
  This is done using the *uniroot* procedure in S-plus. A starting point for the search is determined by the linear Taylor expansion of $K'(t)$ about $t = t_0$; evaluating the result at $t_0 = 0$ yields

  $$\hat{K}'(t) = X_{k+1} + \bar{d}S_{1:k} + t\left[\frac{1}{k}\sum_{r=1}^{k}d_r^2 - (\bar{d})^2\right]S_{1:k}^{(2)}.$$

  Solving $\hat{K}'(t) = x$ for t yields an initial closed-form approximation to $\hat{t}_x$.

- SOLVE $G(w_\xi) - G'(w_\xi)w_\xi = K(\hat{t}_x) - x\hat{t}_x$ AND $G'(w_\xi) = \xi$ FOR $(w_{\hat{\xi}}, \hat{\xi})$:
  Earlier, we described how to solve this system of equations by first finding $\hat{\xi}$, and then $w_{\hat{\xi}}$. It is often easier to do this in reverse. From (18), we first note that

  $$G(w) - G'(w)w = \log\Gamma(1-\tilde{a}w) + w\tilde{a}\Psi(1-\tilde{a}w).$$

  As a function of w, this function is negative and concave, undefined for $w > \tilde{a}^{-1}$, and reaches a maximum of zero at $w = 0$. Since $K(\hat{t}_x) - x\hat{t}_x$ is negative for every $x \ne E(X_1^*|\mathbf{d}, X_{k+1})$, this implies that there will be two solutions to (15) as a function of w. Denote these as $w^- \in (-\infty, 0)$ and $w^+ \in (0, \tilde{a}^{-1})$. In addition, since $\mu_1^* = K'(0) = G'(0)$ and $K'(t)$ and $G'(w)$ are both increasing functions, it follows that $G'(w_\xi^-) < \mu_1^*$ and $G'(w_\xi^+) > \mu_1^*$. The solution $(w_{\hat{\xi}}, \hat{\xi})$ may then be determined as:

  if $x < \mu_1^*$ then set $w_{\hat{\xi}} = w^-$ and $\hat{\xi} = G'(w^-)$
  if $x > \mu_1^*$ then set $w_{\hat{\xi}} = w^+$ and $\hat{\xi} = G'(w^+)$.

For a suitably chosen value of $M > 0$, we restrict the range of the root-finding algorithm to $(-M, 0)$ or $(0, \tilde{a}^{-1})$ according to whether $x < K'(0)$ or $x > K'(0)$. We then employ *uniroot* to numerically solve (15) for $w_{\hat{\xi}}$, and consequently set

$$\hat{\xi} = G'(w_{\hat{\xi}}) = \tilde{b} - \tilde{a}\Psi(1 - \tilde{a}w_{\hat{\xi}}) .$$

- CALCULATE $P_W(x)$

We have calculated $\tilde{a} = \overline{X}_k - X_{k+1}$, $\tilde{b} = \tilde{a}\Psi(k+2) + X_{k+1}$, $\hat{t}_x$, $w_{\hat{\xi}}$, and $\hat{\xi}$. From (16),

$$\hat{u} = \hat{t}_x \left(\frac{K''(\hat{t}_x)}{\tilde{a}^2 \Psi'(1 - \tilde{a}w_{\hat{\xi}})} \right)^{1/2} ,$$

where $\Psi'(s)$ is the trigamma function. The resulting approximation to $P\{X_1^* \leq x | \mathbf{d}, X_{k+1}\}$ based on (14) is now

$$P_W(x) = \Upsilon(\hat{\xi}) - v(\hat{\xi})\left\{ \hat{u}^{-1} - w_{\hat{\xi}}^{-1} \right\} .$$

5. Simulation results

As in Section 4, we concentrate on approximating the distribution of the largest order statistic X_1. We want to examine the accuracy of $P_W(x)$ as an approximation to the true bootstrap CDF $P\{X_1^* \leq x | \mathbf{d}, X_{k+1}\}$, and also the accuracy of $P\{X_1^* \leq x | \mathbf{d}, X_{k+1}\}$ (or $P_W(x)$) as an approximation to $P\{X_1 \leq x\}$ for suitably chosen k.

The bootstrap results throughout this section are based on two randomly generated datasets. The first dataset consists of a single random sample of $n = 1000$ observations from a Weibull distribution with scale parameter equal to one and shape (or index) parameter equal to four. The second dataset consists of a single random sample of $n = 75$ observations from the standard normal distribution. It is well known that the convergence of the tail of the normal distribution to the Gumbel distribution is slow (Hall, 1979). The small sample size, combined with the slow rate of convergence, lead us to expect that the bootstrap procedure might not perform very well. All simulations were done in S-plus on a SPARCSTATION 20.

How well does $P_W(x)$ approximate $P\{X_1^* \leq x | \mathbf{d}, X_{k+1}\}$?

To generate the bootstrap distribution $P\{X_1^* \leq x | \mathbf{d}, X_{k+1}\}$ for the Weibull dataset, we first resampled the spacings $\mathbf{d}$ 100,000 times, each time generating a bootstrap replicate

$$X_1^* = X_{k+1} + \sum_{i=1}^{k} i^{-1} d_i^* .$$

The same procedure was done for the normal dataset. The largest 150 observations in the Weibull data were arbitrarily chosen to construct the d'_is. In this particular dataset, $X_{151} = 1.19$. For the normal dataset, the tail was chosen (again, arbitrarily) to begin at $k = 15$, with $X_{16} = 0.558$.

We then linearly transformed the bootstrap variates so that the empirical mean and variance of the 100,000 bootstrap replicates matched the theoretical bootstrap mean and variance given in (9) and (10). This was done solely to improve the Monte Carlo approximation to the true bootstrap distribution, the latter of which $P_W(x)$ aims to approximate. The resampling procedure took approximately thirty minutes for each dataset. Finally, the approximation to the CDF was obtained by calculating the empirical CDF of the transformed bootstrap variates. The approximation $P_W(x)$ is based on the original sample spacings, and was calculated exactly as described in Section 3. Generating $P_W(x)$ on a grid of 100 points took approximately two minutes. The resulting bootstrap CDF's and the corresponding saddlepoint approximations are given Figures 1a and 1b. In both cases, the saddlepoint approximation is an extremely close approximation to the bootstrap CDF. Similar investigations for other arbitrarily chosen value of k yielded equally encouraging results.

The significance levels for the test of exponentiality based on the Gini statistic described in Section 2 are 0.089 and 0.796 respectively for the two distributions. That is, the tail of the Weibull data using $k = 150$ is not very close to being exponentially distributed, while such an assumption seems very reasonable for the normal data. From the plots, one can tell that the saddlepoint approximation in Figure 1b (normal data) is somewhat better than in Figure 1a (Weibull data), although the differences are small.

How well does $P\{X_1^* \leq x | \mathbf{d}, X_{k+1}\}$ approximate $P\{X_1 \leq x\}$?

We restrict attention here to the four methods of determining k summarized in Table 1. As a rule of thumb, we also required that $\log(n)^{4/5} < k < n^{4/5}$ so that the conditions in Section 2 for strong consistency of the extremal index estimators are satisfied.

Table 2 lists the choice of k as determined by each method for the two datasets being considered. Estimates of a_n and b_n in (1) (i.e., $\tilde{a}$ and $\tilde{b}$) are compared with their true values, which are $b_n = F^{-1}(1 - 1/n)$ and $a_n = F^{-1}(1 - e^{-1}/n) - b_n$ (Gnedenko, 1943). For these two datasets, it is apparent that the approximation based on method IV yields estimates that are closest to their respective true values. A small simulation study (not presented here) indicates that method IV also minimizes the mean square difference

$$E[(\tilde{a} - a_n)^2 + (\tilde{b} - b_n)^2]$$

among the four methods considered here.

Figures 2a and 2b give the saddlepoint approximation to the bootstrap distribution corresponding to the four choices of k as well as the true distribution of

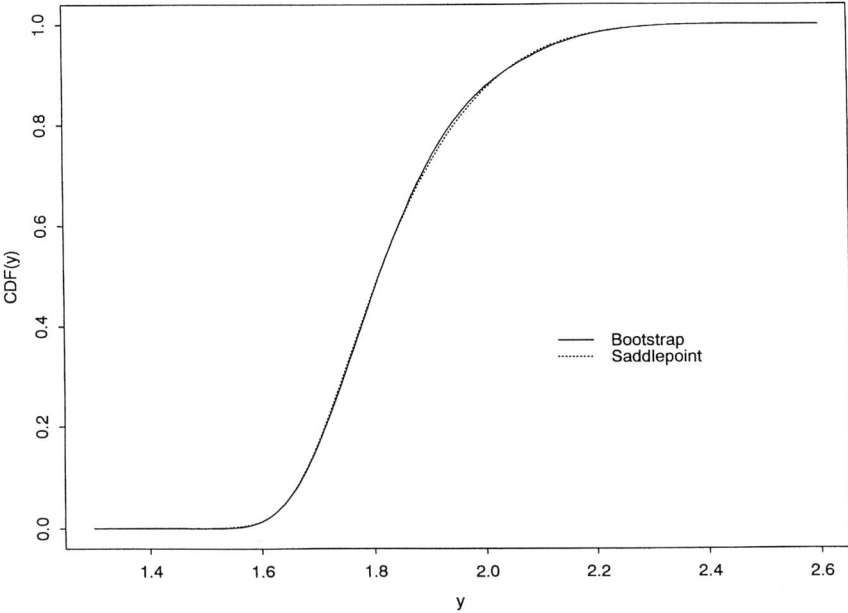

Fig. 1a. Saddlepoint and bootstrap CDF's for Weibull data.

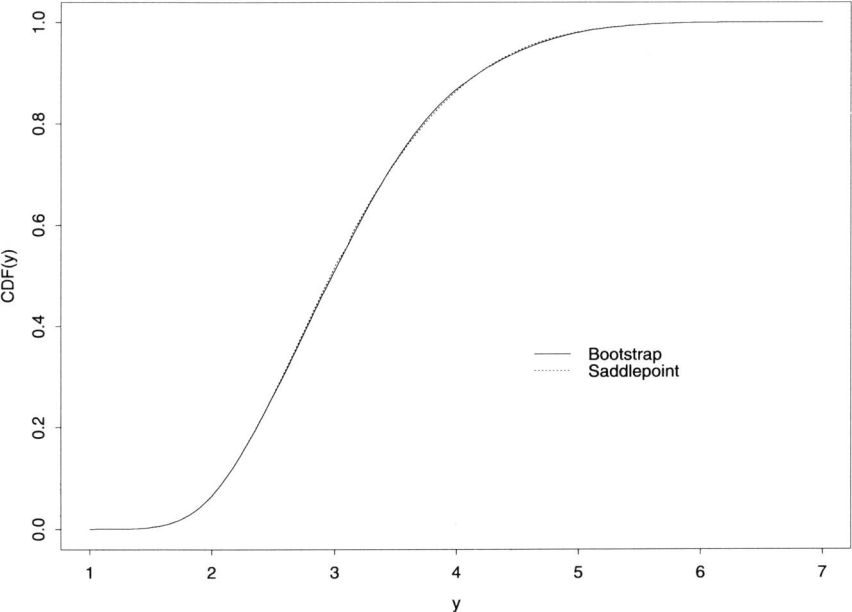

Fig. 1b. Saddlepoint and bootstrap CDF's for Normal data.

Table 2
Comparison of parameter estimates for optimal k

Method	Weibull(4), $n = 1000$			$N(0,1)$, $n = 75$		
	k	b_n	a_n	k	b_n	a_n
True	–	1.6211	0.0557	–	2.2164	0.3661
I	70	1.7212	0.1023	19	2.5084	0.6752
II	9	1.6472	0.0811	14	2.4747	0.6403
III	29	1.6571	0.0803	13	2.3688	0.5778
IV	24	1.6109	0.0607	10	2.1776	0.4487

the largest order statistic. Keep in mind that the bootstrap approximations for the Weibull and normal data are each based on one random sample only. In both cases, it is clear that the choice of k based on method IV produces a superior approximation to the other methods.

The bootstrap approximation which minimized the Cramér–Von Mises distance (cf Durbin, 1973) between $P\{X_1 \leq x\}$ and $P\{X_1^* \leq x | \mathbf{d}, X_{k+1}\}$, or

$$\int_{-\infty}^{\infty} \left(P\{X_1 \leq u\} - P\{X_1^* \leq u | \mathbf{d}, X_{k+1}\}\right)^2 d(P\{X_1 \leq u\}) \;,$$

is also given. This nonnegative discrepancy measure is small when the two distributions are close to each other over the support of $P\{X_1 \leq u\}$. The optimal values of k under this criterion for the Weibull and normal data are respectively $k = 22$ and $k = 10$. As implemented here, this criterion requires knowledge of $P\{X_1 \leq x\}$ and is therefore useless from a practical standpoint.

We also compared the best bootstrap estimates for the two datasets to their corresponding approximations based on (2) using various estimators of γ. The estimators of γ used and corresponding estimated values are summarized in Table 3. Two of the three estimators are discussed in Section 2; the third, due to Pickands (1975), is given by

$$\hat{\gamma}_k = (\log 2)^{-1} \log \frac{X_k - X_{2k}}{X_{2k} - X_{4k}} \;.$$

The estimators used in the approximations based on (2) are based on the same value of k as the bootstrap approximation. Plots of the bootstrap distribution and GPD-based estimates are given in Figures 3a and 3b. The bootstrap approximation outperforms the GPD estimator in each case, especially in the upper tail

Table 3
Generalized Pareto Distribution (GPD) parameter estimator

Author	Estimator	Plot Label	Weibull data	Normal data
Davis and Resnick	$R_k^{(1)}$	GPD-DR	0.041195	0.303438
Dekkers et al.	$\tilde{\gamma}_k$	GPD-DDH	0.039678	0.211518
Pickands	$\hat{\gamma}_k$	GPD-P	0.110514	0.073435

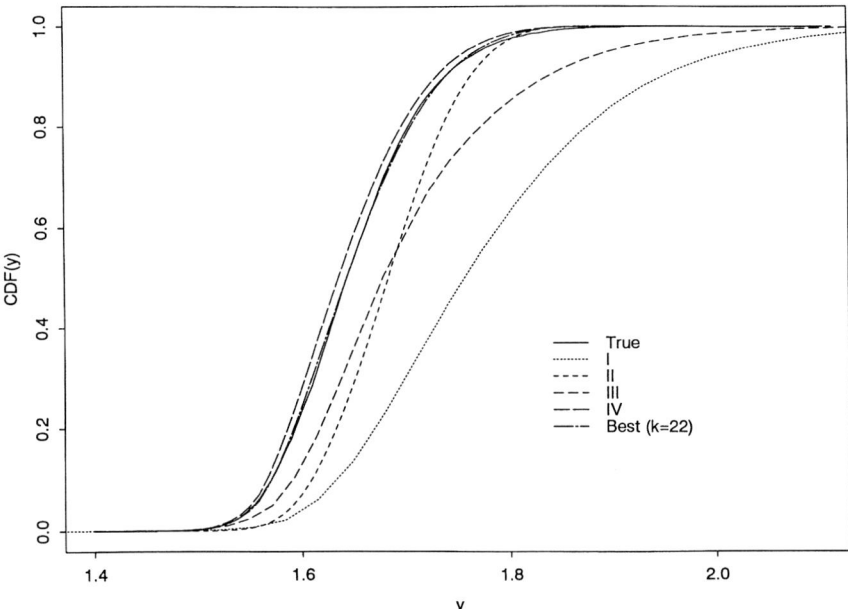

Fig. 2a. Comparison of bootstrap CDF's to true CDF for Weibull data.

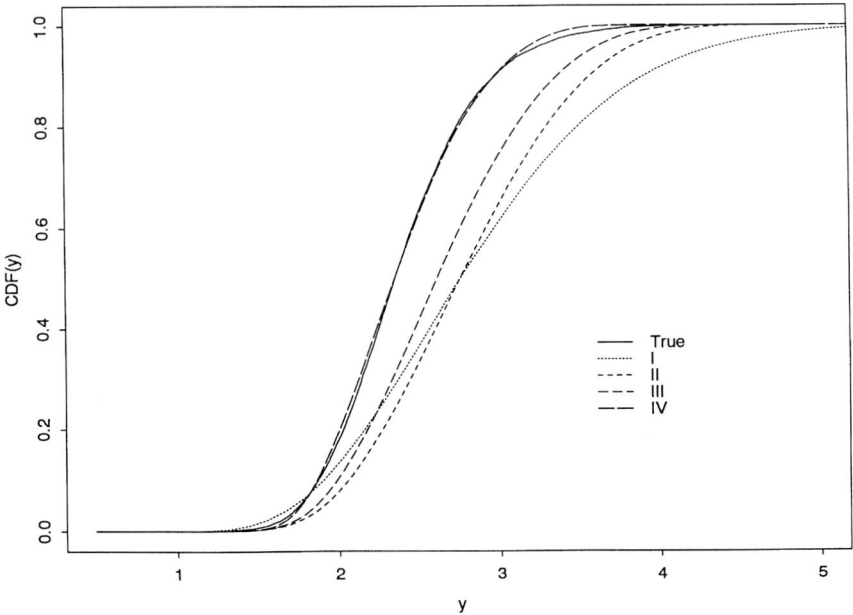

Fig. 2b. Comparison of bootstrap CDF's to true CDF for Normal data.

of the distribution. It is possible that optimizing the choice of k for the GPD-based estimators will lead to comparable results; however, this has not been investigated. Relevant work on this problem can be found in Dekkers and de Haan (1993).

Not shown on these plots is the estimator for $P\{X_1 \leq x\}$ at $\gamma = 0$, or

$$\Upsilon(x) = \exp\left\{-\exp\left\{-\frac{x - \tilde{b}}{\tilde{a}}\right\}\right\} .$$

This is the estimator used by Boos (1984). For the Weibull data, $\Upsilon(x)$ is essentially identical to the best of the 3 GPD estimators; here, "best" is defined as that being closest to the truth for these data. For the normal data, $\Upsilon(x)$ is a definite improvement over the best GPD estimator in the upper tail, and essentially identical everywhere else. In both cases the approximation is still inferior to the bootstrap approximation.

6. Example: The British coal mining data

The British Coal Mining Data (Andrews and Herzberg, 1985, pp. 51–56) consists of the number of days between major coal mining disasters in Britain over the period 1851–1962. A "major disaster" is defined in terms of the number of miner deaths, not the magnitude of the natural event. The data represent inter-accident times and hence spacings between events, and have been previously analyzed by many authors. While Maguire, Pearson, and Wynn (1952) concluded that the exponential model was reasonable, it has since been established that the full data set ($n = 190$) is not consistent with an exponential model (e.g., Simonoff, 1983; Zelterman, 1986, 1993).

Using the Gini statistic G_k defined in Section 2, Zelterman (1993, Figure 2) demonstrates that the spacings between the 50 largest inter-accident times appear close to being exponentially distributed. The same results indicate that reasonable choices of k range somewhere between 20–50. Zelterman (1993) uses $k = 36$, which roughly corresponds to the largest significance level of the Gini statistic and therefore to method III of Section 5. Method IV of Section 5 (see Section 3 and also Table 1) suggests using $k = 41$. Based on the results of Section 5, we expect the latter choice to yield a more accurate approximation to $P\{X_1 \leq x\}$.

Relevant parameter estimates based on the $k = 41$ largest values in the full data set are $\tilde{a} = 327.2$, $\tilde{b} = 1534$, $R_k^{(1)} = 0.551$, $\tilde{\gamma}_k = 0.538$, and $\hat{\gamma}_k = 0.485$. Figure 4 shows the resulting saddlepoint approximation to the bootstrap distribution of the largest spacing between disasters for $k = 41$. Also shown are the GPD-based estimator (using $\tilde{\gamma}_k$ for γ) and the asymptotic approximation based on the extreme value distribution:

$$\Upsilon(x) = \exp\left\{-\exp\left\{-\frac{x - \tilde{b}}{\tilde{a}}\right\}\right\} .$$

A semiparametric bootstrap for simulating extreme order statistics 459

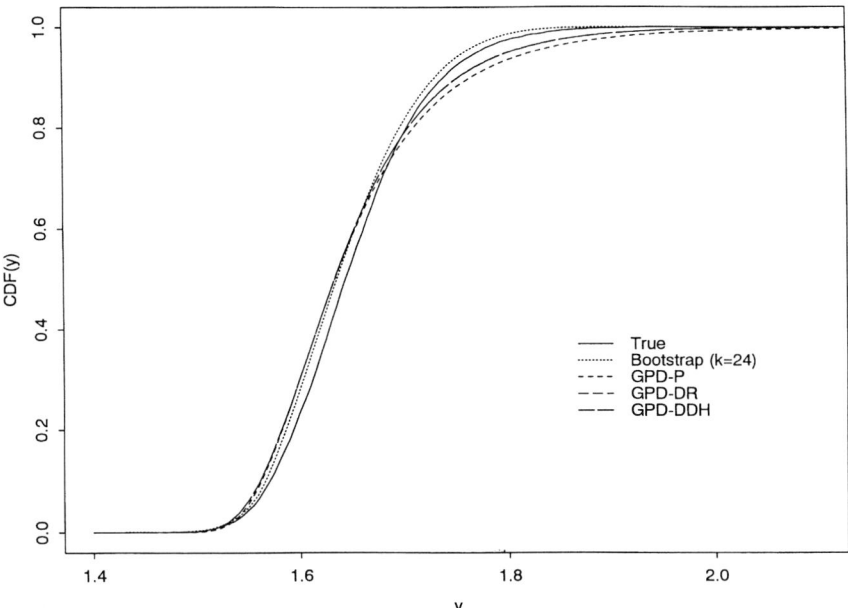

Fig. 3a. Comparison of bootstrap to GPD approximation – Weibull data.

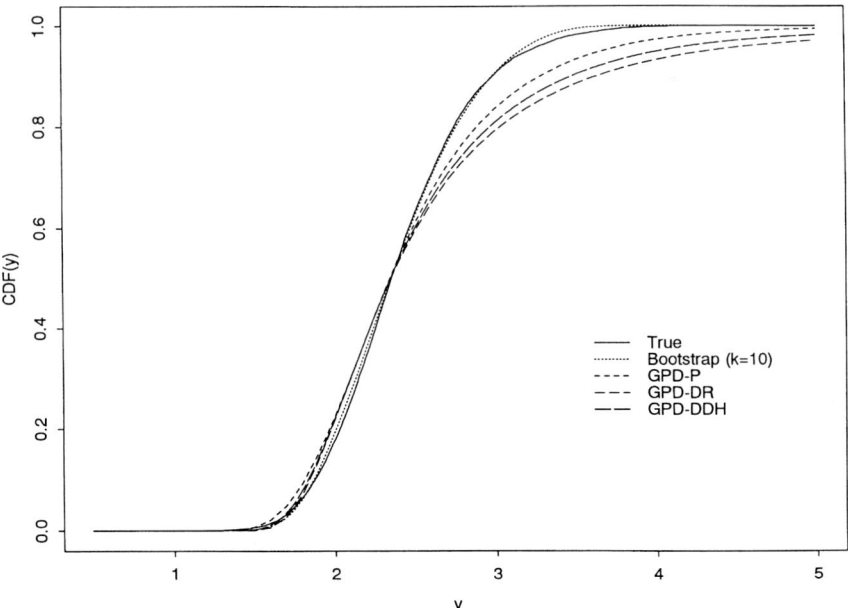

Fig. 3b. Comparison of bootstrap to GPD approximation – Normal data.

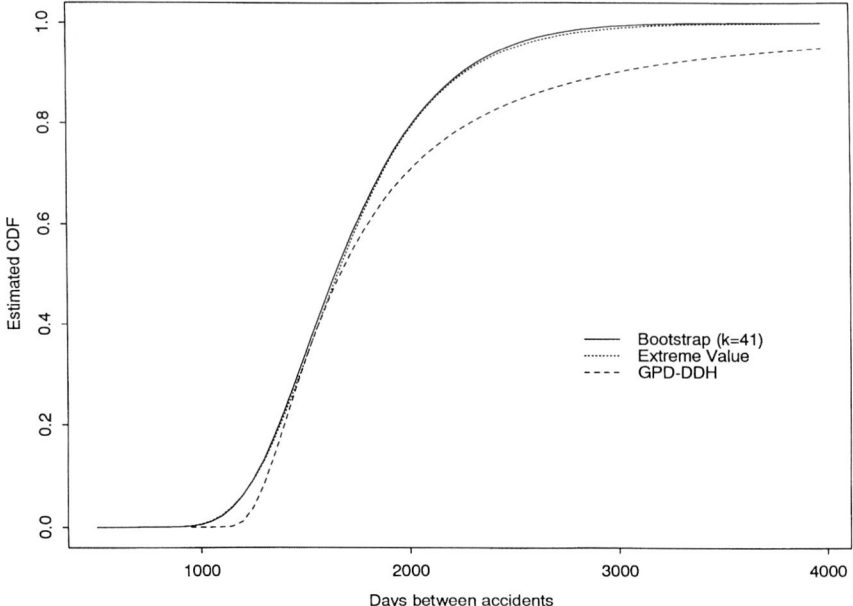

Fig. 4. Distribution of largest spacing between accidents.

The agreement between the bootstrap approximation and $\Upsilon(x)$ is striking. The p-value for the Gini statistic of Section 3 at $k = 41$ is 0.345, and supports the hypothesis that the spacings between the spacings are close to being exponentially distributed. The GPD-based estimator, however, is quite discrepant. The theory suggests that the bootstrap approximation and $\Upsilon(x)$ should be (asymptotically) the same only when the data are from $\mathscr{G}$. Since in finite samples they will not always be close to each other even then (see Figures 3a and 3b), this appears to lend strong support to the answers provided by the bootstrap.

Interestingly, a test of $\gamma = 0$ for $k = 41$ based on the asymptotic distribution of $\sqrt{k}(\tilde{\gamma}_k - \gamma)$ under $H_0 : \gamma = 0$ leads to a highly significant result ($p < 0.00001$). However, it may be difficult to trust the results of this test. The fourth criterion used to choose k is similar in spirit to minimizing the mean squared error of $\sqrt{k}(\tilde{\gamma}_k - \gamma)$; the results of Dekkers and de Haan (1993, Theorem 3.6 and subsequent remarks) imply that there is a potentially significant bias in the asymptotic mean of $\sqrt{k}(\tilde{\gamma}_k - 0)$ when k is chosen as such. We have not attempted to investigate this further here.

Acknowledgments

Some algorithms used for implementing the approximations discussed here are available from *statlib*, an electronic statistical software archive maintained by Michael Meyer at Carnegie Mellon University. For information on how to use *statlib*, send electronic mail to *statlib@lib.stat.cmu.edu* with message body *send index*. The program used by the authors is available upon request from the first named author.

References

Abramowitz, M. A. and I. Stegun (1972). *Handbook of Mathematical Functions*. Dover, NY.
Andrews, D. F. and A. M. Herzberg (1985). *Data*. Springer-Verlag, NY.
Balkema, A. A. and L. de Haan (1974). Residual life time at great age. *Ann. Probab.* **2**, 792–804.
Barndorff-Nielsen, O. E. and D. R. Cox (1989). *Asymptotic Techniques for Use in Statistics*. Chapman-Hall, NY.
Bernardo, J. M. (1976). Psi (Digamma) Function. Algorithm AS 103, *App. Statist.* **25**, 315–317.
Bickel, P. J. and D. A. Freedman (1981). Some asymptotics for the bootstrap. *Ann. Statist.* **9**, 1196–1217.
Boos, D. D. (1984). Using extreme value theory to estimate large percentiles. *Technometrics* **26**, 33–39.
Daniels, H. E. (1954). Saddlepoint approximations in statistics. *Ann. Math. Statist.* **25**, 631–50.
David, H. A. (1981). *Order Statistics* (2nd ed). New York, John Wiley.
Davis, R. and S. Resnick (1984). Tail estimates motivated by extreme value theory. *Ann. Statist.* **12**, 1467–1487.
Davison, A. C. (1988). Approximate conditional inference in generalized linear models. *J. Roy. Statist. Soc. Ser. B* **50**, 445–461.
Davison, A. C. (1988). Discussion of bootstrap methods, by David Hinkley. *J. Roy. Statist. Soc. Ser. B* **50**, 356–357.
Davison, A. C. and D. V. Hinkley (1988). Saddlepoint approximations in resampling methods. *Biometrika* **75**, 417–431.
Davison, A. C. and R. L. Smith (1990). Models for exceedances over high thresholds. *J. Roy. Statist. Soc. Ser. B* **52**, 393–442.
Dekkers, A. L. M. and L. de Haan (1989). On the estimation of the extreme-value index and large quantile estimation. *Ann. Statist.* **17**, 1795–1832.
Dekkers, A. L. M., J. H. J. Einmahl and L. de Haan (1989). A moment estimator for the index of an extreme value distribution. *Ann. Statist.* **17**, 1833–1855.
DiCiccio, T. J. and J. P. Romano (1988). A review of bootstrap confidence intervals. *J. Roy. Statist. Soc. Ser. B* **50**, 338–354.
Do, K.-A. and P. Hall (1991). On importance sampling for the bootstrap. *Biometrika* **78**, 161–167.
Durbin, J. (1973). *Distribution Theory for Tests Based on the Sample Distribution Function*. Philadelphia: SIAM Monograph #9.
Efron, B. (1979). Bootstrap methods: Another look at the jackknife. *Ann. Statist.* **7**, 1–26.
Efron, B. (1982). *The Jackknife, the Bootstrap, and Other Resampling Plans*. Philadelphia: SIAM Monograph #38.
Efron, B. and G. Gong (1983). A leisurely look at the bootstrap, the jackknife, and cross-validations. *Amer. Statist.* **37**, 36–48.
Efron, B. and R. Tibshirani (1993). *An Introduction to the Bootstrap*. Chapman and Hall, NY.
Gail, M. H. and J. L. Gastwirth (1978). A scale-free goodness-of-fit test for the exponential distribution based on the Gini statistic. *J. Roy. Statist. Soc. Ser. B*, **40**, 350–357.
Gnedenko, B. V. (1943). Sur la Distribution Limite du Terme Maximum d'un Serie Aleatoire. *Ann. Math.* **44**, 423–453.
Hall, P. (1979). On the rate of convergence of normal extremes. *J. Appl. Probab.* **12**, 475–490.
Hill, B. M. (1975). A simple general approach to inference about the tail of a distribution. *Ann. Statist.* **3**, 1163–1174.
Hinkley, D. V. (1988). Bootstrap Methods (with discussion). *J. Roy. Statist. Soc. Ser. B* **50**, 321–337.
Johns, M. V. (1988). Importance sampling for bootstrap confidence intervals. *J. Amer. Statist. Assoc.* **83**, 709–714.

Lagakos, S. W., B. J. Wessen and M. Zelen (1986). An analysis of contaminated well water and health effects in Woburn, Massachusetts (with discussion). *J. Amer. Statist. Assoc.* **81**, 583–614.

Lanczos, C (1964). A precision approximation of the gamma function. *SIAM J. Numer. Anal. B* **1**, 86–96.

Loh, W. Y. (1984). Estimating an endpoint of a distribution with resampling methods. *Ann. Statist.* **12**, 1543–1550.

Lugannani, R. and S. Rice (1980). Saddlepoint approximation for the distribution of the sum of independent random variables. *Adv. Appl. Probab.* **12**, 475–490.

Maguire, B. A., E. S. Pearson and A. H. A. Wynn (1952). The time intervals between industrial accidents. *Biometrika* **39**, 168–180.

McCullagh, P. (1987). *Tensor Methods in Statistics*. Chapman and Hall, NY.

Pickands, J. III (1975). Statistical inference using extreme order statistics. *Ann. Statist.* **3**, 119–131.

Pyke, R. (1965). Spacings (with discussion). *J. Roy. Statist. Soc. Ser. B.* **27**, 395–436.

Reiss, R. D. (1989). *Approximate Distributions of Order Statistics*. New York: Springer-Verlag.

Schneider, B. E. (1978). Trigamma function. Algorithm AS 121, *Appl. Statist.* **27**, 97–99.

Serfling, R. J. (1980). *Approximation Theorems of Mathematical Statistics*. New York, John Wiley.

Simonoff, J. S. (1983). A penalty function approach to smoothing large sparse contingency tables. *Ann. Statist.* **11**, 208–218.

Smith, R. L. (1989). Extreme value analysis of environmental time series: An application to trend detection in ground-level ozone. *Statist. Sci.* **4**, 367–393.

Sukhatme, P. V. (1936). On the analysis of k samples from exponential populations with especial reference to the problem of random intervals. *Statist. Res. Mem.* **1**, 94–112.

Weissman, I. (1978). Estimation of parameters and large quantiles based on the k largest observations. *J. Amer. Statist. Assoc.* **73** 812–815.

Wood, A. T. A., J. G., Booth and R. W. Butler (1993). Saddlepoint approximations to the CDF of some statistics with nonnormal limit distributions. *J. Amer. Statist. Assoc.* **88**, 680–686.

Young, G. A. (1992). Bootstrap: More than a stab in the dark? *Statist. Sci.* **9**, 382–415.

Zelterman, D. (1986). The log-likelihood ratio for sparse multinomial mixtures. *Statist. Probab. Lett.*, **4**, 95–99.

Zelterman, D. (1992). A distribution with an unbounded hazard function and its application to a theory from demography. *Biometrics* **48**, 807–818.

Zelterman, D. (1993). A semiparametric bootstrap technique for simulating extreme order statistics. *Jour. Amer. Statist. Assoc.* **88**, 477–485.

Approximations to Distributions of Sample Quantiles

Chunsheng Ma and John Robinson

1. Introduction and definitions

Suppose that F is a cumulative distribution function (d.f.). Its quantile function (q.f.) is defined by

$$F^{-1}(t) = \inf\{x : F(x) \geq t\}, \quad t \in (0,1) .$$

So F^{-1} is the usual inverse of F if F is continuous and strictly increasing. For a given $q \in (0,1)$, $F^{-1}(q)$ is called the qth *quantile* (or *fractile*) of F and alternatively denoted by ξ_q. In particular, $\xi_{\frac{1}{2}}$ is called the *median* of F. Clearly, ξ_q satisfies the equation

$$F(\xi_q-) \leq q \leq F(\xi_q) .$$

A fundamental problem in nonparametric statistical estimation and hypothesis testing deals with inferences about a d.f. F and the qth quantile ξ_q. A natural nonparametric estimate of F is the sample d.f. F_n of a random sample $X_1, \ldots, X_n$ from F, and the sample quantile $F_n^{-1}(q)$ is an appropriate estimate of ξ_q, while in the context of a smooth statistical model, the use of a perturbed (or smoothed) sample d.f. and associated quantile may be more natural and appropriate.

1.1. Sample distribution function and quantile

Assume that $X_1, \ldots, X_n$ is a random sample from d.f. F. The corresponding sample d.f. F_n is constructed by placing at each observation X_j a mass $\frac{1}{n}$ and so represented as

$$F_n(x) = \frac{1}{n} \sum_{j=1}^n I_0(x - X_j), \quad x \in \mathsf{R}^1 \tag{1.1}$$

where I_0 is the d.f. of the unit mass at the origin; namely,

$$I_0(x) = \begin{cases} 0, & x < 0, \\ 1, & x \geq 0 . \end{cases}$$

We can interpret (1.1) from the following two aspects. On the one hand, (1.1) is a distribution function while we consider it as a function of x for each fixed sample. From this view, the sample q^{th} quantile is defined as the qth quantile of the sample d.f. F_n, that is

$$F_n^{-1}(q) = \inf\{x: F_n(x) \geq q\} .$$

We will denote it as $\hat{X}_{nq}$. Obviously,

$$F_n(\hat{X}_{nq}-) \leq q \leq F_n(\hat{X}_{nq}) .$$

On the other hand, for each fixed value of x, (1.1) is a random variable, considered as a function of the sample. Furthermore, $nF_n(x)$ is distributed as *binomial* $(n, F(x))$.

In a view encompassing both features, the distribution function of $\hat{X}_{nq}$ can be written exactly as

$$P(\hat{X}_{nq} \leq x) = P(nF_n(x) \geq nq) = \sum_{j=m}^{n} \binom{n}{j} [F(x)]^j [1 - F(x)]^{n-j} \qquad (1.2)$$

or equivalently,

$$P(\hat{X}_{nq} \leq x) = P(nF_n(x) \geq nq) = m \binom{n}{m} \int_0^{F(x)} t^{m-1}(1-t)^{n-m} dt \qquad (1.2)'$$

where

$$m = \begin{cases} nq, & \text{if } nq \text{ is an integer,} \\ [nq] + 1, & \text{otherwise}, \end{cases}$$

with $[x]$ denoting the largest integer less than or equal to $x \in R^1$. If F has a density f, then the d.f. of $\hat{X}_{nq}$ has also a density

$$\frac{1}{B(m, n-m+1)} F^{m-1}(x)(1 - F(x))^{n-m} f(x) . \qquad (1.3)$$

(1.2) or (1.2)' says that the exact distribution of $\hat{X}_{nq}$ depends only upon F through the incomplete Beta function. Because of (1.2), we can employ known results for binomial r.v.'s when treating the distribution properties of $\hat{X}_{nq}$. Another standard method is firstly to deal with the sample quantile of i.i.d. $(0, 1)$-uniformly distributed r.v.'s, and then use the transformation technique to extend the result to the general case.

Another approach in defining the qth quantile is specified by considering it as a special case of M-estimates. The relevant ψ function is $\psi(x, \theta) = \psi(x - \theta)$, where

$$\psi(x) = \begin{cases} -1, & x < 0, \\ 0, & x = 0, \\ \frac{q}{1-q}, & x > 0 . \end{cases}$$

In this way, ξ_q is defined as a solution θ_0 of the equation

$$\int \psi(x, \theta_0) dF(x) = 0 \ ;$$

and similarly, $\hat{X}_{nq}$ is a solution $\hat{\theta}$ of the equation

$$\sum_{j=1}^{n} \psi(X_j, \hat{\theta}) = 0 \ .$$

This view enables us to formulate the sample quantiles as statistical functions and to use the techniques for the M-estimates.

1.2. Perturbed sample distribution function and quantile

In many statistical models where it is known (or reasonable to assume based on physical considerations under which the data are collected, or empirical evidence of the past) that the underlying d.f. F is smooth enough, it is more natural to use a smoothed version $\hat{F}_n$ of F_n, rather than the step function F_n itself, as the estimate of F. Such a $\hat{F}_n$ is often called the perturbed (or smoothed) sample d.f., and the corresponding q^{th} quantile is called the perturbed (or smoothed) sample q^{th} quantile.

There are essentially two approaches to construct a perturbed sample quantile in common use now. One may first smooth F_n by convolution, then define the qth quantile as in the classical case. Another method is the so-called quantile estimation, which is a certain subclass of L-estimates, where the smoothing is directly applied to the sample q.f. F_n^{-1} rather than F_n itself.

Let $\{I_n\}$ be a sequence of continuous d.f.'s converging weakly to I_0. An intuitively appealing and easily understood competitor to F_n is the perturbed sample d.f. $\hat{F}_n$ constructed by a convolution of the sample d.f. F_n and $\{I_n\}$, $\hat{F}_n = F_n * I_n$; more exactly,

$$\hat{F}_n(x) = \int_{-\infty}^{\infty} I_n(x - t) dF_n(t) = \frac{1}{n} \sum_{j=1}^{n} I_n(x - X_j), \quad x \in \mathsf{R}^1 \ .$$

It differs from F_n in that the mass $\frac{1}{n}$ is no longer concentrated at X_j, but is distributed continuously around X_j, according to I_n. Denote the corresponding q^{th} quantile by $\hat{Z}_{nq}$. Similarly to (1.2), we have

$$P(\hat{Z}_{nq} \leq x) = P(n\hat{F}_n(x) \geq nq) = P\left(\frac{1}{n} \sum_{j=1}^{n} I_n(x - X_j) \geq q\right) \ .$$

This approach is parallel to that used in density estimation, proposed by Parzen (1962) and Rosenblatt (1956), where the density estimate is given by

$$\hat{f}_n(x) = \int_{-\infty}^{\infty} w_n(x-t)\,dF_n(t) = \frac{1}{n}\sum_{j=1}^{n} w_n(x-X_j), \quad x \in \mathbb{R}^1,$$

with a sequence of nonnegative weight functions $\{w_n(\cdot)\}$ satisfying that $\int_{-\infty}^{\infty} w_n(t)\,dt = 1$ $(n \geq 1)$ and that the total mass concentrates in a neighborhood of zero as $n \to +\infty$, that is, for given any $\varepsilon > 0$, $\int_{|t|<\varepsilon} w_n(t)\,dt \to 1$ as $n \to +\infty$.

A typical way of generating sequences $\{I_n\}$ is based on the kernel method. Usually, take

$$I_n(x) = K\left(\frac{x}{\alpha_n}\right),$$

where K is a d.f. and $\{\alpha_n\}$ is a positive sequence with $\alpha_n \to 0$ as $n \to +\infty$; and call K the kernel function, $\{\alpha_n\}$ the bandwidth or window-width, and $\hat{F}_n$ the sample kernel d.f.. In this case,

$$\hat{F}_n(x) = \int_{-\infty}^{\infty} K\left(\frac{t-x}{\alpha_n}\right) dF_n(t) = \frac{1}{n}\sum_{j=1}^{n} K\left(\frac{x-X_j}{\alpha_n}\right), \quad x \in \mathbb{R}^1 \tag{1.4}$$

which is a convolution of F_n and a properly scaled kernel function. For appropriately chosen kernels and sufficiently smooth F's, it has been shown that the asymptotic performance of $\hat{F}_n$ is superior to that of F_n in the sense of relative deficiency and the speed of convergence in the bootstrap; see Reiss (1989, Chapter 8) and the references therein.

Note that $\hat{X}_{nq}$ can be expressed in terms of L-estimates, namely,

$$\hat{X}_{nq} = X_{m:n},$$

where $X_{1:n} \leq \cdots \leq X_{n:n}$ are order statistics pertaining to the sample $X_1,\ldots,X_n$. One might hope that averaging over order statistics close to the sample quantile leads to estimates of better performance. This idea results in considering L-estimates of the form

$$\hat{Z}_{nq} = \sum_{j=1}^{n} s_{j:n}(q) X_{j:n}$$

with certain choice of the scores $\{s_{j:n}(q)\}$ involving q.

A popular subclass of L-estimates related to quantiles is called kernel quantile estimators, introduced by Parzen (1979) and Reiss (1982), where the scores $\{s_{j:n}(q)\}$ are chosen by the kernel method. Taking

$$s_{j:n}(q) = \int_{\frac{j-1}{n}}^{\frac{j}{n}} dK\left(\frac{t-q}{\alpha_n}\right), \quad j=1,\ldots,n$$

where K is a kernel function and $\{\alpha_n\}$ the bandwidth, it gives a kernel quantile estimator

$$\hat{Z}_{nq} = \int_0^1 F_n^{-1}(t) \, dK\left(\frac{t-q}{\alpha_n}\right).$$

The kernel quantile estimators have been studied in the literature; see, for example, Falk (1984, 1985), Sheather and Marron (1990), Yang (1985), Zelterman (1990), and also Reiss (1989, Chapter 8).

The scores of Harrell and Davis (1982) is given by

$$S_{j:n}(q) = \frac{1}{B(m_0, n - m_0 + 1)} \int_{\frac{j-1}{n}}^{\frac{j}{n}} t^{m_0 - 1}(1-t)^{n-m_0} \, dt,$$

where $m_0 = (n+1)q$. In this particular case, the kernel quantile estimator is exactly the bootstrap estimator of $E(X_{[m_0]:n+1})$.

Kaigh and Lachenbruch (1982) proposed a certain U-statistic with representation also an L-estimate, which is the average of qth sample quantiles from all $\binom{n}{r}$ subsamples of size r, chosen without replacement from $X_1, \ldots, X_n$, where the smoothing parameter r is used to regulate the amount of smoothness desired. This idea has been generalized to introduce the so-called O-statistics. See Kaigh (1988) for a survey.

1.3. Summary

This chapter gives a brief review of the asymptotic distribution theory of the (classical and perturbed) sample quantiles. Section 2 presents *Smirnov's Lemma*. Section 3 discusses the normal approximation to distributions of the quantiles. The saddlepoint approximation is studied in Section 4. In Section 5 we review the bootstrap approximation to the quantiles.

Throughout this chapter, denote by ϕ, Φ respectively the density and distribution function of the standard normal distribution $N(0,1)$, and define $\sigma_q = [q(1-q)]^{\frac{1}{2}}$. Unless otherwise specified, limits in all order symbols are taken as $n \to +\infty$.

2. Smirnov's lemma

This section presents *Smirnov's Lemma*, which plays a key role in the study of the asymptotic theory of the sample quantiles as well as the central order statistics.

Assume that $\{a_n\}$ and $\{b_n\}$ are two sequences of constants with $a_n > 0 (n \geq 1)$. Let

$$\lambda_n = \frac{m}{n+1}, \quad \sigma_{\lambda_n} = [\lambda_n(1-\lambda_n)]^{\frac{1}{2}}.$$

Then $|\lambda_n - q| \leq \frac{1}{n}$, and $|\sigma_{\lambda_n} - \sigma_q| = O(\frac{1}{n})$. In view of (1.2)' Smirnov (1949, Part 1, Lemma 2) showed that

$$\sup_{x\in\mathbb{R}^1}\left|P\left(\frac{\hat{X}_{nq}-b_n}{a_n}<x\right)-\Phi\left(\sqrt{n+1}\frac{F(a_nx+b_n)-\lambda_n}{\sigma_{\lambda_n}}\right)\right|\to 0, \quad (2.1)$$

$$n\to+\infty.$$

This result was used by Smirnov (1949) to investigate the types of limit distributions for the sample quantiles as well as the central order statistics. In fact, (2.1) can be improved as the following Berry-Esseen type bound.

Smirnov's Lemma: For every $n\geq 1$,

$$\sup_{x\in\mathbb{R}^1}\left|P\left(\frac{\hat{X}_{nq}-b_n}{a_n}<x\right)-\Phi\left(\sqrt{n}\frac{F(a_nx+b_n)-q}{\sigma_q}\right)\right|\leq C_0 n^{-\frac{1}{2}} \quad (2.2)$$

where the constant C_0 depends only upon q and F.

Clearly, no conditions in (2.2) are imposed on a_n, b_n and the underlying d.f. F. As a result, Smirnov's Lemma provides a useful tool to deal with the asymptotic normality and bootstrap approximation to the sample quantiles.

(2.2) is generalized by Puri and Ralescu (1986, Lemma 4.3) to the random central order statistics (of limiting rank q). More generally, it follows from the Berry–Esseen theorem for the sample quantile of i.i.d. $(0,1)$-uniformly distributed r.v.'s (See Reiss (1989, Theorem 4.2.1)) that for every positive integer r there exists a constant $C_r>0$ such that

$$\sup_{x\in\mathbb{R}^1}\left|P\left(\frac{\hat{X}_{nq}-b_n}{a_n}<x\right)-\int_{-\infty}^{u_n(x)}\left(1+\sum_{k=1}^{r-1}L_{k,n}(t)\right)\phi(t)\,dt\right|\leq C_r n^{-\frac{r}{2}}$$

(2.3)

where $u_n(x)=\sqrt{n}\frac{F(a_nx+b_n)-q}{\sigma_q}$, and $L_{k,n}(t)$ is a polynomial of degree $\leq 3k$, $k=1,\ldots,r-1$.

3. Normal approximation

3.1. Normal approximation to distributions of the quantiles

It is well-known that $\hat{X}_{nq}$ is asymptotically normally distributed under only local assumptions on the underlying d.f. F near ξ_q. As pointed out by Stigler (1973), as far back as 1818 Laplace had shown that the sample median $\hat{X}_{\frac{n}{2}}$ is asymptotically normal. More than a century later Smirnov (1949) described all possible limit distributions for the sample quantiles as well as central order statistics. A more comprehensive treatment was given by Balkema and de Haan (1978). For the asymptotic joint normality of the vector of sample quantiles $(\hat{X}_{nq_1},\ldots,\hat{X}_{nq_r})(0<q_1<\cdots<q_r<1)$, we refer to Serfling (1980, Chapter 2).

By $\hat{X}_{nq}$ being asymptotically normally distributed we mean that a suitably normalized version of $\hat{X}_{nq}$ converges in distribution to $N(0,1)$; more precisely, there exist sequences of constants $\{a_n\}$ and $\{b_n\}$ with $a_n>0(n\geq 1)$ such that

$$\frac{\hat{X}_{nq} - b_n}{a_n} \xrightarrow{\mathscr{D}} N(0,1) \ . \tag{3.1}$$

As a consequence of (2.1) or its improved form (2.2), we obtain (a special case of Smirnov (1949, Theorem 4))

THEOREM 1. *In order that given sequences $\{a_n\}$ and $\{b_n\}$ satisfy (3.1), it is necessary and sufficient that*

$$\sqrt{n}\frac{F(a_n x + b_n) - q}{\sigma_q} - x \to 0, \quad n \to +\infty \ . \tag{3.2}$$

Since Φ is continuous, the convergence in (3.2) holds uniformly in x.

In particular, for some positive constant τ, taking $a_n = \frac{\tau}{\sqrt{n}}, b_n = \xi_q (n \geq 1)$ in (3.1). Then condition (3.2) is equivalent to

$$\sqrt{n}\frac{F(\xi_q + \frac{\tau}{\sqrt{n}}x) - q}{\sigma_q} - x \to 0, \quad n \to +\infty \ . \tag{3.3}$$

Therefore F must be continuous at ξ_q. A sufficient condition for (3.3) is that F is differentiable at ξ_q and $F'(\xi_q) > 0$. With the help of *Smirnov's Lemma* it is found that this condition is also necessary.

THEOREM 2. *For some $\tau > 0$,*

$$\sqrt{n}\frac{\hat{X}_{nq} - \xi_q}{\tau} \xrightarrow{\mathscr{D}} N(0,1)$$

holds if and only if F is differentiable at ξ_q and $F'(\xi_q) > 0$.
In this case, $\tau = \frac{\sigma_q}{F'(\xi_q)}$.

Generally, suppose that (3.1) is true. Then, is F differentiable at ξ_q with $F'(\xi_q) > 0$? A problem of this type was studied by Balkema and de Haan (1978), Lahiri (1992), and Smirnov (1949). It was pointed out by Balkema and de Haan (1978, p 343) that there exists a d.f. F which satisfies (3.1) but $F' = 0$ with probability 1 on R^1.

Assume that $\hat{F}_n$ is the sample kernel d.f. given by (1.4). The asymptotic normality of the perturbed sample quantile $\hat{Z}_{nq}$ has been studied by Mack (1987), Nadaraya (1964), Ralescu and Sun (1993), and Seoh and Puri (1994). Based on certain regularity conditions on the underlying d.f. F and the kernel K, necessary and sufficient conditions for the asymptotic normality of $\hat{Z}_{nq}$ are found by Ralescu and Sun (1993) for the bandwidth $\{\alpha_n\}$.

THEOREM 3. (Ralescu and Sun (1993)): *Let $\{q_n\}$ be a sequence of constants satisfying*

$$0 < q_n < 1, \quad \lim_{n \to +\infty} \sqrt{n}(q_n - q) = c_0 a$$

where a and $c_0 > 0$ are constants.

(1) Suppose that F has a bounded second derivative F'' on its support with $F'(\xi_q) > 0$, and F'' is continuous in a neighborhood of ξ_q with $F''(\xi_q) \neq 0$; K has a density k and

$$\int_{-\infty}^{\infty} tk(t)\,dt = 0 \quad \text{and} \quad \int_{-\infty}^{\infty} t^2 k(t)\,dt < +\infty.$$

Then the condition

$$\lim_{n \to +\infty} n^{\frac{1}{4}} \alpha_n = 0$$

is necessary and sufficient for

$$\sqrt{n}\frac{\hat{Z}_{nq_n} - \xi_{q_n}}{\tau_n} \xrightarrow{\mathscr{D}} N(a, 1) \tag{3.4}$$

where $\tau_n = \frac{\sigma_q}{F'(\xi_{q_n})}$.

(2) Suppose that F has a bounded density f with $f(\xi_q) > 0$, and f is continuous in a neighborhood of ξ_q; K has a density k and

$$\int_{-\infty}^{\infty} |t|\,k(t)\,dt < +\infty \quad \text{but} \quad \int_{-\infty}^{\infty} tk(t)\,dt \neq 0.$$

Then (3.4) holds if and only if

$$\lim_{n \to +\infty} n^{\frac{1}{2}} \alpha_n = 0.$$

Multidimensional asymptotic normality of the kernel quantile estimator was established by Falk (1985), where the conditions on F fit completely with the standard assumptions (see Serfling (1980, p 80)) which are used to prove the asymptotic joint normality of the vector of sample quantiles $(\hat{X}_{nq_1}, \ldots, \hat{X}_{nq_r})$ $(0 < q_1 < \cdots < q_r < 1)$, and the limiting normal distributions are the same. See also Yang (1985).

3.2. Accuracy of the asymptotic normality

When dealing with the asymptotic normality of the quantiles, we are often concerned with its accuracy. The basic tools for this purpose are the Berry-Esseen theorem and *Smirnov's Lemma*.

Under certain regularity conditions on F, it has been shown that the rate of convergence in connection with the asymptotic normality of $\hat{X}_{nq}$ is of order $O(n^{-\frac{1}{2}})$. Assuming that F has a bounded second derivative on R^1 with $F'(\xi_q) > 0$, observing (1.2) and using the Berry–Esseen theorem for independent binomial r.v.'s, Reiss (1974) firstly considered a special case, *i.e.*, F is the uniform distribution on $[0, 1]$, and then made the quantile transformation to derive the general result. A similar result was independently obtained by Serfling (1980) using Hoeffding's inequality and the Berry–Esseen theorem, where the smoothing conditions of F are assumed in a neighborhood of ξ_q rather than on the whole real line R^1. Puri and Ralescu (1986) validated this convergence rate for the generalized class of random central order statistics and further pointed out that the requirements concerning the second derivative may be dropped. In fact, from *Smirnov's Lemma* the following condition is enough to achieve the Berry–Esseen rate $O(n^{-\frac{1}{2}})$

$$\left| F(\xi_q + h) - F(\xi_q) - F'(\xi_q)h \right| \leq O(h^2), \quad h \to 0 \tag{3.5}$$

with $F'(\xi_q) > 0$.

It is natural to ask whether the converse of this assertion is also true. *Smirnov's Lemma* implies that this question can be answered positively. To this end, suppose that for some $\tau > 0$

$$\sup_{x \in R^1} \left| P\left(\sqrt{n} \frac{\hat{X}_{nq} - \xi_q}{\tau} < x \right) - \Phi(x) \right| = O\left(n^{-\frac{1}{2}}\right), \quad n \to +\infty$$

holds. Combining it with (2.2), the triangle inequality gives

$$\sup_{x \in R^1} \left| \Phi\left(\sqrt{n} \frac{F\left(\xi_q + \frac{\tau}{\sqrt{n}}x\right) - q}{\sigma_q} \right) - \Phi(x) \right| = O\left(n^{-\frac{1}{2}}\right), \quad n \to +\infty .$$

By the mean value theorem, we get for all n sufficiently large,

$$\left| \sqrt{n} \frac{F\left(\xi_q + \frac{\tau}{\sqrt{n}}x\right) - q}{\sigma_q} - x \right| \leq O\left(n^{-\frac{1}{2}}\right) ,$$

uniformly in $|x| \leq C_\delta$, where $\delta \in (0, \frac{1}{2})$, and $C_\delta = \Phi^{-1}(1 - \delta)$. As a result, uniformly in $\frac{1}{2}C_\delta \leq |x| \leq C_\delta$,

$$\left| F\left(\xi_q + \frac{\tau}{\sqrt{n}}x\right) - q - \frac{\sigma_q}{\sqrt{n}}x \right| \leq O\left(\left(\frac{\tau x}{\sqrt{n}}\right)^2\right)$$

which implies that $F(\xi_q) = q, \tau = \frac{\sigma_q}{F'(\xi_q)}$, and (3.5) holds.

THEOREM 4. Let $\tau > 0$. Then the following statements are equivalent:

(1) $\quad \displaystyle\sup_{x \in R^1} \left| P\left(\sqrt{n} \frac{\hat{X}_{nq} - \xi_q}{\tau} < x \right) - \Phi(x) \right| = O\left(n^{-\frac{1}{2}}\right), \quad n \to +\infty .$

(2) In a neighborhood of ξ_q, F possesses the property

$$\left| F(\xi_q + h) - F(\xi_q) - \frac{\sigma_q}{\tau} h \right| \leq O(h^2), \quad h \to 0 .$$

In this case, $\tau = \frac{\sigma_q}{F'(\xi_q)}$.

Assume that $\hat{F}_n$ is the sample kernel d.f. given by (1.4). Ralescu (1992a) obtained the rate, $O(n^{-\frac{1}{2}} \log n)$, of convergence in distribution of the perturbed sample quantile $\hat{Z}_{nq}$ to the normal law. In fact, as shown in Seoh and Puri (1994), the Berry–Esseen rate $O(n^{-\frac{1}{2}})$ is still available.

THEOREM 5 (Seoh and Puri (1994)). Let $\{q_n\}$ be a sequence of constants satisfying

$$0 < q_n < 1, \text{ and } |q_n - q| \le \frac{1}{4n} \sigma_q$$

for all sufficiently large n.

(1) Suppose that F has a bounded second derivative F'' on its support with $F'(\xi_q) > 0$; K has a density k and

$$\int_{-\infty}^{\infty} t k(t) \, dt = 0 \quad \text{and} \quad \int_{-\infty}^{\infty} t^2 k(t) \, dt < +\infty \; .$$

Then

$$\sup_{x \in R^1} \left| P\left(\sqrt{n} \frac{\hat{Z}_{nq_n} - \xi_{q_n}}{\tau_n} < x \right) - \Phi(x) \right| \le C_1 \max\{n^{-\frac{1}{2}}, \alpha_n, n^{\frac{1}{2}} \alpha_n^2\}$$

where $\tau_n = \frac{\sigma_q}{F'(\xi_{q_n})}$, and C_1 is an absolute constant (depending only on q, F, K). The Berry–Esseen rate $O(n^{-\frac{1}{2}})$ is obtained by choosing the bandwidth $\alpha_n = O(n^{-\frac{1}{2}})$.

(2) Suppose that F has a bounded density f with $f(\xi_q) > 0$, and f satisfies the Lipschitz condition of order one in a neighborhood of ξ_q; K has a density k with

$$\int_{-\infty}^{\infty} |t| k(t) \, dt < +\infty \; .$$

Then

$$\sup_{x \in R^1} | P\left(\sqrt{n} \frac{\hat{Z}_{nq_n} - \xi_{q_n}}{\tau_n} < x \right) - \Phi(x) | \le C_2 \max\{n^{-\frac{1}{2}}, \alpha_n, n^{\frac{1}{2}} \alpha_n\}$$

where C_2 is an absolute constant (depending only on q, F, K). The Berry–Esseen rate $O(n^{-\frac{1}{2}})$ is obtained by choosing the bandwidth $\alpha_n = O(n^{-1})$.

Theorem 3 and Theorem 5 imply that asymptotic normality and the Berry–Esseen bound for the perturbed sample quantile $\hat{Z}_{nq}$ depend on the selection of the bandwidth $\{\alpha_n\}$. Moreover, in order to obtain the Berry–Esseen rate $O(n^{-\frac{1}{2}})$, the bandwidth need to be smaller than is needed to ensure asymptotic normality.

A bound, $O(n^{-\frac{1}{4}} \log n)$, for the rate at which the distribution of the kernel quantile estimator $\hat{Z}_{nq}$ tends to its limit was presented by Falk (1985). See also Reiss (1989, p 264).

3.3. Equivalence between normal approximation and Bahadur–representation

It is known that $\hat{X}_{nq}$ as well as central order statistics can be asymptotically expressed as sums of i.i.d. random variables via representation as a linear transform of the sample d.f. F_n evaluated at ξ_q. The resulting representation is usually called the Bahadur representation. Ghosh (1971) noted that its use in deriving the asymptotic moments of $\hat{X}_{nq}$ goes back to Karl Pearson. It was Bahadur (1966) who first presented the representation in its own right, with a full view of its significance. Bahadur's work was subsequently refined by Ghosh (1971) and Kiefer (1967), and gave impetus to a number of important additional studies in the i.i.d. case as well as subsequent extensions to nonindependent sequences (see Serfling (1980), Shorack and Wellner (1986)).

Consider the representation

$$\sqrt{n}\frac{\hat{X}_{nq} - \xi_q}{\tau} = \sqrt{n}\frac{q - F_n(\xi_q)}{\sigma_q} + \sqrt{n}R_n$$

where $\tau > 0$, and R_n is the remainder term. Note that $\sqrt{n}\frac{q - F_n(\xi_q)}{\sigma_q}$ is asymptotically normal with the accuracy of order $O(n^{-\frac{1}{2}})$. Thus the key of the problem is the remainder term R_n as well as its precise behavior. Here, two modes of convergence are often involved, that is, with probability 1 and in probability, which give respectively a strong and a weak version of the Bahadur representation.

Assume that F is twice differentiable at ξ_q with $F'(\xi_q) > 0$, Bahadur's original result is with probability 1,

$$R_n = O\left(n^{-\frac{3}{4}}(\log n)^{\frac{1}{2}}(\log \log n)^{\frac{1}{4}}\right), \quad n \to +\infty \ .$$

Kiefer (1967) proved that the exact rate of R_n is $n^{-\frac{3}{4}}(\log \log n)^{\frac{3}{4}}$ and precisely obtained with probability 1,

$$\limsup_{n \to +\infty} \pm \left(\frac{n}{\log \log n}\right)^{\frac{3}{4}} R_n = \frac{2^{\frac{5}{4}}}{3^{\frac{3}{4}}} \sigma_q^{\frac{1}{2}}$$

for either choice of sign.

For many statistical applications, it suffices merely to have

$$\sqrt{n}R_n = o_p(1), \quad n \to +\infty \ .$$

This weak version of the Bahadur representation was obtained by Ghosh (1971) and requires only that F is once differentiable at ξ_q with $F'(\xi_q) > 0$. Indeed, this is also a necessary condition; see Lahiri (1992) for details, where a simpler proof can be given if one utilizes *Smirnov's Lemma*.

THEOREM 6 (Lahiri (1992)). Let $\{q_n\}$ be a sequence of constants satisfying

$$0 < q_n < 1, \quad \lim_{n \to +\infty} \sqrt{n}(q_n - q) = c_0 a$$

where a and $c_0 > 0$ are constants. Then the following statements are equivalent:

(1) For $F(\xi_q)(1 - F(\xi_q)) > 0$,

$$\hat{X}_{nq_n} = \xi_q + \frac{q_n - q}{c_0} + \frac{q - F_n(\xi_q)}{c_0} + R_n$$

holds, with

$$\sqrt{n} R_n = o_p(1), \quad n \to +\infty \ .$$

(2) For some $\tau > 0$

$$\sqrt{n} \frac{\hat{X}_{nq_n} - \xi_{q_n}}{\tau} \xrightarrow{\mathscr{D}} N(a, 1)$$

holds.

(3) F is differentiable at ξ_q and $F'(\xi_q) > 0$.

In this case $c_0 = F'(\xi_q)$, $\tau = \frac{\sigma_q}{F'(\xi_q)}$.

Assume that $\hat{F}_n$ is the sample kernel d.f. given by (1.4). Under appropriate conditions on F and K, proceeding in the same way as Bahadur (1966), Mack (1987) established a pointwise Bahadur type representation for the perturbed sample quantile $\hat{Z}_{nq}$, where the remainder term is of order $O(n^{-\frac{3}{4}}(\log n)^{\frac{3}{4}})$ with probability 1. A stronger result is given by Ralescu (1992b) with the same best order as obtained by Kiefer (1967) for $\hat{X}_{nq}$.

Bahadur representations of the kernel quantile estimator $\hat{Z}_{nq}$ are established by Xiang (1994) and Yang (1985).

3.4. Edgeworth-Type approximation

Consider an Edgeworth expansion for the distribution of a standardized version of $\hat{X}_{nq}$, say $\sqrt{n}\frac{\hat{X}_{nq} - \xi_q}{\tau}$ with $\tau > 0$. Due to Theorem 2, it is natural to choose $\tau = \frac{\sigma_q}{F'(\xi_q)}$.

Reiss (1976) dealt with an Edgeworth-type expansion for the distribution of $\hat{X}_{nq}$. The idea is based on *Smirnov's Lemma* or exactly its extended form (2.3). The leading term in such an expansion is the normal distribution, whereas, the higher order terms are given by integrals of polynomials with respect to the normal distribution. As the sample size n increases, these expansions establish a higher order approximation which holds uniformly in x. If F has $s + 2$ left and right derivatives at ξ_p, the error of the approximation is of the order $O(n^{-(s+1)})$. For a detailed description of Edgeworth-type approximation, we refer to Reiss (1989, Chapter 4).

However, in most cases of practical interest $F'(\xi_q)$ is unknown, we have to consider its estimation, which leads to studentizing a sample quantile.

A simple estimator of $\frac{1}{F'(\xi_q)}$ is the Siddiqui–Bloch–Gastwirth estimator, $S_{r,n}$, which is based on the difference between two order statistics whose indices are $2r$ apart, namely,

$$S_{r,n} = \frac{n}{2r}(X_{m+r:n} - X_{m-r:n})$$

where $r = r(n) \to +\infty$, and $\frac{r}{n} \to 0$ as $n \to +\infty$. In terms of Siddiqui–Bloch–Gastwirth estimator, Hall and Sheather (1988) derived the Edgeworth expansion for the studentized sample quantile $\sqrt{n}\frac{\hat{X}_{nq} - \xi_q}{\sigma_q S_{r,n}}$.

Another estimator of $F'(\xi_q)$, based on a kernel density estimator $\hat{f}_n$ of f, was proposed by Falk and Janas (1992). Letting

$$\tau_n = \frac{\sigma_q}{\hat{f}_n(\xi_q + b(\hat{X}_{nq} - \xi_q))}$$

with b a constant, they established an Edgeworth expansion of length two for the studentized sample quantile $\sqrt{n}\frac{\hat{X}_{nq} - \xi_q}{\tau_n}$.

4. Saddlepoint approximation

4.1. On the saddlepoint approach

Suppose that G is a d.f. and its cumulant generating function $\kappa(t) = \log \int e^{ty} dG(y)$ exists and is finite. The large deviation rate function, which plays a crucial role in the development of large deviation theory, is defined by

$$\rho(y) = \sup_{t \in R^1}[ty - \kappa(t)]$$

where the supremum is usually attained for every value of y with $0 < G(y) < 1$. In fact, we have

$$\rho(y) = \hat{t}y - \kappa(\hat{t})$$

where $\hat{t} = t(y)$, usually called the saddlepoint, is the solution of the equation

$$\kappa'(t) = y \qquad (4.1)$$

Under certain general conditions (see Daniels (1954), Kolassa (1994, Section 4.3)), (4.1) has a single real root $\hat{t}$, with $\kappa''(\hat{t}) > 0$. Hereafter denote

$$\hat{w}^2 = 2\rho(y) \qquad (4.2)$$

Suppose that $Y_1, \ldots, Y_n$ is a random sample from G and $\overline{Y} = \frac{1}{n}\sum_{j=1}^{n} Y_j$ is the sample mean. We want to approximate the tail probability $\overline{G}_n(y) = P(\overline{Y} \geq y)$; and the density or probability function of $\overline{Y}$, $g_n(y)$.

In his pioneering paper, Daniels (1954) derived the following saddlepoint approximation to $g_n(y)$ by using two techniques: the method of steepest descents of asymptotic analysis and the idea of the conjugate density or exponential tilting.

$$g_n(y) = \frac{e^{-n\rho(y)}}{[2\pi\kappa''(\hat{t})/n]^{\frac{1}{2}}} \{1 + O(n^{-1})\} \tag{4.3}$$

This says that the relative error of the saddlepoint approximation is of order $O(n^{-1})$. For a wide class of underlying densities, Daniels (1954) showed that the coefficient of the term of order n^{-1} doesn't depend on y and thus the relative error is of order $O(n^{-1})$ uniformly. This is the most important property of (4.3) and a major advantage with respect to an Edgeworth expansion. Therefore the saddlepoint approximation is often more accurate than the normal or even the one- or two-term Edgeworth series approximation. Another advantage of the saddlepoint approximation is that it is often astonishingly accurate for quite small sample sizes even down to $n = 1$ over the whole range of the variable.

There are essentially three methods of calculating the saddlepoint tail area approximation to $\overline{G}_n(y)$ in common use now; these are the indirect Edgeworth expansion (see e.g. Daniels (1987), Robinson, Hoglund, Holst and Quine (1990)), the numerically integrated saddlepoint density (see Field and Hampel (1982), Field and Ronchetti (1990)), usually renormalized for additional accuracy, and the Lugannani–Rice (1980) formula given by

$$\overline{G}_n(y) = 1 - \Phi(\sqrt{n}\hat{w}) + \phi(\sqrt{n}\hat{w})\left\{\left(\frac{1}{\hat{u}} - \frac{1}{\hat{w}}\right)n^{-\frac{1}{2}} + O\left(n^{-\frac{3}{2}}\right)\right\} \tag{4.4}$$

where the meaning of $\hat{u}$ will be explained below. Daniels (1987) compared the Lugannani-Rice formula and the indirect Edgeworth expansion, where the former is much simpler and easier to use than the latter, with relative error of order $O(n^{-\frac{3}{2}})$ in case of the mean, for exponential and inverse normal distributions, cases where the saddlepoint approximations to the density of the mean are "exact", and found that the latter performs slightly better than the former.

Another form similar to (4.4) was presented by Barndorff–Nielsen (1991), who discussed the relation of this kind of tail area approximation to the modified signed log likelihood ratio. Jensen (1992) showed that the Barndorff–Nielsen and the Lugannani–Rice approximations are equivalent.

Statistical applications of the saddlepoint approximation have been widely developed since the appearance of a discussion paper by Barndorff–Nielsen and Cox (1979). For more detailed summaries, we refer to Barndorff–Nielsen and Cox (1989), Field and Ronchetti (1990), Kolassa (1994) and Reid (1988).

Following the approach of Daniels(1987) and Lugannani and Rice (1980), this subsection will show how to derive Daniels' formula (4.3), the Lugannani–Rice formula (4.4) and Barndorff–Nielsen's formula in a unified way.

First of all, we use the Fourier inversion formula. It leads to

$$g_n(y) = \frac{n}{2\pi i} \int_{c-ic_1}^{c+ic_1} e^{n(\kappa(t)-ty)} \, dt \;,$$

and

$$\overline{G}_n(y) = \begin{cases} \frac{1}{2\pi i} \int_{c-ic_1}^{c+ic_1} e^{n(\kappa(t)-ty)} \frac{dt}{t}, & \text{for the continuous case} \\ \frac{1}{2\pi i} \int_{c-ic_1}^{c+ic_1} e^{n(\kappa(t)-ty)} \frac{dt}{1-e^{-t}}, & \text{for the lattice case} \end{cases}$$

where $c > 0$, and $c_1 = \pi$ (the lattice case) or $+\infty$ (the continuous case). Obviously, these formulas can be written in a unified form as below

$$Q_n(y) = \frac{1}{2\pi i} \int_{c-ic_1}^{c+ic_1} e^{n(\kappa(t)-ty)} l(t) \, dt \tag{4.5}$$

where $l(t) = n$, $\frac{1}{t}$ or $\frac{1}{1-e^{-t}}$.

There are at most two dominant critical points for the integral (4.5). The exponent $\kappa(t) - ty$ has a simple saddlepoint at $t = \hat{t}$. Another possible singularity critical point is the origin; indeed, $t = 0$ is a pole of $l(t)$ in the case $l(t) = \frac{1}{t}$ or $\frac{1}{1-e^{-t}}$.

The basic idea for deriving a saddlepoint expansion of $Q_n(y)$ is usually realized by two steps. Firstly, introduce in (4.5) a new variable of integration, say w, and make a transformation to replace the original exponent $\kappa(t) - ty$ by a simpler exponent, which should be one-to-one over a region containing both $t = 0$ and $t = \hat{t}$ when $\hat{t}$ is small. For given y and $\kappa(\cdot)$, such a transformation with the simplest form would be a polynomial in w of degree 2, say

$$\frac{1}{2}w^2 - \hat{w}w = \kappa(t) - ty \tag{4.6}$$

where we choose $\hat{w}$ so that the minimum of the left side of (4.6) is equal to that of the right side of (4.6), and thus $\hat{w} = w(y)$ is given by (4.2).

It follows from (4.5) and (4.6) that

$$Q_n(y) = \frac{1}{2\pi i} \int_{\hat{w}-i\infty}^{\hat{w}+i\infty} e^{n(\frac{1}{2}w^2-\hat{w}w)} l(t) \frac{dt}{dw} \, dw \tag{4.7}$$

Here, the possible two dominant critical points for (4.7) are $w = 0$ (a possible pole of $l(t)\frac{dt}{dw}$) and $w = \hat{w}$ (a simple saddlepoint of the exponent). In what follows we will write $\frac{1}{\hat{u}} = \left[l(t)\frac{dt}{dw}\right]_{w=\hat{w}}$ to indicate the value of $l(t)\frac{dt}{dw}$ at $w = \hat{w}$, then $\hat{u} = [\kappa''(\hat{t})]^{\frac{1}{2}}/l(\hat{t})$.

If $l(t) = n$, then $w = 0$ is not its pole and thus $l(t)\frac{dt}{dw}$ is analytic. Expanding $\frac{dt}{dw}$ about $\hat{w}$ and integrating (4.7), we obtain a saddlepoint expansion for the density or probability function $g_n(y)$ of the form

$$\sqrt{n}\phi(\sqrt{n}\hat{w})\{b_0 + b_1 n^{-1} + \cdots\}$$

where $b_0 = \frac{1}{n\hat{u}} = [\kappa''(\hat{t})]^{-\frac{1}{2}}$. Thus (4.3) follows.

If $l(t) = 1/t$ or $1/(1 - e^{-t})$, then the origin is its pole. Our next step, the key step of the approach of Daniels(1987) and Lugannani and Rice (1980), is to divide (4.7) into two terms. For a general consideration, let $w^* = \hat{w} + \frac{\zeta(\hat{w})}{n}$, where $\zeta(\hat{w})$ is a function of $\hat{w}$. Noting that $l(t)\frac{dt}{dw} - (e^{\zeta(\hat{w})w}/w)$ is analytic in a neighbourhood of $w = 0$, we rearrange (4.7) as

$$Q_n(y) = \frac{1}{2\pi i}\int_{w^*-i\infty}^{w^*+i\infty} e^{n(\frac{1}{2}w^2 - w^*w)}\frac{dw}{w} + \frac{1}{2\pi i}e^{-\frac{n}{2}\hat{w}^2} \\ \times \int_{\hat{w}-i\infty}^{\hat{w}+i\infty} e^{\frac{n}{2}(w-\hat{w})^2}\left\{l(t)\frac{dt}{dw} - \frac{e^{\zeta(\hat{w})w}}{w}\right\}dw \quad (4.8)$$

The benefit of this division is that the first term of (4.8) is easily evaluated, whose value is exactly $1 - \Phi(\sqrt{n}w^*)$, and simultaneously the dominant critical point for the second integral is reduced to one, namely, the saddlepoint $\hat{w}$. In the second term, using an approach similar to the above case, $l(t)dt/dw - (e^{\zeta(\hat{w})w}/w)$ is expanded about $\hat{w}$ and integrated to give an expansion of the form

$$\phi(\sqrt{n}\hat{w})\left\{c_0 n^{-\frac{1}{2}} + c_1 n^{-\frac{3}{2}} + \cdots\right\}$$

where $c_0 = 1/\hat{u} - (e^{\zeta(\hat{w})\hat{w}}/\hat{w})$. Therefore we get a generalized Lugannani–Rice formula

$$Q_n(y) = 1 - \Phi(\sqrt{n}w^*) + \phi(\sqrt{n}\hat{w})\left\{\left(\frac{1}{\hat{u}} - \frac{e^{\zeta(\hat{w})\hat{w}}}{\hat{w}}\right)n^{-\frac{1}{2}} + O(n^{-\frac{3}{2}})\right\} \quad (4.9)$$

In particular, choosing $\zeta \equiv 0$ in (4.9), $w^* = \hat{w}$ implies $c_0 = \frac{1}{\hat{u}} - \frac{1}{\hat{w}}$. As a result, the Lugannan–Rice formula (4.4) is obtained.

More interestingly, (4.9) becomes Barndorff–Nielsen's formula

$$Q_n(y) = (1 - \Phi(\sqrt{n}w^*))\{1 + O(n^{-1})\} \quad (4.10)$$

if $c_0 = 0$, that is $w^* = \hat{w} + \frac{1}{n\hat{w}}\log\frac{\hat{u}}{\hat{w}}$; where the error holds uniformly for y in a compact set. Furthermore, for any $c > 0$, the error $O(n^{-1})$ can be replaced by $O(n^{-\frac{3}{2}})$ for $|y - EY| < cn^{-\frac{1}{2}}$.

4.2. Saddlepoint approximation to the quantiles

Hampel (1974) introduced a technique, which is an example of what he called "small sample asymptotics" where high accuracy is achieved for quite small sample sizes n, even down to single figures, for approximating to the density of M-estimates. This technique was developed by Field and Hampel (1982) in detail, and its performance compared with that of other approximation methods, such as Edgeworth expansions, large deviation theory, et al.. As Hampel pointed out, this approach is closely related to the saddlepoint method of Daniels (1954); saddlepoint approximations for M-estimates are also given in Daniels (1983). In particular, he obtained the saddlepoint approximation to the density of $\hat{X}_{nq}$ as

$$\left(\frac{n}{2\pi}\right)^{\frac{1}{2}}\left(\frac{n}{m}\right)^{m-\frac{1}{2}}\left(\frac{n}{n-m}\right)^{n-m+\frac{1}{2}} F^{m-1}(x)(1-F(x))^{n-m} f(x) \qquad (4.11)$$

The difference between (1.3) and (4.11) is that the normalizing constant is replaced by its Stirling approximation.

Applying Barndorff–Nielsen formula's (4.10), Ma and Robinson (1994) get a saddlepoint approximation to the binomial tail probability, and observing (1.2) they obtain the saddlepoint approximation of the distribution function of $\hat{X}_{nq}$

$$P(\hat{X}_{nq} \leq x) = \Phi(\sqrt{n} w^*(F(x)))\{1 + O(n^{-1})\} \qquad (4.12)$$

where

$$w^*(t) = \begin{cases} w(t) + \frac{1}{nw(t)} \log \frac{u(t)}{w(t)}, & \text{if } t \neq q \\ -\frac{1}{2n}\sigma_q^{-1}, & \text{if } t = q \end{cases}$$

and $u(t), w(t)$ are given by

$$u(t) = \left(\frac{1-q}{q}\right)^{\frac{1}{2}} \frac{t-q}{1-t},$$

$$w(t) = \operatorname{sgn}\{t-q\}\left\{2\left[q \cdot \log\left(\frac{q}{t}\right) + (1-q) \cdot \log\left(\frac{1-q}{1-t}\right)\right]\right\}^{\frac{1}{2}}.$$

As shown by Ma and Robinson (1994), the asymptotic normal distribution theory of the sample quantiles is equivalent to employing a linear approximation of $w^*(F(x))$ or $w(F(x))$ in the saddlepoint approximation. This saddlepoint approximation is better than the normal approximation in the sense that there are weaker conditions and higher accuracy.

5. Bootstrap approximation

Bootstrap methods have earned an important place in the statistician's toolkit since their systematic introduction by Efron (1979). Recently Beran and Ducharme (1991) and Hall (1992) give a mathematically sophisticated treatment of the bootstrap approach. Efron and Tibshirani (1993) presents an overview of the bootstrap and related methods for assessing statistical accuracy.

In connection with covering probabilities and confidence intervals one is interested in the d.f. of the statistic $\sqrt{n}(\hat{X}_{nq} - \xi_q)$,

$$T_n(F, x) = P_F(\sqrt{n}(\hat{X}_{nq} - \xi_q) \leq x) \ . \tag{5.1}$$

Usually, ξ_q as well as F are unknown. The basic idea of the bootstrap method is to replace ξ_q and F by certain kind of their sample counterpart, and then estimate the d.f. $T_n(F, \cdot)$ by the corresponding sample counterpart.

A basic way of validating a particular bootstrap method is by proving that it is consistent: conditionally on the observed data the bootstrap distribution has the same asymptotic behaviour as the (centered, standardized) sample distribution of the original estimator either with probability 1 or in probability. Singh (1981) established the consistency of the bootstrap approximation of $T_n(F, \cdot)$ and provided the exact rate, which is $O(n^{-\frac{1}{4}}(\log\log n)^{\frac{1}{2}})$ with probability 1, at which the discrepancy converges to zero. Babu and Singh (1984) obtained strong representations of the bootstrap quantiles and L-statistics. Falk (1986) proved that the accuracy of the bootstrap approximation of the joint distribution of the vector of the sample quantiles lies between $O(n^{-\frac{1}{4}})$ and $O(n^{-\frac{1}{4}}c_n)$, where $(\log n)^{\frac{1}{2}} = o(c_n)$. Bootstrapping the sample quantiles is summarized by Falk (1992), see also Babu and Rao (1993).

5.1. The standard bootstrap

The standard bootstrap technique is to estimate ξ_q by the sampling method, with the samples being drawn not from the underlying d.f. F but from the sample d.f. F_n. A bootstrap sample $X_1^*, \ldots, X_n^*$, a random sample from F_n, is then generated by successively selecting uniformly with replacement from the observed data $X_1, \ldots, X_n$. Denote by $\hat{X}_{nq}^*$ the corresponding sample q^{th} quantile. Replacing ξ_q and F in (5.1) by $\hat{X}_{nq}$ and F_n, respectively, we estimate $T_n(F, \cdot)$ by

$$T_n(F_n, x) = P_{F_n}(\sqrt{n}(\hat{X}_{nq}^* - \hat{X}_{nq}) \leq x) \ .$$

Now let's consider the bootstrap error

$$T_n(F_n, x) - T_n(F, x) = P_{F_n}(\sqrt{n}(\hat{X}_{nq}^* - \hat{X}_{nq}) \leq x) - P_F(\sqrt{n}(\hat{X}_{nq} - \xi_q) \leq x) \ .$$

It follows from *Smirnov's Lemma* that

$$\sup_{x \in R^1} \left| T_n(F, x) - \Phi\left(\sqrt{n}\frac{F(\xi_q + \frac{x}{\sqrt{n}}) - q}{\sigma_q}\right) \right| = O(n^{-\frac{1}{2}}) \ ,$$

$$\sup_{x \in R^1} \left| T_n(F_n, x) - \Phi\left(\sqrt{n}\frac{F_n(\hat{X}_{nq} + \frac{x}{\sqrt{n}}) - q}{\sigma_q}\right) \right| = O(n^{-\frac{1}{2}}) \ ,$$

and thus

$$\sup_{x\in R^1}|T_n(F_n,x) - T_n(F,x)| \leq \sup_{x\in R^1}\left|\Phi\left(\sqrt{n}\frac{F\left(\xi_q + \frac{x}{\sqrt{n}}\right) - q}{\sigma_q}\right)\right.$$
$$\left. - \Phi\left(\sqrt{n}\frac{F_n\left(\hat{X}_{nq} + \frac{x}{\sqrt{n}}\right) - q}{\sigma_q}\right)\right| + O(n^{-\frac{1}{2}}) .$$

Consequently the accuracy of the bootstrap approximation is closely related to

$$\sup_{x\in R^1}\left|\Phi\left(\sqrt{n}\frac{F\left(\xi_q + \frac{x}{\sqrt{n}}\right) - q}{\sigma_q}\right) - \Phi\left(\sqrt{n}\frac{F_n\left(\hat{X}_{nq} + \frac{x}{\sqrt{n}}\right) - q}{\sigma_q}\right)\right| .$$

Hence the key to the problem is to compare

$$\sqrt{n}\frac{F_n(\hat{X}_{nq} + \frac{x}{\sqrt{n}}) - q}{\sigma_q} \quad \text{with} \quad \sqrt{n}\frac{F(\xi_q + \frac{x}{\sqrt{n}}) - q}{\sigma_q} .$$

The following ingenious result, which is similar to Bahadur (1966, Lemma 1), was given by Singh (1981, Lemma 3.2)

$$\sup_{|x|\leq \log n} (1+|x|)^{-\frac{1}{2}}\left|F_n\left(\hat{X}_{nq} + \frac{x}{\sqrt{n}}\right) - F_n(\hat{X}_{nq}) - F\left(\hat{X}_{nq} + \frac{x}{\sqrt{n}}\right) + F(\hat{X}_{nq})\right|$$
$$= O(n^{-\frac{3}{4}}(\log\log n)^{\frac{1}{2}})$$

(5.2)

Based on this result, Singh (1981) established

THEOREM 7 (Singh (1981)).: If F has a bounded second derivative in a neighborhood of ξ_q with $F'(\xi_q) > 0$, then with probability 1

$$\limsup_{n\to+\infty} n^{\frac{1}{4}}(\log\log n)^{-\frac{1}{2}}\sup_{x\in R^1}|T_n(F_n,x) - T_n(F,x)| = C_0$$

where the constant C_0 depends only upon q and F.

Ma and Robinson (1994) consider the relative error between $T_n(F,\cdot)$ and $T_n(F_n,\cdot)$ by using (4.12) and (5.2), and obtain

$$T_n(F,x) = T_n(F_n,x)\left(1 + O_p\left\{n^{-\frac{1}{4}}(\log\log n)^{\frac{1}{2}}x(1+|x|)^{\frac{1}{2}}\right\}\right) .$$

5.2. *The smoothed bootstrap*

As a modification to the standard bootstrap procedure, the essential idea of the smoothed bootstrap (Efron (1979)) is to perform the repeated sampling from a smoothed version $\hat{F}_n$ of F_n, rather than sampling from F_n itself. The smoothed bootstrap has been studied by Falk and Reiss (1989a, 1989b), Hall, DiCiccio and Romano (1989), and Silverman and Young (1987), amongst others. For a review, see Angelis and Young (1992).

Assume that $\hat{F}_n$ is a perturbed sample d.f. and $\tilde{Z}_{nq}$ is the associated q^{th} quantile. Let $Z_1^*, \ldots, Z_n^*$ be a random sample from $\hat{F}_n$, and denote by $\hat{Z}_{nq}^*$ the corresponding sample q^{th} quantile. Taking $\tilde{Z}_{nq}$ and $\hat{F}_n$ respectively as an estimate of ξ_q and F in (5.1), we estimate $T_n(F, \cdot)$ by

$$T_n(\hat{F}_n, x) = P_{\hat{F}_n}(\sqrt{n}(\hat{Z}_{nq}^* - \tilde{Z}_{nq}) \leq x) .$$

As above, the key of the problem is to compare

$$\sqrt{n}\frac{\hat{F}_n(\tilde{Z}_{nq} + \frac{x}{\sqrt{n}}) - q}{\sigma_q} \quad \text{with} \quad \sqrt{n}\frac{F(\xi_q + \frac{x}{\sqrt{n}}) - q}{\sigma_q} .$$

Two versions of the smoothed bootstrap for the sample quantiles were described by Falk and Reiss (1989a, 1989b), where $\tilde{Z}_{nq} = \hat{Z}_{nq}$ and $\hat{Z}_{nq}$, respectively.

Falk and Reiss (1989a) took $\hat{F}_n$ as a kernel estimate of F_n, given by (1.4). Under certain regularity conditions of F and K, they showed that uniformly for $|x| \leq \log n$

$$\sqrt{n}\left\{\hat{F}_n\left(\hat{Z}_{nq} + \frac{x}{\sqrt{n}}\right) - F\left(\xi_q + \frac{x}{\sqrt{n}}\right)\right\} = (\hat{f}_n(\xi_q) - f(\xi_q))x + o_p((n\alpha_n)^{-\frac{1}{2}}) .$$

Consequently the accuracy of this smooth bootstrap approximation is roughly $O(n^{-\frac{2}{3}})$ for an appropriate choice of α_n. So in this case the smooth bootstrap estimate outperforms the nonsmoothed one.

Falk and Reiss (1989b) adopted another smoothed version of F_n, where rather than smoothing the sample d.f. F_n directly, they smoothed the sample q.f. F_n^{-1} by the kernel quantile estimate method. The rate of convergence of this smoothed bootstrap estimate is roughly $O(n^{-\frac{1}{3}})$.

References

Angelis, D. D. and G. A. Young (1992). Smoothing the bootstrap. *Inter. Statist. Rev.* **60**, 45–56.
Babu, G. J. and C. R. Rao (1993). Bootstrap methodology. In: C. R. Rao, ed., *Handbook of Statistics, Vol. 9: Comput. Statist.* Elsevier Science Publisher B.V., 627–659.
Babu, G. J. and K. Singh (1984). Asymptotic representations related to jackknifing and bootstrapping L-statistics. *Sankhyā Ser. A.* **46**, 195–206.
Bahadur, R. R. (1966). A note on quantiles in large samples. *Ann. Math. Statist.* **37**, 577–580.
Balkema, A. A. and L. de Haan (1978). Limit distributions for order statistics I, II. *Theory Probab. Appl.* **23**, 77–92, 341–358.
Barndorff-Nielsen, O. E. (1991). Modified signed log likelihood ratio. *Biometrika* **78**, 557–563.
Barndorff-Nielsen, O. E. and D. R. Cox (1979). Edgeworth and saddlepoint approximations with statistical applications (with discussion). *J. Roy. Statist. Soc. Ser. B.* **41**, 279–312.
Barndorff-Nielsen, O. E. and D. R. Cox (1989). *Asymptotic Techniques for Use in Statistics*. Chapman and Hall, London.

Beran, R and G. Ducharme (1991). *Asymptotic Theory for Bootstrap Methods in statistics*. Centere de Reserches Mathematiques, University of Montreal.

Daniels, H. E. (1954). Saddlepoint approximations in statistics. *Ann. Math. Statist.* **25**, 631–650.

Daniels, H. E. (1983). Saddlepoint approximations for estimating equations. *Biometrika* **70**, 89–96.

Daniels, H. E. (1987). Tail probability approximations. *Inter. Statist. Rev.* **55**, 37–48.

Efron, B. (1979). Bootstrap methods: Another look at the jacknife. *Ann. Statist.* **7**, 1–26.

Efron, B. and R. J. Tibshirani (1993). *An Introduction to the Bootstrap*. Chapman and Hall, New York.

Falk, M. (1984). Relative deficiency of kernel type estimators of quantiles. *Ann. Statist.* **12**, 261–268.

Falk, M. (1985). Asymptotic normality of the kernel quantile estimator. *Ann. Statist.* **13**, 428–433.

Falk, M. (1986). On the accuracy of the bootstrap approximation of the joint distribution of sample quantiles. *Commun. Statist.-Theory Meth.* **15**, 2867–2876.

Falk, M. (1992). Bootstrapping the sample quantiles: a survey. In: K.-H. Jochel, G. Rothe, and W. Sendler eds., *Bootstrapping and Related Techniques, proceedings,Trier,FRG*, 1990. Lecture Notes in Economics and Mathematical Systems **376**, Springer-Verlag. 165–172.

Falk, M. and D. Janas (1992). Edgeworth expansions for studentized and prepivoted sample quantiles. *Statist. Probab. Lett.* **14**, 13–24.

Falk, M. and R.-D. Reiss (1989a). Weak convergence of smoothed and nonsmoothed bootstrap quantile estimates. *Ann. Prob.* 362–371.

Falk, M. and R.-D. Reiss (1989b). Bootstrapping the distance between smooth bootstrap and sample quantile function. *Probab. Th. Rel. Fields* **82**, 177–186.

Field, C. A. and F. R. Hampel (1982). Small-sample asymptotic distributions of M-estimators of location. *Biometrika*, **69**, 29–46.

Field, C. A. and E. Ronchetti (1990). *Small Sample Asymptotics*. IMS Notes **13**, Hayward, CA.

Ghosh, J. K. (1971). A new proof of the Bahadur representation of quantiles and an application. *Ann. Math. Statist.* **42**, 1957–1961.

Hall, P. (1992). *The Bootstrap and Edgeworth Expansion*. Springer-Verlag, New York.

Hall, P., T. J. DiCiccio and J. P. Romano (1989). On smoothing and the bootstrap. *Ann. Statist.* **17**, 692–704.

Hall, P. and S. J. Sheather (1988). On the distribution of a studentized quantile. *J. Roy. Statist. Soc. Ser. B.* **50**, 381–391.

Hampel, F. R. (1974). Some small sample asymptotics. In: J. Hajek ed., *Proceedings of the Second Prague Symposium on Asymptotic Statistics*, 1973 **2**. Charles University, Prague. 109–126.

Jensen, J. L. (1992). The modified signed likelihood statistic and saddlepoint approximations. *Biometrika* **79**, 693–703.

Kaigh, W. D. (1988). O-statistics ans their applications. *Commun. Statist.–Theory Meth.* **11**, 2217–2238.

Kiefer, J. (1967). On Bahadur's representation of sample quantiles. *Ann. Math. Statist.* **38**, 1323–1342.

Kolassa, J. E. (1994). *Series Approximation Methods in Statistics*. Lecture Notes in Statistics **88**. Springer-Verlag, New York.

Lahiri, S. N. (1992). On the Bahadur-Ghosh-Kiefer representation of sample quantiles. *Statist. Prob. Lett.*, **15**, 163–168

Lugannani, R. and S. O. Rice (1980). Saddlepoint approximation for the distribution function of the sum of independent random variables. *Adv. Appl. Prob.* **12**, 475–490.

Ma, C. and J. Robinson (1994). Saddlepoint approximation to the sample quantitles. *Submitted for publication*.

Mack, Y. DP. (1987). Bahadur representation of sample quantiles based on smoothed estimates of a distribution function. *Probab. Math. Statist.* **8**, 183–189.

Nadaraya, E. A. (1964). Some new estimate for distribution functions. *Theory Prob. Appl.* **9**, 497–500.

Parzen, E. (1962). On estimation of a probability density function and mode. *Ann. Math. Statist.* **33**, 1065–1076.

Parzen, E. (1979). Nonparametric statistical data modeling. *J. Amer. Statist. Assoc.* **74**, 105–131.

Puri, M. L. and S. S. Ralescu (1986). Limit theorems for random central order statistics. In: J. V.Ryzin ed., *Adaptive Statistical Procedures and Related Topics*. IMS Lecture Notes **8**, 467–475.

Ralescu, S. S. (1992a). A remainder estimate for the normal approximation of perturbed sample quantiles. *Statist. Probab. Lett.* **14**, 293–298.

Ralescu, S. S. (1992b). Asymptotoic derivations between perturbed empirical and quantile processes. *J. Statist. Plann. Infer.* **32**, 243–258.

Ralescu, S. S. and S. Sun (1993). Necessary and sufficient conditions for the asymptotic normality of perturbed sample quantiles. *J. Statist. Plann. Infer.* **35**, 55–64.

Reid, N. (1988). Saddlepoint methods and statistical inference. *Statist. Sci.* **3**, 213–238.

Reiss, R.-D. (1974). On the accuracy of the normal approximation for quantiles. *Ann. Prob.* **2**, 741–744.

Reiss, R.-D. (1976). Asymptotic expansions for sample quantiles. *Ann. Prob.* **4**, 249–258.

Reiss, R.-D. (1982). One-sided test for quantiles in certain nonparametric models. In: B. V. Gnedenko, M. L. Puri and I. Vincze eds., *Nonparametric Statistical Inference, Colloq Math. Soc. János Bolyai 32*, North-Holland Publishing Company, Amsterdam, 759–772.

Reiss, R.-D. (1989). *Approximate Distributions of Order Statistics with Applications to Nonparametric Statistics*. Springer-Verlag, New York.

Robinson, J. (1982). Saddlepoint approximations for permutation tests and confidence intervals. *J. Roy. Statist. Soc. Ser. B.* **44**, 91–101.

Robinson, J., T. Hoglund, L. Holst and M. P. Quine (1990). On approximating probabilities for small and large deviations in R^d. *Ann. Prob.* **18**, 727–753.

Rosenblatt, M. (1956). Remarks on some nonparametric estimates of a density function. *Ann. Math. Statist.* **27**, 832–837.

Seoh, M. and M. L. Puri (1994). Berry-Esséen rate in asymptotic normality for perturbed sample quantiles. *Metrika* **41**, 99–108.

Serfling, R. J. (1980). *Approximation Theorems of Mathematical Statistics*. Wiley, New York.

Sheather, S. J. and J. S. Marron (1990). Kernel quantile estimators. *J. Amer. Statist. Assoc.* 410–416.

Shorack, G. R. and J. A. Wellner (1986). *Empirical Processes with Applications to Statistics*. Wiley, New York.

Silverman, B. W. and G. A. Young (1987). The bootstrap: to smooth or not to smooth? *Biometrika* **74**, 469–479.

Singh, K. (1981). On the asymptotic accuracy of Efrons bootstrap. *Ann. Statist.* **9**, 1187–1195.

Smirnov, N. V. (1949). Limit distributions for the terms of a variational series. *Trudy Mat. Inst. Steklov.* **25**, 1–60. In Russia. (English translation in *Amer. Math. Soc. Transl. Ser.* **1 (11)**, 1962, 82–143.

Stigler, S. M. (1973). Simon Newcomb, Percy Daniell, and the history of robust estimation 1885-1920. *J. Amer. Statist. Assoc.* **68**, 872–879.

Xiang, X. J. (1994). Bahadur representation of kernel quantile estimators. *Scand. J. Statist.* **21**, 169–178.

Yang, S. S. (1985). A smooth nonparametric estimator of a quantile function. *J. Amer. Statist. Assoc.* **80**, 1004–1011.

Zelterman, D. (1990). Smooth nonparametric estimation of the quantile function. *J. Statist. Plann. Infer.* **26**, 339–352.

PART VIII

Related Statistics

Concomitants of Order Statistics

H. A. David and H. N. Nagaraja

1. Introduction and summary

Let (X_i, Y_i), $i = 1, \ldots, n$, be a random sample from a bivariate distribution with cumulative distribution function (cdf) $F(x,y)$. If the sample is ordered by the X_i, then the Y-variate associated with the rth order statistic $X_{r:n}$ will be denoted by $Y_{[r:n]}$ and termed the *concomitant of the rth order statistic* (David, 1973, 1981).

The most important use of concomitants arises in selection procedures when $k(< n)$ individuals are chosen on the basis of their X-values. Then the corresponding Y-values represent performance on an associated characteristic. For example, if the top k out of n rams, as judged by their genetic make-up, are selected for breeding, then $Y_{[n-k+1:n]}, \ldots, Y_{[n:n]}$ might represent the quality of the wool of one of their female offspring. Or X might be the score of a candidate on a screening test and Y the score on a later test. There are related problems dealing with the estimation of parameters from data in which selection has taken place. The study of some aspects of these problems antedates the term concomitant of order statistics (e.g., Watterson, 1959). However, the occurrence of concomitants in a variety of contexts independently prompted also another term, *induced order statistics* (Bhattacharya, 1974). As pointed out by Sen (1981), linear functions of concomitants may also be viewed as mixed rank statistics (Ghosh and Sen, 1971). Egorov and Nevzorov (1982) use the term induced order statistics for the X-values arranged according to ordered $g(X)$ values for a prespecified function g.

Some generalizations should be noted. Associated with X_i there may be ℓ variates $Y_{1i}, \ldots, Y_{\ell i}$, $i = 1, \ldots, n$. There is no clear-cut way of ordering the $(\ell + 1)$-variate measurements $(X_i, Y_{1i}, \ldots, Y_{\ell i})$. One way of doing so is by the order of the X-values (see Barnett, 1976) in which case associated with $X_{r:n}$ will be the vector of concomitants $\left(Y_{1[r:n]}, \ldots, Y_{\ell[r:n]}\right)$. This situation is included in a unified treatment of distribution theory given in Section 2. One could also contemplate selection based on more than a single X. This, however, is awkward and has not been pursued, although selection based on a univariate function of such X's would present no difficulties. An interesting further generalization is due to Egorov and Nevzorov (1984), Reiss (1989, p. 66), and Kaufmann and Reiss (1992). These authors order vectors x_i, $i = 1, \ldots, n$, by the size of a real-valued function $g(x_i)$. Kaufmann and Reiss define the g-ordering

$$x_i \leq_g x_j \quad \text{if } g(x_i) \leq g(x_j) .$$

In particular, if $g(x_i) = g(x_{1i})$, then the vectors are ordered by the first component x_{1i}, and the other components become the concomitants. Concomitants can also be associated with record values (Houchens, 1984; Ahsanullah, 1994) and with generalized order statistics which include order statistics and record values as special cases (Kamps, 1995).

Before outlining the content of the remainder of this article, we mention some general issues. Lo and McKinlay (1990) make a rather basic point, namely that in some practical situations the effect of previous selection of objects by their X-values is ignored and the concomitants for the chosen objects are treated simply as random Y's. The authors study the resulting effects on some standard tests of significance, with specific reference to financial asset pricing models. David and Gunnink (1997) re-examine the paired t-test when $2n$ individuals are paired on the basis of closeness of prior related measurements, i.e., the pairs correspond to the measurements $(x_{2i:2n}, x_{2i-1:2n})$, $i = 1, \ldots, n$. The order within pairs is randomized before two treatments are applied. When a t-test is performed on the signed differences $\pm(y_{[2i:2n]} - y_{[2i-1:2n]})$, of the experimental measurements, the assumptions of the paired t-test no longer hold, but simulation indicates that under a bivariate normal model the test continues to be valid.

In sections 2 and 3 we strive to provide a unified account of the basic finite-sample and asymptotic theory of concomitants, together with multivariate generalizations. Section 4 deals with estimation and hypothesis testing. It is shown how concomitants enter into (a) the estimation of regression and correlation coefficients in a variety of situations, (b) the analysis of censored bivariate data, (c) ranked-set sampling, and (d) double sampling. The rank of $Y_{[r:n]}$ among the Y's, which plays an important role in certain selection procedures, is studied in section 5. Other selection procedures using concomitants are treated in section 6. The asymptotic theory of various functions of concomitants is examined in section 7. Applications include (a) inference on the regression function $E(Y|x)$, (b) the induced selection differential, (c) bootstrapping, and (d) file-matching procedures.

A fine earlier review of concomitants is given in Bhattacharya (1984). Previous reviews by the first author are superseded by the present account which attempts to include a nearly complete set of references for the literature on concomitants.

2. Finite-sample distribution theory and moments

2.1. The simple linear model

We begin with an important special case for which rather explicit results are possible. Suppose that X_i and Y_i ($i = 1, \ldots, n$) have means μ_X, μ_Y, variances σ_X^2, σ_Y^2, and are linked by the linear regression model ($|\rho| < 1$)

$$Y_i = \mu_Y + \rho \frac{\sigma_Y}{\sigma_X}(X_i - \mu_X) + \epsilon_i, \qquad (2.1)$$

where the X_i and the ϵ_i are mutually independent. Then from (2.1) it follows that $E\epsilon_i = 0$, var $\epsilon_i = \sigma_Y^2(1-\rho^2)$, and $\rho = \text{corr}(X,Y)$. In the special case when the X_i and ϵ_i are normal, X_i and Y_i are bivariate normal. Ordering on the X_i we have for $r = 1, 2, \ldots, n$

$$Y_{[r:n]} = \mu_Y + \rho \frac{\sigma_Y}{\sigma_X}(X_{r:n} - \mu_X) + \epsilon_{[r]}, \qquad (2.2)$$

where $\epsilon_{[r]}$ denotes the particular ϵ_i associated with $X_{r:n}$. In view of the independence of the X_i and the ϵ_i we see that the set of $X_{r:n}$ is independent of the $\epsilon_{[r]}$, the latter being mutually independent, each with the same distribution as ϵ_i.

Setting

$$\alpha_{r:n} = E\left(\frac{X_{r:n} - \mu_X}{\sigma_X}\right) \quad \text{and} \quad \beta_{rs:n} = \text{cov}\left(\frac{X_{r:n} - \mu_X}{\sigma_X}, \frac{X_{s:n} - \mu_X}{\sigma_X}\right),$$

$r, s = 1, 2, \ldots, n$, we have from (2.2)

$$\begin{aligned} EY_{[r:n]} &= \mu_Y + \rho \sigma_Y \alpha_{r:n}, \\ \text{var } Y_{[r:n]} &= \sigma_Y^2 (\rho^2 \beta_{rr:n} + 1 - \rho^2), \\ \text{cov}(X_{r:n}, Y_{[s:n]}) &= \rho \sigma_X \sigma_Y \beta_{rs:n}, \\ \text{cov}(Y_{[r:n]}, Y_{[s:n]}) &= \rho^2 \sigma_Y^2 \beta_{rs:n}, \quad r \neq s. \end{aligned} \qquad (2.3\text{a-d})$$

In the bivariate normal case Eqs. (2.3) were given by Watterson (1959). An interesting way of expressing (2.3a,b,d) brings out the relations between the moments of the $Y_{[r:n]}$ and the $Y_{r:n}$ (Sondhauss, 1994):

$$\begin{aligned} EY_{[r:n]} - \mu_Y &= \rho(EY_{r:n} - \mu_Y), \\ \text{var } Y_{[r:n]} - \sigma_Y^2 &= \rho^2 (\text{var } Y_{r:n} - \sigma_Y^2), \\ \text{cov}(Y_{[r:n]}, Y_{[s:n]}) &= \rho^2 \text{cov}(Y_{r:n}, Y_{s:n}), \quad r \neq s. \end{aligned}$$

A generalization of (2.1) may be noted here. Let $Y_i = g(X_i, \epsilon_i)$ represent a general model for the regression of Y on X, where neither the X_i nor the ϵ_i need be identically distributed (but are still independent). Then

$$Y_{[r:n]} = g(X_{r:n}, \epsilon_{[r]}) \quad r = 1, \ldots, n. \qquad (2.4)$$

From the mutual independence of the X_i and the ϵ_i it follows that $\epsilon_{[r]}$ has the same distribution as the ϵ_i accompanying $X_{r:n}$ and that the $\epsilon_{[r]}$ are mutually independent (Kim and David, 1990).

2.2. General results

Dropping now the structural assumption (2.1), we see that quite generally, for $1 \leq r_1 < r_2 < \cdots < r_k \leq n$, the $Y_{[r_h:n]}$ ($h = 1, \ldots, k$) are conditionally independent given $X_{r_h:n} = x_h$ $(h = 1, 2, \ldots, k)$. The joint conditional pdf may be written as $\prod_{h=1}^{k} f(y_h|x_h)$. It follows that

$$f_{Y_{[r_1:n]}, \ldots, Y_{[r_k:n]}}(y_1, \ldots, y_k)$$
$$= \int_{-\infty}^{\infty} \int_{-\infty}^{x_k} \cdots \int_{-\infty}^{x_2} f_{X_{r_1:n}, \ldots, X_{r_k:n}}(x_1, \ldots, x_k) \prod_{h=1}^{k} [f(y_h|x_h) \, dx_h] \; . \quad (2.5)$$

Put $m(x) = E(Y|X = x)$ and $\sigma^2(x) = \text{var}(Y|X = x)$ (Bhattacharya, 1974). It follows (Yang, 1977), in generalization of (2.3), that

$$\begin{aligned} E(Y_{[r:n]}) &= E[m(X_{r:n})] \; , \\ \text{var}(Y_{[r:n]}) &= \text{var}[m(X_{r:n})] + E[\sigma^2(X_{r:n})] \; , \\ \text{cov}(X_{r:n}, Y_{[s:n]}) &= \text{cov}[X_{r:n}, m(X_{s:n})] \; , \\ \text{cov}(Y_{[r:n]}, Y_{[s:n]}) &= \text{cov}[m(X_{r:n}), m(X_{s:n})], \quad r \neq s \end{aligned} \quad (2.6\text{a–d})$$

For example, (2.6b) and (2.6d) are special cases of the general formula (subject to the existence of the quantities involved) for the rv's U, V, W

$$\text{cov}(U, V) = \text{cov}(E(U|W), E(V|W)) + E[\text{cov}(U, V|W)] \quad (2.7)$$

with $U = Y_{r:n}$, $V = Y_{s:n}$, and $W = X_{r:n}$ in (2.6b) and $W = (X_{r:n}, X_{s:n})$ in (2.6d).

Jha and Hossein (1986) note that (2.6) continues to hold when X is absolutely continuous but Y discrete. By straightforward arguments they point out also that for any exchangeable variates (X_i, Y_i), $i = 1, \ldots, n$, familiar recurrence relations for order statistics continue to hold for concomitants.

Specific results when (X, Y) has Gumbel's bivariate exponential distribution are given by Balasubramanian and Beg (1996).

2.3. Multivariate generalizations

Next, suppose that associated with each X there are ℓ variates Y_j ($j = 1, \ldots, \ell$), i.e., we have n independent sets of variates $(X_i, Y_{1i}, \ldots, Y_{\ell i})$. Triggered by a problem in hydrology, this situation has recently been intensively studied with increasing degrees of generality, especially when the $\ell + 1$ variates have a multivariate normal distribution (Song, Buchberger, and Deddens, 1993; Song and Deddens, 1993; Balakrishnan, 1993; Song and Balakrishnan, 1994). See also David and Galambos (1974, p. 765).

We begin without assuming multivariate normality. Setting $m_j(x_i) = E(Y_{ji}|x_i)$ and writing $Y_{j[r:n]}$ for that Y_{ji} paired with $X_{r:n}$, we have

$$EY_{j[r:n]} = E[m_j(X_{r:n})] \; .$$

Also, a slightly different application of (2.7) gives, for $k = 1, \ldots, \ell$,

$$\text{cov}(Y_{j[r:n]}, Y_{k[r:n]}) = \text{cov}(m_j(X_{r:n}), m_k(X_{r:n})) + E\sigma_{jk}(X_{r:n}) \; , \quad (2.8)$$

where $\sigma_{jk}(x_i) = \text{cov}(Y_{ji}, Y_{ki}|x_i)$.

In the multivariate normal case $\sigma_{jk}(x_i)$ does not depend on x_i and may be obtained from standard theory (e.g., Anderson, 1984, p. 35). Let

$$\Sigma = \begin{pmatrix} \Sigma_{11} & \Sigma_{12} \\ \Sigma_{21} & \Sigma_{22} \end{pmatrix}$$

where

$$\Sigma_{11} = \sigma_X^2, \quad \Sigma_{12} = \text{cov}(X, Y_j)_{1 \times \ell} = (\sigma_{xj}), \text{ say}$$

and

$$\Sigma_{22} = \text{cov}(Y_j, Y_k)_{\ell \times \ell} = (\sigma_{jk}) \ .$$

Then from the result

$$\Sigma_{22 \cdot 1} = \Sigma_{22} - \Sigma_{21} \Sigma_{11}^{-1} \Sigma_{12}$$

we have here

$$\Sigma_{22 \cdot 1} = \Sigma_{22} - \Sigma_{21} \Sigma_{12} / \sigma_X^2$$

or

$$\sigma_{jk}(x) = \sigma_{jk} - \sigma_{Xj} \sigma_{Xk} / \sigma_X^2 \ . \tag{2.9}$$

Also, with $\mu_j = EY_j$, $\sigma_j^2 = \text{var } Y_j$, and $\rho_j = \text{corr}(X, Y_j)$, Eq. (2.2) becomes

$$Y_{j[r:n]} = \mu_j + \rho_j \sigma_j (X_{r:n} - \mu_x) / \sigma_X + \epsilon_{j[r]} \ . \tag{2.10}$$

Thus

$$EY_{j[r:n]} = \mu_j + \rho_j \sigma_j \alpha_{r:n} \ . \tag{2.11}$$

Also, noting that all the $\epsilon_{j[r]}$ are independent of $(X_{1:n}, \ldots, X_{n:n})$ and that $\epsilon_{j[r]}$ and $\epsilon_{k[s]}$ are independent unless $r = s$, we have from (2.10) or (2.8) that

$$\text{cov}(Y_{j[r:n]}, X_{k[r:n]}) = \rho_j \sigma_j \rho_k \sigma_k \beta_{rr:n} + \sigma_{jk}(x)$$
$$= \sigma_{jk} - \rho_j \rho_k \sigma_j \sigma_k (1 - \beta_{rr:n}) \tag{2.12}$$

by (2.9). From (2.10) it follows at once that

$$\text{cov}(Y_{j[r:n]}, Y_{k[s:n]}) = \rho_j \rho_k \sigma_j \sigma_k \beta_{rs:n} \ . \tag{2.13}$$

An interesting special case occurs when

$$X_i = \sum_{j=1}^{\ell} Y_{ji} \ ,$$

so that

$$X_{r:n} = \sum_{j=1}^{\ell} Y_{j[r:n]} \ .$$

Eqs. (2.11)–(2.13) now hold with

$$\rho_j = \sum_{k=1}^{\ell} \sigma_{jk}/(\sigma_X \sigma_j) \ .$$

Song and Balakrishnan (1994) treat a multiplicative model of the form $Y_{ji} = X_i \epsilon_{ji}$, $j = 1,\ldots,\ell$, $i = 1,\ldots,n$, and use it to generalize results of Song (1993) on concomitants of gamma order statistics. See also Ma, Yue, and Balakrishnan (1995). Balasubramanian and Balakrishnan (1995) obtain a class of multivariate distributions for which the concomitants are members of the class.

2.4. Dependence structure of concomitants

DEFINITION. The random variables $X_1,\ldots,X_n$ (constituting the vector X) are said to be *associated* if $\text{cov}([h_1(X), h_2(X)]) \geq 0$ for all pairs of increasing functions h_1, h_2 for which the covariance exists.

Let X_i and $\epsilon_i (i = 1,\ldots,n)$ be mutually independent random variables and $Y_i = g(X_i, \epsilon_i)$, leading to (2.4). Then from results on associated random variables given in, e.g., Barlow and Proschan (1975), it is easy to show that the concomitants are associated if g is monotone (Kim and David, 1990). For $(X_{1:n},\ldots,X_{n:n})$ and $(\epsilon_{[1]},\ldots,\epsilon_{[n]})$ are independent sets of associated random variables, so that their union is also associated. Since any monotone functions of associated random variables are associated, $Y_{[1:n]},\ldots,Y_{[n:n]}$ are associated.

Note that association implies positive quadrant dependence, so that for any y_1, y_2

$$P\{Y_{[r:n]} \leq y_1, \ Y_{[s:n]} \leq y_2\} \geq P\{Y_{[r:n]} \leq y_1\} P\{Y_{[s:n]} \leq y_2\} \ .$$

Kim and David (1990) also show that the concomitants satisfy a stronger form of dependence, multivariate total positivity of order two (MTP$_2$) (Karlin and Rinott, 1980) if each $Z_{[r]}$ has a Pólya frequency function of order two (PF$_2$).

We now note some further interesting facts about the dependence structure of order statistics and their concomitants. Bhattacharya (1984) shows that, for a fixed n and arbitrary sequence of constants $(c_i, 1 \leq i \leq n)$, the sequence

$$\{W_j, 1 \leq j \leq n\} \text{ with } W_j = \sum_{i=1}^{j} c_i \{Y_{[i:n]} - m(X_{i:n})\} \text{ forms a martingale } .$$

Bhattacharya and Gangopadhyay (1990) and, independently, Goel and Hall (1994) observe that $F_{Y|X}(Y_{[i:n]}|X_{i:n})$ are i.i.d standard uniform variates independent of the X_i and hence of the X-order statistics. These last results may be seen as follows: The $Y_{[i:n]}$ are clearly conditionally independent given $X_{i:n} = x_i (i = 1,\ldots,n)$. Thus the uniform $(0,1)$ variates $F_{Y|X}(Y_{[i:n]}|x_i)$ are stochastically independent and, since their distribution does not depend on the x_i, are also unconditionally independent and independent of the $X_{i:n}$.

3. Asymptotic theory

3.1. Marginal distributions

The asymptotic distribution of $Y_{[r:n]}$ is affected by how r is related to n and by the dependence structure of (X, Y). In order to streamline our discussion, let us consider three situations regarding the growth pattern of $r(n)$: (a) the quantile case where $r = [np]$, $0 < p < 1$, (b) the extreme case, where either r or $n - r$ is held fixed, and (c) the intermediate case, where $r(n) \to \infty$, $n - r(n) \to \infty$ in such a way that $r(n)/n$ approaches either 0 or 1. (Certainly other possibilities exist, but we refrain from treating these nonstandard situations.) In these cases, the limit properties of $X_{r:n}$ and associated norming constants differ.

Next is the issue of dependence structure. To see how it affects the limit distribution of $Y_{[r:n]}$ and how the norming constants associated with $X_{r:n}$ play a substantial role in the asymptotic distribution, let us look at the simple linear regression model given by (2.1). Without loss of generality we assume that $\mu_X = \mu_Y = 0$, and $\sigma_X = \sigma_Y = 1$. In view of the independence of $\epsilon_{[r]}$ and $X_{[r:n]}$, from (2.2), it is clear that the distribution of $Y_{[r:n]}$ is affected by them in a linear fashion.

If $X_{r:n} - a_n \xrightarrow{P} 0$ as $n \to \infty$, then $Y_{[r:n]} - \rho a_n \xrightarrow{d} \epsilon$, and hence the limit distribution is that of ϵ. This happens in the quantile case where a_n may be chosen as $F_X^{-1}(p)$. Further, in the other two cases related to the growth rate of r, such a situation may also arise. For example, for the bivariate normal parent, when $n - r = k - 1$, corresponding to the upper k^{th} extreme $X_{r:n} - \sqrt{2 \log n} \xrightarrow{P} 0$, and consequently $Y_{[r:n]} - \rho \sqrt{2 \log n} \xrightarrow{d} \sqrt{1 - \rho^2} Z$, where Z is a standard normal variate. A detailed discussion of the consequences is given in David (1994).

Now suppose $X_{r:n}$ fails to converge in probability, but instead there exist constants a_n and $b_n > 0$ such that $(X_{r:n} - a_n)/b_n \xrightarrow{d} W$ as $n \to \infty$, for a nondegenerate random variable W. In this case the scaling constant b_n plays an important role in the limit behavior of $Y_{[r:n]}$. To be precise, in view of (2.2) we may write

$$\frac{Y_{[r:n]} - \rho a_n}{b_n} = \rho \frac{X_{r:n} - a_n}{b_n} + \frac{\epsilon_{[r]}}{b_n} . \tag{3.1}$$

If $b_n \to b$, finite and positive, we may conclude from (3.1) that $(Y_{[r:n]} - \rho a_n)/b_n$ converges in distribution to the convolution of ρW and ϵ. In the upper (lower) extreme case, W will be distributed as the kth lower (upper) record value from one of the three relevant extreme value distributions (e.g., Nagaraja and David, 1994). In the intermediate case, when X is bounded above and $r(n)/n$ approaches 1, under certain smoothness conditions on the tail of F_X, W is normally distributed. Appropriate choices for the norming constants may be found in Reiss (1988, pp. 108–109).

While the brief discussion above indicates that numerous possibilities exist for the limit distribution of $Y_{[r:n]}$ for the model given by (2.1), it also shows the interplay between the conditional pdf $f(y|x)$ and the marginal pdf $f_X(x)$. Of course, this will be true in general. We now state a result due to Galambos (1978), extended in David (1994) and Sondhauss (1994), which gives a representation for the limit distribution of $Y_{[r:n]}$ in the extreme case for an arbitrary absolutely continuous bivariate cdf $F(x,y)$.

THEOREM 3.1. Let $F_X(x)$ satisfy one of the von Mises conditions and assume that the sequences of constants $a_n, b_n > 0$, are such that as $n \to \infty$,

$$\{F_X(a_n + b_n x)\}^n \to G(x) \tag{3.2}$$

for all x. Further, suppose there exist constants A_n and $B_n > 0$ such that

$$\{F_Y(A_n + B_n y | a_n + b_n x)\} \to H(y|x) \tag{3.3}$$

for all x and y. Then

$$P(Y_{[n-k+1:n]} \leq A_n + B_n y) \to \int_{-\infty}^{\infty} H(y|x) \, dG_{(k)}(x), \tag{3.4}$$

where $G_{(k)}$ is the cdf of the kth lower record value from the extreme value cdf G.

If (3.2) holds we say that F_X is in the domain of attraction of G and we write $F_X \in D(G)$. It is well known that G must be one of the three extreme value cdf's which are of the following types (Gnedenko, 1943):

$$G_1(x; \alpha) = \begin{cases} 0, & x \leq 0 \\ \exp\{-x^{-\alpha}\}, & x > 0; \ \alpha > 0 \end{cases} \tag{3.5a}$$

$$G_2(x; \alpha) = \begin{cases} \exp\{-(-x)^{\alpha}\}, & x < 0; \ \alpha > 0 \\ 1, & x \geq 0 \end{cases} \tag{3.5b}$$

$$G_3(x) = \exp\{-\exp(-x)\}, \quad -\infty < x < \infty. \tag{3.5c}$$

The condition (3.3) holds with $H(y|x) = H(y)$ and $A_n = 0$, $B_n = 1$, if the joint distribution of (X, Y) is such that as $x \to F_X^{-1}(1), F_Y(y|x) \to H(y)$. In that situation evidently $P(Y_{[n-k+1:n]} \leq y) \to H(y)$ as is the case with the following example.

EXAMPLE 3.1. Let (X, Y) have Gumbel's bivariate exponential distribution with joint cdf $F(x, y) = (1 - e^{-x})(1 - e^{-y})(1 + \alpha e^{-x-y})$, $x, y \geq 0$, $|\alpha| \leq 1$. Then $F_Y(y|x) = \{1 - \alpha(2e^{-x} - 1)\}(1 - e^{-y}) + \alpha(2e^{-x} - 1)(1 - e^{-2y})$, which converges to $H(y) = (1 - e^{-y})(1 - \alpha e^{-y})$ as $x \to \infty$. Since F_X is an exponential cdf, it satisfies a von Mises condition and (3.2) holds with $G = G_3$. Thus, we may conclude that $P(Y_{[n-k+1:n]} \leq y) \to (1 - e^{-y})(1 - \alpha e^{-y})$ for all $y > 0$. This is in sharp contrast to the asymptotic distribution of the extreme order statistic $Y_{n-k+1:n}$. Since F_Y is also standard exponential, $P(Y_{n-k+1:n} - \log n \leq y) \to G_{3(k)}(y)$ whose pdf is given by

$$g_{3(k)}(y) = \exp(-ky) \frac{\exp\{-\exp(-y)\}}{(k-1)!}, \tag{3.6}$$

for all y (see Arnold et al., 1992, p. 221).

In the above example, since $H(y|x)$ was free of x, k did not influence the distribution of $Y_{[n-k+1:n]}$. This happens to be the case for several common bivariate distributions including the bivariate normal distribution. But this need not be the case always. As an example, in the linear model expressed in (3.1), take X to be standard exponential and Z to be standard normal. Then, $F_X \in D(G_3)$ where G_3 is given by (3.5c) and we can choose $a_n = \log n$, and $b_n = 1$. Consequently $P(Y \leq \rho a_n + y | X = a_n + x) = \Phi(y - \rho x)$ representing $H(y|x)$. Hence we obtain

$$P(Y_{[n-k+1:n]} \leq \rho a_n + y) \to \int_{-\infty}^{\infty} \Phi(y - \rho x) g_{3(k)}(x) \, \mathrm{d}x \ ,$$

where $g_{3(k)}$ is given by (3.6).

See also Coles and Tawn (1994) who, however, are more interested in the distribution of Y given that X is large. Ledford and Tawn (1995) investigate the influence of the joint survival function $P(X > x, Y > y)$ on the nondegenerate limit distribution of $Y_{[n:n]}$.

One can state a limit result similar to Theorem 3.1 for $Y_{[r:n]}$ in the quantile case. With a different set of conditions that involve the assumption of uniform convergence, Suresh (1993) proves such a result.

3.2. Joint distributions

The asymptotic distribution of a finite set of concomitants has also been explored. Under the assumption that $Y - E(Y|X)$ and X are independent, David and Galambos (1974) have shown that $Y_{[r_1:n]}, \ldots, Y_{[r_k:n]}$ are asymptotically independent if $\mathrm{var}(E(Y_{[r_i:n]}|X_{r_i:n}))$ approaches 0 as n increases for all $i = 1, \ldots, k$. In such cases, the asymptotic joint distribution can be obtained from the marginal limit distributions of the $Y_{[r_i:n]}$.

In the quantile case where $r_i/n \to p_i$, $0 < p_i < 1$, for $i = 1, \ldots, k$, and in the extreme case where r_i is either i or $n - i + 1$, interesting results quickly follow from the conditional independence exhibited in (2.5). In the latter case, let $r_i = n - i + 1$, and assume that conditions of Theorem 3.1 hold. It then follows that, as $n \to \infty$,

$$P(Y_{[n:n]} \leq A_n + B_n y_1, \ldots, Y_{[n-k+1:n]} \leq A_n + B_n y_k)$$
$$\to \int_{x_1 > \cdots > x_k} \prod_{i=1}^{k} H(y_i|x_i) \, \mathrm{d}G_k(x_1, \ldots, x_k) \ , \tag{3.7}$$

where G_k is the joint cdf of the first k lower record values from the cdf G.

In the quantile case asymptotic independence prevails if the p_i are distinct. To be precise, if $r_i/n \to p_i$, $0 < p_i < 1$, and $X_{r_i:n} \xrightarrow{P} F_X^{-1}(p_i)$, for $i = 1, \ldots, k$, Yang (1977) proved that

$$P(Y_{[r_1:n]} \leq y_1, \ldots, Y_{[r_k:n]} \leq y_k) \to \prod_{i=1}^{k} F_Y(y_i | F_X^{-1}(p_i))$$

Suresh (1993) has shown that the central concomitants and extreme concomitants are asymptotically independent.

4. Estimation and tests of hypotheses

4.1. Estimation of regression and correlation coefficient

If the regression of Y on the non-stochastic variable x is linear, viz.

$$E(Y|x) = \alpha + \beta x, \tag{4.1}$$

then β may be estimated by the ratio statistic

$$b' = \frac{\overline{Y}'_{[k:n]} - \overline{Y}_{[k:n]}}{\overline{x}'_{k:n} - \overline{x}_{k:n}}, \tag{4.2}$$

where

$$\overline{x}'_{k:n} = \frac{1}{k}\sum_{i=1}^{k} x_{n+1-i:n}, \quad \overline{x}_{k:n} = \frac{1}{k}\sum_{i=1}^{k} x_{i:n},$$

and

$$\overline{Y}'_{[k:n]} = \frac{1}{k}\sum_{i=1}^{k} Y_{[n+1-i:n]}, \quad \overline{Y}_{[k:n]} = \frac{1}{k}\sum_{i=1}^{k} Y_{[i:n]}.$$

If X is stochastic, we may interpret (4.1) as conditional on $X = x$ and have from (4.2)

$$E(b'|x_1, x_2, \ldots, x_n) = \beta. \tag{4.3}$$

Since (4.3) holds whatever the x_i, it also holds unconditionally; that is,

$$B' = \frac{\overline{Y}'_{[k:n]} - \overline{Y}_{[k:n]}}{\overline{X}'_{k:n} - \overline{X}_{k:n}} \tag{4.4}$$

is also an unbiased estimator of β. Note that this result does not require either the X_i or the Y_i to be identically distributed or even to be independent. Barton and Casley (1958) show that B' has an efficiency of 75–80% when (X_i, Y_i), $i = 1, \ldots, n$, is a random sample from a bivariate normal, provided k is chosen as about $0.27n$.

Since $\rho = \beta \sigma_X / \sigma_Y$, (4.4) suggests

$$\hat{\rho}' = B' \cdot \frac{(\overline{X}'_{k:n} - \overline{X}_{k:n})/c_{n,x}}{(\overline{Y}'_{k:n} - \overline{Y}_{k:n})/c_{n,y}} = \frac{(\overline{Y}'_{[k:n]} - \overline{Y}_{[k:n]})/c_{n,x}}{(\overline{Y}'_{k:n} - \overline{Y}_{k:n})/c_{n,y}}$$

as an estimator of ρ, where $c_{n,x} = E(\overline{X}'_{k:n} - \overline{X}_{k:n})/\sigma_X$, etc. If X and Y have the same marginal distributional form (e.g., both normal), $\hat{\rho}'$ simplifies to

$$\hat{\rho}' = \frac{\overline{Y}'_{[k:n]} - \overline{Y}_{[k:n]}}{\overline{Y}'_{k:n} - \overline{Y}_{k:n}}. \tag{4.5}$$

This estimator has been suggested by Tsukibayashi (1962) for $k = 1$ when the denominator is just the range of the Y_i, and also for a mean range denominator. He points out that (4.5) can be calculated even if only the ranks of the X's are available. In order to deal with the distribution of $\hat{\rho}'$ (for $k = 1$) interesting distributional results are developed in Tsukibayashi (1996), such as the joint pdf of $Y_{n:n}$ and $Y_{[n:n]}$. See also Watterson (1959) and Barnett et al. (1976).

An interesting related measure of association, not requiring (4.1) to hold, has been proposed and studied by Schechtman and Yitzhaki (1987). They show that if (X, Y) has a continuous joint cdf with marginals $F_X(x)$, $F_Y(y)$, then

$$G(Y, X) = \frac{\Sigma(2i - 1 - n)Y_{[i:n]}}{\Sigma(2i - 1 - n)Y_{i:n}}$$

is a consistent estimator of

$$\Gamma(Y, X) = \frac{\text{cov}(Y, F_X(Y))}{\text{cov}(Y, F_Y(Y))}.$$

The authors call $\Gamma(Y, X)$ (and $\Gamma(X, Y)$) the *Gini correlation*, since $\text{cov}(Y, F_Y(Y))$ is one-fourth of Gini's mean difference (Stuart, 1954). If (X, Y) is bivariate normal $(\mu_X, \mu_Y, \sigma_X, \sigma_Y, \rho)$, then $\Gamma(Y, X) = \Gamma(X, Y) = \rho$. In general, $\Gamma(Y, X) \neq \Gamma(X, Y)$ and $G(Y, X) \neq G(X, Y)$. This makes explicit a result also applying to (4.5), where the Y's may be replaced by X's if measurements for the latter are available. Likewise, $G(Y, X)$ can be calculated even if only the ranks of the X's are known.

Motivated by confidentiality considerations, Spruill and Gastwirth (1982) have used concomitants to estimate the correlation coefficient between two sensitive rv's X and Y, data on which is kept by separate agencies A and B, respectively. In their method, agency A is asked to divide the $N = mn$ individuals into n groups of size m by ordering on x, and to provide the group identification of each individual as well as the group means and variances, i.e., for group k, $k = 1, \ldots, n$,

$$\bar{x}_k = \sum x_{i:N}/m, \qquad s^2_{x,k} = \sum (x_{i:N} - \bar{x}_k)^2/m,$$

where the sums extend over $i = (k-1)m + 1, \ldots, km$. Given only the group identifications, agency B simply provides

$$\bar{y}_k = \sum y_{[i:N]}/m, \qquad s^2_{y,k} = \sum (y_{[i:N]} - \bar{y}_k)^2/m.$$

The least-squares estimate of ρ can now be obtained from $\hat{\rho} = \hat{\beta}\hat{\sigma}_X/\hat{\sigma}_Y$ as

$$\hat{\rho} = \frac{\sum_{k=1}^{n}(\bar{x}_k - \bar{x})(\bar{y}_k - \bar{y})\left[\sum_{i=1}^{N}(x_i - \bar{x})^2\right]^{1/2}}{\sum_{k=1}^{n}(\bar{x}_k - \bar{x})^2\left[\sum_{i=1}^{N}(y_i - \bar{y})^2\right]^{1/2}},$$

where $\bar{x}$ and $\bar{y}$ are grand means. Note that only the overall sample variances of X and Y occur in $\hat{\rho}$; the $s^2_{x,k}$ and $s^2_{y,k}$ are needed for assessing the efficiency, E, of $\hat{\rho}$. In the bivariate normal case the authors find $\hat{\rho}$ to be nearly unbiased and E to exceed 0.8 for $n = 10$ in the cases studied ($N = 100, 1000$; $\rho = .25, .50, .75, .9$).

Guilbaud (1985) considers related questions of inference in a slightly more general setting. Let $F(x,y)$ be the cdf of (X,Y) and let $0 = \lambda_0 < \lambda_1 < \cdots < \lambda_{k-1} < \lambda_k = 1$. Also let $\xi_j = F_X^{-1}(\lambda_j)$, $j = 1, \ldots, k-1$, $\xi_0 = -\infty$, $\xi_k = \infty$, where $F_X(x)$ has positive derivatives in the neighborhoods of $\lambda_1, \ldots, \lambda_{k-1}$.

Given a random sample $(x_i, y_i) \, i = 1, \ldots, n$, from $F(x,y)$, natural estimates of the class means are

$$\bar{x}_j = \sum_i x_{i:n}/n_j, \quad \bar{y}_j = \sum_i y_{[i:n]}/n_j, \quad j = 1, \ldots, k$$

where the sums now extend over the n_j observations with x-values in (ξ_{j-1}, ξ_j). With $\mu_{X,j} = E(\bar{X}_j)$, $\mu_{Y,j} = E(\bar{Y}_j)$, Guilbaud gives the $2k$-variate asymptotic normal distribution of $n^{1/2}[(\bar{X}_j - \mu_{X,j}), (\bar{Y}_j - \mu_{Y,j})]$, with the help of which large-sample tests and confidence intervals can be constructed. He also treats stratified random sampling.

4.2. Estimation for censored bivariate data

Concomitants of order statistics arise very naturally when multivariate data sets are subject to some form of Type II censoring. In the bivariate case, with data (x_i, y_i), $i = 1, \ldots, n$, three kinds of such censoring may usefully be distinguished (Watterson, 1959): (a) censoring of certain $x_{i:n}$ and of the corresponding $y_{[i:n]}$; (b) censoring of certain $y_{i:n}$ only; and (c) censoring of certain $x_{i:n}$ only. For example, (b) occurs when the $x_{i:n}(i = 1, \ldots, n)$ are entrance scores and the $y_{[i:n]}$ $(i = k+1, \ldots, n)$ later scores of the successful candidates. On the other hand, (c) applies in a life test terminated after $n - k$ failures when measurements on some associated variable are available for all n items.

The bivariate normal case has been considered by Watterson (1959). Some of his findings may be illustrated on case (a) above. Obviously, μ_X and σ_X may be estimated as in the univariate case. The mean and variance of a linear function of the concomitants can be found from (2.3). Since

$$EX_{r:n} = \mu_X + \sigma_X \alpha_{r:n}$$

it follows from (2.3a) that any coefficients a_i making $\Sigma a_i X_{i:n}$ unbiased for μ_X also make $\Sigma a_i Y_{[i:n]}$ unbiased for μ_Y, for any ρ, where Σ ranges over the available observations. However, var$(\Sigma a_i Y_{[i:n]})$ depends on ρ, so that there is no optimal choice of the a_i for the estimation of μ_Y. Two possible choices are to take a_i as in the best linear estimator of μ_X or in Gupta's (1952) simplified linear estimator of μ_X. In cases examined numerically in Watterson (1959) it turns out that the latter choice is the more efficient except when ρ is very close to 1.

Case (a) with censoring on the right has been studied in detail by Harrell and Sen (1979), who use a maximum likelihood approach, modified by replacing the MLE's of μ_X and σ_X by Gupta's estimates. Such a replacement is made because of the considerable biases in the MLE's, especially under heavy censoring. The authors also propose a test for the independence of X and Y based on the correlation coefficient between the order statistics and their concomitants. Gill, Tiku and Vaughan (1990) use Tiku's simplified MLE's to deal with two-sided censoring in the same situation, but allowing for possibly more than one set of concomitants.

Gomes (1981, 1984) has employed the idea of concomitants in her study of estimators based on multivariate samples of order statistics of largest values. Suppose only the top k extreme values in each of n independent large samples are available. For simplicity take $k = 2$ and label the largest sample value as X, and the second largest as Y. Then the data can be modeled as a random sample of size n from the joint pdf $f(x, y) = \frac{g(x)}{G(x)} g(y)$, $y < x$, where G is one of the extreme value cdf's in (3.5) but for an unknown change of location and scale. Gomes compares their linear estimators based on the order statistics of the X-values with and without the information on the concomitants.

4.3. Ranked-set sampling

An ingenious method called ranked-set sampling was introduced by McIntyre (1952) for situations where the primary variable of interest, Y, is difficult or expensive to measure, but where ranking in small subsets is easy. For example, suppose we require an estimate of the mean height μ_Y of a population of trees. Choose a sample of size $n = k^2$ (or a multiple of k^2). Randomly subdivide the sample into subsamples of k. In each subsample rank the trees visually by height and in the ith $(i = 1, \ldots, k)$ subsample measure only the tree of rank i.

Concomitants of order statistics enter when the ranking is subject to error, a situation first studied by Dell and Clutter (1972) who speak of judgment ordering. Such an ordering may be regarded as based on an auxiliary X-variate representing an actual or hypothetical measurement (David and Levine, 1972; Stokes, 1977). Then the Y-value in the ith subsample may be denoted by $Y_{[i:k]}^{(i)}$. The $Y_{[i:k]}^{(i)}$ are independent rv's, $Y_{[i:k]}^{(i)}$ having the same marginal distribution as $Y_{[i:k]}$. Then the ranked-set sample estimator of μ_Y is

$$\hat{\mu}_Y = \frac{1}{k} \sum_{i=1}^{k} Y_{[i:k]}^{(i)} , \qquad (4.5)$$

and

$$\mathrm{E}(\hat{\mu}_Y) = \frac{1}{k} \sum_{i=1}^{k} \mathrm{E}\left(Y_{[i:k]}^{(i)}\right) = \frac{1}{k} \sum_{i=1}^{k} \mathrm{E}(Y_{[i:k]}) .$$

But the $Y_{[i:k]}$ are a permutation of i.i.d variates Y_i having mean μ_Y, so that

$$\mathrm{E}(\hat{\mu}_Y) = \frac{1}{k} \sum_{i=1}^{k} \mathrm{E}(Y_i) = \mu_Y .$$

Thus $\hat{\mu}_Y$ is an unbiased estimator of μ_Y without any distributional assumption other than the existence of μ_Y.

We have also

$$k\sigma_Y^2 = \mathrm{E}\sum_{i=1}^{k}(Y_i - \mu_Y)^2$$

$$= \mathrm{E}\sum_{i=1}^{k}(Y_{[i:k]} - \mu_Y)^2$$

$$= \sum_{i=1}^{k}\mathrm{var}(Y_{[i:k]}) + \sum_{i=1}^{k}(\mathrm{E}Y_{[i:k]} - \mu_Y)^2$$

But (4.5), by the independence of the $Y_{[i:k]}^{(i)}$, gives

$$k^2\mathrm{var}(\hat{\mu}_Y) = \sum_{i=1}^{k}\mathrm{var}(Y_{[i:k]}) ,$$

so that

$$\mathrm{var}(\hat{\mu}_Y) = \frac{1}{k}\left[\sigma_Y^2 - \frac{1}{k}\sum(\mathrm{E}Y_{[i:k]} - \mu_Y)^2\right] .$$

This may be compared with $\mathrm{var}\overline{Y} = \sigma_Y^2/k$, the efficiency, RE, of $\hat{\mu}_Y$ relative to $\overline{Y}$ being

$$\mathrm{RE} = \frac{\mathrm{var}\overline{Y}}{\mathrm{var}(\hat{\mu}_Y)} = \frac{1}{1 - \frac{1}{k}\sum \alpha_{[i:k]}^2} ,$$

where $\alpha_{[i:k]} = (\mathrm{E}Y_{[i:k]} - \mu_Y)/\sigma_Y$.

Although errors in ranking reduce the efficiency of ranked-set sampling, substantial gains in efficiency may remain if ρ is not too small (Dell and Clutter, 1972; Ridout and Cobby, 1987).

These methods have been extended to the estimation of variance (Stokes, 1980a), the correlation coefficient (Stokes, 1980b), and to situations with size biased selection of the X's (Muttlak and McDonald, 1990a,b). There are many other aspects of ranked-set sampling not involving concomitants. See the comprehensive review paper by Patil et al. (1994) and the article in the companion volume by Bimal K. Sinha and Nora Ni Chuiv.

4.4. Double sampling

In this method (O'Connell and David, 1976), which has a similar purpose to ranked-set sampling, $X_1, \ldots, X_n$ represent inexpensive measurements. Based on their ordering $k(<n)$ expensive measurements $Y_{[r_j:n]}$ are made ($j = 1, \ldots, k$; $r_1 < \cdots < r_k$). Then a simple estimator of μ_Y is their average, $\overline{Y}_{[r:n]}$ say, which under the linear model (2.2) is given by

$$\overline{Y}_{[r:n]} = \mu_Y + \rho\frac{\sigma_Y}{\sigma_X}(\overline{X}_{r:n} - \mu_X) + \overline{\epsilon}_{[r]} , \qquad (4.6)$$

where $\overline{X}_{r:n}$ and $\overline{\epsilon}_{[r]}$ are the means of $X_{r_j:n}$ and $\epsilon_{[r_j]}$, $j = 1, \ldots, k$. Evidently, $\overline{Y}_{[r:n]}$ is unbiased for μ_Y if X has a symmetric distribution and the ranks are symmetrically chosen, i.e.,

$$r_{k+1-j} = n + 1 - r_j, \quad j = 1, \ldots, \left[\tfrac{1}{2}(k+1)\right] .$$

From (4.6) we have

$$\mathrm{var}(\overline{Y}_{[r:n]}/\sigma_Y) = \rho^2 \mathrm{var}(\overline{X}_{r:n}) + (1-\rho^2)/k .$$

Thus the ranks r_j minimizing $\mathrm{var}(\overline{X}_{r:n})$ also minimize $\mathrm{var}(\overline{Y}_{[r:n]})$, whatever the value of ρ. The simple form of optimal spacing involved in minimizing $\mathrm{var}(\overline{X}_{r:n})$ was studied already by Mosteller (1946). The (asymptotically) optimal ranks may be taken as the integral parts of $n\lambda_j + 1$, $0 < \lambda_1 < \cdots < \lambda_k < 1$, where the λ_j have been tabulated in the normal case for $k \leq 10$; roughly, $\lambda_j = (j - \tfrac{1}{2})/k$. The following table illustrates that $\overline{Y}_{[r:n]}$ is more efficient than $\hat{\mu}_Y$ under bivariate normality, increasingly so as ρ approaches 1. Although n and k have been chosen to allow a direct comparison, the experimental situations are, of course, different and $\hat{\mu}_Y$ is more robust, remaining unbiased even when the distribution of X is not symmetrical.

5. The rank of $Y_{[r:n]}$

Suppose we have independent measurements (X_i, Y_i), $i = 1, \ldots, n$, with common cdf $F(x, y)$, on individuals or objects $A_1, \ldots, A_n$ and that A_i ranks r^{th} on the x-measurements. In this section we study the following two questions: (a) What is the probability that A_i will rank s^{th} on the y-measurement. (b) What is A_i's expected rank on the y-measurement?

Let $R_{r:n}$ denote the rank of $Y_{[r:n]}$ among the n Y_i, i.e.,

$$R_{r,n} = \sum_{i=1}^{n} I(Y_{[r:n]} - Y_i) , \tag{5.1}$$

Table 1
$\mathrm{var}\hat{\mu}_Y/\mathrm{var}\overline{Y}_{[r:n]}$

ρ	$n = 9, k = 3$	$n = 49, k = 7$	ρ	$n = 9, k = 3$	$n = 49, k = 7$
0	1	1	0.6	1.173	1.068
0.1	1.003	1.001	0.7	1.278	1.110
0.2	1.014	1.005	0.8	1.457	1.185
0.3	1.033	1.013	0.9	1.816	1.343
0.4	1.063	1.024	0.95	2.171	1.512
0.5	1.107	1.042	1	2.864	1.884

where

$$I(x) = 1 \quad \text{if} \quad x \geq 0,$$
$$= 0 \quad \text{if} \quad x < 0.$$

For (a) we require $\pi_{rs} = P\{R_{r,n} = s\}$. Consider first π_{nn}. We have

$$\pi_{nn} = \sum_{i=1}^{n} P\{X_i = X_{n:n}, Y_i = Y_{n:n}\}$$
$$= nP[X_n = X_{n:n}, Y_n = Y_{n:n}]$$
$$= nP\{X_1 < X_n, \ldots, X_{n-1} < X_n; Y_1 < Y_n, \ldots, Y_{n-1} < Y_n\}$$
$$= n \int_{-\infty}^{\infty} \int_{-\infty}^{\infty} [P\{X < x, Y < y\}]^{n-1} dF(x,y)$$

upon conditioning on X_n, Y_n. It is not difficult to extend this argument to finding π_{rs}. Table 2 is an extract from a table in David et al. (1977) for the case when X and Y are bivariate normal.

EXAMPLE 5.1. Suppose that the scores of candidates taking two tests are bivariate normal with $\rho = 0.8$. Out of 9 candidates taking the first (screening) test the top k are selected and given the second test. What is the smallest value of k ensuring with probability at least 0.9 that the best of the 9 candidates, as judged by the second test, is included among the k selected?

We require the smallest k such that

$$\pi_{99} + \pi_{89} + \cdots + \pi_{10-k,9} \geq 0.9$$

i.e.,

$$\pi_{99} + \pi_{98} + \cdots + \pi_{9,10-k} \geq 0.9.$$

Table 2
$\pi_{rs} = P\{R_{r,n} = s\}$ as a function of ρ for $n = 9$

r	s	ρ									
		0.10	0.20	0.30	0.40	0.50	0.60	0.70	0.80	0.90	0.95
9	9	0.1407	0.1746	0.2133	0.2576	0.3087	0.3686	0.4404	0.5306	0.6564	0.7510
	8	0.1285	0.1459	0.1631	0.1797	0.1952	0.2087	0.2185	0.2207	0.2033	0.1725
	7	0.1211	0.1296	0.1363	0.1408	0.1424	0.1401	0.1321	0.1152	0.0817	0.0523
	6	0.1152	0.1173	0.1171	0.1143	0.1085	0.0989	0.0846	0.0640	0.0350	0.0169
	5	0.1100	0.1069	0.1015	0.0938	0.0836	0.0706	0.0546	0.0357	0.0149	0.0053
	4	0.1051	0.0973	0.0877	0.0765	0.0638	0.0497	0.0345	0.0192	0.0059	0.0015
	3	0.0999	0.0877	0.0747	0.0611	0.0472	0.0334	0.0205	0.0095	0.0021	0.0004
	2	0.0939	0.0773	0.0612	0.0462	0.0324	0.0204	0.0107	0.0040	0.0006	0.0001
	1	0.0856	0.0635	0.0451	0.0300	0.0183	0.0097	0.0041	0.0011	0.0001	0.0000

Since, from the column for $\rho = 0.8$, we have

$$0.5306 + 0.2207 + 0.1152 + 0.0640 = 0.9305$$

the required value is $k = 4$.

The expected value of $R_{r,n}$ may be obtained directly by the following characteristic order statistics argument. Let $X_{r:n-1}$ and $X_{r:n}$ denote the rth order statistic for $X_1, X_2, \ldots, X_{n-1}$ and $X_1, X_2, \ldots, X_{n-1}, X_n$, respectively.
Then clearly

$$\begin{aligned} X_{r:n} &= X_{r-1:n-1} & \text{if } X_n < X_{r-1:n-1} \\ &= X_{r:n-1} & \text{if } X_n > X_{r:n-1} \\ &= X_n & \text{if } X_{r-1:n-1} < X_n < X_{r:n-1} \end{aligned} \qquad (5.2)$$

From (5.1) we have

$$E(R_{r,n}) = nP\{Y_{[r:n]} \geq Y_n\}, \qquad (5.3)$$

where corresponding to (5.2)

$$P\{Y_{[r:n]} \geq Y_n\} = P\{X_n < X_{r-1:n-1}, Y_n < Y_{[r-1:n-1]}\} \\ + P\{X_n > X_{r:n-1}, Y_n < Y_{[r:n-1]}\} + \frac{1}{n}.$$

On noting that the joint pdf $f_{X_{r-1:n-1}, Y_{r-1:n-1}}(x, y)$ may be written as $f_{r-1:n-1}(x) \cdot f(y|x)$, we now obtain

$$E(R_{r,n}) = 1 + n \int_{-\infty}^{\infty} \int_{-\infty}^{\infty} f(y|x)[P\{X < x, Y < y\} f_{r-1:n-1}(x) \\ + P\{X > x, Y < y\} f_{r:n-1}(x)] \, dy \, dx.$$

In the bivariate normal case some further simplification is possible. It turns out (David et al., 1977) that, for $r/(n+1) \to \lambda (0 < \lambda < 1)$ as $n \to \infty$, $E[R_{r,n}/(n+1)]$ is quite well approximated even for n as small as 9, by its asymptotic value

$$\lim_{n \to \infty} E\left(\frac{R_{r,n}}{n+1}\right) = \Phi\left(\frac{\rho \Phi^{-1}(\lambda)}{(2-\rho^2)^{1/2}}\right),$$

where Φ denotes the standard normal cdf. It can also be shown (David and Galambos, 1974; David et al., 1977) that for $0 \leq u \leq 1$

$$\lim_{n \to \infty} P\{R_{r,n} \leq nu\} = \Phi\left(\frac{\Phi^{-1}(u) - \rho \Phi^{-1}(\lambda)}{(1-\rho^2)^{1/2}}\right).$$

For the general bivariate case Yang (1977) obtains correspondingly the intuitively appealing result

$$\lim_{n \to \infty} P\{R_{r,n} \leq nu\} = P\{Y \leq F_Y^{-1}(u) | X = F_X^{-1}(\lambda)\}.$$

Probabilities related to those in Table 2 are studied by Spruill and Gastwirth (1996) in connection with employment problems of a professional couple. Ledford and Tawn (1998) investigate the limiting behaviour of $P(R_{n,n} = n)$ for a range of extremal dependence forms.

6. Selection through an associated variable

Yeo and David (1984) consider the problem of choosing the best k objects out of n when, instead of measurements Y_i of primary interest, only associated measurements X_i ($i = 1, \ldots, n$) are available or feasible. For example, Y_i could represent future performance of an individual, with current score X_i, or Y_i might be an expensive measurement on the ith object, perhaps destructive, and X_i an inexpensive measurement. It is assumed that the n pairs (X_i, Y_i) are a random sample from a continuous population. The actual values of the X_i are not required, only their ranks. A general expression is developed for the probability π that the s objects with the largest X-values include the k objects ($k \leq s$) with the largest Y-values. When X and Y are bivariate normal with correlation coefficient ρ, a table of $\pi = {}_n\pi_{s:k}(\rho)$, for selected values of the parameters, gives the smallest s for which $\pi \geq P^*$ is preassigned.

EXAMPLE 6.1. From 10 objects it is desired to select a subset of size s that will contain the k best objects ($k = 1, 2, 3$) with probability at least 0.9. We give a table of s for $\rho = 0.7, 0.8, 0.9$.

ρ	k	1	2	3
0.7		5	7	na
0.8		4	6	7
0.9		3	5	6

Thus if we want to be at least 90% certain that the object with the highest Y-value is in the chosen subset for $\rho = 0.8$, we need to select the four objects with the highest X-value. The full table gives the actual inclusion probability ${}_{10}\pi_{4:1}(0.8)$ as 0.9183 and also shows that the object with the highest X-value has probability 0.5176 of having the highest Y-value. Another table tells us that for the object with the highest X-value to have probability ≥ 0.90 of having the highest Y-value, would require $\rho \geq 0.993$! For a 50:50 chance $\rho = 0.783$.

With the help of a computer program it is also possible to base the selection of the best object on the actual values of the X_i rather than on their ranks (Yeo and David, 1984).

Suppose now that the cost of each Y-measurement is c. Originally unaware of the preceding approach, Feinberg and Huber (1996) have pursued the same aim of using the ordering of the X_i to reduce the number, n_c, of objects for which Y_i needs to be measured. They choose n_c to maximize the difference in expected "utility" and expected cost:

$$E\left[\max\left(Y_{[n-n_c+1:n]}, \ldots, Y_{[n:n]}\right)\right] - cn_c \ . \tag{6.1}$$

Both the finite-sample and asymptotic theory of

$$V_{k,n} = \max(Y_{[n-k+1:n]}, \ldots, Y_{[n:n]}) \quad k = 1, \ldots, n$$

is treated by Nagaraja and David (1994). Note that $V_{k,n}$ is nondecreasing in k and that $V_{n,n} = Y_{n:n}$. Thus the closeness of $V_{k,n}$ to its maximum, a measure of the effectiveness of the auxiliary variable X, may be gauged by $\mathrm{E}(V_{k,n})/\mathrm{E}(Y_{n:n})$. A compact expression for the cdf $F_{k,n}(y)$ of $V_{k,n}$ is

$$F_{k,n}(y) = \int_{-\infty}^{\infty} \left[F_2^{\star}(y|x)\right]^k f_{X_{n-k:n}}(x) \, dx \ , \tag{6.2}$$

where

$$F_2^{\star}(y|x) = \mathrm{P}(Y \le y | X > x) \ .$$

Choice of n_c according to (6.1) is difficult to perform analytically, but has been accomplished by Feinberg and Huber (1996) in the bivariate normal case through extensive simulation. These authors report close agreement between the simulated means of $V_{k,n}$ and values of $\mathrm{E}(V_{k,n})$ obtained by numerical integration based on (6.2).

Using the expression in (6.2), Nagaraja and David (1994) have derived the limit distribution of $V_{k,n}$ in the extreme and the quantile cases. When k is held fixed, if the assumptions made in Theorem 3.1 hold (except that F_Y in Eq. (3.3) is replaced by $F_2^{\star}$), as n increases,

$$F_{k,n}(A_n + B_n y) \to \int_{-\infty}^{\infty} \{H(y|x)\}^k \, dG_{(k+1)}(x) \ .$$

As to be anticipated from (6.2), in the quantile case, where $k = [np]$, $0 < p < 1$, under mild conditions, the limit distribution of $V_{k:n}$ coincides with the limit distribution of the sample maximum from the cdf $F_2^{\star}(y|F_X^{-1}(1-p))$. If this cdf is tail equivalent to F_y (in the right tail), the norming constants associated with $V_{k,n}$ are closely related to those associated with $Y_{k:k}$. See Nagaraja and David (1994) who also discuss the simple linear regression model (2.1) in detail. Let us now present the limit distribution of $V_{k,n}$ in the important case of the bivariate normal population.

EXAMPLE 6.2. Let (X, Y) be bivariate normal with zero means, unit variances and correlation coefficients $\rho (|\rho| < 1)$. When k is held fixed,

$$V_{k,n} - \rho\sqrt{2 \log n} \xrightarrow{d} \sqrt{1 - \rho^2} Z_{k:k} \ , \tag{6.3}$$

where $Z_{k:k}$ is the maximum in a random sample of size k from the standard normal population. While the role of ρ is explicit in (6.3), it works behind the scenes in the quantile case. To be precise, if $\rho > 0$,

$$P(V_{k,n} \le a_n + b_n y) \to G_3(y) \tag{6.4}$$

for all real y, where the norming constants can be chosen as for the sample maximum from a standard normal population. In other words, we can choose

$$a_n = \sqrt{2\log n} - \frac{1}{2}\frac{\log(4\pi \log n)}{\sqrt{2\log n}} \quad \text{and} \quad b_n = 1/\sqrt{2\log n} \tag{6.5}$$

in (6.4).

When $\rho = 0$, while (6.3) holds one has to replace a_n and b_n by a_k and b_k, respectively, in (6.4).

For $\rho < 0$, from Joshi and Nagaraja (1995) it follows that, when k is held fixed, (6.3) holds on replacing ρ by $|\rho|$. But in the quantile case, the norming constants are substantially different. With $k = [np]$, $0 < p < 1$,

$$P\left\{\frac{V_{k,n} - c_n}{\theta b_n} \leq v\right\} \to e^{-p \exp(-v)},$$

where $\theta = \sqrt{1 - \rho^2}$, b_n is given by (6.5), and

$$c_n = \rho \Phi^{-1}(p) + \theta\left\{\sqrt{2\log n} - \frac{\log(4\pi \log n)}{\sqrt{2\log n}}\right\}$$
$$- \theta \frac{\left((\Phi^{-1}(p))^2/2\right) + \log(p|\rho|/\theta)}{\sqrt{2\log n}}.$$

Thus, in the quantile case, the norming constants do not depend on ρ explicitly as long as $\rho > 0$.

Joshi and Nagaraja (1995) explore the joint distribution of $V_{k,n}$ and $V^\star_{k,n}$, where $V^\star_{k,n} = \max(Y_{[1:n]}, \ldots, Y_{[n-k:n]})$. This can be used to study the joint distribution of $V_{k,n}$ and $Y_{n:n}$ since $Y_{n:n} = \max(V_{k:n}, V^\star_{k,n})$. It can be used to choose k such that $V_{k,n}/Y_{n:n}$ is close to 1.

7. Functions of concomitants

The maximum of selected concomitants considered above in section 6 is one of several functions of $Y_{[i:n]}$ studied in the literature. The earliest one was the partial sum $S_n(t) = \sum_{i=1}^{[nt]} Y_{[i:n]}$, for $0 \leq t \leq 1$, discussed extensively by Bhattacharya (1974, 1976). He used its asymptotic properties to carry out inference on the regression function $m(x) = E(Y|x)$ and the integrated regression function. Motivated by genetic selection problems, Nagaraja (1982) introduced and studied the properties of the induced selection differential, a linear function of the upper partial sum. Yang (1981a, b) and Sandström (1987) studied the asymptotic properties of smooth linear functions of the $Y_{[i:n]}$. Stute (1993) and Veraverbeke (1992) have established the asymptotic normality of U-statistic-type functions of concomitants. Guilbaud (1985), Do and Hall (1992), and Goel and Hall (1994) discuss the asymptotic theory for some special functions of concomitants. In this section we briefly discuss the distribution theory (mostly asymptotic) and elaborate on some interesting applications to selection problems, inference on $m(x)$, bootstrapping, and the assessment of the quality of file-matching procedures.

7.1. Partial sum process and some generalizations

Bhattacharya (1974, 1976) has investigated the weak convergence of the process $\{S_n(t),\ 0 \leq t \leq 1\}$ and the closely related process

$$S_n^\star(t) = \sum_{\{i|X_{i:n} \leq F_X^{-1}(t)\}} Y_{[i:n]} \ .$$

The idea behind his approach is a decomposition that uses conditional expectations. For example, for the $S_n(t)$ process we may write

$$S_n(t) = \sum_{i=1}^{[nt]} \left(Y_{[i:n]} - m(X_{i:n})\right) + \sum_{i=1}^{[nt]} m(X_{i:n}) ,$$
$$= S_{1n}(t) + S_{2n}(t), \quad \text{say} \ .$$

On appropriate normalization, $S_{1n}(t)$ converges to a functional of a Brownian motion process and the asymptotic behavior of $S_{2n}(t)$ is associated with an independent Brownian bridge process. From this, the asymptotic properties of $S_n(t)$ (and of $S_n^\star(t)$) follow. For instance, under some regularity conditions, for a fixed t in (0.1),

$$(S_n(t) - S(t))/\sqrt{n} \xrightarrow{d} N\left(0,\ \psi(t) + \text{var}\left\{\int_0^t B(s)\, \mathrm{d}h(s)\right\}\right) . \tag{7.1}$$

where $h(t) = m(F_X^{-1}(t))$, $S(t) = \int_0^t h(s)\, \mathrm{d}s$, $\psi(t) = \int_0^t \text{var}(Y|X = F_X^{-1}(s))\, \mathrm{d}s$, and $B(\cdot)$ is a Brownian bridge. The limiting result in (7.1) and other structural properties of the limiting processes are used to suggest tests on $m(x)$ and a confidence interval for $S(t)$. For further details see the excellent discussion in Bhattacharya (1984, sec. 9) which also elaborates on the work of Sen (1976) establishing some invariance principles for concomitants. Egorov and Nevzorov (1981: Russian version date; 1984: English version date) consider the multidimensional concomitant model (see section 2.3) and show that the vector of partial sum processes of the concomitants converges to a multivariate normal distribution. The rate of this convergence is shown to be of order $1/\sqrt{n}$.

Let us now consider a genetic selection problem where X and Y represent the measurement of a certain characteristic associated with the parent and offspring populations, respectively. Suppose k parents, ranked highest on X, are selected and the average $\overline{Y}'_{[k:n]}$ (introduced in (4.2)) of the Y-values associated with the offspring of the selected parents is recorded. A measure of improvement in the offspring group due to the selection is the *induced selection differential* $D_{[k:n]} = (\overline{Y}'_{[k:n]} - \mu_Y)/\sigma_Y$, also known as the *response to selection* in the genetics literature. Nagaraja (1982) has investigated the finite-sample and asymptotic properties of $D_{[k:n]}$. In the extreme case, its limit distribution can be obtained from the limiting joint cdf given in (3.7). In the quantile case, (7.1) can be used to establish the asymptotic normality of $D_{[k:n]}$. Nagaraja has also established the asymptotic bivariate normality of appropriately normalized $D_{[k:n]}$ and $D_{k:n} = (X'_{k:n} - \mu_X)/\sigma_X$. While the data here can be viewed as a Type II censored bivariate sample, Guilbaud's (1985) work discussed in section 4 refers to a certain kind of Type I censored sample.

For the linear regression model (2.1), asymptotic properties of $D_{[k:n]}$ have been thoroughly investigated. Let $\mu_X(p)$ and $\sigma_X^2(p)$ be the conditional mean and variance of the distribution of X given $X > F_X^{-1}(q)$, where $k = [np]$, $0 < p < 1$, and $q = 1 - p$. For simplicity, let us take $\mu_X = 0$ and $\sigma_X = 1$. Then as $n \to \infty$,

$$\sqrt{k}(D_{[k:n]} - \rho\mu_X(p)) \xrightarrow{d} N\left(0, \frac{\sigma_\epsilon^2}{\sigma_Y^2} + \rho^2\left(\sigma_X^2(p) + q\{\mu_X(p) - F_X^{-1}(q)\}^2\right)\right) .$$

7.2. Smooth functions of $X_{i:n}$ and $Y_{[i:n]}$

Yang (1981a) has considered general linear functions of the form

$$L_{1n} = \frac{1}{n}\sum_{i=1}^{n} J\left(\frac{i}{n}\right) Y_{[i:n]} \quad \text{and} \quad L_{2n} = \frac{1}{n}\sum_{i=1}^{n} J\left(\frac{i}{n}\right) \eta(X_{i:n}, Y_{[i:n]}) , \qquad (7.2)$$

where J is a bounded smooth function which may depend on n, and η is a real-valued function. Using the idea of Hájek's projection lemma, he has established the asymptotic normality of these statistics. These results are used to construct consistent estimators of quantities associated with the conditional distribution of Y given $X = x$. For example, suppose our interest is in the regression function $m(x)$. Yang presents an estimator of the form of L_{1n} where the weight function J depends on x and on an auxiliary density function f_0. He shows that the estimator

$$L_{3n}(x) = \sum_{i=1}^{n} \frac{1}{\delta(n)} f_0\left(\frac{t - F_n(x)}{\delta(n)}\right)$$

is a mean square consistent estimator of $m(x)$, where $F_n(x)$ is the empirical cdf of the X sample, and $\delta(n) \to 0$ as $n \to \infty$. Johnston (1982) has derived the asymptotic distribution of the maximal deviation $\sup|L_{3n}(x) - m(x)|$ and has obtained a large-sample uniform confidence interval for the regression function.

Yang (1981b) considers the limiting properties of functions of the form $L_{4n} = \sum_{i=1}^{n} J(t_{ni}) Y_{[i:n]}$ with t_{ni} being close to i/n. Such statistics are used to test univariate and bivariate normality and to test independence of the X and Y samples. Mehra and Upadrasta (1992) establish the asymptotic normality of a class of linear functions of concomitants with random weights. They take the weight associated with $Y_{[i:n]}$ to be $J(i/n, R_{i,n})$, where $R_{i,n}$ is the rank of $Y_{[i:n]}$ (see (5.1)) and J is a smooth bivariate bounded score function.

The statistic L_{2n} in (7.2) may also be written as

$$T(F_n) = \int_{-\infty}^{\infty} \int_{-\infty}^{\infty} J(F_n(x)) H(x,y) \, dF_n(x,y) \ .$$

Sandström (1987) has proved the asymptotic normality of $\sqrt{n}(T(F_n) - T(F))$ using a stochastic differential while Yang's (1981a) result had used $E(T(F_n))$ in place of $T(F)$ as the location-shift parameter. See also Chanda and Ruymgaart (1992). Sandström also establishes, under certain assumptions, the asymptotic normality of $\sqrt{n}(T(F_n) - T(F_N))$, where F_N is the cdf of a finite population of size N.

7.3. Other functions

Stute (1993, preprint 1989) has generalized Bhattacharya's limit result on $S_n(t)$ to U-statistic type functions of degree 2. More precisely, for a symmetric function $\eta(x,y)$, consider the process $\{S_{3n}(t), 0 \le t \le 1\}$ with

$$S_{3n}(t) = \binom{n}{2}^{-1} \sum_{i<j \le [nt]} \eta(Y_{[i:n]}, Y_{[j:n]}) \ .$$

Using Hájek's projection lemma – a classical tool in the asymptotic theory of U-statistics – Stute proves weak convergence results for the process $S_{3n}(t)$. Such functions arise in the inference for quantities related to the conditional distribution of Y given $X \le x$. For instance, with $\eta(y_1, y_2) = (y_1 - y_2)^2/2$, $S_{3n}(t)/t^2$ provides a consistent asymptotically normal estimator of $\mathrm{var}(Y|X \le F_X^{-1}(t))$. Veraverbeke (1992) expands the study to U-functions of higher orders by considering trimmed versions of U-functions of the form

$$S_{4n}(s,t) = \frac{1}{\binom{n}{m}} \sum \eta(Y_{[i_1:n]}, \ldots, Y_{[i_m:n]})$$

where the summation is over $[ns] < i_1 < \cdots < i_m < n - [nt]$ with $0 \le s < t \le 1$.

To estimate the percentiles of the bootstrap distribution of a statistic of interest, Efron (1990) suggested a method that represents the statistic as the sum of two random variables one of which has a known distribution. Do and Hall (1992) relate it to concomitants of order statistics. To elaborate, let $Y = X + \epsilon$, where F_X is completely known and $F_Y(y)$ is to be estimated. We observe $(X_{i:n}, \epsilon_{[i:n]})$, $1 \le i \le n$, where $\epsilon_{[i:n]} = Y_{[i:n]} - X_{i:n}$. Do and Hall compare the asymptotic properties of the empirical cdf $F_{n,Y}(y)$ with the concomitant based estimator

$$F_{n,Y}^\star(y) = \frac{1}{n} \sum_{i=1}^n I\left(F_X^{-1}\left(\frac{i}{n}\right) + \epsilon_{[i:n]} \le y\right) \ ,$$

where $I(\cdot)$ represents the indicator function. If the ϵ's are sufficiently small, they show that $F_{n,y}^\star$ outperforms the classical estimator $F_{n,Y}$.

Concomitants appear naturally in the study of file-matching techniques. Suppose the linkage between the X-values and Y-values is unavailable and the goal is to match the X value with its concomitant Y. If X and Y are positively correlated, one might use $Y_{i:n}$ as the predictor of $Y_{[i:n]}$. The total cost of mismatches can be expressed as the sum $S_{5n} = \sum_{i=1}^{n} \eta\left(Y_{i:n} - Y_{[i:n]}\right)$ where η represents a penalty function. Assuming η is smooth, Goel and Hall (1994) establish the strong convergence of S_{5n} and show that it is asymptotically normal. Their approach uses the interesting fact (Section 2.4) that $F_{Y|X}(Y|X)$ is standard uniform and is independent of X.

References

Ahsanullah, M. (1994). Record values, random record models and concomitants. *J. Statist. Res.* **28**, 89–109.

Anderson, T. W. (1984). *Introduction to Multivariate Analysis.* 2nd edn. Wiley, New York.

Arnold, B. C., N. Balakrishnan and H. N. Nagaraja (1992). *A First Course in Order Statistics*. Wiley, New York.

Balakrishnan, N. (1993). Multivariate normal distribution and multivariate order statistics induced by ordering linear combinations. *Statist. Prob. Letters* **17**, 343–350.

Balasubramanian, K. and N. Balakrishnan (1995). On a class of multivariate distributions closed under concomitance of order statistics. *Statist. Prob. Letters* **23**, 239–242.

Balasubramanian, K. and M. I. Beg (1996). Concomitant of order statistics in Gumbel's bivariate exponential distribution (submitted).

Barlow, R. E. and F. Proschan (1975). *Statistical Theory of Reliability and Life Testing*. Holt, Rinehart and Winston, New York.

Barnett, V. (1976). The ordering of multivariate data (with Discussion). *J. Roy. Statist. Soc. A* **139**, 318–354.

Barnett, V., P. J. Green and A. Robinson (1976). Concomitants and correlation estimates. *Biometrika* **63**, 323–328.

Barton, D. E. and D. J. Casley (1958). A quick estimate of the regression coefficient. *Biometrika* **45**, 431–435.

Bhattacharya, P. K. (1974). Convergence of sample paths of normalized sums of induced order statistics. *Ann. Statist.* **2**, 1034–1039.

Bhattacharya, P. K. (1976). An invariance principle in regression analysis. *Ann. Statist.* **4**, 621–624.

Bhattacharya, P. K. (1984). Induced order statistics: Theory and applications. In: P. R. Krishnaiah and P. K. Sen, eds, *Handbook of Statistics* Vol. 4. Elsevier, Amsterdam, 383–403.

Bhattacharya, P. K. and A. K. Gangopadhyay (1990). Kernel and nearest-neighbor estimation of a conditional quantile. *Ann. Statist.* **18**, 1400–1415.

Chanda, K. C. and F. H. Ruymgaart (1992). Asymptotic normality of linear combinations of functions of the concomitant order statistics. *Commun. Statist. – Theory Meth.* **21**, 3247–3254.

Coles, S. G. and J. A. Tawn (1994). Statistical methods for multivariate extremes: An application to structural design. *Appl. Statist.* **43**, 1–48.

David, H. A. (1973). Concomitants of order statistics. *Bull. Internat. Statist. Inst.* **45**, 295–300.

David, H. A. (1981). *Order Statistics.* 2nd edn. Wiley, New York.

David, H. A. (1994). Concomitants of extreme order statistics. In: J. Galambos et al., eds, *Extreme Value Theory and Applications*, Kluwer, Dordrecht, 211–224.

David, H. A. and J. Galambos (1974). The asymptotic theory of concomitants of order statistics. *J. Appl. Probab.* **11**, 762–770.

David, H. A. and J. L. Gunnink (1997). The paired t-test under artificial pairing. *Amer. Statist.* **51**, 9–12.

David, H. A. and D. N. Levine (1972). Ranked set sampling in the presence of judgement error. *Biometrics* **28**, 553–555.

David, H. A., M. J. O'Connell and S. S. Yang (1977). Distribution and expected value of the rank of a concomitant of an order statistic. *Ann. Statist.* **5**, 216–223.

Dell, T. R. and J. L. Clutter (1972). Ranked set sampling theory with order statistics background. *Biometrics* **28**, 545–555.

Do, K.-A. and P. Hall (1992). Distribution estimation using concomitants of order statistics, with applications to Monte Carlo simulation for the bootstrap. *J. Roy. Statist. Soc.* **B 54**, 595–607.

Efron, B. (1990). More efficient bootstrap computations. *J. Amer. Statist. Assn.* **85**, 79–89.

Egorov, V. A. and V. B. Nevzorov (1982). Some theorems on induced order statistics. *Theory Prob. Appl.* **27**, 633–639.

Egorov, V. A. and V. B. Nevzorov (1984). Rate of convergence to the normal law of sums of induced order statistics. *J. Soviet Math.* **25**, 1139–1146. Russian version, 1981.

Feinberg, F. M. and J. Huber (1996). A theory of cutoff formation under imperfect information. *Mgmt. Sci.* **42**, 65–84.

Galambos, J. (1978, 1987). *The Asymptotic Theory of Extreme Order Statistics.* 1st, 2nd edn. Wiley, New York, Krieger, Florida.

Ghosh, M. and P. K. Sen (1971). On a class of rank order tests for regression with partially informed stochastic predictors. *Ann. Math. Statist.* **42**, 650–661.

Gill, P. S., M. L. Tiku and D. C. Vaughan (1990). Inference problems in life testing under multivariate normality. *J. Appl. Statist.* **17**, 133–147.

Gnedenko, B. (1943). Sur la distribution limite du terme maximum d' une série aléatoire. *Ann. Math.* **45**, 423–453.

Goel, P. K. and P. Hall (1994). On the average difference between concomitants and order statistics. *Ann. Probab.* **22**, 126–144.

Gomes, M. I. (1981). An i-dimensional limiting distribution function of largest values and its relevance to the statistical theory of extremes. In: C. Tailie et al., eds. *Statistical Distributions in Scientific Work* **6**. Reidel, Holland, 389–410.

Gomes, M. I. (1984). Concomitants in a multidimensional extreme model. In: J. Tiago de Oliveira, ed., *Statistical Extremes and Applications.* Reidel, Holland, 353–364.

Guilbaud, O. (1985). Statistical inference about quantile class means with simple and stratified random sampling. *Sankhyā Ser.* **B 47**, 272–279.

Gupta, A. K. (1952). Estimation of the mean and standard deviation of a normal population from a censored sample. *Biometrika* **39**, 260–273.

Harrell, F. E. and P. K. Sen (1979). Statistical inference for censored bivariate normal distributions based on induced order statistics. *Biometrika* **66**, 293–298.

Houchens, R. L. (1984). Record value theory and inference. Ph.D Dissertation, University of California, Riverside.

Jha, V. D. and M. G. Hosssein (1986). A note on concomitants of order statistics. *J. Ind. Soc. Agric. Statist.* **38**, 417–420.

Johnston, G. J. (1982). Probabilities of maximal deviations for nonparametric regression function estimates. *J. Multivar. Anal.* **12**, 402–414.

Joshi, S. N. and H. N. Nagaraja (1995). Joint distribution of maxima of concomitants of selected order statistics. *Bernoulli* **1**, 245–255.

Kamps, U. (1995). *A Concept of Generalized Order Statistics.* Teubner, Stuttgart.

Karlin, S. and Y. Rinott (1980). Classes of orderings of measures and related correlation inequalities. 1. Multivariate totally positive distributions. *J. Multivar. Anal.* **10**, 467–498.

Kaufmann, E. and R.-D. Reiss (1992). On conditional distributions of nearest neighbors. *J. Multivar. Anal.* **42**, 67–76.

Kim, S. H. and H. A. David (1990). On the dependence structure of order statistics and concomitants of order statistics. *J. Statist. Plann. Infer.* **24**, 363–368.

Ledford, A. W. and J. A. Tawn (1998). Concomitant tail behavior for extremes. *Advances in Appl. Probab.* (to appear).

Lo, A. W. and A. C. Mckinlay (1990). Data-snooping biases in tests of financial asset pricing models. *Rev. Financ. Stud.* **3**, 431–467.

Ma, C., X. Yue and N. Balakrishnan (1994). Multivariate *p*-order Liouville distributions: Definition, properties, and multivariate order statistics induced by ordering ℓ_p-norm (submitted).

McIntyre, G. A. (1952). A method of unbiased selective sampling using ranked sets. *Austral. J. Agri. Res.* **3**, 385–390.

Mehra, K. L. and S. P. Upadrasta (1992). Asymptotic normality of linear combinations of induced order statistics with double weights. *Sankhyā Ser. A* **54**, 332–350.

Mosteller, F. (1946). On some useful "inefficient" statistics. *Ann. Math. Statist.* **17**, 377–408.

Muttlak, H. A. and L. I. McDonald (1990a). Ranked set sampling with respect to concomitant variables and with size biased probability of selection. *Commun. Statist. – Theory Meth.* **19**, 205–219.

Muttlak, H. A. and L. I. McDonald (1990b). Ranked set sampling with size-biased probability of selection. *Biometrics* **46**, 435–445.

Nagaraja, H. N. (1982). Some asymptotic results for the induced selection differential. *J. Appl. Probab.* **19**, 253–261.

Nagaraja, H. N. and H. A. David (1994). Distribution of the maximum of concomitants of selected order statistics. *Ann. Statist.* **22**, 478–494.

O'Connell, M. J. and H. A. David (1976). Order statistics and their concomitants in some double sampling situations. In: S. Ikeda et al., eds, *Essays in Probability and Statistics*. Shinko Tsusho, Tokyo, 451–466.

Patil, G. P., A. K. Sinha and C. Taillie (1994). Ranked set sampling. In: G. P. Patil and C. R. Rao, eds, *Handbook of Statistics, Vol. 12*. Elsevier, Amsterdam, 167–200.

Pinhas, M. (1983). Variables concomitantes et information qualitative *Metron* **41**, 147–153.

Reiss, R.-D. (1989). *Approximate Distributions of Order Statistics*. Springer, New York.

Ridout, M. S. and J. M. Cobby (1987). Ranked set sampling with non-random selection of sets and errors in ranking. *Appl. Statist.* **38**, 145–152.

Sandström, A. (1987). Asymptotic normality of linear functions of concomitants of order statistics. *Metrika* **34**, 129–142.

Schechtman, E. and S. Yitzhaki (1987). A measure of association based on Gini's mean difference. *Commun. Statist. – Theory Meth.* **16**, 207–231.

Sen, P. K. (1976). A note on invariance principles for induced order statistics. *Ann. Prob.* **4**, 474–479.

Sen, P. K. (1981). Some invariance principles for mixed rank statistics and induced order statistics and some applications. *Commun. Statist. – Theory Meth.* **10**, 1691–1718.

Sondhauss, U. (1994). Asymptotische Eigenschaften intermediärer Ordnungsstatistiken und ihrer Konkomitanten. Diplomarbeit, Department of Statistics, Dortmund University.

Song, R. (1993). Moments of variables summing to gamma order statistics. *Commun. Statist. – Theory Meth.* **22**, 797–803.

Song, R. and N. Balakrishnan (1994). On order statistics induced by the additive and multiplicative models (submitted).

Song, R., S. G. Buchberger and J. A. Deddens (1992). Moments of variables summing to normal order statistics. *Statist. Prob. Letters.* **15**, 203–208.

Song. R. and J. A. Deddens (1993). A note on moments of variables summing to normal order statistics. *Statist. Prob. Letters.* **17**, 337–341.

Spruill, N. L. and J. Gastwirth (1982). On the estimation of the correlation coefficient from grouped data. *J. Amer. Statist. Assn.* **77**, 614–620.

Spruill, N. L. and J. Gastwirth (1996). Probability models for an employment problem. In: H. N. Nagaraja, P. K. Sen and D. F. Morrison, eds, *Statistical Theory and Applications: Papers in Honor of Herbert A. David*. Springer, New York, 199–213.

Stokes, S. L. (1977). Ranked set sampling with concomitant variables. *Commun. Statist. – Theory Meth.* **6**, 1207–1212.

Stokes, S. L. (1980a). Estimation of variance using judgment ordered ranked set samples. *Biometrics* **36**, 35–42.

Stokes, S. L. (1980b). Inference on the correlation coefficient in bivariate normal populations from ranked set samples. *J. Amer. Statist. Assoc.* **75**, 989–995.

Stuart, A. (1954). The correlation between variate-values and ranks in samples from a continuous distribution. *Brit. J. Statist. Psychol.* **7**, 37–44.

Stute, W. (1993). U-functions of concomitants of order statistics. *Probab. Math. Statist.* **14**, 143–155. Preprint, University of Giessen, Germany, 1989.

Suresh, R. P. (1993). Some asymptotic results for the induced percentile selection differential. *Sankhyā Ser. A* **55**, 120–129.

Tsukibayashi, S. (1962). Estimation of bivariate parameters based on range. *Rep. Statist. Appl. Res. JUSE* **9**, 10–23.

Tsukibayashi, S. (1996). The joint distribution and moments of an extreme of the dependent variable and the concomitant of an extreme of the independent variable (submitted).

Veraverbeke, N. (1992). Asymptotic results for U-functions of concomitants of order statistics. *Statistics* **23**, 257–264.

Watterson, G. A. (1959). Linear estimation in censored samples from multivariate normal populations. *Ann. Math. Statist.* **30**, 814–824.

Yang, S. S. (1977). General distribution theory of the concomitants of order statistics *Ann. Statist.* **5**, 996–1002.

Yang, S. S. (1981a). Linear functions of concomitants of order statistics with application to non-parametric estimation of a regression function. *J. Amer. Statist. Assoc.* **76**, 658–662.

Yang, S. S. (1981b). Linear combinations of concomitants of order statistics with application to testing and estimation. *Ann. Inst. Statist. Math.* **33**, 463–470.

Yeo, W. B. and H. A. David (1984). Selection through an associated characteristic with applications to the random effects model. *J. Amer. Statist. Assoc.* **79**, 399–405.

A Record of Records

Valery B. Nevzorov and N. Balakrishnan

1. Introduction

In his pioneering paper, Chandler (1952) defined *records* and laid the groundwork for a mathematical study of records. Since then, around 350 papers have been published on records. Elaborate review articles by Glick (1978), Nevzorov (1987a) and Nagaraja (1988a) have highlighted many of these advances. The books by Galambos (1978, 1987), Arnold and Balakrishnan (1989), and Arnold, Balakrishnan, and Nagaraja (1992) also contain some limited discussions on the theory of records in the framework of order statistics. Books by Ahsanullah (1995) and Arnold, Balakrishnan and Nagaraja (1998) provide an elaborate treatment to records. Though only ten years have passed since the publication of the review articles by Nevzorov (1987a) and Nagaraja (1988a), numerous results have been obtained on different issues concerning records during this period. This increased activity in the area of records is clearly evident from the fact that our list of all articles dealing with records has around 230 in number (of which more than 210 are listed in this article) while Nevzorov's (1987a) review article lists 166 references and Nagaraja's (1988a) article contains 75 references. It naturally prompted us to prepare this updated review of the theory and applications of records; furthermore, the above mentioned review articles by Nevzorov (1987a) and Nagaraja (1988a) also allow us to concentrate more on the developments during the last decade and peripherally on the earlier work. The first part of our review is devoted to the classical records (based on sequences of independent and identically distributed variables), while the second part deals with non-classical record models and schemes and some new generalizations of records.

2. Classical records

The work *records* is bound to bring a smile or two in any one as one constantly hears of new records being created in natural events such as rainfall, temperature, flood-level, snowfall, etc. or in different sports events. It may even bring a picture of Wayne Gretzky or Sergey Bubka in his/her mind should that individual be an avid sports fan (like both of us). Interestingly enough, a similar reaction may be observed even among the statistical community as *Records* has not become a mainstream area of Statistics and some may even question the usefulness of the study of records. While it is true that modelling the occurrence of records in sports events by means of a simple stochastic model may often be very difficult (if not impossible), the theory of records nonetheless provides a basis for some very interesting, intriguing and neat theoretical results.

In many respects, the theory of records is very closely connected with that of order statistics and in particular with extreme order statistics. The phenomenal progress in the theory of order statistics is one more reason for the increased activity in the study of records. We shall consider the period 1952–1979 first during which time the classical theory of records was initiated as well as consolidated considerably. Of the various works that appeared in this period, we want to mention specifically the following:

- Chandler (1952) – presented the basic definitions and the first theoretical results on records
- Foster and Stuart (1954), Foster and Teichroew (1955), and Stuart (1954, 1956) – discussed the use of records in statistical procedures
- Rényi (1962) – presented many remarkable results including representations of distributions of record times in terms of distributions of independent random variables
- Tata (1969) – gave representations similar to those of Rényi for record values. It needs to mentioned here that the representations presented by Rényi and Tata enabled the application of the well-developed theory of sums of independent random variables to the study of records
- Neuts (1967) and Holmes and Strawderman (1969) – discussed inter-record times
- Dwass (1964, 1966), Lamperti (1964) and Tiago de Oliveira (1968) – investigated extremal processes which are very closely related to records
- Shorrock (1972a,b, 1973, 1974, 1975) – made elaborate discussions on extremal processes, record values and record times, and also included a nice representation of record values from discrete distributions as a sum of independent terms
- Resnick (1973a,b,c, 1974, 1975) and Resnick and Rubinovitch (1973) – discussed relationships between maxima, record values and extremal processes, and also presented the set of all possible asymptotic distributions for record values and the associated normalizing constants
- Vervaat (1973) – presented limit theorems for records from discrete populations and also discussed weak records
- Williams (1973) – gave another interesting representation for record times
- Dziubdziela and Kopocinsky (1976) and Dziubdziela (1977) – introduced the generalizations of record values and record times through k^{th} record values and k^{th} record times
- Ahsanullah (1978, 1979), Nagaraja (1977) and Srivastava (1978, 1979) – presented some characterizations of distributions, like exponential and geometric, using different properties of record values

- Biondini and Siddiqui (1975) and Guthrie and Holmes (1975) – investigated records arising from sequences of dependent random variables.

Of the 70 or so papers that appeared in the period 1952–1977, we have mentioned here only about half of them. Most of these papers deal with the classical records model and these developments have been ably reviewed by Glick (1979), Nevzorov (1987a) and Nagaraja (1988a). We will, therefore, focus our attention more on the "nonclassical" records models than the "classical" records model.

3. Definitions

Let $X_1, X_2, \ldots$ be a sequence of random variables, $X_{1:n} \leq \cdots \leq X_{n:n}$ ($n = 1, 2, \ldots$) be the corresponding order statistics, $M(n) = X_{n:n} = \max\{X_1, X_2, \ldots, X_n\}$, and $m(n) = X_{1:n} = \min\{X_1, X_2, \ldots, X_n\}$. For any $k = 1, 2, \ldots$, we define k^{th} record times $L(n, k)$ and k^{th} record values $X(n, k)$ as follows:

$$L(0, k) = 0, \quad L(1, k) = k, \quad L(n+1, k) = \min\{j > L(n, k) : X_j > X_{j-k, j-1}\} \tag{3.1}$$

and

$$X(n, k) = X_{L(n,k)-k+1:L(n,k)} . \tag{3.2}$$

The more simple notation $L(n) = L(n, 1)$ and $X(n) = X(n, 1)$ will be used for the most important case $k = 1$. In this case, we can also define the upper record times $L(n)$ and the upper record values $X(n)$ in the following way:

$$L(0) = 0, \quad L(1) = 1, \quad L(n+1) = \min\{j : X_j > M(L(n))\} \tag{3.3}$$

and

$$X(n) = M(L(n)), \quad n = 1, 2, \ldots \tag{3.4}$$

If we replace $>$ and $M(L(n))$ by $<$ and $m(L(n))$ in (3.3) and (3.4), we get the definitions of the lower record times and the lower record values. In an analogous way, we can define the kth lower record times and the kth lower record values. We will consider only the upper records because the theories of lower and upper records practically coincide in all their details. In fact, we can obtain the lower records from the upper records by changing the sequence $X_1, X_2, \ldots$ to $-X_1, -X_2, \ldots$, or (in the case then X's are positive) to the new sequence $1/X_1, 1/X_2, \ldots$. Note that for discrete distributions we can introduce the so-called weak records. For it, we have to use the sign $\geq$ in (3.3) instead of $>$. In this case, any repetition of a record value is also a record.

We will use $N(n, k)$ ($N(n) = N(n, 1)$) and $\Delta(n, k) = L(n, k) - L(n-1, k)$ ($\Delta(n) = L(n, 1) - L(n-1, 1)$), $n = 1, 2, \ldots$, to denote the numbers of kth records in a sequence $X_1, X_2, \ldots$ and k^{th} inter-record times correspondingly.

Finally, let $\Phi(x)$ denote the cumulative distribution function of the standard normal distribution and $E(a, \sigma)$ denote the exponential distribution with the density function

$$f(x) = (1/\sigma) \exp\{-(x-a)/\sigma\}, \quad x > a ,$$

where a is the location (or threshold) parameter and σ is the scale parameter.

4. Representations of record times and record values using sums of independent terms

In many respects, the intensive development of the theory of records was due to results of Dwass, Rényi, Tata and Shorrock, which have allowed to the excess record statistics in terms of sums of independent random variables. The first result in this series was established independently by Dwass in 1960 and Rényi (1962).

THEOREM 4.1. Let $X_1, X_2, \ldots$ be a sequence of independent random variables with a common continuous distribution function, and let indicators $\xi_1, \xi_2, \ldots$ be defined as follows: $\xi_n = 1$ if X_n is a record value and $\xi_n = 0$ otherwise. Then, the random variables $\xi_1, \xi_2, \ldots$ are independent and $P\{\xi_n = 1\} = 1 - P\{\xi_n = 0\} = 1/n$, $n = 1, 2, \ldots$.

Note that $\xi_n = \mathbf{1}_{\{X_n > M(n-1)\}}$, $n = 2, 3, \ldots$, and $P\{\xi_1 = 1\} = 1$.

Theorem 4.1 is a corollary of the following result for sequential ranks R_n.

LEMMA 4.1. [Rényi (1962), Barndorff–Nielsen (1963)] Let $X_1, X_2, \ldots$ be a sequence of independent random variables with a common continuous distribution function and R_n, $n = 1, 2, \ldots$, be a rank of X_n in a sequence $X_1, X_2, \ldots, X_n$, that is $R_n = \sum_{k=1}^{n} \mathbf{1}_{\{X_n \geq X_k\}}$. Then, the random variables $R_1, R_2, \ldots$ are independent and $P\{R_n = r\} = 1/n$, $1 \leq r \leq n$.

As a matter of fact, another form of Lemma 4.1 was given by Wilks (1959).

Theorem 4.1 presents a very important representation for the random variables $N(n)$ and $L(n)$.

REPRESENTATION 4.1. If $X_1, X_2, \ldots$ are independent random variables with a continuous distribution function, then

$$N(n) = \xi_1 + \xi_2 + \cdots + \xi_n, \quad n = 1, 2, \ldots ,$$

and

$$P\{L(m) \geq n\} = P\{N(n) \leq m\} = P\{\xi_1 + \xi_2 + \cdots + \xi_n \leq m\} .$$

COROLLARY 4.1. If $X_1, X_2, \ldots$ are independent and have a common continuous distribution function F, then the distributions of $N(n)$ and $L(n)$ do not depend on F.

Lemma 4.1 helps us to obtain more general results than Theorem 4.1 [see, for example, Nevzorov's (1986a) paper on k_n-records]. The following results are valid for the kth records.

THEOREM 4.2. Let $X_1, X_2, \ldots$ be a sequence of independent random variables with a common continuous distribution function, and let indicators $\xi_n^{(k)}$, $n \geq k$, be defined as follows: $\xi_n^{(k)} = 1$ if X_n is a kth record value and $\xi_n^{(k)} = 0$ otherwise. Then, for any fixed $k = 1, 2, \ldots$, the random variables $\xi_k^{(k)}, \xi_{k+1}^{(k)}, \ldots$ are independent and $P\{\xi_n^{(k)} = 1\} = k/n$, $n \geq k$.

REPRESENTATION 4.2. Under the conditions of Theorem 4.2, the following equalities hold for any $k = 1, 2, \ldots$:

$$N(n, k) = \xi_k^{(k)} + \xi_{k+1}^{(k)} + \cdots + \xi_n^{(k)}, \quad n = k, k+1, \ldots,$$

and

$$P\{L(m, k) \geq n\} = P\{N(n, k) \leq m\} = P\{\xi_k^{(k)} + \xi_{k+1}^{(k)} + \cdots + \xi_n^{(k)} \leq m\}.$$

A number of such representations were obtained for record values $X(n)$ and the kth record values $X(n, k)$.

THEOREM 4.3. [Tata (1969)] Let $X_1, X_2, \ldots$ be a sequence of independent $E(0, 1)$ random variables. Then, the record spacings $X(1), X(2) - X(1), \ldots, X(n) - X(n-1), \ldots$ are independently distributed as $E(0, 1)$.

This theorem implies the following interesting representation for the record values from the $E(0, 1)$ distribution.

REPRESENTATION 4.3. For the $E(0, 1)$ distribution,

$$X(n) \stackrel{d}{=} X_1 + X_2 + \cdots + X_n, \quad n = 1, 2, \ldots.$$

The results of Dziubdziela and Kopocinsky (1976) and Deheuvels (1984a) give the following generalization of the last representation.

REPRESENTATION 4.4. When $X_1, X_2, \ldots$ is a sequence of independent $E(0, 1)$ random variables, then for any $k = 1, 2, \ldots$,

$$\{kX(n, k)\}_{n=1}^{\infty} \stackrel{d}{=} \{X_1 + X_2 + \cdots + X_n\}_{n=1}^{\infty}.$$

Now taking into account Representation 4.4 and using Smirnov's transformation, one can obtain the following result.

REPRESENTATION 4.5. Let $X_1, X_2, \ldots$ be a sequence of independent random variables with a common continuous distribution function F. Then, for any $k = 1, 2, \ldots$ and $n = 1, 2, \ldots$,

$$\{X(n,k)\}_{n=1}^{\infty} \stackrel{d}{=} \{H((\omega_1 + \omega_2 + \cdots + \omega_n))/k\}_{n=1}^{\infty} ,$$

where $H(x) = G(1 - e^{-x})$ with G being the inverse function of F, and the random variables $\omega_1, \omega_2, \ldots$ are independently distributed as $E(0,1)$.

A very important and interesting property of records from discrete sequences was established by Shorrock (1972a). Without loss of generality, let us consider non-negative integer values for X's and suppose that $P\{X = n\} > 0$ for any $n = 0, 1, 2, \ldots$.

THEOREM 4.4. Let $X, X_1, X_2, \ldots$ be a sequence of independent random variables taking values $0, 1, 2, \ldots$, and let indicators $\eta_0, \eta_1, \ldots$ be defined as follows: $\eta_n = 1$ if n is a record value in the sequence $X_1, X_2, \ldots$; that is, $X(m) = n$ for some m, and $\eta_n = 0$ otherwise. Then, the random variables $\eta_0, \eta_1, \ldots$ are independent and $P\{\eta_n = 1\} = 1 - P\{\eta_n = 0\} = P\{X = n\}/P\{X \geq n\}$, $n = 0, 1, 2, \ldots$.

This theorem allows us to obtain distributions of record values $X(n)$ in the discrete case using the following representation.

REPRESENTATION 4.6. Under the conditions of Theorem 4.4,

$$P\{X(n) > m\} = P\{\eta_0 + \eta_1 + \cdots + \eta_m < n\}, \quad n \geq 1, \quad m \geq 0 .$$

Nevzorov (1986a, 1987b) generalized Theorem 4.4 for the case of the kth records using the indicators $\eta_n^{(k)}$ defined as follows: $\eta_n^{(k)} = 1$ if n is a kth record value in the sequence $X_1, X_2, \ldots$, and $\eta_n^{(k)} = 0$ otherwise.

THEOREM 4.5. Under the conditions of Theorem 4.4, for any $k = 1, 2, \ldots$, the indicators $\eta_0^{(k)}, \eta_1^{(k)}, \ldots$ are independent and

$$P\{\eta_n^{(k)} = 1\} = \left(\frac{P\{X = n\}}{P\{X \geq n\}}\right)^k, \quad n = 0, 1, 2, \ldots .$$

REPRESENTATION 4.7. Under the conditions of Theorem 4.4, for any $k = 1, 2, \ldots$,

$$P\{X(n,k) > m\} = P\{\eta_0^{(k)} + \cdots + \eta_m^{(k)} < n\}, \quad m \geq 0, \quad n \geq 1 .$$

5. Distributions and probability structure of record times

The main results on distributions of record times were given in the early papers of Chandler (1952) and Foster and Stuart (1954). All their results were rediscovered later by Rényi (1962) with the help of Representation 4.1. He also obtained a lot of new formulae for records. We list below some of the most important results connected with distributions of record times in the case when a parent distribution function F is continuous.

Recall first of all (see Corollary 4.1) that distributions of random variables $N(n)$ and $L(n)$ in this situation do not depend on the population distribution F.

THEOREM 5.1. [Chandler (1952), Rényi (1962)] The joint and marginal distributions of record times are as follows:

(a) if $1 < \ell_2 < \cdots < \ell_n$, then

$$P\{L(2) = \ell_2, \ldots, L(n) = \ell_n\} = \frac{1}{\ell_n(\ell_n - 1)(\ell_{n-1} - 1) \cdots (\ell_2 - 1)} ;$$

(b) for any $n \geq 2$ and $\ell \geq n$,

$$P\{L(n) = \ell\} = \sum_{1 < \ell_2 < \cdots < \ell_{n-1} < \ell} \frac{1}{\ell_n(\ell_n - 1) \cdots (\ell_2 - 1)}$$

and, in particular, $P\{L(2) = \ell\} = \frac{1}{\ell(\ell-1)}$, $\ell \geq 2$.

To connect the distributions of $L(n)$ and $N(n)$, we have to use the equalities

$$P\{L(n) \geq m\} = P\{N(m) \leq n\}$$

and

$$P\{L(n) = m\} = P\{N(m - 1) = n - 1, \xi_m = 1\}$$
$$= P\{N(m - 1) = n - 1\}/m .$$

These relationships between $L(n)$ and $N(n)$ were used to prove the second part of the following theorem.

THEOREM 5.2. [Rényi (1962), Shorrock (1972a), Westcott (1977a)] Let S_n^k denote the Stirling number of the first kind, defined by

$$x(x - 1) \cdots (x - n + 1) = \sum_{k=0}^{\infty} S_n^k x^k .$$

Then:

(a) $P\{N(n) = m\} = \frac{|S_n^m|}{n!}$ and $P\{N(n) = m\} \sim \frac{(\log n)^{m-1}}{n(m-1)!}$ as $n \to \infty$;

(b) $P\{L(n) = m\} = \frac{|S_{m-1}^{n-1}|}{m!}$ and $P\{L(n) = m\} \sim \frac{(\log m)^{n-2}}{m^2(n-2)!}$ as $m \to \infty$.

Rényi has discovered the generating functions of $L(n)$.

THEOREM 5.3. [Rényi (1962)] For any $n = 1, 2, \ldots$, the probability generating function of $L(n)$ is given by

$$Es^{L(n)} = 1 + (s - 1) \sum_{\ell=0}^{n-1} (-\log(1 - s))^\ell / \ell! .$$

Note that the above expression of the probability generating function of $L(n)$ can be rewritten as

$$\frac{1}{(n-1)!} \int_0^{-\log(1-s)} x^{n-1} e^{-x} dx \ .$$

The statements of Theorem 5.1 imply the Markov structure of record times $L(n)$.

THEOREM 5.4. The sequence $L(1), L(2), \ldots$ is a Markov chain, where

$$P\{L(n) = k | L(n-1) = m\} = \frac{m}{k(k-1)}, \text{ if } k > m \geq n-1 \ ,$$

and

$$P\{L(n) > k | L(n-1) = m\} = \frac{m}{k} \text{ if } k \geq m \ .$$

One can also prove Theorem 5.4 through Representation 4.1. For example, it follows from the independence of indicators $\xi_1, \xi_2, \ldots$ that

$$\begin{aligned}
P\{L(n) &= k | L(n-1) = m\} \\
&= P\{\xi_1 + \xi_2 + \cdots + \xi_{k-1} = n-1, \ \xi_k = 1 | \xi_1 + \xi_2 + \cdots + \xi_{m-1} \\
&= n-2, \ \xi_m = 1\} \\
&= P\{\xi_{m+1} + \cdots + \xi_{k-1} = 0, \ \xi_k = 1\} \\
&= P\{\xi_{m+1} = 0\} \cdots P\{\xi_{k-1} = 0\} P\{\xi_k = 1\} \\
&= \frac{m}{m+1} \cdots \frac{k-2}{k-1} \frac{1}{k} = \frac{m}{k(k-1)} \ .
\end{aligned}$$

The probability structure of the record times $L(n)$ can be also seen from the following result.

THEOREM 5.5. [Williams (1973)] Let $\omega_1, \omega_2, \ldots$ be independent $E(0,1)$ random variables and $[x]$ denote the integer part of x. Then, $L(n+1) \stackrel{d}{=} [L(n) \exp\{\omega_n\}] + 1$, $n = 1, 2, \ldots$.

Note that the Markov property of the sequence $L(n)$ is seen easily from Theorem 5.5.

One more interesting result which we give below describes the probability structure of the quotients $L(n+1)/L(n)$.

THEOREM 5.6. [Galambos and Seneta (1975)] Let integer-valued random variables $T(1), T(2), \ldots$ be defined by the following inequalities:

$$T(n) - 1 < \frac{L(n+1)}{L(n)} \leq T(n), \quad n = 1, 2, \ldots \ .$$

Then, $T(1), T(2), \ldots$ are independent and identically distributed random variables and

$$P\{T(n) = j\} = \frac{1}{j(j-1)}, \quad j = 2, 3, \ldots, \quad n = 1, 2, \ldots .$$

The following result can be considered as a corollary of Theorem 5.4 or Theorem 5.6.

COROLLARY 5.1. [Tata (1969), Galambos (1978, 1987)]

(a) For any $m = 1, 2, \ldots$ and $n = 1, 2, \ldots,$ $P\left\{\frac{L(n+1)}{L(n)} > m\right\} = \frac{1}{m}$;

(b) For any $x \geq 1$, $\lim P\left\{\frac{L(n+1)}{L(n)} > x\right\} = \frac{1}{x}$ as $n \to \infty$.

All the results given in this section are valid for any continuous parent distribution. Unfortunately, in the discrete case the distributions of record times depend on the distribution of X's. It is possible in this situation to write the necessary formulae for distributions of random variables $L(n)$, but the corresponding expressions are rather complicated. For example, if X's take on values $0, 1, 2, \ldots$, then

$$P\{L(2) = m\} = \sum_{n=0}^{\infty} P\{X = n\}(P\{X \leq n\})^{m-2} P\{X > n\}, \quad m = 2, 3, \ldots ,$$

and, in particular,

$$P\{L(2) = 2\} = P\{X_2 > X_1\} = \left(1 - \sum_{n=0}^{\infty} (P\{X = n\})^2\right) \bigg/ 2 .$$

It is known that any record time $L(n)$ exists with probability one for continuous distributions. In the discrete case, X's have to satisfy the following condition, which guarantees the existence of any record time: if $\beta \equiv \sup\{x : F(x) < 1\}$, then either $\beta = \infty$ or $\beta < \infty$ and $P\{X = \beta\} = 0$.

In the discrete case, there are many more interesting results for record values (we will discuss them below) than for record times. One can find some information about random variables $L(n)$ for discrete sequences in Vervaat (1973).

6. Moments of records times and numbers of records

Representation 4.1 allows us to calculate the moments of the random variable $N(n)$. It is not difficult to see that

$$EN(n) = A(n) = 1 + \frac{1}{2} + \frac{1}{3} + \cdots + \frac{1}{n}$$

and

$$\mathrm{Var}(N(n)) = B(n) = A(n) - \sum_{k=1}^{n}\left(\frac{1}{k}\right)^2, \qquad n = 1, 2, \ldots .$$

Note also that $\frac{A(n)}{\log n} \to 1$ and $B(n) \to 1$ as $n \to \infty$.

Now consider record times $L(n)$. We know that $P\{L(2) = k\} = \frac{1}{k(k-1)}$, $k > 1$. It means that $EL(2) = \infty$, and hence $EL(n) = \infty$ for any $n > 1$. This fact, may be, was the main reason why classical records have not been very popular amongst statisticians. Note that k^{th} record times $L(n,k)$, beginning from $k = 2$, have finite expectations. Meanwhile, some moments (factorial moments of the negative order and logarithmic moments) of record times $L(n)$ have been derived which have a rather simple form.

In the following section, it will be shown that random variables $\frac{L(n)-n}{\sqrt{(n)}}$ is asymptotically normally distributed. Hence, it is important to have formulae for the logarithmic moments of the random variable $L(n)$. Pfeifer (1984a) used Williams' representation (Theorem 5.5) to prove that

$$\mathrm{E}\log L(n) = n - C + O(n^2/2^n),$$
$$\mathrm{Var}(L(n)) = n - \pi^2/6 + O(n^3/2^n) \text{ as } n \to \infty,$$

where $C = 0.5772\ldots$ is Euler's constant.

More precise results were obtained by employing the martingale approach to finding the moments of record times [see Nevzorov (1986d, 1987c, 1989) and Nevzorov and Stepanov (1988).] It was shown, in particular, that

$$\mathrm{E}\log L(n) = n - C - 2^{-(n+1)} - \frac{5}{24}3^{-n} - \frac{1}{8}4^{-n} - A5^{-n},$$

where $0 < A < 49/96$, and

$$\mathrm{Var}(L(n)) = n - \pi^2/6 + n/2^{(n+1)} + O(2^{-n}) \text{ as } n \to \infty .$$

Let us now consider the factorial moments of random variable ξ of negative order in the following way: $m_r(\xi) = \mathrm{E}\{(\xi+1)(\xi+2)\cdots(\xi-r)\}^{-1}$, $r = -1, -2, \ldots$. Among other results, Nevzorov (1986d, 1990b) has proved that

$$m_r(L(n)) = \frac{1}{(1-r)!(1-r)^{n-1}}, \qquad r = -1, -2, \ldots ,$$

and in particular, $\mathrm{E}\left(\frac{1}{L(n)+1}\right) = 2^{-n}$, $\mathrm{E}\left(\frac{1}{(L(n)+1)(L(n)+2)}\right) = 3^{-n}/2$, $n = 1, 2, \ldots$.

We should also mention the following useful result due to Nevzorov and Stepanov (1988): for any $\alpha > 0$,

$$\mathrm{E}(L(n))^{1-\alpha} = \frac{1}{\Gamma(\alpha)}\left(\left(\frac{1}{\alpha}\right)^n + \frac{\alpha-1}{2}\left(\frac{1}{\alpha+1}\right)^n + O((\alpha+2)^{-n})\right)$$

as $n \to \infty$,

and, in particular,

$$E(L(n))^{1/2} = \frac{2^n}{\sqrt{\pi}} - \frac{(2/3)^n}{4\sqrt{\pi}} + O\left(\frac{2}{5}\right)^n,$$

$$E\left(\frac{1}{L(n)}\right) = 2^{-n} + \frac{3^{-n}}{2} + O(4^{-n}),$$

$$E\left(\frac{1}{L^2(n)}\right) = 3^{-n} + \frac{4^{-n}}{2} + O(5^{-n}) \text{ as } n \to \infty.$$

7. Limit theorems for record times

Representations of record times and numbers of records via sums of independent indicators presented earlier in Section 4 are convenient and simple tools for obtaining different limit theorems for random variables $N(n)$ and $L(n)$. Most of the classical limit theorems for records were obtained by Rényi (1962). It was shown that $\frac{N(n) - \log n}{(\log n)^{1/2}}$ asymptotically has the standard normal distribution. The laws of large numbers and the law of the iterated logarithm were also obtained for $N(n)$. These theorems, with the use of the relation of equality $P\{L(m) \geq n\} = P\{N(n) \leq m\}$, have been transformed to obtain the corresponding asymptotic results for the record times $L(n)$. In particular, the following results were established by Rényi (1962).

THEOREM 7.1. As $n \to \infty$,

$$P\left\{\frac{\log L(n)}{n} \to 1\right\} = 1,$$

$$P\left\{\frac{\log L(n) - n}{\sqrt{n}} < x\right\} \to \Phi(x),$$

$$P\left\{\limsup \frac{\log L(n) - n}{(2n \log \log n)^{1/2}} = 1\right\}$$

$$= P\left\{\liminf \frac{\log L(n) - n}{(2n \log \log n)^{1/2}} = -1\right\} = 1.$$

Another interesting result is due to Shorrock (1972a) which is as follows.

THEOREM 7.2. For any fixed $r > 1$, the ratios $\frac{L(n+1)}{L(n)}, \ldots, \frac{L(n+r)}{L(n+r-1)}$ are asymptotically independent as $n \to \infty$.

A survey on strong limit theorems and various techniques for record times as well as for record values has been made by Pfeifer and Zhang (1989).

8. Inter-Record times

Although inter-record times $\Delta(n) = L(n) - L(n-1)$, $n = 1, 2, \ldots$, are expressed very simply in terms of record times, there are many interesting results for these random variables and there also exist some special methods to investigate them. Below we give some useful formulae for these inter-record times, $\Delta(n)$.

Distributions of inter-record times for sequences of independent random variables $X_1, X_2, \ldots$, having a common continuous distribution function F, do not depend on F. These distributions are determined by the following, as given by Neuts (1967):

$$P\{\Delta(n) = r\} = \int_0^\infty (1 - e^{-x})^{r-1} \frac{x^{n-2} e^{-2x}}{(n-2)!} \, dx, \quad r = 1, 2, \ldots,$$

$$P\{\Delta(n) > r\} = \int_0^\infty (1 - e^{-x})^r \frac{x^{n-2} e^{-x}}{(n-2)!} \, dx = \int_0^1 v^r \frac{-(\log(1-v))^{n-2}}{(n-2)!} \, dv$$

$$= \sum_{m=0}^r \binom{r}{m} (-1)^m (1+m)^{1-n}, \quad r = 0, 1, \ldots, \ n = 2, 3, \ldots,$$

from which it follows immediately that $E\Delta(n) = \infty$ for any $n = 2, 3, \ldots$.

The logarithmic moments of inter-record times were given by Nevzorov and Stepanov (1988). For example, they have shown that

$$E \log \Delta(n) = n - 1 - C + \frac{n - 1 + \theta_1}{2^{n+1}} + \frac{n - 1 + \theta_2}{2 \cdot 3^n}, \quad n = 2, 3, \ldots,$$

where C is Euler's constant, $0 \leq \theta_1 \leq 2/3$ and $0 \leq \theta_2 \leq 6$.

Some limit theorems for inter-record times were proved by Neuts (1967), Holmes and Strawderman (1969), and Strawderman and Holmes (1970), among others. Some of these results are as follows: as $n \to \infty$,

$$P\left\{\frac{\log \Delta(n)}{n} \to 1\right\} = 1,$$

$$P\left\{\frac{\log \Delta(n) - n}{\sqrt{n}} < x\right\} \to \Phi(x),$$

$$P\left\{\limsup \frac{\log \Delta(n) - n}{(2n \log \log n)^{1/2}} = 1\right\}$$

$$= P\left\{\liminf \frac{\log \Delta(n) - n}{(2n \log \log n)^{1/2}} = -1\right\} = 1.$$

The joint asymptotic distribution of two successive inter-record times were obtained by Tata (1969) as

$$\lim P\{\log \Delta(n) - n > x\sqrt{n}, \ \log \Delta(n-1) - n < y\sqrt{n}\}$$
$$= \max\{0, \Phi(y) - \Phi(x)\} \text{ as } n \to \infty.$$

Strong limit theorems for the joint distribution of $L(n)$ and $\Delta(n)$ were obtained by Galambos and Seneta (1975) and Pfeifer (1987).

9. Distributions and probability structure of record values in sequences of continuous random variables

Tata's result (Representation 4.3) is a tool to obtain the distribution function for record values $X(n)$ if independent random variables $X_1, X_2, \ldots$ are distributed as $E(0, 1)$. In this case, $X(n)$ can be represented as a sum of n independent terms with the same $E(0, 1)$ distribution, and hence

$$P\{X(n) < x\} = \frac{1}{(n-1)!} \int_0^x v^{n-1} e^{-v} dv, \quad x > 0 \; .$$

Note also that in this case the joint distribution density function f_n of random variables $X(1), X(2), \ldots, X(n)$ has the following form:

$$f_n(x_1, x_2, \ldots, x_n) = e^{-x_n} \text{ if } 0 < x_1 < x_2 < \cdots < x_n$$
$$= 0 \text{ otherwise } . \tag{9.5}$$

One can now apply Representation 4.5 (with $k = 1$) to obtain an analogous formula for record values in the general case, when i.i.d. random variables $X_1, X_2, \ldots$ have any continuous distribution function F. In this situation,

$$P\{X(n) < x\} = \frac{1}{(n-1)!} \int_0^{-\log(1-F(x))} v^{n-1} e^{-v} dv \; .$$

It is of interest to compare this expression with the following one:

$$P\{X(n) < x\} = E\left\{(F(x))^{L(n)}\right\} \; .$$

If X_i's have a common density function f, then

$$f_n(x_1, x_2, \ldots, x_n) = r(x_1)r(x_2)\cdots r(x_n)(1 - F(x_n)),$$
$$\text{for } x_1 < x_2 < \cdots < x_n \; ,$$

where $r(x) = \frac{f(x)}{1-F(x)}$ is the hazard function, and the density function of $X(n)$ has the form

$$f_n(x) = \{-\log(1 - F(x))\}^{n-1} f(x)/(n-1)! \; .$$

Representation 4.5 also implies the Markov structure of record values. Moreover, Shorrock (1972b) proved that the equalities

$$P\{X(n+1) \geq x | X(1), X(2), \ldots, X(n) = v\} = P\{X(n+1) \geq x | X(n) = v\}$$
$$= \frac{P\{X \geq x\}}{P\{X > v\}}, \quad x > v \; , \tag{9.6}$$

are valid not only for continuous distribution function F. The only restriction to provide this result is the existence with probability one of any record value $X(n)$.

For continuous distributions, the following relationship between order statistics and record values exists. Deheuvels (1984b) and Gupta (1984) showed that for any $n > 1$ and $m > 1$,

$$P\{X(n) > y | X(n-1) = x\} = P\{X_{m:m} > y | X_{m-1:m} = x\} \text{ a.s.}$$

10. Limit theorems for record values from continuous distributions

We know that $X(n) = M(L(n))$, where $M(n) = \max\{X_1, X_2, \ldots, X_n\}$. Hence, existence of some relationships between the asymptotic behavior of record values and maximal order statistics can not be surprising to us. It is well-known that there are three types of nondegenerate asymptotic distributions for centered and normalized maxima $M(n)$. The corresponding limit distribution functions may be written as $\exp\{-\exp(-g(x))\}$, where $g(x)$ is one of the following functions [see Galambos (1978, 1985)]:

(i) $g(x) = x$, $-\infty < x < \infty$;
(ii) $g(x) = \alpha \log x$, $\alpha > 0$, $x > 0$, and $g(x) = -\infty$, $x < 0$;
(iii) $g(x) = \infty$, $x > 0$ and $g(x) = -\alpha \log(-x)$, $x < 0$, $\alpha > 0$.

Tata (1969) and Resnick (1973b) posed and solved the problem of describing the set of all possible asymptotic distributions of record values $X(n)$ under suitable normalization. It appears that these limit distribution functions have the form $\Phi(g(x))$, where $g(x)$ is any of the functions given above. Resnick (1973b) also gave a description of the domains of attraction of the corresponding limit laws along with a form of appropriate centering and normalizing constants.

Some other limit theorems (laws of large numbers, law of the iterated logarithm, etc.) for record values and their differences and quotients were obtained by Resnick (1973a,b), de Haan and Resnick (1973), Goldie (1982), and Freudenberg and Szynal (1976).

11. Record values from discrete distributions

In this section, without loss of generality, we confine ourselves to i.i.d. random variables $X, X_1, X_2, \ldots$ taking on values $0, 1, 2, \ldots$ and such that $P\{X < n\} < 1$ for any integer n. The simplest way to investigate record values $X(n)$ in this situation is to apply Theorem 4.4 and Representation 4.6.

The joint distribution of record values is given as follows:

$$P\{X(1) = j_1, X(2) = j_2, \ldots, X(n) = j_n\}$$
$$= P\{X = j_n\} \prod_{r=1}^{n-1} \left(\frac{P\{X = j_r\}}{P\{X > j_r\}} \right), \quad 0 \leq j_1 < \cdots < j_n.$$

It follows from the latter formula that the sequence $X(1), X(2), \ldots$ forms a Markov chain and

$$P\{X(n+1) = j | X(n) = i\} = \frac{P\{X = j\}}{P\{X > i\}}, \quad j > i \geq n - 1 \;.$$

For the case of the geometric distribution, results for record values can be simplified considerably. For example, if we take the geometric distribution with probability mass function $P\{X = n\} = (1-p)p^{n-1}$, $0 < p < 1$, $n = 1, 2, \ldots$, then the record spacings $X(1), X(2) - X(1), X(3) - X(2), \ldots$ are all independent and $X(n)$ has the same distribution as the sum $X_1 + \cdots + X_n$.

Some limit theorems for $X(n)$ in this case were also obtained by Vervaat (1973).

12. Weak records

Sometimes, a repetition of a record value can be counted as a new record. Vervaat (1973) and Stepanov (1992) considered the corresponding record model. Essentially, this model is new only in the case when two random variables can coincide with a positive probability. That is the reason why weak records are connected with sequences of discrete distributions. Definitions of weak record times $L^w(n)$ and weak record values $X^w(n)$ are given as follows:

$$L^w(1) = 1, \quad L^w(n+1) = \min\{j > L^w(n) : X_j \geq \max(X_1, X_2, \ldots, X_{j-1})\} \;,$$
$$X^w(n) = X_{L^w(n)}, \quad n \geq 1 \;.$$

Note that all weak records, unlike classical ("strong") records, exist with probability one. The joint distribution of weak records is given by the equality

$$P\{X^w(1) = j_1, X^w(2) = j_2, \ldots, X^w(n) = j_n\}$$
$$= P\{X = j_n\} \prod_{r=1}^{n-1} \left(\frac{P\{X = j_r\}}{P\{X \geq j_r\}} \right), \quad 0 \leq j_1 \leq \cdots \leq j_n \;.$$

This formula implies the Markov property of weak record values.

THEOREM 12.1. [Vervaat (1973)] The sequence $X^w(1), X^w(2), \ldots$ forms a Markov chain with probabilities

$$P\{X^w(n+1) = j | X^w(n) = i\} = \frac{P\{X = j\}}{P\{X \geq i\}}, \quad j \geq i \;.$$

One more useful result is based on the formula for the joint distribution of weak records which was given above. If X has a geometric distribution with probability mass function $P\{X = n\} = (1-p)p^n$, $n = 0, 1, 2, \ldots$, then the record spacings $X^w(1), X^w(2) - X^w(1), X^w(3) - X^w(2), \ldots$ are all independent and have the same geometric distribution as X.

There are specific representations for weak records. The next result is presented in the form as given by Stepanov (1992), eventhough it was already present in an implicit form in Vervaat (1973). Now, let us introduce new random variables $\mu_0, \mu_1, \mu_2, \ldots$, where μ_n is a number of records in the sequence $X^w(1), X^w(2), \ldots$ taking value n, that is, μ_n coincides with a number of repetitions of a record value n.

THEOREM 12.2. Random variables $\mu_0, \mu_1, \mu_2, \ldots$ are all independent and $P\{\mu_0 = m\} = (1 - p_n)p_n^m$, $n = 1, 2, \ldots$, $m = 0, 1, 2, \ldots$, where $p_n = \frac{P\{X=n\}}{P\{X \geq n\}}$.

Theorem 12.2 implies the following relationships.

REPRESENTATION 12.1. For any $n = 1, 2, \ldots$ and $m = 0, 1, \ldots$,

$$P\{X^w(n) > m\} = P\{\mu_0 + \mu_1 + \cdots + \mu_m < n\}$$

and

$$P\{X^w(n) = m\} = P\{\mu_0 + \mu_1 + \cdots + \mu_{m-1} < n, \ \mu_0 + \mu_1 + \cdots + \mu_m \geq n\} \ .$$

Vervaat (1973) and Stepanov (1992) used this result to express the distributions of weak record values as sums of independent random variables, and applied them in turn to establish limit theorems for $X^w(n)$.

13. Bounds and approximations for moments of record values

Along the lines of the derivations of bounds and approximations for moments of order statistics, some bounds and approximations have also been developed for the moments of record values. For example, by employing Cauchy-Schwarz inequality, Nagaraja (1978) has shown that the mean of the nth upper record value, $E\{X(n)\}$, for any arbitrary continuous distribution (with mean 0 and variance 1), satisfies the inequality

$$E\{X(n)\} \leq \left\{ \binom{2n-2}{n-1} - 1 \right\}^{1/2}$$

and that this bound is sharp; it is achieved for the population with its inverse cumulative distribution function as

$$F^{-1}(u) = \frac{1}{\{\binom{2n-2}{n-1} - 1\}^{1/2}} \left[\frac{1}{(n-1)!} \{-\log(1-u)\}^{n-1} - 1 \right]$$

for $0 < u < 1$.

Nagaraja (1978) has also derived an improved bound for the case when the population distribution is symmetric. Bounds and approximations can also be developed for the moments of record values using an orthogonal inverse expansion; for details, one may refer to Nagaraja (1978) and Arnold, Balakrishnan and Nagaraja (1998).

Upon noting that these bounds (like the one presented above) become very large very quickly even though they are universally sharp, Arnold, Balakrishnan and Nagaraja (1998) have developed some "extrapolation-type bounds", along the lines of Balakrishnan (1990) in the order statistics context, which give good improvements over the bounds as they are distribution-based. Nagaraja (1978) also followed the work of van Zwet (1964) in order to derive some bounds for the moments of record values based on c- and s-comparisons.

Generalizing the work of Nagaraja (1978), Grudzien and Szynal (1983) derived the Cauchy-Schwarz bounds for the moments of the kth record values, $X(n,k)$. Though these bounds are simple and taken on a similar form (to the one given above), unfortunately these bounds are not sharp. In fact, they are sharp only for the case when $k = 1$ (in which case it coincides with the result presented above for the usual record values). For this purpose, following the lines of Moriguti (1953) and Balakrishnan (1993), Raqab and Balakrishnan (1997) used the greatest convex minorant principle in order to derive sharp bounds for the moments of the kth record values.

14. Recurrence relations for moments of record values

Numerous recurrence relations and identities exist in the literature for the moments of order statistics; see, for example, David (1981), Arnold and Balakrishnan (1989), and Arnold, Balakrishnan and Nagaraja (1992). Along the same lines, recurrence relations may be established for the single and product moments of record values as well.

For example, in the case of the standard exponential distribution, upon making use of the differential equation $f(x) = 1 - F(x)$, Balakrishnan and Ahsanullah (1995) established the following recurrence relations for the moments of upper record values:

$$E\{X(n)\}^{a+1} = E\{X(n-1)\}^{a+1} + (a+1)E\{X(n)\}^a \text{ for } n \geq 1, a \geq 0 ,$$

$$E\{X^a(m)X^{b+1}(m+1)\} = E\{X^{a+b+1}(m)\} + (b+1)E\{X^a(m)X^b(m+1)\}$$
$$\text{for } m \geq 1, a, b \geq 0 ,$$

$$E\{X^a(m)X^{b+1}(n)\} = E\{X^a(m)X^{b+1}(n-1)\} + (b+1)E\{X^a(m)X^b(n)\}$$
$$\text{for } 1 \leq m < n, a, b \geq 0 .$$

Balakrishnan and Ahsanullah (1995) have also presented more general results for the higher order product moments of record values, as well as some recurrence relations for the moments of record values from the non-identical exponential model.

Proceeding along the same lines and exploiting the underlying differential equation of the population distribution assumed, recurrence relations for single and product moments of record values have been established for a number of different distributions including Rayleigh, Weibull, Gumbel, generalized extreme value, Lomax, generalized Pareto, normal and logistic distributions. Interested readers may refer to the papers by Balakrishnan and Chan (1994, 1995), Balakrishnan, Ahsanullah and Chan (1992, 1995), Balakrishnan, Chan and Ahsanullah (1993), and Balakrishnan and Ahsanullah (1994a,b). Reference also be made to Chapter 3 of Arnold, Balakrishnan and Nagaraja (1998) for a review of all these results.

15. Joint distributions of record times and record values

Earlier, it was pointed out that

$$P\{X(n) < x\} = E\{(F(x))^{L(n)}\} \ .$$

This formula connects the record values and their corresponding record times. The easiest way of establishing this relationship between $X(n)$ and $L(n)$ is through the following result which was proved by Ballerini and Resnick (1987b) in a more general situation.

THEOREM 15.1. Let $X_1, X_2, \ldots, X_n$ be independent random variables with a common distribution function. Then for any n, the indicators of records $\xi_1, \xi_2, \ldots, \xi_n$ (as defined in Representation 4.1) do not depend on $M(n) = \max\{X_1, X_2, \ldots, X_n\}$.

From Theorem 15.1, one can write

$$\begin{aligned} P\{X(n) < x\} &= P\{M(L(n)) < x\} \\ &= \sum_{m=n}^{\infty} P\{M(L(n)) < x | L(n) = m\} P\{L(n) = m\} \\ &= \sum_{m=n}^{\infty} P\{M(m) < x | L(n) = m\} P\{L(n) = m\} \\ &= \sum_{m=n}^{\infty} P\{M(m) < x\} P\{L(n) = m\} \\ &= \sum_{m=n}^{\infty} \{F(x)\}^m P\{L(n) = m\} \\ &= E\{(F(x))^{L(n)}\} \ . \end{aligned}$$

Note that in the above proof, we have used the fact that the event $\{L(n) = m\}$ for any $m \geq n$ coincides with the event $\{\xi_1 + \cdots + \xi_{m-1} = n - 1, \xi_m = 1\}$ and, therefore, it does not depend on $M(m)$.

The joint distribution of record times and record values has been discussed by Rényi (1962). Let $X_1, X_2, \ldots$ be i.i.d. random variables having a common continuous distribution function F. Then, for any $1 < k(2) < \cdots < k(n)$,

$$P\{L(1) = 1, L(2) = k(2), \ldots, L(n) = k(n), X(1) < x_1, X(2) < x_2, \ldots, X(n) < x_n\}$$
$$= \int \cdots \int u_1^{k(2)-1} u_2^{k(3)-k(2)-1} \cdots u_{n-1}^{k(n)-k(n-1)-1} \, du_1 \, du_2 \ldots du_n ,$$

where the integrals are taken over the set A of points $\mathbf{u} = (u_1, u_2, \ldots, u_n)$ given by

$$A = \{\mathbf{u} : -\infty < u_1 < \cdots < u_n, \, F(u_j) < x_j, \, j = 1, \ldots, n\} .$$

It is more convenient to consider the joint behavior of record values $X(n)$ and inter-record times $\Delta(n)$. Shorrock (1972b) showed that the sequence of two-dimensional vectors $(X(n), \Delta(n))$, $n = 1, 2, \ldots$, forms a Markov chain with probabilities

$$P\{X(n) > x, \Delta(n) = m | \Delta(1), X(1), \ldots, \Delta(n-2), X(n-2), \Delta(n-1), X(n-1) = y\}$$
$$= \{F(y)\}^{m-1} \{1 - F(x)\}, \quad x > y .$$

It was also proved [see Strawderman and Holmes (1970) and Shorrock (1972b)] that the inter-record times $\Delta(n)$, $n = 1, 2, \ldots$ are conditionally independent under fixed values of random variables $X(1), X(2), \ldots$, and

$$P\{\Delta(n) = m | X(1), X(2), \ldots\} = \{1 - F(X(n))\} \{F(X(n))\}^{m-1},$$
$$m = 1, 2, \ldots .$$

There also exists some results on the rates of closeness of the random variables $\Delta(n)$, $L(n)$ and $\tau_n = -\log\{1 - F(X(n))\}$. For example, Shorrock (1972b) showed that, almost surely as $n \to \infty$,

$$\limsup \frac{|\log \Delta(n+1) - \tau_n|}{\log n} = 1$$

and

$$\limsup \frac{|\log L(n) - \tau_n|}{\log n} = 1 .$$

Note that an analogous result is also valid for random variables $\frac{|\log \Delta(n+1) - \log L(n)|}{\log n}$, as shown by Galambos and Seneta (1975). On these lines, the following result of Nevzorov (1995) is worth mentioning here.

THEOREM 15.2. Let $H(x) = e^{-e^{-x}}$. Then for any q, $0 < q < 1$, the following inequality is valid:

$$|P\{\tau_n - \log L(n) < x\} - H(x)| \le r_n(x),$$

where

$$r_n(x) = H(x)\left(e^{-2x} 2^{2-n} + \frac{2}{1-q} e^{-4x} 3^{1-n}\right) \quad \text{if } x > x_n$$

$$= \exp\{-(3qn/2)^{1/2}\}\left(1 + 3qn2^{1-n} + \frac{q^2}{2(1-q)} n^2 3^{3-n}\right) \quad \text{if } x < x_n,$$

with $x_n = -\frac{1}{2}\log\left(\frac{3nq}{2}\right)$.

Siddiqui and Biondini (1975) investigated the asymptotic behavior of random variables $\Delta(n+1)e^{-\tau_n}$ and $\Delta(n)e^{-\tau_n}$. The first sequence has asymptotically a standard exponential distribution while the limiting distribution function for the second sequence is $1 - \int_1^\infty z^{-2} e^{-zx} \, dz$, $x > 0$.

16. Generalizations of the classical record model

Until now, we have discussed only the classical records arising from sequences of i.i.d. random variables. Nearly twenty years after the mathematical foundation of records was made by Chandler, only in the mid-seventies were the first attempts made in order to generalize the record scheme of Chandler. Three different directions of generalizations of records were considered during this time. One of them [see Yang (1975)] was based on relaxing the assumption of identicaly distributed part of the initial random variables. The record times and record values from Markov sequences were then studied. Another direction taken by Dziubdziela and Kopocinsky (1976) and Dziubdziela (1977) kept the i.i.d. structure of the initial X's but they, unlike their predecessors, chose to investigate a more general random variable than the classical records – the so-called k^{th} records. Hence, we begin our review here of the non-classical record models starting with k^{th} record times and k^{th} record values.

17. k^{th} record times

Classical records simply a particular case of the kth records (when $k = 1$). It was precisely the reason why we gave in the very beginning of this chapter the definitions of kth record times (3.1) and k^{th} record values (3.2) and some general results (Theorems 4.2 and 4.5 and Representations 4.2, 4.4, 4.5 and 4.7). One can see that the theory of kth records relates to the theory of classical records in the same way as the theory of the kth extremes relates to the theory of maxima. We will explain in this section what new properties and advantages were obtained due to this generalization.

We may note first of all that there is another definition of the kth record times, which in the case of continuous distributions coincides with (3.1). Let R_n,

$n = 1, 2, \ldots$, (as in Lemma 4.1) be sequential ranks of random variables $X_1, X_2, \ldots$. Then, the kth record times for any fixed $k = 1, 2, \ldots$ can be defined as follows:

$$L(0, k) = 0, \quad L(1, k) = k \text{ and } L(n+1, k)$$
$$= \min\{j > L(n, k) : R_j \geq j - k + 1\}, \quad n \geq 1. \quad (17.7)$$

As was already mentioned in Section 2, Definitions (3.1) and (3.2) and some first results on the kth records were obtained by Dziubdziela and Kopocinsky (1976).

THEOREM 17.1. For any $k = 1, 2, \ldots$ and $k < l_2 < \cdots < l_n$,

$$P\{L(1, k) = k, \ L(2, k) = l_2, \ldots, \ L(n, k) = l_n\}$$
$$= k!(l_n - k)! k^{n-1} / \{l_n!(l_2 - k) \cdots (l_n - k)\}.$$

THEOREM 17.2. For any $k = 1, 2, \ldots$, random variables $L(1, k), L(2, k), \ldots$ form a Markov chain and $P\{L(n, k) > j | L(n-1, k) = i\} = (j - k)! i! / \{(i - k)! j!\}$.

The following properties of the kth record times were established by Nevzorov (1990a,b).

THEOREM 17.3. (Compare with Theorem 5.3). For any $k = 1, 2, \ldots$,

$$E\{s^{L(n,k)}\} = H(-\log(1-s)), \quad n = 1, 2, \ldots,$$

where $H(x) = \frac{k^n}{(n-1)!} \int_0^x u^{n-1} e^{-ku} (1 - e^{-(x-u)})^{k-1} du$.

THEOREM 17.4. For any $k = 2, 3, \ldots$ and $n = 1, 2, \ldots$, $E\{L(n, k)\} = k^n / (k-1)^{n-1}$.

The latter result, if compared with equality $E\{L(n, 1)\} = \infty$, $n = 2, 3, \ldots$ for the classical records, "rehabilitates" records in statisticians' opinion. Note that this new property of the kth record times for $k \geq 2$ agree with the results of Wilks (1959) and Gumbel (1961), who studied the moments of the number of additional observations that are needed to surpass the kth largest of n existing elements of a sample.

THEOREM 17.5. For any $k = 1, 2, \ldots$ and $\alpha > 0$, as $n \to \infty$,

$$E\{L(n, k)\}^{k-\alpha} = \frac{(k-1)!}{\Gamma(\alpha)} \left\{ \left(\frac{k}{\alpha}\right)^n + \frac{(\alpha - k)(\alpha - k + 1)}{2\alpha} \left(\frac{k}{\alpha + 1}\right)^n \right.$$
$$\left. + O\left(\left(\frac{k}{\alpha + 2}\right)^n\right) \right\}.$$

Now, let $m_r(\xi) = E\{\xi(\xi - 1) \cdots (\xi - r + 1)\}$ for $r = 1, 2, \ldots, m_0(\xi) = 1$ and $m_r(\xi) = E[1/\{(\xi + 1)(\xi + 2) \cdots (\xi - r)\}]$ for $r = -1, -2, \ldots$, be the factorial moments of a random variable ξ, of both positive and negative orders.

THEOREM 17.6. For any $k = 1, 2, \ldots$ and $r = k - 1, k - 2, \ldots, 1, 0, -1, -2, \ldots,$

$$m_r(L(n,k)) = k^{n-1} k! / \left\{ (k-r)^{n-1} (k-r)! \right\} \quad \text{for} \quad n \geq 1 \ .$$

Of course, Theorem 17.4 is a corollary of Theorem 17.6. Amongst other corollaries of Theorem 17.6, there are the following equalities:

$$E\{L(n,k)(L(n,k) - 1)\} = k^n (k-1)/(k-2)^{n-1}, \quad k > 2 \ ,$$
$$E\{1/(L(n,k) + 1)\} = k^{n-1}/(k+1)^n \ ,$$
$$E[1/\{(L(n,k) + 1)(L(n,k) + 2)\}] = k^{n-1}/\{(k+2)^n(k+1)\}, \quad n \geq 1 \ .$$

Theorems 17.5 and 17.6 are based on the next result [see Nevzorov (1990a,b)] in which $\mathscr{T}(n,k)$ denotes the σ-algebra of events generated by random variables $L(1,k), L(2,k), \ldots, L(n,k)$.

THEOREM 17.7. For any fixed $k = 1, 2, \ldots$ and $\gamma < k$, the sequence $T_n(\gamma)$ is a martingale with respect to σ-algebras $\mathscr{T}(n,k)$, and

$$E\{T_n(\gamma)\} = (k - \gamma) \Gamma(k+1) / \{k \Gamma(k - \gamma + 1)\} \ .$$

There are also some results for logarithmic moments of the kth record times [see Nevzorov and Stepanov (1988)]. For example, it has been shown by these authors that for any $k = 1, 2, \ldots,$

$$\begin{aligned}
E\{\log L(n,k)\} &= 1 + 1/2 + \cdots + 1/(n-1) + n/k \\
&\quad - C - k^{n-1}/\{2(k+1)^n\} - 5k^{n-1}/\{12(k+1)(k+2)^n\} \\
&\quad - 3k^{n-1}/\{4(k+1)(k+2)(k+3)^n\} \\
&\quad - Ak^{n-1}/\{(k+1)(k+2)(k+3)(k+4)^n\}, \quad n \geq 1 \ ,
\end{aligned}$$

where $0 < A < 49/4$ and $C = 0.5772\ldots$ is Euler's constant, and

$$\text{Var}(L(n,k)) = \frac{n}{k^2} - \frac{\pi^2}{6} + \sum_{\ell=1}^{k-1} (1/\ell^2) + \frac{nk^{n-2}}{(k+1)^{n+1}} + O((k/(k+1))^{n+1}), \quad n \to \infty \ .$$

Representation 4.2 can be used to obtain a central limit theorem as well as some other asymptotic results for random variables $L(n,k)$ and $N(n,k)$ [see Deheuvels (1981, 1982a, 1983a, 1984a,c,d)]. For example, it has been proved that for any $k = 1, 2, \ldots$, almost surely as $n \to \infty$,

$$\lim \frac{N(n,k)}{k \log n} = 1, \quad \lim \frac{k \log L(n,k)}{n} = 1 \ ,$$

and

$$\limsup \frac{k \log L(n,k) - n}{(2n \log \log \log n)^{1/2}} = 1 \ .$$

Note also that $\frac{k \log L(n,k) - n}{n^{1/2}}$ asymptotically has a standard normal distribution. Deheuvels also obtained some strong approximation results for the kth record times by Wiener and Poisson processes.

Sometimes [see, for example, Resnick (1987) or Deheuvels (1988)] the notion "k^{th} record times" is used for random variable $L^{(k)}(n)$, which are defined as follows:

$$L^{(k)}(0) = k - 1 \text{ and } L^{(k)}(n)$$
$$= \min\{j > L^{(k)}(n-1) : R_j = j - k + 1\}, \quad n \geq 1. \quad (17.8)$$

This definition almost coincides with the definition of the classical record times. As a matter of fact, the corresponding indicators of records, say $\xi_1(k), \xi_2(k), \ldots$, in this case are also independent, $\xi_1(k) = 0, \xi_2(k) = 0, \ldots, \xi_{k-1}(k) = 0$ and $\xi_n(k)$ for any $n \geq k$, has the same distribution as indicators ξ_n from Theorem 4.1 and Representation 4.1. Then, $\{N^{(k)}(m)\}_{m=k}^{\infty} \stackrel{d}{=} \{N(m) - N(k-1)\}_{m=k}^{\infty}$, where $N^{(k)}(m) = \xi_1(k) + \xi_2(k) + \cdots + \xi_m(k)$, and it means (since $1 \leq N(k-1) \leq k-1$) that limiting distributions of $N^{(k)}(m)$ and $N(m)$ as well as asymptotic properties of $L^{(k)}(n)$ and $L(n)$ coincide.

18. k^{th} inter-record times

A lot of properties for k^{th} inter-record times $\Delta(n, k) = L(n, k) - L(n - 1, k)$ come out as corollaries of theorems for k^{th} record times. We list here some results for these k^{th} inter-record times $\Delta(n, k)$. Distributions and moments of these random variables have been discussed by Nevzorov (1987a,c), Nevzorov and Stepanov (1988), and Stepanov (1987). In particular, it has been shown that

$$P\{\Delta(n,k) = m\} = \int_0^\infty \frac{1}{(n-2)!} k^{n-1} x^{n-2} e^{-x(k+1)} (1 - e^{-x})^{m-1} dx,$$
$$k \geq 1, \ n \geq 2 \ ;$$

$$P\{\Delta(n,k) > r\} = \int_0^\infty \frac{1}{(n-2)!} k^{n-1} x^{n-2} e^{-xk} (1 - e^{-x})^r dx$$
$$= \sum_{m=0}^r (-1)^m \binom{m}{r} \left(1 + \frac{m}{k}\right)^{1-n}, \quad r \geq 0, \ n \geq 2 \ ;$$

$$E\{\Delta(n,k)\} = E\{L(n,k)\} - E\{L(n-1,k)\} = \frac{k^n}{(k-1)^{n-1}} - \frac{k^{n-1}}{(k-1)^{n-2}}$$
$$= \frac{k^{n-1}}{(k-1)^{n-1}}, \quad k \geq 2 \ ;$$

and

$$E\{\log \Delta(n,k)\} = \frac{n-1}{k} - C + \frac{n-1+\theta_1}{2k}\left(\frac{k}{k+1}\right)^n$$
$$+ \frac{n-1+\theta_2}{2k}\left(\frac{k}{k+2}\right)^n,$$

where $|\theta_1| < (k+1)/3$, $0 \leq \theta_2 \leq 2(k+2)$.

Several limit theorems for k^{th} inter-record times have been established by Deheuvels (1983b, 1984a,c); see also Nevzorov (1987a). Amongst many other results, it has been proved that as $n \to \infty$,

$$P\{k(\log \Delta(n,k)/n < x\sqrt{n}\} \to \Phi(x) \quad \text{and} \quad P\{(k/n)\log \Delta(n,k) \to 1\} = 1 .$$

The closeness of $L(n,k)$ and $\Delta(n,k)$ is emphasized by the following two theorems of Deheuvels.

THEOREM 18.1. For any $k = 1, 2, \ldots$,

$$1/k \leq \limsup_{n \to \infty}(|\log \Delta(n+1,k) - \log L(n,k)| \leq 1 .$$

THEOREM 18.2. For any $k = 1, 2, \ldots$ it is possible to define a standard Wiener process $W^{(k)}(t)$, $t \geq 0$, such that at the same time almost surely, as $n \to \infty$,

$$|\log L(n,k) - n/k - W^{(k)}/k| = O(\log n)$$

and

$$|\log \Delta(n+1,k) - n/k - W^{(k)}(n)/k| = O(\log n) .$$

Note that there is a possibility of obtaining a new type of result (as compared to the classical records) for the kth records. One can investigate the asymptotic behavior of $L(n,k)$, $\Delta(n,k)$ and $X(n,k)$ in the scheme of series as $k \to \infty$. In this connection, we mention the paper of Gajek (1985) wherein it has been shown that the distributions of random variables $k\Delta(n,k)$, under some general conditions, converge to the exponential distribution $E(0,\sigma)$ as $k \to \infty$.

19. k^{th} record values for the continuous case

Some interesting results for the kth record values in the continuous case have been obtained by Deheuvels (1984d, 1988). It follows from Representation 4.5 that distributions of the kth record values $X(n,k)$ can be expressed via distributions of the classical record values. Let $X_1, X_2, \ldots$ be independent random variables having a continuous distribution function F, and $Y_1 = \min\{X_1, \ldots, X_k\}$, $Y_2 = \min\{X_{k+1}, \ldots, X_{2k}\}, \ldots$ and so on. Further, let $X(n,k)$ be the kth record values based on X's and $Y(n)$ be the classical record values based on Y's.

THEOREM 19.1. *For any* $k = 1, 2, \ldots,$ $\{X(n,k)\}_{n=1}^{\infty} \stackrel{d}{=} \{Y(n)\}_{n=1}^{\infty}.$

This result, together with theorems for the classical record values, helps us to make several statements for the kth record values. For example, it can be immediately obtained [see Dziubdziela (1977)] that

$$P\{X(n,k) < x\} = P\{X(n,k) < x\} = \frac{1}{(n-1)!} \int_0^{-k\log\{1-F(x)\}} u^{n-1} e^{-u} du.$$

Note also that $X(1,k), X(2,k), \ldots$ forms a Markov chain and

$$P\{X(n+1,k) > x | X(n,k) = y\} = \{(1 - F(x))/(1 - F(y))\}^k, \quad x > y.$$

The last equality can be rewritten in another form, giving a curious relation between the kth record values and order statistics.

THEOREM 19.2. *For any* $k = 1, 2, \ldots, n = 2, 3, \ldots$ *and* $m = k+1, k+2, \ldots,$ *almost surely,*

$$P\{X(n,k) > x | X(n-1,k) = y\} = P\{X_{m-k+1:m} > x | X_{m-k:m} = y\}$$
$$\text{for} \quad x > y.$$

Random variables $Y_1, Y_2, \ldots$ in Theorem 19.1 have the joint distribution function $G(x) = 1 - \{1 - F(x)\}^k$. Accordingly, $F(x) = 1 - \{1 - G(x)\}^{1/k}$, and it means that F and G are distribution functions simultaneously. Therefore, it follows from Theorem 19.1 that the set of possible limit distribution functions for suitably normalized random variables $X(n,k)$ is the same as for $X(n)$; hence, it consists of three types of functions $H(x) = \exp\{-\exp(-g(x))\}$, where $g(x)$ is one of the following functions:

(i) $g(x) = x,\ -\infty < x < \infty$;
(ii) $g(x) = \alpha \log x$ for $\alpha > 0,\ x > 0$, and $g(x) = -\infty$ for $x < 0$;
(iii) $g(x) = -\alpha \log(-x)$ for $x < 0,\ \alpha > 0$, and $g(x) = \infty$ for $x > 0$.

This and some related problems have been solved by Dziubdziela and Kopocinsky (1976), Dziubdziela (1977), Grudzien (1979), and Nevzorov (1988). Nevzorov (1986a) also showed that even for a more general set of random variable $X_{L(n,k)-\ell+1:L(n,k)}$, where ℓ is fixed or increases with increasing values of n, as well as for random variables $X_{L(n,k)}$, all possible limit distribution functions include only three types of functions H, as presented above.

The following alternate definition of the kth record values $X^{(k)}(n)$ corresponds to the definition of the kth record times $L^{(k)}(n)$ given in (17.8):

$$X^{(k)}(n) = X_{L^{(k)}(n)}, \quad n = 1, 2, \ldots.$$

It means that we select from the initial sequences $X_1, X_2, \ldots$ only those X_m such that $m = L^{(k)}(1), L^{(k)}(2), \ldots,$ that is, $\sum_{i=1}^{m} \mathbf{1}_{\{X_i \geq X_m\}} = k$. There are some results available for $X^{(k)}(n)$. For example, Deheuvels (1988) has obtained several limit theorems for these random variables. It appears that the most curious result on these record values $X^{(k)}(n)$ is the so-called *Ignatov's theorem*. In order to state this result, we introduce for any $k = 1, 2, \ldots$ a counting process $N^{(k)}(x) = \sum_{n=1}^{\infty} \mathbf{1}_{\{X^{(k)}(n) \leq x\}}$. It appears that $N^{(k)}$, $k = 1, 2, \ldots$, are i.i.d. point processes. This statement is due to Ignatov (1981), but the proof of this theorem was given only in Ignatov (1986). Goldie (1982) and Stam (1982, 1985) independently suggested their own proofs of Ignatov's assertion. Meanwhile, Deheuvels (1983a), knowing nothing of the papers of Ignatov, Goldie and Stam, also discovered this result. Mention should also be made here to the papers of Goldie and Rogers (1984), Deheuvels (1988), Engelen, Tommassen and Vervaat (1988), Rogers (1989) and the book of Resnick (1987), where Ignatov's theorem and related problems have been discussed in great detail.

20. k^{th} record values for the discrete case

Without loss of generality, we consider here i.i.d. random variables $X, X_1, X_2, \ldots$, taking on values $0, 1, 2, \ldots$ such that $P\{X < n\} < 1$ for any n. In this case, all k^{th} record values $X(n, k)$ exist with probability one. From Theorem 4.5, one can then express the distribution functions of random variables $X(n, k)$ via distributions of independent indicators as follows:

$$P\{X(n, k) \leq m\} = P\{\eta_0^{(k)} + \cdots + \eta_m^{(k)} \geq n\} , \qquad (20.9)$$

where $P\{\eta_n^{(k)} = 1\} = 1 - P\{\eta_n^{(k)} = 0\} = \left(\frac{P\{X=n\}}{P\{X \geq n\}}\right)^k$, $n = 0, 1, 2, \ldots$.

As in Theorem 19.1, let us once again consider two sequences of independent random variables, $X_1, X_2, \ldots$ with distribution function F, and $Y_1, Y_2, \ldots$ such that $Y_1 = \min\{X_1, \ldots, X_k\}$, $Y_2 = \min\{X_{k+1}, \ldots, X_{2k}\}$, and so on. Let $X(n, k)$ be the kth record values based on $X_1, X_2, \ldots$ and $Y(n)$ be the classical record values constructed using the sequence $Y_1, Y_2, \ldots$. Then, comparing (20.9) with the analogous result for record values $Y(n)$, we observe that the result in Theorem 19.1 is also valid in the discrete case. Moreover, by combining the statements of Theorem 19.1 for continuous and discrete distributions and the fact that the number of records taking values in any interval $(a, b]$ depends only on values $F(x)$, $a \leq x \leq b$, leads to the validity of Theorem 19.1 for any distribution function F. The only restriction on F is the existence of record values with probability one. Therefore, the statement of Theorem 19.1 is true if $P\{X = \beta\} = 0$, where $\beta = \sup\{x : F(x) < 1\}$.

21. Weak k^{th} record values

Stepanov (1992) introduced weak k^{th} record times $L^w(n, k)$ and weak k^{th} record values $X^w(n, k)$ in the following manner:

$$L^w(1,k) = k, \quad L^w(n+1,k) = \min\{j > L^w(n,k) : X_j \geq X_{j-k,j-1}\},$$
$$X^w(n,k) = X_{L^w(n,k)-k+1:L^w(n,k)}, \quad n \geq 1.$$

For continuous distributions, the weak k^{th} records coincide with the strong k^{th} records. Therefore, they need to be considered only in the case of discrete distributions. Note that weak k^{th} records exist with probability one in all situations.

Let us consider a sequence of i.i.d. random variables $X, X_1, X_2, \ldots$, taking on values $0, 1, 2 \ldots$, and let $P\{X < n\} < 1$ for any n. Define random variables $\mu_i(k)$, $i = 0, 1, \ldots$, as follows: $\mu_i(k) = m$, $m = 0, 1, \ldots$ if in the sequence $X_1, X_2, \ldots$ we have exactly m weak k^{th} records taking a value i. Then, the following results due to Stepanov (1992), are generalizations of Theorem 12.2 and Representation 12.1.

THEOREM 21.1. For any $k = 1, 2, \ldots$, the random variables $\mu_0(k), \mu_1(k), \ldots$ are independent and have negative binomial distributions with probabilities

$$P\{\mu_i(k) = m\} = \binom{k-1}{k+m-1}(1-q_i)^k q_i^m, \quad i = 0, 1, \ldots, \; m = 0, 1, \ldots,$$

where $q_i = P\{X = i\}/P\{X \geq i\}$.

REPRESENTATION 21.1. For any $k = 1, 2, \ldots$, $m = 0, 1, \ldots$, and $n = 1, 2, \ldots$,

$$P\{X^w(n,k) > m\} = P\{\mu_0(k) + \mu_1(k) + \cdots + \mu_m(k) < n\}.$$

Comparing Representations 21.1 and 12.1, one can construct examples which will reveal that Theorem 19.1 is not true for weak records.

Stepanov (1992) also applied Representation 21.1 to establish some limit theorems for weak k^{th} record values.

22. k_n-records

The definition of kth record times given in (17.7) allows us to introduce further generalizations of records. Nevzorov (1986a) considered the so-called k_n-record times and k_n-record values. Sometimes, they are called **K**-record times and **K**-record values. This section is based on results of Nevzorov (1986a) and Deheuvels and Nevzorov (1994a).

Let $X_1, X_2, \ldots$ be i.i.d. random variables with a continuous distribution function F. As in Lemma 4.1, we consider sequential ranks $R_1, R_2, \ldots$ of $X_1, X_2, \ldots$. Let $\mathbf{K} = \{k_n, n \geq 1\}$ denote a sequence of integers such that $0 \leq k_n \leq n$. Then k_n-record times $L_k(n)$ and k_n-record values $X_k(n)$ are defined in the following manner:

$$L_k(0) = 0, \quad L_k(n) = \inf\{m > L_k(n-1) : R_m \geq m - k_m + 1\}$$

and

$$X_k(n) = X_{L_k(n)-k_n+1:L_k(n)}, \quad n = 1, 2, \ldots.$$

Note that the case when $k_n = 0$ for $n = 1, \ldots, k-1$, and $k_n = k$ for $n > k-1$, corresponds to the kth records.

In this section, we define indicators $\xi(1), \xi(2), \ldots$ as follows: $\xi(n) = 1$ if $R_n \geq n - k_n + 1$, and $\xi(n) = 0$ otherwise. The following results then look rather natural when compared with Theorem 4.2 and Representation 4.2.

THEOREM 22.1. *Indicators $\xi(1), \xi(2), \ldots$ are independent and $P\{\xi(n) = 1\} = k_n/n$, $n = 1, 2, \ldots$.*

REPRESENTATION 22.1. For any $n = 1, 2, \ldots$ and $m \geq n$,

$$P\{L_k(n) > m\} = P\{\xi(1) + \xi(2) + \cdots + \xi(m) < n\},$$
$$P\{L_k(n) = m\} = P\{\xi(1) + \xi(2) + \cdots + \xi(m-1) = n-1, \xi(m) = 1\}.$$

These equalities help us to write the joint distributions of k_n-record times. Note that if $k_n = n$ or $k_n = 0$, then the corresponding indicator has a degenerate distribution. Therefore, we only consider the case when $0 < k_n < n$. Then for any $r = 1, 2, \ldots$ and $1 \leq m(1) < m(2) < \cdots < m(r)$,

$$P\{L_k(1) = m(1), L_k(2) = m(2), \ldots L_k(r) = m(r)\}$$
$$= \prod_{\ell=1}^{m(r)} \left(1 - \frac{k_\ell}{\ell}\right) \prod_{s=1}^{r} \left\{\frac{k_{m(s)}}{m(s) - k_{m(s)}}\right\}.$$

THEOREM 22.2. *The sequence $L_k(n)$, $n = 1, 2, \ldots$, forms a Markov chain and*

$$P\{L_k(n+1) = r | L_k(1), L_k(2), \ldots, L_k(n-1), L_k(n) = s\}$$
$$= \frac{k_r}{r} \prod_{j=s+1}^{r-1} \left(1 - \frac{k_j}{j}\right).$$

Let $N_k(n) = \xi(1) + \xi(2) + \cdots + \xi(n)$ be the number of records in the sequence $X_1, X_2, \ldots, X_n$. Denote $A(n) = E\{N_k(n)\} = \sum_{m=1}^{n} \frac{k_m}{m}$ and $B(n) = \text{Var}(N_k(n)) = A(n) - D(n)$, where $D(n) = \sum_{m=1}^{n} \left(\frac{k_m}{m}\right)^2$. Also, let $\mathcal{T}_k(n)$ be a σ-algebra of events generated by random variables $L_k(m)$, $m = 1, 2, \ldots, n$. It was shown in Deheuvels and Nevzorov (1994a) that the sequences $V(n) = A(L_k(n)) - n$ and $W(n) = \{A(L_k(n)) - n\}^2 + D(L_k(n)) - n$, $n = 1, 2, \ldots$, are martingales with respect to the sequence of σ-algebras $\mathcal{T}_k(n)$, $n = 1, 2, \ldots$. These results imply that for any $n = 1, 2, \ldots$,

$$E\{A(L_k(n))\} = n$$

and

$$\text{Var}(A(L_k(n))) = n - E\{D(L_k(n))\}.$$

If $D(\infty) = \sum_{m=1}^{\infty} \left(\frac{k_m}{m}\right)^2 < \infty$, then the latter equality can be rewritten as

$$\operatorname{Var}(A(L_k(n))) = n - D(\infty) + o(1) \text{ as } n \to \infty .$$

Some limit theorems were obtained for $L_k(n)$. For example, it was shown that if $\limsup(k_n/n) < 1$, then as $n \to \infty$

$$|P\{A(L_k(n)) - n < x(C(n))^{1/2}\} - \Phi(x)| = O(n^{-1/2}) ,$$

$$A(L_k(n))/n \to 1 ,$$

and

$$\limsup \pm \frac{A(L_k(n)) - n}{(2C(n) \log \log C(n))^{1/2}} = 1 ,$$

where $C(n) = B(A \leftarrow (n))$ and $A \leftarrow (n) = \inf\{m : A(m) \geq n\}$.

It needs to be mentioned here that Berred (1994a) has used some statistics based on k_n-records to estimate the shape parameter γ of the generalized extreme value distribution with cdf $G_\gamma(x) = \exp\{-(1 + \gamma x)^{1/\gamma}\}$.

23. Records in sequences of dependent random variables

As we have already seen, the classical record model requires independence of the original X's. There are only a few papers in which records in sequences of dependent random variables $X_1, X_2, \ldots$ have been investigated.

The simplest case in this direction is connected with sequences of exchangeable random variables. It is evident [see Rényi (1962), for example] that Theorem 4.1 and Lemma 4.1 are valid for exchangeable X's. Of course, Representations 4.1 and 4.2, Theorem 4.2 and Corollary 4.1 are also true if we consider sequences of exchangeable random variables $X_1, X_2, \ldots$. In this situation, all the results for record times as well as those for k^{th} and k_n-record times which are based on the property of independence of indicators of records remain unchanged if one considers exchangeable random variables instead of independent random variables. Of course, de Finetti's representation of exchangeable random variables as a mixture of independent random variables is a tool to investigate the record values in this case. An example that showed how the distributions of record values $X(n)$ can be found for some stationary Gaussian sequences $X_1, X_2, \ldots$ was given by Nevzorov (1987a). Haiman (1987a) and Haiman and Puri (1993) used another method to study records from more general stationary Gaussian sequences. They obtained the following very curious result. Let X_n, $-\infty < n < \infty$, be a stationary Gaussian sequence with zero means and covariance function $\Gamma(n) = E(X_i X_{i+n})$ and Y_n, $n = 1, 2, \ldots$ be a sequence of independent random variables each having the standard normal distribution. Further, let $L(n, k)$ and $X(n, k)$ denote the kth record values and the kth record times as defined in (3.1) and (3.2) with the following exception: Haiman and Puri defined $L(n, 1)$ as the first n, $n \geq k$, such that $X_{n-k+1:n} > \Omega$, where Ω is a fixed real number. They defined the same way the kth record times $S(n,k)$ and the kth record values $R(n,k)$ based on Ω and a sequence Y_n, $n = 1, 2, \ldots$. It appears that under some conditions, for example, if $\sum_{n=1}^{\infty} |\Gamma(n)| < 1/2$, then for some Ω, there exist almost surely an n_0 and a q such that for all $n > n_0$, $S(n,k) = L(n-q,k)$ and $R(n,k) = X(n-q,k)$. The main point of this *almost sure invariance principle* is the fact that under weak dependence conditions the distributions of some functionals in the case of dependent random variables behave almost the same way as in the case of independent random variables. Haiman (1987b, 1992) proved an analogous result (but only for $k = 1$) for m-dependent sequences of random variables. He considered a stationary m-dependent sequence of random variables X_n, $n = 1, 2, \ldots$, which satisfy the following condition:

$$\limsup_{u \to \omega} \left\{ \sup_{u < v < z < \omega} \frac{P\{X_1 > u, v < X_k < z\}}{\{-\log P(X_1 > u)\}^{2+\beta} P(v < X_1 < z)} \right\}$$
$$< \infty, \quad 2 \le k \le m, \qquad (23.10)$$

where $\omega = \sup\{x : P(X_1 < x) < 1\}$ and β is some positive constant.

THEOREM 23.1. *Let a stationary m-dependent sequence X_n, $n = 1, 2, \ldots$, satisfy condition (23.10). Then, there exists a probability space which carries, in addition to $\{X_n\}$, an i.i.d. sequence $\{Y_n, n \ge 1\}$ having the same marginal distribution and such that almost surely there exist an n_0 and a q such that for all $n > n_0$, $S(n) = L(n-q)$ and $R(n) = X(n-q)$, where record times $L(n)$, $S(n)$ and record values $X(n)$, $R(n)$ are defined respectively on the sequences $\{X_n\}$ and $\{Y_n\}$, as in (3.3) and (3.4).*

Biondini and Siddiqui (1975) [see also Pfeifer (1984b)] investigated Markov sequences $X_1, X_2, \ldots$. They proved that record values $X(1), X(2), \ldots$ inherit the Markov property of original random variables $X_1, X_2, \ldots$. Moreover, for stationary Markov sequences, they demonstrated the process of obtaining the transition density $h(x,y)$ of the sequence $X(1), X(2), \ldots$, knowing a transition density $p(x,y)$ of X's. It has been shown that

$$h(x,y) = p(x,y) + \int_{-\infty}^{x} k(x,t)p(t,y)dt \text{ if } x < y$$
$$= 0 \text{ if } x \ge y, \qquad (23.11)$$

where $k(x,y)$ can be found from the equation

$$k(x,y) = p(x,y) + \int_{-\infty}^{x} k(x,t)p(t,y)dt \text{ if } y \le x$$
$$= 0 \text{ if } y > x. \qquad (23.12)$$

Note that if $X_1, X_2, \ldots$ are independent and have a joint density function f, then $p(x,y) = f(y)$ and it follows from (23.12) that $k(x,y) = f(y)\left(1 + \int_{-\infty}^{x} k(x,t)dt\right)$ for $y \le x$. Evidently, $k(x,y) = g(x)f(y)$, where $g(x) = 1/\{1 - F(x)\}$. Now, (23.11) corresponds to the result for the classical records. Biondini and Siddiqui (1975) also considered discrete Markov sequences. Investigations of records in Markov sequences under more mild restrictions were continued by Adke (1993). It has been shown by him that $\{(X(n), L(n)), n \geq 1\}$ generated by the Markov sequence $X_1, X_2, \ldots$ is itself a bivariate Markov sequence, and that $\{X(n), n \geq 1\}$ is also a Markov sequence. As for $\{L(n), n \geq 1\}$, in general, it is not a Markov chain. Some properties of record times were obtained for the Markov sequences Y_n and Z_n, $n \geq 1$, defined as follows:

$$Y_1 = Z_1 = X_1, \quad Y_{n+1} = k \max(Y_n, X_{n+1}), \quad 0 < k < 1,$$
$$Z_{n+1} = \max(Z_n, X_{n+1}) - c, \quad c > 0, \quad n = 1, 2, \ldots.$$

Andel (1990) investigated the probabilities $P\{\xi_n = 1\}$ that a record occurs at time n for autoregressive processes with an exponentially distributed white noise. In his scheme X_1 has an exponential $E(0, a/(1-b))$ distribution, $X_{n+1} = bX_n + Y_{n+1}$, $n = 1, 2, \ldots$, and $Y_2, Y_3, \ldots$ are independent exponential $E(0, a)$ random variables, where $a > 0$ and $0 \leq b < 1$ are arbitrary constants.

24. Random record models

A large number of papers starting with Pickands (1971) dealt with the so-called [following Gaver (1976)] *random record models*. Let P be a point process (usually, a homogeneous or nonhomogeneous Poisson process, renewal process, or birth process) with arrival times $0 = \tau_0 < \tau_1 < \tau_2 < \ldots$. Further, let $X_0, X_1, X_2, \ldots$ (as a rule, they are i.i.d. random variables, but sometimes one may consider dependent or nonstationary sequences) be associated with points $\tau_0, \tau_1, \tau_2, \ldots$, respectively. Then, the record times $L(n)$ based on the sequence of X's generate a new point process $P^*(t)$ with arrival times $\tau_n^* = \tau_{L(n)}$. Now let $\Delta^*(n) = \tau_n^* - \tau_{n-1}^*$ be the inter-record times for the process $P^*(t)$. Distributions of random variables τ_n^* and $\Delta^*(n)$ as well as several characteristics of the process $P^*(t)$ and some related problems have been discussed by Pickands (1971), Gaver (1976), Westcott (1977b, 1979), Gaver and Jacobs (1978), Embrechts and Omey (1983), Yakymiv (1986), Bruss (1988), Bruss and Rogers (1991), and Bunge and Nagaraja (1991, 1992a,b).

25. Nonstationary record models

In this section, we consider sequences of independent X's, which can have different distributions. Yang (1975) was the first one to consider records in a nonstationary scheme. We present here a review of some nonstationary record models, beginning with some models which can be regarded as a certain mixture of stationary and nonstationary schemes.

25.1. Pfeifer's model

Pfeifer (1982, 1984c) considered records in a scheme of series of i.i.d. random variables $X_{n1}, X_{n2}, \ldots$, $n = 0, 1, 2, \ldots$. Let F_n be the distribution function of random variables $X_{n1}, X_{n2}, \ldots$. The sequence $\Delta(n)$, $n = 0, 1, 2, \ldots$, of inter-record times is recursively defined as follows:

$$\Delta(0) = 0 \text{ and } \Delta(n) = \min\{j : X_{nj} > X_{n-1:\Delta(n-1)}\}, \quad n = 1, 2, \ldots.$$

Then, following Pfeifer, $L(n) = 1 + \Delta(1) + \Delta(2) + \cdots + \Delta(n)$ and $X(n) = X_{u,\Delta(n)}$, $n = 0, 1, 2, \ldots$, form sequences of record times and record values. It has been shown that under some additional restrictions, $(\Delta(n), X(n))$, $n = 0, 1, \ldots$ is a Markov chain. Pfeifer has also showed that if $F_n(x) = 1 - \exp(-\lambda_n x)$, $x > 0$, where $\lambda_n > 0$, $n = 1, 2, \ldots$ then the random variables $X(n-1)$ and $\delta(n) = X(n) - X(n-1)$ are independent, and the distribution function of $\delta(n)$ coincides with F_n. Pfeifer (1984c) proved in addition some limit theorems for $\log \Delta(n)$.

The record model described above reflects situations where conditions of an experiment change after the occurrence of a new record value. For example, after the destruction of an old component, a modified one has to be used. Pfeifer's model was further discussed by Deheuvels (1984b) and Gupta (1984). Reliability properties of record values in Pfeifer's scheme have been studied by Kamps (1994). Kamps (1995, Section 1.7) has also considered k_n-records in Pfeifer's record model.

25.2. Balabekyan–Nevzorov's record model

The next scheme [see, for example, Balabekyan and Nevzorov (1986) and Rannen (1991)] to be described in this subsection combines elements of the i.i.d. model and a nonstationary scheme. It can be well illustrated by the following example. Let m athletes have in succession n starts each. Then, the distribution functions which correspond to their results $X_1, \ldots, X_m, \ldots, X_{nm}$ form a sequence $F_1, \ldots, F_{nm}$ such that $F_{rm+k} = F_k$, $1 \leq k \leq m$, $0 \leq r \leq n-1$, that is, we have m different (nonstationarity!) distribution functions $F_1, \ldots, F_m$, and this group of distribution functions is repeated (an element of stationarity!) n times. A value m can be fixed or $m = m(n)$ may be permitted to increase to a certain degree with n, like $m(n) = o((\log n)^{1/2})$. It turns out that the number of records $N(nm)$ among random variables $X_1, \ldots, X_{nm}$ has the same asymptotic distribution as in the i.i.d. scheme. It has been shown that

$$|P\{N(nm) - \log n < x(\log n)^{1/2}\} - \Phi(x)| \leq Cg(m^2/\log n),$$

where C is an absolute constant, $g(x) = x^{1/4}$ (Balabekyan and Nevzorov), and $g(x) = \max\{x^{1/2}, x^{1/2}\log(1/x)\}$ (the recent improvement due to Rannen). It follows from this estimate that the random variable $(\log L(n) - n)/n^{1/2}$ also asymptotically has a standard normal distribution. The main idea of the method is to compare the original sequence of X's and the sequence $Y_r =$

$\max\{X_{(r-1)m+1}, \ldots, X_{rm}\}$, $r = 1, 2, \ldots, n$. Evidently, $Y_1, \ldots, Y_n$ are independent random variables with the joint distribution function $G = \prod_{k=1}^{m} F_k$; consequently, we have the asymptotical normality of $N_1(n)$, where $N_1(n)$ is the number of records in the sequence $Y_1, \ldots, Y_n$. The second step involves estimating the difference $N_2(n) = N(nm) - N_1(n)$. Balabekyan and Nevzorov (1986) have proved that $0 \leq \mathrm{E}\{N_2(n)\} \leq 2(m-1)$. Rannen (1991) managed to show that for any $m = 2, 3, \ldots, n = 1, 2, \ldots$, and $k = 1, 2, \ldots$,

$$\mathrm{E}\{N_2(n)\}^k \leq (m-1)^k k^{k+1} \ .$$

This implies that the second term in the sum $N(nm) = N_1(n) + N_2(m)$ is negligible with respect to $N_1(n)$. Therefore, the asymptotic behavior of $N(nm)$ is determined by $N_1(n)$.

25.3. Records in sequences with trend

Sports competitions give a remarkable material to illustrate the theory of records and to introduce new record models. We have a lot of data covering sometimes a century or two or even more longer periods in some instances. For example, Smith (1988) discussed athletic records for mile races beginning from 1860. No doubt, the assumptions of the classical record model can not be applied to the results of different athletes over such a long period of time. Changes in competition rules, a great progress in athletic equipments, new methods of training, and some other reasons force us to find more suitable schemes. Investigation of models with different trends seems to be very natural in this context. Note that the first results for records in sequences with a trend were given by Foster and Stuart (1954) and Foster and Teichroew (1955), where such models were introduced as an alternative to a null hypothesis considering the i.i.d. case. Recent results of this kind are due to Ballerini and Resnick (1985, 1987a), Smith and Miller (1986), de Haan and Verkade (1987), and Smith (1988); see also Ballerini (1987) and Nevzorov (1987a).

Smith (1988) considered a best result Y_n in the nth year, $n = 1, 2, \ldots$, where $Y_n = X_n + c_n$ with $X_1, X_2, \ldots$ being an i.i.d. sequence, and c_n being a nonrandom trend. Note that only the latest record in each year is presented in the sequence $\{Y_n\}$. Three types of distributions of X's (normal, Gumbel and generalized extreme value distributions) as well as three types of trends (linear, quadratic and exponential) were considered by Smith. He applied the maximum likelihood method to estimate parameters of the distribution of X's and also the unknown coefficients in the trend term. Smith used this method to analyze the record results for mile (1860–1985) and marathon (1909–1985) races and to determine the corresponding predictions of the future records in these two events.

The analogous model $Y_n = X_n + c_n$, $n = 1, 2, \ldots$, but only with the linear trend $c_n = cn$, $c > 0$, was the subject of investigation of Ballerini and Resnick (1985). They gave a formula which will help one to compute explicitly or approximately the asymptotic record rate p, which is defined as $\lim \mathrm{P}\{Y_n \text{ is a record}\}$ as $n \to \infty$. For example, for the standard normal distribution and $c = 1$, $p = 0.72506\ldots$.

The strong law of large numbers and central limit theorems were proved for $N(n)$ and $L(n)$. Ballerini and Resnick (1987a) considered next the model with a linear trend after relaxing the assumption of independence of X's. They took $\{X_n\}$ to be a doubly infinite, strictly stationary sequence. They then proved that, almost surely and in L_p for $p > 0$, $N(n)/n \to p$ as $n \to \infty$. One more result of Ballerini and Resnick is the following theorem.

THEOREM 25.1. Let $\{X_n\}$ be strictly stationary, strong mixing sequence with mixing coefficients $\{\alpha_n\}$, $E\{X^n\} < \infty$ and $\sum_{n=1}^{\infty} \alpha_n < \infty$. Then, as $n \to \infty$,

$$\sup_x P\{n^{1/2}(N(n)/n - p) < x\} \to \Phi(x) .$$

The standard technique based on the equality $P\{L(n) \geq m\} = P\{N(m) \leq n\}$ will allow one to prove the corresponding limit theorems for the random variables $L(n)$ under this model.

25.4. F^α-scheme

25.4.1. Yang's model

One more nonstationary record model was initiated by Yang (1975) who reasonably supposed that the breaking of sports records (for example, records in Olympic games) was due in some degree to the increase in the population of the world. Therefore, he considered a sequence of independent random variables X_n with distribution functions $F_n = F^{m(n)}$, $n = 1, 2, \ldots$. Here, F_n corresponds to the random variable $\max\{Y_1, Y_2, \ldots, Y_{m(n)}\}$, where Y's are independent random variables with a joint continuous distribution function F. The form of the coefficients $m(1), m(2), \ldots$ is due to the assumption that the population increases geometrically and we have $m(n) = \lambda^{n-1} m$, where $m = m(1)$ is the initial population size. Yang got the exact formula for $P\{\Delta(n) > j\}$ and proved that as $n \to \infty$, $\lim P\{\Delta(n) = j\} = (\lambda - 1)/\lambda^j$, $j = 1, 2, \ldots$. These results were then applied to analyze records of nine consecutive Olympic games. Of course, it has been shown that the increasing population is not the main reason for the rapid breaking of Olympic records. Alpuim (1985) showed that $\Delta(n)$ asymptotically has the geometric distribution for more general sequences of coefficients $m(n)$ and even, under certain conditions, when these coefficients themselves are random variables. Meanwhile, Nevzorov (1981, 1984, 1985, 1986b,c) observed that indicators of records $\xi_1, \xi_2, \ldots$ stay independent in Yang's model. Moreover, these indicators are independent for any sequence of positive coefficients $m(1), m(2), \ldots$. This independence property of record indicators led to an intensive study of a new model.

25.4.2. Definition of F^α-scheme

We say that a sequence of independent random variables $X_1, X_2, \ldots$ with distribution functions $F_1, F_2, \ldots$ obeys the F^α-scheme if $F_n = F^{\alpha_n}$, $n = 1, 2, \ldots$, where F is a continuous distribution function and $\alpha_1, \alpha_2, \ldots$ are arbitrary positive numbers. Note that equal values of α's lead to the classical record model. In addition to the above mentioned papers of Nevzorov, a number of other results connected with the F^α-scheme were given by Pfeifer (1989, 1991), Deheuvels and Nevzorov (1993), and Nevzorov (1990a, 1993, 1995). Further generalizations of the F^α-scheme were considered by Ballerini and Resnick (1987b), Pfeifer (1989) (the same marginal distributions as in the F^α-scheme, but a certain dependence structure of X's), and Deheuvels and Nevzorov (1994b) (mixtures of F^α-schemes).

25.4.3. Indicators of records

Let indicators of records $\xi_1, \xi_2, \ldots$ be as defined in Theorem 1. Nevzorov (1981, 1985) proved the independence of the indicators for any sequence of positive $\alpha_1, \alpha_2, \ldots$ and showed that $p_n = P\{\xi_n = 1\} = \alpha_n/(\alpha_1 + \cdots + \alpha_n)$. It seems that this independence of indicators in a certain sense characterizes the F^α-scheme.

THEOREM 25.2. [Nevzorov (1986b, 1993).] Let $X_1, X_2, \ldots, X_n$ be independent random variables with continuous distribution functions $F_1, F_2, \ldots, F_n$ and $0 < F_i(a) < F_i(b) < 1$ ($1 \le i \le n-1$) for some a and b, $-\infty < a < b < \infty$. If the indicator ξ_n and the vector $(\xi_1, \ldots, \xi_{n-1})$ are independent for any distribution function F_n, then there exist positive constants $\alpha_2, \ldots, \alpha_{n-1}$ such that $F_i = F_1^{\alpha_i}$, $1 \le i \le n-1$, and $\xi_1, \ldots, \xi_{n-1}$ are also independent.

From a result of Ballerini and Resnick (1987b) dealing with a more general situation, it follows that the vector $(\xi_1, \ldots, \xi_n)$ and $M(n) = \max\{X_1, \ldots, X_n\}$ are independent for the F^α-scheme. This property also characterizes the F^α-scheme.

THEOREM 25.3. [Nevzorov (1990b, 1993).] Let the distribution functions of independent random variables $X_1, X_2, \ldots, X_n$ be continuous and $0 < F_i(a) < F_i(b) < 1$ ($1 \le i \le n$) for some a and b, $-\infty < a < b < \infty$. If $M(n) = \max\{X_1, \ldots, X_n\}$ and the vector $(\xi_1, \ldots, \xi_n)$ are independent, then there exist positive constants $\alpha_2, \ldots, \alpha_n$ such that $F_i = F_1^{\alpha_i}$, $2 \le i \le n$, and the random indicators $\xi_1, \ldots, \xi_n$ are also independent.

Since $N(n) = \xi_1 + \cdots + \xi_n$, we have

$$A(n) = E\{N(n)\} = \sum_{k=1}^{n} p_k = \sum_{k=1}^{n} \frac{\alpha_k}{\alpha_1 + \cdots + \alpha_k} \qquad (25.13)$$

and

$$B(n) = \text{Var}(N(n)) = A(n) - \sum_{k=1}^{n} \left(\frac{\alpha_k}{\alpha_1 + \cdots + \alpha_k}\right)^2. \qquad (25.14)$$

In the following, we take $S(n) = \alpha_1 + \cdots + \alpha_n \to \infty$ and $n \to \infty$. It follows from Dini's test that only when $A(n) \to \infty$, $P\{N(n) < \infty\} = 1$. It means that in this case any record time $L(n)$ exists with probability one. Without loss of generality, we can take $\alpha_1 = 1$.

25.4.4. Distributions of records times

Note that distributions of $N(n)$ and $L(n)$ in the F^α-scheme as well as in the classical model do not depend on the continuous distribution function F. The independence of indicators $\xi_1, \xi_2, \ldots$ leads us to the following equalities [Nevzorov (1985)]:

$$P\{L(1) = 1, L(2) = m_2, \ldots, L(n) = m_n\}$$
$$= \left(\prod_{r=2}^{n} \frac{\alpha_{m_r}}{S(m_r - 1)}\right) \Big/ S(m_n) \quad \text{for } 1 < m_2 < \cdots < m_n ;$$

and

$$P\{L(n) = j | L(1), L(2), \ldots, L(n-2), L(n-1) = i\}$$
$$= S(i)\left(\frac{1}{S(j-1)} - \frac{1}{S(j)}\right) \quad \text{and } j > i \geq n - 1 .$$

One can now see that the sequence $L(n), n = 1, 2, \ldots$, forms a Markov chain, and

$$P\{L(n) = j | L(n-1) = i\} = S(i)\left(\frac{1}{S(j-1)} - \frac{1}{S(j)}\right)$$

and

$$P\{L(n) > j | L(n-1) = i\} = S(i)/S(j) \quad \text{for } j > i .$$

25.4.5. Relationship between F^α- and other record schemes

Shorrock (1972a) showed that distributions of record times $\{L(n+1), n \geq 1\}$ in the i.i.d. model coincide with distributions of record values $\{Y(n), n \geq 1\}$ in a sequence of independent integer-valued random variables $Y_1, Y_2, \ldots$ such that $P\{Y_k > n\} = 1/n$ for $k = 1, 2, \ldots$ and $n = 1, 2, \ldots$. Nevzorov (1985) proved that the distributions of $Y(n)$ for all sequences of i.i.d. random variables $Y_1, Y_2, \ldots$ taking values $2, 3, \ldots$ with positive probabilities can be embedded in the F^α-scheme with suitably chosen coefficients $\alpha_1, \alpha_2, \ldots$ As a matter of fact, if $\alpha_1 = 1$, $\alpha_r = (1/P\{Y_1 > r\}) - (1/P\{Y_1 > r - 1\})$, $r = 2, 3, \ldots$, and $1 = L(1) < L(2) < \cdots$ are the corresponding record times for this F^α-scheme, then for any $n = 1, 2, \ldots$,

$$\{L(1), L(2), \ldots, L(n)\} \stackrel{d}{=} \{Y(0), Y(1), \ldots, Y(n-1)\} ,$$

where $1 = Y(0) < Y(1) < \cdots$ are the record values based on $Y_1, Y_2, \ldots$. Deheuvels and Nevzorov (1993) gave an analogous result for the kth record values from discrete sequences. In this case, one has to take $\alpha_1 = 1$ and $\alpha_r = (P\{Y_1 > r\})^{-k} - (P\{Y_1 > r - 1\})^{-k}$, $r = 2, 3, \ldots$, to obtain the result

$$\{L(1), L(2), \ldots, L(n)\} \stackrel{d}{=} \{Y(0, k), Y(1, k), \ldots Y(n-1, k)\} ,$$

where $1 = Y(0, k) < Y(1, k) < \cdots$ are the kth record values based on the sequence $Y_1, Y_2, \ldots$.

Of course, the classical record times corresponds to the case of F^α-scheme with equal coefficients. As for the kth record times $L(n,k)$, Deheuvels and Nevzorov proved that

$$\{L(1,k),\ldots,L(n,k)\} \stackrel{d}{=} \{L(1)+k-1,\ldots,L(n)+k-1\}, \quad n=1,2,\ldots ,$$

where $1 = L(1) < L(2) < \cdots$ are the record times for the F^α-scheme with the coefficients $\alpha_n = \binom{k-1}{k+n-2}$, $n = 1, 2, \ldots$. They also showed that the k_n-record times $L_k(n)$, under the condition $k_1 = 1$, $0 < k_n < n$, $n = 2, 3, \ldots$, have the same distribution as the record times $L(n)$ in the F^α-scheme with the coefficients $\alpha_1 = 1$ and $\alpha_n = k_n(n-1)!/\prod_{i=2}^{n}(i-k_i)$.

25.4.6. Martingale properties of record times

There are three known classical sequences of martingale sequences generated by sums $S_n = X_1 + \cdots + X_n$, $n = 0, 1, 2, \ldots$ of independent random variables. They are: $T_1(n) = S_n - \mathrm{E}\{S_n\}$, $T_2(n) = (S_n - \mathrm{E}\{S_n\})^2 - \mathrm{Var}(S_n)$ and $T_3(n) = \lambda^{S_n}/\mathrm{E}\{\lambda^{S_n}\}$, $n = 1, 2, \ldots$. Substitution of sequences of Markov times $\tau_1 < \tau_2 < \cdots$ usually leads to new martingale sequences $V_i(n) = T_i(\tau_n)$, $n = 1, 2, \ldots$, $i = 1, 2, 3$. One may take $N(n) = \xi_1 + \cdots + \xi_n$ and $L(n)$ as S_n and τ_n, respectively, guess the form of new martingale sequence, and then prove the martingale property of the obtained sequences. To simplify the form of these sequences, we can use the identity $N(L(n)) = n$. This, in fact, was utilized by Deheuvels and Nevzorov (1993) to obtain the following results. Let $A(n)$, $B(n)$ and p_n be defined as in (25.13) and (25.14), and $\mathscr{F}_n$ be a σ-algebra of events generated by random variables $L(1), L(2), \ldots, L(n)$, $n \geq 1$.

THEOREM 25.4. Let $S(n) = \alpha_1 + \cdots + \alpha_n \to \infty$ as $n \to \infty$. Then, the sequences of random variables

$$V_1(n) = A(L(n)) - n ,$$
$$V_2(n) = \{A(L(n)) - n\}^2 - B(L(n)) ,$$
$$V_3(n) = s^n / \prod_{j=1}^{L(n)}(1 - p_j + p_j s), \quad n = 1, 2, \ldots ,$$

form martingales with respect to σ-algebras $\mathscr{F}_n$, $n = 1, 2, \ldots$, and $\mathrm{E}\{V_k(n)\} = 0$ for any $k = 1, 2, 3$, and $n \geq 1$.

Note that, in the above definition of $V_3(n)$, the parameter s can not be equal to $1 - 1/p_j$. For example, we can suppose that $s > 0$.

25.4.7. Moments of record times

A number of moment properties of record times $L(n)$ can be obtained as corollaries of Theorem 25.4.

COROLLARY 25.1. For any $n = 1, 2, \ldots,$

(i) $E\{A(L(n))\} = n$;

(ii) $\mathrm{Var}(A(L(n))) = n - E\{D(L(n))\}$, where $D(n) = A(n) - B(n) = \sum_{k=1}^{n} \left(\frac{\alpha_k}{\alpha_1 + \cdots + \alpha_k}\right)^2$, and if $D = D(\infty) < \infty$, then $D(A(L(n))) = n - D + o(1)$ as $n \to \infty$,

(iii) $\exp(-cn) \leq E\{\exp(-cA(L(n)))\} \leq \exp(-n \log(1 + c))$ if $c > 0$;

(iv) $E\{(S(L(n)))^{-c}\} \leq \alpha_1^{-c}(1 + c)^{-n}$, $c > 0$.

Nevzorov (1985) obtained some other inequalities for record times $L(n)$ and inter-record times $\Delta(n) = L(n) - L(n-1)$. Let us now choose any continuous strictly increasing function $S(x)$, $x \geq 1$, and introduce coefficients $\alpha_1, \alpha_2, \ldots$ as $\alpha_n = S(n) - S(n-1)$ for $n = 1, 2, \ldots$. Without loss of generality, we suppose that $S(1) = \alpha_1 = 1$ and $S(x) \to \infty$ as $x \to \infty$. Then, the following inequalities are valid:

$$\int_1^\infty \frac{1}{S(x)} dx \leq E\{\Delta(2)\} = E\{L(2)\} - 1 = \sum_{k=2}^\infty \frac{1}{S(k)} \leq \int_1^\infty \frac{1}{S(x)} dx + 1,$$

$$c_1(c_1 + 1)^{n-2} \leq E\{\Delta(n)\} \leq (c_2 + 1)^{n-1}$$

and

$$(c_1 + 1)^{n-1} \leq E\{L(n)\} \leq \{(c_2 + 1)^n - 1\}/c_2, \quad n = 3, 4, \ldots,$$

where

$$c_1 = \inf_{x \geq 1} \frac{S(x)}{x} \int_x^\infty \frac{1}{S(t)} dt \quad \text{and} \quad c_2 = \sup_{x \geq 1} \frac{S(x)}{x} \int_x^\infty \frac{1}{S(t)} dt.$$

In particular, $c_1 = c_2 = 1/(\gamma - 1)$ if $S(x) = x^\gamma$, $\gamma > 1$. It is easy to see that $E\{L(n)\} < \infty$ and $E\{\Delta(n)\} < \infty$ for any $n \geq 2$ if, for example, $S(x)$ is a regularly varying function with index $\gamma > 1$.

25.4.8. Limit theorems for record times

Some asymptotic results for record times were given by Nevzorov (1985, 1986c, 1995).

THEOREM 25.5. Let $S(x) \to \infty$ and $\frac{S(x)}{S(x+1)} \to 1$ as $x \to \infty$. Then, for any $k > 1$, as $n \to \infty$, the random variables

$$R_1(n) = \frac{S(L(n))}{S(L(n+1))}, \ldots, R_k(n) = \frac{S(L(n+k-1))}{S(L(n+k))}$$

are asymptotically independent and

$$\lim P\left\{\frac{S(L(n))}{S(L(n+1))} < x\right\} = x \quad \text{for} \quad 0 < x < 1.$$

THEOREM 25.6. Let $S(x)$ be a regularly varying function with positive index γ. Then, for any fixed $k > 1$, the random variables $\left(\frac{L(n+m-1)}{L(n+m)}\right)^{\gamma}$, $m = 1, 2, \ldots, k$, are asymptotically independent and $\lim P\left\{\frac{L(n)}{L(n+1)} < x^{1/\gamma}\right\} = x$ for $0 < x < 1$, as $n \to \infty$.

COROLLARY 25.2. Under the assumptions of Theorem 25.6, as $n \to \infty$,

$$\lim P\left\{\frac{\Delta(n)}{L(n+1)} > x\right\} = (1+x)^{-\gamma}, \quad x > 0,$$

$$\lim P\left\{\frac{\Delta(n)}{L(n)} > x\right\} = (1-x)^{-\gamma}, \quad 0 < x < 1,$$

$$\lim P\left\{\frac{\Delta(n+1)}{\Delta(n)} > x\right\} = H_{\gamma}(x) = \gamma x^{-\gamma} \int_0^x \frac{y^{\gamma-1}}{1+y} dy, \quad x > 0,$$

where, in particular, $H_1(x) = \frac{\log(1+x)}{x}$ and $H_2(x) = \frac{2}{x} - \frac{2\log(1+x)}{x^2}$.

Central limit theorems for random variables $N(n), A(L(n))$ and $\log S(L(n))$ have been obtained by Nevzorov (1986c, 1995).

25.4.9. Limit theorems for record values

The independence of $M(n)$ and the indicators $\xi_1, \ldots, \xi_n$ implies that $P\{X(n) < x\} = E\{(F(x))^{L(n)}\}$. Since we know the asymptotic properties of $L(n)$, this result can be used to obtain limit theorems for record values [see Nevzorov (1995)]. The main result obtained there is that the set of all possible asymptotic distributions of suitably normalized record values $X(n)$, under some conditions on sequences of the coefficients $\alpha_1, \alpha_2, \ldots$, in the F^{α}-scheme coincides with the corresponding asymptotic distributions in the stationary scheme.

25.4.10. Applications of the F^{α}-scheme

We mentioned earlier that the record models with a trend were applied for forecasting records in sports events. Quite often, in this situation one takes an extreme value distribution as the distribution of initial X's. Note that if $X_1, X_2, \ldots$ have a common distribution function $F(x) = \exp\{-\exp\{-(x-a)/b\}\}$, then the distribution functions $F_j(x)$ of random variables $Y_j = X_j + cj$, $c > 0$, $j = 1, 2, \ldots$, have the form $F_j(x) = (F(x))^{\alpha_j}$, where $\alpha_j = \exp\{cj/b\}$. In this case, the results for the F^{α}-scheme can be used to simplify arguments [see also Nevzorov (1987a), Examples 9.2 and 9.3]. If we have the i.i.d. scheme, all $n!$ permutations of ranks $(1, 2, \ldots, n)$ of random variables $X_1, X_2, \ldots, X_n$ as well as all orderings of these random variables have equal probabilities. It means that $\max\{X_1, X_2, \ldots, X_n\}$ has equal probabilities $1/n$ to coincide with the first sample element, with the second element,..., and with the last of X's. Note also that this maximum has to be the last record value in a sequence $X_1, X_2, \ldots, X_n$. This is the reason why we can apply the results on the classical record model to solve the so-called *secretary problem*. Pfeifer (1989, 1991) considered some situations with non-equiprobable orderings or ranks. In his case, $P\{\max(X_1, X_2, \ldots, X_n) = X_j\} = \alpha_j/(\alpha_j + \cdots + \alpha_n)$,

$j = 1, 2, \ldots, n$. Pfeifer showed that one can get this case with the help of the F^α-scheme. He used the independence of record indicators and some other results for the F^α-scheme to solve non-homogeneous "secretary problem" and to analyze the probabilistic behavior of some searching algorithms (the linear search for the maximum of n elements by comparisons) in models with non-equiprobable orderings of elements.

25.4.11. Mixture of F^α-schemes

We mentioned earlier that the independence of record indicators characterizes in some sense F^α-schemes and one can not expect this property to be true for sequences $X_1, X_2, \ldots$ with arbitrary distribution functions $F_1, F_2, \ldots$. It seems that we can, however, combine sequences $F_1, F_2, \ldots$. The first results in this direction were obtained by Nevzorov (1987a) and Deheuvels and Nevzorov (1994b).

For any continuous distribution function F, we consider a set $\mathscr{T}(F)$ of distribution functions defined as follows: $G \in \mathscr{T}(F)$ if there exists a distribution function H ($H(0) = 0$) such that

$$G(x) = \int_0^\infty (F(x))^\alpha H\{d\alpha\}, \quad -\infty < x < \infty.$$

One can present $G(x)$ as $\mathrm{E}\{(F(x))^v\}$, where a random variable, v, has the distribution function H. In the following, we will suppose that $\mathrm{P}\{v > 0\} = 1$. Then, we say that a sequence of independent random variables $X_1, X_2, \ldots$ with distribution functions $F_1, F_2, \ldots$ belongs to a set $\mathscr{T}(F, v_1, v_2, \ldots)$, where positive random variables $v_1, v_2, \ldots$ have distribution functions $H_1, H_2, \ldots$, respectively, if

$$F_k(x) = \int_0^\infty (F(x))^\alpha H_k\{d\alpha\}, \quad k = 1, 2, \ldots. \qquad (25.15)$$

We will now discuss records from sequences of random variables $X_1, X_2, \ldots$ having distribution functions of the form (25.15). Smirnov's transformation implies that the distributions of record times are preserved if we consider a set $\mathscr{T}(F_0, v_1, v_2, \ldots)$ instead of any set $\mathscr{T}(F, v_1, v_2, \ldots)$, where $F_0(x) = x$, $0 < x < 1$.

Let us consider a sequence of X's with distribution functions $F_k(x)$, $k = 1, 2, \ldots$, as defined in (25.15). Conditionally, if $v_1 = \alpha_1, v_2 = \alpha_2, \ldots$, we have the F^α-scheme with coefficients $\alpha_1, \alpha_2, \ldots$ and $N(n)$ can be represented as a sum of independent indicators $\xi_1(\alpha_1) + \cdots + \xi_n(\alpha_n)$. Therefore,

$$\mathrm{P}\{N(n) < x\} = \int_0^\infty \cdots \int_0^\infty \mathrm{P}\{\xi_1(\alpha_1) + \cdots + \xi_n(\alpha_n) < x\}$$
$$H_1\{d\alpha_1\} \cdots H_n\{d\alpha_n\}.$$

The normal approximation of $\mathrm{P}\{\xi_1(\alpha_1) + \cdots + \xi_n(\alpha_n) < x\}$ leads to the fact that asymptotic distributions of $N(n)$ in this model are expressible as mixtures of normal distributions. Deheuvels and Nevzorov then obtained a number of estimates of the difference

$$\delta_n = \left| P\{N(n) < x\} - \mathrm{E}\Phi\left(\frac{x - \log((v_1 + \cdots + v_n)/v_1)}{(\log((v_1 + \cdots + v_n)/v_1))^{1/2}}\right) \right|.$$

For example, they showed that if the sequence $X_1, X_2, \ldots$ belongs to a set $\mathscr{T}(F, v_1, v_2, \ldots)$, then

$$\delta_n \leq c\mathrm{E}\left\{\left(1 + \sum_{k=2}^{n} \frac{v_k^2}{(v_1 + \cdots + v_{k-1})^2}\right) \Big/ \left(\log\frac{v_1 + \cdots + v_n}{v_1}\right)^{1/2}\right\},$$

where c is an absolute constant. Denote now $e_n = \mathrm{E}\{v_n\}$ and $d_n = \mathrm{Var}(v_n)$, $n = 1, 2, \ldots$. As $n \to \infty$, let $e_1 + e_2 + \cdots + e_n \to \infty$ and $(d_1 + \cdots + d_n)/(e_1 + \cdots + e_n)^2 \to 0$. Also let $0 < v_k \leq A < \infty$ for $k = 1, 2, \ldots$. Then,

$$\Delta_n = \left| P\{N(n) < x\} - \Phi\left(\frac{x - \log(e_1 + \cdots + e_n)}{(\log(e_1 + \cdots + e_n))^{1/2}}\right) \right| \to 0$$

as $n \to \infty$, and, moreover, there exists a constant $c(A)$, depending on n only, such that

$$\Delta_n \leq c(A)\left\{(\log(e_1 + \cdots + e_n))^{-1/2} + (d_1 + \cdots + d_n)/(e_1 + \cdots + e_n)^2\right\}.$$

The latter result implies the asymptotic normality of the random variables $\{\log e(L(n)) - n\}/n^{1/2}$, where $e(n) = e_1 + e_2 + \cdots + e_n$. It needs to be mentioned that this method also works in the case of dependent X's.

25.5. Records in Archimedean copula processes

Recently, Ballerini (1994), Nevzorova, Nevzorov and Balakrishnan (1997), and Bagdonavicius, Malov and Nikulin (1997) all discussed some properties of records from *Archimedean copula processes* (AC processes). In this model, initial random variables $X_1, X_2, \ldots$ are dependent in general and can have different distributions. A sequence $X_1, X_2, \ldots$ with marginal distribution functions $F_1, F_2, \ldots$ is said to be an *AC process* if for any $n = 1, 2, \ldots$, one has

$$P\{X_1 < t_1, \ldots, X_n < t_n\} = B\left(\sum_{i=1}^{n} A(F_i(t_i))\right), \tag{25.17}$$

where B is a dependence function such that $B(0) = 1$ and B is completely monotone, and A is the inverse of the function B. Ballerini (1994) studied in detail the special case of AC process with $B(s) = \exp\{-s^{1/\gamma}\}$, $\gamma \geq 1$, and $F_i(x) = (F(x))^{\alpha_i}$, $i = 1, 2, \ldots$, where $F(x)$ is a continuous distribution function and $\alpha_1, \alpha_2, \ldots$ are positive constants. In this case,

$$P\{X_1 < t_1,\ldots, X_n < t_n\} = \exp\left\{-\left(\sum_{i=1}^{n}\{-\log(F(t_i))^{\alpha_i}\}^{\gamma}\right)^{1/\gamma}\right\}, \quad (25.18)$$

and if $\gamma = 1$ the distributions in (25.18) coincide with the ones in the F^{α}-scheme. For $\gamma > 1$, the marginal distributions of $X_1, X_2, \ldots$ are the same as in the F^{α}-scheme while these random variables are dependent. Ballerini incidentally called his model as a *dependent F^{α}-scheme*. Under this model, he proved that record indicators $\xi_1, \xi_2, \ldots, \xi_n$ and maximal value $M(n) = \max\{X_1, X_2, \ldots, X_n\}$ are independent for any $n = 1, 2, \ldots$.

Nevzorova, Nevzorov and Balakrishnan (1997) considered more general AC processes. In their construction, $B(s)$ is any complete monotone function such that $B(0) = 1$. It means that $B(s)$ coincides with a Laplace transform of some proper distribution. Marginal distribution functions $F_1, F_2, \ldots$ are taken to be

$$F_n(x) = B(c_n A(F(x))), \quad n = 1, 2, \ldots, \quad (25.19)$$

where $F(x)$ is any continuous distribution function and $c_1, c_2, \ldots$ are positive constants. Then

$$P\{X_1 < t_1, \ldots, X_n < t_n\} = H(t_1, \ldots, t_n) = B\left(\sum_{i=1}^{n} A(F_i(t_i))\right)$$

$$= B\left(\sum_{i=1}^{n} c_i R(t_i)\right) \quad (25.20)$$

where $R(t) = A(F(t))$.

It is easy to see that this construction includes the case considered by Ballerini ($B(s) = \exp\{-s^{1/\gamma}\}$, $\gamma \geq 1$) and gives some new families of marginal distributions as well. For example, if we take $B(s) = \frac{1}{1+s}$ which is the Laplace transform of the standard exponential distribution, then

$$F_n(x) = \frac{F(x)}{c_n + (1-c_n)F(x)}, \quad n = 1, 2, \ldots. \quad (25.21)$$

THEOREM 25.7. Let $X_1, X_2, \ldots$ be an AC process with joint distributions $H(t_1, \ldots, t_n)$ as given in (25.20). Then for any $n = 1, 2, \ldots$, the record indicators $\xi_1, \xi_2, \ldots, \xi_n$ and the maxima $M(n)$ are independent.

Under some natural restrictions on the marginal distribution functions $F_1, F_2, \ldots$ (as in Theorem 27.2), Nevzorova, Nevzorov and Balakrishnan (1997) proved that if

$$H(t_1, \ldots, t_n) = B\left(\sum_{k=1}^{n} A(F_k(t_k))\right), \quad n = 1, 2, \ldots,$$

and random variables $\xi_1, \xi_2, \ldots, \xi_n$ and $M(n)$ are independent for any $n = 1, 2, \ldots$, then the multivariate distributions $H(t_1, \ldots, t_n)$ are of the form

$$H(t_1, \ldots, t_n) = B\left(\sum_{k=1}^{n} c_k R(t_k)\right) . \tag{25.22}$$

Thus, this independence property provides a characterization for the class of dependent distributions in (25.22).

25.6. BR-scheme

We discuss here a model which was suggested by Ballerini and Resnick (1987b). This BR (Ballerini-Resnick)-scheme is a generalization of the F^α-scheme for sequences of dependent random variables $X_1, X_2, \ldots$.

Let $Y(t)$, $0 < t < \infty$, be an extremal-F process with finite dimensional distribution functions

$$F_{t_1, \ldots, t_n}(x_1, \ldots, x_n) = F^{t_1}(\min(x_1, \ldots, x_n)) F^{(t_2 - t_1)}(\min(x_2, \ldots, x_n))$$
$$\cdots F^{(t_n - t_{n-1})}(x_n), 0 < t_1 < t_2 < \cdots < t_n .$$

Consider a sequence of random variables $X_1, X_2, \ldots$ for which the sequence of successive maxima $M(n) = \max\{X_1, \ldots, X_n\}$, $n = 1, 2, \ldots$, can be embedded in some extremal-F process $Y(t)$, where F is continuous. It means that there exists an increasing sequence of numbers $0 = a_0 < a_1 < \cdots$ (we will suppose that $a_n \to \infty$ as $n \to \infty$) such that $\{M(n)\}_{n=1}^{\infty} \stackrel{d}{=} \{Y(a_n)\}_{n=1}^{\infty}$. Note that if $X_1, X_2, \ldots$ are independent, then their distribution functions $F_1, F_2, \ldots$ have the form $F_n = F^{\alpha_n}$, $n = 1, 2, \ldots$, where $\alpha_n = a_n - a_{n-1}$. It means that the successive maxima $M(n)$, $n = 1, 2, \ldots$, for initial X's has the same distributions as the corresponding maxima for the F^α-scheme with coefficients $\alpha_n = a_n - a_{n-1}, n \geq 1$. Therefore, the distributions of record times and record values (which undoubtedly are determined by sequences of maxima) for both sequences coincide and so one can have for the BR-scheme all the results which are proved for the F^α-scheme. Ballerini and Resnick showed that for any $n = 2, 3, \ldots$, the record indicators $\xi_1, \xi_2, \ldots, \xi_n$ and $M(n)$ are independent in the BR-scheme. They also proved some traditional limit theorems for random variables $N(n)$ and $L(n)$. It has also been shown that if the maxima $M(n)$ suitably normalized have a limit law, then the vectors $\{M(n), N(n), L(n)\}$ have a joint limit law. The following natural result was given for the record values: as $n \to \infty$, if $P\{M(n) - b(n) < xa(n)\} \to G(x)$ where G is one of the three extreme value distributions, then $P\{X(n) - b(L(n)) < xa(L(n))\} \to G(x)$. It should be mentioned that limit theorems for record values $X(n)$ under nonrandom normalization obtained by Nevzorov (1995) as well as other results for the records in the F^α-scheme are valid for the BR-scheme.

26. Multivariate records

The definition of records is clearly closely connected with the ordering of random variables. Therefore, it will be natural that multidimensional generalizations of records require the existence of some order in the corresponding set. Goldie and Resnick (1989, 1994) [see also Kinoshita and Resnick (1989)] introduced multivariate records in a partially ordered set. They obtained some results for general partial orders, but the most interesting theorems are formulated for independent identically distributed R^2-valued random vectors $\mathbf{X}_n = (X_n^{(1)}, X_n^{(2)})$, $n = 1, 2, \ldots$, with a common joint distribution function F. Several definitions of records in R^2 have been suggested. For example, $\mathbf{X}_n$ is a record if simultaneously $X_n^{(1)} > \max(X_1^{(1)}, \ldots, X_{n-1}^{(1)})$ and $X_n^{(2)} > \max(X_1^{(2)}, \ldots, X_{n-1}^{(2)})$, or $\mathbf{X}_n$ is a record if it falls outside the convex hull of $\mathbf{X}_1, \ldots, \mathbf{X}_{n-1}$. Let us consider the first of these two definitions of records. If vectors $\mathbf{X}_n$ have independent coordinates and marginal distributions of components $X^{(1)}$ and $X^{(2)}$ are continuous, then $\mathrm{P}\{\mathbf{X}_n \text{ is a record}\} = 1/n^2$ and the total number N of records in a sequence $\mathbf{X}_1, \mathbf{X}_2, \ldots$ is almost surely finite. The following result is a rather curious one.

THEOREM 26.1. *Let F be continuous and in the domain of attraction of the bivariate extreme-value distribution G. Then $\mathrm{P}(N < \infty) = 1$ or $\mathrm{P}(N = \infty) = 1$ according as G is or is not a product measure.*

It follows from this theorem that $\mathrm{P}(N < \infty) = 1$ for the bivariate normal distribution with correlation $\rho < 1$. An analogous property for the number of records $N(A)$ in some rectangle A is governed by the hazard measure H defined as follows: $H(\mathrm{d}x) = \mathrm{P}(\mathbf{X}_1 \in \mathrm{d}x)\mathrm{P}(\{\mathbf{X}_1 < x\}^c)$. It turns out that $\mathrm{P}(N(A) < \infty) = 1$ or $\mathrm{P}(N(A) = \infty) = 1$ according as $H(A) < \infty$ or $H(A) = \infty$.

Since it is typical that the total number of records in a sequence $\mathbf{X}_1, \mathbf{X}_2, \ldots$ is finite, Goldie and Resnick (1994) suggested studying the behavior of the records in a fixed rectangle A conditional on the fact there exists a large number of records $N(A)$ in the rectangle. They have proved some limit theorems connected with this situation.

27. Relations between records and other probabilistic and statistical problems

We have already mentioned that records are used for tests of some statistical hypotheses (the hypothesis of absence of a trend, constancy of a variance against natural alternatives, and the hypothesis of randomness against normal regression) – see, for example, Foster and Stuart (1954), Foster and Teichroew (1955), Stuart (1956, 1957), and Barton and Mallows (1961). Estimation of some parameters of distributions using records was the subject of discussion by Samaniego and Whitaker (1986, 1988), Ahsanullah (1989, 1990a), Berred (1991, 1992, 1994a,b), Balakrishnan and Chan (1993, 1998), Balakrishnan, Ahsanullah and Chan (1995), Chan (1998), and Sultan and Balakrishnan (1997a,b,c). There are a lot of papers dealing with the estimation of some characteristics of distributions or parameters of record models based on the observed values of existing records, and also with the prediction of future records. Nonparametric inference has also been discussed in the context of record values by Samaniego and Whitaker (1988) and Gulati and Padgett (1992, 1994a,b,c,d). Several record models for analyzing sports records and their prediction have been suggested by Yang (1975), Ahsanullah (1980, 1992), Dunsmore (1983), Nagaraja (1984), Tryfos and Blackmore (1985), Ballerini and Resnick (1985, 1987a,b), Smith and Miller (1986), Smith (1988), Basak and Bagchi (1990), and Sibuya and Nishimura (1997).

Teugels (1984) showed some applications of record statistics in insurance mathematics. Pfeifer (1985, 1991) used the theory of records to investigate several properties of searching algorithms. The so-called *secretary problem* in connection with records has been considered by Bruss (1988) and Pfeifer (1989). The relation between inter-record times and sequences of cycles lengths for the symmetric group of permutations of a set $\{1, 2, \ldots, n\}$ has been discussed by De Laurentis and Pittel (1985) and Goldie (1989). Devroye (1988) applied the theory of records to study random trees. There exists a curious connection between lower record values and the speeds of groups of vehicles that are formed in a long single-lane traffic; see, for example, Haghighi-Taleb and Wright (1973) and Shorrock (1973).

28. Nonclassical characterizations based on records

Characterizations of exponential and geometric distributions by properties of record values were very popular in the seventies and eighties. A number of references on this topic can be found in Nevzorov (1987a), Nagaraja (1988a), and Rao and Shanbhag (1994). Some characterization theorems have been obtained recently by Lin and Huang (1987), Nagaraja (1988b), Nagaraja, Sen and Srivastava (1989), Stepanov (1989), Too and Lin (1989), Ahsanullah (1990b, 1991), Witte (1988, 1990), Ahsanullah and Kirmani (1991), Nevzorov (1992), and Huang and Li (1993). It has been shown that there is a close connection between characterizations associated with record values $X(n)$ and those related to order statistics $X_{m:n}$. As we know, for any $n > 1$ and $m > 1$,

$$P\{X(n) > y | X(n-1) = x\} = P\{X_{m:m} > y \mid X_{m-1:m} = x\} \text{ a.s. }.$$

This equality accounts for the similarity between several characterization theorems for record values and for order statistics. There are some parallel characterizations of the uniform distribution (based on certain properties of order statistics) and of the exponential distribution (based on the analogous properties of record values). For example, Nagaraja (1988b) proved that the underlying distribution is uniform iff $E(X_{m-1:m}|X_{m:n})$ is linear for some $m \leq n$, and it is exponential iff $E(X(n-1)|X(n))$ is linear. Then, Szekely and Mori (1985) showed that for any $1 \leq i \leq j \leq n$, the correlation coefficient $\rho(X_{i:n}, X_{j:n}) \leq \{i(n+1-j)/j(n+1-i)\}^{1/2}$ where the equality is attained only for the uniform distribution. Nevzorov (1992) established that $\rho(X(m), X(n)) \leq (m/n)^{1/2}$, $m < n$, and its maximal value is attained for the exponential distribution (see also

Nevzorov (1997)). Huang and Su (1994) investigated this duality by comparing the sample processes and nonhomogeneous Poisson processes which are processes with the order statistics property, and generalized several existing characterizations. The characterization theorem for order statistics and for records do not always coincide, however. For example, it is evident that the conditional distributions of X_k, given $X_{m:n}$ are the same for any $1 \leq k \leq n$. This symmetry fails for record values and a new (as compared to order statistics) type of characterization is suggested by Nagaraja and Nevzorov (1997). They have solved the problem of finding the underlying distribution function F_0 such that $E(X_1|X(2) = x) = E(X_2|X(2) = x)$ for all $x \geq x_0 \geq -\infty$.

Theorems 25.2 and 25.3 presented above can also be regarded as nonclassical characterizations of certain sequences of random variables. The analogous results for discrete distribution were proved by Nevzorov and Rannen (1992) wherein a certain sequence of distributions were characterized by the property of independence of record indicators. Characterizations of distributions in several nonclassical record models were also given by Pfeifer (1982) and Nevzorov (1986b).

29. Processes associated with records

Record values are closely related to extreme order statistics. Therefore the theory of extremal, extremal-F and related processes [see, for example, Dwass (1964, 1966, 1974), Lamperti (1964) and Resnick (1987)] can be applied to study records. The corresponding results in this direction are given in Tiago de Oliveira (1968), Pickands (1971), Resnick (1973c, 1974, 1975), Resnick and Rubinovitch (1973), Deheuvels (1973, 1974, 1981, 1982a,b, 1983b), Shorrock (1974, 1975), de Haan (1984), and Pfeifer (1986). There are many relations between records and Wiener, homogeneous and nonhomogeneous Poisson processes. Various approximations of record times and record values using these processes and related results were obtained by Vervaat (1973), Deheuvels (1983a, 1984a,b, 1988), Pfeifer (1986), and Gupta and Kirmani (1988). See also the papers mentioned above that are connected with Ignatov's theorem. Counting processes associated with record times were studied by Gut (1990). Zahle (1989) investigated the structure of the set of points where a random process with continuous time takes its record values.

30. Diverse results

Mentioned here are some other work dealing with record values and record times. Katzenbeisser (1990) obtained the joint distribution of the numbers of inversions, upper and lower records. Haas (1992) applied some properties of record times to obtain the joint asymptotic distribution of the sample mean and the maximum value of a random number of i.i.d. random variables. Nayak and Wali (1992) investigated the number of occurrence of events $X(n) > r_1(n)$ and $X(n) < r_2(n)$ for some classes of sequences $r_1(n)$ and $r_2(n)$.

Gupta and Kirmani (1988) and Kochar (1990) compared distribution functions of successive record values $X(n)$ and $X(n+1)$. Amongst other results, it has been proved that if $X(n)$ has an increasing failure rate (IFR) distribution then $X(n+1)$ also has an IFR distribution. If $X(n+1)$ has a decreasing failure rate (DFR) distribution, then $X(n)$ also has a DFR distribution.

Haiman and Nevzorov (1995) investigated a stochastic ordering of the numbers of records among random variables $X_{k(1)}, X_{k(2)}, \ldots, X_{k(n)}$, where $(k(1), k(2), \ldots, k(n))$ covers all $n!$ permutations of $1, 2, \ldots, n$. They have supposed that the original random variables $X_1, X_2, \ldots, X_n$ are stochastically ordered and four types of stochastic ordering of X's are considered.

A δ-exceedance record model, in which a value will be considered a new record if it exceeds the previous record by at least δ (a pre-fixed quantity), has been discussed by Balakrishnan, Balasubramanian and Panchapakesan (1996) with special emphasis on exponential and extreme value distributions.

Nagaraja and Nevzorov (1996) proved that for a continuous parent distribution, while the correlation between functions of successive record values $X(n)$ and $X(n+1)$ is always nonnegative, it can be negative for nonconsecutive records. As for discrete parent distributions, one can construct for any $n > m \geq 1$ a function g such that the covariance between $g(X(m))$ and $g(X(n))$ is negative.

Mention should also be made here to the papers by Cheng (1987), Blom (1988), Bruss, Mahiat and Pierard (1988), Blom, Thorburn and Vessey (1990), and Dziubdziela (1990), which are all of a survey character.

Acknowledgement

This work was partially supported by the Russian Foundation of Fundamental Research grants (Nos. 95-01-01260 and 96-01-00547) and the Natural Sciences and Engineering Research Council of Canada grant

References

Adke, S. R. (1993). Records generated by Markov sequences. *Statist. Probab. Lett.* **18**, 257–263.
Ahsanullah, M. (1978). Record values and the exponential distribution. *Ann. Inst. Statist. Math.* **30**, 429–433.
Ahsanullah, M. (1979). Characterization of the exponential distribution by record values. *Sankhyā, Series B.* **41**, 116–121.
Ahsanullah, M. (1980). Linear prediction of record values for the two-parameter exponential distribution. *Ann. Inst. Statist. Math.* **32**, 363–368.
Ahsanullah, M. (1989). Estimation of the parameters of a power function distribution by record values. *Pak. J. Statist.* **5**, 189–194.
Ahsanullah, M. (1990a). Estimation of the parameters of the Gumbel distribution based on the m record values. *Comput. Statist. Quart.* **5**, 231–239.
Ahsanullah, M. (1990b). Some characterizations of the exponential distribution by the first moment of record values. *Pak. J. Statist.* **6**, 183–188.
Ahsanullah, M. (1991). Some characteristic properties of the record values from the exponential distribution. *Sankhyā, Series B.* **53**, 403–408.

Ahsanullah, M. (1992). Inference and prediction problems of the generalized Pareto distribution based on record values. In: *Order Statistics and Nonparametrics: Theory and Applications* (Eds., P. K. Sen and I. A. Salama), pp. 47–57, Elsevier, Amsterdam.

Ahsanullah, M. (1995). *Record Statistics*. Commack, NY: Nova Science Publishers.

Ahsanullah, M. and S. N. U. A. Kirmani (1991). Characterizations of the exponential distribution through a lower record. *Commun. Statist. – Theory Meth.* **20**, 1293–1299

Ahsanullah, M. and V. B. Nevzorov (1996). Distributions of order statistics generated by records. Zapiski Nauchn. Semin. POMI **228**, 24–30 (in Russian).

Albeverio, S., S. A. Molchanov, and D. Surgailis (1994). Stratified structure of the Universe and Burger's equation–a probabilitistic approach. *Probability Theory and Related Fields* **100**, 457–484.

Alpuim, M. T. (1985). Record values in populations with increasing or random dimension. *Metron* **43**, 145–155.

Andel, J. (1990). Records in an AR(1) process. *Ricerche Mat.* **39**, 327–332.

Arnold, B. C. and N. Balakrishnan (1989). *Relations, Bounds and Approximations for Order Statistics*. Lecture Notes in Statistics – **53**, Springer-Verlag, New York.

Arnold, B. C., N. Balakrishnan, and H. N. Nagaraja (1992). *A First Course in Order Statistics*. John Wiley & Sons, New York.

Arnold, B. C., N. Balakrishnan, and H. N. Nagaraja (1998). *Records*. John Wiley & Sons, New York (to appear).

Bagdonavicius, V., S. Malov, and M. Nikulin (1997). On characterizations and semi-parametric regression estimation in Archimedean copula. *Rapport Interne de l'Unité Mathématiques Appliquées de Bordeaux*, N. 97007.

Balabekyan, V. A. and V. B. Nevzorov (1986). On the number of records in a sequence of series of non-identically distributed random variables. In: *Rings and Modules – Limit Theorems of Probability Theory 1* (Eds., Z. I. Borevich and V. V. Petrov), pp. 147–153, Leningrad State University, Leningrad.

Balakrishnan, N. (1990). Improving the Hartley-David-Gumbel bound for the mean of extreme order statistics. *Statist. Probab. Lett.* **9**, 291–294.

Balakrishnan, N. (1993). A simple application of binomial-negative binomial relationship in the derivation of bounds for moments of order statistics based on greatest convex minorants. *Statist. Probab. Lett.* **18**, 301–305.

Balakrishnan, N. and M. Ahsanullah (1994a). Recurrence relations for single and product moments of record values from generalized Pareto distribution. *Commun. Statist. – Theory Meth.* **23**, 2841–2852.

Balakrishnan, N. and M. Ahsanullah (1994b). Relations for single and product moments of record values from Lomax distribution. *Sankhyā, Series B.* **56**, 140–146.

Balakrishnan, N. and M. Ahsanullah (1995). Relations for single and product moments of record values from exponential distribution. *J. Appl. Statist. Sci.* **2**, 73–87.

Balakrishnan, N., M. Ahsanullah and P. S. Chan (1992). Relations for single and product moments of record values from Gumbel distribution. *Statist. Probab. Lett.* **15**, 223–227.

Balakrishnan, N., M. Ahsanullah and P. S. Chan (1995). On the logistic record values and associated inference. *J. Appl. Statist. Sci.* **2**, 233–248.

Balakrishnan, N. and K. Balasubramanian (1995). A characterization of geometric distribution based on record values. *J. Appl. Statist. Sci.* **2**, 277–282.

Balakrishnan, N., K. Balasubramanian and S. Panchapakesan (1996). δ-exceedance records. *J. Appl. Statist. Sci.* **4**, 123–132.

Balakrishnan, N. and P. S. Chan (1994). Record values from Rayleigh and Weibull distributions and associated inference. *Extreme Value Theory and Applications – Vol. 3, NIST Special Publication* (Eds., J. Galambos, J. Lechner and E. Simiu) **866**, pp. 41–51.

Balakrishnan, N. and P. S. Chan (1995). On the normal record values and associated inference. *Technical Report*, McMaster University, Hamilton, Ontario, Canada.

Balakrishnan, N. and P. S. Chan (1998). On the normal record values and associated inference. *Statist. Probab. Lett.* (to appear).

Balakrishnan, N., P. S. Chan and M. Ahsanullah (1993). Recurrence relations for moments of record values from generalized extreme value distribution. *Commun. Statist. – Theory Meth.* **22**, 1471–1482.

Balakrishnan, N. and V. B. Nevzorov (1997). Stirling numbers and records. In: *Advances in Combinatorial Methods and Applications to Probability and Statistics* (Ed., N. Balakrishnan), pp. 189–200, Birkhäuser, Boston.

Ballerini, R. (1987). Another characterization of the type I extreme value distribution. *Statist. Probab. Lett.* **5**, 87–93.

Ballerini, R. (1994). A dependent F^α-scheme. *Statist. Probab. Lett.* **21**, 21–25.

Ballerini, R. and S. Resnick (1985). Records from improving populations. *J. Appl. Probab.* **22**, 487–502.

Ballerini, R. and S. Resnick (1987a). Records in the presence of a linear trend. *Adv. Appl. Probab.* **19**, 801–828.

Ballerini, R. and S. Resnick (1987b). Embedding sequences of successive maxima in extremal processes, with applications. *J. Appl. Probab.* **24**, 827–837.

Barndorff-Nielsen, O. (1963). On the limit behavior of extreme order statistics. *Ann. Math. Statist.* **34**, 992–1002.

Barton, D. E. and C. L. Mallows (1961). The randomization bases of the problems of the amalgamation of weighted means. *J. Roy. Statist. Soc., Series B* **23**, 423–433.

Basak, P. and P. Bagchi (1990). Application of Laplace approximation to record values. *Commun. Statist. – Theory Meth.* **19**, 1875–1888.

Berred, M. (1991). Record values and the estimation of the Weibull tail-coefficient. *Comptes Rendus, Academy of Sciences of Paris* **312**, 943–946.

Berred, M. (1992). On record values and the exponent of a distribution with regularly varying upper tail. *J. Appl. Probab.* **29**, 575–586.

Berred, M. (1994a). K-record values and the extreme-value index. *Technical Report 9425*. Centre de Recherche en Economie et Statistique.

Berred, M. (1994b). On the estimation of the Pareto tail-index using k-record values. *Technical Report 9426*. Centre de Recherche en Economie et Statistique.

Biondini, R. and M. M. Siddiqui (1975). Record values in Markov sequences. In: *Statistical Inference and Related Topics – 2* (Ed., M. L. Puri), pp. 291–352, Academic Press, New York.

Blom, G. (1988). Om rekord. *Elementa.* **71**, 67–69 (in Swedish).

Blom, G., D. Thorburn and T. Vessey (1990). The distribution of the record position and its applications. *Amer. Statist.* **44**, 151–153.

Borovkov, K. and D. Pfeifer (1995). On record indices and record times. *J. Stat. Plann. Infer.* **45**, 65–80.

Bruss, F. T. (1988). Invariant record processes and applications to best choice modelling. *Stoch. Proc. Appl.* **30**, 303–316.

Bruss, F. T., H. Mahiat and M. Pierard (19bb). Sur une fonction generatrice du nombre de records d'une suite de variables aleatoires de longueur aleatoire. *Ann. Soc. Sci. Bruxelles, Series 1*, **100**, 139–149 (in French).

Bruss, F. T. and B. Rogers (1991). Pascal processes and their characterization. *Stoch. Proc. Appl.* **37**, 331–338.

Bunge, J. A. and H. N. Nagaraja (1991). The distributions of certain record statistics from a random number of observations. *Stoch. Proc. Appl.* **38**, 167–183.

Bunge, J. A. and H. N. Nagaraja (1992a). Dependence structure of Poisson-paced records. *J. Appl. Probab.* **29**, 587–596.

Bunge, J. A. and H. N. Nagaraja (1992b). Exact distribution theory for some point process record models. *Adv. Appl. Probab.* **24**, 20–44.

Chan, P. S. (1998). Interval estimation of parameters of life based on record values. *Statist. Probab. Lett.* (to appear).

Chandler, K. N. (1952). The distribution and frequency of record values. *J. Roy. Statist. Soc. Series B*, **14**, 220–228.

Cheng, S.-H. (1987). Records of exchangeable sequences. *Acta Math. Appl. Sin.* **10**, 464–471 (in Chinese).

De Haan, L. (1984). Extremal processes. In: *Statistical Extremes and Applications* (Ed., J. Tiago de Oliveira), pp. 297–309, D. Reidel, Dordrecht, The Netherlands.

De Haan, L. and S. I. Resnick (1973). Almost sure limit points of record values. *J. Appl. Probab.* **10**, 528–542.

De Haan, L. and E. Verkade (1987). On extreme-value theory in the presence of a trend. *J. Appl. Probab.* **24**, 62–76.

Deheuvels, P. (1981). The strong approximation of extremal processes. *Z. Wahrsch. verw. Geb.* **58**, 1–6.

Deheuvels, P. (1982a). Spacings, record times and extremal processes. In: *Exchangeability in Probability and Statistics*. pp. 233–243, North-Holland, Amsterdam.

Deheuvels, P. (1982b). A construction of extremal processes. In: *Probability and Statistical Inference* (Eds., W. Grossmann, G. Pflug and W. Wertz), pp. 53–58, D. Reidel, Dordrecht, The Netherlands.

Deheuvels, P. (1983a). The strong approximation of extremal processes. II. *Z. Wahrsch. verw. Geb.* **62**, 7–15.

Deheuvels, P. (1983b). The complete characterization of the upper and lower class of the record and inter-record times of an i.i.d. sequence. *Z. Wahrsch. verw. Geb.* **62**, 1–6.

Deheuvels, P. (1984a). On record times associated with k^{th} extremes. In: *Proceedings of the 3rd Pannonian Symposium on Mathematical Statistics at Visegrad, Hungary* pp. 43–51, Budapest, Hungary.

Deheuvels, P. (1984b). The characterization of distributions by order statistics and record values – a unified approach. *J. Appl. Probab.* **21**, 326–334. Correction, **22**, 997.

Deheuvels, P. (1984c). Strong approximation in extreme value theory and applications. In: *Colloquia Mathematica Societatis Janos Bolya, 36: Limit Theorems in Probability and Statistics* Vol. 1, pp. 326–404, North-Holland, Amsterdam.

Deheuvels, P. (1984d). Strong approximations of records and record times. In: *Statistical Extremes and Applications* (Ed., J. Tiago de Oliveira) pp. 491–496, D. Reidel, Dordrecht, The Netherlands.

Deheuvels, P. (1988). Strong appoximations of kth records and j^{th} record times by Wiener processes. *Probability Theory and Related Fields*. **77**, 195–209.

Deheuvels, P. and V. B. Nevzorov (1993). Records in F^{α}-scheme. I. Martingale properties. *Zapiski Naucnyh Seminarov Leningrad. Otdel. Mat. Inst. Steklova.* **207**, 19–36 (in Russian).

Deheuvels, P. and V. B. Nevzorov (1994a). Limit laws for K-record times. *J. Statist. Plann. Infer.* **38**, 279–308.

Deheuvels, P. and V. B. Nevzorov (1994b). Records in F^{α}-scheme. II. Limit theorems. *Zapiski Naucnyh Seminarov Leningrad. Otdel. Mat. Inst. Steklova* **216**, 42–51 (in Russian).

De Laurentis, J. M. and B. G. Pittel (1985). Random permutations and Brownian motion. *Pac. J. Math.* **119**, 287–301.

Devroye, L. (1988). Applications of the theory of records in the study of random trees. *Acta Informatica.* **26**, 123–130.

Diersen, J. and G. Trenkler (1996). Records tests for trend in location. *Statistics.* **28**, 1–12.

Dunsmore, I. R. (1983). The future occurrence of records. *Ann. Inst. Statist. Math.* **35**, 267–277.

Dwass, M. (1964). Extremal processes. *Ann. Math. Statist.* **35**, 1718–1725.

Dwass, M. (1966). Extremal processes. II. *Illin. J. Math.* **10**, 381–395.

Dwass, M. (1974). Extremal processes. III. *Bull. Inst. Math. Academia Sinica*, **2**, 255–265.

Dziubdziela, W. (1977). Rozklady graniczne ekstremalnych statystyk pozycyjnych (Limit distributions of the extreme order statistics). *Roczniki Polsk. Tow. Mat, Series 3*, **9**, 45–71 (in Polish).

Dziubdziela, W. (1990). O czasach rekordowych i liczbie rekordow w ciagu zmiennych losowych (On record times and numbers of records in sequences of random variables). *Roczniki Polsk. Tow. Mat. Series 2*, **29**, 57–70 (in Polish).

Dziubdziela, W. and B. Kopocinsky (1976). Limiting properties of the kth record values. *Zastos. Mat.* **15**, 187–190.

Embrechts, P. and E. Omey (1983). On subordinated distributions and random record processes. *Math. Proc. Camb. Philos. Soc.* **93**, 339–353.

Engelen, R., P. Tommassen and W. Vervaat (1988). Ignatov's theorem: a new and short proof. *J. Appl. Probab.* **25**, 229–236.

Foster, F. G. and D. Teichroew (1955). A sampling experiment on the powers of the records tests for trend in a time series. *J. Roy. Statist. Soc. Series B*, **17**, 115–121.

Foster, F. G. and A. Stuart (1954). Distribution free tests in time-series band on the breaking of records. *J. Roy. Statist. Soc. Series B*, **16**, 1–22.

Freudenberg, W. and D. Szynal (1976). Limit laws for a random number of record values. *Bull. Acad. Polon. Sci. Ser. Sci. Math. Astr. Phys.* **24**, 193–199.

Freudenberg, W. and D. Szynal (1977). On domains of attraction of record value distributions. *Colloquium Math.* **38**, 129–139.

Gajek, L. (1985). Limiting properties of difference between the successive k^{th} record values. *Probab. Math. Statist.* **5**, 221–224.

Galambos, J. (1978, 1985). *The Asymptotic Theory of Extreme Order Statistics*. First edition, John Wiley & Sons, New York; Second edition, Krieger, Malabar, Florida.

Galambos, J. and E. Seneta (1975). Record times. *Proc. Amer. Math. Soc.* **50**, 383–387.

Gaver, D. P. (1976). Random record models. *J. Appl. Probab.* **13**, 538–547.

Gaver, D. P. and P. A. Jacobs (1978). Non-homogeneously paced random records and associated extremal processes. *J. Appl. Probab.* **15**, 552–559.

Glick, N. (1978). Breaking records and breaking boards. *Amer. Math. Monthly* **85**, 2–26.

Goldie, Ch. M. (1982). Differences and quotients of record values. *Stoch. Proc. Appl.* **12**, 162.

Goldie, Ch. M. (1989). Records, permutations and greatest convex minorants. *Math. Proc. Camb. Philos. Soc.* **106**, 177–189.

Goldie, Ch. M. and S. I. Resnick (1987). Records in a partially ordered set. *Ann. Probab.* **17**, 678–699.

Goldie, Ch. M. and S. I. Resnick (1994). Multivariate records and ordered random scattering. Preprint.

Goldie, Ch. M. and L. C. G. Rogers (1984). The k-record processes are i.i.d. *Z. Wahr. verw. Geb.* **67**, 197–211.

Grudzien, Z. (1979). On distribution and moments of ith record statistic with random index. *Ann. Univ. Mariae Curie Sklodowska*, **33**, 89–108.

Gulati, S. and W. J. Padgett (1992). Kernel density estimation from record-breaking data. In: *Survival Analysis: State of the Art* (Eds., J. P. Klein and P. K. Goel), pp. 197–210, Kluwer Academic Publishers, The Netherlands.

Gulati, S. and W. J. Padgett (1994a). Nonparametric quantile estimation from record-breaking data. *Austral. J. Statist.* **36**, 211–223.

Gulati, S. and W. J. Padgett (1994b). Estimation of nonlinear statistical functions from record-breaking data: A review-*Nonlinear Times and Digest.* **1**, 97–112.

Gulati, S. and W. J. Padgett (1994c). Smooth nonparametric estimation of the hazard rate functions from record-breaking data. *J. Statist. Plann. Infer.* **42**, 331–341.

Gulati, S. and W. J. Padgett (1994d). Smooth nonparametric estimation of the distribution and density functions from record-breaking data. *Commun. Statist. – Theory Meth.* **23**, 1259–1274.

Gumbel, E. J. (1961). The return period of order statistics. *Ann. Inst. Statist. Math.* **12**, 249–256.

Gupta, R. C. (1984). Relationships between order statistics and record values and some characterization results. *J. Appl. Probab.* **21**, 425–430.

Gupta, R. C. and S. N. U. A. Kirmani (1988). Closure and monotonicity properties of non-homogeneous Poisson processes and record values. *Probability in the Engineering and Informational Sciences*, **2**, 475–484.

Gut, A. (1990). Convergence rates for record times and the associated counting process. *Stoch. Proc. Appl.* **36**, 135–152.

Guthrie, G. L. and P. T. Holmes (1975). On record and inter-record times for a sequence of random variables defined on a Markov chain. *Adv. Appl. Probab.* **7**, 195–214.

Haas, P. J. (1992). The maximum and mean of a random length sequences. *J. Appl. Probab.* **29**, 460–466.

Haghighi-Taleb, D. and C. Wright (1973). On the distribution of records in a finite sequence of observations with an application to a road traffic problem. *J. Appl. Probab.* **10**, 556–571.

Haiman, G. (1987a). Almost sure asymptotic behavior of the record and record time sequences of a stationary Gaussian process. In: *Mathematical Statistics and Probability Theory*, Vol. A (Eds., M. L. Puri, P. Revesz and W. Wertz), pp. 105–120, D. Reidel, Dordrecht, The Netherlands.

Haiman, G. (1987b). Etude des extremes d'une suite stationnaire m-dependante avec une application relative aux accroissements due processus de Wiener. *Ann. Inst. Henri Poincare*, **23**, 425–458 (in French).

Haiman, G. (1992). A strong invariance principle for the extremes of multivariate stationary m-dependent sequences. *J. Statist. Plann. Infer.* **32**, 147–163.

Haiman, G. and M. L. Puri (1993). A strong invariance principle concerning the J-upper order statistics for stationary Gaussian sequences. *Ann. Probab.* **21**, 86–135.

Haiman, G. and V. B. Nevzorov (1995). Stochastic orderings of number of records. Statistical Theory and Applications: Papers in Honor of Herbert A. David (Eds., H. N. Nagaraja, P. K. Sen and D. F. Morrison), pp. 105–116, Springer Verlag.

Holmes, P. T. and W. Strawderman (1969). A note on the waiting times between record observations. *J. Appl. Probab.* **6**, 711–714.

Huang, W. J. and S. H. Li (1993). Characterization results based on record values. *Statistica Sinica*, **3**, 583–599.

Huang, W. J. and J. C. Su (1994). On certain problems involving order statistics – A unified approach through order statistics property of point processes. *Preprint*. Abstract. *IMS Bulletin*, **23**, 400–401.

Ignatov, Z. (1981). Point processes generated by order statistics and their applications. *Colloquia Mathematica Societatis Janos Bolyai 24 – Point Processes and Queueing Problems* (Eds., P. Bartfai and J. Tomko), pp. 109–116, North-Holland, Amsterdam.

Ignatov, Z. (1986). Ein von der variationsreihe erzeugter poissonscher punktprozess. *Annuaire Univ. Sofia Fac. Math. Mec.* **71**, 79–94.

Kamps, U. (1994). Reliability properties of record values from non-identically distributed random variables. *Commun. Statist. – Theory Meth.* **23**, 2101–2112.

Kamps, U. (1995). *A Concept of Generalized Order Statistics*. Teubner, Stuttgart, Germany.

Katzenbeisser, W. (1990). On the joint distribution of the number of upper and lower records and the number of inversions in a random sequence. *Adv. Appl. Probab.* **22**, 957–960.

Kinoshita, K. and S. I. Resnick (1989). Multivariate records and shape. In: *Extreme Value Theory* (Eds., J. Husler and R.D. Reiss), Lecture Notes in Statistics – **51**, pp. 222–233, Springer-Verlag, Berlin.

Kochar, S. C. (1990). Some partial ordering results on record values. *Commun. Statist. – Theory Meth.* **19**, 299–306.

Lamperti, J. (1964). On extreme order statistics. *Ann. Math. Statist.* **35**, 1726–1737.

Lin, G. D. (1987). On characterizations of distributions via moments of record values. *Probability Theory and Related Fields*, **74**, 479–483.

Lin, G. D. and J. S. Huang (1987). A note on the sequence of expectations of maxima and of record values. *Sankhyā, Series A*, **49**, 272–273.

Malov, S. V. and V. B. Nevzorov (1997). Characterizations using ranks and order statistics. In: *Advances in the Theory and Practice of Statistics: A Volume in Honor of Samuel Kotz* (Eds., N.L. Johnson and N. Balakrishnan), pp. 479–489, John Wiley & Sons, New York.

Moreno Rebollo J. L., I. Barranco Chamoor, F. Lopez Blazquez and T. Gomez Gomez (1996). On the estimation of the unknown sample size from the number of records. *Statist. Probab. Lett.* **31**, 7–12.

Moriguti, S. (1953). A modification of Schwarz's inequality with applications to distributions. *Ann. Math. Statist.* **24**, 107–113.

Nagaraja, H. N. (1977). On a characterization based on record values. *Austral. J. Statist.* **19**, 70–73.

Nagaraja, H. N. (1978). On the expected values of record values. *Austral. J. Statist.* **20**, 176–182.

Nagaraja, H. N. (1984). Asymptotic linear prediction of extreme order statistics. *Ann. Inst. Statist. Math.* **36**, 289–299.

Nagaraja, H. N. (1986). Comparison of estimators from two-parameter exponential distribution. *Sankhyā Series B*, **48**, 10–18.

Nagaraja, H. N. (1988a). Record values and related statistics – a review. *Commun. Statist. – Theory Meth.* **17**, 2223–2238.

Nagaraja, H. N. (1988b). Some characterizations of continuous distributions based on regressions of adjacent order statistics and record values. *Sankhyā, Series A*, **50**, 70–73.

Nagaraja, H. N. and V. B. Nevzorov (1997). On characterizations based on record values and order statistics. *J. Statist. Plann. Infer.* **63**, 271–284.

Nagaraja, H. N. and V. B. Nevzorov (1996). Correlations between functions of records can be negative. *Statist. Probab. Lett.* **29**, 95–100.

Nagaraja, H. N., P. Sen and R. C. Srivastava (1989). Some characterizations of geometric tail distributions based on record values. *Statistical Papers* **30**, 147–159.

Nayak, S. S. and K. S. Wali (1992). On the number of boundary crossings related to LIL and SLLN for record values and partial maxima of i.i.d. sequences and extremes of uniform spacings. *Stoch. Proc. Appl.* **43**, 317–329.

Neuts, M. F. (1967). Waiting times between record observations. *J. Appl. Probab.* **4**, 206–208.

Nevzorov, V. B. (1981). Limit theorems for order statistics and record values. In: *Abstracts of Third International Vilnius Conference on Probability Theory and Mathematical Statistics*, Vol. 2, pp. 86–87, Vilnius (in Russian).

Nevzorov, V. B. (1984). Record times in the case of nonidentically distributed random variables. *Teoriya veroyatnostey i ee Primenenija* **29**, 808–809 (in Russian). Translated version in *Theory of Probability and its Applications*.

Nevzorov, V. B. (1985). Record and inter-record times for sequences of nonidentically distributed random variables. *Zapiski Nauchn. Semin. LOMI*, **142**, 109–118 (in Russian). Translated version in *J. Soviet Math.* **36** (1987), 510–516.

Nevzorov, V. B. (1986a). K-th record times and their generalizations. *Zapiski Nauchn. Semin. LOMI*, **153**, 115–121 (in Russian). Translated version in *J. Soviet Math.* **44** (1989), 510–515.

Nevzorov, V. B. (1986b). Two characterizations using records. In: *Stability Problems for Stochastic Models* (Eds., V.V. Kalashnikov, B. Penkov and V.M. Zolotarev), Lecture Notes in Mathematics – **1233**, pp. 79–85, Springer-Verlag, Berlin.

Nevzorov, V. B. (1986c). The number of records in a sequence of nonidentically distributed random variables. *Veroyatnostnye Raspredeleniya i Matematicheskaja Statistika*, Part 2, pp. 373–388 (in Russian). Translated version in *J. Soviet Math.* **38** (1987), 2375–2382.

Nevzorov, V. B. (1986d). Record times and their generalizations. *Teoriya Veroyatnostey i ee Primenenija*, **31**, 629–630 (in Russian). Translated version in *Theory Probab. Appl.*

Nevzorov, V. B. (1987a). Records, *Teoriya Veroyatnostey i ee Primenenija* **32**, 219–251 (in Russian). Translated version in *Theory Probab Appl.* **32** (1988), 201–228.

Nevzorov, V. B. (1987b). Distribution of kth record values in the discrete case. *Zapiski Nauchn. Semin. LOMI*, **158**, 133–137 (in Russian). Translated version in *J. Soviet Math.* **43** (1988), 2830–2833.

Nevzorov, V. B. (1987c). Moments of some random variables connected with records. *Vestnik of Leningrad Univ.* **8**, 33–37 (in Russian).

Nevzorov, V. B. (1988). Centering and normalizing constants for extrema and for records. *Zapiski Nauchn. Semin. LOMI*, **166**, 103–111 (in Russian). Translated version in *J. Soviet Math.* **52** (1990), 2935–2941.

Nevzorov, V. B. (1989). Martingale methods of investigation of records. *Statistics and Control Random Processes* 156–160 (in Russian).

Nevzorov, V. B. (1990a). Generating functions for k^{th} record values – a martingale approach. *Zapiski Nauchn. Semin. LOMI*, **184**, 208–214 (in Russian). Translated version in *J. Math. Sci.* **68** (1994), 545–550.

Nevzorov, V. B. (1990b). Records for nonidentically distributed random variables. In: *Proceedings of the Fifth Vilnius Conference on Probability and Statistics* Vol. 2, pp. 227–233, VSP Publishers, Mokslas.

Nevzorov, V. B. (1992). A characterization of exponential distributions by correlations between records. In: *Mathematical Methods of Statistics 1*, pp. 49–54, Allerton Press.

Nevzorov, V. B. (1993). Characterizations of certain nonstationary sequences by properties of maxima and records. In: *Rings and Modules. Limit Theorems of Probability Theory* (Eds., Z.I. Borevich and V.V. Petrov), Vol. 3, pp. 188–197, St.-Petersburg State University, St.-Petersburg (in Russian).

Nevzorov, V. B. (1995). Asymptotic distributions of records in nonstationary schemes. *J. Statist. Plann. Infer.* **44**, 261–273.

Nevzorov, V. B. (1997). One limit relation between order statistics and records. Zapiski Nauchn. Semin. POMI, 244, 218–226 (in Russian).

Nevzorov, V. B. and M. Rannen (1992). On record times in sequences of nonidentically distributed discrete random variables. *Zapiski Nauchn. Semin. LOMI*, **194**, 124–133 (in Russian). Translated version in *J. Math. Sci.*.

Nevzorov, V. B. and A. V. Stepanov (1988). Records: martingale approach to finding of moments. In: *Rings and Modules. Limit Theorems of Probability Theory* (Eds., Z.I. Borevich and V.V. Petrov), Vol. 2, pp. 171–181, St.-Petersburg State University, St.-Petersburg (in Russian).

Nevzorova, L. N., V. B. Nevzorov and N. Balakrishnan (1997). Characterizations of distributions by extremes and records in Archimedean copula process. In: *Advances in the Theory and Practice of Statistics: A Volume in Honor of Samuel Kotz* (Eds., N.L. Johnson and N. Balakrishnan), pp. 469–478, John Wiley & Sons, New York.

Nevzorova, L. N. and V. N. Nikoulina (1997). Intervalles de confiance pour la taille de l'échantillon, basées sur le nombre de records. *Rapport Interne de l'Unité Mathématiques Appliquées de Bordeaux* N. 97008.

Pfeifer, D. (1981). Asymptotic expansions for the mean and variance of logarithmic inter-record times. *Meth. Oper. Res.* **39**, 113–121.

Pfeifer, D. (1982). Characterizations of exponential distributions by independent non-stationary record increments. *J. Appl. Probab.* **19**, 127–135. Correction, **19**, 906.

Pfeifer, D. (1984a). A note on moments of certain record statistics. *Z. Wahr. verw. Geb.* **66**, 293–296.

Pfeifer, D. (1984b). A note on random time changes of Markov chains. *Scand. Actuar. J.* 127–129.

Pfeifer, D. (1984c). Limit laws for inter-record times from non-homogeneous record values. *J. Organ. Behav. Statist.* **1**, 69–74.

Pfeifer, D. (1985). On a relationship between record values and Ross's model of algorithm efficiency. *Adv. Appl. Probab.* **27**, 470–471.

Pfeifer, D. (1986). Extremal processes, record times and strong approximation. *Publ. Inst. Statist. Univ. Paris* **31**, 47–65.

Pfeifer, D. (1987). On a joint srong approximation theorem for record and inter-record times. *Probab. Theory Rel. Fields* **75**, 213–221.

Pfeifer, D. (1989). Extremal processes, secretary problems and the 1/e law. *J. Appl. Probab.* **26**, 722–733.

Pfeifer, D. (1991). Some remarks on Nevzorov's record model. *Adv. Appl. Probab.* **23**, 823–834.

Pfeifer, D. and Y. C. Zhang (1989). A survey on strong approximation techniques in connection with records. In: *Extreme Value Theory* (Eds., J. Husler and R.D. Reiss), Lecture Notes in Statistics – **51**, pp. 50–58, Springer-Verlag, Berlin.

Pickands, J. (1971). The two-dimensional Poisson process and extremal processes. *J. Appl. Probab.* **8**, 745–756.

Rannen, M. M. (1991). Records in sequences of series of nonidentically distributed random variables. *Vestnik of the Leningrad State University, Series 1*, 62–66.

Rao, C. R. and D. N. Shanbhag (1994). *Choquet-Deny Type Functional Equations with Applications to Stochastic Models*. John Wiley & Sons, Chichester, England.

Raqab, M. Z. and N. Balakrishnan (1997). Bounds based on the greatest convex minorants for moments of record values. *Submitted for publication*.

Renyi, A. (1962). Theorie des elements saillants d'une suite d'observations. In: *Colloquium on Combinatorial methods in Probability Theory* (August 1–10, 1962), pp. 104–117, Mathematical Institute,

Aarhus University, Aarhus, Denmark (in French). See also *Selected Papers of Alfred Renyi*, Vol. 3 (1976), pp. 50–65, Akademiai Kiado, Budapest, Hungary.

Resnick, S. I. (1973a). Limit laws for record values. *Stoch. Proc. Appl.* **1**, 67–82.

Resnick, S. I. (1973b). Extremal processes and record value times. *J. Appl. Probab.* **10**, 864–868.

Resnick, S. I. (1973c). Record values and maxima. *Annal. Probab.* **1**, 650–662.

Resnick, S. I. (1974). Inverses of extremal processes. *Adv. Appl. Probab.* **6**, 392–406.

Resnick, S. I. (1975). Weak convergence to extremal processes. *Ann. Probab.* **3**, 951–960.

Resnick, S. I. and M. Rubinovitch (1973). The structure of extremal processes. *Adv. Appl. Probab.* **5**, 287–307.

Rogers, L. C. G. (1989). Ignatov's theorem: an abbreviation of the proof of Engelen, Tommassen and Vervaat. *Adv. Appl. Probab.* **21**, 933–934.

Samaniego, F. J. and L. R. Whitaker (1986). On estimating population characteristics from record-breaking observations. I. Parametric results. *Naval Res. Log. Quart.* **33**, 531–543.

Samaniego, F. J. and L. R. Whitaker (1988). On estimating population characteristics from record-breaking observations. II. Nonparametric results. *Naval Res. Log. Quart.* **33**, 221–236.

Shorrock, R. W. (1972a). On record values and record times. *J. Appl. Probab.* **9**, 316–326.

Shorrock, R. W. (1972b). A limit theorem for inter-record times. *J. Appl. Probab.* **9**, 219–223.

Shorrock, R. W. (1973). Record values and inter-record times. *J. Appl. Probab.* **10**, 543–555.

Shorrock, R. W. (1974). On discrete time extremal processes. *Adv. Appl. Probab.* **6**, 580–592.

Shorrock, R. W. (1975). Extremal processes and random measures. *J. Appl. Probab.* **12**, 316–323.

Sibuya, M. and K. Nishimura (1997). Prediction of record-breakings. *Statistica Sinica* **7**, 893–906.

Siddiqui, M. M. and R. W. Biondini (1975). The joint distribution of record values and inter-record times. *Ann. Probab.* **3**, 1012–1013.

Smith, R. L. (1988). Forecasting records by maximum likelihood. *J. Amer. Statist. Assoc.* **83**, 331–338.

Smith, R. L. and J. E. Miller (1986). A non-Gaussian state space model and application to prediction of records. *J. Roy. Statist. Soc. Series B*, **48**, 79–88.

Srivastava, R. C. (1979). Two characterizations of the geometric distribution by record values. *Sankhyā Series B*, **40**, 276–278.

Stam, A. I. (1985). Independent Poisson processes generated by record values and inter-record times. *Stoch. Proc. Appl.* **19**, 315–325.

Stepanov, A. V. (1987). On logarithmic moments for inter-record times. *Theory Probab. Appl.* **32**, 708–710.

Stepanov, A. V. (1989). Characterizations of geometric class of distributions. *Teoriya Veroyatnostey i Matematicheskaya Statistika* **41**, 133–136 (in Russian). Translated version in *Theory of Probability and Mathematical Statistics* **41** (1990).

Stepanov, A. V. (1992). Limit theorems for weak records. *Theory Probab. Appl.* **37**, 586–590.

Stuart, A. (1954). Asymptotic relative efficiencies of distribution-free tests of randomness against normal alternatives. *J. Amer. Statist. Assoc.* **49**, 147–157.

Stuart, A. (1956). The efficiencies of tests of randomness against normal regression. *J. Amer. Statist. Assoc.* **51**, 285–287.

Stuart, A. (1957). The efficiency of the records test for trend in normal regression. *J. Roy. Statist. Soc. Series B*, **19**, 149–153.

Sultan, K. S. and N. Balakrishnan (1997a). Higher order moments of record values from Rayleigh distribution and Edgeworth approximate inference. *Submitted for publication*.

Sultan, K. S. and N. Balakrishnan (1997b). Higher order moments of record values from Weibull distribution and Edgeworth approximate inference. *Submitted for publication*.

Sultan, K. S. and N. Balakrishnan (1997c). Higher order moments of record values from Gumbel distribution with applications to inference. *Submitted for publication*.

Szekely, G. J. and T. F. Mori (1985). An extremal property of rectangular distributions. *Statist. Probab. Lett.* **3**, 107–109.

Tata, M. N. (1969). On outstanding values in a sequence of random variables. *Z. Wahr. verw. Geb.* **12**, 9–20.

Teugels, J. L. (1984). On successive record values in a sequence of independent identically distributed random variables. In: *Statistical Extremes and Applications* (Ed., J. Tiago de Oliveira), pp. 639–650, D. Reidel, Dordrecht, The Netherlands.

Tiago de Oliveira, J. (1968). Extremal processes: definitions and properties. *Publ. Inst. Statist. Univ. Paris* **17**, 25–36.

Too, Y. H. and G. D. Lin (1989). Characterizations of uniform and exponential distributions. *Statist. Probab. Lett.* **7**, 357–359.

Tryfos, P. and R. Blackmore (1985). Forecasting records. *J. Amer. Statist. Assoc.* **80**, 46–50.

Van Zwet, W. R. (1964). *Convex Transformations of Random Variables*. Mathematical Centre Tracts 7, Mathematisch Centrum, Amsterdam.

Vervaat, W. (1973). Limit theorems for records from discrete distributions. *Stoch. Proc. Appl.* **1**, 317–334.

Westcott, M. (1977a). A note on record times. *J. Appl. Probab.* **14**, 637–639.

Westcott, M. (1977b). The random record model. *Proc. Roy. Soc. London, Series A*, **356**, 529–547.

Westcott, M. (1979). On the tail behavior of record-time distributions in a random record process. *Ann. Probab.* **7**, 868–873.

Wilks, S. S. (1959). Recurrence of extreme observations. *J. Amer. Math. Soc.* **1**, 106–112.

Williams, D. (1973). On Renyi's record problem and Engel's series. *Bull. London Math. Soc.* **5**, 235–237.

Witte, H. J. (1988). Some characterizations of distributions based on the integrated Cauchy functional equation. *Sankhyā, Series A*, **50**, 59–63.

Witte, H. J. (1990). Characterizations of distributions of exponential or geometric type by the integrated lack of memory property and record values. *Comput. Statist. Data Anal.* **10**, 283–288.

Yakymiv, A. L. (1986). Asymptotic properties of changing states times in random record process. *Theory Probab. Appl.* **31**, 577–581.

Yang, M. C. K. (1975). On the distribution of the inter-record times in an increasing population. *J. Appl. Probab.* **12**, 148–154.

Zahle, U. (1989). Self-similar random measures, their carrying dimension and application to records. In: *Extreme Value Theory* (Eds., J. Husler and R.D. Reiss), pp. 59–68, Lecture Notes in Statistics – **51**, Springer-Verlag, Berlin.

PART IX

Related Processes

Weighted Sequential Empirical Type Processes with Applications to Change-Point Problems

Barbara Szyszkowicz

1. Introduction

Let $X_1, X_2, \ldots$ be independent random variables with the same continuous distribution function F. We consider the two-time parameter, frequently called sequential, empirical processes

$$\zeta_n(x,t) = n^{-1/2} \sum_{i=1}^{[nt]} \left(\mathbf{1}\{X_i \leq x\} - F(x) \right), \quad x \in \mathbb{R}, \ 0 \leq t \leq 1, \quad (1.1)$$

$$\beta_n(s,t) = n^{-1/2} \sum_{i=1}^{[nt]} \left(\mathbf{1}\{R_{in} \leq s\} - \frac{[ns]}{n} \right), \quad 0 \leq s,t \leq 1, \quad (1.2)$$

$$\gamma_n(s,t) = n^{-1/2} \sum_{i=1}^{[nt]} \left(\mathbf{1}\{\xi_i \leq s\} - \frac{[is]}{i} \right), \quad 0 \leq s,t \leq 1, \quad (1.3)$$

where $R_{1n}, \ldots, R_{nn}$ denote the normalized ranks

$$R_{in} = n^{-1} \sum_{k=1}^{n} \mathbf{1}\{X_k \leq X_i\}, \quad i = 1, \ldots, n,$$

and $\xi_i, \ldots, \xi_n$ are the normalized sequential ranks

$$\xi_i = i^{-1} \sum_{k=1}^{i} \mathbf{1}\{X_k < X_i\}, \quad i = 1, \ldots, n,$$

of the first n of the chronologically ordered random variables $X_1, X_2, \ldots$. Thus, the rank of X_i, in our notation nR_{in}, is the integer among the numbers $1, \ldots, n$, which corresponds to the position of X_i in the order statistics $X_{1n} \leq \cdots \leq X_{nn}$ of a random sample of size $n \geq 1$, while the sequential rank of X_i, here $i\xi_i$, is the position of X_i in the first $i-1$ ordered observations $X_{0,i-1} \leq \cdots \leq X_{i-1,i-1}$, $1 \leq i \leq n$, of a random sample of size $n \geq 1$, where $X_{0,0} \equiv 0$ by convention, and $\xi_1 = 0$ by definition. We note that $\lambda_n(s) := [ns]/n$, $0 \leq s \leq 1$, and $\lambda_i(s) := [is]/i$,

$0 \leq s \leq 1$, $1 \leq i \leq n$, are the distribution functions of the normalized ranks R_{in} and the normalized sequential ranks ξ_i, respectively, under the assumption that $X_1, \ldots, X_n$ are independent, identically distributed random variables (i.i.d.r.v.'s) with the same continuous distribution function F.

The asymptotic behaviour of the empirical process in (1.1), which is of practical interest, may be concluded from that of its uniform version $\alpha_n(s,t)$, where

$$\alpha_n(s,t) = n^{-1/2} \sum_{i=1}^{[nt]} (\mathbf{1}\{F(X_i) \leq s\} - s), \quad 0 \leq s, t \leq 1,$$

due to the fact that with a continuous distribution function F, we have

$$\{\zeta_n(F^{-1}(s),t); \ 0 \leq s,t \leq 1, \ n \geq 1\} = \{\alpha_n(s,t); \ 0 \leq s,t \leq 1, \ n \geq 1\},$$

i.e., these two empirical processes are identical. Hence all results proved for $\alpha_n(s,t)$ will hold automatically also for $\zeta_n(x,t)$. Indeed, for every $\omega \in \Omega$, $\alpha_n(F(x),t) = \zeta_n(x,t)$, $x \in \mathbb{R}^1$, $0 \leq t \leq 1$, $n \geq 1$.

The limiting distributions of all these processes are well known. Here we present their asymptotics in $D[0,1]^2$ in weighted supremum and L_p-metrics for the optimal classes of weights q, which are functions of the time variable $t \in [0,1]$.

We note that the process $\alpha_n(s,t)$ is a two-time parameter version of the usual (uniform) empirical process

$$\alpha_n(s) = n^{-1/2} \sum_{i=1}^{n} (\mathbf{1}\{U_i \leq s\} - s), \quad 0 \leq s \leq 1,$$

where $U_1, \ldots, U_n$ are uniform-$(0,1)$ random variables. Starting with Anderson and Darling (1952), Rényi (1953), Chibisov (1964) and O'Reilly (1974), there has been considerable interest in the asymptotic behaviour of weighted uniform empirical and quantile processes. For an insightful treatise of this subject we refer to Csörgő, Csörgő, Horváth and Mason ([CsCsHM]) (1986), Shorack and Wellner (1986), and to Csörgő and Horváth (1993), as well as to the references in these works. Due to CsCsHM (1986) and Csörgő, Horváth and Shao (1993), there are now complete characterizations available for describing the asymptotic behaviour of the weighted uniform empirical and quantile processes in supremum and L_p-metrics. For a treatize on recent advances on weighted approximations in probability and statistics in general, we refer to Csörgő and Horváth (1993).

In this paper we are interested in asymptotics of weighted two-time parameter (sequential) uniform empirical process, namely $\alpha_n(s,t)/q(t)$, $0 \leq s,t \leq 1$, with weights in the time parameter $t \in (0,1]$ instead of the "space" parameter $s \in (0,1)$, where $q(t)$ is a nonnegative function on $(0,1]$. In a similar way we consider two-time parameter (sequential) empirical processes of ranks and sequential ranks.

In particular, for the process $\alpha_n(\cdot,\cdot)$ we have (cf Theorem 2.1(a)) that with $q(t)$ which is a positive function on $(0,1]$ and nondecreasing near zero, as $n \to \infty$,

$$\sup_{0<t\leq 1}\sup_{0\leq s\leq 1}|\alpha_n(s,t)-n^{-1/2}K(s,nt)|/q(t)=o_P(1) \qquad (1.4)$$

if and only if $I(q,c)<\infty$ for all $c>0$, where

$$I(q,c)=\int_0^1 t^{-1}\exp\left(-ct^{-1}q^2(t)\right)dt,\quad c>0,$$

and $\{K(s,t);\ 0\leq s\leq 1,\ t\geq 0\}$ is a Kiefer process, i.e., a two-time parameter separable Gaussian process with mean zero and covariance function

$$EK(s_1,t_1)K(s_2,t_2)=(s_1\wedge s_2-s_1s_2)(t_1\wedge t_2).$$

Obviously, via (1.4) we conclude convergence in distribution of any continuous in sup-norm functional of $\alpha_n(s,t)/q(t)$ to the corresponding functional of $K(s,t)/q(t)$ with weight function $q(t)$ as above, and with $\{K(s,t);\ 0\leq s,t\leq 1\}$ being a Kiefer process.

Moreover, when considering the supremum functional on its own we have (cf Theorem 2.1 (c)) that the class of admissible weight functions $q(t)$ for the convergence in distribution of $\sup_{0<t\leq 1}\sup_{0\leq s\leq 1}|\alpha_n(s,t)|/q(t)$ is bigger than for the weak convergence of the whole process $\alpha_n(s,t)/q(t)$ in supremum norm. Similar results hold also for the processes $\beta_n(s,t)$ and $\gamma_n(s,t)$. Such a phenomenon was first noted and proved by CsCsHM (1986) for the empirical and quantile processes with weight functions in the space parameter $s\in(0,1)$. For instance, in case of the $t\in(0,1)$ variable, we have

$$\sup_{0<t\leq 1}\sup_{0\leq s\leq 1}|\alpha_n(s,t)|/(t\log\log((1/t)\vee 3))^{1/2}$$
$$\xrightarrow{\mathcal{D}}\sup_{0<t\leq 1}\sup_{0\leq s\leq 1}|K(s,t)|/(t\log\log((1/t)\vee 3))^{1/2}, \qquad (1.5)$$

where $K(\cdot,\cdot)$ is a Kiefer process, *though weak convergence in supremum norm with the weight function* $q(t)=(t\log\log((1/t)\vee 3))^{1/2}$ *is impossible*.

Since $\limsup_{t\downarrow 0}\sup_{0\leq s\leq 1}|K(s,t)|/t^{1/2}=\infty$ a.s., it is clear that the statement as in (1.5) is impossible with $q(t)=t^{1/2}$. Such a weight function is, however, an immediate candidate for a weighted L_p-functional of $\alpha_n(s,t)$ to converge, due to the almost sure finiteness of the integral $\int_0^1\int_0^1|K(s,t)|/t^{1/2}\,ds\,dt$ or, in general, that of $\int_0^1\int_0^1|K(s,t)|^p/t^{p/2}\,ds\,dt$, $0<p<\infty$ (cf Lemma 2.B). Indeed, the class of admissible weight functions when considering L_p-norms is bigger than that in the case of supremum norms. For example, it will follow from Theorem 2.2 that, as $n\to\infty$,

$$\int_0^1\int_0^1|\alpha_n(s,t)|/t\,ds\,dt\xrightarrow{\mathcal{D}}\int_0^1\int_0^1|K(s,t)|/t\,ds\,dt, \qquad (1.6)$$

where $\{K(s,t);\ 0\leq s,t\leq 1\}$ is a Kiefer process. Moreover, in (1.6) we can even have t^v instead of t with any $v<3/2$.

The empirical processes defined in (1.1)–(1.3) play an important role in the so-called change-point problem. In particular, Csörgő and Horváth (1987a) propose nonparametric procedures for the change-point problem based on the $\beta_n(s,t)$ process. Leipus (1988, 1989) studies the processes $\alpha_n(s,t)$ and $\beta_n(s,t)$, as well as some functionals of $\beta_n(s,t)$, in a similar vein. The process $\gamma_n(s,t)$ was used by Csörgő and Horváth (1987b) to construct a sequential procedure for detecting a possible change-point in a random sequence. The three processes (1.1)–(1.3) were studied by Pardzhanadze and Khmaladze (1986) under a class of contiguous alternatives which accommodate the possible occurrence of a changepoint in a series of measurements. Szyszkowicz (1991b, 1994) studied these processes, as well as the "bridge type" (tied down at $t=1$) versions of $\alpha_n(s,t)$ and $\gamma_n(s,t)$, in weighted supremum metrics under the null assumption of no change and also under a sequence of contiguous alternatives as parametrized by Pardzhanadze and Khmaladze (1986). In the same vein Correa (1995) studies weighted sequential empirical processes, where the parameters of the underlying distribution function are estimated.

As illustrated in Csörgő and Horváth (1988a) and argued in Brodsky and Darkhovsky (1993), a large number of nonparametric as well as parametric modelling of change-point problems result in the same test statistic, namely in (cf also references in the just mentioned works)

$$\max_{1 \leq k < n} \left\{ |S_k - kS_n/n|/(k(1-k/n))^{1/2} \right\} ,$$

where $S_k = X_1 + \cdots + X_k$, $k = 1, \ldots, n$ and $X_1, \ldots, X_n$ are independent observations. This is the standardized difference between the mean of the first k observations (before the change), S_k/k, $1 \leq k < n$, and the overall mean, S_n/n, or, *equivalently*, the standardized difference between the mean of the first k observations (before the change) and the mean of the remaining $(n-k)$ observations (after the change), $k = 1, 2, \ldots, n-1$, where $\max_{1 \leq k < n}$ accounts for the fact that the time of change is actually unknown. *Namely*, we arrive at considering the sequence (in n) of stochastic processes in k

$$\frac{S_k}{k} - \frac{S_n - S_k}{n-k} = n\left(S_k - \frac{k}{n}S_n\right)/k(n-k), \quad k = 1, \ldots, n-1 ,$$

and consequently, hoping for the convergence in distribution, as $n \to \infty$, of the statistics

$$n^{1/2} \max_{1 \leq k < n} \left| \frac{S_k}{k} - \frac{S_n - S_k}{n-k} \right| = n^{-1/2} \max_{1 \leq k < n} \frac{|S_k - kS_n/n|}{\left(\frac{k}{n}\left(1-\frac{k}{n}\right)\right)} \tag{1.7}$$

so that we should be able to reject the null assumption of having no change in the mean if the latter were too large. Unfortunately, these statistics, and even the statistics resulting from replacing the weight function $k/n(1-k/n)$ in (1.7) by $(k/n(1-k/n))^{1/2}$, converge in probability to ∞, as $n \to \infty$, even if the null assumption of having no change in the mean were true. This, in turn, leads to studying the problem of weighted asymptotics of partial sum processes tied down at $t=1$ (cf Csörgő and Horváth (1988a) for sup-norm results for i.i.d.r.v.'s under the assumption of $E|X_1|^v < \infty$ for some $v > 2$, and Szyszkowicz (1991c, 1992, 1993a, 1996a, 1997) for weighted asymptotics in supremum and L_p-norms for i.i.d.r.v.'s, as well as for weighted weak convergence under contiguous measures, when only two moments for X_1 are assumed to be finite).

The general idea of studying change in distribution of a random sequence of chronologically ordered observations via comparing the mean of the first k of them to that of the remaining $(n-k)$ of them appears to be even more appropriate when expressed in terms of comparing the empirical distribution function of the first k observations to that of the last $(n-k)$ observations. Consequently, we are interested in studying the asymptotic distribution of the sequence of processes

$$n^{1/2} \left| \frac{1}{k} \sum_{i=1}^{k} \mathbf{1}\{X_i \leq x\} - \frac{1}{n-k} \sum_{i=k+1}^{n} \mathbf{1}\{X_i \leq x\} \right|$$

$$= \left| \sum_{i=1}^{k} \mathbf{1}\{X_i \leq x\} - \frac{k}{n} \sum_{i=1}^{n} \mathbf{1}\{X_i \leq x\} \right| \bigg/ n^{1/2} \left(\frac{k}{n} \left(1 - \frac{k}{n}\right) \right), \quad (1.8)$$

$$x \in \mathbb{R}, \ 1 \leq k < n, \ n = 1, 2, \ldots .$$

Considering their $\sup_{1 \leq k < n} \sup_{x \in \mathbb{R}}$ functionals we note that the resulting sequence of random variables, and even the one where $\frac{k}{n}(1 - \frac{k}{n})$ is replaced by $\left(\frac{k}{n}(1 - \frac{k}{n})\right)^{1/2}$, namely

$$\sup_{1 \leq k < n} \sup_{x \in \mathbb{R}} \left| \sum_{i=1}^{k} \mathbf{1}\{X_i \leq x\} - \frac{k}{n} \sum_{i=1}^{n} \mathbf{1}\{X_i \leq x\} \right| \bigg/ n^{1/2} \left(\frac{k}{n} \left(1 - \frac{k}{n}\right) \right)^{1/2},$$

(1.9)

converges to ∞ in probability, as $n \to \infty$, even if the null assumption of no change in distribution were true. Consequently, just like in case of partial sum processes, in order to have non-degenerate limits as $n \to \infty$, in supremum norm we are led to considering weight functions $q(k/n) = ((k/n)(1 - k/n))^{1/2} h(k/n)$, where the function $h(k/n)$ neccessarily goes to ∞ as $k/n \to 0$ or $k/n \to 1$ (cf Picard (1985), Deshayes and Picard (1986), and Szyszkowicz (1991b, 1994)).

Just like tests based on the classical Kolmogorov–Smirnov statistic, the ones based on

$$\sup_{1 \leq k < n} \sup_{x \in \mathbb{R}} \frac{k}{n} \left(1 - \frac{k}{n}\right) n^{1/2} \left| \frac{1}{k} \sum_{i=1}^{k} \mathbf{1}\{X_i \leq x\} - \frac{1}{n-k} \sum_{i=k+1}^{n} \mathbf{1}\{X_i \leq x\} \right|$$

$$= \sup_{1 \leq k < n} \sup_{x \in \mathbb{R}} n^{-1/2} \left| \sum_{i=1}^{k} \mathbf{1}\{X_i \leq x\} - \frac{k}{n} \sum_{i=1}^{n} \mathbf{1}\{X_i \leq x\} \right|$$

(1.10)

should be more powerful for detecting changes that occur in the middle, namely near $n/2$, where $\frac{k}{n}(1 - \frac{k}{n})$ has its maximum, than for noticing the ones occurring near the endpoints 0 and n. Thus, a weighted version of (1.10) should emphasize changes which may have occurred near the endpoints, while retaining sensitivity to possible changes in the middle as well. To see what kind of weight functions are possible for statistics based on the processes in (1.8), we will study the asymptotic behavior of the processes

$$(\alpha_n(s,t) - t\alpha_n(s,1))/q(t)$$

for a wide class of functions q. For example, as a result of Theorem 3.1 (c) we arrive at the following modification of (1.9) and (1.10)

$$\sup_{1\leq k<n} \sup_{x\in\mathbb{R}} \frac{\left|\sum_{i=1}^{k} \mathbf{1}\{X_i \leq x\} - \frac{k}{n}\sum_{i=1}^{n} \mathbf{1}\{X_i \leq x\}\right|}{n^{1/2}\left(\frac{k}{n}\left(1-\frac{k}{n}\right)\log\log\frac{1}{\frac{k}{n}\left(1-\frac{k}{n}\right)}\right)^{1/2}},$$

which has a nondegenerate limiting distribution, namely that of (cf (3.5))

$$\sup_{0<t<1} \sup_{0\leq s\leq 1} |K(s,t) - tK(s,1)| \Big/ \left(t(1-t)\log\log\frac{1}{t(1-t)}\right)^{1/2}, \qquad (1.11)$$

where $\{K(s,t);\ 0\leq s,t\leq 1\}$ is a Kiefer process.

On the other hand, if we were to study the problem of the asymptotic behaviour of the processes in (1.8) in L_1 say, then, for having a non-degenerate limit, there is no need to replace the naturally arrived at weight function $((k/n)(1-k/n))$ in there by any other function that would be milder on the tails. For example, as a result of Theorem 3.2, we obtain that the very L_1-functional of our process as in (1.8) converges in distribution to the corresponding functional of $\{K(s,t) - tK(s,1);\ 0\leq s,t\leq 1\}$ process, namely to $\int_0^1 \int_0^1 |K(s,t) - tK(s,1)|/(t(1-t))\,ds\,dt$ (cf (3.3), (3.6) and (3.7)) where $\{K(s,t);\ 0\leq s,t\leq 1\}$ is a Kiefer process. Thus, the latter L_1-functional, which does not require any modification of the naturally arrived at weights for convergence, results in a natural asymptotic solution to the change-point problem as posed in (1.8).

After studying, in Section 2, approximations of $\alpha_n(s,t)/q(t)$ in probability in supremum and L_p-metrics for the optimal classes of weight functions, we use these results to obtain the same kind of approximations for the weighted tied down processes $(\alpha_n(s,t) - t\alpha_n(s,1))/q(t)$ in Section 3. In Section 4 we consider the two-time parameter empirical process of normalized ranks as defined by (1.2). Noting first that we have

$$\beta_n(s,t) = \alpha_n(U_n(s),t) - \frac{[nt]}{n}\alpha_n(U_n(s),1),$$

where $U_n(\cdot)$ is the empirical quantile function of (4.1), which already shows that $\beta_n(s,t)$ has, as it is, a two-time parameter bridge type structure, though random in s, we then prove approximations for the $\beta_n(s,t)/q(t)$ process which are like those of the $(\alpha_n(s,t) - t\alpha_n(s,1))/q(t)$ process.

The two-time parameter empirical process $\gamma_n(s,t)$ of sequential ranks defined by (1.3) is ready-made for analyzing chronological observations. In Section 5 we study approximations of the weighted versions $\gamma_n(s,t)/q(t)$ with continuous F, and those of their tied down at $t=1$ forms $(\gamma_n(s,t) - \gamma_n(s,1))/q(t)$ in Section 6, for the sake of constructing tools, statistics for sequential studies of chronologically ordered finite samples of size n, as $n \to \infty$. In the light of the representations $\xi_k \stackrel{\mathcal{D}}{=} [kU_k]/k$, $k=1,2,\ldots$, of sequential ranks in terms of independent uniform-(0,1) random variables $U_1, U_2, \ldots$, (cf (5.1)) it is reasonable that we get the same Gaussian weighted approximations like in case of the processes in (1.1) and their tied down at $t=1$ versions, respectively.

All in all, our considerations and results in Sections 2–6 amount to a unified treatment, under the assumption that $X_1, \ldots, X_n$ are i.i.d.r.v.'s, of the two-time parameter empirical processes of these observations, their ranks and sequential ranks, respectively, in weighted supremum and L_p-metrics. The same can be said about our study of these processes under contiguous alternatives in Section 7, where we succeed in transforming our i.i.d. based asymptotics into contiguous asymptotics with their respective optimal classes of weight functions.

In Section 8 we summarize similarly optimal weighted asymptotics of multi-time parameter empirical processes of arbitrary distributions in $\mathbb{R}^d$, $d \geq 1$, in supremum and L_p-metrics. As a consequence of these results we should note, of course, that all the results of Sections 2 and 3 can be restated in terms of the $\zeta_n(x,t)$ process of (1.1) with an arbitrary distribution function F on $\mathbb{R} \equiv \mathbb{R}^1$. Indeed, the assumption of continuity of F in Sections 2–7 is only for the sake of a unified treatment of the three sequential empirical processes of (1.1), (1.2) and (1.3), respectively.

2. Weighted empirical processes based on observations

Let $X_1, X_2, \ldots$ be i.i.d.r.v.'s with distribution function F and define *the empirical distribution function* of the sample $X_1, \ldots, X_n$ by

$$F_n(x) = \frac{1}{n} \sum_{i=1}^n \mathbf{1}\{X_i \leq x\} .$$

An equivalent definition of the empirical distribution function F_n can be given in terms of the *order statistics* $X_{1,n} \leq X_{2,n} \leq \cdots \leq X_{n,n}$ of the random sample $X_1, \ldots, X_n$ as follows:

$$F_n(x) = \begin{cases} 0, & X_{1,n} > x \\ \frac{k}{n}, & X_{k,n} \leq x < X_{k+1,n}, \; k=1,2,\ldots,n-1, \\ 1, & X_{n,n} \leq x . \end{cases}$$

Clearly, for every fixed x, $F_n(x)$ is the relative frequency of successes in a Bernoulli sequence of trials with $EF_n(x) = F(x)$ and Var $F_n(x) = \frac{1}{n}F(x)(1-F(x))$. Consequently, by the classical strong law of large numbers,

$$F_n(x) \stackrel{a.s.}{\to} F(x), \quad \text{for } x \text{ fixed}.$$

Hence, using the language of statistics, $F_n(x)$ is an unbiased and strongly consistent estimator of $F(x)$ for each fixed x. Viewing $\{F_n(x); -\infty < x < \infty\}$ as a stochastic process, its sample functions are distribution functions, and it is of great importance that $F(x)$ can be uniformly estimated by this process with probability one, as proved by Cantelli (1933) and Glivenko (1933). Namely, for any F, we have

$$\sup_{-\infty < x < \infty} |F_n(x) - F(x)| \stackrel{a.s.}{\to} 0, \tag{2.1}$$

which tells us that, sampling ad infinitum, $F(x)$ can be uniquely determined with probability one.

From a practical point of view it is also of interest to study the rate of convergence in (2.1). Towards this end we define *the empirical process*

$$\zeta_n(x) = n^{1/2}(F_n(x) - F(x)), \quad -\infty < x < \infty.$$

For each fixed x, one has immediately the central limit theorem:

$$\zeta_n(x) \stackrel{\mathscr{D}}{\to} N(0, F(x)(1-F(x))).$$

As to the rate of convergence in (2.1), we have the following result of Kolmogorov (1933) and Smirnov (1939).

THEOREM 2.A. *If $F(x)$ is a continuous distribution function, then*

$$P\left\{\sup_{-\infty < x < \infty} |\zeta_n(x)| \leq y\right\} \to K(y),$$

where

$$K(y) = \begin{cases} \sum_{k=-\infty}^{\infty} (-1)^k e^{-2k^2 y^2}, & y > 0 \\ 0, & \text{otherwise}, \end{cases}$$

and

$$P\left\{\sup_{-\infty < x < \infty} \zeta_n(x) \leq y\right\} \to S(y),$$

where

$$S(y) = \begin{cases} 1 - e^{-2y^2}, & y > 0 \\ 0, & \text{otherwise}. \end{cases}$$

Let $U_i = F(X_i)$, $i = 1, 2, \ldots$. Then U_i are uniform-$(0,1)$ random variables, provided that $F(\cdot)$ is continuous. Let now $E_n(s)$ be the empirical distribution function of the sample $U_1, \ldots, U_n$, and denote the resulting empirical process in this case by

$$\alpha_n(s) = \sqrt{n}(E_n(s) - s), \quad 0 \leq s \leq 1 ,$$

that is to say

$$\alpha_n(s) = n^{-1/2} \sum_{i=1}^{n} (\mathbf{1}\{U_i \leq s\} - s), \quad 0 \leq s \leq 1 .$$

Then $E\alpha_n(s) = 0$, and the covariance function of the process $\{\alpha_n(s); 0 \leq s \leq 1\}$ is

$$\rho(s_1, s_2) = E\alpha_n(s_1)\alpha_n(s_2) = s_1 \wedge s_2 - s_1 s_2 ,$$

which coincides with that of a Brownian bridge $\{B(s); 0 \leq s \leq 1\}$. The weak convergence of $\alpha_n(\cdot)$ to a Brownian bridge was proved by Donsker (1952). The first strong approximation of the empirical process by a sequence of Brownian bridges is due to Brillinger (1969). The following theorem, which gives the optimal rate of approximation, is due to Komlós, Major and Tusnády (1975), and then to Bretagnolle and Massart (1989) with the herewith given constants.

THEOREM 2.B. *Given independent uniform-$(0, 1)$ random variables $U_1, U_2, \ldots$, there exists a sequence of Brownian bridges $\{B_n(s); 0 \leq s \leq 1\}$ such that for all n and x we have*

$$P\left\{n^{-1/2} \sup_{0 \leq s \leq 1} |\alpha_n(s) - B_n(s)| > x + 12 \log n\right\} \leq 2 \exp(-x/6) .$$

Consequently

$$\sup_{0 \leq s \leq 1} |\alpha_n(s) - B_n(s)| \stackrel{a.s.}{=} O(n^{-1/2} \log n) .$$

Rényi (1953) studied the asymptotic distributions of statistics like

$$\sup_{a \leq s \leq b} \alpha_n(s)/s, \quad 0 < a < b \leq 1, \text{ and } \sup_{a \leq s \leq b} \alpha_n(s)/(1-s), \quad 0 \leq a < b < 1 ,$$

as well as those of their two-sided versions. His idea of introducing these modifications of the classical Kolmogorov–Smirnov statistics was to make them more sensitive on the tails from a hypothesized continuous distribution function $s = F(x)$. In the classical Kolmogorov–Smirnov formulation the difference $|E_n(F(x)) - F(x)| = |F_n(x) - F(x)|$, where $E_n(F(x)) = F_n(x)$ is the empirical distribution function of a random sample $X_1, \ldots, X_n$ with distribution function F, is considered without taking into account the value of $F(x)$ at x. For example, the difference $|F_n(x) - F(x)| = 0.01$ at some point x where $F(x) = 0.5$ (a difference that is only 2% of the value of $F(x)$) is considered just as significant as the same difference at the point x where $F(x) = 0.01$, though the difference now is 100% of the value of $F(x)$. Rényi (1953) therefore proposed that, instead of the absolute deviations $|F_n(x) - F(x)|$, the relative values $|F_n(x) - F(x)|/F(x)$ should be considered. In the same spirit, Anderson and Darling (1952) studied the asymptotics of statistics like

$$\sup_{a \le s \le b} \alpha_n(s)/(s(1-s))^{1/2}, \quad 0 < a < b < 1,$$

and their two-sided versions, measuring distances in terms of their standard deviation.

Considerations like these weighted statistics have led to posing the problem of weak convergence of the empirical processes α_n in $D[0,1]$ in the so-called $\|\cdot/q\|$-metrics (weighted supremum metrics) to a Brownian bridge process B. Namely, if we let $\mathscr{Q}^*$ denote the class of positive functions on $(0,1)$ that are bounded away from zero on $(\delta, 1-\delta)$ for all $0 < \delta < 1/2$, nondecreasing in a neighbourhood of 0 and nonincreasing in a neighbourhood of 1, then one would want to characterize those functions $q \in \mathscr{Q}^*$ for which we have

$$\alpha_n(s)/q(s) \xrightarrow{\mathscr{D}} B(s)/q(s), \quad 0 < s < 1. \tag{2.2}$$

Assuming that $q \in \mathscr{Q}^*$ and is continuous, such a class of functions was first characterized by Chibisov (1964) for $q(s)$ resp. $q(1-s)$ regularly varying at 0 of order $0 < \alpha \le 1/2$, and then by O'Reilly (1974), assuming only the continuity of $q \in \mathscr{Q}^*$. From their theorem one can conclude that we have (2.2) if and only if for a standard Wiener process W we have

$$\limsup_{s \downarrow 0} |W(s)|/q(s) = \limsup_{s \uparrow 1} |W(s)|/q(1-s) = 0 \quad \text{a.s.} \tag{2.3}$$

This characterization of the weighted weak convergence in (2.2) is somewhat surprising at the first sight, and there have been unsuccessful attempts in the literature to reprove this Chibisov–O'Reilly theorem. For comments on these, we refer to Section 2 in Csörgő, Csörgő and Horváth (1986).

CsCsHM (1986) obtained a weighted approximation of the uniform empirical and quantile processes by a sequence of Brownian bridges in supremum norm, for the optimal class of weight functions. Weighted approximations in L_p-metrics, $0 < p < \infty$ for such processes were proven by Csörgő, Horváth and Shao (1993). For further discussions on this subject we refer to CsCsHM (1986), Shorack and Wellner (1986), and to Csörgő and Horváth (1993), as well as to the references in these works. The proof of the following theorem can be found in CsCsHM (1986).

Let $\mathscr{Q}^*$ be the class of positive functions on $(0,1)$, i.e., such that $\inf_{\delta \le s \le 1-\delta} q(s) > 0$ for all $\delta \in (0, 1/2)$, which are nondecreasing near zero and nonincreasing near one. We define also the integrals

$$E^*(q,c) = \int_0^1 (s(1-s))^{-3/2} q(s) \exp(-c(s(1-s))^{-1} q^2(s)) \, ds$$

and

$$I^*(q,c) = \int_0^1 (s(1-s))^{-1} \exp(-cq^2(s)/(s(1-s))) \, ds, \quad c > 0 .$$

THEOREM 2.C. Let $U_1, U_2, \ldots$ be independent uniform-$(0,1)$ random variables and $q \in \mathcal{Q}^*$.

(a) We can define a sequence of Brownian bridges $\{B_n(s); 0 \leq s \leq 1\}$ such that, as $n \to \infty$,

$$\sup_{0 < s < 1} |\alpha_n(s) - B_n(s)|/q(s) = o_P(1)$$

if and only if $I^*(q,c) < \infty$ for all $c > 0$.

(b) With $\{B(s); 0 \leq s \leq 1\}$ being a Brownian bridge, as $n \to \infty$, we have

$$\sup_{0 < s < 1} |\alpha_n(s)|/q(s) \xrightarrow{\mathcal{D}} \sup_{0 < s < 1} |B(s)|/q(s)$$

if and only if $I^*(q,c) < \infty$ for some $c > 0$.

We note that, according to Theorems 3.3 and 3.4 of CsCsHM (1986), Theorem 2.C can be stated equivalently in terms of the integral $E^*(q,c)$ as well.

The following description for the L_p-functionals of weighted empirical processes is due to Csörgő, Horváth and Shao (1993).

THEOREM 2.D. We assume that $0 < p < \infty$ and q is a positive function on $(0,1)$. Then the following statements are equivalent.

(i) We have

$$\int_0^1 (s(1-s))^{p/2}/q(s) \, ds < \infty .$$

(ii) There is a sequence of Brownian bridges $\{B_n(s); 0 \leq s \leq 1\}$ such that, as $n \to \infty$,

$$\int_0^1 |\alpha_n(s) - B_n(s)|^p/q(s) \, ds = o_P(1) .$$

(iii) We have, as $n \to \infty$,

$$\int_0^1 |\alpha_n(s)|^p/q(s) \, ds \xrightarrow{\mathcal{D}} \int_0^1 |B(s)|^p/q(s) \, ds$$

where $\{B(s); 0 \leq s \leq 1\}$ is a Brownian bridge.

Similar results hold true for uniform quantile processes (cf CsCsHM (1986) and Csörgő, Horváth and Shao (1993), and for immediate reference we refer to Theorems 3.A and 3.B in Csörgő and Szyszkowicz (1998) in this volume).

Kiefer (1970) was the first one to call attention to the fact that the empirical process should be viewed as a two-time parameter process and that it should be approximated almost surely by an appropriate two-time parameter Gaussian process. He also gave a solution to this problem and proved the first two-time parameter strong approximation theorem for $\alpha_n(s)$, (cf Kiefer (1972)) in terms of a two-time parameter Gaussian process.

THEOREM 2.E. *Given independent uniform-$(0,1)$ random variables $U_1, U_2, \ldots$, there exists a two-time parameter Gaussian process $\{K(s,t);\ 0 \leq s \leq 1,\ t \geq 0\}$ with mean zero and covariance function*

$$EK(s_1, t_1)K(s_2, t_2) = (s_1 \wedge s_2 - s_1 s_2)(t_1 \wedge t_2) \tag{2.4}$$

such that

$$\sup_{0 \leq s \leq 1} |n^{1/2}\alpha_n(s) - K(s,n)| \stackrel{a.s.}{=} O(n^{1/3}(\log n)^{2/3})\ .$$

Let $X_1, X_2, \ldots$ be independent identically distributed random variables with distribution function F. We consider the two-time parameter empirical process

$$\zeta_n(x,t) = n^{-1/2} \sum_{i=1}^{[nt]} \left(\mathbf{1}\{X_i \leq x\} - F(x)\right), \quad x \in \mathbb{R},\ 0 \leq t \leq 1\ .$$

We assume throughout that F is continuous and write, without loss of generality, in the statements and proofs of our theorems

$$\zeta_n\left(F^{-1}(s), t\right) = n^{-1/2} \sum_{i=1}^{[nt]} \left(\mathbf{1}\{F(X_i) \leq s\} - s\right) =: \alpha_n(s, t), \quad 0 \leq s, t \leq 1\ .$$

The limiting distribution of the process $\alpha_n(s,t)$, as $n \to \infty$, is that of a Kiefer process $K(\cdot,\cdot)$, i.e., a separable two-time parameter Gaussian process $\{K(s,t);\ 0 \leq s \leq 1,\ t \geq 0\}$ with mean zero and covariance function as in (2.4). For weak convergence of $\alpha_n(s,t)$, we refer to Bickel and Wichura (1971) and Müller (1970). The best available strong approximation of $\alpha_n(s,t)$ is due to Komlós, Major, and Tusnády [KMT] (1975), and then to Bonvalot and Castelle (1991) with the herewith given constants.

THEOREM 2.F. *Given independent uniform-$(0,1)$ random variables $U_1, U_2, \ldots$, there exists a Kiefer process $\{K(s,t);\ 0 \leq s \leq 1,\ t \geq 0\}$ such that*

$$P\left\{\sup_{1 \leq k \leq n} \sup_{0 \leq s \leq 1} |k^{1/2}\alpha_k(s) - K(s,k)| > (x + 76\mathrm{LOG}n)\mathrm{LOG}n\right\}$$
$$\leq 2.028 \exp(-x/41)$$

with $\mathrm{LOG}n = \max(\log n, \log 4)$. *Consequently*

$$\sup_{0 \leq s \leq 1} |n^{1/2}\alpha_n(s) - K(s,n)| \stackrel{a.s.}{=} O(\log^2 n)\ .$$

Picard (1985), Deshayes and Picard (1986), and Szyszkowicz (1991b, 1994) study the problem under what conditions does weak convergence continue to hold for the weighted two-time parameter empirical processes $\alpha_n(s,t)/q(t)$, where $q(t)$ is a nonnegative function on $(0,1]$ approaching zero as $t \to 0$.

Here we present the optimal conditions for weighted asymptotics of the processes $\alpha_n(s,t)$ in supremum and L_p-metrics.

We will assume throughout, without loss of generality, that all random variables and stochastic processes introduced so far and later on, are defined on the same probability space (cf Lemma 4.4.4 of Csörgő and Révész (1981) and Section A.2 in Csörgő and Horváth (1993)).

Let $\mathscr{Q}$ be the class of positive functions q on $(0,1]$, i.e., such that $\inf_{\delta \leq t \leq 1} q(t) > 0$ for all $0 < \delta < 1$, which are non-decreasing in a neighborhood of zero, and let

$$I(q,c) = \int_0^1 t^{-1} \exp(-ct^{-1}q^2(t)) \, dt, \quad c > 0.$$

THEOREM 2.1. Let $q \in \mathscr{Q}$. Then one can construct a Kiefer process $\{K(s,t);\ 0 \leq s \leq 1,\ t \geq 0\}$ such that, as $n \to \infty$, we have

(a) $\quad \sup_{0 < t \leq 1} \sup_{0 \leq s \leq 1} |\alpha_n(s,t) - n^{-1/2} K(s,nt)|/q(t) = o_P(1) \qquad (2.5)$

if and only if $I(q,c) < \infty$ for all $c > 0$,

(b) $\quad \sup_{0 < t \leq 1} \sup_{0 \leq s \leq 1} |\alpha_n(s,t) - n^{-1/2} K(s,nt)|/q(t) = O_P(1) \qquad (2.6)$

if and only if $I(q,c) < \infty$ for some $c > 0$,

(c) $\quad \sup_{0 < t \leq 1} \sup_{0 \leq s \leq 1} |\alpha_n(s,t)|/q(t) \xrightarrow{\mathscr{D}} \sup_{0 < t \leq 1} \sup_{0 \leq s \leq 1} |K(s,t)|/q(t) \qquad (2.7)$

if and only if $I(q,c) < \infty$ for some $c > 0$, where $\{K(s,t);\ 0 \leq s,t \leq 1\}$ is a Kiefer process.

COROLLARY 2.1. Let $q \in \mathscr{Q}$ and $\{K(s,t);\ 0 \leq s,t \leq 1\}$ be a Kiefer process. Then, as $n \to \infty$, we have

$$\alpha_n(s,t)/q(t) \xrightarrow{\mathscr{D}} K(s,t)/q(t) \quad \text{in} \quad D[0,1]^2 \qquad (2.8)$$

if and only if $I(q,c) < \infty$ for all $c > 0$.

REMARK 2.1. Throughout this paper weak convergence statements on Skorohod spaces are stated as corollaries to approximations in probability. Naturally, when talking about weighted weak convergence on such spaces, we will always assume that the weights are c.d.l.g. functions.

REMARK 2.2. Obviously, Corollary 2.1 implies convergence in distribution of any continuous in sup-norm functional of $\alpha_n(s,t)/q(t)$ to the corresponding functional of $K(s,t)/q(t)$ with $q \in \mathcal{Q}$ and such that $I(q,c) < \infty$ for all $c > 0$. However, for the sup-functional itself, by Theorem 2.1(c), the class of admissible weight functions $q(t)$ for the convergence in distribution of $\sup_{0<t\leq 1} \sup_{0\leq s\leq 1} |\alpha_n(s,t)|/q(t)$ is bigger than that for the weak convergence of the whole process $\alpha_n(s,t)/q(t)$ in supremum norm. For example, as $n \to \infty$, we have

$$\sup_{0<t\leq 1} \sup_{0\leq s\leq 1} |\alpha_n(s,t)|/(t\log\log((1/t) \vee 3))^{1/2}$$
$$\xrightarrow{\mathcal{D}} \sup_{0<t\leq 1} \sup_{0\leq s\leq 1} |K(s,t)|/(t\log\log((1/t) \vee 3))^{1/2},$$

though weak convergence in supremum norm with this weight function is impossible.

Theorem 2.1 and Corollary 2.1 were proved by Szyszkowicz (1994) under stronger conditions. Namely, (2.5) and (2.8) were obtained for weight functions q which are positive on $(0,1]$ and such that

$$\lim_{t\downarrow 0} (t\log\log 1/t)^{1/2}/q(t) = 0 \tag{2.A}$$

while (2.6) and (2.7) were proved for positive weight functions q and such that

$$\lim_{t\downarrow 0} (t\log\log 1/t)^{1/2}/q(t) < \infty. \tag{2.B}$$

As mentioned in Szyszkowicz (1994), using upper-lower class results (tests for upper and lower functions) for suprema of Kiefer processes, we could also state (2.5), (2.6) and (2.7) of Theorem 2.1 under seemingly different conditions. Namely, assuming that $q \in C(0,1]$ and $q(t)/t^{1/2}$ is nonincreasing near zero, by Theorem 5.2 of Adler and Brown (1986) (cf also Chung (1949) and Kiefer (1961)), we can obtain (2.5) under the condition

$$\mathcal{I}(q,c) < \infty \quad \text{for all} \quad c > 0, \tag{2.C}$$

while (2.6) and (2.7) may be obtained under the condition

$$\mathcal{I}(q,c) < \infty \quad \text{for some} \quad c > 0, \tag{2.D}$$

where

$$\mathcal{I}(q,c) = \int_0^1 t^{-2} q^2(t) \exp(-ct^{-1} q^2(t)) \, dt, \quad c > 0.$$

It follows, however, from Lemma 2.1 and Theorem 2.2 of Csörgő, Horváth and Szyszkowicz (1994) that, assuming the above mentioned monotonicity of $q(t)/t^{1/2}$, (2.A) is equivalent to (2.C), and (2.B) is equivalent to (2.D) (cf the lines right after Theorem 2.2 of the just mentioned paper).

Theorem 2.1 here is an improvement of Theorem 2.1 in Szyszkowicz (1994) in that we obtain (2.5), (2.6) and (2.7) for their largest possible respective classes of weight functions. This improvement is due to the following integral test for suprema of Kiefer processes which follows from Theorem 2.1 of Csörgő, Horváth and Szyszkowicz (1994), where the additional conditions of continuity of q and monotonicity of $q(t)/t^{1/2}$ near zero, inherited from the Adler and Brown (1986) test for upper and lower functions for suprema of Kiefer processes, are dropped.

LEMMA 2.A. *Let $q \in \mathcal{Q}$ and $\{K(s,t); 0 \leq s \leq 1, t \geq 0\}$ be a Kiefer process. Then we have*

(a) $\quad \limsup_{t \downarrow 0} \sup_{0 \leq s \leq 1} |K(s,t)|/q(t) < \infty \quad$ a.s.

if and only if $I(q,c) < \infty$ for some $c > 0$,

(b) $\quad \lim_{t \downarrow 0} \sup_{0 \leq s \leq 1} |K(s,t)|/q(t) = 0 \quad$ a.s.

if and only if $I(q,c) < \infty$ for all $c > 0$.

REMARK 2.3. We note that requiring that $q \in \mathcal{Q}$ be such that $q(t)/t^{1/2}$ is nonincreasing near zero is a considerable restriction on the class of possible weight functions $q \in \mathcal{Q}$ for which we have $I(q,c) < \infty$ for all $c > 0$. For examples and further discussion along these lines we refer to Csörgő, Csörgő and Horváth (1986) (cf also Proposition 4.1.1 and Examples 4.1.1 and 4.1.2 in Csörgő and Horváth (1993)). We note also that assuming q to be nondecreasing near zero, i.e., that $q \in \mathcal{Q}$, is not a restriction at all, since if q is decreasing near zero, then

$$\lim_{t \downarrow 0} \sup_{0 \leq s \leq 1} |K(s,t)|/q(t) = 0 \quad \text{a.s.}$$

Consequently, by assuming that $q \in \mathcal{Q}$, we consider the non-trivial cases of weight functions when a test for the latter statement to be true is indeed required.

For a review of the integral $I(q,c)$ being used for characterizing the local (near zero) behaviour of a standard Wiener process, we refer to CsCsHM (1986), Csörgő, Shao and Szyszkowicz (1991) and Csörgő and Horváth (1993). We wish to emphasize that the classes of weight functions characterizing the local behaviour of a standard Wiener process and the suprema of Kiefer processes are found to be the same. Namely, on account of Lemma 2.A and Theorems 3.3 and 3.4 of CsCsHM (1986), the following results hold true.

COROLLARY 2.A. *Let $\{W(t); t \geq 0\}$ be a standard Wiener process and $\{K(s,t); 0 \leq s \leq 1, t \geq 0\}$ be a Kiefer process. Then, with $q \in \mathcal{Q}$, the following three statements are equivalent.*

(a) $I(q,c) < \infty$ for some $c > 0$,
(b) $\limsup_{t \downarrow 0} |W(t)|/q(t) < \infty$ a.s.,
(c) $\limsup_{t \downarrow 0} \sup_{0 \leq s \leq 1} |K(s,t)|/q(t) < \infty$ a.s.

COROLLARY 2.B. Let $\{W(t); t \geq 0\}$ be a standard Wiener process and $\{K(s,t); 0 \leq s \leq 1, t \geq 0\}$ be a Kiefer process. Then, with $q \in \mathscr{Q}$, the following three statements are equivalent.

(a) $I(q,c) < \infty$ for all $c > 0$,
(b) $\lim_{t \downarrow 0} |W(t)|/q(t) = 0$ a.s.,
(c) $\lim_{t \downarrow 0} \sup_{0 \leq s \leq 1} |K(s,t)|/q(t) = 0$ a.s.

PROOF OF THEOREM 2.1. Using KMT (1975) and arguing as in the proof of Theorem 2.1 of Szyszkowicz (1994), one can construct a Kiefer process $K(\cdot,\cdot)$ such that with a positive function q on $(0,1]$ and such that $\lim_{t \downarrow 0} t^{1/2}/q(t) = 0$, we have, as $n \to \infty$ (cf (2.15) in the just mentioned paper),

$$\sup_{0 < t \leq 1} \sup_{0 \leq s \leq 1} \left| \sum_{i=1}^{[nt]} (1\{F(X_i) \leq s\} - s) - \overline{K}(s, nt) \right| / n^{1/2} q(t) \stackrel{\text{a.s.}}{=} o(1), \qquad (2.9)$$

where

$$\overline{K}(s, nt) = \begin{cases} K(s, nt) & \text{for } t \in [1/n, 1], \\ 0 & \text{elsewhere} . \end{cases}$$

Since $I(q,c) < \infty$ for some $c > 0$ with $q \in \mathscr{Q}$ implies $\lim_{t \downarrow 0} t^{1/2}/q(t) = 0$ (cf Theorem 3.3 of CsCsHM (1986)), and we have for each $n \geq 1$,

$$\sup_{0 < t < 1/n} \sup_{0 \leq s \leq 1} |K(s, nt)|/n^{1/2} q(t) \stackrel{\mathscr{D}}{=} \sup_{0 < t < 1/n} \sup_{0 \leq s \leq 1} |K(s,t)|/q(t) ,$$

it is clear that the class of weight functions for obtaining (2.5) or (2.6) is determined by the local (near zero) behaviour of the supremum (in s) of a Kiefer process $\{K(s,t), 0 \leq s, t \leq 1\}$. Using now Lemma 2.A, we get

$$\sup_{0 < t < 1/n} \sup_{0 \leq s \leq 1} |K(s, nt)|/n^{1/2} q(t) = \begin{cases} o_P(1) & \text{if and only if } I(q,c) < \infty \\ & \text{for all } c > 0, \\ O_P(1) & \text{if and only if } I(q,c) < \infty \\ & \text{for some } c > 0 , \end{cases}$$

which together with (2.9) gives the "if" parts of (a) and (b) respectively. The converse parts of (a) and (b) follow from Lemma 2.A, since

$$\sup_{0 < t \leq 1} \sup_{0 \leq s \leq 1} |\alpha_n(s,t) - n^{-1/2} K(s, nt)|/q(t)$$

$$\geq \sup_{0 < t < 1/n} \sup_{0 \leq s \leq 1} |n^{-1/2} K(s, nt)|/q(t) .$$

The proof of the "if" part of (c) is similar to the proof of Theorem 2.1 (c) in Szyszkowicz (1994). For the converse part we assume that with $q \in \mathscr{Q}$ we have

$$\sup_{0 < t \leq 1} \sup_{0 \leq s \leq 1} |\alpha_n(s,t)|/q(t) \stackrel{\mathscr{D}}{\to} \sup_{0 < t \leq 1} \sup_{0 \leq s \leq 1} |K(s,t)|/q(t) ,$$

where $K(\cdot,\cdot)$ is a Kiefer process. Hence the limiting random variable is a.s. finite. Consequently Lemma 2.A implies $I(q,c) < \infty$ for some $c > 0$. □

PROOF OF COROLLARY 2.1. Obviously, with $q \in \mathcal{Q}$ and $I(q,c) < \infty$ for all $c > 0$, by Theorem 2.1(a) we have (2.8). Assuming now (2.8) to hold with $q \in \mathcal{Q}$, by Skorohod–Dudley–Wichura theorem (cf Shorack and Wellner (1986), p. 47) there exists α_n^*, $n \geq 1$, and K^* such that

$$\left(\frac{\alpha_n}{q}\right)^* \stackrel{\mathcal{D}}{=} \frac{\alpha_n}{q}, \ n \geq 1, \ \left(\frac{K}{q}\right)^* \stackrel{\mathcal{D}}{=} \frac{K}{q}$$

and

$$\sup_{0<t\leq 1}\left|\left(\frac{\alpha_n}{q}\right)^*(t) - \left(\frac{K}{q}\right)^*(t)\right| = o(1) \quad \text{a.s.}$$

Hence

$$\sup_{0<t<1/n}\left|\left(\frac{K}{q}\right)^*(t)\right| = o(1) \quad \text{a.s.} ,$$

which, by Lemma 2.A, implies $I(q,c) < \infty$ for all $c > 0$. □

Considering L_p-functionals, we obtain that the class of admissible weight functions is even bigger than for supremum functionals, via the following best possible asymptotic result.

THEOREM 2.2. Let $0 < p < \infty$ and q be a positive function on $(0,1]$. Then the following three statements are equivalent.

(a) We have

$$\int_0^1 t^{p/2}/q(t) \, dt < \infty \ . \tag{2.10}$$

(b) One can construct a Kiefer process $\{K(s,t); 0 \leq s \leq 1, t \geq 0\}$ such that, as $n \to \infty$, we have

$$\int_0^1 \int_0^1 |\alpha_n(s,t) - n^{-1/2}K(s,nt)|^p/q(t) \, ds \, dt = o_P(1) \ . \tag{2.11}$$

(c) We have, as $n \to \infty$,

$$\int_0^1 \int_0^1 |\alpha_n(s,t)|^p/q(t) \, ds \, dt \stackrel{\mathcal{D}}{\to} \int_0^1 \int_0^1 |K(s,t)|^p/q(t) \, ds \, dt \ , \tag{2.12}$$

where $\{K(s,t); 0 \leq s, t \leq 1\}$ is a Kiefer process.

The proof of Theorem 2.2 is based on KMT (1975) approximation and the test for the integrals of Gaussian processes of Csörgő, Horváth and Shao (1993).

Namely, with q being a positive function on $(0, 1]$ and such that (2.10) holds, using KMT (1975) approximation, a Kiefer process $K(\cdot, \cdot)$ can be so constructed that we obtain, as $n \to \infty$,

$$\int_{1/n}^{1} \int_{0}^{1} |\alpha_n(s,t) - n^{-1/2}K(s,nt)|^p/q(t) \, ds \, dt = o_P(1) \ .$$

Since $\alpha_n(s,t) \equiv 0$ for $0 < t < 1/n$, it remains to show that

$$\int_{0}^{1/n} \int_{0}^{1} |n^{1/2}K(s,nt)|^p/q(t) \, ds \, dt = o_P(1) \ ,$$

which holds true by the following result, which is a consequence of Theorem 2.1 of Csörgő, Horváth and Shao (1993).

LEMMA 2.B. *Let $\{K(s,t); \ 0 \leq s, t \leq 1\}$ be a Kiefer process and q be a positive function on $(0,1]$. We assume that $0 < p < \infty$. Then the statements*

$$\int_{0}^{1} t^{p/2}/q(t) dt < \infty$$

and

$$\int_{0}^{1} \int_{0}^{1} |K(s,t)|^p/q(t) \, ds \, dt < \infty \quad \text{a.s.}$$

are equivalent.

For detailed proofs of Theorem 2.2 and Lemma 2.B we refer to Szyszkowicz (1993) and Csörgő and Szyszkowicz (1994).

3. "Bridge-type" two-time parameter empirical processes

Let $X_1, X_2, \ldots$ be i.i.d.r.v.'s with a continuous distribution function F. In this section we study two-time parameter empirical process tied down at $t = 1$. We define

$$\hat{\alpha}_n(s,t) = \begin{cases} n^{-1/2} \left(\sum_{i=1}^{[(n+1)t]} \mathbf{1}\{F(X_i) \leq s\} - \frac{[(n+1)t]}{n} \sum_{i=1}^{n} \mathbf{1}\{F(X_i) \leq s\} \right), \\ \qquad\qquad\qquad\qquad\qquad\qquad\qquad 0 \leq s \leq 1, \ 0 \leq t < 1 \\ 0, \qquad\qquad\qquad\qquad\qquad\qquad 0 \leq s \leq 1, \ t = 1 \ . \end{cases}$$

The approximating process of $\hat{\alpha}_n(s,t)$ is, in terms of the Kiefer process $K(s,nt)$, given by

$$\{n^{1/2}(K(s, nt) - tK(s, n)); \ 0 \leq s, \ t \leq 1\}$$
$$\stackrel{\mathcal{D}}{=} \{K(s, t) - tK(s, 1); \ 0 \leq s, \ t \leq 1\} \quad \text{for each} \ n \geq 1$$
$$\stackrel{\mathcal{D}}{=} \{\Gamma(s, t); \ 0 \leq s, \ t \leq 1\} \ ,$$

that is to say $\{\Gamma(s, t); \ 0 \leq s, t \leq 1\}$ is a mean zero separable Gaussian process with covariance function

$$E\Gamma(s_1, t_1)\Gamma(s_2, t_2) = (s_1 \wedge s_2 - s_1 s_2)(t_1 \wedge t_2 - t_1 t_2) \ , \tag{3.1}$$

which Parzen (1992) calls a pinned Brownian sheet (two-time parameter Brownian bridge). This limiting Gaussian process Γ also appeared in Hoeffding (1948) and Blum, Kiefer and Rosenblatt (1961) in the context of testing for independence, as well as in Csörgő and Horváth (1987a), proposing nonparametric procedures for changepoint problems based on the empirical process of full sample ranks.

Let $\mathcal{Q}^*$ be the class of functions $q: (0, 1) \to (0, \infty)$ which are positive, non-decreasing in a neighborhood of zero and non-increasing in a neighborhood of one, and let

$$I^*(q, c) = \int_0^1 (t(1-t))^{-1} \exp(-c(t(1-t))^{-1} q^2(t)) \, dt, \quad c > 0 \ .$$

THEOREM 3.1. Let $q \in \mathcal{Q}^*$. One can construct a Kiefer process $\{K(s, t); \ 0 \leq s \leq 1, \ t \geq 0\}$ such that with the sequence of stochastic processes $\Gamma_n(\cdot, \cdot)$,

$$\{\Gamma_n(s, t); \ 0 \leq s, t \leq 1\} = \{n^{-1/2}(K(s, nt) - tK(s, n)); \ 0 \leq s, t \leq 1\}$$
$$\stackrel{\mathcal{D}}{=} \{\Gamma(s, t); \ 0 \leq s, t \leq 1\} \quad \text{for each} \ n \geq 1 \ ,$$

as $n \to \infty$, we have

(a) $$\sup_{0 < t < 1} \sup_{0 \leq s \leq 1} |\hat{\alpha}_n(s, t) - \Gamma_n(s, t)|/q(t) = o_P(1)$$

if and only if $I^*(q, c) < \infty$ for all $c > 0$,

(b) $$\sup_{0 < t < 1} \sup_{0 \leq s \leq 1} |\hat{\alpha}_n(s, t) - \Gamma_n(s, t)|/q(t) = O_P(1)$$

if and only if $I^*(q, c) < \infty$ for some $c > 0$,

(c) $$\sup_{0 < t < 1} \sup_{0 \leq s \leq 1} |\hat{\alpha}_n(s, t)|/q(t) \stackrel{\mathcal{D}}{\to} \sup_{0 < t < 1} \sup_{0 \leq s \leq 1} |\Gamma(s, t)|/q(t)$$

if and only if $I^*(q, c) < \infty$ for some $c > 0$, where $\{\Gamma(s, t); \ 0 \leq s, t \leq 1\}$ is Gaussian process as defined in (3.1).

COROLLARY 3.1. Let $q \in \mathcal{Q}^*$ and $\{\Gamma(s,t); 0 \leq s,t \leq 1\}$ be a Gaussian process as in (3.1). Then, as $n \to \infty$, we have

$$\hat{\alpha}_n(s,t)/q(t) \xrightarrow{\mathcal{D}} \Gamma(s,t)/q(t) \quad \text{in} \quad D[0,1]^2$$

if and only if $I^*(q,c) < \infty$ for all $c > 0$.

REMARK 3.1. Remarks 2.1 and 2.2 are also valid here.

We note that Theorem 3.1 does not follow directly from Theorem 2.1 since now we consider also weight functions q such that $\lim_{t\uparrow 1} q(t) = 0$. The proof of Theorem 3.1 can be done similarly to that of Theorem 2.2 in Szyszkowicz (1994) when using Theorem 2.1 here instead of Theorem 2.1 of Szyszkowicz (1994), and the following integral test of Csörgő, Horváth and Szyszkowicz (1994).

LEMMA 3.A. Let $q \in \mathcal{Q}^*$ and $\{\Gamma(s,t); 0 \leq s,t \leq 1\}$ be a Gaussian process as in (3.1). Then we have

(a) $\quad \limsup_{t\downarrow 0} \sup_{0\leq s\leq 1} |\Gamma(s,t)|/q(t) < \infty \quad \text{a.s.}$

and

$\quad \limsup_{t\uparrow 1} \sup_{0\leq s\leq 1} |\Gamma(s,t)|/q(t) < \infty \quad \text{a.s.}$

if and only if $I^*(q,c) < \infty$ for some $c > 0$,

(b) $\quad \lim_{t\downarrow 0} \sup_{0\leq s\leq 1} |\Gamma(s,t)|/q(t) = 0 \quad \text{a.s.}$

and

$\quad \lim_{t\uparrow 1} \sup_{0\leq s\leq 1} |\Gamma(s,t)|/q(t) = 0 \quad \text{a.s.}$

if and only if $I^*(q,c) < \infty$ for all $c > 0$.

Theorem 3.1 was proved by Szyszkowicz (1994) under stronger conditions. Namely, part (a) was obtained for weight functions q which are positive on $(0,1)$ and such that

$$\begin{cases} \lim_{t\downarrow 0} \left(t(1-t)\log\log \frac{1}{t(1-t)}\right)^{1/2}\!\!\!\Big/ q(t) = 0 \\ \text{and} \\ \lim_{t\uparrow 1} \left(t(1-t)\log\log \frac{1}{t(1-t)}\right)^{1/2}\!\!\!\Big/ q(t) = 0 \end{cases} \quad (3.A^*)$$

while parts (b) and (c) were proved for positive functions q such that

$$\begin{cases} \lim_{t\downarrow 0} \left(t(1-t)\log\log \frac{1}{t(1-t)}\right)^{1/2}\!\!\!\Big/ q(t) < \infty \\ \text{and} \\ \lim_{t\uparrow 1} \left(t(1-t)\log\log \frac{1}{t(1-t)}\right)^{1/2}\!\!\!\Big/ q(t) < \infty \end{cases} \quad (3.B^*)$$

Equivalent conditions resulting from upper-lower class results (tests for upper and lower functions) for suprema of Kiefer processes as given in Theorem 5.2 of Adler and Brown (1986) (cf Chung (1949) and Kiefer (1961)) were also discussed there (cf also the lines of discussion following Remark 2.2 here, concerning conditions on weight functions). Deshayes and Picard (1986) concluded the convergence in distribution of the test statistic

$$\sup_{0<t<1} \sup_{0\leq s\leq 1} \psi(t) \frac{|\hat{\alpha}_n(s,t)|}{t(1-t)} \xrightarrow{\mathscr{D}} \sup_{0<t<1} \sup_{0\leq s\leq 1} \psi(t) \frac{|\Gamma(s,t)|}{t(1-t)}$$

with ψ, a non-negative, piecewise continuous function on the interval $(0, 1)$, under the condition $\int_0^1 \left(\frac{\psi(t)}{t(1-t)}\right)^2 dt < \infty$. Theorem 3.1 here gives the largest possible classes of weight functions for asymptotics of $\hat{\alpha}_n(s,t)$ in weighted supremum metrics. In particular, it follows from Theorem 3.1(c) that

$$\sup_{0<t<1} \sup_{0\leq s\leq 1} |\hat{\alpha}_n(s,t)| \Big/ \left(t(1-t)\log\log\frac{1}{t(1-t)}\right)^{1/2}$$

$$\xrightarrow{\mathscr{D}} \sup_{0<t<1} \sup_{0\leq s\leq 1} |\Gamma(s,t)| \Big/ \left(t(1-t)\log\log\frac{1}{t(1-t)}\right)^{1/2} . \qquad (3.2)$$

Next we present L_p-approximation and the convergence of L_p-functionals for the optimal class of weight functions, which is bigger than that in the case of sup-norm asymptotics.

THEOREM 3.2. Let $0 < p < \infty$ and q be a positive function on $(0,1)$. Then the following three statements are equivalent.

(a) We have

$$\int_0^1 (t(1-t))^{p/2}/q(t) \, dt < \infty \ .$$

(b) There exists a Kiefer process $\{K(s,t); 0 \leq s \leq 1, t \geq 0\}$ such that with the sequence of stochastic processes $\Gamma_n(\cdot,\cdot)$

$$\{\Gamma_n(s,t); 0 \leq s, t \leq 1\} = \left\{n^{1/2}(K(s,nt) - tK(s,n)); 0 \leq s, t \leq 1\right\}$$

$$\stackrel{\mathscr{D}}{=} \{\Gamma(s,t); 0 \leq s, t \leq 1\} \quad \text{for each } n \geq 1 \ ,$$

as $n \to \infty$, we have

$$\int_0^1 \int_0^1 |\hat{\alpha}_n(s,t) - \Gamma_n(s,t)|^p/q(t) \, ds \, dt = o_P(1) \ .$$

(c) We have, as $n \to \infty$,

$$\int_0^1 \int_0^1 |\hat{\alpha}_n(s,t)|^p/q(t) \, ds \, dt \xrightarrow{\mathscr{D}} \int_0^1 \int_0^1 |\Gamma(s,t)|^p/q(t) \, ds \, dt \ ,$$

where $\{\Gamma(s,t);\ 0 \leq s, t \leq 1\}$ is a Gaussian process as defined in (3.1).

The proof of Theorem 3.2 is based on Theorem 2.2 and the following lemma which is a consequence of Theorem 2.1 of Csörgő, Horváth and Shao (1993). For details we refer to Szyszkowicz (1993b) and Csörgő and Szyszkowicz (1994).

LEMMA 3.B. *Let $\{\Gamma(s,t); 0 \leq s, t \leq 1\}$ be a stochastic process as in (3.1) and q be a positive function on $(0,1)$. We assume that $0 < p < \infty$. Then the statements*

$$\int_0^1 (t(1-t))^{p/2}/q(t)\, dt < \infty$$

and

$$\int_0^1 \int_0^1 |\Gamma(s,t)|^p/q(t)\, ds\, dt < \infty \quad \text{a.s.}$$

are equivalent.

We note that the class of admissible weight functions for the convergence in distribution of L_p-functionals of the processes $\hat{\alpha}_n(s,t)/q(t)$ is considerably bigger than that for the convergence in distribution of supremum functionals of this process. For example, considering L_1-functionals it follows from Theorem 3.2 that, as $n \to \infty$,

$$\int_0^1 \int_0^1 |\hat{\alpha}_n(s,t)|/(t(1-t))\, ds\, dt \tag{3.3}$$
$$\xrightarrow{\mathscr{L}} \int_0^1 \int_0^1 |K(s,t) - tK(s,1)|/(t(1-t))\, ds\, dt ,$$

where $\{K(s,t);\ 0 \leq s, t \leq 1\}$ is a Kiefer process (cf (2.4)). Moreover, according to Theorem 3.2, in (3.3) we can even have $(t(1-t))^\nu$ with any $\nu < 3/2$ instead of $t(1-t)$.

We note that $|\hat{\alpha}_n(s,t)|/t(1-t)$ is essentially the sequence of stochastic processes of (1.8), rewritten a bit for the sake of theorem proving.

In order to better relate the result in (3.3) to the change-point problem as summarized by the sequence of stochastic processes in (1.8), we put $s = F(x)$ and consider $\hat{\alpha}_n(F(x), t)$. Consequently, we obtain the sequence of stochastic processes

$$\{\hat{\alpha}_n(F(x), t);\ x \in \mathbb{R},\ 0 \leq t \leq 1\}$$
$$= \{\hat{\zeta}_n(x,t);\ x \in \mathbb{R},\ 0 \leq t \leq 1\}$$
$$:= \begin{cases} n^{-1/2}\left(\sum_{i=1}^{[(n+1)t]} \mathbf{1}\{X_i \leq x\} - \frac{[(n+1)t]}{n}\sum_{i=1}^{n}\mathbf{1}\{X_i \leq x\}\right), & x \in \mathbb{R},\ 0 \leq t < 1 , \\ 0, & x \in \mathbb{R},\ t = 1,\ n = 1, 2, \ldots , \end{cases}$$

which, unlike $\zeta_n(x,t)$ (cf (1.1)), does not depend on the possibly unknown distribution function F. Indeed, the definition of $\hat{\alpha}_n(s,t)$ is for the convenience of theorem proving only; all the weighted supremum norm results presented here (and those of Szyszkowicz (1991, 1994)) remain true of course under the transformation $s \to F(x)$. In particular, based on the processes in (1.8), in terms of $\hat{\alpha}_n(F(x),t) = \hat{\zeta}_n(x,t)$, as $n \to \infty$, we have, for example (cf (3.2)),

$$\sup_{0<t<1} \sup_{x \in \mathbb{R}} |\hat{\zeta}_n(x,t)| / \left(t(1-t)\log\log\frac{1}{t(1-t)}\right)^{1/2}$$
$$\xrightarrow{\mathscr{L}} \sup_{0<t<1} \sup_{0\leq s\leq 1} |K(s,t) - tK(s,1)| / \left(t(1-t)\log\log\frac{1}{t(1-t)}\right)^{1/2},$$
(3.5)

i.e., a sequence of statistics (computable) converges in distribution to a distribution free (not a function of F) random variable.

On the other hand, (3.3) via $s \to F(x)$ translates into

$$\int_0^1 \int_{-\infty}^{\infty} |\hat{\zeta}_n(x,t)| / (t(1-t)) dF(x) \, dt$$
$$\xrightarrow{\mathscr{L}} \int_0^1 \int_0^1 |K(s,t) - tK(s,1)| / (t(1-t)) \, ds \, dt ,$$
(3.6)

i.e., now we have a sequence of "statistics" (incomputable, unless F is specified) converging to a distribution free (of F) random variable. Hence, in order to make our optimal asymptotic results on $\hat{\zeta}_n$ in weighted L_p-norms more relevant to change-point analysis, it is desirable to replace the dF measure of integration in the empirical parts of the theorems, like, for example, on the left hand side of (3.6), by the empirical dF_n. This would make the appropriate empirical left hand side L_p-functionals computable statistics, while keeping the right hand side Gaussian ones distribution free (of F). This program of replacing dF by dF_n, when desirable for computational purposes, can be carried out by combining appropriate parts of our proofs with that of Corollary 5.6.4 of Csörgő and Révész (1981, pp. 186–188).

Another way to deal with $\hat{\alpha}_n(F(x),t) = \hat{\zeta}_n(x,t)$ in weighted L_p-norms is to replace the measure of integration dF on the left hand side of (3.6) by dx, which will result in having (cf Corollary 3.2 below)

$$\int_0^1 \int_{-\infty}^{\infty} |\hat{\zeta}_n(x,t)| / (t(1-t)) \, dx \, dt$$
$$\xrightarrow{\mathscr{L}} \int_0^1 \int_{-\infty}^{\infty} |K(F(x),t) - tK(F(x),1)| / (t(1-t)) \, dx \, dt ,$$
(3.7)

on assuming a bit more than two moments for F. That is to say, we now have a sequence of statistics (computable) converging to a non distribution free (a function of F) Gaussian random variable whose distribution can be simulated for each unknown F via repeated large samples, or by bootstrapping a given one.

In order to obtain convergence of statistics like those in (3.7), first we need to formulate results for the $\alpha_n(F(x), t)$ process.

Let $\zeta_n(x, t)$ be as in (1.1), i.e.,

$$\{\zeta_n(x, t); x \in \mathbb{R}, \ 0 \leq t \leq 1\} = \{\alpha_n(F(x), t); \ x \in \mathbb{R}, \ 0 \leq t \leq 1\},$$

where F is a continuous distribution function. We note that $\{K(F(x), t); x \in \mathbb{R}, 0 \leq t \leq 1\}$ is a Kiefer process, i.e., a two-time parameter separable Gaussian process with mean zero and covariance function

$$\mathrm{E}K(F(x_1), t_1)K(F(x_2), t_2) = (F(x_1) \wedge F(x_2) - F(x_1)F(x_2))(t_1 \wedge t_2).$$

The analog of Theorem 2.2 for $\zeta_n(\cdot, \cdot)$ process can be formulated as follows.

THEOREM 3.3. *Let q be a positive function on $(0, 1]$ and F be a continuous distribution function. Then, with $0 < p < \infty$, the following three conditions are equivalent.*

(a) *We have*

$$\int_0^1 t^{p/2}/q(t) \, dt < \infty \quad \text{and} \quad \int_{-\infty}^\infty (F(x)(1 - F(x)))^{p/2} \, dx < \infty.$$

(b) *One can construct a Kiefer process $\{K(F(x), t); x \in \mathbb{R}, t \geq 0\}$ such that, as $n \to \infty$, we have*

$$\int_0^1 \int_{-\infty}^\infty |\zeta_n(x, t) - n^{-1/2}K(F(x), nt)|^p/q(t) \, dx \, dt = o_P(1).$$

(c) *We have, as $n \to \infty$,*

$$\int_0^1 \int_{-\infty}^\infty |\zeta_n(x, t)|^p/q(t) \, dx \, dt \xrightarrow{\mathscr{L}} \int_0^1 \int_{-\infty}^\infty |K(F(x), t)|^p/q(t) \, dx \, dt,$$

where $\{K(F(x), t); x \in \mathbb{R}, \ 0 \leq t \leq 1\}$ is a Kiefer process.

The proof of Theorem 3.3 is like that of Theorem 2.2, where instead of Lemma 2.B we use the following result.

LEMMA 3.1. *Let $\{K(F(x), t); x \in \mathbb{R}, \ 0 \leq t \leq 1\}$ be a Kiefer process and q be a positive function on $(0, 1]$. We assume that $0 < p < \infty$ and F is a continuous distribution function. Then the statements*

$$\int_0^1 t^{p/2}/q(t) \int_{-\infty}^\infty (F(x)(1 - F(x)))^{p/2} \, dx \, dt < \infty$$

and

$$\int_0^1 \int_{-\infty}^{\infty} |K(F(x),t)|^p/q(t)\, dx\, dt < \infty \quad \text{a.s.}$$

are equivalent.

PROOF. Arguing similarily as in the proof of Lemma 2.B (cf Szyszkowicz (1993b) or Csörgő and Szyszkowicz (1994)), we have

$$E\left(\int_{-\infty}^{\infty} |K(F(x),t)|\, dx\right)^2 = t^p E|N(0,1)|^{2p} \left(\int_{-\infty}^{\infty} (F(x)(1-F(x)))^{p/2}\, dx\right)^2$$

$$= \left(E\int_{-\infty}^{\infty} |K(F(x),t)|^p\, dx\right)^2$$

and, consequently, the condition of Theorem 2.1 of Csörgő, Horváth and Shao (1993) is satisfied with $r=2$. Hence

$$\int_0^1 \int_{-\infty}^{\infty} |K(F(x),t)|^p/q(t)\, dx\, dt < \infty \quad \text{a.s.}$$

if and only if

$$\int_0^1 \int_{-\infty}^{\infty} E|N(0,1)|^p (tF(x)(1-F(x)))^{p/2}/q(t)\, dx\, dt < \infty \;,$$

i.e., if and only if

$$\int_0^1 \int_{-\infty}^{\infty} (tF(x)(1-F(x)))^{p/2}/q(t)\, dx\, dt < \infty \;. \quad \square$$

Considering the sequence of stochastic processes as in (3.4), we have the following result.

Let $\Gamma(\cdot,\cdot)$ be a Gaussian process as in (3.1) and $F(\cdot)$ be a continuous distribution function. We note that $\{\Gamma(F(x),t); x \in \mathbb{R}, 0 \le t \le 1\}$ is a separable Gaussian process with mean zero and covariance function

$$E\Gamma(F(x_1),t_1)\Gamma(F(x_2),t_2) = (F(x_1) \wedge F(x_2) - F(x_1)F(x_2))(t_1 \wedge t_2 - t_1 t_2) \;. \tag{3.8}$$

THEOREM 3.4. Let q be a positive function on $(0,1)$ and F be a continuous distribution function. Then, with $0 < p < \infty$, the following three conditions are equivalent.

(a) We have

$$\int_0^1 (t(1-t))^{p/2}/q(t)\, dt < \infty \quad \text{and} \quad \int_{-\infty}^{\infty} (F(x)(1-F(x)))^{p/2}\, dx < \infty \;.$$

(b) One can construct a Kiefer process $\{K(F(x),t); x \in \mathbb{R}, t \ge 0\}$ such that with the sequence of stochastic processes $\Gamma_n(\cdot,\cdot)$

$$\{\Gamma_n(F(x),t); x \in \mathbb{R}, 0 \leq t \leq 1\}$$
$$= \left\{n^{-1/2}(K(F(x),t) - tK(F(x),1)); x \in \mathbb{R}, 0 \leq t \leq 1\right\}$$
$$\stackrel{\mathscr{D}}{=} \{\Gamma(F(x),t); x \in \mathbb{R}, 0 \leq t \leq 1\} \quad \text{for each } n \geq 1,$$

as $n \to \infty$, we have

$$\int_0^1 \int_{-\infty}^{\infty} |\hat{\zeta}_n(x,t) - \Gamma_n(F(x),t)|^p / q(t) \, dx \, dt = o_P(1).$$

(c) We have, as $n \to \infty$,

$$\int_0^1 \int_{-\infty}^{\infty} |\hat{\zeta}_n(x,t)|^p / q(t) \, dx \, dt \stackrel{\mathscr{D}}{\to} \int_0^1 \int_{-\infty}^{\infty} |\Gamma(F(x),t)|^p / q(t) \, dx \, dt,$$

where $\{\Gamma(F(x),t); x \in \mathbb{R}, 0 \leq t \leq 1\}$ is a Gaussian process as in (3.8).

In order to obtain Theorem 3.4 the following result is needed.

LEMMA 3.2. Let $\{\Gamma(F(x),t); x \in \mathbb{R}, 0 \leq t \leq 1\}$ be a separable Gaussian process with mean zero and covariance function as in (3.8), where F is a continuous distribution function. We assume that $0 < p < \infty$ and q is a positive function on $(0,1)$. Then the statements

$$\int_0^1 (t(1-t))^{p/2} / q(t) \int_{-\infty}^{\infty} (F(x)(1-F(x)))^{p/2} \, dx \, dt < \infty$$

and

$$\int_0^1 \int_{-\infty}^{\infty} |\Gamma(F(x),t)|^p / q(t) \, dx \, dt < \infty \quad \text{a.s.}$$

are equivalent.

PROOF. Similar to that of Lemma 3.1. □

PROOF OF THEOREM 3.4. Similar to that of Theorem 3.2 when using Theorem 3.3 and Lemma 3.2 in lieu of Theorem 2.2 and Lemma 3.B, respectively. □

When considering L_1-functionals we note that the condition $\int_{-\infty}^{\infty} (F(x)(1-F(x)))^{1/2} \, dx < \infty$ is slightly stronger than the existence of the 2nd moment (cf Appendix of Hoeffding (1973), and Csörgő, Csörgő and Horváth (1986, p. 34)). Consequently, we have the following result.

COROLLARY 3.2. Let q be a positive function on $(0,1)$ and F be a continuous distribution function such that

$$\int_{-\infty}^{\infty} x^2 (\log(1+x))^{1+\delta} \, dF(x) < \infty, \quad \text{with any } \delta > 0.$$

Then the following three statements are equivalent.

(a) We have
$$\int_0^1 (t(1-t))^{1/2}/q(t)\, dt < \infty .$$

(b) With the sequence of stochastic processes $\Gamma_n(\cdot,\cdot)$ as in Theorem 3.4, as $n \to \infty$, we have
$$\int_0^1 \int_{-\infty}^{\infty} |\hat{\zeta}_n(x,t) - \Gamma_n(F(x),t)|/q(t)\, dx\, dt = o_P(1),$$

(c) We have, as $n \to \infty$,
$$\int_0^1 \int_{-\infty}^{\infty} |\hat{\zeta}_n(x,t)|/q(t)\, dx\, dt \xrightarrow{\mathcal{D}} \int_0^1 \int_{-\infty}^{\infty} |\Gamma(F(x),t)|/q(t)\, dx\, dt ,$$

where $\{\Gamma(F(x),t);\ x \in \mathbb{R},\ 0 \le t \le 1\}$ is a Gaussian process as in (3.8).

For example, under the conditions of Corollary 3.2 we have (3.7). Consequently, in this weighted L_1-convergence in distribution, we obtain a most natural statistic (computable) for the change-point problem as formulated in (1.8) (cf also the discussion of this problem in the Introduction).

Similar applicable statistics-type results can be formulated also in terms of the other stochastic processes considered in this paper.

4. Weighted empirical processes based on ranks

Let $X_1, X_2, \ldots$ be independent identically distributed random variables with an unknown continuous distribution function F. For each $n \ge 1$, let $R_{1n}, \ldots, R_{nn}$ denote the normalized ranks
$$R_{in} = n^{-1} \sum_{k=1}^{n} \mathbf{1}\{X_k \le X_i\}, \quad i = 1, \ldots, n ,$$

of the first n of the random variables $X_1, X_2, \ldots$. Thus, the rank of X_i, in our notation nR_{in}, is the integer among the numbers $1, \ldots, n$ which corresponds to the position of X_i in the order statistics $X_{1n} \le \cdots \le X_{nn}$ of a random sample of size $n \ge 1$. In this section we consider the asymptotic behaviour of the empirical process based on ranks with weight functions q, namely $\beta_n(s,t)/q(t)$, where
$$\beta_n(s,t) = n^{-1/2} \sum_{i=1}^{[nt]} \left(\mathbf{1}\{R_{in} \le s\} - \frac{[ns]}{n} \right), \quad 0 \le s,t \le 1 .$$

As in Section 3, let $\{\Gamma(s,t);\ 0 \le s,t \le 1\}$ be a stochastic process as defined in (3.1), and let $B(s)$ be a Brownian bridge. The following test statistics for the change-point problem were proposed by Csörgő and Horváth (1987a), where they derive their limiting distributions under the assumption of $X_1,\ldots,X_n$ being i.i.d.r.v.'s. Namely, as $n \to \infty$

$$\sup_{0\leq t\leq 1}\sup_{0\leq s\leq 1} |\beta_n(s,t)| \xrightarrow{\mathscr{D}} \sup_{0\leq t\leq 1}\sup_{0\leq s\leq 1} |\Gamma(s,t)|,$$

$$\int_0^1\int_0^1 \beta_n^2(s,t)\,ds\,dt \xrightarrow{\mathscr{D}} \int_0^1\int_0^1 \Gamma^2(s,t)\,ds\,dt,$$

$$(s_0(1-s_0))^{-1/2}\sup_{0\leq s\leq 1} |\beta_n(s_0,t)| \xrightarrow{\mathscr{D}} \sup_{0\leq t\leq 1} |B(t)|,$$

$$(t_0(1-t_0))^{-1/2}\sup_{0\leq s\leq 1} |\beta_n(s,t_0)| \xrightarrow{\mathscr{D}} \sup_{0\leq s\leq 1} |B(s)|,$$

$$12\int_0^1\int_0^1 \beta_n(s,t)\,ds\,dt \xrightarrow{\mathscr{D}} N(0,1),$$

$$\left(\frac{12}{s_0(1-s_0)}\right)^{1/2}\int_0^1 \beta_n(s_0,t)\,dt \xrightarrow{\mathscr{D}} N(0,1),$$

$$(48)^{1/2}\int_0^1 \left(-\beta_n(\tfrac{1}{2},t)\right)\,dt \xrightarrow{\mathscr{D}} N(0,1),$$

$$\left(\frac{12}{t_0(1-t_0)}\right)^{1/2}\int_0^1 \beta_n(s,t_0)\,ds \xrightarrow{\mathscr{D}} N(0,1),$$

where $0 < s_0, t_0 < 1$, and $N(0,1)$ stands for the standard normal random variable.

Theorems 4.1 and 4.2 below give the asymptotics of weighted versions of such functionals for the optimal classes of weight functions. In order to present our results we define

$$\hat{\beta}_n(s,t) = \begin{cases} n^{-1/2}\sum_{i=1}^{[(n+1)t]}\left(\mathbf{1}\{R_{in}\leq s\} - \frac{[ns]}{n}\right), & 0\leq s,\ t < 1 \\ 0, & 0\leq s\leq 1,\ t = 1 \end{cases}.$$

THEOREM 4.1. Let $q \in \mathscr{Q}^*$. One can construct a Kiefer process $\{K(s,t);\ 0 \leq s \leq 1,\ t \geq 0\}$ such that with the sequence of stochastic processes $\Gamma_n(\cdot,\cdot)$,

$$\{\Gamma_n(s,t);\ 0 \le s, t \le 1\} = \left\{n^{-1/2}(K(s,nt) - tK(s,n));\ 0 \le s, t \le 1\right\}$$
$$\stackrel{\mathscr{D}}{=} \{\Gamma(s,t);\ 0 \le s, t \le 1\} \quad \text{for each}\quad n \ge 1,$$

as $n \to \infty$, we have

(a) $\quad\sup\limits_{0 < t < 1}\sup\limits_{0 \le s \le 1} |\hat{\beta}_n(s,t) - \Gamma_n(s,t)|/q(t) = o_P(1)$

if and only if $I^*(q,c) < \infty$ for all $c > 0$,

(b) $\quad\sup\limits_{0 < t < 1}\sup\limits_{0 \le s \le 1} |\hat{\beta}_n(s,t) - \Gamma_n(s,t)|/q(t) = O_P(1)$

if and only if $I^*(q,c) < \infty$ for some $c > 0$,

(c) $\quad\sup\limits_{0 < t < 1}\sup\limits_{0 \le s \le 1} |\hat{\beta}_n(s,t)|/q(t) \stackrel{\mathscr{D}}{\to} \sup\limits_{0 < t < 1}\sup\limits_{0 \le s \le 1} |\Gamma(s,t)|/q(t)$

if and only if $I^*(q,c) < \infty$ for some $c > 0$, where $\{\Gamma(s,t);\ 0 \le s, t \le 1\}$ is a Gaussian process as in (3.1).

COROLLARY 4.1. *Let* $q \in \mathscr{Q}^*$ *and* $\{\Gamma(s,t);\ 0 \le s, t \le 1\}$ *be a Gaussian process as in* (3.1). *Then, as* $n \to \infty$, *we have*

$$\hat{\beta}_n(s,t)/q(t) \stackrel{\mathscr{D}}{\to} \Gamma(s,t)/q(t) \quad \text{in}\ D[0,1]^2$$

if and only if $I^*(q,c) < \infty$ *for all* $c > 0$.

We note that from Theorem 4.1 (c) (but not as a consequence of Corollary 4.1) we have, for example, as $n \to \infty$,

$$\sup_{0 < t < 1}\sup_{0 \le s \le 1} |\hat{\beta}_n(s,t)| \Big/ \left(t(1-t) \log\log \frac{1}{t(1-t)}\right)^{1/2}$$
$$\stackrel{\mathscr{D}}{\to} \sup_{0 < t < 1}\sup_{0 \le s \le 1} |\Gamma(s,t)| \Big/ \left(t(1-t) \log\log \frac{1}{t(1-t)}\right)^{1/2}.$$

As an immediate consequence of Theorem 4.2 below we obtain, for example, as $n \to \infty$,

$$\int_0^1\int_0^1 |\hat{\beta}_n(s,t)|/(t(1-t))\ ds\ dt \stackrel{\mathscr{D}}{\to} \int_0^1\int_0^1 |\Gamma(s,t)|/(t(1-t))\ ds\ dt.$$

THEOREM 4.2. *Let* $0 < p < \infty$ *and* q *be a positive function on* $(0,1)$. *Then the following three statements are equivalent.*

(a) *We have*
$$\int_0^1 (t(1-t))^{p/2}/q(t)\ dt < \infty.$$

(b) With the sequence of stochastic processes $\Gamma_n(\cdot,\cdot)$ as in Theorem 4.1 we have, as $n \to \infty$,

$$\int_0^1 \int_0^1 |\hat{\beta}_n(s,t) - \Gamma_n(s,t)|^p / q(t) \, ds \, dt = o_P(1) \ .$$

(c) We have, as $n \to \infty$,

$$\int_0^1 \int_0^1 |\hat{\beta}_n(s,t)|^p / q(t) \, ds \, dt \xrightarrow{\mathscr{D}} \int_0^1 \int_0^1 |\Gamma(s,t)|^p / q(t) \, ds \, dt \ ,$$

where $\{\Gamma(s,t);\ 0 \le s,t \le 1\}$ is a Gaussian process as in (3.1).

Proofs of Theorems 4.1 and 4.2 are based on the following observation.

OBSERVATION 4.1. We note that

$$\beta_n(s,t) = n^{-1/2} \sum_{i=1}^{[nt]} \left(\mathbf{1}\{R_{in} \le s\} - \frac{[ns]}{n} \right)$$

$$= n^{-1/2} \sum_{i=1}^{[nt]} \left(\mathbf{1}\{F(X_i) \le U_n(s)\} - \frac{[ns]}{n} \right) \ ,$$

where $U_n(s)$ is defined by

$$U_n(s) = \begin{cases} 0 & \text{for } s \in [0, 1/n) \\ U_{k:n} = F(X_{k:n}) & \text{for } s \in [k/n, (k+1)/n),\ k = 1, \ldots, n-1 \\ U_{n:n} & \text{for } s = 1 \ , \end{cases}$$

(4.1)

and $X_{1:n} \le \cdots \le X_{n:n}$ denote the order statistics of $X_1, \ldots, X_n$. We note also that the process $\beta_n(s,t)$ already has a "bridge" type structure as it is, namely we have

$$\beta_n(s,t) = \alpha_n(U_n(s), t) - \frac{[nt]}{n} \alpha_n(U_n(s), 1) \ .$$

Theorem 4.1 was proved by Szyszkowicz (1994) under stronger conditions, namely (3.A*) and (3.B*) in lieu of $I^*(q,c) < \infty$ for all $c > 0$ and $I^*(q,c) < \infty$ for some $c > 0$, respectively (cf also discussion after Corollary 3.1 here).

PROOF OF THEOREM 4.1. Arguing as in the proof of Lemma 3.1 of Szyszkowicz (1994) and using Lemma 2.A, with any $q \in \mathscr{Q}$, we have

$$\sup_{0<t\le 1} \sup_{0\le s\le 1} |K(U_n(s), nt) - K(s, nt)| / (n^{1/2} q(t))$$

$$= \begin{cases} o_P(1) & \text{if } I(q,c) < \infty \text{ for all } c > 0 \\ O_P(1) & \text{if } I(q,c) < \infty \text{ for some } c > 0 \ , \end{cases}$$

(4.2)

where $K(\cdot,\cdot)$ is a Kiefer process and $U_n(s)$ is defined as in (4.1). Consequently, the proof of Theorem 3.1 can be repeated here, since Theorem 2.1 and (4.2) imply that there exists a Kiefer process $K(\cdot,\cdot)$ such that, as $n \to \infty$,

$$\sup_{0<t\leq 1}\sup_{0\leq s\leq 1}|\alpha(U_n(s),nt)-n^{-1/2}K(s,nt)|/q(t)$$
$$=\begin{cases}o_P(1) & \text{if } I(q,c)<\infty \text{ for all } c>0\\ O_P(1) & \text{if } I(q,c)<\infty \text{ for some } c>0\end{cases}\qquad\square$$

PROOF OF COROLLARY 4.1. Similar to that of Corollary 2.1. $\square$

In order to obtain Theorem 4.2 we will need the following result.

LEMMA 4.1. Let q be a positive function on $(0,1]$ and such that

$$\int_0^1 t^{p/2}/q(t)\,dt<\infty\ .\tag{4.3}$$

Then, as $n\to\infty$, we have

$$\int_0^1\int_0^1 |n^{-1/2}(K(U_n(s),nt)-K(s,nt))|^p/q(t)\,ds\,dt=o_P(1)\ ,$$

where $\{K(s,t);0\leq s\leq 1, t\geq 0\}$ is a Kiefer process and $U_n(s)$ is defined as in (4.1).

PROOF OF LEMMA 4.1. By Lemma 4.5.1 of Csörgő and Révész (1981), we have

$$\sup_{1/n\leq s\leq 1}|K(U_n(s),nt)-K(s,nt)|\stackrel{a.s.}{=}O\left((nt)^{1/4}(\log\log nt)^{1/4}(\log nt)^{1/2}\right)\ ,$$

as $nt\to\infty$. Hence for any $\lambda\geq 1$

$$\int_{\lambda/n}^1\int_{1/n}^1 |n^{-1/2}(K(U_n(s),nt)-K(s,nt))|^p/q(t)\,ds\,dt$$
$$\leq\int_{\lambda/n}^1\left(\sup_{1/n\leq s\leq 1}|n^{-1/2}(K(U_n(s),nt)-K(s,nt))|\right)^p/q(t)\,dt$$
$$\stackrel{a.s.}{=}O\left(\int_{\lambda/n}^1\left((nt)^{1/4}(\log\log nt)^{1/4}(\log nt)^{1/2}/(nt)^{1/2}\right)^p(t^{p/2}/q(t))\,dt\right)$$
$$\leq O\left(\sup_{\lambda/n\leq t<\delta}\left((nt)^{-1/4}(\log nt)^{3/4}\right)^p\int_{\lambda/n}^\delta t^{p/2}/q(t)\,dt\right.$$
$$\left.+\sup_{\delta\leq t\leq 1}\left((nt)^{-1/4}(\log nt)^{3/4}\right)^p\int_\delta^1 t^{p/2}/q(t)\,dt\right)$$
$$=O(O(1)o(1)+o(1)O(1))$$
$$\stackrel{a.s.}{=}o(1)\ ,$$

as $n\to\infty$, and taking $\delta>0$ arbitrarily small, for any function q which is positive on $(0,1]$ and such that (4.3) holds. Also, if (4.3) holds, we have

$$\int_0^{\lambda/n} \int_{1/n}^1 |n^{-1/2}(K(U_n(s), nt) - K(s, nt))|^p / q(t) \, ds \, dt$$

$$\leq 2^{p+1} \int_0^{\lambda/n} \int_0^1 |n^{-1/2} K(s, nt)|^p / q(t) \, ds \, dt$$

$$= o_P(1)$$

as $n \to \infty$, by Lemma 2.B. Hence, by the last two results, for any function q which is positive on $(0, 1]$ and such that (4.3) holds, as $n \to \infty$, we have

$$\int_0^1 \int_{1/n}^1 |n^{-1/2}(K(U_n(s), nt) - K(s, nt))|^p / q(t) \, ds \, dt = o_P(1) \, .$$

Also, on account of $U_n(s) = 0$ for $s \in [0, 1/n)$ and using again Lemma 2.B, as $n \to \infty$, we have

$$\int_0^1 \int_0^{1/n} |n^{-1/2}(K(U_n(s), nt) - K(s, nt))|^p / q(t) \, ds \, dt$$

$$= \int_0^1 \int_0^{1/n} |n^{-1/2} K(s, nt)|^p / q(t) \, ds \, dt$$

$$\stackrel{\mathscr{L}}{=} \int_0^1 \int_0^{1/n} |K(s, t)|^p / q(t) \, ds \, dt$$

$$\stackrel{\text{a.s.}}{=} o(1)$$

for any function q which is positive on $(0, 1]$ and such that (4.3) holds. This concludes the proof of Lemma 4.1. $\square$

As to the proof of Theorem 4.2, we can simply say that the proof of Theorem 3.2 can be repeated here since Theorem 2.2 and Lemma 4.1 imply that there exists a Kiefer process $\{K(s,t); 0 \leq s \leq 1, t \geq 0\}$ such that, as $n \to \infty$,

$$\int_0^1 \int_0^1 |\alpha_n(U_n(s), t) - n^{-1/2} K(s, nt)|^p / q(t) \, ds \, dt = o_P(1)$$

if

$$\int_0^1 t^{p/2} / q(t) \, dt < \infty \, .$$

5. Weighted empirical processes based on sequential ranks

Let $X_1, X_2, \ldots$ be independent random variables with a continuous distribution function F. For each $n \geq 1$, let $\xi_1, \ldots, \xi_n$ denote the normalized sequential ranks

$$\xi_i = i^{-1} \sum_{k=1}^i \mathbf{1}\{X_k < X_i\}, \quad i = 1, \ldots, n \, ,$$

of the first n of the random variables $X_1, X_2, \ldots$. Thus the sequential rank of X_i, here $i\xi_i$, is the position of X_i in the first $i-1$ ordered observations $X_{0,i-1} \leq \cdots \leq X_{i-1,i-1}$, $1 \leq i \leq n$ of a random sample of size $n \geq 1$, where $X_{0,0} \equiv 0$ by convention, and $\xi_1 = 0$ by definition.

When observations are available sequentially and to be kept in their chronological order, then, for computing nR_{in}, one has to recompute the ranks of all the available observations at each stage. This is often quite tedious as well as time consuming. However, computing only the rank of the last observation at each stage gives the sequential ranks $i\xi_i$ immediately. Hence, it is, comparatively, also easy to calculate statistics of sequential ranks. In particular, one can calculate these statistics successively as the sequential ranks are calculated, in contrast to statistics based on ranks that require the entire sequence $R_{1n}, \ldots, R_{nn}$ to be calculated. Moreover, sequential ranks have the important property that, under the null hypothesis of all the independent X_i's being also identically distributed, they themselves are independent (in contrast to ranks). The following result is due to Barndorff–Nielsen (1963).

THEOREM 5.A. *If the random variables $X_1, \ldots, X_n$ are independent and identically continuously distributed, then the sequential ranks $\xi_1, \ldots, \xi_n$ are independent and*

$$P\left\{\xi_i = \frac{k}{i}\right\} = \frac{1}{i}, \quad k = 0, \ldots, i-1; \; i = 1, \ldots, n \ .$$

Statistical tests based on sequential ranks are frequently used because of these convenient properties. Sequential ranks are ready made, of course, for sequential procedures. However, they are also very convenient to use for sequential studies of chronologically ordered observations of finite size samples.

For related works on statistics based on sequential ranks we refer to Parent (1965), Reynolds (1975), Mason (1981), Lombard and Mason (1985). Bhattacharya and Frierson (1981) propose a control chart based on partial sums of sequential ranks for detecting small changes in the distribution of a given random sample. They consider $\max_{1 \leq k \leq n} \sum_{i=1}^{k} \xi_i / n^{1/2}$ type statistics while Gombay (1995) studies their sequentially weighted versions $\max_{1 \leq k \leq n} \sum_{i=1}^{k} \xi_i / k^{1/2}$ instead, via a theorem of Darling and Erdős (1956). Using different type of weight functions Szyszkowicz (1996b) obtains weighted asymptotics of partial sums of functionals of sequential ranks in supremum and L_p-norms. Csörgő and Horváth (1987b) propose a sequential procedure based on empirical process of sequential ranks for detecting a possible changepoint in a random sequence. Asymptotic properties of CUSUM versions of these procedures are studied by Huse (1989). Test statistics based on sequential ranks under contiguous alternatives and in connection with changepoint problems are considered by Lombard (1981, 1983).

Here, we consider the two-time parameter empirical process

$$\gamma_n(s,t) = n^{-1/2} \sum_{i=1}^{[nt]} \mathbf{1}\left(\{\xi_i \leq s\} - \frac{[is]}{i}\right), \quad 0 \leq s, t \leq 1 \ ,$$

of the normalized sequential ranks ξ_i, $i = 1, \ldots, n$, and prove the optimal asymptotic characterization of $\gamma_n(s,t)$ in weighted supremum and L_p-metrics.

THEOREM 5.1. *Let* $q \in \mathcal{Q}$. *There exists a sequence of Kiefer processes* $\{K_n(s,t); 0 \leq s, t \leq 1\}$ *such that, as* $n \to \infty$, *we have*

(a) $$\sup_{0<t\leq 1} \sup_{0\leq s\leq 1} |\gamma_n(s,t) - K_n(s,t)|/q(t) = o_P(1)$$

if and only if $I(q,c) < \infty$ *for all* $c > 0$,

(b) $$\sup_{0<t\leq 1} \sup_{0\leq s\leq 1} |\gamma_n(s,t) - K_n(s,t)|/q(t) = O_P(1)$$

if and only if $I(q,c) < \infty$ *for some* $c > 0$,

(c) $$\sup_{0<t\leq 1} \sup_{0\leq s\leq 1} |\gamma_n(s,t)|/q(t) \xrightarrow{\mathcal{D}} \sup_{0<t\leq 1} \sup_{0\leq s\leq 1} |K(s,t)|/q(t)$$

if and only if $I(q,c) < \infty$ *for some* $c > 0$, *where* $\{K(s,t); 0 \leq s, t \leq 1\}$ *is a Kiefer process.*

COROLLARY 5.1. *Let* $q \in \mathcal{Q}$ *and* $\{K(s,t); 0 \leq s, t \leq 1\}$ *be a Kiefer process. Then, as* $n \to \infty$, *we have*

$$\gamma_n(s,t)/q(t) \xrightarrow{\mathcal{D}} K(s,t)/q(t) \quad \text{in } D[0,1]^2$$

if and only if $I(q,c) < \infty$ *for all* $c > 0$.

Considering L_p-functionals, we obtain here the following result.

THEOREM 5.2. *Let* $0 < p < \infty$ *and* q *be a positive function on* $(0,1]$. *Then the following three statements are equivalent.*

(a) *We have*

$$\int_0^1 t^{p/2}/q(t)\, dt < \infty \ .$$

(b) *There exists a sequence of Kiefer processes* $\{K_n(s,t); 0 \leq s, t \leq 1\}$ *such that, as* $n \to \infty$, *we have*

$$\int_0^1 \int_0^1 |\gamma_n(s,t) - K_n(s,t)|^p/q(t)\, ds\, dt = o_P(1) \ .$$

(c) *We have, as* $n \to \infty$,

$$\int_0^1 \int_0^1 |\gamma_n(s,t)|^p/q(t)\, ds\, dt \xrightarrow{\mathcal{D}} \int_0^1 \int_0^1 |K(s,t)|^p/q(t)\, ds\, dt \ ,$$

where $\{K(s,t); 0 \leq s, t \leq 1\}$ *is a Kiefer process.*

In order to prove our results, we will represent the normalized sequential ranks ξ_i in terms of uniformly distributed random variables. By Theorem 5.A, sequential ranks generated by i.i.d. observations, i.e., $i\xi_i$ in our terminology, are independent, the ith being uniformly distributed on $\{0, \ldots, i-1\}$. Hence, with independent random variables U_i, uniformly distributed on $[0, 1]$, we have

$$\xi_i \stackrel{\mathscr{D}}{=} \frac{[iU_i]}{i}, \quad i = 1, 2, \ldots. \tag{5.1}$$

Such a representation of sequential ranks was used by Csörgő and Horváth (1987b).

PROOF OF THEOREM 5.1. Let $U_i = F(X_i)$, and

$$\tilde{\gamma}_n(s, t) = n^{-1/2} \sum_{i=1}^{[nt]} \left(\mathbf{1}\left\{ \frac{[iU_i]}{i} \leq s \right\} - s \right), \quad 0 \leq s, t \leq 1.$$

By Csörgő and Horváth (1987b), one can construct a Kiefer process $K(\cdot, \cdot)$ such that

$$\sup_{0 \leq s \leq 1} |n^{1/2} \tilde{\gamma}_n(s, t) - K(s, nt)| = O(\log^2 nt), \tag{5.2}$$

as $nt \to \infty$. Hence, imitating the proof of Theorem 2.1, we get with $q \in \mathscr{Q}$

$$\sup_{0 < t \leq 1} \sup_{0 \leq s \leq 1} |\tilde{\gamma}_n(s, t) - n^{-1/2} K(s, nt)|/q(t)$$
$$= \begin{cases} o_P(1) & \text{if } I(q, c) < \infty \text{ for all } c > 0 \\ O_P(1) & \text{if } I(q, c) < \infty \text{ for some } c > 0 \end{cases},$$

and

$$\sup_{0 < t \leq 1} \sup_{0 \leq s \leq 1} |\tilde{\gamma}_n(s, t)|/q(t) \stackrel{\mathscr{D}}{\to} \sup_{0 < t \leq 1} \sup_{0 \leq s \leq 1} |K(s, t)|/q(t)$$

if $I(q, c) < \infty$ for some $c > 0$. The rest of the proof concerning the "if" parts of (a), (b) and (c) is exactly the same as the corresponding part of the proof of Theorem 4.1 of Szyszkowicz (1994), where the results of Theorem 5.1 were proved under stronger conditions resulting from the law of the iterated logarithm or from tests for upper and lower functions for the suprema of Kiefer processes (cf our discussion after Remark 2.2 here). The converse parts of (a), (b) and (c) follow the same way as the corresponding parts in the proof of Theorem 2.1 here. □

PROOF OF COROLLARY 5.1. Similar to that of Corollary 2.1. □

PROOF OF THEOREM 5.2. First we show that (a) implies (b). Assume that q is a positive function on $(0, 1]$ such that (2.10) holds, i.e., (a) is satisfied. Let $U_i = F(X_i)$, and $\tilde{\gamma}_n(s, t)$ be defined as in the proof of Theorem 5.1. Recall that, by

Csörgő and Horváth (1987b), one can construct a Kiefer process $\{K(s,t); 0 \le s \le 1, t \ge 0\}$ such that (5.2) holds. Hence, with a sequence of Kiefer processes $\widetilde{K}_n(s,t)$, namely $\widetilde{K}_n(s,t) = n^{-1/2}K(s,nt)$, similarly as when proving Theorem 2.2, we obtain

$$\int_0^1 \int_0^1 |\tilde{\gamma}_n(s,t) - \widetilde{K}_n(s,t)|^p/q(t)\,\mathrm{d}s\,\mathrm{d}t = o_P(1)\ . \tag{5.3}$$

Since, by (5.1),

$$\{\bar{\gamma}_n(s,t);\ 0 \le s,t \le 1\} \stackrel{\mathscr{D}}{=} \{\tilde{\gamma}_n(s,t);\ 0 \le s,t \le 1\}\ , \tag{5.4}$$

where

$$\bar{\gamma}_n(s,t) = n^{-1/2} \sum_{i=1}^{[nt]} (\mathbf{1}\{\xi_i \le s\} - s),\quad 0 \le s,t \le 1\ ,$$

we can use Lemma 3.1.2 in Csörgő (1983) to conclude (5.3) with $\bar{\gamma}_n(s,t)$ instead of $\tilde{\gamma}_n(s,t)$ and a possibly different sequence of Kiefer processes $K_n(s,t)$. Since, for $nt \to \infty$, we have

$$\sum_{i=1}^{[nt]} \left|\frac{[is] - is}{i}\right| = O(\log nt) \quad \text{for any } s \in [0,1]\ , \tag{5.5}$$

consequently, we obtain

$$\int_0^1 \int_0^1 |\bar{\gamma}_n(s,t) - \gamma_n(s,t)|^p/q(t)\,\mathrm{d}s\,\mathrm{d}t = o(1)$$

for any positive q such that $\int_0^1 t^{p/2}/q(t)\,\mathrm{d}t < \infty$, which implies part (b).

Next we assume that (b) holds true. Then we have

$$\int_0^{1/n} \int_0^1 |K_n(s,t)|^p/q(t)\,\mathrm{d}s\,\mathrm{d}t \stackrel{\mathscr{D}}{=} \int_0^{1/n} \int_0^1 |K(s,t)|^p/q(t)\,\mathrm{d}s\,\mathrm{d}t = o_P(1)\ ,$$

as $n \to \infty$, where $\{K(s,t); 0 \le s,t \le 1\}$ is a Kiefer process. Now (a) follows from Lemma 2.B in the same way as in the proof of Theorem 2.2.

Finally we show that (a) and (c) are equivalent. Assume that (a) holds. Using (5.2), (5.4), Lemma 3.1.2 in Csörgő (1983) and (5.5), we conclude that there exists a sequence of Kiefer processes $\{K_n(s,t); 0 \le s,t \le 1\}$ such that, as $n \to \infty$,

$$\sup_{0 \le t \le 1} \sup_{0 \le s \le 1} |\gamma_n(s,t) - K_n(s,t)| = o_P(1)\ .$$

Consequently, due to the positivity of q, as $n \to \infty$,

$$\int_\delta^1 \int_0^1 |\gamma_n(s,t)|^p/q(t)\,\mathrm{d}s\,\mathrm{d}t \stackrel{\mathscr{D}}{\to} \int_\delta^1 \int_0^1 |K(s,t)|^p/q(t)\,\mathrm{d}s\,\mathrm{d}t$$

for any $\delta \in (0, 1]$, where $\{K(s,t); 0 \leq s, t \leq 1\}$ is a Kiefer process. By Lemma 2.B we have

$$\lim_{\delta \to 0} \int_0^\delta \int_0^1 |K(s,t)|^p/q(t)\,ds\,dt = 0 \quad \text{a.s.} \ ,$$

which, combined with the previous statement, implies (c). Assuming now that (c) holds, we have

$$\int_0^1 \int_0^1 |K(s,t)|^p/q(t)\,ds\,dt < \infty \quad \text{a.s.} \ ,$$

which yields (a) by Lemma 2.B. Details of this proof are similar to those of the corresponding parts of the proof of Theorem 2.2 and can be found in Szyszkowicz (1993b). □

6. "Bridge-type" empirical processes of sequential ranks

As in Section 5, let $X_1, X_2, \ldots$ be i.i.d.r.v.'s with a continuous distribution function F and let $\xi_1, \xi_2, \ldots$ be their normalized sequential ranks. With (1.8) and (3.4) in mind, we study the asymptotic behaviour of the "bridge" type processes

$$(\gamma_n(s,t) - t\gamma_n(s,1))/q(t) \ .$$

We introduce

$$\hat{\gamma}_n(s,t) = \begin{cases} n^{-1/2}\left(\sum_{i=1}^{[(n+1)t]} \mathbf{1}\{\xi_i \leq s\} - \frac{[(n+1)t]}{n} \sum_{i=1}^{n} \mathbf{1}\{\xi_i \leq s\} \right), \\ \qquad\qquad\qquad\qquad\qquad\qquad 0 \leq s \leq 1, \ 0 \leq t \leq 1, \\ 0, \qquad\qquad\qquad\qquad\qquad\qquad 0 \leq s \leq 1, \ t = 1. \end{cases}$$

Again, let $\{\Gamma(s,t); 0 \leq s, t \leq 1\}$ be a stochastic process as defined in (3.1).

Here we establish the optimal asymptotics for the processes $\hat{\gamma}_n(s,t)$ in weighted supremum and L_p-metrics.

THEOREM 6.1. Let $q \in \mathscr{Q}^*$. There exists a sequence of stochastic processes $\Gamma_n(\cdot, \cdot)$, $n \geq 1$,

$$\{\Gamma_n(s,t); 0 \leq s, t \leq 1\} \stackrel{\mathscr{D}}{=} \{\Gamma(s,t); 0 \leq s, t \leq 1\}$$

such that, as $n \to \infty$, we have

(a) $\quad \sup_{0 < t < 1} \sup_{0 \leq s \leq 1} |\hat{\gamma}_n(s,t) - \Gamma_n(s,t)|/q(t) = o_P(1)$

if and only if $I^*(q, c) < \infty$ for all $c > 0$,

(b)
$$\sup_{0<t<1}\sup_{0\leq s\leq 1}|\gamma_n(s,t)-\Gamma_n(s,t)|/q(t)=O_P(1)$$

if and only if $I^*(q,c)<\infty$ for some $c>0$,

(c)
$$\sup_{0<t<1}\sup_{0\leq s\leq 1}|\hat{\gamma}_n(s,t)|/q(t)\xrightarrow{\mathcal{D}}\sup_{0<t<1}\sup_{0\leq s\leq 1}|\Gamma(s,t)|/q(t)$$

if and only if $I^*(q,c)<\infty$ for some $c>0$, where $\{\Gamma(s,t);\ 0\leq s,t\leq 1\}$ is a Gaussian process as in (3.1).

COROLLARY 6.1. *Let $q\in\mathcal{Q}^*$ and $\{\Gamma(s,t);\ 0\leq s,t\leq 1\}$ be a Gaussian process as in (3.1). Then, as $n\to\infty$, we have*

$$\hat{\gamma}_n(s,t)/q(t)\xrightarrow{\mathcal{D}}\Gamma(s,t)/q(t)\quad\text{in }D[0,1]^2$$

if and only if $I^*(q,c)<\infty$ for all $c>0$.

THEOREM 6.2. *Let $0<p<\infty$ and q be a positive function on $(0,1)$. Let $\{\Gamma(s,t);\ 0\leq s,t\leq 1\}$ denote a Gaussian process as in (3.1). Then the following three statements are equivalent.*

(a) We have

$$\int_0^1 (t(1-t))^{p/2}/q(t)\,dt<\infty\ .$$

(b) There exists a sequence of stochastic processes $\Gamma_n(\cdot,\cdot)$, $n\geq 1$,

$$\{\Gamma_n(s,t);\ 0\leq s,t\leq 1\}\stackrel{\mathcal{D}}{=}\{\Gamma(s,t);\ 0\leq s,t\leq 1\}$$

such that, as $n\to\infty$, we have

$$\int_0^1\int_0^1 |\hat{\gamma}_n(s,t)-\Gamma_n(s,t)|^p/q(t)\,ds\,dt=o_P(1)\ .$$

(c) We have, as $n\to\infty$,

$$\int_0^1 |\hat{\gamma}_n(s,t)|^p/q(t)\,ds\,dt\xrightarrow{\mathcal{D}}\int_0^1\int_0^1 |\Gamma(s,t)|^p/q(t)\,ds\,dt\ ,$$

where $\{\Gamma(s,t);\ 0\leq s,t\leq 1\}$ is a Gaussian process as in (3.1).

Theorem 6.1 (and Corollary 6.1) is an improvement of Theorem 4.2 (and Corollary 4.2) of Szyszkowicz (1994), where stronger conditions on weight functions were used. Namely, part (a) and Corollary 6.1 were obtained for weight functions q which are positive on $(0,1)$ and such that $(3.A^*)$ holds, while parts (b) and (c) were proved for positive functions q such that $(3.B^*)$ holds (cf also discussion following Remark 3.1 here). The proof of Theorem 6.1 can be done along the lines of the proof of Theorem 4.2 in Szyszkowicz (1994), now making use of

Lemma 2.B in order to get the optimal conditions on the weight functions. On account of Lemma 2.B, Corollary 6.1 follows from Theorem 6.1 in the very same way as Corollary 2.1 follows from Theorem 2.1.

PROOF OF THEOREM 6.2. Let

$$\widetilde{S}_{[(n+1)t]}(s) = \sum_{i=1}^{[(n+1)t]} \left(1\left\{\frac{[iU_i]}{i} \leq s\right\} - s\right).$$

We introduce also

$$\widetilde{S}^{(1)}_{[(n+1)t]}(s) = \sum_{i=1}^{[(n+1)t]} \left(1\left\{\frac{[iU_i]}{i} \leq s\right\} - s\right), \quad 0 \leq t \leq 1/2, \; 0 \leq s \leq 1,$$

$$\widetilde{S}^{(2)}_{[(n+1)t]}(s) = \sum_{i=[(n+1)t]+1}^{n} \left(1\left\{\frac{[iU_i]}{i} \leq s\right\} - s\right), \quad 1/2 \leq t < 1, \; 0 \leq s \leq 1.$$

Then

$$\widetilde{S}_{[(n+1)t]}(s) = \begin{cases} \widetilde{S}^{(1)}_{[(n+1)t]}(s), & 0 \leq t \leq 1/2, \; 0 \leq s \leq 1 \\ \widetilde{S}^{(1)}_{[\frac{n+1}{2}]}(s) + \widetilde{S}^{(2)}_{[\frac{n+1}{2}]}(s) - \widetilde{S}^{(2)}_{[(n+1)t]}(s), & 1/2 \leq t < 1, \; 0 \leq s \leq 1. \end{cases}$$

Using the representation of sequential ranks $\xi_i \stackrel{\mathcal{D}}{=} [iU_i]/i$ (cf (5.1)) and the approximation of (5.2), one can construct two independent Kiefer processes $K^{(1)}(\cdot,\cdot)$ and $K^{(2)}(\cdot,\cdot)$ such that

$$\int_0^{1/2} \int_0^1 \left| n^{-1/2} \left(\widetilde{S}^{(1)}_{[(n+1)t]}(s) - K^{(1)}(s, (n+1)t) \right) \right|^p \Big/ q(t) \, ds \, dt = o_P(1) \tag{6.1}$$

if

$$\int_0^{1/2} t^{p/2}/q(t) \, dt < \infty,$$

and

$$\int_0^{1/2} \int_0^1 \left| n^{-1/2} \left(\widetilde{S}^{(2)}_{[(n+1)t]}(s) - K^{(2)}(s, (n+1)t) \right) \right|^p \Big/ q(1-t) \, ds \, dt = o_P(1) \tag{6.2}$$

if

$$\int_0^{1/2} t^{p/2}/q(1-t) \, dt < \infty.$$

In order to get (6.2), we note that, along the lines of the proof of Theorem 2.1 in Csörgő and Horváth (1987b), we get, as $nt \to \infty$,

$$\sup_{0 \le s \le 1} \left| \sum_{i=1}^{[(n+1)t]} \mathbf{1}\left\{ \frac{[(n-i+1)U_{n-i+1}]}{n-i+1} \le s \right\} - \sum_{i=1}^{[(n+1)t]} \mathbf{1}\{U_{n-i+1} \le s\} \right|$$
$$= O(\log nt) .$$

By this and (5.2), we have, as $nt \to \infty$,

$$\sup_{0 \le s \le 1} \left| \sum_{i=1}^{[(n+1)t]} \mathbf{1}\left\{ \frac{[(n-i+1)U_{n-i+1}]}{n-i+1} \le s \right\} - K^{(2)}(s, (n+1)t) \right|$$
$$= O(\log^2 nt) \quad \text{a.s.}$$

Hence, with function q positive on $(0,1)$, we have

$$\int_{1/2}^{1} \int_{0}^{1} \left| n^{-1/2} \widetilde{S}^{(2)}_{[(n+1)t]}(s) - n^{-1/2} K^{(2)}(s, (n+1)(1-t)) \right|^{p} / q(t) \, ds \, dt$$
$$= \int_{0}^{1/2} \int_{0}^{1} \left| n^{-1/2} \widetilde{S}^{(2)}_{[(n+1)t]}(s) - n^{-1/2} K^{(2)}(s, (n+1)t) \right|^{p} / q(1-t) \, ds \, dt$$
$$= o_P(1)$$

if

$$\int_{1/2}^{1} (1-t)^{p/2} / q(t) \, dt < \infty .$$

Let $\hat{\tilde{\gamma}}_n$ be a version of the process $\hat{\gamma}_n$ when replacing ξ_i by $[iU_i]/i$ (cf (5.1)). Using (6.1) and (6.2), as well as the method of proof of Theorem 3.2 (for details of the proof of Theorem 3.2 we refer to Szyszkowicz (1993b) or Csörgő and Szyszkowicz (1994)), with the Kiefer process

$$\{K(s, nt), \ 0 \le t \le 1, \ 0 \le s \le 1\}$$
$$= \begin{cases} K^{(1)}(s, nt), & 0 \le t \le 1/2, \ 0 \le s \le 1 \\ K^{(1)}(s, \tfrac{n}{2}) + K^{(2)}(s, \tfrac{n}{2}) - K^{(2)}(s, n - nt), & 1/2 < t \le 1, \ 0 \le s \le 1 , \end{cases}$$

we have

$$\int_{0}^{1/2} \int_{0}^{1} \left| \hat{\tilde{\gamma}}_n(s,t) - n^{-1/2}(K(s,(n+1)t) - tK(s, n+1)) \right|^{p} / q(t) \, ds \, dt$$
$$= \int_{0}^{1/2} \int_{0}^{1} \left| n^{-1/2} \left(\widetilde{S}_{[(n+1)t]}(s) - \frac{[(n+1)t]}{n} \widetilde{S}_n(s) \right) \right.$$
$$\left. - n^{-1/2}(K(s,(n+1)t) - tK(s, n+1)) \right|^{p} / q(t) \, ds \, dt$$
$$= o_P(1)$$

if
$$\int_0^1 (t(1-t))^{p/2}/q(t)\,dt < \infty .$$

Also
$$\int_{1/2}^1 \int_0^1 \left|\hat{\tilde{\gamma}}_n(s,t) - n^{-1/2}(K(s,(n+1)t) - tK(s,n+1))\right|^p \Big/ q(t)\,ds\,dt$$
$$= \int_{1/2}^1 \int_0^1 \left|n^{-1/2}\left(\widetilde{S}_{[(n+1)t]}(s) - \frac{[(n+1)t]}{n}\widetilde{S}_n(s)\right)\right.$$
$$\left. - n^{-1/2}(K(s,(n+1)t) - tK(s,n+1))\right|^p \Big/ q(t)\,ds\,dt$$
$$= o_P(1)$$

if
$$\int_0^1 (t(1-t))^{p/2}/q(t)\,dt < \infty .$$

Consequently, assuming (a), with the sequence of stochastic processes Γ_n, $n \geq 1$,
$$\{\Gamma_n(s,t);\ 0 \leq s,t \leq 1\} = \left\{n^{-1/2}(K(s,nt) - tK(s,n));\ 0 \leq s,t \leq 1\right\}$$

we have, as $n \to \infty$
$$\int_0^1 \int_0^1 \left|\hat{\tilde{\gamma}}_n(s,t) - \Gamma_n(s,t)\right|^p \Big/ q(t)\,ds\,dt = o_P(1) .$$

We note again that for each $n \geq 1$
$$\{\Gamma_n(s,t);\ 0 \leq s,t \leq 1\} \stackrel{\mathcal{D}}{=} \{\Gamma(s,t);\ 0 \leq s,t \leq 1\} .$$

Since by (5.1)
$$\{\hat{\gamma}_n(s,t);\ 0 \leq s,t \leq 1\} \stackrel{\mathcal{D}}{=} \left\{\hat{\tilde{\gamma}}_n(s,t);\ 0 \leq s,t \leq 1\right\} ,$$

we can use Lemma 3.1.2 in Csörgő (1983) to conclude that, with a possibly different sequence of stochastic processes $\hat{\Gamma}_n(\cdot,\cdot)$, but still such that
$$\{\hat{\Gamma}_n(s,t);\ 0 \leq s,t \leq 1\} \stackrel{\mathcal{D}}{=} \{\Gamma(s,t);\ 0 \leq s,t \leq 1\},\quad n \geq 1 ,$$

we have, as $n \to \infty$,
$$\int_0^1 \int_0^1 \left|\hat{\gamma}_n(s,t) - \hat{\Gamma}_n(s,t)\right|^p / q(t)\,ds\,dt = o_P(1) \quad \text{if (a) holds} .$$

This concludes the proof that (a) implies (b). The proof that (a) follows from (b) and the proof of the equivalence of (a) and (c) can be done along the lines of the proof of Theorem 3.2 (for details of the proof of Theorem 3.2 we refer to Szyszkowicz (1993b) or Csörgő and Szyszkowicz (1994)). □

7. Contiguous alternatives

Let $X_1, X_2, \ldots$ be independent random variables. We wish to test the null hypothesis

$$H_0: X_i, \ 1 \leq i \leq n, \ \text{have the same distribution} \ F \ ,$$

versus the alternative hypothesis

$$H_1: X_i, \ 1 \leq i \leq n, \ \text{have the respective distribution functions} \ F_{in} \ ,$$

where we assume that all F_{in} are absolutely continuous with respect to the distribution function F and

$$\left[\frac{dF_{in}}{dF}(F^{-1}(u))\right]^{1/2} = 1 + \frac{1}{2n^{1/2}} h_n(t, F^{-1}(u)),$$

$$\equiv 1 + \frac{1}{2n^{1/2}} g_n(t, u), \quad \frac{i-1}{n} < t \leq \frac{i}{n} \ , \quad (7.1)$$

where $F^{-1}(u) = \inf\{x : F(x) \geq u\}$, $0 \leq u \leq 1$, $F^{-1}(0) = F^{-1}(0^+)$. We assume also that there exists a function $g \in L^2[0,1]^2$ such that

$$\int_0^1 g(t, u) \, du = 0 \quad \text{for almost all} \ t \in [0, 1] \quad (7.2)$$

and

$$\int_0^1 \int_0^1 \left[g_n(t, u) - g\left(\frac{[nt]}{n}, u\right)\right]^2 du \, dt \to 0, \quad n \to \infty \ . \quad (7.3)$$

It is known that the sequence of direct products $F_{1n} \times \cdots \times F_{nn}$, $n = 1, 2, \ldots$ is contiguous (for the notion of contiguity see, e.g., Le Cam (1960, 1986), Greenwood and Shiryayev (1985), Hájek and Šidák (1967), Roussas (1972)) to the sequence $F \times \cdots \times F$ (cf Oosterhoff and van Zwet (1975), Szyszkowicz (1991a)). In particular, for the so-called change-point problem we assume that there exists $\lambda \in (0, 1)$ such that

$$g(t, u) = \mathbf{1}\{t \geq \lambda\} \, g(u) \quad (7.4)$$

for some square integrable function g.

This description of alternatives was used by Khmaladze and Parjanadze (1986) in the context of studying changepoint problems using linear statistics of sequential ranks. Statistics based on functionals of the empirical process based on sequential ranks are considered by Pardzhanadze and Khmaladze (1986), where they also discuss some merits of the use of statistics of sequential ranks.

In this section we present asymptotic distributions (weak convergence) of the weighted empirical processes $\alpha_n(s,t)/q(t)$, $\hat{\beta}_n(s,t)/q(t)$ and $\gamma_n(s,t)/q(t)$, as well as of the "bridge type" processes $\hat{\alpha}_n(s,t)/q(t)$ and $\hat{\gamma}_n(s,t)/q(t)$ under the sequence of contiguous alternatives of H_1. We show that the weak convergence of the just mentioned processes, which we have established under the null hypothesis of no change for the optimal classes of weight functions $q(t)$ (cf Corollaries 2.1, 3.1, 4.1, 5.1 and 6.1), continues to hold for the same classes of weight functions under a sequence of contiguous alternatives as in (7.1)–(7.3), which accommodates, for example, the occurrence of a change-point in a series of chronologically ordered data (cf (7.4)).

Let

$$c(s,t) = \int_0^t \int_0^s g(\tau, u)\, du\, d\tau \tag{7.5}$$

and

$$r(s,t) = -\int_0^t \frac{c(s,y)}{y}\, dy + c(s,t), \tag{7.6}$$

where $g(\cdot, \cdot)$ is the function defined by (7.1)–(7.3).

THEOREM 7.1. We assume that under H_0, $X_1, \ldots, X_n$, $n \geq 1$, are i.i.d.r.v.'s with a continuous distribution function F. Let $q \in \mathcal{Q}$ and $I(q, c) < \infty$ for all $c > 0$. Then, under the altnerative H_1, as $n \to \infty$, we have

$$\alpha_n(s,t)/q(t) \xrightarrow{\mathcal{D}} (K(s,t) + c(s,t))/q(t) \tag{7.7}$$

and

$$\gamma_n(s,t)/q(t) \xrightarrow{\mathcal{D}} (K(s,t) + r(s,t))/q(t) \tag{7.8}$$

in $D[0,1]^2$, where $\{K(s,t);\ 0 \leq s, t \leq 1\}$ is a Kiefer process.

COROLLARY 7.1. We assume that under H_0, $X_1, \ldots, X_n$, $n \geq 1$, are i.i.d.r.v.'s with a continuous distribution function F. Let $q \in \mathcal{Q}$ and $I(q,c) < \infty$ for all $c > 0$. Then, under the change-point alternatives of (7.4), as $n \to \infty$, we have

$$\alpha_n(s,t)/q(t) \xrightarrow{\mathcal{D}} \left(K(s,t) + (t-\lambda)\mathbf{1}\{t \geq \lambda\} \int_0^s g(u)\, du\right)\Big/q(t) \tag{7.9}$$

and

$$\gamma_n(s,t)/q(t) \xrightarrow{\mathcal{D}} \left(K(s,t) + \lambda \log(t/\lambda)\mathbf{1}\{t \geq \lambda\} \int_0^s g(u)\, du\right)\Big/q(t) \tag{7.10}$$

in $D[0,1]^2$, where $\{K(s,t);\ 0 \leq s, t \leq 1\}$ is a Kiefer process.

Let
$$d(s,t) = c(s,t) - tc(s,1) \; , \qquad (7.11)$$

where $c(s,t)$ is defined by (7.5), i.e.,

$$d(s,t) = \int_0^t \int_0^s g(\tau,u) \, du \, d\tau - t \int_0^1 \int_0^s g(\tau,u) \, du \, d\tau \; ,$$

and let

$$e(s,t) = r(s,t) - tr(s,1) \; , \qquad (7.12)$$

where $r(s,t)$ is defined by (7.6), i.e.,

$$e(s,t) = -\int_0^t \frac{c(s,y)}{y} \, dy + t \int_0^1 \frac{c(s,y)}{y} \, dy + c(s,t) - tc(s,1) \; .$$

THEOREM 7.2. We assume that under H_0, $X_1, \ldots, X_n$, $n \geq 1$, are i.i.d.r.v.'s with a continuous distribution function F. Let $q \in \mathcal{Q}^*$ and $I^*(q,c) < \infty$ for all $c > 0$. Then, under the alternative H_1, as $n \to \infty$, we have

$$\hat{\alpha}_n(s,t)/q(t) \xrightarrow{\mathcal{D}} (\Gamma(s,t) + d(s,t))/q(t) \; , \qquad (7.13)$$

$$\hat{\beta}_n(s,t)/q(t) \xrightarrow{\mathcal{D}} (\Gamma(s,t) + d(s,t))/q(t) \; , \qquad (7.14)$$

$$\hat{\gamma}_n(s,t)/q(t) \xrightarrow{\mathcal{D}} (\Gamma(s,t) + e(s,t))/q(t) \; , \qquad (7.15)$$

in $D[0,1]^2$, where $\{\Gamma(s,t);\ 0 \leq s, t \leq 1\}$ is a Gaussian process as in (3.1).

COROLLARY 7.2. We assume that under H_0, $X_1, \ldots, X_n$, $n \geq 1$, are i.i.d.r.v.'s with the continuous distribution function F. Let $q \in \mathcal{Q}^*$ and $I^*(q,c) < \infty$ for all $c > 0$. Then, under the change-point alternatives of (7.4), as $n \to \infty$, we have

$$\frac{\hat{\alpha}_n(s,t)}{q(t)} \xrightarrow{\mathcal{D}} \frac{\Gamma(s,t) - t(1-\lambda) \int_0^s g(u) \, du \mathbf{1}\{t < \lambda\} - (1-t)\lambda \int_0^s g(u) \, du \mathbf{1}\{t \geq \lambda\}}{q(t)} \; ,$$
$$(7.16)$$

$$\frac{\hat{\beta}_n(s,t)}{q(t)} \xrightarrow{\mathcal{D}} \frac{\Gamma(s,t) - t(1-\lambda) \int_0^s g(u) \, du \mathbf{1}\{t < \lambda\} - (1-t)\lambda \int_0^s g(u) \, du \mathbf{1}\{t \geq \lambda\}}{q(t)} \; ,$$
$$(7.17)$$

$$\frac{\hat{\gamma}_n(s,t)}{q(t)} \xrightarrow{\mathcal{D}} \frac{\Gamma(s,t) + (t\lambda \log \lambda \mathbf{1}\{t < \lambda\} + (\lambda \log(t/\lambda) + t\lambda \log \lambda)\mathbf{1}\{t \geq \lambda\}) \int_0^1 g(u) \, du}{q(t)}$$
$$(7.18)$$

in $D[0,1]^2$, where $\{\Gamma(s,t);\ 0 \leq s, t \leq 1\}$ is a Gaussian process as in (3.1).

Theorems 7.1 and 7.2 (and Corollaries 7.1 and 7.2) are improvements of the corresponding results of Szyszkowicz (1994) in that the weak convergence statements of (7.7)–(7.10) and (7.13)–(7.18) are obtained here for larger classes of weight functions.

The proofs of Theorems 7.1 and 7.2 are exactly the same as those of Theorems 5.1 and 5.2 of Szyszkowicz (1994) when using Corollaries 2.1, 3.1, 4.1, 5.1 and 6.1 instead of the corresponding statements in the just mentioned paper.

In order to outline the proofs of our results, we assume without loss of generality that under H_0 the X_i are of the form $F^{-1}(\overline{U}_i)$, where the $\overline{U}_i$ are independent, identically distributed uniform-$(0,1)$ random variables, and so $X_i \stackrel{\mathcal{D}}{=} F^{-1}(\overline{U}_i)$, $i = 1, 2, \ldots$.

Define the centered log-likelihood ratio process $L_n(\cdot)$ by

$$L_n(t) = \sum_{i=1}^{[nt]} \left\{ \log \frac{dF_{in}}{dF}(F^{-1}(\overline{U}_i)) - E \log \frac{dF_{in}}{dF}(F^{-1}(\overline{U}_i)) \right\}, \tag{7.19}$$

and let $L(t)$ be a Gaussian process with mean zero and covariance function $Q_L(t_1 \wedge t_2)$, where

$$Q_L(t) = \int_0^t \int_0^1 g^2(s, u) \, du \, ds . \tag{7.20}$$

Then, by (7.1)–(7.3), Theorem 2 of Oosterhoff and van Zwet (1975), under H_0, yields (cf., e.g., Szyszkowicz (1991a))

$$L_n(t) \xrightarrow{\mathcal{D}} L(t) \quad \text{in } D[0,1] . \tag{7.21}$$

Let

$$a(u) \in L^2(0,1), \quad \text{where} \quad \int_0^1 a(u) \, du = 0 \quad \text{and} \quad \int_0^1 a^2(u) \, du = 1 . \tag{7.22}$$

Then, under H_0, $a(F(X_i))$ are i.i.d.r.v.'s with mean 0 and variance 1, and

$$E\left(n^{-1/2} \sum_{i=1}^{[nt]} a(F(X_i))\right) L_n(z) \longrightarrow \int_0^{t \wedge z} \int_0^1 g(\tau, u) a(u) \, du \, d\tau . \tag{7.23}$$

Let (W, L) be a two-dimensional Gaussian process, where W is a standard Wiener process, L is as in (7.20), and the covariance function of the process (W, L) is

$$EW(t)L(z) = \int_0^{t \wedge z} \int_0^1 g(\tau, u) a(u) \, du \, d\tau .$$

Using Donsker's Theorem, (7.21) and (7.23), under H_0 as $n \to \infty$, we obtain

$$\left(n^{-1/2} \sum_{i=1}^{[nt]} a(F(X_i)), L_n(z)\right) \xrightarrow{\mathcal{D}} (W(t), L(z)) . \tag{7.24}$$

In particular, the function $a(u)$ can be taken as

$$a(u) = a_s(u) = (\mathbf{1}\{u \leq s\} - s)/(s(1-s))^{1/2},$$

for any fixed $0 < s < 1$, and we note that for any fixed $0 < s < 1$

$$W(t) = \frac{K(s,t)}{(s(1-s))^{1/2}}, \qquad t \geq 0$$

is a Wiener process. Consequently, we have

$$E\alpha_n(s,t)L_n(z) \longrightarrow \int_0^{t\wedge z}\!\!\int_0^s g(\tau,u)\,du\,d\tau$$

and

$$EK(s,t)L(z) = (s(1-s))^{1/2} E\frac{K(s,t)}{(s(1-s))^{1/2}}L(z)$$

$$= (s(1-s))^{1/2}\int_0^{t\wedge z}\!\!\int_0^1 g(\tau,u)\frac{(\mathbf{1}\{u\leq s\}-s)}{(s(1-s))^{1/2}}\,du\,d\tau$$

$$= \int_0^{t\wedge z}\!\!\int_0^s g(\tau,u)\,du\,d\tau - s\int_0^{t\wedge z}\!\!\int_0^1 g(\tau,u)\,du\,d\tau$$

$$= \int_0^{t\wedge z}\!\!\int_0^s g(\tau,u)\,du\,d\tau,$$

due to (7.2). Hence, we conclude that all finite-dimensional distributions of the process $(\alpha_n(s,t), L_n(z))$ converge to those of the process $(K(s,t), L(z))$ with covariance function

$$EK(s,t)L(z) = \int_0^{t\wedge z}\!\!\int_0^s g(\tau,u)\,du\,d\tau.$$

Tightness of $(\alpha_n(s,t), L_n(z))$ follows from the weak convergence of $\alpha_n(s,t)$ (cf., e.g., Theorem 2.1 (a), or Corollary 2.1 with $q \equiv 1$) and that of $L_n(z)$ (cf (7.21)). Consequently, under H_0 as $n \to \infty$, we obtain

$$(\alpha_n(s,t), L_n(z)) \xrightarrow{\mathcal{D}} (K(s,t), L(z)) \quad \text{in} \quad D[0,1]^2 \times D[0,1]. \tag{7.25}$$

We introduce the following Gaussian processes:

(i) $(K(s,t), L(z))$, where $K(s,t)$ is a Kiefer process, i.e., the process defined by (2.1), and the covariance function of the process (K,L) is

$$Q_{K,L}(s,t,z) = EK(s,t)L(z) = c(s, t \wedge z), \tag{7.26}$$

where $c(s,t)$ is defined by (7.5),

(ii) $(K^*(s,t), L(z))$, where $K^*(s,t)$ is again a Kiefer process, but the covariance function of the process (K^*, L) is

$$Q_{K^*,L}(s,t,z) = \mathrm{E}K^*(s,t)L(z) = -\int_0^t \frac{c(s,y\wedge z)}{y}\,dy + c(s,t\wedge z) \ . \qquad (7.27)$$

LEMMA 7.1. We assume that H_0 holds. Then, as $n \to \infty$, we have

(a) $\quad (\alpha_n, L_n) \xrightarrow{\mathcal{D}} (K, L),$

(b) $\quad (\gamma_n, L_n) \xrightarrow{\mathcal{D}} (K^*, L),$

in $D[0,1]^2 \times D[0,1]$.

PROOF. Part (a) is just the statement (7.25). Considering now the processes $\gamma_n(s,t)$, we note that although the limiting Gaussian process of the sequence $\gamma_n(s,t)$ is again a Kiefer process, the joint weak convergence of γ_n and L_n results in a different limiting process than in the case of α_n and L_n. Khmaladze and Parjanadze (1986) consider the asymptotic behaviour of the sequence of the partial sum processes

$$V_n(t) = n^{-1/2} \sum_{i=1}^{[nt]} a(\xi_i), \quad 0 \le t \le 1 \ ,$$

with $a(u)$ as in (7.22), under H_0, as well as under the class of contiguous alternatives of H_1. Using their Theorem 3, again with the function $a(u) = a_s(u) = (\mathbf{1}\{u \le s\} - s)/(s(1-s))^{1/2}$, and Theorem 4.1 (a), or Corollary 4.1 with $q \equiv 1$, the proof is similar to that of part (a). $\square$

PROOF OF THEOREM 7.1. Since (α_n, L_n) converges weakly to (K, L) when we assume H_0 to hold, Le Cam's third lemma implies that, under contiguous alternatives H_1,

$$\alpha_n(s,t) \xrightarrow{\mathcal{D}} K(s,t) + c(s,t) \quad \text{in} \quad D[0,1]^2 \ . \qquad (7.28)$$

In order to conclude a similar statement under H_1 for the processes $\alpha_n(s,t)/q(t)$, first we need to verify the joint weak convergence of $(\alpha_n/q, L_n)$ under H_0 to the appropriate Gaussian process. The joint weak convergence of $(\alpha_n/q, L_n)$ under H_0 follows, via the Cramér-Wold device, from Corollary 2.1, (7.21) and the fact that

$$\mathrm{E}\frac{\alpha_n(s,t)}{q(t)}L_n(z) = \frac{1}{q(t)}\mathrm{E}\alpha_n(s,t)L_n(z) \to \frac{1}{q(t)}c(s,t\wedge z) \ .$$

That is to say, under H_0, as $n \to \infty$

$$(\alpha_n(s,t)/q(t), L_n(z)) \xrightarrow{\mathcal{D}} (K(s,t)/q(t), L(z)) \quad \text{in } D[0,1]^2 \times D[0,1] \ , \qquad (7.29)$$

provided the covariance function $c(s,t\wedge z)/q(t)$ of the mean zero Gaussian process (cf (7.26))

$(K(s,t)/q(t), L(z))$, $0 \le s, t, z \le 1$

is finite uniformly in $t \in (0, 1]$. This, in turn, is true since $g \in L^2[0,1]^2$, and hence we have

$$|c(s,t)| = \left|\int_0^t \int_0^s g(\tau, u) \, du \, d\tau\right| \le s^{1/2} t^{1/2} \left(\int_0^t \int_0^s g^2(\tau, u) \, du \, d\tau\right)^{1/2}, \tag{7.30}$$

which implies

$$\sup_{0<t\le 1} |c(s,t)|/q(t) \le s^{1/2} \sup_{0<t\le 1} \left(\int_0^t \int_0^s g^2(\tau, u) \, d u d\tau\right)^{1/2}$$
$$\times \sup_{0<t\le 1} t^{1/2}/q(t) < \infty,$$

due to the fact that q is positive on $(0, 1]$ and $\lim_{t\downarrow 0} t^{1/2}/q(t) = 0$. Hence we have (7.29), and Le Cam's third lemma implies (7.7).

The proof of (7.8) is similar, when using part (b) of Lemma 7.1 and Corollary 5.1. In order to prove the finiteness of the covariance function of $\gamma_n(s,t)/q(t)$ and $L_n(z)$, we note that

$$\left|\int_0^t \frac{c(s,y)}{y} dy\right| \le \int_0^t \left|\frac{c(s,y)}{y}\right| dy,$$

and, since due to (7.30)

$$\left|\frac{c(s,y)}{y}\right| \le s^{1/2} y^{-1/2} \left(\int_0^1 \int_0^1 g^2(\tau, u) \, du \, d\tau\right)^{1/2},$$

we get

$$\left|\int_0^t \frac{c(s,y)}{y} dy\right| \le 2 s^{1/2} t^{1/2} \left(\int_0^1 \int_0^1 g^2(\tau, u) \, du \, d\tau\right)^{1/2}. \tag{7.31}$$

Since q is positive on $(0, 1]$ and $\lim_{t\downarrow 0} t^{1/2}/q(t) = 0$, using (7.30) and (7.31), we obtain

$$\sup_{0<t\le 1} |r(s,t)|/q(t) < \infty,$$

where $r(s,t)$ is defined by (7.6). $\square$

In order to prove Theorem 7.2, we need to introduce the following Gaussian processes:

(i) $(\Gamma(s,t), L(z))$, where $\Gamma(s,t)$ is a Gaussian process with mean zero and covariance function as in (3.1), and the covariance function of the process (Γ, L) is

$$Q_{\Gamma,L}(s,t,z) = E\Gamma(s,t)L(z) = c(s, t \wedge z) - tc(s,z),$$

where $c(s,t)$ is defined by (7.5).

(ii) $(\Gamma^*(s,t), L(z))$, where $\Gamma^*(s,t)$ is again a Gaussian process with mean zero and covariance function as in (3.1), and the covariance function of the process (Γ^*, L) is

$$Q_{\Gamma^*,L}(s,t,z) = E\Gamma^*(s,t)L(z)$$
$$= -\int_0^t \frac{c(s, y \wedge z)}{y}\,dy + c(s, t \wedge z) + t\int_0^1 \frac{c(s, y \wedge z)}{y}\,dy - tc(s,z) ,$$

where $c(s,t)$ is defined by (7.5).

LEMMA 7.2. We assume that H_0 holds. Then, as $n \to \infty$, we have

$$(\hat{\alpha}_n, L_n) \to (\Gamma, L) , \tag{7.32}$$
$$(\hat{\beta}_n, L_n) \to (\Gamma, L) , \tag{7.33}$$
$$(\hat{\gamma}_n, L_n) \to (\Gamma^*, L) \tag{7.34}$$

in $D[0,1]^2 \times D[0,1]$.

PROOF. The weak convergence statements of (7.32) and (7.34) follow from Lemma 7.1. As to (7.33), we note that (cf Observation 4.1 and Lemma 4.1 with $q \equiv 1$)

$$\sup_{0 \leq t \leq 1} \sup_{0 \leq s \leq 1} |\hat{\beta}_n(s,t) - \hat{\alpha}_n(s,t)|$$
$$= \sup_{0 \leq t \leq 1} \sup_{0 \leq s \leq 1} |\hat{\alpha}_n(U_n(s),t) - \hat{\alpha}_n(s,t)|$$
$$= o_P(1). \qquad \square$$

PROOF OF THEOREM 7.2. Arguing similarly as in the proof of Theorem 7.1, in order to prove (7.13), (7.14) and (7.15), we use Lemma 7.2 and Corollaries 3.1, 4.1 and 6.1, respectively. The finiteness of the limiting covariance functions is shown similarly to that in the proof of Theorem 7.1. $\qquad \square$

8. Weighted multi-time parameter empirical processes

Let $X = (X^{(1)}, \ldots, X^{(d)})$, $X_i = (X_i^{(1)}, \ldots, X_i^{(d)})$, $i = 1, 2, \ldots$, be independent random vectors in $\mathbb{R}^d$, $d \geq 1$, with a distribution function F. Define the $(d+1)$-time parameter empirical process $\zeta_n(\mathbf{x}, t)$ by

$$\zeta_n(\mathbf{x}, t) = n^{-1/2} \sum_{i=1}^{[nt]} (\mathbf{1}\{X_i \leq \mathbf{x}\} - F(\mathbf{x})), \quad \mathbf{x} \in \mathbb{R}^d, \quad 0 \leq t \leq 1 .$$

A Kiefer process $K_F(\mathbf{x}, t)$ on $\mathbb{R}^d \times [0, \infty)$ associated with a distribution function F on $\mathbb{R}^d$, $d \geq 1$, is a separable $(d+1)$-parameter real-valued Gaussian process with $K_F(\mathbf{x}, 0) = 0$, $EK_F(\mathbf{x}, t) = 0$ and covariance function

$$EK_F(\mathbf{x}, s)K_F(\mathbf{y}, t) = (F(\mathbf{x} \wedge \mathbf{y}) - F(\mathbf{x})F(\mathbf{y}))(s \wedge t)$$

for all $\mathbf{x}, \mathbf{y} \in \mathbb{R}^d$ and $s, t \geq 0$, where, and throughout, the symbol $\wedge$ means minimum, component-wise in higher dimensions.

Let F be an arbitrary distribution function on $\mathbb{R}^d$, $d \geq 1$, and let $\{K_F(\mathbf{x}, t); \mathbf{x} \in \mathbb{R}^d, t \geq 0\}$ be a Kiefer process associated with F.

Csörgő and Horváth (1988b) proved the following strong approximation result.

THEOREM 8.A. *Assume that $X_1, X_2, \ldots$ are independent random vectors with an arbitrary distribution function F on $\mathbb{R}^d$. Then there exists a Kiefer process $\{K_F(\mathbf{x}, t); \mathbf{x} \in \mathbb{R}^d, t \geq 0\}$, associated with F on $\mathbb{R}^d$, such that*

$$\sup_{0 \leq t \leq 1} \sup_{\mathbf{x} \in \mathbb{R}^d} |n^{1/2} \zeta_n(\mathbf{x}, t) - K_F(\mathbf{x}, nt)| \stackrel{a.s.}{=} O(n^{1/2 - 1/(4d)} (\log n)^{3/2}).$$

Here we give characterizations of the asymptotics of weighted multi-time parameter empirical processes and their tied-down at $t = 1$ versions, in supremum and L_p-metrics. Our weighted asymptotics in supremum metrics are based on Theorem 8.A and the following integral test which follows immediately from Theorem 2.1 of Csörgő, Horváth and Szyszkowicz (1994).

LEMMA 8.A. *Let $q \in \mathcal{Q}$ and $\{K_F(\mathbf{x}, t); x \in \mathbb{R}^d, t \geq 0\}$ be a Kiefer process. Then we have*

(a) $\quad \limsup_{t \downarrow 0} \sup_{\mathbf{x} \in \mathbb{R}^d} |K(\mathbf{x}, t)|/q(t) < \infty \quad$ a.s.

if and only if $I(q, c) < \infty$ for some $c > 0$,

(b) $\quad \lim_{t \downarrow 0} \sup_{\mathbf{x} \in \mathbb{R}^d} |K(\mathbf{x}, t)|/q(t) = 0 \quad$ a.s.

if and only if $I(q, c) < \infty$ for all $c > 0$.

THEOREM 8.1. *Assume that $X_1, X_2, \ldots$ are independent random vectors with an arbitrary distribution function F on $\mathbb{R}^d$ and let $q \in \mathcal{Q}$. Then, there exists a Kiefer process $\{K_F(\mathbf{x}, t); \mathbf{x} \in \mathbb{R}^d, t \geq 0\}$ associated with F such that, as $n \to \infty$, we have*

(a) $\quad \sup_{0 < t \leq 1} \sup_{\mathbf{x} \in \mathbb{R}^d} |\zeta_n(\mathbf{x}, t) - n^{-1/2} K_F(\mathbf{x}, nt)|/q(t) = o_P(1) \quad$ (8.1)

if and only if $I(q, c) < \infty$ for all $c > 0$,

(b) $\quad \sup_{0 < t \leq 1} \sup_{\mathbf{x} \in \mathbb{R}^d} |\zeta_n(\mathbf{x}, t) - n^{-1/2} K_F(\mathbf{x}, nt)|/q(t) = O_P(1) \quad$ (8.2)

if and only if $I(q, c) < \infty$ for some $c > 0$,

(c) $$\sup_{0<t\leq 1}\sup_{\mathbf{x}\in\mathbb{R}^d}|\zeta_n(\mathbf{x},t)|/q(t) \xrightarrow{\mathcal{D}} \sup_{0<t\leq 1}\sup_{\mathbf{x}\in\mathbb{R}^d}|K_F(\mathbf{x},t)|/q(t) ,$$ (8.3)

if and only if $I(q,c) < \infty$ for some $c > 0$, where $\{K_F(\mathbf{x},t); \mathbf{x} \in \mathbb{R}^d, 0 \leq t \leq 1\}$ is a Kiefer process associated with F on $\mathbb{R}^d$.

Theorem 8.1 was proved by Csörgő and Szyszkowicz (1994) under stronger conditions. Namely, (8.1) was obtained for weight functions q which are positive on $(0, 1]$ and such that (2.A) holds or, equivalently, on assuming that

$$q \in C(0,1], \quad q(t)/t^{1/2} \quad \text{is nonincreasing near zero},$$ (8.4)

and such that for a given integer $d \geq 1$ and all $c > 0$

$$I_d(q,c) < \infty ,$$ (8.5)

where

$$I_d(q,c) := \int_0^1 \frac{q^{2d}(t)}{t^{d+1}} \exp\left(-c\frac{q^2(t)}{t}\right) dt .$$

Statements as in (8.2) and (8.3) were proved for positive weight functions q and such that (2.B) holds, or equivalently, on assuming (8.4) and that for a given $d \geq 1$ and some $c > 0$ we have (8.5). Conditions (2.A) and (2.B) resulted from the law of the iterated logarithm for the suprema of Kiefer processes, while those which involved the integral $I_d(q,c)$ were arrived at as a result of the Adler and Brown (1986) test for upper and lower functions for suprema of Kiefer processes (cf also the lines of discussion following Remark 2.2 in our Section 2). The proof of Theorem 8.1 is similar to that of Theorem 3.1 in Csörgő and Szyszkowicz (1994), when using Lemma 8.A in lieu of the just discussed conditions on weight functions. When $d = 1$, Theorem 8.1 reduces to restating Theorem 2.1 with an arbitrary distribution function F on $\mathbb{R}^1$.

When testing for the possibility of having a change in distribution of a sequence of chronologically ordered d-dimensional observations $X_i = (X_i^{(1)}, \ldots, X_i^{(d)})$, $i = 1, \ldots, n$, $d \geq 1$, at an unknown time $1 \leq k < n$, it is natural to compare the empirical distributions "before" to those "after" (for $d = 1$, see (1.8) as well as our discussion in the Introduction and Section 3).

Thus, similarly, as in the case of $d = 1$, we are led to considering "tied down" multi-time parameter empirical processes with weights which would continue emphasizing the possibility of having a change in distribution on the tails, but in a non-degenerate way (cf Csörgő and Szyszkowicz (1994) and Csörgő, Horváth and Szyszkowicz (1994)).

We define the "tied down in $t = 1$" multi-time parameter empirical bridge process $\{\hat{\zeta}_n(\mathbf{x},t); \mathbf{x} \in \mathbb{R}^d, 0 \leq t \leq 1\}$, $n \geq 1$, by

$$\hat{\zeta}_n(\mathbf{x}, t) = \begin{cases} n^{-1/2}\left(\sum_{i=1}^{[(n+1)t]}\mathbf{1}\{X_i \leq \mathbf{x}\} - \frac{[(n+1)t]}{n}\sum_{i=1}^{n}\mathbf{1}\{X_i \leq \mathbf{x}\}\right), \\ \qquad\qquad\qquad \mathbf{x} \in \mathbb{R}^d, \ 0 \leq t < 1, \\ 0, \qquad\qquad\qquad \mathbf{x} \in \mathbb{R}^d, \ t = 1 \ . \end{cases}$$

Let $\{K_F(\mathbf{x}, t); \mathbf{x} \in \mathbb{R}^d, t \geq 0\}$ be a Kiefer process associated with an arbitrary distribution function F on $\mathbb{R}^d$. Define the Gaussian process $\{\Gamma_F(\mathbf{x}, t); \mathbf{x} \in \mathbb{R}^d, 0 \leq t \leq 1\}$ by

$$\Gamma_F(\mathbf{x}, t) = K_F(\mathbf{x}, t) - tK_F(\mathbf{x}, 1), \quad \mathbf{x} \in \mathbb{R}^d, \quad 0 \leq t \leq 1 \ . \tag{8.6}$$

Consequently, $\Gamma_F(\cdot, \cdot)$ is a separable Gaussian process with mean zero and the covariance function

$$E\Gamma_F(\mathbf{x}_1, t_1)\Gamma_F(\mathbf{x}_2, t_2) = (F(\mathbf{x}_1 \wedge \mathbf{x}_2) - F(\mathbf{x}_1)F(\mathbf{x}_2))(t_1 \wedge t_2 - t_1 t_2) \ . \tag{8.7}$$

Along the lines of Theorem 8.1, we have the following result.

THEOREM 8.2. Assume that $X_1, X_2, \ldots$ are independent random vectors with an arbitrary distribution function F on $\mathbb{R}^d$ and let $q \in \mathcal{Q}^*$. Then there exists a Kiefer process $\{K_F(\mathbf{x}, t); \mathbf{x} \in \mathbb{R}^d, t \geq 0\}$, associated with F on $\mathbb{R}^d$, such that with

$$\{\Gamma_{F,n}(\mathbf{x}, t); \mathbf{x} \in \mathbb{R}^d, 0 \leq t \leq 1, n \geq 1\}$$
$$:= \{n^{-1/2}(K_F(\mathbf{x}, nt) - tK_F(\mathbf{x}, n)); \mathbf{x} \in \mathbb{R}^d, 0 \leq t \leq 1, n \geq 1\} \tag{8.8}$$
$$\stackrel{\mathcal{D}}{=} \{\Gamma_F(\mathbf{x}, t); \mathbf{x} \in \mathbb{R}^d, 0 \leq t \leq 1\} \ ,$$

as $n \to \infty$, we have

(a) $\quad \sup_{0<t<1} \sup_{\mathbf{x}\in\mathbb{R}^d} |\hat{\zeta}_n(\mathbf{x}, t) - \Gamma_{F,n}(\mathbf{x}, t)|/q(t) = o_P(1)$

if and only if $I^*(q, c) < \infty$ for all $c > 0$,

(b) $\quad \sup_{0<t<1} \sup_{\mathbf{x}\in\mathbb{R}^d} |\hat{\zeta}_n(\mathbf{x}, t) - \Gamma_{F,n}(\mathbf{x}, t)|/q(t) = O_P(1)$

if and only if $I^*(q, c) < \infty$ for some $c > 0$,

(c) $\quad \sup_{0<t<1} \sup_{\mathbf{x}\in\mathbb{R}^d} |\hat{\zeta}_n(\mathbf{x}, t)|/q(t) \stackrel{\mathcal{D}}{\to} \sup_{0<t<1} \sup_{\mathbf{x}\in\mathbb{R}^d} |\Gamma_F(\mathbf{x}, t)|/q(t)$

if and only if $I^*(q, c) < \infty$ for some $c > 0$, where $\{\Gamma_F(\mathbf{x}, t); \mathbf{x} \in \mathbb{R}^d, 0 \leq t \leq 1\}$ is a separable Gaussian process as in (8.7), associated with F.

Theorem 8.2 was proved by Csörgő and Szyszkowicz (1994) under stronger conditions that are parallel to those summarized after Theorem 8.1 here (cf also our discussion after Remark 3.1). Again, the proof of Theorem 8.2 under the present optimal conditions on weight functions is similar to the proof of Theorem 3.2 in Csörgő and Szyszkowicz (1994) when making use now of the following integral test of Csörgő, Horváth and Szyszkowicz (1994).

LEMMA 8.B. Let $q \in \mathscr{Q}^*$ and $\{\Gamma_F(\mathbf{x},t); \mathbf{x} \in \mathbb{R}^d, 0 \leq t \leq 1\}$ be a Gaussian process as in (8.7). Then we have

(a) $\qquad \limsup_{t \downarrow 0} \sup_{\mathbf{x} \in \mathbb{R}^d} |\Gamma_F(\mathbf{x},t)|/q(t) < \infty \quad \text{a.s.}$

and

$$\limsup_{t \uparrow 1} \sup_{\mathbf{x} \in \mathbb{R}^d} |\Gamma_F(\mathbf{x},t)|/q(t) < \infty \quad \text{a.s.}$$

if and only if $I^*(q,c) < \infty$ for some $c > 0$,

(b) $\qquad \limsup_{t \downarrow 0} \sup_{\mathbf{x} \in \mathbb{R}^d} |\Gamma_F(\mathbf{x},t)|/q(t) = 0 \quad \text{a.s.}$

and

$$\limsup_{t \uparrow 1} \sup_{\mathbf{x} \in \mathbb{R}^d} |\Gamma_F(\mathbf{x},t)|/q(t) = 0 \quad \text{a.s.}$$

if and only if $I^*(q,c) < \infty$ for all $c > 0$.

Part (c) of Theorem 8.2 was proven by Csörgő, Horváth and Szyszkowicz (1994). When $d = 1$, Theorem 8.2 reduces to restating Theorem 3.1 with an arbitrary distribution function F on $\mathbb{R}^1$.

While it is true that appropriate weighted L_p-approximations for $\zeta_n(\mathbf{x},t)$ and $\hat{\zeta}_n(\mathbf{x},t)$ follow from (a) of Theorem 8.1 and (a) of Theorem 8.2 respectively, nevertheless, such approximations for a larger class of weight functions are of interest on their own. Indeed, the weighted L_p-approximations of empirical processes we state here are best possible in that they hold true for the optimal class of weight functions as characterized by (8.9).

THEOREM 8.3. Assume that $X_1, X_2, \ldots$ are independent random vectors with an arbitrary distribution function F on $\mathbb{R}^d$, $d \geq 1$. Assume also that q is a positive weight function on $(0,1]$ and let $0 < p < \infty$. Then the following statements are equivalent:

(a) We have

$$\int_0^1 t^{p/2}/q(t)\,dt < \infty \ . \tag{8.9}$$

(b) There exists a Kiefer process $\{K_F(\mathbf{x},t); \mathbf{x} \in \mathbb{R}^d, t \geq 0\}$, associated with F, such that, as $n \to \infty$, we have

$$\int_0^1 \int_{\mathbb{R}^d} |\zeta_n(\mathbf{x},t) - n^{-1/2} K_F(\mathbf{x},nt)|^p/q(t) \, dF(\mathbf{x}) \, dt = o_P(1) \ .$$

(c) We have, as $n \to \infty$

$$\int_0^1 \int_{\mathbb{R}^d} |\zeta_n(\mathbf{x},t)|^p/q(t) dF(\mathbf{x}) \, dt \xrightarrow{\mathscr{D}} \int_0^1 \int_{\mathbb{R}^d} |K_F(\mathbf{x},t)|^p/q(t) \, dF(\mathbf{x}) dt \ ,$$

where $\{K_F(\mathbf{x},t); \mathbf{x} \in \mathbb{R}^d, \ 0 \leq t \leq 1\}$ is a Kiefer process associated with F on $\mathbb{R}^d$.

The L_p-version of Theorem 8.2 is also best possible. Namely we have the following optimal results.

THEOREM 8.4. Assume that $X_1, X_2, \ldots$ are independent random vectors with an arbitrary distribution function F on $\mathbb{R}^d$. Assume also that q is a positive weight function on $(0,1)$ and let $0 < p < \infty$. Then the following statements are equivalent:

(a) We have

$$\int_0^1 (t(1-t))^{p/2}/q(t) \, dt < \infty \ .$$

(b) There exists a Kiefer process $\{K_F(\mathbf{x},t); \mathbf{x} \in \mathbb{R}^d, t \geq 0\}$, associated with F, such that for the Gaussian process $\{\Gamma_{F,n}(\mathbf{x},t); \mathbf{x} \in \mathbb{R}^d, \ 0 \leq t \leq 1, \ n \geq 1\}$ as in (8.8), as $n \to \infty$, we have

$$\int_0^1 \int_{\mathbb{R}^d} |\hat{\zeta}_n(\mathbf{x},t) - \Gamma_{F,n}(\mathbf{x},t)|^p/q(t) \, dF(\mathbf{x}) \, dt = o_P(1) \ .$$

(c) We have, as $n \to \infty$,

$$\int_0^1 \int_{\mathbb{R}^d} |\hat{\zeta}_n(\mathbf{x},t)|^p/q(t) dF(\mathbf{x}) \, dt \xrightarrow{\mathscr{D}} \int_0^1 \int_{\mathbb{R}^d} |\Gamma_F(\mathbf{x},t)|^p/q(t) dF(\mathbf{x}) \, dt \ ,$$

(8.10)

where $\{\Gamma_F(\mathbf{x},t); \mathbf{x} \in \mathbb{R}^d; \ 0 \leq t \leq 1\}$ is a Gaussian process as in (8.7), associated with F.

It is of interest to note here that a first step towards the applicability of (8.10) in statistics is to replace $dF(\mathbf{x})$ by $dF_n(\mathbf{x})$ on the left hand side of (8.10) for the sake of computability, and then prove that the same convergence in distribution remains true. This would also open the door for tackling the problem of tabulating critical values by bootstrap methods.

The proof of Theorem 8.3 is based on Theorem 8.A and the following result which is a consequence of Theorem 2.1 of Csörgő, Horváth and Shao (1993).

LEMMA 8.C. Let $\{K_F(\mathbf{x}, t); \mathbf{x} \in \mathbb{R}^d, 0 \leq t \leq 1\}$ be a Kiefer process associated with an arbitrary distribution function F on $\mathbb{R}^d$, and let q be a positive function on $(0, 1]$. Then, with $0 < p < \infty$, we have

$$\int_0^1 \int_{\mathbb{R}^d} |K_F(\mathbf{x}, t)|^p / q(t) \, dF(\mathbf{x}) \, dt < \infty \quad \text{a.s.}$$

if and only if

$$\int_0^1 t^{p/2} / q(t) \, dt < \infty \ .$$

The proof of Theorem 8.4 is based on Theorem 8.3 and the following analogue of Lemma 8.C.

LEMMA 8.D. Let $\{\Gamma_F(\mathbf{x}, t); \mathbf{x} \in \mathbb{R}^d, 0 \leq t \leq 1\}$ be a separable Gaussian process with mean zero and covariance function as in (8.7) and q be a positive function on $(0, 1)$. We assume that $0 < p < \infty$. Then the statements

$$\int_0^1 (t(1-t))^{p/2} / q(t) \, dt < \infty$$

and

$$\int_0^1 \int_{\mathbb{R}^d} |\Gamma_F(\mathbf{x}, t)|^p / q(t) \, dF(\mathbf{x}) \, dt < \infty \quad \text{a.s.}$$

are equivalent.

For detailed proofs of Theorems 8.3 and 8.4 and those of Lemmas 8.C and 8.D respectively, we refer to Csörgő and Szyszkowicz (1994). When $d = 1$, Theorems 8.3 and 8.4 reduce to restating Theorems 2.2, 3.3, and 3.2, 3.4, respectively, with an arbitrary distribution function F on $\mathbb{R}^1$.

REMARK 8.1. We note that, due to Theorem 8.1(a), we of course have also weak convergence of $\zeta_n(\mathbf{x}, t)/q(t)$ to $K_F(\mathbf{x}, t)/q(t)$ in $D(\mathbb{R}^d \times [0, 1])$ for $q \in \mathcal{Q}$ and such that $I(q, c) < \infty$ for all $c > 0$. Consequently, for such a class of weight functions (arguing similarly as in Section 7), we obtain Theorem 7.1 of Csörgő and Szyszkowicz (1994) which gives weak convergence of $\zeta_n(\mathbf{x}, t)/q(t)$ under a sequence of contiguous measures. Also, weak convergence of $\hat{\zeta}_n(\mathbf{x}, t)/q(t)$ to $\Gamma_F(\mathbf{x}, t)/q(t)$ in $D(\mathbb{R}^d \times [0, 1])$ with $q \in \mathcal{Q}^*$ and such that $I^*(q, c) < \infty$ for all $c > 0$, which follows from Theorem 8.2(a), allows one to improve Theorem 7.2 of Csörgő and Szyszkowicz (1994) in the very same way.

Acknowledgement

The author wishes to thank Miklós Csörgő for reading this manuscript and for his helpful comments and suggestions. Research supported by an NSERC Canada grant at Carleton University, Ottawa

References

Adler, R. J. and L. D. Brown (1986). Tail behaviour for suprema of empirical processes. *Ann. Probab.* **14**, 1–30.

Anderson, T. W. and D. A. Darling (1952). Asymptotic theory of certain "goodness of fit" criteria based on stochastic processes. *Ann. Math. Statist.* **23**, 193–212.

Barndorff-Nielsen, O. (1963). On the limit behaviour of extreme order statistics. *Ann. Math. Statist.* **34**, 992–1002.

Bhattacharya, P. K. and D. Jr. Frierson (1981). A nonparametric control chart for detecting small disorders. *Ann. Statist.* **9**, 544–554.

Bickel, P. J. and M. J. Wichura (1971). Convergence criteria for multiparameter stochastic processes and some applications. *Ann. Math. Statist.* **42**, 1656–1670.

Blum, J. R., J. Kiefer and M. Rosenblatt (1961). Distribution free tests of independence based on the sample distribution function. *Ann. Math. Statist.* **32**, 485–498.

Bonvalot, F. and N. Castelle (1991). Strong approximation of uniform empirical process by Kiefer process. Prépublications 91-41, Université de Paris-Sud Mathématiques, Bât 425, Orsey, France.

Bretagnolle, J. and P. Massart (1989). Hungarian constructions from the nonasymptotic viewpoint. *Ann. Prob.* **17**, 239–256.

Brillinger, D. R. (1969). An asymptotic representation of the sample distribution function. *Bull. Amer. Math. Soc.* **75**, 545–547.

Brodsky, B. E. and B. S. Darkhovsky (1993). Nonparametric methods in Change-Point Problems. Kluwer, Dordrecht.

Cantelli, F. P. (1933). Sulla determinazione empirica delle leggi di probabilità. *Giorn. Ist. Ital. Attuari* **4**, 421–424.

Chibisov, D. (1964). Some theorems on the limiting behaviour of empirical distribution functions. *Selected Transl. Math. Statist. Prob.* **6**, 147–156.

Chung, K. L. (1949). An estimate concerning the Kolmogoroff limit distribution. *Trans. Amer. Math. Soc.* **67**, 36–50.

Correa, J. A. (1995). Weighted Approximations and Contiguous Weak Convergence of Parameter-Estimated Empirical Processes with Applications to Changepoint Analysis. Ph.D. Thesis, Carleton University.

Csörgő, M. (1983). Quantile Processes with Statistical Applications. SIAM 42, Philadelphia, Pennsylvania.

Csörgő, M., S. Csörgő and L. Horváth, L. (1986). An Asymptotic Theory for Empirical Reliability and Concentration Processes. *Lecture Notes in Statistics* **33**. Springer-Verlag, Berlin.

Csörgő, M., S. Csörgő, L. Horváth and D. Mason (1986). Weighted empirical and quantile processes. *Ann. Probab.* **14**, 31–85.

Csörgő, M. and L. Horváth (1987a). Nonparametric tests for the changepoint problem. *J. Statist. Plann. Infer.* **17**, 1–9.

Csörgő, M. and L. Horváth (1987b). Detecting change in a random sequence. *J. Multivar. Anal.* **23**, 119–130.

Csörgő, M. and L. Horváth (1988a). Nonparametric methods for changepoint problems. *Handbook of Statistics*, Vol. 7 403–425, Elsevier Science Publishers B.V. (North-Holland).

Csörgő, M. and L. Horváth (1988b). A note on strong approximations of multivariate empirical processes. *Stoch. Process. Appl.* **28**, 101–109.

Csörgő, M. and L. Horváth (1993). *Weighted Approximations in Probability and Statistics*. Wiley, Chichester.

Csörgő, M., L. Horváth and Q. M. Shao (1993). Convergence of integrals of uniform empirical and quantile processes. *Stochastic Process. Appl.* **45**, 283–294.

Csörgő, M., L. Horváth and Szyszkowicz, B. (1994). Integral tests for suprema of Kiefer processes with application. In: Tech. Rep. Ser. Lab. Res. Statist. and Probab. No. **257**, Carleton U.-U. of Ottawa.

Csörgő, M., Q. M. Shao and B. Szyszkowicz (1991). A note on local and global functions of a Wiener process and some Rényi-type statistics. *Studia Sci. Math. Hungar.* **26**, 239–259.

Csörgő, M. and B. Szyszkowicz (1994). Applications of multi-time parameter processes to change-point analysis. In: *Probability Theory and Mathematical Statistics*–Proceedings of the Sixth Vilnius Conference, pp. 159–222, B. Grigelionis et al. (Eds.), VSP/TEV 1994.

Csörgő, M. and B. Szyszkowicz (1998). Sequential quantile and Bahadur-Kiefer processes. This volume.

Csörgő, M. and P. Révész (1981). *Strong Approximations in Probability and Statistics*. Academic Press, New York.

Darling, D.A. and P. Erdős (1956). A limit theorem for the maximum of normalized sums of independent random variables. *Duke Math. J.* **23**, 143–145.

Deshayes, J. and D. Picard (1986). Off-line statistical analysis of change-point models using non parametric and likelihood methods. Lecture Notes in Control and Information Sciences, M. Thoma and A. Wyner, eds., **77**: Detection of Abrupt Changes in Signals and Dynamical Systems, M. Basseville and A. Benveniste, eds. Springer-Verlag, Berlin 103–168.

Donsker, M. (1952). Justification and extension of Doob's heuristic approach to the Kolmogorov-Smirnov theorems. *Ann. Math. Statist.* **23**, 277–283.

Glivenko, V. (1933). Sulla determinazione empirica delle leggi di probabilità. *Giorn. Ist. Ital. Attuari* **4**, 92–99.

Gombay, E. (1995). Nonparametric truncated sequential change-point detection. *Statistics and Decisions* **13**, 71–82.

Greenwood, P. E. and A. N. Shiryayev (1985), *Contiguity and the Statistical Invariance Principle*. Gordon and Breach Science Publishers.

Hájek, J. and Z. Šidák (1967). *Theory of Rank Test*. Academic Press, New York.

Hoeffding, W. (1948). A nonparametric test of independence. *Ann. Math. Statist.* **19**, 546–557.

Hoeffding, W. (1973). On the centering of a simple linear rank statistic. *Ann. Statist.* **1**, 54–66.

Huse, V. R. (1989). Asymptotic properties for the sequential CUSUM procedure. *Statist. Prob. Lett.* **7**, 73–80.

Khmaladze, E. V. and A. M. Parzhanadze (1986). Functional limit theorems for linear statistics of sequential ranks. *Prob. Theory Related Fields* **73**, 322–334.

Kiefer, J. (1961). On the large deviations of the empiric distribution function of vector chance variables and a law of the iterated logarithm. *Pacific J. Math.* **11**, 649–660.

Kiefer, J. (1970). Old and new methods for studying order statistics and sample quantiles. In: *Nonparametric Techniques in Stat. Inference*. Cambridge Univ. Press, London, 349–357.

Kiefer, J. (1972). Skorohod embedding of multivariate RV's and the sample D.F. *Z. Wahrsh. Verw. Gebiete* **24**, 1–35.

Kolmogorov, A. N. (1933). Sulla determinazione empirica di une legge di distribuzione. *Giorn. Ist. Ital. Attuari* **4**, 83–91.

Komlós, J., P. Major and G. Tusnády (1975). An approximation of partial sums of independent R.V.'s and the sample D.F.I. *Z. Wahrsch. verw. Gebiete* **32**, 111–131.

Le Cam, L. (1960). Locally asymptotically normal families of distributions. University of California Publications in Statistics **3**, 37–98.

Le Cam, L. (1986). *Asymptotic Methods in Statistical Decision Theory*. Springer-Verlag, New York.

Leipus, R. (1988). Weak convergence of two parameter empirical fields in the change-point problem (in Russian). *Lietuvos Matematikos Rinkinys* **28**, no. 4. 716–723.

Leipus, R. (1989). Functional limit theorems for the rank tests in change-point problem (in Russian). *Lietuvos matematikos Rinkinys* **29**, no. 4, 733–744.

Lombard, F. (1981). An invariance principle for sequential nonparametric test statistics under contiguous alternatives. *South African Statist. J.* **15**, 107–111.

Lombard, F. (1983). Asymptotic distributions of rank statistics in the change-point problem. *South African Statist. J.* **17**, 83–105.

Lombard, F. and D. M. Mason (1985). Limit theorems for generalized sequential rank statistics. *Z. Wahrsch. verw. Gebiete* **70**, 395–410.

Mason, D. M. (1981). On the use of a statistic based on sequential ranks to prove limit theorems for simple linear rank statistics. *Ann. Statist.* **9**, 424–436.

Müller, D.W. (1970). On Glivenko-Cantelli convergence. *Z. Wahrsch. Verw. Gebiete* **16**, 195–210.

Oosterhoff, J. and W. R. van Zwet (1975). A note on contiguity and Hellinger distance. In: J. Jurechkova, ed., *Contributions to Statistics*. Reidel, Dordrecht 157–166.

O'Reilly, N. (1974). On the weak convergence of empirical processes in sup-norm metrics. *Ann. Probab.* **2**, 642–651.

Pardzhanadze, A. M. and E. V. Khmaladze (1986). On the asymptotic theory of statistics of sequential ranks. *Theory Prob. Appl.* **31**, 669–682.

Parent, E. A. Jr. (1965). Sequential ranking procedures. Doctoral Dissertation, Stanford Univ.

Parzen, E. (1979). Nonparametric statistical data modeling. *J. Amer. Statist. Assoc.* **74**, 105–131.

Picard, D. (1985). Testing and estimating change-points in time series. *Adv. Appl. Prob.* **17**, 841–867.

Rényi, A. (1953). On the theory of order statistics. *Acta. Math. Acad. Sci. Hung.* **4**, 191–232.

Reynolds, M. R., Jr. (1975). A sequential signed-rank test for symmetry. *Ann. Statist.* **3**, 382–400.

Roussas, G. G. (1972). *Contiguity of Probability Measures: Some Applications in Statistics*. Cambridge University Press.

Shorack, G. R. and J. A. Wellner (1986). *Empirical Processes with Applications to Statistics*. Wiley, New York.

Smirnov, N. V. (1939). On the estimation of the discrepancy between empirical curves of distribution for two independent samples. *Bull. Math. de l'Université de Moscow* **2** (Fasc. 2).

Szyszkowicz, B. (1991a). Changepoint problem and contiguous alternatives. *Statist. and Prob. Lett.* **11**, 299–308.

Szyszkowicz, B. (1991b). Empirical type processes and contiguity. *C.R. Math. Rep. Acad. Sci. Canada* **13**, 161–166.

Szyszkowicz, B. (1991c). Weighted stochastic processes under contiguous alternatives. *C.R. Math. Rep. Acad. Sci. Canada* **13**, 211–216.

Szyszkowicz, B. (1992). On $\|\cdot/q\|$-metric convergence and contiguous alternatives. In: *Tech. Rep. Ser. Lab. Res. Stat. Prob.* No. **191**, Carleton U.-U. of Ottawa.

Szyszkowicz, B. (1993a). L_p-approximations of partial sum processes. *Stochastic Process. Appl.* **45**, 295–308.

Szyszkowicz, B. (1993b). L_p-approximations of weighted empirical type processes. In: *Tech. Rep. Ser. Lab. Res. Stat. Prob.* No. **231**, Carleton U.-U. of Ottawa.

Szyszkowicz, B. (1994). Weak convergence of weighted empirical type processes under contiguous and changepoint alternatives. *Stochastic Process. Appl.* **50**, 281–313.

Szyszkowicz, B. (1996a). Weighted approximations of partial sum processes in $D[0, \infty)$. I. *Studia Sci. Math. Hungar.* **31**, 323–353.

Szyszkowicz, B. (1996b). Asymptotic results for weighted partial sums of sequential ranks. *Statistics and Decisions.* **14**, 53–72.

Szyszkowicz, B. (1997). Weighted approximations of partial sum processes in $D[0, \infty)$. II. *Studia Sci. Math. Hungar.* **33**, 305–320.

Sequential Quantile and Bahadur–Kiefer Processes

Miklós Csörgő and Barbara Szyszkowicz

1. Introduction: Basic notions, definitions and some preliminary results

Our initial aim in writing this exposition was to develop a more or less complete theory of weighted asymptotics for sequential quantile processes along the lines of sequential empirical processes as in Szyszkowicz (1998) in this volume. We soon realized the inevitability, as well as the desirability, of having to deal simultaneously also with the Bahadur (1966) and Kiefer (1967, 1970) theory of quantiles and that of their extensions, as initiated by Csörgő and Révész (1975, 1978), in the same sequential spirit. Though this made our work more enjoyable, it became also harder and slower as well. One of the outcomes of this effort is that, while in addition to Bahadur-Kiefer elements throughout, there are three full sections of this work that are entirely devoted to the "explanation" of the Bahadur-Kiefer theory of quantiles (cf Sections 4, 5 and 6), there is only one section, Section 3, that is on weighted asymptotics for sequential quantiles. The latter, of course, is not the "complete" theory we have hoped for initially. In our present Section 3 we state only a few immediate results of our forthcoming paper, Csörgő and Szyszkowicz (1995/96), on weighted sequential quantile and Bahadur-Kiefer processes and their applications, along lines that are similar to those of Szyskowicz (1998) in this volume on sequential empirical processes.

Our global *approach to the Bahadur-Kiefer theory of quantiles and quantile processes is based on one unifying invariance principle*, namely only on the notion, first expressed by Csörgő and Révész (1981, Theorem 4.5.3 and Lemma 4.5.1), that *the uniform Bahadur-Kiefer process* is, essentially, *a process of random increments of a Kiefer process* that is *iterated* on the very empirical uniform quantiles (cf (1.13), (1.14), (1.18) and (1.19), as well as (1.26), (1.27) and (1.30) in this exposition) which, as well as their extensions to more general quantiles, we want to study via the better understood empirical distribution function as first proposed by Bahadur (1966). This approach evolves as our major guideline throughout, first mainly for uniform quantiles (cf. Sections 1 and 4) and then, when combined with the Csörgő and Révész (1978) study of deviations between the general and uniform quantile processes (cf. Sections 2 and 3), it leads also to an appropriate extension of the Bahadur-Kiefer approach to studying quantiles via empiricals (cf. Sections 5 and 6).

The theory and practice of quantiles, and quantile and Bahadur-Kiefer processes are dispersed in a vast literature all over the world. This exposition is not concerned with the impossible task of reviewing this literature. We are guided mainly, if not only, by the above described unifying notion and the tools of strong approximations (invariance principles) in probability and statistics, as summarized, for example, in the books by Csörgő and Révész (1981), Csörgő (1983), Shorack and Wellner (1986), Csörgő and Horváth (1993), and in their references, as well as by similar advances since. The choice of material for inclusion and further development in this exposition reflects only our own predilections.

1.1. Sequential uniform empirical and quantile processes

Let $U_1, U_2, \ldots$ be independent identically distributed random variables (i.i.d.r.v.'s) that are uniformly distributed on $[0,1]$. Given a chronologically ordered random sample $U_1, U_2, \ldots, U_n$, $n \geq 1$, their *sequential uniform empirical distribution function* is defined by

$$\hat{E}_{[nt]}(y) = \begin{cases} 0, & 0 \leq t < 1/n, \\ \frac{1}{[nt]} \sum_{i=1}^{[nt]} \mathbf{1}\{U_i \leq y\}, & 0 \leq y \leq 1, \ 1/n \leq t \leq 1, \end{cases} \quad (1.1)$$

where $\mathbf{1}\{A\}$ is the indicator function of the set A, and $[x]$ denotes the integer part of the number x. In terms of the *sequential uniform order statistics* $U_{1,[nt]} \leq U_{2,[nt]} \leq \cdots \leq U_{[nt],[nt]}$, $1/n \leq t \leq 1$, of the random sample $U_1, U_2, \ldots, U_n$, $\hat{E}_{[nt]}$ reads as follows:

$$\hat{E}_{[nt]}(y) = \begin{cases} 0, & \text{if } 0 \leq y < U_{1,[nt]}, \ 0 \leq t \leq 1, \\ k/[nt], & \text{if } U_{k,[nt]} \leq y < U_{k+1,[nt]}, \ 1 \leq k \leq [nt]-1, 0 \leq t \leq 1, \\ 1, & \text{if } U_{[nt],[nt]} \leq y \leq 1, \ 0 \leq t \leq 1, \end{cases} \quad (1.2)$$

with $U_{1,[nt]} = U_{[nt],[nt]} \equiv 0$ for $0 \leq t < 1/n$. Now the *sequential uniform empirical process* is defined by

$$\alpha_n(y,t) = n^{-1/2} \sum_{i=1}^{[nt]} (\mathbf{1}\{U_i \leq y\} - y) \quad (1.3)$$
$$= n^{-1/2}[nt](\hat{E}_{[nt]}(y) - y), \ 0 \leq y \leq 1, \ 0 \leq t \leq 1 \ .$$

This is the so-called two-time parameter empirical process in Szyszkowicz (1998) in this volume.

Next we define the *sequential uniform empirical quantile function*

$$\hat{U}_{[nt]}(y) = \begin{cases} U_{1,[nt]}, & \text{if } y = 0, \ 0 \leq t \leq 1, \\ U_{k,[nt]}, & \text{if } (k-1)/[nt] < y \leq k/[nt], \ 1 \leq k \leq [nt], \\ & \hspace{4cm} 0 \leq t \leq 1, \end{cases} \quad (1.4)$$

with $\hat{U}_{[nt]}(y) = 0$ for $0 \leq t < 1/n$, that is to say, we have

$$\hat{U}_{[nt]}(y) = \inf\{s : \hat{E}_{[nt]}(s) \geq y\}, \quad 0 \leq y \leq 1, \quad (1.5)$$
$$\hat{U}_{[nt]}(0) = \hat{U}_{[nt]}(0+), \quad 0 \leq t \leq 1.$$

The *sequential uniform quantile process* is defined by

$$u_n(y, t) = n^{-1/2}[nt]\big(y - \hat{U}_{[nt]}(y)\big), \quad 0 \leq y \leq 1, \ 0 \leq t \leq 1 \quad (1.6)$$

which, in the vein of Szyszkowicz (1998) in this volume, could be called the two-time parameter uniform quantile process.

We note that, by definition, $\alpha_n(y, t) = u_n(y, t) = 0$ for $0 \leq t < 1/n$.

A *Kiefer process* $\{K(y, x); 0 \leq y \leq 1, 0 \leq x < \infty\}$ is a separable 2-time parameter real-valued Gaussian process with $K(y, 0) = 0$, $EK(y, x) = 0$ and

$$EK(y_1, x_1)K(y_2, x_2) = (x_1 \wedge x_1)(y_1 \wedge y_2 - y_1 y_2) \quad (1.7)$$

for all $(y_i, x_i) \in [0, 1] \times [0, \infty)$, $i = 1, 2$, where, and throughout, the symbol $\wedge$ means minimum and, for later use, $\vee$ means maximum. For the existence and properties of a Kiefer process we refer to Csörgő and Révész (1981, Section 1.15).

Komlós, Major and Tusnády (1975) proved the following important embedding inequality for the sequential empirical process $\alpha_n(y, t)$.

THEOREM 1.A. *There exists a Kiefer process* $\{K(y, x), 0 \leq y \leq 1, 0 \leq x < \infty\}$ *such that we have*

$$P\left\{\sup_{0 \leq t \leq 1} \sup_{0 \leq y \leq 1} |n^{1/2}\alpha_n(y, t) - K(y, nt)| > (C_1 \log n + x) \log n\right\} \quad (1.8)$$
$$\leq C_2 \exp(-C_3 x)$$

for all $x > 0$, *where* C_1, C_2 *and* C_3 *are positive constants.*

Bonvalot and Castelle (1991) gave a more detailed proof of this theorem than that of Komlós, Major and Tusnády (1975), and established this result with $C_1 = 76$, $C_2 = 2.028$ and $C_3 = 1/41$. This and the Borel–Cantelli lemma give that, on the probability space of Theorem 1.A, we have with any $\epsilon > 0$

$$\limsup_{n \to \infty} \frac{n^{1/2}}{(\log n)^2} \sup_{0 \leq t \leq 1} \sup_{0 \leq y \leq 1} |\alpha_n(y, t) - n^{-1/2}K(y, nt)| \leq 117 + \epsilon \quad \text{a.s.}$$
$$(1.9)$$

Extending the Skorohod embedding scheme to multivariate random variables, Kiefer (1972) obtained the first similar strong approximation of $n^{1/2}\alpha_n(\cdot, \cdot)$ at the rate of $\mathcal{O}(n^{1/3}(\log n)^{2/3})$.

The very same Kiefer process *that was constructed for approximating* $\alpha_n(y, t)$ *in Theorem 1.A can be used also to approximate the sequential uniform quantile process* $u_n(y, t)$ *on the same probability space.* Namely we have

THEOREM 1.B. On the probability space of Theorem 1.A, with the Kiefer process $\{K(y,x), 0 \leq y \leq 1, 0 \leq x < \infty\}$ of (1.8) and an absolute positive constant C, we have

$$\limsup_{n \to \infty} \frac{n^{1/4}}{(\log n)^{1/2}(\log \log n)^{1/4}} \times \sup_{0 \leq t \leq 1} \sup_{0 \leq y \leq 1} |u_n(y,t) - n^{-1/2}K(y,nt)| \leq C \quad \text{a.s.} \tag{1.10}$$

Theorem 1.B. is an immediate corollary to Theorem 3 of Csörgő and Révész (1975) (cf Theorem 4.5.3 of Csörgő and Révész (1981)) where (1.10) is established with $t = 1$. Our outline of the proof of Theorem 1.B first concludes the latter theorem (cf (1.20)) and then (1.10) follows as a consequence.

AN OUTLINE OF THE PROOF OF THEOREM 1.B. The two sequential processes $\alpha_n(y,t)$ and $u_n(y,t)$ live on the same probability space. Hence, both of them can be viewed as living on that of Theorem 1.A as well, on which $\alpha_n(y,t)$ lives almost surely as near to $K(y,nt)$ as given in (1.9). The essence of this outline is to show that $u_n(y,t)$ can, at best, live only as near to the same Kiefer process as given in (1.10). In order to see this, we write $\|\cdot\|$ for $\sup_{0 \leq y \leq 1}|\cdot|$, and consider

$$\|u_n(y,1) - \alpha_n(y,1)\| - \|\alpha_n(y,1) - n^{-1/2}K(y,n)\|$$
$$\leq \|u_n(y,1) - n^{-1/2}K(y,n)\| \tag{1.11}$$
$$\leq \|u_n(y,1) - \alpha_n(y,1)\| + \|\alpha_n(y,1) - n^{-1/2}K(y,n)\| \ .$$

By (1.9) we have

$$\|\alpha_n(y,1) - n^{-1/2}K(y,n)\| = \mathcal{O}(n^{-1/2}(\log n)^2) \quad \text{a.s.} \ , \tag{1.12}$$

and, on account of $\alpha_n(U_{k,n},1) = u_n(\frac{k}{n},1)$, we have

$$\|u_n(y,1) - \alpha_n(y,1)\| = \|\alpha_n(\hat{U}_n(y),1) - \alpha_n(y,1)\| + \mathcal{O}(n^{-1/2}) \ . \tag{1.13}$$

Applying (1.9) again, estimating the right hand side of (1.13) we obtain

$$\|\alpha_n(\hat{U}_n(y),1) - \alpha_n(y,1)\|$$
$$= n^{-1/2}\|K(\hat{U}_n(y),n) - K(y,n)\| + \mathcal{O}(n^{-1/2}(\log n)^2) \quad \text{a.s.} \tag{1.14}$$

Consequently, we now need to know only the size of the indicated random increments of a Kiefer process. Let a_n be a sequence of positive numbers for which we have

$$\lim_{n \to \infty} \frac{\log 1/a_n}{\log \log n} = \infty \ . \tag{1.15}$$

Then (cf Theorem 1.15.2 in Csörgő and Révész (1981))

$$\lim_{n\to\infty} \sup_{0\le y\le 1-a_n} \sup_{0\le s\le a_n} \gamma(n)|K(y+s,n) - K(y,n)| = 1 \quad \text{a.s.}, \tag{1.16}$$

where $\gamma(n) = (2na_n \log 1/a_n)^{-1/2}$. The latter combined with the law of the iterated logarithm

$$\limsup_{n\to\infty} ||u_n(y,1)||/(\log\log n)^{1/2} = 2^{-1/2} \quad \text{a.s.} \tag{1.17}$$

yields (cf Lemma 4.5.1 of Csörgő and Révész (1981) and (3.1.10) of Csörgő (1983))

$$\limsup_{n\to\infty} n^{-1/4}(\log n)^{-1/2}(\log\log n)^{-1/4} \\ \times ||K(y - u_n(y,1)/n^{1/2}, n) - K(y,n)|| \le 2^{-1/4} \quad \text{a.s.} \tag{1.18}$$

Hence by (1.13), (1.14) and (1.18) we arrive at

$$\limsup_{n\to\infty} n^{1/4}(\log n)^{-1/2}(\log\log n)^{-1/4}||u_n(y,1) - \alpha_n(y,1)|| \le 2^{-1/4} \quad \text{a.s.} \tag{1.19}$$

Consequently, by (1.19) via (1.11) and (1.12), we have as well

$$\limsup_{n\to\infty} n^{1/4}(\log n)^{-1/2}(\log\log n)^{-1/4}||u_n(y,1) - n^{-1/2}K(y,n)|| \\ \le 2^{-1/4} \quad \text{a.s.} \tag{1.20}$$

The latter, in turn, implies that, with $t \in (0,1]$ and $(nt) \to \infty$ as $n \to \infty$, for n large we have

$$||n^{1/2}u_n(y,t) - K(y,nt)|| \\ \le (2^{-1/4} + o(1))(nt)^{1/4}(\log(nt))^{1/2}(\log\log(nt))^{1/4} \quad \text{a.s.}, \tag{1.21}$$

and hence, on dividing both sides by $n^{1/4}(\log n)^{1/2}(\log\log n)^{1/4}$ and then taking $\sup_{0\le t\le 1}$ on both sides, we obtain (1.10) as well, as $n \to \infty$. □

COROLLARY 1.A. *Let* $\{K(y,t), 0 \le y \le 1, 0 \le t \le 1\}$ *be a Kiefer process. As* $n \to \infty$, *we have*

(a) $\alpha_n(y,t) \xrightarrow{\mathcal{D}} K(y,t)$ *in* $D[0,1]^2$.
(b) $u_n(y,t) \xrightarrow{\mathcal{D}} K(y,t)$ *in* $D[0,1]^2$.

PROOF. The statement (a) follows from (1.9) and (b) is a consequence of (1.10), both on account of having

$$\{n^{-1/2}K(y,nt), 0 \le y \le 1, 0 \le t \le 1\} \stackrel{\mathcal{D}}{=} \{K(y,t); 0 \le y \le 1, 0 \le t \le 1\}$$

for each $n \ge 1$, which in turn implies that, as $n \to \infty$,

$$n^{-1/2}K(y,nt) \xrightarrow{\mathscr{D}} K(y,t) \text{ in } D[0,1]^2 .$$

REMARK 1.1. Müller (1970) was first to prove (a) of Corollary 1.A. The covariance function of the Gaussian process $\{K(y,1),\ 0 \leq y \leq 1\}$ (cf (1.7)) is equal to that of a *Brownian bridge* $\{B(y),\ 0 \leq y \leq 1\}$, a Gaussian process with $EB(y) = 0$ and $EB(y_1)B(y_2) = y_1 \wedge y_2 - y_1 y_2$. Hence, (a) and (b) of Corollary 1.A in this case reduce to

(a) $\alpha_n(y,1) \xrightarrow{\mathscr{D}} B(y)$ in $D[0,1]$,

(b) $u_n(y,1) \xrightarrow{\mathscr{D}} B(y)$ in $D[0,1]$,

respectively, as $n \to \infty$. The latter (a) statement is one of the famous theorems of Donsker (cf Donsker (1952)). For some historical notes we refer to Csörgő (1987).

1.2. The classical uniform Bahadur–Kiefer process and its strong approximation

Consider

$$\left\{K(y,x)/(2x \log \log x)^{1/2},\ 0 \leq y \leq 1,\ x \geq 3\right\}, \tag{1.22}$$

a net of functions in x, taking values in $C[0,1]$. This net of functions is relatively compact in $C[0,1]$ with probability one, and the set of its limit points (as $x \to \infty$) is $\mathscr{F}$, where $\mathscr{F} \subset C[0,1]$ is the set of absolutely continuous functions f for which

$$f(0) = f(1) = 0 \quad \text{and} \quad \int_0^1 (f'(t))^2 \, dt \leq 1 \tag{1.23}$$

(cf Theorem 1.15.1 of Csörgő and Révész (1981)). This, on account of (1.20), implies that the sequence of functions

$$\left\{u_n(y,1)/(2 \log \log n)^{1/2},\ 0 \leq y \leq 1, n \geq 3\right\} \tag{1.24}$$

must be also relatively compact with probability one and the set of its limit points must be $\mathscr{F}$ as well. Consequently, following Shorack (1982), as in proving Theorem 3.3.1 in Csörgő and Horváth (1993) for example, we obtain that, on the probability space of Theorem 1.A we have

$$\limsup_{n \to \infty} n^{-1/4}(\log n)^{-1/2}(\log \log n)^{-1/4}\|K(\hat{U}_n(y),n) - K(y,n)\|$$
$$\geq 2^{-1/4} \quad \text{a.s.} \tag{1.25}$$

Now a combination of (1.18) and (1.25) yields

PROPOSITION 1.A. *On the probability space of Theorem 1.A we have*

$$\limsup_{n\to\infty} n^{-1/4}(\log n)^{-1/2}(\log\log n)^{-1/4}$$
$$\times \sup_{0\leq y\leq 1} |K(\hat{U}_n(y),n) - K(y,n)| = 2^{-1/4} \quad \text{a.s.} \tag{1.26}$$

This, in turn, by (1.13) and (1.14), leads to (cf Kiefer (1970))

PROPOSITION 1.B. We have

$$\limsup_{n\to\infty} n^{-1/4}(\log n)^{-1/2}(\log\log n)^{-1/4}$$
$$\times \sup_{0\leq y\leq 1} |n^{1/2}\alpha_n(y,1) - n^{1/2}u_n(y,1)| = 2^{-1/4} \quad \text{a.s.} \tag{1.27}$$

This result is Kiefer's celebrated exact rate uniform distance version of the Bahadur (1966) representation of quantiles in the uniform-[0, 1] case, which initiated the study of uniform distance, asymptotically exact rate Bahadur–Kiefer representations of general quantiles (cf Kiefer (1970), Csörgő and Révész (1978)). Now (1.27), in turn, via (1.11) and (1.12) yields

PROPOSITION 1.C. On the probability space of Theorem 1.A we have

$$\limsup_{n\to\infty} n^{1/4}(\log n)^{-1/2}(\log\log n)^{-1/4}$$
$$\times \sup_{0\leq y\leq 1} |u_n(y,1) - n^{-1/2}K(y,n)| = 2^{-1/4} \quad \text{a.s.} \tag{1.28}$$

Kiefer (1970) used a direct method of proof for proving the result of (1.27). Whichever way, it implies that the best possible rate for the joint approximation of $u_n(y,t)$ and $\alpha_n(y,t)$ by the same Kiefer process is that given in (1.10) of Theorem 1.B. *The exact rate and notion of* (1.27) *is frequently called the Bahadur–Kiefer principle.* For summaries and further results along these lines we refer to Csörgő and Révész (1981), Csörgő (1983), Shorack and Wellner (1986), and Csörgő and Horváth (1993).

REMARK 1.2. Due to the so-called Erdős-Rényi (1970) law of large numbers, the almost sure $\mathcal{O}((\log n)^2/n^{1/2})$ rate of convergence in (1.9) is at most $\mathcal{O}(\log n)$ away from the best possible $\mathcal{O}((\log n)/n^{1/2})$ rate of convergence, i.e., it cannot be improved beyond the latter rate (cf Komlós, Major and Tusnády (1975), and/or Theorem 4.4.2 in Csörgő and Révész (1981)). Via constructing a different Kiefer process than that of (1.9) for approximating the sequential uniform quantile process directly, the rate of convergence of such an approximation is much improvable as compared to that of (1.10). Indeed, in principle, up to being as good as $\mathcal{O}((\log n)/n^{1/2})$, again the best possible rate of convergence for such a construction. This has been an *open problem* since posed in Csörgő and Révész (1975), where the first version of (1.20) was proved. Should one succeed in carrying out this new construction of a Kiefer process directly for approximating

$u_n(y,t)$, it is clear from the above proof of the Bahadur–Kiefer principle of (1.27) or, equivalently, from that of (1.26) that, as a consequence, one would have also an exact analogue of (1.28) for $\alpha_n(y,1)$ in terms of this Kiefer process, a *manifestation, again, of the Bahadur–Kiefer principle*.

The stochastic process

$$\{R_n^*(y,1),\ 0 \leq y \leq 1,\ n = 1,2,\ldots\}$$
$$:= \left\{n^{1/2}(\alpha_n(y,1) - u_n(y,1)),\ 0 \leq y \leq 1,\ n = 1,2,\ldots\right\} \quad (1.29)$$

is called the *uniform Bahadur–Kiefer process*. A direct application of (1.12) and (1.13) leads to (cf Remark 6.2.3 of Csörgő (1983))

COROLLARY 1.B. *On the probability space of Theorem 1.A, with the Kiefer process $\{K(y,x),\ 0 \leq y \leq 1,\ 0 \leq x < \infty\}$ of (1.8), we have, as $n \to \infty$,*

$$\sup_{0 \leq y \leq 1} n^{-1/2}|R_n^*(y,1) - (K(y,n) - K(\hat{U}_n(y),n))|$$
$$= \sup_{0 \leq y \leq 1} |(\alpha_n(y,1) - u_n(y,1)) - n^{-1/2}(K(y,n) - K(\hat{U}_n(y),n))| \quad (1.30)$$
$$= \mathcal{O}\left((\log n)^2/n^{1/2}\right) \quad \text{a.s.}$$

Corollary 1.B *shows the equivalence of* (1.26) *and* (1.27). The statement of (1.30) *is a strong invariance principle for the Bahadur–Kiefer process $R_n^*(y,1)$*. Namely from (1.30) we conclude again that via (1.26) we have (1.27) as well. That is to say, the sup-distance of $\alpha_n(y,1)$ from $u_n(y,1)$ is exactly the same as that of $n^{-1/2}K(y,n)$ from $n^{-1/2}K(\hat{U}_n(y),n)$, which is now seen to be just *another way of putting* again *the Bahadur–Kiefer principle*. For further discussions along these lines we refer to Csörgő and Révész (1978, 1981), Csörgő (1983, Chapter 6), Deheuvels and Mason (1990), and Csörgő and Horváth (1993, Section 3.3).

1.3. The sequential uniform Bahadur–Kiefer process and its strong approximation

It is now natural to define the *sequential uniform Bahadur–Kiefer process* $R_n^*(y,t)$ (cf (1.29)) by

$$\{R_n^*(y,t),\ 0 \leq y \leq 1,\ 0 \leq nt \leq n,\ n = 1,2,\ldots\}$$
$$:= \{n^{1/2}(\alpha_n(y,t) - u_n(y,t)),\ 0 \leq y \leq 1,\ 0 \leq nt \leq n,\ n = 1,2,\ldots\}. \quad (1.31)$$

Due to (1.26) we have, as $(nt) \to \infty$,

$$\sup_{0 \leq y \leq 1} \left|K(\hat{U}_{[nt]}(y), nt) - K(y,nt)\right|$$
$$\leq \left(2^{-1/4} + o(1)\right)(nt)^{1/4}(\log(nt))^{1/2}(\log\log(nt))^{1/4} \quad \text{a.s.}, \quad (1.32)$$

and hence, on dividing both sides by $n^{1/4}(\log n)^{1/2}(\log\log n)^{1/4}$ and then taking $\sup_{0\leq t\leq 1}$ on both sides, with an absolute positive constant C we obtain

$$\limsup_{n\to\infty} n^{-1/4}(\log n)^{-1/2}(\log\log n)^{-1/4}$$
$$\times \sup_{0\leq t\leq 1}\sup_{0\leq y\leq 1}\left|K(\hat{U}_{[nt]}(y),nt) - K(y,nt)\right| \leq C \quad \text{a.s.} \tag{1.33}$$

By definition, $\alpha_n(U_{k,[nt]},t) = u_n\left(\frac{k}{[nt]},t\right)$ and

$$\sup_{0\leq t\leq 1}\sup_{1\leq k\leq [nt]}\sup_{(k-1)/[nt]<y\leq k/[nt]}\left|u_n(y,t) - u_n\left(\frac{k}{[nt]},t\right)\right| = \mathcal{O}(n^{-1/2}). \tag{1.34}$$

Consequently, we have

$$\sup_{0\leq t\leq 1}\sup_{0\leq y\leq 1}|u_n(y,t) - \alpha_n(y,t)|$$
$$= \sup_{0\leq t\leq 1}\sup_{0\leq y\leq 1}\left|\alpha_n(\hat{U}_{[nt]}(y),t) - \alpha_n(y,t)\right| + \mathcal{O}(n^{-1/2}), \tag{1.35}$$

and now, an application of (1.9) and (1.35) yields

PROPOSITION 1.1. *On the probability space of Theorem 1.A, with the Kiefer process* $\{K(y,x), 0\leq y\leq 1, 0\leq x<\infty\}$ *of (1.8) we have, as* $n\to\infty$,

$$\sup_{0\leq t\leq 1}\sup_{0\leq y\leq 1} n^{-1/2}\left|R_n^*(y,t) - \left(K(y,nt) - K(\hat{U}_{[nt]}(y),nt)\right)\right|$$
$$= \sup_{0\leq t\leq 1}\sup_{0\leq y\leq 1}\left|(\alpha_n(y,t) - u_n(y,t)) - n^{-1/2}\left(K(y,nt) - K(\hat{U}_{[nt]}(y),nt)\right)\right|$$
$$= \mathcal{O}\left((\log n)^2/n^{1/2}\right) \quad \text{a.s.} \tag{1.36}$$

Proposition 1.1 is a strong invariance principle for the sequential uniform Bahadur–Kiefer process $R_n^*(y,nt)$ of (1.31). Namely, on account of (1.33) and (1.36) we conclude the following sequential uniform Bahadur–Kiefer principle.

PROPOSITION 1.2. *With an absolute positive constant C we have*

$$\limsup_{n\to\infty} n^{-1/4}(\log n)^{-1/2}(\log\log n)^{-1/4}\sup_{0\leq t\leq 1}\sup_{0\leq y\leq 1}|R_n^*(y,t)|$$
$$= \limsup_{n\to\infty} n^{-1/4}(\log n)^{-1/2}(\log\log n)^{-1/4} \tag{1.37}$$
$$\times \sup_{0\leq t\leq 1}\sup_{0\leq y\leq 1}\left|n^{1/2}\alpha_n(y,t) - n^{1/2}u_n(y,t)\right|$$
$$\leq C \quad \text{a.s.}$$

1.4. Definitions of sequential general empirical and quantile processes

Now we are to study sequential processes in terms of more general distributions than the uniform one. Let $X_1, X_2, \ldots$ be i.i.d.r.v.'s with distribution function F

that is defined to be right continuous. Given a chronologically ordered random sample $X_1, X_2, \ldots, X_n$, $n \geq 1$, their *sequential empirical distribution function* is defined by

$$\hat{F}_{[nt]}(x) = \begin{cases} 0, & 0 \leq t < 1/n, \\ \frac{1}{[nt]} \sum_{i=1}^{[nt]} \mathbf{1}\{X_i \leq x\}, & -\infty < x < \infty, \ 1/n \leq t \leq 1, \end{cases} \tag{1.38}$$

which in terms of the *sequential order statistics* $X_{1,[nt]} \leq X_{2,[nt]} \leq \cdots \leq X_{[nt],[nt]}$, $1/n \leq t \leq 1$, of the random sample $X_1, X_2, \ldots, X_n$, reads as follows:

$$\hat{F}_{[nt]}(x) = \begin{cases} 0, & \text{if } -\infty < x < X_{1,[nt]}, \ 0 \leq t \leq 1, \\ k/[nt], & \text{if } X_{k,[nt]} \leq x < X_{k+1,[nt]}, \ 1 \leq k \leq [nt]-1, \ 0 \leq t \leq 1, \\ 1, & \text{if } X_{[nt],[nt]} \leq x < \infty, \ 0 \leq t \leq 1, \end{cases} \tag{1.39}$$

with $X_{1,[nt]} = X_{[nt],[nt]} \equiv 0$ for $0 \leq t < 1/n$. The *sequential empirical process* is defined by

$$\beta_n(x,t) = n^{-1/2} \sum_{i=1}^{[nt]} (\mathbf{1}\{X_i \leq x\} - F(x)) \tag{1.40}$$

$$= n^{-1/2}[nt](\hat{F}_{[nt]}(x) - F(x)), \quad -\infty < x < \infty, \ 0 \leq t \leq 1.$$

Let Q be the *quantile function* of F, defined by

$$Q(y) = F^{-1}(y) = \inf\{x \colon F(x) \geq y\}, \ 0 < y \leq 1, \ Q(0) = Q(0+), \tag{1.41}$$

i.e., Q is defined to be the left continuous inverse of the right continuously defined distribution function F. Thus, for any right continuous distribution function G on the real line we have for any $0 < y < 1$

$$G(x) \geq y \quad \text{if and only if} \quad G^{-1}(y) \leq x, \tag{1.42}$$

and

$$G(x) < y \quad \text{if and only if} \quad G^{-1}(y) > x. \tag{1.43}$$

Consequently, *a random variable X with distribution function F has the same distribution as the random variable $Q(U)$*, i.e.

$$X \stackrel{\mathscr{D}}{=} Q(U), \tag{1.44}$$

where U is a uniform-$[0,1]$ random variable, since by (1.42) we have $P\{Q(U) \leq x\} = P\{U \leq F(x)\} = F(x)$.

À la (1.5), the *sequential empirical quantile function* $\hat{Q}_{[nt]}$ of $\hat{F}_{[nt]}$ is defined by

$$\hat{Q}_{[nt]}(y) = \hat{F}_{[nt]}^{-1}(y) = \inf\{x : \hat{F}_{[nt]}(x) \geq y\}, \ 0 < y \leq 1,$$
$$\hat{Q}_{[nt]}(0) = \hat{Q}_{[nt]}(0+), \ 0 \leq t \leq 1 \ , \tag{1.45}$$

i.e., we have

$$\hat{Q}_{[nt]}(y) = \hat{F}_{[nt]}^{-1}(y) = \begin{cases} X_{1,[nt]} & \text{if } y = 0, \ 0 \leq t \leq 1, \\ X_{k,[nt]} & \text{if } (k-1)/[nt] < y \leq k/[nt], \\ & 1 \leq k \leq [nt], \ 0 \leq t \leq 1 \ , \end{cases} \tag{1.46}$$

with $\hat{Q}_{[nt]}(y) = 0$ for $0 \leq t < 1/n$. Imitating (1.40), and the definition of the sequential uniform quantile process (cf (1.6)), it seems that a reasonable definition of the sequential general quantile process should be

$$\gamma_n(y,t) = n^{-1/2}[nt]\big(Q(y) - \hat{Q}_{[nt]}(y)\big), \quad 0 < y < 1, \ 0 \leq t \leq 1 \ . \tag{1.47}$$

By the Dvoretzky, Kiefer and Wolfowitz (1956) inequality there is a constant C such that

$$P\bigg\{\sup_{x \in \mathbb{R}} |\beta_n(x,1)| > z\bigg\} \leq C \exp(-2z^2) \tag{1.48}$$

for all $z > 0$. Massart (1990) obtained the optimal choice of $C = 2$. In passing, for a short history of (1.48) we refer to Proof of Lemma 3.1.4 on p. 119 of Csörgő and Horváth (1993). By (1.48) we have

$$P\bigg\{\sup_{0 \leq t \leq 1} \sup_{x \in \mathbb{R}} |\beta_n(x,t)| > \epsilon n^{1/2}\bigg\} \leq 2n \exp(-2\epsilon^2 n) \tag{1.49}$$

for all $\epsilon > 0$ and $n \geq 1$ and, on account of

$$\sup_{0 \leq t \leq 1} \sup_{0 \leq y \leq 1} |\beta_n(Q(y),t)| = \sup_{0 \leq t \leq 1} \sup_{0 \leq y \leq 1} |\alpha_n(y,t)| = \sup_{0 \leq t \leq 1} \sup_{0 \leq y \leq 1} |u_n(y,t)| \tag{1.50}$$

if F is continuous, the same inequality holds true also for the sequential uniform quantile process $u_n(y,t)$. Hence, by the Borel–Cantelli Lemma, as $n \to \infty$, we have also the *Glivenko–Cantelli theorems*

$$\sup_{0 \leq t \leq 1} \sup_{0 \leq y \leq 1} \frac{[nt]}{n}\Big|y - F\big(\hat{Q}_{[nt]}(y)\big)\Big| \to 0 \quad \text{a.s.} \tag{1.51}$$

if F is continuous, and

$$\sup_{0 \leq t \leq 1} \sup_{x \in \mathbb{R}} \frac{[nt]}{n}\big|\hat{F}_{[nt]}(x) - F(x)\big| \to 0 \quad \text{a.s.} \ , \tag{1.52}$$

with any F. While (1.51) "justifies" the definition of the sequential uniform quantile process (cf (1.6)), we note that, in contrast to (1.51), we have

$$P\left\{\lim_{n\to\infty}\sup_{0\leq t\leq 1}\sup_{0\leq y\leq 1}\frac{[nt]}{n}\left|Q(y)-\hat{Q}_{[nt]}(y)\right|=\infty\right\}=1\ ,\tag{1.53}$$

unless F has finite support. Also, only if Q is continuous at $y=y_0$, do we have that

$$\lim_{n\to\infty}Q_n(y_0)=Q(y_0)\quad\text{a.s.}\tag{1.54}$$

Otherwise this statement cannot be true. For further results along these lines we refer to Parzen (1980). These remarks already show that, *at best, the process $\gamma_n(y,t)$ can be well behaved only at its points of continuity*.

From now on we assume that F is a continuous distribution function. Then Q satisfies

$$Q(y)=\inf\{x\colon F(x)=y\},\quad F(Q(y))=y,\quad 0\leq y\leq 1\ .\tag{1.55}$$

Consequently, if X is a random variable with a continuous distribution function F, then, on account of $P\{F(X)\geq y\}=P\{X\geq Q(y)\}=1-F(Q(y))=1-y$, we have that

$$F(X)\stackrel{\mathscr{D}}{=}U\ ,\tag{1.56}$$

where U is a uniform-$[0,1]$ random variable. Hence, in this case, $U_1=F(X_1)$, $U_2=F(X_2),\ldots$ are independent uniform-$[0,1]$ random variables, and the sequential order statistics $X_{1,[nt]}\leq X_{2,[nt]}\leq\cdots\leq X_{[nt],[nt]}$, $1/n\leq t\leq 1$, of the random sample $X_1,X_2,\ldots,X_n$ *induce the sequential uniform order statistics* $U_{1,[nt]}=F(X_{1,[nt]})\leq U_{2,[nt]}=F(U_{2,[nt]})\leq\cdots\leq U_{[nt],[nt]}=F(X_{[nt],[nt]})$ of the *induced* uniform-$[0,1]$ random sample $U_1=F(X_1)$, $U_2=F(X_2),\ldots,U_n=F(X_n)$, with $U_{1,[nt]}=U_{[nt],[nt]}\equiv 0$ for $0\leq t<1/n$. Then, the thus *induced sequential uniform empirical distribution function* $\hat{E}_{[nt]}$ of this uniform random sample is given by

$$\hat{E}_{[nt]}(y)=\begin{cases}0,&\text{if }0\leq y<F(X_{1,[nt]}),\ 0\leq t\leq 1,\\ k/[nt],&\text{if }F(X_{k,[nt]})\leq y<F(X_{k+1,[nt]}),\\ &\quad 1\leq k\leq [nt]-1,\ 0\leq t\leq 1,\\ 1,&\text{if }F(X_{[nt],[nt]})\leq y\leq 1\end{cases}\tag{1.57}$$
$$=\hat{F}_{[nt]}(Q(y)),\quad 0\leq y\leq 1,\ 0\leq t\leq 1$$

or, equivalently, by

$$\hat{E}_{[nt]}(y)=\begin{cases}0,&\text{if }0\leq t\leq 1/n,\\ (1/[nt])\sum_{i=1}^{[nt]}\mathbf{1}\{F(X_i)\leq y\},&\text{if }0\leq y\leq 1,\ 1/n\leq t\leq 1,\end{cases}$$
$$=\hat{F}_{[nt]}(Q(y)),\quad 0\leq y\leq 1,\ 0\leq t\leq 1\ .\tag{1.58}$$

The similarly *induced sequential uniform empirical quantile function* is given by

$$\hat{U}_{[nt]}(y) = \begin{cases} F(X_{1,[nt]}), & \text{if } y = 0, \ 0 \leq t \leq 1, \\ F(X_{k,[nt]}), & \text{if } (k-1)/[nt] < y \leq k/[nt], \\ & 1 \leq k \leq [nt], \ 0 \leq t \leq 1, \end{cases} \quad (1.59)$$

$$= F(\hat{Q}_{[nt]}(y)), \quad 0 \leq y \leq 1, \ 0 \leq t \leq 1 \ .$$

Thus, in terms of $U_i = F(X_i)$, $i = 1, \ldots, n$, we have for any continuous distribution function F

$$\beta_n(Q(y), t) = \alpha_n(y, t), \quad 0 \leq y \leq 1, \ 0 \leq t \leq 1 \ , \quad (1.60)$$

where $\alpha_n(y, t)$ is now the *F induced version of the sequential uniform empirical process*, defined in (1.3). Hence, all theorems proved for $\alpha_n(y, t)$ will hold automatically for $\beta_n(x, t)$ as well, simply by letting $y = F(x)$ in (1.60). In particular, Theorem 1.A and (1.9) hold true automatically for $\beta_n(x, t) = \alpha_n(F(x), t)$ with any continuous distribution function F. We note in passing that, implicitly, we already used (1.60) in stating (1.50) and, similarly, the notion of (1.59) in turn was already utilized in (1.51).

Unfortunately, there is no such immediate simple route as that of (1.60) for transforming $\gamma_n(y, t)$ of (1.47) into its own *corresponding sequential uniform quantile process*

$$\begin{aligned} u_n(y, t) &= n^{-1/2}[nt](y - \hat{U}_{[nt]}(y)) \\ &= n^{-1/2}[nt](y - F(\hat{Q}_{[nt]}(y))), \quad 0 \leq y \leq 1, \ 0 \leq t \leq 1 \ . \end{aligned} \quad (1.61)$$

The transformation $y \to F(x)$ does not work directly for tying up $u_n(y, t)$ with $\gamma_n(y, t)$. Using, however, the mean value theorem, we can write

$$\begin{aligned} \gamma_n(y, t) &= n^{-1/2}[nt](Q(y) - \hat{Q}_{[nt]}(y)) \\ &= n^{-1/2}[nt](Q(y) - Q(F(\hat{Q}_{[nt]}(y)))) \\ &= n^{-1/2}[nt](y - F(\hat{Q}_{[nt]}(y)))/f(Q(\theta_n(y, t))) \\ &= u_n(y, t)/f(Q(\theta_n(y, t))), \ 0 < y < 1, \ 0 \leq t \leq 1 \ , \end{aligned} \quad (1.62)$$

where $\hat{U}_{[nt]}(y) \wedge y < \theta_n(y, t) < \hat{U}_{[nt]}(y) \vee y$, $y \in (0, 1)$, $t \in [0, 1]$, $n = 1, 2, \ldots$, provided of course that we have $Q'(y) = 1/f(Q(y)) < \infty$ for $y \in (0, 1)$, i.e., provided that F is an absolutely continuous distribution function (with respect to Lebesgue measure) with a strictly positive density function $f = F'$ on the real line. The function $f(Q(y))$ is called the *density-quantile function*, and $Q'(y) = 1/f(Q(y))$ the *quantile density function* by Parzen (1979a,b).

The relationship (1.62) shows that, for the sake of comparing $\gamma_n(y, t)$ with $u_n(y, t)$, one should first of all multiply the former by $f(Q(y))$. Hence we *assume* that $f = F'$ exists on the real line, and define the *general sequential quantile process* $\rho_n(y, t)$ by

$$\rho_n(y,t) = f(Q(y))\gamma_n(y,t)$$
$$= n^{-1/2}[nt]f(Q(y))(Q(y) - \hat{Q}_{[nt]}(y)), \quad 0 \leq y \leq 1, \ 0 \leq t \leq 1. \tag{1.63}$$

We wish to emphasize that the introduction of the density quantile function $f(Q(y))$ into the definition of the general sequential quantile process $\{\rho_n(y,t), 0 \leq y \leq 1, 0 \leq t \leq 1, n = 1,2,\ldots\}$ is for the sake of making it 'look like' the process $\{u_n(y,t), 0 \leq y \leq 1, 0 \leq t \leq 1, n = 1,2,\ldots\}$ of (1.61), or like $\{\alpha_n(y,t) = \beta_n(Q(y),t), 0 \leq y \leq 1, 0 \leq t \leq 1, n = 1,2,\ldots\}$, asymptotically. In the rest of this section we will see that the idea of studying $\rho_n(y,t)$ via its own uniform version $u_n(y,t)$ of (1.61) is a fruitful one that will also lead to conveniently comparing $\rho_n(y,t)$ with $\alpha_n(y,t)$ of (1.60) as well. On the other hand, owing to the presence of the density-quantile function in its definition, the form of ρ_n does not lend itself easily to constructing confidence bands for the quantile function $Q(y)$ (cf Csörgő and Révész (1984), and Chapter 4 of Csörgő (1983)). Indeed, having confidence bands for Q in mind, it is better to start with the *weak convergence* of $\alpha_n(y,1)$ of (1.60) to a Brownian bridge $\{B(y), 0 \leq y \leq 1\}$ (cf (a) of Remark 1.1). Assuming only that F is continuous, via $\alpha_n(y,1)$ of (1.60) one can easily arrive, for example, at (cf Csörgő and Horváth (1989))

$$\lim_{n\to\infty} P\left\{Q_n\left(y - n^{-1/2}c(\alpha)\right) \leq Q(y) < Q_n(y + n^{-1/2}c(\alpha)), \ \epsilon_n \leq y \leq 1-\epsilon_n\right\}$$
$$= P\left\{\sup_{0\leq y\leq 1} |B(y)| \leq c(\alpha)\right\} = 1 - \alpha, \tag{1.64}$$

where $B(\cdot)$ is a Brownian bridge, $c(\alpha)$ is a positive real number for which we have the latter equality holding true for a given $\alpha \in (0,1)$, and $\{\epsilon_n, n \geq 1\}$ is any sequence of positive real numbers such that $\epsilon_n \to 0$ and $n^{1/2}\epsilon_n \to \infty$ as $n \to \infty$.

1.5. Comparing the general sequential quantile process to its uniform version: Preliminary notions

Returning now to the problem of comparing the general sequential quantile process $\rho_n(y,t)$ of (1.63) to its uniform version of (1.61), with $\theta_n(y,t)$ as in (1.62), we have

$$\rho_n(y,t) = u_n(y,t)(f(Q(y))/f(Q(\theta_n(y,t)))), \quad 0 < y < 1, \ 0 \leq t \leq 1. \tag{1.65}$$

Hence one would expect $\rho_n(y,t)$ to have an asymptotic theory like that of $u_n(y,t)$, provided only that we could 'regulate' the ratio $f(Q(y))/f(Q(\theta_n(y,t)))$ uniformly in y and t over the unit interval. The following lemma and its conditions constitute a sufficient background for achieving this goal.

LEMMA 1.A. Let F be a continuous distribution function F and assume

(i) F is twice differentiable on (a, b), where

$$a = \sup\{x: F(x) = 0\}, \quad b = \inf\{x: F(x) = 1\}, \quad -\infty \leq a < b \leq \infty, \tag{1.66}$$

(ii) $F'(x) = f(x) > 0$, $x \in (a, b)$,
(iii) for some $\gamma > 0$ we have

$$\sup_{0 < y < 1} y(1-y)|f'(Q(y))|/f^2(Q(y)) \leq \gamma.$$

Then we have

$$\frac{f(Q(y_1))}{f(Q(y_2))} \leq \left\{\frac{y_1 \vee y_2}{y_1 \wedge y_2} \frac{1 - (y_1 \wedge y_2)}{1 - (y_1 \vee y_2)}\right\}^{\gamma} \tag{1.67}$$

for every pair $y_1, y_2 \in (0, 1)$.

This is Lemma 1 of Csörgő and Révész (1978). Proof can be also found in Csörgő and Révész (1981, Lemma 4.5.2), Csörgő (1983, Lemma 1.4.1), and Csörgő and Horváth (1993, Lemma 6.1.1).

In the literature on non-parametric statistics, it is customary to define the so-called *score function* (cf for example, Hájek and Šidák (1967, p. 19)):

$$J(y) = -\frac{d}{dy}f(Q(y)) = -f'(Q(y))/f(Q(y)). \tag{1.68}$$

Thus, our condition (1.66) (iii) can be written as

$$\sup_{0 < y < 1} y(1-y)|J(y)|/f(Q(y)) \leq \gamma. \tag{1.69}$$

For example,

$$J(y) = -1, \quad \text{if } F(x) = 1 - \exp(-x), \ x \geq 0,$$
$$J(y) = 1, \quad \text{if } F(x) = 1 - \exp(x), \ x \leq 0,$$

and

$$f(Q(y)) = 1 - y, \quad \text{if } F(x) = 1 - \exp(-x), \ x \geq 0$$
$$f(Q(y)) = y, \quad \text{if } F(x) = 1 - \exp(x), \ x \leq 0.$$

Hence, in these two cases of exponential distributions, γ of (1.66) (iii) is equal to 1. For further examples, and a discussion of tail monotonicity assumptions of extreme value theory as related to (1.69), we refer to Parzen (1979a, 1980).

We note in passing that the score function J of (1.68) plays an important role in non-parametric and robust statistical analysis (cf, for example, Hájek and Šidák (1967) and Huber (1981)). Owing to its importance, and because of our lack of knowledge of f in most practical situations, it is desirable to estimate J, given a random sample $X_1,\ldots,X_n$ on F. For results on estimating J, and for further discussions along these lines, we refer to Hájek and Šidák (1967, p. 259), Parzen (1979a), Csörgő (1983, Chapter 10), Csörgő and Révész (1986) and Burke and Horváth (1989), as well as to references in these works.

Back to the problem of the general sequential quantile process $\rho_n(y,t)$ *versus* its uniform version $u_n(y,t)$, by (a) of Remark 1.1 we have

$$\alpha_n(y,1) \xrightarrow{\mathscr{D}} N(0, y(1-y)) \quad \text{for every } \textit{fixed } y \in (0,1), \tag{1.70}$$

and one has also (cf, e.g., Rényi (1970, p. 490))

$$\gamma_n(y,1) \xrightarrow{\mathscr{D}} N(0, y(1-y)/f^2(Q(y))), \quad y \textit{ fixed in } (0,1), \tag{1.71}$$

provided that F is absolutely continuous in an interval around $Q(y)$ and $f(Q(y))$ is positive and continuous at y. (Continuity of the latter at y can be dropped).

1.6. Preliminary notions on the weak convergence of the general sequential quantile process

Given the conditions of (1.71), an equivalent way of putting it is to write

$$\rho_n(y,1) = f(Q(y))\gamma_n(y,1) \xrightarrow{\mathscr{D}} N(0, y(1-y)), \quad y \textit{ fixed in } (0,1). \tag{1.72}$$

The latter (1.72)-version of (1.71) underlines again the rationale behind defining the general sequential quantile process $\rho_n(y,t)$ the way we did in (1.63). In the light of (1.72) and (b) of Corollary 1.A, it is only natural to ask for a similar result for $\rho_n(y,t)$. The next result (cf Theorem 1.5.1 of Csörgő (1983)) will illustrate the immediate usefulness of Lemma 1.A in this direction.

THEOREM 1.C. *Given the conditions of Lemma 1.A on F, we have for all $\epsilon > 0$, $0 < c < 1$ and all $n \geq 1$*

$$\begin{aligned}P\bigg\{\sup_{c \leq y \leq 1-c} &\left|\frac{f(Q(y))}{f(Q(\theta_n(y,1)))} - 1\right| > \epsilon\bigg\} \\ &\leq 4([\gamma]+1)\bigg\{\exp\left(-nch((1+\epsilon)^{(1/2)([\gamma]+1)})\right) \\ &\quad + \exp\left(-nch(1/(1+\epsilon)^{(1/2)([\gamma]+1)})\right)\bigg\} \\ &=: H_n(\gamma, c, h, \epsilon), \end{aligned} \tag{1.73}$$

where $\theta_n(y,1)$ is as in (1.62), $\gamma > 0$ is as in Lemma 1.A, and $h(x) = x + \log(1/x) - 1$, $x \geq 1$.

As a consequence of (1.73), we have also

$$P\left\{\sup_{0\leq t\leq 1}\sup_{c\leq y\leq 1-c}\left|\frac{f(Q(y))}{f(Q(\theta_n(y,t)))}-1\right|>\epsilon\right\} \leq nH_n(\gamma,c,h,\epsilon) \tag{1.74}$$

for all $\epsilon > 0$, $0 < c < 1$ and all $n \geq 1$, where $\theta_n(y,t)$ is as in (1.62).

COROLLARY 1.1. With $h(\cdot)$ as in Theorem 1.C, let

$$c_n := 2(\log n)\bigg/\left(n\left(h\left((1+\epsilon)^{(1/2)([\gamma]+1)}\right)\wedge h\left((1+\epsilon)^{-(1/2)([\gamma]+1)}\right)\right)\right). \tag{1.75}$$

Assume that F satisfies the conditions of Lemma 1.A on F. Then, as $n \to \infty$, we have

$$\sup_{0\leq t\leq 1}\sup_{c_n\leq y\leq 1-c_n}|\rho_n(y,t)-u_n(y,t)|=o_P(1), \tag{1.76}$$

where $\rho_n(\cdot,\cdot)$ and $u_n(\cdot,\cdot)$ are as in (1.63) and (1.61) respectively.

PROOF. By (1.65) we have

$$\sup_{0\leq t\leq 1}\sup_{c_n\leq y\leq 1-c_n}|\rho_n(y,t)-u_n(y,t)|$$
$$\leq \sup_{0\leq t\leq 1}\sup_{0\leq y\leq 1}|u_n(y,t)|\sup_{0\leq t\leq 1}\sup_{c_n\leq y\leq 1-c_n}\left|\frac{f(Q(y))}{f(Q(\theta_n(y,t)))}-1\right| \tag{1.77}$$
$$= O_P(1)o_P(1),$$

where $O_P(1)$ is on account of (b) of Corollary 1.A, and $o_P(1)$ is via (1.74) on noting that $h(\cdot)$ and $h(1/\cdot)$ are positive. □

Combining Theorem 1.B with Corollary 1.1, we conclude also

COROLLARY 1.2. Assume that F satisfies the conditions of Lemma 1.A on F. Let c_n be as in (1.75). Then, on the probability space of Theorem 1.A we have, as $n \to \infty$,

$$\sup_{0\leq t\leq 1}\sup_{c_n\leq y\leq 1-c_n}|\rho_n(y,t)-n^{-1/2}K(y,nt)|=o_P(1), \tag{1.78}$$

where $\{K(y,x),\ 0\leq y\leq 1,\ 0\leq x<\infty\}$ is the Kiefer process of Theorem 1.A.

PROOF. We have

$$\sup_{0\le t\le 1}\sup_{c_n\le y\le 1-c_n}|\rho_n(y,t)-n^{-1/2}K(y,nt)|$$
$$\le \sup_{0\le t\le 1}\sup_{c_n\le y\le 1-c_n}|\rho_n(y,t)-u_n(y,t)|$$
$$+\sup_{0\le t\le 1}\sup_{c_n\le y\le 1-c_n}|u_n(y,t)-n^{-1/2}K(y,nt)| \qquad (1.79)$$
$$= o_P(1)+o_P(1), \text{ as } n\to\infty,$$

by (1.76) and (1.10) respectively. □

Let $\{K(y,t),\ 0\le y\le 1, 0\le t\le 1\}$ be a Kiefer process. Then

$$\left\{n^{-1/2}K(y,nt),\ 0\le y\le 1,\ 0\le t\le 1\right\}\stackrel{\mathscr{D}}{=}\{K(y,t),\ 0\le y\le 1, 0\le t\le 1\} \qquad (1.80)$$

for each $n\ge 1$. Also, on account of almost sure continuity of a Kiefer process $K(\cdot,\cdot)$ (cf, e.g., Theorem 1.13.1 in Csörgő and Révész (1981)), as $n\to\infty$ we have

$$\sup_{0\le t\le 1}\sup_{c_n\le y\le 1-c_n}|K(y,t)| \longrightarrow \sup_{0\le t\le 1}\sup_{0\le y\le 1}|K(y,t)| \text{ a.s.} \qquad (1.81)$$

Consequently, by combining (1.78), (1.80) and (1.81), we obtain that, under the conditions of Lemma 1.A on F, as $n\to\infty$

$$\sup_{0\le t\le 1}\sup_{c_n\le y\le 1-c_n}|\rho_n(y,t)|\stackrel{\mathscr{D}}{\to}\sup_{0\le t\le 1}\sup_{0\le y\le 1}|K(y,t)|. \qquad (1.82)$$

Convergence in distribution of some other functionals of interest of the truncated sequential quantile process $\{\rho_n(y,t)\mathbf{1}\{c_n\le y\le 1-c_n\},\ 0\le t\le 1\}$ can be similarly established. Our eventual aim, however, is to study the weak convergence of $\rho_n(\cdot,\cdot)$ of (1.63) in $D[0,1]^2$. Hence, we are to extend the respective results of (1.76) and (1.78) with such a goal in mind. We note in passing that, under the conditions of Lemma 1.A, Csörgő (1983, Chapter 2) proved that, as $n\to\infty$,

$$\sup_{1/(n+1)\le y\le n/(n+1)}|\rho_n(y,1)-u_n(y,1)|=o_P(1) \qquad (1.83)$$

and that, with an appropriately constructed sequence of Brownian bridges $\{B_n(y),\ 0\le y\le 1\}$, we have as well

$$\sup_{1/(n+1)\le y\le n/(n+1)}|\rho_n(y,1)-B_n(y)|=o_P(1), \qquad (1.84)$$

where $\rho_n(y,1)$ and $u_n(y,1)$ are as in (1.63) and (1.61) respectively.

Indeed, (1.83) is an immediate consequence of Theorem 1.C (cf Chapter 2 of Csörgő (1983)), and then there we can have (1.84) as well, via the triangular inequality and (1.10) of Theorem 1.B. While (1.84) just about takes care of the weak convergence of the general quantile process $\rho_n(y,1)$ to a Brownian bridge in

$D[0,1]$, it is not strong enough for what one has to do for establishing the weak convergence of $\rho_n(y,t)$ to a Kiefer process $K(y,t)$ in $D[0,1]^2$.

2. Deviations between the general and uniform quantile processes and their sequential versions

In this section we quote, and outline the proofs of, some known results on the deviations between the general and uniform quantile processes (cf Csörgő and Révész (1978)) that have led to an extension of the Bahadur (1966) and Kiefer (1970) theory of deviations between sample quantiles and the empirical distribution function (cf Csörgő and Révész (1978; 1981, Sections 5.2 and 5.3) and Csörgő (1983, Chapter 6), as well as Sections 4 and 5 of this exposition). The results to be quoted here will also play a fundamental role in what we intend to say about the sup-norm distance of $\rho_n(y,t)$ and its uniform version $u_n(y,t)$ (cf (1.63), (1.61) respectively and Propositions 2.1, 2.2 below), about the approximations of the general sequential quantile process by a Kiefer process (cf Section 3 below), and about the corresponding sequential versions of the Bahadur–Kiefer deviations of Section 6.

Csörgő and Révész (1975, 1978) initiated the study of the general quantile process $\rho_n(y,1)$ via its own uniform quantile process $u_n(y,1)$. It has turned out (cf e.g., Csörgő and Révész (1981), Csörgő (1983), Csörgő and Horváth (1993), and the related references in these works) that the assumptions (1.66) (i), (ii) and (iii) of Lemma 1.A are frequently convenient in dealing with problems that are inherently based on joint distributions of empirical quantiles (cf, e.g., Rao and Zhao (1995)). In particular, Lemma 1.A plays a crucial role, in combination with a Csáki-type law of the iterated logarithm for the uniform quantile process, in comparing the two processes $\rho_n(y,1)$ and $u_n(y,1)$.

Csáki (1977) investigated the almost sure behaviour of the upper limits of the sequence

$$\sup_{\epsilon_n \leq y \leq 1-\epsilon_n} (y(1-y)\log\log n)^{-1/2}|\alpha_n(y,1)|$$

for a wide class of sequences $\{\epsilon_n\}$, $\epsilon_n \downarrow 0$. One special case of his many results reads as follows:

THEOREM 2.A. With $\epsilon_n = dn^{-1}\log\log n$ and $d \geq 0.236\ldots$, we have

$$\limsup_{n\to\infty} \sup_{\epsilon_n \leq y \leq 1-\epsilon_n} (y(1-y)\log\log n)^{-1/2}|\alpha_n(y,1)| = 2 \quad \text{a.s.} \tag{2.1}$$

For further details on this theorem we refer to Theorem 5.1.6 and Remark 5.1.1 in Csörgő and Révész (1981). In the theory of quantiles we use an analogue of this theorem for $u_n(y,1)$ (cf Csörgő and Révész (1978, Theorem 2) and/or (1981, Theorem 4.5.5)).

THEOREM 2.B. With $\delta_n = 25n^{-1}\log\log n$ we have

$$\limsup_{n\to\infty} \sup_{\delta_n\leq y\leq 1-\delta_n} (y(1-y)\log\log n)^{-1/2}|u_n(y,1)| \leq 4 \quad \text{a.s.} \tag{2.2}$$

The basic results of Theorems 2.C and 2.D below on the distance between $\rho_n(y,1)$ (cf (1.63)) and $u_n(y,1)$ (cf (1.61)) are essentially due to Csörgő and Révész (1978, Theorem 3). M. Csörgő, S. Csörgő, Horváth and Révész (1985), and Csörgő and Horváth (1993, Chapter 6) contain the present, somewhat modified versions, as well as the details of their proofs.

THEOREM 2.C. Given the conditions (1.66)(i), (ii) and (iii) on F, as $n \to \infty$, we have

$$\sup_{1/(n+1)\leq y\leq n/(n+1)} |\rho_n(y,1) - u_n(y,1)|$$

$$\stackrel{\text{a.s.}}{=} \begin{cases} \mathcal{O}\left(n^{-1/2}(\log\log n)^{1+\gamma}\right) & \text{if } \gamma \leq 1, \\ \mathcal{O}\left(n^{-1/2}(\log\log n)^{\gamma}(\log n)^{(1+\epsilon)(\gamma-1)}\right) & \text{if } \gamma > 1, \end{cases} \tag{2.3}$$

for all $\epsilon > 0$.

AN OUTLINE OF THE PROOF OF THEOREM 2.C. First, with δ_n as in Theorem 2.B, one shows that

$$\sup_{\delta_n\leq y\leq 1-\delta_n} |\rho_n(y,1) - u_n(y,1)| = \mathcal{O}(n^{-1/2}\log\log n) \quad \text{a.s.} \tag{2.4}$$

This is accomplished along the following lines. A two-term Taylor expansion gives

$$\rho_n(y,1) = u_n(y,1) - \frac{n^{-1/2}}{2} u_n^2(y,1) \frac{f'(Q(\xi))}{f^3(Q(\xi))} f(Q(y)), \tag{2.5}$$

where $\xi = \xi(y,n)$ and $|y - \xi| \leq n^{-1/2}|u_n(y,1)|$. By Theorem 2.B

$$\sup_{\delta_n\leq y\leq 1-\delta_n} |u_n^2(y,1)|/(y(1-y))| = \mathcal{O}(\log\log n) \quad \text{a.s.}, \tag{2.6}$$

by (1.66) (iii) we have

$$\sup_{0<y<1} \xi(1-\xi) \frac{|f'(Q(\xi))|}{f^2(Q(\xi))} \leq \gamma, \tag{2.7}$$

and by Lemma 1.A we obtain

$$\frac{f(Q(y))}{f(Q(\xi))} \leq \left(\frac{y \vee \xi}{y \wedge \xi}\right)^\gamma \left(\frac{1-(y\wedge\xi)}{1-(y\vee\xi)}\right)^\gamma. \tag{2.8}$$

Now calculations over the interval $\delta_n \leq y \leq 1 - \delta_n$ for estimating the right hand side of (2.8) combined with (2.5)–(2.7) yield (2.4).

Next, using again Theorem 2.B, one shows that, as $n \to \infty$, we have

$$\sup_{0 \leq y \leq \delta_n} |u_n(y, 1)| = \mathcal{O}(n^{-1/2} \log \log n) \quad \text{a.s.} \tag{2.9}$$

as well as

$$\sup_{1-\delta_n \leq y \leq 1} |u_n(y, 1)| = \mathcal{O}(n^{-1/2} \log \log n) \quad \text{a.s.} \tag{2.10}$$

Somewhat more difficult calculations yield also that, as $n \to \infty$,

$$\sup_{1/(n+1) \leq y \leq \delta_n} |\rho_n(y, 1)|$$
$$\stackrel{\text{a.s.}}{=} \begin{cases} \mathcal{O}\left(n^{-1/2} (\log \log n)^{1+\gamma}\right), & \text{if } \gamma \leq 1, \\ \mathcal{O}\left(n^{-1/2} (\log \log n)^{\gamma} (\log n)^{(1+\epsilon)(\gamma-1)}\right), & \text{if } \gamma > 1, \end{cases} \tag{2.11}$$

as well as that $\sup_{1-\delta_n \leq y \leq 1} |\rho_n(y, 1)|$ can be estimated exactly the same way. Consequently, Theorem 2.C follows by (2.4), (2.9), (2.10), (2.11) and what we have just said right after, concerning the estimation of $\sup_{1-\delta_n \leq y \leq 1} |\rho_n(y, 1)|$. □

In the light of Theorem 2.C it is natural to wonder what happens to (2.3) when taking sup over $[0, 1]$ instead of $[1/(n+1), 1 - 1/(n+1)]$. Somewhat surprisingly, still as in Csörgő and Révész (1978), we need extra conditions for comparing $\rho_n(y, 1)$ and $u_n(y, 1)$ on the whole unit interval. Namely we have

THEOREM 2.D. *Assume that* (1.66)(i)–(iii) *hold true and that we have also*

(i) $A = \lim_{y \downarrow 0} f(Q(y)) < \infty$, $B = \lim_{y \uparrow 1} f(Q(y)) < \infty$ \hfill (2.12)

and

(ii) $A \wedge B > 0$,

or

(iii) *if* $\lim_{y \downarrow 0} f(Q(y)) = 0$, *then* f *is non-decreasing in a right-neighbourhood of* $Q(0) = Q(0+)$, *and if* $\lim_{y \uparrow 1} f(Q(y)) = 0$, *then* f *is non-increasing in a left-neighbourhood of* $Q(1)$.

Then, as $n \to \infty$, *we have*

$$\sup_{0 \leq t \leq 1} |\rho_n(y, 1) - u_n(y, 1)|$$
$$\stackrel{\text{a.s.}}{=} \begin{cases} \mathcal{O}(n^{-1/2} \log \log n) & \text{if } \gamma < 1, \\ \mathcal{O}\left(n^{-1/2} (\log \log n)^2\right) & \text{if } \gamma = 1, \\ \mathcal{O}\left(n^{-1/2} (\log \log n)^{\gamma} (\log n)^{(1+\epsilon)(\gamma-1)}\right) & \text{if } \gamma > 1, \end{cases} \tag{2.13}$$

for all $\epsilon > 0$.

AN OUTLINE OF THE PROOF OF THEOREM 2.D. Under the conditions (1.66) (i)–(iii) we have (2.4). Assuming now (2.12) (i) and (ii), the one term Taylor series expansion of (1.65) combined with (2.9) and (2.10) respectively, yields

$$\sup_{0 \leq y \leq \delta_n} |\rho_n(y, 1) - u_n(y, 1)| = \mathcal{O}(n^{-1/2} \log \log n) \quad \text{a.s.} \tag{2.14}$$

and

$$\sup_{1 - \delta_n \leq y \leq 1} |\rho_n(y, 1) - u_n(y, 1)| = \mathcal{O}(n^{-1/2} \log \log n) \quad \text{a.s.} \tag{2.15}$$

Consequently, under the conditions (1.66) (i)–(iii), and (2.12) (i), (ii) we have

$$\sup_{0 \leq y \leq 1} |\rho_n(y, 1) - u_n(y, 1)| = \mathcal{O}(n^{-1/2} \log \log n) \quad \text{a.s.} \tag{2.16}$$

Next we assume (2.12) (i) and (iii), the latter with $\lim_{y \downarrow 0} f(Q(y)) = 0$ and $f(\cdot)$ non-decreasing on the right of $Q(0) = Q(0+)$. Restricting attention to the region $0 < y \leq \delta_n$ and considering first the case of $\hat{U}_n(y) \geq y$, condition (2.12) (iii) gives the trivial estimate

$$|\rho_n(y, 1)| = n^{1/2} \int_y^{\hat{U}_n(y)} \frac{f(Q(y))}{f(Q(s))} ds \leq -u_n(y, 1) . \tag{2.17}$$

Hence, applying (2.9), we obtain

$$\sup_{0 \leq y \leq \delta_n} |\rho_n(y, 1) \mathbf{1}\{\hat{U}_n(y) \geq y\}| = \mathcal{O}(n^{-1/2} \log \log n) \quad \text{a.s.} \tag{2.18}$$

Still in the region $0 < y \leq \delta_n$, but assuming now that $\hat{U}_n(y) < y$, calculations yield (cf Csörgő and Révész (1978, Proof of Theorem 3, or 1981, Proof of Theorem 4.5.6))

$$\sup_{0 \leq y \leq \delta_n} |\rho_n(y, 1) \mathbf{1}\{\hat{U}_n(y) < y\}|$$
$$\stackrel{\text{a.s.}}{=} \begin{cases} \mathcal{O}(n^{-1/2} \log \log n) & \text{if } \gamma < 1, \\ \mathcal{O}(n^{-1/2} (\log \log n)^2) & \text{if } \gamma = 1, \\ \mathcal{O}\left(n^{-1/2} (\log \log n)^\gamma (\log n)^{(1+\epsilon)(\gamma - 1)}\right) & \text{if } \gamma > 1 , \end{cases} \tag{2.19}$$

for all $\epsilon > 0$. Combining now (2.9), (2.18) and (2.19) we obtain that $\sup_{0 \leq y \leq \delta_n} |\rho_n(y, 1) - u_n(y, 1)|$ is estimated at the almost sure rate of (2.19). This combined with (2.4) yields that we now have the desired bounds for estimating $\sup_{0 \leq y \leq 1 - \delta_n} |\rho_n(y, 1) - u_n(y, 1)|$. Namely, as $n \to \infty$, we have

$$\sup_{0 \le y \le 1-\delta_n} |\rho_n(y,1) - u_n(y,1)|$$

$$\stackrel{a.s.}{=} \begin{cases} \mathcal{O}(n^{-1/2} \log \log n) & \text{if } \gamma < 1, \\ \mathcal{O}(n^{-1/2} (\log \log n)^2) & \text{if } \gamma = 1, \\ \mathcal{O}\left(n^{-1/2} (\log \log n)^\gamma (\log n)^{(1+\epsilon)(\gamma-1)}\right) & \text{if } \gamma > 1, \end{cases} \quad (2.20)$$

for all $\epsilon > 0$.

Assuming now (2.12) (i) and (iii) with $\lim_{y \uparrow 1} f(Q(y)) = 0$ and non-increasing f on the left of $Q(1)$, by (2.10) and calculations over the region $1 - \delta_n \le y \le 1$ that are similar to those yielding (2.18) and (2.19), we arrive at estimating $\sup_{1-\delta_n \le y \le 1} |\rho_n(y,1) - u_n(y,1)|$ at the almost sure rate of (2.20) as well. Hence we conclude also (2.13). □

As *an immediate consequence* of Theorem 2.C, we get its *sequential version* which reads as follows.

PROPOSITION 2.1. Given the conditions (1.66) (i), (ii) and (iii) on F, as $n \to \infty$, we have

$$\sup_{0 \le t \le 1} \sup_{1/(n+1) \le y \le n/(n+1)} |\rho_n(y,t) - u_n(y,t)|$$

$$\stackrel{a.s.}{=} \begin{cases} \mathcal{O}\left(n^{-1/2} (\log \log n)^{1+\gamma}\right) & \text{if } \gamma \le 1, \\ \mathcal{O}\left(n^{-1/2} (\log \log n)^\gamma (\log n)^{(1+\epsilon)(\gamma-1)}\right) & \text{if } \gamma > 1, \end{cases} \quad (2.21)$$

for all $\epsilon > 0$.

PROOF. With $t \in (0,1]$ and $(nt) \to \infty$ as $n \to \infty$, by (2.3) for n large we have

$$\sup_{1/(n+1) \le y \le n/(n+1)} |n^{1/2} \rho_n(y,t) - n^{1/2} u_n(y,t)| \le (C + o(1)) r_n(t, \gamma, \epsilon) \quad \text{a.s.} \quad (2.22)$$

where C is an absolute constant, and

$$r_n(t, \gamma, \epsilon) = \begin{cases} (\log \log(nt))^{1+\gamma} & \text{if } \gamma \le 1, \\ \left((\log \log(nt))^\gamma (\log(nt))^{(1+\epsilon)(\gamma-1)}\right) & \text{if } \gamma > 1. \end{cases} \quad (2.23)$$

Dividing both sides of (2.22) by $r_n(1, \gamma, \epsilon)$ and then taking $\sup_{0 \le t \le 1}$ on both sides, we obtain (2.21) as $n \to \infty$. □

In a similar fashion, we also get the following sequential version of Theorem 2.D.

PROPOSITION 2.2. Assume (1.66) (i)–(iii), (2.12) (i) and (2.12) (ii) or (iii). Then, as $n \to \infty$, we have

$$\sup_{0\le t\le 1}\sup_{0\le y\le 1}|\rho_n(y,t)-u_n(y,t)|$$
$$\stackrel{a.s.}{=}\begin{cases} \mathcal{O}(n^{-1/2}\log\log n) & \text{if } \gamma<1, \\ \mathcal{O}(n^{-1/2}(\log\log n)^2) & \text{if } \gamma=1, \\ \mathcal{O}\left(n^{-1/2}(\log\log n)^{\gamma}(\log n)^{(1+\epsilon)(\gamma-1)}\right) & \text{if } \gamma>1, \end{cases} \quad (2.24)$$

for all $\epsilon>0$.

3. Weighted sequential quantile processes in supremum and L_p-metrics

3.1. Some weighted asymptotics for sequential uniform quantile processes

We note and recall that the sequential uniform quantile process $u_n(y,t)$ is a sequential version of the usually considered uniform quantile process

$$u_n(y):=u_n(y,1)=n^{1/2}(y-\hat{U}_n(y)), \quad 0\le y\le 1 .$$

Starting with Rényi (1953), Chibisov (1964) and O'Reilly (1974), there has been considerable interest in the asymptotic behaviour of weighted uniform empirical and quantile processes. For an insightful treatize of this subject we refer to Csörgő, Csörgő, Horváth and Mason (1986), Shorack and Wellner (1986) and to Csörgő and Horváth (1993), as well as to the references in these works. There are now complete characterizations available for describing the asymptotic behaviour of the weighted uniform empirical and quantile processes in supremum and L_p-metrics.

Let $\mathcal{Q}^*$ be the class of positive functions q on $(0,1)$, i.e., such that $\inf_{\delta\le y\le 1-\delta}q(y)>0$ for all $0<\delta<1$, which are nondecreasing near zero and nonincreasing near one. Let

$$I^*(q,c)=\int_0^1 \frac{1}{y(1-y)}\exp\left(-cq^2(y)/(y(1-y))\right)dy, \quad c>0 .$$

Let $U_{1,n}\le U_{2,n}\le\cdots\le U_{n,n}$ denote the order statistics of a uniform-$[0,1]$ random sample and define the modified uniform quantile function $\tilde{U}_n(y)$ by

$$\tilde{U}_n(y)=U_{k,n}, \ k/(n+2)<y\le(k+1)/(n+2), \quad k=0,\ldots,n+1 ,$$

where $U_{0,n}=0$ and $U_{n+1,n}=1$, and the modified uniform quantile process $\tilde{u}_n(y)$ by

$$\tilde{u}_n(y)=n^{1/2}(y-\tilde{U}_n(y)), \quad 0\le y\le 1 .$$

For the proof of the following theorem we refer to Csörgő, Csörgő, Horváth and Mason (1986).

THEOREM 3.A. *Let* $q \in \mathscr{Q}^*$.

(i) We can define a sequence of Brownian bridges $\{B_n(y), 0 \leq y \leq 1\}$ such that, as $n \to \infty$

$$\sup_{0<y<1} |\tilde{u}_n(y) - B_n(y)|/q(y) = o_P(1)$$

if and only if $I^*(q,c) < \infty$ for all $c > 0$.

(ii) As $n \to \infty$ we have

$$\sup_{0<y<1} |\tilde{u}_n(y)|/q(y) \xrightarrow{\mathscr{D}} \sup_{0<y<1} |B(y)|/q(y) ,$$

where $\{B(y), 0 \leq y \leq 1\}$ is a Brownian bridge, if and only if $I^*(q,c) < \infty$ for some $c > 0$.

Concerning the problem of various possible modifications of the uniform quantile process $u_n(y) := u_n(y, 1)$ for the sake of a weighted approximation as in Theorem 3.A (i), and hence also for its weighted weak convergence in $D[0,1]$ with appropriate c.d.l.g. weight functions $q(y)$, we refer to O'Reilly (1974), Csörgő, Csörgő, Horváth and Mason (1986), and Csörgő and Horváth (1993, Chapters 4 and 5).

It is of interest to note that (i) of Theorem 3.A does not imply its statement (ii) for all possible weight functions $q(y)$ of interest. For example, choosing

$$q(y) = \left(y(1-y)\log\log\frac{1}{y(1-y)}\right)^{1/2} ,$$

we have (ii), but not (i), of Theorem 3.A. Thus, as $n \to \infty$, convergence in distribution of $\sup_{0<y<1} |\tilde{u}_n(y)|/q(y)$ to $\sup_{0<y<1} |B(y)|/q(y)$ holds for a larger subclass of functions $q \in \mathscr{Q}^*$ than for having $\tilde{u}_n(\cdot)/q(\cdot) \xrightarrow{\mathscr{D}} B(\cdot)/q(\cdot)$ in $D[0,1]$. Moreover, the class of functions q for the convergence of L_p-functionals of $\tilde{u}_n(\cdot)/q(\cdot)$ is even larger than in the latter case, as demonstrated by the next trichotomy theorem of Csörgő, Horváth and Shao (1993).

THEOREM 3.B. *Let q be a positive function on $(0,1)$, and assume that $0 < p < \infty$. Then the following statements are equivalent:*

(i) We have

$$\int_0^1 (y(1-y))^{p/2}/q(y)\,dy < \infty ,$$

(ii) There is a sequence of Brownian bridges $\{B_n(y), 0 \leq y \leq 1\}$ such that

$$\int_0^1 |\tilde{u}_n(y) - B_n(y)|^p/q(y)\,dy = o_P(1) ,$$

(iii) As $n \to \infty$, we have

$$\int_0^1 |\tilde{u}_n(y)|^p / q(y) \, dy \overset{\mathscr{D}}{\to} \int_0^1 |B(y)|^p / q(y) \, dy \,,$$

where $\{B(y),\ 0 \leq y \leq 1\}$ is a Brownian bridge.

We note in passing that, naturally, Theorems 3.A and 3.B hold true also for the uniform empirical process $\alpha_n(y) := \alpha_n(y, 1)$. We quoted these two versions, for here we are concentrating mainly on quantile processes.

In this subsection of Section 3 we are interested in studying asymptotics of weighted sequential uniform quantile process, namely $u_n(y,t)/q(t)$, $0 \leq y,\ t \leq 1$ with weights in the time parameter $t \in (0,1)$, instead of the "space" parameter $y \in [0,1]$, where $q(t)$ is a nonnegative function on $(0,1]$. For the complete and optimal characterization of sequential empirical processes of observations, their ranks and sequential ranks in weighted supremum and L_p-metrics we refer to Szyszkowicz (1996) in this volume.

Let $\mathscr{Q}$ be the class of positive functions q on $(0,1]$, i.e., such that $\inf_{\delta \leq t \leq 1} q(t) > 0$ for all $0 < \delta < 1$, which are nondecreasing near zero. Let also

$$I(q,c) = \int_0^1 t^{-1} \exp\left(-ct^{-1}q^2(t)\right) dt, \quad c > 0 \,.$$

THEOREM 3.1. Let $q \in \mathscr{Q}$. Then, on the probability space of Theorem 1.A, with the Kiefer process $\{K(y,x),\ 0 \leq y \leq 1,\ 0 \leq x < \infty\}$ of (1.8), as $n \to \infty$, we have

(a) $\sup\limits_{0 < t \leq 1} \sup\limits_{0 \leq y \leq 1} |u_n(y,t) - n^{-1/2} K(y, nt)|/q(t) = o_P(1)$

if and only if $I(q,c) < \infty$ for all $c > 0$,

(b) $\sup\limits_{0 < t \leq 1} \sup\limits_{0 \leq y \leq 1} |u_n(y,t) - n^{-1/2} K(y, nt)|/q(t) = O_P(1)$

if and only if $I(q,c) < \infty$ for some $c > 0$, and

(c) $\sup\limits_{0 < t \leq 1} \sup\limits_{0 \leq y \leq 1} |u_n(y,t)|/q(t) \overset{\mathscr{D}}{\to} \sup\limits_{0 < t \leq 1} \sup\limits_{0 \leq y \leq 1} |K(y,t)|/q(t)$

if and only if $I(q,c) < \infty$ for some $c > 0$, where $\{K(y,t),\ 0 \leq y \leq 1,\ 0 \leq t \leq 1\}$ is a Kiefer process.

The proof of Theorem 3.1 is similar to that of Theorem 2.1 of Szyszkowicz (1998) in this volume (cf also Szyszkowicz (1994)), and it is based on Theorem 1.B of this paper and on the integral test of Csörgő, Horváth and Szyszkowicz (1994) for suprema of Kiefer processes, spelled out as Lemma 2.A in Szyszkowicz (1998). The results of Theorem 3.1 themselves are exact analogues of those of Theorem 2.1 of Szyszkowicz (1998) for the sequential uniform empirical process $\alpha_n(y,t)$. Clearly, Theorem 3.1 is also parallel to Theorem 3.A, only here the weight function $q(t)$ relates to the question of finiteness of $\limsup_{t \downarrow 0} \sup_{0 \leq y \leq 1} |K(y,t)|/q(t)$, while in Theorem 3.A. $q(y)$ relates to those of $\limsup_{y \downarrow 0} |B(y)|/q(y)$ and $\limsup_{y \uparrow 1} |B(y)|/q(y)$.

On account of Theorem 3.1 (a) we obtain

COROLLARY 3.1. *Let $q \in \mathcal{Q}$. Then as $n \to \infty$, we have*

$$u_n(y,t)/q(t) \xrightarrow{\mathcal{D}} K(y,t)/q(t) \quad \text{in} \quad D[0,1]^2$$

if and only if $I(q,c) < \infty$ for all $c > 0$, where $\{K(y,t), 0 \le y \le 1, 0 \le t \le 1\}$ is a Kiefer process.

REMARK 3.1. Throughout this paper weak convergence statements on Skorohod spaces are stated as corollaries to approximations in probability. Naturally, when talking about weighted weak convergence on such spaces, we will always assume that the weights are c.d.l.g. functions.

REMARK 3.2. Obviously Corollary 3.1 implies convergence in distribution of any continuous in sup-norm functional of $u_n(y,t)/q(t)$ to the corresponding functional of $K(y,t)/q(t)$ with $q \in \mathcal{Q}$ and such that $I(q,c) < \infty$ for all $c > 0$. However, for the sup-functional itself, by Theorem 3.1 (c), the class of possible weight functions is bigger. For example, as $n \to \infty$, we have

$$\sup_{0<t\le 1} \sup_{0\le y\le 1} |u_n(y,t)|/(t\log\log((1/t) \vee 3))^{1/2}$$
$$\xrightarrow{\mathcal{D}} \sup_{0<t\le 1} \sup_{0\le y\le 1} |K(y,t)|/(t\log\log((1/t) \vee 3))^{1/2},$$

while the *weak convergence of the process* $\{u_n(y,t)/(t\log\log((1/t) \vee 3))^{1/2}, 0 \le y, t \le 1\}$ *is impossible*.

Such a phenomenon was first noted and proved by Csörgő, Csörgő, Horváth and Mason (1986) for the uniform empirical and quantile processes (cf Theorem 3.A (ii) and our discussion that follows after, there with the example of $q(y) = (y(1-y)\log\log\frac{1}{y(1-y)})^{1/2}$. We note also that statement (b) of Theorem 3.1 does not imply its statement (c).

Considering convergence of L_p-functionals, we obtain here that the optimal class of weight functions is even bigger than for convergence of supremum functionals.

THEOREM 3.2. *Let $0 < p < \infty$ and q be a positive function on $(0,1]$. Then the following three statements are equivalent:*

(a) *We have*

$$\int_0^1 t^{p/2}/q(t)\,dt < \infty \ .$$

(b) *On the probability space of Theorem 1.A, with the Kiefer process $\{K(y,x), 0 \le y \le 1, 0 \le x < \infty\}$ of (1.8), as $n \to \infty$, we have*

$$\int_0^1 \int_0^1 |u_n(y,t) - n^{-1/2}K(y,nt)|^p/q(t)\,dy\,dt = o_P(1) \ .$$

(c) We have, as $n \to \infty$,

$$\int_0^1 \int_0^1 |u_n(y,t)|^p/q(t)\,dy\,dt \xrightarrow{\mathscr{D}} \int_0^1 \int_0^1 |K(y,t)|^p/q(t)\,dy\,dt \ ,$$

where $\{K(y,t),\ 0 \leq y \leq 1,\ 0 \leq t \leq 1\}$ is a Kiefer process.

The proof of Theorem 3.2 is similar to that of Theorem 2.2 of Szyszkowicz (1998) in this volume, and it is based on Theorem 1.B of this paper and on the integral test of Csörgő, Horváth and Shao (1993) for weighted L_p-functionals of a Kiefer process, spelled out as Lemma 2.B in Szyszkowicz (1998). The results of Theorem 3.2, which parallel those of Theorem 3.B for the modified uniform quantile process $\tilde{u}_n(y)$, are exact analogues of those of Theorem 2.2 of Szyszkowicz (1998) for the sequential uniform empirical process $\alpha_n(y,t)$.

3.2. Approximations of the sequential general quantile process by a Kiefer process

A combination of Theorem 1.B with Proposition 2.1 yields the following result immediately.

PROPOSITION 3.1. Assume the conditions (1.66) (i)–(iii) on F. Then, on the probability space of Theorem 1.A, with the Kiefer process $\{K(y,x),\ 0 \leq y \leq 1,\ 0 \leq x < \infty\}$ of (1.8), as $n \to \infty$, we have

$$\sup_{0 \leq t \leq 1} \sup_{1/(n+1) \leq y \leq n/(n+1)} |\rho_n(y,t) - n^{-1/2}K(y,nt)|$$
$$= \mathcal{O}\left(n^{-1/4}(\log n)^{1/2}(\log \log n)^{1/4}\right) \quad \text{a.s.} \tag{3.1}$$

PROOF. We have

$$\sup_{0 \leq t \leq 1} \sup_{1/(n+1) \leq y \leq n/(n+1)} |\rho_n(y,t) - n^{-1/2}K(y,nt)|$$

$$\leq \sup_{0 \leq t \leq 1} \sup_{1/(n+1) \leq y \leq n/(n+1)} |\rho_n(y,t) - u_n(y,t)|$$

$$+ \sup_{0 \leq t \leq 1} \sup_{1/(n+1) \leq y \leq n/(n+1)} |u_n(y,t) - n^{-1/2}K(y,nt)| \ .$$

Now (3.1) follows by (2.21) and (1.10) combined. □

Let

$$\hat{\rho}_n(y,t) = \rho_n(y,t)\mathbf{1}\{1/(n+1) \leq y \leq n/(n+1)\},\ 0 \leq t \leq 1 \ . \tag{3.2}$$

We note that $\hat{\rho}_n$ is based on all the observations of a random sample of size n. Hence the following weak convergence of $\hat{\rho}_n(y,t)$ as a two-time parameter process is of interest.

COROLLARY 3.1. Assume the conditions (1.66) (i)–(iii) on F. Then, as $n \to \infty$, we have

$$\hat{\rho}_n(y,t) \xrightarrow{\mathscr{D}} K(y,t) \quad \text{in } D[0,1]^2, \tag{3.3}$$

where $\{K(y,t),\ 0 \le y \le 1,\ 0 \le t \le 1\}$ is a Kiefer process.

PROOF. On account of (3.1) and the almost sure continuity of a Kiefer process (cf e.g., Theorem 1.13.1 in Csörgő and Révész (1981) combined with their definition of a Kiefer process in their Section 1.15; cf also (4.28) below), as $n \to \infty$, we have

$$\sup_{0 \le t \le 1} \sup_{0 \le y \le 1} |\hat{\rho}_n(y,t) - n^{-1/2} K(y,nt)| = o_P(1), \tag{3.4}$$

which in turn implies (3.3) as $n \to \infty$ via having (cf (1.80))

$$n^{-1/2} K(y,nt) \xrightarrow{\mathscr{D}} K(y,t) \text{ in } D[0,1]^2. \tag{3.5}$$

□

Corollary 3.1 practically takes care of the problem of weak convergence of the general sequential quantile process $\rho_n(y,t)$ via that of $\hat{\rho}_n(y,t)$. Nevertheless, the next two results are also of interest.

PROPOSITION 3.2. Assume the conditions (1.66) (i)–(iii), (2.12) (i) and (2.12) (ii) or (iii) on F. Then, on the probability space of Theorem 1.A, with the Kiefer process $\{K(y,x),\ 0 \le y \le 1,\ 0 \le x < \infty\}$ of (1.8), as $n \to \infty$, we have

$$\sup_{0 \le t \le 1} \sup_{0 \le y \le 1} |\rho_n(y,t) - n^{-1/2} K(y,nt)|$$
$$= \mathcal{O}\left(n^{-1/4} (\log n)^{1/2} (\log \log n)^{1/4}\right) \quad \text{a.s.} \tag{3.6}$$

PROOF. Similarly to the proof of Proposition 3.1, (3.6) follows by (2.24) and (1.10) combined. □

COROLLARY 3.2. Assume the conditions (1.66) (i)–(iii), (2.12) (i) and (2.12) (ii) or (iii). Then, as $n \to \infty$, we have

$$\rho_n(y,t) \xrightarrow{\mathscr{D}} K(y,t) \quad \text{in } D[0,1]^2, \tag{3.7}$$

where $\{K(y,t),\ 0 \le y \le 1,\ 0 \le t \le 1\}$ is a Kiefer process.

PROOF. Immediate by (3.6). □

3.3. Weighted general sequential quantile processes in supremum and L_p-metrics

There are many results available along the global lines of Theorems 3.A and 3.B also for the general quantile process $\rho_n(y) := \rho_n(y, 1)$, and for its appropriate modifications, under various conditions on the underlying distribution function F. For information on these – weighted in the "space" parameter $y \in (0, 1)$ via $q(y)$ – type studies, we refer to Csörgő (1986), Csörgő and Horváth (1990a,b), and Csörgő and Horváth (1993, Chapter 6), and to the references in these works.

In this subsection of Section 3, we spell out some results along the lines of Theorems 3.1 and 3.2, which can be immediately based on Propositions 3.1 and 3.2, and on the already hinted at methods of proof of Theorems 3.1 and 3.2 respectively.

THEOREM 3.3. Assume the conditions (1.66) (i)–(iii) on F. Let $q \in \mathcal{Q}$. Then, on the probability space of Theorem 1.A, with the Kiefer process $\{K(y,x), 0 \leq y \leq 1, 0 \leq x < \infty\}$ of (1.8), as $n \to \infty$, we have

(a) $\sup\limits_{0<t\leq 1} \sup\limits_{1/(n+1)\leq y\leq n/(n+1)} |\rho_n(y,t) - n^{-1/2}K(y,nt)|/q(t) = o_P(1)$

if and only if $I(q,c) < \infty$ for all $c > 0$,

(b) $\sup\limits_{0<t\leq 1} \sup\limits_{1/(n+1)\leq y\leq n/(n+1)} |\rho_n(y,t) - n^{-1/2}K(y,nt)|/q(t) = O_P(1)$

if and only if $I(q,c) < \infty$ for some $c > 0$,

(c) $\sup\limits_{0<t\leq 1} \sup\limits_{1/(n+1)\leq y\leq n/(n+1)} |\rho_n(y,t)|/q(t) \xrightarrow{\mathcal{D}} \sup\limits_{0<t\leq 1} \sup\limits_{0\leq y\leq 1} |K(y,t)|/q(t)$

if and only if $I(q,c) < \infty$ for some $c > 0$, where $\{K(y,t), 0 \leq y \leq 1, 0 \leq t \leq 1\}$ is Kiefer process.

THEOREM 3.4. Assume the conditions (1.66) (i)–(iii) on F. Let $0 < p < \infty$, and q be a positive function on $(0, 1]$. Then the following three statements are equivalent:

(a) We have

$$\int_0^1 t^{p/2}/q(t)\, dt < \infty ,$$

(b) On the probability space of Theorem 1.A, with the Kiefer process $\{K(y,x), 0 \leq y \leq 1, 0 \leq x < \infty\}$ of (1.8), as $n \to \infty$ we have

$$\int_0^1 \int_{1/(n+1)}^{n/(n+1)} |\rho_n(y,t) - n^{-1/2}K(y,nt)|^p/q(t)\, dy\, dt = o_P(1) ,$$

(c) We have, as $n \to \infty$,

$$\int_0^1 \int_{1/(n+1)}^{n/(n+1)} |\rho_n(y,t)|^p/q(t)\,dy\,dt \xrightarrow{\mathcal{D}} \int_0^1 \int_0^1 |K(y,t)|^p/q(t)\,dy\,dt,$$

where $\{K(y,t),\ 0 \leq y \leq 1,\ 0 \leq t \leq 1\}$ is a Kiefer process.

THEOREM 3.5. Assume the conditions (1.66) (i)–(iii), (2.12) (i) and (2.12) (ii) or (iii) on F. Let $q \in \mathcal{Q}$. Then, on the probability space of Theorem 1.A, with the Kiefer process $\{K(y,x),\ 0 \leq y \leq 1,\ 0 \leq x < \infty\}$ of (1.8), as $n \to \infty$, we have

(a) $\sup_{0<t\leq 1} \sup_{0\leq y\leq 1} |\rho_n(y,t) - n^{-1/2}K(y,nt)|/q(t) = o_P(1)$

if and only if $I(q,c) < \infty$ for all $c > 0$,

(b) $\sup_{0<t\leq 1} \sup_{0\leq y\leq 1} |\rho_n(y,t) - n^{-1/2}K(y,nt)|/q(t) = O_P(1)$

if and only if $I(q,c) < \infty$ for some $c > 0$,

(c) $\sup_{0<t\leq 1} \sup_{0\leq y\leq 1} |\rho_n(y,t)|/q(t) \xrightarrow{\mathcal{D}} \sup_{0<t\leq 1} \sup_{0\leq y\leq 1} |K(y,t)|/q(t)$

if and only if $I(q,c) < \infty$ for some $c > 0$, where $\{K(y,t),\ 0 \leq y \leq 1,\ 0 \leq t \leq 1\}$ is Kiefer process.

THEOREM 3.6. Assume the conditions (1.66) (i)–(iii), (2.12) (i) and (2.12) (ii) or (iii) on F. Let $0 < p < \infty$, and q be a positive function on $(0,1]$. Then the following three statements are equivalent:

(a) We have

$$\int_0^1 t^{p/2}/q(t)\,dt < \infty,$$

(b) On the probability space of Theorem 1.A, with the Kiefer process $\{K(y,x),\ 0 \leq y \leq 1,\ 0 \leq x < \infty\}$ of (1.8), as $n \to \infty$, we have

$$\int_0^1 \int_0^1 |\rho_n(y,t) - n^{-1/2}K(y,nt)|^p/q(t)\,dy\,dt = o_P(1),$$

(c) We have, as $n \to \infty$,

$$\int_0^1 \int_0^1 |\rho_n(y,t)|^p/q(t)\,dy\,dt \xrightarrow{\mathcal{D}} \int_0^1 \int_0^1 |K(y,t)|^p/q(t)\,dy\,dt,$$

where $\{K(y,t),\ 0 \leq y \leq 1,\ 0 \leq t \leq 1\}$ is a Kiefer process.

We conclude this section by saying that Theorem 1.B, Propositions 2.1 and 2.2, as well as the results of this Section 3 constitute only the initial steps of our projected study of sequential quantile and Bahadur–Kiefer processes in weighted metrics in Csörgő and Szyszkowicz (1995/96), along the lines of Szyszkowicz (1998) in this volume.

4. A summary of the classical Bahadur-Kiefer process theory via strong invariance principles

The modern theory of sample quantiles was initiated by Bahadur (1966), who, in terms of our notations, studied the following representation of the yth sample quantile $\hat{Q}_n(y)$ (cf (1.45)) in terms of the empirical distribution function $\hat{F}_n(\cdot)$ (cf (1.38)):

$$\hat{Q}_n(y) = Q(y) + \left(1 - \hat{F}_n(Q(y)) - (1-y)\right)/f(Q(y)) + R_n(y) \tag{4.1}$$

for $y \in (0,1)$ *fixed* and for the *whole sequence in n*, i.e., for the stochastic process $\{\hat{Q}_n(y),\ n=1,2,\ldots\}$ in n for *fixed* $y \in (0,1)$. Using again our notations, we have (cf (1.40), (1.47), (1.60) and (1.63))

$$\begin{aligned}R_n(y) &= \left(\hat{F}_n(Q(y)) - y\right)/f(Q(y)) + \hat{Q}_n(y) - Q(y) \\ &= \frac{B_n(Q(y), 1)}{n^{1/2} f(Q(y))} - \frac{\gamma_n(y, 1)}{n^{1/2}} \\ &= \frac{\alpha_n(y, 1)}{n^{1/2} f(Q(y))} - \frac{\rho_n(y, 1)}{n^{1/2} f(Q(y))} .\end{aligned} \tag{4.2}$$

Bahadur (1966) proved

THEOREM 4.A. *For any distribution function F on the real line which, for $y \in (0,1)$ fixed and x such that $y = F(x)$, has the property that it is twice differentiable in a neighbourhood of our fixed y, and that $F''(x) = f'(x) = f'(Q(y))$ is bounded in that neighbourhood and $F'(x) = f(x) = f(Q(y)) > 0$, we have, as $n \to \infty$*

$$R_n(y) = \mathcal{O}\left(n^{-3/4} (\log n)^{1/2} (\log \log n)^{1/4}\right), \quad \text{a.s.} \tag{4.3}$$

Consequently, via (4.2) and (4.3), under the conditions of Theorem 4.A, for $y \in (0,1)$ *fixed* the quantile process in n

$$\{\gamma_n(y,t),\ n=1,2,\ldots\} = \{\rho_n(y,1)/f(Q(y)),\ n=1,2,\ldots\} \tag{4.4}$$

behaves like the uniform empirical process

$$\{B_n(Q(y), 1)/f(Q(y)),\ n=1,2,\ldots\} = \{\alpha_n(y,1)/f(Q(y)),\ n=1,2,\ldots,\} \tag{4.5}$$

at least at the a.s. rate of convergence of (4.3). This amounts to saying that Bahadur's result of (4.3) is a pointwise (in $y \in (0,1)$), strong (almost sure) approximation (invariance principle) for $\rho_n(y,1)$ in terms of the pointwise (in $y \in (0,1)$) much better known binomial process $\alpha_n(y,1)$. Indeed, Bahadur (1966) notes that the above representation of sample quantiles in (4.1) gives new insight into the well known result that, for $y \in (0,1)$ fixed, $\gamma_n(y,1)$ of (1.62) is asymptotically normally distributed with mean zero and variance $y(1-y)/f^2(Q(y))$ (cf

(1.71)) and that it gives easy access, via the multivariate central limit theorem for zero-one variables, to the asymptotic joint distribution of several quantiles in the same vein. Moreover, he mentions also that his representation shows that the following law of the iterated logarithm (LIL) holds *for fixed sample quantiles*

$$\limsup_{n\to\infty} |\gamma_n(y, 1)|/(2\log\log n)^{1/2} = (y(1-y))^{1/2}/f(Q(y)) \quad \text{a.s.}, \quad (4.6)$$

due to the same LIL for the binomial random variables $\{n\hat{F}_n(Q(y)), n = 1, 2, \ldots\}$. He also raised the question of finding the exact order of $R_n(y)$, which was, in turn, established by Kiefer (1967) as follows.

THEOREM 4.B. *Under the conditions of Theorem 4.A on F for $y \in (0, 1)$ fixed, we have*

$$\limsup_{n\to\infty} n^{3/4}|f(Q(y))R_n(y)|/(\log\log n)^{3/4} \stackrel{\text{a.s.}}{=} \frac{2^{5/4}}{3^{3/4}}\sigma_y^{1/2}, \quad (4.7)$$

where $\sigma_y = (y(1-y))^{1/2}$.

Kiefer's proof of this particularly difficult result of (4.7) is based on showing that, under the conditions of Theorem 4.A on F for fixed $y \in (0, 1)$, we have (cf (1.9) of Kiefer (1967))

$$|\tilde{R}_n(y, 1) - R_n^*(y, 1)| \stackrel{\text{a.s.}}{=} o(n^{1/4}), \quad (4.8)$$

where $R_n^*(y, 1)$ is as defined in (1.29), and

$$\begin{aligned}\{\tilde{R}_n(y, 1), \ 0 \le y \le 1, \ n = 1, 2, \ldots\} \\ := \{nf(Q(y))R_n(y), \ 0 \le y \le 1, \ n = 1, 2, \ldots\}.\end{aligned} \quad (4.9)$$

We note that in the case of the uniform-(0,1) distribution, i.e., $F(y) = y$, we have (cf (4.2))

$$\tilde{R}_n(y, 1) = R_n^*(y, 1) = n^{1/2}(\alpha_n(y, 1) - u_n(y, 1)). \quad (4.10)$$

In this case the conditions of Theorem 4.A on F for fixed $y \in (0, 1)$ are, of course, satisfied and (4.7) then reads as follows

$$\limsup_{n\to\infty} |R_n^*(y, 1)|/\left(n^{1/4}(\log\log n)^{3/4}\right) \stackrel{\text{a.s.}}{=} \frac{2^{5/4}}{3^{3/4}}\sigma_y^{1/2}. \quad (4.11)$$

Consequently, (4.8) *is a strong invariance principle* (an almost sure approximation) *for the now exact order Bahadur result of* (4.7), i.e., (4.11) implies (4.7) by (4.8), provided of course that one can establish (4.11). Indeed, Kiefer (1967, Lemma 1) observed the same and proved the exact a.s. rate of (4.7) via finding it first for $R_n^*(y, 1)$, as it is stated in (4.11). Kiefer's direct proof of the latter result is an example of true virtuosity.

Using the approach of Csörgő and Révész (1975, 1978, 1981, Section 4.5)), and Csörgő (1983, Chapter 6) to studying quantile processes, a further insight can be obtained into these theorems. By (4.2) and (4.9) we have

$$\tilde{R}_n(y, 1) = nf(Q(y))R_n(y)$$
$$= n^{1/2}\alpha_n(y, 1) - n^{1/2}\rho_n(y, 1) \qquad (4.12)$$
$$= n^{1/2}(\alpha_n(y, 1) - u_n(y, 1)) - n^{1/2}(\rho_n(y, 1) - u_n(y, 1)) ,$$

where $u_n(y, 1)$ is as defined in (1.61). Hence, using the notations of (1.29) and (4.9) respectively, we can write

$$\tilde{R}_n(y, 1) - R_n^*(y, 1) = -n^{1/2}(\rho_n(y, 1) - u_n(y, 1)) \qquad (4.13)$$

and, consequently, *the Bahadur–Kiefer deviations can be studied via those of the general quantile process and its corresponding uniform version.* Indeed, on assuming the conditions of Theorem 4.A on F for fixed $y \in (0, 1)$, and expanding as in (2.5), from (4.13) we obtain

$$\tilde{R}_n(y, 1) - R_n^*(y, 1) = \frac{1}{2}u_n^2(y, 1)\frac{f'(Q(\xi))}{f^3(Q(\xi))}f(Q(y)) , \qquad (4.14)$$

where $\xi = \xi(y, n)$ and $|y - \xi| \leq n^{-1/2}|u_n(y, 1)|$. Thus, *under the conditions of Theorem 4.A on F for fixed* $y \in (0, 1)$, we obtain

$$|\tilde{R}_n(y, 1) - R_n^*(y, 1)|$$
$$\overset{\text{a.s.}}{=} \mathcal{O}(\log \log n) , \qquad (4.15)$$

where the latter a.s. $\mathcal{O}(\log \log n)$ rate is due to applying a pointwise version of the law of the iterated logarithm for $u_n(y, 1)$ (cf (1.17)). Thus (4.15) is a stronger version of (4.8) and, indeed, essentially this is also the very way Kiefer (1967) obtained the latter result.

The proof of (4.11) *is more difficult.* Kiefer (1967) obtained it by establishing direct upper and lower class estimates for $R_n^*(y, 1)$. A bit of *insight* into what is going on can be *gained by looking at* (1.30) *again.* Naturally the latter obtains for each *fixed* $y \in (0, 1)$, namely, as $n \to \infty$, we have

$$n^{-1/2}|R_n^*(y, 1) - (K(y, n) - K(\hat{U}_n(y), n))| \overset{\text{a.s.}}{=} \mathcal{O}((\log n)^2/n^{1/2}) . \qquad (4.16)$$

Hence, on account of (4.11), we have also

$$\limsup_{n \to \infty} |K(\hat{U}_n(y), n) - K(y, n)|/(n^{1/4}(\log \log n)^{3/4}) \overset{\text{a.s.}}{=} \frac{2^{5/4}}{3^{3/4}}\sigma_y^{1/2} . \qquad (4.17)$$

This, of course, *invites the question of proving* (4.17) *directly, and then deducing Kiefer's result of* (4.11) *via the strong invariance principle of* (4.16).

In order to *establish* (4.17), *we study* now the stochastic process

$$\{K(\hat{U}_n(y),n) - K(y,n),\ n \geq 1\}$$
$$= \{K(y - u_n(y,1)/n^{1/2},\ n) - K(y,n), n \geq 1\} \tag{4.18}$$

for $y \in (0,1)$ *fixed*, via first *proving a strong invariance principle for the randomly perturbed Kiefer process* $\{K(\hat{U}_n(y),n),\ n \geq 1\}$.

PROPOSITION 4.1. The probability space of Theorem 1.A can be extended in such a way that, in addition to the original sequence of independent uniform-$[0,1]$ random variables and the Kiefer process $\{K(y,x),\ 0 \leq y \leq 1,\ 0 \leq x < \infty\}$ as in (1.8), for any given $y \in (0,1)$ there lives also a Wiener process $\{W(x),\ 0 \leq x < \infty\}$ on it which is independent of $\{K(y,x),\ 0 \leq y \leq 1,\ 0 \leq x < \infty\}$ and, as $n \to \infty$, we have

$$|K(\hat{U}_n(y),n) - K(y - (y(1-y))^{1/2}n^{-1}W(n),n)|$$
$$\stackrel{a.s.}{=} \mathcal{O}(n^{1/8}(\log n)^{3/4}(\log\log n)^{1/8})\ . \tag{4.19}$$

PROOF. With any fixed value of $y \in (0,1)$ we have (cf (1.3))

$$n^{-1/2}\alpha_n(y,1) = (y(1-y))^{1/2}n^{-1}\sum_{i=1}^{n}(\mathbf{1}\{U_i \leq y\} - y)/(y(1-y))^{1/2}$$
$$:= (y(1-y))^{1/2}n^{-1}S(n)\ , \tag{4.20}$$

where, by definition now, $\{S(n), n \geq 1\}$ is the partial sum process of independent standardized, i.e., mean 0 and variance 1 binomial random variables. Hence, by Komlós, Major and Tusnády (1975, 1976) (cf Theorems 2.6.1 and 2.6.2 in Csörgő and Révész (1981)), we can define the original sequence $\{U_n,\ n \geq 1\}$ of i.i.d. uniform-$(0,1)$ r.v.'s on a probability space together with a Wiener process $\{W(x),\ x \geq 0\}$ such that, almost surely, as $n \to \infty$

$$|S(n) - W(n)| = \mathcal{O}(\log n)\ . \tag{4.21}$$

Consequently, on extending the probability space of Theorem 1.A accordingly, we can define the original sequence $\{U_n,\ n \geq 1\}$, the Kiefer process of (1.8) and the Wiener process of (4.21) such that (1.8) and (4.21) hold true simultaneously, while the two approximating Gaussian processes are constructed independently from each other. On this probability space for $y \in (0,1)$ *fixed* we have

$$K(\hat{U}_n(y),n) = K(y - n^{-1/2}u_n(y,1),n)$$
$$= K(y + n^{-1/2}(\alpha_n(y,1) - u_n(y,1)) - n^{-1/2}\alpha_n(y,1),n)$$
$$= K\left(y + n^{-1/2}(\alpha_n(y,1) - u_n(y,1)) + (y(1-y))^{1/2}\right.$$
$$\left.\times\left(\frac{W(n)}{n} - \frac{S(n)}{n}\right) - (y(1-y))^{1/2}\frac{W(n)}{n},n\right) \tag{4.22}$$

where $S(\cdot)$ is defined in (4.20) and $W(\cdot)$ is as in (4.21).
Hence

$$\left|K(\hat{U}_n(y), n) - K\left(y - (y(1-y))^{1/2}\frac{W(n)}{n}, n\right)\right|$$
$$= \left|K\left(y - (y(1-y))^{1/2}\frac{W(n)}{n} + \Delta_n(y), n\right)\right. \quad (4.23)$$
$$\left. - K\left(y - (y(1-y))^{1/2}\frac{W(n)}{n}, n\right)\right|,$$

where

$$\Delta_n(y) := n^{-1/2}(\alpha_n(y, 1) - u_n(y, 1)) + (y(1-y))^{1/2}\left(\frac{W(n)}{n} - \frac{S(n)}{n}\right). \quad (4.24)$$

By (1.27) and (4.21) combined, for $y \in (0, 1)$ fixed and with some positive constant C, we have

$$|\Delta_n(y)| \leq (2^{-1/4} + o(1))n^{-3/4}(\log n)^{1/2}(\log \log n)^{1/4} + (C + o(1))n^{-1}\log n$$
$$\leq (2^{-1/4} + o(1))n^{-3/4}(\log n)^{1/2}(\log \log n)^{1/4} \quad \text{a.s.} \quad (4.25)$$

if n is large. Consequently, if n is large and $y \in (0, 1)$ is *fixed*, by (1.16) and (4.25) combined via (4.23), we obtain

$$\left|K(\hat{U}_n(y), n) - K(y - (y(1-y))^{1/2}n^{-1}W(n), n)\right|$$
$$\leq (1 + o(1))(2n(2^{-1/4} + o(1))n^{-3/4}(\log n)^{1/2}(\log \log n)^{1/4}$$
$$\times (\log n)(3/4 + o(1)))^{1/2}$$
$$\leq (1 + o(1))(2^{-5/4}3 + o(1))^{1/2}n^{1/8}(\log n)^{3/4}(\log \log n)^{1/8} \quad \text{a.s.}$$
$$(4.26)$$

This also concludes the proof of Proposition 4.1. □

Since we study the stochastic process of (4.18) for the sake of proving (4.17), we now restate (4.19) accordingly.

COROLLARY 4.1. In the context of Proposition 4.1 we have, as $n \to \infty$,

$$\left|(K(\hat{U}_n(y), n) - K(y, n)) - (K(y - (y(1-y))^{1/2}n^{-1}W(n), n) - K(y, n))\right|$$
$$= \left|(K(y - u_n(y, 1)/n^{1/2}, n) - K(y, n))\right.$$
$$\left. - (K(y - (y(1-y))^{1/2}n^{-1}W(n), n) - K(y, n))\right| \quad (4.27)$$
$$\stackrel{\text{a.s.}}{=} \mathcal{O}(n^{1/8}(\log n)^{3/4}(\log \log n)^{1/8}).$$

By definition of a Kiefer process $\{K(y,x),\ 0 \leq y \leq 1,\ 0 \leq x < \infty\}$ we have

$$K(y,n) = W(y,n) - yW(1,n)$$
$$= \sum_{i=1}^{n}(W(y,i) - W(y,i-1)) - y\sum_{i=1}^{n}(W(1,i) - W(1,i-1))$$
$$:= \sum_{i=1}^{n}W_i(y) - y\sum_{i=1}^{n}W_i(1),\qquad (4.28)$$

where $\{W(y,x),\ 0 \leq y < \infty,\ 0 \leq x < \infty\}$ is a two-parameter Wiener process, i.e., $W(y,x)$ is a Gaussian process with $EW(y,x) = 0$ and $EW(y_1,x_1)W(y_2,x_2) = (y_1 \wedge y_2)(x_1 \wedge x_2)$, and, consequently, the stochastic processes $\{W_i(y),\ 0 \leq y \leq 1\}$, $i = 1,2,\ldots$, form a sequence of independent standard Wiener processes.

Conversely, let $\{r_n\}$ be the sequence of positive dyadic rational numbers and let $\{W_{r_n}(y),\ 0 \leq y < \infty\}$ be independent standard Wiener processes. Then, using these standard Wiener processes, we can construct a two-parameter Wiener process $\{W(y,x),\ 0 \leq y < \infty,\ 0 \leq x < \infty\}$ as in Section 1.11 of Csörgő and Révész (1981). Let $\{\tilde{W}_{r_n}(u),\ 0 \leq u < \infty\}$ be another sequence of independent standard Wiener processes indexed by the sequence of positive dyadic rational numbers $\{r_n\}$ that is also independent of $\{W_{r_n}(y),\ 0 \leq y < \infty\}$. Then, in the above construction of $W(\cdot,\cdot)$, each of the Wiener processes $\{W_{r_n}(y),\ 0 \leq y < \infty\}$ can be extended to the real line $\mathbb{R} = (-\infty,\infty)$ by letting $W_{r_n}(y) = \tilde{W}_{r_n}(-y)$ for $y < 0$ for each r_n of the sequence $\{r_n\}$. Doing this yields a construction of a two-parameter Wiener process $\{W(y,x),\ -\infty < y < \infty,\ 0 \leq x < \infty\}$ on $\mathbb{R} \times \mathbb{R}_+$. Consequently, given any Kiefer process $\{K(y,x),\ 0 \leq y \leq 1,\ 0 \leq x < \infty\}$, we can extend its definition and write (cf (4.28)), without loss of generality,

$$K(y,n) = W(y,n) - yW(1,n)$$
$$= \sum_{i=1}^{n}W_i(y) - y\sum_{i=1}^{n}W_i(1),\qquad (4.29)$$

where $\{W(y,x),\ -\infty < y < \infty,\ 0 \leq x < \infty\}$ is an extended two-parameter Wiener process (Wiener sheet extended to $\mathbb{R} \times \mathbb{R}_+$), and the thus resulting $\{W_i(y),\ -\infty < y < \infty\} = \{W(y,i) - W(y,i-1),\ -\infty < y < \infty\}$, $i = 1,2,\ldots$, are independent Wiener processes that are now extended to the real line $\mathbb{R} = (-\infty,\infty)$.

PROPOSITION 4.2. Let $\{W(y,x),\ -\infty < y < \infty,\ 0 \leq x < \infty\}$ be an extended two parameter Wiener process and let $\{W(x),\ 0 \leq x < \infty\}$ be a standard Wiener process that is independent of the given extended Wiener sheet. Then, almost surely as $x \to \infty$, the limit set of

$$(V_1(x), V_2(x)) := \left(\frac{W(x^{-1}|W(x)|,x)}{(2|W(x)|\log\log x)^{1/2}},\ \frac{|W(x)|}{(2x\log\log x)^{1/2}} \right) \qquad (4.30)$$

is the semicircle

$$D(u,v) = \{(u,v) : u^2 + v^2 \leq 1, \ v > 0\} \ . \tag{4.31}$$

Here and throughout, we will use the convention $0/0 = 1$.

AN OUTLINE OF THE PROOF OF PROPOSITION 3.2. The set of limit points of $V_1(x)$ is the interval $[-1, 1]$ a.s. The set of limit points of $V_2(x)$ is the interval $[0, 1]$ a.s. The set of the limit points of $(V_1(x), V_2(x))$ is the semicircle $D(u,v)$.

REMARK 4.1. The details of the proof of Proposition 4.2 are somewhat lengthy. They can be constructed along the lines of Csáki, Csörgő, Földes and Révész (1989, 1994), and/or those of Deheuvels and Mason (1992).

PROPOSITION 4.3. Let $V_1(x)$ and $V_2(x)$ be as in Proposition 4.2. Then, almost surely as $x \to \infty$, the set of limit points of

$$V_1(x)(V_2(x))^{1/2} = \frac{W(x^{-1}|W(x)|, x)}{2^{3/4} x^{1/4} (\log \log x)^{3/4}} \tag{4.32}$$

is the interval $[0, 2^{1/2} 3^{-3/4}]$ and, consequently, we have

$$\limsup_{x \to \infty} \frac{W(x^{-1}|W(x)|, x)}{x^{1/4} (\log \log x)^{3/4}} \stackrel{\text{a.s.}}{=} 2^{5/4} 3^{-3/4} \ , \tag{4.33}$$

as well as

$$\limsup_{x \to \infty} \frac{W(x^{-1} W(x), x)}{x^{1/4} (\log \log x)^{3/4}} \stackrel{\text{a.s.}}{=} 2^{5/4} 3^{-3/4} \ . \tag{4.34}$$

PROOF. In the light of Proposition 4.2 we are to maximize the function: $(u,v) \to uv^{1/2}$, subject to the constraint $u^2 + v^2 - 1 = 0$, $v > 0$. Using, for example, the method of Lagrange multipliers, we obtain

$$\sup_{(u,v) \in D(u,v)} uv^{1/2} = 2^{1/2} 3^{-3/4} \ , \tag{4.35}$$

and hence also (4.33).

For the sake of proving now (4.34), let

$$W^+(y,x) = \begin{cases} W(y,x) & \text{if } y \geq 0, \\ 0 & \text{if } y \leq 0 \ , \end{cases} \tag{4.36}$$

$$W^-(y,x) = \begin{cases} W(y,x) & \text{if } y \leq 0, \\ 0 & \text{if } y \geq 0 \ , \end{cases} \tag{4.37}$$

on recalling that $W(\cdot, \cdot)$ is a Wiener sheet on $\mathbb{R} \times \mathbb{R}_+$. Then, clearly,

$$W(W(x),x) = W^+(W(x),x) + W^-(W(x),x)$$
$$= \begin{cases} W^+(W(x),x) & \text{if } W(x) \geq 0, \\ W^-(W(x),x) & \text{if } W(x) \leq 0, \end{cases} \quad (4.38)$$

$$W(|W(x)|,x) = W^+(W(x),x) + W^+(-W(x),x)$$
$$= \begin{cases} W^+(W(x),x) & \text{if } W(x) \geq 0, \\ W^+(-W(x),x) & \text{if } W(x) \leq 0, \end{cases} \quad (4.39)$$

where $0 \leq x < \infty$. By (4.33) and (4.39) we have

$$2^{5/4} 3^{-3/4} \stackrel{\text{a.s.}}{=} \limsup_{x \to \infty} \frac{W(x^{-1}|W(x)|, x)}{x^{1/4}(\log\log x)^{3/4}}$$

$$\stackrel{\text{a.s.}}{=} \max\left(\limsup_{\substack{x \to \infty \\ x \in A}} \frac{W^+(x^{-1}W(x), x)}{x^{1/4}(\log\log x)^{3/4}},\ \limsup_{\substack{x \to \infty \\ x \in B}} \frac{W^+(-x^{-1}W(x), x)}{x^{1/4}(\log\log x)^{3/4}} \right),$$
$$(4.40)$$

where

$$A := \{x : x \geq 0, W(x) > 0\}, \quad B := \{x : x \geq 0, W(x) < 0\}.$$

Since, due to symmetry, we have

$$\limsup_{\substack{x \to \infty \\ x \in A}} \frac{W^+(x^{-1}W(x), x)}{x^{1/4}(\log\log x)^{3/4}} = \limsup_{\substack{x \to \infty \\ x \in B}} \frac{W^+(-x^{-1}W(x), x)}{x^{1/4}(\log\log x)^{3/4}}$$
$$= \limsup_{\substack{x \to \infty \\ x \in B}} \frac{W^-(x^{-1}W(x), x)}{x^{1/4}(\log\log x)^{3/4}}, \quad (4.41)$$

on combining (4.38)–(4.41), we obtain (4.34) as well. $\square$

PROPOSITION 4.4. *Let $\{K(y,x),\ 0 \leq y \leq 1,\ 0 \leq x < \infty\}$ be a Kiefer process defined in terms of an extended two-parameter Wiener process $\{W(y,x),\ -\infty < y < \infty,\ 0 \leq x < \infty\}$. Let $\{W(x),\ 0 \leq x < \infty\}$ be a standard Wiener process that is independent of the given Kiefer process $K(\cdot,\cdot)$. Then, for each fixed $0 \leq y \leq 1$, almost surely as $x \to \infty$, the limit set of*

$$(K_y(x), V_2(x)) := \left(\frac{K(y + x^{-1}|W(x)|, x) - K(y, x)}{(2|W(x)|\log\log x)^{1/2}},\ \frac{|W(x)|}{(2x\log\log x)^{1/2}} \right)$$
$$(4.42)$$

is the semicircle $D(u,v)$ of (4.31).

PROOF. By definition of a Kiefer process $K(\cdot,\cdot)$, now in terms of an extended Wiener sheet $W(\cdot,\cdot)$ on $\mathbb{R} \times \mathbb{R}_+$, we have

$$K(y,x) = W(y,x) - yW(1,n) \ . \tag{4.43}$$

Hence, and because of the independence of $W(\cdot)$ and $W(\cdot,\cdot)$, we obtain

$$\begin{aligned} &K(y + x^{-1}W(x), x) - K(y,x) \\ &= W(y + x^{-1}W(x), x) - (y + x^{-1}W(x))W(1,x) - (W(y,x) - yW(1,x)) \\ &= W(x^{-1}W(x), x) - x^{-1}W(x)W(1,x) \ . \end{aligned} \tag{4.44}$$

Consequently, for each *fixed* $0 \leq y \leq 1$,

$$\frac{|K(y + x^{-1}|W(x)|, x) - W(x^{-1}|W(x)|, x)|}{(2|W(x)| \log \log x)^{1/2}} \leq \frac{|W(1,x)|}{x^{1/2}(2x \log \log x)^{1/2}}$$
$$= \mathcal{O}(1/x^{1/2}) \quad \text{a.s.} \ , \tag{4.45}$$

by the law of the iterated logarithm, as $x \to \infty$. Hence, by Proposition 4.2 and (4.45) we conclude Proposition 4.4. $\square$

As a consequence of Proposition 4.4, along the lines of Proposition 4.3, the next corollary is immediate.

COROLLARY 4.2. *Let $K_y(x)$ and $V_2(x)$ be as in Proposition 4.4. Then, for each fixed $0 \leq y \leq 1$, almost surely as $x \to \infty$, the set of limit points of*

$$K_y(x)(V_2(x))^{1/2} = \frac{K(y + x^{-1}|W(x)|, x) - K(y,x)}{2^{3/4} x^{1/4} (\log \log x)^{3/4}} \tag{4.46}$$

is the interval $[0, 2^{1/2} 3^{-3/4}]$ *and, consequently, we have also*

$$\limsup_{x \to \infty} \frac{K(y + x^{-1}|W(x)|, x) - K(y,x)}{x^{1/4} (\log \log x)^{3/4}} \stackrel{a.s.}{=} 2^{5/4} 3^{-3/4} \ , \tag{4.47}$$

as well as

$$\limsup_{x \to \infty} \frac{K(y + x^{-1}W(x), x) - K(y,x)}{x^{1/4} (\log \log x)^{3/4}} \stackrel{a.s.}{=} 2^{5/4} 3^{-3/4} \ . \tag{4.48}$$

REMARK 4.2. Let $\{K(y,x), \ 0 \leq y \leq 1, \ 0 \leq x < \infty\}$ be the Kiefer process of Theorem 1.A. We write, without loss of generality,

$$K(y,x) = W(y,x) - yW(1,x) \ , \tag{4.49}$$

where $\{W(y,x), \ 0 \leq y < \infty, \ 0 \leq x < \infty\}$ is a Wiener sheet. Let $\{\widetilde{W}(u,x), \ 0 \leq u < \infty, \ 0 \leq x < \infty\}$ be another Wiener sheet that is independent of $W(\cdot,\cdot)$, and extend the latter to $\mathbb{R} \times \mathbb{R}_+$ by letting $W(y,x) = \widetilde{W}(-y,x)$ for $y < 0$. Thus, the probability space of Theorem 1.A is also extended accordingly, *and* (1.8) *remains true*, of course. Also, given the latter Kiefer process $\{K(y,x), \ 0 \leq y \leq 1, \ 0 \leq x < \infty\}$, we can and, indeed, do write whenever needed from now on, without loss of generality,

$$K(y,x) = W(y,x) - yW(1,x) \tag{4.50}$$

where $\{W(y,x), -\infty < y < \infty, 0 \leq x < \infty\}$ is the Wiener sheet of (4.49) extended to $\mathbb{R} \times \mathbb{R}_+$.

We are now ready to establish (4.17) that will now be concluded in Corollary 4.4 below. First we are to revisit Corollary 4.1 in the context of Remark 4.2 (cf Proposition 4.6 below) via Proposition 4.5 as follows.

PROPOSITION 4.5. *Let* $\{K(y,x), 0 \leq y \leq 1, 0 \leq x < \infty\}$ *be the Kiefer process of Theorem* 1.A, *and let* $\{W(x), 0 \leq x < \infty\}$ *be the Wiener process defined in Proposition* 4.1 *that is independent of* $K(\cdot,\cdot)$. *Then, for each fixed* $0 < y < 1$, *as* $n \to \infty$ *we have*

$$\begin{aligned}
|(K(y - (y(1-y))^{1/2}&n^{-1}W(n), n) - K(y,n)) \\
&- (y(1-y))^{1/4} W(-n^{-1}W(n), n)| \\
&\leq (y(1-y))^{1/2}(n^{-1}|W(n)|)|W(1,n)| \\
&\leq (y(1-y))^{1/2}(2 + o(1))(\log \log n) \quad \text{a.s.} \\
&\stackrel{\text{a.s.}}{=} \mathcal{O}(\log \log n) ,
\end{aligned} \tag{4.51}$$

where $\{W(y,x), -\infty < y < \infty, 0 \leq x < \infty\}$ *is the extended Wiener sheet of* (4.50).

PROOF. Via (4.50), and because of the independence of $W(\cdot)$ and $W(\cdot,\cdot)$, we have for each fixed $0 < y < 1$

$$\begin{aligned}
K(y - &(y(1-y))^{1/2}n^{-1}W(n), n) - K(y, n) \\
&= W(-(y(1-y))^{1/2}n^{-1}W(n), n) + (y(1-y))^{1/2}n^{-1}W(n)W(1, n) \\
&= (y(1-y))^{1/4} W(-n^{-1}W(n), n) + (y(1-y))^{1/2}n^{-1}W(n)W(1, n) ,
\end{aligned} \tag{4.52}$$

which, in turn, implies (4.51). □

COROLLARY 4.3. *Let* $\{K(y,x), 0 \leq y \leq 1, 0 \leq x < \infty\}$ *be the Kiefer process of Theorem* 1.A, *and let* $\{W(x), 0 \leq x < \infty\}$ *be the Wiener process defined in Proposition* 4.1 *that is independent of* $K(\cdot,\cdot)$. *Then, for each fixed* $0 < y < 1$, *we have*

$$\begin{aligned}
\limsup_{n \to \infty} &\frac{K(y - (y(1-y))^{1/2}n^{-1}W(n), n) - K(y,n)}{(y(1-y))^{1/4} n^{1/4} (\log \log n)^{3/4}} \\
&\stackrel{\text{a.s.}}{=} \limsup_{n \to \infty} \frac{W(-n^{-1}W(n), n)}{n^{1/4}(\log \log n)^{3/4}} \stackrel{\text{a.s.}}{=} 2^{5/4} 3^{-3/4} .
\end{aligned} \tag{4.53}$$

PROOF. The first almost sure equality follows by (4.51), while the second one is implied by (4.34), due to symmetry of $W(\cdot)$. □

In the light of Proposition 4.5, Proposition 4.1 á la Corollary 4.1 can be rewritten as follows.

PROPOSITION 4.6. The probability space of Theorem 1.A can be extended in such a way that, in addition to the original sequence of independent uniform-$[0,1]$ random variables and the Kiefer process $\{K(y,x), 0 \leq y \leq 1, 0 \leq x < \infty\}$ as in (1.8) and written as in (4.50) in terms of a Wiener sheet extended to $\mathbb{R} \times \mathbb{R}_+$, for each fixed $0 < y < 1$ there lives also a Wiener process $\{W(x),\ 0 \leq x < \infty\}$ on it which is independent of $\{K(y,x),\ 0 \leq y \leq 1,\ 0 \leq x < \infty\}$, and hence also from the extended Wiener sheet $\{W(y,x),\ -\infty < y < \infty,\ 0 \leq x < \infty\}$ of (4.50), and, as $n \to \infty$, we have

$$\left|\left(K(\hat{U}_n(y),n) - K(y,n)\right) - (y(1-y))^{1/4} W(-n^{-1} W(n), n)\right|$$
$$\stackrel{a.s.}{=} \left|\left(K(y - u_n(y,1)/n^{1/2}, n) - K(y,n)\right)\right.$$
$$\left. - \left(K(y - (y(1-y))^{1/2} n^{-1} W(n), n) - K(y,n)\right)\right| + \mathcal{O}(\log\log n)$$
$$\stackrel{a.s.}{=} \mathcal{O}\left(n^{1/8} (\log n)^{3/4} (\log\log n)^{1/8}\right). \qquad (4.54)$$

PROOF. A combination of (4.51) with (4.57) yields (4.54). □

COROLLARY 4.4. On the probability space of Theorem 1.A extended as in Proposition 4.6, we have for each fixed $0 < y < 1$

$$\limsup_{n \to \infty} \frac{K(\hat{U}_n(y), n) - K(y, n)}{n^{1/4} (\log\log n)^{3/4}} \stackrel{a.s.}{=} 2^{5/4} 3^{-3/4} (y(1-y))^{1/4}. \qquad (4.55)$$

PROOF. A combination of (4.53) with (4.54) gives (4.55). □

REMARK 4.3. Corollary 4.4 amounts to saying that (4.17) is now proven directly. Hence Kiefer's original result of (4.11) now follows also from (4.16), which, in turn, via (4.15) and (4.9), implies Theorem 4.B as well. For related details on, and the first explanation of, (4.11) along somewhat similar lines, we refer to Deheuvels and Mason (1992) and their references. The aim and essence of our approach here to proving (1.27) and (4.11) in a unified way amount to saying that we have demonstrated that both these results follow from the same invariance principle (cf (1.30) and (4.16)) for the uniform Bahadur–Kiefer process (cf (1.29)) in terms of the randomly perturbed increments $\{K(\hat{U}_n(y), n) - K(y, n),\ n \geq 1\}$, of a Kiefer process, via establishing (1.26) and (4.55), respectively, for the latter randomly stopped approximating Gaussian process.

Kiefer also studied the asymptotic distribution of $R_n^*(y,1)$ for $y \in (0,1)$ fixed (cf Kiefer (1967)), as well as that of $\sup_{0 \leq y \leq 1} |R_n^*(y,1)|$ (cf Kiefer (1970)). We summarize these results of Kiefer in the following theorem (cf Kiefer (1967, 1970)).

THEOREM 3.C. For every fixed $0 < y < 1$ we have

$$\lim_{n \to \infty} P\{R_n^*(y,1)/n^{1/4} \leq x\} = 2 \int_0^\infty \Phi(x/v^{1/2}) \, d_v \Phi(v/\sigma_y) \,, \quad (4.56)$$

$$\lim_{n \to \infty} P\{|R_n^*(y,1)|/n^{1/4} \leq x\} = 2 \int_0^\infty \left(2\Phi(x/v^{1/2}) - 1\right) d_v \Phi(v/\sigma_y) \,, \quad (4.57)$$

where Φ is the standard normal distribution function, $\sigma_y = (y(1-y))^{1/2}$, and, as $n \to \infty$,

$$n^{-1/4}(\log n)^{-1/2} \sup_{0 \leq y \leq 1} |R_n^*(y,1)| \xrightarrow{\mathscr{L}} \left(\sup_{0 \leq y \leq 1} |B(y)|\right)^{1/2}, \quad (4.58)$$

where $\{B(y),\ 0 \leq y \leq 1\}$ is a Brownian bridge.

We are to show now that (4.56) and (4.57) follow *via* (4.16), while (4.58) can be established *via* (1.30), that is to say we show that, just like (1.27) and (4.11), *these results also follow from the same invariance principle.*

OBSERVATION 4.1. Let $\{W(y,x),\ -\infty < y < \infty,\ 0 \leq x < \infty\}$ be an extended Wiener sheet, and let $\{W(x),\ 0 \leq x < \infty\}$ be a standard Wiener process that is independent of the given Wiener sheet. Then, for each $x \in (0, \infty)$, we have

$$W(-x^{-1}W(x), x)/x^{1/4} \stackrel{\mathscr{L}}{=} N_1 (|N_2|)^{1/2} \,, \quad (4.59)$$

where N_1 and N_2 are independent standard normal random variables.

PROOF. We have (cf (4.36)–(4.38))

$$W(-x^{-1}W(x), x) = W^+(-x^{-1}W(x), x) + W^-(-x^{-1}W(x), x)$$

$$= \frac{W^+(-x^{-1}W(x), x)}{(x(-W(x))/x)^{1/2}}(-W(x))^{1/2}\mathbf{1}\{W(x) \leq 0\}$$

$$+ \frac{W^-(-x^{-1}W(x), x)}{(xW(x)/x)^{1/2}}(W(x))^{1/2}\mathbf{1}\{W(x) \geq 0\} \quad (4.60)$$

$$\stackrel{\mathscr{L}}{=} W^+(1,1)(-W(x))^{1/2}\mathbf{1}\{W(x) \leq 0\}$$

$$+ W^-(-1,1)(W(x))^{1/2}\mathbf{1}\{W(x) \geq 0\}$$

$$\stackrel{\mathscr{L}}{=} N_1(|W(x)|)^{1/2} \,,$$

where N_1 is a standard normal random variable that is independent of $W(\cdot)$. Hence we have also (4.59). $\square$

As a consequence of this observation, we have

PROPOSITION 4.7. Let $\{K(y,x), 0 \leq y \leq 1, 0 \leq x < \infty\}$ be the Kiefer process of Theorem 1.A, $\{W(y,x), -\infty < y < \infty, 0 \leq x < \infty\}$ be the extended Wiener sheet of (4.50) and let $\{W(x), 0 \leq x < \infty\}$ be the Wiener process defined in Proposition 4.1 that is independent of $K(\cdot,\cdot)$. Then, for each fixed $0 < y < 1$ and $n \geq 1$, we have

$$\begin{aligned}&\left(K(y - (y(1-y))^{1/2}n^{-1}W(n), n) - K(y, n)\right)/n^{1/4} \\ &\stackrel{\mathscr{D}}{=} (y(1-y))^{1/4} N_1(|N_2|)^{1/2} + (y(1-y))^{1/2} N_1 N_2/n^{1/4} ,\end{aligned} \quad (4.61)$$

where N_1 and N_2 are independent standard normal random variables.

PROOF. A combination of (4.52) with (4.59) yields (4.61). □

PROOF OF (4.56) AND (4.57). Due to (4.61), as $n \to \infty$, we have for each fixed $0 < y < 1$

$$\begin{aligned}&\left(K(y - (y(1-y))^{1/2}n^{-1}W(n), n) - K(y, n)\right)/n^{1/4} \\ &\stackrel{\mathscr{D}}{\to} (y(1-y))^{1/4} N_1(|N_2|)^{1/2} .\end{aligned} \quad (4.62)$$

Consequently, by (4.27), as $n \to \infty$ we have also

$$\left(K(\hat{U}_n(y), n) - K(y, n)\right)/n^{1/4} \stackrel{\mathscr{D}}{\to} (y(1-y))^{1/4} N_1(|N_2|)^{1/2} \quad (4.63)$$

for every fixed $0 < y < 1$. Now, as $n \to \infty$, (4.16) yields

$$R_n^*(y, 1)/n^{1/4} \stackrel{\mathscr{D}}{\to} (y(1-y))^{1/4} N_1(|N_2|)^{1/2} \quad (4.64)$$

for each fixed $0 < y < 1$, which coincides with (4.56). On repeating the same proof for absolute values, we arrive at

$$|R_n^*(y, 1)|/n^{1/4} \stackrel{\mathscr{D}}{\to} (y(1-y))^{1/4} |N_1|(|N_2|)^{1/2} , \quad (4.65)$$

i.e., (4.57) is proved as well. □

COROLLARY 4.5. Under the conditions of Theorem 4.A on F for $y \in (0,1)$ fixed, as $n \to \infty$ we have

$$\tilde{R}_n(y, 1)/n^{1/4} \stackrel{\mathscr{D}}{\to} (y(1-y))^{1/4} N_1(|N_2|)^{1/2} \quad (4.66)$$

and

$$|\tilde{R}_n(y, 1)|/n^{1/4} \stackrel{\mathscr{D}}{\to} (y(1-y))^{1/4} |N_1|(|N_2|)^{1/2} , \quad (4.67)$$

where N_1 and N_2 are independent standard normal random variables.

PROOF. On account of (4.15), we have (4.66) and (4.67) via (4.64) and (4.65) respectively. □

We note in passing that Kiefer (1967) proved (4.66) and (4.67) via (4.8), by establishing (4.56) and (4.57) directly.

REMARK 4.4. The distribution function appearing in (4.56) is the same as that in Dobrushin's theorem (1955). In Csáki, Csörgő, Földes and Révész (1992, p. 681) it appears as the asymptotic distribution of certain additive functionals. The connection of these additive functionals and the distribution of the random variable $N_1(|N_2|)^{1/2}$ is explained in the same paper. For further connections along these lines we refer to Csáki, Csörgő, Földes and Révész (1989), where Brownian local time is studied as a two-time parameter stochastic process via approximating it almost surely by a randomly perturbed Wiener sheet. When viewed via strong invariance principles, the very nature of the Kiefer theorems we have studied so far in this section, *as well as that of the ones coming up*, is very much like that of the results concluded in *4. Applications* of the latter paper.

Kiefer (1970) deduces the uniform version (4.58) of (4.57) from proving directly the following convergence in probability, as $n \to \infty$:

$$n^{-1/4}(\log n)^{-1/2} \sup_{0 \leq y \leq 1} |R_n^*(y,1)| / \left(\sup_{0 \leq y \leq 1} |\alpha_n(y,1)| \right)^{1/2} \xrightarrow{P} 1 . \qquad (4.68)$$

Kiefer (1970) noted also that (4.68) was actually true with probability one, and hence also that not only (4.58), but (1.27) as well followed at once from it and the law of the iterated logarithm for $\sup_{0 \leq y \leq 1} |\alpha_n(y,1)|$. Namely, Kiefer (1970) announced

THEOREM 4.D. We have

$$\lim_{n \to \infty} n^{-1/4}(\log n)^{-1/2} \sup_{0 \leq y \leq 1} |R_n^*(y,1)| / \left(\sup_{0 \leq y \leq 1} |\alpha_n(y,1)| \right)^{1/2} \stackrel{a.s.}{=} 1 . \qquad (4.69)$$

He, however, proved (4.68) and (1.27) directly and did not publish his proof of (4.69) due to its length and tediousness at that time. Deheuvels and Mason (1990) gave the first published proof of (4.69). Here *we prove* Theorem 4.D via streamlining the proof of Theorem 3.3.3 for (4.69) in Csörgő and Horváth (1993). The proof is based on the strong invariance principle of (1.30) and on the following randomized version of (1.16) (cf (A.1.11) of Csörgő and Horváth (1993)).

THEOREM 4.E. Let $\{K(y,x),\ 0 \leq y \leq 1,\ 0 \leq x < \infty\}$ be a Kiefer process written as in (4.50) in terms of a Wiener sheet extended to $\mathbb{R} \times \mathbb{R}_+$. Then

$$\lim_{n\to\infty} (\log n)^{-1/2} \sup_{0\le y\le 1} |K(y - n^{-1}K(y,n), n)$$
$$- K(y,n)|/\left(\sup_{0\le y\le 1} |K(y,n)|\right)^{1/2} \stackrel{\text{a.s.}}{=} 1 . \tag{4.70}$$

Thinking via Theorem 4.E instead of (1.16), the *original approach* of Csörgő and Révész (1975, 1978, 1981) *to quantile* and *Bahadur–Kiefer processes* as formulated in Proposition 1.1 can now be *reformulated* as follows.

THEOREM 4.1. Let $\{K(y,x),\ 0 \le y \le 1,\ 0 \le x < \infty\}$ be the Kiefer process of Theorem 1.A as in (1.8), written as in (4.50) in terms of a Wiener sheet extended to $\mathbb{R} \times \mathbb{R}_+$, and let $\{R_n^*(y,1),\ 0 \le y \le 1,\ n=1,2,\ldots\}$ be the uniform Bahadur–Kiefer process as in (1.29). Then, as $n \to \infty$, we have

$$\sup_{0\le y\le 1} |R_n^*(y,1) - (K(y,n) - K(y - n^{-1}K(y,n), n))|$$
$$\stackrel{\text{a.s.}}{=} \mathcal{O}\left(n^{1/8}(\log n)^{3/4}(\log\log n)^{1/8}\right) . \tag{4.71}$$

PROOF. On the probability space of Theorem 1.A, we have (1.10) of Theorem 1.B as well. Hence, along the lines of the proof of Proposition 4.1, by (1.10) and (1.16) combined, as $n \to \infty$, we obtain

$$\sup_{0\le y\le 1} |(K(\hat{U}_n(y),n) - K(y,n)) - (K(y - n^{-1}K(y,n),n) - K(y,n))|$$
$$= \sup_{0\le y\le 1} |(K(y - n^{-1}(n^{1/2}u_n(y,1)),n) - K(y,n))$$
$$- (K(y - n^{-1}K(y,n),n) - K(y,n))| \tag{4.72}$$
$$\stackrel{\text{a.s.}}{=} \mathcal{O}\left(n^{1/8}(\log n)^{3/4}(\log\log n)^{1/8}\right) .$$

As a consequence of (1.30) and (4.72), we conclude (4.71). □

PROOF OF THEOREM 4.D. We assume, without loss of generality, that we are on the probability space of Theorem 1.A, as described in Theorem 4.1. Mogul'skiĭ (1980) (cf Theorem 5.1.7 in Csörgő and Révész (1981)) proved

$$\liminf_{n\to\infty} (\log\log n)^{1/2} \sup_{0\le y\le 1} |\alpha_n(y,1)| \stackrel{\text{a.s.}}{=} \pi/8^{1/2} , \tag{4.73}$$

the so-called other law of the iterated logarithm for the uniform empirical process $\{\alpha_n(y,1),\ 0 \le y \le 1,\ n=1,2,\ldots\}$ (cf Chung (1948) for the first such law, and Chapter 3 of Csörgő and Révész (1981) for some further related material). Naturally, via (1.12), the scaled Kiefer process $\{n^{-1/2}K(y,n),\ 0 \le y \le 1,\ n = 1,2,\ldots\}$ inherits the same law. Namely we have

$$\liminf_{n\to\infty} (\log\log n)^{1/2} \sup_{0\le y\le 1} |K(y,n)|/n^{1/2} \stackrel{\text{a.s.}}{=} \pi/8^{1/2} , \qquad (4.74)$$

on account of

$$\#\left(\liminf_{n\to\infty} (\log\log n)^{1/2} \sup_{0\le y\le 1} |\alpha_n(y,1)| - \liminf_{n\to\infty} (\log\log n)^{1/2} \sup_{0\le y\le 1} |K(y,n)|/n^{1/2}\right)$$

$$\le \limsup_{n\to\infty} (\log\log n)^{1/2} \#\left(\sup_{0\le y\le 1} |\alpha_n(y,1)| - \sup_{0\le y\le 1} |K(y,n)|/n^{1/2}\right)$$

$$\le \limsup_{n\to\infty} \sup_{0\le y\le 1} |\alpha_n(y,1) - K(y,n)/n^{1/2}|(\log\log n)^{1/2}$$

$$\stackrel{\text{a.s.}}{=} \limsup_{n\to\infty} \mathcal{O}(n^{-1/2}(\log n)^2)(\log\log n)^{1/2} = 0 , \qquad (4.75)$$

where $\# = +$ or $-$, and in the last line we utilized (1.12).

Combining now (4.73) with (4.71), we obtain

$$\lim_{n\to\infty} n^{-1/4}(\log n)^{-1/2} \frac{\sup_{0\le y\le 1} |R_n^*(y,1) - (K(y,n) - K(y - n^{-1}K(y,n),n))|}{(\sup_{0\le y\le 1} |\alpha_n(y,1)|)^{1/2}}$$

$$\le \frac{\limsup_{n\to\infty} \sup_{0\le y\le 1} |R_n^*(y,1) - (K(y,n) - K(y - n^{-1}K(y,n),n))|}{\liminf_{n\to\infty} \left((\log\log n)^{1/2} \sup_{0\le y\le 1} |\alpha_n(y,1)|\right)^{1/2}}$$

$$\times \frac{(\log\log n)^{1/4}}{n^{1/4}(\log n)^{1/2}} = 0 \quad \text{a.s.} , \qquad (4.76)$$

and, note in passing that, using (4.74) instead of (4.73), we obtain similarly that we have

$$\lim_{n\to\infty} n^{-1/4}(\log n)^{-1/2} \frac{\sup_{0\le y\le 1} |R_n^*(y,1) - (K(y,n) - K(y - n^{-1}K(y,n),n))|}{\left(n^{-1/2} \sup_{0\le y\le 1} |K(y,n)|\right)^{1/2}}$$

$$\stackrel{\text{a.s.}}{=} 0 \qquad (4.77)$$

as well. On account of (1.26) combined with (4.72), we have also

$$\limsup_{n\to\infty} n^{-1/4}(\log n)^{-1/2}(\log\log n)^{-1/4}$$
$$\times \sup_{0\le y\le 1} |K(y,n) - K(y - n^{-1}K(y,n),n)| \stackrel{\text{a.s.}}{=} 2^{-1/4} , \qquad (4.78)$$

yet another manifestation of Kiefer's result in (1.27). Hence

$$\lim_{n\to\infty} n^{-1/4}(\log n)^{-1/2} \sup_{0\leq y\leq 1} |K(y,n) - K(y - n^{-1}K(y,n), n)|$$

$$\times \left| \frac{1}{\left(n^{-1/2} \sup_{0\leq y\leq 1} |K(y,n)|\right)^{1/2}} - \frac{1}{\left(\sup_{0\leq y\leq 1} |\alpha_n(y,1)|\right)^{1/2}} \right|$$

$$\leq \lim_{n\to\infty} \frac{\left(\sup_{0\leq y\leq 1} |K(y,n) - K(y - n^{-1}K(y,n), n)|\right)\left(\sup_{0\leq y\leq 1} |\alpha_n(y,1) - n^{-1/2}K(y,n)|\right)}{(a_n)^{1/2}(b_n)^{1/2}\left((a_n)^{1/2} + (b_n)^{1/2}\right)}$$

$$\times \frac{1}{n^{1/4}(\log n)^{1/2}}$$

$$\leq \frac{\limsup_{n\to\infty}\left(\sup_{0\leq y\leq 1} |K(y,n) - K(y - n^{-1}K(y,n), n)|\right)\left(\sup_{0\leq y\leq 1} |\alpha_n(y,1) - n^{-1/2}K(y,n)|\right)}{\liminf_{n\to\infty} ((\log\log n)^{1/2}a_n)^{1/2}((\log\log n)^{1/2}b_n)^{1/2}(((\log\log n)^{1/2}a_n)^{1/2} + ((\log\log n)^{1/2}b_n)^{1/2})}$$

$$\times \frac{(\log\log n)^{3/4}}{n^{1/4}(\log n)^{1/2}}$$

$$= 0 \quad \text{a.s.}, \tag{4.79}$$

where $a_n := n^{-1/2}\sup_{0\leq y\leq 1}|K(y,n)|$, $b_n := \sup_{0\leq y\leq 1}|\alpha_n(y,1)|$ and, in deducing the latter conclusion, in addition to (4.78), we also made use of (1.12), (4.73) and (4.74). Now Theorem 4.E combined with (4.79) yields

$$\lim_{n\to\infty} n^{-1/4}(\log n)^{-1/2} \sup_{0\leq y\leq 1} |K(y - n^{-1}K(y,n), n) - K(y,n)| / \left(\sup_{0\leq y\leq 1} |\alpha_n(y,1)|\right)^{1/2} \stackrel{\text{a.s.}}{=} 1$$

$$\tag{4.80}$$

which, in turn via (4.76), also completes the proof of Theorem 4.D. □

REMARK 4.5. In the light of (4.70) and (4.77) we have also

$$\lim_{n\to\infty} (\log n)^{-1/2} n^{-1/4} \sup_{0\leq y\leq 1} |R_n^*(y,1)| / \left(n^{-1/2} \sup_{0\leq y\leq 1} |K(y,n)|\right)^{1/2} \stackrel{\text{a.s.}}{=} 1. \tag{4.81}$$

Since, for each $n \geq 1$ and elementary outcome $\omega \in \Omega$, we have

$$\sup_{0\leq y\leq 1} |\alpha_n(y,1)| = \sup_{0\leq y\leq 1} |u_n(y,1)|, \tag{4.82}$$

by (4.73) we conclude

$$\liminf_{n\to\infty} (\log\log n)^{1/2} \sup_{0\leq y\leq 1} |u_n(y,1)| \stackrel{\text{a.s.}}{=} \pi/8^{1/2}, \tag{4.83}$$

as well as

$$\lim_{n\to\infty} n^{-1/4}(\log n)^{-1/2} \sup_{0\le y\le 1} |R_n^*(y,1)| \Big/ \Big(\sup_{0\le y\le 1} |u_n(y,1)|\Big)^{1/2} \stackrel{\text{a.s.}}{=} 1 \ . \quad (4.84)$$

COROLLARY 4.A. We have, as $n \to \infty$,

$$n^{-1/4}(\log n)^{-1/2} \sup_{0\le y\le 1} |R_n^*(y,1)| \stackrel{\mathscr{D}}{\to} \Big(\sup_{0\le y\le 1} |B(y)|\Big)^{1/2} , \quad (4.85)$$

where $\{B(y), \ 0 \le y \le 1\}$ is a Brownian bridge, as well as

$$\limsup_{n\to\infty} n^{-1/4}(\log n)^{-1/2}(\log\log n)^{-1/4} \sup_{0\le y\le 1} |R_n^*(y,1)| \stackrel{\text{a.s.}}{=} 2^{-1/4} , \quad (4.86)$$

and

$$\liminf_{n\to\infty} n^{-1/4}(\log n)^{-1/2}(\log\log n)^{1/4} \sup_{0\le y\le 1} |R_n^*(y,1)| \stackrel{\text{a.s.}}{=} 8^{-1/4}\pi^{1/2} \ .$$

(4.87)

PROOF. (4.85) is a restatement of (4.58) of Theorem 4.C whose proof now follows from any one of the respective statements of (4.69), (4.81) and (4.84).

As to (4.86), it is a restatement of (1.27), which we have already proved via (1.26), (1.13) and (1.14). This time around we can deduce it again from (4.78) via (4.71), or from any one of the respective statements of (4.69), (4.81) and (4.84), each combined with the corresponding law of the iterated logarithm for $\sup_{0\le y\le 1} |\alpha_n(y,1)|$, $n^{-1/2}\sup_{0\le y\le 1} |K(y,n)|$ and $\sup_{0\le y\le 1}|u_n(y,1)|$, respectively.

Concerning (4.87), it follows from any one of the respective statements of (4.69), (4.81) and (4.84), each combined with the corresponding *other* law of the iterated logarithm for $\sup_{0\le y\le 1} |\alpha_n(y,1)|$, $n^{-1/2}\sup_{0\le y\le 1} |K(y,n)|$ and $\sup_{0\le y\le 1}|u_n(y,1)|$ as in (4.73), (4.74) and (4.83), respectively. This also completes the proof of Corollary 4.A. □

For further results along these lines we refer to Section 3.3 of Csörgő and Horváth (1993) and references therein.

Another route to take to having (4.87) is via

COROLLARY 4.6. Let $\{K(y,x), \ 0 \le y \le 1, \ 0 \le x < \infty\}$ be a Kiefer process written as in (4.50) in terms of a Wiener sheet extended to $\mathbb{R} \times \mathbb{R}_+$. Then

$$\liminf_{n\to\infty} n^{-1/4}(\log n)^{-1/2}(\log\log n)^{1/4} \sup_{0\le y\le 1} |K(y - n^{-1}K(y,n), n) - K(y,n)|$$
$$= 8^{-1/4}\pi^{1/2} \quad \text{a.s.} \quad (4.88)$$

PROOF. A combination of Theorem 4.E with (4.74) yields the result. □

REMARK 4.6. *An alternative proof of* (4.87) now follows from (4.88) via (4.71) on the probability space of Theorem 1.A.

5. An extension of the classical Bahadur–Kiefer process theory via strong invariance principles

Kiefer (1970), in addition to proving (1.27) directly, as well as (4.58) via (4.68), actually carried out his ingenious calculations in terms of a random sample with a common, twice differentiable distribution function F, having finite support on $\mathbb{R}$, and assuming that $\inf_{0 \leq y \leq 1} f(Q(y)) > 0$ and $\sup_{0 \leq y \leq 1} |f'(Q(y))| < \infty$. Naturally, he noted also the desirability of extending his results for F not necessarily satisfying the just mentioned assumptions. A glance at the relationship of (4.13) clearly indicates that the *results* we have summarized and proved *in* Section 2 can be immediately *restated* as *strong invariance principles* for the *general Bahadur–Kiefer process* (cf (4.2) and (4.9))

$$\{\tilde{R}_n(y, 1),\ 0 \leq y \leq 1,\ n = 1, 2, \ldots\} := \{n^{1/2}(\alpha_n(y, 1) \\ - \rho_n(y, 1)),\ 0 \leq y \leq 1,\ n = 1, 2 \ldots\} \tag{5.1}$$

in terms of the *uniform Bahadur–Kiefer process* (cf (1.29))

$$\{R_n^*(y, 1),\ 0 \leq y \leq 1,\ n = 1, 2, \ldots\} := \{n^{1/2}(\alpha_n(y, 1) \\ - u_n(y, 1)),\ 0 \leq y \leq 1,\ n = 1, 2 \ldots\} \tag{5.2}$$

where $\alpha_n(y, 1) = \beta_n(Q(y), 1)$ and $u_n(y, 1) = n^{1/2}(y - F(\hat{Q}_n(y)))$ (cf (1.60) and (1.61) respectively). *As a consequence* of doing this, we *obtain an extension of Kiefer's theory of uniform deviations for* $\tilde{R}_n(\cdot, \cdot)$ *under the milder conditions on F* of Csörgő and Révész (1978) as spelled out and summarized also in Section 2 above. For related details of these extensions we refer to Csörgő and Révész (1978; 1981, Sections 5.2 and 5.3), Csörgő (1983, Chapter 6), and Csörgő and Horváth (1993, Section 6.5). For immediate use in stating some of the consequent extensions of the results of Kiefer (1970) here, *we restate* (4.13): for each elementary outcome $\omega \in \Omega$, we have (cf (5.1) and (5.2))

$$\{\tilde{R}_n(y, 1) - R_n^*(y, 1),\ 0 \leq y \leq 1,\ n = 1, 2, \ldots\} \\ = \{-n^{1/2}(\rho_n(y, 1) - u_n(y, 1)),\ 0 \leq y \leq 1,\ n = 1, 2, \ldots\}, \tag{5.3}$$

where $u_n(y, 1) = n^{1/2}(y - F(\hat{Q}_n(y)))$.

THEOREM 5.1. *Given the conditions* (1.66) (i), (ii) *and* (iii) *on F, we have*

$$\lim_{n\to\infty} n^{-1/4}(\log n)^{-1/2} \sup_{1/(n+1)\leq y\leq n/(n+1)} |\tilde{R}_n(y,1)| \Big/ \Big(\sup_{0\leq y\leq 1} |\alpha_n(y,1)|\Big)^{1/2} \overset{\text{a.s.}}{=} 1 \ . \tag{5.4}$$

PROOF. The result follows immediately via (5.3) by Theorem 2.C, (4.73) and Theorem 4.D. □

COROLLARY 5.A. Given the conditions (1.66) (i), (ii) and (iii) on F, as $n \to \infty$, we have

$$n^{-1/4}(\log n)^{-1/2} \sup_{1/(n+1)\leq y\leq n/(n+1)} |\tilde{R}_n(y,1)| \overset{\mathcal{L}}{\to} \Big(\sup_{0\leq y\leq 1} |B(y)|\Big)^{1/2} , \tag{5.5}$$

where $\{B(y),\ 0 \leq y \leq 1\}$ is a Brownian bridge, as well as

$$\limsup_{n\to\infty} n^{-1/4}(\log n)^{-1/2}(\log\log n)^{-1/4} \sup_{1/(n+1)\leq y\leq n/(n+1)} |\tilde{R}_n(y,1)| \overset{\text{a.s.}}{=} 2^{-1/4} , \tag{5.6}$$

and

$$\liminf_{n\to\infty} n^{-1/4}(\log n)^{-1/2}(\log\log n)^{1/4} \sup_{1/(n+1)\leq y\leq n/(n+1)} |\tilde{R}_n(y,1)| \overset{\text{a.s.}}{=} 8^{-1/4}\pi^{1/2} \ . \tag{5.7}$$

PROOF. (5.5) follows immediately from (5.4). So does also (5.6), if we combine (5.4) with the law of the iterated logarithm for $\sup_{0\leq y\leq 1}|\alpha_n(y,1)|$. We note also that (5.6) also follows via (5.3) from Theorem 2.C and (4.86). As to (5.7), it follows from (5.4) and (4.73), as well as via (5.3) from Theorem 2.C and (4.87).

A more direct approach to studying the general Bahadur–Kiefer process $\tilde{R}_n(y,1)$ is in terms of the iterated Kiefer process increments of Theorem 4.E.

THEOREM 5.2. Let $\{K(y,x),\ 0 \leq y \leq 1,\ 0 \leq x < \infty\}$ be the Kiefer process of Theorem 1.A as in (1.8), written as in (4.50) in terms of a Wiener sheet extended to $\mathbb{R} \times \mathbb{R}_+$, and let $\{\tilde{R}_n(y,1),\ 0 \leq y \leq 1,\ n=1,2,\ldots\}$ be the general Bahadur–Kiefer process as in (5.1). Given the conditions (1.66) (i), (ii) and (iii) on F, as $n \to \infty$, we have

$$\sup_{0\leq y\leq 1} |\tilde{R}_n(y,1)\mathbf{1}\{1/(n+1) \leq y \leq n/(n+1)\} - (K(y,n)$$
$$- K(y - n^{-1}K(y,n),n))| \overset{\text{a.s.}}{=} \mathcal{O}(n^{1/8}(\log n)^{3/4}(\log\log n)^{1/8}) \ . \tag{5.8}$$

PROOF. We have

$$\sup_{0\leq y\leq 1} |\widetilde{R}_n(y,1)\mathbf{1}\{1/(n+1) \leq y \leq n/(n+1)\}$$
$$- (K(y,n) - K(y - n^{-1}K(y,n), n))|$$
$$\leq \sup_{0\leq y\leq 1} |(\widetilde{R}_n(y,1) - R_n^*(y,1))\mathbf{1}\{1/(n+1) \leq y \leq n/(n+1)\}|$$
$$+ \sup_{0\leq y\leq 1} |R_n^*(y,1) - (K(y,n) - K(y - n^{-1}K(y,n), n))| \quad (5.9)$$
$$\stackrel{\text{a.s.}}{=} \sup_{1/(n+1)\leq y\leq n/(n+1)} |n^{1/2}(\rho_n(y,1) - u_n(y,1))|$$
$$+ \mathcal{O}\left(n^{1/8}(\log n)^{3/4}(\log\log n)^{1/8}\right)$$
$$\stackrel{\text{a.s.}}{=} \mathcal{O}\left(n^{1/8}(\log n)^{3/4}(\log\log n)^{1/8}\right) ,$$

where, just before the last line of (5.9), we utilized (5.3) and (4.71), and then applied Theorem 2.C as well. □

ANOTHER PROOF OF THEOREM 5.1 AND COROLLARY 5.A. Clearly, Theorem 5.1 and Corollary 5.A are corollaries of Theorem 5.1. Indeed, (5.4) is implied by (5.8), (4.80) and (4.73). Also, (5.6) and (5.7) follow from (5.8) combined with (4.78) and (4.88) respectively. □

Our next result is along the lines of Theorem 5.2.

THEOREM 5.3. Let $\{K(y,x), 0 \leq y \leq 1, 0 \leq x < \infty\}$ be the Kiefer process of Theorem 1.A as in (1.8), written as in (4.50) in terms of a Wiener sheet extended to $\mathbb{R} \times \mathbb{R}_+$, and let $\{\widetilde{R}_n(y,1), 0 \leq y \leq 1, n = 1, 2, \ldots\}$ be the general Bahadur–Kiefer process as in (5.1). Given the conditions (1.66) (i), (ii) and (iii), as well as (2.12) (i) and (ii) or (2.12) (iii) on F, as $n \to \infty$, we have

$$\sup_{0\leq y\leq 1} |\widetilde{R}_n(y,1) - (K(y,n) - K(y - n^{-1}K(y,n), n))|$$
$$\stackrel{\text{a.s.}}{=} \mathcal{O}\left(n^{1/8}(\log n)^{3/4}(\log\log n)^{1/8}\right) . \quad (5.10)$$

PROOF. We follow the lines of the proof of Theorem 5.1, only now use Theorem 2.D instead of Theorem 2.C. □

COROLLARY 5.B. Given the conditions (1.66) (i),(ii) and (iii), as well as (2.12) (i) and (ii) or (2.12) (iii) on F, we have

$$\lim_{n\to\infty} n^{-1/4}(\log n)^{-1/2} \sup_{0\leq y\leq 1} |\widetilde{R}_n(y,1)| / \left(\sup_{0\leq y\leq 1} |\alpha_n(y,1)|\right)^{1/2} \stackrel{\text{a.s.}}{=} 1 , \quad (5.11)$$

$$n^{-1/4}(\log n)^{-1/2} \sup_{0\leq y\leq 1} |\widetilde{R}_n(y,1)| \xrightarrow{\mathcal{D}} \left(\sup_{0\leq y\leq 1} |B(y)|\right)^{1/2}, \quad (n\to\infty), \quad (5.12)$$

where $\{B(y),\ 0\leq y\leq 1\}$ is a Brownian bridge,

$$\limsup_{n\to\infty} n^{-1/4}(\log n)^{-1/2}(\log\log n)^{-1/4} \sup_{0\leq y\leq 1} |\widetilde{R}_n(y,1)| \stackrel{\text{a.s.}}{=} 2^{-1/4} \quad (5.13)$$

and

$$\liminf_{n\to\infty} n^{-1/4}(\log n)^{-1/2}(\log\log n)^{1/4} \sup_{0\leq y\leq 1} |\widetilde{R}_n(y,1)| \stackrel{\text{a.s.}}{=} 8^{-1/4}\pi^{1/2}. \quad (5.14)$$

PROOF. (5.11) is implied by (5.10), (4.80) and (4.73). Now (5.12) follows from (5.11), while (5.13) and (5.14) follow from (5.10) combined with (4.78) and (4.88), respectively.

In concluding this section we underline again that Corollaries 5.A and 5.B essentially come from Csörgő and Révész (1975, 1978) as augmented somewhat in Csörgő and Révész (1981, Section 5.2), Csörgő (1983, Chapter 6), Csörgő and Horváth (1993, Section 6.5), and as stated herewith. Theorems 5.1, 5.2 and 5.3 constitute new convenient formulations of these extensions of the Bahadur–Kiefer process theory of quantiles.

6. An outline of a sequential version of the extended Bahadur-Kiefer process theory via strong invariance principles

This section is built on many of our results so far. In the light of (1.31) and (5.1), it is natural to define the *sequential general Bahadur–Kiefer process* $\widetilde{R}_n(y,t)$ by

$$\begin{aligned}
&\{\widetilde{R}_n(y,t),\ 0\leq y\leq 1,\ 0\leq nt\leq n,\ n=1,2,\ldots\} \\
&:= \{n^{1/2}(\alpha_n(y,t) - \rho_n(y,t)),\ 0\leq y\leq 1,\ 0\leq nt\leq n,\ n=1,2,\ldots\} \\
&= \{n^{1/2}(\beta_n(Q(y),t) - \rho_n(y,t)),\ 0\leq y\leq 1,\ 0\leq nt\leq n,\ n=1,2,\ldots\}.
\end{aligned} \quad (6.1)$$

Consequently, the sequential version of (5.3) is

$$\begin{aligned}
&\{\widetilde{R}_n(y,t) - R_n^*(y,t),\ 0\leq y\leq 1,\ 0\leq nt\leq n,\ n=1,2,\ldots\} \\
&= \{-n^{1/2}(\rho_n(y,t) - u_n(y,t)),\ 0\leq y\leq 1,\ 0\leq nt\leq n,\ n=1,2,\ldots\} \\
&= \{-n^{1/2}(\rho_n(y,t) - n^{-1/2}[nt](y - F(\hat{Q}_{[nt]}(y)))), \\
&\quad 0\leq y\leq 1,\ 0\leq nt\leq n,\ n=1,2,\ldots\}.
\end{aligned} \quad (6.2)$$

As an immediate consequence of (6.2) and Proposition 2.1, we have

PROPOSITION 6.1. Given the conditions (1.66) (i), (ii) and (iii) on F, as $n \to \infty$, we have

$$\sup_{0 \leq t \leq 1} \sup_{1/(n+1) \leq y \leq n/(n+1)} |\widetilde{R}_n(y,t) - R_n^*(y,t)|$$
$$\stackrel{a.s.}{=} \begin{cases} \mathcal{O}\big((\log\log n)^{1+\gamma}\big) & \text{if } \gamma \leq 1, \\ \mathcal{O}\big((\log\log n)^\gamma (\log n)^{(1+\epsilon)(\gamma-1)}\big) & \text{if } \gamma > 1, \end{cases} \quad (6.3)$$

for all $\epsilon > 0$.

Similarly, (6.2) and Proposition 2.2 yield

PROPOSITION 6.2. Assume (1.66) (i)–(iii), (2.12) (i) and (2.12) (ii) or (iii). Then, as $n \to \infty$, we have

$$\sup_{0 \leq t \leq 1} \sup_{0 \leq y \leq 1} |\widetilde{R}_n(y,t) - R_n^*(y,t)|$$
$$\stackrel{a.s.}{=} \begin{cases} \mathcal{O}(\log\log n) & \text{if } \gamma < 1 \\ \mathcal{O}\big((\log\log n)^2\big) & \text{if } \gamma = 1, \\ \mathcal{O}\big((\log\log n)^\gamma (\log n)^{(1+\epsilon)(\gamma-1)}\big) & \text{if } \gamma > 1, \end{cases} \quad (6.4)$$

for all $\epsilon > 0$.

Combining now Proposition 1.2 with Proposition 6.1, we obtain

PROPOSITION 6.3. Given the conditions (1.66) (i), (ii) and (iii) on F, with an absolute positive constant C we have

$$\limsup_{n \to \infty} n^{-1/4} (\log n)^{-1/2} (\log\log n)^{-1/4}$$
$$\times \sup_{0 \leq t \leq 1} \sup_{1/(n+1) \leq y \leq n/(n+1)} |\widetilde{R}_n(y,t)| \leq C \text{ a.s.} \quad (6.5)$$

Similarly, Proposition 1.2 combined with Proposition 6.2 implies

PROPOSITION 6.4. Assume (1.66) (i)–(iii), (2.12) (i) and (2.12) (ii) or (iii). Then, with an absolute positive constant C we have

$$\limsup_{n \to \infty} n^{-1/4} (\log n)^{-1/2} (\log\log n)^{-1/4} \sup_{0 \leq t \leq 1} \sup_{0 \leq y \leq 1} |\widetilde{R}_n(y,t)| \leq C \quad \text{a.s.}$$
(6.6)

Proposition 1.1 itself can be reformulated along the lines of Theorem 4.1. First we prove

THEOREM 6.1. Let $\{K(y,x),\ 0 \leq y \leq 1,\ 0 \leq x < \infty\}$ be the Kiefer process of Theorem 1.A as in (1.8), written as in (4.50) in terms of a Wiener sheet extended to $\mathbb{R} \times \mathbb{R}_+$. Then, as $n \to \infty$, we have

$$\sup_{1/n \leq t \leq 1} \sup_{0 \leq y \leq 1} \left| \left(K(\hat{U}_{[nt]}(y), nt) - K(y, nt) \right) \right.$$
$$\left. - \left(K(y - [nt]^{-1} K(y, nt), nt) - K(y, nt) \right) \right|$$
$$= \sup_{1/n \leq t \leq 1} \sup_{0 \leq y \leq 1} \left| \left(K(y - [nt]^{-1} (n^{1/2} u_n(y, t)), nt) - K(y, nt) \right) \right. \quad (6.7)$$
$$\left. - \left(K(y - [nt]^{-1} K(y, nt), nt) - K(y, nt) \right) \right|$$
$$\stackrel{\text{a.s.}}{=} \mathcal{O}\left(n^{1/8} (\log n)^{3/4} (\log \log n)^{1/8}\right).$$

PROOF. We first recall that, by definition, $u_n(y,t) = 0$ for $0 \leq t < 1/n$ (cf (1.6)). On the probability space of Theorem 1.A, we have (1.10) of Theorem 1.B as well. With $t \in (1/n, 1]$ and $(nt) \to \infty$ as $n \to \infty$, by (4.72) with some positive constant $C > 0$ and for n large we have that the left hand side of (6.7) is almost surely bounded above by

$$(C + o(1))((nt)^{1/8} (\log(nt))^{3/4} (\log \log(nt))^{1/8}). \quad (6.8)$$

Hence, on dividing the left hand side of (6.7), as well as the expression in (6.8), by $n^{1/8} (\log n)^{3/4} (\log \log n)^{1/8}$ and then taking $\sup_{1/n \leq t \leq 1}$ of both expressions, we arrive at (6.7). $\square$

Now *the promised reformulation* of Proposition 1.1 *á la Theorem 4.1 reads as follows.*

THEOREM 6.2. Let $\{K(y,x),\ 0 \leq y \leq 1,\ 0 \leq x < \infty\}$ be the Kiefer process of Theorem 1.A as in (1.8), written as in (4.50) in terms of a Wiener sheet extended to $\mathbb{R} \times \mathbb{R}_+$, and let $\{R_n^*(y,t),\ 0 \leq y \leq 1,\ 0 \leq nt \leq n,\ n = 1, 2, \ldots\}$ be the sequential uniform Bahadur-Kiefer process as in (1.31). Then, as $n \to \infty$, we have

$$\sup_{1/n \leq t \leq 1} \sup_{0 \leq y \leq 1} |R_n^*(y,t) - (K(y,nt) - K(y - [nt]^{-1} K(y,nt), nt))| \quad (6.9)$$
$$\stackrel{\text{a.s.}}{=} \mathcal{O}(n^{1/8} (\log n)^{3/4} (\log \log n)^{1/8}).$$

PROOF. The left hand side of (6.9) is bounded above by

$$\sup_{1/n \leq t \leq 1} \sup_{0 \leq y \leq 1} |R_n^*(y,t) - (K(y,nt) - K(\hat{U}_{[nt]}(y), nt))|$$
$$+ \sup_{1/n \leq t \leq 1} \sup_{0 \leq y \leq 1} |K(y - [nt]^{-1} K(y,nt), nt) - K(\hat{U}_{[nt]}(y), nt)|$$
$$\stackrel{\text{a.s.}}{=} \mathcal{O}(n^{1/8} (\log n)^{3/4} (\log \log n)^{1/8}), \quad (6.10)$$

where the latter almost sure upper bound results follow from (1.36) and (6.7) combined. □

If at this stage we had an *exact* analogue of (4.70) of Theorem 4.E for the stochastic process

$$\{K(y, nt) - K(y - [nt]^{-1}K(y, nt), nt), \ 0 \le y \le 1,$$
$$1 \le nt \le n, \ n = 1, 2, \ldots\} \ , \tag{6.11}$$

then Theorem 6.2 would, of course, imply the same for $R_n^*(y, t)$ that would amount to an appropriate version of (4.69), which should read as follows:

$$\lim_{n \to \infty} n^{-1/4}(\log n)^{-1/2} \sup_{1/n \le t \le 1} \sup_{0 \le y \le 1} |R_n^*(y,t)| \Big/ \left(\sup_{1/n \le t \le 1} \sup_{0 \le y \le 1} |\alpha_n(y,t)| \right)^{1/2}$$
$$\stackrel{a.s.}{=} 1 \ , \tag{6.12}$$

an analogue of Theorem 4.D. As a consequence of such a result we would also have appropriate analogues of (4.85), (4.86) and (4.87) as well. Naturally, (6.12) along with the just mentioned analogues for $R_n^*(y, t)$ would be inherited by $\widetilde{R}_n(y, t)$ via the strong invariance principles of Proposition 6.1 and Proposition 6.2, respectively. The thus inherited results by $\widetilde{R}_n(y, t)$ would, of course, also include the exact versions of Propositions 6.3 and 6.4, respectively. For now we can say that we have here Propositions 6.1 and 6.2, as well as Theorem 6.2, for future considerations. (cf our discussion at the end of Section 3.) We note also that, using Proposition 6.1 (respectively, Proposition 6.2) in combination with Theorem 6.2, one can also formulate a sequential analogue of Theorem 5.2 (respectively that of Theorem 5.3) for similar future considerations.

As to the exact analogue of (4.70) of Theorem 4.E we seek for the stochastic process of (6.12), it can be obtained via using Theorem S.1.15.1 in Csörgő and Révész (1981) in combination with an argument like that of Proposition 1 of Deheuvels and Mason (1990).

Acknowledgement

This research was supported by NSERC Canada grants at Carleton University, Ottawa.

References

Bahadur, R. R. (1966). A note on quantiles in large samples. *Ann. Math. Statist.* **37**, 577–580.
Bonvalot, F. and N. Castelle (1991). Strong approximation of uniform empirical process by Kiefer process. Prépublications 91-41, Université de Paris–Sud Mathématiques, Bât 425, Orsey, France.
Burke, M. D. and L. Horváth (1989). Large sample properties of kernel-type score function estimators. *J. Statist. Plann. Infer.* **22**, 307–321.

Chibisov, D. (1964). Some theorems on the limiting behaviour of empirical distribution functions. *Selected Transl. Math. Statist. Probab.* **6**, 147–156.

Chung, K. L. (1948). On the maximum partial sums of sequences of independent random variables. *Trans. Amer. Math. Soc.* **64**, 205–233.

Csáki, E. (1977). The law of iterated logarithm for normalized empirical distribution functions. *Z. Wahrsch. verw. Gebiete* **38**, 147–167.

Csáki, E., M. Csörgő, A. Földes and P. Révész (1989). Brownian local time approximated by a Wiener sheet. *Ann. Probab.* **17**, 516–537.

Csáki, E., M. Csörgő, A. Földes and P. Révész (1992). Strong approximation of additive functionals. *J. Theor. Probab.* **5**, 679–706.

Csáki, E., M. Csörgő, A. Földes and P. Révész (1994). Global Strassen-type theorems for iterated Brownian motions. In: *Tech. Rep. Ser. Lab. Res. Stat. Probab.* No. 250-1994, Carleton U.-U. of Ottawa. In: *Stoch. Process. Appl.* **59** (1995), 321–341.

Csörgő, M. (1983). *Quantile Processes with Statistical Applications*. SIAM, Philadelphia.

Csörgő, M. (1986). Quantile Processes. In: *Encyclopedia of Statistical Sciences*, Volume 7, pp. 412–424, Wiley-Interscience.

Csörgő, M. (1987). *Empirical processes with applications to statistics*, by Galen R. Shorack and John A. Wellner, Wiley 1986. Review in *Bull. Amer. Math. Soc.* (New Series) **17**, 189–200.

Csörgő, M., S. Csörgő, L. Horváth and D. M. Mason (1986). Weighted empirical and quantile processes. *Ann. Prob.* **14**, 31–85.

Csörgő, M., S. Csörgő, L. Horváth and P. Révész (1985). On weak and strong approximations of the quantile process. In: *Proc. Seventh Conf. Probability Theory*, 81–95. Editura Academiiei, Bucuresti and Nuscience Press, Utrecht.

Csörgő, M. and L. Horváth (1990a). On the distributions of the supremum of weighted quantile processes. *Studia Sci. Math. Hung.* **25**, 353–375.

Csörgő, M. and L. Horváth (1990b). On the distributions of L_p norms of weighted quantile processes. *Ann. Inst. Henri Poincaré Prob. Stat.* **26**, 65–90.

Csörgő, M. and L. Horváth (1989). On confidence bands for the quantile function of a continuous distribution function. In: *Coll. Math. Soc. J. Bolyai* **57**, *Limit Theorems of Prob. Theory and Statist.*, Pécs, Hungary, 95–106. North-Holland, Amsterdam.

Csörgő, M. and L. Horváth (1993). *Weighted Approximations in Probability and Statistics*. Wiley, Chichester.

Csörgő, M., L. Horváth and Q.-M. Shao (1993). Convergence of integrals of uniform empirical and quantile processes. *Stochastic Process. Appl.* **45**, 283–294.

Csörgő, M., L. Horváth and B. Szyszkowicz (1994). Integral tests for suprema of Kiefer processes with application. In: Tech. Rep. Ser. Lab. Res. Statist. and Probab. No. **257**, Carleton U.-U. of Ottawa.

Csörgő, M. and P. Révész (1975). Some notes on the empirical distribution function and the quantile process. In: *Coll. Math. Soc. J. Bolyai* **11**, *Limit Theorems of Prob. Theory*. Keszthely, Hungary, 59–71. North-Holland, Amsterdam.

Csörgő, M. and P. Révész (1978). Strong approximations of the quantile process. *Ann. Statist.* **6**, 882–894.

Csörgő, M. and P. Révész (1981). *Strong Approximations in Probability and Statistics*. Academic-Press, New York.

Csörgő, M. and P. Révész (1984). Two approaches to constructing simultaneous confidence bounds for quantiles. *Prob. and Math. Statist.* **4**, 221–236.

Csörgő, M. and P. Révész (1986). A nearest neighbour-estimator for the score function. *Probab. Th. Rel. Fields* **71**, 293–305.

Csörgő, M. and B. Szyszkowicz (1995/96). Sequential quantile and Bahadur–Kiefer processes in weighted supremum and L_p-metrics. In progress.

Deheuvels, P. and D. M. Mason (1990). Bahadur–Kiefer-type processes. *Ann. Prob.* **18**, 669–697.

Deheuvels, P. and D. M. Mason (1992). A functional LIL approach to pointwise Bahadur–Kiefer theorems. In: R. M. Dudley et al., eds., *Probability in Banach Spaces*. Birkhäuser, Boston, pp. 255–266.

Dobrushin, R. L. (1955). Two limit theorems for the simplest random walk on a line (in Russian). *Uspehi Mat. Nauk* (N.S.) **10**, (3) (65), 139–146.

Donsker, M. (1951). An invariance principle for certain probability limit theorems. *Four Papers on Probability. Mem. Amer. Math. Soc. No.* **6**.

Donsker, M. (1952). Justification and extension of Doob's heuristic approach to the Kolmogorov-Smirnov theorems. *Ann. Math. Statist.* **23**, 277–283.

Dvoretzky, A., J. Kiefer and J. Wolfowitz (1956). Asymptotic minimax character of the sample distribution function and of the classical multinomial estimator. *Ann. Math. Statist.* **27**, 642–669.

Erdős, P. and A. Rényi (1970). On a new law of large numbers. *J. Analyse Math.* **23**, 103–111.

Hájek, J. and Z. Šidák (1967). *Theory of Rank Tests*. Academic Press, New York.

Huber, P. J. (1981). *Robust Statistics*. John Wiley, New York.

Kiefer, J. (1967). On Bahadur's representation of sample quantiles. *Ann. Math. Statist.* **38**, 1323–1342.

Kiefer, J. (1970). Deviations between the sample quantile process and the sample D.F. In: *Nonparametaric Techniques in Stat. Inference*. Cambridge Univ. Press, London, 299–319.

Kiefer, J. (1972). Skorohod embedding of multivariate R.V.'s and the sample DF. *Z. Wahrsch. verw. Gebiete* **13**, 321–332.

Komlós, J., P. Major and G. Tusnády (1975). An approximation of partial sums of independent R.V.'s and the sample DF. I. *Z. Wahrsch. verw. Gebiete* **32**, 111–131.

Komlós, J., P. Major and G. Tusnády (1976). An approximation of partial sums of independent R.V.'s and the sample DF. II. *Z. Wahrsch. verw. Gebiete* **34**, 33–58.

Massart, P. (1990). The tight constant in the Dvoretzky–Kiefer–Wolfowitz inequality. *Ann. Prob.* **18**, 1269–1283.

Mogul'skii, A. A. (1980). On the law of the iterated logarithm in Chung's form for functional spaces. *Theor. Prob. Appl.* **24**, 405–413.

Müller, D. W. (1970). On Glivenko-Cantelli convergence. *Z. Wahrsch. Verw. Gebiete* **16**, 195–210.

O'Reilly, N. (1974). On the weak convergence of empirical processes in sup-norm metrics. *Ann. Prob.* **2**, 642–651.

Parzen, E. (1979a). Nonparametric statistical data modelling. *J. Amer. Statist. Assoc.* **74**, 105–131.

Parzen, E. (1979b). A density–quantile function perspective on robust estimation. In: *Robust Estimation Workshop Proceedings*, Academic Press, New York.

Parzen, E. (1980). Quantile functions, convergence in quantile, and extreme value distribution theory. *Technical Report*, B-3, Statistical Inst., Texas A & M Univ., College Station, TX.

Rao, C. R. and L. C. Zhao (1995). Strassen's law of the iterated logarithm for the Lorenz curves. *J. Multiv. Analysis.* **54**, 239–252.

Rényi, A. (1953). On the theory of order statistics. *Acta. Math. Acad. Sci. Hung.* **4**, 191–232.

Rényi, A. (1970). *Probability Theory*. North-Holland, Amsterdam.

Shorack, G. R. (1982). Kiefer's theorem via the Hungarian construction. *Z. Wahrsch. verw. Gebiete* **61**, 369–373.

Shorack, G. R. and J. A. Wellner (1986). *Empirical Processes with Applications to Statistics*. Wiley, New York.

Szyszkowicz, B. (1994). Weak convergence of weighted empirical type processes under contiguous and changepoint alternatives. *Stoch. Process. Appl.* **50**, 281–313.

Szyszkowicz, B. (1998). Weighted sequential empirical type processes with applications to changepoint problems. This volume.

Author Index

Abdel-Aty, S. H. 155, 222
Abramowitz, M. 176, 105, 222, 443, 461
Abu-Salih, M. S.293, 310
Abu-Youssef, S. E. 184, 190, 202, 211, 219, 226, 227, 293, 307, 310
Acton, S. T. 46, 48
Adke, S. R. 545, 561
Adler, R. J. 586, 587, 593, 623, 628
Adrain, R. 28, 48
Aggarwala, R. 8, 13, 21, 69, 70
Ahmed, S. E. 32, 49, 58, 153, 221, 223, 226, 292, 297, 305, 308, 310
Ahsanullah, M. 19, 21, 47, 48, 231, 253, 264–267, 269, 271, 274, 285, 296, 304, 308, 487, 515, 516, 531, 532, 558, 559, 561, 562
Aitchison, J. 35, 48
Aitken, A. C. 41, 48
Alam, S. N. 245, 255
Albeverio, S. 562
Ali, M. M. 138, 142
Ali, M. Masoom, 14, 21, 31, 48, 209, 222
Alimoradi, S. 14, 21
Alpium, M. T. 548, 562
Aly, M. A. H. 231, 254, 285, 286
Alzaid, A. A. 232, 245, 254, 262, 286
Andel, J. 545, 561
Anderson, C. L. 47, 48
Anderson, T. W. 42, 43, 48, 256, 491, 510, 574, 582, 628
Andrews, D. F. 18, 21, 29, 48, 150, 222, 444, 458, 461
Angelis, D. D. 482
Arce, G. R. 45, 46, 48, 49
Arnold, B. C., 3, 4, 7, 9–11, 17, 19, 21, 26, 27, 32, 33, 47, 48, 65, 70, 75, 76, 78–80, 82–87, 106, 113, 117–119, 124, 125, 137, 138, 142, 151, 153–155, 222, 231, 240, 245, 254, 257, 263, 270, 271, 275, 279, 282, 286, 288, 292, 295, 297, 302, 305, 308, 404, 414, 416, 420, 438, 494, 510, 515, 531, 532, 562
Athreya, K. B. 256

Aven, T. 113, 142
Azlarov, T. A. 217, 218, 222, 231, 254, 257, 286, 296, 308

Babu, A. J. G. 68, 71
Babu, G. J. 480, 482
Bagai, I. 97, 103
Bagchi, P. 559, 563
Bagchi, S. N. 262, 284–286
Bagdonavicius, V. 555, 562
Bahadur, R. R. 341, 371, 473, 474, 481, 482, 636, 648, 661, 685
Bai, Z. D. 47, 64
Bain, L. J. 16, 21, 30, 48, 150, 222
Balabekyan, V. A. 546, 547, 562
Balakrishnan, A. V. 120, 126, 142
Balakrishnan, N. 3, 4, 6–14, 16–34, 36–38, 40–42, 44–50, 52, 54–58, 60, 62–72, 106, 117, 121, 122, 137, 138, 142, 149, 150–160, 162, 164, 166, 168–174, 176–180, 182–184, 186, 188–192, 194–196, 198, 200-206, 208, 210, 212–216, 218, 220, 221–228, 231, 241, 242, 253–255, 257, 263, 279, 282, 286–287, 292, 297, 301, 302, 305, 308, 310, 403, 404, 406, 408, 410–416, 418, 420, 422, 424, 426, 428, 430, 432, 434, 436, 438, 490, 492, 494, 510, 512, 515, 516, 518, 522, 524, 526, 528, 530–532, 534, 536, 538, 540, 542, 544, 546, 548, 550, 552, 554–556, 558, 560–563, 566, 568–570
Balasubramanian, K. 7, 9, 22, 46, 49, 172, 213–215, 222, 226, 292, 293, 308, 404, 490, 492, 510, 561, 562
Baldessari, B. A. 287, 511
Balkema, A. A. 448, 461, 468, 469, 482
Ballerini, R. 532, 547–549, 555–557, 559, 563
Bapat, R. B. 46, 49, 292, 308
Bapat, R. B. 99, 103
Barbera, F. 36, 62
Bargmann, R. E. 42, 61

Barlow, R. E. 26, 33, 49, 92–94, 102, 103, 134, 136, 137, 143, 364, 371, 492, 510
Barndorff-Nielson, O. 376, 379, 380–382, 384, 449, 461, 476, 482, 518, 563, 605, 628
Barner, K. E. 46, 48, 49
Barnett, V. 4, 17, 22, 29, 41, 44, 47, 49, 150, 152, 175, 223, 412, 438, 487, 497, 510
Barranco Chamoor, I. 566
Bartfai, P. 566
Bartlett, M. S. 41, 42, 49
Barton, D. E. 496, 510, 558, 563
Bartoszewicz, J. 103
Basak, P. 559, 563
Bassett, G. 350, 371
Basseville, M. 629
Basu, A. P. 8, 17, 22, 72, 372
Basu, S. K. 372
Bechenbach, E. F. 303, 308
Bechhofer, R. E. 38, 49, 50
Becker, A. 266, 280, 282, 283, 286
Beckman, R. J. 31, 63
Beesack, P. R. 118, 143
Beg, M. I. 46, 49, 245, 255, 293, 308, 310, 490, 510
Bell, C. B. 261, 286
Bellman, R. 303, 308
Belsley, D. A. 387, 401
Belson, J. 35, 50
Bendre, S. M. 46, 49, 62, 122, 142, 292, 308, 411, 438
Bennet, S. 204, 223
Benveniste, A. 629
Beran, R. 479, 483
Bereanu, B. 144
Berenson, M. L. 70, 71
Berk, R. H. 275, 286
Berman, S. M. 330, 331
Bernardo, J. M. 451, 461
Bernoulli, D. 29, 50
Bernoulli, N. 39
Berred, M. 543, 558, 563
Bhattacharjee, M. C. 364, 368, 369–372
Bhattacharya, P. K. 47, 50, 340, 352, 354, 355, 371, 487, 488, 490, 492, 506, 507, 510, 605, 628
Bhattacharyya, B. B. 46, 50, 361, 373
Bickel, P. J. 18, 21, 29, 48, 150, 222, 289, 444, 461, 584, 628
Billingsley, P. 342, 371
Biondini, R. 517, 534, 544, 545, 563, 569
Birnbaum, A. 178, 223
Blackmore, R. 559, 569
Block, H. W. 104, 204, 223

Blom, G. 40, 44, 50, 135, 143, 149, 223, 561, 563
Blum, J. R. 591, 628
Boas, R. P. 294, 308
Boland, P. J. 7, 22, 27, 33, 50, 89, 90, 92, 94–96, 98–104
Boncelet, C. G. Jr. 46, 60
Bonvalot, F. 584, 628, 632, 685
Boos, D. D. 443, 458, 461
Booth, J. D. 444, 449, 450, 452, 462
Borevich, Z. I. 562, 567, 568
Borthakur, A. C. 372
Bortkiewicz, L. von, 39, 50, 319, 331
Bosch, K. 263, 286
Boscovich, R. J. 28, 50
Bose, R. C. 42, 61
Bovik, A. C. 46, 48, 50
Boyd, A. V. 117, 118, 143
Boyd, D. W. 389, 402
Braun, H. J. 63
Bray, T. A. 67, 71
Breiman, L. 375, 384
Breiter, M. C. 184, 223
Bretangolle, J. 581, 628
Brillinger, D. R. 581, 628
Brodsky, B. E. 576, 628
Brown, L. D. 586, 587, 593, 623, 628
Brunk, H. D. 44, 50
Bruss, F. T. 545, 559, 561, 563
Buchberger, S. G. 490, 512
Bunge, J. A. 545, 563
Burke, M. D. 645, 685
Burr, I. W. 203, 204, 223
Burrows, P. M. 174, 223
Butler, R. W. 444, 449, 450, 452, 462

Cadwell, J. H. 159, 223
Cantelli, F. P. 580, 628
Caraux, G. 109, 113, 143
Casley, D. J. 496, 510
Castelle, N. 584, 628, 632, 685
Castillo, E. 4, 12, 22, 40, 50, 150, 223, 326, 327, 328, 329, 331
Chakraborty, S. 149, 176, 177, 224, 225
Chan, L. K. 11, 22, 31, 50, 138, 142, 248, 254, 262, 286, 293, 308
Chan, P. S. 8, 13, 21, 532, 558, 562, 563
Chan, W. 96, 102, 104
Chanda, K. C. 509, 519
Chandler, K. N. 47, 50, 515, 516, 520, 521, 534, 563
Chaplin, W. S. 39, 40, 50
Chatterjee, S. 387, 401

Chaudhury, H. 372
Chauvenet, W. 29, 50
Chen, H. J. (Mrs.), 45, 62
Cheng, S. W. 31, 50
Cheng, S.-H. 561, 563
Chernoff, H. 41, 50
Chibisov, D. M. 40, 50, 574, 582, 628, 653, 686
Childs, A. 11, 18, 22, 403, 404, 406, 408, 410, 412, 414, 418, 420, 422, 424, 426, 428, 430, 432, 434, 436, 438
Chow, Y. S. 376, 384
Chung, K. L. 586, 593, 628, 675, 686
Cislak, P. J. 203, 204, 223
Clark, C. E. 33, 50
Clemm, D. S. 37, 55
Clutter, J. L. 499, 500, 511
Cobby, J. M. 500, 512
Cohen, A. C. 4, 14, 16, 22, 30, 31, 40–51, 65, 71, 150, 222, 223, 292, 301, 308
Cole, R. H. 151, 155, 223, 291, 308
Coles, S. G. 495, 510
Constantine, A. G. 42, 51
Cook, R. D. 387, 401
Correa, J. A. 576, 628
Cournot, A. A. 36, 51
Cox, D. R. 354, 355, 356, 371, 449, 461, 476, 482
Cramér, H. 43, 51
Crawford, G. B. 231, 236, 254
Csáki, E. 648, 667, 674, 686
Csörgő, M. 45, 51, 255, 261, 272, 273–276, 278, 286, 290, 574, 575–577, 582, 583, 585–592, 594, 595, 597, 598, 600, 603, 605, 607, 608, 612–614, 622–629, 631-638, 604, 642–660, 662–664, 666–668, 670, 672, 674–680, 682, 684–686
Csörgő, S. 574, 575, 582, 583, 587, 588, 598, 628, 649, 653, 654, 656, 686
Curry, T. F. 29, 56

D'Agostino, R. B. 18, 22, 45, 51, 150, 223
Dallas, A. C. 231, 249, 251, 254, 263, 287
Daniell, P. J. 30, 40, 51
Daniels, H. A. 361, 371
Daniels, H. E. 444, 449, 461, 475–479, 483
Darkhovsky, B. S. 576, 628
Darling, D. A. 43, 48, 574, 582, 605, 628, 629
David, F. N. 10, 11, 22, 33, 51, 173, 224
David, H. A. 3, 10, 11, 17, 19, 20, 22, 23, 25, 26, 32, 33, 45–47, 51, 56, 65, 71, 106, 114, 118–120, 143, 149, 150, 153–155, 168, 184, 222, 224, 225, 231, 254, 258, 287, 291, 292, 302, 305, 308, 309, 340, 353, 371, 403, 412, 438, 441, 445, 461, 487–490, 492–496, 498–500, 502–506, 508, 510–513, 531
Davies, L. 277, 290
Davies, P. L. 235, 254, 262, 287
Davis, C. S. 151, 153, 173, 224
Davis, R. 442, 447, 456, 461
Davison, A. C. 441, 443, 449, 461
Dawson, D. A. 255
de Haan, L. 40, 51, 321, 322, 328, 332, 441, 442, 447, 448, 456, 458, 460, 461, 468, 469, 482, 528, 547, 560, 563, 564
De Laurentis, J. M. 559, 564
Deddens, J. A. 490, 512
Deheuvels, P. 375–377, 382, 384, 519, 528, 536–542, 546, 549–551, 554, 560, 564, 637, 667, 671, 674, 685, 686
Dekker, A. L. M. 441, 442, 447, 448, 456, 458, 460, 461
Dell, T. R. 499, 500, 511
Deny, J. 234, 254
Deshayes, J. 577, 585, 593, 629
Deshpande, J. V. 371, 372
Deus, M. M. 38, 45, 51, 52, 261, 263, 266, 269, 270, 275, 287
Devroye, L. 41, 52, 66, 69, 71, 559, 564
DiCiccio, T. J. 443, 461, 482, 483
Diersen, J. 564
Dimaki, C. 270, 287, 301, 308
Dixit, U. J. 292, 311
Dixit, U. J. 46, 62
Dixon, W. J. 29, 52, 149, 150, 224
Do, K.-A. 443, 461, 506, 509, 511
Dobrushin, R. L. 674, 687
Dodd, E. L. 39, 52, 319, 331
Dodge, Y. 289, 310, 372
Doganaksoy, N. 14, 22
Donsker, M. 581, 629, 635, 687
Doornbos, R. 29, 52
Downton, F. 32, 52, 151, 156, 194, 224, 251, 254
Drake, S. 53
Droste, W. 92, 104
Dubey, S. D. 276, 287
Ducharme, G. 479, 483
Dudley, R. M. 686
Dudman, J. 178, 223
Dufour, R. 247, 254, 274, 287
Duncan, D. B. 37, 52
Dunnett, C. W. 38, 49
Dunsmore, I. R. 35, 48, 559, 564
Durban, J. 465, 461
Durbin, J. 42, 44, 52

Dvoretzky, A. 640, 687
Dwass, M. 516, 518, 560, 564
Dyer, D. D. 199, 224
Dykstra, R. L. 46, 52
Dyson, F. J. 42, 52
Dziubdziela, W. 516, 519, 534, 535, 539, 561, 564

Edgeworth, F. Y. 28, 52
Efron, B. 348, 371, 443, 461, 479, 481, 483, 509, 511
Egorov, V. A. 487, 507, 511
Einmahl, J. H. J. 447, 448, 465, 461
Einstein, A. 53
Eisenberger, I. 150, 224
El-Neweihi, E. 95, 98, 100–103
Embrechts, P. 545, 564
Engelen, R. 540, 565
Engelhardt, M. 16, 21, 30, 48, 150, 222
Epstein, B. 258, 287
Erdős, P. 605, 629, 636, 687
Erds, P. 379, 384
Euler, L. 28, 52

Földes, A. 667, 674, 686
Fahmy, S. 118, 143
Falk, M. 329, 331, 467, 470, 472, 475, 480, 482, 483
Fang, B. Q. 261, 287
Fang, K. T. 261, 287
Feinberg, F. M. 504, 505, 511
Ferguson, T. S. 45, 52, 231, 236, 248, 254, 293, 308
Field, C. A. 475, 479, 483
Fisher, R. A. 12, 22, 29, 39, 42, 52, 319, 331
Fishman, G. S. 70, 71
Fisz, M. 45, 52, 238, 254
Fosam, E. B. 234, 240–242, 251, 254, 287
Foster, F. G. 516, 520, 547, 558, 564, 565
Fourier, J. B. J. 28, 52
Fowlkes, E. B. 41, 52
Fox, J. 387, 401
Fréchet, M. 12, 22, 39, 52, 319, 331
Frankel, M. 376, 380, 384
Fraser, D. A. S. 35, 52
Freedman, D. A. 444, 461
Freudenberg, W. 528, 565
Frierson, D. Jr. 605, 628

Gail, M. H. 446, 461
Gajek, L. 123, 126, 128, 130, 133, 143, 264–267, 287, 303, 304, 308, 309, 538, 565

Galambos, J. 4, 11, 12, 22, 40, 45, 47, 51, 53, 150, 224, 231, 254, 255, 257, 262, 263, 272, 275, 276, 278, 280, 283, 284, 287, 293, 295, 306, 309, 315, 316, 318–333, 337, 371, 372, 380, 384, 490, 493, 495, 503, 510, 511, 515, 522, 523, 526, 528, 533, 562, 565
Galilei, G. 28, 53
Gallagher, N. C. 45, 48, 53
Galton, F. 39, 53, 160, 224
Gangopadhyay, A. K. 354, 371, 492, 510
Gascuel, O. 109, 113, 143
Gastwirth, J. L. 75, 87, 446, 461, 497, 504, 512
Gather, U. 11, 23, 45, 53, 123, 143, 255, 257, 258, 260, 262, 264–270, 272, 274, 276, 278, 280, 282, 284, 286–288, 290, 292, 296, 303, 304, 306, 308, 309
Gauss, C. F. 28, 30, 53
Gaver, D. P. 545, 565
Geary, R. C. 42, 53
Geffroy, J. 375–377, 384
Geisser, S. 29, 53
Gentle, J. E. 41, 57, 70, 71
George, E. O. 278, 279, 287
Gerontidis, I. 41, 53, 70, 71
Gessaman, M. P. 35, 60
Ghosh, J. K. 341, 371, 473, 483
Ghosh, M. 240, 254, 270, 271, 275, 286, 288, 353, 371, 487, 511
Ghurye, S. G. 272, 275, 279, 280, 288
Gibbons, J. D. 39, 53, 149, 224
Gill, P. S. 499, 511
Gilstein, C. Z. 119, 143
Girshick, M. A. 41, 42, 53
Glick, N. 19, 23, 47, 53, 515, 517, 565
Glivenko, V. 580, 629
Gnanadesikan, R. 44, 64
Gnedenko, B. 12, 23, 39, 53, 319, 320, 322, 331, 332, 441, 454, 461, 484, 494, 511
Godbole, A. P. 380, 384
Godwin, H. J. 30, 31, 53, 151, 224
Goel, P. K. 492, 506, 510, 511
Goldberger, A. S. 14, 23
Goldie, Ch. M. 528, 540, 558, 559, 565
Gombay, E. 605, 629
Gomes, M. I. 324, 326, 332, 333, 499, 511
Gomez Gomez, T. 566
Gong, G. 443, 461
Govindarajulu, Z. 32, 35, 45, 47, 53, 151, 152, 161–164, 171–173, 222, 224, 231, 255, 292, 295-297, 299, 305–307, 309
Gravey, A. 11, 143
Gray, J. B. 388, 394, 401

Grechanovsky, E.   4, 24
Green, J. R.   43, 53
Green, P. J.   497, 510
Greenberg, B. G.   3, 24, 26, 31, 60, 62, 149, 223, 225, 227, 343, 372
Greenwood, P. E.   614, 629
Grigelionis, B.   629
Groeneveld, R. A.   33, 48, 113, 118, 142, 143
Grossmann, W.   564
Grubbs, F. E.   29, 53
Grudzien, Z.   19, 23, 531, 539, 565
Guilbaud, O.   498, 506, 508, 511
Gulati, S.   559, 565
Gumbel, E. J.   10, 23, 26, 32, 39, 40, 41, 53, 118, 143, 150, 224, 302, 309, 326, 329, 332, 535, 565
Gunnink, J. L.   488, 510
Gupta, A. K.   30, 54, 498, 511
Gupta, R. C.   231, 249, 255, 263, 269, 277, 288, 528, 546, 560, 561, 565
Gupta, S. S.   6, 8, 22, 33, 38, 39, 49, 54, 178, 184, 224
Gurenther, W. C.   35, 53
Gurland, J.   389, 402
Gut, A.   560, 565
Gutenbrunner, C.   352, 371
Guthrie, E. H.   37, 55
Guthrie, G. L.   517, 565
Guttman, I.   35, 52, 54

Hájek, J.   483, 614, 629, 644, 645, 687
Hüsler, J.   287, 308, 329, 331, 332, 566, 568, 570
Haas, P. J.   560, 565
Hadi, A. S.   387, 401
Haghighi-Taleb, D.   559, 565
Hagwood, Ch.   321, 331
Hahn, G. J.   35, 54
Haiman, G.   543, 544, 561, 565, 566
Haines, J.   36, 60
Hajós, G.   270, 288
Hall, P.   323, 332, 443, 453, 461, 475, 479, 482, 483, 492, 506, 509–511
Hall, W. J.   248, 255, 323, 332
Halperin, M.   30, 54
Hamedani, G. G.   274, 286
Hampel, F. R.   18, 21, 29, 48, 54, 150, 222, 476, 478, 478, 483
Hannan, E. J.   42, 54, 64
Hardy, G. H.   75, 87
Harrell, F. E.   498, 511

Harter, H. L.   3, 8, 16, 23, 25, 26, 28, 29–32, 34, 36–38, 40–44, 46, 48, 50, 52, 54–56, 58, 60, 62, 64, 150, 224
Hartley, H. O.   10, 23, 32, 41, 56, 58, 66, 68, 71, 118, 119, 120, 143, 302, 309
Hawkins, D. M.   29, 56, 63, 117, 118, 143, 150, 224
Hazen, A.   40, 56
Hegazy, Y. A. S.   43, 53
Henshaw, R. C. Jr.   42, 56
Herzberg, A. M.   444, 458, 461
Hewett, J. E.   46, 52
Hill, B. M.   442, 447, 461
Hill, I. D.   205, 227
Hinkley, D. V.   443, 449, 461
Ho, F. C. M.   70, 71
Hoaglin, D. C.   29, 56
Hoeffding, W.   11, 23, 45, 56, 154, 225, 257, 288, 309, 338, 363, 371, 591, 598, 629
Hogg, R. V.   29, 56, 150, 225
Hoglund, T.   476, 484
Holland, B.   254
Hollander, M.   99, 100, 104, 149, 225, 368, 371
Holmes, P. T.   516, 517, 526, 533, 565, 566
Holst, L.   476, 484
Hoppe, F. M.   51
Horn, P. S.   41, 56, 68, 71
Horváth, L.   574–577, 582, 583, 585–592, 594, 597, 598, 600, 605, 607, 608, 612, 622, 623, 625, 626, 628, 629, 635–637, 640, 643–645, 648, 649, 653–657, 659, 674, 678, 679, 682, 685, 686
Hosking, J. R. M.   14, 23
Hossein, M. G.   490, 511
Hotelling, H.   41, 42, 56
Houchens, R. L.   487, 511
Hsu, P. L.   42, 56
Huang, J. S.   11, 23, 194, 225, 255, 257, 266, 270, 271, 275, 288, 290, 294–296, 309, 420, 438, 559, 566
Huang, W. J.   559, 566
Huang, W. L.   249, 255
Huber, J.   504, 505, 511
Huber, P. J.   18, 21, 23, 29, 48, 56, 150, 222, 225, 644, 687
Huse, V. R.   605, 629
Hwang, J. S.   257, 288, 294, 295, 309

Igantov, Z.   540, 566
Iglehart, D. L.   256
Ikeda, S.   512
Inagaki, N.   45, 56

Iosifescu, M. 144
Isaacson, D. 263, 286
Iwińska, M. 267, 288, 304, 309

Jackson, O. A. Y. 45, 56
Jacobs, P. A. 545, 565
James, A. T. 42, 51, 56, 57
Janardan, K. G. 269, 288
Janas, D. 475, 483
Janssen, A. 326, 332
Jense, D. R. 18, 23
Jensen, D. R. 387–390, 392, 394, 396, 398, 400, 402
Jensen, J. L. 476, 483
Jha, V. D. 197, 116, 490, 511
Jlek, M. 35, 57
Joag-Dev, K. 99, 100, 104
Jochel, K.-H. 483
Jogdeo, K. 93, 104
Johnson, L. G. 40, 57
Johnson, M. E. 65, 70, 72
Johnson, M. V. 443, 461
Johnson, N. L. 10, 11, 22, 29, 38, 51, 55, 57, 69, 70, 71, 104, 173, 177, 204, 224, 225, 257, 288, 389, 402, 404, 438, 566, 568
Johnston, G. J. 508, 511
Jones, H. L. 151, 155, 225
Joshi, P. C. 32, 33, 46, 49, 57, 122, 143, 151–153, 157, 159, 163, 164, 166, 168, 170, 173, 174, 176, 177, 182, 184, 188–191, 194–196, 217, 218, 222, 224, 225, 291, 292, 308, 404, 406, 412, 415, 438
Joshi, S. M. 506, 511
Jung, J. 149, 225
Junkins, D. B. 190, 227
Jurekov, J. 336, 344, 345, 347, 351, 352, 371

Kabe, D G. 194, 225, 269, 290
Kabir, A. B. M. L. 138, 143
Kadane, J. B. 293 , 309
Kagan, A. M. 11, 23, 45, 57, 247, 255, 279, 288
Kahaner, D. K. 31, 63
Kaigh, W. D. 467, 483
Kakosyan, A. V. 45, 57, 264, 277, 288
Kalashnikov, V. V. 287, 308, 567
Kalbfleisch, J. D. 30, 57
Kallianpur, G. 289
Kaminsky, K. S. 14, 23, 36, 57
Kamps, U. 11, 20, 23, 45, 53, 57, 123, 135, 143, 216, 218, 225, 231, 255, 257, 258, 260, 262–264, 266–268, 270, 272, 274, 276, 278, 280, 282, 284, 286, 288, 290–294, 296, 298–304, 306, 308–310, 487, 511, 546, 566
Kaplan, E. L. 349, 371
Karlin, S. 492, 511
Katai, I. 331
Katzenbeisser, W. 560, 566
Kaufmann, E. 487, 511
Keilson, J. 90, 91, 104
Kemperman, J. H. B. 35, 57, 111, 143
Kendall, M. G. 43, 57
Kennedy, W. J. 41, 57, 70, 71
Keuls, M. 37, 57
Khamis, H. J. 44, 55, 57
Khan, A. H. 175, 190, 209, 210, 217, 218, 221, 222, 225, 226, 293, 297, 300, 307, 310
Khan, I. A. 210, 217, 221, 225, 293, 300, 310
Khatri, C. G. 152, 226
Khmaladze, E. V. 576, 614, 619, 629, 630
Khurana, A. P. 197, 226
Kiefer, J. 377, 384, 473, 474, 483, 584, 586, 591, 593, 628, 629, 632, 636, 640, 648, 662, 663, 672, 674, 679, 687
Kim, S. H. 489, 492, 511
Kim, Y-T. 46, 48
Kimball, B. F. 41, 57
Kinoshita, K. 558, 566
Kipnis, V. 4, 24
Kirmani, S. N. U. A. 245, 254, 255, 293, 308, 559, 560–562, 565
Klass, M. J. 376, 380, 381, 383, 384
Klebanov, L. B. 45, 57, 233, 255, 264, 273, 277, 288
Klincsek, T. 66, 71
Knuth, D. E. 66, 71
Kochar, S. C. 97, 99, 100, 103, 104, 561, 566
Kocherlakota, S. 172, 180, 222, 223
Koenker, R. 350, 371
Kolassa, J. E. 475, 476, 483
Kolmogorov, A. N. 43, 58, 580, 629
Komlós, J. 581, 584, 590, 629, 632, 636, 664, 687
Konecny, F. 23
Konheim, A. G. 293, 310
Kopocinsky, B. 516, 519, 534, 535, 539, 564
Kotlarski, I. I. 255, 272, 280, 288, 289
Kotz, S. 11, 22, 29, 45, 51, 53, 55, 57, 60, 68, 70, 71, 104, 177, 204, 225, 231, 236, 253–255, 257, 263, 272, 275, 285, 287–289, 293, 295, 296, 306, 309, 310, 324, 331, 389, 402, 404, 438
Koul, H. L. 336, 368, 371, 372
Krishnaiah, P. R. 32, 42, 43, 50, 54, 55, 57–59, 62–64, 151, 184, 223, 226, 371, 510

Kuh, E.   387, 401
Kuhlmann, F.   45, 58
Kuiper, N. H.   43, 58
Kulldorff, G.   416, 419, 438

Lagakos, S. w.   441, 462
Lahiri, S. N.   469, 473, 483
Lai, T. L.   109, 111, 143
Lamb, R. E.   44, 55
Lambert, J. H.   28, 29, 58
Lamperti, J.   516, 560, 566
Lanczos, C.   451, 462
Lange, K.   156, 226
Laplace, P. S.   28, 58, 468
Lau, K. S.   45, 60, 231, 232–234, 239, 245, 249, 254, 255, 256, 259, 262, 286, 289
Launer, R. L.   29, 58, 224, 438
Lawless, J. F.   16, 23, 30, 58, 150, 226
Lawley, D. N.   42, 43, 58
Lawrence, M. J.   132, 134, 138, 139, 143
Le Cam, L.   144, 614, 629
Leadbetter, M. R.   12, 23, 40, 58, 330, 332
Lebedev, V. V.   40, 58
Lechner, J.   329, 332, 562
Ledford, A. W.   495, 504, 511
Lee, S. K.   8, 13, 22
Lefevre, C.   114, 143
Legendre, A. M.   28, 58
Lehmann, E. L.   94, 104, 149, 226
Leipus, R.   576, 629, 630
Lenic, E.   123, 143
Leroy, A. M.   29, 61, 387, 402
Leslie, J. R.   247, 248, 255, 274, 275, 289
Levine, D. N.   499, 510
Lewis, T.   4, 17, 22, 29, 44, 49, 150, 22, 412, 438
Lewis, T.   92, 104
Li, S. H.   249, 255, 559, 566
Liang, T. C.   241, 242, 253, 255
Liberman, U.   255
Lieberman, G. J.   41, 50
Lien, D. D. 213–215, 226
Likeš, J.   416, 438
Lilliefors, H. W.   43, 58
Lin, G. D.   119, 122, 144, 217, 218, 226, 257, 288, 289, 294, 295, 297, 299, 301–307, 309–311, 559, 566, 569
Lindgren, G.   12, 23, 40, 58, 330, 332
Linnik, Yu. V.   11, 12, 45, 57, 247, 255, 279, 288
Littlewood, J. E.   75, 87
Lka, O.   35, 57
Lloyd, E. H.   13, 23, 30, 58, 149, 226

Lo, A. W.   487, 511
Lockhart, R. A.   18, 23, 45, 58
Loh, W. Y.   444, 462
Lombard, F.   605, 630
Lopez Blazquez, F.   566
Ludwig, O.   32, 58, 119, 121, 144
Lugannani, R.   449, 462, 476–478, 483
Lurie, D.   41, 58, 66–68, 71
Lynch, J.   92, 101, 104

Müller, D. W.   584, 630, 635, 687
Müntz, C. H.   294, 310
Mărgăritescu, E.   118, 144
Ma, C.   463, 464, 466, 468, 470, 472, 474, 476, 478–484, 492, 512
Maag, U. R.   274, 287
Mack, Y. D. P.   469, 474, 483
MacLaren, M. D.   67, 71
Maddala, G. S.   29, 58
Madreimov, I.   271, 272, 289, 295, 310
Maguire, B. A.   458, 462
Mahalanobis, P. C.   387, 402
Mahiat, H.   561, 563
Mahmoud, M. A. W.   184, 190, 192, 101, 211, 226, 227, 293, 310
Mahmoud, M. R.   212, 227
Major, P.   58Ĭ, 584, 590, 629, 632, 636, 664, 687
Malik, H. J.   32, 46, 49, 58, 151–153, 155–157, 159, 162, 170, 174, 191, 194, 200, 201, 203–205, 218, 221, 223, 226, 292, 297, 301, 305, 308, 310, 411, 438
Mallows, C. L.   109, 111, 117, 144, 293, 310, 558, 563
Malmquist, S.   17, 23, 69, 258, 269, 289
Malov, S.   555, 562, 566
Mann, N. R.   16, 23, 30, 31, 59, 150, 226
Marcus, M. B.   321, 332
Mardia, K. V.   45, 59
Margolin, B. H.   46, 59
Marohn, F.   329, 331, 332
Marron, J. S.   467, 484
Marsaglia, G.   67, 71, 251, 255
Marshall, A. W.   101, 104, 108, 134, 144, 263, 289
Mason, D. M.   574, 575, 582, 583, 587, 588, 605, 628, 630, 637, 653, 654, 656, 667, 671, 674, 685, 686
Mason, R. L.   41, 58, 67, 71
Massart, P.   581, 628, 640, 687
Mathai, A. M.   33, 45, 59, 389, 402
Mattner, L.   218, 225, 300, 310
Matusita, K.   56

Maurer, W.   46, 59
Maxwell, A. E.   43, 58
McCabe, G. P.   27, 59
McCullagh, P.   449, 462
McDonald, L. I.   500, 512
McIntyre, G. A.   499, 512
McKinlay, A. C.   488, 511
McLaughlin, D. H.   29, 63, 149, 228
Meeden, G.   295, 308
Meeker, W. Q.   35, 54
Mehra, K. L.   508, 512
Mehta, M. L.   42, 59
Meier, P.   349, 371
Mejzler, D. G.   329, 332
Melamed, J. A.   45, 57, 264, 277, 288
Melnick, E. L.   151, 155, 226, 291, 310
Menon, M. V.   274, 289
Miller, J. E.   547, 559, 569
Miller, L. H.   44, 59
Mimmach, G.   92, 101, 104
Mitra, S. K.   402
Mitrinović, D. S.   303, 310
Mogul'skiĭ, A. A.   675, 687
Mogyorodi, J.   23
Mohan, N. R.   255
Mohie El-Din, M. M.   184, 190, 192, 202, 211, 216, 219, 226, 227, 293, 307, 310
Molchanov, S. A.   562
Moore, A. H.   29, 30, 43, 55, 56, 64
Moore, D. S.   27, 59
Moothathu, T. S. K.   185, 227
Moreno Rebollo, J. L.   566
Mori, T.   375, 377, 384, 559, 569
Moriguti, S.   32, 50, 120, 121, 125, 144, 302, 310, 531, 566
Morrison, D. F.   43, 59, 225, 512
Mosler, K.   86
Mosteller, F.   29, 30, 56, 59, 150, 226, 501, 512
Mudholkar, G. S.   278, 279, 287
Murphy, R. B.   34, 35, 59
Muttlak, H. A.   500, 512

Nadaraya, E. A.   469, 483
Nagar, A. L.   42, 63
Nagaraja, H. N.   3, 5, 7, 11, 17, 19–23, 26, 27, 47, 48, 51, 59, 65, 70, 71, 80, 86, 106, 119, 137, 138, 142, 144, 153, 154, 222, 225, 231, 254, 255, 257, 263, 279, 282, 285, 286, 289, 292, 305, 308, 353, 416, 420, 438, 487, 488, 490, 492–494, 496, 498, 500, 502, 504–506, 508, 510–512, 515–517, 530–532, 545, 559, 560–563, 566
Nair, K. R.   117, 144

Nakano, K.   35, 50
Nanda, D. N.   42, 59, 60
Nassar, M. M.   212, 227
Natanson, I. P.   294, 310
Nayak, S. S.   231, 251, 255, 560, 567
Nelson, W.   16, 23, 30, 60, 150, 227
Nelson, L. S.   35, 60
Nelson, P. I.   14, 23, 36, 57
Neuts, M. F.   516, 526, 567
Nevzorov, V. B.   19, 23, 45, 47, 60, 487, 507, 511, 515–520, 522, 524, 526, 528, 530, 532–534, 536–544, 546–564, 566–568, 570
Nevzorova, L. N.   555, 556, 568
Newby, M. J. H.   70, 71
Newcomb, S.   29, 60
Newman, D.   37, 60
Neyman, J.   64, 144
Ni Chuiv, N.   500
Nicolae, T.   118, 144
Nikoulina, V. N.   568
Nikulin, K.   559, 569
Nodes, T. A.   45, 48

Obretenov, A.   322, 331
O'Connell, M. J.   500–503, 511, 512
Ogawa, J.   30, 31, 60
Oja, H.   127, 144
Olkin, I.   39, 47, 53, 54, 60, 101, 104, 108, 117, 144, 263, 289
Omey, E.   545, 564
Oosterhoff, J.   614, 617, 630
O'Quigley, J.   204, 227
Ord, J. K.   45, 51, 60, 253–255, 285–287, 309
O'Reilly, N.   574, 582, 630, 653, 654, 687
Owen, D. B.   35, 60

Padgett, W. J.   559, 565
Palmieri, F.   46, 60
Panchapakesan, S.   33, 38, 39, 54, 561, 562
Pancheva, E.   324, 332
Pardzhanadze, A. M.   576, 614, 619, 629, 630
Parent, E. A. Jr.   605, 630
Parrish, R. S.   31, 60
Parvez, S.   175, 190, 217, 218, 221, 226, 297, 300, 310
Parzen, E.   465, 466, 483, 591, 630, 641, 642, 644, 645, 687
Patel, J. K.   35, 60
Patil, G. P.   45, 51, 60, 253–255, 275, 285–287, 289, 309, 500, 511, 512
Pearson, E. S.   36, 60, 118, 143, 458, 462
Pearson K.   30, 39, 43, 60, 160, 208, 227, 473
Pederzoli, G.   45, 59

Peirce, B. 29, 60
Penkov, B. 567
Petrov, V. V. 562, 567, 568
Petunin, Y. I. 271, 272, 289, 295, 310
Pfeifer, D. 524–526, 544, 546, 549, 553, 559, 560, 568
Pflug, G. 564
Picard, D. 577, 585, 593, 629, 630
Pickands, J. 322, 332, 442, 456, 462, 545, 560, 568
Pierard, M. 561, 563
Pike, M. C. 205, 227
Pillai, K. C. S. 43, 60
Pinhas, M. 512
Pinsker, I. Sh. 4, 24
Pinsky, M. 321, 332
Pittel, B. G. 559, 564
Plackett, R. L. 119, 125, 144, 302, 310
Pledger, G. 46, 60, 101, 104
Pollak, M. 293, 310
Pólya, G. 75, 87, 310
Posner, E. C. 150, 224
Postelnicu, T. 144
Prentice, R. L. 30, 57
Proschan, F. 26, 33, 46, 49, 54, 60, 92–96, 98, 101–104, 118, 134, 136, 137, 143, 144, 364, 368, 371, 492, 510
Provost, S. B. 389, 402
Pudeg, A. 260, 289
Puri, M. L. 62, 339, 359, 372, 468, 469, 471, 472, 483, 484, 543, 563, 565, 566
Puri, P. S. 239, 255, 266, 283, 284, 289
Puthenpura, S. 179, 223
Pyke, R. 44, 60, 442, 446, 462

Quesenberry, C. P. 35, 60
Quine, M. P. 476, 484
Qureishi, A. S. 178, 224

Rabonomwitz, M. 70, 71
Raghavarao, D. 38, 52
Rahman, M. 138, 143, 266, 286, 296, 308
Rainville, E. D. 197, 227
Ralescu, S. S. 468, 469, 471, 472, 474, 483, 484
Ramachandran, B. 45, 60, 231, 232, 234, 235, 239, 256, 259, 262, 266, 283, 289
Ramberg, J. S. 41, 60, 68, 71
Ramirez, D. E. 18, 23, 381–390, 392, 394, 396, 398, 400, 402
Rannen, M. M. 546, 547, 560, 568
Rao, C. R. 3, 4, 6, 8, 10–12, 14, 16, 18, 20, 22–24, 27, 29, 42, 43, 45, 49, 54, 55, 57, 58, 61, 64, 231–240, 242, 244–257, 259, 262, 266, 279, 283, 286–289, 292, 402, 480, 482, 512, 559, 568, 648, 687
Rao, B. R. 204, 223
Rao, J. N. K. 255
Raqab, M. Z. 531, 568
Rauhut, B. 293, 310
Reeder, H. A. 41, 61, 67, 71
Reid, N. 476, 484
Reiss, R.-D. 12, 24, 33, 40, 61, 287, 308, 329, 331, 332, 441, 462, 466–468, 470, 472, 474, 482–484, 487, 493, 511, 512, 566, 568, 570
Rényi, A. 258, 269, 270, 288, 289, 330, 332, 516, 518, 520, 521, 525, 583, 543, 568, 574, 581, 582, 630, 636, 645, 653, 687
Resnick, S. I. 12, 24, 40, 47, 61, 442, 447, 456, 461, 516, 528, 532, 537, 540, 547–549, 557–560, 563–566, 568, 569
Restrepo, A. 46, 50
Révsz, P. 565, 585, 595, 603, 629, 632–637, 643–645, 647–651, 658, 663, 664, 666, 667, 674, 675, 679, 682, 685, 686
Reynolds, M. R. Jr. 605, 630
Richter, D. 117, 144
Rider, P. R. 29, 61
Ridout, M. S. 500, 512
Riedel, M. 261, 263, 267, 289
Rinott, Y. 47, 61, 492, 511
Riordan, J. 152, 227
Rizvi, M. H. 32, 58, 151, 184, 226
Robbins, H. 109, 111, 143, 380, 384
Robinson, J. 463, 464, 466, 468, 470, 472, 474, 476, 478–484
Robinson, A. 497, 510
Rogers, G. S. 256
Rogers, W. H. 18, 21, 29, 48, 150, 222
Rogers, M. P. 45, 51
Rogers, L. C. G. 540, 565, 569
Rogers, B. 545, 563
Romano, J. P. 43, 461, 482, 483
Romanovsky, V. 160, 227
Ronchetti, E. M. 29, 54, 476, 483
Rootzén, H. 12, 23, 40, 58, 330, 332
Rosenblatt, M. 465, 484, 591, 628
Ross, S. 100, 104
Rossberg, H. J. 45, 61, 239, 256, 260, 261, 266, 267, 269, 283, 289
Rothe, G. 483
Rothmann, M. D. 377, 380, 384
Roussas, G. G. 614, 630
Rousseeuw, P. J. 29, 54, 61, 387, 402
Roy, J. 42, 61
Roy, S. N. 42, 43, 61

Rubin, H.   239, 255, 266, 283, 289
Rubinovitch, M.   516, 560, 569
Russo, R. P.   377, 380, 384
Rüschendorf, L.   7, 24
Rustagi, J. S.   54, 60, 104, 119, 144
Ruymgaart, F. H.   509, 510
Rychlik, T.   10, 24, 33, 61, 105, 106, 108–118, 120, 122–124, 126, 128, 130, 132–134, 136, 138, 140–144, 257, 292

Sackrowitz, H. E.   253, 256
Salama, I. A.   62, 144, 562
Saleh, A. K. Md. E.   14, 21, 190, 227, 255, 295, 309, 311, 371
Samaniego, F. J.   558, 559, 569
Sampson, A. R.   104
Samuel, P.   166, 227
Samuel-Cahn, E.   47, 61, 253, 256
Samuelson, P. A.   117, 144
Sandhu, R. A.   41, 49, 69, 71, 183, 184, 223
Sanström, A.   506, 509, 512
Sapatinas, T.   264, 289
Sarabia, J. M.   326, 327, 331
Sarhan, A. E.   3, 24, 26, 31, 60, 62, 149, 223, 225, 227, 343, 372
Sarkar, S. K.   14, 24, 31, 47, 62, 64
Sarma, Y. R. K.   261, 286
Sasvári, Z.   272, 289
Sathe, Y. S.   46, 62, 292, 311
Savits, T. H.   104
Scarsini, M.   86
Schaeffer, L. R.   174, 227
Schafter, R. E.   16, 23, 30, 59, 150, 226
Schechtman, E.   497, 512
Scheffé, H.   34, 62, 149, 227
Schlipf, J. S.   41, 56, 68, 71
Schmeiser, B. W.   41, 62, 68, 69, 71
Schneider, B. E.   451, 462
Schneider, H.   36, 62
Schucany, W. R.   41, 62, 67, 68, 71
Schweitzer, N.   11, 23, 45, 53, 257, 258, 260, 262, 264, 266, 268, 270, 272, 274, 276, 278, 280, 282, 284, 286, 288, 290, 292, 296, 304, 306
Scott, C.   190, 227
Scott, J. M. C.   117, 144
Scott, E. L.   144
Seal, K. C.   38, 62, 173
Sen, P.   559, 567
Sen, P. K.   12, 24, 40, 46, 50, 62, 101, 104, 144, 225, 287, 335, 337–340, 342, 344–348, 350–362, 364, 366, 368–373, 487, 498, 507, 510–512, 562

Sendler, W.   483
Seneta, E.   522, 526, 533, 565
Seoh, M.   469, 472, 484
Serfling, R. J.   12, 24, 40, 54, 62, 344, 373, 441, 462, 468, 470, 471, 473, 484
Seshadri, V.   45, 51, 261, 272–276, 278, 286, 289, 290
Sethuraman, J.   46, 60, 96, 101, 102, 104, 263, 276, 290
Shah, B. K.   32, 62, 178, 217, 218, 224, 227
Shah, S. M.   269, 290
Shaked, M.   7, 22, 27, 33, 50, 80, 81, 87, 89–92, 94–98, 100–104
Shanbhag, D. N.   11, 24, 45, 61, 231–242, 244–257, 259, 262, 283, 286, 287, 289, 290, 292, 559, 568
Shanthikumar, J. G.   7, 22, 27, 33, 50, 80, 81, 87, 89–92, 94–98, 100–104
Shao, Q. M.   574, 582, 583, 587, 589, 590, 595, 597, 626, 629, 654, 657, 686
Shapiro, S. S.   18, 24, 44, 45, 62
Sheather, S. J.   467, 475, 483, 484
Shewhart, W. A.   36, 62
Sheynin, O. B.   30, 62
Shimizu, R.   234, 247, 256, 260, 262, 263, 270, 271, 277, 290
Shiryayev, A. N.   614, 629
Shorack, G. R.   12, 24, 40, 62, 336, 350, 373, 376, 381, 384, 473, 484, 574, 582, 589, 630, 635, 636, 653, 686, 687
Shorrock, R. W.   47, 62, 231, 256, 516, 518, 520, 521, 525, 527, 533, 550, 560, 569
Shu, V. S.   150, 224, 227
Shubha, 159, 225
Sibuya, M.   559, 569
Šidák,   614, 629, 644, 645, 687
Siddiqui, M. M.   45, 62, 517, 534, 544, 545, 563, 569
Siegel, A. F.   47, 62
Siegmund, D.   380, 384
Sillitto, G. P.   32, 62, 151, 156, 159, 160, 227
Silverman, B. W.   482, 484
Simiu, E.   329, 332, 562
Simonoff, J. S.   458, 462
Singer, J. M.   336, 373
Singh, B.   8, 17, 22
Singh, K.   480–482, 484
Sinha, A. K.   500, 512
Sinha, B. K.   372, 500
Smirnov, N.   43, 62, 467–469, 484, 580, 630
Smith, R. L.   41, 53, 70, 71, 327, 332, 441, 442, 461, 462, 547, 559, 569
Sobel, M.   38, 39, 49, 50, 53, 54, 258, 287

Somerville, P. R.   34, 35, 62
Sondhauss, U.   489, 494, 512
Song, R.   490, 492, 512
Spruill, N. L.   497, 504, 512
Sreehari, M.   245, 256
Srikantan, K. S.   32, 62, 151, 156, 227
Srivastava, J. N.   42, 62, 63
Srivastava, R. C.   255, 256, 280, 282, 285, 289, 290, 516, 559, 567, 569
Stadje, W.   239, 240, 256
Stahel, W. A.   29, 54
Stam, A. I.   540, 569
Stegun, I. A.   176, 205, 222, 443, 461
Stepanov, A. V.   524, 526, 529, 530, 536, 537, 540, 541, 559, 568, 569
Stephens, M. A.   18, 22, 23, 45, 51, 58, 150, 151, 153, 173, 223, 224, 261, 274, 275, 290
Stigler, S. M.   40, 63, 468, 484
Stokes, S. L.   499, 500, 512
Strassen, V.   76, 87
Strawderman, R. L.   441, 442, 444, 446, 448, 450, 452, 454, 456, 458, 460, 462
Strawderman, W.   516, 526, 533, 566
Struthers, L.   204, 227
Stuart, A.   43, 57, 497, 513, 516, 520, 547, 558, 565, 569
Stute, W.   506, 509, 513
Styan, G. P. H.   117, 145
Su, J. C.   560, 566
Sugiura, N.   33, 63, 125, 144
Suh, M. W.   361, 373
Sukhatme, P. V.   16, 24, 70, 71, 258, 290, 442, 462
Sultan, K. S.   9, 22, 32, 49, 149, 150, 152, 154, 156, 158, 160, 162, 164, 166, 169, 170, 172, 174, 176, 178, 180, 182, 184, 186, 188, 190, 192, 194, 196, 198, 200, 202, 204, 108, 210, 212, 214, 216, 218–220, 222, 224, 226–228, 307, 310, 558, 569
Sumita, U.   90, 91, 104
Sun, S.   469, 484
Suresh, R. P.   495, 513
Surgailis, D.   562
Sweeting, T. J.   322, 333
Swetits, J. J.   128, 145
Szász, O.   294, 311
Szegö, G.   310
Szekely, G.   266, 287, 559, 569
Szynal, D.   19, 23, 528, 531, 565
Szyszkowicz, B.   573, 574, 576–578, 580, 582–588, 590, 592, 594–598, 600, 602, 604–610, 612, 614, 616–618, 620, 622–632, 634, 636, 638, 640, 642, 644, 646, 648, 650, 652, 654–658, 660, 662, 664, 666, 668, 670, 672, 674, 676, 678, 680, 682, 684, 686, 687

Tadikamalla, P. R.   17, 24, 41, 60, 63, 65–68, 70–72, 208, 227
Taillie, C.   287, 500, 511, 512
Takahasi, K.   27, 63
Tan, W. Y.   18, 24, 29, 63, 150, 227
Taneja, V. S.   269, 288
Tarter, M. E.   183, 227
Tata, M. N.   516, 518, 519, 523, 526, 528, 569
Tăutu, P.   144
Tawn, J. A.   495, 504, 510, 511
Taylor, H. M.   329, 333
Tchen, A.   111, 145
Teicher, H.   376, 384
Teichroew, D.   31, 63, 516, 547, 558, 564
Teugels, J. L.   559, 569
Theil, H.   42, 63
Thoma, M.   629
Thomas, P. Y.   166, 185, 186,227
Thompson, G. W.   118, 145
Thompson, J. W.   92, 104
Thompson, W. A. Jr.   46, 52
Thompson, W. R.   117, 145
Thomson, G. H.   41, 63
Thorburn, D.   561, 563
Tiago de Oliveira, J.   40, 47, 63, 325, 326, 327, 329, 332, 333, 511, 516, 560, 563, 564, 569
Tibshirani, R.   443, 461, 479, 483
Tietjen, G. L.   31, 63
Tiku, M. L.   18, 24, 29, 44, 63, 150, 227, 499, 511
Tintner, G.   42, 63
Tippett, L. H. C.   12, 22, 39, 52, 63, 319, 331
Todorovic, P.   327, 333
Tomkins, R. J. 375–382, 384
Tomko, J.   566
Tommassen, P.   540, 564
Tong, Y. L.   93, 104
Too, Y. H.   302–304, 311, 559, 569
Trenkler, G.   564
Tryfos, P.   559, 569
Tsukibayashi, S.   497, 513
Tubilla, A.   251, 255
Tukey, J. W.   18, 21, 29, 34, 35, 37, 45, 48, 52, 56, 62, 63, 149, 150, 222, 224, 226, 228
Tusnády, G.   581, 584, 590, 629, 632, 636, 664, 687

Ubhaya, V. A.   128, 145
Umbach, D.   14, 21, 31, 48
Upadrasta, S. P.   508, 512

van Eeden, C.   247, 248, 255, 274, 275, 287, 289, 290
Vännman, K.   416, 419, 438
Van Vleck, L. D.   174, 227
van Zwet, W. R.   33, 63, 92, 104, 127, 132, 135, 136, 138, 145, 531, 570, 614, 617
Vaughan, D. C.   175, 228, 499, 511
Vaughan, R. J.   20, 24, 46, 63, 403, 438
Velasco, J. A.   174, 227
Venables, W. N.   20, 24, 46, 63, 403, 438
Veraverbeke, N.   506, 509, 513
Verkade, E.   547, 564
Vervaat, W.   516, 523, 529, 530, 540, 560, 564, 570
Vessey, T.   561, 563
Viana, M. A. G.   47, 60
Villaseñor, J. A.   7, 21, 27, 48, 75, 76, 78–80, 82–87
Vincze, I.   331, 484
Viswanathan, R.   46, 63
Volodin, N. A.   217, 218, 222, 231, 254, 257, 286, 296, 308
von Mises, R.   12, 24, 39, 43, 59, 319, 321, 332, 373

Wald, A.   35, 63
Wali, K. S.   560, 567
Walsh, J. E.   35, 64
Wang, H.   375, 376, 378–384
Wang, W.   14, 24, 31, 47, 62, 64
Watson, G. S.   42, 43, 52, 64
Watterson, G. A.   487, 489, 497, 498, 513
Wefelmeyer, W.   92, 104
Weibull, W.   40, 64
Weinstein, S. E.   128, 145
Weisberg, S.   387, 401
Weissman, I.   329, 333, 442, 443, 445, 462
Wellner, J. A.   12, 24, 40, 62, 248, 255, 323, 332, 336, 350, 373, 376, 381, 384, 473, 484, 574, 82, 289, 630, 636, 653, 686, 687
Welsch, R. E.   387, 401
Wertz, W.   23, 564, 565
Wessen, B. J.   441, 462
Westcott, M.   521, 545, 570
Whisenand, C. W.   199, 224
Whitaker, L. R.   558, 559, 569
Whitten, B. J.   16, 22, 30, 51, 150, 223
Wichura, M. J.   584, 628
Wiegand, R. P.   41, 56
Wigner, E. P.   42, 64

Wilk, M. B.   18, 24, 44, 45, 62, 64
Wilkinson, G. N.   29, 58, 224, 438
Wilks, S. S.   30, 33–35, 42, 64, 518, 535, 570
Williams, D.   516, 522, 570
Williams, G. T.   33, 50
Wise, G. L.   45, 53, 58
Witte, H.-J.   249, 256, 559, 570
Wolfe, D. A.   149, 225
Wolfowtiz, J.   35, 64, 640, 687
Wolkowicz, H.   117, 145
Wood, A. T. A.   444, 449, 450, 452, 462
Woodruff, B. W.   43, 64
Wormleighton, R.   35, 52
Wright, C.   559, 565
Wyner, A.   629
Wynn, A. H. A.   458, 462

Xekalaki, E.   270, 287, 301, 308
Xiang, X. J.   474, 484
Xu, J. L.   275, 290
Xu, Y.   128, 145, 322, 332

Yakymiv, A. L.   545, 570
Yalovsky, M.   45, 51, 286
Yamauti, Z.   31, 64
Yang, G. L.   275, 290
Yang, M. C. K.   534, 545, 548, 559, 570
Yang, S. S.   467, 470, 474, 484, 495, 502, 503, 506, 508, 509, 511, 513
Yao, D. D.   91, 95, 104
Yaqub, M.   175, 190, 217, 218, 221, 226, 297, 300, 310
Yeo, W. B.   504, 513
Yitzhaki, S.   497, 512
Young, D. H.   46, 64, 151, 185, 228, 292, 311
Young, G. A.   443, 462, 482, 484
Yue, X.   492, 512
Yuen, K. K.   150, 228

Zahedi, H.   286
Zahle, U.   560, 570
Zaretzki, 294
Zelen, M.   441, 462
Zelterman, D.   441, 442, 444–446, 448, 450–452, 454, 456, 458, 460, 462, 467, 484
Zhang, Y. C.   525, 568
Zhao, L. C.   648, 687
Zijlstra, M.   240, 256, 283, 290
Zolotarev, V. M.   287, 308, 332, 567

Subject Index

Adaptive procedures 29
Admissible weight functions 589, 594
Affine invariance property 340
Almost sure invariance principle 544
Analysis of covariance (ANOCOVA) 357
Analysis of variance 44
Anderson-Darling test 43
Antiranks 339, 340
Applications
– birth defects 441
– breaking strength of materials 39, 40
– breeding 487
– childhood leukemia 441
– concentration of airborne pollutants 441
– duration of human life 39
– econometrics 25
– employment problems of a professional couple 504
– energy levels in nuclear spectra 42
– engineering 40
– environmental issues 441
– estimation theory 253
– exposure to radioactivity 441
– extinction times for bacteria 39
– file-matching procedures 488, 510
– financial asset pricing models 488
– floods and droughts 39, 515
– genetic selection problems 506, 507
– gust loads 39
– hydrology 25, 40, 327, 328, 490
– insurance mathematics 559
– life-testing 14, 15, 30, 337, 338, 355, 361
– major flood 441
– meterology 25, 39
– model building 253, 362
– Olympic games 548
– paired t-test 488
– public health issues 441
– quality control 25, 36, 39
– queuing theory 253
– radioactive decay 39
– rainfall 515
– rate of cancer 441
– reliability 30, 89, 90, 253, 335, 336, 337, 338, 361, 362
– searching algorithms 559
– secretary problem 559
– seismology 25
– selection procedures 487, 488, 507
– snowfall 515
– speeds of groups of vehicles 559
– sports records 559
– sports events 515, 553
– stock market 40
– strength of materials 25, 329
– strength of bundle of filaments 336, 361
– survival analysis 89, 90, 335, 336, 354–356
– system availability 336, 361
– temperature 515
Approximations 4, 32, 530
Archimedean copula process 555, 556
Ascending method 41, 66, 68
Associated random variables 94, 492
Asymmetric censoring 31
Asymptotic theory 4
Asymptotically best linear estimators 149, 343
Autoregressive process 545
Auxiliary variable 505
Average availability 364
Average absolute deviation 27

Bahadur representation 341, 342, 473, 474, 636
Bahadur-Kiefer principle 636, 637
Bahadur-Kiefer process 631, 637, 660, 661, 675, 679–682
Balabekyan-Nevzorov's records 546
Ballerini-Resnick scheme 557
Bandwidth 466
Behrens-Fisher problem 273
Bernoulli distribution 38, 284, 580

Bernstein basis 112
Berry-Esseen theorem 470, 471
Berry-Esseen type bound 468, 472
Bessel process 356
Best linear unbiased estimation 12, 13, 18, 25, 28, 30, 31, 179, 343, 404, 415, 416, 421, 426, 498
Best linear invariant estimation 31
Best linear unbiased prediction 12, 14
Beta distribution 26, 68, 77
Beta function 154, 203
Bias 18, 416–419, 421–425, 443
Binomial distribution 7, 26, 38, 252, 337, 341, 464, 470, 479, 662, 664
Binomial expansion 9
Binomial form 4
Birth defects 441
Birth process 545
Bivariate dependence 89
Bivariate extreme value distribution 558
Bivariate Markov sequence 545
Bivariate normal distribution 489, 493, 495–498, 502, 504, 505, 508, 558
Block-diagonal dispersion matrix 393, 399
Block-diagonal form 388
Bonferroni-type inequalities 315, 380
Bootstrap 348, 362, 368, 441, 443–445, 453–456, 466, 480, 488, 507, 595
Bootstrap approximation 446, 455–457, 458–460, 468, 479, 481
Bootstrap distribution 444, 448, 449, 454, 458, 509
Bootstrap error 480
Bootstrap estimator 467
Bootstrap quantiles 480
Bootstrap sample 480
Borel-Cantelli lemma 379, 632, 640
Borel measurable function 232, 234, 236, 237
Borel σ-field 232, 250
Borel subset 250
Bounds 4, 32, 105, 303, 337, 530
Break points 107
Breaking strength of materials 39
Breeding 487
Bridge type process 576, 609
British Coal Mining data 444, 458
Brownian bridge 342, 507, 581–583, 599, 635, 643, 644, 654, 655, 672, 678, 680
Brownian bridge process 582
Brownian motion 507
Bucket sorting 70
Bundle of parallel filaments 336, 361

Burr distribution 153, 204, 208, 209, 216, 218–221, 268, 269, 298, 300

c-comparison 136, 531
c-ordering 127, 369
Canonical correlation 42
Canonical form 390–392
Canonical leverages 397
Cauchy distribution 26, 33, 152, 175, 274, 301
Cauchy-Schwarz inequality 33, 113, 118, 120, 122, 126, 302, 530, 531
Censored bivariate data 488, 498
Censored BLUE 416, 419, 426
Censored sample 25, 150, 336, 343
Centered log-likelihood ratio 617
Central concomitants 495
Central distributions 389
Central limit theorem 341, 536, 548, 553, 580
Changepoint 605, 615
Change-point alternatives 615, 616
Change-point problems 573, 576, 578, 591, 595, 599, 614
Characteristic function 278, 279, 301
Characteristic roots 41
Characterizations 11, 45, 90, 231–253, 257–285, 291–307, 516, 559
Chi (see Half normal) distribution
Chi-square distribution 274, 388
Chi-square test 43
Chibisov-O'Reilly theorem 582
Childhood leukemia 441
Choquet-Deny type functional equations 45
Closure property 91
Combinatorial identities 151, 152, 157
Compact differentiability 345
Compatible 276
Complete covariance matrix dominance 14
Complete statistic 339
Complete sufficient statistic 416
Concave function 108, 139
Concentration of airborne pollutants 441
Concomitant order statistics 4, 19, 20, 336, 340, 352, 353, 355, 356, 487, 492
Concomitant variate 359
Conditional L-functionals 336, 359
Conditional moments 257
Conditional ordering 89
Conditional quantiles 358
Conditional residual moment 321
Confidence bands 343
Confidence bounds 328, 643
Confidence intervals 25, 480, 498, 507, 508
Confidence procedures 37, 42

Confidence region 400
Conjugate density 476
Contamination 150
Contiguous alternatives 576, 579, 614, 615, 619
Contiguous functions 192, 197
Control chart 605
Convex cones 105, 106, 120, 126–128, 134
Convex function 75, 93
Convex hull 558
Convex transform ordering 89, 92
Convolutions 91, 95, 97
Cook's D_1 statistics 378, 400
Cook's distance 18
Correlation coefficient 488, 496, 497, 500, 504, 505, 559
Counting process 540, 560
Covariance function 543, 581, 584, 591, 596–598, 617, 618, 620–622, 624, 627, 635
Covariance matrix 416, 419–421
Cox regression model 354
Cramér-Von Mises distance 456
Cramér-Von Mises test 43
Cramér-Wold device 619
Criterion series 375
Cross-cumulants 11
Csáki-type law of the iterated logarithm 648
Cumulant generating function 449, 450, 475
Cumulants 11
Cumulative hazard function 320, 365
Cumulative rounding error 152
CUSUM chart 605
Cutting function 340
Cycles lengths 559
δ-exceedance records 561

Data compression 150
Decomposition principle 325, 326
Decreasing Mean Residual life (DMRL) 260, 261, 367
Decreasing failure probability 139
Decreasing Failure Rate (DFR) 26, 91, 97, 128, 130, 136, 140, 260, 261, 264, 265, 267, 282,
Decreasing Failure Rate on Average (DFRA) 27, 367
de Finetti's representation 543
de Finetti's theorem 235, 236, 248, 330
Delete k-jackknifing 347
Density-quantile function 642, 643
Deny's theorem 235
Dependence structure 26, 493
Dependent F^α-scheme 556

Descending method 41, 67, 68
Determinant-efficient estimators 14
Determinantal equations 42
Differential equation 189, 191, 404, 414, 531, 532
Digamma function 443
Direct products 614
Discrete Pareto distribution 284
Discriminant analysis 42
Dispersion matrix 394
Dispersive ordering 89, 91, 92, 96, 97
Dispersive random variable 92
Distributions
– Bernoulli 38, 284, 580
– beta 26, 68, 77
– binomial 4, 7, 26, 38, 252, 284, 337, 341, 464, 470, 479, 662, 664
– bivariate extreme value 558
– bivariate normal 489, 493, 495–498, 502, 504, 505, 508, 558
– bootstrap 444, 448, 449, 454, 458
– Burr 153, 204, 208, 209, 216, 218–221, 268, 269, 298, 300
– Cauchy 26, 33, 152, 175, 274, 301
– central 389
– chi-square 274, 388
– closed under maxima 413
– DFR 26, 128, 130, 140, 260, 261, 264, 265, 267, 282, 304, 367, 561
– DFRA 27, 367
– discrete Pareto 284
– DMRL 260, 261, 367
– double Weibull 172
– doubly truncated Burr 209
– doubly truncated Cauchy 301
– doubly truncated exponential 188, 190, 404
– doubly truncated Laplace 212, 213, 216
– doubly truncated linear-exponential 201
– doubly truncated logistic 180, 183, 184
– doubly truncated log-logistic 205
– doubly truncated parabolic 211
– doubly truncated Pareto 194, 404, 413, 414, 421
– doubly truncated power function 191, 192
– doubly truncated skewed 211
– elliptically contoured 47
– exponential 8, 13, 14, 16, 26, 31–33, 38, 43, 45, 46, 67, 70, 77–80, 82, 86, 91, 92, 95–97, 99, 101, 102, 128, 130, 134, 152, 153, 188, 201, 216, 221, 231, 233, 234, 236–242, 245, 248, 249, 251, 253, 258–272, 174, 175, 277–279, 282, 284, 295–299, 302, 304, 306, 307, 321- -325, 327, 337, 338, 369, 403, 413,

442–447, 454, 458, 460, 476, 494, 516, 518–520, 525, 527, 531, 534, 538, 545, 556, 559, 561, 644
- exponential family 339, 367
- extended normal 307
- extreme value
- F 388, 396, 398, 400, 401
- folded 151, 171, 172
- Frechét
- gamma 26, 31, 32, 38, 70, 184–186, 278, 321, 442, 492
- generalized extreme value 442, 532, 543, 547
- generalized gamma 186, 187
- generalized half logistic 32, 183
- generalized logistic 32
- generalized Pareto 442, 447, 456, 459, 532
- geometric 26, 231, 234, 236–242, 245, 249, 251, 253, 259, 264, 265, 280, 282–285, 516, 529, 548, 559
- Gompertz 442
- Gumbel 441, 442, 453, 532, 547
- Gumbel's bivariate exponential 490, 494
- half normal (chi) 173
- HNBUE 368
- IFR 26, 131, 132, 137, 260, 261, 264, 265, 267, 282, 304, 367, 561
- IFRA 26, 367
- IMRL 260, 261, 367
- inverse multinomial 185
- inverse normal 476
- J-shaped 131, 132, 135, 137, 139
- k-dimensional extremal 442, 443
- Laplace (or Double exponential) 26, 29, 152, 153, 172, 276, 279
- left-truncated Pareto 415
- life 325
- linear-exponential 32, 200
- ℓ_1-normal symmetric 261
- location-scale family 13, 343, 350
- logarithmic 300
- logistic 8, 26, 32, 138, 152, 153, 177, 216–218, 268, 269, 272, 275, 277–279, 297–301, 404, 442, 532
- log-logistic 32, 153, 204, 205, 208, 210, 218
- lognormal 26, 272, 273, 442
- log-Weibull 31
- Lomax 153, 203, 210, 216–218, 532
- mixture 100, 130, 134, 543
- mixture of exponential 153, 212
- mixture of normal 554
- modified geometric type 285
- multinomial 26, 70
- multivariate beta 370
- multivariate exponential 261
- multivariate normal 42, 45, 47, 341, 342, 351, 490, 491, 507
- NBRUE 367–369
- NBU 27, 260, 264, 267, 269, 275, 282, 304, 366
- NBUE 367, 368
- negative binomial 4, 26, 541
- negative Pareto 136
- noncentral 389
- non-lattice 278
- normal 17, 26, 28, 30–32, 37–39, 41, 43–45, 99, 137, 138, 151–153, 155, 173, 174, 272–274, 292, 297, 305–307, 321–323, 325, 348, 354, 362, 388, 394, 442, 444, 447, 449, 453–459, 467, 468, 470, 472–474, 479, 489, 493, 495, 498, 503, 505, 506, 509, 510, 518, 524, 532, 537, 547, 554, 555, 595, 600, 661, 672, 673
- NWRUE 367
- NWU 27, 260, 264, 267, 269, 270, 275, 282, 304, 366, 368
- NWUE 367
- parabolic 153, 211
- Pareto 8, 13, 14, 18, 26, 32, 79, 82, 84, 86, 136, 152, 153, 193, 194, 198, 217–221, 259, 268–270, 297–301, 307, 403, 404, 406, 411–413, 415, 416, 421, 426, 442
- Poisson 26, 284
- power function
- Rayleigh 199–201, 218–221, 532
- reflected Weibull 276, 300
- restricted family 105, 106
- right truncated exponential 132, 190, 403
- right-truncated Pareto 415
- scaled F 400
- scale-family 13
- shifted exponential 243, 244, 250
- shifted geometric 243, 245
- singular Gaussian 394
- skewed 153, 211
- Snedecor-Fisher 387
- stable 45
- symmetric 105
- symmetric unimodal 105
- symmetric U-shaped 138, 141
- t 26
- truncated 32, 153
- truncated Burr 209
- truncated Cauchy 175
- truncated normal 307
- truncated Pareto 403, 413, 414

- type I Gumbel 441, 442, 447, 448, 450
- type II Gumbel 442, 448
- uniform 8, 10, 13, 14, 17, 26, 29, 30, 39, 41, 45, 65–69, 77, 78, 84, 86, 91, 97, 107, 110, 112, 123, 130, 133, 138, 248, 252, 253, 258, 269–272, 275, 278–280, 284, 295, 297, 299, 302, 303, 306, 339, 343, 382, 464, 468, 470, 492, 510, 559, 574, 581, 584, 607, 617, 636, 653, 662, 664, 671
- U-shaped 133, 134
- Weibull 8, 12, 26, 31, 40, 83, 86, 92, 190, 218–221, 247, 259, 268–270, 272, 274, 276, 277, 300, 321, 328, 329, 442, 453–457, 459, 532
- Weibull-exponential 210
- Weibull-gamma 210

Distribution-free 25, 149, 357
Distribution-free tolerance procedures 33
Distributions closed under maxima 413
Domain of (maximal) attraction 12, 318, 319, 322, 323, 327, 441, 442, 448, 494, 528
Donsker's theorem 617, 635
Double exponential (*see* Laplace) distribution
Double sampling 488, 500
Double Weibull distribution 172
Doubly stochastic matrix 17
Doubly truncated Burr distribution 209
Doubly truncated Cauchy distribution 301
Doubly truncated exponential distribution 188, 190, 404
Doubly truncated Laplace distribution 212, 213, 216
Doubly truncated linear-exponential distribution 201
Doubly truncated logistic distribution 180, 183, 184
Doubly truncated log-logistic distribution 205
Doubly truncated parabolic distribution 211
Doubly truncated Pareto distribution 194, 404, 413, 414, 421
Doubly truncated power function distribution 191, 192
Doubly truncated skewed distribution 211
Duality result 9, 21, 46
Duncan's multiple range test 37
Duration of human life 39

Econometrics 25
Edge set 330
Edgeworth approximate inference 6
Edgeworth-type expansion 33, 474–476, 479
Efficient estimation 336

Ellipsoid 400
Elliptically contoured distributions 47
Empirical distribution function 18, 25, 43, 327, 335, 336, 339, 342, 344, 349, 354, 362, 448, 454, 508, 509, 577, 579, 581, 648, 661
Empirical mean residual life 447
Empirical process 362, 380, 574–576, 579, 580, 582, 584, 585, 590, 591, 599, 605, 614, 625
Empirical process of normalized ranks 578
Empirical process of ranks 574
Empirical process of sequential ranks 605
Empirical quantile function 578
Empirical quantiles 648
Employment problems of a professional couple 504
Energy levels in nuclear spectra 42
Engineering applications 40
Environmental issues 441
Error mean square 392
Estimable parameter 338
Estimation theory 253
Euclidean k-space 388
Euler's constant 524, 536
Exchangeability 232, 235, 248
Exchangeable variables 46, 111, 151, 261, 291, 330, 340, 348, 490, 543
Expected residual life 320
Exponential distribution 8, 13, 14, 16, 26, 31–33, 38, 43, 45, 46, 67, 70, 77–80, 82, 86, 91, 92, 95–97, 99, 101, 102, 128, 130, 134, 152, 153, 188, 201, 216–221, 231, 233, 234, 236–242, 249, 251, 253, 258–272, 274, 275, 277–279, 282, 284, 295–299, 302, 304, 306, 307, 321–325, 327, 337, 338, 369, 403, 413, 442–447, 454, 458, 460, 476, 494, 516, 518–520, 527, 531, 534, 538, 545, 556, 559, 561, 644
Exponential family 339, 367
Exponential spacings 66
Exponential tilting 476
Exponential trend 547
Exposure to radioactivity 441
Extended Bahadur-Kiefer process 682
Extended normal distribution 307
Extended Wiener sheet 666, 668, 671, 673
Extinction times for bacteria 39
Extrapolation-type bounds 531
Extremal-F process 557, 560
Extremal processes 516, 560
Extreme concomitants 495
Extreme order statistics 4, 17
Extreme quantiles 443, 445
Extreme value theory 4, 39, 261, 315–317, 319, 320, 323, 325, 329, 644

Extreme value distribution 8, 12, 26, 317, 318, 325, 444, 449, 458, 493, 494, 499, 553, 561
Extreme values 337

F distribution 388, 396–398, 400, 401
F^α-scheme 548, 556
F-scheme 548–551, 553, 554, 557
F test 36
Factorial moment generating function 184, 190, 216
Factorial moments 524, 535
Failure rate 89, 101, 259, 260, 264, 267, 320
Fatigue failure 315
File-matching procedures 488, 510
Financial asset pricing models 488
Finite intersection test 42
First-order asymptotic distributional representation (FOADR) 347, 348
Floods and droughts 39, 515
Folded distribution 151, 171, 172
Fourier inversion formula 477
Fractiles 463
Fréchet differentiability 345
Fréchet distribution 12, 39
Functional jackknifing 336, 345, 347

G-dependence 329
g-ordering 488
Galton's problem 39
Gamma distribution 26, 31, 32, 38, 70, 184–186, 278, 321, 442, 492
Gamma function 194, 205
Gauss-Markov theorem 30, 149
Gaussian function 342
Gaussian process 356, 575, 584, 589, 591, 592, 594, 598, 599, 601, 602, 610, 616–622, 624–626, 632, 635, 664, 666, 671
General linear models 351
Generalized beta function 153
Generalized distance 387
Generalized extreme value distributions 442, 532, 543, 547
Generalized gamma distribution 186, 187
Generalized half logistic distribution 32, 183
Generalized inverse 351, 355, 401
Generalized L-estimators 363
Generalized least squares 343
Generalized logistic distribution 32
Generalized order statistics 20, 258, 293, 298, 304, 488
Generalized Pareto distributions 442, 447, 456, 459, 532
Generating function approach 21

Generating functions 521
Generating sequences 466
Genetic selection problems 506, 507
Geometric distribution 26, 231, 234, 236–242, 249, 251, 253, 259, 264, 280, 282–285, 516, 529, 548, 559
Gini correlation 497
Gini statistic 446, 454, 458, 460
Gini's mean difference 497
Glivenko-Cantelli theorem 640
Glivenko-Cantelli type convergence 363
Gompertz 442
Goodness-of-fit tests 18, 25, 43, 45, 150, 261, 272, 274
Graph dependent models 330
Greatest convex function 106, 120
Greatest convex minorant principle 531
Grouping method 41, 70
Gumbel distributions 441, 442, 453, 532, 547
Gumbel probability paper 327
Gumbel's bivariate exponential distribution 490, 494
Gust loads 39

Hadamard derivative 345
Hadamard differentiability 345, 347, 348, 350, 361
Hàjek's projection lemma 471
Half logistic distribution 32, 178, 179, 184
Half normal (chi) distribution 173
Harmonic New Better than Used in Expectation (HNBUE) 368
Hartley-David-Gumbel bounds 33
Hazard function 321, 354, 355, 365, 366, 527
Hazard measure 558
Hazard rate 89, 90, 95, 96, 101, 102, 320
Hazard rate ordering 90, 91, 94, 95, 98, 101, 102
High level exceedances 442
High leverage 387
Hilbert spaces 126
Hoeffding-decomposition 363
Hoeffding's inequality 471
Hölder's inequality 33, 112, 113, 119, 123, 303
Homogeneous Poisson process 545, 560
Homoscedasticity 336
Hybrid censoring 15
Hydrology 25, 40, 327, 328, 490
Hypergeometric function 192, 197
Hypothesis of randomness 588

Idempotent matrix 390
Identity 8, 32, 149–228, 291

Ignatov's theorem 540, 560
Imputational-type bounds 33
Incomplete beta function 464
Incomplete beta ratio 154, 208
Increasing binary functions 94
Increasing Failure Rate (IFR) 26, 91, 92, 131, 132, 136, 137, 140, 200, 260, 261, 264, 265, 267, 282, 304, 367, 561
Increasing Failure Rate on Average (IFRA) 26, 92, 367
Increasing Mean Residual Life (IMRL) 260, 261, 267
Indicator function 341, 509
Indicator method 21, 46
Indifference-zone approach 38
Induced linear order statistics 357
Induced order statistics 47, 336, 340, 353, 354, 487
Induced selection differential 488, 506, 507
Induced sequential uniform empirical distribution function 641
Induced sequential uniform empirical process 642
Induced sequential uniform empirical quantile function 641
Inference 253
Infinitely often 375
Influential 387, 397
Inliers 388, 416, 421
Insurance mathematics 559
Integrated Cauchy functional equation 231, 259, 262, 283
Integrated regression function 506
Inter-accident times 458
Inter-record times 516, 517, 525, 526, 533, 545, 546, 552, 559
Interval estimation 31
Invariance principle 672
Inverse distribution function 75, 77
Inverse multinomial distribution 185
Inverse normal distribution 476
Inverse transformation method 65
Inversion method 41, 70
ISD test 37
Iterated Kiefer process 680

J-shaped distributions 131, 132, 135, 137, 139
Jackknifed variance estimator 346, 348
Jackknifing 336, 345, 347, 348, 362, 365, 368
Jensen's inequality 105, 135, 267, 282, 304
Joint survival function 495

k-dimensional extremal distribution 442, 443

k-out of-K multicomponent system 364
K-component system-in parallel 361, 364
K-component system-in series 361
K-record times 541
K-record values 541
k_n-records 293, 519, 541, 543, 546
k_n-record times 541–543, 551
k^{th} inter-record times 537, 538
k^{th} record values
k^{th} lower record times 517
k^{th} lower record values 517
k^{th} order rank weighted average 348
k^{th} record times 516, 517, 524, 534–537, 539, 540, 541, 543, 544, 551
Kernel 338, 344, 364, 474, 482
Kernel density estimator 475
Kernel function 466
Kernel method 466
Kernel quantile estimators 466, 467, 470, 472, 474, 482
Kernel smoothing method 353
Kiefer process 575, 578, 584–591, 593, 594, 596, 597, 600, 602–604, 606–609, 611, 612, 615, 618, 619, 622–627, 632–634, 636–638, 646–648, 655–660, 664, 666, 668–671, 673–675, 678, 680, 681, 684
Kolmogorov-Smirnov test 43, 44
– δ-corrected 44
Kolmogorov-Smirnov type statistics 342, 577, 581
Kuiper test 43
Kurtosis 132

ℓ_1-norm symmetric distributions 261
ℓ_2-projection 106
L-estimates 105, 106, 111, 112, 114, 116–118, 122, 126–129, 131, 133–135, 138, 140, 142, 344, 350, 360, 363, 465, 466, 467, 480
L-functionals 345, 348, 349, 354, 358, 361
L-statistic 336, 338, 339, 344, 348, 358, 368
L_1-functional 578, 594, 598
L_p approximation 593
L_p functional 583, 589, 593–595, 654, 656
L_p-metric 574, 578, 579, 582, 585, 606, 609, 622, 653, 659
L_p-norm 577, 605, 655
Lack of memory 78
Lagrange multipliers 667
Laplace transform 260, 556
Large deviation rate function 475
Large deviation theory 479
Lau-Rao theorem 235
Law of large numbers 525, 528, 636

Law of the iterated logarithm   525, 528, 607, 623, 634, 662, 663, 669, 674, 675, 678, 680
Le Cam's third lemma   619, 620
Least significant difference (LSD) test   36, 37
Least squares   25, 28, 30, 41, 149, 328, 392, 393
Least-squares estimator   387
Lebesgue measure   232, 234, 642
Left tail increasing   94
Left-truncated Pareto distribution   415
Legendre orthonormal basis   122
Legendre-Fenchel transformation   449
Leverage   395
Life distributions   325
Life-testing experiments   14, 15, 30, 150, 337, 338, 355, 361
Life-time distributions   3
Lightly trimmed mean   168
Likelihood ratio ordering   89, 91, 95, 96, 98, 99, 102
Likelihood function   16
Linear discriminant analysis   42
Linear-exponential distribution   32, 200
Linear estimators   149, 499
Linear functional   344, 369, 448
Linear invariant estimation   31
Linear models   350, 360, 387, 401, 500
Linear programming   28, 107, 350
Linear regression model   488, 505, 508
Linear trend   547, 548
Linkage   510
Lipschitz condition   472
Location parameter   30, 105, 112, 115, 149, 179, 295, 404, 416, 419, 426, 441, 449, 499, 518
Location-scale family   13, 336, 343, 350
Location-scale transformation   135
Log concave density   95, 97
Log-concavity   7, 21, 46, 91, 92
Log-gamma function   451
Log-logistic distribution   32, 153, 204, 205, 208, 210, 218
Log-Weibull distribution   31
Logarithmic moments   524, 526, 536
Logarithmic distributions   300
Logistic distribution   8, 26, 32, 138, 152, 153, 177, 216–218, 268, 269, 272, 275, 277–279, 297–301, 404, 442, 532
Loglikelihood function   325
Lognormal distribution   26, 29, 152, 153, 172, 276, 279
Lomax distribution   153, 203, 210, 216–218, 532
Lorenz ordering   75, 76, 78, 79–83, 85, 86
Lower record times   517

Lower records   19, 77, 78, 81, 82, 493–495, 517, 559
Lower bounds   109
Lower-case probability   375, 376, 379, 382

m-component mixed random variable   215
M-estimates   18, 448, 464, 465, 479
M-statistics   279
Major flood   441
Majorization   100, 101
Maple program   11
Markov chain   7, 26, 522, 529, 533, 535, 539, 542, 545, 546, 550
Markov process   47
Markov property   522, 529, 544
Markov sequences   534, 544, 545
Markov structure   522, 527
Markov times   551
Markovian relationship   445
Martingale   355, 492, 524, 536, 542
Martingale argument   380
Martingale properties   551
Martingale sequences   551
Maximal deviation   508
Maximal order statistics   529
Maximum likelihood estimates   16, 25, 28–31, 404, 415, 416, 426, 447, 547
Maximum sequence   375
Mean range   39
Mean absolute deviation from median   29
Mean remaining life function   236
Mean residual life   248, 365, 367
Mean square error   18, 345, 354, 362, 416–419, 421–426, 460
Mean value theorem   471, 642
Measures of central tendency   27, 28, 348, 358
Measures of dispersion   27, 28
Median   27, 29, 79, 150, 155, 203, 204, 279, 348, 463, 468
Median filter   45
Meteorology   25, 39
Method of induction   21
Midrange   27, 29, 279
Minimal sufficient statistic   339
Minimax method   28
Minimum mean-square-error estimators   31
Minimum variance unbiased estimator   416, 443
Mismatches   510
Mixed-effects models   336, 357, 359
Mixed-rank statistics   352, 357, 487
Mixture   86, 122, 151, 543
Mixture distribution   100, 130, 134

Mixture of two exponential distributions 153, 212
Mixture of normal distributions 554
Mixture of F^α-schemes 549, 554
Model building 253
Modified geometric type distributions 285
Modified signed log likelihood ratio 476
Modified uniform quantile function 653
Modified uniform quantile process 653, 657
Moment generating function 184, 190, 445
Monitoring sequences 380
Monotone failure probability 105
Monotone failure rate 105, 329
Monotone likelihood ratio property 91, 100
Monotonic stochastic process 248
Monotonicity hypotheses 380
Monte Carlo approximation 454
Monte Carlo simulation 65, 443, 448
Moving medians 45
Moving averages 45
Moving maxima 45
Moving minima 45
Moving order statistics 45, 46
Multi-time parameter empirical bridge process 623
Multidimensional concomitant model 507
Multinomial distribution 26, 70
Multiple analysis of variance 42
Multiple comparison 25, 36, 37
Multiple decision procedure 38
Multiple factor analysis 41
Multiple outlier model 18, 404, 412, 413, 415, 416, 421, 426
Multiple range procedure 38
Multiple shape outliers 416, 417, 419, 422, 424
Multiplicative model 492
Multivariate beta distribution 370
Multivariate central limit theorem 662
Multivariate exponential distributions 261
Multivariate normal distribution 42, 45, 47, 341, 342, 351, 490, 491, 507
Multivariate order statistics 47
Multivariate records 558
Multivariate tolerance regions 35
Multivariate total positivity of order 2 (MTP_2) 492
Müntz's theorem 266
Müntz-Szász lemma 294, 299, 306

Nearest neighbor (NN) method 353
Nearly best linear unbiased estimation 31, 149
Negative binomial distribution 26, 541
Negative binomial form 4

Negative Pareto distribution 136
New Better than Used (NBU) 27, 93, 260, 264, 267, 269, 275, 282, 304, 366
New Worse than Used (NWU) 27, 260, 264, 267, 269, 270, 275, 282, 304, 366, 368
New Better than Renewal Used in Expectation (NBRUE) 367-369
New Better than Used in Expectation (NBUE) 367, 368
New Worse than Renewal Used in Expectation (NWRUE) 367
New Worse than Used in Expectation (NWUE) 367
Newman-Keuls test 37
Neyman-Pearson theory 326
Non-lattice distribution 278
Non-linear regression 28
Non-overlapping mixture model 214
Noncentral distributions 389
Nonclassical records 517, 534
Nonhomogeneous Poisson process 545, 560
Nonparametric bootstrap 443, 444
Nonparametric estimator 354
Nonparametric procedures 576
Nonparametric regression function 353
Nonparametric regression 336, 358
Nonparametric statistics 25, 149, 336, 644
Nonstationary records 545
Normal distribution 17, 26, 28, 30–32, 37–39, 41, 43–45, 99, 137, 138, 151–153, 155, 173, 174, 272–274, 292, 297, 305–307, 321–323, 325, 348, 354, 362, 388, 394, 442, 447, 449, 453–459, 467, 468, 470, 472–474, 479, 489, 493, 495, 498, 503, 505, 506, 509, 510, 518, 524, 525, 532, 537, 543, 547, 554, 555, 595, 600, 661, 672, 673
Normalized extremes 329
Normalized moments 151
Normalized ranks 573, 574, 599
Normalized sequential ranks 573, 574, 604, 606, 607, 609
Normalized spacings 16, 337, 366, 442, 444–446
Nuisance functionals 336
Numerical error propagation 151

O-statistics 467
Olympic games 548
Operator method 21
Optimum quantiles 31
Ordered roots 42
Orthogonal designs 397
Orthogonal inverse expansion 33, 530

Orthogonal matrices 388, 390
Orthogonal transformation 273
Outliers 4, 17, 18, 25, 28, 29, 39, 41, 46, 150, 168, 316, 325, 343, 387, 388, 395–398, 401, 413, 416, 421
Outlier tests 44

p-norm bounds 33, 124
Paired t-test 488
Parabolic distribution 153
Parallel system 329
Pareto distribution 8, 13, 14, 18, 26, 32, 79, 82, 86, 136, 152, 153, 193, 194, 197, 198, 217–221, 259, 268–270, 297–301, 307, 403, 404, 406, 411-413, 415, 416, 421, 426, 442
Partial likelihood function 355, 356
Partial likelihood score test statistics 356
Partial sum processes 577, 578, 619, 664
Partial sums 353
Partitioning method 69
Penalty function 510
Percentile 442, 509
Percentile estimator 44
Percentile points 349
Permanent 20, 21, 46, 403
Permutation 116, 561
Perturbed sample quantile 472, 474
Pfeifer's records 293, 546
Pinned Brownian sheet 591
Pivot 358
Plotting positions 40, 41
Point process 47
Point prediction 25, 35
Poisson distribution 26, 284
Poisson process 251–253, 274, 537, 545
Pólya frequency function of order 2 (PF_2) 492
Positive dependence 93, 97
Positively quadrant dependent 94, 492
Positively regression dependent 94
Power function distribution 8, 13, 14, 32, 79, 82, 84, 86, 123, 135, 152, 153, 190, 216–218, 259, 268–271, 298, 300–302, 304, 307, 404
Prediction intervals 25, 35, 36
Prediction of records 559
Principal components 41, 42
Probabilistic approach 21
Probability generating function 521, 522
Probability paper 25, 31, 40, 44, 327, 328
Product-limit (PL) estimator 349, 368
Progressive censoring 15, 16, 356
Progressive Type-I censoring 16
Progressive Type-II censoring 16, 41, 68, 69

Projection method 126, 127
Proportional hazard model (PHM) 354
Pseudo-samples 444
Pseudovariables 354
Public health issues 441

Q-Q plots 18, 41
Quadratic forms 395, 398
Quadratic trend 547
Quality control 25, 36, 39
Quantile density function 642
Quantile estimation 443
Quantile function 75, 106, 113, 116, 125, 127, 338, 341, 358, 463, 639, 643
Quantile process 575, 575, 582, 631, 638, 647, 648, 653, 655, 656, 659, 661, 663, 675
Quantile transformation 470
Quantiles 33, 91, 97, 105, 106, 126, 127, 135, 136, 326, 337, 339, 341, 342, 344, 347, 350, 442, 463–469, 472, 475, 478–480, 482, 495, 505, 508, 636, 662
Quartile deviation 27
Quasi-median 27
Quasi-midrange 27
Quasi-parametric model 359
Quasi-range 33, 124, 125, 161
Quasidiagonal forms 390
Queuing theory 253

R-estimators 352
r-ordering 134
Radioactive decay 39
Rainfall 515
Random censoring 15, 349, 350, 368
Random records 545
Random trees 559
Randomly disturbed Wiener sheet 674
Randomly perturbed Kiefer process 664
Range 25, 27, 33, 36, 39, 113, 119, 120, 124, 125, 161, 203, 204, 273, 504
Rank order statistics 25
Rank sequence 376, 382, 383
Rank-weighted mean 348
Ranked-set sampling 488, 499, 500
Ranking and selection 25, 38, 39
Ranks 339, 340, 375, 376, 553, 579, 599, 605, 655
Rao-Shanbhag result 249
Rate of cancer 441
Rayleigh distribution 199–201, 218–221, 532
Reach statistic 174
Record indicators 532, 537, 543, 548, 549, 554, 556, 560

Record spacings 519
Record times 249, 516, 518, 520–525, 532–534, 537, 544–546, 550–552, 554, 557, 560
Record values 4, 20, 45, 75, 77, 81, 86, 231, 248, 249, 251, 293, 302, 304, 488, 516–520, 523, 525, 527–529, 531, 533, 534, 538–540, 543, 54, 546, 550, 553, 557, 559–561
Records 11, 19, 47, 86, 326, 515–561
Records in sequences with trend 547, 553
Recurrence relations 4, 9, 21, 32, 46, 149–228, 257, 291, 403, 412, 531
Reflected Weibull distribution 276, 300
Reflexive symmetric g-inverse 399
Regression analysis 42
Regression coefficients 25, 28, 488, 496
Regression diagnostics 387
Regression-equivariance property 350
Regression function 488, 506, 508
Regression functionals 339, 341, 359
Regression quantiles 336, 350–352, 360
Regression rank scores estimators 352
Regression-scale family 336
Regular functional 338
Regular variation 322
Regularly varying function 552, 553
Reliability 30, 89, 90, 253, 335–338, 361, 362
Reliability function 363, 366
Reliability properties 546
Renewal process 252, 545
Repair times 364
Repeated significance testing 356
Response to selection 507
Restricted family of distributions 105, 106
Reverse martingale 346
Reverse sub-martingale property 362
Riemann integration 320
Riemann zeta functions 176
Right corner set increasing (RCSI) 93
Right tail increasing 94, 99
Right truncated exponential distribution 132, 190, 403
Right-truncated Pareto distribution 415
Robust estimates 18, 29, 45, 46, 150, 336, 345, 350
Robust inference 17, 18, 41, 150, 343, 347, 348, 351, 403, 404, 415, 416, 419, 644
Rounding error 161, 190
σ-algebra 536, 542, 551
σ-field 235

s-comparisons 133, 138, 531
s-ordering 132–134
Saddlepoint 475

Saddlepoint approximation 444, 448, 449, 451, 454, 455, 458, 467, 475, 476, 478
Saddlepoint density 476
Saddlepoint expansion 477, 478
Saddlepoint methods 443, 479
Sample differential 118
Sampling with replacement 444, 480
Sampling without replacement 94, 100, 116, 338, 467
Samuelson inequality 117
Scale parameter 30, 31, 105, 112, 149, 179, 208, 404, 415, 416, 419, 426, 441, 449, 453, 499, 518
Scale-parameter family 13
Scaled Kiefer process 675
Scaled F distribution 400
Score function 344, 347, 644
Scores 355, 467
Searching algorithms 554, 559
Secretary problem 553, 554, 559
Seismology 25
Selection differentials 118, 119, 125, 138
Selection procedures 487, 488, 507
Semi-interquartile range 27
Semiparametric inference 336, 354, 442, 444
Semirange 29
Separable Gaussian process 591, 596–598, 624, 627
Sequential Bahadur-Kiefer process 682
Sequential empirical process 632, 638, 639, 655
Sequential empirical distribution function 639
Sequential empirical quantile function 639
Sequential order statistics 293, 639
Sequential procedure 576, 605
Sequential process 633
Sequential quantile process 631, 640, 642, 543, 645, 648, 657, 658, 660
Sequential ranks 518, 535, 574, 579, 604, 605, 607, 609, 611, 614, 655
Sequential testing 356
Sequential uniform Bahadur-Kiefer principle 638
Sequential uniform Bahadur-Kiefer process 637, 638, 684
Sequential uniform empirical distribution function 631
Sequential uniform empirical process 631, 632, 657
Sequential uniform empirical quantile function 631
Sequential uniform order statistics 631, 641

Sequential uniform quantile process 633, 636, 640, 642
Serial correlation 42
Series system 329
Shanbhag's lemma 236, 283
Shape parameter 329, 415, 416, 453
Shapiro-Wilk's test 18
Sharp bounds 105, 119, 121, 126, 127, 531
Shifted exponential distribution 243, 244, 250
Shifted geometric distribution 243, 245
Shifts 91
Siddiqui-Bloch-Gastwirth estimator 475
Significance level 443, 447, 454
Simple random sampling 26
Simplified linear estimator 498
Simplified MLEs 499
Simulation algorithms 17, 41, 65–72, 441
Singular decomposition 390
Singular Gaussian distribution 394
Size biased selection 500
Skewed distribution 153
Skewness 127
Skorohod embedding 632
Skorohod-Dudley-Wichura theorem 589
Skorohod space 585, 656
Slippage 412, 421
Slowly varying function 322
Small sample asymptotic 478
Smirnov's lemma 467–471, 473, 474, 480
Smirnov's transformation 519, 554
Smooth L-functionals 361
Smooth bootstrap approximation 482
Smooth quantile estimator 344
Smoothed bootstrap 481, 482
Snedecor-Fisher distribution 387
Snowfall 515
Sorting 66
Spacings 11, 25, 44, 70, 258, 259, 261, 265, 270, 271, 275, 278, 295, 298, 337, 338, 442, 445, 446, 454, 458
Span 232, 234, 237, 239, 245, 250
Speeds of groups of vehicles 559
Sports events 515, 553
Sports records 559
SSD test 37
Stable distributions 45
Standardized selection differential 153, 174
Star-ordering 76, 81, 92
Star-relation 134
Star-shaped 76, 93
Starshaped ordering 89, 134
Stationary m-dependent sequence 544
Stationary Gaussian sequences 543

Stationary, strong mixing sequence 548
Statistical modeling 362
Step-down procedure 42
Stirling number of the first kind 521
Stirling's formula 383, 479
Stochastic comparisons 103
Stochastic orderings, 7, 27, 89–91, 93, 94, 98, 99, 101, 107, 120, 561
Stochastic predictors 353
Stochastic process 338, 358, 585, 591, 593, 54, 597, 599, 600, 602, 609, 610, 613, 637, 663 655, 666, 674, 685
Stochastically increasing 94
Stock market 40
Strassen's theorem 79–82
Stratified random sampling 26, 498
Strength of materials 25, 329
Strong approximation 622, 635, 637
Strong invariance principle 363, 638, 661–664, 674, 679, 682, 685
Strong law of large numbers 548, 580
Strong limit theorems 525, 526
Strong memoryless property 231, 234, 237, 239, 241, 253
Strong records 529
Strongly unimodal 92
Structural design 150
Structural parameters 388
Studentized range 25, 36, 37
Studentized selection differential 153, 174
Sub-additivity 270, 271
Sub-diagonal product moments 152
Sub-sigma fields 3346
Subset-selection approach 38
Substochastic density 141
Successive maxima 557
Sufficient statistic 339
Sup-norm asymptotics 593
Sup-norm functional 369, 586, 656
Sup-norm test statistic 369, 370
Super-additivity 270, 271
Superadditive 134
Superadditive ordering 89, 92
Supremum functional 575, 589, 594, 656
Supremum metric 622
Supremum norm 583, 605, 648, 653, 659
Survival analysis 89, 90, 335, 336, 354–356
Survival function 90, 98, 247, 261, 267, 282, 349, 357
Symmetric U-shaped distribution 138, 141
Symmetric distribution 105
Symmetric group of permutations 559
Symmetric Luxeburg norms 118

Symmetric norms   117
Symmetric unimodal distribution   105
System availability   336, 361

t distribution   26
t test   36, 37
Taylor series   10, 11, 323, 345, 452, 649, 651
Temperature   515
Tests of hypotheses   3, 149, 488
Threshold method   327
Tied down multi-time parameter empirical process   623
Tied-down Wiener process   342, 362
Tilted Edgeworth expansion   449
Time-sequential tests   356
Tolerance intervals   25, 34, 35
Tolerance limits   33, 34
Total time on testing (TTT)   336, 337, 365, 366, 370
Totally positive or order 2   91, 93, 97–100
Trace-efficient estimators   14
Transition density   544
Translation-invariant functional   359
Triangle rule   9, 21
Triangular array   257, 293
Triangular inequality   647
Trigamma function   453
Trimmed L-estimators   349
Trimmed least squares   336
Trimmed LSE   350, 351, 360
Trimmed means   18, 29, 108, 113, 149, 168, 348, 351
Trimmed procedures   29
Trimmed U-functions   509
Truncated Burr distribution   209
Truncated Cauchy distribution   175
Truncated forms   32, 153, 336
Truncated normal distribution   307
Truncated Pareto distribution   403, 413, 414
Truncated sequential quantile process   647
Truncation point   348
TTT-asymptotics   337
TTT transformation   337
Tukey's studentized range test   37
Two-sided censoring   499
Type-I censoring   15, 348, 356, 508
Type I Gumbel family   441, 447, 448, 451
Type-I right censoring   344, 348
Type II censored bivariate sample   508
Type-II censored samples   3, 12, 15, 30, 31, 356, 498
Type II Gumbel family   442, 448

Type-II right censoring   16, 344, 349

U-shaped distribution   133, 134
U-statistic   339, 348, 363, 364, 467, 506, 509
U-statistic type functions of degree 2   509
Unbiased and strongly consistent estimator   580
Unbiased estimator   335
Uniform Bahadur-Kiefer process   635, 637, 671, 675, 679
Uniform distribution   8, 10, 13, 14, 17, 26, 29, 30, 39, 41, 45, 65–69, 77, 78, 84, 86, 91, 97, 107, 110–112, 123, 128, 130, 133, 138, 248, 252, 253, 258, 269–272, 275, 278–280, 284, 295, 297, 299, 302, 303, 306, 339, 343, 382, 464, 468, 470, 492, 510, 559, 574, 581, 583, 584, 607, 617, 636, 641, 653, 662, 664, 671
Uniform empirical distributional process   342
Uniform empirical process   582, 655, 656, 661, 675
Uniform empirical quantile process   343
Uniform-Gaussian weighted approximations   579
Uniform quantile process   583, 648, 653, 654
Uniform stochastic order in the positive direction   90
Union-intersection principle   42
Universal bound   10, 32
Upper bounds   109
Upper-case probability   375, 377
Upper-diagonal product moments   162
Upper record times   517
Upper records   18, 77, 78, 81, 82, 84, 85, 493, 517, 530
Urn sampling   155

Variable ranks   382
Variance-covariance matrix   17, 151, 153, 173, 343
Variational methods   119
von Mises conditions   494

Waiting time problems   156
Warranty period   316
Watson test   43
Weak k^{th} record values   540, 541
Weak convergence   507, 615, 616, 618, 619, 627, 643, 647, 648, 656, 658
Weak record times   529
Weak records   516, 517, 529, 530
Weakest-link theory   40
Weibull distribution   8, 12, 26, 31, 40, 83, 86, 92, 190, 218–221, 247, 259, 268–270, 272,

274, 276, 277, 300, 321, 328, 329, 442, 453–457, 459, 532
Weibull-Exponential distribution 210
Weibull-Gamma distribution 210
Weight functions 578, 587, 592, 593, 599, 600, 605, 610, 611, 617, 623, 625, 627, 654, 656
Weighted L_1-convergence in distribution 599
Weighted L_p approximation 625
Weighted L_p functional 575, 657
Weighted L_p-norm 595
Weighted asymptotic 577, 605
Weighted empirical process 583, 599, 604, 615
Weighted least squares 361
Weighted multi-time parameter empirical process 621
Weighted regression 42
Weighted sequential empirical type processes 573, 576
Weighted sequential quantile process 653, 659
Weighted sequential uniform quantile process 655

Weighted supremum 574, 579, 606, 609
Weighted supremum norm 595, 655
Weighted uniform empirical process 574, 653
Weighted weak convergence 577, 582, 654, 656
Wiener-Hopf technique 283
Wiener process 537, 538, 560, 582, 587, 588, 617, 618, 664, 666, 668, 672, 673
Wiener sheet 667, 669–672, 674, 675, 678, 680, 681, 684
Williams' representation 524
Window-width 466
Winsorized means 18, 108, 149, 348
Winsorized procedures 29
WSD test 37

Yang's model 548
Yule process 251, 252

Zero-one laws 375, 376, 378–380, 382
Zonal polynomials 42

Handbook of Statistics
Contents of Previous Volumes

Volume 1. Analysis of Variance
Edited by P. R. Krishnaiah
1980 xviii + 1002 pp.

1. Estimation of Variance Components by C. R. Rao and J. Kleffe
2. Multivariate Analysis of Variance of Repeated Measurements by N. H. Timm
3. Growth Curve Analysis by S. Geisser
4. Bayesian Inference in MANOVA by S. J. Press
5. Graphical Methods for Internal Comparisons in ANOVA and MANOVA by R. Gnanadesikan
6. Monotonicity and Unbiasedness Properties of ANOVA and MANOVA Tests by S. Das Gupta
7. Robustness of ANOVA and MANOVA Test Procedures by P. K. Ito
8. Analysis of Variance and Problem under Time Series Models by D. R. Brillinger
9. Tests of Univariate and Multivariate Normality by K. V. Mardia
10. Transformations to Normality by G. Kaskey, B. Kolman, P. R. Krishnaiah and L. Steinberg
11. ANOVA and MANOVA: Models for Categorical Data by V. P. Bhapkar
12. Inference and the Structural Model for ANOVA and MANOVA by D. A. S. Fraser
13. Inference Based on Conditionally Specified ANOVA Models Incorporating Preliminary Testing by T. A. Bancroft and C.-P. Han
14. Quadratic Forms in Normal Variables by C. G. Khatri
15. Generalized Inverse of Matrices and Applications to Linear Models by S. K. Mitra
16. Likelihood Ratio Tests for Mean Vectors and Covariance Matrices by P. R. Krishnaiah and J. C. Lee

17. Assessing Dimensionality in Multivariate Regression by A. J. Izenman
18. Parameter Estimation in Nonlinear Regression Models by H. Bunke
19. Early History of Multiple Comparison Tests by H. L. Harter
20. Representations of Simultaneous Pairwise Comparisons by A. R. Sampson
21. Simultaneous Test Procedures for Mean Vectors and Covariance Matrices by P. R. Krishnaiah, G. S. Mudholkar and P. Subbiah
22. Nonparametric Simultaneous Inference for Some MANOVA Models by P. K. Sen
23. Comparison of Some Computer Programs for Univariate and Multivariate Analysis of Variance by R. D. Bock and D. Brandt
24. Computations of Some Multivariate Distributions by P. R. Krishnaiah
25. Inference on the Structure of Interaction in Two-Way Classification Model by P. R. Krishnaiah and M. Yochmowitz

Volume 2. Classification, Pattern Recognition and Reduction of Dimensionality
Edited by P. R. Krishnaiah and L. N. Kanal
1982 xxii + 903 pp.

1. Discriminant Analysis for Time Series by R. H. Shumway
2. Optimum Rules for Classification into Two Multivariate Normal Populations with the Same Covariance Matrix by S. Das Gupta
3. Large Sample Approximations and Asymptotic Expansions of Classification Statistics by M. Siotani
4. Bayesian Discrimination by S. Geisser
5. Classification of Growth Curves by J. C. Lee
6. Nonparametric Classification by J. D. Broffitt
7. Logistic Discrimination by J. A. Anderson
8. Nearest Neighbor Methods in Discrimination by L. Devroye and T. J. Wagner
9. The Classification and Mixture Maximum Likelihood Approaches to Cluster Analysis by G. J. McLachlan
10. Graphical Techniques for Multivariate Data and for Clustering by J. M. Chambers and B. Kleiner
11. Cluster Analysis Software by R. K. Blashfield, M. S. Aldenderfer and L. C. Morey
12. Single-link Clustering Algorithms by F. J. Rohlf
13. Theory of Multidimensional Scaling by J. de Leeuw and W. Heiser
14. Multidimensional Scaling and its Application by M. Wish and J. D. Carroll
15. Intrinsic Dimensionality Extraction by K. Fukunaga

16. Structural Methods in Image Analysis and Recognition by L. N. Kanal, B. A. Lambird and D. Lavine
17. Image Models by N. Ahuja and A. Rosenfeld
18. Image Texture Survey by R. M. Haralick
19. Applications of Stochastic Languages by K. S. Fu
20. A Unifying Viewpoint on Pattern Recognition by J. C. Simon, E. Backer and J. Sallentin
21. Logical Functions in the Problems of Empirical Prediction by G. S. Lbov
22. Inference and Data Tables and Missing Values by N. G. Zagoruiko and V. N. Yolkina
23. Recognition of Electrocardiographic Patterns by J. H. van Bemmel
24. Waveform Parsing Systems by G. C. Stockman
25. Continuous Speech Recognition: Statistical Methods by F. Jelinek, R. L. Mercer and L. R. Bahl
26. Applications of Pattern Recognition in Radar by A. A. Grometstein and W. H. Schoendorf
27. White Blood Cell Recognition by E. S. Gelsema and G. H. Landweerd
28. Pattern Recognition Techniques for Remote Sensing Applications by P. H. Swain
29. Optical Character Recognition – Theory and Practice by G. Nagy
30. Computer and Statistical Considerations for Oil Spill Identification by Y. T. Chinen and T. J. Killeen
31. Pattern Recognition in Chemistry by B. R. Kowalski and S. Wold
32. Covariance Matrix Representation and Object-Predicate Symmetry by T. Kaminuma, S. Tomita and S. Watanabe
33. Multivariate Morphometrics by R. A. Reyment
34. Multivariate Analysis with Latent Variables by P. M. Bentler and D. G. Weeks
35. Use of Distance Measures, Information Measures and Error Bounds in Feature Evaluation by M. Ben-Bassat
36. Topics in Measurement Selection by J. M. Van Campenhout
37. Selection of Variables Under Univariate Regression Models by P. R. Krishnaiah
38. On the Selection of Variables Under Regression Models Using Krishnaiah's Finite Intersection Tests by J. L Schmidhammer
39. Dimensionality and Sample Size Considerations in Pattern Recognition Practice by A. K. Jain and B. Chandrasekaran
40. Selecting Variables in Discriminant Analysis for Improving upon Classical Procedures by W. Schaafsma
41. Selection of Variables in Discriminant Analysis by P. R. Krishnaiah

Volume 3. Time Series in the Frequency Domain
Edited by D. R. Brillinger and P. R. Krishnaiah
1983 xiv + 485 pp.

1. Wiener Filtering (with emphasis on frequency-domain approaches) by R. J. Bhansali and D. Karavellas
2. The Finite Fourier Transform of a Stationary Process by D. R. Brillinger
3. Seasonal and Calender Adjustment by W. S. Cleveland
4. Optimal Inference in the Frequency Domain by R. B. Davies
5. Applications of Spectral Analysis in Econometrics by C. W. J. Granger and R. Engle
6. Signal Estimation by E. J. Hannan
7. Complex Demodulation: Some Theory and Applications by T. Hasan
8. Estimating the Gain of a Linear Filter from Noisy Data by M. J. Hinich
9. A Spectral Analysis Primer by L. H. Koopmans
10. Robust-Resistant Spectral Analysis by R. D. Martin
11. Autoregressive Spectral Estimation by E. Parzen
12. Threshold Autoregression and Some Frequency-Domain Characteristics by J. Pemberton and H. Tong
13. The Frequency-Domain Approach to the Analysis of Closed-Loop Systems by M. B. Priestley
14. The Bispectral Analysis of Nonlinear Stationary Time Series with Reference to Bilinear Time-Series Models by T. Subba Rao
15. Frequency-Domain Analysis of Multidimensional Time-Series Data by E. A. Robinson
16. Review of Various Approaches to Power Spectrum Estimation by P. M. Robinson
17. Cumulants and Cumulant Spectral Spectra by M. Rosenblatt
18. Replicated Time-Series Regression: An Approach to Signal Estimation and Detection by R. H. Shumway
19. Computer Programming of Spectrum Estimation by T. Thrall
20. Likelihood Ratio Tests on Covariance Matrices and Mean Vectors of Complex Multivariate Normal Populations and their Applications in Time Series by P. R. Krishnaiah, J. C. Lee and T. C. Chang

Volume 4. Nonparametric Methods
Edited by P. R. Krishnaiah and P. K. Sen
1984 xx + 968 pp.

1. Randomization Procedures by C. B. Bell and P. K. Sen
2. Univariate and Multivariate Mutisample Location and Scale Tests by V. P. Bhapkar
3. Hypothesis of Symmetry by M. Hušková
4. Measures of Dependence by K. Joag-Dev
5. Tests of Randomness against Trend or Serial Correlations by G. K. Bhattacharyya
6. Combination of Independent Tests by J. L. Folks
7. Combinatorics by L. Takács
8. Rank Statistics and Limit Theorems by M. Ghosh
9. Asymptotic Comparison of Tests–A Review by K. Singh
10. Nonparametric Methods in Two-Way Layouts by D. Quade
11. Rank Tests in Linear Models by J. N. Adichie
12. On the Use of Rank Tests and Estimates in the Linear Model by J. C. Aubuchon and T. P. Hettmansperger
13. Nonparametric Preliminary Test Inference by A. K. Md. E. Saleh and P. K. Sen
14. Paired Comparisons: Some Basic Procedures and Examples by R. A. Bradley
15. Restricted Alternatives by S. K. Chatterjee
16. Adaptive Methods by M. Hušková
17. Order Statistics by J. Galambos
18. Induced Order Statistics: Theory and Applications by P. K. Bhattacharya
19. Empirical Distribution Function by E. Csáki
20. Invariance Principles for Empirical Processes by M. Csörgo
21. M-, L- and R-estimators by J. Jurečková
22. Nonparametric Sequantial Estimation by P. K. Sen
23. Stochastic Approximation by V. Dupač
24. Density Estimation by P. Révész
25. Censored Data by A. P. Basu
26. Tests for Exponentiality by K. A. Doksum and B. S. Yandell
27. Nonparametric Concepts and Methods in Reliability by M. Hollander and F. Proschan
28. Sequential Nonparametric Tests by U. Müller-Funk
29. Nonparametric Procedures for some Miscellaneous Problems by P. K. Sen
30. Minimum Distance Procedures by R. Beran
31. Nonparametric Methods in Directional Data Analysis by S. R. Jammalamadaka
32. Application of Nonparametric Statistics to Cancer Data by H. S. Wieand

33. Nonparametric Frequentist Proposals for Monitoring Comparative Survival Studies by M. Gail
34. Meterological Applications of Permutation Techniques based on Distance Functions by P. W. Mielke, Jr.
35. Categorical Data Problems Using Information Theoretic Approach by S. Kullback and J. C. Keegel
36. Tables for Order Statistics by P. R. Krishnaiah and P. K. Sen
37. Selected Tables for Nonparametric Statistics by P. K. Sen and P. R. Krishnaiah

Volume 5. Time Series in the Time Domain
Edited by E. J. Hannan, P. R. Krishnaiah and M. M. Rao
1985 xiv + 490 pp.

1. Nonstationary Autoregressive Time Series by W. A. Fuller
2. Non-Linear Time Series Models and Dynamical Systems by T. Ozaki
3. Autoregressive Moving Average Models, Intervention Problems and Outlier Detection in Time Series by G. C. Tiao
4. Robustness in Time Series and Estimating ARMA Models by R. D. Martin and V. J. Yohai
5. Time Series Analysis with Unequally Spaced Data by R. H. Jones
6. Various Model Selection Techniques in Time Series Analysis by R. Shibata
7. Estimation of Parameters in Dynamical Systems by L. Ljung
8. Recursive Identification, Estimation and Control by P. Young
9. General Structure and Parametrization of ARMA and State-Space Systems and its Relation to Statistical Problems by M. Deistler
10. Harmonizable, Cramér, and Karhunen Classes of Processes by M. M. Rao
11. On Non-Stationary Time Series by C. S. K. Bhagavan
12. Harmonizable Filtering and Sampling of Time Series by D. K. Chang
13. Sampling Designs for Time Series by S. Cambanis
14. Measuring Attenuation by M. A. Cameron and P. J. Thomson
15. Speech Recognition Using LPC Distance Measures by P. J. Thomson and P. de Souza
16. Varying Coefficient Regression by D. F. Nicholls and A. R. Pagan
17. Small Samples and Large Equation Systems by H. Theil and D. G. Fiebig

Volume 6. Sampling
Edited by P. R. Krishnaiah and C. R. Rao
1988 xvi + 594 pp.

1. A Brief History of Random Sampling Methods by D. R. Bellhouse
2. A First Course in Survey Sampling by T. Dalenius
3. Optimality of Sampling Strategies by A. Chaudhuri
4. Simple Random Sampling by P. K. Pathak
5. On Single Stage Unequal Probability Sampling by V. P. Godambe and M. E. Thompson
6. Systematic Sampling by D. R. Bellhouse
7. Systematic Sampling with Illustrative Examples by M. N. Murthy and T. J. Rao
8. Sampling in Time by D. A. Binder and M. A. Hidiroglou
9. Bayesian Inference in Finite Populations by W. A. Ericson
10. Inference Based on Data from Complex Sample Designs by G. Nathan
11. Inference for Finite Population Quantiles by J. Sedransk and P. J. Smith
12. Asymptotics in Finite Population Sampling by P. K. Sen
13. The Technique of Replicated or Interpenetrating Samples by J. C. Koop
14. On the Use of Models in Sampling from Finite Populations by I. Thomsen and D. Tesfu
15. The Prediction Approach to Sampling theory by R. M. Royall
16. Sample Survey Analysis: Analysis of Variance and Contingency Tables by D. H. Freeman, Jr.
17. Variance Estimation in Sample Surveys by J. N. K. Rao
18. Ratio and Regression Estimators by P. S. R. S. Rao
19. Role and Use of Composite Sampling and Capture-Recapture Sampling in Ecological Studies by M. T. Boswell, K. P. Burnham and G. P. Patil
20. Data-based Sampling and Model-based Estimation for Environmental Resources by G. P. Patil, G. J. Babu, R. c. Hennemuth, W. L. Meyers, M. B. Rajarshi and C. Taillie
21. On Transect Sampling to Assess Wildlife Populations and Marine Resources by F. L. Ramsey, C. E. Gates, G. P. Patil and C. Taillie
22. A Review of Current Survey Sampling Methods in Marketing Research (Telephone, Mall Intercept and Panel Surveys) by R. Velu and G. M. Naidu
23. Observational Errors in Behavioural Traits of Man and their Implications for Genetics by P. V. Sukhatme
24. Designs in Survey Sampling Avoiding Contiguous Units by A. S. Hedayat, C. R. Rao and J. Stufken

Volume 7. Quality Control and Reliability
Edited by P. R. Krishnaiah and C. R. Rao
1988 xiv + 503 pp.

1. Transformation of Western Style of Management by W. Edwards Deming
2. Software Reliability by F. B. Bastani and C. V. Ramamoorthy
3. Stress–Strength Models for Reliability by R. A. Johnson
4. Approximate Computation of Power Generating System Reliability Indexes by M. Mazumdar
5. Software Reliability Models by T. A. Mazzuchi and N. D. Singpurwalla
6. Dependence Notions in Reliability Theory by N. R. Chaganty and K. Joag-dev
7. Application of Goodness-of-Fit Tests in Reliability by H. W. Block and A. H. Moore
8. Multivariate Nonparametric Classes in Reliability by H. W. Block and T. H. Savits
9. Selection and Ranking Procedures in Reliability Models by S. S. Gupta and S. Panchapakesan
10. The Impact of Reliability Theory on Some Branches of Mathematics and Statistics by P. J. Boland and F. Proschan
11. Reliability Ideas and Applications in Economics and Social Sciences by M. C. Bhattacharjee
12. Mean Residual Life: Theory and Applications by F. Guess and F. Proschan
13. Life Distribution Models and Incomplete Data by R. E. Barlow and F. Proschan
14. Piecewise Geometric Estimation of a Survival Function by G. M. Mimmack and F. Proschan
15. Applications of Pattern Recognition in Failure Diagnosis and Quality Control by L. F. Pau
16. Nonparametric Estimation of Density and Hazard Rate Functions when Samples are Censored by W. J. Padgett
17. Multivariate Process Control by F. B. Alt and N. D. Smith
18. QMP/USP–A Modern Approach to Statistical Quality Auditing by B. Hoadley
19. Review About Estimation of Change Points by P. R. Krishnaiah and B. Q. Miao
20. Nonparametric Methods for Changepoint Problems by M. Csörgo and L. Horváth
21. Optimal Allocation of Multistate Components by E. El-Neweihi, F. Proschan and J. Sethuraman
22. Weibull, Log-Weibull and Gamma Order Statistics by H. L. Herter
23. Multivariate Exponential Distributions and their Applications in Reliability by A. P. Basu

24. Recent Developments in the Inverse Gaussian Distribution by S. Iyengar and G. Patwardhan

Volume 8. Statistical Methods in Biological and Medical Sciences
Edited by C. R. Rao and R. Chakraborty
1991 xvi + 554 pp.

1. Methods for the Inheritance of Qualitative Traits by J. Rice, R. Neuman and S. O. Moldin
2. Ascertainment Biases and their Resolution in Biological Surveys by W. J. Ewens
3. Statistical Considerations in Applications of Path Analytical in Genetic Epidemiology by D. C. Rao
4. Statistical Methods for Linkage Analysis by G. M. Lathrop and J. M. Lalouel
5. Statistical Design and Analysis of Epidemiologic Studies: Some Directions of Current Research by N. Breslow
6. Robust Classification Procedures and Their Applications to Anthropometry by N. Balakrishnan and R. S. Ambagaspitiya
7. Analysis of Population Structure: A Comparative Analysis of Different Estimators of Wright's Fixation Indices by R. Chakraborty and H. Danker-Hopfe
8. Estimation of Relationships from Genetic Data by E. A. Thompson
9. Measurement of Genetic Variation for Evolutionary Studies by R. Chakraborty and C. R. Rao
10. Statistical Methods for Phylogenetic Tree Reconstruction by N. Saitou
11. Statistical Models for Sex-Ratio Evolution by S. Lessard
12. Stochastic Models of Carcinogenesis by S. H. Moolgavkar
13. An Application of Score Methodology: Confidence Intervals and Tests of Fit for One-Hit-Curves by J. J. Gart
14. Kidney-Survival Analysis of IgA Nephropathy Patients: A Case Study by O. J. W. F. Kardaun
15. Confidence Bands and the Relation with Decision Analysis: Theory by O. J. W. F. Kardaun
16. Sample Size Determination in Clinical Research by J. Bock and H. Toutenburg

Volume 9. Computational Statistics
Edited by C. R. Rao
1993 xix + 1045 pp.

1. Algorithms by B. Kalyanasundaram
2. Steady State Analysis of Stochastic Systems by K. Kant
3. Parallel Computer Architectures by R. Krishnamurti and B. Narahari
4. Database Systems by S. Lanka and S. Pal
5. Programming Languages and Systems by S. Purushothaman and J. Seaman
6. Algorithms and Complexity for Markov Processes by R. Varadarajan
7. Mathematical Programming: A Computational Perspective by W. W. Hager, R. Horst and P. M. Pardalos
8. Integer Programming by P. M. Pardalos and Y. Li
9. Numerical Aspects of Solving Linear Lease Squares Problems by J. L. Barlow
10. The Total Least Squares Problem by S. Van Huffel and H. Zha
11. Construction of Reliable Maximum-Likelihood-Algorithms with Applications to Logistic and Cox Regression by D. Böhning
12. Nonparametric Function Estimation by T. Gasser, J. Engel and B. Seifert
13. Computation Using the QR Decomposition by C. R. Goodall
14. The EM Algorithm by N. Laird
15. Analysis of Ordered Categorial Data through Appropriate Scaling by C. R. Rao and P. M. Caligiuri
16. Statistical Applications of Artificial Intelligence by W. A. Gale, D. J. Hand and A. E. Kelly
17. Some Aspects of Natural Language Processes by A. K. Joshi
18. Gibbs Sampling by S. F. Arnold
19. Bootstrap Methodology by G. J. Babu and C. R. Rao
20. The Art of Computer Generation of Random Variables by M. T. Boswell, S. D. Gore, G. P. Patil and C. Taillie
21. Jackknife Variance Estimation and Bias Reduction by S. Das Peddada
22. Designing Effective Statistical Graphs by D. A. Burn
23. Graphical Methods for Linear Models by A. S. Hadi
24. Graphics for Time Series Analysis by H. J. Newton
25. Graphics as Visual Language by T. Selker and A. Appel
26. Statistical Graphics and Visualization by E. J. Wegman and D. B. Carr
27. Multivariate Statistical Visualization by F. W. Young, R. A. Faldowski and M. M. McFarlane
28. Graphical Methods for Process Control by T. L. Ziemer

Volume 10. Signal Processing and its Applications
Edited by N. K. Bose and C. R. Rao
1993 xvii + 992 pp.

1. Signal Processing for Linear Instrumental Systems with Noise: A General Theory with Illustrations for Optical Imaging and Light Scattering Problems by M. Bertero and E. R. Pike
2. Boundary Implication Rights in Parameter Space by N. K. Bose
3. Sampling of Bandlimited Signals: Fundamental Results and Some Extensions by J. L. Brown, Jr.
4. Localization of Sources in a Sector: Algorithms and Statistical Analysis by K. Buckley and X.-L. Xu
5. The Signal Subspace Direction-of-Arrival Algorithm by J. A. Cadzow
6. Digital Differentiators by S. C. Dutta Roy and B. Kumar
7. Orthogonal Decompositions of 2D Random Fields and their Applications for 2D Spectral Estimation by J. M. Francos
8. VLSI in Signal Processing by A. Ghouse
9. Constrained Beamforming and Adaptive Algorithms by L. C. Godara
10. Bispectral Speckle Interferometry to Reconstruct Extended Objects from Turbulence-Degraded Telescope Images by D. M. Goodman, T. W. Lawrence, E. M. Johansson and J. P. Fitch
11. Multi-Dimensional Signal Processing by K. Hirano and T. Nomura
12. On the Assessment of Visual Communication by F. O. Huck, C. L. Fales, R. Alter-Gartenberg and Z. Rahman
13. VLSI Implementations of Number Theoretic Concepts with Applications in Signal Processing by G. A. Jullien, N. M. Wigley and J. Reilly
14. Decision-level Neural Net Sensor Fusion by R. Y. Levine and T. S. Khuon
15. Statistical Algorithms for Noncausal Gauss Markov Fields by J. M. F. Moura and N. Balram
16. Subspace Methods for Directions-of-Arrival Estimation by A. Paulraj, B. Ottersten, R. Roy, A. Swindlehurst, G. Xu and T. Kailath
17. Closed Form Solution to the Estimates of Directions of Arrival Using Data from an Array of Sensors by C. R. Rao and B. Zhou
18. High-Resolution Direction Finding by S. V. Schell and W. A. Gardner
19. Multiscale Signal Processing Techniques: A Review by A. H. Tewfik, M. Kim and M. Deriche
20. Sampling Theorems and Wavelets by G. G. Walter
21. Image and Video Coding Research by J. W. Woods
22. Fast Algorithms for Structured Matrices in Signal Processing by A. E. Yagle

Volume 11. Econometrics
Edited by G. S. Maddala, C. R. Rao and H. D. Vinod
1993 xx + 783 pp.

1. Estimation from Endogenously Stratified Samples by S. R. Cosslett
2. Semiparametric and Nonparametric Estimation of Quantal Response Models by J. L. Horowitz
3. The Selection Problem in Econometrics and Statistics by C. F. Manski
4. General Nonparametric Regression Estimation and Testing in Econometrics by A. Ullah and H. D. Vinod
5. Simultaneous Microeconometric Models with Censored or Qualitative Dependent Variables by R. Blundell and R. J. Smith
6. Multivariate Tobit Models in Econometrics by L.-F. Lee
7. Estimation of Limited Dependent Variable Models under Rational Expectations by G. S. Maddala
8. Nonlinear Time Series and Macroeconometrics by W. A. Brock and S. M. Potter
9. Estimation, Inference and Forecasting of Time Series Subject to Changes in Time by J. D. Hamilton
10. Structural Time Series Models by A. C. Harvey and N. Shephard
11. Bayesian Testing and Testing Bayesians by J.-P. Florens and M. Mouchart
12. Pseudo-Likelihood Methods by C. Gourieroux and A. Monfort
13. Rao's Score Test: Recent Asymptotic Results by R. Mukerjee
14. On the Strong Consistency of M-Estimates in Linear Models under a General Discrepancy Function by Z. D. Bai, Z. J. Liu and C. R. Rao
15. Some Aspects of Generalized Method of Moments Estimation by A. Hall
16. Efficient Estimation of Models with Conditional Moment Restrictions by W. K. Newey
17. Generalized Method of Moments: Econometric Applications by M. Ogaki
18. Testing for Heteroskedasticity by A. R. Pagan and Y. Pak
19. Simulation Estimation Methods for Limited Dependent Variable Models by V. A. Hajivassiliou
20. Simulation Estimation for Panel Data Models with Limited Dependent Variable by M. P. Keane
21. A Perspective on Application of Bootstrap methods in Econometrics by J. Jeong and G. S. Maddala
22. Stochastic Simulations for Inference in Nonlinear Errors-in-Variables Models by R. S. Mariano and B. W. Brown
23. Bootstrap Methods: Applications in Econometrics by H. D. Vinod
24. Identifying outliers and Influential Observations in Econometric Models by S. G. Donald and G. S. Maddala
25. Statistical Aspects of Calibration in Macroeconomics by A. W. Gregory and G. W. Smith

26. Panel Data Models with Rational Expectations by K. Lahiri
27. Continuous Time Financial Models: Statistical Applications of Stochastic Processes by K. R. Sawyer

Volume 12. Environmental Statistics
Edited by G. P. Patil and C. R. Rao
1994 xix + 927 pp.

1. Environmetrics: An Emerging Science by J. S. Hunter
2. A National Center for Statistical Ecology and Environmental Statistics: A Center Without Walls by G. P. Patil
3. Replicate Measurements for Data Quality and Environmental Modeling by W. Liggett
4. Design and Analysis of Composite Sampling Procedures: A Review by G. Lovison, S. D. Gore and G. P. Patil
5. Ranked Set Sampling by G. P. Patil, A. K. Sinha and C. Taillie
6. Environmental Adaptive Sampling by G. A. F. Seber and S. K. Thompson
7. Statistical Analysis of Censored Environmental Data by M. Akritas, T. Ruscitti and G. P. Patil
8. Biological Monitoring: Statistical Issues and Models by E. P. Smith
9. Environmental Sampling and Monitoring by S. V. Stehman and W. Scott Overton
10. Ecological Statistics by B. F. J. Manly
11. Forest Biometrics by H. E. Burkhart and T. G. Gregoire
12. Ecological Diversity and Forest Management by J. H. Gove, G. P. Patil, B. F. Swindel and C. Taillie
13. Ornithological Statistics by P. M. North
14. Statistical Methods in Developmental Toxicology by P. J. Catalano and L. M. Ryan
15. Environmental Biometry: Assessing Impacts of Environmental Stimuli Via Animal and Microbial Laboratory Studies by W. W. Piegorsch
16. Stochasticity in Deterministic Models by J. J. M. Bedaux and S. A. L. M. Kooijman
17. Compartmental Models of Ecological and Environmental Systems by J. H. Matis and T. E. Wehrly
18. Environmental Remote Sensing and Geographic Information Systems-Based Modeling by W. L. Myers
19. Regression Analysis of Spatially Correlated Data: The Kanawha County Health Study by C. A. Donnelly, J. H. Ware and N. M. Laird
20. Methods for Estimating Heterogeneous Spatial Covariance Functions with Environmental Applications by P. Guttorp and P. D. Sampson

21. Meta-analysis in Environmental Statistics by V. Hasselblad
22. Statistical Methods in Atmospheric Science by A. R. Solow
23. Statistics with Agricultural Pests and Environmental Impacts by L. J. Young and J. H. Young
24. A Crystal Cube for Coastal and Estuarine Degradation: Selection of Endpoints and Development of Indices for Use in Decision Making by M. T. Boswell, J. S. O'Connor and G. P. Patil
25. How Does Scientific Information in General and Statistical Information in Particular Input to the Environmental Regulatory Process? by C. R. Cothern
26. Environmental Regulatory Statistics by C. B. Davis
27. An Overview of Statistical Issues Related to Environmental Cleanup by R. Gilbert
28. Environmental Risk Estimation and Policy Decisions by H. Lacayo Jr.

Volume 13. Design and Analysis of Experiments
Edited by S. Ghosh and C. R. Rao
1996 xviii + 1230 pp.

1. The Design and Analysis of Clinical Trials by P. Armitage
2. Clinical Trials in Drug Development: Some Statistical Issues by H. I. Patel
3. Optimal Crossover Designs by J. Stufken
4. Design and Analysis of Experiments: Nonparametric Methods with Applications to Clinical Trials by P. K. Sen
5. Adaptive Designs for Parametric Models by S. Zacks
6. Observational Studies and Nonrandomized Experiments by P. R. Rosenbaum
7. Robust Design: Experiments for Improving Quality by D. M. Steinberg
8. Analysis of Location and Dispersion Effects from Factorial Experiments with a Circular Response by C. M. Anderson
9. Computer Experiments by J. R. Koehler and A. B. Owen
10. A Critique of Some Aspects of Experimental Design by J. N. Srivastava
11. Response Surface Designs by N. R. Draper and D. K. J. Lin
12. Multiresponse Surface Methodology by A. I. Khuri
13. Sequential Assembly of Fractions in Factorial Experiments by S. Ghosh
14. Designs for Nonlinear and Generalized Linear Models by A. C. Atkinson and L. M. Haines
15. Spatial Experimental Design by R. J. Martin
16. Design of Spatial Experiments: Model Fitting and Prediction by V. V. Fedorov
17. Design of Experiments with Selection and Ranking Goals by S. S. Gupta and S. Panchapakesan

18. Multiple Comparisons by A. C. Tamhane
19. Nonparametric Methods in Design and Analysis of Experiments by E. Brunner and M. L. Puri
20. Nonparametric Analysis of Experiments by A. M. Dean and D. A. Wolfe
21. Block and Other Designs in Agriculture by D. J. Street
22. Block Designs: Their Combinatorial and Statistical Properties by T. Calinski and S. Kageyama
23. Developments in Incomplete Block Designs for Parallel Line Bioassays by S. Gupta and R. Mukerjee
24. Row-Column Designs by K. R. Shah and B. K. Sinha
25. Nested Designs by J. P. Morgan
26. Optimal Design: Exact Theory by C. S. Cheng
27. Optimal and Efficient Treatment – Control Designs by D. Majumdar
28. Model Robust Designs by Y-J. Chang and W. I. Notz
29. Review of Optimal Bayes Designs by A. DasGupta
30. Approximate Designs for Polynomial Regression: Invariance, Admissibility, and Optimality by N. Gaffke and B. Heiligers

Volume 14. Statistical Methods in Finance
Edited by G. S. Maddala and C. R. Rao
1996 xvi + 733 pp.

1. Econometric Evaluation of Asset Pricing Models by W. E. Ferson and R. Jegannathan
2. Instrumental Variables Estimation of Conditional Beta Pricing Models by C. R. Harvey and C. M. Kirby
3. Semiparametric Methods for Asset Pricing Models by B. N. Lehmann
4. Modeling the Term Structure by A. R. Pagan, A. D. Hall, and V. Martin
5. Stochastic Volatility by E. Ghysels, A. C. Harvey and E. Renault
6. Stock Price Volatility by S. F. LeRoy
7. GARCH Models of Volatility by F. C. Palm
8. Forecast Evaluation and Combination by F. X. Diebold and J. A. Lopez
9. Predictable Components in Stock Returns by G. Kaul
10. Interset Rate Spreads as Predictors of Business Cycles by K. Lahiri and J. G. Wang
11. Nonlinear Time Series, Complexity Theory, and Finance by W. A. Brock and P. J. F. deLima
12. Count Data Models for Financial Data by A. C. Cameron and P. K. Trivedi
13. Financial Applications of Stable Distributions by J. H. McCulloch
14. Probability Distributions for Financial Models by J. B. McDonald
15. Bootstrap Based Tests in Financial Models by G. S. Maddala and H. Li

16. Principal Component and Factor Analyses by C. R. Rao
17. Errors in Variables Problems in Finance by G. S. Maddala and M. Nimalendran
18. Financial Applications of Artificial Neural Networks by M. Qi
19. Applications of Limited Dependent Variable Models in Finance by G. S. Maddala
20. Testing Option Pricing Models by D. S. Bates
21. Peso Problems: Their Theoretical and Empirical Implications by M. D. D. Evans
22. Modeling Market Microstructure Time Series by J. Hasbrouck
23. Statistical Methods in Tests of Portfolio Efficiency: A Synthesis by J. Shanken

Volume 15. Robust Inference
Edited by G. S. Maddala and C. R. Rao
1997 xviii + 698 pp.

1. Robust Inference in Multivariate Linear Regression Using Difference of Two Convex Functions as the Discrepancy Measure by Z. D. Bai, C. R. Rao and Y. H. Wu
2. Minimum Distance Estimation: The Approach Using Density-Based Distances by A. Basu, I. R. Harris and S. Basu
3. Robust Inference: The Approach Based on Influence Functions by M. Markatou and E. Ronchetti
4. Practical Applications of Bounded-Influence Tests by S. Heritier and M-P. Victoria-Feser
5. Introduction to Positive-Breakdown Methods by P. J. Rousseeuw
6. Outlier Identification and Robust Methods by U. Gather and C. Becker
7. Rank-Based Analysis of Linear Models by T. P. Hettmansperger, J. W. McKean and S. J. Sheather
8. Rank Tests for Linear Models by R. Koenker
9. Some Extensions in the Robust Estimation of Parameters of Exponential and Double Exponential Distributions in the Presence of Multiple Outliers by A. Childs and N. Balakrishnan
10. Outliers, Unit Roots and Robust Estimation of Nonstationary Time Series by G. S. Maddala and Y. Yin
11. Autocorrelation-Robust Inference by P. M. Robinson and C. Velasco
12. A Practitioner's Guide to Robust Covariance Matrix Estimation by W. J. den Haan and A. Levin
13. Approaches to the Robust Estimation of Mixed Models by A. H. Welsh and A. M. Richardson

14. Nonparametric Maximum Likelihood Methods by S. R. Cosslett
15. A Guide to Censored Quantile Regressions by B. Fitzenberger
16. What Can Be Learned About Population Parameters When the Data Are Contaminated by J. L. Horowitz and C. F. Manski
17. Asymptotic Representations and Interrelations of Robust Estimators and Their Applications by J. Jurečková and P. K. Sen
18. Small Sample Asymptotics: Applications in Robustness by C. A. Field and M. A. Tingley
19. On the Fundamentals of Data Robustness by G. Maguluri and K. Singh
20. Statistical Analysis With Incomplete Data: A Selective Review by M. G. Akritas and M. P. LaValley
21. On Contamination Level and Sensitivity of Robust Tests by J. Á. Visšek
22. Finite Sample Robustness of Tests: An Overview by T. Kariya and P. Kim
23. Future Directions by G. S. Maddala and C. R. Rao